Concise Encyclopedia of the Sciences

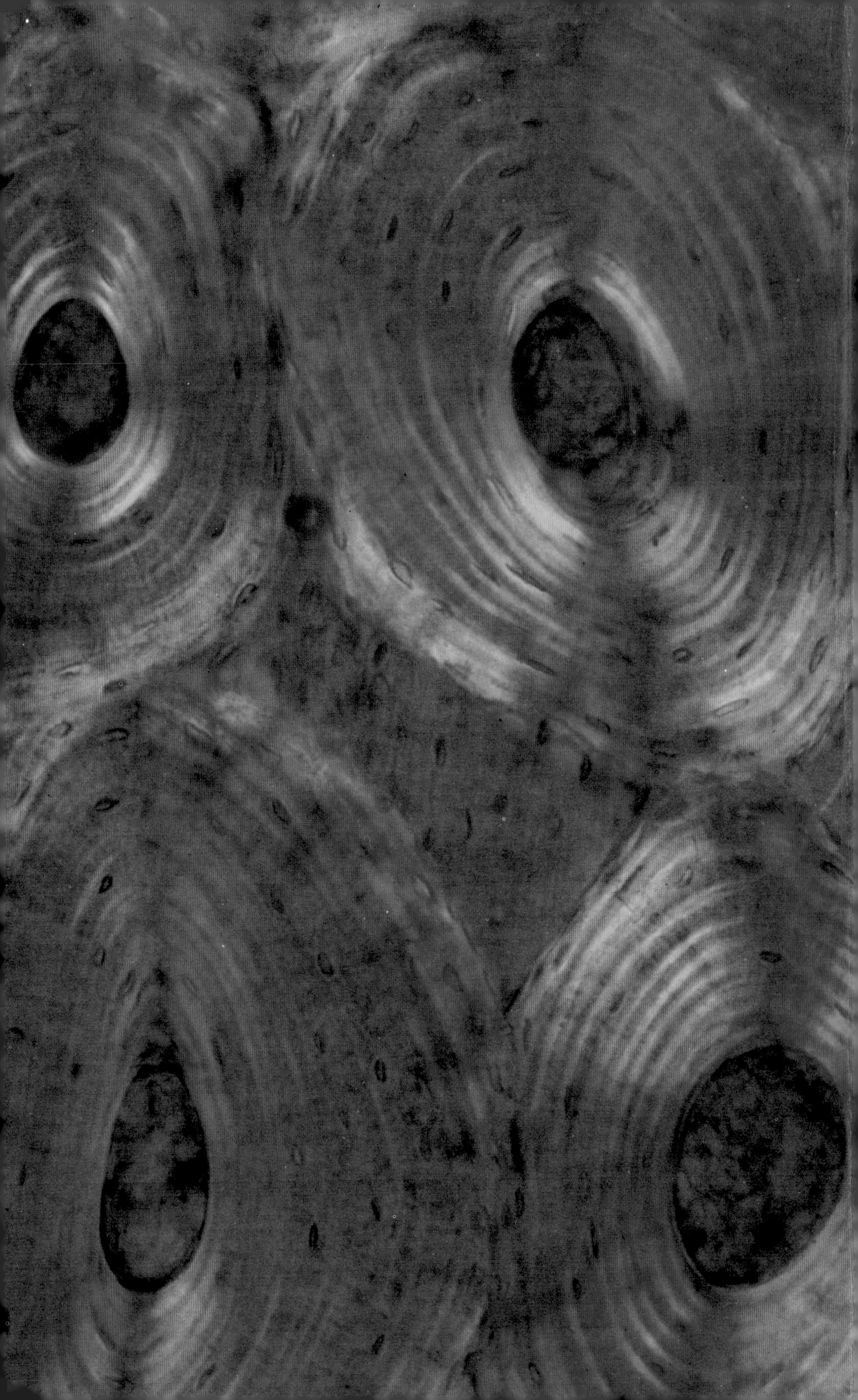

Concise Encyclopedia of the Sciences

John-David Yule,
editor

Facts On File
119 West 57th Street, New York, N.Y. 10019

Contributors

P. le P. Barnett
J. G. Bateman, BSc, PhD
M. G. Desebrock, BA
P. C. Gardner, MA
B. Gibbs, BSc
G. S. Harbinson, BSc, DPhil
P. Hutchinson, BSc, PhD
A. J. Pinching, BM, BCh, MA, DPhil
J. P. Revel, BSc
P. R. Robinson, BA, BLitt
W. O. Saxton, MA, PhD
M. Scott Rohan, MA

Frontispiece: Microscopic section through bone tissue viewed in polarized light.

Facts On File, Inc.
119 West 57th Street
New York, N.Y. 10019

Library of Congress Cataloging in Publication Data
Main entry under title:

Concise encyclopedia of the sciences.

First published in 1978 under title: Phaidon concise encyclopedia of science and technology.
Includes index.
1. Science—Dictionaries. I. Yule, John-David.
II. Facts on File, inc., New York.
Q121.P48 1980 503'.21 80-21173
ISBN 0-87196-491-0

Typeset by Keyspools Ltd., Golborne, Lancashire
Printed in Hong Kong

Preface

As the technicalities of science intrude ever more deeply into every aspect of man's daily life, affecting the work he does, the food he eats, the energy he consumes and the environment of the cities he inhabits, so a need has become apparent for a compact, easy-to-use reference work which will interpret this new empire of science to its reader. The present volume is designed to meet this need by supplying both a dictionary of the most commonly encountered words of science and an encyclopedia of the background material necessary for understanding their use in a wider context. Prepared with the reference requirements of young people, parents, students, educators, professional workers and businessmen in mind, it gives instant access to the facts, theories, inventions and formulas vital for mastering the complexities of the science and technology which underpin 20th-century industrial society.

Contents

So that it can function as an efficient and yet comprehensive science dictionary, the book has been organized around some 5 500 carefully selected keywords. The subjects treated include mathematics; the physical, earth and life sciences; technology; the philosophy of science, and the sciences of man—medicine, anthropology and psychology. In addition there are more than 1 000 brief biographical notices of the men and women who have contributed the most to the development of modern science and technology. In botany and zoology, the descriptive articles included are confined to the most general taxonomic groupings; individual species are the subject of separate entries only when they are of particular scientific interest.

Since it is impossible to tailor a reference work of this kind so that it will immediately and completely answer all the many questions that different users (having varying degrees of prior knowledge) will bring to it, and since in a volume of this size there is little room for the duplication of information, most articles have been generously cross-referenced (SMALL CAPITALS). To make the most efficient use of the reference facility thus provided, the reader is urged to read an entry straight through, without consulting any cross-referenced article, before selecting for further study only those cross-references which will assist him with his immediate problem.

Illustrations

Only two-thirds of the space in this volume is occupied by the text articles. The remaining third is devoted to diagrams and photographs (mainly in full color) which illustrate and enlarge upon the information presented in the text, and to attractively designed tables which summarize a wealth of useful additional detail. Particularly noteworthy are the visual history panels which review the development of several key sciences, and the context panels which, for others, both reveal their internal structure and display their connections with related disciplines.

Units and Nomenclature

The scientific units employed in this book in general follow the Système International d'Unités (SI units) in the form recommended by the US National Bureau of Standards (NBS). Other conventions followed in this volume include the naming of numbers in excess of one million according to the customary American system (see below) and the general adoption of the system of chemical nomenclature recommended by the International Union of Pure and Applied Chemistry (IUPAC).

SI Units

The International System of Units (SI units) is founded upon seven empirically defined *base units* (e.g., ampere), which can be combined, sometimes with the assistance of two geometric *supplementary units* (e.g., radian) to yield the *derived units* (e.g., cubic metre) which together with the base units constitute a coherent set of units capable of application to all measurable physical phenomena. Some of the derived units have special names (e.g. volt). For the sake of convenience smaller and larger units, the multiples and submultiples of the SI units, can be formed by adding certain prefixes (e.g., milli) to the names of the SI units. In any instance only one prefix can be added to the name of a unit (thus nanometre, not millimicrometre). Of the other units in common scientific use, it is recognized that several (e.g., hour) will continue to be used alongside the SI units, although combinations of these units with SI units (as in kilowatt hour) are discouraged. However, other units (e.g., angstrom unit) are redundant if the International System is fully utilized, and it is intended that these should drop out of use.

Base units

Quantity	Unit	Symbol
length	metre	m
mass	kilogram	kg
time	second	s
electric current	ampere	A
temperature	kelvin	K
luminous intensity	candela	cd
amount of substance	mole	mol

Supplementary units

Quantity	Name	Symbol
plane angle	radian	rad
solid angle	steradian	sr

SI Prefixes

Multiplying factor	Prefix	Symbol
10^{12}	tera	T
10^{9}	giga	G
10^{6}	mega	M
10^{3}	kilo	k
10^{2}	hecto	h
10^{1}	deka	da
10^{-1}	deci	d
10^{-2}	centi	c
10^{-3}	milli	m
10^{-6}	micro	μ
10^{-9}	nano	n
10^{-12}	pico	p
10^{-15}	femto	f
10^{-18}	atto	a

Derived units with special names

Quantity	Name	Symbol	Equivalent in other units	As expressed in base units
frequency	hertz	Hz	—	s^{-1}
force	newton	N	—	$m.kg.s^{-2}$
work, energy	joule	J	N.m	$m^2.kg.s^{-2}$
power	watt	W	J/s	$m^2.kg.s^{-3}$
pressure	pascal	Pa	N/m^2	$m^{-1}.kg.s^{-2}$
quantity of electricity	coulomb	C	—	s.A
potential difference	volt	V	W/A	$m^2.kg.s^{-3}.A^{-1}$
electric resistance	ohm	Ω	V/A	$m^2.kg.s^{-3}.A^{-2}$
capacitance	farad	F	C/V	$m^{-2}.kg^{-1}.s^4.A^2$
conductance	siemens	S	A/V	$m^{-2}.kg^{-1}.s^3.A^2$
magnetic flux	weber	Wb	V.s	$m^2.kg.s^{-2}.A^{-1}$
flux density	tesla	T	Wb/m^2	$kg.s^{-2}.A^{-1}$
inductance	henry	H	Wb/A	$m^2.kg.s^{-2}.A^{-2}$
luminous flux	lumen	lm	—	cd.sr
illuminance	lux	lx	lm/m^2	$m^{-2}.cd.sr$

Non-SI units in continuing use

(a) defined in terms of SI units

Quantity	Name	Symbol	Equivalent
plane angle	degree	°	$\frac{\pi}{180}$ rad
plane angle	minute	′	$\frac{1}{60}$°
plane angle	second	″	$\frac{1}{60}$′
time	minute	min	60s
time	hour	h	3 600s = 60min
time	day	d	86 400s = 24h
volume	litre	l	$10^{-3}m^3$
mass	tonne	t	10^3kg

(b) defined empirically

Quantity	Name	Symbol	Approximate value
energy	electronvolt	eV	1.602×10^{-19}J
mass	atomic mass unit	u	$1.660\,531 \times 10^{-27}$kg
length	astronomical unit	AU	$149\,600 \times 10^6$m
length	parsec	pc	$30\,857 \times 10^{12}$m

Temperature

The kelvin (K) as a unit of temperature difference is defined as 1/273.16 of the thermodynamic temperature of the triple point of water. It is thus identical to the old "centrigrade degree" (C°). As an alternative to thermodynamic temperatures expressed in kelvins, it is often convenient to express commonly encountered temperatures using the Celsius temperature scale. Temperatures thus expressed in "degrees Celsius" (°C) are identical to thermodynamic temperatures given in kelvins (K) *less* 273.15K.

Numbers greater than one million

In this volume the American system (which names large numbers according to the number of groups of three zeros which follow 1000 when they are expressed in numerals) is followed in preference to the European system (in which they are named for the power of 1 000 which they represent).

Number expressed as a power of 10	American name	European name	Number expressed as a power of 1 000 000
$10^6 = 10^{(3+3)}$	million	million	$1\,000\,000^1$
$10^9 = 10^{(3+2\times3)}$	billion	milliard	
$10^{12} = 10^{(3+3\times3)}$	trillion	billion	$1\,000\,000^2$
$10^{15} = 10^{(3+4\times3)}$	quadrillion	—	
$10^{18} = 10^{(3+5\times3)}$	quintillion	trillion	$1\,000\,000^3$
$10^{33} = 10^{(3+10\times3)}$	decillion	—	
10^{60}	—	decillion	$1\,000\,000^{10}$
$10^{303} = 10^{(3+100\times3)}$	centillion	—	
10^{600}	—	centillion	$1\,000\,000^{100}$

Abbreviations

A ampere
A mass number
Å angstrom unit
a are, atto-
AC alternating current
AD *anno domini* (in the year of our Lord)
AF audio frequency
Ala. Alabama
AM amplitude modulation
Ariz. Arizona
Ark. Arkansas
asb apostilb
atm atmosphere
AU astronomical unit
Aug. August
AW atomic weight

b barn
b. born
bar bar
BC before Christ
bhp brake horse power
bp boiling point
Btu British thermal unit

C coulomb
circa, centi-
c electromagnetic constant (speed of light)
C° centrigrade degree
°C degrees Celsius
Cal SEE kcal
Cal. California
cal calorie
ccp cubic close-packed
cd candela
CGS centimetre-gram-second (system)
Ci curie
Col. Colorado
Conn. Connecticut
cos cosine
cosec SEE csc
cosech SEE csch
cosh hyperbolic cosine
cot cotangent
coth hyperbolic cotangent
csc cosecant
csch hyperbolic cosecant
cu ft cubic foot
cwt hundredweight

d day, deci-
d. died
da deka-
dB decibel
DC direct current
Dec. December
Del. Delaware
DIN Deutsche Industrie Norm
dp decimal place
dr dram
dyn dyne

E east
e electron charge, base of natural logarithms
e.g. *exempli gratia* (for example)
EHF extremely high frequency
emf electromotive force
emu electromagnetic unit
erg erg
esu electrostatic unit
etc. *et cetera* (and others)
eV electron volt

F farad
f femto-
f/ aperture ratio

°F	degrees Fahrenheit
fcc	face-centered cubic
Feb.	February
Fla.	Florida
FM	frequency modulation
ft	foot
G	universal constant of gravitation, giga-
g	gram
g	acceleration due to gravity
Ga.	Georgia
Gal	gal (or galileo)
gal	gallon
Gb	gilbert
gr	grain
Gs	gauss
H	henry
h	hour, hecto-
h	Planck constant
ha	hectare
hcf	highest common factor
hcp	hexagonal close-packed
HF	high frequency
hp	horse power
Hz	hertz
i	imaginary operator
Ia.	Iowa
Ida.	Idaho
i.e.	*id est* (that is)
IF	intermediate frequency
iff	if and only if
Ill.	Illinois
in	inch
Ind.	Indiana
ir	infrared
J	joule
Jan.	January
K	kelvin
k	kilo-
k	Boltzmann constant
Kan.	Kansas
kcal	kilocalorie
kg	kilogram
kgf	kilogram-force
kn	knot
Ky.	Kentucky
L	lambert
l	litre
La.	Louisiana
lb	pound
lbf	pound-force
lcm	lowest common multiple
LF	low frequency
lm	lumen
ln	natural logarithm
log	common logarithm
LW	long wave
lx	lux
ly	light year
M	mega-
m	metre, milli-
Mass.	Massachusetts
mbar	millibar
Md.	Maryland
Me.	Maine
MF	medium frequency
mi	mile
Mich.	Michigan
min	minute (time)
Minn.	Minnesota
MKSA	metre-kilogram-second-ampere (system)
mmHg	millimetres of mercury
Mo.	Missouri
Mont.	Montana
mp	melting point
mph	miles per hour
MW	medium wave, molecular weight
Mx	maxwell
N	newton, north
N	Avogadro number, neutron number
n	nano-
N.B.	*nota bene* (note)
N.C.	North Carolina
N.D.	North Dakota
Neb.	Nebraska
Nev.	Nevada
N.H.	New Hampshire
N.J.	New Jersey
N.M.	New Mexico
nmr	nuclear magnetic resonance
Nov.	November
NTP	SEE STP
N.Y.	New York
Oct.	October
Oe	oersted
Okla.	Oklahoma
O.N.	oxidation number
Ore.	Oregon
oz	ounce
P	poise
p	pico-
Pa	pascal
Pa.	Pennsylvania
pc	parsec
pH	hydrogen ion concentration
ph	phot
Pl	poiseuille
ppm	parts per million
psi	pounds per square inch
pt	pint
qt	quart
R	röntgen
R	universal gas constant
°R	degrees Rankine (or Réaumur)
rad	radian, SEE also rd
rd	rad (dose)
RF	radio frequency
R.I.	Rhode Island
rms	root mean square
rpm	revolutions per minute
S	south
s	second
sb	stilb
S.C.	South Carolina
S.D.	South Dakota
sec	secant
sech	hyperbolic secant
Sept.	September
sf	significant figure
sg	specific gravity
SHF	superhigh frequency
sin	sine
sinh	hyperbolic sine
sp(p)	species
sq mi	square mile
sr	steradian
St	stokes
St.	Saint
STP	standard temperature and pressure
subl	sublimation point
SW	short wave
T	tesla, tera-
t	tonne
tan	tangent
tanh	hyperbolic tangent
Tenn.	Tennessee
Tex.	Texas
Torr	torr
U.	University
u	atomic mass unit
UHF	ultrahigh frequency
UK	United Kingdom
UN	United Nations
US	United States
USSR	Union of Soviet Socialist Republics
Ut.	Utah
uv	ultraviolet
V	volt
Va.	Virginia
VHF	very high frequency
Vt.	Vermont
W	watt, west
Wash.	Washington
Wb	weber
Wis.	Wisconsin
W.Va.	West Virginia
WWI	World War I
WWII	World War II
Wyo.	Wyoming
yd	yard
yr	year
Z	atomic number
/	(virgule) per
%	percent
‰	per mile (per thousand)
°	degree (angle)
′	minute (angle)
″	second (angle)
ε_0	permittivity of free space
μ	micron, micro-
μ_0	permeability of free space
Ω	ohm

Credits

Unless otherwise stated all the illustrations on a given page are credited to the same source.
AGA Infrared Systems 310/Allard Design Group 85, 252, 255/Amsterdam Harbor Authority 175/Amsterdam Universiteits Bibliotheek 464/Ann Ronan Picture Library 96, 157t, 172b, 354, 491, 589/Ardea Photographics 534/ASEA B.V. 215b/F. B. Atkins 379/A-Z Botanical Collections Limited 35/Barnaby Picture Library 510/R. Barnard 17, 23, 39, 49, 58/Bayer *frontispiece*/D. Bayliss 377/B.I.C.C. Limited 97t/Bowens of London 227r/B. Bracegirdle 78, 123, 308t/J. Brennan 1, 81b, 185, 341, 393t, 411, 458, 478b, 533, 575/P. Brierley 8, 97b/E. J. Brill 124t/British Aerospace Company 567br/British Leyland 192b/British Museum of Natural History 25/California Institute of Technology and Carnegie Institute of Washington 244/Central Electricity Generating Board 6/M.-Th. Cerceau-Larrival 422/C.E.R.N. 92/S.Cocking 72, 81t, 82, 109t, 120, 132, 137, 138, 140, 152, 188, 204, 211, 235, 237, 238, 280, 281, 320, 323, 330, 337t, 337bl, 345b, 346, 358, 386, 417/Coles Cranes Limited 149/P. Cook 425/Courtauld Limited 222/Culham Laboratories 241/Diagram Limited 51, 231, 293, 296, 301, 361, 497t, 577/Deutsches Hygiene Museum 324/J. M. Dinwoodie 332/J. A. Dixon 474/D. Ellis 33, 442/Elsevier Amsterdam 13b, 31, 57b, 60, 62, 66, 67, 79l, 83, 87, 91, 94, 98, 101, 103, 117b, 124b, 127, 128, 134, 136, 142, 143, 153, 164, 168, 180, 183, 187, 189, 196, 197, 200, 202t, 220t, 223, 225t, 229b, 233, 236, 245b, 247, 254, 257, 262, 270, 272, 275, 283, 287, 303, 311, 314t, 325, 329, 333, 335, 337br, 338, 345t, 348, 352, 356, 359, 366, 381, 400, 414, 415, 428, 431, 434, 439, 466, 470/1, 473, 478t, 484, 485, 488, 492, 508r, 512/3, 539, 541, 548, 550, 563t, 567t, 570b, 572r, 582, 586, 590/R. Estall 84/R. Fransen 86, 112, 125, 154, 155, 166, 174, 193b, 206, 224b, 263, 266/R. Geiger 498/Prof. Dr. H. H. van Gelderen 327b/General Dynamics (Shipbuilding Division) 397/Gist Brocades 340/R. Gorringe 43, 36, 157b, 171, 219, 239t, 327t, 339/B. Harvey 412/Heineken Brewery 221/Heurtier 547/A. A. M. van der Heyden 145/P. Hinchcliffe (Bruce Coleman Limited) 336t/E. Hosking 460/Houtvoorlichtings Institut 582/Hoverlloyd Limited 13t/Hydraulics Research Station (Crown copyright, reproduced by permission Controller H.M.S.O.) 170, 572l/Imperial Chemical Industries Limited 322/Institute of Geological Sciences 24, 279/International Wool Secretariat 584/W. A. Jensen 396/K. L. M. Aerocarto 217/Koninklyke Luchtmacht 277/Leiden Universiteit, Pathologisch Laboratory 568/B. Lenthall 515/R. Lewington 113/London Transport 558/Lovell Johns 423, 570t, 448/The Mansell Collection 42b, 44t/T. McArthur 158, 176, 186b, 251, 265, 268, 286, 306, 307, 418, 420, 456, 475/P.J. McCabe 207/M. McGregor 186t, 215t, 248, 291/Massey-Ferguson (U.K.) Limited 213/J. Midderigh 10b/Ministry of Defence, P.E.R.M.E. (Crown copyright) 214/D. M. Moore 426/P. Morris 234/Mount Wilson & Palomar Observatories (Hale Observatory) 398, 494, 561/Museum of the History of Science 48/G. Myatt 38t/N.A.S.A. 363, 375, 388/Natural Science Photos 161/A. van den Nieuwenhuizen 376, 399/Nereid N.V. 63/Nederlands Scheepsbouwkundig Proefstation 111/Oxford Illustrators Limited 3t, 9, 10t, 28, 29, 30, 38b, 42t, 42c, 44b, 50, 55, 66, 141, 271b, 278, 295b, 302, 309, 231, 342, 357, 384, 450, 497b, 522, 540/Oxford Scientific Films Limited 135, 312/Perkins-Elmer Nederland B.V. 517/Philips N.V. 198/Philips Research Laboratories, photo D. Washington 347/Polaroid Europa B.V. 451/R. D. Preston 57/National Physical Laboratory (Crown copyright) and Radiation Dynamics Limited 3b/Rapho 512/Renold Limited 117t/A. W. Robards 536/R.T.Z. Services Limited 229t/P. Scott 73/A Shell photograph 433, 444/Servicio Editoriale Fotografico 107b/N. J. Severs & E. G. Jordon 269/Space Frontiers Limited 243, 300, 326, 490, 588/Visual Art Productions 75, 107t, 109b, 151, 172t, 192t, 193t, 202b, 220b, 224t, 227l, 271t, 273, 285, 298, 299, 308b, 315t, 314b, 317, 364, 365, 370, 371, 378, 393b, 385, 394, 401, 403, 404, 413, 419, 435, 436, 441, 445t, 452, 453, 454, 463, 467, 472, 477, 505, 508l, 519, 521, 551, 553, 556, 563b, 565, 566, 567bl, 571, 576, 580/Ward's Natural Science Establishment Incorporated 181, 245t/K. Waters 156, 560/J. Way 12, 27, 41, 54, 59, 61, 64, 70, 76, 79r, 80, 88, 115, 148, 163, 169, 179, 194, 218, 225b, 230, 239b, 246, 249, 250, 261, 295t, 297, 315b, 336b, 351, 380, 391, 392, 407, 424, 437, 445b, 486 (*after Pains-Wessex Limited*), 501, 502, 518, 545, 554, 573/Josiah Wedgwood & Sons Limited 455/Westminster Dredging Company Limited 178/W. van Rijn 516/D. Wyllie 53/T. W. K. Young & D. N. Pegler 0, 216, 469, 585/Zuid-Nederlandse Melkindustrie B.V. 95.
The Publishers have attempted to observe the legal requirements with respect to the rights of the suppliers of photographic material. Nevertheless, persons who have claims are invited to apply to the Publishers.

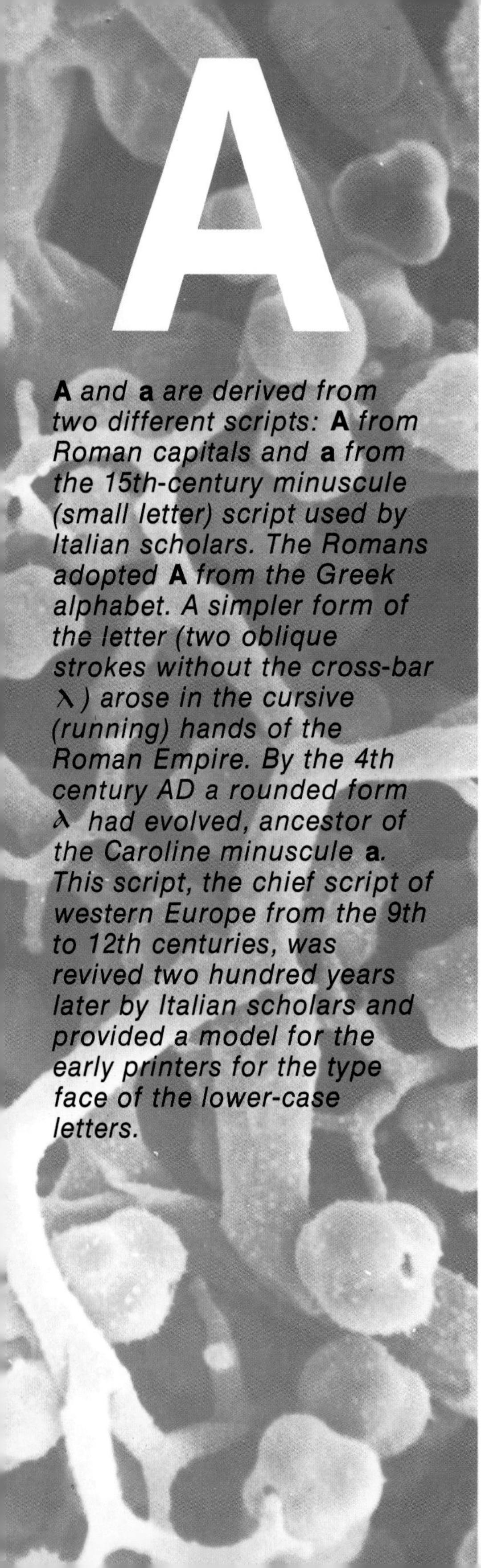

A

A *and* **a** *are derived from two different scripts:* **A** *from Roman capitals and* **a** *from the 15th-century minuscule (small letter) script used by Italian scholars. The Romans adopted* **A** *from the Greek alphabet. A simpler form of the letter (two oblique strokes without the cross-bar λ) arose in the cursive (running) hands of the Roman Empire. By the 4th century AD a rounded form λ had evolved, ancestor of the Caroline minuscule* **a**. *This script, the chief script of western Europe from the 9th to 12th centuries, was revived two hundred years later by Italian scholars and provided a model for the early printers for the type face of the lower-case letters.*

AA, (pronounced *ah-ah*), block lava. (See LAVA.)

ABACUS, or counting frame, a simple calculating instrument still widely used in Asia. It comprises a wooden frame containing a series of parallel rods divided into upper and lower portions. The rods represent the powers of 10, with each of the five beads on their lower portion counting 1 and the two on their upper portion each counting 5. In the hands of a skilled operator it allows addition, subtraction, multiplication and division problems to be solved with great rapidity.

ABAMPERE, the unit of current in the CGS electromagnetic system of units (emu—see CGS UNITS). Other **ab-units** include the abcoulomb, abohm and abvolt. They derive from equations in which the PERMEABILITY of free space is set dimensionless (see DIMENSIONS) and equal to UNITY.

Ab-units

quantity	CGS emu system	SI units
charge	1 abcoulomb	10 coulomb (10 C)
current	1 abampere	10 ampere (10 A)
potential	1 abvolt	10 nanovolt $(10^{-8}\,V)$
resistance	1 abohm	1 nanoohm $(10^{-9}\,\Omega)$
capacitance	1 abfarad	1 gigafarad $(10^{9}\,F)$
inductance	1 abhenry	1 nanohenry $(10^{-9}\,H)$

ABBE, Cleveland (1838–1916), nicknamed "Old Probabilities," US astronomer and meteorologist. In 1869, as director of the Cincinnati Observatory, he published the first daily weather forecasts in the US, becoming in 1871 the first chief meteorologist of the US Weather Service.

ABBE, Ernst (1840–1905), German physicist, research director of and partner in the optical firm of Carl ZEISS and founder of the Carl Zeiss Foundation (1891). He invented an apochromatic condenser LENS for use in a MICROSCOPE.

ABBOT, Charles Greeley (1872–1973), US astrophysicist, noted for his studies of solar radiation, and director of the Smithsonian Astrophysical Observatory from 1907 to 1944.

ABDOMEN, in VERTEBRATES, the part of the body between the CHEST and the PELVIS. In man, it contains most of the GASTROINTESTINAL TRACT (from the stomach to the colon) together with the LIVER, GALL BLADDER and SPLEEN in a potential cavity lined by PERITONEUM, while the KIDNEYS, ADRENAL GLANDS and PANCREAS lie behind this cavity, with the abdominal AORTA and inferior VENA CAVA. It is surrounded and protected by a muscular abdominal wall attached to the spine, ribs and pelvic bones and is separated from the chest by the DIAPHRAGM. In the ARTHROPODA, the abdomen is the rear division of the body.

ABEL, John Jacob (1857–1938), US pharmacologist who first isolated the hormone ADRENALINE (1897) and prepared INSULIN in crystalline form (1926).

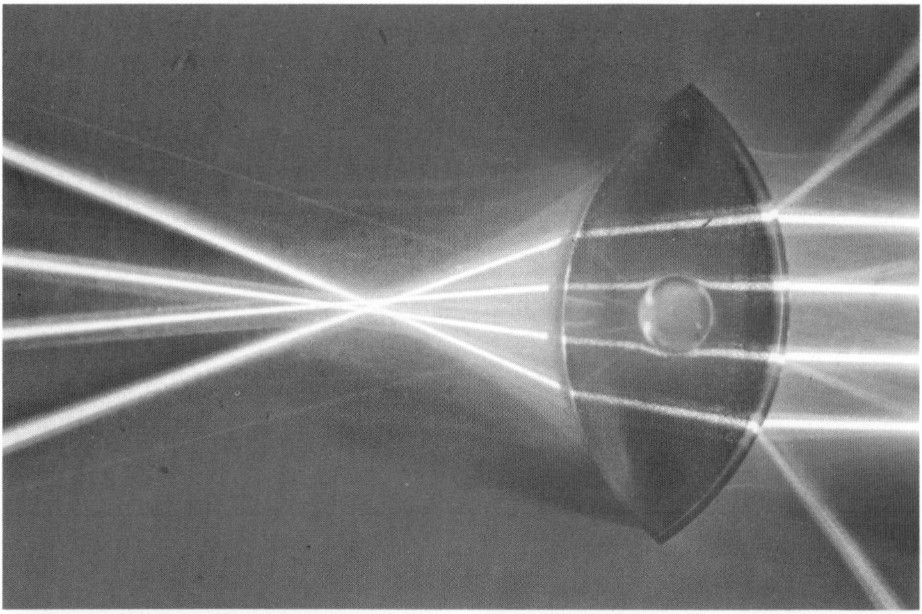

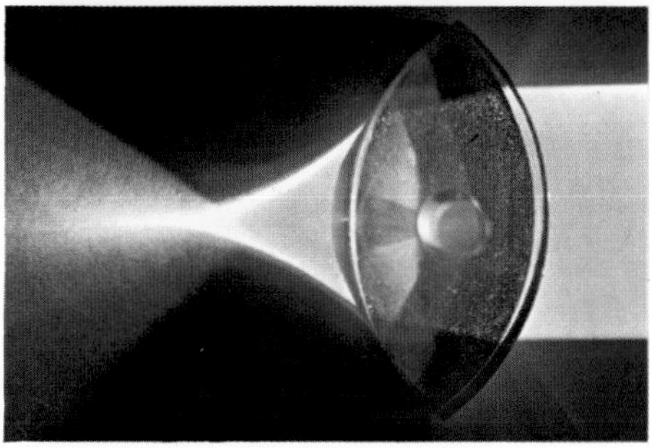

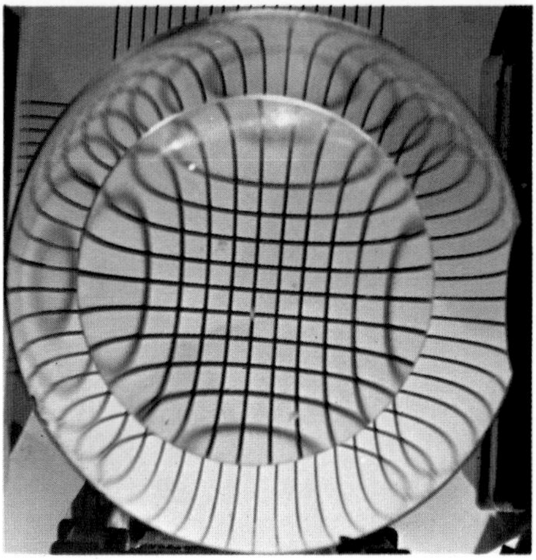

Optical aberration. In spherical aberration, (top), light rays passing through the outer regions of the lens are brought to a shorter focus than rays passing through the center. A broad beam of light passing through the same perspex lens (above). Chromatic aberration (right) demonstrated using a combination of two large converging lenses; pinchushion distortion is also evident.

ABEL, Niels Henrik (1802–1829), Norwegian mathematician who proved (1824) that the general EQUATION of the fifth degree cannot be solved algebraically. A pioneering memoir on transcendental functions (1826) was published posthumously in 1841.

ABELIAN GROUP, a GROUP in which, for every pair of elements a and b under an operation *, the commutative law (see ALGEBRA) holds: a*b=b*a. For example, the set of all integers under multiplication is an infinite abelian group. The name commemorates the work of Niels ABEL.

ABERRATION OF LIGHT, in astronomy, a displacement between a star's observed and true position caused by the earth's motion about the sun and the finite nature of the velocity of light. The effect is similar to that observed by a man walking in the rain: though the rain is in fact falling vertically, because of his motion it appears to be falling at an angle. The maximum aberrational displacement is 20.5″ of arc; stars on the ECLIPTIC appear to move to and fro along a line of 41″; stars 90° from the ecliptic appear to trace out a circle of radius 20.5″; and stars in intermediate positions ellipses of major axis 41″.

ABERRATION, Optical, the failure of a lens to form a perfect image of an object. The commonest types are chromatic aberration, where DISPERSION causes colored fringes to appear around the image; and spherical aberration, where blurring occurs because light from the outer parts of the lens is brought to a focus at a shorter distance from the lens than that passing through the center. Chromatic aberration can be reduced by using an ACHROMATIC LENS and spherical aberration by separating the elements of a compound lens.

ABIOGENESIS. See SPONTANEOUS GENERATION.

ABLATION, in aerospace technology, the FUSION and EVAPORATION of the surface layers of an object heated by frictional contact with air (e.g. a spacecraft on reentry into the atmosphere). HEAT SHIELDS incorporate outer layers which ablate, thus preventing overheating of the spacecraft's interior.

ABLATION, in glaciology, the loss of snow and ice from the surface of a GLACIER by melting, EVAPORATION or SUBLIMATION; also, the quantity so lost.

ABORIGINES, a mythical Italian tribe held to have been the first inhabitants of that country (Latin: *ab origine*). The term is now applied to the earliest known occupants of any region; hence its application to the AUSTRALIAN ABORIGINES, although it appears that the TASMANIANS were Australia's first inhabitants.

ABORTION, ending of PREGNANCY before the fetus is able to survive outside the womb. It can occur spontaneously (in which case it is often termed **miscarriage**) or it can be artificially induced. Spontaneous abortion may occur as a result of maternal or fetal disease and faulty implantation in the WOMB. Induction may be mechanical, chemical or using HORMONES, the maternal risk varying with fetal age, the method used and the skill of the physician. In most countries, and until recently throughout the US, the practice was considered criminal unless the mother's life was at risk. In recent years, despite continuing moral controversy, abortion has become widely regarded as a means of BIRTH CONTROL.

ABRAHAM, Karl (1877–1925), German psychoanalyst whose most important work concerned the development of the LIBIDO, particularly in infancy. He suggested that various PSYCHOSES should be interpreted in terms of the interruption of this development.

ABRASION, the wearing down of a surface when a harder surface rubs over it. Abrasion results from small chips of the softer material being sheared off when they obstruct the passage of irregularities in the surface of the harder material. Unwanted abrasion in BEARINGS can be prevented by LUBRICATION.

ABRASIVE, any material used to cut, grind or polish a softer material by ABRASION. Mild abrasives such as CHALK are incorporated in toothpaste, and others, SILICA, PUMICE or ALUMINUM oxide, are used in household cleansers; but various industrial applications demand even harder abrasives (see HARDNESS) such as CARBORUNDUM, BORAZON or DIAMOND. Some abrasives are used in solid blocks (as with knife-grinding stones), but **coated abrasives** such as sandpaper in which abrasive granules are stuck onto a carrier make more economic use of the material. **Sandblasting** exemplifies a third technique in which abrasive particles are thrown against the workpiece in a stream of compressed air or steam. Sandblasting is used for cleaning buildings and engraving glass.

ABREACTION. See CATHARSIS.

ABSCESS, a localized accumulation of PUS, usually representing one response of the body to bacterial infection. Abscesses, which may occur in any tissue or organ of the body, often show themselves in pain, redness and swelling. They may drain spontaneously, otherwise they should be incised.

ABSCISIC ACID, formerly known as abscisin II or dormin, a plant HORMONE that inhibits plant growth, induces bud DORMANCY and promotes leaf ABSCISSION (shedding).

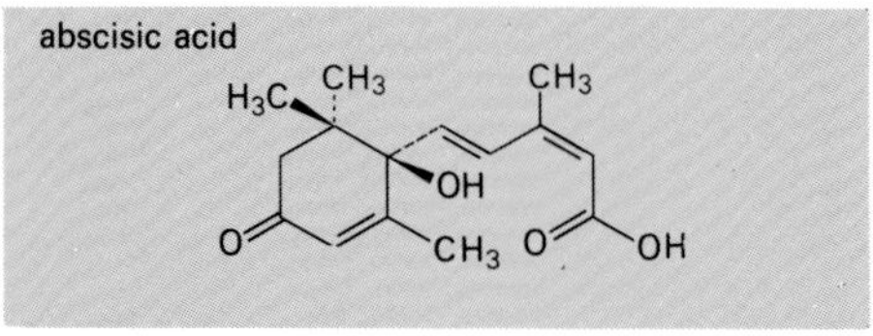

ABSCISSA, in CARTESIAN COORDINATES, the distance (measured parallel to the *x*-axis) of a point from the *y*-axis, positive in sign to the right of the *y*-axis, negative to its left. The abscissa of the point (a, b) is thus a.

ABSCISSION, in botany, the process whereby plants shed leaves, flowers and fruits. Controlled by plant HORMONES (abscisins) such as ABSCISIC ACID, leaf drop occurs in many plants through the formation of an intermediate **abscission layer** of cells which constricts sap flow to the leaf and is then broken.

ABSOLUTE, an adjective frequently encountered in scientific terminology, used to imply a fundamental theoretic or physical significance, as opposed to one merely empirical or practical (as in absolute SPACE-TIME, TEMPERATURE scale, UNITS); to define an actual, as opposed to an apparent, relative or comparative, measurement (as in absolute HUMIDITY; MAGNITUDE), or to describe a limiting case (as in ABSOLUTE ZERO or, for an aircraft, absolute ceiling). In philosophy the term refers to what is unconditional, noncontingent, self-existent or even arbitrary. In 19th-century IDEALISM, the Absolute (Idea) came to refer to the ultimate cosmic totality.

ABSOLUTE ZERO, the TEMPERATURE at which all substances have zero thermal ENERGY and thus, it is believed, the lowest possible temperature. Although many substances retain some nonthermal ZERO-POINT ENERGY at absolute zero, this cannot be eliminated and so the temperature cannot be reduced further. Originally conceived as the temperature at which an ideal GAS at constant pressure would contract to zero volume, absolute zero is of great significance in THERMODYNAMICS, and is used as the fixed point for ABSOLUTE temperature scales. In practice the absolute zero of temperature is unattainable, although temperatures within a few millionths of a KELVIN of it have been achieved in CRYOGENICS laboratories.

$$0\,K = -273.16°C = -459.69°F$$

ABSORPTION, any process by which a substance incorporates another substance into itself, or takes in radiant or sound ENERGY. In chemistry it is distinguished from ADSORPTION in which the adsorbed substance merely adheres to the surface of the adsorbent. In AIR-POLLUTION control and the chemical industries gas absorption is a key process, while the varying extent to which different surfaces absorb SOUND is an important factor in ACOUSTIC design. According to QUANTUM THEORY, an atom or MOLECULE can absorb a PHOTON of ELECTROMAGNETIC RADIATION only if the quantum of energy so received raises it exactly from its present ENERGY LEVEL to a higher one, a fact of great significance for SPECTROSCOPY and for the theory of DYES and COLOR vision.

ABUNDANCE RATIO. See ISOTOPES.

AB-UNITS. See ABAMPERE.

ABYSSAL FAUNA, the animals inhabiting the oceans at depths greater than 1km. Here, temperatures range between 5°C and 1°C, the pressure approaches 600atm (at 6km depth) and daylight is absent. Abyssal animals are all highly specialized, some being parasites, others scavengers or predators. Some of these animals are blind; some sport bioluminescent (see LUMINESCENCE) lures and body panels.

ABYSSAL HILLS, small hills on the OCEAN floor, commonly flanking or projecting through the ABYSSAL PLAINS. They are thought to represent once volcanic seamounts now largely buried by sediment.

ABYSSAL PLAIN, a large flat area of the OCEAN floor, lying usually between 4km and 6km below the surface. These plains, which together cover some 40% of the earth's surface, are formed of thick layers of mud and other sediments.

ACANTHOCEPHALA, or spiny-headed worms, a phylum of worms parasitic (see PARASITE) on vertebrates such as fish and birds. Named for the proboscis, which bears tiny hooks that anchor them to the intestinal walls of their hosts, these worms are so degenerate that most have little more than a reproductive system and simple brain. They often cause fatal infections.

ACANTHODII, a subclass of the OSTEICHTHYES known only as fossils. They occur in rocks of the Silurian period, 440–400 million years ago, and are characterized by spines that supported the front edges of the fins.

ACCELERATION, the rate at which the VELOCITY of a moving body changes. Since velocity, a VECTOR quantity, is speed in a given direction, a body can accelerate both by changing its speed and by changing its direction. The units of acceleration, itself a vector quantity, are those of velocity per unit time—e.g., metres per second, per second (m/s^2). In calculus notation, acceleration **a** is the first differential of velocity **v** with respect to time t:

$$\mathbf{a}=\frac{d\mathbf{v}}{dt}$$

According to NEWTON's second law of MOTION, acceleration is always the result of a force acting on a body; the acceleration **a** produced in a body of mass m by a force **F** is given by $\mathbf{a}=\mathbf{F}/m$. The **acceleration due to gravity** (g) of a body falling freely near the earth's surface is about $9.81 m/s^2$. In the aerospace industry the accelerations experienced by men and machines are often expressed as multiples of g. Headward (vertical) accelerations of as little as $3g$ can cause pilots to black out (see SPACE MEDICINE).

ACCELERATORS, Particle, atomic research tools used to accelerate SUBATOMIC PARTICLES to high velocities. Their power is rated according to the kinetic energy they impart, measured in ELECTRON VOLTS (eV). Linear accelerators use electrostatic and electromagnetic fields to accelerate particles in a straight line, but greater energies are obtained by leading the particles around a spiral or circular path (see BETATRON; BEVATRON; CYCLOTRON; SYNCHROCYCLOTRON; SYNCHROTRON).

ACCELEROMETER, a device used to measure ACCELERATION, usually consisting of a heavy body of known mass free to move in only one dimension, in

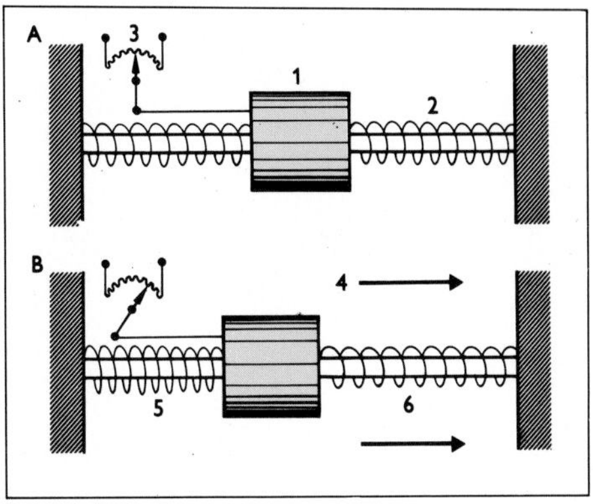

A linear accelerometer. If the device is moving at constant velocity the mass (1), supported on a bar by springs (2), remains static and an intermediate reading is registered on the potentiometer (3). On acceleration in the direction of the bar (4), i.e., along the accelerometer's sensitive axis, inertia causes the mass to "lag behind", compressing the spring behind it (5) and stretching the spring ahead of it (6): a high voltage is registered. On deceleration, inertia causes the mass to compress the spring ahead of it and stretch that behind it, and thus a low voltage is registered on the potentiometer.

The "Linac" electron linear accelerator at the UK's National Physical Laboratory, used in determining safety standard for X-rays and other ionizing radiations. Electrons produced in the electron gun assembly (*rear left*) are accelerated through the accelerating wave guides in the center of the picture.

which it is supported by springs. Any acceleration experienced along that line is computed from electrical measurements of the resulting distortion in the springs. Three such instruments set at mutual right angles are needed to measure accelerations in three dimensions.

ACCLIMATIZATION, the process of adjustment that allows an individual organism to survive under changed conditions. In a hot, sunny climate man acclimatizes by eating less, drinking more and wearing lighter clothes; furthermore, his skin may darken. At higher altitudes he can adjust to the diminished oxygen by increased production of red blood corpuscles. (See also ADAPTATION.)

ACCUMULATOR, or storage battery. See BATTERY.

ACETALDEHYDE (CH_3CHO), or ethanal, a colorless, flammable liquid, an ALDEHYDE, made by catalytic oxidation of ETHANOL. An important reagent, it is used in the manufacture of dyes, plastics and many other organic chemicals. In the presence of acids it forms the cyclic polymers paraldehyde, $(CH_3CHO)_3$, and metaldehyde, $(CH_3CHO)_4$. The former is used as a hypnotic, and the latter as a solid fuel for portable stoves and as a poison for snails and slugs.

ACETATES, compounds in which the acid hydrogen atom in ACETIC ACID is replaced either by a metal (forming soluble acetate SALTS containing the ion CH_3COO^-) or by an organic radical (giving a covalent ESTER such as ethyl acetate, $CH_3COOC_2H_5$). Acetate esters are of great commercial importance, being used as solvents and for the manufacture of plastics and fibers. The various cellulose acetates are used in sheet form as a base for photographic films and also as the well-known fibers, Celanese and Arnel.

Acetates

sodium acetate
MW 82.03, mp 324°C

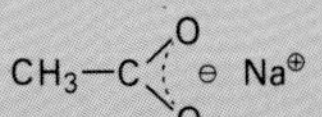

methyl acetate
MW 74.08,
mp −98°C, bp 57°C

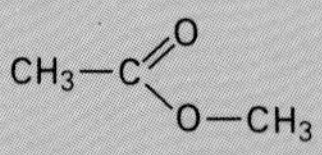

ethyl acetate
MW 88.12,
mp −84°C, bp 77°C

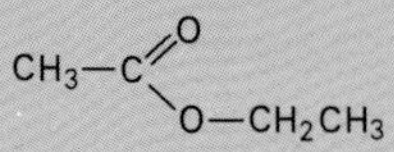

phenyl acetate
MW 136.16, bp 196°C

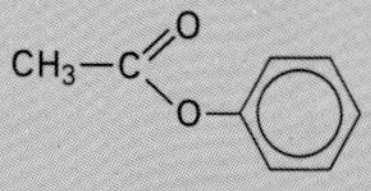

benzyl acetate
MW 150.18,
mp −52°C, bp 216°C

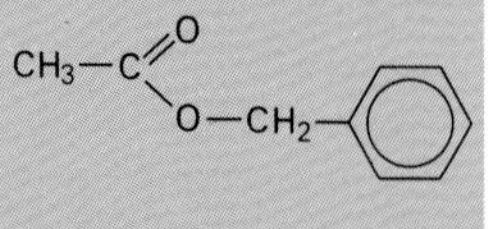

polyvinyl
acetate (PVA)

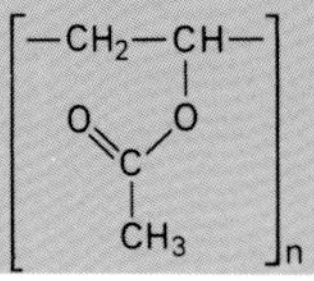

ACETIC ACID, or ethanoic acid (CH_3COOH), most important of the CARBOXYLIC ACIDS, is a pungent, colorless liquid used to make ACETATES. It is important in BIOSYNTHESIS. Acetic acid is produced by bacterial action on alcohol in air (yielding VINEGAR), and industrially by the oxidation of ACETALDEHYDE or ETHANOL. Pure ("glacial") acetic acid solidifies to ice-like crystals at 17°C; it is corrosive. MW 60.85, bp 118°C.

ACETONE (CH_3COCH_3), or 2-propanone, the simplest KETONE, a fragrant, colorless liquid used in industry as a solvent and for organic synthesis. It is prepared by fermentation of starch or dehydrogenation of 2-propanol (see ALCOHOLS). Large amounts of acetone occur in diabetics (see DIABETES; ACIDOSIS). MW 58.08, mp −95°C, bp 56°C.

ACETYLCHOLINE (ACh), a substance playing an important role in the transmission of nerve impulses within the NERVOUS SYSTEM. On release from the end of one nerve fiber, it stimulates the adjoining one before being rapidly broken down by the enzyme cholinesterase to choline and ACETIC ACID, of which it is the ester.

acetylcholine

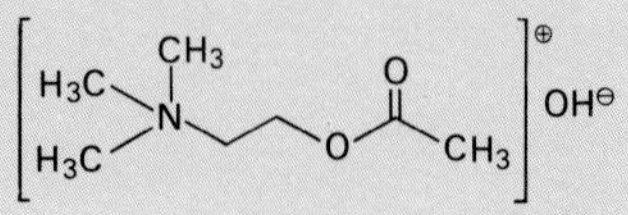

ACETYLENE, or ethyne (CHCH), the simplest ALKYNE; a colorless, flammable gas prepared by reaction of water and calcium carbide (or acetylide; see CARBON); it is a very weak ACID. Acetylene may explode when under pressure, so is stored dissolved in acetone. It is used in the oxyacetylene torch for cutting and WELDING metals, in lamps, and in the synthesis of ACETALDEHYDE, VINYL compounds, neoprene rubbers and various solvents and insecticides. MW 26.04, subl −84°C.

ACETYLSALICYLIC ACID. See ASPIRIN.

ACHENE, dry indehiscent FRUIT with one seed, as in the buttercup.

ACHESON, Edward Goodrich (1856–1931), US inventor who discovered the powerful abrasive, CARBORUNDUM, and devised a method for producing high-quality GRAPHITE. Earlier he had assisted EDISON in developing the incandescent filament lamp.

ACHEULIAN, a lower Paleolithic (see STONE AGE) culture, prevalent about 430 000–130 000 years ago, whose remains were first discovered near St. Acheul (France). The type-tool is a flint hand ax.

ACHILLES TENDON, the TENDON from the calf muscles to the heel bone, important in standing, walking and running. In the Achilles myth, it was the warrior's only vulnerable part.

ACHROMATIC LENS, or achromat, a compound LENS designed to minimize the effects of chromatic ABERRATION, which is due to the dispersion of light in REFRACTION. This is done by employing lens elements made of glasses of different dispersive powers, traditionally crown glass and flint glass. In practice a given lens can be correctly balanced only for a few wavelengths.

ACID, a substance capable of providing HYDROGEN

ions (H^+) for chemical reaction. In an important class of chemical reactions (acid-base reactions) a hydrogen ION (identical to the physicist's PROTON) is transferred from an acid to a BASE, this being defined as any substance which can accept hydrogen ions. The strength of an acid is a function of the availability of its acid protons (see PH). Free hydrogen ions are available only in SOLUTION where the minute proton is stabilized by association with a solvent molecule. In aqueous solution it exists as the hydronium ion (H_3O^+).

Chemists use several different definitions of acids and bases simultaneously. In the Lewis theory, an alternative to the Brönsted-Lowry theory outlined above, species which can accept ELECTRON pairs from bases are defined as acids.

Many chemical reactions are speeded up in acid solution, giving rise to important industrial applications (acid-base CATALYSIS). Mineral acids including SULFURIC ACID, NITRIC ACID and HYDROCHLORIC ACID find widespread use in industry. Organic acids, which occur widely in nature, tend to be weaker. CARBOXYLIC ACIDS (including ACETIC ACID and OXALIC ACID) contain the acidic group –COOH; aromatic systems with attached hydroxyl group (PHENOLS) are often also acidic. AMINO ACIDS, constitutive of proteins, are essential components of all living systems.

Acid anhydrides

ethanoic anhydride (acetic anhydride) MW 102.09, mp −73°C, bp 140°C	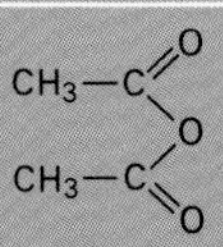
propanoic anhydride (propionic anhydride) MW 130.15, mp −45°C, bp 168°C	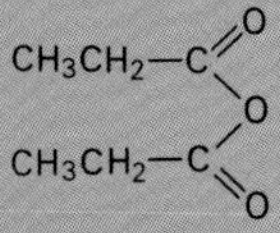
butanoic anhydride (butyric anhydride) MW 158.20, mp −75°C, bp 200°C	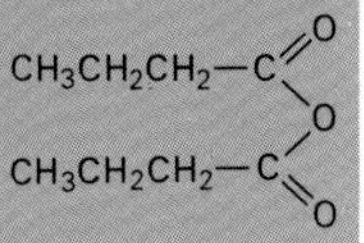
succinic anhydride MW 100.08, mp 120°C, bp 261°C	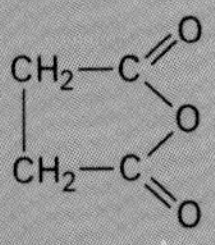
glutaric anhydride MW 114.10, mp 57°C	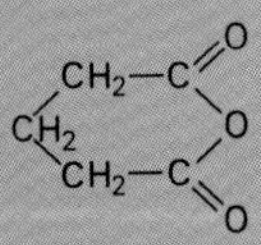
benzoic anhydride MW 226.24, mp 42°C, bp 360°C	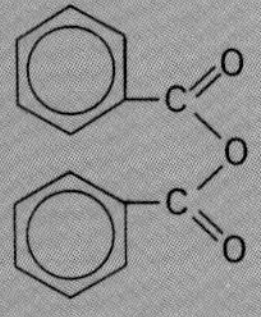
phthalic anhydride MW 148.12, mp 132°C, bp 285°C	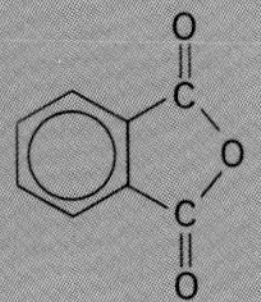

ACID ANHYDRIDES, class of organic compounds derived formally (but not in practice) from CARBOXYLIC ACIDS by elimination of water, of general formula RCOOCOR′; prepared by reaction of an ACID CHLORIDE with a carboxylate. Chemically they resemble acid chlorides, but are less violently reactive. They are used in the FRIEDEL-CRAFTS and DIELS-ALDER reactions, and to make ESTERS.

ACID CHLORIDES, or acyl chlorides, class of organic compounds of general formula RCOCl; prepared by reacting CARBOXYLIC ACIDS with phosphorus (III or V) chloride or thionyl chloride ($SOCl_2$). They are volatile, fuming, pungent liquids, corrosive and very reactive. They react with ALCOHOLS to give ESTERS, with AMMONIA and AMINES to give AMIDES, and with water to give carboxylic acids. They are reduced to KETONES by GRIGNARD REAGENTS and in the FRIEDEL-CRAFTS REACTION. Other acid HALIDES are similar. (See also ALDEHYDES; ACID ANHYDRIDES.)

Acid chlorides

ethanoyl chloride (acetyl chloride) MW 78.50, mp −112°C, bp 51°C	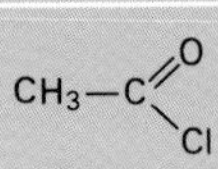
propanoyl chloride (propionyl chloride) MW 92.53, mp −94°C, bp 80°C	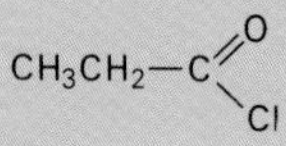
butanoyl chloride (*n*-butyl chloride) MW 106.55, mp −89°C, bp 102°C	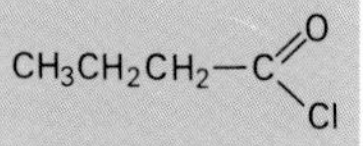
benzoyl chloride MW 140.57, mp 0°C, bp 197°C	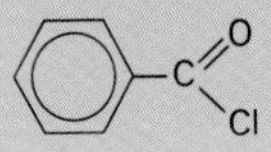

ACIDOSIS, medical condition in which the acid-base balance in the blood PLASMA is disturbed in the direction of excess acidity, the pH falling below 7.35. It may cause deep sighing breathing and drowsiness or coma. Respiratory acidosis, associated with lung disease, heart failure and central respiratory depression, results from underbreathing and a consequent buildup of plasma CARBON dioxide. Alternative metabolic causes include the ingestion of excess acids (as in ASPIRIN overdose), KETOSIS (resulting from malnutrition or diabetes), heavy alkali loss (as from a FISTULA) and the inability to excrete acid which occurs in some KIDNEY disorders. (See also ALKALOSIS.)

ACNE, a common pustular SKIN disease of the face

An anechoic chamber used by the UK's Central Electricity Generating Board for studies into the effects of industrial noise and fundamental acoustical research. The rubber wedges on all the internal surfaces absorb all the sound energy reaching them.

and upper trunk, most prominent in ADOLESCENCE. BLACKHEADS become secondarily inflamed due either to local production of irritant FATTY ACIDS by BACTERIA or to bacterial infection itself. In severe cases, with secondary infection and picking of spots, scarring may occur. Acne may be aggravated by diet (chocolate and nuts being worst offenders), by HORMONE imbalance, by greasy skin or by poor hygiene. Methods of treatment include degreasing the skin, removing the blackheads, controlling diet or hormones, and exposure to ULTRAVIOLET RADIATION. TETRACYCLINES may be used to decrease fatty acid formation.

ACORN WORMS. See HEMICHORDATA.

ACOUSTICS, the science of SOUND, dealing with its production, transmission and effects. Engineering acoustics deals with the design of sound-systems and their components, such as MICROPHONES, headphones and LOUDSPEAKERS; musical acoustics is concerned with the construction of musical instruments, and ULTRASONICS studies sounds having frequencies too high for men to hear them. Architectural acoustics gives design principles of rooms and buildings having optimum acoustic properties. This is particularly important for auditoriums, where the whole audience must be able to hear the speaker or performers clearly and without ECHOES. Also, the **reverberation time** (the time taken for the sound to decay to one millionth of its original intensity) must be matched to the intended uses of the hall; for speech it should be less than 1s; for chamber music between 1s and 2s; for larger scale works, from 2s to 3.5s. All this is achieved by attending to the geometry and furnishings of the hall and incorporating the appropriate sound-absorbing, diffusing and reflecting surfaces. **Anechoic chambers,** used for testing acoustic equipment, are completely surfaced with diffusing and absorbing materials so that reverberation is eliminated. **Noise insulation engineering** is a further increasingly important branch of acoustics.

ACQUIRED CHARACTERISTICS, modifications in an organism resulting from interaction with its environment. In 1801 LAMARCK proposed an evolutionary theory in which the assumption that acquired characteristics could be inherited provided the mechanism for species divergence. In later editions of *The Origin of Species,* DARWIN moved towards accepting this explanation in parallel to that of NATURAL SELECTION, but eventually the Lamarckian mechanism was entirely discounted. It is now thought, however, that organisms which reproduce asexually (see REPRODUCTION) can pass on acquired characteristics. (See also ADAPTATION; EVOLUTION.)

ACROPHOBIA, a morbid fear of heights, sometimes associated with physical symptoms; often an isolated PHOBIA in otherwise normal people.

ACTH (Adrenocorticotrophic Hormone), or corticotropin, a HORMONE secreted by the PITUITARY GLAND which stimulates the secretion of various STEROID hormones from the cortex of the ADRENAL GLANDS. ACTH has been used in the treatment of a number of diseases including MULTIPLE SCLEROSIS.

ACTINIDES, the 15 elements with atomic numbers (see ATOM) 89–103, beginning with ACTINIUM, analogous to the LANTHANUM SERIES, though rather more diverse in properties. They are separated by ion-exchange CHROMATOGRAPHY. The elements through uranium occur in nature; except actinium, they show higher valencies than +3. The TRANSURANIUM ELEMENTS are synthetic; the +3 valence state becomes progressively more stable, and higher valencies less stable. (See also PERIODIC TABLE.)

Actinides

element	Z	outermost electrons	valencies
actinium (Ac)	89	$6d^1 7s^2$	3
thorium (Th)	90	$6d^2 7s^2$	4
protactinium (Pa)	91	$5f^2 6d^1 7s^2$	4,5
uranium (U)	92	$5f^3 6d^1 7s^2$	3,4,5,6
neptunium (Np)	93	$5f^4 6d^1 7s^2$	3,4,5,6,7
plutonium (Pu)	94	$5f^6 7s^2$	3,4,5,6,7
americium (Am)	95	$5f^7 7s^2$	3,4,5,6
curium (Cm)	96	$5f^7 6d^1 7s^2$	3,4
berkelium (Bk)	97	$5f^9 7s^2$	3,4
californium (Cf)	98	$5f^{10} 7s^2$	2,3
einsteinium (Es)	99	$5f^{11} 7s^2$	2,3
fermium (Fm)	100	$5f^{12} 7s^2$	2,3
mendelevium (Md)	101	$5f^{13} 7s^2$	2,3
nobelium (No)	102	$5f^{14} 7s^2$	2,3
lawrencium (Lr)	103	$5f^{14} 6d^1 7s^2$	3

ACTINIUM (Ac), radioactive TRANSITION ELEMENT in Group IIIB of the PERIODIC TABLE, resembling LANTHANUM; it occurs in URANIUM ores, and Ac^{227} (half-life 22yr) is formed by irradiation of RADIUM. It is the prototypical member of the ACTINIDES. (For the **Actinium Series**, see RADIOACTIVITY.) AW 227, mp 1050°C, sg 10.

ACTINOMYCETES, a large group of filamentous mold-like bacteria found in all types of soil. They help maintain soil fertility by their action in breaking down organic matter, and they are valuable as a source of ANTIBIOTICS such as STREPTOMYCIN.

ACTINOPTERYGII, or ray-finned fishes, a subclass of the OSTEICHTHYES that have fins supported by bony rays which lack lobes in the bases of the paired fins. Primitive actinopterygians are included in the **Chondrostei** and have asymmetrical tails, heavy scales and an upper jawbone that is fused to the bones of the cheek. They were abundant during the Carboniferous period, 345–280 million years ago. **Holosteans** (the Holostei) replaced chondrosteans during the Triassic period, about 220 million years ago, and show characteristics intermediate between those of chondrosteans and **teleosts** (the Teleostei). The latter group make up the vast majority of present-day fish populations in lakes, rivers and oceans. Teleosts have symmetrical tails, thin scales and an upper jawbone that is free from the cheek and adapted to a great variety of feeding habits. They number about 25000 species, many of which are important sources of food for man.

ACTIVATION ENERGY, the ENERGY which must be supplied to molecules to enable them to react together, usually obtained from their kinetic (thermal) energy. It represents the energy above the ground state of a transient activation complex which decomposes to give the products. (See also KINETICS, CHEMICAL.)

ACUPUNCTURE, an ancient Chinese medical practice in which fine needles are inserted into the body at specified points, used for relieving pain and in treating a variety of conditions including MALARIA and RHEUMATISM. It was formerly believed that this would correct the imbalance between the opposing forces of yin and yang in the body which lay behind the symptoms of sickness. Although it is not yet understood how acupuncture works, it is still widely practiced in China and increasingly in the West, mainly as a form of ANESTHESIA.

ACUTE, term descriptive of an ANGLE of less than 90°.

ADAMS, John Couch (1819–1892), British astronomer who, independently of LEVERRIER, inferred the existence of the planet NEPTUNE from the PERTURBATIONS it induced in the orbit of URANUS.

ADAPTATION, the process of modification of the form or functions of a part of an organism, to fit it for its environment and so to achieve efficiency in life and reproduction. Adaptation of individual organisms is called ACCLIMATIZATION, and is temporary since it involves ACQUIRED CHARACTERISTICS; the permanent adaptation of species arises from transmitted genetic variations preserved by NATURAL SELECTION (see also EVOLUTION). Successful and versatile adaptation in an organism usually leads to widespread distribution and long-term survival. Examples include the development of lungs in amphibians, and of wings in birds and insects. The term is sometimes also used for the modified forms of the organism.

ADAPTIVE RADIATION, a sequence of EVOLUTION in which an unspecialized group of organisms gives rise to various differentiated types adapted to specific modes of life. Early placental mammals, for example, gave rise to modern burrowing, climbing, flying, running and swimming forms.

ADDEND, in ALGEBRA, one of two or more terms undergoing ADDITION. For example, in the expression $x+y+z$, each of x, y, z is an addend.

ADDICTION. See DRUG ADDICTION.

ADDISON'S DISEASE, failure of STEROID production by the ADRENAL GLAND cortex, first described by English physician **Thomas Addison** (1793–1860). Its features include brownish skin pigmentation, loss of appetite, nausea and vomiting, weakness, and malaise and faintness on standing. The stress associated with an infection or an operation can lead to sudden collapse. Autoimmune disease (see IMMUNITY), TUBERCULOSIS and disseminated CANCER may damage the adrenals and long-term steroid therapy may suppress normal production. Treatment is normally by steroid replacement.

ADDITION, one of the basic operations of ALGEBRA and ARITHMETIC, denoted by the sign "+". The addition of positive numbers can best be defined in terms of SET THEORY: if one considers set A to contain 4 elements, set B to contain 5 elements, then $A \cup B$ contains 9 elements; i.e., $4+5=9$. The addition of negative numbers is equivalent to SUBTRACTION in that $a+(-b) \equiv a-b$.

ADENOIDS, lymphoid tissue (see LYMPH) draining the nose, situated at the back of the throat. They are normally largest in the first five years and by adult life have undergone ATROPHY. Excessive size resulting from repeated nasal infection may lead to mouthbreathing, middle-ear diseases, sinusitis and chest infection. If these are prominent or persistent complications, surgical removal of the adenoids may be needed.

ADENOSINE TRIPHOSPHATE (ATP). See NUCLEOTIDES.

ADHESION, the force of attraction between contacting surfaces of unlike substances, such as glue and wood or water and glass. Adhesion is due to intermolecular forces of the same kind as those causing COHESION. Thus the force depends on the nature of the materials, temperature and the pressure between the surfaces. A liquid in contact with a solid surface will "wet" it if the adhesive force is greater than the cohesive force within the liquid. (See also ADHESIVES; SOLDERING; SOLUTION.)

ADHESIVES, substances that bond surfaces to each other by mechanical ADHESION (the adhesive filling the pores of the substrate) and in some cases by chemical reaction. Thermoplastic adhesives (including most animal and vegetable glues) set on cooling or evaporation of the solvent. Thermosetting adhesives (including the epoxy resins) set on heating or when mixed with a catalyst. There are now many strong, long-lasting adhesives designed for use in such varied fields as electronics, medicine, house-building and bookbinding, and for bonding plastics, wood and rubber. (See also CEMENT, GLUE, SOLDERING.)

ADIABATIC PROCESS, in THERMODYNAMICS, a

The permanent tack of a new adhesive under test.

change in a system without transfer of HEAT to or from the environment. An example of an adiabatic process is the vertical flow of air in the atmosphere; air expands and cools as it rises, and contracts and grows warmer as it descends. The generation of heat when a gas is rapidly compressed, as in a piston engine or SOUND waves, is approximately adiabatic.

ADIPOSE TISSUE, specialized fat-containing connective TISSUE, mainly lying under the skin and within the ABDOMEN, whose functions include FAT storage, energy release and insulation. In individuals its distribution varies with age, sex and OBESITY.

ADLER, Alfred (1870–1937), Austrian psychiatrist who broke away from FREUD to found his own psychoanalytic school, "individual psychology," which saw AGGRESSION as the basic drive. Adler emphasized the importance of feelings of inferiority in individual maladjustments to society.

ADOBE (from Spanish *adobar*, to plaster), sun-dried brick made from clay soil mixed with grass or straw. The earliest building material of Egypt and Assyria, it is still used in China, Japan and particularly Mexico, Central America and the SW states of the US.

ADOLESCENCE, in humans, the transitional period between childhood and adulthood. The term has no precise biological meaning, but adolescence is generally considered to start with the onset of PUBERTY and to end at the age of about 20. In primitive societies the period is marked by RITES OF PASSAGE such as that at puberty and that on MARRIAGE. These formal rites are reflected less overtly in more sophisticated societies, and this lack of formalization, where the individual is expected to adjust to standards which he does not fully appreciate, is believed to be responsible, as much as physiological (see PHYSIOLOGY) and hormonal (see HORMONES) changes, for adolescent emotional stresses. The frustration by society, through parental or other disapproval, of the adolescent's sex drive, which during this period is exceptionally strong, causes further stresses. Partial outlets are found through masturbation (sexual stimulation by oneself), diary-writing, artistic creativity, political militancy and, frequently, vandalism. These emotional stresses, coupled with those caused by differences in the ages at which physical developments take place in different individuals (see INFERIORITY COMPLEX), create enormous educational and social problems that are often underestimated.

ADONIS, asteroid about one mile in diameter with a highly eccentric ORBIT. Its perihelion is within the orbit of Venus, its aphelion beyond that of Mars. It was discovered in 1936.

ADRENAL GLANDS, or **Suprarenal Glands,** two ENDOCRINE GLANDS, one above each kidney. The inner portion (medulla) produces the hormones ADRENALINE and noradrenaline and is part of the autonomic NERVOUS SYSTEM. The outer portion (cortex), which is regulated by ACTH, produces a number of STEROID HORMONES which control sexual development and function, glucose metabolism and electrolyte balance. Adrenal cortex damage causes ADDISON'S DISEASE.

ADRENALINE, or **Epinephrine,** a HORMONE secreted by the ADRENAL GLANDS, together with smaller quantities of **Noradrenaline.** The nerve endings of the sympathetic NERVOUS SYSTEM also secrete both hormones, noradrenaline in greater quantities. They are similar chemically and in their pharmacological effects. These constitute the "fight or flight" response to stress situations: blood pressure is raised, smaller blood-vessels are constricted, heart rate is increased, METABOLISM is accelerated, and levels of blood glucose and FATTY ACIDS are raised. Adrenaline is used as a heart stimulant, and to treat serious acute ALLERGIES.

adrenaline

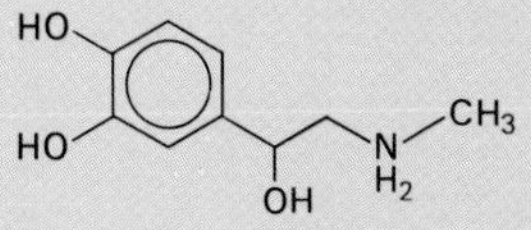

ADRIAN, Edgar Douglas. 1st Baron Adrian of Cambridge (1889–), English physiologist who shared the 1932 Nobel Prize for Physiology or Medicine with Charles SHERRINGTON for work elucidating the functioning of the neurons of the NERVOUS SYSTEM.

ADSORPTION, the ADHESION of molecules of a fluid (the adsorbate) to a solid surface (the adsorbent); the degree of adsorption depends on temperature, pressure and the surface area—porous solids such as CHARCOAL being especially suitable. The forces binding the adsorbate may be physical or chemical; chemical adsorption is specific, and is used to separate mixtures (see CHROMATOGRAPHY). Adsorption is used in GAS MASKS and to purify and decolorize liquids. (See also ABSORPTION.)

AEDES, a genus of MOSQUITO that includes the species that carries YELLOW FEVER, *Aedes aegypti*. Other species live in the Arctic and one in hot volcanic pools. Family: Culicidae.

AEGEAN CIVILIZATION, a collective term for the BRONZE AGE civilizations surrounding the Aegean Sea, usually extended to include the preceding STONE AGE cultures there. Early archaeological work in the area was performed by Heinrich SCHLIEMANN in the 1870s–80s, whose successes included the location of Troy, and early in this century by Sir Arthur EVANS. The Bronze Age cultures of the Aegean have been identified as follows: **Helladic**, the cultures of the Greek mainland, including subdivisions such as Macedonian; **Cycladic**, the cultures not only of the Cyclades but of all the Aegean Islands except Crete; and **Minoan**, the cultures of Crete, so named by Evans for Minos, in legend the most powerful of Cretan kings. The Late Helladic cultures are often termed **Mycenaean.**

Around 3000 BC the region was invaded by Chalcolithic (i.e. bronze- and stone-using) peoples, displacing the previous Neolithic inhabitants. This population appears to have remained static until around 2000 BC, when the Greek tribes arrived on the mainland, overpowering and submerging the previous cultures. Around the same time Crete established a powerful seafaring empire, and throughout the area there were rapid and substantial advances in the arts, technology and social organization. Around 1550 BC it would appear that the Mycenaeans occupied Crete, and certainly by this time the Greeks were established as the dominant culture in the area. The Cretan civilization seems to have been eclipsed about 1400 BC. During the 17th century BC there emerged on the mainland a wealthy and powerful aristocracy, whose riches have been discovered in many of their tombs. It would appear that for several hundred years there was a period of stability, since fortifications were not added to the aristocrats' palaces until the 13th century BC. The artistry of this era is exquisite, as evidenced by archaeological discoveries in the tombs: gold cups superbly wrought, small sculptures, jewelry, dagger blades inlaid with precious metals, and delicate frescoes. During the 13th century BC there probably was a war with Troy, ending with the destruction of that city around 1260 BC and a general decline of the civilizations as a whole into the so-called Dark Ages. (See also MINOAN LINEAR SCRIPTS.)

AERIAL PHOTOGRAPHY. See PHOTOGRAMMETRY.

AEROBE, an organism that needs oxygen for its survival. The term is usually applied to certain kinds of BACTERIA. (See also ANAEROBE.)

AERODYNAMIC EFFICIENCY, the EFFICIENCY with which an AIRFOIL uses the AERODYNAMIC forces acting on it: in particular the ratio of lift to drag.

AERODYNAMICS, the branch of physics dealing with the flow of air or other gas around a body in motion relative to it. Aerodynamic forces depend on the body's size, shape and velocity; and on the density, compressibility, VISCOSITY, temperature and pressure of the gas. At low velocities, flow around the body is streamlined or laminar, and causes low drag; at higher velocities TURBULENCE occurs, with fluctuating eddies, and drag is much greater. "Streamlined" objects, such as AIRFOILS, are designed to maintain laminar flow even at relatively high velocities. Pressure impulses radiate at the speed of SOUND ahead of a moving body; at SUPERSONIC velocities these impulses pile up, producing a shock wave—the "sonic boom" (see DOPPLER EFFECT). In AIRPLANE design all of these factors must be considered. In normal cruising flight all the forces acting on an airplane must balance. The lift provided by the wings must equal the aircraft's weight; the forward thrust of the engine must balance the forces of drag. Lift occurs because the wing's upper surface is more convex, and therefore longer, than the lower surface. Air must therefore travel faster past the upper surface than past the lower, which leads to reduced pressure above the wing. (See also WIND TUNNEL; REYNOLDS NUMBER.)

AEROEMBOLISM, presence of air in the blood circulation. Direct entry of air into veins may occur through trauma, cannulation or surgery, and a large air embolus reaching the HEART may cause death. In acute decompression (as with flying to high altitude or sudden surfacing after deep diving) bubbles of air come out of solution. These may block small blood vessels causing severe muscle pains ("**bends**"), tingling and choking sensations and occasionally PARALYSIS or COMA. Recompression and slow decompression is the correct treatment.

AERONAUTICS, the technology of aircraft design, manufacture and performance. See AERODYNAMICS;

Of fundamental significance in aerodynamics is the airfoil shape. The arrows (1) show the relative motion of the air through which the airfoil (2) is moving. Air travels past the upper edge faster than past the lower edge, so that there is a region of low pressure (6) above the airfoil and one of high pressure (7) below. This results in the airfoil experiencing vertical *lift* (3) at right angles to the *drag* force (4). The resultant of the lift and the drag forces is the *aerodynamic force* (5).

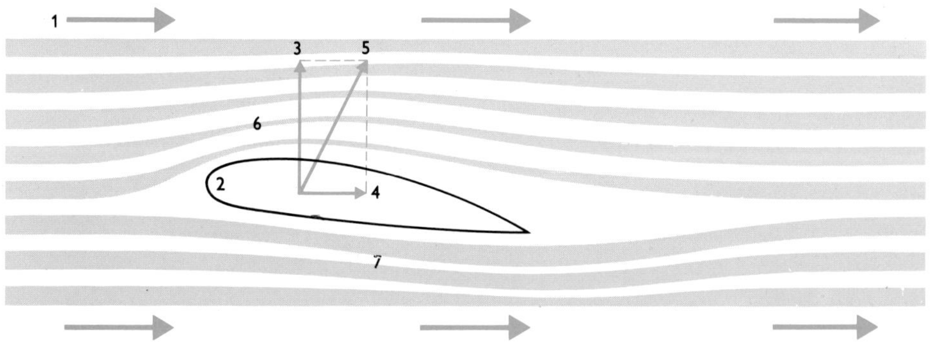

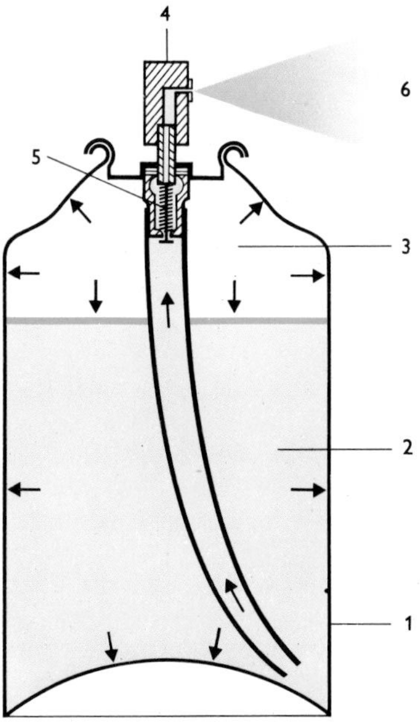

A commercial aerosol can. Pressure is exerted on all walls of the strengthened metal container (1) by both the charge, (2) in which is dissolved some of the propellant, and the propellant in its vapor phase (3). On depression of the button (4) the valve (5) opens, and the liquid is forced up the standpipe and through a fine nozzle in the button.

AIR-CUSHION VEHICLE; AIRPLANE; AIRSHIP; BALLOON; FLIGHT, HISTORY OF; GLIDER; HELICOPTER.

AEROSOL, a suspension of small liquid or solid particles (0.1–100μm diameter) in a gas. Examples include smoke (solid particles in air), FOG and CLOUDS. Aerosol particles can remain in suspension for hours, or even indefinitely. Commercial aerosol sprays are widely used for insecticides, air fresheners, paints, cosmetics, etc. (See also COLLOID; ATMOSPHERE.)

AEROSPACE MEDICINE. See SPACE MEDICINE.

AESCULAPIUS, the Roman god of healing and medicine. In Greek myth he was known as Asclepius (or Asklepios), the son of Apollo and Coronis, and learned the art of healing from Chiron the centaur. His symbol was a snake entwined around a staff.

AESTIVATION. See HIBERNATION.

AFFECT, in psychology. See EMOTION.

AFTERBIRTH, the material, primarily the PLACENTA, expelled from the mother's body after child delivery.

AFTERBURNER, device used in AIRPLANE turbojet engines (see JET PROPULSION) during periods of takeoff, climb or dash to increase thrust by as much as 100% or more by the burning of additional fuel.

AFTERGLOW, the radiant coloration of the western sky following sunset. It arises from the SCATTERING of sunlight by dust particles (see COLLOID) in the upper atmosphere. The term is also used as an alternative name for phosphorescence (see LUMINESCENCE).

AFTERIMAGE, the persistence of a visual image after the original image has gone. It may be positive (the same COLOR or shade as the original) or negative (the complementary color) depending on the background color. It is a perceptual illusion due to the differential stimulation of the EYE's retinal receptors.

AGAR, or agar-agar, gelatinous product prepared from the ALGAE *Gracilaria* and *Gelidium*. It dissolves in hot water, the solution gelling on cooling (see COLLOID). Its main use is as a thickener in bacteria culture mediums and in cooking.

AGASSIZ, Jean Louis Rodolphe (1807–1873), Swiss-American naturalist, geologist and educator, who first proposed (1840) that large areas of the northern continents had been covered by ice sheets (see ICE AGE) in the geologically recent past. He is also noted for his studies of fishes. Becoming natural history professor at Harvard in 1848, he founded the Museum of Comparative Zoology there in 1859. On his death he was succeeded as its curator by his son, **Alexander Agassiz** (1835–1910).

AGASSIZ, Lake, a large prehistoric lake which covered parts of N.D., Minn., Manitoba, Ontario and Saskatchewan in the PLEISTOCENE epoch, named for Louis AGASSIZ. It was formed by the melting ice sheet as it retreated (see ICE AGE). When all the ice had melted, the lake drained northward, leaving fertile silt.

AGATE, a gemstone, a variety of CHALCEDONY streaked with bands of color, formed by intermittent depositions of mineral or organic matter, usually SILICA, on the walls of cavities in volcanic rock. It is often used in jewelry and ornamental work, usually after having been dyed.

AGE OF REASON. See ENLIGHTENMENT, THE.

AGE SET, a compulsory grouping by age practiced by certain peoples of N and NE Africa. Each male of the TRIBE is enlisted either at birth or at a particular age in an age set, in which he will remain for the rest of his life. As he and the rest of his set grow older, the set adopts different functions within the tribe. The number of sets within a tribe is limited, their names recurring in cycles.

AGGLOMERATE, rock made up of angular

A cross-section of an agate (center) showing parallel bands, and two smaller agates, cut and polished.

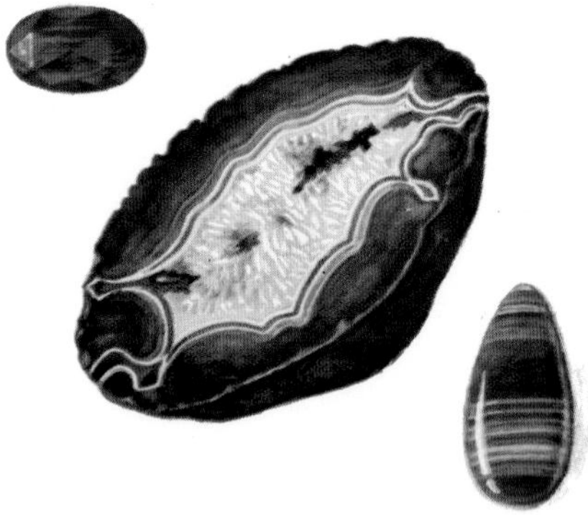

fragments of lava in a matrix of smaller, often ashy, particles. It is a result of volcanic activity (see VOLCANO). (See also CONGLOMERATE.)

AGGLUTINATIVE LANGUAGES (from Latin *gluten,* glue), LANGUAGES (e.g. Turkish) in which words are formed by joining together groups of MORPHEMES (individual meaning elements), so that a single word may convey the sense of a complete English clause. Their words do not undergo INFLECTION.

AGGLUTININS, ANTIBODIES found in BLOOD plasma which cause the agglutination (sticking together) of antigens such as foreign red blood cells and bacteria. Each agglutinin acts on a specific antigen, removing it from the blood. An agglutinin is produced in large quantities after immunization with its particular antigen. Agglutinins which agglutinate red blood cells are called isohemagglutinins, and the blood group of an individual is determined by which of these are present in his blood. Group O blood contains isohemagglutinins anti-A and anti-B; group A contains anti-B; group B contains anti-A, and group AB contains neither.

AGGREGATE, building material which is mixed with an adhesive such as CEMENT to make CONCRETE and MORTAR. Fine aggregate, usually sand or crushed stone, is used for smaller structural members where a smooth surface is required; coarse aggregate, such as pebbles, for larger members.

AGGREGATES, Theory of. See SET THEORY.

AGGREGATION, in ECOLOGY, the grouping together of plants or animals in response to environmental as opposed to behavioral stimuli.

AGGREGATION, in physics, the clustering of particles into larger groups or aggregates such as those forming rigid bodies.

AGGRESSION, behavior adopted by animals, especially vertebrates, in the defense of their territories and in the establishment of social hierarchies. An animal's aggressive behavior is usually directed towards members of its own species, but it is possible that the behavior of predators, although not generally regarded as aggression, may be controlled by the same mechanism. Aggressive behavior is commonly ritualized, the combatants rarely inflicting serious wounds upon one another. Ritual fighting has become established by the evolution of a language of signs, such as the threat posture, by which animals make known their intentions. Equally as important are submission or appeasement postures, which signal that one combatant acknowledges defeat.

It has been claimed in recent years that such signs are particularly well developed in man, and that he is unique in having aggressive tendencies which have led to the extermination of large numbers of his own species. Detractors from such claims point out that comparisons between social and political situations and those occurring in animal populations are invalid or, at best, misleading.

AGING, the process of progressive degeneration that occurs in an organism by which it becomes more liable to die. The rate of aging is more or less constant for a given species so that the life span of any individual is determined within limits.

AGNATHA, members of the phylum CHORDATA that are distinguished from the GNATHOSTOMATA by their lack of jaws. They include two living groups, the lampreys and hagfishes, and a number of fossil forms that were common during the Ordovician, Silurian and Devonian periods, 500–350 million years ago. The modern lampreys are eel-like and feed on fish by attaching themselves to their prey and sucking their body fluids. Hagfishes live by feeding on dead or dying fishes and are, unlike the lampreys, exclusively marine.

AGONISTIC BEHAVIOR, behavior resulting from conflict between an animal's instinct to fight and its instinct to flee when it is confronted by a member of its own species at the border of its territory. Agonistic behavior often results in would-be combatants presenting their flanks to one another, thus avoiding both threat and submissive postures.

AGORAPHOBIA, a morbid fear of public places, a pathological PHOBIA for the unfamiliar. The sufferers, commonly young women, may be unable to leave the home. Behavior therapy may be successful.

AGRAMONTE, Aristides (1869–1931), Cuban physician and bacteriologist, a member of the US Army Yellow Fever Commission headed by Walter REED which in 1900 proved that YELLOW FEVER can be transmitted only by certain mosquitos.

AGRANULOCYTOSIS, a BLOOD condition in which there are inadequate numbers of granulocytes, white cells that eliminate BACTERIA. It is a rare complication of certain drugs and systemic diseases. Severe throat infections and SEPTICEMIA are common; ANTIBIOTICS are the mainstay of treatment.

AGRICOLA, Georgius, Georg Bauer (1494–1555), German physician and scholar, "the father of mineralogy." His pioneering studies in geology, metallurgy and mining feature in his *De natura fossilium* (1546) and *De re metallica* (1556).

AGRONOMY, the branch of agricultural science dealing with production of field crops and management of the SOIL. The agronomist studies crop diseases, selective breeding, crop rotation and climatic factors. He also tests and analyzes the soil, investigates SOIL EROSION and designs LAND RECLAMATION and IRRIGATION schemes.

AILERONS, the control surfaces on the outer trailing edges of an AIRPLANE's wings, moved in opposite senses (one up, the other down) to cause or correct roll, particularly in banking (tilting the plane into a turn). When combined with the ELEVATORS (as in *Concorde*) they become "elevons."

AINU, the primitive hunting and fishing AUSTRALOID Japanese ABORIGINES. They are distinguished by stockiness, pale skins and profuse body hair, hence their frequent description as "the Hairy Ainu." Ainu speech, little used now, bears no relation to any other language. They are now few in numbers, many having been absorbed into ordinary Japanese society.

AIR. See ATMOSPHERE.

AIR BLADDER, or swim bladder, organ found in modern bony fishes that has evolved from the lungs of earlier forms. Control of the gas pressure inside the bladder enables the fish to remain buoyant in water at any depth.

AIR BRAKE, on trains, the fail-safe brake patented by George WESTINGHOUSE in 1872, which is released when the engineer allows compressed air to enter the train air line and applied when the pressure is released. On many trucks and buses, the air brake is

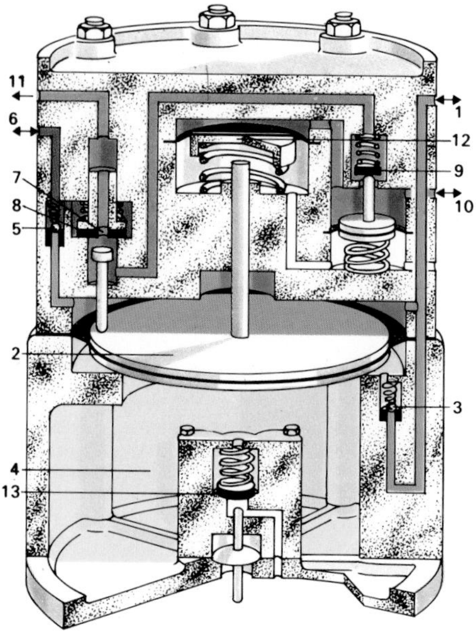

Greatly simplified diagram of an air brake distributor as used on railroad rolling stock, based on a design by the Westinghouse Brake and Signal Co. Ltd. Brake distributors are the devices which control the pressure in the brake cylinders along the train according to the way in which the engineer controls the pressure in the train brake pipe. In order to charge the system, the engineer allows the normal maximum pressure (about $5kg/cm^2$) to prevail in the train pipe (1). Air thus enters the area (blue) above the main diaphragm (2) and flows through non-return valve (3) into the control chamber (4—thus providing a reference pressure). It also flows through non-return valve (5) to charge the auxiliary reservoir (6—red). To apply the brakes, the engineer allows the pressure in the train pipe to drop (to about $3.4kg/cm^2$). The pressure difference across the main diaphragm moves it upward, closing the exhaust port (7) of the inlet/exhaust valve (8) but allowing air from the auxiliary reservoir (red) to pass through this valve and the limiting valve (9) to the brake cylinder (10). To release the brake again, the engineer again increases the pressure in the train pipe, allowing the exhaust port on the inlet exhaust valve again to open, and the brake cylinder is vented to the atmosphere (11—green), while the auxiliary reservoir is recharged through valve (5). Additionally, the upper diaphragm (12) and limiting valve system allow the brake to be applied partially when intermediate pressures prevail in the train pipe. Should the train pipe be broken, or in an emergency brake application, the pressure in the train pipe drops to atmospheric pressure and the brake is fully applied. When the vehicle is parked, the control chamber can be vented manually through valve (13). As described here, there is only one train pipe. Often a "two-pipe" system is employed in which there is a separate continuous air supply to the auxiliary reservoir. The air-brake system has in recent years been adopted on many railroads formerly using vacuum brakes.

applied, using compressed air, on depressing the brake pedal.

AIR COMPRESSOR. See JET PROPULSION; PUMP.

AIR CONDITIONING, the regulation of the temperature, humidity, circulation and composition of the air in a building, room or vehicle. In warm weather an air-conditioning plant, working like a refrigerator (see REFRIGERATION), cools, dehumidifies (see also DEHYDRATION) and filters the air. In colder weather it may be reversed to run as a HEAT PUMP.

The first commercial air-conditioning installation dates from 1902, when W. H. CARRIER designed a cooling and humidifying system for a New York printing plant. During the 1920s, motion-picture theaters and then office buildings, department stores and hospitals began to install air-conditioning equipment. After WWII, home units became available, resulting in the rapid growth of the industry manufacturing the equipment. Room air conditioners (window units) are the most widely used domestic equipment, though the installation of the more versatile central air-conditioning equipment (unitary equipment) is becoming more widespread.

AIR-CUSHION VEHICLE (ACV), or **Hovercraft,** a versatile marine, land or amphibious vehicle which supports its weight on a high-pressure air cushion maintained by a system of fans. Because this minimizes the friction between the craft and the ground, the auxiliary propulsion equipment can maintain speeds up to 100 knots, even over difficult surfaces. Although the air-cushion principle was rediscovered by the UK engineer Christopher Cockerell in the early 1950s, technical difficulties have so far limited the use of ACVs to military applications and ferry services on a few short sea crossings. **Ground-effect machines** (GEMs) are sometimes distinguished from ACVs as, like AIRPLANES, they derive most of their lift from their aerodynamic design and forward motion.

AIRFOIL, any surface designed to have a mechanical interaction with the air through which it passes. In particular the term refers to the cross-sectional shape of an AIRPLANE wing, though the tailfin and propeller blades are also airfoils.

AIRGLOW, a faint reddish or greenish light, of similar nature to the AURORA, visible in night skies at low and middle latitudes. It is caused by the reforming of molecules split by the sun's ultraviolet light.

AIR GUN, a weapon using compressed air to fire a dart or pellet with a maximum range of about 90m. The charge of air released on pressing the trigger is produced either by prior compression or instantaneously by releasing a spring-loaded piston. In the similar **gas gun,** a replaceable carbon dioxide reservoir provides several hundred charges.

AIR LOCK, an airtight chamber with two doors used to allow men and materials to pass between environments having different air pressures. Air locks are used on caissons facilitating underwater excavation, on spacecraft, as submarine escape hatches and in industrial high-vacuum installations.

AIR MASS. See METEOROLOGY.

AIRPLANE, a powered heavier-than-air craft which obtains lift from the aerodynamic effect of the air rushing over its wings (see AERODYNAMICS). The typical airplane has a cigar-shaped fuselage which carries the pilot and payload; wings to provide lift; a

This giant SRN4 MKII hovercraft carries 280 passengers and 37 automobiles across the English Channel in only 40 minutes.

power unit to provide forward thrust; stabilizers and a tail fin for controlling the plane in flight, and landing gear for supporting it on the ground. The plane is piloted using the throttle and the three basic control surfaces: the ELEVATORS on the stabilizers which determine "pitch" (whether the plane is climbing, diving or flying horizontally); the rudder on the tail fin which governs "yaw" (the rotation of the plane about a vertical axis), and the AILERONS on the wings which control "roll" (the rotation of the plane about the long axis through the fuselage). In turning the plane, both the rudder and the ailerons must be used to "bank" the plane into the turn. The airplane's control surfaces are operated by moving a control stick or steering column (elevators and ailerons) in conjunction with a pair of footpedals (rudder).

The pilot has many instruments to guide him. Chief among these are the air-speed indicator, altimeter, compass, fuel gauge and engine-monitoring instruments. Large modern aircraft also have flight directors, artificial horizons, course indicators, slip and turn indicators, instruments which interact with ground-based navigation systems and radar. In case any individual instrument fails, most are duplicated. (See also FLIGHT, HISTORY OF.)

AIR PLANT. See EPIPHYTE.

AIR POLLUTION, the contamination of the atmosphere by harmful vapors, AEROSOLS and dust particles, resulting principally from the activities of man but to a lesser extent from natural processes. Natural pollutants include pollen particles, salt-water spray, wind-blown dust and fine debris from volcanic eruptions. Most man-made pollution involves the products of COMBUSTION—smoke (from burning wood, coal, and oil in municipal, industrial and domestic furnaces); carbon monoxide and lead (from automobiles), and oxides of nitrogen and sulfur dioxide (mainly from burning coal)—though other industrial processes, crop-spraying and atmospheric

The controls of a typical light plane. The control column (1) moves both the ailerons (2) and the elevators (3); the rudder (4) is moved by pedals (5).

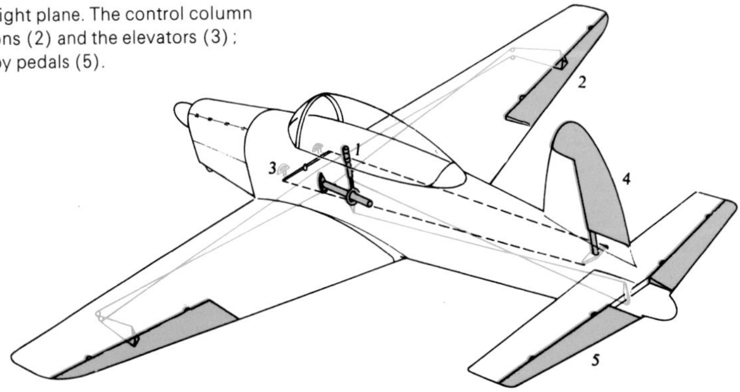

nuclear explosions also contribute. Most air pollution arises in the urban environment, with a large portion of that coming from the AUTOMOBILE. **Pollution control** involves identifying the sources of contamination; developing improved or alternative technologies and sources of raw materials, and persuading industries and individuals to adopt these, if need be under the sanction of legislation. Automobile emission control is a key area for current research, exploring avenues such as the RECYCLING and thorough OXIDATION of exhaust gases; the production of lead-free GASOLINE, and the development of alternatives to the conventional INTERNAL COMBUSTION ENGINE. On the industrial front, flue-gas cleansing using catalytic conversion (see CATALYSIS) or centrifugal, water-spray or electrostatic precipitators is becoming increasingly widespread. The matching of smokestack design to local meteorological and topographic conditions is important for the efficient dispersal of remaining pollutants. Domestic pollution can be reduced by restricting the use of high-pollution fuels as in the UK's "smokeless zones." In the short term, the community must be prepared to pay the often high prices of such pollution-control measures, but bearing in mind the continuing economic rewards ensuing and the vital necessity of preserving the purity of the air we breathe, the sacrifice must be worthwhile. (See also POLLUTION).

AIR PRESSURE. See ATMOSPHERE.

AIR PUMP. See PUMP.

AIR SACS, small respiratory cavities: in birds, leading off the lungs and often entering into the bones; in many insects, expansions in the TRACHEAE.

AIRSCREW. See PROPELLER.

AIRSHIP, or dirigible, a lighter-than-air, self-propelled aircraft whose buoyancy is provided by gasbags containing hydrogen or helium. The first successful airship was designed by Henri Giffard, a French engineer, and flew over Paris in 1852, though it was only with the development of the INTERNAL COMBUSTION ENGINE that the airship became truly practical. From 1900 Germany led the world in airship design, as Count Ferdinand von ZEPPELIN began to construct his famous "Zeppelins." Most of the large airships built during the next 40 years were of the "rigid" type, with a metal-lattice frame, and used hydrogen as the lifting gas. Their vulnerability in storms and a series of spectacular fire disasters brought an abrupt end to their use in about 1937. During WWII much use was made of small "nonrigid" patrol airships ("blimps") in which the gasbag formed the outer skin and altitude was controlled by inflating and venting air "ballonets" inside the main gas bag. Most existing craft are of this type, with engines slung beneath the gasbag either on the cabin or in separate "nacelles." "Semi-rigid" airships are similar to blimps but, being larger, usually have a longitudinal metal keel. Airship enthusiasts envisage a great future for airships filled with nonflammable helium, noiselessly transporting freight right into the heart of large cities.

AIRSICKNESS. See MOTION SICKNESS.

AIR TURBULENCE, irregular eddying in the ATMOSPHERE, such as that encountered in gusts of wind. Turbulence disperses water vapor, dust, smoke and other pollutants through the atmosphere, and is important in transferring heat energy upward from the ground. There is little turbulence in the upper atmosphere except in developing thunderclouds. **Clear-air turbulence** (CAT), which is often found around the margins of JET STREAMS, can be hazardous to high-flying jet aircraft.

ALABASTER, fine-grained, massive form of GYPSUM, usually translucent and white; used ornamentally for centuries, being easily carved. Ancient oriental alabaster was a yellowish MARBLE.

ALBATEGNIUS, or **Albatenius** or **al-Battani.** See BATTANI, ABU-ABDULLAH MUHAMMAD IBN-JABIR AL-.

ALBEDO, the ratio between the amount of light reflected from a surface and the amount of light incident upon it. The term is usually applied to celestial objects within the SOLAR SYSTEM: the moon reflects about 7% of the sunlight falling upon it, and hence has an albedo of 0.07.

ALBERTUS MAGNUS, Saint (c1200–1280), German scholastic philosopher and scientist; the teacher of St. Thomas Aquinas. Albert's main significance was in promoting the study of ARISTOTLE and in helping to establish Aristotelianism and the study of the natural sciences within Christian thought. In science he did important work in botany and was possibly the first to isolate ARSENIC.

ALBINO, an organism lacking the pigmentation normal to its kind. The skin and hair of albino animals (including man) is uncolored while the irises of their eyes appear pink. Albinism, which may be total or only partial, is generally inherited. Albino plants contain no CHLOROPHYLL and thus, being unable to perform PHOTOSYNTHESIS, rapidly die.

ALBITE, common mineral occurring in igneous rocks, consisting of sodium aluminum silicate ($NaAlSi_3O_8$); often forms vitreous crystals of various colors. It is one of the three end-members (pure compounds) of the FELDSPAR group.

ALBUMIN, group of PROTEINS soluble in water and in 50% saturated ammonium sulfate solution; present in animals and plants. Ovalbumin is the chief protein in egg white; serum albumin occurs in blood PLASMA, where it controls osmotic pressure.

ALCHEMY, a blend of philosophy, mysticism and chemical technology, originating before the Christian era, seeking variously the conversion of base metals into gold, the prolongation of life and the secret of immortality. In the Classical world alchemy began in Hellenistic Egypt and passed through the writings of the great Arab alchemists such as Al-Razi (RHAZES) to the Latin West. The late medieval period saw the discovery of NITRIC, SULFURIC and HYDROCHLORIC acids and ETHANOL (*aqua vitae*, the water of life) in the alchemists' pursuit of the "philosopher's stone" or *elixir* which would transmute base metals into gold.

In the early 16th century PARACELSUS set alchemy on a new course, towards a chemical pharmacy (IATROCHEMISTRY), although other alchemists—including John DEE and even Isaac NEWTON—continued to work along mystical, quasireligious lines. Having strong ties with ASTROLOGY, interest in alchemy, particularly in the Hermetic writings (see HERMES TRISMEGISTUS), has never quite died out, though without any further benefit to medical or chemical science. (See also CHEMISTRY.)

ALCOHOLIC BEVERAGES, drinks containing ETHANOL, the only variety of ALCOHOL that may be consumed in moderation without damaging effects. Popular alcoholic beverages include BEER, WINE, WHISKEY, BRANDY, RUM and compounded liquors such as liqueurs and GIN. They vary widely in alcoholic content, ranging from 2% or 3% in light beers to more than 60% in some VODKAS and distilled fruit brandy. (See also DISTILLED LIQUOR.)

Intoxicating beverages were known to the ancient Egyptians and Babylonians, and in the past many people used them instead of impure water. Today their use is general for conviviality; excessive drinking may be due to ALCOHOLISM. (See also INTOXICATION.)

ALCOHOLISM, compulsive drinking of alcohol in excess, one of the most serious problems in modern society. Many people drink for relaxation and can stop drinking without ill effects; the alcoholic cannot give up drinking without great discomfort: he is dependent on alcohol, physically and psychologically.

Alcohol is a DEPRESSANT that acts initially by reducing activity in the higher centers of the BRAIN. The drinker loses judgment and inhibitions; he feels free of his responsibilities and anxieties. This is the basis for initial psychological dependence. With further alcohol intake, thought and body control are impaired (see also INTOXICATION). The alcoholic starts by drinking more and longer than his fellows. He then finds that the unpleasant symptoms of withdrawal—"hangover," tremor, weakness and hallucinations—are relieved by alcohol. In this way his drinking extends through the greater part of the day and physical dependence is established. The alcoholic often has a reduced tolerance to the effects of alcohol and may suffer from AMNESIA after a few drinks. Social pressures soon lead to secretive drinking, work is neglected and financial difficulties add to the disintegration of personality; denial and pathological jealousy hasten social isolation. Alcohol depresses the appetite and the alcoholic may stop or reduce eating. Many of the diseases associated with alcoholism are in part due to MALNUTRITION and VITAMIN deficiency: CIRRHOSIS, NEURITIS, dementia, and KORSAKOV'S PSYCHOSIS. Prolonged alcohol withdrawal leads to DELIRIUM TREMENS. Treatment of alcoholism is very difficult. SEDATIVES and ANTABUSE may help to counteract dependence. Reconciliation of the patient to society is crucial; he must understand the reasons for his drinking and learn to approach his problems and fears realistically. Psychotherapy and Alcoholics Anonymous are valuable in this. Total abstinence is essential to avoid relapse.

ALCOHOLS, class of ALIPHATIC COMPOUNDS, of general formula ROH, containing a hydroxyl group bonded to a carbon atom. They are classified as monohydric, dihydric, etc., according to the number of hydroxyl groups; and as primary, secondary or tertiary according to the number of hydrogen atoms adjacent to the hydroxyl group. Alcohols occur widely in nature, and are used as solvents and antifreezes and in chemical manufacture. They are obtained by fermentation, oxidation or hydration of ALKENES from petroleum and natural gas, and by reduction of fats and oils.

ALDEBARAN (Alpha Tauri), the 14th brightest star in the night sky and the brightest star in TAURUS. At a distance of 21pc, it has an absolute magnitude varying about −0.8.

ALDEHYDES, class of organic compounds of general formula RCHO, containing a carbonyl group (see also KETONES). They are highly reactive, and find many uses in industry in the preparation of solvents, dyes, resins and other compounds. Many aldehydes

Alcohols

methanol (methyl alcohol) MW 32.04, mp −94°C, bp 65°C	CH_3-OH
ethanol (ethyl alcohol) MW 46.07, mp −117°C, bp 79°C	CH_3CH_2-OH
1-propanol (*n*-propyl alcohol) MW 60.11, mp −127°C, bp 97°C	$CH_3CH_2CH_2-OH$
2-propanol (isopropyl alcohol) MW 60.11, mp −90°C, bp 82°C	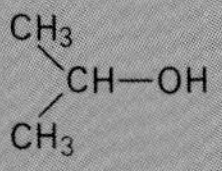
1-butanol (*n*-butyl alcohol) MW 74.12, mp −90°C, bp 117°C	$CH_3CH_2CH_2CH_2-OH$
1-pentanol (*n*-amyl alcohol) MW 88.15, mp −79°C, bp 137°C	$CH_3(CH_2)_3CH_2-OH$
1,2-ethanediol (ethylene glycol) MW 62.07, mp −12°C, bp 198°C	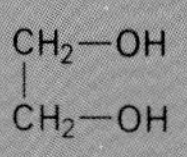 CH_2-OH | CH_2-OH
1,2,3-propanetriol (glycerol) MW 92.11, mp 20°C	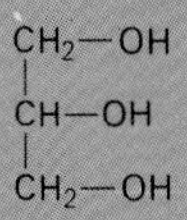 CH_2-OH | $CH-OH$ | CH_2-OH
benzyl alcohol MW 108.15, mp −15°C, bp 205°C	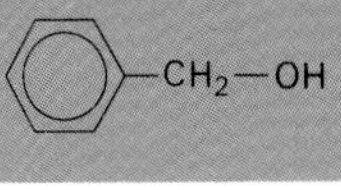

Aldehydes

methanal (formaldehyde) MW 30.03, mp −92°C, bp −21°C	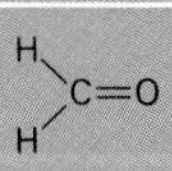
ethanal (acetaldehyde) MW 44.05, mp −121°C, bp 21°C	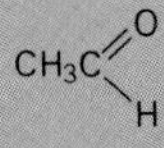
propanal (propionaldehyde) MW 58.08, mp −81°C, bp 49°C	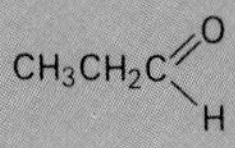
propenal (acrolein) MW 56.07, mp −87°C, bp 53°C	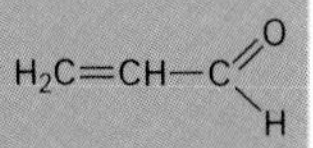
benzaldehyde MW 106.13, mp −26°C, bp 178°C	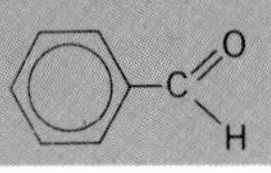

occur in nature and are often responsible for the flavor and scent of animals and plants. The simplest aldehydes are FORMALDEHYDE and ACETALDEHYDE. Aromatic aldehydes, such as BENZALDEHYDE and vanillin, are used in dyes and as perfumes and food flavorings. Aldehydes can be prepared by dehydrogenation or oxidation of primary ALCOHOLS, or by reduction of ACID CHLORIDES. Aldehydes may be reduced to primary alcohols, or oxidized to CARBOXYLIC ACIDS (if ammoniacal silver nitrate is used, a silver MIRROR is formed). They undergo addition reactions with BASES such as AMMONIA and CYANIDES; and condensation reactions with HYDRAZINE, alcohols, ACID ANHYDRIDES and other reactive compounds.

ALDER, Kurt (1902–1958), German organic chemist who shared the 1950 Nobel Prize for Chemistry with Otto DIELS for demonstrating the usefulness of the diene synthesis (DIELS–ALDER REACTION) in forming ALICYCLIC COMPOUNDS.

ALEMBERT, Jean Le Rond d' (1717–1783), French philosopher, physicist and mathematician, a leading figure in the French ENLIGHTENMENT and coeditor with DIDEROT of the renowned *Encyclopedia*. His early fame rested on his formulation of D'ALEMBERT'S PRINCIPLE in mechanics (1743). His other works treat calculus, music, philosophy and astronomy.

ALEMBIC, an early type of still, popularly associated with the experiments of alchemists (see ALCHEMY); the term strictly refers only to a particular form of DISTILLATION head.

ALEPH NULL ($\aleph_0$). See TRANSFINITE CARDINAL NUMBER.

ALEXANDRIAN LIBRARY, the greatest collection of books in antiquity, containing perhaps 400 000 manuscripts, in Alexandria, Egypt. Commenced under Ptolemy Soter, it came to be housed mainly in the Museum (see ALEXANDRIAN SCHOOL). Portions were destroyed by fires between 47 BC and the final fall of the city to the Arabs in 646 AD.

ALEXANDRIAN SCHOOL, or *Museum* (place dedicated to the Muses), founded c300 BC, the foremost center of learning in the ancient world during the HELLENISTIC AGE, and which housed the ALEXANDRIAN LIBRARY. The school was renowned from the first, its teachers including the mathematicians APOLLONIUS OF PERGA, EUCLID and HERO; the physicians ERASISTRATUS, EUDEMUS and HEROPHILUS; the geographer ERATOSTHENES and the astronomer HIPPARCHUS. The last great Alexandrian scientist was Claudius PTOLEMY, who worked in the city between 127 AD and 151 AD. With the decline of Hellenistic culture, activity in the school turned away from original research towards compilation and criticism, the study of mystical philosophy and theology assuming an increasingly significant role.

ALEXANDRITE, rare variety of the mineral CHRYSOBERYL, found in the Urals. A valuable GEM, it has a brilliant luster, and appears dark green in daylight but red in artificial or transmitted light.

ALEXIA, a disorder of language, the complete inability to comprehend the written word. (See SPEECH AND SPEECH DISORDERS; DYSLEXIA.)

ALFVÉN, Hannes Olof Gösta (1908–), Swedish physicist who shared the 1970 Nobel Prize for Physics with Louis NÉEL for contributing to the development of PLASMA physics. Alfvén himself introduced the study of MAGNETOHYDRODYNAMICS.

ALGAE, a large and extremely diverse group of plants, including some of the simplest organisms known to man. They are mostly aquatic, and range in size from microscopic single-celled organisms living on trees, in snow, ponds and the surface waters of oceans to strands of seaweed several metres long in the deep oceans. Some algae are free-floating, some are motile (see LOCOMOTION) and some grow attached to a substrate.

Algae are separated into seven major divisions, primarily on the basis of pigmentation. Blue–green algae have also been grouped in the algae by some authorities but differ from other algae in that they are prokaryotic organisms (see PROKARYOTE). Green algae (division Chlorophyta) are found mainly in freshwater and may be single-celled, form long filaments (like *Spirogyra*) or a flat leaf-like mass of cells called a thallus (like the sea lettuce, *Ulva lactuca*). Golden-brown algae (division Chrysophyta) also include the DIATOMS. Brown algae (division Phaeophyta) include the familiar seaweeds found on rocky shores. The largest, the kelps, can grow to enormous lengths. Red algae (division Rhodophyta) are found mostly in warmer seas and include several species of economic importance. Desmids and dinoflagellates (both in division Pyrrophyta) are single-celled algae and are important constituents of marine PLANKTON. Yellow–green algae and chloromonads (division Xanthophyta) are mainly freshwater forms, mostly unicellular and nonmotile. Motile unicellular algae such as *Euglena* (division Euglenophyta) are classified by some biologists as PROTOZOA, but most contain CHLOROPHYLL and can synthesize their own food.

Algae in both marine and freshwater plankton are important as the basis of food chains (see ECOLOGY). Many of the larger algae are important to man; for example, the red algae *Porphyra* and *Chondrus crispus* are used as foodstuffs. *Gelidium*, another red alga, is a source of AGAR, and the kelps (such as the giant kelp *Macrocystis*) produce alginates, one use of which is in the manufacture of ice cream. Other uses of algae are in medicine and as manure. (See also PLANT KINGDOM.)

ALGEBRA, that part of mathematics dealing with the relationships and properties of number systems by use of general symbols (such as a, b, x, y) to represent mathematical quantities. These are combined by addition ($x+y$), subtraction ($x-y$), multiplication ($x \times y$, $x.y$ or most usually xy) and division ($x \div y$, x/y or most usually $\frac{x}{y}$). The relationships between them are expressed by symbols such as $=$ ("is equal to"), $\neq$ ("is not equal to") $\simeq$ (is approximately equal to"), $>$ ("is greater than"), and $<$ ("is less than"). These symbols are also used in ARITHMETIC. Should a number be multiplied by itself one or more times it is said to be raised to a power:

$$x.x = x^2,$$
$$x.x.x = x^3, \text{ etc.}$$

x^2 is termed "x to the power of two" or more usually "x squared"; x^3 is termed "x to the power of three" or more often "x cubed." From this emerges the concept of the ROOT: if $x^2 = y$ then x is the square root of y or $\sqrt[2]{y}$ (or $\sqrt{y}$).

Algebraic operations are described by the

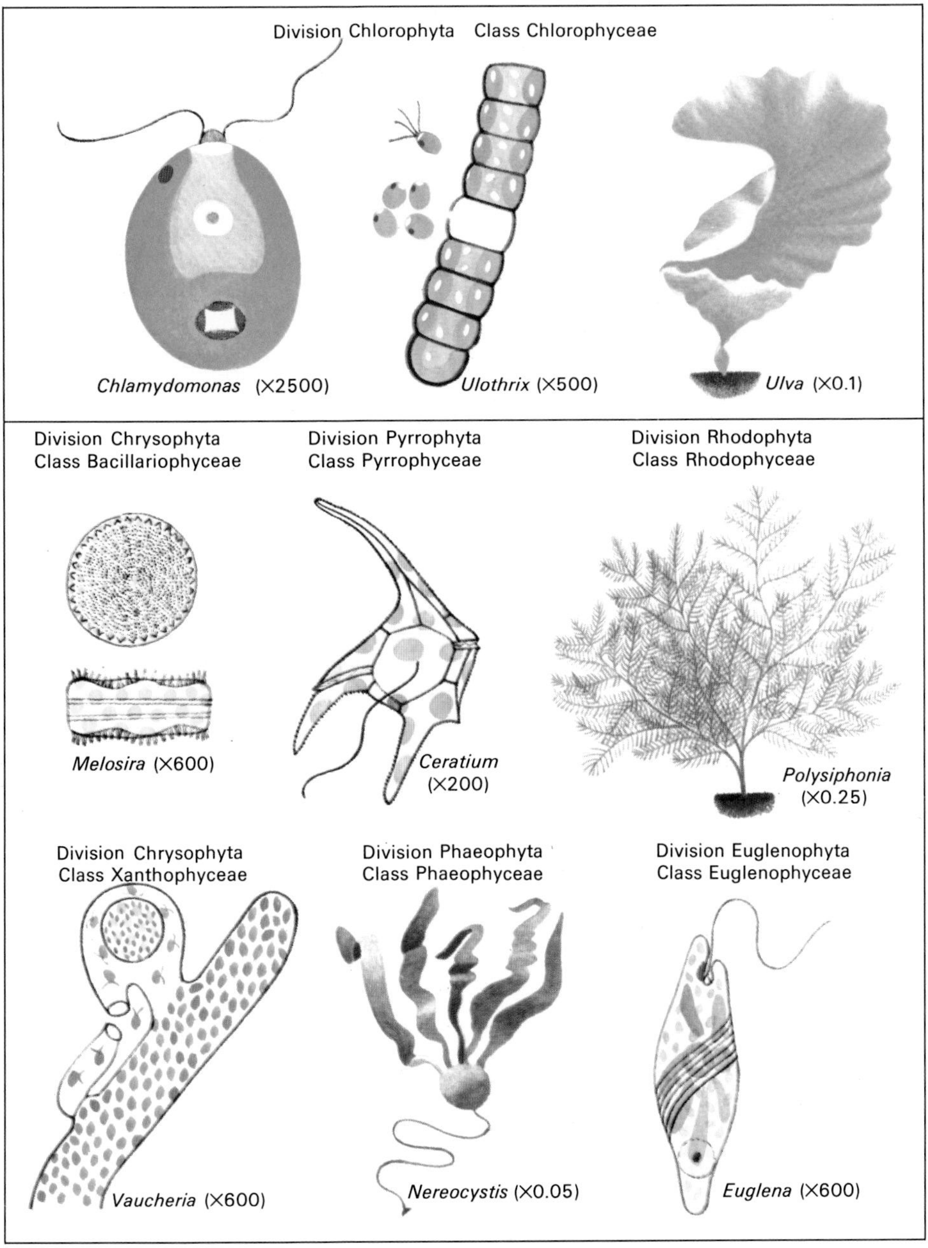

Representatives of the principal divisions of algae.

Associative, Commutative and Distributive Laws (see table).

Statements such as $x+y=z$ or $x^2=y$ are termed identities. Should one of the terms in such an identity be of unknown value, then we may discover its value through examination of the identity: if in the first example $x=3$, y is unknown and $z=7$, then $3+y=7$ and $y+4$. This is known as solution of an EQUATION.

Expressions containing two or more terms, such as $x+y+z$, are called polynomials. A special case is an expression containing two terms, such as $x+y$, which is termed a binomial.

Not all quantities combine in the ways described above. In vector algebra (see VECTOR ANALYSIS), for example, the operation equivalent to multiplication is not commutative: $\mathbf{a}\times\mathbf{b}=-\mathbf{b}\times\mathbf{a}$. There are thus different algebras to cope with different types of quantities. A special example of this is the application of algebraic techniques to LOGIC.

In higher algebra, a set of items combinable by an operation "x" is called a group if, for any a and b, $a\,x\,b$ is a unique element of the set (if, say, $a=3$ and $b=\sqrt{4}$, $a\,x\,b$ would not be unique if x implied multiplication, since $\sqrt{4}=+2$ or -2, and hence $a \times b$ equals $+6$ or -6), if $a\,x\,y=b$ and $z\,x\,b=a$ have unique solutions, and if the operation $a\,x\,b\,x\,c$ is associative. If the operation is commutative the group is called a commutative group: the set of all INTEGERS, for example, is commutative with respect to addition; the set of all RATIONAL NUMBERS is commutative with respect to multiplication. Should the set of mathematical quantities be distributive for, say, the two operations "$\times$" and "$+$" the set is called a ring; and should the multiplication be commutative the set is a commutative ring: the set of all integers is an example. If for any non-zero element a, there is an element a', known as its inverse, such that $a.a'=1$, the set is termed a field: the set of all rational numbers is such a field.

Algebraic methods are used throughout mathematics. (See also ALGEBRAIC GEOMETRY; BINOMIAL THEOREM; CALCULUS; GROUPS; MATHEMATICS; SET THEORY.)

Commutation, association and distribution

Addition and multiplication in the field of real numbers are said to be *commutative* since

$$a+b=b+a$$

and $$a \cdot b=b \cdot a.$$

Division and subtraction are not commutative

since $$a \div b \neq b \div a$$

and $$a-b \neq b-a.$$

Addition and multiplication are also *associative*

since $$a+(b+c)=(a+b)+c$$

and $$a \cdot (b \cdot c)=(a \cdot b) \cdot c$$

but division and subtraction are not associative

since $$a \div (b \div c) \neq (a \div b) \div c$$

and $$a-(b-c) \neq (a-b)-c.$$

Addition and multiplication are together *distributive* since

$$a \cdot (b+c)=a \cdot b+a \cdot c.$$

ALGEBRAIC GEOMETRY, that branch of ALGEBRA concerned with the visual realization of algebraic FUNCTIONS, whether such realization is practically possible or not. An extension of ANALYTIC GEOMETRY, it is now used primarily as an intuitive aid in discovering or understanding THEOREMS.

ALGIN, polysaccharide (see CARBOHYDRATES) extracted from brown seaweeds. A derivative, sodium alginate, is a thickening agent and emulsion stabilizer used in the food industry, especially ice cream manufacture, and as sizing in paper and textiles.

ALGOL, Beta Persei, second-brightest star in the constellation PERSEUS. It is a multiple star of at least three but probably four components, two of which form an eclipsing binary (see VARIABLE STAR) causing a 10h diminution of brightness every 59h.

ALGOL (*Algo*rithmic *L*anguage, sometimes given as *Alg*ebraic *O*riented *L*anguage), universal COMPUTER language devised 1958, adapted 1960. Similar to FORTRAN, but with several important advantages, it is used more in Europe than in the US.

ALGORITHM, a set of simple mathematical operations which together, in the right order, constitute a complex mathematical operation. Algorithms are used extensively in COMPUTER science (see ALGOL).

A simple algorithm

The object is to find the square root of 9 using only simple arithmetical operations. An initial guess is made – in this case 4 – and the calculator then proceeds:

try out guess: $9 \div 4 = 2.25$

find the mean of guess and result:

$$\frac{2.25+4}{2} = 3.12$$

try this result: $9 \div 3.125 = 2.88$

find the mean of this result and the previous one:

$$\frac{2.88+3.125}{2} = 3.0025$$

and so on. Already the calculator is close to the right answer – 3 – and he will know that it has been attained when two successive averaging operations yield the same result to the accuracy that he requires.

ALICYLIC COMPOUNDS, class of organic compounds in which carbon atoms are linked to form one or more rings. AROMATIC COMPOUNDS are excluded because of their special properties. In general, alicyclic compounds resemble analogous ALIPHATIC COMPOUNDS. However, strain occurs in small rings (with three, four or five members) because the angles between adjacent bonds are less than the preferred angle of 109° 28′, and these compounds are less stable and more reactive. Larger rings are nonplanar and unstrained. Many TERPENES, such as MENTHOL, are alicyclic. (See also HETEROCYCLIC COMPOUNDS.)

ALIMENTARY CANAL. See GASTROINTESTINAL TRACT.

ALIPHATIC COMPOUNDS, major class of organic compounds that includes all those with carbon atoms linked in straight or branched open chains. The other classes are ALICYCLIC, HETEROCYCLIC and AROMATIC compounds.

ALIZARIN, 1,2-dihydroxyanthraquinone, a once-important orange-red DYE, originally extracted from the root of the MADDER (*Rubia tinctorum*) but after 1871 made synthetically from ANTHRAQUINONE by a process developed by PERKIN (1869). An earlier German synthesis (1868) was the first laboratory preparation of a natural dyestuff. Alizarin is now little used.

alizarin

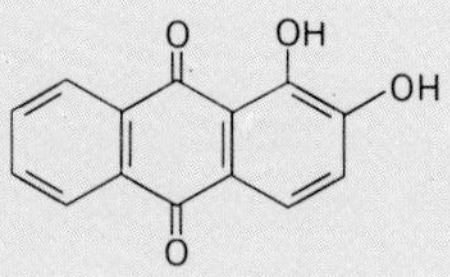

ALKALI, a water-soluble compound of the ALKALI METALS (or ammonia) which acts as a strong BASE producing a high concentration of hydroxyl ions in aqueous solution. Alkalis neutralize acids to form salts and turn red litmus paper blue. Common alkalis are sodium hydroxide (NaOH), ammonia (NH_3), sodium carbonate (Na_2CO_3) and potassium carbonate (K_2CO_3). They have important industrial applications in the manufacture of glass, soap, paper and textiles. Caustic alkalis are corrosive and can cause severe burns.

ALKALI FLATS, level, barren areas in dry regions covered with EVAPORITES, mainly salts of the ALKALI METALS and ALKALINE-EARTH METALS. Alkali flats are formed by the repeated periodic evaporation of shallow lakes lacking outlets.

ALKALI METALS, highly reactive metals in Group IA of the PERIODIC TABLE, comprising LITHIUM, SODIUM, POTASSIUM, RUBIDIUM, CESIUM and FRANCIUM. They are soft and silvery-white with low melting points. Alkali metals react with water to give off hydrogen, and so much heat is generated (except by lithium and sodium) that spontaneous combustion may occur. Because of this extreme reactivity they never occur naturally as the metals, but are always found as monovalent ionic salts.

Alkali metals

element	atomic number	atomic weight
lithium (Li)	3	6.94
sodium (Na)	11	23.00
potassium (K)	19	39.09
rubidium (Rb)	37	85.47
cesium (Cs)	55	132.91
francium (Fr)	87	223*

* isotope of greatest stability.

ALKALINE-EARTH METALS, gray-white metals in Group IIA of the PERIODIC TABLE, comprising BERYLLIUM, MAGNESIUM, CALCIUM, STRONTIUM, BARIUM and RADIUM. They never occur in an uncombined state, but are usually found as carbonates or sulfates. Except for beryllium, they are highly reactive and inflammable, readily dissolving in acids to form divalent ionic salts. The hydroxides of the four heaviest elements are alkalis.

Alkaline earth metals

element	atomic number	atomic weight
beryllium (Be)	4	9.01
magnesium (Mg)	12	24.31
calcium (Ca)	20	40.08
strontium (Sr)	38	87.62
barium (Ba)	56	137.34
radium (Ra)	88	226.03

ALKALOIDS, narcotic poisons found in certain plants and fungi. They have complex molecular structures and are usually heterocyclic nitrogen-containing BASES. Many, such as coniine (from hemlock) or atropine (deadly nightshade), are extremely poisonous. Others, such as morphine, nicotine and cocaine, can be highly addictive, and some, such as mescaline, are psychedelics. But in small doses alkaloids are often powerful medicines, and are used as analgesics, tranquilizers, and cardiac and respiratory stimulants. Other examples are quinine, reserpine and ephedrine. Caffeine (found in coffee and tea) is a stimulant. Although alkaloids may be found in any part of the plant, they are usually contained in the seeds, seed capsules, bark or roots. One plant, the opium poppy, contains about 30 alkaloids. Alkaloids are extracted from plants and separated by chromatography; synthetic alkaloids are seldom economically competitive.

ALKALOSIS, medical condition in which the blood PLASMA becomes excessively alkaline (i.e., the PH rises above 7.45) with resulting nausea, anorexia or TETANY. Respiratory alkalosis is due to over-ventilation with loss of plasma CARBON dioxide. Metabolic causes can include the consumption of excess ALKALI or the large acid loss involved in severe vomiting. (See also ACIDOSIS.)

ALKANES, or **Paraffins,** the homologous series of saturated HYDROCARBONS of general formula C_nH_{2n+2}. The lowest members, which are gases, are METHANE, ETHANE, PROPANE and BUTANE; higher members are named for the number of carbon atoms in the molecule. From pentane (C_5H_{12}) to heptadecane ($C_{17}H_{36}$) they are liquids, and above that waxy solids. Alkanes with four or more carbon atoms have several ISOMERS, the straight-chain isomers being called normal alkanes (*n*-alkanes). Branched alkanes are named as derivatives of the longest straight chain in the molecule. Alkanes are obtained from petroleum and natural gas; they may be synthesized by hydrogenation of ALKENES or from carbon monoxide and hydrogen. They are soluble in most organic solvents, but not in water. The lower alkanes are less reactive than the higher alkanes. Typical reactions include combustion in air, decomposition and rearrangement on heating, isomerization and condensation with alkenes (with acid catalyst), nitration, sulfonation, and halogenation by fluorine, chlorine and bromine (with heat or light). The monovalent radicals C_nH_{2n+1} derived from alkanes by loss of one hydrogen atom are called alkyl groups (methyl, ethyl, etc.), usually denoted by the symbol R. (See also OCTANE.)

ALKENES, or **Olefins,** the homologous series of unsaturated HYDROCARBONS having one or more double bonds (see BOND, CHEMICAL) between adjacent carbon atoms. The monoalkenes (one double bond) have general formula C_nH_{2n}; their systematic names are derived from those of the corresponding ALKANES by replacing the suffix -ane by -ene. They are prepared by thermal cracking of alkanes (from petroleum or natural gas), dehydration of ALCOHOLS, or base-catalyzed elimination of hydrogen halides from ALKYL HALIDES. Alkenes physically resemble the corresponding alkanes, but chemically their properties are due mainly to the double bond. Many reagents add across the double bond: hydrogen (with nickel or platinum catalyst), halogens, hydrogen halides, and sulfuric acid. Alkenes are oxidized by permanganate or hypochlorite to GLYCOLS, and by OZONE to ozonides which readily decompose to ALDEHYDES and KETONES. Alkenes may readily be polymerized (by catalysts) to give plastics and resins. (See also ETHYLENE; BUTADIENE; ISOPRENE.)

AL-KHWARIZMI. See KHWARIZMI, MUHAMMAD IBN-MUSA AL-.

ALKYLATION, the introduction of an alkyl group (see ALKANES) into a compound by substitution or addition, usually done by reacting the compound with an ALKENE or an ALKYL HALIDE. (See FRIEDEL-CRAFTS REACTION.) Specifically in petroleum refining, alkylation refers to the thermal or catalytic process in which branched alkanes are reacted with alkenes to yield highly branched products of high OCTANE rating.

ALKYL HALIDES, organic compounds consisting of an alkyl group (see ALKANES) bonded to a HALOGEN atom; polyhalogen derivatives of alkanes are similar. They are prepared by direct halogenation of alkanes (except for the iodides), addition of hydrogen halides to ALKENES, or halogenation of ALCOHOLS by hydrogen halides, phosphorus (III) halides, etc. Alkyl halides are used as solvents and as intermediates in chemical manufacture. The halogen atom is readily replaced by other NUCLEOPHILES such as hydroxide or cyanide. Elimination of hydrogen halides (base-catalyzed) yields alkenes.

ALKYNES, or acetylenes, the homologous series of unsaturated HYDROCARBONS having one or more triple bonds (see BOND, CHEMICAL) between adjacent carbon atoms. The monoalkynes (one triple bond) have general formula C_nH_{2n-2}; their systematic names are derived from those of the corresponding ALKANES by replacing the suffix -ane by -yne. They are prepared by elimination of two hydrogen halide molecules from a dihaloalkane. Alkynes physically resemble the corresponding alkanes, but chemically their properties are due mainly to the triple bond, and are similar to those of the ALKENES. Addition reactions take place in two stages, forming first a substituted alkene and then a substituted alkane. The triple bond being nucleophilic (see NUCLEOPHILES), alkynes add to unsaturated compounds such as aldehydes or ketones. Alkynes readily polymerize to various products, including AROMATIC and ALICYCLIC compounds.

ALLELE, or allelomorph, one of the two or more genes that can and do occupy particular loci on homologous chromosomes. Different alleles are responsible for the different though similar effects of genetic variation (e.g., whether an individual has brown, green or blue eyes) and are interconvertible by mutation. (See HEREDITY.)

ALLERGY, a state of abnormal sensitivity to foreign material (allergen) in susceptible individuals. It is essentially the inappropriate reaction of ANTIBODY AND ANTIGEN defense responses to environmental substances. Susceptibility is often inherited but manifestations vary with age. Exposure to allergen induces the formation of antibodies; when, at a later

Aliphatic hydrocarbons

alkanes		alkenes	
methane mp −182°C, bp −164°C	CH_4		
ethane mp −183°C, bp −88°C	$CH_3{-}CH_3$	ethene (ethylene) mp −169°C, bp −104°C	$CH_2{=}CH_2$
propane mp −190°C, bp −42°C	$CH_3{-}CH_2{-}CH_3$	propene (propylene) mp −185°C, bp −47°C	$CH_2{=}CH{-}CH_3$
2-methylpropane (isobutane) mp −138°C, bp −50°C	$CH_3{-}CH(CH_3){-}CH_3$	2-methylpropene (isobutylene) mp −140°C, bp −7°C	$CH_2{=}C(CH_3)_2$
butane mp −138°C, bp −1°C	$CH_3{-}CH_2{-}CH_2{-}CH_3$	1-butene (butylene) mp −185°C, bp −6°C	$CH_2{=}CH{-}CH_2{-}CH_3$
		2-butene mp −139°C, bp 4°C (*cis*-) mp −106°C, bp 1°C (*trans*-)	$CH_3{-}CH{=}CH{-}CH_3$

dienes			
conjugated		**allenic**	
		propadiene (allene) mp −136°C, bp −35°C	$H_2C{=}C{=}CH_2$
1,3-butadiene (bivinyl) mp −109°C, bp −4°C	$H_2C{=}CH{-}CH{=}CH_2$	1,2-butadiene (methylallene) mp −136°C, bp 11°C	$H_2C{=}C{=}CH{-}CH_3$

date, the material is again encountered, it reacts with the antibodies causing release of HISTAMINE from mast cells in the tissues. INFLAMMATION follows, with local irritation, redness and swelling, which in skin appear as ECZEMA or urticaria (see HIVES). In the nose and eyes HAY FEVER results, and in the GASTROINTESTINAL tract diarrhea may occur. In the LUNGS a specific effect leads to spasm of bronchi, which gives rise to the wheeze and breathlessness of ASTHMA. In most cases, the route of entry determines the site of the response; but skin rashes may occur regardless of route and asthma may follow eating allergenic material. If the allergen is injected, ANAPHYLAXIS may occur. Localized allergic reactions in skin following chronic exposure to chemicals (e.g., nickel, poison ivy) are the basis of contact dermatitis. Common allergens include drugs (PENICILLIN, ASPIRIN), foods (shellfish), plant pollens, animal furs or feathers, insect stings and the house dust mite. Treatment includes ANTIHISTAMINES, cromoglycate, STEROIDS and desensitizing INJECTIONS; ADRENALINE may be life-saving in severe allergic reactions. (See also IMMUNITY.)

ALLOMORPH. See MORPHEME.

ALLOPATHY (from German *Allopathie*, term coined 1842 by HAHNEMANN), the cure of a disease by the induction of symptoms differing from those of the disease; the opposite of HOMEOPATHY. A variation, **enantiopathy**, the countering of an overabundance of one HUMOR by the overabundance of another, was used in medieval times.

ALLOPHONE. See PHONEME.

ALLOTROPY, the occurrence of some elements in more than one form (known as allotropes) which differ in their crystalline or molecular structure. Allotropes may have strikingly different physical or chemical properties. Allotropy in which the various forms are stable under different conditions and are reversibly interconvertible at certain temperatures and pressures, is called enantiotropy. Notable examples of allotropy include DIAMOND and GRAPHITE, OXYGEN and OZONE, and SULFUR. (See also POLYMORPHISM.)

ALLOY, a combination of metals with each other or with nonmetals such as carbon or phosphorus. They are useful because their properties can be adjusted as desired by varying the proportions of the constituents. Very few metals are used today in a pure state. Alloys are formed by mixing their molten components. The structures of alloys consisting mainly of one component may be substitutional or interstitial, depending on the relative sizes of the atoms. The study of alloy structures in general is complex. (See also PHASE EQUILIBRIA.)

The commonest alloys are the different forms of STEEL, which all contain a large proportion of iron and small amounts of carbon and other elements. BRASS and BRONZE, two well-known and ancient metals, are alloys of copper, while PEWTER is an alloy of tin and lead. The very light but strong alloys used in aircraft construction are frequently alloys of aluminum with magnesium, copper or silicon. SOLDERS contain tin with lead and bismuth; type metal is an alloy of lead,

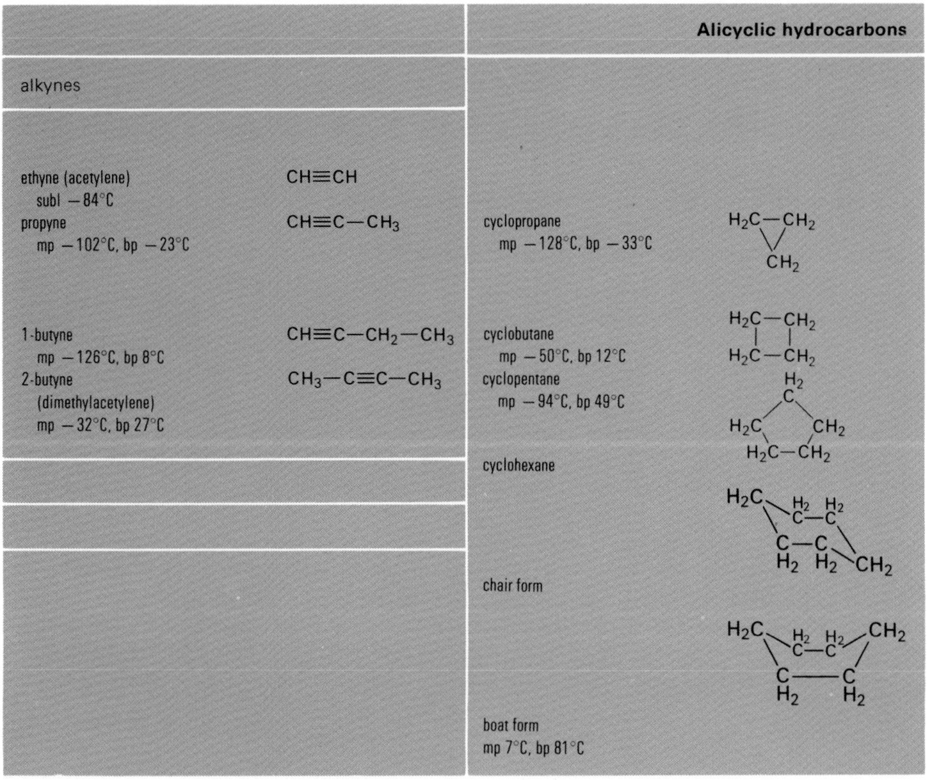
Alicyclic hydrocarbons
alkynes
ethyne (acetylene)
subl −84°C
CH≡CH
propyne
mp −102°C, bp −23°C
CH≡C—CH3
1-butyne
mp −126°C, bp 8°C
CH≡C—CH2—CH3
2-butyne
(dimethylacetylene)
mp −32°C, bp 27°C
CH3—C≡C—CH3
cyclopropane
mp −128°C, bp −33°C
cyclobutane
mp −50°C, bp 12°C
cyclopentane
mp −94°C, bp 49°C
cyclohexane
chair form
boat form
mp 7°C, bp 81°C

Some common alloys

name	normal or typical composition	uses or properties
alnico-4	55% Fe, 28% Ni, 12% Al, 5% Co	magnets
babbitt metal	91% Sn, 4.5% Sb, 4.5% Cu	bearings
brass	60% Cu, 40% Zn	wide general use
bronze	92% Cu, 8% Sn	wide general use
coinage metal	95% Cu, 4% Sn, 1% Zn	"copper" coins
coinage metal	75% Cu, 25% Ni	"silver" coins
constantan	55% Cu, 45% Ni	thermocouples
dental amalgam	52% Hg, 33% Ag, 12.5% Sn, 2% Cu, 0.5% Zn	dental fillings
elektron	86.5% Mg, 11% Al, 1.5% Zn, 1% Mn	very light aircraft parts
german silver	56% Cu, 24% Zn, 20% Ni	base for electroplating
gun metal	88% Cu, 10% Sn, 2% Zn	strong and tough
invar	64% Fe, 36% Ni	zero temperature coefficient of expansion
monel metal	67% Ni, 33% Cu	corrosion-resistant
nichrome	60% Ni, 25% Fe, 15% Cr	electrical heating elements
pewter	65% Sn, 30% Pb, 5% Sb	drinking vessels
solder	60% Pb, 35% Sn, 5% Bi	electrical connections
stainless steel	73% Fe, 18% Cr, 8% Ni, 1% C	corrosion-resistant
Wood's metal	50% Bi, 25% Pb, 12.5% Sn, 12.5% Cd	low mp

tin and antimony. Among familiar alloys are those used in coins: modern "silver" coinage in most countries is an alloy of nickel and copper. Special alloys are used for such purposes as die-casting, dentistry, high-temperature use, and for making thermocouples, magnets and low-expansion materials. (See also AMALGAM; BABBITT METAL; GERMAN SILVER; GUN METAL; INVAR; MONEL METAL.)

ALLPORT, Gordon Willard (1897–1967). US psychologist, important figure in the study of personality, who stressed the "functional autonomy of motives." Among his many works, *The Nature of Prejudice* (1954) has become a classic in its field.

ALLUVIUM, material such as GRAVEL, SILT and SAND deposited, mainly near their mouths, by streams and rivers. Alluvium makes rich agricultural soil, and the earliest civilizations originated as farming communities centered on alluvial flood plains.

ALMAGEST. See PTOLEMY, CLAUDIUS.

ALMANAC, originally a calendar giving the positions of the planets, the phases of the moon, etc., particularly as used by navigators (nautical almanacs), but now any yearbook of miscellaneous information, often containing abstracts of annual statistics.

ALOPECIA. See BALDNESS.

ALPHABET (from Greek *alpha* and *beta*), a set of characters intended to represent the sounds of spoken language. Because of this intention (which in practice is never realized) written languages employing alphabets are quite distinct from those using characters which represent whole words (see IDEOGRAM; HIEROGLYPHICS). The word alphabet is, however, usually extended to describe syllabaries, languages in which characters represent syllables. The chief alphabets of the world are Roman (Latin), Greek, Hebrew, Cyrillic (Slavic), Arabic and Devanagari.

Alphabets probably originated around 2000 BC. Hebrew, Arabic and other written languages sprang from a linear alphabet which had appeared c1500 BC. From the Phoenician alphabet, which appeared around 1700 BC, was derived the Greek. Roman letters were derived from Greek and from the rather similar Etruscan, also a descendant of the Greek. Most of the letters we now use are from the Latin alphabet, U and W being distinguished from V, and J from I, in the early Middle Ages. The Cyrillic alphabet, used with the Slavic languages, derives from the Greek. It is thought that Devanagari was possibly invented to represent Sanskrit.

Chinese and Japanese are the only major languages that function without alphabets, although Japanese has syllabary elements. (See also CUNEIFORM and WRITING, HISTORY OF; and, for the evolution of the letters of our alphabet, the headings to each alphabetical section.)

ALPHA CENTAURI, multiple star in the constellation CENTAURUS, comprising a DOUBLE STAR around which orbits at a distance of 10000AU a red dwarf, Proxima Centauri, which is the nearest star to the solar system, being 1.33parsecs distant.

ALPHA PARTICLES, HELIUM nuclei ($_2He^4$) emitted at velocities of about 1.6Mm/s from radioactive materials undergoing alpha disintegration (see RADIOACTIVITY). Alpha particles, discovered by RUTHERFORD in 1899, carry a double positive charge and are strongly absorbed by air, thin paper and metal foils.

ALTAIR, brightest star in the constellation AQUILA and the eleventh brightest in the night sky (apparent magnitude +0.89). It has an extremely rapid rotation and is 4.9 parsecs from the earth.

ALTED (∇). See DEL.

ALTERNATING CURRENT (AC). See ELECTRICITY.

ALTERNATION OF GENERATIONS, a feature of the life cycle of most plants and many lower animals by which successive generations reproduce alternately sexually and asexually. In animals the feature is exhibited by FLUKES, TAPEWORMS and some CNIDARIA including the common jellyfish (*Aurelia aurita*) and

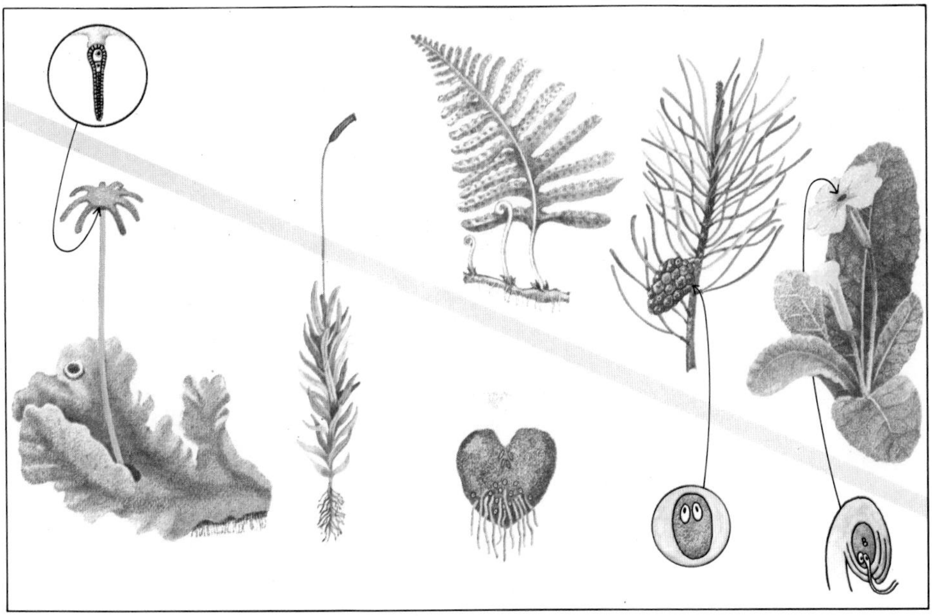

Many plants have two distinct phases in their life cycles. The gametophyte which produces the haploid male and female gametes; and the sporophyte which produces diploid spores. During the course of evolution there has been a gradual shift from the gametophyte being the dominant form, i.e. the plant we recognize, to the sporophyte being the dominant form. This illustration shows the gametophytes of various plant types below the diagonal line and the sporophytes above the line. From left to right the examples are: a liverwort in which the green thallus is the gametophyte, while the sporophyte is represented by the fertilized egg and the capsule that develops from it; a moss in which the upright leafy plant is the gametophyte and the capsule the sporophyte that lives upon it; in the ferns the dominant plant is the sporophyte, while the gametophyte is a small, free-living flattened prothallus that produces the gametes; in conifers and flowering plants the sporophyte is the only conspicuous phase, the male gametophyte being contained in the pollen grains and the female gametophyte represented by the egg and surrounding tissues.

the sea fir *Obelia*. In plants the sexually reproducing or **gametophyte** generation gives rise to HAPLOID male and female sex cells (GAMETES) which, on FERTILIZATION, produce a diploid ZYGOTE which in turn germinates into the asexually reproducing **sporophyte** generation. This reproduces by forming SPORES which germinate to give the gametophyte generation again. In lower plants such as the LIVERWORTS and MOSSES, the gametophyte generation is dominant but in flowering plants the gametophytes are reduced to microscopic proportions, the plant itself being the sporophyte generation. (See also REPRODUCTION.)

ALTIMETER, an instrument used for estimating the height of an aircraft above sea level. Most are modified aneroid BAROMETERS and work on the principle that air pressure decreases with increased altitude, but these must be constantly recalibrated throughout the flight to take account of changing meteorological conditions (local ground temperature and air pressure reduced to sea level). **Radar altimeters**, which compute ABSOLUTE altitudes (the height of the aircraft above the ground surface immediately below) from the time taken for RADAR waves to be reflected to the aircraft from the ground, although essential for blind landings, are as yet too expensive for general installation.

ALTITUDE SICKNESS, a condition of OXYGEN lack in blood and tissues due to low atmospheric PRESSURE. Night vision is impaired, followed by breathlessness, headache, and faintness. At 5000m mental changes include indifference, euphoria and faulty judgment but complete ACCLIMATIZATION is possible up to those heights. At very high altitude (6000m to 7000m), CYANOSIS, COMA and death rapidly supervene. Treatment is by oxygen and descent. The use of pressurized cabins prevents the occurrence of the condition.

ALTOCUMULUS. See CLOUDS.

ALTOSTRATUS. See CLOUDS.

ALUM, a double salt comprising sulfates of two metals (one monovalent, one trivalent) combined with 12 molecules of water of crystallization: $M^IM^{III}(SO_4)_2.12H_2O$. The monovalent metal is commonly potassium, sodium or ammonium; the trivalent metal may be aluminum, chromium or ferric iron. Alums are soluble in water and are usually acid. They are used as astringents (styptic pencils), as a mordant in dyes, and in the manufacture of baking powder, antiperspirants and fire extinguishers. Potash alum (potassium aluminum sulfate,

Crystals of common alum (potassium aluminum sulfate). Left: a polycrystalline aggregate formed naturally. Right: a single octahedral crystal grown artificially from a seed crystal.

$KAl(SO_4)_2 . 12H_2O$) is used in the sizing of paper and in water purification.

ALUMINA, or aluminum oxide. See ALUMINUM.

ALUMINUM (Al), silvery-white metal in Group IIIA of the PERIODIC TABLE, the most abundant metal, comprising 8% of the earth's crust. It occurs naturally as BAUXITE, CRYOLITE, FELDSPAR, clay and many other minerals, and is smelted by the HALL-HÉROULT PROCESS, chiefly in the US, USSR and Canada. It is a reactive metal, but in air is covered with a protective layer of the oxide. Aluminum is light and strong when alloyed, so that aluminum ALLOYS are used very widely in the construction of machinery, and domestic appliances. It is also a good conductor of electricity and is often used in overhead transmission cables where lightness is crucial. AW 27.0, mp 660°C, bp 2467°C, sg 2.6989 (20°C).

Aluminum compounds are trivalent and mainly cationic (see CATION), though with strong bases aluminates are formed. (See also ALUM.) **Aluminum Oxide** (Al_2O_3), or **Alumina,** is a colorless or white solid occurring in several crystalline forms, and is found naturally as CORUNDUM, EMERY and BAUXITE. Solubility in acid and alkali increases with hydration. mp 2045°C, bp 2980°C. **Aluminum Chloride** ($AlCl_3$) is a colorless crystalline solid, used as a catalyst (see FRIEDEL-CRAFTS REACTION). The hexahydrate is used in deodorants and as an astringent.

ALVAREZ, Luis Walter (1911–), US physicist awarded the 1968 Nobel Prize for Physics for work on SUBATOMIC PARTICLES, including the discovery of the transient resonance particles. He helped develop much of the hardware of NUCLEAR PHYSICS.

AM (Amplitude Modulation). See RADIO.

AMALGAM, an ALLOY of MERCURY with other metals. Most metals except iron will form amalgams; those with high mercury content are liquid, but most are solid. Amalgams of some NOBLE METALS occur naturally: SILVER and GOLD are extracted from their ores by forming amalgams. Dental amalgam, containing silver, copper, zinc and tin, is used to fill TEETH. Various amalgams may be used as ELECTRODES. (See also MIRROR.)

AMBER, fossilized RESIN from prehistoric EVERGREENS. Brownish-yellow and translucent, it is highly valued and can be easily cut and polished for ornamental purposes. Its chief importance is that FOSSIL insects up to 20 million years old have been found embedded in it. The main source of amber is along the shores of the Baltic Sea.

AMBERGRIS, waxy solid formed in the intestines of Sperm whales, perhaps to protect them from the bony parts of their squid diet. When obtained from dead whales, it is soft, black and evil-smelling, but on weathering (as when found as flotsam) it becomes hard, gray and fragrant, and is used as a perfume fixative and in the East as a spice.

AMBIDEXTERITY. See HANDEDNESS.

AMBIVALENCE, contrasting and alternating EMOTIONS toward a person or object, typified by the "love–hate" relationship. FREUD suggested that ambivalence was basic to many NEUROSES.

AMERICAN ASSOCIATION FOR THE ADVANCEMENT OF SCIENCE (AAS), the largest US organization for the promotion of scientific understanding. Founded in Boston in 1848 but now centered in Washington, it has over 100 000 individual and 300 corporate members. Its publications include the weekly, *Science*.

AMERICAN MUSEUM OF NATURAL HISTORY, an institution in New York City founded in 1869 and dedicated to research and popular education in anthropology, astronomy, mineralogy and natural history. Its public museums include the Hayden Planetarium and it publishes several technical and popular periodicals.

AMERICAN PHILOSOPHICAL SOCIETY, the oldest surviving US learned society, based in Philadelphia where it was founded by Benjamin FRANKLIN in 1743. The US counterpart of the ROYAL SOCIETY OF LONDON (1660), it currently has approaching 600 US and foreign members. It has an extensive library, much relating to early American science, its own regular publications commencing in 1769 with its *Transactions*.

AMERICIUM (Am), silvery-white radioactive TRANSURANIUM ELEMENT, one of the ACTINIDES. It is prepared by NEUTRON irradiation of PLUTONIUM. Am^{241}, the most readily available ISOTOPE (half-life 458yr), emits GAMMA RAYS and is used in industrial density and thickness gauges.

AMERINDS, term coined (1897–98) by John Wesley POWELL to denote the American Indians. It is believed that the Amerinds originated in NE Asia, having crossed the Bering Strait (perhaps by land bridge) before the 10th millennium BC. However, there are marked blood-type differences between them and the Asian Mongoloid peoples.

AMETHYST, transparent violet or purple variety of QUARTZ, colored by iron or manganese impurities. The color changes to yellow on heating. Amethysts are semiprecious GEMS. The best come from Brazil, Uruguay, Ariz. and the USSR.

AMIA, or bowfin, a North American fish that is a survivor of the fossil group HOLOSTEI.

AMIDES, class of ALIPHATIC COMPOUNDS, of general formula $RCONH_2$, derived from CARBOXYLIC ACIDS and AMMONIA by replacing the acid hydroxyl group by the amino group (NH_2). N-substituted amides are derived from primary or secondary AMINES instead of ammonia. Other amides are derived analogously from inorganic OXY-ACIDS or from SULFONIC ACIDS (see also SULFA DRUGS). Amides are prepared by reaction of ACID CHLORIDES, ACID ANHYDRIDES or ESTERS with AMMONIA or AMINES, or by partial HYDROLYSIS of NITRILES. Most simple amides are low-melting solids with strong HYDROGEN BONDING, soluble in water. Formamide and N-substituted amides are liquids widely used as solvents. Amides are both weak ACIDS and weak BASES. They may be hydrolyzed to CARBOXYLIC ACIDS, and dehydrated to NITRILES. Metallic HYDRIDES convert them to AMINES, and treatment with bromine and sodium hydroxide (the Hofmann degradation) yields amines with one fewer carbon atom. Polymeric amides (such as NYLON) are used as SYNTHETIC FIBERS, and similar amide linkages join AMINO ACIDS in PROTEINS and PEPTIDES. (See also UREA; IMIDES.)

Amides

methanamide (formamide) MW 45.04, mp 3°C, bp 193°C	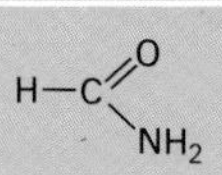
ethanamide (acetamide) MW 59.07, mp 83°C, bp 221°C	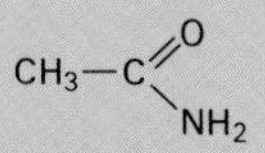
ethanediamide (oxamide) MW 88.07	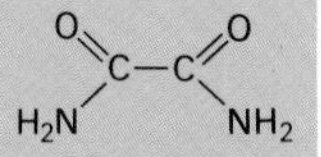
benzamide MW 121.14, mp 133°C, bp 290°C	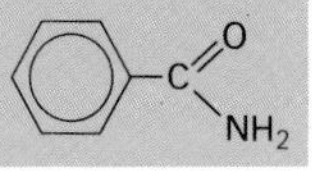

AMINES, class of organic compounds derived from AMMONIA by replacing one or more hydrogen atoms by alkyl groups (see ALKANES) or aryl groups (see AROMATIC COMPOUNDS). Primary amines have general

A piece of amber containing fossil insects and mounted as an ornament.

formula RNH_2; secondary R_2NH; and tertiary R_3N. HETEROCYCLIC nitrogen bases (including the ALKALOIDS and PYRIDINE) are tertiary amines. Amines may be formed by reduction of AMIDES, NITRILES or nitro compounds, or by reaction of ammonia with organic HALIDES, ALCOHOLS or SULFONIC ACIDS. Simple amines are pungent liquids which are strong BASES and LIGANDS; many occur naturally in decaying organic matter. They give AMIDES with acid derivatives. Amines have many uses, including the manufacture of dyes, drugs and SYNTHETIC FIBERS. (See also ANILINE; AMINO ACIDS.)

Amines

primary	
aminomethane (methylamine) MW 31.06, mp −94°C, bp −6°C	$H_3C—NH_2$
aminoethane (ethylamine) MW 45.09, mp −81°C, bp 17°C	$H_3C—CH_2—NH_2$
aminobenzene (aniline, phenylamine) MW 93.13, mp −6°C, bp 184°C	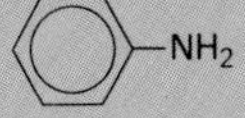
α-aminotoluene (benzylamine) MW 107.16, bp 185°C	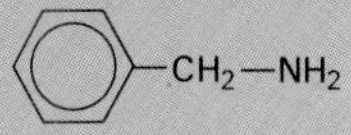
secondary	
dimethylamine MW 45.09, mp −93°C, bp 7°C	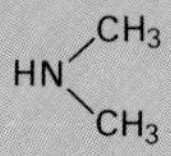
tertiary	
trimethylamine MW 59.11, mp −117°C, bp 3°C	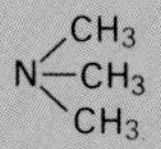
quaternary ammonium compound	
methylamine hydrochloride MW 67.52, subl 225°C	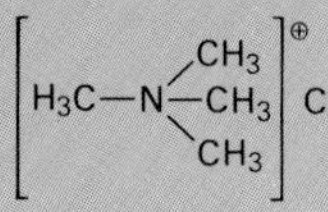

AMINO ACIDS, an important class of CARBOXYLIC ACIDS containing one or more amino ($-NH_2$) groups (see AMINES). Twenty or so α-amino acids ($RCH[NH_2]COOH$) are the building blocks of the PROTEINS found in all living matter. They are also found and synthesized in cells. Amino acids are white, crystalline solids, soluble in water; they can act as ACIDS or BASES depending on the chemical environment (see pH). In neutral solution they exist as ZWITTERIONS. An amino acid mixture may be analyzed by CHROMATOGRAPHY. All α-amino acids (except glycine) contain at least one asymmetric carbon atom to which are attached the carboxyl group, the amino group, a hydrogen atom and a fourth group (R) that differs for each amino acid and determines its character. Thus amino acids can exist in two mirror-image forms (see STEREOISOMERS). Generally only L-isomers occur in nature, but a few bacteria contain D-isomers. Humans synthesize most of the amino acids needed for NUTRITION, but depend on protein foods for eight "essential amino acids" which they cannot produce. Inside the body, amino acids derived from food are metabolized (see METABOLISM) in various ways. As each amino acid contains both an acid and an amino group, they can form a long chain of amino acids bridged by AMIDE links and called PEPTIDES. Peptide synthesis from constituent amino acids is a stage in PROTEIN SYNTHESIS. Thus some are converted into HORMONES, ENZYMES and NUCLEIC ACIDS. Proteins may be broken down again by HYDROLYSIS into their constituent amino acids, as in digestion. When amino acids are deaminated (the amino group removed), the nitrogen passes out as UREA. The remainder of the molecule enters the CITRIC ACID CYCLE, being broken down to provide energy.

Scientists have produced amino acids and simple peptide chains by combining carbon dioxide, ammonia and water vapor under the sort of conditions (including electric discharges) thought to exist on earth millions of years ago. This may provide a clue to the origin of LIFE.

Amino acids

L-isomer of a generalized amino acid	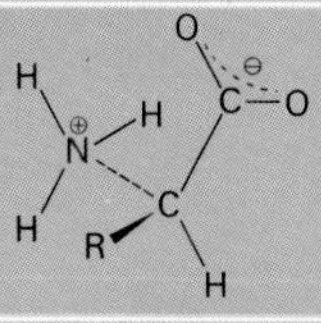
name	R
glycine	$—H$
α-alanine	$—CH_3$
serine	$—CH_2—OH$
tyrosine	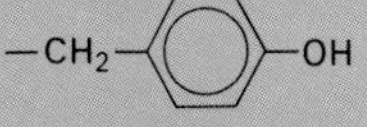
lysine	$—CH_2CH_2CH_2CH_2—NH_2$
β-asparagine	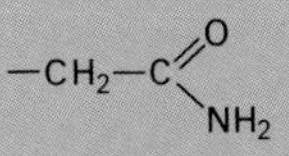
cysteine	$—CH_2—HS$
proline (complete molecule)	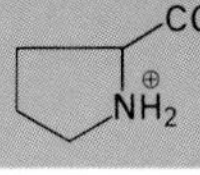

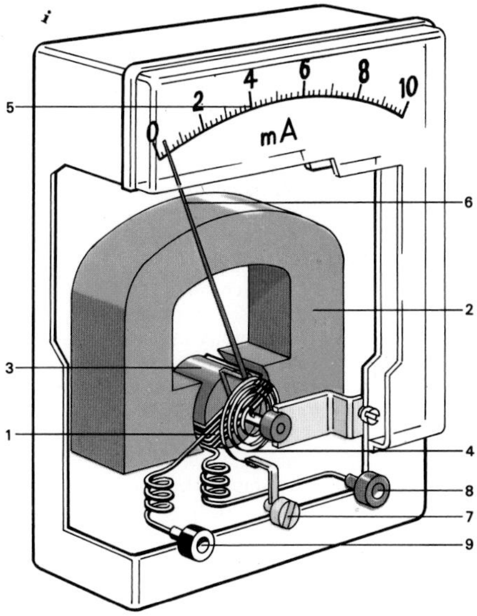

A moving-coil milliammeter. This consists of a rectangular coil wound on an aluminum former (1) mounted so that it can rotate between the poles of a permanent magnet (2). A soft iron core (3) ensures that a uniform radial magnetic field acts on the coil. When a current flows in the coil it experiences a torque which tries to turn it against the tension in a hair spring (4). The extent to which the coil turns is proportional to the current flowing and this is read off on a scale (5) with the aid of a pointer (6) attached to the coil assembly. Adjustment (7) is provided for zeroing the pointer and the instrument is connected into a circuit through terminals (8) and (9).

AMMETER, an instrument used to measure electric currents greater than 1 μA. Most direct-current ammeters are similar in design to the moving-coil GALVANOMETERS used for smaller currents, though they differ in passing most of the test current through a low "shunt" RESISTANCE (thus bypassing the coil) and in using a pointer fixed to the coil assembly to indicate the reading on the linearly calibrated scale. For alternating currents either a rectifier can be used with a moving-coil instrument or the less sensitive hot-wire or moving-iron instruments can be used. (See also ELECTRICITY; VOLTMETER.)

AMMONIA (NH_3), colorless acrid gas, made by the HABER PROCESS; a covalent HYDRIDE. The pyramidal molecule turns inside out very rapidly, which is the basis of the ammonia clock (see ATOMIC CLOCK). Ammonia's properties have typical anomalies due to HYDROGEN BONDING; liquid ammonia is a good solvent. Ammonia is a BASE; its aqueous solution contains ammonium hydroxide, and is used as a household cleaning fluid. It forms ammine (NH_3) LIGAND complexes with transition metal ions, and yields AMIDES and AMINES with many organic compounds. Ammonia is used as a fertilizer, a refrigerant, in the OSTWALD PROCESS, and to make ammonium salts, UREA, and many drugs, dyes and plastics. mp −78°C, bp −33°C.

On reaction with acids, ammonia gives ammonium salts, containing the NH_4^+ ion, which resemble ALKALI METAL salts. They are mainly used as fertilizers. The analogous quaternary ammonium salts, NR_4^+, are made by alkylation of tertiary AMINES and are used as ANTISEPTICS. **Ammonium Chloride** (NH_4Cl), or **Sal Ammoniac,** a colorless crystalline solid used in dry cells and as a flux, formed as a by-product in the SOLVAY PROCESS. subl 340°C. **Ammonium Nitrate** (NH_4NO_3), a colorless crystalline solid, used as a fertilizer and in explosives. mp 170°C. (See also HYDRAZINE.)

AMMONITES, extinct order of mollusks (Class: CEPHALOPODA), extant between 200 and 70 million years ago. Typically spiral-shelled, of diameter 0.01–2m (0.4in–6.6ft), (though helical—see HELIX—shells have been found), they evolved rapidly and their FOSSILS are thus of use in dating geological strata.

AMMUNITION, any material designed to be used with destructive effect against a target or an enemy. It includes mines, self-propelled missiles, BOMBS, TORPEDOES and grenades, together with gun ammunition. Sufficient materials to operate a weapon a single time constitute a round of ammunition. **Gun ammunition** usually comprises a bullet, shell or shot, a propellant charge and a primer which fires the propellant. High-explosive shells also include a fuse which detonates the charge either upon impact or a fixed time after firing. All this is usually supplied as a single unit (fixed ammunition) although large artillery often uses separate-loading ammunition for ease of handling. The propellant is usually a mixture of relatively slow burning explosives which liberates a large volume of gas on firing, thus propelling the projectile up the gun barrel. The size of ammunition depends on the **caliber** of gun used. This is the diameter of the barrel, usually expressed in mm or in decimal fractions of an inch. (See also BALLISTICS.)

AMNESIA, the total loss of MEMORY for a period of time or for events. In cases of CONCUSSION, **retrograde amnesia** is the permanent loss of memory for events just preceding a head injury while **post-traumatic** amnesia applies to a period after injury during which the patient may be conscious but incapable of recall, both at the time and later. Similar behavior to the latter, termed **fugue**, occurs as a psychiatric phenomenon.

AMNION, a tough membrane surrounding the EMBRYO of reptiles, birds and mammals and containing the AMNIOTIC FLUID. All land-laid EGGS contain amnions; those of fishes and amphibians do not, and thus must be laid in moist surroundings or water. (See also AMNIOTES; PLACENTA.)

AMNIOTES, those VERTEBRATES (mammals, reptiles, birds) characterized by the development of an AMNION to protect the EMBRYO.

AMNIOTIC FLUID, the fluid contained within the AMNION of AMNIOTES which provides a moist, aquatic environment for the EMBRYO. (See also PLACENTA.)

AMOEBAS, a large order (Amoebida) of the class Sarcodina (Rhizopodea) of PROTOZOA. They are unicellular (see CELL), a relatively rigid outer layer of ectoplasm surrounding a more fluid mass of endoplasm, in which lie one or more nuclei. They move by extending PSEUDOPODIA, into which they flow; and feed by surrounding and absorbing organic particles. REPRODUCTION is almost always asexual, generally by binary FISSION, though sometimes by multiple fission of the nucleus; a tough wall of CYTOPLASM forms about each of these small nuclei to create cysts. These can survive considerable rigors, returning to normal amoeboid form when circumstances are more clement. (Some species of amoeba may form a single cyst to survive adversity.) Certain amoebas can reproduce sexually. Amoebas are found wherever there is moisture, some parasitic (see PARASITE) forms living within other animals: *Entamoeba histolytica*, for example, causes amoebic DYSENTERY in man. The type-species is *Amoeba proteus*, which has a single nucleus and can form only one pseudopodium at a time.

AMPERE (A), the SI base unit of electric current, named for A.M. AMPÈRE and defined as the constant current which, if maintained in two straight parallel conductors of infinite length, of negligible circular cross-section, and placed 1 metre apart in vacuum, would produce between these conductors a force equal to 2×10^{-7} newton per metre of length. (See ELECTRICITY; SI UNITS.)

AMPÈRE, André Marie (1775–1836), French mathematician, physicist and philosopher best remembered for many discoveries in electrodynamics and electromagnetism. In the early 1820s he developed OERSTED's experiments on the interaction between magnets and electric currents and investigated the forces set up between current-carrying conductors.

AMPHETAMINES, a group of STIMULANT drugs, including **benzedrine** and **methedrine,** now in medical disfavor following widespread abuse and addiction. They counteract fatigue, suppress appetite, speed up performance (hence **"Speed"**) and give confidence, but pronounced DEPRESSION often follows; thus psychological and then physical addiction are encouraged. A paranoid PSYCHOSIS (resembling SCHIZOPHRENIA) may result from prolonged use, although it may be that amphetamine abuse is rather an early symptom of the psychosis. While no longer acceptable in treatment of OBESITY, they are useful in **narcolepsy**, a rare condition of abnormal sleepiness.

AMPHIBIA, a class of the CHORDATA that includes two fossil groups, the LEPOSPONDYLA and LABYRINTHODONTIA and three living groups, the ANURA (frogs and toads), URODELA (newts and salamanders) and the APODA (caecilians). Amphibians typically spend part of their life in water and part on land. Living members have soft moist skin through which they breathe, but also have gills and/or lungs. The group is composed of about 3 000 species and is widespread throughout the world. Amphibians of temperate regions commonly hibernate because they are cold-blooded and become sluggish at low temperatures.

AMPHIBOLES, a class of SILICATE minerals found in igneous rocks and metamorphic SCHISTS and GNEISSES. They contain infinite double-chains of SiO_4 tetrahedra, and have a cleavage of about 56°. Amphiboles include HORNBLENDE, JADE and certain ASBESTOS minerals.

AMPHIOXUS, genus of the CEPHALOCHORDATA thought to have many of the characteristics that one would expect in the ancestor to the higher members of the CHORDATA.

AMPHIPODA, one of the largest orders of CRUSTACEA (phylum ARTHROPODA) with over 3 600 species, found in both fresh and salt water. Typical amphipods include freshwater shrimps and sandhoppers.

AMPLIFIER, any device which increases the strength of an input signal. Amplifiers play a vital role in most electronic devices: RADIO and TELEVISION receivers, PHONOGRAPHS, TAPE-RECORDERS and COMPUTERS: but nonelectronic devices such as the horn of a windup phonograph or the PANTOGRAPH used for enlarging drawings are also amplifiers of a kind. Electronic amplifiers, usually based on

Typical amphipod showing the main divisions of the body: head (red), thorax (yellow) and abdomen (brown). The first two pairs of thoracic legs (purple), the gnathopods, assist feeding and are used by males during copulation. The first three pairs of abdominal legs (blue) create a water current which aerates the gills for respiration, and provide the main propulsive force in swimming. The rear three abdominal appendages are known as uropods, and two of these (blue) are usually modified to act as rudders.

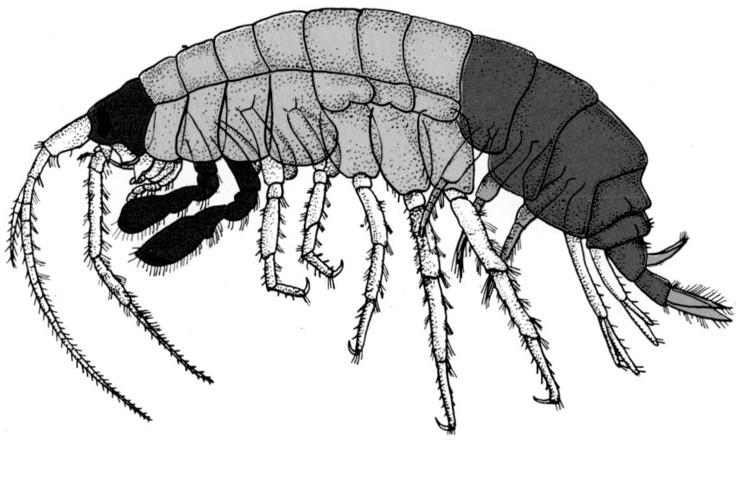

The symbol for an amplifier in circuit diagrams.

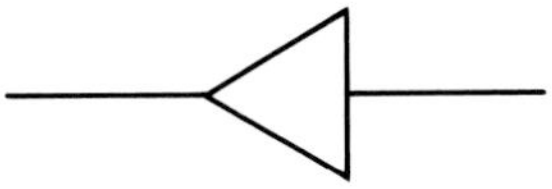

TRANSISTORS or ELECTRON TUBES, can be thought of as a sort of variable switch in which the output from a power source is controlled (modulated) by a weak input signal. An important factor is the fidelity (see HIGH-FIDELITY) with which the waveform of the output signal reproduces that of the input over the desired BANDWIDTH.

AMPLITUDE, in WAVE MOTIONS, the maximum displacement from its MEAN value of the oscillating property; thus for a PENDULUM, half the extent of its swing. **Amplitude modulation** (AM) is a common method of encoding a carrier wave in RADIO.

AMPUTATION, the surgical or traumatic removal of a part or the whole of a limb or other structure. It is necessary for severe limb damage, infective GANGRENE, loss of BLOOD supply and certain types of CANCER. Healthy tissue is molded to form a stump as a base for artificial limb prosthesis (see PROSTHETICS).

AMULET, a natural or artificial object believed to bring good luck (in which case it may be termed a **talisman**) or ward off evil. It is either placed at the focus of its desired sphere of influence (e.g., over a doorway) or carried or worn.

AMYL COMPOUNDS, organic compounds containing the amyl group C_5H_{11} (see ALKANES), which has eight ISOMERS. They include the amyl ALCOHOLS, synthesized from HYDROCARBONS or extracted from FUSEL OIL, and used as a solvent. **Isoamyl nitrite** ($C_5H_{11}ONO$) is a vasodilator used to give relief in ANGINA PECTORIS. **Amyl acetate** (which has a pleasant, fruity smell and is known as banana oil) is used as a solvent for NITROCELLULOSE and as a flavoring for candy.

ANAEROBE, any organism whose RESPIRATION does not make use of OXYGEN. Many BACTERIA and PARASITES are **facultative anaerobes** (that is, they can survive without oxygen for short or long periods), and a few are **obligate anaerobes** (unable to use oxygen in respiration). (See also AEROBE.)

ANALGESICS, drugs used for relief of pain. They mainly impair perception of or emotional response to pain by action on the higher BRAIN centers. ASPIRIN and paracetamol are mild but effective. Phenylbutazone, indomethacin and ibuprofen are, like aspirin, useful in treating RHEUMATOID ARTHRITIS by reducing INFLAMMATION as well as relieving pain. NARCOTIC analgesics derived from OPIUM ALKALOIDS range from the milder CODEINE and dextropropoxyphene, suitable for general use, to the highly effective euphoriant and addictive MORPHINE and HEROIN. These are reserved for severe acute pain and terminal disease, where addiction is either unlikely or unimportant. Pethidine (demerol) is an intermediate narcotic.

ANALOGUE, in a plant or animal, an organ performing the same function as one in another species but which differs from it in origin and structure (i.e., is not a HOMOLOGUE). Thus a bird's wing and a bee's wing are in this sense merely analogous.

ANALYSIS, the branch of MATHEMATICS concerned particularly with the concepts of FUNCTION and LIMIT. Its important divisions are CALCULUS, ANALYTIC GEOMETRY and the study of DIFFERENTIAL EQUATIONS.

The eight isomers of amyl alcohol

1-pentanol (*n*-amyl alcohol) MW 88.15, mp −79°C, bp 137°C	$CH_3CH_2CH_2CH_2CH_2OH$
2-pentanol bp 119°C	$CH_3CH_2CH_2CH(OH)CH_3$
3-pentanol bp 116°C	$CH_3CH_2CH(OH)CH_2CH_3$
2-methyl-1-butanol (*act*-amyl alcohol) bp 128°C	$CH_3CH_2CH(CH_3)CH_2OH$
2-methyl-2-butanol (*tert*-amyl alcohol) mp −8°C, bp 102°C	$CH_3CH_2C(CH_3)_2OH$
3-methyl-1-butanol (isopentyl alcohol) bp 129°C	$CH_3CH(CH_3)CH_2CH_2OH$
3-methyl-2-butanol bp 112°C	$CH_3CH(CH_3)CH(OH)CH_3$
2,2-dimethyl-1-propanol (neopentyl alcohol) mp 52°C, bp 113°C	$CH_3C(CH_3)_2CH_2OH$

ANALYSIS, in psychology, determination of the individual components of a complex experience or mental process: often also used for PSYCHOANALYSIS.

ANALYSIS, Chemical, determination of the compounds or elements comprising a chemical substance. Qualitative analysis deals with what a sample contains; quantitative analysis finds the amounts. The methods available depend on the size of the sample: macro (>100mg), semimicro (1–100mg), micro (1μg–1mg), or submicro (<1μg). Chemical analysis is valuable in chemical research, industry, archaeology, medicine and many other fields. A representative sample must first be taken (see STATISTICS) and prepared for analysis. Preliminary separation is often carried out by CHROMATOGRAPHY, ION-EXCHANGE, DISTILLATION or precipitation.

In qualitative analysis, classical methods involve characteristic reactions of substances. After preliminary tests—inspection, heating, and FLAME TESTS—systematic schemes are followed which separate the various IONS into groups according to their reactions with standard reagents, and which then identify them individually. Cations and anions are analyzed separately. For organic compounds, carbon and hydrogen are identified by heating with copper (II) oxide, carbon dioxide and water being formed; nitrogen, halogens and sulfur are identified by heating with molten sodium and testing the residue for CYANIDE, HALIDES and SULFIDE respectively. Classical quantitative analysis is performed by GRAVIMETRIC ANALYSIS and VOLUMETRIC ANALYSIS.

Modern chemical analysis employs instrumental methods to give faster, more accurate assessments than do classical methods. Many modern methods have the additional advantage of being nondestructive. They include COLORIMETRY, SPECTROPHOTOMETRY, POLAROGRAPHY, MASS SPECTROSCOPY, differential THERMAL ANALYSIS, potentiometric titration (see POTENTIOMETER), and methods for determining MOLECULAR WEIGHT. Neutron activation analysis subjects a sample to NEUTRON irradiation and measures the strength of induced radioactivity and its rate of decay. In X-ray analysis, a sample is irradiated with X RAYS and emits X rays of different, characteristic wavelengths (see also X-RAY DIFFRACTION).

ANALYTIC GEOMETRY, that branch of GEOMETRY based on the idea that a POINT may be defined relative to another point or to AXES by a set of numbers. In plane geometry, there are usually two axes, commonly designated the x- and y-axes, at right ANGLES. The position of a point in the plane of the axes may then be defined by a pair of numbers (x, y), its coordinates, which give its distance in units in the x- and y-direction from the **origin** (the point of INTERSECTION of the two axes). In three dimensions there are three axes, usually at mutual right angles, commonly designated the x-, y- and z-axes. (See also ABSCISSA; ORDINATE; and CARTESIAN COORDINATES.) In the coordinates (x, y, z), consider the situation when two of these have fixed values: there is a set of points, called a coordinate LINE, corresponding to all values of the third coordinate. Repeating this for each of the three coordinates, it can be seen that through each point defined by this coordinate system there are three coordinate lines. For all points, all three of these are straight (the system is rectilinear) and at mutual right angles (the system is rectangular). In plane polar coordinates there are two coordinated lines through each point: these are at right angles and one is curved (the system is rectangular and curvilinear).

Equation of a curve. A CURVE may be defined as a set of points. A relationship may be established between the coordinates of every point of the set, and this relationship is known as the EQUATION of the curve. The simplest form of plane curve is the straight LINE, which in the system we have described has an equation of the form $y=ax+b$, where a and b are CONSTANTS. Set $a=2$ and $b=3$: then, if $x=1$, $y=2+3=5$, if $x=2$, $y=4+3=7$, and so on; and conversely if $y=1$, $x=(1-3)/2=-1$, and so on. All points whose coordinates satisfy the relationship $y=2x+3$ will lie on this line. Equations of curves may involve higher POWERS of x or y: a parabola (see CONIC SECTIONS) may be expressed as $y=ax^2+b$. Since

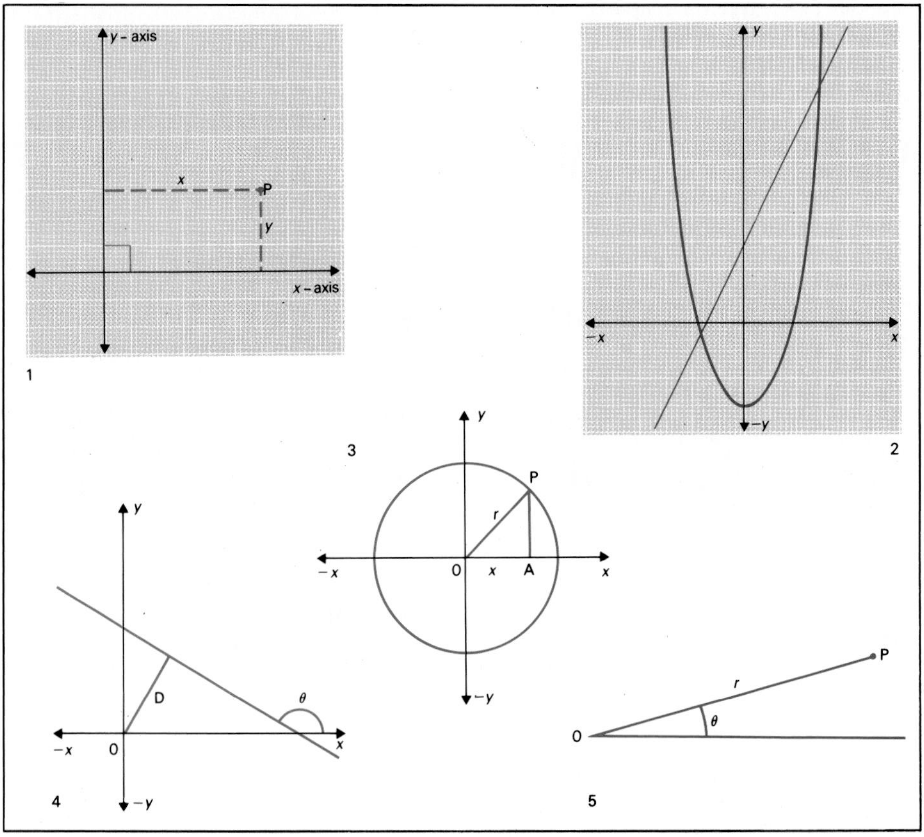

$x^2 = (-x)^2$ for all values of x, the curve is symmetrical (see SYMMETRY) about the y-axis. Similarly, the curve $x = ay^2 + b$ is symmetrical about the x-axis.

These principles may be applied to different coordinate systems, and to figures in more than two dimensions.

(See also ALGEBRAIC GEOMETRY; ANALYSIS; CALCULUS; CYLINDRICAL COORDINATES; FUNCTION; SPHERICAL COORDINATES.)

ANAPHYLAXIS, a severe allergic reaction (see ALLERGY) due to injection of foreign material, mediated by HISTAMINE and KININS. In man, sudden severe breathlessness—due to spasm in bronchi and larynx—and circulatory collapse (SHOCK) occur. ADRENALINE and ANTIHISTAMINES should be given.

ANASAZI (Navaho: ancient ones), a prehistoric culture of the American Southwest, whose modern manifestation is the Pueblo Indians. The earliest Anasazi remains are dated c50 AD. They were skilled basket weavers, in later times farmers and outstanding potters. Between about 1000 and 1300 AD they built the famous cliff dwellings.

ANATOMY, the structure and form of biological organisms (see BIOLOGY) and its study (morphology). The subject has three main divisions: gross anatomy, dealing with components visible to the naked eye; microscopic anatomy, dealing with microstructures seen only with the aid of an optical MICROSCOPE, and submicroscopic anatomy, dealing with still smaller ultrastructures. Since structure is closely related to function, anatomy is related to PHYSIOLOGY. (See also HUMAN BODY; EMBRYOLOGY.)

The study of anatomy is as old as that of MEDICINE, though for many centuries physicians' knowledge of anatomy left much to be desired. ANAXAGORAS had studied the anatomy of animals and anatomical observations can be found in the Hippocratic writings (see HIPPOCRATES), but it was ARISTOTLE who was the true father of comparative anatomy, and human dissection (the basis of all systematic human anatomy) was rarely practiced before the era of the ALEXANDRIAN SCHOOL and the work of HEROPHILUS and ERASISTRATUS. The last great experimental anatomist of antiquity was GALEN. His theories, as transmitted through the writings of the Arab scholars RHAZES and AVICENNA, held sway throughout the medieval period. Further progress had to await the revival of the practice of human dissection by SERVETUS and VESALIUS in the 16th century. The latter founded the famous Paduan school of anatomy which also included FALLOPIUS and FABRICIUS, whose pupil William HARVEY reunited the studies of anatomy and physiology in postulating the circulation of the BLOOD in *de Motu Cordis* (1628). This theory was confirmed some years later when MALPIGHI discovered the capillaries linking the arteries with the veins. Since the 17th century many important anatomical schools have been founded and the study of anatomy has become an essential part of medical training. Important developments in the late 18th century included the foundation of HISTOLOGY by BICHAT and that of modern comparative anatomy by CUVIER.

The rise of microscopic anatomy has of course depended on the development of the microscope; it found its greatest success in the announcement of SCHWANN's cell theory in 1839.

(1) A Cartesian coordinate system whose origin is at 0 where the x- and y-axes intersect at right angles. The position of point P is determined by its coordinates, (x, y). (2) On this coordinate system, the set of all points that satisfy a particular relationship between x and y may be plotted. Here the set of all points satisfying the equation $y = 2x + 3$ forms a straight line cutting the y-axis at $y = 3$ and the x-axis at $x = -1\frac{1}{2}$. The set of all points satisfying $y = x^2 - 3$ forms a parabola cutting the y-axis at $y = -3$ and $x = +\sqrt{3}$ and $x = -\sqrt{3}$. Note the symmetry of the curve about the y-axis. (3) Again using these axes, we can find the equation for the set of all points lying on the circumference of a circle. Here the circle has radius r, and its center lies at the origin, 0. Applying Pythagoras' Theorem, we see that for any point P on the circumference, $x^2 + y^2 = r^2$. This relationship is the equation of the circle. (4) A straight line may be determined using line coordinates, where the angle θ that the line makes with the x-axis and the perpendicular distance D from the line to the origin are specified. (5) Polar coordinates employ the distance r of a point P from the origin O and the angle θ which the line OP makes with a fixed line called the polar axis. The coordinates of P are therefore (r, θ).

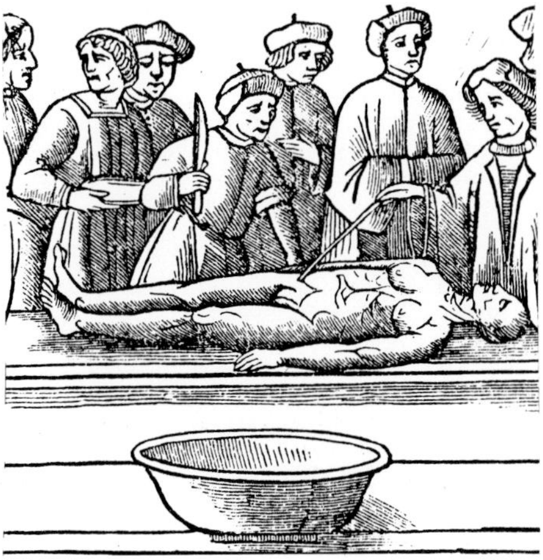

16th-century woodcut showing a dissection in progress.

ANAXAGORAS (c500–c428 BC), Greek philosopher of the Ionian school, resident in Athens, who taught that the elements were infinite in number and that every thing contained a portion of every other thing. He also discovered the true cause of ECLIPSES, thought of the sun as a blazing rock and showed that air has substance.

ANAXIMANDER (c610–c545 BC), Greek philosopher of the Ionian school who taught that the cosmos was all derived from one primordial substance by a process of the separating out of opposites. He was probably the first Greek to attempt a map of the whole known world and thought of the earth as a stubby cylinder situated at the center of all things. Animal life, he thought, had begun in the sea.

ANAXIMENES OF MILETUS (6th century BC), Greek philosopher of the Ionian school who held that all things were derived from air; this becoming, for instance, fire on rarefaction, water, and finally earth on condensation.

ANDERSON, Carl David (1905–), US physicist who shared the 1936 Nobel Prize for Physics for his discovery of the positron (1932). Later he was codiscoverer of the first meson (see SUBATOMIC PARTICLES).

ANDERSON, Philip Warren (1923–), US physicist who shared the 1977 Nobel Prize for Physics with N. F. MOTT and J. H. VAN VLECK, for their contributions to solid state physics.

ANDREWS, Roy Chapman (1884–1960), US naturalist, explorer and author. From 1906 he worked for the AMERICAN MUSEUM OF NATURAL HISTORY (later becoming its director, 1935–41) and made important expeditions to Alaska, the Far East and Central Asia.

ANDROECIUM. See FLOWER.

ANDROGENS, STEROID HORMONES which produce secondary male characteristics such as facial and body hair and a deep voice. They also develop the male reproductive organs. The main androgen is TESTOSTERONE, produced in the TESTES; others are produced in small quantities in the cortex of the ADRENAL GLANDS. Small amounts occur in women in addition to the ESTROGENS and may produce some male characteristics. (See also ENDOCRINE GLANDS; PUBERTY.)

ANDROMEDA, constellation in the N Hemisphere. The Great Andromeda Nebula (M31), seen near the DOUBLE STAR Gamma Andromedae, is the most distant object visible to the naked eye in N skies. It is the nearest external GALAXY to our own, like it a spiral but larger (49kpc across), and about 600kpc away.

ANECHOIC CHAMBER. See ACOUSTICS.

ANEMIA, condition in which the amount of HEMOGLOBIN in the BLOOD is abnormally low, thus reducing the blood's oxygen-carrying capacity. Anemic people may feel weak, tired, faint and breathless, have a rapid pulse and appear pale.

Of the many types of anemia, five groups can be described. In **iron-deficiency anemia**, the red blood cells are smaller and paler than normal. Usual causes include inadequate diet (especially in PREGNANCY), failure of iron absorption, and chronic blood loss (as from heavy MENSTRUATION HEMORRHAGE or disease of the GASTROINTESTINAL TRACT). Iron replacement is essential and in severe cases blood transfusion may be needed. In **megaloblastic anemia**, the red cells are larger than normal. This may be due to nutritional lack of VITAMIN B_{12} or folate, but the most important cause is **pernicious anemia**. B_{12} is needed for red-cell formation and patients with pernicious anemia cannot absorb B_{12} because they lack a factor in the stomach essential for absorbtion. Regular B_{12} injections are needed for life. In **aplastic anemia**, inadequate numbers of red cells are produced by the bone MARROW owing to damage by certain poisons, drugs or irradiation. Treatment includes transfusions and androgens. In **hemolytic anemia**, the normal life span of red cells is reduced, either because of ANTIBODY reactions or because they are abnormally fragile. A form of the latter is **sickle-cell anemia**, a hereditary disease, common among NEGROIDS in which abnormal hemoglobin is made. Finally, many chronic diseases including RHEUMATOID ARTHRITIS, chronic infection and UREMIA suppress red-cell formation and thus cause anemia.

ANEMOMETER, any instrument for measuring wind speed. The rotation type, which estimates wind speed from the rotation of cups mounted on a vertical shaft, is the most common of mechanical anemometers. The pressure-tube instrument utilizes the PITOT-TUBE effect, while the sonic or acoustic anemometer depends on the velocity of sound in the wind. For laboratory work a hot-wire instrument is used: here air flow is estimated from the change in RESISTANCE it causes by cooling an electrically heated wire.

ANESTHESIA, or absence of sensation, may be of three types: general, local or pathological. **General anesthesia** is a reversible state of drug-induced unconsciousness with muscle relaxation and suppression of REFLEXES; this facilitates many surgical procedures and avoids distress. An anesthesiologist attends to ensure stable anesthesia and to protect vital functions. While ETHANOL and NARCOTICS have been used for their anesthetic properties for centuries, modern anesthesia dates from the use of diethyl ETHER by William MORTON in 1846, and of CHLOROFORM by Sir James SIMPSON in 1847. Nowadays injections of short-acting BARBITURATES, such as PENTOTHAL SODIUM, are frequently used to induce anesthesia rapidly; inhaled agents, including halothane, ether, nitrous oxide, trichlorethylene and cyclopropane, are used for induction and maintenance. **Local** and **regional anesthesia** are the reversible blocking of pain impulses by chemical action of COCAINE derivatives (e.g., PROCAINE, lignocaine). Nerve trunks are blocked for minor SURGERY and DENTISTRY, and more widespread anesthesia may be achieved by blocking spinal nerve roots, useful in obstetrics and patients unfit for general anesthesia. **Pathological anesthesia** describes loss of sensation following trauma or disease.

ANEURYSM, a pathological enlargement of, or defect in, a blood vessel. These may occur in the HEART after CORONARY THROMBOSIS, or in the AORTA and ARTERIES due to ARTERIOSCLEROSIS, high blood pressure, congenital defect, trauma or infection (specifically syphilis). They may rupture, causing HEMORRHAGE, which in the heart or aorta is rapidly fatal. Again, their enlargement may cause pain, swelling or pressure on nearby organs; these complications are most serious in the arteries of the BRAIN. SURGERY for aneurysm includes tying off and removal; larger vessels may be repaired by grafting.

ANFINSEN, Christian Boehmer (1916–), US biochemist, corecipient of the 1972 Nobel Prize for Chemistry for research into the structure of the ENZYME ribonuclease (see also NUCLEIC ACIDS).

ANGINA PECTORIS, severe, short-lasting CHEST pain caused by inadequate blood supply to the myocardium (see HEART), often due to coronary artery disease such as ARTERIOSCLEROSIS. It is precipitated by exertion or other stresses which demand increased heart work. Pain may spread to nearby areas, often the arms; sweating and breathlessness may occur. It is rapidly relieved or prevented by sucking NITROGLYCERIN tablets or inhaling AMYL nitrite.

ANGIOSPERMS, or **flowering plants,** large and very important class of seed-bearing plants, characterized by having seeds that develop

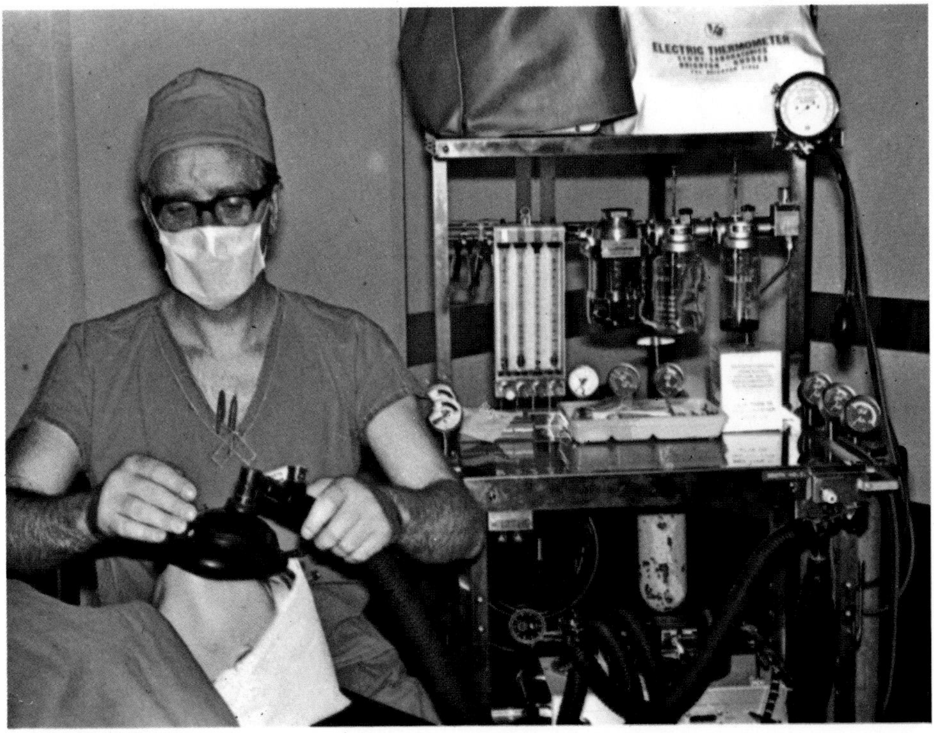

An anesthetic machine in use at University College Hospital, London.

completely enclosed in the tissue of the parent plant, rather than unprotected as in the only other seed-bearing group, the GYMNOSPERMS. Containing about 250 000 species distributed throughout the world, and ranging in size from tiny herbs to huge trees, angiosperms are the dominant land flora of the present day. They have sophisticated mechanisms to ensure that pollination and fertilization take place and that the resulting seeds are readily dispersed and able to germinate. There are two subclasses: MONOCOTYLEDONS (with one seed-leaf) and DICOTYLEDONS (with two).

ANGLE, in plane GEOMETRY, the figure formed by the intersection of two straight lines. The point of intersection is known as the vertex.

Consider the two lines to be radii of a CIRCLE of unit radius. There is then a direct way of defining the magnitudes of angles in terms of the proportion of the circle's circumference cut off by their two sides: as the length of the circle's circumference is given by $2\pi r$, and $r=1$, the **radian** (rad) is defined as the magnitude of an angle whose two sides cut off $1/2\pi$ of the circumference. A **degree** (°) is defined as an angle whose two sides cut off 1/360 of the circumference.

An angle of $\pi/2$ rad (90°), whose sides cut off one quarter of the circumference, is a **right angle,** the two lines being said to be PERPENDICULAR. Should the two sides cut off one half of the circumference (πrad or 180°), the angle is a straight angle or straight line. Angles less than $\pi/2$ rad are termed acute; those greater than $\pi/2$ but less than πrad, obtuse; and those greater than πrad but less than 2πrad, reflex. Pairs of angles that add up to $\pi/2$ rad are termed complementary; those that add up to πrad, supplementary. Further properties of angles lie in the province of TRIGONOMETRY.

In SPHERICAL GEOMETRY a **spherical angle** is that formed by intersecting arcs of two great circles: its magnitude is equal to that of the angle between the PLANES (see DIHEDRAL ANGLE) of the great circles.

A **solid angle** is formed by a conical surface (see CONE). Considering its vertex to lie at the center of a SPHERE, then a measure of its magnitude may be obtained from the ratio between the area (L^2) of the surface of the sphere cut off by the angle, and the square (R^2) of the sphere's radius. Solid angles are measured in steradians (sr), an angle of one steradian subtending an area of R^2 at distance R.

ÅNGSTRÖM, Anders Jonas (1814–1874), Swedish physicist who was one of the founders of SPECTROSCOPY and was the first to identify hydrogen in the solar spectrum (1862). The ANGSTROM UNIT is named in his honor.

ANGSTROM UNIT (Å), the CGS UNIT used to express optical wavelengths (see LIGHT), and equal to 0.1nm.

ANHYDRIDES, compounds derived from others by reversible DEHYDRATION. Most inorganic anhydrides are soluble OXIDES which dissolve in water to give ALKALIS or OXYACIDS. (See also ACID ANHYDRIDES.)

ANHYDRITE, mineral form of anhydrous calcium sulfate (see CALCIUM), occurring worldwide as white-gray masses of orthorhombic CRYSTALS. Thick beds of anhydrite are formed where old seawater lagoons have evaporated. It is used in producing sulfuric acid and ammonium sulfate and as a drying agent.

ANILINE, or aminobenzene ($C_6H_5NH_2$), a primary aromatic AMINE. It is a toxic, oily, colorless liquid, readily oxidized to various products, and a weak BASE. It is made by the reduction of nitrobenzene or by reaction of chlorobenzene with AMMONIA. Aniline is used in making synthetic dyes, rubber, pharmaceuticals, explosives and resins. (See also AROMATIC COMPOUNDS; DIAZONIUM COMPOUNDS.)

ANIMA AND ANIMUS, terms used by JUNG to denote, respectively, the unconscious female part of the male personality and the unconscious male part of the female personality. (See also PERSONA.)

ANIMAL, any organism that is classified in the ANIMAL KINGDOM. There is no single criterion that defines an animal, and it is not possible to distinguish all animals from all plants by simply listing characters found in one group and not the other. This is because animals and plants have evolved from a common ancestor, and organisms that have changed little since the separation of these two groups display intermediate characteristics. Apart from such intermediate forms, the vast majority of animals are able to respond to stimulation by rapid, controlled movement; they are composed of cells contained within a membrane; and they are unable to manufacture foodstuffs from inorganic chemicals such as are found in soil and water.

Animals are unable to manufacture foodstuffs because they do not possess a chemical called CHLOROPHYLL which is found in almost all plants. For this reason, animals are ultimately dependent on plants for their food.

The ability to move in response to external stimulae has led to the evolution of increasingly refined sense organs with which to monitor the surrounding environment, of muscles that make movement possible, and a nervous system that links them together. The development of a co-ordinating center, the brain, makes possible a greater variety of responses in "higher" animals. (See also ANIMAL BEHAVIOR; ECOLOGY; EVOLUTION; FOSSIL; TAXONOMY; ZOOLOGY.)

ANIMAL BEHAVIOR, the responses of animals to internal and external stimuli. Study of these responses can enable advances to be made in our understanding of human PSYCHOLOGY and behavior. Animal responses may be learned by the animal during its lifetime or may be instinctive or inherited (see HEREDITY; INSTINCT).

Even the simplest animals are capable of learning—to associate a particular stimulus with pain or pleasure, to negotiate mazes, etc. Moreover, there are critical periods in an animal's life when it is capable of learning a great deal in a very short time. Thus baby geese hatched in the absence of the mother will follow the first moving object they see, another animal or a human being. If, later, they must choose between this other animal and the mother, they prefer the other animal. This rapid early learning is called **imprinting.**

Among even the most intelligent animals much behavior is instinctive: the shape of a baby's head, for example, evokes an instinctive parental response in man. The complicated dance of the BEES, by which they inform the hive of the whereabouts of food, each species of bee having its own dance "dialect," is an example of more complex instinctive behavior. Instinctive ritual, too, plays its part (see APPEASEMENT BEHAVIOR; MATING RITUALS). Instinct can determine the behavior of a single animal, or of a whole animal society (see HIBERNATION; MIGRATION). (See also AGGRESSION: ETHOLOGY; TERRITORIALITY.)

ANIMAL KINGDOM, Animalia, one of the two kingdoms into which all living organisms are classified (see also PLANT KINGDOM). In practice, the border-line between the two is not clearly defined (see ANIMAL) but most organisms fall easily into one group or the other. Although there are more than a million different species of animal, and these may be broadly arranged within a taxonomic system, many details of animal TAXONOMY remain the subject of debate among zoologists.

The Animalia are often divided into three subkingdoms, the Protozoa (unicellular animals), the Parazoa (sponges—loosely organized animals) and the Metazoa (multicellular animals). The first two of these contain only a single phylum each, but zoologists recognize about 25 phyla within the Metazoa. Within these phyla, the animals are divided into classes—primarily according to their bodily structure, but also taking into account their evolutionary history (see EVOLUTION). Classes may be grouped into subphyla and divided into subclasses, superorders and orders. Within the orders are distinguished families, genera, species and subspecies (varieties). MAN forms the species *Homo sapiens*, within the genus *Homo*, within the family Hominidae, within the suborder Anthropoidea, within the order PRIMATES, within the subclass EUTHERIA of the class MAMMALIA, within the phylum CHORDATA.

ANIMISM, a term first used by E. B. TYLOR to designate a general belief in spiritual beings, which belief he held to be the origin of all religions. A common corruption of Tylor's sense is to refer to a belief that all natural objects possess spirits as animism. PIAGET has proposed that the growing child characteristically passes through an animistic phase.

ANIMUS. See ANIMA AND ANIMUS.

ANION, a negatively-charged ION which moves to the ANODE in ELECTROLYSIS; often a BASE.

ANKYLOSIS, fusion or stiffness of a JOINT, restricting movement. It may be caused by injury, SURGERY, inflammatory ARTHRITIS and TUBERCULOSIS.

ANNEALING, the slow heating and cooling of METALS and GLASS to remove stresses which have arisen in CASTING, cold working or machining (see MACHINE TOOLS). The annealed material is tougher and easier to process further. (See also METALLURGY.)

ANNELIDA, or true worms, a phylum containing worms that have segmented bodies, coelomic body cavities, a central nervous system composed of cerebral ganglia from which branches extend along the lower part of the body and, typically, bristles called chaetae that project from the body wall. Annelids are classified into three classes, the Polychaeta, Oligochaeta, and Hirudinea. The polychaetes include the ringworms and are marine and have groups of long chaetae extending in bundles from extensions of the body wall called parapodia. The oligochaetes include the earthworms, which are terrestrial forms, and some small aquatic species. They have relatively few chaetae per segment and lack parapodia—features that distinguish them from the polychaetes. Members of the Hirudinea, otherwise known as leeches, are found mainly in freshwater. They lack chaetae and parapodia, but possess suckers at each end of the body.

ANNUAL, plant that completes its life cycle in one growing season and then dies. Annuals propagate themselves only by seeds. They include such garden flowers and food plants as marigolds, cornflowers, cereals, peas and tomatoes. Preventing seeding may convert an annual, e.g. mignonette, to a BIENNIAL or a PERENNIAL.

ANNUAL RINGS, concentric rings each representing one year's growth, visible in cross sections of woody plants. Each ring is usually composed of two growth layers, a broad, large-celled layer representing spring growth and a narrow, denser layer showing summer growth. The relative amounts of these layers are affected by the environment, a fact that forms the basis of the science of DENDROCHRONOLOGY.

ANODE, the positive ELECTRODE of a BATTERY, electric CELL or ELECTRON TUBE. ELECTRONS, which conventionally carry negative charge, enter the device at the CATHODE and leave by the anode.

ANODIZING, a process for building up a corrosion-resistant or decorative oxide layer on the surface of metal (usually ALUMINUM) objects. The item to be coated is made the ANODE in a CELL containing an aqueous solution of sulfuric, chromic or oxalic acid as electrolyte. The desired oxide coating is formed when a current is passed through the cell (see ELECTROLYSIS). Further treatment can render this oxide layer waterproof, electrically insulating or brightly colored.

ANOPHELES, a genus of MOSQUITOES. The only known MALARIA carrier, it also transmits ENCEPHALITIS and FILARIASIS.

ANOREXIA NERVOSA, pathological loss of appetite with secondary MALNUTRITION and HORMONE changes. It often affects young women with diet obsession and may reflect underlying psychiatric disease.

ANORTHITE, mineral occurring in igneous rocks, consisting of calcium aluminum silicate ($CaAl_2Si_2O_8$); vitreous white or gray crystals. It is one of the three end-members (pure compounds) of the FELDSPAR group.

ANOXIA, or hypoxia, lack of oxygen in BLOOD and body TISSUES. ASPHYXIA, LUNG disease, PARALYSIS of respiratory muscles and some forms of COMA prevent enough oxygen reaching the blood. Disease of HEART or circulation may also lead to tissue anoxia. Irreversible BRAIN damage follows prolonged anoxia.

ANTABUSE, or disulfiram (tetraethylthiuram disulfide), drug used in the treatment of ALCOHOLISM. Though nontoxic, it prevents the breakdown of ACETALDEHYDE, a highly toxic product of ETHANOL metabolism. Thus if alcohol is drunk after Antabuse has been taken, unpleasant symptoms occur, including palpitations and vomiting.

ANTACIDS, mild ALKALIS or BASES taken by mouth to neutralize excess STOMACH acidity for relief of DYSPEPSIA, including peptic ULCER and HEARTBURN. MILK OF MAGNESIA, aluminum hydroxide and sodium bicarbonate are common antacids.

ANTARES, Alpha Scorpii, a DOUBLE STAR comprising a red supergiant 480 times larger than the sun and a blue star of unknown type 3 times larger than the sun (apparent magnitudes +1.23 and +5.5). It is 52pc from the earth.

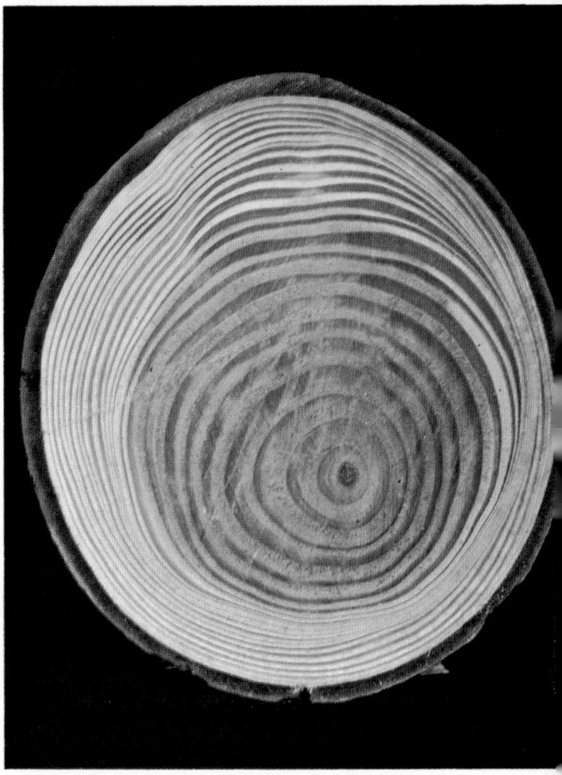

Cross section of a log of Douglas Fir (*Pseudotsuga menziesii*) showing annual rings. Faster growth occurs when the stem is young, and hence the inner rings are broader than the outer ones. The ring system is eccentric in this example because the stem was bent.

ANTENNA, or aerial, a component in an electrical circuit which radiates or receives RADIO waves. In essence a transmitting antenna is a combination of conductors which converts AC electrical ENERGY into ELECTROMAGNETIC RADIATION. The simple **dipole** consists of two straight conductors aligned end on and energized at the small gap which separates them. The length of the dipole determines the frequency for which this configuration is most efficient. It can be made directional by adding electrically isolated director and reflector conductors in front and behind. Other configurations include the folded dipole, the highly-directional loop antenna and the dish type used for MICROWAVE links. Receiving antennas can consist merely of a short DIELECTRIC rod or a length of wire for low-frequency signals. For VHF and microwave signals, complex antenna configurations similar to those used for transmissions must be used. (See also RADIO TELESCOPE.)

ANTHER. See FLOWER.

ANTHOCEROTAE. See HORNWORTS.

ANTHOCYANINS, pigments producing most red, purple and blue colors in higher plants. Their main function is to provide flowers and fruit with bright colors to attract insects and other animals for purposes

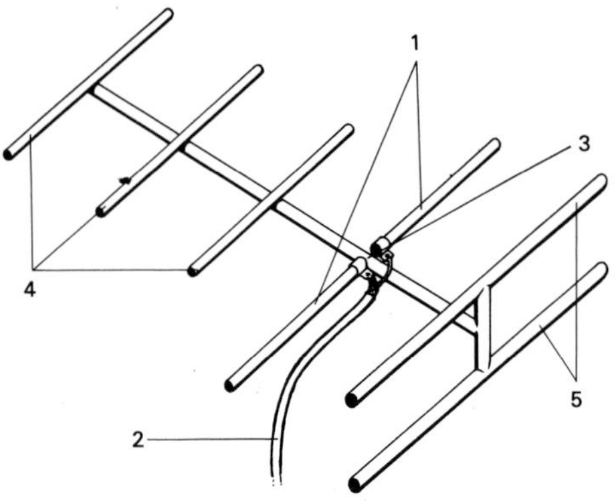

A simple TV antenna (Yagi-Uda type) consisting of a half-wave dipole (1) which converts the electromagnetic radiation received into high-frequency electrical oscillations. These are coupled to the TV via a coaxial cable lead (2) connected across the air gap (3). The antenna is made more efficient and directional by placing several passive directors (4) in front of the dipole and a two-element passive reflector (5) behind it.

of POLLINATION and seed dispersal.

ANTHOZOA, or Actinozoa, the class of the CNIDARIA including sea anemones and CORALS.

ANTHRACITE. See COAL.

ANTHRAQUINONE, a yellow, crystalline KETONE made from anthracene or by condensation of BENZENE and phthalic anhydride (see ACID ANHYDRIDES). It is the parent of numerous DYES of all colors, which are bright, fast and suitable for natural and synthetic fibers: they include ALIZARIN and COCHINEAL.

ANTHRAX, a rare BACTERIAL DISEASE causing characteristic SKIN pustules and LUNG disease; it may progress to SEPTICEMIA and death. Anthrax spores, which can survive for years, may be picked up from infected animals (such as sheep or cattle), or bone meal. Treatment is with PENICILLIN and people at risk are vaccinated; the isolation of animal cases and disinfection of spore-bearing material is essential. It was the first disease in which bacteria were shown (by KOCH) to be causative and it had one of the earliest effective vaccines, developed by PASTEUR.

ANTHROPOID APES, the animals (genus *Pan*) most closely resembling MAN (genus *Homo*) and probably sharing with him a common evolutionary ancestor (see EVOLUTION). Together the genera *Pan* and *Homo* form the family Hominidae in the suborder Anthropoidea (order PRIMATES). The apes concerned, the gorilla and chimpanzee, have a far higher intelligence than the other primates and greater manual dexterity. The term "anthropoid apes" is often also used to embrace the orang-utan (family Pongidae) and gibbon (family Hylobatidae), the other members of the superfamily Hominoidea.

ANTHROPOLOGY, the study of man from biological, cultural and social viewpoints. HERODOTUS may perhaps be called the father of anthropology, but it was not until the 14th and 15th centuries AD, with the mercantilist expansion of the Old World into new regions, that contact with other peoples kindled a scientific interest in the subject. In the modern age there are two main disciplines, physical anthropology and cultural anthropology, the latter embracing social anthropology. **Physical anthropology** is the study of man as a biological species, his past EVOLUTION and his contemporary physical characteristics. In its study of PREHISTORIC MAN it has many links with ARCHAEOLOGY, the difference being that anthropology is concerned with the remains or fossils of man himself while archaeology is concerned with the remains of his material culture. The physical anthropologist studies also the difference between RACES and groups, relying to a great extent on techniques of ANTHROPOMETRY and, more recently, genetic studies. **Cultural anthropology** is divided into several classes. ETHNOGRAPHY is the study of the culture of a single group, either primitive (see PRIMITIVE MAN) or civilized. Fieldwork is the key to

The different branches of anthropology (red) and their links with allied sciences (yellow).

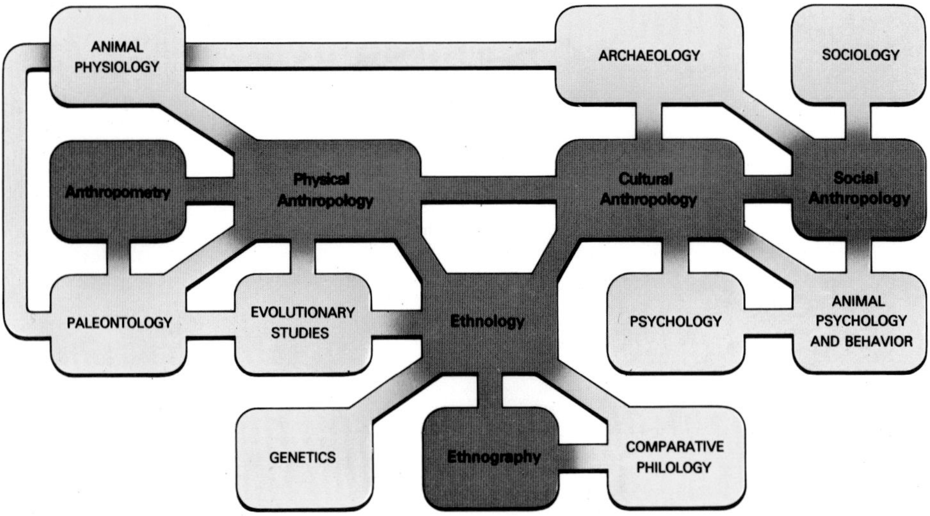

ethnographical studies, which are themselves the key to cultural anthropology. ETHNOLOGY is the comparative study of the cultures of two or more groups. Cultural anthropology is also concerned with cultures of the past, and the borderline in this case between it and archaeology is vague. **Social anthropology** is concerned primarily with social relationships and their significance and consequences in primitive societies. In recent years its field has been extended to cover more civilized societies, though these are still more generally considered the domain of SOCIOLOGY.

Though it might appear that anthropology could amplify the differences between races or groups, its results in fact indicate the opposite: physical anthropology has shown that divisions between races are at best dubious; and, with cultural anthropology, that tensions between races are of cultural, rather than biological, origin. (See also CEPHALIC INDEX; CLAN; ENDOGAMY AND EXOGAMY; FAMILY; MARRIAGE; TRIBE.)

ANTHROPOMETRY, the anthropological study of the physical characteristics of man; originally restricted to measurements of parts of the body, it now includes blood-typing, biostatistics, etc. Anthropometry has contributed considerably to modern ideas of human evolution.

ANTIBIOTICS, substances produced by microorganisms that kill or prevent growth of other microorganisms; their properties are made use of in the treatment of bacterial and fungal infection. PASTEUR noted the effect and Alexander FLEMING in 1929 first showed that the mold *Penicillium notatum* produced PENICILLIN, a substance able to destroy certain bacteria. It was not until 1940 that FLOREY and CHAIN were able to manufacture sufficient penicillin for clinical use. The isolation of STREPTOMYCIN by WAKSMAN, of Gramicidin (from tyrothricin) by DUBOS, and of the Cephalosporins were among early discoveries of antibiotics useful in human infection. Numerous varieties of antibiotics now exist and the search continues for new ones. Semi-synthetic antibiotics, in which the basic molecule is chemically modified, have increased the range of naturally occurring substances.

Each antibiotic is effective against a wider or narrower range of bacteria at a given dosage; their mode of action ranges from preventing cell-wall synthesis to interference with PROTEIN and NUCLEIC ACID metabolism. Bacteria resistant to antibiotics either inherently lack susceptibility to their mode of action or have acquired resistance by ADAPTATION (e.g., by learning to make substances which inactivate an antibiotic). Among the more important antibiotics are the PENICILLINS, Cephalosporins, TETRACYCLINES, STREPTOMYCIN, Gentamicin and Rifampicin. Each group has its own particular value and side effect, and antibiotics may induce ALLERGY. Many antibiotics are effective by mouth but INJECTION may be more suitable; topical application can also be used.

ANTIBODIES AND ANTIGENS. As one of the body's defense mechanisms, PROTEINS called antibodies are made by specialized white cells to counter foreign proteins known as antigens. Common antigens are VIRUSES, bacterial products (including TOXINS) and allergens (see ALLERGY). A specific antibody is made for each antigen. Antibody reacts with antigen in the body, leading to a number of effects including enhanced phagocytosis by white cells, activation of complement (a substance capable of damaging cell membranes) and HISTAMINE release. Antibodies are produced faster and in greater numbers if the body has previously encountered the particular antigen. IMMUNITY to second attacks of diseases such as MEASLES and CHICKENPOX, and VACCINATION against diseases not yet contracted are based on this principle. Antibody detection in blood samples may show AGGLUTININS, precipitins or complement fixation, according to the technique used and the antibody involved.

ANTICOAGULANTS, drugs that interfere with blood CLOTTING, used to treat or prevent THROMBOSIS and clot EMBOLISM. The two main types are heparin, which is injected and has an immediate but short-lived effect, and the coumarins (including WARFARIN) which are taken by mouth and are longer-lasting. They affect different parts of the clotting mechanism, coumarins depleting factors made in the LIVER.

ANTIDEPRESSANTS, drugs used in the treatment of DEPRESSION; they are of two types: tricyclic compounds and monoamine oxidase inhibitors. Although their mode of action is obscure, they have revolutionized the treatment of depression.

ANTIFREEZE, a substance added to water, particularly that in AUTOMOBILE cooling systems, to prevent ice forming in cold weather. The additive most commonly used is ETHYLENE GLYCOL; METHANOL and ETHANOL, although cheaper alternatives, tend to need more frequent replacement, being much more volatile.

ANTIGENS. See ANTIBODIES AND ANTIGENS.

ANTIHISTAMINES, drugs that counteract HISTAMINE action; they are useful in HAY FEVER and HIVES (in which ALLERGY causes histamine release) and in some insect bites. They also act as SEDATIVES and may relieve MOTION SICKNESS.

ANTIKNOCK ADDITIVES, substances added to GASOLINE to slow the burning of the fuel and thus prevent "knocking," the premature ignition of the combustion mixture in the cylinder head. Most widely used is LEAD tetraethyl $[Pb(C_2H_5)_4]$. This is usually mixed with 1,2-dibromo- and 1,2-dichloroethane, which prevent the formation of lead deposits in the engine.

ANTILOGARITHM. See LOGARITHMS.

ANTIMATTER, a variety of MATTER differing from the matter which predominates in our part of the UNIVERSE in that it is composed of antiparticles rather than particles. Individual antiparticles, many of which have been found in COSMIC RAY showers or produced using particle ACCELERATORS, differ from their particle counterparts in that they are oppositely charged (as with the antiproton-PROTON pair) or in that their magnetic moment is orientated in the opposite sense with respect to their SPIN (as with the antineutrino and neutrino). In our part of the universe antiparticles are very short-lived, being rapidly annihilated in collisions with their corresponding particles, their mass-energy reappearing as a gamma-ray PHOTON. (The reverse is also true; a high-energy GAMMA RAY sometimes spontaneously forms itself into a positron-ELECTRON pair.) However, it is by no means inconceivable that regions of the universe exist in which all the matter is antimatter,

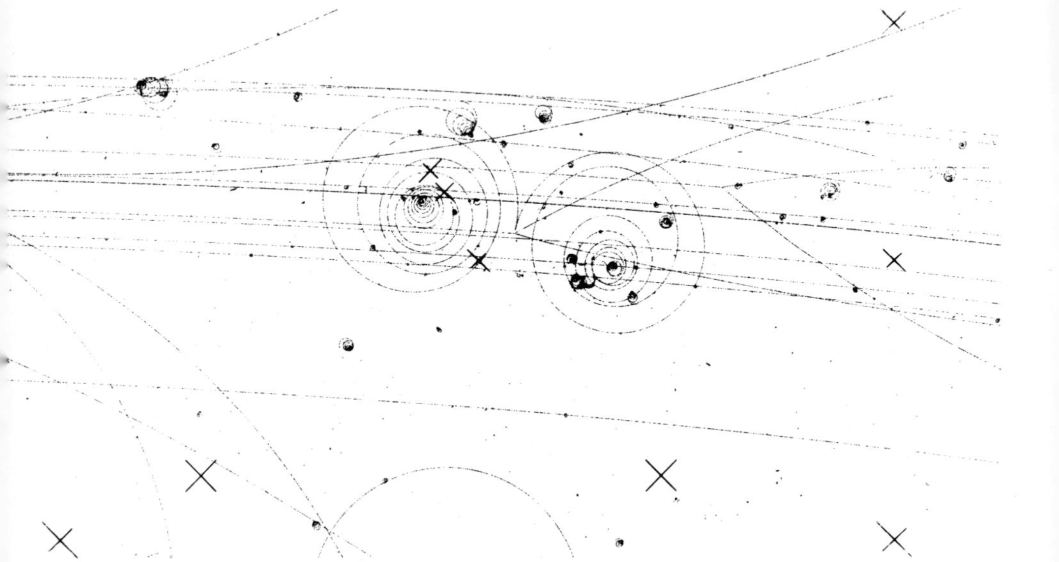

The production of a positron-electron pair from a high-energy gamma ray. In this bubble-chamber photograph, a beam of 4.2 GeV/c K^- mesons enters from the left. In an event near the middle of the picture one of these mesons gives rise to two new charged particles (forking tracks) and a gamma ray which is uncharged and hence invisible in the bubble chamber. This gamma ray almost immediately converts into a positron-electron pair (tight spirals). The spiral curving the same way as the main beam (clockwise) is the negative electron; the other (curving counterclockwise) belongs to the positron, the corresponding antiparticle.

composed of what are to us antiparticles. The first antiparticle, the positron (i.e., the antielectron), was discovered by C. D. ANDERSON in 1932, only four years after DIRAC had theoretically predicted the existence of antiparticles. (See also SUBATOMIC PARTICLES.)

ANTIMONY (Sb), brittle, silver-white metal in Group VA of the PERIODIC TABLE, occurring mainly as STIBNITE, which is roasted to give antimony (III) oxide. Its ALLOTROPY resembles that of ARSENIC. Antimony, though rather unreactive, forms trivalent and pentavalent oxides, halides and oxyanions. It is used in SEMICONDUCTORS and in lead ALLOYS, chiefly BABBITT METAL, PEWTER, type metal and in lead storage batteries. Certain antimony compounds are used in the manufacture of medicines, paints, matches, explosives and fireproofing materials. AW 121.75, mp 631°C, bp 1750°C, sg 6.691 (20°C).

ANTIPARTICLES. See ANTIMATTER; SUBATOMIC PARTICLES.

ANTIPRISM, a right PRISM, one of whose regular (see POLYGON) bases has been rotated parallel to the PLANE in which it lies through an ANGLE equal to half that subtended by each of the sides of the base at its center. The faces of an antiprism are congruent isosceles TRIANGLES.

ANTISEPTICS, or germicides, substances that kill or prevent the growth of microorganisms (particularly BACTERIA and FUNGI); they are used to avoid SEPSIS

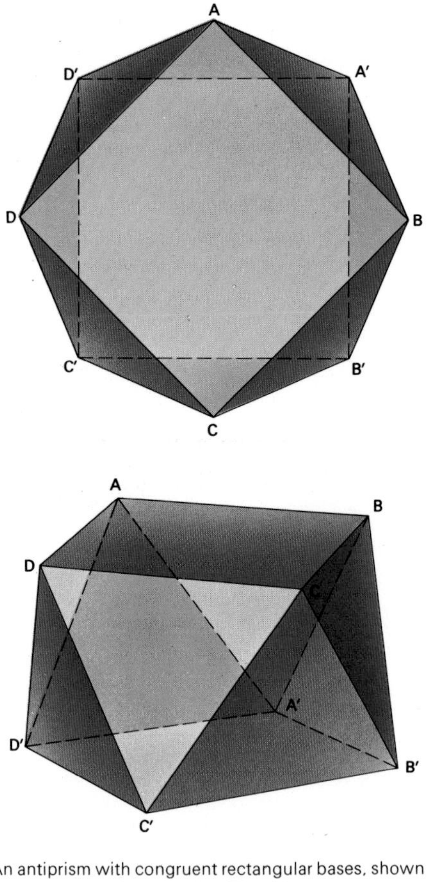

An antiprism with congruent rectangular bases, shown from above and from the side. The sides are congruent isosceles triangles: AA′B ≡ A′BB′ ≡ BB′C, etc.

from contamination of body surfaces and surgical instruments. Some antiseptics are used as disinfectants to make places or objects GERM-free. VINEGAR and cedar oil have been used from earliest times to treat wounds and for EMBALMING. Modern antisepsis was pioneered by SEMMELWEIS, LISTER and KOCH, and dramatically reduced deaths from childbirth and surgery. Commonly used antiseptics and disinfectants include IODINE, CHLORINE, hypochlorous acid, ETHANOL, isopropanol (see ALCOHOLS), PHENOLS (including hexachlorophene), quaternary ammonium salts (see AMMONIA), FORMALDEHYDE, hydrogen PEROXIDE, potassium permanganate, and acriflavine (an acridine dye). Heat, ULTRAVIOLET and ionizing radiations also have antiseptic effects. (See also STERILIZATION; ASEPSIS.)

ANTITOXINS, antibodies produced in the body against the TOXINS of some bacteria. They are also formed after INOCULATION of toxoid, chemically inactivated toxin that can still confer IMMUNITY.

ANTLERITE, green mineral consisting of basic copper sulfate ($Cu_3[OH]_4SO_4$). A minor COPPER ore of widespread occurrence, it is formed by oxidation of primary copper minerals.

ANTLERS, the horns of deer. Unlike other animal HORNS, they are made of BONE rather than modified EPIDERMIS.

ANTS, social insects of the family Formicidae of the order Hymenoptera, recognizable through the petiole or "waist" between abdomen and THORAX. There are some 3500 species of ant, each species containing three distinct castes: male, female and worker. **Males** can be found only at certain times of year: winged, they are not readmitted to the nest after the mating flight. The **queen** is likewise winged, but she rubs her wings off after mating; she may survive for as long as 15 years, still laying eggs fertilized during the original mating flight. The **workers** are sterile females, sometimes falling into two distinct size categories, the larger ones (soldiers) defending the nest and assisting with heavier work. The most primitive ants (*Ponerinae*) may form nests with only a few individuals; nests of wood ants (*Formica rufa*), however, may contain more than 100000 individuals. *Dorylinae,* the so-called Army Ants, do not build nests at all but are nomadic, traveling in "armies" up to 150000 strong: like *Ponerinae* (but unlike the more sophisticated species, which are vegetarian) they are carnivorous. Nesting ants welcome some insects, mainly beetles, to their nests, and often "farm" APHIDS for honeydew.

ANURA, an order of the AMPHIBIA that includes tailless forms, the frogs and toads. There are about 2600 species widely distributed in temperate and tropical regions. All have long hindlimbs, a modification associated with the jumping habit. Anurans breed in water and the larva, or tadpole, has a tail which is lost as adult features develop.

ANXIETY, one of two reactions of the EGO to outside threat. **Signal anxiety** warns the ego of impending threat so that **primary anxiety,** the EMOTION connected with dissolution of the ego, may be avoided (see DEFENSE MECHANISM). Primary anxiety, however, does occur in nightmares (see DREAMS).

AORTA, the chief systemic ARTERY, distributing oxygenated blood from the heart to the whole body except the LUNGS via its branches. (See also BLOOD CIRCULATION; HEART.)

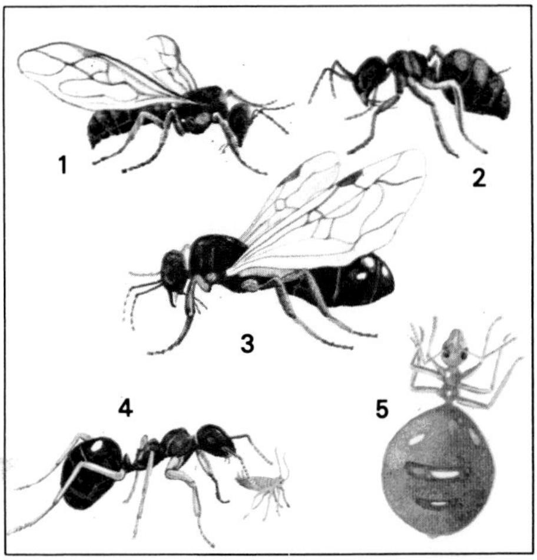

Examples of typical queen (3), worker (2) and male (1) ants. Many ants are fond of "honeydew" excreted by aphids. They "milk" the aphids by gently stroking the creatures' backs (4). "Honeypot" ants (5), found in some species, are fed so much sugary liquid by the workers that their abdomens swell grossly. They hang like storage tanks from the roof of a chamber of the nest, regurgitating the liquid food on demand.

APATITE, the chief PHOSPHATE mineral, found in the Kola peninsula, USSR, N Africa, Mont. and Fla., and mined for FERTILIZER and as the major ore of PHOSPHORUS. Its chemical composition is $Ca_5(PO_4)_3X$, where $X=F$ (fluorapatite, the most common), Cl (chlorapatite), OH (hydroxyapatite) or a mixture of all three; it forms hexagonal crystals.

APEX, that ANGLE of a POLYGON or POLYHEDRON farthest from the side or plane designated the base of the figure. The apex of a CONE is its vertex.

APHASIA, a speech defect resulting from injury to certain areas of the brain and causing inability to use or comprehend words; it may be partial (dysphasia) or total. Common causes are cerebral THROMBOSIS, HEMORRHAGE and brain TUMORS. (See SPEECH AND SPEECH DISORDERS.)

APHELION. See ORBIT.

APHIDS (or greenflies or plant-lice), some 4000 species of sap-feeding insects, comprising the family Aphididae of the order Homoptera. They have needle-like mouthparts with which they pierce the plant tissue, the pressure within this forcing the sap into the insect's gut. Because of the damage caused by their feeding and because many species carry harmful VIRUSES, aphids are one of the world's greatest crop pests. The life cycle is a complex one, so that within a species there may at any one time be a diversity of forms; winged and wingless, reproducing sexually or parthenogenetically (see PARTHENOGENESIS; REPRODUCTION; also ALTERNATION OF GENERATIONS). Aphids excrete (see EXCRETION) a substance known as honeydew, a major food source for ANTS and other insects.

APHRODISIAC, anything contributing to sexual excitement. Aphrodisiacs may be external (touch, sight, etc.) or internal (foods, drugs, etc.), the latter working usually by SUGGESTION or, like alcohol (see ETHANOL) and MARIJUANA, by lowering INHIBITIONS. Most foods traditionally regarded as aphrodisiac depend for their efficacy merely on their chance genital shape.

APODA, an order of the AMPHIBIA that includes forms in which the limbs have been lost, the caecilians. They superficially resemble large earthworms and are often blind. Some 158 species are known and are distributed in warm temperate regions of the world.

APOGEE. See ORBIT.

APOLLONIUS OF PERGA (b. c262 BC), the "Great Geometer," who, building on the foundation of EUCLID, went on to investigate the properties of CONIC SECTIONS, introducing the terms ellipse, hyperbola and parabola and exploring the properties of tangents.

APOLLO PROGRAM, US SPACE EXPLORATION program. Its prime objective was not only to land men on the moon, but to carry out research into its nature and origins. Of 17 missions, 6 were unmanned, 2 earth orbital, 2 lunar orbital, 1 was aborted by in-flight accident, and 6 made lunar landings.

APOPLEXY, obsolete term for STROKE due to cerebral HEMORRHAGE.

APOSTILB (asb), in PHOTOMETRY, a unit of LUMINANCE, being that of a uniformly diffusing surface reflecting or emitting one lumen per square metre.

APOTHECARIES' WEIGHTS, a system of weights formerly used by pharmacists in preparing medicines.

Apothecaries' weights

unit	symbol	equivalent	approximate SI equivalent
pound	lb (lb ap)	12 ounces	0.373 242kg
ounce	℥ (oz ap)	8 drams†	31.103g
dram†	ʒ (dr ap)	3 scruples	3.888g
scruple	℈ (s ap)	20 grains	1.296g
grain	gr	–	*64.798 91mg

†drachm in UK *exact equivalent

1oz ap = 1oz troy = 1.0971oz avoirdupois = 31.103g. (See also PHARMACY; WEIGHTS AND MEASURES.)

APPEASEMENT BEHAVIOR, action (usually ritual) taken by one animal to allay AGGRESSION toward it by another member of its species. There are a vast number of appeasement (or submission) rituals, the most common among mammals being the adoption of the submissive sexual posture, even by one male towards another. (See also MATING RITUALS; TERRITORIALITY.)

APPENDICITIS, inflammation of the APPENDIX, often caused by obstruction to its narrow opening, followed by swelling and bacterial infection. Acute appendicitis may lead to rupture of the organ, formation of an ABSCESS or PERITONITIS. Symptoms include abdominal pain, usually in the right lower ABDOMEN, nausea, vomiting and FEVER. Early surgical removal of the appendix is essential; any abscess requires drainage of PUS and delayed excision.

APPENDIX, Vermiform, narrow tubular structure opening into the cecum (see GASTROINTESTINAL TRACT) found in some vertebrates, including man. The human appendix contains lymphoid tissue (see LYMPH) and is probably vestigial. (See APPENDICITIS.)

APPLETON, Sir Edward Victor (1892–1965), English physicist who discovered the Appleton layer (since resolved as two layers termed F_1 and F_2) of ionized gas molecules in the IONOSPHERE. His work in atmospheric physics won him the 1947 Nobel Prize for Physics and contributed to the development of RADAR. During WWII he helped develop the ATOMIC BOMB.

APPROXIMATION, the setting of an approximate value V_a in place of a true but imprecisely known value V where $V - V_a$ lies within known limits. There are several different ways of expressing approximations and their limits of accuracy.

If we approximate 2.365 420 2 by 2.365 420 we say that this is correct to 6 decimal places, in that the six figures after the decimal point are correct (see also DECIMAL SYSTEM). It is common practice, when approximating to $(n-1)$ decimal places a number that has n decimal places, to round *up* if the nth figure after the point is 5 or greater, *down* if it is less than 5. Thus 3.65 can be written as 3.7 to 1 decimal place; 3.64 as 3.6.

Similarly, 2.365 420 2 can be expressed as 2.365 420 to 7 significant figures, since the first 7 figures are correct, rounding down.

An alternative way of writing approximations is by use of the sign $\pm$ (read "plus or minus"). Thus $2.365\,420 \pm 0.000\,000\,4$ is an approximation stating that the correct value lies between 2.365 419 6 and 2.365 420 4. This can be expressed as a percentage (see PERCENT): $2.365\,420 \pm 0.000\,016\,9\%$.

Approximation is required in almost all computations, due either to inherent inaccuracies in the calculating device or to the technique of calculation, or because greater accuracy is unnecessary. The techniques of approximation to a FUNCTION are of paramount importance in CALCULUS.

A PRIORI AND A POSTERIORI, terms descriptive of knowledge or reasoning reflecting whether or not it is the result of our experience of the real world. Alleged knowledge attained solely through reasoning from arbitrary principles is *a priori* (Latin: from earlier things); that gained empirically, from observation or experience, is *a posteriori* (Latin: from later things). This modern usage of the terms derives from the philosopher KANT.

APSIDES, Line of, the imaginary straight line connecting the two points of an elliptical ORBIT which represent the orbiting body's greatest (higher apsis) and least (lower apsis) distances from the body around which it is revolving.

AQUALUNG or **SCUBA** (*s*elf-*c*ontained *u*nderwater *b*reathing *a*pparatus), a device allowing divers to breath and move about freely underwater. It comprises a mouthpiece, a connecting tube, a valve and at least one compressed-air cylinder. The key component is the "demand valve" which allows the diver to breathe air at the PRESSURE prevailing in the surrounding water however great the pressure in his supply cylinder. Used air is vented into the water.

AQUAMARINE, transparent pale blue or blue-green semiprecious GEM stone; a variety of BERYL.

AQUA REGIA, caustic mixture of one part NITRIC ACID and three parts HYDROCHLORIC ACID, which dissolves gold and platinum. It is used in analysis of minerals and alloys, and as a powerful cleaning agent.

AQUARIUM (or aquavivarium), tank or tanks for the display of fish and marine plants and animals. Aquaria provide an environment as close to the original habitat as possible: they may be fresh- or salt-water, tropical or cold. Although pet fish were kept as long ago as 2500 BC, it was not until 1853 that the first public aquarium was established (London), the first in the US in 1856 (New York). Today there are many public and private aquaria.

AQUARIUS (the Water Bearer), a large but faint constellation on the ECLIPTIC; the 11th sign of the ZODIAC.

AQUATINT, an ENGRAVING process used with particular success by Goya, in which the plate is repeatedly etched through a porous (usually resin) ground. What are to produce the white areas of the finished print are stopped out with acid-resisting varnish before the first ETCHING, the other areas being stopped out in order of increasing darkness between subsequent etchings. The color-wash effect of aquatint is particularly striking when used in combination with drypoint.

AQUEOUS HUMOR, the clear, watery fluid between the cornea and the lens of the EYE.

AQUIFER, an underground rock formation through which GROUNDWATER can easily percolate. SANDSTONES, GRAVEL beds and jointed LIMESTONES make good aquifers.

AQUILA (the Eagle), large autumn constellation in the Northern Hemisphere, lying in the plane of the Milky Way. (See also ALTAIR.)

ARACHNIDA, a class of ARTHROPODA which includes scorpions, spiders, harvestmen, ticks and mites. Some 60000 species are known, most being terrestrial and living in soil, leaf litter and low vegetation. Some species are blood-sucking parasites.

ARAGO, Dominique François Jean (1786–1853), French physicist and mathematician whose work helped establish the wave theory of LIGHT. He discovered the polarization of light in quartz crystals (1811) and was awarded the Royal Society's Copley Medal in 1825 for demonstrating the magnetic effect of a rotating copper disk.

ARBORETUM (from Latin *arbor*, tree), a collection of clearly labeled trees and shrubs kept for educational, scientific and ornamental purposes. Notable examples are the US National Arboretum, Wash., D.C.; the Arnold Arboretum, Harvard; and in the Royal Botanic Gardens, Kew, England.

ARC. See CIRCLE.

ARCH, structural device to span openings and support loads. In architecture the simplest form of arch is the round (semicircular): here, as in most

An aqualung demand value (regulator). The flexible hoses to right (1) and left (2) are connected to the diver's mouthpiece: air from the supply cylinder enters from the top (3). When the diver breathes in, this reduces the pressure in the upper half of the regulator and the diaphragm (4) flexes upward, operating a spring-loaded valve (5) via a chain of levers (6) and thus admitting air from the cylinder through a filter (7) and a venturi (8). On breathing out, waste air is vented to the water through a one-way "duckbill" valve (9). Admitting water to the underside of the diaphragm through the vent holes (10) ensures that the diver breathes air at the prevailing water pressure.

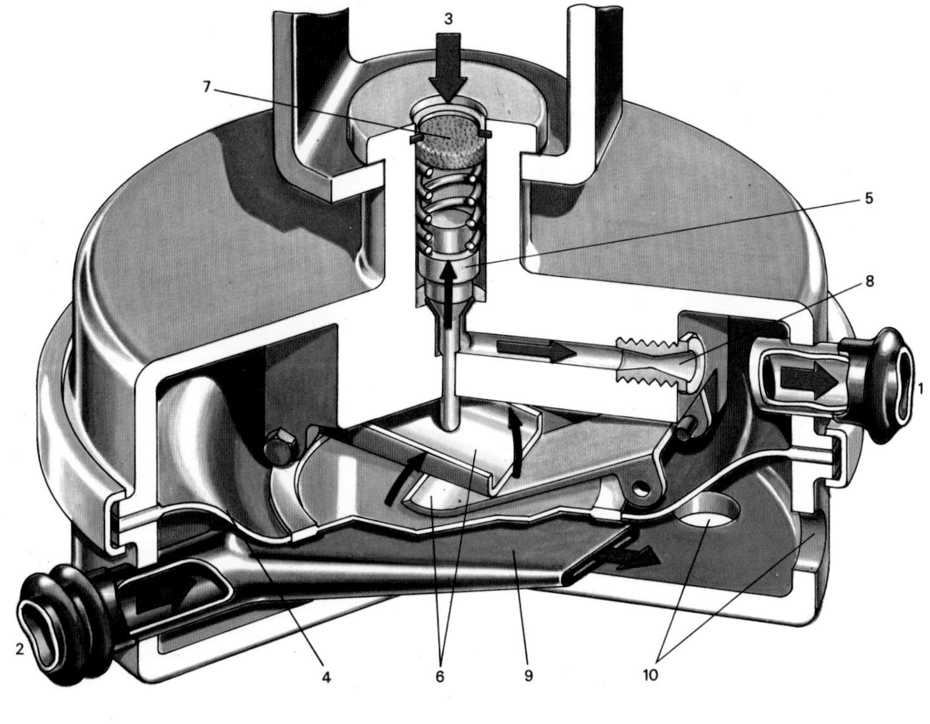

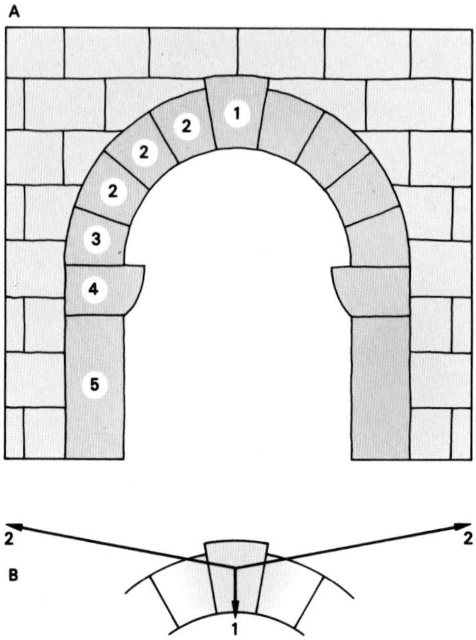

(A) Round arch with (1) keystone, (2) voussoir, (3) springer, (4) impost, (5) pier. (B) The forces acting on the keystone: the downward load (1) is balanced by the upward thrusts (2) from the two adjacent voussoirs. Ultimately, of course, the sideways components of these forces are borne by the wall into which the arch is built.

arches, wedge-shaped stones (**voussoirs**) are fitted together so that stresses in the arch exert outward forces on them; downward forces from the load combine with these to produce a diagonal resultant termed the THRUST. The voussoirs at each end of the arch are termed **springers**; that in the center, usually the last to be inserted, is the **keystone**. Although the arch was known in Ancient Egypt and Greece, it was not until Roman times that its use became popular.

ARCHAEOLOGY, the study of the past through identification and interpretation of the material remains of human cultures. A comparatively new science, involving many academic and scientific disciplines, including ANTHROPOLOGY, history, PALEOGRAPHY and PHILOLOGY, it makes use of numerous scientific techniques. Its keystone is fieldwork.

Archaeology was born in the early 18th century. There were some excavations of Roman and other sites, and the famous ROSETTA STONE, which provided the key to Egyptian HIEROGLYPHICS, was discovered in 1799 and deciphered in 1818. In 1832 archaeological time was classified into three divisions: STONE AGE, BRONZE AGE and IRON AGE; though this system is now more commonly used for cultures of PRIMITIVE MAN.

However, it was not until the 19th century that archaeology graduated from its amateur status to become a systematized science. SCHLIEMANN, Arthur EVANS, WOOLLEY, CARTER and others adopted an increasingly scientific approach in their researches.

Excavation is a painstaking procedure, as great care must be taken not to damage any object or fragment of an object, and each of the different levels of excavation must be carefully documented and photographed. The location of suitable sites for excavation is assisted by historical accounts, topographical surveys and aerial photography.

Dating is accomplished in several ways. First, of course, is comparison of the relative depths of objects that are discovered. Analysis of the types of pollen in an object can provide an indication of its date. The most widespread dating technique is RADIOCARBON DATING, incorporating the corrections formulated through discoveries in DENDROCHRONOLOGY.

ARCHAEOPTERYX, a fossil bird that is thought to be the ancestor to all other birds. (See ARCHAEORNITHES.)

ARCHAEORNITHES, a group of the AVES that includes the fossil link between birds and the REPTILIA, *Archaeopteryx*. This animal is known from rocks of the Jurassic period, about 160 million years old, and has a long tail like most reptiles, but feathers like all birds.

ARCHEAN, synonym for ARCHEOZOIC.

ARCHEOZOIC, the portion of the PRECAMBRIAN

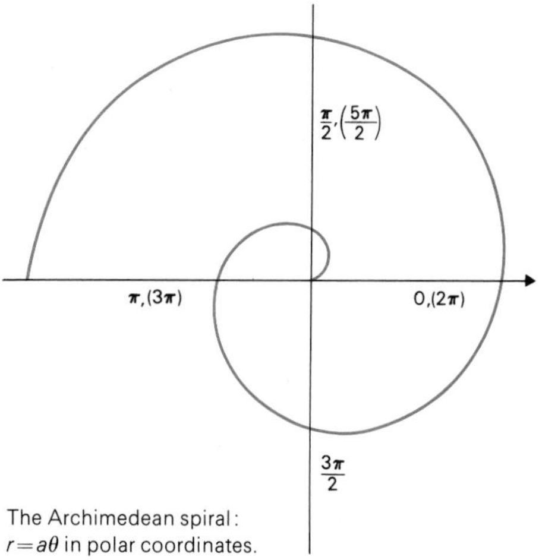

The Archimedean spiral: $r = a\theta$ in polar coordinates.

An Archimedes screw of the cylindrical type.

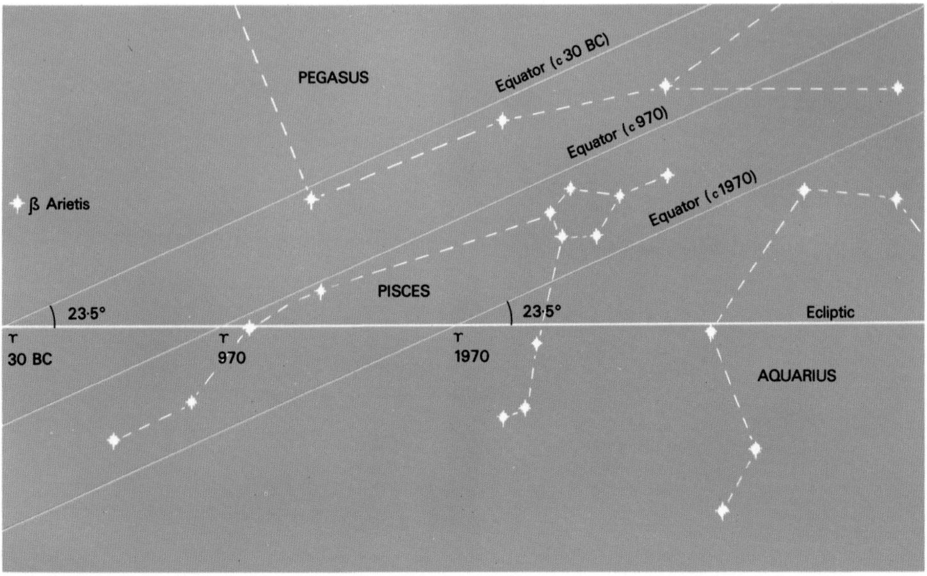

The region of the sky near the First Point of Aries, or Vernal Equinox, symbolized by ϒ, the point of intersection of the celestial equator and the ecliptic. Because of precession of the Earth's poles, ϒ moves backward along the ecliptic at a rate of about 0.014° annually. As can be seen, the point lies no longer in Aries (one star of Aries, β Arietis, can be seen) but is moving through Pisces toward Aquarius.

prior to about 2390 million years ago. (See also GEOLOGY; PROTEROZOIC.)

ARCHETYPE, term used by JUNG to refer to the COLLECTIVE UNCONSCIOUS.

ARCHIMEDES (c287–212 BC), Greek mathematician and physicist who spent most of his life at his birthplace, Syracuse (Sicily). In mathematics he worked on the areas and volumes associated with CONIC SECTIONS, fixed the value of PI (π) between $3\frac{10}{70}$ and $3\frac{10}{71}$ and defined the **Archimedean Spiral** ($r=a\theta$). He founded the science of HYDROSTATICS with his enunciation of **Archimedes' Principle**. This states that the force acting to buoy up a body partially or totally immersed in a fluid is equal to the weight of the fluid displaced. In MECHANICS he studied the properties of the LEVER and applied his experience in the construction of military catapults and grappling irons. He is also said to have invented the **Archimedes Screw**, a machine for raising water still used to irrigate fields in Egypt. This consists of a helical tube or a cylindrical tube containing a close-fitting screw with the lower end dipping in the water. When the tube (or screw) is rotated, water is moved up the tube and is discharged from the top.

ARC LAMP, an intensely bright and comparatively efficient form of LIGHTING used for lighthouses, floodlights and spotlights, invented by DAVY in 1809. An arc discharge is set up when two carbon ELECTRODES at a moderate POTENTIAL difference (typically 40V) are "struck" (touched together then drawn apart). The light is emitted from vaporized carbon IONS in the discharge. In modern lamps the arc is enclosed in an atmosphere of high-pressure XENON.

ARCTURUS, Alpha Boötis, a red giant star 11pc distant. It has a very high PROPER MOTION and an apparent magnitude of -0.04.

AREA, in plane GEOMETRY, the measure of the extent of an enclosed surface or region in terms of the number of squares with sides of unit length, or fractions of those squares, that could be fitted exactly into it. Areas are usually measured in square metres (m^2). The area of an enclosed region on a curved surface is given by the area it would cover if spread out on a PLANE surface.

ARGAND BURNER, an improved oil lamp invented in 1784 by the Genevan physicist, Amié Argand. A hollow wick and cylindrical glass chimney increased the supply of air to the flame, thus improving light output and reducing the smell, smoke and flickering that were characteristic of earlier forms of oil LIGHTING.

ARGAND DIAGRAM, a way of representing complex numbers graphically (see IMAGINARY NUMBERS), named for the French mathematician Jean Robert Argand (1768–1822) by whom it was devised.

ARGENTITE, a soft, dark SULFIDE ore of SILVER (Ag_2S), related to CHALCOCITE. It occurs in Norway, Czechoslovakia, South America, Mexico and Nev.

ARGOL. See TARTARIC ACID.

ARGON (Ar), the commonest of the NOBLE GASES, comprising 0.934% of the ATMOSPHERE. It is used as an inert shield for arc welding and for the production of silicon and germanium crystals, to fill electric light bulbs and fluorescent lamps, and in argon–ion LASERS. AW 39.9, mp $-189°C$, bp $-186°C$.

ARIES (the Ram), second constellation of the ZODIAC. In N skies it is a winter constellation.

ARIES, First Point of, the vernal equinox, a point of intersection of the EQUATOR and the ECLIPTIC used as the zero of celestial longitude. Owing to PRECESSION, the First Point of Aries is currently in PISCES.

ARISTARCHUS OF SAMOS (c310–230 BC), Alexandrian Greek astronomer who realized that the sun is larger than the earth and who is reported by ARCHIMEDES to have taught that the earth orbited a motionless sun.

ARISTOTLE (384–322 BC), Greek philosopher and scientist, a pupil of PLATO, the foremost systematizer of the knowledge of the ancient world and founder of the Peripatetic School of philosophy. In science his best work concerned biology. He proposed new principles for the classification of animals; he effectively founded the science of comparative ANATOMY, and contributed greatly to EMBRYOLOGY. His physics and cosmology were less successful although none the less influential: his views in this area, in the guise of Aristotelianism, dominated the mind of science down to the renaissance. He rejected the atomic theory and held the essence of matter to reside in the four qualities—hot, cold, wet and dry—the combination of these constituting the four Aristotelian elements—earth, air, fire and water. The heavens, kept in motion by the Unmoved Mover, are composed of a fifth "quintessential" element—aether—and circle a fixed, central and spherical earth. In logic, Aristotle originated the formal method of the SYLLOGISM; however, it was the application of the Aristotelian logic to scientific problems which in great measure led to the sterile RATIONALISM of medieval science.

ARITHMETIC (from Greek *arithmos*, number), the science of NUMBER. Until the 16th century arithmetic was viewed as the study of all the properties and relations of all numbers; in modern times, the term usually denotes the study of the positive REAL NUMBERS and ZERO under the operations of ADDITION, SUBTRACTION, MULTIPLICATION and DIVISION. Arithmetic can therefore be viewed as merely a special case of ALGEBRA, although it is of importance in considerations of the history of MATHEMATICS.

ARKWRIGHT, Sir Richard (1732–1792), English industrialist and inventor of cotton carding and SPINNING machinery. In 1769 he patented a spinning frame which was the first machine able to produce

The derivation of the four elements from the Aristotelian qualities: for example, fire represents the union of the hot with the dry.

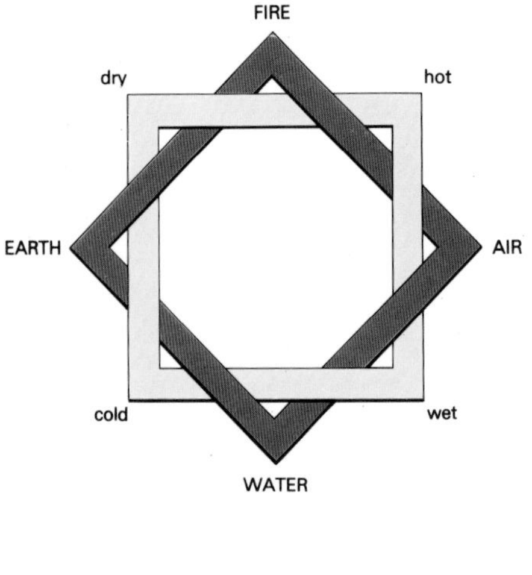

A typical armillary sphere.

cotton thread strong enough to use in the warp. He was a pioneer of the factory system of production, building several water- and later steam-powered mills. (See also CROMPTON, SAMUEL; HARGREAVES, JAMES.)

ARM, the part of the human forelimb between WRIST and SHOULDER. The upper arm bone (humerus) is attached at the shoulder by a ball and socket JOINT and to the lower arm by a hinge joint at the elbow. The lower arm consists of the ulna and the radius, which rotates over the ulna to turn the wrist. The chief arm muscles are the BICEPS and TRICEPS and the strong hand muscles.

ARMATURE, the part of an electric MOTOR or GENERATOR which includes the principal current-carrying windings. In small motors it usually comprises several coils of wire wound on a soft iron core and mounted on the drive shaft, though on larger AC motors the armature is often the stationary component. When the current flows in the armature winding of a motor, it interacts with the magnetic field produced by the field windings giving rise to a TORQUE between the rotor and stator. In the generator the armature is rotated in a magnetic field giving rise to an ELECTROMOTIVE FORCE in the windings.

ARMILLARY SPHERE, a model displaying the mutual dispositions of the imaginary circles of classical astronomy in which metal circles were used to represent the celestial EQUATOR, the ECLIPTIC, the TROPICS, the arctic and antarctic circles, the hours of the day, the HORIZON and a MERIDIAN. Derived from ancient astronomical instruments, armillary spheres became particularly popular in the 17th and 18th centuries.

ARMORPLATE, metal protective covering consisting usually, but not always, of several layers of differently heat-treated metals (see METALLURGY) whose combined properties provide the best possible protection. Case-hardened (see CASE-HARDENING) body armor appeared as early as the 15th century; but it was not until the 19th century that armorplate was applied to ships, playing an important part in the Crimean War. Since the 19th century, armorplate has been made either of several laminae, typically carbon STEEL with a backing of WROUGHT IRON, or as a single thick layer of metal. Modern armorplate is applied to land vehicles such as tanks as well as to sea-going craft.

ARMSTRONG, Edwin Howard (1890–1954), US electronics engineer who developed the FEEDBACK concept for AMPLIFIERS (1912), invented the superheterodyne circuit used in radio receivers (1918) and perfected FM RADIO (1925–39).

AROMATIC COMPOUNDS, major class of organic compounds containing one or more planar rings of atoms having special stability due to their electronic structure. This includes TORUS-shaped ORBITALS above and below the plane of the ring, known as a 1-electron system, containing (4n+2) electrons (i.e. 6, 10, 14 and so on). Such systems can be represented by RESONANCE structures of alternate single and double bonds (see BOND, CHEMICAL) round the ring. Typical properties of aromatic compounds include ease of formation, tendency to react by substitution rather than addition, and modification of the properties of attached groups. The most important aromatic compounds are BENZENE and its derivatives, including PHENOLS, TOLUENE, BENZALDEHYDE, BENZOIC ACID, BENZYL ALCOHOL, ANILINE and SALICYLIC ACID. Compounds with more than one benzene ring (polycyclic) include NAPHTHALENE and anthracene. Non-benzenoid aromatics include cyclopentadienyl and cycloheptatrienylium ions (see FERROCENE) and azulene. Many HETEROCYCLIC COMPOUNDS are aromatic.

Aromatic compounds

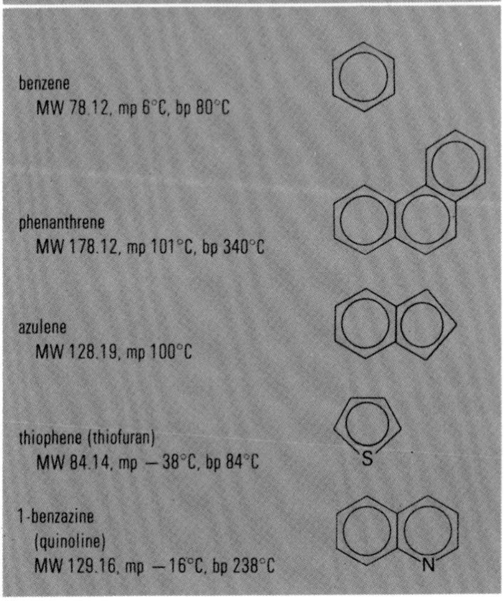

ARRHENIUS, Svante August (1859–1927), Swedish physical chemist whose theory concerning the DISSOCIATION of SALTS in solution (see CONDUCTIVITY) earned him the 1903 Nobel Prize for Chemistry and laid the foundations for the study of ELECTROCHEMISTRY.

ARROW-WORMS. See CHAETOGNATHA.

ARSENIC (As), metalloid in Group VA of the PERIODIC TABLE. Its chief ore is ARSENOPYRITE, which is roasted to give arsenic (III) oxide, or white arsenic, used as a poison. Arsenic has two main allotropes (see ALLOTROPY): yellow arsenic, As_4, resembling white PHOSPHORUS; and gray (metallic) arsenic. It burns in air and reacts with most other elements, forming trivalent and pentavalent compounds, all highly toxic. It is used as a doping agent in TRANSISTORS; gallium arsenide is used in LASERS. AW 74.9, subl 613°C, sg 1.97 (yellow), 5.73 (gray).

ARSENOPYRITE, or mispickel, silvery-white mineral with metallic luster, crystallizing in the monoclinic system, iron sulfarsenide (FeAsS). It is the chief ore of ARSENIC, and is found in the US, Canada, Germany, England and Scandinavia.

ARTERIOSCLEROSIS, disease of arteries in which the wall becomes thickened and rigid, and blood flow is hindered. **Atherosclerosis** is the formation of fatty deposits (containing CHOLESTEROL) in the inner lining of an ARTERY, followed by scarring and calcification. It is commoner in older age groups, but in DIABETES, disorders of fat METABOLISM and high-blood pressure, its appearance may be earlier. Excess saturated fats in the blood may play a role in its formation. A rarer form, **medial sclerosis,** is caused by degeneration and calcification of the middle muscular layer of the artery. Narrowing or obstruction of cerebral arteries may lead to STROKE, while that of coronary arteries causes ANGINA PECTORIS and CORONARY THROMBOSIS. Reduced blood flow to the limbs may cause CRAMP on exertion, ULCERS and GANGRENE. Established arteriosclerosis cannot be reversed, but a low fat diet, exercise and the avoidance of smoking help in prevention. Surgery by artery-replacement or removal of deposits is occasionally indicated.

ARTERY, blood vessel which carries BLOOD from the HEART to the TISSUES (see BLOOD CIRCULATION). The arteries are elastic and expand with each PULSE. In most vertebrates, the two main arteries leaving the heart are the pulmonary artery, which carries blood from the body to the LUNGS to be reoxygenated, and the AORTA which supplies the body with oxygenated blood. Major arteries supply each limb and organ and within each they divide repeatedly until arterioles and CAPILLARIES are reached. Fish have only one arterial system, which leads from the heart via the GILLS to the body. (See also ARTERIOSCLEROSIS.)

ARTESIAN WELL, a well in which water rises under hydrostatic pressure above the level of the AQUIFER in which it has been confined by overlying impervious strata. Often pumping is necessary to bring the water to the surface, but true artesian wells (named for the French province of Artois where they were first constructed) flow without assistance.

ARTHRITIS, INFLAMMATION, with pain and swelling, of JOINTS. **Osteoarthritis** is most common, though there is no true inflammation; it is a wear-and-tear arthritis, causing pain and limitation of movement. OBESITY, previous trauma and inflam-

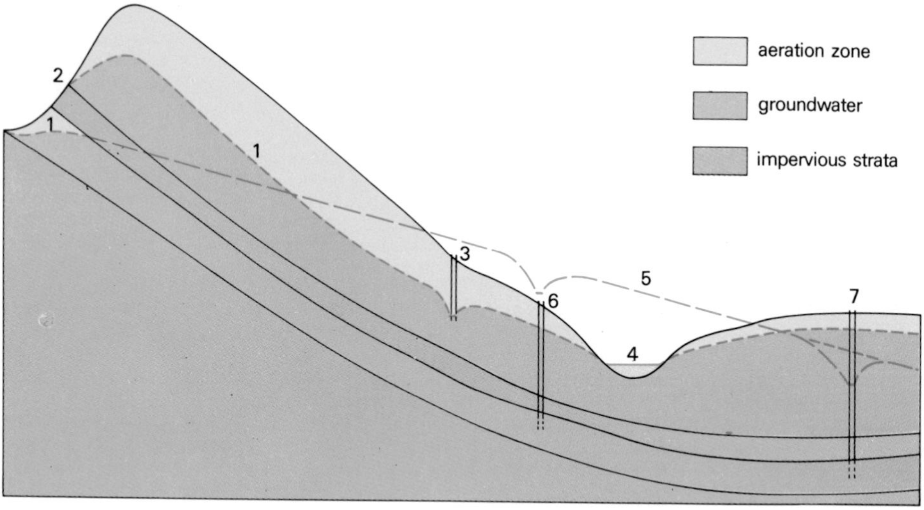

Two aquifers, the lower one artesian being confined by impervious strata above and below. In the upper stratum, the water table (1)—i.e., the upper limit of the groundwater zone—is drawn down by a spring at (2), a pumped well (3) and seepage into a stream (4). Only a small section of water table (1) occurs in the lower stratum, but the piezometric surface (5) represents the height of water which would produce the observed hydrostatic pressure distribution. Like the water table, it is drawn down by the two artesian wells (6 and 7) sunk into the artesian aquifer. One well (6) flows naturally, since the piezometric surface is above ground level at that point; the other (7), where this is not so, requires to be pumped.

matory arthritis predispose. Bacterial infection (e.g., by STAPHYLOCOCCI or TUBERCULOSIS), with PUS in the joint, and GOUT, due to deposition of crystals in SYNOVIAL FLUID, may lead to serious joint destruction. **Rheumatoid arthritis** is a systemic disease manifested mainly in joints, with inflammation of synovial membranes and secondary destruction. In the hands, tendons may be disrupted and extreme deformity can result. Arthritis also occurs in many other systemic diseases including RHEUMATIC FEVER, LUPUS ERYTHEMATOSUS, PSORIASIS and some VENEREAL DISEASES. Treatment of arthritis includes anti-inflammatory ANALGESICS (e.g., ASPIRIN), rest, local heat and PHYSIOTHERAPY. STEROIDS are sometimes helpful but their long-term use is now discouraged. Badly damaged joints may need surgical treatment or replacement.

ARTHROPODA, the largest and most diverse phylum of the animal kingdom. Most members are characterized by a segmented EXOSKELETON with jointed limbs, a heart that lies in the upper part of the body and a nerve cord that runs along the lower part of the body. Members of the phylum are classified into five classes, the CRUSTACEA, ONYCHOPHORA, MYRIOPODA, INSECTA and ARACHNIDA.

ARTIFICIAL INSEMINATION, introduction of SPERM into the vagina by means other than copulation. The technique is widely used for breeding livestock as it produces many offspring from one selected male (see HEREDITY). It has a limited use in treating human impotence and STERILITY.

ARTIFICIAL LIMBS. See PROSTHETICS.

ARTIFICIAL ORGANS, mechanical devices that can perform the functions of bodily organs. The **heart-lung machine** can maintain BLOOD CIRCULATION and oxygenation and has enabled much new cardiac SURGERY. **Artificial kidneys** clear waste products from the blood (see UREMIA) by DIALYSIS and may take over KIDNEY function for life. Both machines require ANTICOAGULANTS during use.

ARTIFICIAL RESPIRATION, the means of inducing RESPIRATION when it has ceased, as after DROWNING, ASPHYXIA, in COMA or respiratory PARALYSIS. It must be continued until natural breathing returns and ensuring a clear airway via the mouth to the lungs is essential. The most common first aid methods are: "mouth-to-mouth", in which air is breathed via the mouth into the lungs and is then allowed to escape, and the less effective Holger Nielsen technique where rhythmic movements of the CHEST force air out and encourage its entry alternately. If prolonged artificial respiration is needed, mechanical pumps are used and these may support respiration for months or even years.

ARTIFICIAL SELECTION, the method by which man has determined the evolution of certain animals and plants by selecting for breeding those individuals which display desired characters. These include fast growth rates in cattle, heavy crop yields and disease resistance in plants. (See also DOMESTICATION.)

ARTILLERY, once the term for all military machinery, it now refers to guns too heavy to be carried by one or two men. The branch of the army involved is also known as the artillery. Modern artillery may be said to have its origins in the 14th century when weapons that used gunpowder were first developed. The importance of artillery in battle increased as it became more mobile, and scientific advances improved its accuracy and effectiveness. WWII saw the development of specialized antitank and antiaircraft guns, and the first really effective use of rockets. Since then the guided MISSILE has been produced, with its long ranges and high accuracy.

ARTIODACTYLA, an order of the MAMMALIA that includes the hippos, camels, deer, giraffe, pronghorns and the buffalo. All are even-toed; that is, they have cloven hooves, a specialization that originally evolved for fast running. Among the artiodactyls are to be found some of the most important domestic animals such as swine, cattle, sheep and goats.

ASBESTOS, name of various fibrous minerals, chiefly CHRYSOTILE and AMPHIBOLE. Canada and the USSR are the chief producers. It is a valuable industrial material because it is refractory, alkali- and acid-resistant and an electrical insulator. It can be spun to make fireproof fabrics for protective clothing and safety curtains, or molded to make tiles, bricks and automobile brake linings. Asbestos particles may cause PNEUMOCONIOSIS and lung CANCER if inhaled.

ASCHELMINTHES, a phylum of small or microscopic worm-like animals. The members of the phylum form a rather heterogeneous group of animals previously classified in separate phyla. They include the rotifers, microscopic forms found mainly in freshwater, and the nematodes, or roundworms, which are parasitic or free-living in freshwater or soil.

ASCLEPIUS. See AESCULAPIUS.

ASCOMYCETES. See FUNGI.

ASCORBIC ACID, or Vitamin C. See VITAMIN.

ASEPSIS, the principle in modern SURGERY of excluding GERMS. Means include the STERILIZATION of instruments, dressings, gowns and gloves, and the use of ANTISEPTICS for cleaning the skin of patient and surgeon.

ASH, the inorganic residue left after organic substances are burned. Some ash is commercially useful: plant ash is a FERTILIZER and seaweed ash yields IODINE. Conglomerated ash may be called clinker.

ASPHALT, a tough black material used in road paving, roofing and canal and reservoir lining. Now obtained mainly from PETROLEUM refinery residues (although natural deposits are still worked), it consists mainly of heavy HYDROCARBONS.

ASPHYXIA, the complex of symptoms due to inability to take oxygen into or excrete carbon dioxide from the LUNGS. The commonest causes are DROWNING, suffocation or strangling; inhalation of toxic gases, obstruction of LARYNX, TRACHEA or BRONCHI (which can occur in severe cases of CROUP, ASTHMA and DIPHTHERIA). Early ARTIFICIAL RESPIRATION is essential.

ASPIRIN, or **acetylsalicylic acid,** an effective analgesic, which also reduces FEVER and INFLAMMATION and also affects BLOOD platelets. It is useful in HEADACHE, minor feverish illness, MENSTRUATION pain, RHEUMATIC FEVER, inflammatory ARTHRITIS, and may also be used to prevent THROMBOSIS. Aspirin may cause gastrointestinal irritation and HEMORRHAGE, and should be avoided in cases of peptic ULCER.

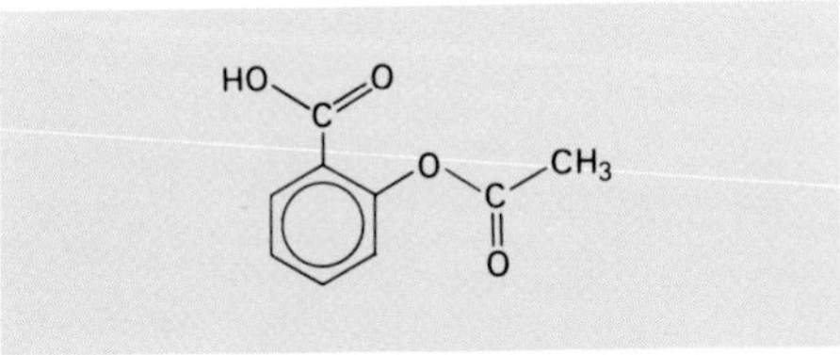

The structure of acetylsalicylic acid (aspirin).

ASSAYING, a method of chemical ANALYSIS for determining NOBLE METALS in ores or alloys, used since the 2nd millennium BC. The sample is fused with a flux containing LEAD (II) oxide. This produces a lead button containing the noble metals, which is heated in oxygen to oxidize the lead and other impurities, leaving a bead of the noble metals which is weighed and separated chemically.

ASSOCIATION, in PSYCHOLOGY, the mental linking of one item with others: e.g., black and white, Tom with Dick and Harry, etc. The connections are described by the primary (similarity and contiguity) and secondary (frequency, recency, vividness and primacy) laws of association. In **association tests,** subjects are presented with one word and asked to respond either with a specifically related word, such as a rhyme or antonym, or merely with the first word that comes to mind.

ASSOCIATIONISM, a psychological school which held that the sole mechanism of human learning consisted in the permanent association in the intellect of impressions which had been repeatedly presented to the senses. Originating in the philosophy of John LOCKE and developed through the work of John Gay, David HARTLEY, James and John Stuart MILL and Alexander Bain, the "association of ideas" was the dominant theme in British PSYCHOLOGY for 200 years.

ASSOCIATIVE LAW. See ALGEBRA.

ASTATINE (At), radioactive HALOGEN, occurring naturally in minute quantities, and prepared by bombarding BISMUTH with ALPHA PARTICLES. The most stable isotope, At^{210}, has half-life 8.3h. Tracer studies show that astatine closely resembles IODINE. AW 210, mp 302°C, bp 337°C.

ASTEROIDEA, a class of ECHINODERMATA which includes the starfishes. They are star-shaped with arms—five or multiples of five—there being no clear anatomical distinction between the arms and the central body.

ASTEROIDS, the thousands of planetoids or minor planets, ranging in diameter from a few metres to 760km (CERES), most of whose orbits lie in the Asteroid Belt between the orbits of Mars and Jupiter. Vesta is the only asteroid visible to the naked eye, though Ceres was the first to be discovered (1801 by PIAZZI). Their total mass is estimated to be 0.001 that of the earth. A second asteroid belt beyond the orbit of Pluto has been postulated. (See also METEORITE; SOLAR SYSTEM.)

ASTHMA, chronic respiratory disease marked by recurrent attacks of wheezing and acute breathlessness. It is due to abnormal bronchial sensitivity and is usually associated with ALLERGY to house dust mite, pollen, FUNGI, furs and other substances which may precipitate an attack. Chest infection, exercise or emotional upset may also provoke an attack. The symptoms are caused by spasm of bronchioles (see BRONCHI) and the accumulation of thick MUCUS. Cyanosis may occur in severe attacks. Desensitization INJECTIONS, cromoglycate, STEROIDS and drugs that dilate bronchi are used in prevention; acute attacks may require OXYGEN, aminophylline or ADRENALINE, and steroids.

ASTIGMATISM, a defect of VISION in which the LENS of the EYE exhibits different curvatures in different planes, corrected using cylindrical lenses. Also, an ABERRATION of lenses having spherical surfaces.

A Moorish brass astrolabe made in Toledo, Spain in 1068 AD. In addition to the usual degree scales and alidade, it has six plates, five of which are inscribed (in Arabic) with the names of various cities, their latitude and the duration of their longest day.

ASTON, Francis William (1877–1945), British physicist who designed the first mass spectrograph (see MASS SPECTROSCOPY) and used it to identify and separate the ISOTOPES of the nonradioactive elements. This work earned him the Nobel Prize for Chemistry in 1922 and led to the formulation of the whole-number rule for isotopic weights (see ATOMIC WEIGHT).

ASTRINGENT, agent used to shrink mucous membranes and to dry up secretions or wet lesions. They may act by vasoconstriction, dehydration (as with ETHANOL) or by denaturing proteins (as with TANNIN).

ASTROLABE, an astronomical instrument dating from the Hellenic Period, used to measure the altitude of celestial bodies and, before the introduction of the SEXTANT, as a navigational aid. It consisted of a vertical disk with an engraved scale across which was mounted a sighting rule or "alidade" pivoted at its center.

ASTROLOGY, the art and science of divining the future from the study of the heavens. Originating in ancient Mesopotamia as a means for predicting the fate of states and their rulers, the astrology which found its way into Hellenistic culture applied itself also to the destinies of individuals. Together with the desire to devise accurate CALENDARS, astrology provided a key incentive leading to the earliest systematic ASTRONOMY and was a continuing spur to the development of astronomical techniques until the 17th century. The majority of classical and medieval astronomers, PTOLEMY and KEPLER among them, practiced astrology, often earning their livelihoods thus. Astrology exercised its greatest influence in the Graeco-Roman world and again in renaissance Europe (despite the opposition of the Church) and, although generally abandoned after the 17th century, it has continued to excite a fluctuating interest down to the present. The key datum in Western astrology is the position of the stars and planets, described relative to the 12 divisions of the ZODIAC, at the moment of an individual's birth.

ASTRONOMICAL UNIT (AU), a unit of distance equal to 149.6Gm—approximately the mean distance of the earth from the sun—used for describing distances within the solar system. (See SI UNITS.)

ASTRONOMY, the study of the heavens. Born at the crossroads of agriculture and religion, astronomy, the earliest of the sciences, was of great practical importance in ancient civilization. Before 2000 BC, Babylonians, Chinese and Egyptians all sowed their crops according to calendars computed from the regular motions of the sun and moon.

Although early Greek philosophers were more concerned with the physical nature of the heavens than with precise observation, later Greek scientists (see ARISTARCHUS; HIPPARCHUS) returned to the problems of positional astronomy. The vast achievement of Greek astronomy was epitomized in the writings of Claudius PTOLEMY. His *Almagest*, passing through Arabic translations, was eventually transmitted to medieval Europe and remained the chief authority among astronomers for over 1 400 years.

Throughout this period the main purpose of positional astronomy had been to assist in the casting of accurate horoscopes, the twin sciences of astronomy and ASTROLOGY having not yet parted company. The structure of the universe meanwhile remained the preserve of (Aristotelian) physics. The work of COPERNICUS represented an early attempt to harmonize an improved positional astronomy with a true physical theory of planetary motion. Against the judgment of antiquity that sun, moon and planets circled the earth as lanterns set in a series of concentric transparent shells, in his *de Revolutionibus* (1543) Copernicus argued that the sun lay motionless at the center of the planetary system.

Although the Copernican (or heliocentric) hypothesis proved to be a sound basis for the computation of navigators' tables (the needs for which were stimulating renewed interest in astronomy), it did not become unassailably established in astronomical theory until NEWTON published his mathematical derivation of KEPLER'S LAWS in 1687. In the meanwhile KEPLER, working on the superb observational data of Tycho BRAHE, had shown the orbit of Mars to be elliptical and not circular and GALILEO had used the newly invented TELESCOPE to discover SUNSPOTS, the phases of Venus and four moons of Jupiter.

Since the 17th century the development of astronomy has followed on successive improvements in the design of telescopes. In 1781 William HERSCHEL discovered Uranus, the first discovery of a new PLANET to be made in historical times. Measurement of the PARALLAX of a few stars in 1838 first allowed the estimation of interstellar distances. Analysis of the FRAUNHOFER LINES in the spectrum of the sun gave scientists their first indication of the chemical composition of the STARS.

In the present century the scope of observational astronomy has extended as radio and X-ray telescopes (see X-RAY ASTRONOMY) have come into use, leading to

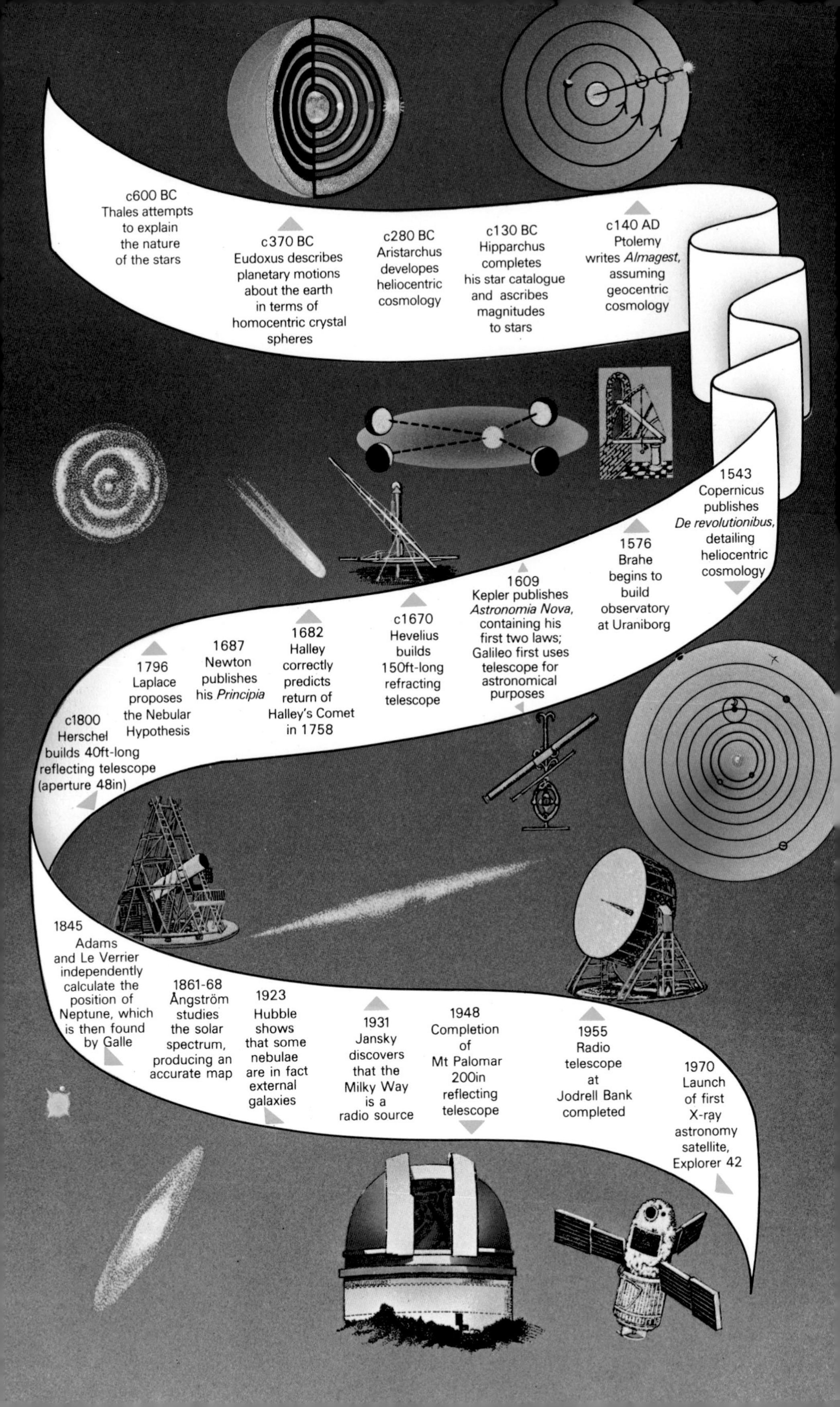
c600 BC
Thales attempts to explain the nature of the stars
c370 BC
Eudoxus describes planetary motions about the earth in terms of homocentric crystal spheres
c280 BC
Aristarchus developes heliocentric cosmology
c130 BC
Hipparchus completes his star catalogue and ascribes magnitudes to stars
c140 AD
Ptolemy writes *Almagest*, assuming geocentric cosmology
1543
Copernicus publishes *De revolutionibus*, detailing heliocentric cosmology
1576
Brahe begins to build observatory at Uraniborg
1609
Kepler publishes *Astronomia Nova*, containing his first two laws; Galileo first uses telescope for astronomical purposes
c1670
Hevelius builds 150ft-long refracting telescope
1682
Halley correctly predicts return of Halley's Comet in 1758
1687
Newton publishes his *Principia*
1796
Laplace proposes the Nebular Hypothesis
c1800
Herschel builds 40ft-long reflecting telescope (aperture 48in)
1845
Adams and Le Verrier independently calculate the position of Neptune, which is then found by Galle
1861-68
Ångström studies the solar spectrum, producing an accurate map
1923
Hubble shows that some nebulae are in fact external galaxies
1931
Jansky discovers that the Milky Way is a radio source
1948
Completion of Mt Palomar 200in reflecting telescope
1955
Radio telescope at Jodrell Bank completed
1970
Launch of first X-ray astronomy satellite, Explorer 42

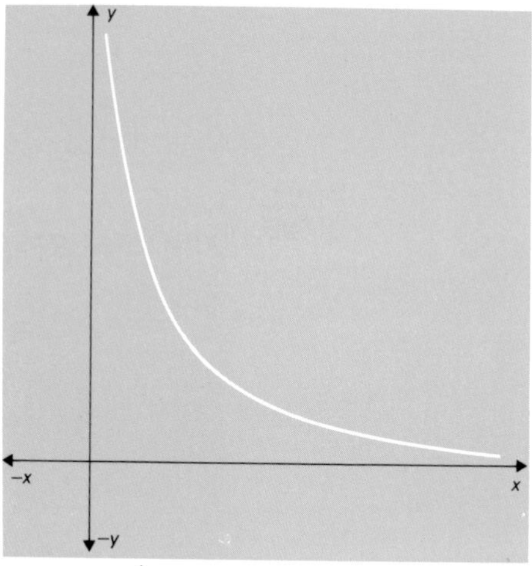

The curve $y=\frac{1}{x}(x>0)$, whose asymptotes are the x- and y-axes.

the discovery of QUASARS, PULSARS and neutron stars. In their turn these discoveries have enabled cosmologists to develop ever more self-consistent models of the UNIVERSE. (See also COSMOLOGY; OBSERVATORY.)

ASTROPHOTOGRAPHY, photography of celestial objects, usually by focusing a TELESCOPE onto a photographic plate. Of paramount importance in, e.g., measurement of stellar PARALLAX, astrophotography has almost entirely replaced direct observation.

ASTROPHYSICS, deals with the physical and chemical nature of celestial objects and events, using data produced by RADIO ASTRONOMY and SPECTROSCOPY. By investigating the laws of the universe as they currently operate, astronomers can formulate theories of stellar evolution and behavior (see COSMOLOGY).

ASYMPTOTE, term used in ANALYTIC GEOMETRY. If a CURVE is such that a straight LINE may be drawn which the curve approaches ever more closely (but never meets) with increasing distance from the origin, then the straight line is termed an asymptote and the curve asymptotic. Consider the curve with equation $y=1/x$, where x is always greater than zero. As x tends to infinity, y tends to, but never equals, zero. One asymptote of the curve is therefore the x-axis. In the same way, as x tends to zero, y tends to infinity; and therefore the other asymptote of the curve is the y-axis. All hyperbolas (see CONIC SECTIONS) are asymptotic.

ATABRINE (quinacrine hydrochloride), a synthetic ALKALOID drug for treating MALARIA and TAPEWORM, introduced in WWII when the Japanese captured natural QUININE supplies, and now rarely used.

ATAVISM, the inheritance (see HEREDITY) by an individual organism of characteristics not shown by its parental generation. Once thought to be throwbacks to an ancestral form, atavisms are now known to be primarily the result of the random reappearance of recessive traits (see GENETICS), though they may result also from aberrations in the development of the embryo or from disease.

ATAXIA, impaired coordination of body movements resulting in unsteady gait, difficulty in fine movements and speech disorder. Caused by disease of the cerebellum or spinal cord, ataxia occurs with MULTIPLE SCLEROSIS, certain hereditary conditions and in the late stages of syphilis (see VENEREAL DISEASES).

ATHEROSCLEROSIS. See ARTERIOSCLEROSIS.

ATHLETE'S FOOT, a common form of RINGWORM, a contagious fungal infection of the feet, causing inflammation and scaling or maceration of the skin, especially between the toes. It may be contracted in swimming pools or from shared towels or footwear. Treatment consists in foot hygiene, dusting powder, certain CARBOXYLIC ACIDS and antifungal ANTIBIOTICS.

ATLANTIC CABLES, the telegraph and telephone cables on the bed of the North Atlantic Ocean linking North America with Europe. The first successful cable was laid by promoter Cyrus West Field (1819–1892) under the direction of William Thomson (later Lord KELVIN) at the third attempt, in 1858, but this soon failed and was replaced by a more permanent one in 1866. The first Atlantic telephone cable was not laid until 1956 and the method is at present suffering strong competition from communications SATELLITES.

ATLAS, the uppermost VERTEBRA of the spinal column, supporting the skull, and forming a pivot joint with the **axis** vertebra below it, thus allowing the head to turn. (See also SKELETON.)

ATMOSPHERE, the roughly spheroidal envelope of GAS, VAPOR and AEROSOL particles surrounding the EARTH, retained by gravity and forming a major constituent in the environment of most forms of terrestrial life, protecting it from the impact of METEORS, COSMIC RAY particles and harmful solar radiation. The composition of the atmosphere and most of its physical properties vary with ALTITUDE, certain key properties being used to divide the whole into several zones, the upper and lower boundaries of which change with LATITUDE, the time of day and the season of the year. About 75% of the total MASS of atmosphere and 90% of its water vapor and aerosols are contained in the **troposphere,** the lowest zone. Excluding water vapor,the **air** of the troposphere contains 78% NITROGEN; 20% OXYGEN; 0.9% ARGON; and 0.03% CARBON dioxide, together with traces of the other NOBLE GASES; and METHANE, HYDROGEN and

The physical properties (left) of the atmosphere, and the phenomena which occur in it (center), are strikingly dependent on the altitude. The regions into which the atmosphere is divided are shown in the right-hand column, with the characteristic molecules, atoms and ions found in each. The atmosphere is an effective shield against particles—all meteors except the very largest are burned up in the lower ionosphere and mesosphere—and also against radiation. X rays and most ultraviolet radiation are absorbed by the ionosphere, creating the layers of ions for which it is named. Most infrared radiation is absorbed in the mesosphere and stratosphere, where also the remaining harmful ultraviolet radiation is absorbed by the ozone layer. Only visible light, a narrow band of infrared radiation, and radio waves reach the earth's surface. This is illustrated in the center column, where the horizontal scale (top) shows the wavelength of the radiation in three scales: angstrom, centimetre and metre. Radio waves transmitted from the surface are reflected (except VHF) by the various layers (D, E, F_1 and F_2) of the ionosphere where ionization is greatest; the shorter wavelengths are reflected from the highest layers.

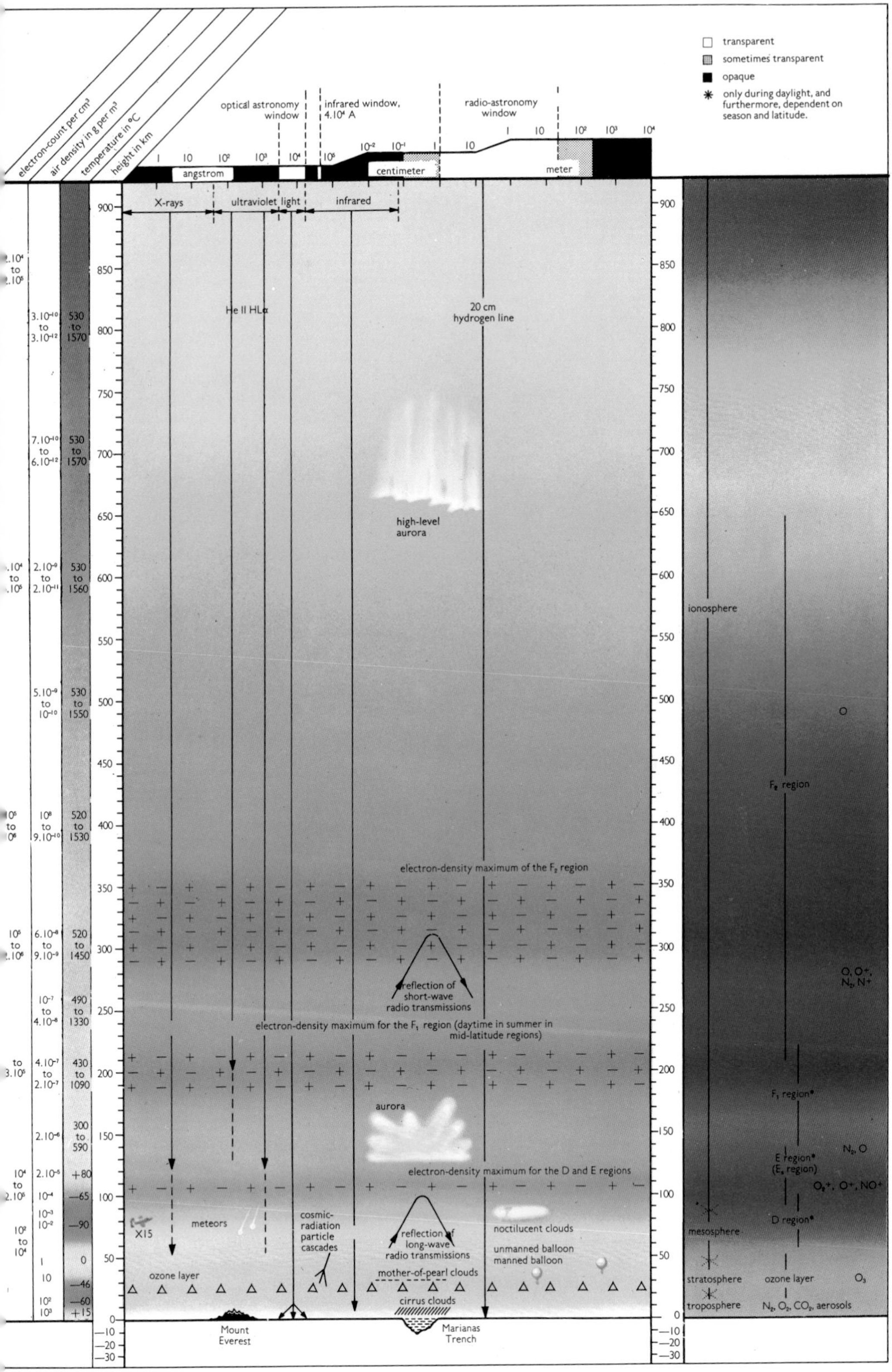

transparent
sometimes transparent
opaque
only during daylight, and furthermore, dependent on season and latitude.
electron-count per cm³
air density in g per m³
temperature in °C
height in km
optical astronomy window
infrared window, 4.10⁴ A
radio-astronomy window
angstrom
centimeter
meter
X-rays
ultraviolet light
infrared
He II HLα
20 cm hydrogen line
high-level aurora
ionosphere
F₂ region
electron-density maximum of the F₂ region
reflection of short-wave radio transmissions
electron-density maximum for the F₁ region (daytime in summer in mid-latitude regions)
F₁ region*
aurora
E region* (Eₛ region)
electron-density maximum for the D and E regions
D region*
X15
meteors
cosmic-radiation particle cascades
reflection of long-wave radio transmissions
noctilucent clouds
unmanned balloon
manned balloon
mother-of-pearl clouds
ozone layer
cirrus clouds
mesosphere
stratosphere
troposphere
Mount Everest
Marianas Trench
O
O, O⁺, N₂, N⁺
N₂, O
O₂⁺, O⁺, NO⁺
O₃
N₂, O₂, CO₂, aerosols

nitrous oxide. The water vapor content fluctuates within wide margins as water is evaporated from the OCEANS, carried in CLOUDS and precipitated upon the continents. The air flows in meandering currents, transferring ENERGY from the warm equatorial regions to the colder poles (see also GREENHOUSE EFFECT). The troposphere is thus the zone in which weather occurs (see METEOROLOGY), as well as that in which most air-dependent life exists. Apart from occasional INVERSIONS, the TEMPERATURE falls with increasing altitude through the troposphere until at the tropopause (altitude 7km at the poles; 16km on the equator) it becomes constant (about 217K), and then slowly increases again into the **stratosphere** (up to about 48km). The upper stratosphere contains the OZONE layer which filters out the dangerous ULTRAVIOLET RADIATION incident from the SUN. Above the stratosphere, the **mesosphere** merges into the IONOSPHERE, a region containing various layers of charged particles (IONS) of immense importance in the propagation of RADIO waves, being used to reflect signals between distant ground stations. At greater altitudes still, the ionosphere passes into the **exosphere,** a region of rarefied HELIUM and hydrogen gases, in turn merging into the interplanetary medium. In all, the atmosphere has a mass of about 5.2×10^{18}kg, its DENSITY being about 1.23kg/m^3 at sea level. Its WEIGHT results in its exerting an average **air pressure** of 101.3kPa (1013mbar) near the surface, this fluctuating greatly with the weather and falling off rapidly with height (see also PRESSURE). The other PLANETS of the SOLAR SYSTEM (with the possible exception of PLUTO), though only two of their SATELLITES, all have distinctive atmospheres, though none of these contains as much life-supporting oxygen as does that of the earth.

ATMOSPHERE (atm), CGS UNIT of PRESSURE. See BAR.

ATMOSPHERIC REFRACTION, the REFRACTION of light rays passing through the ATMOSPHERE, due to variations in its density and temperature which produce corresponding variations in its refractive index. Under standard conditions, slight curvature results in the case of rays with a horizontal component, and thus the apparent position of celestial bodies is altered. Unusual density variations may produce MIRAGES, shimmer and other deceptive effects.

ATOLL, a typically circular CORAL reef enclosing a LAGOON. Many atolls, often supporting low arcuate islands, are found in the Pacific Ocean.

ATOM, classically one of the minute, indivisible, homogeneous material particles of which material objects are composed (see ATOMISM), and in 20th-century science the name given to a relatively stable package of MATTER, typically about 0.1nm across, and itself made up of at least two SUBATOMIC PARTICLES. Every atom consists of a tiny nucleus (containing positively charged PROTONS and electrically neutral NEUTRONS) with which is associated a number of negatively charged ELECTRONS. These, although individually much smaller than the nucleus, occupy a hierarchy of ORBITALS which represent the atom's electronic ENERGY LEVELS, and fill most of the space taken up by the atom. The number of protons in the nucleus of an atom (the atomic number, Z) defines of which chemical ELEMENT the atom is an example. In an isolated neutral atom the number of electrons equals the atomic number, but in an electrically charged ION of the same atom there is either a surfeit or a deficit of electrons. The number of neutrons in the nucleus (the neutron number, N) can vary between different atoms of the same element, the resulting species being called the ISOTOPES of the element. Most stable isotopes have slightly more neutrons than protons. Although the nucleus is very small, it contains nearly all the MASS of the atom—protons and neutrons having very similar masses, and the mass of the electron (about 0.05% of the proton mass) being almost negligible. Counting the proton mass as one, this means that the mass of the atom is roughly equal to the total number of its protons and neutrons. This number $Z+N$, is known as the mass number of the atom, A. In equations representing nuclear reactions the atomic number of an atom is often written as a subscript preceding the chemical symbol for the element and the mass number as a superscript following it. Thus an atomic nucleus with mass number 16 and containing 8 protons belongs to an atom of "oxygen-16", written ${}_8O^{16}$. The average of the mass numbers of the various naturally occurring isotopes of an element, weighted according to their relative abundance, gives the chemical ATOMIC WEIGHT of the element. Subatomic particles fired into atomic nuclei can cause nuclear reactions giving rise either to new isotopes of the original element or to atoms of a different element, and emitting ALPHA PARTICLES, BETA RAYS or GAMMA RAYS.

The earliest atomistic concept, regarding the atom as that which could not be subdivided, was implicit in the first modern, chemical atomic theory, that of John DALTON (1808). Although it survives in the once-common chemical definition of atom—the smallest fragment of a chemical element which retains the properties of that element and can take part in chemical reactions—chemists now recognize that it is the MOLECULE and not the atom which is the natural chemical unit of matter. The atomic nuclei form the vertebra of the molecules; but it is the interaction of the VALENCE electrons associated with these nuclei, rather than the properties of the individual component atoms, which is responsible for the chemical behavior of matter.

ATOMIC BOMB, a weapon of mass destruction deriving its energy from nuclear FISSION. The first atomic bomb was exploded at Alamogordo, N.M., on July 16, 1945. As in the bomb dropped over Hiroshima, Japan a few weeks later (August 6), the fissionable material was uranium-235, but when Nagasaki was destroyed by another bomb three days after that, plutonium-239 was used. Together the Hiroshima and Nagasaki bombs killed more than 100000 people. Since the early 1950s, the power of the fission bomb (equivalent to some 20000 tons of TNT in the case of the Hiroshima bomb) has been vastly exceeded by that of the HYDROGEN BOMB which depends on nuclear FUSION. (See also NUCLEAR WARFARE.)

ATOMIC CLOCK, a device which utilizes the exceptional constancy of the FREQUENCIES associated with certain electron SPIN reversals (as in the CESIUM clock) or the inversion of AMMONIA molecules (the ammonia clock) to define an accurately reproducible TIME scale.

ATOMIC ENERGY. See NUCLEAR ENERGY.
ATOMIC NUMBER, (*Z*). See ATOM.
ATOMIC REACTOR. See NUCLEAR REACTOR.
ATOMIC WASTE DISPOSAL. See NUCLEAR ENERGY.
ATOMIC WEIGHT, the MEAN MASS of the ATOMS of an ELEMENT weighted according to the relative abundance of its naturally occurring ISOTOPES and measured relative to some standard. Since 1961 this standard has been provided by the CARBON isotope C^{12} whose atomic mass is defined to be exactly 12. On this scale atomic weights for the naturally occurring elements range from 1.008 (HYDROGEN) to 238.03 (URANIUM).
ATOMISM, the theory that all matter consists of atoms—minute indestructible particles, homogeneous in substance but varied in shape. Developed in the 5th century BC by LEUCIPPUS and DEMOCRITUS and adopted by EPICURUS, it was expounded in detail by the Roman poet LUCRETIUS.
ATOM SMASHER. See ACCELERATORS, PARTICLE.
ATP (adenosine triphosphate). See NUCLEOTIDES.
ATROPHY, wasting away of bodily TISSUES or organs because of disease, MALNUTRITION, disuse or old age. (See also MUSCULAR DYSTROPHY.)
ATROPINE, an ALKALOID derived from HYOSCYAMINE obtained from Belladonna. It decreases the effects of the parasympathetic NERVOUS SYSTEM, and is used to dilate the pupils of the eyes, to increase heart-rate, to reduce the secretion of mucus and saliva, and to relax spasm.

Structural formula of atropine

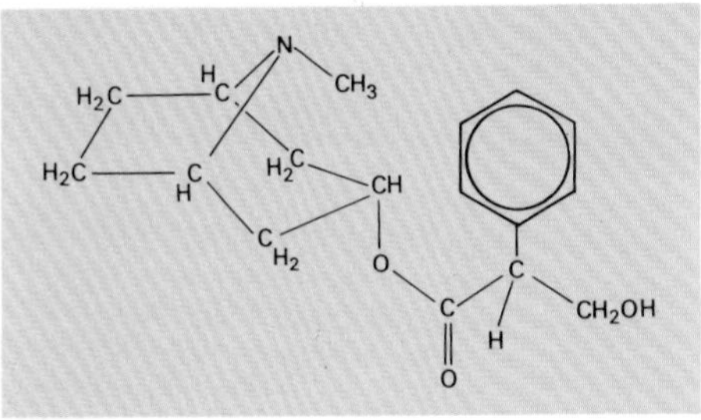

ATTENTION, the process of mental selection in which the individual concentrates on certain elements by considering them apart their environment. Active attention is a voluntary focusing of the mind; passive attention is involuntary reaction to outside stimuli. ELECTROENCEPHALOGRAPHS show that attention is characterized by fast, low-amplitude BRAIN waves.
AUDUBON, John James (1785–1851),US artist and naturalist famous for his bird paintings, born in Santo Domingo (Haiti) of French parents and brought up in France. Some years after emigrating to the US in 1803, he embarked on what was to become his major achievement: the painting of all the then-known birds of North America. His *Birds of America* (London: 1827–38), was followed by a US edition (1840–44) and other illustrated works on American natural history.
AUDUBON SOCIETY, National, US nature conservancy organization, named for J. J. AUDUBON. Itself running many nature sanctuaries, it works closely with government departments in the fight to preserve American wildlife.

A striking display of *aurora borealis*.

AUREOMYCIN. See TETRACYCLINES; ANTIBIOTICS.
AURIGA (the Charioteer), winter constellation of N skies, containing CAPELLA, Alpha Aurigae, the fifth brightest star in the night sky.
AURIGNACIAN, an Upper Paleolithic (see STONE AGE) European culture named for Aurignac in S France. It is marked by fine cave paintings and bone, horn and stone-flake tool-making.
AURORA, or **polar lights,** striking display of lights seen in night skies near the earth's geomagnetic poles. The *aurora borealis* (northern lights) is seen in Canada, Alaska and N Scandinavia; the *aurora australis* (southern lights) is seen in Antarctic regions. The auroras are caused by the collision of air molecules in the upper atmosphere with charged particles from the sun that have been accelerated and "funneled" by the earth's magnetic field. Particularly intense auroras are associated with high solar activity. Nighttime AIRGLOW is termed the permanent aurora.
AUSTRALIAN ABORIGINES, aboriginal population (see ABORIGINES) of Australia. They have dark wavy hair (except in childhood), medium stance, broad noses and narrow heads—typical AUSTRALOID features. Before white encroachment in the 18th and 19th centuries they lived by well-organized nomadic food-gathering and hunting and numbered about

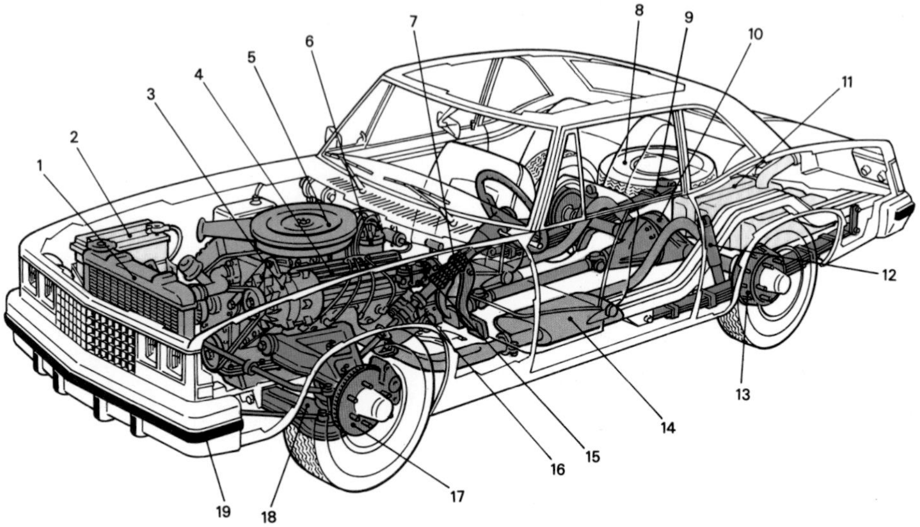

300 000. Since the enfranchisement of the remaining 40 000 full-blooded aborigines in 1962, varyingly successful attempts at integration have been made. (See also TASMANIANS.) Recent researches have suggested that they may be the result of interbreeding between an original population of *Homo erectus* and the insurgent earliest members of *Homo sapiens* (see PREHISTORIC MAN).

AUSTRALOIDS, an ethnic group including the AUSTRALIAN ABORIGINES, the AINU, the DRAVIDIANS, the population of the Vedda of Sri Lanka and, debatably, Melanesians, Negritos and Papuans.

AUSTRALOPITHECUS. See PREHISTORIC MAN.

AUTISM, withdrawal from reality and relations with others. Pathological autism occurs in various psychoses, especially SCHIZOPHRENIA. Certain children who fail to establish normal communication with others or social responses are termed autistic; it may represent a juvenile form of schizophrenia.

AUTOCLAVE, a strong-walled pressure vessel suitable for heating liquids above their BOILING POINTS, used in the study of high-pressure chemical reactions, for sterilizing and cooking and for impregnating wood. The industrial autoclave derives from Denis Papin's "steam digester" of 1679, the prototype for the modern PRESSURE COOKER.

AUTOGIRO, a short-takeoff, heavier-than-air flying machine that derives its lift from a rotating wing which turns in response to the aerodynamic forces (see AERODYNAMICS) acting on it as the aircraft is driven forward through the air by a conventional PROPELLER. It is potentially faster and mechanically far simpler than the HELICOPTER, capable of flying more slowly than the AIRPLANE, and impossible to stall.

AUTOMATIC PILOT. See GYROPILOT.

AUTOMATIC TRANSMISSION. See TRANSMISSION.

AUTOMATION, the detailed control of a production process without recourse to human decision-making at every point, typically involving a negative-FEEDBACK system. (See MECHANIZATION AND AUTOMATION.)

The layout of a typical modern automobile with automatic transmission. The engine, radiator and exhaust system are colored brown, the transmission green, the braking system red, the steering and suspension blue, and the storage battery and fuel tank yellow. (1) radiator, (2) battery, (3) AC generator, (4) V8 engine, (5) air cleaner and carburetor, (6) distributor, (7) telescopic steering column, (8) spare wheel, (9) rear spring, (10) differential and rear axle, (11) fuel tank, (12) shock absorber, (13) drum brake (rear), (14) exhaust system, (15) brake master cylinder, (16) automatic transmission, (17) disk brake (front), (18) front suspension, (19) energy-absorbing bumpers.

AUTOMOBILE, or passenger car, a small self-propelled passenger-carrying vehicle designed to operate on ordinary highways and usually supported on four wheels. Power is provided in most modern automobiles by an INTERNAL COMBUSTION ENGINE which uses GASOLINE (vaporized and premixed with a suitable quantity of air in the CARBURETOR) as FUEL. This is ignited in the (usually 4, 6 or 8) cylinders of the engine by SPARK PLUGS, fired from the DISTRIBUTOR in the appropriate sequence. The gas supply and thus the engine speed is controlled from the accelerator pedal. The driving power is communicated to the road wheels through the TRANSMISSION which includes a clutch (enabling the driver to disengage the engine without stopping it), a gearbox (allowing the most efficient use to be made of the engine power), various drive-shafts (with universal joints), and a DIFFERENTIAL which allows the driving wheels to turn at marginally different rates in cornering. Steering is controlled from a hand wheel which moves a transverse tie rod mounted between the independently-pivoted front wheels. Service BRAKES of various types are mounted on all wheels, an additional parking-brake mechanism being used when stationary. In modern automobiles service brakes and steering may be power-assisted and the transmission automatic rather than manually controlled with a gearshift. Although the first propelled steam vehicles were built by the French army officer Nicholas-Joseph Cugnot in the 1760s, it was not until Karl BENZ and Gottlieb DAIMLER began to build gasoline-powered carriages in the mid-1880s that the day of the modern automobile dawned. The DURYEA brothers built the first US automobile in 1893 and within a few years several automobile manufacturers, including Henry Ford, had started into business. The Ford Motor Company itself was founded in 1903, pioneering the cheap mass-market auto with the Model T of 1908. The automobile industry expanded fitfully until the 1960s, improving automobile performance, comfort and styling, but more recently has been forced by economic considerations and the activity of consumer groups (led by Ralph Nader) to pay more attention to safety and environmental factors.

AUTOMOBILE EMISSION CONTROL, the reduction of the AIR POLLUTION caused by AUTOMOBILES by modification of the fuel and careful design. The principal pollutants are unburnt HYDROCARBONS, CARBON monoxide, oxides of NITROGEN and LEAD halide particles. The last can be eliminated if alternatives to lead-based ANTIKNOCK ADDITIVES in the GASOLINE are used but the others require redesigned cylinder heads, recycling and afterburning of exhaust gases and better metering of the GASOLINE supply through FUEL INJECTION. An alternative approach investigates alternatives to the conventional INTERNAL COMBUSTION ENGINE—BATTERY- and FUEL-CELL-powered vehicles, GAS-TURBINE and even steam-powered units.

AUTOPSY, or postmortem examination, the dissection of a corpse to determine the cause of death and the nature and progress of the prior disease by the recognition of abnormal ANATOMY (see PATHOLOGY). It enables assessment of diagnosis and treatment; in FORENSIC MEDICINE, it may determine identity and the manner and time of death.

AUTOSUGGESTION. See SUGGESTION.

AUTOTROPH, an organism which requires only inorganic compounds and solar energy (see PHOTOSYNTHESIS). Green plants and some bacteria are autotrophs; animals, fungi and most bacteria are **heterotrophs,** requiring also organic compounds. Autotrophs form the primary link in the food chain (see ECOLOGY).

AUXINS, HORMONES which promote lengthwise plant growth, and control ABSCISSION and the plant's responses to light and gravity (see TROPISMS). Natural auxins are derivatives of indole (see HETEROCYCLIC COMPOUNDS). Synthetic auxins are used for crop control and as WEEDKILLERS.

AVALANCHE, mass of snow, ice and mixed rubble moving down a mountainside or over a precipice, usually caused by loud noises or other shock waves acting on snow already unstable from subsurface thawing. They can be major factors in EROSION.

AVERAGE. See MEAN, MEDIAN AND MODE.

AVERROËS, Latin name of abu-al-Walid Muhammad ibn-Ahmad **ibn-Rushd** (1126–1198), Spanish/N African Arab philosopher, a commentator on Aristotle and Plato who exerted a great influence on the development of the later Latin scholastic philosophy.

AVES, a class of the CHORDATA that includes all birds. Birds are warm-blooded, lay eggs and display numerous modifications that enable them to fly, although some species have lost this ability. There are about 8600 living species. To make flight possible, the skeleton is light; the body cavity is filled by a series of air-sacs; the fore limbs are modified as wings, with large flight muscles for their operation, and the body is covered with feathers. Feathers, which are not found in any other group of animals, provide the wing and tail flight surfaces, as well as an insulating layer over the entire body.

AVICENNA, Latin name of abu-Ali al-Husayn **ibn-Sina** (980–1037), the greatest of the Arab scientists of the medieval period. His *Canon of Medicine* remained a standard text in Europe until the Renaissance.

AVOGADRO, Count Amedeo (1776–1856), Italian physicist who first realized that gaseous ELEMENTS might exist as MOLECULES which contain more than one ATOM, thus distinguishing molecules from atoms. In 1811 he published **Avogadro's hypothesis**—that equal volumes of all GASES under the same conditions of TEMPERATURE and PRESSURE contain the same number of molecules—but his work in this area was ignored by chemists for over 50 years. The **Avogadro Number** ($\mathcal{N}$), the number of molecules in one MOLE of substance, 6.02×10^{23}, is named for him.

AVOIRDUPOIS (corruption of French: property of weight), the system of weights customarily used in the

Axes used in rectangular coordinate systems. Those below are used in plane geometry, the *x*-axis being at right angles to the *y*-axis. Those at the bottom are for use in three dimensions, the *x*-, *y*- and *z*-axes being at mutual right angles.

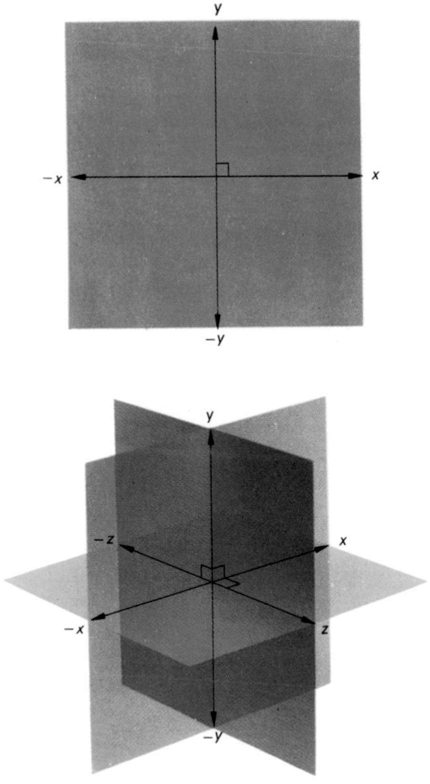

US (and formerly in the UK) for most goods except gems and drugs for which are employed respectively TROY WEIGHTS and APOTHECARIES WEIGHTS. There are 7000 grains or 16 ounces in the pound avoirdupois (see WEIGHTS AND MEASURES).

AXELROD, Julius (1912–   ), US biochemist who shared the 1970 Nobel Prize for Medicine or Physiology with Bernard KATZ and Ulf von EULER for their independent contributions toward elucidating the chemistry of the transmission of nerve impulses (see NERVOUS SYSTEM). Axelrod identified a key ENZYME in this mechanism.

AXES, in mathematics, straight lines used as reference lines. In plane ANALYTIC GEOMETRY (and GRAPHS) two axes, usually at right angles, are most commonly used. In three-dimensional geometry, three axes, usually at mutual right angles, are most common. Their point of INTERSECTION is the **origin.**

AXIOM, one of the fundamental propositions which must be assumed true without proof in the compilation of a logical system (see LOGIC). An axiom need not be self-evident but should be consistent with the other axioms of the system. Closely related is the **postulate** which is a less arbitrary or basic assumption, provisionally accepted for some particular purpose but more freely open to substitution.

AXIS. See ATLAS; VERTEBRAE.

AXIS OF SYMMETRY, a LINE drawn through a geometric figure such that the figure is symmetrical (see SYMMETRY) about it. The line may be considered as an axis of rotation: if the figure is rotated about it, there will be two or more positions which are indistinguishable from each other. If the letter Z, for example, is rotated about an axis drawn perpendicularly into the paper through the center of the diagonal stroke, there will be two correspondent positions: the letter has 2-fold rotational symmetry about that axis. In general, if a figure has n correspondent positions on rotation about an axis, it is said to have n-fold symmetry about that axis. (See also PLANE OF SYMMETRY.)

AXOLOTL, neotenic LARVA (see NEOTENY) of the salamander *Ambystoma mexicanum.* Perhaps owing to iodine deficiency in the lakes near Mexico City where they are found, they usually never become adult; but in a controlled environment they may develop into normal adult salamanders. They are 100–175mm (4–7in) in length. The term is also applied to neotenic larvae of other salamanders.

AXON, the fiber of a nerve cell (NEURON) that conducts impulses away from the cell body. (See NERVOUS SYSTEM.)

AYER, Sir Alfred Jules (1910–   ), English philosopher whose *Language, Truth and Logic* (1936) was influential in founding the "Oxford school" of philosophy, with its emphasis on the careful analysis of the use of words. The LOGICAL POSITIVISM of the Vienna Circle was an important formative influence in his philosophy.

AZEOTROPIC MIXTURE, or constant-boiling-point mixture, a SOLUTION of two or more liquids which on DISTILLATION behaves as a pure liquid: its boiling-point is invariable at a given pressure, and the composition of the vapor phase is the same as that of the liquid. (See also PHASE EQUILIBRIA.)

AZIMUTH, in navigation and astronomy, the angular distance measured from 0–360° along the horizon eastward from an observer's north point to the point of intersection of the horizon and a great circle (see CELESTIAL SPHERE; SPHERICAL GEOMETRY) passing through the observer's ZENITH and a star or planet.

AZO COMPOUNDS, class of organic compounds of general formula $R{-}N{=}N{-}R'$. The most important azo compounds have AROMATIC groups for R and R′, and are made by coupling of DIAZONIUM COMPOUNDS with NUCLEOPHILES such as PHENOLS or aromatic AMINES. They comprise more than half the DYES commercially available.

AZOIC, the portion of geological time prior to the CRYPTOZOIC. (See also GEOLOGY; PRECAMBRIAN.)

AZURITE, blue mineral consisting of basic COPPER carbonate ($Cu_3[OH]_2[CO_3]_2$), occurring with MALACHITE, notably in France, SW Africa and Ariz. It forms monoclinic crystals used as GEMS, and was formerly used as a pigment.

*Capital **B** was adopted into the Roman alphabet from the Greek. Originally the Greeks wrote from right to left, hence a. Later they wrote boustrophedon (alternating the direction of the line of writing from right to left, left to right). After 500 BC Greek writing invariably ran from left to right and the letter was reversed in mirror fashion. When written with a running pen the component strokes of the capital coalesced into cl in Roman cursive hands. The form **b** is not found until the 3rd century AD when scribes began to reverse the bow to differentiate between **b** and **d**; **b** did not become standard until the 6th century and appears in all the later minuscule scripts.*

BABBAGE, Charles (1792–1871), English mathematician and inventor who devoted much labor and expense to an unsuccessful attempt to devise mechanical calculating engines (see CALCULATING MACHINE). More significant was the part he played with J. HERSCHEL and G. Peacock in introducing the Leibnizian "d" notation for CALCULUS into British mathematical use in place of the less flexible "dot" notation devised by NEWTON.

BABBITT METAL, an ALLOY containing 89% TIN, 9% ANTIMONY and 2% COPPER, devised in 1839 by US inventor **Isaac Babbitt** (1799–1862) for lining BEARINGS. Today the term babbitt metal is also applied to other high-tin and high-LEAD bearing alloys.

BABCOCK, Stephen Moulton (1843–1931), US agricultural chemist who devised the **Babcock test** for determining the butterfat content of milk (1890).

BABY. See BIRTH; OBSTETRICS; PEDIATRICS.

BACILLUS, genus of rod-shaped BACTERIA of the family Bacillaceae.

BACKBONE. In animals, see VERTEBRATES; INVERTEBRATES; in man, see SKELETON; VERTEBRAE.

BACON, Francis; Lord Verulam, Viscount St. Albans (1561–1626), English philosopher and statesman who rose to become Lord Chancellor (1618–21) to James I but is chiefly remembered for the stimulus he gave to scientific research in England. Although his name is indelibly associated with the method of INDUCTION and the rejection of A PRIORI reasoning in science, the painstaking collection of miscellaneous facts without any recourse to prior theory which he advocated in the *Novum Organum* (1620) has never been adopted as a practical method of research. The application of the Baconian method was, however, an important object in the foundation

A micrograph of a tetanus bacillus, taken with an electron microscope (magnification ×100000 approx.). The bacillus has a large number of long, threadlike flagella.

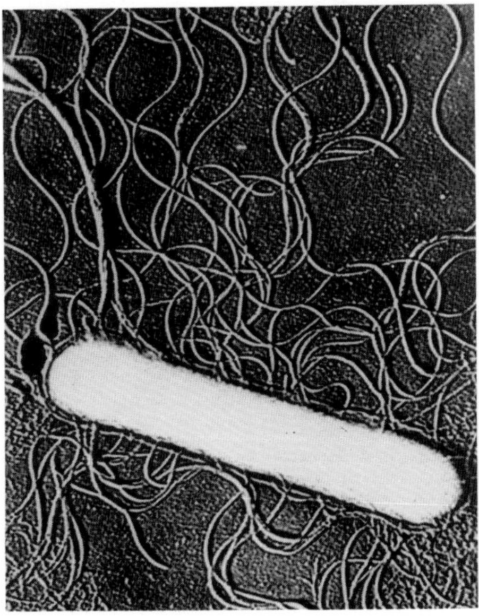

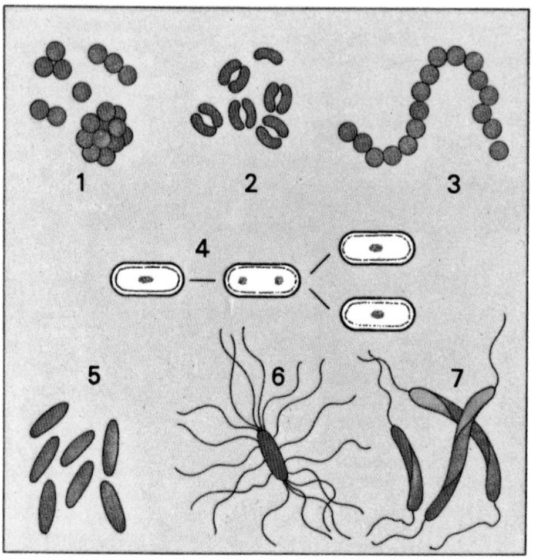

Bacteria: (1) cocci, (2) diplococci, (3) streptococci, (5) bacilli. Some bacteria possess hairlike flagella; for example, (6) flagellate rods or (7) flagellate spirilla. At (4) a bacillus is shown undergoing reproduction by binary fission.

of the ROYAL SOCIETY OF LONDON some 40 years later.

BACON, Roger (c1214–1292?), English scholar renowned in his own day for his great knowledge of science and remembered today for allegedly prophesying many of the inventions of later centuries: aircraft; telescopes; steam engines; microscopes. In fact he was a wealthy lecturer in the schools of Oxford and Paris with a passion for alchemical and other experiments, whose later life was overshadowed by disputes with the Franciscan Order, of which he had become a member in 1257. His principal writings (in aid of an encyclopedia of knowledge) were the *Opus Majus*, *Opus Minor* and *Opus Tertium*.

BACTERIA, unicellular microorganisms between 0.3 and 2 μm in diameter. They differ from plant and animal cells in that their nucleus is not a distinct organelle surrounded by a membrane: they are usually placed in a separate kingdom, the Protista.

The majority of bacteria are saprophytic: they exist independently of living hosts and are involved in processes of decomposition of dead animal and plant material. As such they are essential to the natural economy of living things.

Some bacteria are parasitic, and their survival depends on their presence in or on other living cells. They may be commensals (see COMMENSALISM), which coexist harmlessly with host cells, or pathogens, which damage the host organism by producing toxins, which may cause tissue damage (see BACTERIAL DISEASES). This distinction is not absolute: *Escherichia coli* is a commensal in the human intestine, but may cause infection in the urinary tract.

Bacteria are like plant cells in that they are surrounded by a rigid cell wall. Most species are incapable of movement, but certain types can swim using hairlike (single-fibril) flagella. Bacteria vary in their food requirements: AUTOTROPHS can obtain energy by oxidizing substances which they have built up from simple inorganic matter; heterotrophs need organic substances for nutrition. Aerobic (see AEROBE) bacteria need oxygen to survive, whereas anaerobic (see ANAEROBE) species do not. Included in the latter group are the putrefactive bacteria, which aid decomposition. Bacteria generally reproduce asexually by binary FISSION, but some species reproduce sexually. Some can survive adverse conditions by forming highly resistant spores.

Bacteria are important to man in many ways. Commensal bacteria in the human intestine aid digestion of food; industrially they are used in the manufacture of, for example, acetone, citric acid and butyl alcohol and in many dairy products. Some bacteria, especially the ACTINOMYCETES, produce ANTIBIOTICS, used in destroying pathogenic bacteria.

Classification. There is no standard way of classifying bacteria. The higher bacteria are filamentous and the cells may be interdependent—they include the family Actinomycetes. The lower bacteria are subdivided according to shape: cocci (round), BACILLI (cylindrical), vibrios (curved), and spirilla (spiral). Cocci live singly, in pairs (DIPLOCOCCI), in clusters (STAPHYLOCOCCI) or in chains (STREPTOCOCCI)—as a group they are of great medical importance. Spirochetes form a separate group from the above: although spiral they are able to move. Bacteria are also classified medically in terms of their response to GRAM'S STAIN: those absorbing it are termed Gram-positive, those not, Gram-negative. (See also BACTERIOLOGY; MICROBIOLOGY.)

BACTERIAL DISEASES, diseases caused by BACTERIA or their products. Many bacteria have no effect and some are beneficial, while only a small number lead to disease. This may be a result of bacterial growth, the INFLAMMATION in response to it or of TOXINS (e.g., TETANUS, BOTULISM and CHOLERA). Bacteria may be contracted from the environment, other animals or humans, or from other parts of a single individual. Infection of SKIN and soft tissues with STAPHYLOCOCCUS or STREPTOCOCCUS leads to BOILS, carbuncles, IMPETIGO, cellulitis, SCARLET FEVER and ERYSIPELAS. ABSCESS represents the localization of bacteria, while SEPTICEMIA is infection circulating in the BLOOD. Sometimes a specific bacteria causes a specific disease (e.g., ANTHRAX, DIPHTHERIA, TYPHOID FEVER), but any bacteria in some organs cause a similar disease: in LUNGS, PNEUMONIA occurs; in urinary tract, CYSTITIS or pyelonephritis, and in the BRAIN coverings, MENINGITIS. Many VENEREAL DISEASES are due to bacteria. In some diseases (e.g., TUBERCULOSIS, LEPROSY, RHEUMATIC FEVER), many manifestations are due to hypersensitivity (see IMMUNITY) to the bacteria. While ANTIBIOTICS have greatly reduced death and ill-health from bacteria, and VACCINATION against specific diseases (e.g., WHOOPING COUGH) has limited the number of cases, bacteria remain an important factor in disease.

BACTERIOLOGICAL WARFARE. See CHEMICAL AND BIOLOGICAL WARFARE.

BACTERIOLOGY, the science that deals with BACTERIA, their characteristics and their activities as related to medicine, industry and agriculture. Bacteria were discovered in 1676 by Anton von LEEUWENHOEK. Modern techniques of study originate from about 1870 with the use of stains and the discovery of culture methods using plates of nutrient AGAR media. Much pioneering work was done by

Louis PASTEUR and Robert KOCH. (See also BACTERIAL DISEASES; NITROGEN FIXATION; SPONTANEOUS GENERATION.)

BACTERIOPHAGE, or **phage**, a VIRUS which attacks BACTERIA. They have a thin PROTEIN coat surrounding a central core of DNA (or occasionally RNA), and a small protein tail. The phage attaches itself to the bacterium and injects the NUCLEIC ACID into the cell. This genetic material (see GENETICS) alters the metabolism of the bacterium, and several hundred phages develop inside it: eventually the cell bursts, releasing the new, mature phages. Study of phages has revealed much about PROTEIN SYNTHESIS and nucleic acids.

BADLANDS, arid to semiarid areas of pinnacles, ridges and gullies, usually lacking in vegetation, formed by heavy EROSION of non-uniform rock. The Big Badlands of S.D. are particularly notable.

BAEKELAND, Leo Hendrik (1863–1944), Belgian-born chemist who, after emigrating to the US in 1889, devised Velox photographic printing paper (selling the process to EASTMAN in 1899) and went on to discover BAKELITE, the first modern synthetic PLASTIC.

BAER, Karl Ernst von (1792–1876), Estonian-born German embryologist who discovered the mammalian egg (see REPRODUCTION) and the NOTOCHORD of the vertebrate embryo. He is considered to have been one of the founders of comparative EMBRYOLOGY.

BAEYER, Johann Friedrich Wilhelm Adolf von (1835–1917), German organic chemist who proposed a "strain theory" to account for the relative stabilities of ring compounds (see ALICYCLIC COMPOUNDS). He was awarded the 1905 Nobel Prize for Chemistry for his research on dyestuffs: in 1878 he had become the first to synthesize INDIGO.

BAILEY, Liberty Hyde (1858–1954), US horticulturalist and botanist who, as professor of HORTICULTURE at Cornell U. (1888–1913), did much to put US horticulture on a scientific footing.

BAILEY BRIDGE, a strong temporary bridge built to a design devised in 1941 for military use by British engineer **Sir Donald Coleman Bailey** (1901–). The design, a variant of which, resting on plywood floats called "pontoons", is used for crossing broad rivers, has proved of lasting value in peacetime emergencies.

BAILLY, Jean Sylvain (1736–1793), French astronomer and politician. After studying the satellites of Jupiter and writing a five-volume history of astronomy (1775–87), he turned to politics, becoming mayor of Paris 1789–91; he was executed in the Terror.

BAILY'S BEADS, named for Francis Baily (1774–1844), the apparent fragmentation of the thin crescent of the sun just before totality in a solar ECLIPSE, caused by sunlight shining through mountains at the edge of the lunar disk.

BAIRD, John Logie (1888–1946), Scottish inventor who, by first transmitting moving pictorial images (1925), inaugurated the TELEVISION era. Although the Baird system was tested when the British public television service began in 1936, it was dropped in favor of a rival which used electronic rather than mechanical scanning.

BAKELITE, synthetic RESIN discovered by Leo BAEKELAND, made by chemical reaction of FORMALDEHYDE and PHENOL. It is a thermosetting PLASTIC (see also POLYMERS). A hard, strong material, it is used as an electrical insulator, an adhesive and a paint binder.

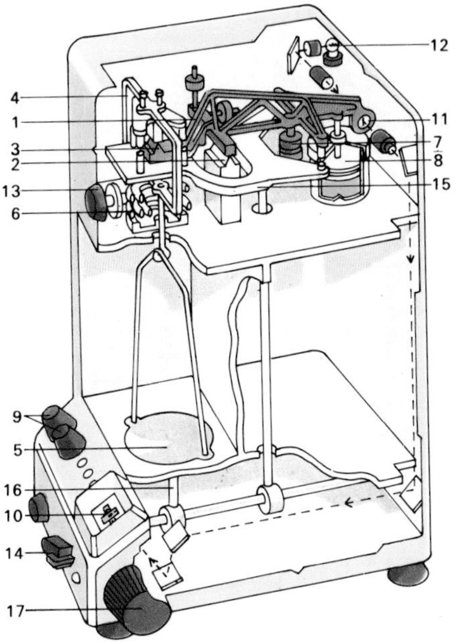

A modern single-pan substitution balance, much simplified. A substitution balance comprises a beam (1) supported on a knife edge (2), with, on the shorter beam arm, a further knife edge (3) supporting a stirrup (4) carrying the weighing pan (5) and a set of weights (6), and, on the longer arm, a fixed weight (7). When the beam is allowed to swing with the pan empty, the moments of the pan and weights about the main knife edge exactly balance that due to the fixed weight. When an object is placed on the pan and the beam is released, the weight of the object can be determined from its equalling the total weight (from the set of weights) that must be removed from the stirrup to return the beam to its equilibrium position. Any actual balance includes many refinements such as an air-damping system (8) for the beam, rapid taring and rough weighing-in facilities (not shown) and the use of carefully matched sets of ring weights. In the balance in the diagram (much simplified from a design by Mettler Instrumente AG) the weight in whole grams of a sample is determined by turning the ring-weight control knobs (9) until the beam is roughly balanced. Decimal fractions of a gram can then be read off an optical scale (10) (a simplified representation of the optical system is indicated on the diagram) which magnifies the image of a graticule (11) showing the extent to which the beam is out of equilibrium. Other features include: lamp (12), zero adjustment knob (13), arrestment lever (14), arrestment system (15), pan brake (16), digital counter knob (for high-precision weighing) (17).

BAKEWELL, Robert (1725–1795), English agriculturalist who pioneered the selective BREEDING of sheep and cattle for meat and introduced the technique of repeated INBREEDING.

BAKING. See BREAD.

BAKING POWDER, white YEAST substitute which causes dough to rise by giving off carbon dioxide bubbles when moistened. It is composed of SODIUM bicarbonate; an acid (usually TARTARIC ACID or ALUM) which reacts with sodium bicarbonate in water to release carbon dioxide; and an inert substance such as starch to retard the reaction while the powder is dry.

BAKING SODA, or sodium bicarbonate. See SODIUM.

BALANCE, instrument used for measuring the WEIGHT of an object, typically by comparison with objects of known weight. The equal-arm balance, known to ancient Egyptians and Mesopotamians, consists of two identical pans hung from either end of a centrally suspended beam. When objects of equal weight are placed in each pan, the beam swings level because the MOMENTS of the gravitational FORCES acting on each object and pan about the central pivot or fulcrum are equal in magnitude and opposite in sense. Other types of beam balance involve fixed weights sliding along or hung below unequal beam arms, but the principle of the balancing of equal and opposite gravitational moments remains the same. The relatively inaccurate spring balance utilizes HOOKE's law to determine the weight of the specimen from the extension it produces in a coiled spring, and the much finer TORSION balance utilizes the resistance of a wire to being twisted. Modern chemical microbalances can measure weights as small as $1\mu g$.

BALANCE OF NATURE, late 19th- and early 20th-century concept of Nature existing in an EQUILIBRIUM maintained by interdependencies between different animals and plants. In fact, this balance is unstable, since many factors (e.g., climate, POLLUTION) may cause dynamic change in large or small natural populations. (See also ECOLOGY.)

BALDNESS, or **alopecia,** loss of hair, usually from the scalp, due to disease of hair FOLLICLES. **Male-pattern baldness** is an inherited tendency, often starting in the twenties. **Alopecia areata** is a disease of unknown cause producing patchy baldness,though it may be total. Prolonged FEVER, LUPUS ERYTHEMATOSUS and RINGWORM may lead to temporary baldness, as may certain drugs and poisons.

BALEEN, material found as fibrous plates hanging in rows from the roof of the mouth in whalebone whales and often known as whalebone. Its function is to strain PLANKTON, on which these whales feed, from the water. Strong and elastic, it was used for many purposes before the advent of PLASTIC and spring STEEL.

BALLISTIC MISSILE. See MISSILE.

BALLISTICS, the science concerned with the behavior of projectiles, traditionally divided into three parts. **Interior ballistics** is concerned with the progress of the projectile before it is released from the launching device. In the case of a gun this involves determining the propellant charge, barrel design and firing mechanism needed to give the desired muzzle velocity and stabilizing spin to the projectile. **External ballistics** is concerned with the free flight of the projectile. At the beginning of the 17th century GALILEO determined that the trajectory (flight path) of a projectile should be parabolic (see CONIC SECTIONS), as indeed it would be if the effects of air resistance, the rotation and curvature of the earth, the variation of air density and gravity with height, and the rotational INERTIA of the projectile could be ignored. The shockwaves accompanying projectiles moving faster than the speed of sound (see SUPERSONICS) are also the concern of this branch.

Terminal or **penetration ballistics** deals with the behavior of projectiles on impacting at the end of their trajectory. The velocity-to-mass ratio of the impact particle is an important factor and results are of equal interest to the designers of AMMUNITION and of ARMORPLATE. A relatively recent development in the science is **forensic ballistics**, which now plays an important role in the investigation of gun crimes.

BALLOON, a nonpowered, nonrigid lighter-than-air craft comprising a bulbous envelope containing the lifting medium and a payload-carrying basket or "gondola" suspended below. Balloons may be captive (secured to the ground by a cable, as in the barrage balloons used during WWII to protect key installations and cities from low-level air bombing) or free-flying (blown along and steered at the mercy of the WIND). Lift may be provided either by GAS (usually HYDROGEN or noninflammable HELIUM) or by heating the air in the envelope. A balloon rises or descends through the air until it reaches a level at which it is in

Diagram of a modern gas balloon (1) landing-run line; (2) valve cord; (3) rip cord; (4) equator; (5) rip panel; (6) safety device; (7) valve; (8 & 9) net; (10) rain deflector; (11) appendix; (12) load ring; (13) basket ropes; (14) anchor cable.

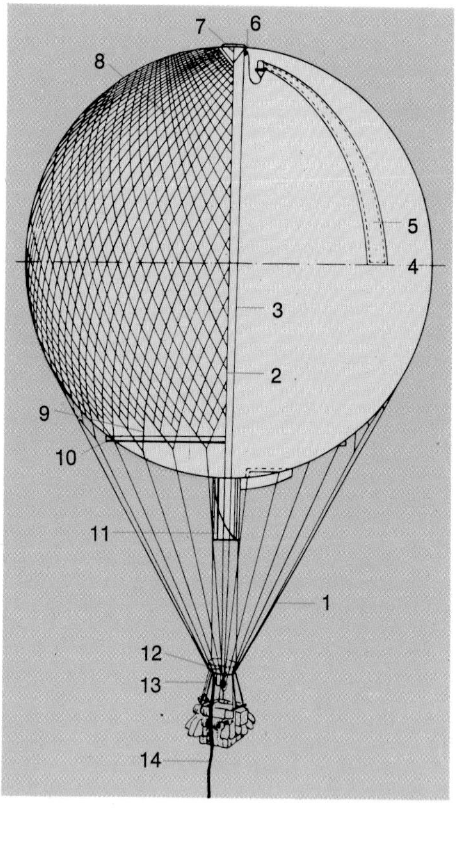

EQUILIBRIUM in accordance with ARCHIMEDES' principle. In this situation the total WEIGHT of the balloon and payload is equal to that of the volume of air which it is displacing.

If the pilot of a gas balloon wishes to ascend, he throws ballast (usually sand) over the side, thus reducing the overall DENSITY of the craft; to descend he releases some of the lifting gas through a small valve in the envelope. The ALTITUDE of a hot-air balloon is controlled using the PROPANE burner which heats the air; increased heat causes the craft to rise; turning off the burner gives a period of level flight followed by a slow descent as the trapped air cools.

The MONTGOLFIER brothers' hot-air balloon became the first manned aircraft in 1783, and in the same year the first gas balloon was flown by Jacques CHARLES. In 1785 Jean BLANCHARD piloted a balloon across the English Channel. In due time the powered balloon, or AIRSHIP, was developed, though free balloons have remained popular for sporting, military and scientific purposes. The upper atmosphere is explored using unmanned gas balloons, and RADIOSONDE balloons are in regular meteorological use. In 1931 Auguste PICCARD pioneered high-altitude manned flights. Many modern sporting balloons are built on the hot-air principle.

BALLPOINT PEN, pen designed to minimize ink leakage. At one end of a narrow, cylindrical ink reservoir a freely-rotating metal ball is held in a socket. The viscous (see VISCOSITY) ink is drawn through internal ducts in the socket by capillary action (see CAPILLARITY). Ballpoint pens came into general use in the late 1930s.

BALTIMORE, David (1938–), US virologist who shared the 1975 Nobel Prize for Physiology or Medicine with R. DULBECCO and H. M. TEMIN in recognition of their work linking VIRUSES with the development of some cancerous tumors.

BALUCHITHERIUM, extinct genus of rhinoceros known from FOSSILS of the late OLIGOCENE and early MIOCENE found in Asia. Probably the world's largest land mammal, it stood some 6m (20ft) at the shoulders.

BANDWIDTH, in telecommunications, the difference (expressed in HERTZ) between the upper and lower limits of the band range of FREQUENCIES either needed or available to transmit a given signal, or which can be adequately passed by a component device.

BANG'S DISEASE. See BRUCELLOSIS.

BANKS, Sir Joseph (1743–1820), British botanist and president of the ROYAL SOCIETY OF LONDON (1778–1820), the foremost British man of science of his time. He accompanied COOK as naturalist on his first expedition (1768–71) and was a key figure in the establishment of the Botanic Gardens at Kew, London.

BANNEKER, Benjamin (1731–1806), American mathematician and astronomer, notable as the first American Negro to gain distinction in science and the author of celebrated ALMANACS (1791–1802).

BANTING, Sir Frederick Grant (1891–1941), Canadian physiologist who, with C. H. BEST, first isolated the hormone INSULIN from the pancreases of dogs (1922). For this he shared the 1923 Nobel Prize for Physiology or Medicine with J. J. R. MACLEOD who had provided the experimental facilities.

BAR, a unit of pressure in the CGS system equal to 100kPa. The **millibar** (mbar or mb)— 100Pa—is commonly used in meteorology. The standard **atmosphere** is 1013.25mbar.

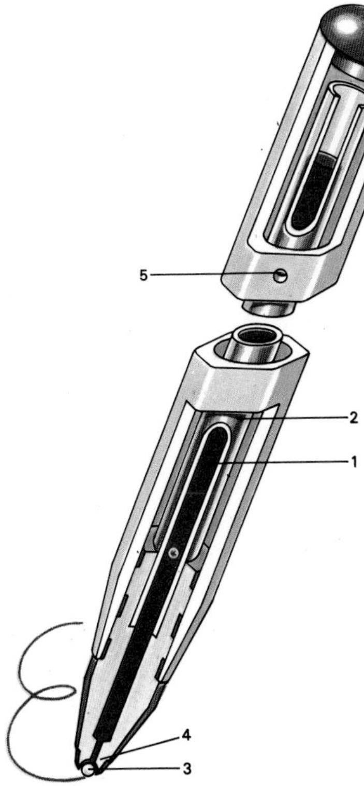

A typical mass-produced ballpoint pen. The ink (1) is contained in a plastic reservoir (2), at one end of which is a small ball (3)—usually about 1 mm in diameter—in a brass socket (4). When the pen is in use, the ball rotates because of friction with the writing surface, and the ink is drawn out. A hole (5) in the outer casing permits air to enter, thus preventing the formation of a partial vacuum as the ink is used up. In pens with larger reservoirs the risk of leakage from the open end of the reservoir is minimized by use of a "follower", a small quantity of extremely viscous liquid which follows the ink down the tube toward the socket.

BÁRÁNY, Robert (1876–1936), Austrian-born physiologist who received the 1914 Nobel Prize for Physiology or Medicine for research on the functioning of the organs of the inner EAR. From 1916 he continued his research in Sweden.

BARBED WIRE, fencing material comprising a single or double strand of steel or aluminum WIRE with sharply-pointed wire "barbs" twisted round it at short intervals. The introduction of barbed-wire fencing in the American West (following the invention of an efficient manufacturing technique by Joseph E. Glidden in 1873) marked an important epoch in the development of US agriculture.

BARBELS, tactile organs protruding about the mouth of certain fishes, notably the barbels (*Barbus*) of the family Cyprinidae.

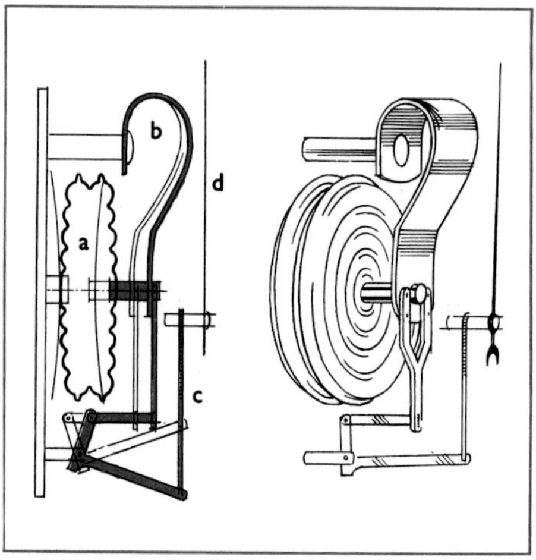

The aneroid barometer comprises a partially evacuated corrugated metal box (a), prevented from collapsing by a spring (b). The strain in this spring, proportional to the difference in pressures between the air inside and that outside the box, is amplified by a train of levers that operate a pointer (d) that moves over a calibrated scale. Aneroid barometers are convenient to use but require regular calibration against an accurate mercury barometer.

BARBITURATES, a class of drugs acting on the central NERVOUS SYSTEM which may be SEDATIVES, anesthetics or anticonvulsants. They depress nerve cell activity, the degree of depression and thus clinical effect varying in different members of the class. Although widely used in the past for insomnia, their use is now discouraged in view of high rates of addiction and their danger in over-dosage; safer alternatives are now available. Short-acting barbiturates are useful in ANESTHESIA; phenobarbitone is used in treatment of CONVULSIONS, often in combination with other drugs.

BARDEEN, John (1908–), US physicist who shared the 1956 Nobel Prize for Physics with SHOCKLEY and BRATTAIN for their development of the TRANSISTOR. In 1972 he became the first person to win the physics prize a second time, sharing the award with COOPER and SCHRIEFFER for their development of a comprehensive theory of SUPERCONDUCTIVITY.

BARGE, a flat-bottomed freight vessel which may be pushed, towed or self-propelled and is used mainly on inland waterways. The term is also applied to small vessels used to convey important personages.

BARITE, commonest barium mineral, consisting of barium sulfate (see BARIUM); white or yellow orthorhombic crystals. It is very dense and is mined for use in oil-well drilling muds and as the chief source of barium compounds. Largest US deposits are in Mo. and Ark.

BARIUM (Ba), silvery-white ALKALINE-EARTH METAL resembling CALCIUM; chief ores BARITE and witherite ($BaCO_3$). Barium is used to remove traces of gases from vacuum tubes; its compounds are used in making flares, fireworks, paint pigments and poisons. AW 137.3, mp 725°C, bp 1640°C, sg 3.5 (20°C). **Barium Sulfate** ($BaSO_4$), highly insoluble and opaque to X-rays, can be safely ingested for X-ray examination of the GASTROINTESTINAL TRACT.

BARIUM ENEMA; Barium Meal. See GASTROINTESTINAL SERIES.

BARK, general term for the covering of stems of woody plants, comprising the secondary phloem, cork cambium and CORK. The bark is impervious to water and protects the stem from excessive evaporation; it also protects the more delicate tissues within. Extracts of bark may have medicinal uses, e.g., QUININE from chinchona bark. (See TREE.)

BARKLA, Charles Glover (1877–1944), British physicist who was awarded the 1917 Nobel Prize for Physics for research on the scattering of X RAYS by GASES, in particular for his discovery of the "characteristic radiation" scattered by different elements.

BARN (b), unit of nuclear CROSS-SECTION (area) equal to $10^{-28}\,m^2$.

BARNARD, Christiaan Neethling (1922–), South African surgeon who performed the first successful human heart TRANSPLANT operation in 1967.

BARNARD, Edward Emerson (1857–1923), US astronomer who discovered the fifth satellite of JUPITER (1892). In 1916 he discovered **Barnard's star**, a red dwarf STAR only 6ly from the earth and which has the largest known stellar PROPER MOTION.

BAROMETER, an instrument for measuring air pressure (see ATMOSPHERE), used in WEATHER FORECASTING and for determining altitude. Most commonly encountered is the **aneroid barometer** in which the effect of the air in compressing an evacuated thin cylindrical corrugated metal box is amplified mechanically and read off on a scale or, in the **barograph**, used to draw a trace on a slowly rotating drum, thus giving a continuous record of the barometric pressure. The aneroid instrument is that used for aircraft ALTIMETERS. The earliest barometers, as invented by TORRICELLI in 1643, consisted simply of a glass tube about 800mm long closed at one end and filled with MERCURY before being inverted over a pool of mercury. Air pressure acting on the surface of the pool held up a column of mercury about 760mm tall in the tube, a "Torricellian" vacuum appearing in the closed end of the tube. The height of the column was read as a measure of the pressure. In the **Fortin barometer**, devised by Jean Fortin (1750–1831) and still used for accurate scientific work, the lower mercury level can be finely adjusted and the column height is read off with the aid of a VERNIER SCALE.

BARRIER REEF, CORAL reef, lying roughly parallel to a shore. (See also LAGOON).

BARROW, also termed a tumulus or MOUND (in the US), a large mound of earth and stones containing or covering an ancient burial place. Long barrows are generally Neolithic, round ones Early Bronze Age.

BARROW, Isaac (1630–1677), English mathematician and theologian. The first Lucasian professor of mathematics in the University of Cambridge (1663), Barrow resigned in favor of his pupil Isaac NEWTON in 1669. His work on tangents and areas was influential in Newton's development of the CALCULUS.

BARTHOLIN (Latin: Bartholinus), family of Danish physicians. **Caspar Berthelsen Bartholin**

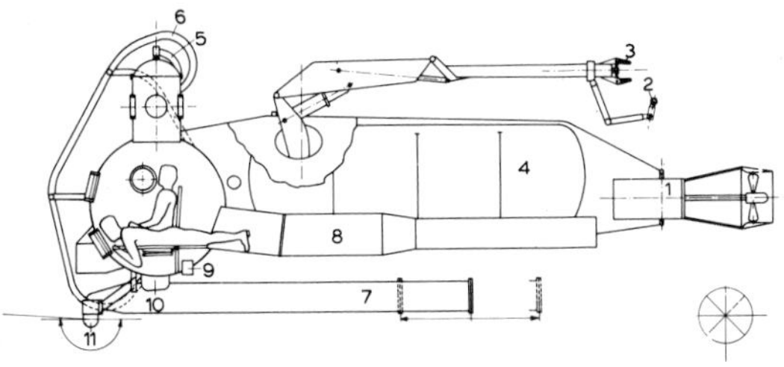

Right: The *Nereid 330* bathyscaphe is winched aboard its mother ship.
Above: Section through *Nereid 330*: (1) main propulsion unit; (2) secondary manipulator; (3) main manipulator hand; (4) flotation tanks; (5) hatch entrance; (6) fender; (7) replaceable battery cylinder; (8) auxiliary engines; (9) echo sounder; (10) emergency ballast; (11) retractable sonar dome.

(1585–1629) was author of the much-used *Institutiones Anatomicae* (1611). This was enlarged by his son, **Thomas Bartholin** (1616–1680), noted also for his study of the human lymphatic system (1652). **Erasmus Bartholin** (1625–1698), brother of Thomas, in 1669 discovered the phenomenon of DOUBLE REFRACTION in Iceland spar crystals. **Caspar Bartholin** (1655–1738), son of Thomas, was the discoverer of Bartholin's vaginal glands.

BARTON, Sir Derek Harold Richard (1918–), British organic chemist who shared the 1969 Nobel Prize for Chemistry with Odd HASSEL for the development of CONFORMATIONAL ANALYSIS.

BARYONS, in particle physics, a class of SUBATOMIC PARTICLES (comprising the nucleons and hyperons) distinct from the mesons.

BASAL METABOLIC RATE (BMR), a measure of the rate at which an animal at rest uses energy. Human BMR is a measure of the heat output per unit time from a given area of body surface, the subject being at rest under certain standard conditions. It is usually estimated from the amounts of oxygen and carbon dioxide exchanged in a certain time. (See METABOLISM.)

BASALT, a dense igneous rock, mainly plagioclase FELDSPAR, fine-grained and dark gray to black in color; volcanic in origin, it is widespread as lava flows or intrusions. Basalt can assume a striking columnar structure, as exhibited in the Palisades along the Hudson R, or in Devils Postpile in Cal., and it can also form vast plateaus, such as the 200 000sq mi Deccan of India. Most oceanic islands of volcanic origin, such as Hawaii, are basaltic.

BASE, in chemistry, the complement of an acid. Bases used to be defined as substances which react with acids to form SALTS, or as substances which give rise to hydroxyl ions (see HYDROXIDE) in aqueous solution. Some such inorganic strong bases are known as ALKALIS. In modern terms, bases are species which can accept a HYDROGEN ion from an acid, or which can donate an electron-pair to a Lewis ACID.

BASIDIOMYCETES. See FUNGI.

BASOV, Nikolai Gennadievich (1922–), Soviet physicist who shared the 1964 Nobel Prize for Physics with TOWNES and PROKHOROV for research in quantum physics which led to the development of LASERS and MASERS.

BATES, Henry Walter (1825–1892), English entomologist who first drew attention to the phenomenon of MIMICRY. One form, **Batesian mimicry**, is named for him.

BATESON, William (1861–1926), English biologist, known as the "father of GENETICS." In 1900 he translated MENDEL's classic HEREDITY paper into English and thereafter his work did much to promote general acceptance of the Mendelian theory.

BATHOLITH, large subterranean mass of IGNEOUS ROCK formed by the intrusion of MAGMA across the enclosing rock beds, and its subsequent cooling. Batholiths have an extent of over 100km², and frequently form the cores of mountain ranges.

BATHYSCAPHE, submersible deep-sea research vessel, invented by Auguste PICCARD in the late 1940s, comprising a small, spherical, pressurized passenger cabin suspended beneath a cigar-shaped flotation hull. On the surface most of the flotation tanks in the hull are filled with GASOLINE, the rest, sufficient to float the vessel, with air. To dive the air is vented and seawater takes its place. During descent, sea water is allowed to enter the gasoline-filled tanks from the bottom, compressing the gasoline and thus increasing the DENSITY of the vessel. The rate of descent is checked by releasing iron ballast. To begin ascent, the remaining ballast is jettisoned. As the vessel rises, the gasoline expands, expelling water from the flotation tanks, thus lightening the vessel further and accelerating the ascent. Battery-powered motors provide the vessel with a degree of submarine mobility.

BATHYSPHERE, a hollow steel sphere suspended by cables from a surface ship and used for deep-sea research before the development of the BATHYSCAPHE. The first bathysphere was built and used by engineer Otis Barton and naturalist William BEEBE in 1930.

BATTANI, abu-Abdullah Muhammad ibn-Jabir, al-, or Latin **Albategnius** or **Albatenius** (c858–929), Arab mathematician and astronomer who improved on the results of Claudius PTOLEMY by applying TRIGONOMETRY to astronomical computations. He had a powerful influence on medieval European ASTRONOMY.

BATTERY, a device for converting internally-stored chemical ENERGY into direct-current ELECTRICITY. The term is also applied to various other electricity

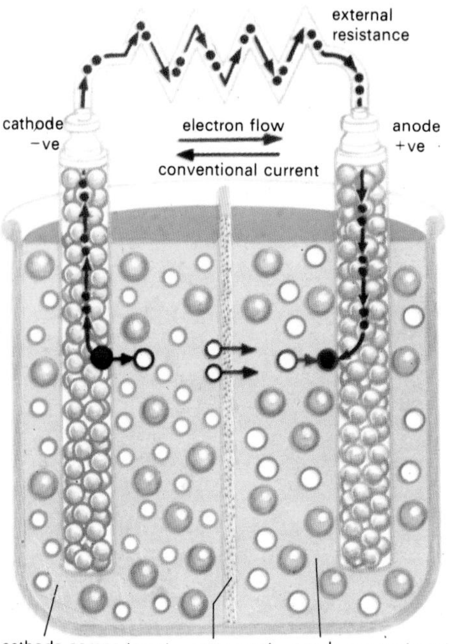

Battery: Schematized diagram of a Daniell Cell with zinc cathode (blue spheres), copper anode (green spheres), dilute sulfuric acid (grey circles—H_3O^+—and yellow spheres—SO_4^{2-}) in the cathode compartment, copper sulfate (green cirlces—Cu^{2+}—and yellow spheres) solution in the cathode compartment and porous pot separating the two compartments. The fundamental reactions of the cell are demonstrated by the "atoms" picked out in bolder colors. At the cathode, a zinc atom (Zn—blue ring with red center) gives up two electrons (e^-—red dot) and, as a zinc ion (Zn^{2+}—blue circle with a white center) enters the cathode solution. The cathode compartment now has a charge excess +2 and so two hydrogen ions (H_3O^+—black circle) pass through the porous pot barrier to the anode compartment. This allows a copper ion (Cu^{2+}—green circle with white center) to accept two electrons from the anode and deposit itself as a copper atom (Cu—green circle with red center) on the anode. The electrons are made available at the anode because electrons from the cathode are allowed to flow through an external circuit to the anode. The energy for the process comes from the overall reaction:

$$Zn + Cu^{2+} \rightarrow Zn^{2+} + Cu$$

and by means of the cell is made available as electricity to do work against a resistance in the external circuit. The conventional current flows in the opposite direction from the actual flow of electrons.

sources, including the SOLAR CELL and the nuclear cell, but is usually taken to exclude the FUEL CELL, which requires the continuous input of a chemical fuel for its operation. Chemical batteries consist of one or more electrochemical (voltaic) cells (comprising two ELECTRODES immersed in a conducting electrolyte) in which a chemical reaction occurs when an external circuit is completed between the electrodes. Most of the energy liberated in this reaction can be tapped if a suitable load is placed in the external circuit, impeding the flow of ELECTRONS from CATHODE to ANODE. (The conventional current, of course, flows in the opposite sense.) Batteries are classified in two main divisions. In **primary cells**, the chemical reaction is ordinarily irreversible and the battery can yield only a finite quantity of electricity. Single primary-cell batteries are used in flashlights, shavers, LIGHT METERS, etc. The most common type is the dry **Leclanché cell,** which has a ZINC cathode, a CARBON anode and uses ammonium chloride paste as electrolyte. MANGANESE dioxide "depolarizer" is distributed around the anode (mixed with powdered GRAPHITE) to prevent the accumulation of the HYDROGEN gas which would otherwise stop the operation of the cell. The dry Leclanché cell gives a nominal 1.54V. For the higher voltages necessary to power transistor radios, batteries containing several thin laminar cells are used. **Secondary cells**, known also as storage batteries or **accumulators**, can be recharged and reused at will provided too much electricity has not been abstracted from them. The most common type, as used in AUTOMOBILES, is the lead-acid type, in which both electrodes are made of LEAD (the positive covered with lead (IV) oxide when charged) and the electrolyte is dilute SULFURIC ACID. Its voltage is about 2V, depending on the state of charge. The robust yet light nickel-iron battery (having a POTASSIUM hydroxide solution electrolyte) is widely used in telephone exchanges and other heavy-duty situations but is being partly displaced by the nickel-cadmium type. They give about 1.3V.

The first battery was the voltaic pile invented c1800 by VOLTA. This comprised a stack of pairs of silver and zinc disks, each pair separated by a brine-soaked board. For many years from 1836 the standard form of battery was the **Daniell cell**, with a zinc cathode, a copper cathode and a porous-pot barrier separating the anode electrolyte (copper(II) sulfate) from the cathode electrolyte (sulfuric acid). The lead-acid storage battery was invented by Gaston Planté in 1859 and the wet Leclanché cell, the prototype for the modern dry cell, by Georges Leclanché in 1865.

BAUER, George. See AGRICOLA, GEORGIUS.

BAUMÉ, Antoine (1728–1804), French chemist, inventor of the Baumé HYDROMETER and remembered in the Baumé hydrometer scales, used to describe the DENSITIES of liquids.

BAUXITE, the main ore of ALUMINUM, consisting of hydrated aluminum oxide, usually with iron oxide impurity. It is a claylike, amorphous material formed by the weathering of silicate rocks, especially under tropical conditions. High-grade bauxite, being highly refractory, is used as a lining for furnaces. Synthetic corundum is made from it, and it is an ingredient in some quick-setting cements. Leading bauxite-producing countries include Jamaica, Australia, the USSR, Surinam, Guyana, France, Guinea and the US (especially Ark.).

BAYLISS, Sir William Maddock (1860–1924), English physiologist who, in collaboration with **Ernest Henry Starling** (1866–1927), introduced the term HORMONE to describe substances which, when secreted from one part of the body, have a specific effect on another. In 1902 they discovered SECRETIN, one of the first hormones to be identified. Bayliss also introduced the saline injection for cases of surgical SHOCK during WWI.

BAYOU (from Choctaw *bayuk*, a creek), a minor creek or river tributary to a larger body of water, especially in La. By extension, the term is applied to muddy or sluggish bodies of water in general.

BCG VACCINE (*B*acillus *C*almette-*G*uérin), anti-TUBERCULOSIS vaccine (see VACCINATION) developed by CALMETTE and GUÉRIN.

BEACH, stretch of SAND, shingle, GRAVEL and other material along the shore of a lake, river or sea. Caused by erosion and deposition of sediment by waves, it usually extends from the furthest point reached by waves out to a water depth of around 10m.

BEADLE, George Wells (1903–), US geneticist who shared part of the 1958 Nobel Prize for Physiology or Medicine with E. L. TATUM (and J. LEDERBERG) for work showing that individual GENES controlled the production of particular ENZYMES (1937–40).

BEAGLE, H.M.S., British survey ship which, under the command of **Captain Robert Fitzroy** (1805–1865), and carrying Charles DARWIN as naturalist, sailed around the world between 1831 and 1836. Darwin's observations on this voyage, particularly those of the fauna of the Galápagos Islands, helped bring him to his theory of EVOLUTION.

BEAK, or **bill,** general term for a rigid, projecting oral structure. All monotremes, birds (for both of which the term "bill" is preferred) and turtles, as well as some fish, cephalopod and insects, are beaked, as were many dinosaurs. Beak shapes and sizes are usually highly specialized.

BEARINGS, components of MACHINES which support and direct loads while reducing FRICTION where moving parts are in contact. The simplest type is the journal bearing in which a rotating shaft is supported in a hole in a fixed frame. The inner surface of the hole is usually lined with a bearing metal such as BABBITT METAL to reduce wear. Friction and wear are also reduced by suitable LUBRICATION. Lubricants, which include greases, oils, water and even air, form a thin FLUID film between the moving parts of the bearing. Usually the motion itself is sufficient to form the film (hydrodynamic bearings) but sometimes the lubricant must be applied under PRESSURE (hydrostatic bearings) as in air-lubricated dental drills. For many applications roller bearings and ball bearings are used. In these a separator holds a series of short cylinders or balls between the inner and outer rings of the bearing. In recent years dry (unlubricated) plastic bearings and self-lubricating bearings have been developed for applications where lubrication is difficult or undesirable.

BEATING, a phenomenon of importance in RADIO and ACOUSTICS resulting from the INTERFERENCE of two wave-trains of similar frequency (see WAVE MOTION) in which a new periodicity is set up in the aggregate AMPLITUDE having frequency equal to the difference of the two constituent frequencies. Beats between two musical notes of similar pitch can often be heard as an unpleasant throbbing; beating between two ULTRASONIC tones may result in an audible tone.

BEAUFORT SCALE, method of measuring wind force, developed in 1806 by the British Admiral Sir Francis Beaufort. Wind strength is measured on a scale ranging from 0–12 (0–17 in Britain and the US). Internationally, the scale has now been superseded by measurement in knots.

BEAUMONT, William (1785–1853), US army physician noted for his researches into the human DIGESTIVE SYSTEM. While on assignment in northern Mich. in 1822 he treated a trapper with a serious stomach wound; when the wound healed, an opening (or FISTULA) into the victim's stomach remained, through which Beaumont was able to extract gastric juices for analysis.

BECHER, Johann Joachim (1635–1682), German chemist, physician and economist whose conception of an active principle of combustion was developed by his pupil STAHL into the PHLOGISTON theory.

BECKMANN THERMOMETER, a mercury-in-glass THERMOMETER used in CALORIMETRY which offers an accuracy of up to ± 0.001K but which has a range of only 5K. This is achieved through its having a large bulb and fine bore. It was devised by the German organic chemist **Ernst Otto Beckmann** (1853–1923), who is also remembered for his discovery (1886) of the Beckmann rearrangement of ketoximes (see OXIMES) into AMIDES under acid CATALYSIS.

BECQUEREL, Antoine Henri (1852–1908), French physicist who, having discovered natural RADIOACTIVITY in a URANIUM salt in 1896, shared the 1903 Nobel physics prize with Pierre and Marie CURIE.

BEDDOES, Thomas (1760–1808), English chemist and physician, who pioneered the inhalation of various gases in medicine. While in Bristol (from 1792) he gathered around him an important group of scientists and men of letters, including DAVY, WATT, S. T. Coleridge, Southey and Wordsworth.

BEDSORES, sores and ULCERS occurring in bedridden patients when pressure and friction restrict skin blood supply. They may be prevented by frequent change of position and bathing; treatment includes ASTRINGENTS, SILICONE creams and ultraviolet light.

BEEBE, Charles William (1877–1962), US naturalist remembered for the descents into the ocean depths he made with Otis Barton in their BATHYSPHERE. Diving off Bermuda in 1934 they reached a then-record depth of 3028ft (923m).

BEER, an ALCOHOLIC BEVERAGE made by fermenting cereals (see BREWING). Known since ancient times, beer became common where the climate was unsuited to WINE production. Beer includes all the malt liquors variously called ale, stout, porter (drunk in the UK and Ireland) and lager. The alcohol content is 3–7%.

BEES, superfamily (Apoidea) of insects which convert nectar and pollen into HONEY for use as food. There are about 20 000 species. Bees and flowering plants are largely interdependent; plants are pollinated (or fertilized) as the bees gather their pollen. Many farmers keep bees specially for this purpose.

Most bees are solitary and each female builds her own nest, although many bees may occupy a single site. Eggs are laid in cells provided with enough pollen-nectar paste to feed the larva until it becomes a flying, adult bee. Social bees (honeybees and bumblebees) live in a complex society of 10 000–50 000 members. Headed by the queen, whose

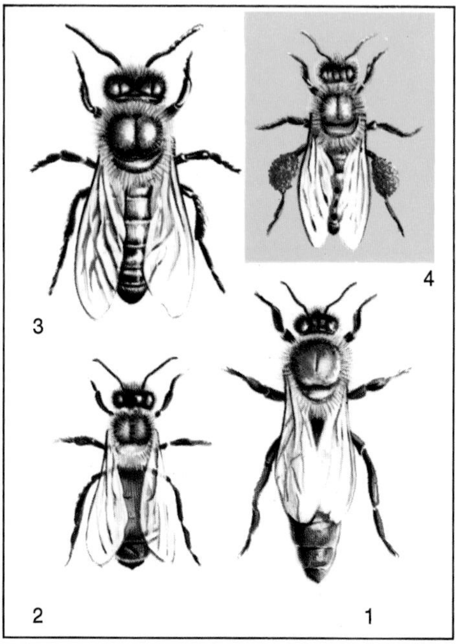

The bees in a hive are of three kinds: the queen (1), the male drone (2) and the female worker (3). Pollen is the stuff of life to all of them; it is gathered by the worker, in "baskets" on each hindleg (4).

function is to lay eggs (up to 2000 a day), the community comprises female workers which collect pollen and build cells, and male bees, or drones, which fertilize the few young queens that appear each fall. Parasitic bees, not equipped to build hives, develop in the cells of the host working bees.

BEESWAX, substance secreted by worker BEES and used to build the cell walls of the honeycomb. It contains cerotic acid, myricin and long-chain ALKANES, and melts at around 65°C. The purified wax is used for candles and in furniture waxes, cosmetics, some printing inks and elsewhere.

BEHAVIORAL SCIENCES, those sciences dealing with human activity, individually or socially. The term, which is sometimes treated synonymously with SOCIAL SCIENCES, embraces such fields as physical and, in particular, cultural and social ANTHROPOLOGY, PSYCHOLOGY and SOCIOLOGY.

BEHAVIORISM, school of PSYCHOLOGY based on the proposal that behavior should be studied empirically—by objective observations of reactions—(see EMPIRICISM) rather than speculatively. It had its roots in ANIMAL BEHAVIOR studies, defining behavior as the actions and reactions of a living organism (and, by extension, man) in its environment; and more specifically in the work of PAVLOV in such fields as conditioned REFLEXES. Behaviorism developed as an effective factor in US psychology following the work of J. B. WATSON just before WWI; and since then it has influenced most schools of psychological thought.

BEHRING, Emil Adolf von (1854–1917), German bacteriologist who was awarded the first Nobel Prize for Physiology or Medicine (1901) in recognition of his part in the development of an ANTITOXIN giving protection against DIPHTHERIA.

BÉKÉSY, Georg von (1899–), Hungarian-born US physicist who was awarded the 1961 Nobel Prize for Physiology or Medicine for his development of a new theory of the physical mechanism of hearing (see EAR).

BEL, a unit used to express power level relative to an arbitrary reference level defined as the common LOGARITHM of the ratio of the powers. Named for Alexander Graham BELL, the bel is seldom used, the DECIBEL (1dB = 0.1 bel) and NEPER being preferred by acoustics and telecommunications engineers.

BELL, a resonant metal object in the form of a cup hung from its closed end (the head), which rings when struck with a "clapper" near its rim. It is used ritually, to give audible time signals and as a warning device. Modern bells are cast in BELL METAL to carefully computed designs so that the correct mix of fundamentals and overtones is produced when the bell is rung. As well as large bells hung in peals (tuned sets) in church towers and campaniles (freestanding bell towers), there are also the smaller handbells and tubular chimes, which are musical instruments of increasing popularity.

BELL, Alexander Graham (1847–1922), Scottish-born US scientist and educator who invented the TELEPHONE (1876), founded the Bell Telephone Company and devised the wax-cylinder PHONOGRAPH and various aids for teaching the deaf. In later life he helped perfect the AILERON for airplanes.

BELL, Sir Charles (1774–1842), Scottish anatomist, a pioneer investigator of the working of the NERVOUS SYSTEM, whose most important discovery was to distinguish the functions of sensory and motor nerves.

BELL, Electric, audible warning device in which a small clapper is made to vibrate rapidly against a resonant metal gong. This is usually achieved by using a direct-current ELECTROMAGNET to attract the clapper arm toward the gong, while a make-and-break contact cuts off the current just before the clapper strikes, so that the spring-loaded clapper arm recoils—remaking the circuit and commencing another cycle. In the single-action bell, there is no repeater mechanism.

BELL METAL, a BRONZE with a high tin content (15–25%), used for casting bells because of its sonority.

BELLOWS, device used to produce a blast of air to speed the burning of a fire, or for musical instruments (accordions and organs). Bellows usually consist of two variously-shaped boards hinged at the nozzle end, and elsewhere joined by pleated leather sides. As the boards are opened, air is drawn in through a valve; as they are shut it is forced out through the nozzle. A blacksmith's bellows has two such chambers working out-of-phase, thus ensuring a continuous flow of air.

BELL TELEPHONE LABORATORIES, research organization set up in 1925 by the American Telephone and Telegraph Company. The Bell Laboratories have been responsible for many important developments in telecommunications technology, notably the TRANSISTOR.

BENDIX, Vincent (1882–1945), US inventor and industrialist who devised the Bendix self-starter for AUTOMOBILES and developed a four-wheeled brake system.

BENDS. See AEROEMBOLISM.

Alexander Graham Bell, the inventor of the telephone, making the first call from New York to Chicago in 1892.

BENEDICT, Ruth (née Fulton; 1887–1948), US cultural anthropologist, whose extensive fieldwork helped illustrate the theory of cultural relativism—that what is deemed deviant in one culture may be normal in another. (See ANTHROPOLOGY.)

BENTHOS, plants and animals living on the sea bottom, as distinct from NEKTON (creatures which swim freely) and PLANKTON (creatures which drift with the current). Benthos include sea anemones and sea cucumbers. (See also OCEAN.)

BENTONITE, fine-grained CLAY formed from volcanic ASH (see VOLCANISM; VOLCANO) by the hydration (formation of HYDRATES or ABSORPTION of water) of, and loss of BASES and perhaps SILICA from, tiny particles of volcanic GLASS present in the ash. SODIUM bentonites, which expand considerably when saturated with water, are used diversely, as in the making of PAPER and sealing of DAMS. CALCIUM bentonites are used in the making of FULLER'S EARTH.

BENZ, Karl (1844–1929), German engineer who built the first commercially successful AUTOMOBILE (1885). His earliest autos were tricycle carriages powered by a small INTERNAL-COMBUSTION ENGINE.

BENZALDEHYDE (C_6H_5CHO), colorless liquid with a smell of almonds. It is an aromatic ALDEHYDE, found in bitter almonds but synthesized from TOLUENE, and is used as a flavoring and in chemical synthesis. It is oxidized to BENZOIC ACID. mp −26°C, bp 179°C.

BENZEDRINE, or **amphetamine,** the prototypical member of the AMPHETAMINES.

BENZENE (C_6H_6), colorless toxic liquid HYDROCARBON produced from PETROLEUM by reforming, and from COAL GAS and COAL TAR. It is the prototypical AROMATIC COMPOUND: its molecular structure, first proposed by KEKULÉ, is based on a regular planar hexagon of carbon atoms. Stable and not very reactive, benzene forms many substitution products, and also reacts with the HALOGENS to give addition products—including γ-benzene hexachloride, a powerful insecticide. It is used as a solvent, in motor fuel, and as the starting material for the manufacture of a vast variety of other aromatic compounds, especially PHENOL, STYRENE, ANILINE and maleic anhydride. mp 5°C, bp 80°C.

BENZINE, volatile, inflammable liquid obtained from PETROLEUM. It is a mixture of aliphatic HYDROCARBONS, used as a solvent in DRY CLEANING.

BENZOIC ACID (C_6H_5COOH), white crystalline solid, an aromatic CARBOXYLIC ACID. It occurs naturally in many plants, and is made by oxidation of TOLUENE. It is mainly used as a food preservative. mp 122°C, bp 249°C. Compounds containing the benzoyl group (C_6H_5CO-) were studied by von LIEBIG and WÖHLER.

BENZYL ALCOHOL ($C_6H_5CH_2OH$), or phenylcarbinol, colorless liquid, an aromatic ALCOHOL whose esters are found in flowers and used in perfumes. It is used in color-film developing and in dyeing. mp −15°C, bp 205°C.

BERGIUS, Friedrich (1884–1949), German industrial chemist who, during WWI, developed a process for making GASOLINE by the high-pressure HYDROGENATION of COAL. He shared the 1931 Nobel chemistry prize with Karl BOSCH for work on high-pressure reactions.

BERGSON, Henri Louis (1859–1941), French philosopher, the first exponent of PROCESS PHILOSOPHY. Reacting against the physicists' definition of TIME and substituting a notion of experienced duration; rejecting the psychophysical parallelism of the day and asserting the independence of mind, and viewing EVOLUTION not as a mechanistic but as a creative process energized by an *élan vital* (vital impulse), Bergson was perhaps the most original philosopher of the early 20th century. He was awarded the Nobel Prize for Literature in 1927.

BERIBERI, deficiency disease caused by lack of VITAMIN B_1 (thiamine); it may occur in MALNUTRITION, ALCOHOLISM or as an isolated deficiency. NEURITIS leading to sensory changes, and foot or wrist drop, palpitations, EDEMA and HEART failure are features; there may be associated dementia. Onset may be insidious or acute. Treatment is thiamine replacement; thiamine enrichment of common foods prevents beriberi.

BERKELEY, George (1685–1753), Irish philosopher and bishop who, rejecting the views of LOCKE as to the nature of material substance, substituted the *esse-percipi* principle: to be is to be perceived (or to be capable of perception). His visit to Rhode Island (1728–31) is commemorated in the name of Berkeley, Cal.

BERKELIUM (Bk), a TRANSURANIUM ELEMENT in the ACTINIDE series. Bk^{249} is prepared by bombarding curium-244 with neutrons.

BERNARD, Claude (1813–1887), French physiologist regarded as the father of experimental medicine. Following the work of BEAUMONT he opened artificial FISTULAS in animals to study their DIGESTIVE SYSTEMS. He demonstrated the role of the PANCREAS in digestion, discussed the presence and function of GLYCOGEN in the LIVER (1856) and in 1851 reported the existence of the vasomotor nerves.

BERNOULLI, family of Swiss mathematicians important in establishing CALCULUS as a mathematical tool of widespread application. **Jacques (Jakob) Bernoulli** (1654–1705), who applied calculus to many geometrical problems, is best remembered in the Bernoulli numbers and the Theorem of Bernoulli that appeared in a posthumous work on PROBABILITY. **Jean (Johann) Bernoulli** (1667–1748), brother of Jacques, also a propagandist on behalf of the Leibnitzian calculus, assisted his brother in founding the calculus of variations. **Daniel Bernoulli** (1700–1782), son of Jean, anatomist, botanist and mathematician—perhaps the family's most famous member—published his *Hydrodynamics* in 1738, applying calculus to that science. In it he proposed **Bernoulli's principle,** which states that in any small volume of space through which a fluid is flowing steadily, the total ENERGY, comprising the pressure, potential and kinetic energies, is constant. This means that the PRESSURE is inversely related to the VELOCITY. This principle is applied in the design of the AIRFOIL, the key component in making possible all heavier-than-air craft, where the faster flow of air over the longer upper surface results in reduction of pressure there and hence a lifting force acting on the airfoil (see also AERODYNAMICS).

BERRY, a fleshy FRUIT normally with many seeds, although occasionally only one, as in the date. The tomato, melon, orange and grape are examples of berries. The name is often given to "false fruits" such as the strawberry, and aggregate fruits such as the raspberry. (See also DRUPE.)

BERTHELOT, Pierre Eugène Marcel(l)in (1827–1907), French chemist and statesman, who pioneered the synthesis of organic compounds not found in nature and later introduced the terms exothermic and endothermic (descriptive of chemical reactions) to THERMOCHEMISTRY. His public career was crowned in 1895 when he became foreign secretary.

BERTHOLLET, Claude Louis, Count (1748–1822), Savoyard-born French chemist noted for his work on CHLORINE (first using it in BLEACHING) but best remembered for his generally erroneous belief that the components in a chemical compound might be present in any of a continuous range of proportions (see COMPOSITION, CHEMICAL).

BERTILLON, Alphonse (1853–1914), French criminologist who devised a system (*Bertillonage*) for identifying criminals based on anthropometric measurements (see ANTHROPOMETRY), adopted by the French police in 1888 and used until the adoption of fingerprinting (see FINGERPRINTS).

BERYL, aluminum beryllium silicate ($Al_2[Be_3(SiO_3)_6]$), the commonest ore of BERYLLIUM, mainly found as hexagonal crystals in granite throughout the world. EMERALD is a deep-green beryl with some chromium; AQUAMARINE is a pale-blue beryl. mp 1400°C.

BERYLLIUM (Be), a gray ALKALINE-EARTH METAL, found mainly as BERYL, prepared by reducing beryllium fluoride with magnesium. It is strong, hard and very light, and has a high melting point and high heat absorption—all useful properties. Combined with copper it makes a very hard alloy resistant to corrosion and fatigue. It is also used in NUCLEAR REACTOR construction to moderate neutrons, and in X-RAY tube windows. Beryllium is relatively unreactive; it forms divalent, tetracoordinate compounds which are poisonous, causing the disease berylliosis. The refractory **Beryllium Oxide** (BeO) is used in ceramics and in electronics. AW 9.01, mp 1278°C, bp 2970°C, sg 1.848 (20°C).

BERZELIUS, Jöns Jakob, Baron (1779–1848), Swedish chemist who determined the ATOMIC WEIGHTS of nearly 40 elements before 1818, discovered CERIUM (1803), SELENIUM (1818) and THORIUM (1829), introduced the terms PROTEIN, ISOMERISM and CATALYSIS and devised the modern method of writing empirical formulas (1813).

BESSEL, Friedrich Wilhelm (1784–1846), German astronomer who first observed stellar PARALLAX (1838) and set new standards of accuracy for positional astronomers. From the parallax observation, which was of 61 Cygni, he calculated the star to be about 6 ly distant, setting a new lower limit for the scale of the universe. In applied mathematics he was the first to employ **Bessel functions** in a systematic way.

BESSEMER, Sir Henry (1813–1898), British inventor of the BESSEMER PROCESS for the manufacture of steel, patented in 1856.

BESSEMER PROCESS, the first cheap, large-scale method of making STEEL from PIG IRON, invented in the 1850s by Henry BESSEMER. The Bessemer converter is a pivoting, pear-shaped BLAST FURNACE lined with refractory bricks. The furnace is tilted, loaded with molten pig iron, then righted. Compressed air blown through the tuyeres burns off most of the carbon and converts silicon and manganese to slag as the temperature rises. If lime is added, an afterblow removes phosphorus. The process is largely superseded by the OPEN-HEARTH PROCESS.

BEST, Charles Herbert (1899–), US–Canadian physiologist who assisted F. G. BANTING in the isolation of INSULIN but, to Banting's annoyance, did not share in the Nobel prize that Banting shared with J. J. R. MACLEOD.

BESTIARY, medieval or other collection of prose or poetry detailing characteristics and habits of real or imagined animals and drawing morals therefrom. The earliest such compilation, *Physiologus* (The Naturalist), appeared in Greek before 200 AD.

BETA RAY, a stream of beta particles (i.e., ELECTRONS or POSITRONS) emitted from radioactive nuclei undergoing beta disintegration (see RADIOACTIVITY). Beta particles are emitted with velocities approaching that of light and can penetrate up to 1mm of lead. Positive beta rays are not emitted from any naturally occurring material.

BETATRON, an electron ACCELERATOR in which the particles are accelerated in a circular path of constant radius by a steadily increasing magnetic field produced by an alternating-current ELECTROMAGNET. The maximum energy attained by the ELECTRONS is about 300MeV.

BETELGEUSE, Alpha Orionis, second brightest star in ORION. An irregularly variable red supergiant (see VARIABLE STAR) with a variable radius some 300 times that of the sun, it is over 150pc from earth.

BETHE, Hans Albrecht (1906–), German-born US theoretical physicist who proposed the nuclear CARBON CYCLE to account for the sun's energy output (1938). During WWII he worked on the Manhattan Project. He was awarded the 1967 Nobel physics prize for his work on the source of stellar energy.

BETTELHEIM, Bruno (1903–), Austrian-born US psychologist who drew on his prewar experience as an inmate of Nazi concentration camps to describe men's behavior in extreme situations (1943). His subsequent work has mainly concerned the treatment of autistic (see AUTISM) and disturbed children.

BEVATRON, contraction for *b*illion *e*lectron-*v*olt synchro*tron*, a name used to describe several high-energy particle ACCELERATORS, first applied to a 6-GeV proton SYNCHROTRON at the University of California, Berkeley.

BICARBONATE OF SODA, or sodium bicarbonate. See SODIUM.

BICARBONATES, or hydrogen carbonates, acid salts of carbonic acid (see CARBON), containing the ion HCO_3^-. Bicarbonates are formed by the action of carbon dioxide on carbonates in aqueous solution; this reaction is reversed on heating. Dissolved calcium and magnesium bicarbonates give rise to HARD WATER.

BICEPS, either of two MUSCLES that are split in two in their upper part to form a Y-shape. *Biceps brachii* is the chief upper ARM muscle, attached to the shoulder blade and the radius. *Biceps femoris* is a thigh muscle (see LEG), attached to the PELVIS, the FEMUR and the FIBULA.

BICHAT, Marie François Xavier (1771–1802), French anatomist and pathologist, the founder of HISTOLOGY. Although working without the MICROSCOPE, Bichat distinguished 21 types of elementary TISSUES from which the organs of the body are composed.

BICHIR. See POLYPTERUS.

BIENNIAL, plant that completes its LIFE CYCLE in two years. During the first year leaves are produced and food is stored (as in the carrot and cabbage) for use in the second year when the plant bears flowers and fruit, then dies. (See ANNUAL; PERENNIAL.)

BIG BANG THEORY. See COSMOLOGY; HUBBLE; LEMAÎTRE.

BIG DIPPER. See GREAT BEAR.

BILE, a yellow-brown fluid secreted by the liver and containing salts derived from CHOLESTEROL. Stored and concentrated in the GALL BLADDER and released into the DUODENUM after a meal, the bile emulsifies fats and aids absorption of fat-soluble vitamins A, D, E and K. Other constituents of bile are in fact waste products. Yellow bile and black bile were two of the HUMORS of Hippocratic medicine.

BILHARZIA, chronic PARASITIC DISEASE of the BLADDER, intestine or LIVER, caused by *Schistosoma* species; often contracted by swimming in infected water.

BILL. See BEAK.

BILLROTH, Albert Christian Theodor (1829–1894), German surgeon and founder of modern abdominal SURGERY. A keen pianist, he is also remembered for his friendship with Brahms.

BIMETALLIC STRIP. See THERMOSTAT.

BINARY NUMBER SYSTEM, a number system which uses the POWERS of 2. Thus the number which in our everyday system, the DECIMAL SYSTEM, would be represented as 25 ($=(2\times10^1)+(5\times10^0)$) is in binary notation 11001 ($=(1\times2^4)+(1\times2^3)+(0\times2^2)+(0\times2^1)+(1\times2^0)$), which is equivalent to $(1\times16)+(1\times8)+(1\times1)$ or $(16+8+1)$.
The system is of particular note since digital COMPUTERS use binary numbers for calculation.

BINARY OPERATION, for any ordered pair of elements (a, b) in a set S (see SET THEORY), an operation * such that $a*b$ is a unique element of S. Examples of binary operations in the FIELD of REAL NUMBERS include ADDITION, MULTIPLICATION and SUBTRACTION. (See also ALGEBRA.)

BINARY STAR. See DOUBLE STAR.

BINET, Alfred (1857–1911), French psychologist who pioneered methods of mental testing. He collaborated with Théodore Simon in devising the Binet-Simon tests, widely used to estimate INTELLIGENCE.

BINOCULARS, an optical instrument comprising two compact TELESCOPES mounted parallel, used to obtain magnified stereoscopic views of distant scenes. The opera glass employs Galilean telescopes and the field glass uses low-power nautical telescopes, but for greater magnifications, reflecting PRISMS must be used to allow an objective lens of long focal length to be incorporated without making the instrument too elongate to be convenient in use. The arrangement of the prisms in the prismatic binocular also allows the objectives to be set farther apart than the eyes of the user, thus allowing the stereoscopic effect to be enhanced. The separation of the eyepiece mountings and the focusing of the individual eyepieces is adjustable in most models.

BINOCULAR VISION, the use of two EYES, set a small distance apart in the head and aligned approximately parallel, to view a single object. Owing to PARALLAX, the images in the two eyes are slightly different, which enables the observer to perceive what is seen in three dimensions and so to judge distance, size and shape. Only man and some higher animals possess binocular vision. (See also STEREOSCOPE.)

BINOMIAL. See POLYNOMIAL.

BINOMIAL COEFFICIENTS, in the expansion (see BINOMIAL THEOREM) of $(a+b)^n$, the COEFFICIENTS of the POWERS of a and b. The coefficient of a typical term, $a^{(n-r)}b^r$, may be written $\binom{n}{r}$.

BINOMIAL THEOREM, the theorem that a binomial (see POLYNOMIAL) $(a+b)$ may be raised to the POWER n by application of the formula

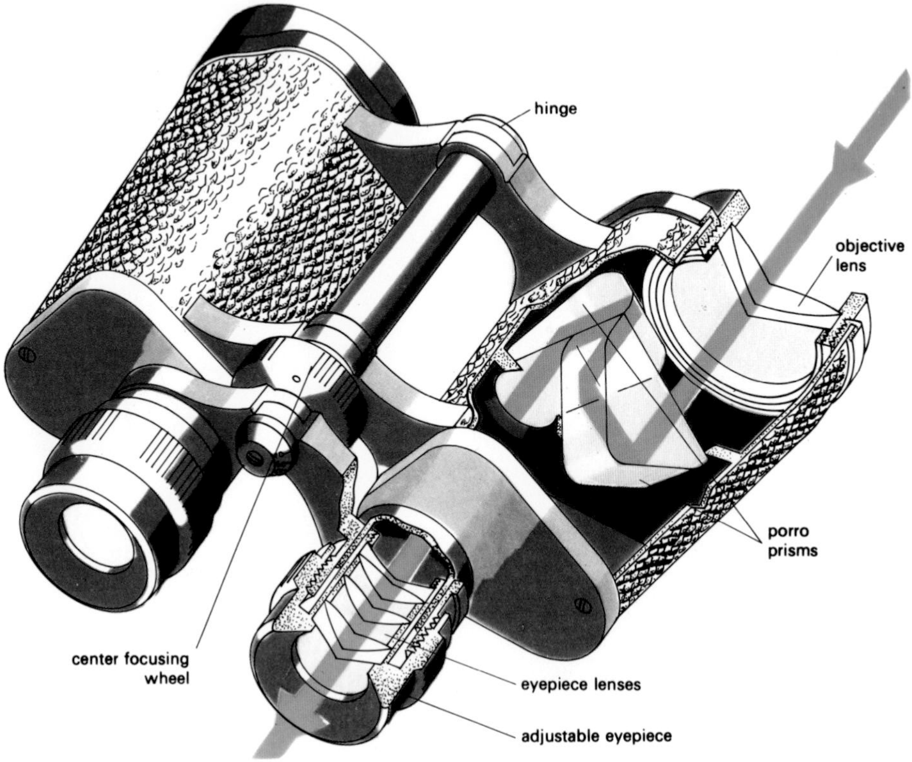

Cutaway of a set of binoculars incorporating the reflecting prism system devised by Italian, Ignazo Porro, in 1851.

$$(a+b)^n = a^n + na^{(n-1)}b + \frac{(n-1).n}{2} \cdot a^{(n-2)}b^2 + \ldots + an.b^{(n-1)} + b^n$$

(for evaluation of COEFFICIENTS see PASCAL'S TRIANGLE). Thus, for example,

$$(a+b)^4 = a^4+4a^3b+6a^2b^2+4ab^3+b^4.$$

Expansion of $(a-b)^n$ is equivalent to expansion of $(a+(-b))^n$. Thus, for example,

$$(a-b)^4 = a^4+4a^3(-b)+6a^2(-b)^2+4a(-b)^3+(-b)^4$$
$$= a^4-4a^3b+6a^2b^2-4ab^3+b^4.$$

Note that the expansion of $(a+b)^n$ has $(n+1)$ terms.

BIOCHEMISTRY, study of the substances occurring in living organisms and the reactions in which they are involved. It is a science on the border between BIOLOGY and ORGANIC CHEMISTRY. The main constituents of living matter are water, CARBOHYDRATES, LIPIDS and PROTEINS. The total chemical activity of the organism is known as its METABOLISM. Plants use sunlight as an energy source to produce carbohydrates from carbon dioxide and water (see PHOTOSYNTHESIS). The carbohydrates are then stored as starch; used for structural purposes, as in the CELLULOSE of plant cell walls; or oxidized through a series of reactions including the CITRIC ACID CYCLE, the energy released being stored as adenosine triphosphate (see NUCLEOTIDES). In animals energy is stored mainly as lipids, which as well as forming fat deposits are components of all cell membranes. Proteins have many functions, of which metabolic regulation is perhaps the most important. ENZYMES, which control almost all biochemical reactions, and some HORMONES are proteins. Plants synthesize proteins using simpler nitrogenous compounds from the soil. Animals obtain proteins from food and break them down by HYDROLYSIS to AMINO ACIDS. New proteins are made according to the pattern determined by the sequence of NUCLEIC ACIDS in the GENES. Many reactions occur in all CELLS and may be studied in simple systems. Methods used by biochemists and chemists are similar and include labelling with radioactive ISOTOPES and separation techniques such as CHROMATOGRAPHY, used to analyze very small amounts of substances, and the high-speed CENTRIFUGE. Molecular structures may be determined by X-RAY DIFFRACTION. Landmarks in biochemistry include the synthesis of urea by WÖHLER (1828), the pioneering research of von LIEBIG, PASTEUR and BERNARD, and more recently the elucidation of the structure of DNA by James WATSON and Francis CRICK in 1953.

BIOGENESIS, theory that all living organisms are derived from other living organisms. It is the opposite of the theory of SPONTANEOUS GENERATION. (See also LIFE.)

BIOLOGICAL CLOCKS, the mechanisms which control the rhythm of various activities of plants and animals. Some activities, such as mating, migration and hibernation, have a yearly cycle; others, chiefly reproductive functions (including human menstruation) follow the lunar month. The majority, however, have a period of roughly 24 hours, called a **circadian rhythm**. As well as obvious rhythms such as the patterns of leaf movement in plants and the activity/sleep cycle in animals, many other features such as body temperature and cell growth oscillate daily. Although related to the day/night cycle, circadian rhythms are not directly controlled by it. Organisms in unvarying environments will continue to show 24-hr rhythms, but the pattern can be changed—the clock reset. Scientists in the Arctic, with 6 months of daylight, used watches which kept a 21-hr day, and gradually their body rhythms changed to a 21-hr period. The delay in adjustment is important in modern travel. After moving from one time zone to another, it takes some time for the body to adjust to the newly imposed cycle. Biological clocks are important in animal navigation. Many animals, such as migrating birds or bees returning to the hive, navigate using the sun. They can only do this if they have some means of knowing what time of day it is. (See also MIGRATION.) Biological clocks are apparently inborn, not learned, but need to be triggered. An animal kept in the light from birth shows no circadian rhythms, but if placed in the dark for an hour or so immediately starts rhythms based on a 24-hr cycle. Once started, the cycles are almost independent of external changes, indicating that they cannot be based on a simple rhythm of chemical reactions, which would be affected by temperature. The biological clock may be somehow linked to external rhythms in geophysical forces, or may be an independent and slightly adjustable biochemical oscillator. In either case the mechanism is unknown. Not all biological rhythms are controlled by a "clock": in many cases they are determined simply by the time taken to complete a certain sequence of actions. For example, the heart rate depends on the time taken for the heart muscles to contract and relax. Unlike those controlled by biological clocks, such rhythms are easily influenced by drugs and temperature.

BIOLOGICAL CONTROL, the control of pests by the introduction of natural predators, parasites or disease, or by modifying the environment so as to encourage those already present. This first took place in Cal. in 1888, when Australian Ladybird beetles were introduced to eliminate the damage to citrus trees by the cottony-cushion scale insect. Another example is the introduction of myxomatosis to combat crop damage by rabbits. The "sterile male" technique is used to control many insect pests. Large numbers of males are bred, sterilized with X-rays and released. Since the females mate only once, many of them with the sterile males, the population rapidly decreases.

BIOLOGICAL WARFARE. See CHEMICAL AND BIOLOGICAL WARFARE.

BIOLOGY, the study of living things, i.e. the science of plants and animals, including humans. Broadly speaking there are two main branches of biology, the study of ANIMALS (ZOOLOGY) and the study of PLANTS (BOTANY). Within each of these main branches are a number of traditional divisions dealing with structure (ANATOMY, CYTOLOGY), development and function PHYSIOLOGY, EMBRYOLOGY), inheritance GENETICS, EVOLUTION), classification (TAXONOMY) and interrelations of organisms with each other and with their environment (ECOLOGY). These branches are also split into a number of specialist fields, such as mycology, ENTOMOLOGY, HERPETOLOGY.

However, the traditional division into zoology and botany no longer applies since groups of biosciences have developed which span their limits, e.g. MICROBIOLOGY, BACTERIOLOGY, virology, OCEANOGRAPHY, MARINE BIOLOGY, LIMNOLOGY. There are also biosciences that bridge the gap between the physical sciences of chemistry, physics and geology, e.g., BIOCHEMISTRY, BIOPHYSICS and PALEONTOLOGY. Similarly there are those that relate to areas of human behavior, e.g., PSYCHOLOGY and SOCIOLOGY.

Disciplines such as MEDICINE, VETERINARY MEDICINE, AGRONOMY and HORTICULTURE also have a strong basis in biology.

To a large extent the history of biology is the history of its constituent sciences. Since the impetus to investigate the living world generally arose in a desire to improve the techniques of medicine or of agriculture, most early biologists were in the first instance physicians or landowners. An exception is provided by ARISTOTLE, the earliest systematist of biological knowledge and himself an outstanding biologist—he founded the science of comparative anatomy—but most other classical authors, as GALEN, CELSUS and the members of the Hippocratic school, were primarily physicians. In the medieval period much biological knowledge became entangled in legend and allegory. The classical texts continued to be the principal sources of knowledge although new compilations, such as AVICENNA's *Canon* of medicine, were produced by Muslim philosophers. In 16th-century Europe interest revived in descriptive natural history, the work of GESNER being notable; physicians such as PARACELSUS began to develop a chemical pharmacology (see IATROCHEMISTRY) and experimental anatomy revived in the work of VESALIUS, FABRICIUS and FALLOPIUS. The discoveries of SERVETUS, HARVEY and MALPIGHI followed. Quantitative plant physiology began with the work of van HELMONT and was taken to spectacular ends in the work of Stephen HALES. In the 17th century, microscopic investigations began with the work of HOOKE and van LEEUWENHOEK; GREW advanced the study of plant organs and RAY laid the foundation for LINNAEUS' classic 18th-century formulation of the classification of plants. This same era saw BUFFON devise a systematic classification of animals and von HALLER lay the groundwork for the modern study of physiology.

The 17th century had seen controversies over the role of mechanism in biological explanation—LA METTRIE had even developed the theories of DESCARTES to embrace the mind of man; the 19th century saw similar disputes, now couched in the form of the mechanist-vitalist controversy concerning the possible chemical nature of life (see BICHAT; MAGENDIE; Claude BERNARD). Development biology, foreshadowed by LAMARCK, was thoroughly established following the work of DARWIN; in anatomy, SCHWANN and others developed the cell concept; in histology Bichat's pioneering work was continued; in physiology, organic and even physical chemists began to play a greater role, and medical theory was

The History of Biology
medicine
HIPPOCRATES
general biology
ARISTOTLE
400 BC
300 BC
medicine
CELSUS
100 AD
GALEN
200 AD
medicine
AVICENNA
1000
medieval herbals and bestiaries
1500
botany
GESNER
VAN HELMONT
iatrochemistry
PARACELSUS
physiology
SERVETUS
anatomy
VERSALIUS
FALLOPIUS
FABRICIUS
1600
RAY
GREW
microscopy
LEEUWENHOEK
HOOKE
HARVEY
MALPIGHI
pathology
REDI
1700
zoology
BUFFON
LAMARCK
HALES
LINNAEUS
VON HALLER
LAVOISIER
BICHAT
evolution
C. DARWIN
WEISMANN
genetics
MENDEL
DE VRIES
cell theory
BROWN
SCHWANN
MAGENDIE
BERNARD
germ theory of disease
PASTEUR
KOCH
ethology
LORENZ
MORGAN
molecular biology
CRICK
WATSON
BANTING

revolutionized by the advent of bacteriology (see PASTEUR; KOCH). The impact of MENDEL's discoveries in genetics was not felt until the early 1900s. Possibly the high point of 20th-century biology came with the proposal of the double-helix model for DNA (see NUCLEIC ACIDS), the chemical carrier of genetic information, by CRICK and WATSON in 1953.

BIOLUMINESCENCE, the production of nonthermal light by living organisms such as fireflies, many marine animals, bacteria and fungi. The effect is an example of CHEMILUMINESCENCE. In some cases its utility to the organism is not apparent, though in others its use is clear. Thus, in the firefly, the ABDOMEN of the female glows, enabling the male to find her. Similarly, LUMINESCENCE enables many deep-sea fish to locate each other or to attract their prey. The glow in a ship's wake at night is due to luminescent microorganisms.

BIOME, ecological region characterized by the predominant vegetation type, such as savanna. The biome is the largest biogeographical unit. (See ECOLOGY.)

BIOMEDICAL ENGINEERING, development and application of mechanical electrical, electronic and nuclear devices in medicine. The many recent advances in biomedical engineering have occurred in four main areas: ARTIFICIAL ORGANS; new surgical techniques involving the use of LASERS, cryosurgery and ULTRASONICS; diagnosis and monitoring using thermography and computers; and PROSTHETICS.

BIONICS, the science of designing artificial systems which have the desirable characteristics of living organisms. These may be simply imitations of nature, such as military vehicles with jointed legs, or, more profitably, systems which embody a principle learned from nature. Examples of the latter include RADAR, inspired by the echolocation system of bats, or the development of associative memories in COMPUTERS as in the human brain.

BIOPHYSICS, a branch of BIOLOGY in which the methods and principles of PHYSICS are applied to the study of living things. It has grown up in the 20th century alongside the development of ELECTRONICS. Its tools include the ELECTROENCEPHALOGRAPH and the ELECTRON MICROSCOPE, its techniques those of SPECTROSCOPY and X-RAY DIFFRACTION and its problems the study of nerve transmission, BIOLUMINESCENCE and materials transfer in RESPIRATION and secretion.

BIOPSY, removal and microscopic examination of tissue from a living patient for purposes of diagnosis. The tissue is removed by needle, suction, swabbing, scraping or excision.

BIOSPHERE, the region inhabited by living things. It forms a thin layer around the earth, including the surface of the LITHOSPHERE, the HYDROSPHERE and the lower ATMOSPHERE. The importance of the concept was first pointed out by LAMARCK.

BIOSYNTHESIS, or **anabolism,** the biochemical reactions by which living cells build up simple molecules into complex ones. These reactions require energy, which is obtained from light (see PHOTOSYNTHESIS) or from ATP which is produced in degradation reactions. (See also METABOLISM; PROTEIN SYNTHESIS.)

BIOT, Jean Baptiste (1774–1862), French physicist who first demonstrated the extraterrestrial origin of and hence the actual existence of METEORITES (1803); who accompanied GAY-LUSSAC on his pioneering BALLOON ascent to collect data concerning the upper ATMOSPHERE (1804); and who, having shown that some organic substances show OPTICAL ACTIVITY (1815), first developed the methods of POLARIMETRY—despite his rejection of the wave theory of LIGHT.

A Berwick's swan cygnet clearly displaying the numbered band on its leg.

BIOTIN. See VITAMINS.

BIOTITE, a range of iron-rich varieties of MICA, grading into PHLOGOPITE. It is a constituent of most igneous and many metamorphic rocks.

BIRD BANDING, placing numbered metal or plastic bands on the legs, wings or necks of birds for identification. Birds are banded as nestlings or when trapped in nets. All relevant data is recorded at a national agency whose address is on the band. Birds found can then be reported, an important aid to studies of migration and distribution.

BIRD MIGRATION. See MIGRATION.

BIRDS. See AVES.

BIRD SANCTUARIES, areas set aside for the protection of birds, either as reserves or as "hostels" for the injured. Reserves can help to preserve rare species. The first US official state sanctuary was established at Lake Merritt, Cal., in 1870.

BIRDSEYE, Clarence (1886–1956), US inventor and industrialist who, having observed during fur-trading expeditions to Labrador (1912–16) that many foods keep indefinitely if frozen, developed a process for the rapid commercial freezing of foodstuffs. In 1924 he organized the company later known as General Foods to market frozen produce.

BIRDSONG, the pattern of notes, often musical and complex, with which birds attract a mate and proclaim their territory. Ornithologists call all such sounds songs, though those that are harsh and unmusical are often referred to simply as the "voice."

BIRD WATCHING, observation of birds in their natural surroundings. Observers may study such things as courtship and nesting, migration patterns, and the occurrence of rare or threatened species. Bird-watching societies in the US include the Nuttall Ornithological Club and the American Ornithologists' Union.

BIREFRINGENCE. See DOUBLE REFRACTION.

BIRKHOFF, George David (1884–1944), US mathematician who made important contributions to the analysis of dynamical systems.

BIRTH, emergence from the mother's WOMB, or, in the case of most lower animals, from the EGG, marking the beginning of an independent life. The birth process is triggered by HORMONE changes in the mother's bloodstream. Birth may be induced, if required, by oxytocin. Mild labor pains (contractions of the womb) are the first sign that a woman is about to give birth. Initially occurring about every 20 minutes, in a few hours they become stronger and occur every few minutes. This is the first stage of labor, usually lasting about 14 hours. The contractions push the baby downward, usually head first, which breaks the membranes surrounding the baby, and the AMNIOTIC FLUID escapes.

In the second stage of labor, stronger contractions push the baby through the cervix and vagina. This is the most painful part and lasts less than 2 hours. Anesthetics (see ANESTHESIA) or ANALGESICS are usually given, and delivery aided by hand or obstetric forceps. A CESARIAN SECTION may be performed if great difficulty occurs. Some women choose "natural childbirth," in which no anesthetic is used, but pain is minimized by prior relaxation exercises.

As soon as the baby is born, its nose and mouth are cleared of fluid and breathing starts, whereupon the UMBILICAL CORD is cut and tied. In the third stage of labor the PLACENTA is expelled from the womb and bleeding is stopped by further contractions. Birth normally occurs 38 weeks after conception. Premature births are those occurring after less than 35 weeks. Most premature babies develop normally with medical care, but if born before 28 weeks the chances of survival are poor. (See also EMBRYO; GESTATION; OBSTETRICS; PREGNANCY.)

BIRTH CONTROL, prevention of unwanted births, by means of CONTRACEPTION, ABORTION, STERILIZATION and formerly infanticide. It is medically advisable if the child would be likely to be defective. At family level, birth control can help to prevent poverty, while globally it could help prevent mass starvation.

BIRTHMARKS, skin blemishes, usually congenital. There are two main types: pigmented nevuses, or moles, which are usually brown or black and may be raised or flat; and vascular nevuses, local growths of small blood vessels, such as the "strawberry mark" and the "port-wine stain." Although harmless, they are sometimes removed for cosmetic reasons or if they show malignant tendencies. (See also TUMOR.)

BIRTH RATE, ratio of the annual number of births in a population to the total midyear population, usually expressed in births per 1000 persons. Being uncorrected for sex and age distribution, it is a crude measure of fertility. Birth rates range from around 50 per 1000 in some developing countries to under 20 per 1000 in advanced nations. (See also DEMOGRAPHY.)

BISMUTH (Bi), metal in Group VA of the PERIODIC TABLE, brittle and silvery-gray with a red tinge. It occurs naturally as the metal, and as the sulfide and oxide, from which it is obtained by roasting and reduction with carbon. In the US it is obtained as a byproduct of the refining of copper and lead ores. Bismuth is rather unreactive; it forms trivalent and some pentavalent compounds. Physically and chemically it is similar to LEAD and ANTIMONY. Bismuth is used in low-melting-point alloys in fire-detection safety devices. Since bismuth expands on solidification, it is used in alloys for casting dies and type metal. Bismuth (III) oxide is used in GLASS and CERAMICS; various bismuth salts are used in medicine. AW 209.0, mp 271°C, bp 1560°C, sg 9.747 (20°C).

BISULFATES. See SULFATES.

BIT, abbreviation for *bi*nary digi*t*, in computer technology and information theory, the smallest conceivable unit of information, representing a choice between only two possible states: the presence or absence of a signal pulse; + or −; 0 or 1; a switch being off or on. The capacity of an information storage or handling device is measured in bits.

BITUMEN, any naturally-occurring HYDROCARBON, including PETROLEUM, but referring especially to the solid hydrocarbons such as WAX, PITCH, ASPHALT and GILSONITE. These are fusible and soluble in organic solvents, unlike the **pyrobitumens**.

BITUMINOUS COAL. See COAL.

BITUMINOUS SANDS, sands which contain natural BITUMEN. The largest deposit is in the Athabasca region of N Alberta, and there are substantial deposits in Cal. and Ut. The heavy tar is extracted and synthetic crude oil produced.

BIVALVIA, a class of the phylum MOLLUSCA.

BLACK, Davidson (1884–1934), Canadian anthropologist who inferred the existence of "Peking Man," an early species of hominid, from the discovery of a single tooth (1927), but later reinterpreted his discovery as representing a variety of *Pithecanthropus erectus* (see PREHISTORIC MAN).

BLACK, Joseph (1728–1799), Scottish physician and chemist who investigated the properties of CARBON dioxide, discovered the phenomena of LATENT and SPECIFIC HEATS, distinguished HEAT from TEMPERATURE and pioneered the techniques used in the quantitative study of CHEMISTRY.

BLACK BODY, in theoretical physics, an object which absorbs all the ELECTROMAGNETIC RADIATION which falls on it. In practice, no object acts as a perfect black body, though a closed box admitting radiation only through a small hole is a good approximation. Black bodies are also ideal thermal radiators.

BLACKBODY RADIATION, the ELECTROMAGNETIC RADIATION emitted from a BLACK BODY in virtue of its thermal energy. The derivation of its properties by PLANCK in 1901 was the occasion of the proposal and first success of QUANTUM THEORY. The energy emitted from a black body is proportional to the fourth power of its (absolute) temperature (Stefan-BOLTZMANN Law). The intensity SPECTRUM of radiation from a black body takes the form of a skew hump which tails off in its longer-wavelength branch.

Blackbody radiation laws

Stefan–Boltzmann law

$$E = \sigma T^4$$

where E is the total energy radiated, T is the absolute temperature, and σ is the Stefan-Boltzmann constant.

Wien's displacement law

$$\lambda_m T = A$$

where λ_m is the wavelength of most intense emission, T is the absolute temperature, and A is a constant.

Wien's second law

$$dE_{\lambda(max)} = BT^5 \, d\lambda$$

where $E_{\lambda(max)}$ is the emissive power in the waveband of maximum emission, λ is the wavelength and B is a constant.

Wien's third law

$$E_\lambda \, . \, d\lambda = \frac{c_1}{\lambda^5} \cdot e^{c_2/\lambda T} \, . \, d\lambda$$

where E_λ is the emissive power in the waveband dλ and c_1 and c_2 are constants.

All the above can be derived from the

Planck radiation formula

$$E_\lambda \, . \, d\lambda = \frac{hc^3}{\lambda^5} \cdot \frac{1}{(e^{hc/k\lambda T} - 1)} \cdot d\lambda$$

where additionally h is the Planck constant, k is the Boltzmann constant and c is the electromagnetic constant.

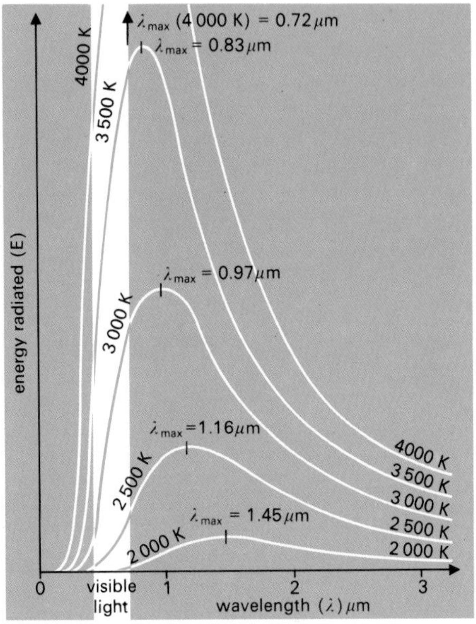

A graph illustrating the intensity spectra for thermal radiation from blackbody sources at a range of temperatures between 2000 and 4000K. The area under each curve represents the total energy radiated at each temperature; as specified in the Stefan-Boltzmann law, this increases with the fourth power of the absolute temperature. The hump in successive curves (the wavelength of greater emission) shifts steadily to the left (shorter wavelengths) in accordance with Wien's displacement law. In practical terms this means that the color of hot bodies progresses from straw through red to white and blue as the source temperature increases and greater proportions of their thermal radiation are emitted in the blue region of the visible spectrum. These colors are indicated on the illustration for Color.

Its precise shape is described in WIEN's laws, the first of which (Wien's displacement law) states that the greatest emission occurs at a wavelength which is inversely proportional to the absolute temperature.

BLACK DEATH. See PLAGUE.

BLACK EARTH. See CHERNOZEM.

BLACKETT, Patrick Maynard Stuart, Baron (1897–1974), British physicist who, having developed the Wilson CLOUD CHAMBER into an instrument for observing COSMIC RAYS, won the 1948 Nobel physics prize for the results he obtained using it.

BLACKHEAD, or **comedo**, a plug of dried sebum in the duct of a SEBACEOUS GLAND, often obstructing flow of sebum, so producing a small pimple. (See ACNE.)

BLACK HOLE, the final stage of evolution for very massive stars, following total gravitational collapse. At the center of the black hole are the infinitely densely packed remains of the star, perhaps only a few km across, if not crushed entirely out of existence. The gravitational field of a black hole is so intense that nothing, not even ELECTROMAGNETIC RADIATION (including light), can escape. For this reason black holes can only be detected through their gravitational effects on other bodies and through the emission of X- and gamma-rays by matter falling into them. It has been suggested that the end of the universe will be its becoming a single black hole.

BLACK LUNG DISEASE, a lay term for PNEUMOCONIOSIS, which affects coal miners.

BLADDER, a hollow muscular sac; especially the urinary bladder (see also AIR BLADDER; GALL BLADDER), found in most vertebrates except birds. In humans it lies in the front of the PELVIS. URINE trickles continually into the bladder from the KIDNEYS through two tubes called ureters, and the bladder stretches until it contains about 500ml, causing desire to urinate. The bladder empties through the urethra, a tube which issues from its base, being normally closed by the external sphincter muscle. The female urethra is about 30mm long: the male urethra, which runs through the PROSTATE GLAND and the PENIS , is about 200mm long. The bladder is liable to CYSTITIS and to the formation of CALCULI.

BLANCHARD, Jean Pierre François, (1783–1809), French balloonist and inventor who made the first aeronautical crossing of the English Channel

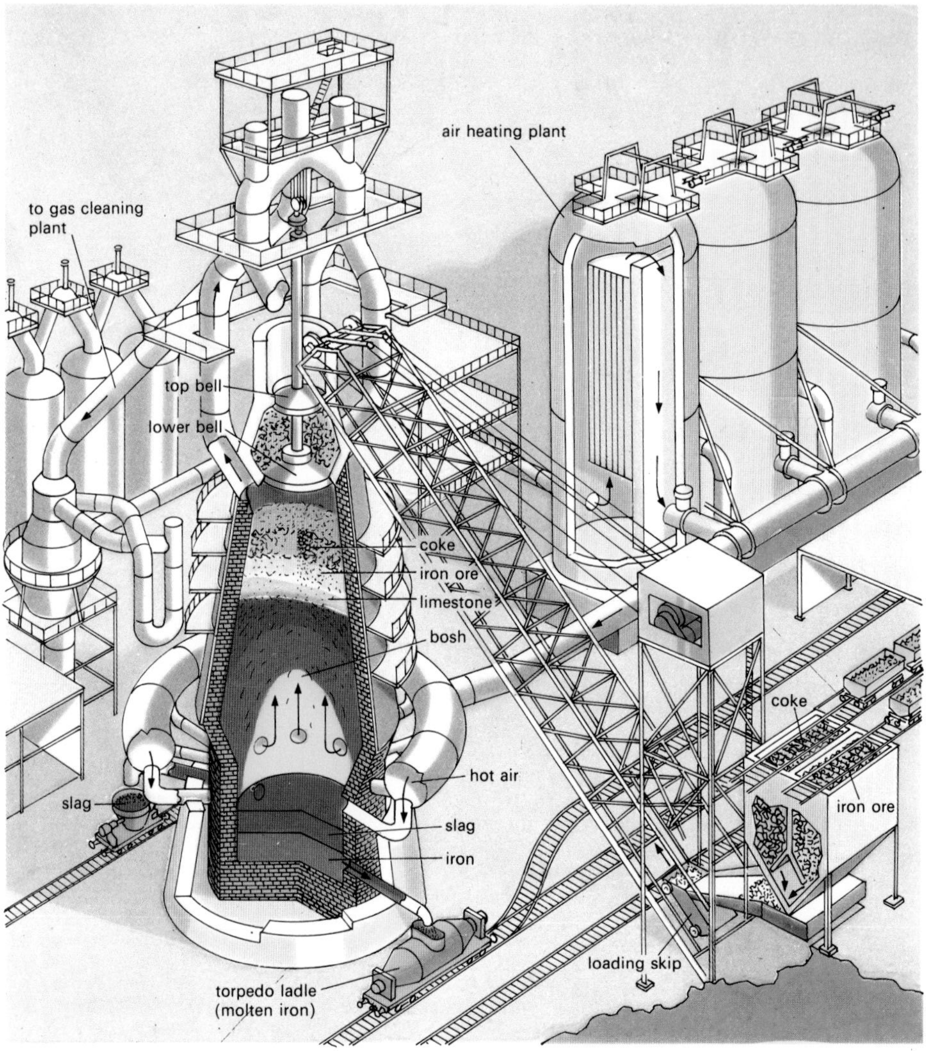

The principal components of a medium-scale blast furnace complex.

(1785) and the first BALLOON ascent in America (1793). He also invented the PARACHUTE (1785).

BLAST FURNACE, furnace in which a blast of hot, high-pressure air is used to force combustion; used mainly to reduce IRON ore to PIG IRON, and also for lead, tin and copper. It consists of a vertical, cylindrical stack surmounting the bosh (the combustion zone) and the hearth from which the molten iron and slag are tapped off. Modern blast furnaces are about 30m high and 10m in diameter, and can produce more than 1800 tonnes per day. Layers of iron oxide ore, COKE and LIMESTONE are loaded alternately into the top of the stack. The burning coke heats the mass and produces CARBON monoxide, which reduces the ore to iron; the limestone decomposes and combines with ash and impurities to form a SLAG, which floats on the molten iron. The hot gases from the top of the stack are burned to preheat the air blast.

BLASTOMYCOSIS, rare FUNGAL DISEASE caused by infection of lungs, skin and viscera by *Blastomyces* fungi. Characteristic SKIN lesions, ABSCESSES and FEVER occur. Treatment is with ANTIBIOTICS.

BLASTULA, a hollow sphere composed of a single layer of cells, formed by cleavage of a fertilized OVUM; the first stage in the development of the EMBRYO. In mammals a similar cluster, the blastocyst, is formed, with an inner cell mass and a spherical envelope that develops into the PLACENTA. (See also EMBRYOLOGY.)

BLEACHING, process of whitening materials by sunlight, ULTRAVIOLET RADIATION or chemicals that reduce or oxidize DYES into a colorless form. Hydrogen PEROXIDE is used to bleach wool, silk and cotton; HYPOCHLORITES, including BLEACHING POWDER, are used for cotton; and sodium chlorite for synthetic fibers. Sulfur dioxide bleaches are impermanent.

Careful control of pH is essential.

BLEACHING POWDER, white powder consisting of calcium HYPOCHLORITE and basic calcium chloride, made by reacting CALCIUM hydroxide with CHLORINE. It is used for BLEACHING and as a disinfectant, but in time loses its strength.

BLEEDING. See HEMORRHAGE.

BLENDE. See SPHALERITE.

BLEULER, Eugen (1857–1939), Swiss psychiatrist who introduced the term SCHIZOPHRENIA (1908) as a generic term for a group of mental illnesses which he had learned to differentiate in a classic research project. He was an early supporter of FREUD but later criticized his dogmatism.

BLIGHT. See PLANT DISEASES.

BLINDNESS, severe loss or absence of VISION, caused by injury to the EYES, congenital defects, or diseases including CATARACT, DIABETES, GLAUCOMA, LEPROSY, TRACHOMA and VASCULAR disease. MALNUTRITION (especially VITAMIN A deficiency) may cause blindness in children. Infant blindness can result if the mother had GERMAN MEASLES early in PREGNANCY; it was also formerly caused by gonorrheal infection of eyes at birth, but routine use of silver nitrate reduced this risk. Transient blindness may occur if one is exposed to a vertical ACCELERATION of more than 5*g*. Cortical blindness is a disease of the higher perceptive centers in the BRAIN concerned with vision: the patient may even deny blindness despite severe disability. Blindness due to cataract may be relieved by removal of the eye lens and the use of GLASSES. Prevention or early recognition and treatment of predisposing conditions is essential to save sight, as established blindness is rarely recoverable.

Many special books (using braille), instruments, utensils and games have been designed for the blind. With the help of guide dogs or long canes, many blind persons can move about freely. They can detect obstacles around them by the change of pitch of high-frequency sound from the feet or a cane, a skill acquired by training, and by using other senses.

BLIND SPOT, the area of the retina of each EYE where the optic nerve and blood vessels enter, about 2mm in diameter. It has no light-sensitive receptors. In binocular vision the two spots do not receive corresponding images, and so are not noticeable.

BLINK COMPARATOR, or Blink Microscope, astronomical instrument used to detect differences between apparently similar star pictures, viewed as in a STEREOSCOPE, by rapidly obscuring each alternately. Anything that has moved flickers. The planet PLUTO was discovered this way. It can also detect variable stars and those of large PROPER MOTION.

BLISTER, a swelling filled with serum or BLOOD formed between two layers of skin following BURNS, friction or contact with certain corrosive chemicals (vesicants). Blisters also occur in certain skin diseases.

BLIZZARD, snowstorm in which wind velocity reaches 32mph (14m/s) or more, the temperature is below freezing and visibility poor. Blizzards are common in polar regions.

BLOCH, Felix (1905–), Swiss-born US physicist who shared the 1952 Nobel Prize for Physics with E. M. PURCELL for developing a method for determining the magnetic fields of NEUTRONS in atomic nuclei (see ATOM). This was developed into the nuclear magnetic resonance (nmr) method of determining chemical structures (see SPECTROSCOPY).

BLOCH, Konrad Emil (1912–), German-born US biochemist who shared the 1964 Nobel Prize for Physiology or Medicine with F. LYNEN for developing an isotopic labeling technique (see ISOTOPES), which he used to elucidate the path by which CHOLESTEROL is synthesized in the body.

BLOCH, Markus Elieser (1723–1799), German ichthyologist whose 12-volume *General Natural History of Fishes* (1782–95) remained for many years a standard work in the field.

BLONDEL, a little-used photometric unit for describing the helios of a source, numerically equivalent to the APOSTILB. It was named for the French physicist André Eugène Blondel (1863–1938).

BLOOD, the body fluid pumped by the heart through the vessels of those animals (all vertebrates and many invertebrates) in which diffusion alone is not adequate for transport of materials, and which therefore require BLOOD CIRCULATION systems. Blood plays a part in every major bodily activity. As the body's main transport medium it carries a variety of materials: oxygen and nutrients (such as glucose) to the tissues for growth and repair (see METABOLISM); carbon dioxide and wastes from the tissues for excretion; HORMONES to various tissues and organs for chemical signaling; digested food from the gut to the LIVER; immune bodies for prevention of infection and clotting factors to help stop bleeding to all parts of the body. Blood also plays a major role in HOMEOSTASIS, as it contains BUFFERS which keep the acidity (pH) of the body fluids constant and, by carrying heat from one part of the body to another, tends to equalize body temperature.

The adult human has about 5 litres of blood, half PLASMA and half blood cells (erythrocytes or red cells, leukocytes or white cells, and thrombocytes or platelets). The formation of blood cells (hemopoiesis) occurs in bone MARROW, lymphoid tissue and the RETICULOENDOTHELIAL SYSTEM. Red cells (about 5 million per mm^3) are produced at a rate of over 100 million per minute and live only about 120 days. They have no nucleus, but contain a large amount of the red pigment HEMOGLOBIN, responsible for oxygen transfer from lungs to tissues and carbon dioxide transfer from tissues to lungs. (Some lower animals employ copper-based HEMOCYANINS instead of hemoglobin. Others, e.g., cockroaches, have no respiratory pigments.) White cells (about 6 000 per mm^3) are concerned with defense against infection and poisons. There are three types of white cells: granulocytes (about 70%), which digest bacteria and greatly increase in number during acute infection; lymphocytes (20–25%), which participate in immune reactions (see IMMUNITY; ANTIBODIES); and monocytes (3–8%), which digest nonbacterial particles, usually during chronic infection. (See also HODGKIN'S DISEASE; LEUKEMIA.) Platelets, which live for about 8 days and which are much smaller than white cells and about 40 times as numerous, assist in the initial stages of blood CLOTTING together with at least 12 plasma clotting factors and fibrinogen. This occurs when blood vessels are damaged, causing THROMBOSIS, and when HEMORRHAGE occurs (see also HEMOPHILIA).

Blood from different individuals may differ in the type of antigen on the surface of its red cells and the type of ANTIBODY in its plasma. Consequently, in a

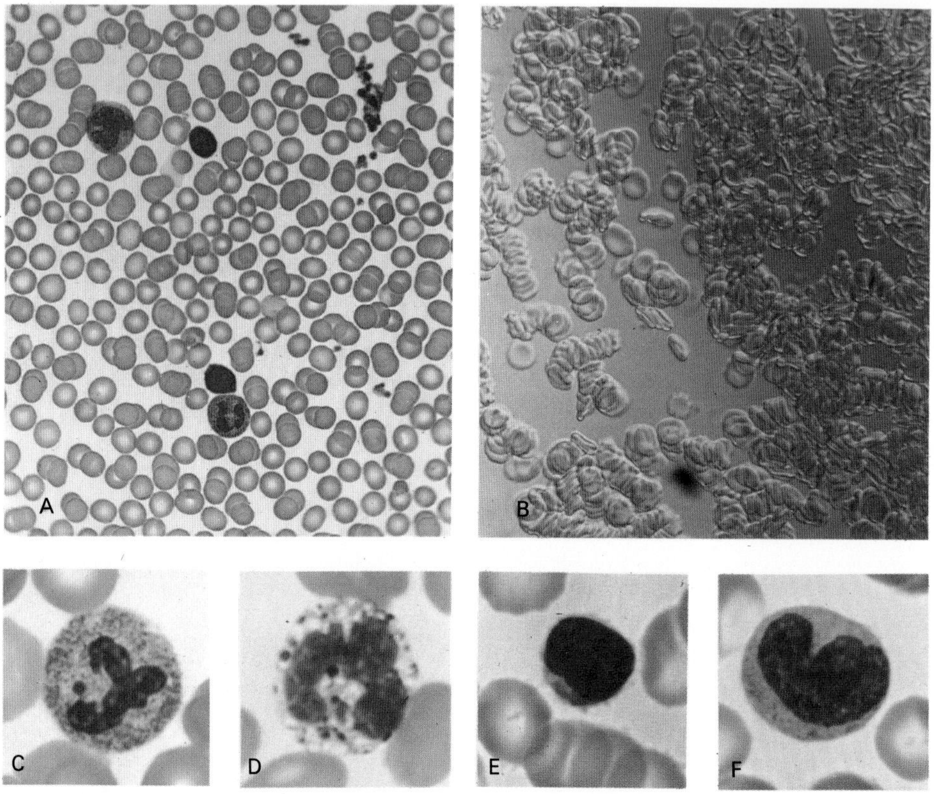

(A) Blood film viewed at ×550, red cells (erythrocytes) predominating. (B) Fresh red cells in a thick layer viewed at ×950 using an interference microscope. Of the white cells (leukocytes), the majority are granular (granulocytes): most of these are neutrophils (C—this one showing a drumstick); a few are basophils (D). Other white cells include small lymphocytes (E) and monocytes (F). Illustrations (C) to (F) are at ×2100 magnification.

blood TRANSFUSION, if the blood groups of the donor and recipient are incompatible with respect to antigens and antibodies present, a dangerous reaction occurs, involving aggregation or clumping of the red cells of the donor in the recipient's circulation. Many blood group systems have been discovered, the first and most important being the ABO system by Karl LANDSTEINER in 1900. In this system, blood is classified by whether the red cells have antigens A (blood group A), B (group B), A and B (group AB), or neither A nor B antigens (group O). Another important antigen is the Rhesus antigen (or Rh factor). People who have the Rh factor (84%) are designated Rh+, those who do not, Rh−. Rhesus antibodies do not occur naturally but may develop in unusual circumstances. In a few cases, where Rh− women are pregnant with Rh+ babies, blood leakage from baby to mother causes production of antibodies by the mother which may progressively destroy the blood of any subsequent baby. (See also EDEMA; POLYCYTHEMIA; SEPTICEMIA; SERUM.

BLOOD CIRCULATION, the movement of BLOOD from the HEART through the ARTERIES, CAPILLARIES and VEINS and back to the heart. The circulatory system has two distinct parts in animals with lungs: the pulmonary circulation, in which blood is pumped from the right ventricle to the left atrium via the blood vessels of the lungs (where the blood is oxygenated and carbon dioxide is eliminated); and the systemic circulation, in which the oxygenated blood is pumped from the left ventricle to the right atrium via the blood vessels of the body tissues (where—in the capillaries—the blood is deoxygenated and carbon dioxide is taken up). As it leaves the heart, the blood is under considerable pressure—about 120mmHg maximum (systolic pressure) and 80mmHg minimum (diastolic pressure). Sustained high blood pressure, or **hypertension**, occurs in kidney and hormone diseases and in old age, but generally its cause is unknown. It may lead to ARTERIOSCLEROSIS and heart, brain and kidney damage. Low blood pressure occurs in SHOCK, TRAUMA and ADDISON'S DISEASE.

BLOOD POISONING. See SEPTICEMIA.

BLOODSTONE, a dark-green variety of CHALCEDONY containing nodules of red JASPER, used in medieval sculptures of martyrdom or flagellation.

BLOWPIPE, narrow tapered tube through which air is blown into a flame to increase the temperature and to direct part of the flame—the oxidizing or reducing zone—onto a substance undergoing chemical ANALYSIS. (See FLAME TEST.)

BLOWTORCH, portable burner used for melting

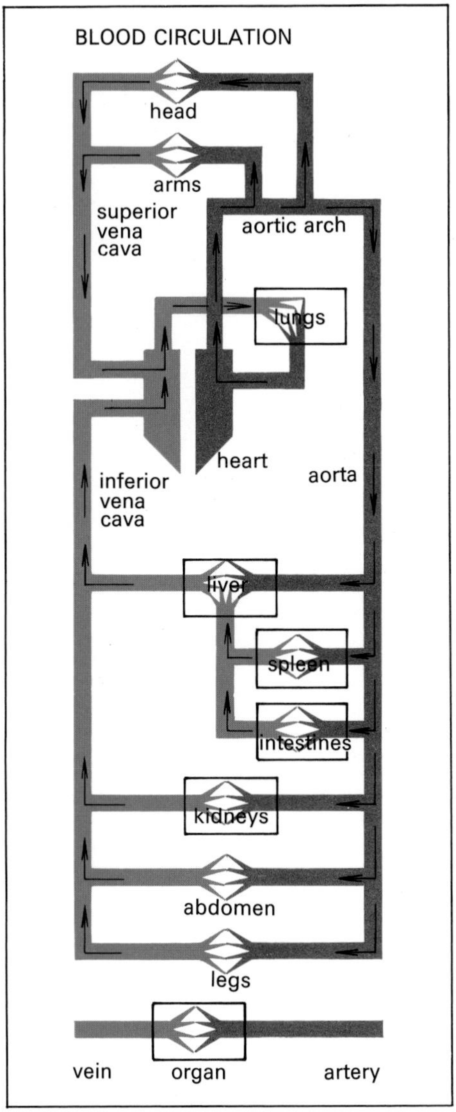

solder and removing paint. The fuel—liquid or pressurized gas—is mixed with compressed air in order to obtain a hot flame.

BLUBBER, the layer of fat below the skin of the whale and some other marine mammals; it is several inches thick and provides buoyancy, insulation and energy reserves. It yields oil which was once used for lighting, and more recently for making soap and margarine. A blue whale can yield more than 20 tonnes of oil.

BLUE BABY, infant born with a HEART defect (a hole between the right and left sides, or malformation of the arteries) that permits much of the BLOOD to bypass the LUNGS. The resulting lack of oxygen causes CYANOSIS. These conditions used to be fatal but can now often be corrected by surgery.

BLUE-GREEN ALGAE, widely distributed photosynthetic prokaryotic (see CELL), microrganisms forming the class Schizophyceae of the division Schizophyta of the PLANT KINGDOM. An important property of blue-green algae is NITROGEN FIXATION.

BLUEPRINT, a process used for copying architectural and engineering plans. Paper sensitized with iron (III) ammonium citrate and potassium ferricyanide is exposed to ULTRAVIOLET LIGHT through a transparent original. Where the light strikes the paper, a blue coloration results, which is often intensified with potassium dichromate solution. The result is a white copy of the original on a blue background. The OZALID PROCESS is now more widely used.

BLUE VITRIOL, or copper (II) sulfate. See COPPER.

BLUE WHALE, *Balaenoptera musculus*, a member of

A handy domestic blowtorch utilizing disposable bottled-gas cartridges. For use, the gas cartridge is first fitted into the cartridge holder. The valve unit is then screwed into the cartridge holder, the pin it contains piercing the gas cartridge. When the needle valve is opened by turning the control knob, gas is allowed to pass through the filter to the burner head. As the gas flows through the burner head, air is drawn in through the air holes provided, allowing a hot flame to be ignited outside the mouth of the burner head. Often a variety of burner-head attachments are supplied, producing different flame configurations suitable for specific jobs.

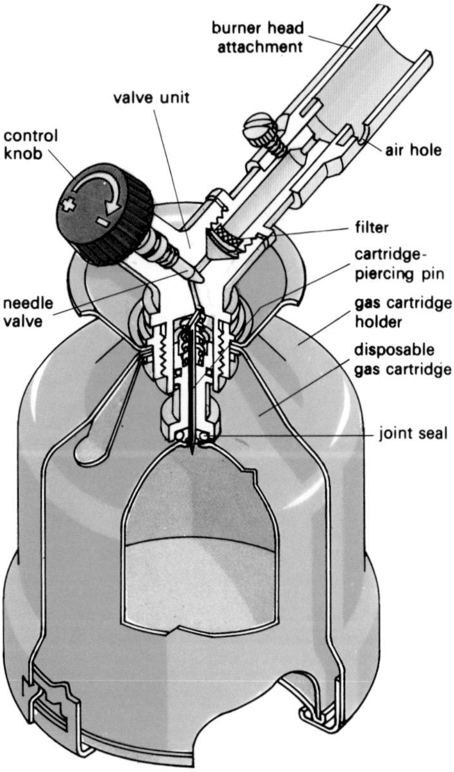

the CETACEA, the largest living animal, attaining a length of 30m (100ft) and weighing up to 130 tonnes. A BALEEN whale, it lives mainly in the Antarctic Ocean.

BLUING, process used in laundering to brighten white fabrics. The fabric is immersed briefly in a dilute blue dye solution at about 45°C, to counteract the yellowish tint which washing generally produces. A blue dye is often added to detergents.

BLUMBERG, Samuel Baruch (1925–), US physiologist who shared the 1976 Nobel Prize for Physiology or Medicine with D. C. GAJDUSEK for research into the nature of infection. He discovered the "Australia antigen" which is associated with serum HEPATITIS.

BLUMENBACH, Johann Friedrich (1752–1840), German physiologist generally regarded as the father of physical ANTHROPOLOGY. As a result of careful measurement of a large collection of skulls, he divided mankind into five racial groups: Caucasian, Mongolian, Malayan, Ethiopian and American.

BOAS, Franz (1858–1942), German-born US anthropologist who played a leading part in the establishment of the cultural-relativist school of ANTHROPOLOGY in the English-speaking world.

BODE, Johann Elert (1747–1826), German astronomer remembered for promoting **Bode's Law**, a numerical relationship which was found to hold between the radii of the ORBITS of the then-known

A compact modern industrial steam generator (boiler). Fuel (various fuel gases or fuel oil) is mixed with air in the burners and burned in the furnace area. The combustion gases are passed through the superheater to the convection bank, through which they are directed by baffles and screens, before being vented to the stack. The main convection bank consists of water tubes running between the steam drum and the water drum. Other similar but finned tubes form the baffle screens and the internal walls of the boiler. Water feed and steam extraction occur at the upper drum which also contains separators and driers to dry the steam before it is passed to the superheater or application. This drawing is based on a design by Foster Wheeler Power Products Ltd.

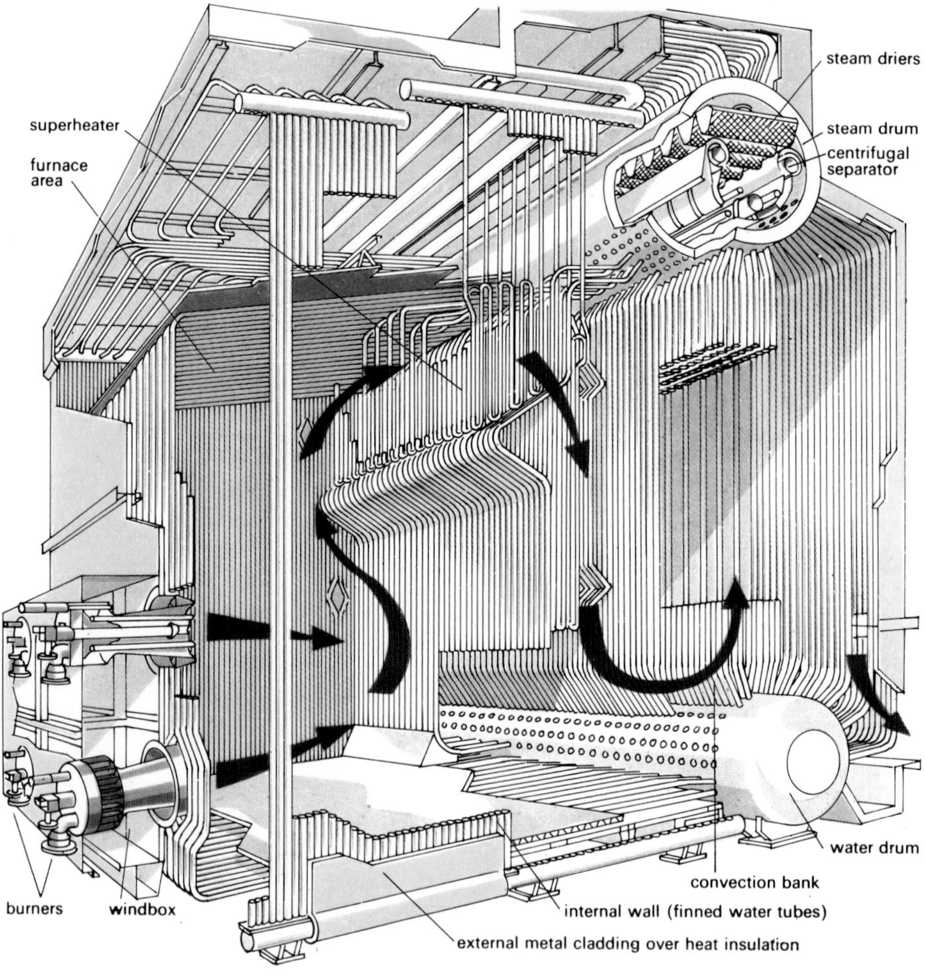

PLANETS. Bode started from the sequence 0, 3, 6, 12, 24, 48, 96, 192, . . . and added 4 to each member: 4, 7, 10, 16, 28, 52, 100, 196, . . . ; then, if the radius of the orbit of the earth was taken to be 10, he found that MERCURY fell into place at 4, VENUS at ~ 7, MARS at 16, JUPITER at 52 and SATURN at ~ 100. The discovery of URANUS at ~ 196 initiated a hunt for a missing planet at 28, which was supplied by the asteroid CERES but when NEPTUNE was discovered in 1846, it failed to satisfy the "law," which immediately declined in importance.

BODY. See HUMAN BODY; SOMATOTYPES.

BOERHAAVE, Hermann (1668–1738), Dutch chemist and physician, renowned in his day as a leading man of science, although he made no discoveries of lasting importance. His teaching did much to establish Leiden as a medical center of international repute.

BOG, commonly, any marsh or SWAMP; specifically, a low-lying area, usually formed by the action of GLACIERS, which is poorly drained, perhaps containing shallow water, and in which organic matter is accumulating.

BOHR, Niels Henrik David (1885–1962), Danish physicist who proposed the Bohr model of the ATOM while working with RUTHERFORD in Manchester, England, in 1913. Bohr suggested that a HYDROGEN atom consisted of a single electron performing a circular orbit around a central PROTON (the nucleus), the energy of the ELECTRON being quantized (i.e., the electron could only carry certain well-defined quantities of energy—see QUANTUM THEORY). At one stroke this accounted both for the properties of the atom and for the nature of its characteristic radiation (a SPECTRUM comprising several series of discrete sharp lines). In 1927 Bohr proposed the COMPLEMENTARITY PRINCIPLE to account for the apparent paradoxes which arose on comparing the wave and particle approaches to describing SUBATOMIC PARTICLES. After escaping from Copenhagen in 1943 he went to the UK and then to the USA, where he helped develop the ATOMIC BOMB, but he was always deeply concerned about the graver implications for humanity of this development. In 1922 he received the Nobel Prize for Physics in recognition of his contributions to atomic theory. His son, **Aage Niels Bohr** (1922–), shared the 1975 Nobel Prize for Physics with B. MOTTELSON and J. RAINWATER for contributions made to the physics of the atomic nucleus.

BOIL, an ABSCESS in a hair FOLLICLE, usually caused by infection with STAPHYLOCOCCUS. A **sty** is a boil on the eyelid; a **carbuncle** is a group of contiguous boils. Small boils may heal spontaneously, but most cannot until PUS has escaped, by thinning and rupture of overlying SKIN. This is hastened by local application of heat. In severe cases lancing and ANTIBIOTICS may be required.

BOILER, device used to convert WATER into STEAM by the action of HEAT (see also BOILING POINT), usually to drive a STEAM ENGINE. A boiler requires a heat source (i.e., a FURNACE), a surface whereby the heat may be conveyed to the water, and enough space for steam to form. The two main types of boiler are the fire-tube, where the hot gases are passed through tubes surrounded by water; and the water-tube, where the water is passed through tubes surrounded by hot gases. Fuels include COAL, OIL and fuel GAS; NUCLEAR ENERGY is also used. HERO designed boilers, but used them only in toys. Steam power proper was barely considered until the 17th century, and little used before the 18th.

BOILING POINT, the temperature at which the VAPOR PRESSURE of a liquid becomes equal to the external pressure, so that boiling occurs; the temperature at which a liquid and its vapor are at equilibrium. Measurement of boiling point is important in chemical ANALYSIS and the determination of MOLECULAR WEIGHTS. (See also EVAPORATION; PHASE EQUILIBRIUM; PRESSURE COOKER.)

BOLOMETER, an instrument used to measure radiant ENERGY, usually in the infrared and microwave regions of the spectrum of ELECTROMAGNETIC RADIATION. It comprises a lens or stop system which focuses the test radiation on a thermoconductive device (usually a THERMISTOR), which is set in a WHEATSTONE BRIDGE circuit with another nonilluminated reference thermistor. Sensitive bolometers are used in conjunction with spectroscopes to measure the intensities of spectral lines (see SPECTROSCOPY).

BOLTS AND SCREWS, devices in which the principle of the screw thread, which may be traced back as far as ARCHIMEDES, is applied to the fastening

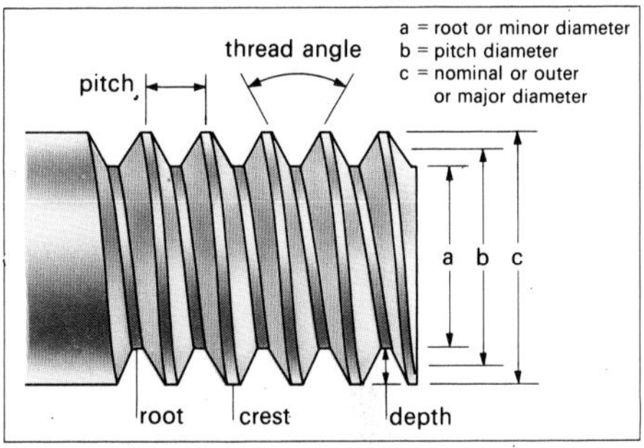

Bolts and screws: (top) screw thread nomenclature; (above) some of the many types of bolts and screws in common industrial and domestic use. Materials used range from brass and various steels to polymer plastics.

together of objects. A screw is essentially conical, with a sharp point and widening toward the head—which is usually shaped to take a screwdriver—with a helical ridge (see CONE; HELIX). If the point is pressed into the material (usually wood) and the screw longitudinally rotated by means of a screwdriver, the screw will be driven into the wood and will be held in place by FRICTION. A bolt is essentially cylindrical, again with a helical ridge (see CYLINDER), and has a broad head usually shaped to take a spanner or wrench. It is used in conjunction with a nut, a member containing a prethreaded hole into which the bolt fits. The objects to be fastened are held together by the pressure of the bolthead on one side, the nut on the other. The distance between consecutive turns of a screw thread is termed the **pitch** of the thread.

BOLTZMANN, Ludwig (1844–1906), Austrian physicist who made fundamental contributions to THERMODYNAMICS, classical statistical mechanics and KINETIC THEORY. The **Boltzmann constant** (k), the quotient of the universal gas constant R and the AVOGADRO number ($\mathcal{N}$), is used in statistical mechanics.

BOLYAI FARKAS (1775–1856), Hungarian mathematician who expended much effort in trying to prove the Euclidean parallel lines postulate. His son, **Bolyai János** (1802–1860), renowned as a duelist and violinist, was a codiscoverer of NON-EUCLIDEAN GEOMETRY, having worked out his theory before 1823, thus preceding Lobachevski.

BOMB, device designed to explode, with the aim of destroying property or killing and maiming human beings. A bomb may be dropped from an aircraft, incorporated in a warhead, or "planted" in position. Essential to all bombs is a fuze (see AMMUNITION), in effect a miniature bomb whose explosion precipitates the explosion of the bomb proper. In particular, fuzes of **timebombs** incorporate devices such that the fuze, and hence the bomb, may be set to explode after a determined elapse of time. Types of conventional bombs include **fragmentation bombs**, with cases designed to disintegrate into shrapnel; **fire bombs** and **incendiary bombs**, whose purpose is to destroy by fire (see NAPALM); **gas bombs**, whose explosion distributes poisonous gas (see CHEMICAL AND BIOLOGICAL WARFARE); **smoke bombs**, which create a smokescreen; and **photoflash bombs**, used in night photography. Underwater bombs, designed to explode at a specific depth, are usually termed **depth charges**. (See also ATOMIC BOMB; EXPLOSIVES; HYDROGEN BOMB; GRENADE.)

BOND, Chemical, the links which hold ATOMS together in compounds. In the 19th century it was found that many substances, known as **covalent compounds**, could be represented by structural FORMULAS in which lines represented bonds. By using double and triple bonds, most organic compounds could be formulated with constant VALENCES of the constituent atoms. STEREOISOMERISM showed that the bonds must be localized in fixed directions in space. **Electrovalent compounds** (see ELECTROCHEMISTRY) consist of oppositely charged IONS arranged in a lattice; here the bonds are nondirectional electrostatic interactions. The theory that atoms consist of electrons orbiting in shells around the nucleus (see PERIODIC TABLE) led to a simple explanation of both kinds of bonding: atoms

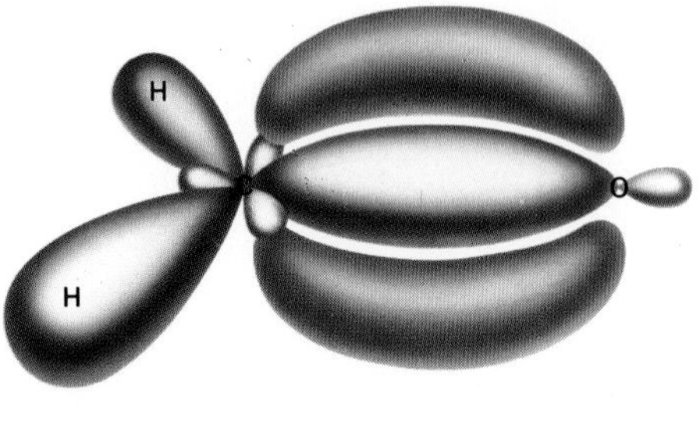

A representation of the bonding molecular orbitals in formaldehyde (CH_2O). The bonding orbitals of the carbon atom are hybridized in a planar sp^2 configuration (blue). Above and below the plane of the sp^2 σ orbitals lie the sausage-shaped high-probability regions of the C—O π bond (red).

combine to achieve highly-stable filled outer shells containing 2, 8 or 18 electrons, either by transfer of electrons from one atom to the other **(ionic bond)**, or by the sharing of one electron from each atom so that both electrons orbit around both nuclei **(covalent bond)**. In the **coordinate bond**, a variant of the covalent bond, both shared electrons are provided by one atom. QUANTUM THEORY has now shown that electrons occupy ORBITALS having certain shapes and energies, and that, when atoms combine, the outer atomic orbitals are mixed to form molecular orbitals. The energy difference constitutes the bond energy—the energy required to break the bond by separating the atoms. Molecular orbitals are classified as σ if symmetric when rotated through 180° about the line joining the nuclei, or π if antisymmetric. The energy and length of chemical bonds, and the angles between them, may be investigated by SPECTROSCOPY and X-RAY DIFFRACTION. (See also HYDROGEN BONDING.)

BOND, William Cranch (1789–1859), US astronomer, first director of the Harvard Observatory and a pioneer of ASTROPHOTOGRAPHY. In 1850 he made the first DAGUERREOTYPE of a celestial object and discovered the third, "crape," ring of SATURN.

BONDERIZING, chemical process to help prevent CORROSION of iron, steel and other metals. The surface is sprayed with or immersed in a hot PHOSPHATE solution, and a superficial insoluble phosphate layer is formed. The metal is then usually painted or lacquered.

BONDI, Sir Hermann (1919–   ), Austrian-born British cosmologist who with T. GOLD in 1948 formulated the steady-state theory (see COSMOLOGY).

BONE, the hard tissue that forms the SKELETON of vertebrates. Bones support the body, protect its organs, act as anchors for MUSCLES and as levers for the movement of limbs, and are the main reserve of calcium and phosphate in the body. Bone consists of living cells (osteocytes) embedded in a matrix of COLLAGEN fibers with calcium salts similar in composition to hydroxyapatite (see APATITE) deposited between them. Some carbonates are also present. All bones have a shell of compact bone in concentric layers (lamellae) around the blood vessels, which run in small channels (Haversian canals). Within this shell is porous or spongy bone, and in the case of "long" bones (see below) there is a hollow

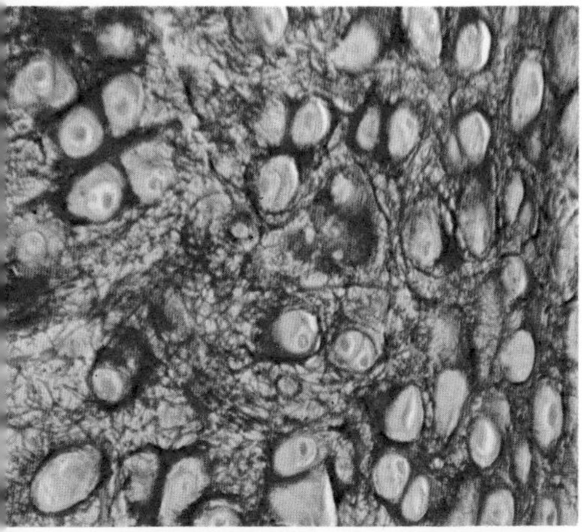

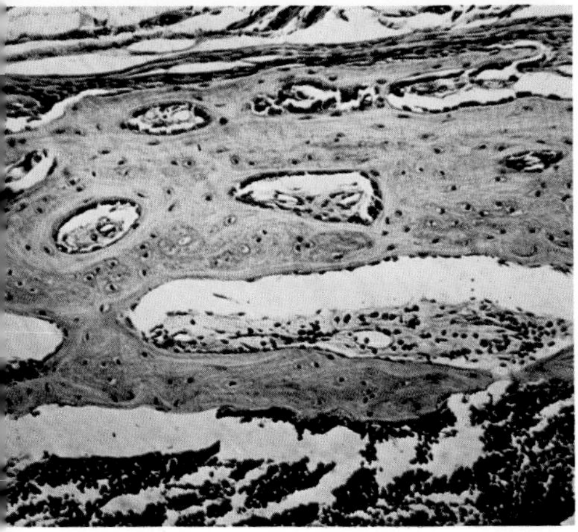

Top: The structure of compact bone.
Above: The structure of spongy bone.

cavity containing MARROW. The bone is enveloped by a fibrous membrane, the periosteum, which is sensitive to pain, unlike the bone itself, and which has a network of nerves and blood vessels which penetrate the bone surface. After primary growth has ended, bone formation (ossification) occurs where the periosteum joins the bone, where there are many bone-forming cells (osteoblasts). Ossification begins in the embryo at the end of the second month, mostly by transformation of CARTILAGE: some cartilage cells become osteoblasts and secrete collagen and a hormone which causes calcium salts to be deposited. Vitamin D makes calcium available from the food to the blood, and its deficiency leads to RICKETS. The two ends of a "long" bone (the epiphyses) ossify separately from the shaft, and are attached to it by cartilaginous plates, at which lengthwise growth takes place. Radical growth is controlled by the periosteum, and at the same time the core of the bone is eroded by osteoclast cells to make it hollow. Primary growth is stimulated by the pituitary and sex HOROMONES; it is completed in adolescence, when the epiphyses fuse to the shaft. Bones are classified anatomically as "long," cylindrical and usually hollow, with a knob at each end; "short," spongy blocks with a thin shell; and "flat" two parallel layers of compact bone with a spongy layer in between. Some hand and foot bones are short; the ribs, sternum, skull and shoulder-blades are flat; and most other bones are long. The shape and structure of bones are quickly modified if the forces on them alter. Disorders of bone include OSTEOMYELITIS and various TUMORS and CANCER. Dead bone is not readily absorbed and can be a focus of infection. In old age thinning and weakening of the bones by loss of calcium (osteoporosis) is common.

BONE CHINA, fine porcelain first introduced c1800 (see POTTERY). It is made of china clay mixed with bone-ash and china stone, and is similar to hard porcelain but more workable and less easily chipped.

BONE MEAL, ground BONE used as fertilizer, providing essential nitrogen, phosphorus and calcium. Sterilized, it is used as an animal feed.

BONSAI, the ancient oriental art of growing trees in dwarf form. The modern enthusiast may spend three years cultivating the "miniature" trees, mainly by root pruning and shoot trimming. Plants that can be "dwarfed" include the cedars, myrtles, junipers, oaks, cypresses, pyracanthas and pines. Bonsai has spread worldwide, and is a fast-growing hobby in North America, where there are many "bonsai" clubs.

BOOLE, George (1815–1864), British mathematician and logician, chiefly remembered for devising **Boolean algebra**, which allowed mathematical methods to be applied to nonquantifiable entities such as logical propositions. In the 20th century Boolean algebra has become important in the design of telecommunications systems and electronic logic circuits, and hence in COMPUTER technology. An example of a Boolean algebra is the algebra of sets (see SET THEORY). (See also LOGIC.)

BOÖTES (the Herdsman), a constellation of the N Hemisphere, containing the star ARCTURUS.

BORANES, covalent HYDRIDES of boron, of unusual molecular structure: they have hydrogen-bridge bonding, and the boron atoms form the vertices of polyhedra. Boranes are volatile, reactive and often flammable in air. They may be used as high-energy fuels for rockets and jet planes.

BORAX, mineral name for sodium tetraborate, $Na_2B_4O_7 . 10H_2O$, found mainly in Cal. (For sodium borates see SODIUM; KERNITE.)

BORAZON, cubic form of boron nitride (BN), resembling DIAMOND in structure and properties. It is made by heating hexagonal boron nitride (a white powder resembling GRAPHITE) to 1500°C at 65 000atm pressure. Borazon is as hard as diamond, and more useful industrially because it is more resistant to oxidation and heat.

BORDEAUX MIXTURE, FUNGICIDE made from COPPER sulfate, CALCIUM hydroxide and water. It was once widely used on crops, but has generally been replaced by fungicides less harmful to fruit and foliage.

BORDET, Jules Jean-Baptiste Vincent (1870–1961), Belgian bacteriologist and immunologist awarded the 1919 Nobel Prize for Physiology or Medicine for his discovery of the substances later named "complement" and the process of complement fixation (see ANTIBODIES AND ANTIGENS). He also

A tidal bore sweeps up the Petitcodiac River near Moncton, New Brunswick, Canada. Bores occur on several of the rivers which empty into the Bay of Fundy, largely because of the great tidal range experienced around its shores.

discovered the BACILLUS responsible for WHOOPING-COUGH.

BORE, or **eagre**, tidal phenomenon of rivers that widen gradually toward broad mouths, and that are subject to high TIDES. During spring flood, larger quantities of water from the sea than can normally flow upriver are driven into the rivermouth, resulting in a high wave that travels upriver at great speed. Perhaps the best-known bores are those of the Ganges, the Severn and the Bay of Fundy.

BORELLI, Giovanni Alfonso (1608–1679), Italian astronomer, physicist and physiologist, the founder of iatrophysics. After making contributions to astronomy, including the proposal that COMETS travel along elliptical paths, he turned his attention to the working of the living body and successfully explained MUSCLE action on mechanical principles.

BORIC ACID (H_3BO_3), or boracic acid, colorless crystalline solid, a weak inorganic acid. It gives boric oxide (B_2O_3) when strongly heated; SODIUM borate typifies its salts. Boric acid is used as an external antiseptic, in the production of glass and as a welding flux.

BORLAUG, Norman Ernest (1914–   ), US agricultural scientist who was awarded the 1970 Nobel Peace Prize for his part in the development of improved varieties of CEREAL CROPS, important in the green revolution.

BORN, Max (1872–1970), German theoretical physicist active in the development of quantum physics, whose particular contribution was the probabilistic interpretation of the SCHRÖDINGER wave equation, thus providing a link between WAVE MECHANICS and the QUANTUM THEORY. Sharing the Nobel physics prize with BOTHE in 1954, he devoted his later years to the philosophy of physics.

BORNITE, reddish-brown SULFIDE mineral of iron and copper (Cu_5FeS_4); a COPPER ore, called "peacock ore" because of its tarnish when fractured. Its occurrence is widespread, especially in Chile, Peru and Tasmania. It alters to CHALCOCITE.

BORON (B), nonmetallic element in Group IIIA of the PERIODIC TABLE, occurring as KERNITE and BORAX in Cal. and Turkey. Boron has three black crystalline allotropes and an amorphous form, and is best prepared by reduction of the halides with hydrogen. Normally inert, it becomes reactive at high temperatures; it is trivalent. Boron is a trace element vital to plant growth; it is used to produce heat-resistant alloy steels, and boron fibers are used as a high-strength construction material. Boron-steel rods absorb neutrons in nuclear reactors. The borates are salts of BORIC ACID; the SODIUM salts are the most important. AW 10.8, mp 2 300°C, bp 2 550°C, sg 2.34 (See also BORANES; BORAZON.)

BOSCH, Karl (1874–1940), German industrial chemist who adapted the HABER PROCESS for AMMONIA manufacture for large-scale industrial use. In 1931 he shared the Nobel chemistry prize with F. BERGIUS for their work on high-pressure synthesis.

BOSE, Sir Jagadis Chandra (1858–1937), Indian biologist who devised and used sensitive instruments for measuring the growth and response to stimuli of plants.

BOSE-EINSTEIN STATISTICS, in QUANTUM MECHANICS, the statistical behavior of a system of indistinguishable particles with a number of discrete states, each of which may be occupied at any one time by any number of particles. SUBATOMIC PARTICLES that show this behavior are termed bosons. (See also FERMI-DIRAC STATISTICS.)

BOSON. See SUBATOMIC PARTICLES.

BOTANICAL GARDENS, collections of living plants made for scientific and educational purposes, often also providing recreation. They may contain flower beds, greenhouses, pools, herbaria and laboratories. Among their principal functions are the preservation of rare plants and the development of new varieties.

BOTANY, the study of plant life. Botany and ZOOLOGY are the major divisions of BIOLOGY. There are many specialized disciplines within botany, the classical ones being morphology, physiology, GENETICS, ECOLOGY and TAXONOMY. Although the presentday botanist often specializes in a single discipline, he frequently draws upon techniques and information obtained from others.

The plant morphologist studies the form and structure of plants, particularly the whole plant and its major components, while the plant anatomist concentrates upon the cellular and subcellular structure, perhaps using the ELECTRON MICROSCOPE. The behavior and functioning of plants is studied by the plant physiologist, though since he frequently uses biochemical techniques, he is often called a plant biochemist. A plant geneticist uses biochemical and biophysical techniques to study the mechanism of inheritance and may relate this to the EVOLUTION of an individual. An important practical branch of genetics is plant BREEDING. The plant ecologist relates the form (morphology and anatomy), function (physiology) and evolution of plants to their environment. The plant taxonomist or systematic botanist specializes in the science of classification, which involves cataloging, indentifying and naming plants using their morphological, physiological and genetic characters. CYTOLOGY, the study of the individual cell, necessarily involves techniques used in morphology, physiology and genetics.

Within these broad divisions there are many specialist fields of research. The plant physiologist may, for instance, be particularly interested in PHOTOSYNTHESIS or RESPIRATION. Similarly, the systematic botanist may specialize in the study of ALGAE (algology), FUNGI (mycology) or MOSSES (bryology). Other specialists study the plant in relation to its uses (economic botany), PLANT DISEASES (plant pathology) or the agricultural importance of plants (agricultural botany). BACTERIOLOGY is often considered to be a division of botany since bacteria are often classified as plants. (See also AGRONOMY; BIOCHEMISTRY; BIOPHYSICS; HORTICULTURE; PLANT; PLANT KINGDOM.)

The forerunners of the botanists were men who collected herbs for medical use long before philosophers turned to the scientific study of nature. However, the title of "father of botany" goes to

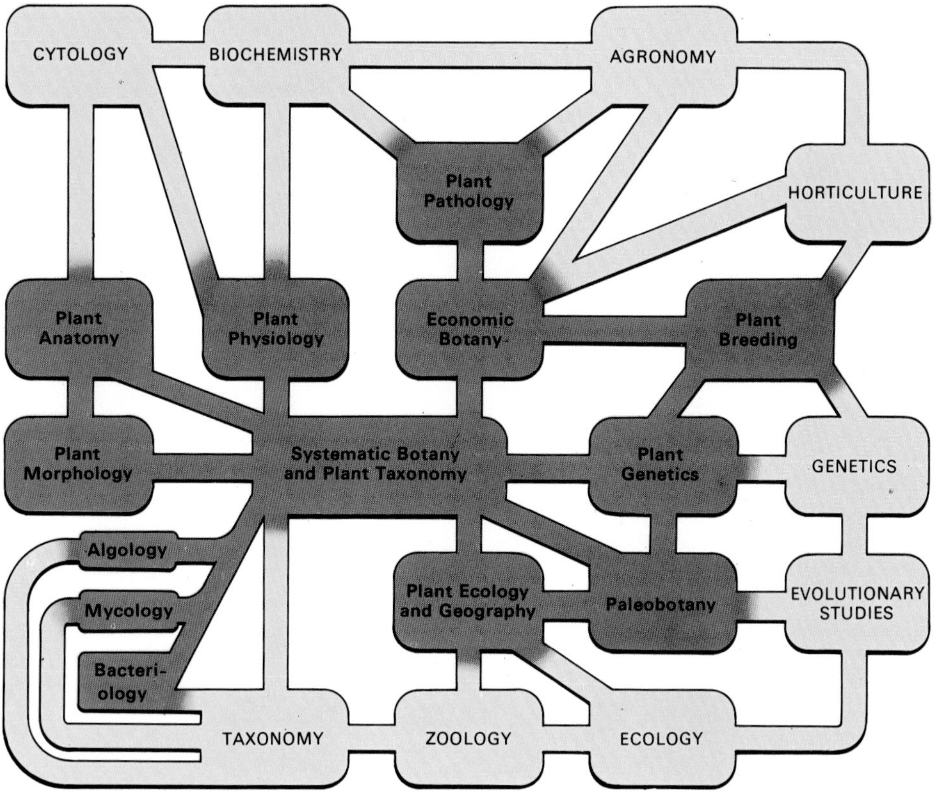

The different branches of botany (red) and their links with allied sciences (yellow).

THEOPHRASTUS, a pupil of ARISTOTLE, whose *Inquiry into Plants* sought to classify the types, parts and uses of the members of the plant kingdom. Passing over the work of the elder Pliny and that of his contemporary, Dioscorides, botany received few further lasting contributions until the Renaissance, the intervening period making do with the more or less fabulous "herbals" of the medical herbalists. The most famous pre-Darwinian classification of the plant kingdom was that of LINNAEUS, in which modern binomial names first appeared (1753). While Nehemiah GREW and John RAY had laid the foundations for plant anatomy and physiology in the 17th and 18th centuries, and HOOKE had even identified the cell (1665) with the aid of the MICROSCOPE, these subjects were not actively pursued until the 19th century when R. BROWN identified the nucleus and SCHWANN proposed his comprehensive cell theory. The work of DARWIN revolutionized the theory of classification, while that of MENDEL pointed the way to a true science of plant breeding.

BOTHE, Walter Wilhelm Georg Franz (1891–1957), German experimental physicist who devised the coincidence method for the detection of COSMIC RAY showers (1929). This work won him a share (with BORN) in the 1954 Nobel physics prize.

BOTTLED GAS, liquefied PETROLEUM gas (LPG) kept under pressure in steel cylinders and used for fuel by campers etc., and for tractors and buses. It is PROPANE, BUTANE or a mixture of the two.

BOTULISM, usually fatal type of FOOD POISONING caused by a toxin produced by the anaerobic bacteria *Clostridium botulinum* and *C. parabotulinum*, which normally live in soil but may infect badly canned food. The toxin paralyzes the nervous system. Thorough cooking destroys both bacteria and toxin.

BOULTON, Matthew (1728–1809), English industrial innovator and a founder member of the LUNAR SOCIETY OF BIRMINGHAM. In 1775 he went into partnership with James WATT to manufacture the latter's improved STEAM ENGINE. Boulton became England's foremost manufacturer and it is often said that his engines powered the industrial revolution.

BOUNDARY LAYER, the portion of a FLUID near to a surface in motion relative to it: specifically, the layers of air nearest to the wing of an aircraft in flight (see AERODYNAMICS). Because of the air's VISCOSITY, these layers are subject to SHEARING, which reduces their velocity relative to the wing; thus lift is reduced and drag increased. Turbulence may also occur. (See also REYNOLDS NUMBER.)

BOURBAKI, Nicolas, pen-name adopted by a group of French mathematicians under which (since 1939) they have published a momentous survey of mathematics along original but strictly formal lines. The work of the Bourbaki authors has been of considerable importance in the development of 20th-century mathematics.

BOUSSINGAULT, Jean-Baptiste Joseph Dieudonné (1802–1887), French agricultural chemist who studied the GERMINATION of SEEDS and promoted the use of inorganic fertilizers containing NITROGEN and PHOSPHORUS compounds.

BOVET, Daniel (1907–), Swiss-born Italian pharmacologist who discovered the first ANTIHISTAMINE and later developed CURARE and curare-like compounds for use as muscle-relaxants during surgery. He was awarded the 1957 Nobel Prize for Physiology or Medicine.

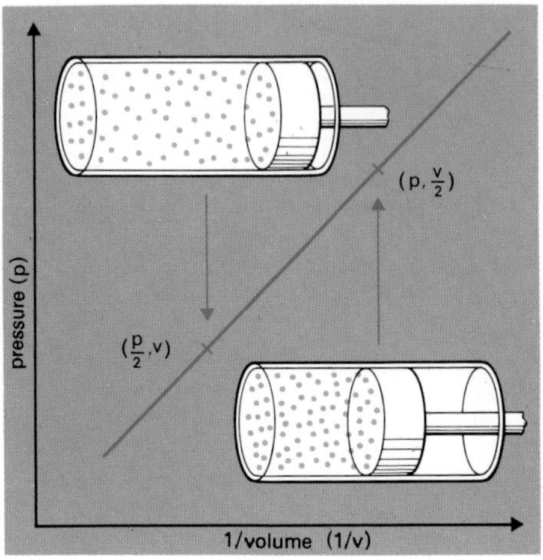

Boyle's law implies that if a quantity of an ideal gas is to be compressed at constant temperature so that its volume is reduced by one half (from v to $\frac{v}{2}$ its pressure must be doubled (from $\frac{p}{2}$ to p).

BOWDITCH, Nathaniel (1773–1838), self-taught US mathematician and astronomer remembered for his *New American Practical Navigator* (1802), "the seaman's bible," later made standard in the US navy. He was the first to describe the LISSAJOUS' FIGURES (Bowditch curves), later studied in detail by Lissajous.

BOYD-ORR, John, Baron (1880–1971), British agricultural scientist and nutritionist who was awarded the 1949 Nobel Peace Prize for his services to the Food and Agriculture Organization (FAO) of which he was director 1945–48.

BOYLE, Robert (1627–1691), British natural philosopher often called the father of modern CHEMISTRY for his rejection of the theories of the alchemists and his espousal of ATOMISM. A founder member of the ROYAL SOCIETY OF LONDON, he was noted for his pneumatic experiments.

BOYLE'S LAW, or Mariotte's Law, an empirical relation reported by BOYLE (1662) and MARIOTTE (1676) but actually discovered by Boyle's assistant R. Townely, which states that given a fixed MASS of GAS at constant TEMPERATURE, its VOLUME is inversely proportional to its PRESSURE. Real gases deviate considerably from this law.

BRACHIOPODA, or **lampshells**, a phylum of marine mollusk-like animals once classified with the MOLLUSCA. Lampshells are enclosed in bivalved shells and are attached to the sea bed by a long flexible stalk. There are about 260 living species, but at least 30 000 are known as fossils from rocks dating back to the Lower CAMBRIAN period.

BRADYODONTI, a subclass of the class CHONDRICHTHYES that contains the bradyodonts, fish that have small mouths with teeth that are composed of large plates firmly attached to the jaws, and an upper jaw that is fused to the skull.

BRADLEY, James (1693–1762), English astronomer who discovered the ABERRATION OF LIGHT (1728) and the earth's nutation (see NUTATION, ASTRONOMICAL).

BRAGG, Sir William Henry (1862–1942), British physicist who shared the 1915 Nobel Prize for Physics with his son, **Sir William Lawrence Bragg** (1890–1971), for learning how to deduce the atomic structure of CRYSTALS from their X-RAY DIFFRACTION patterns (1912).

BRAHE, Tycho (1546–1601), Danish astronomer, the greatest exponent of naked-eye positional ASTRONOMY. KEPLER became his assistant in 1601 and was driven to postulate an elliptical orbit for MARS only because of his absolute confidence in the accuracy of Tycho's data. Brahe is also remembered for the "Tychonic system," in which the planets circled the sun, which in turn orbited a stationary earth, this being the principal 17th-century rival of the Copernican hypothesis.

BRAHMAGUPTA (c598–660), Hindu mathematician and astronomer whose writings influenced Arab and hence medieval European scholars.

BRAIN, complex organ which, together with the SPINAL CORD, comprises the central NERVOUS SYSTEM and coordinates all nerve-cell activity. In INVERTEBRATES the brain is no more than a GANGLION; in VERTEBRATES it is more developed—tubular in lower vertebrates and larger, more differentiated and more rounded in higher ones. In higher mammals, including man, the brain is dominated by the highly developed cerebral cortex. The brain is composed of many billions of interconnecting nerve cells (see NEURONS) and supporting cells (neuroglia). The BLOOD CIRCULATION, in particular the regulation of blood pressure is designed to ensure an adequate supply of oxygen to these cells: if this supply is cut off, neurons die in only a few minutes. The brain is well protected inside the SKULL and is surrounded, like the spinal cord, by three membranes, the meninges. Between the two inner meninges lies the CEREBROSPINAL FLUID (CSF), an aqueous solution of salts and GLUCOSE. CSF also fills the four ventricles (cavities) of the brain and the central canal of the spinal cord. If the circulation of CSF between ventricles and meninges becomes blocked, HYDROCEPHALUS results. Relief of this may involve draining CSF to the atrium of the heart.

The human brain may be divided structurally into three parts: (1) the **hindbrain** consisting of the *medulla oblongata*, which contains vital centers to control heartbeat and breathing; the *pons* which, like the *medulla oblongata*, contains certain cranial nerve nuclei and numerous fibers passing between the higher brain centers and the spinal cord; and the *cerebellum*, which regulates balance, posture and coordination. (2) The **midbrain**, a small but important center for REFLEXES in the brain stem, also containing nuclei of the cranial nerves and the *reticular formation*, a diffuse network of neurons involved in

Section through the human brain: (1) cerebrum (a) frontal lobe, (b) parietal lobe, (c) occipital lobe, (d) temporal lobe; (2) cerebellum; (3) brain stem; (4) corpus callosum; (5) diencephalon; (6) mesencephalon; (7) pons; (8) medulla oblongata.

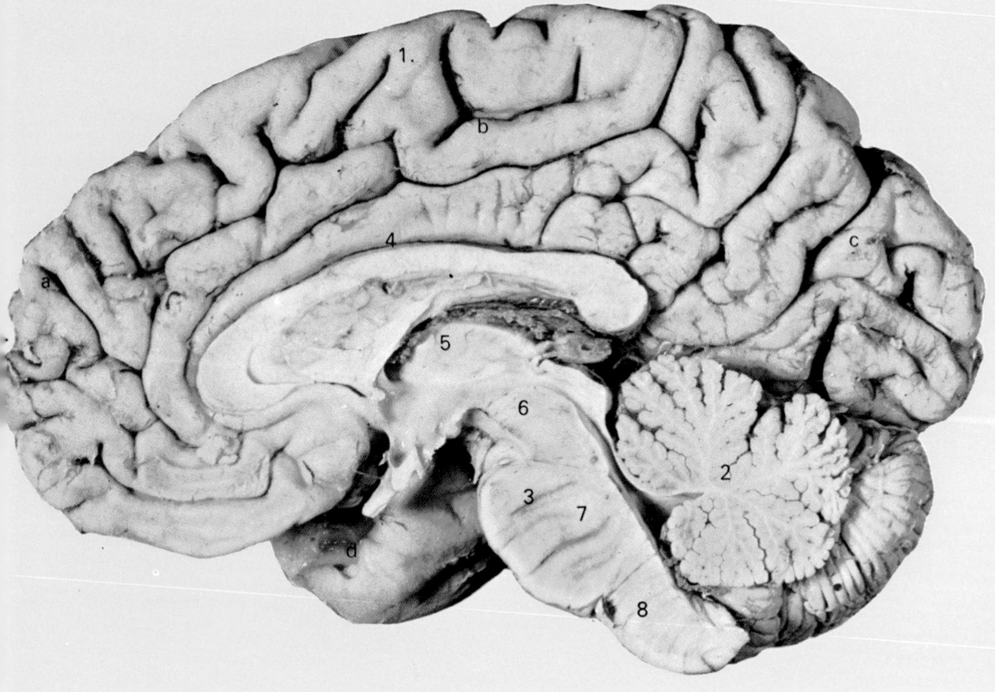

regulating arousal: SLEEP and alertness. (3) The **forebrain**, consisting of *thalamus*, which relays sensory impulses to the cortex; the *hypothalamus*, which controls the autonomic nervous system, food and water intake and temperature regulation, and to which the PITUITARY GLAND is closely related (see also PINEAL BODY); and the *cerebrum*. The cerebrum makes up two-thirds of the entire brain and has a deeply convoluted surface; it is divided into two interconnected halves or hemispheres. The main functional zones of the cerebrum are the surface layers of gray matter, the cortex, below which is a broad white layer of nerve fiber connections, and the *basal ganglia*, concerned with muscle control. (Disease of the basal ganglia causes PARKINSON'S DISEASE.) Each hemisphere has a motor cortex, controlling voluntary movement, and a sensory cortex, receiving cutaneous sensation, both relating to the opposite side of the body. Other areas of cortex are con cerned with language (see APHASIA, SPEECH AND SPEECH DISORDERS), memory, and perception of the special senses (sight, smell, sound); higher functions such as abstract thought may also be a cortical function (see also INTELLIGENCE, LEARNING). Diseases of the brain include infections—specifically MENINGITIS, ENCEPHALITIS, syphilis (see VENEREAL DISEASES) and ABSCESSES; also trauma, TUMORS, STROKES, MULTIPLE SCLEROSIS, and degenerative diseases with early ATROPHY, either generalized or localized. Investigation of brain diseases includes X RAYS using various contrast methods, SPINAL TAP (lumbar puncture)—to study CSF abnormalities—and the use of the ELECTROENCEPHALOGRAM. Treatments range from a variety of drugs, including ANTIBIOTICS and STEROIDS, to SURGERY.

BRAINWASHING, the manipulation of an individual's will, generally without his knowledge and against his wishes. Most commonly, it consists of a combination of isolation, personal humiliation, disorientation, systematic indoctrination and alternating punishment and reward.

BRAKES, devices for slowing or halting motion, usually by conversion of kinetic ENERGY into HEAT energy via the medium of FRICTION. Perhaps most common are **drum brakes**, where a stationary member is brought into contact with the wheel or a drum that rotates with it. They may be either *band brakes*, where a band of suitable material encircling the drum is pulled tightly against its circumference; or *shoe brakes*, where one or more shoes (shaped blocks of suitable material) are applied to the inner or outer circumference of the drum. Similar in principle are **disk brakes**, where the frictional force is applied to the sides of the wheel or a disk that rotates with it. The simplest form is the *caliper brake*, as used on bicycles, in which rubber blocks are pressed against the rim of the wheel. Almost all aircraft, AUTOMOBILE and RAILROAD brakes are of drum or disk type.

Mechanically operated brakes cannot always be used; as when a single control must operate on a number of wheels, thus involving problems in simultaneity and equality of braking action. In such cases, pressure is applied to a HYDRAULIC system (usually oil-filled), and hence equally to the brakes. Similar in principle are vacuum brakes, where creation of a partial VACUUM operates a PISTON which applies the braking action; and AIR BRAKES. **Fluid brakes**, used mainly in trucks to restrict speed in downhill travel, must be used in combination with mechanical brakes if it is desired to halt the vehicle.

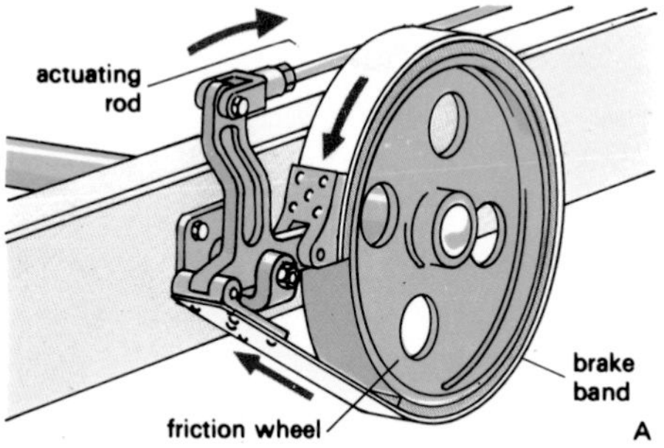

A

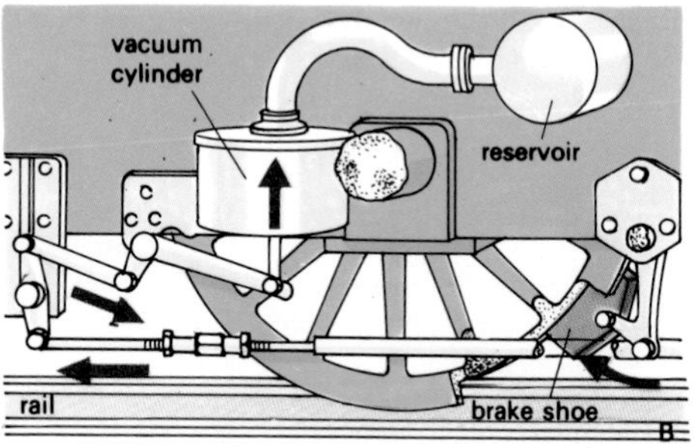

B

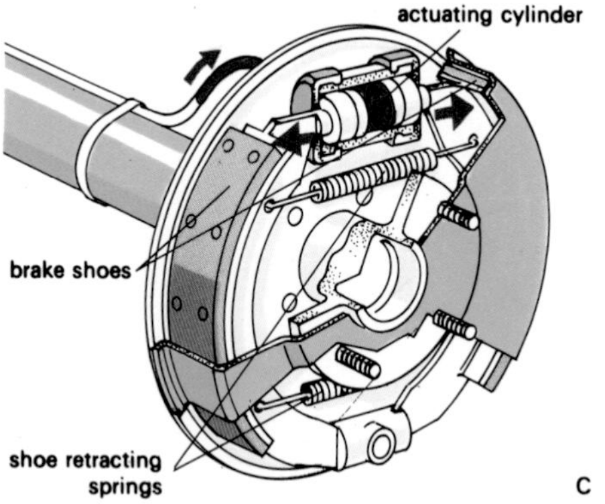

C

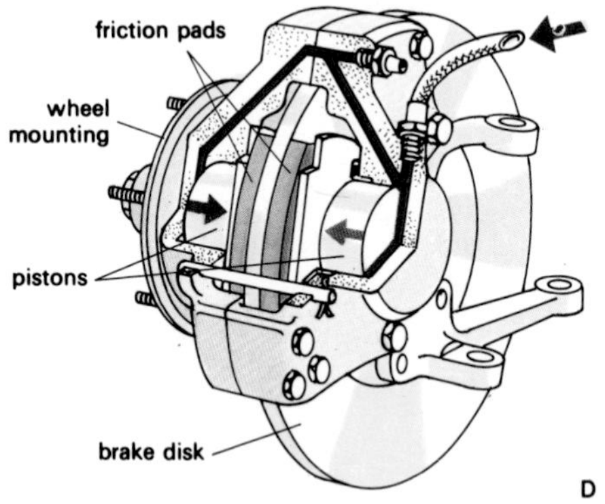

D

They consist of a rotating and a stationary element, between which a liquid (usually water) is introduced. Here it is FLUID, rather than mechanical, friction that converts the kinetic energy. (Cooling is usually performed by circulation through the radiator.) **Electric brakes**, similarly, may only restrict motion. The most common, used on electric trains on downhill runs, consists merely of a GENERATOR driven by the axle (the electricity generated may be used by the train).

In SPACE EXPLORATION there are clearly unique braking problems. In space or when landing on planets with little atmosphere (as on the MOON), ROCKETS are used. During descent to earth (and eventually other planets) friction between ATMOSPHERE and craft is used, supplemented by PARACHUTES (see ABLATION; HEAT SHIELD).

BRAMAH, Joseph (1749–1814), English engineer who invented the hydraulic press, a machine which made possible many of the 19th century's greatest constructed feats. (See also LOCKS AND KEYS.)

BRAN, husk removed from grains of wheat, rye, etc., during FLOUR milling. Wheat bran largely comprises carbohydrate, protein and fiber. Bran is used as cattle feeds and is added to some breads as roughage.

BRANDY, alcoholic drink of distilled grape or other wine, usually matured in wood. Brandies include cognac, from French wines of the Cognac area, kirsch (made from cherries) and slivovitz (made from plums). (See also ALCOHOLIC BEVERAGES.)

BRASS, an ALLOY of COPPER and ZINC, known since Roman times, and widely used in industry and for ornament and decoration. Up to 36% zinc forms α-brass, which can be worked cold; with more zinc a mixture of α- and β-brass is formed, which is less ductile but stronger. Brasses containing more than 45% zinc (white brasses) are unworkable and have few uses. Some brasses also contain other metals: lead to improve machinability, aluminum or tin for greater corrosion-resistance, and nickel, manganese or iron for higher strength.

Four common types of friction brake. (A) Band brake as used to control machinery. When the actuating rod is pulled, the brake band is tightened around a friction wheel on the shaft to be controlled. (B) Shoe brake as commonly used on railroad stock. Here a vacuum cylinder actuates a brake rod which presses brake shoes against the running surfaces of several wheels. Part of the flange is here shown cut away. Although on many railroad systems vacuum brake-activation systems have been replaced by air-brake systems (see illustration page 12), the use of shoe brakes remains common, particularly on low-speed stock. Railroad stock designed for high-speed running uses drum and, increasingly, disk brakes (see below). (C) Drum brake, hydraulically actuated. Increased pressure in the brake fluid forces curved brake shoes outward against a brake drum attached to the wheel mounting. Often a single brake-fluid circuit operates the brakes on all a vehicle's wheels, although for greater safety duplicate circuits can be employed on the same wheels. Alternatively, separate circuits can be used to activate the brakes on different sets of wheels. (D) Disk brake as used in airplanes, many autos and some railroad stock. Increased pressure in the brake fluid forces opposed friction pads against a disk rotating with the wheel.

BRATTAIN, Walter Houser (1902–   ), US physicist who shared the 1956 Nobel physics prize with SHOCKLEY and BARDEEN for their development of the TRANSISTOR.

BRAUN, Karl Ferdinand (1850–1918), German physicist who shared the 1909 Nobel physics prize with MARCONI for his discovery that certain crystals could act as RECTIFIERS, and for his proposal that these could be used in (crystal-set) RADIOS.

BRAUN, Wernher Magnus Maximillian von (1912–   ), German ROCKET engineer who designed the first self-contained missile, the V-2, which was used against the UK in 1944. In 1945 he went to America, where he led the team that put the first US artificial SATELLITE in ORBIT (1958).

BRAVAIS LATTICES, in crystallography, the 14 different unit cells (arrangements of structurally-significant points) using which regular three-dimensional structures can be built. (See CRYSTALS.)

BRAZING, technique in METALLURGY whereby two pieces of metal are joined using a nonferrous ALLOY (usually of COPPER) of lower MELTING POINT. The process is akin to SOLDERING but is performed at higher temperatures (about 1000 K).

BREAD, one of humanity's earliest and most important foods, basically comprising baked "dough"—a mixture of FLOUR and water. In developed western societies, wheat flour is most commonly used and the dough is "leavened" (i.e., increased in volume by introducing small bubbles of CARBON dioxide throughout) using YEAST. In making bread, the chosen blend of flours is mixed with water, yeast, shortening and salt (and sometimes sugar and milk) to form the dough. This is then kneaded to distribute the GLUTEN throughout the mix, left to rise, kneaded again, molded into shape and left to rise a second time before baking. Bread is generally high in CARBOHYDRATES though low in PROTEIN. The vitamin and mineral content depends on the ingredients and additives used.

BREASTS, or mammary glands, the milk-secreting glands in mammals. The breasts develop alike in both sexes, about 20 ducts being formed leading to the nipples, till puberty when the female breasts develop in response to sex HORMONES. In PREGNANCY the breasts enlarge and milk-forming tissue grows around multiplied ducts; later milk secretion and release in response to suckling occur under the control of specific pituitary hormones. Disorders of the breast include mastitis, breast CANCER (see also MASTECTOMY) and adenosis. In humans, the breasts are erogenous zones in both males and females.

BREATHING. See RESPIRATION.

BRECCIA. See CONGLOMERATE; TALUS.

BREEDER REACTOR, a NUCLEAR REACTOR that produces more nuclear fuel than it consumes, used to convert material that does not readily undergo FISSION into material that does. Commonly, nonfissile URANIUM-238 is converted into PLUTONIUM-239. (See also NUCLEAR ENERGY.)

BREEDING, the development of new strains of plants and animals with more desirable characteristics, such as higher yields or greater resistance to disease and suitability to the climate. Breeding has been practiced since prehistoric times— producing our modern domestic animals—but without firm scientific basis until MENDEL's theory of GENETICS. The breeder first decides which traits he wishes to develop, and observes the range of PHENOTYPES in the breeding population. Discounting variants due to environmental differences, he selects those individuals of superior GENOTYPE. This genetic variation may occur naturally, or may be produced by HYBRIDIZATION or MUTATIONS induced by radiation or certain chemicals. The selected individuals are used as parent stock for INBREEDING to purify the strain.

BREEDING BEHAVIOR, any behavior by which animals attract members of the opposite sex for the purposes of reproduction. Such behavior includes visual displays, calls and song and the production of scent for attraction and stimulation. Breeding behavior is found throughout the animal kingdom, but is particularly well developed in birds and mammals. (See also MATING RITUALS.)

BRENTANO, Franz Clemens (1838–1917), German philosopher and psychologist, a Roman Catholic priest from 1864 to 1873, who founded the school of intentionalism and taught both FREUD and HUSSERL.

BREUER, Josef (1842–1925), Austrian physician who pioneered the methods of PSYCHOANALYSIS and collaborated with FREUD in writing *Studies in Hysteria* (1895). He also discovered the role of the semicircular canals of the inner EAR in maintaining balance (1873).

BREWING, the process of making ALCOHOLIC BEVERAGES—generally BEER, but also sake and pulque—from starchy cereal grains. Brewing has been practiced for more than five millennia. The cereal (generally barley) is malted (steeped in moisture and germinated). The malt is then dried, cured at 100°C, and mashed—ground and infused with hot water; in mashing, the ENZYMES produced during malting break down the starch to fermentable SUGARS. The wort (aqueous solution) is filtered and boiled with hops, and then fermented in large vessels with YEAST (see FERMENTATION). When this has almost ceased, the beer is run off and stored to mature; finally, it is filtered, carbon dioxide is added, and the beer is packaged. The strain of yeast used helps to determine the type of beer made.

BREWSTER, Sir David (1781–1868), Scottish man of science who discovered Brewster's law in optics and did much to popularize science in 19th-century Britain. **Brewster's law** states that the maximum polarization of a ray of light reflected from a transparent surface occurs when the reflected and refracted rays are at right angles. (See POLARIZED LIGHT.)

BRIDGE, any device that spans an obstacle and permits traffic of some kind (usually vehicular, bridges that carry canals being more generally termed aqueducts) across it.

The most primitive form is the **beam** (or girder) **bridge**, consisting of a rigid beam resting at either end on piers. The span may be increased by use of intermediate piers, possibly bearing more than one beam. A development of this is the **truss bridge,** a truss being a metal framework specifically designed for greatest strength at those points where the load has greatest MOMENT about the piers. Where piers are impracticable, **cantilever bridges** may be built: from each side extends a beam (cantilever), firmly anchored at its inshore end. The gap between the two outer ends may be closed by a third beam. Another form of bridge is the **arch bridge**, essentially an ARCH built across the gap: a succession of arches supported by intermediate piers may be used for wider gaps. A **suspension bridge** comprises two towers that carry one or more flexible cables that are firmly anchored at each end. From these is suspended the roadway by means of vertical cables. **Movable bridges** take many forms, the most common being the **swing bridge**, pivoted on a central pier; the **bascule** (a descendant of the medieval drawbridge), whose cantilevers are pivoted inshore so that they may be swung upward; the **vertical-lift bridge**, comprising a pair of towers between which runs a beam that may be winched vertically upward; and the less common **retractable bridge**, whose cantilevers may be run inshore on wheels. The most common temporary bridges are the **pontoon**, or floating bridge, comprising a number of floating members that support a continuous roadway; and the BAILEY BRIDGE.

BRIDGMAN, Percy Williams (1882–1961), US physicist who won the 1946 Nobel Prize for Physics for his investigation of substances at very high pressures. His work led to the production of synthetic DIAMONDS (1955). In the philosophy of science he championed the view that scientific terms are only meaningful if they can be given "operational definitions."

BRIGGS, Henry (1561–1631), English mathematician who proposed that common (base 10) LOGARITHMS would prove of greater practical use than NAPIER's original natural (base *e*) ones and subsequently calculated and published the appropriate tables (1617).

BRIGHT'S DISEASE, a form of acute NEPHRITIS that may follow infections with certain STREPTOCOCCUS types. Blood and protein are lost in the urine; there may be EDEMA and raised blood pressure. Recovery is usually complete but a few patients progress to chronic KIDNEY disease.

BRILL, Abraham Arden (1874–1948), Austrian-born US psychiatrist, the "father of American PSYCHOANALYSIS," who introduced the Freudian method to the US and translated many of FREUD's works into English.

BRINDLEY, James (1716–1772), self-taught English canal builder who inaugurated the canal age by constructing the Bridgewater Canal. Brindley laid out more than 360mi (580km) of canals before his death, thus providing the transportation system which made possible the Industrial Revolution.

BRINE, a concentrated solution of SALT and other compounds. Natural brine occurs as seawater and as underground deposits. It is a major source of salt, BROMINE, IODINE and potassium compounds, and is also used as a food preservative and in refrigeration. (See also EVAPORITES.)

BRITANNIA METAL, an ALLOY consisting of 5–10% ANTIMONY, about 1% COPPER and the remainder TIN. It resembles PEWTER, is hard and workable, and is used as a base for electroplated silverware.

BRITISH THERMAL UNIT (Btu), the quantity of ENERGY required to raise one pound of water through one Fahrenheit degree. The International steam tables define the Btu_{IT} as $251.9958cal_{IT}$; this is equivalent to 1055.056 joules.

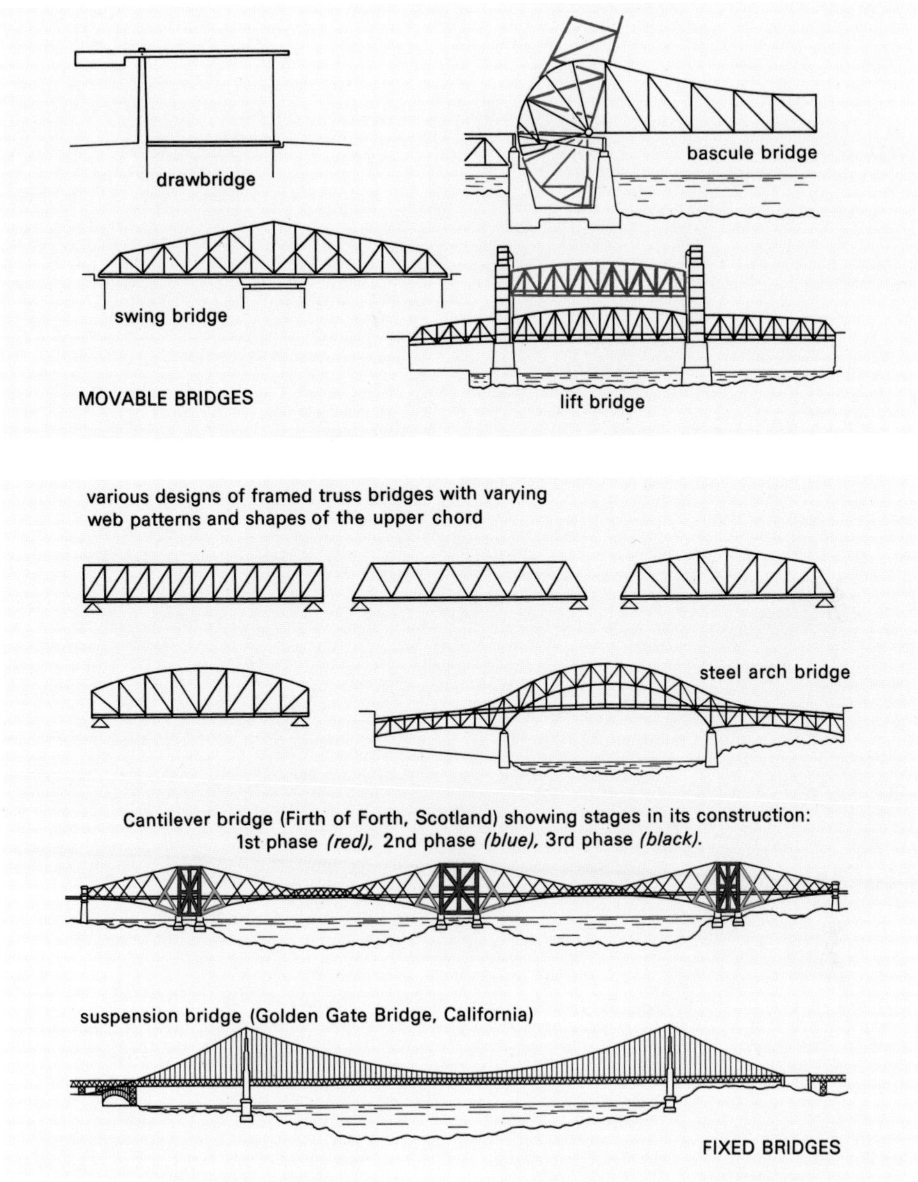

Various types of movable (*top*) and fixed (*above*) bridges.

BROCKEN SPECTER, phenomenon observed when SHADOWS of aircraft, or of people on a high mountain, are cast on cloud below. An observer's own shadow appears vastly magnified: this is an optical ILLUSION. Those of others appear to him surrounded by rings of color, due to DIFFRACTION by water droplets in the cloud.

BROGLIE, Louis Victor Pierre Raymond de. See DE BROGLIE, LOUIS VICTOR PIERRE RAYMOND.

BROMINE (Br), dark-red fuming liquid, toxic and caustic, with a pungent odor; one of the HALOGENS, intermediate in properties between CHLORINE and IODINE. It occurs as bromides, mainly in seawater, from which it is extracted by oxidation with chlorine. Soluble metal bromides (see HALIDES) are used as SEDATIVES; silver bromide, being light-sensitive, is used in PHOTOGRAPHY. Ethylene dibromide, the chief bromine product, is used as a lead scavenger in ANTIKNOCK ADDITIVES. Alkyl bromides (see ALKYL HALIDES) are used as fumigants and solvents. AW 79.9, mp −7°C, bp 59°C, sg 3.12 (20°C).

BRONCHI, tubes through which air passes from the TRACHEA to the LUNGS. The trachea divides into the two primary bronchi, one to each lung, which divide into smaller branches and finally into the narrow

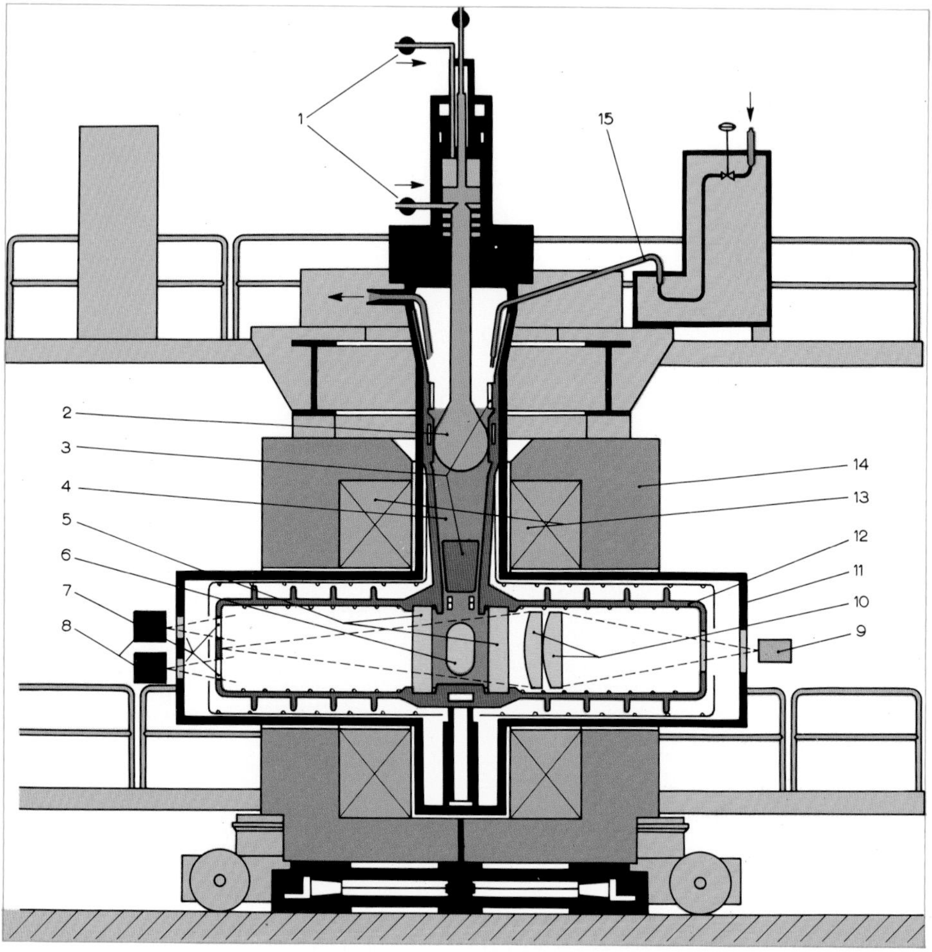

Section through the CERN two-metre hydrogen bubble chamber: (1) pneumatic apparatus actuating the expansion-compression cylinder (2); (3) thermometer sites; (4) liquid hydrogen chamber; (5) glass windows; (6) window through which the ionizing particles enter (the particle pathway is perpendicular to the plane of the section); (7) windows for the photographic cameras (8); (9) light source; (10) condenser lenses; (11) external casing of vacuum chamber (12); (13) electromagnet windings; (14) laminated iron yoke of the electromagnet; (15) cooling hydrogen supply.

bronchioles connecting with the alveolar sacs. The bronchi are lined with a mucus membrane which has motile CILIA to remove dust, etc.

BRONCHITIS, inflammation of BRONCHI. **Acute bronchitis**, often due to VIRUS infection, is accompanied by COUGH and FEVER and is short-lived; ANTIBIOTICS are only needed if there is bacterial infection. **Chronic bronchitis** is a more serious, often disabling and finally fatal disease. The main cause is SMOKING which irritates the LUNGS and causes overproduction of MUCUS. The CILIA fail, and sputum has to be coughed up. Bronchi thus become liable to recurrent bacterial infection, sometimes progressing to PNEUMONIA. Areas of lung become non-functional, and ultimately CYANOSIS and HEART failure may result. Treatment includes PHYSIOTHERAPY, antibiotics and bronchial dilator drugs. Stopping smoking limits damage and may improve early cases.

BRONCHOSCOPE, a tube with a light and lens system, used to examine the TRACHEA and BRONCHI, and also to perform a BIOPSY.

BRØNSTED, Johannes Nicolaus (1879–1947), Danish physical chemist principally remembered for formulating (independently of T. M. Lowry) the Brønsted-Lowry theory of ACIDS and bases.

BRONTOSAURUS, a vegetarian DINOSAUR whose fossilized skeleton has been found in the western US. About 20m (66ft) long and calculated to be 35 tonnes in weight, it had a long neck and tail, small head and brain, ponderous body and thick legs.

BRONZE, an ALLOY of COPPER and TIN, known since the 4th millennium BC (see BRONZE AGE), and used then for tools and weapons, now for machine parts and marine hardware. Statues are often cast in bronze. It is a hard, strong alloy with good corrosion-resistance (the patina formed in air is protective). Various other components are added to bronze to improve hardness or machinability, such as aluminum, iron, lead, zinc and phosphorus. Aluminum bronzes, and some others, contain no tin. (See also BELL METAL; GUN METAL.)

BRONZE AGE, the phase of man's material cultural development following the STONE AGE, and the first phase in which metal was used. The start of the bronze age varies from region to region, but certainly the use of copper was known as early as 6500 BC in Asia Minor, and its use was widespread shortly thereafter. By about 3000 BC BRONZE was widely used, to be replaced around 1000 BC by iron.

BROWN, Robert (1773–1858), Scottish botanist who first observed BROWNIAN MOTION (1827) and who identified and named the plant CELL nucleus (1831).

BROWNIAN MOTION, frequent, random fluctuation, illustrated by the motion of particles of the dispersed phase of a fluid COLLOID; first described by Robert BROWN (1827) after observation of a SUSPENSION of pollen grains in water. It is a result of the bombardment of the colloidal particles by the MOLECULES of the continuous phase (see KINETIC THEORY): a chance greater number of impacts in one direction changes the direction of motion of the particle. It is believed that all molecules of FLUIDS undergo Brownian motion. Observation of Brownian motion in colloids is of value in studies of DIFFUSION.

BROWNSTONE, reddish-brown variety of SANDSTONE, used for building. Also a house of brownstone: best known are those of New York.

BRUCE, Sir David (1855–1931), Australian-born British microbiologist who discovered the organisms responsible for undulant fever, nagana and African SLEEPING SICKNESS. The BACILLUS causing the first of these (now known as BRUCELLOSIS) is named *Brucella* in his honor.

BRUCELLOSIS, or **Bang's disease**, a BACTERIAL DISEASE of cattle, goats and swine, caused by *Brucella*. It causes ABORTION and affected animals have to be slaughtered to prevent spread. The disease in man (once known as **undulant fever**) is contracted from milk or by contact with infected animals; it is a variable illness, often causing FEVER, malaise and DEPRESSION, and may be treated with ANTIBIOTICS.

BRUISE, or contusion, a lesion of the SKIN, usually caused by a blow, in which CAPILLARY damage allows blood to leak into the dermis where it is slowly broken down and absorbed. A **hematoma**, a larger blood-filled cavity, may require draining. Bruising without injury may indicate BLOOD disease.

BRUNEL, Sir Marc Isambard (1769–1849), French-born British engineer and inventor who built the world's first underwater tunnel (under the River Thames) and devised machines for the mass production of pulley blocks and army boots. His son, **Isambard Kingdom Brunel** (1806–1859), pioneered many important construction techniques, designing the Clifton suspension bridge at Bristol, England laying the Great Western Railway with a controversial 7ft (2.13m) gauge and building iron-hulled steamships, including the giant *Great Eastern*.

BRUNO, Giordano (1548–1600), Italian pantheist philosopher, poet and cosmologist, an apostate Dominican, who taught the plurality of inhabited worlds, the infinity of the universe and the truth of the Copernican hypothesis. Burned at the stake for heresy, he became renowned as a martyr to science.

BRYOPHYTA, division of the PLANT KINGDOM that contains the most primitive of the green land plants. The Bryophyta are normally divided into three classes: Hepaticae (LIVERWORTS), Anthoceratae (HORNWORTS) and Musci (MOSSES). Bryophytes have a characteristic life cycle in which the GAMETOPHYTE is dominant and the SPOROPHYTE is attached to and dependent on the gametophyte for nutrition. (See also ALTERNATION OF GENERATIONS.)

BRYOZOA. See POLYZOA.

BUBBLE CHAMBER, device invented by GLASER (1952) to observe the paths of SUBATOMIC PARTICLES with energies too high for a CLOUD CHAMBER to be used. A liquid (e.g. liquid HYDROGEN or OXYGEN) is held under PRESSURE just below its BOILING POINT. Sudden reduction in pressure lowers this boiling point: boiling starts along the paths of energetic subatomic particles, whose passage creates local heating. At the instant of reduction, their paths may thus be photographed as a chain of bubbles.

BUBONIC PLAGUE. See PLAGUE.

BUCHNER, Eduard (1860–1917), German organic chemist, awarded the 1907 Nobel Prize for Chemistry for his discovery that FERMENTATION did not require the presence of complete YEAST cells but only an extract containing the enzyme ZYMASE. This discovery inaugurated ENZYME chemistry.

BUD, a condensed shoot in which the stem is very short, the inner leaves are closely packed and the outer scale leaves form a protective covering. In the spring, the stem elongates rapidly and the leaves unfold. The term bud is also used in zoology for a point from which new growth develops.

BUDDING, a form of GRAFTING particularly used for roses and fruit trees. The grafted portion, or scion, in this case is a bud which is inserted under the bark of a stock that is usually not more than one year old.

BUFFER, a solution in which pH is maintained at a nearly constant value. It consists of a relatively concentrated solution of a weak ACID and its conjugate BASE, and works best if their concentrations are roughly equal, in which case the hydrogen ion concentration equals the DISSOCIATION constant of the acid. When a small amount of a different acid or base is added, the buffer equilibrium shifts so that the pH value hardly changes (see LE CHATELIER). One common buffer is acetic acid/sodium acetate. Buffers are used in many chemical and biochemical experiments. Biochemical processes in the body are controlled by natural buffer systems.

BUFFON, Georges Louis Leclerc, Comte de (1707–1788), French naturalist who was the first modern taxonomist of the ANIMAL KINGDOM and who led the team which produced the 44-volume *Histoire naturelle* (1749–1804).

BULB, a short, underground storage stem composed of many fleshy scale leaves that are swollen with stored food and an outer layer of protective scale leaves. Bulbs are a means of overwintering; in the spring, flowers and foliage leaves are rapidly produced when growing conditions are suitable. Examples of plants producing bulbs are daffodil, tulip, snowdrop and onion.

BUNION, deformity of the joint of the big toe, generally caused by ill-fitting shoes. Pressure on the tissues at this point creates a BURSA which becomes painfully inflamed. Treatment entails wearing wide shoes, PODIATRY and in some cases surgery.

BUNSEN, Robert Wilhelm Eberhard (1811–1899), German chemist who, after important work on organo-arsenic compounds went on (with G. R. KIRCHHOFF) to pioneer chemical SPECTROSCOPY, discovering the elements CESIUM (1860) and RUBIDIUM (1861). He also helped to popularize the gas burner known by his name.

BUNSEN BURNER, burner, promoted by BUNSEN, used widely in laboratories and sometimes in metal heat-treatment (see METALLURGY). Through a nozzle at the base is introduced slightly pressurized fuel GAS (usually COAL GAS) which is mixed with air (primary air) which the flow induces through adjustable inlets: the usual gas:air ratio is 1:2. The mixture proceeds up a short metal tube and is ignited at the top, giving a flame temperature of the order of 2000K. Outer portions of the flame mix with further air (secondary air) to give a cooler, more luminous flame, where the gas:air ratio is about 1:4. The burner is now used mostly in highschools.

BURBANK, Luther (1849–1926), US horticulturalist who developed more than 800 varieties of plants, including the Burbank potato.

BURETTE, graduated glass tube with a stopcock, used in VOLUMETRIC ANALYSIS to measure the volume of a liquid, especially of one of the reagents in a TITRATION. A gas burette measures gas volume by the volume of liquid displaced.

BURIDAN, Jean (c1300–c1385), French philosopher and critic of ARISTOTLE who proposed the impetus theory, often considered to foreshadow NEWTON's first law of motion.

BURNET, Sir Frank Macfarlane (1899–   ), Australian physician and virologist who shared the 1970 Nobel Prize for Physiology or Medicine with P. B. MEDAWAR for his suggestion that the ability of organisms to form ANTIBODIES in response to foreign tissues was acquired and not inborn. Medawar followed up the suggestion and performed successful skin transplants in mice.

BURNHAM, Sherbourne Wesley (1838–1921), US astronomer noted for his studies of DOUBLE STARS, reported in his *General Catalogue of Double Stars* (1906).

BURNING GLASS. See MAGNIFYING GLASS.

BURNS AND SCALDS, injuries caused by heat, electricity, radiation or caustic substances, in which protein denaturation causes death of tissues. (Scalds are burns due to boiling water or steam.) Burns cause PLASMA to leak from blood vessels into the tissues and in severe burns substantial leakage leads to SHOCK. In **first-degree burns**, such as mild SUNBURN, damage is superficial. **Second-degree burns** destroy only the epidermis so that regeneration is possible. **Third-degree burns** destroy all layers of SKIN, which cannot then regenerate, so skin-grafting is required. Infection, ulceration, hemolysis, KIDNEY failure and severe scarring may complicate burns. Treatment includes ANALGESICS, dressings and ANTISEPTICS and fluids for shock. Immediate FIRST AID measures include cold water cooling to minimize continuing damage.

BURR, or bur, a seed with barbed spines which catch onto animal fur. Some common burr-producing plants are the burdock, bur marigold, burcock and cocklebur.

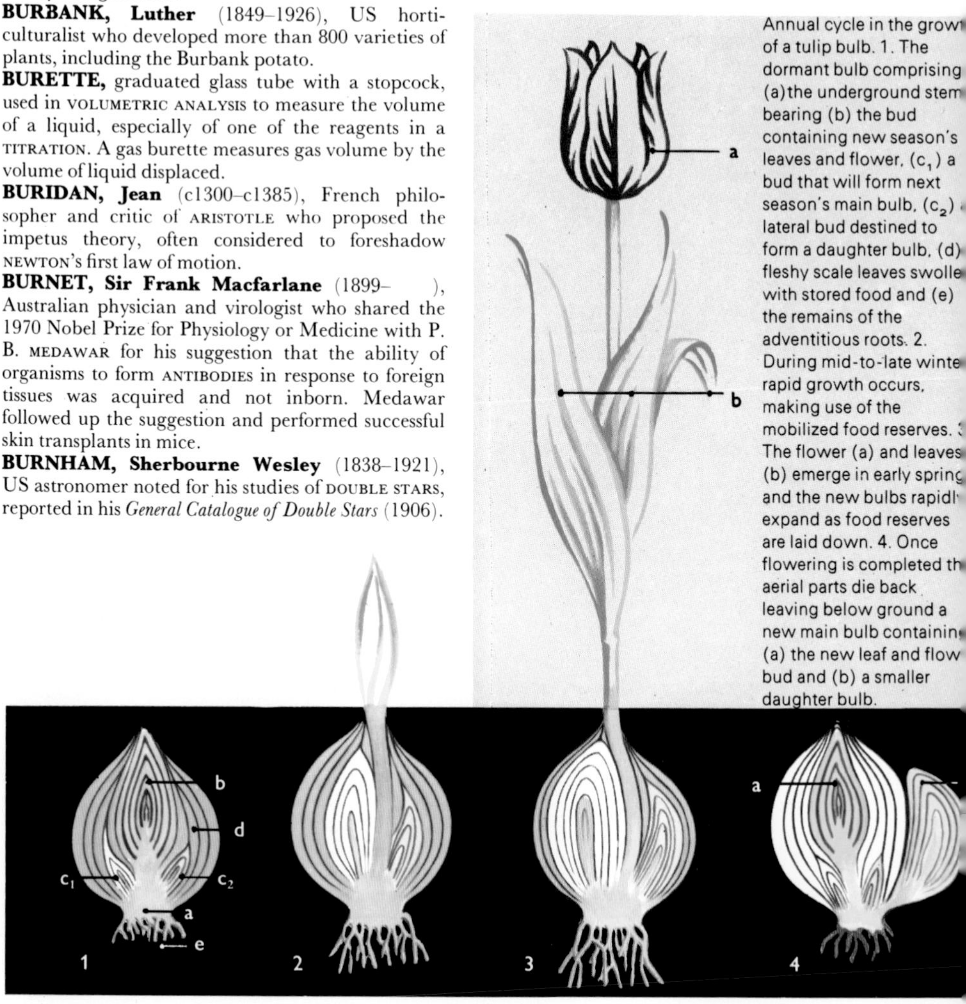

Annual cycle in the grow[th] of a tulip bulb. 1. The dormant bulb comprising (a) the underground stem bearing (b) the bud containing new season's leaves and flower, (c_1) a bud that will form next season's main bulb, (c_2) [a] lateral bud destined to form a daughter bulb, (d) fleshy scale leaves swolle[n] with stored food and (e) the remains of the adventitious roots. 2. During mid-to-late winte[r] rapid growth occurs, making use of the mobilized food reserves. [3.] The flower (a) and leaves (b) emerge in early spring and the new bulbs rapidl[y] expand as food reserves are laid down. 4. Once flowering is completed th[e] aerial parts die back leaving below ground a new main bulb containin[g] (a) the new leaf and flow[er] bud and (b) a smaller daughter bulb.

BURROUGHS, William (1855–1898), US inventor of the first practical adding machine (1885). (See also CALCULATING MACHINE.)

BURSA, fibrous sac containing SYNOVIAL FLUID which reduces friction where TENDONS move over bones. Extra bursae may develop where there is abnormal pressure or friction.

BURSITIS, inflammation of a BURSA, commonly caused by excessive wear and tear (as in housemaid's knee) or by rheumatoid ARTHRITIS, GOUT or various bacteria. It causes pain and stiffness of the affected part, and may require CORTISONE injections and, if infected, surgical drainage.

BURTON, Robert (1577–1640), English clergyman, author of the *Anatomy of Melancholy* (1621), a compendious study of the causes and symptoms of melancholy, which, from the frankness and perception of its section "On Love Melancholy," has led to his being regarded as a precursor of FREUD.

BUSH, Vannevar (1890–1974), US electrical engineer, director of the Office of Scientific Research and Development in WWII. In the 1930s he developed a "differential analyzer"—in effect the first analog COMPUTER.

BUSHEL, a unit of dry measure equal to 64 pints. (See WEIGHTS AND MEASURES.)

BUSHMEN, a people of South Africa related to the pygmies, living around the Kalahari Desert. They average about 5ft in height and have yellowish-brown skin, broad noses and closely curled hair. They are nomadic hunters, living in bands of 25–60. Their language, related to Hottentot and belonging to the Khoisan group, employs a series of "clicks." Bushmen are a musical people, and are also noted for their vivid painting.

BUSHNELL, David (1742–1824), American inventor whose one-man SUBMARINE, *Turtle*, was used against British warships during the American Revolutionary War, but to little effect.

BUTADIENE, or 1,3-Butadiene ($CH_2{=}CHCH{=}CH_2$), gaseous HYDROCARBON made by dehydrogenation of BUTANE and butene. It is mostly polymerized with STYRENE to make synthetic rubber. mp −109°C, bp −4°C. (See also ALKENES; DIELS-ALDER REACTION.)

BUTANE (C_4H_{10}), gaseous ALKANE found in NATURAL GAS and also made by the cracking of PETROLEUM. It is used in BOTTLED GAS and in the manufacture of 1,3-BUTADIENE and high-octane GASOLINE. It has two isomers.

BUTENANDT, Adolf Friedrich Johann (1903–), German chemist who shared the 1939 Nobel Prize for Chemistry (with L. RUZICKA) for his work in isolating and determining the chemical structures of the human sex HORMONES estrone, androsterone and progesterone.

BUTTE, small, flat-topped hill formed when EROSION dissects a MESA.

BUTTER, a soft solid dairy product made by churning MILK or cream, containing fat, protein and water. Made in some countries from the milk of goats, sheep or yaks, it is made in the US from cows' milk only. Continuous mechanized production has been general since the 1940s. After skimming, the cream is ripened with a bacterial culture, pasteurized (see PASTEURIZATION), cooled to 4°C and then churned (see CHURNING), causing the butterfat to separate from the liquid residue, buttermilk. The butter is then washed, worked, colored and salted. World butter production in 1971 was 5.28 million tonnes, of which the US accounted for 10%.

BUYS-BALLOT'S LAW, in meteorology, states that when an observer stands with his back to the WIND in the N Hemisphere, there is high pressure to his right and low pressure to his left. In the S Hemisphere the reverse holds true. It was named for the Dutch metorologist, **Christoph Hendrik Didericus Buys Ballot** (1817–1890), who published it in 1875. (See also FERREL'S LAW.)

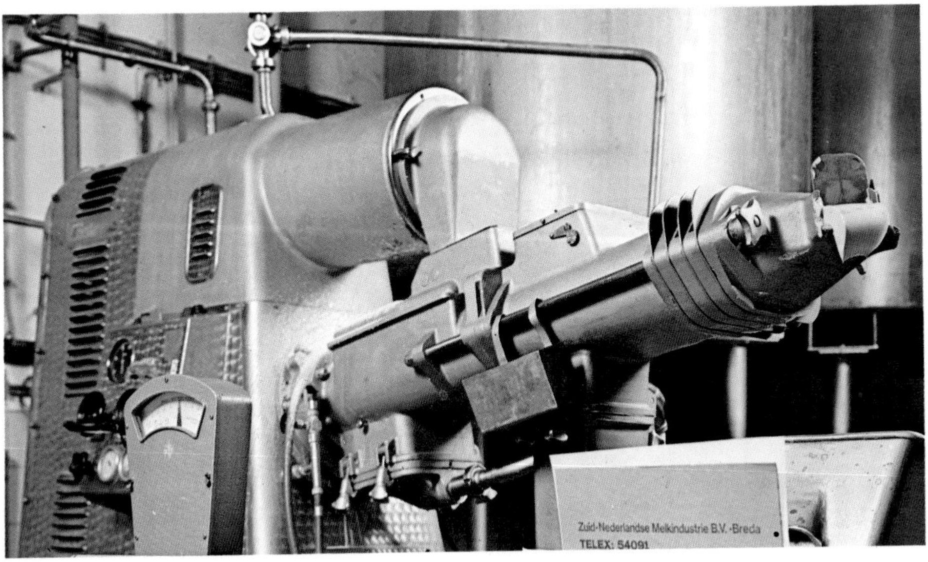

A butter cannon: a machine used in the continuous butter-making process.

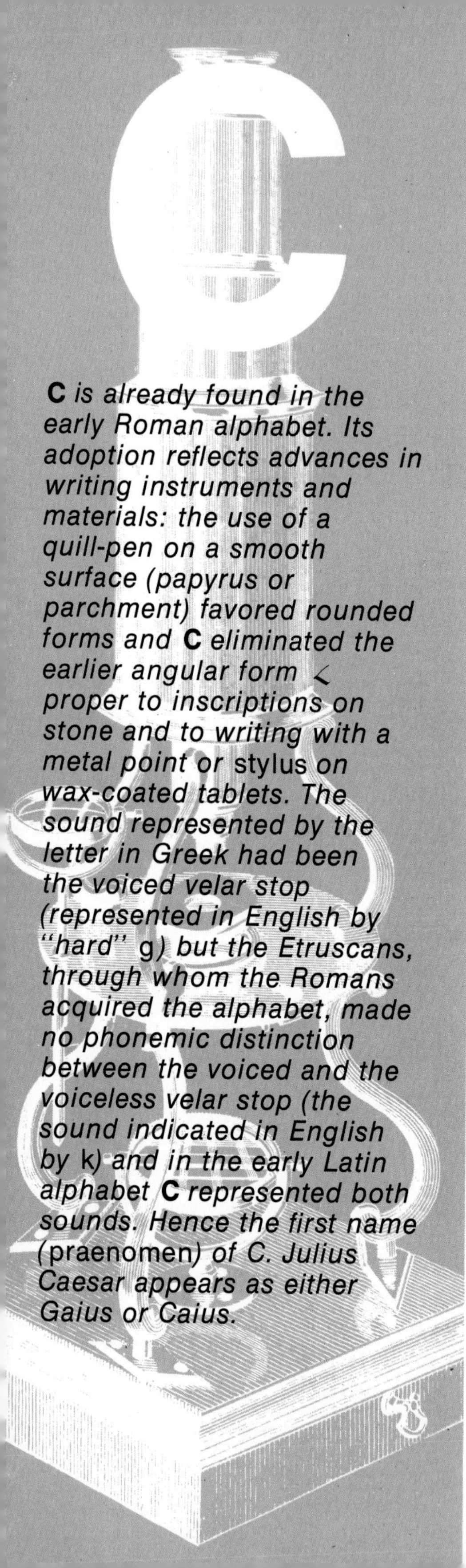

C *is already found in the early Roman alphabet. Its adoption reflects advances in writing instruments and materials: the use of a quill-pen on a smooth surface (papyrus or parchment) favored rounded forms and* **C** *eliminated the earlier angular form* < *proper to inscriptions on stone and to writing with a metal point or stylus on wax-coated tablets. The sound represented by the letter in Greek had been the voiced velar stop (represented in English by "hard" g) but the Etruscans, through whom the Romans acquired the alphabet, made no phonemic distinction between the voiced and the voiceless velar stop (the sound indicated in English by k) and in the early Latin alphabet* **C** *represented both sounds. Hence the first name (praenomen) of C. Julius Caesar appears as either Gaius or Caius.*

CABLE, Electric, insulated electric CONDUCTOR used to carry power or signals. In simplest form, a cable has a core of conducting metal (e.g., COPPER), usually several wires stranded together, surrounded by an insulating sheath of PLASTIC or RUBBER. Cable for VHF signals, as used in RADIO, is **coaxial cable**. Here a central core of wire is surrounded by insulation, then by a sheath of wire braid or an ALUMINUM or copper tube, then by a final layer of insulation. For electric power (see ELECTRICITY) simple cables are used in the home. Overhead cables, used to transmit power from power stations, are insulated from the pylons that support them, but between pylons there is generally no insulation around the core. Cables with a number of cores (multicore cables) are used for telephone cables and especially for the great submarine cables that connect countries as far apart as Australia and Canada. The first submarine cable, laid between Dover and Calais in 1850, had a single core insulated with GUTTA PERCHA.

CABLE CAR, passenger vehicle drawn by a moving cable driven by an exterior power source. The most famous, those of San Francisco, are driven by a continuous cable running beneath the street surface: from each car a grip descends through a slot in the roadway; it can be attached to or detached from the cable by the driver's controls. **Aerial tramways** are used in mountainous regions to span valleys and ascend mountains. On one cable the car is supported by means of PULLEY wheels, which run along the cable; a separate cable provides the motive power. **Ski lifts** are similar, though the car is usually supported by the moving cable alone. The cable-car principle is used by rail transportation on steep gradients in the form of a **funicular**; here an ascending car is usually counterbalanced by a descending car attached to the same cable.

CABLE TELEVISION, or **CATV** (*c*ommunity *a*ntenna *t*ele*v*ision), system used primarily in areas where mountains or tall buildings make TELEVISION reception poor or impossible. Subscribers' sets are connected by coaxial CABLE to a single ANTENNA erected in a suitably exposed position.

CADMIUM (Cd), soft, silver-white metal in Group IIB of the PERIODIC TABLE; an anomalous TRANSITION ELEMENT. It is found as greenockite (CdS) and in ZINC ores, from which it is extracted as a by-product. Cadmium is intermediate in chemical properties between ZINC and MERCURY, forming mainly Cd^{2+} compounds. Cadmium is used for electroplating to give corrosion-resistance; in storage BATTERIES; in low-melting ALLOYS; and as a moderator in nuclear reactors. Some compounds are used in pigments. AW 112.4, mp 321°C, bp 765°C, sg 8.65 (20°C).

CAESARIAN SECTION. See CESARIAN SECTION.

CAFFEINE, or trimethylxanthine ($C_8H_{10}N_4O_2$), an

caffeine

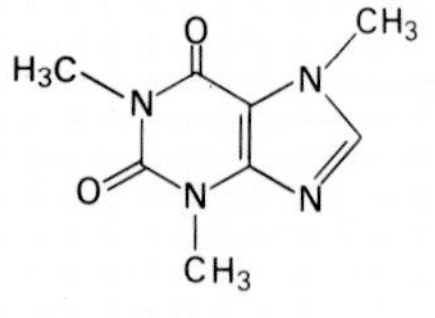

A group of BICC telecommunications cables. Groups of individual conductors are bundled together and the bundles collected within a protective jacket.

ALKALOID extracted from coffee, and also found in tea, cocoa and cola. Caffeine stimulates the central NERVOUS SYSTEM and HEART, and is a DIURETIC. It increases alertness, in excess causing insomnia (see SLEEP), and is mildly addictive. mp 238°C.

CAHOKIA MOUNDS, a group of prehistoric MOUNDS, mostly in the form of truncated pyramids, near East St. Louis, Ill. The largest of these, Monks Mound, is about 350m by 200m at base and some 30m high, and is the largest mound in the US. More than 300 of the mounds have in recent years been bulldozed to make way for agricultural and municipal expansion, but the 18 largest remain.

CAINOZOIC. See CENOZOIC.

CAIRN (from Gaelic *carn*, heap), a mound of stones, usually conical in form, used as a marker (as on a mountaintop), a memorial or to cover an ancient burial place. Found all over the world, cairns are sometimes regarded as a form of BARROWS.

CAISSON, boxlike structure used primarily in BRIDGE building. **Box caissons** are used where the existing bed is firm. They are open at the top, usually being built on land, floated into position, filled with CONCRETE and sunk. **Open caissons** have neither top nor bottom, their bottom edge being sharp. They are placed on the bed and excavation inside them proceeds. The caisson sinks, and as it does so its walls are built up, the increased weight assisting the sinking. When a firm bed is reached, the caisson is filled with concrete. **Pneumatic caissons** have no bottom, and are filled with air under PRESSURE to prevent the entry of water. Workmen may enter via an AIR LOCK and excavate from within the caisson, which on completion of their task is filled with concrete.

CAISSON DISEASE, or **bends.** See AEROEMBOLISM.

CALAMINE, former name of two ZINC minerals, SMITHSONITE and HEMIMORPHITE. In medicine, calamine lotion (zinc carbonate) is a mild ASTRINGENT used to treat ECZEMA and other itchy skin diseases.

CALCIMINE. See WHITEWASH.

CALCINATION, the process of heating materials in air to drive off moisture, carbon dioxide or other volatile compounds, and sometimes to oxidize them. Ores are often calcined after grinding, and plaster, cement and some pigments are made by calcination.

CALCITE, mineral form of calcium carbonate (see CALCIUM) of widespread occurrence; hexagonal symmetry. LIMESTONE, MARBLE and CHALK are types of calcite. Iceland spar, a very pure calcite, is transparent and exhibits DOUBLE REFRACTION of light, so being useful in optical instruments. Lime, CEMENT and fertilizers are manufactured from calcite.

CALCIUM (Ca), a fairly soft, silvery-white ALKALINE-EARTH METAL, the fifth most abundant element. It occurs naturally as CALCITE, GYPSUM and FLUORITE. The metal is prepared by ELECTROLYSIS of fused calcium chloride. Calcium is very reactive, reacting with water to give a surface layer of calcium hydroxide, and burning in air to give the nitride and oxide. Calcium metal is used as a reducing agent to prepare other metals, as a getter in vacuum tubes, and in alloys. AW 40.1, mp 839°C, bp 1484°C, sg 1.55 (20°C).

Calcium compounds are important constituents of animal skeletons: calcium phosphate forms the bones and teeth of vertebrates, and many seashells are made of the carbonate. **Calcium Carbonate** ($CaCO_3$), colorless crystalline solid, occurring naturally as CALCITE and aragonite, which loses carbon dioxide on heating above 900°C. It is an insoluble BASE. **Calcium Chloride** ($CaCl_2$), colorless crystalline solid, a by-product of the Solvay process. Being very deliquescent, it is used as an industrial drying agent. mp 782°C. **Calcium Fluoride** (CaF_2), or FLUORITE,

Double refraction displayed by a crystal of Iceland spar (calcite). Certain crystals can split a ray of unpolarized light into two rays plane-polarized at right-angles to each other. One is refracted in the normal way, the other with a refractive index that depends on the original direction of the ray.

colorless phosphorescent crystalline solid, used as windows in ultraviolet and infrared SPECTROSCOPY. mp 1423°C, bp c2500°C. **Calcium Hydroxide** ($Ca(OH)_2$), or **Slaked Lime**, colorless crystalline solid, slightly soluble in water, prepared by hydrating calcium oxide and used in industry and agriculture as an ALKALI, in mortar and in glass manufacture. **Calcium Oxide** (CaO), or **Quicklime**, white crystalline powder, made by calcination of calcium carbonate minerals, which reacts violently with water to give calcium hydroxide and is used in arc lights and as an industrial dehydrating agent. mp 2580°C, bp 2850°C **Calcium Sulfate** ($CaSO_4$), colorless crystalline solid, occurring naturally as GYPSUM and ANHYDRITE. When the dihydrate is heated to 128°C, it loses water, forming the hemihydrate, **plaster of paris**. This re-forms the dihydrate as a hard mass when mixed with water, and is used for casts.

CALCULATING MACHINE, device that performs simple ARITHMETIC operations (see also ALGEBRA). There are two main classes: **adding machines**, for ADDITION and SUBTRACTION only; and **calculators**, able also to perform MULTIPLICATION and DIVISION. They may be mechanical, electromechanical, or electronic.

The forerunner of the calculating machine was perhaps the ABACUS. The first adding machine, invented by PASCAL (1642), was able to add and carry. A few decades later (1671), LEIBNIZ designed a device that multiplied by repeated addition (the device was built in 1694). BABBAGE built a small adding machine (1822): in 1833 he conceived his Difference Engine, a predecessor of the digital COMPUTER, but his device was never completed. (See also SLIDE RULE.)

CALCULI, or stones, solid concretions of calcium salts or organic compounds, formed in the KIDNEY, BLADDER or GALL BLADDER. They are often associated with infection. There may be no symptoms but they may pass down or block tubes, causing COLIC, or obstruction in an organ. They may pass on unaided with antispasm drugs and ANALGESICS, but may require surgical removal to prevent damage to kidney or liver.

CALCULUS, the branch of MATHEMATICS dealing with continually varying quantities. It can be seen as an extension of ANALYTIC GEOMETRY, much of whose terminology it shares.

Differential Calculus. Consider the FUNCTION $y=f(x)$. This may be plotted as a CURVE on a set of CARTESIAN COORDINATES. Assuming that the curve is not a straight LINE, tangents (see TANGENT OF A CURVE) to it at different points will have different GRADIENTS. Two points close together on the curve, (x,y) and $(x+\delta x, y+\delta y)$, where δx and δy mean very small distances in the x- and y-directions, will usually have tangents of similar, though not identical, gradients. The gradient of the line passing through these two points is given by

$$\frac{(y+\delta y)-y}{(x+\delta x)-x},$$

and is the same as that of a tangent to the curve somewhere between the two points. The smaller δx is, the closer the two gradients will be; and if δx is infinitely small, they will be identical. This LIMIT as $x \to 0$ is the derivative

$$\frac{dy}{dx} \text{ or } f'(x)$$

of the function and is given by (putting $f(x)$ in place of y):

$$f'(x) = \lim_{\delta x \to 0} \frac{f(x+\delta x)-f(x)}{\delta x}$$

This formula will give the gradient of the curve for $f(x)$ at any value of x. For example, in the parabola (see CONIC SECTIONS) $y=x^2$ the derivative at $x=a$ is given by

(1) Graph of function $f(x)$, showing geometrical interpretation of a derivative. As δx tends to zero, AB and $f(x)$ coincide. (2) The construction of thin strips under the curve $y=f'(x)$. Each of the strips has an area of approximately $\delta x.f'(n\delta x)$, where n gives the position of the strip. As can be seen, the approximation is more accurate the smaller the value of δx. (3) Integration of a function between bounds $x=a$ and $x=b$.

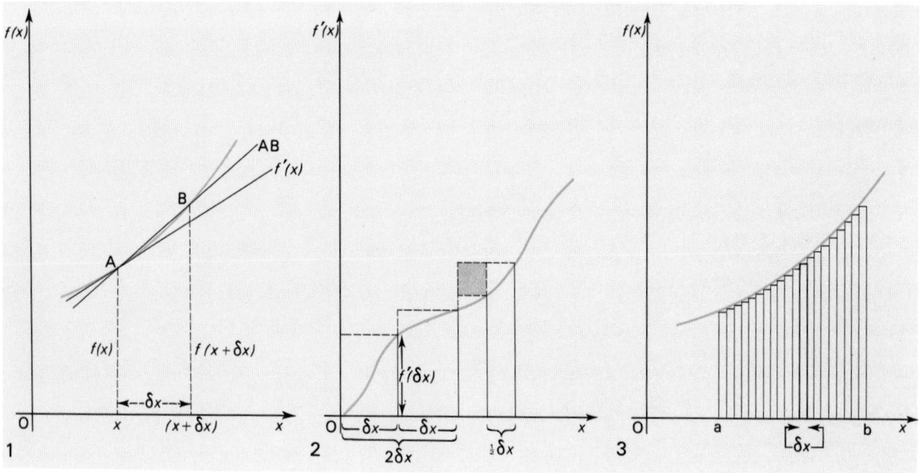

$$f'(a) = \lim_{x\to a}\frac{(x^2-a^2)}{(x-a)} = \lim_{x\to a}\frac{(x+a)(x-a)}{(x-a)}$$
$$= \lim_{x\to a}(x+a) = 2a.$$

Hence we can say that $f'(x)=2x$ for $f(x)=x^2$. This process is termed differentiation. In general if $f(x)=x^z$ then $f'(x)=z.x^{z-1}$, and the second derivative,

$$f''(x) \text{ or } \frac{d^2y}{dx^2},$$

the result of differentiating again, is $(z-1).z.x^{z-2}$ — and so on.

Integral Calculus. The derivative of $f(x)$ gives the instantaneous rate of change of $f(x)$ for a particular value of x. Now, consider the plot of $y=f'(x)$, and assume that $f'(x)=0$, $f(x)=0$ when $x=0$. A very thin strip with one vertical side the $f'(x)$-axis and the other the line joining the points $(\delta x, 0)$ and $(\delta x, f'(\delta x))$, will have an area of approximately $\delta x.f'(\delta x)$. A second thin strip drawn next to it will have an area of roughly $\delta x.f'(2\delta x)$, and so on. The area between the curve and the x-axis from $x=0$ to $x=a$ will therefore be roughly $\delta x.f'(\delta x)+\delta x.f'(2\delta x)+\ldots+\delta x.f'(a-\delta x)+\delta x.f'(a)$. If one plots on a different graph x against (area of strip 1), (area of strips 1+2), (area of strips 1+2+3), etc., one finds that one is plotting a close approximation to $y=f(x)$. This reverse of differentiation is called integration, and $f(x)$ is the integral of $f'(x)$. We find, too, that the integral of x^z is

$$\frac{x^{z+1}}{z+1},$$

which is what we would expect. This is the indefinite integral of x^z since we have not specified how much of the curve we wish to consider, and we must add a constant, c, since the derivative of any constant is 0. As integration is the sum of the areas of the strips described, we symbolize it by an extended S: $\int$. The definite integral of $y=x^z$ between $x=a$ and $x=b$ is therefore expressed as

$$\int_a^b x^z.dx = \int_0^b x^z.dx - \int_0^a x^z.dx.$$

Calculus is one of the most powerful mathematical tools, and is of fundamental importance in many branches of science. (See also EXTREMUM.)

CALCULUS OF VARIATIONS, an extension of CALCULUS concerned with the examination of definite integrals and the calculation of their maximum or minimum values. One of its most famous problems, proposed by BERNOULLI, was the brachistochrone problem, in which it was required to find out the least time taken for a particle to fall, under the influence of gravity alone, between two points at different heights though not vertically above each other. The path is in fact a CYCLOID, the solution involving minimization of the integral that expresses this path.

CALDERA, extremely large crater of volcanic (see VOLCANISM) origin, caused by repeated or massive explosion, collapse, or the amalgamation of a number of smaller craters.

CALENDAR, a system for reckoning the passing of time. The principal problem in drawing up calendars arises from the fact that the solar DAY, the lunar MONTH and the tropical YEAR—the most immediate natural time units—are not simple multiples of each other. In practice a solution is found in basing the system either on the phases of the moon (lunar calendar) or on the changing of the SEASONS (solar calendar). The difficulty that the days eventually get out of step with the moon or the seasons is got over by adding in (intercalating) one or more extra days or months at regular intervals in an extended cycle of months or years. The earliest Egyptian calendar had a year of 12 months with 30 days each, though later 5 extra days were added at the end of each year so that it approximated the tropical year of 365¼ days. In classical times, the Greeks came to use a lunar calendar in which three extra months were intercalated every eight years (the octennial cycle), though, about 432 BC, the astronomer Meton discovered that 235 lunar months fitted exactly into 19 years (the Metonic cycle), this becoming the basis of the modern Jewish and ecclesiastical calendars. The Roman calendar was reformed under Julius Caesar in 46 BC, fixing the year at 365 days but intercalating an additional day every fourth year (thus giving an average 365¼-day year). The 366-day year is known as a leap year. This Julian calendar continued in use until the 16th century when it had become about 10 days out of step with the seasons, the tropical year in fact being a little less than 365¼ days. In 1582, therefore, Pope Gregory XIII ordered that 10 days be omitted from that year. Furthermore, century years would no longer be leap years unless divisible by 400, so that there would be no recurrence of any discrepancy. This Gregorian calendar was only slowly adopted, particularly in non-Catholic countries—the reform waiting until 1752 in England and its American colonies, by which time 11 days had to be dropped. But today it is in civil use throughout the world. Various proposals for further reform have come to nothing.

Years are commonly numbered in Western societies from the birth of Christ—as computed by a 6th-century monk. Years since that epoch are labeled AD, years before, BC. There is no year 0, 1 AD following directly from 1 BC. Astronomers, on the other hand, figure years BC as negative numbers one less than the date BC and include a year 0 (=1 BC). The astronomers' year −10 is thus the same as 11 BC. (See also CHRONOLOGY.)

CALENDERING, process use in the manufacture of TEXTILES, RUBBER, some PLASTICS and especially high-quality PAPER. The substance concerned is passed between a series of pairs of heated rollers, which squeeze it to form a smooth or textured sheet.

CALIBER. See AMMUNITION.

CALIBRATION, a most important step in preparing a scientific instrument or experimental apparatus for use, in which it is provided with a numerical scale in accordance with an internationally agreed procedure or by comparison with a standard instrument or measure. Thus an ordinary thermometer may be calibrated simply by noting its reading first in freezing and then in boiling water (other conditions being specified), but for greater accuracy, complex calibration procedures must be used to within carefully-controlled tolerances.

CALIFORNIUM (Cf), a TRANSURANIUM ELEMENT in the ACTINIDE series. Numerous isotopes have been synthesized by various bombardment methods; most undergo spontaneous nuclear FISSION.

CALIPERS, devices used in measuring. Simple calipers comprise a pair of metal legs, pivoted about a shared screw at one end, the other ends being turned inward (for outer dimensions) or outward (for inner dimensions). They are usually used in conjunction with a rule. **Vernier calipers** resemble a sliding wrench and incorporate a VERNIER SCALE.

CALLUS, connective tissue initially formed around a FRACTURE, and slowly ossified as repair proceeds (see BONE).

CALLUS, or Callosity. See CORNS AND CALLUSES.

CALMETTE, Albert Léon Charles (1863–1933), French bacteriologist who with C. GUÉRIN discovered the BCG (bacillus Calmette-Guérin) vaccine, which has greatly reduced the incidence of TUBERCULOSIS.

CALMS, Regions of, areas where the sea is of mirror-like calmness and where the wind-speed is less than 1 knot. Such wind-speed is denoted 0 on the BEAUFORT SCALE. These conditions are characteristic of the HORSE LATITUDES.

CALOMEL, or **mercury (I) chloride.** See MERCURY.

CALORIC THEORY OF HEAT, the view, formalized by LAVOISIER toward the end of the 18th century, that heat consists of particles of a weightless, invisible fluid, caloric, which resides between the atoms of material substances. The theory fell from favor as physicists began to appreciate the equivalence of WORK and HEAT.

CALORIE, the name of various units of HEAT. The calorie or gram calorie (c or cal), originally defined as the quantity of heat required to raise 1g of water through 1C° at 1 atm pressure, is still widely used in chemical THERMODYNAMICS. The large calorie, kilogram calorie or kilocalorie (Cal or kcal), 1000 times as large, is the "calorie" of dietitians. The 15° calorie (defined in terms of the 1C° difference between 14.5°C and 15.5°C) is 4.184 joules; the International Steam Table calorie (cal_{IT}) of 1929, originally defined as 1/860 watt-hour, is now set equal to 4.1868J in SI UNITS.

CALORIMETRY, the measurement of the HEAT changes associated with chemical reactions and physical processes. This is done using various types of calorimeter. The water calorimeter is a thermally insulated metal cup of known thermal properties containing a known mass of water. When a reaction is carried out in the vessel, any heat liberated or absorbed is taken or given up by the water, the heat changes occurring being monitored by means of a thermometer dipped in the water. Heats of combustion are measured using a bomb calorimeter in which the reaction is carried out in an enclosed, pressurized chamber immersed in the water. Other types of apparatus used to measure SPECIFIC and LATENT HEATS include ice and steam calorimeters. (See also THERMOCHEMISTRY.)

CALOTYPE. See TALBOT, WILLIAM HENRY FOX.

CALVIN, Melvin (1911–   ), US biochemist who gained the 1961 Nobel Prize for Chemistry after having led the team that unraveled the details of the chemistry of PHOTOSYNTHESIS.

CALX, an old term for the product of calcining a metal or its ore (see CALCINATION). The term thus usually refers to the metal's oxide.

CALYX. See FLOWER.

CAM, mechanical device which, on rotation, imparts to another member (the follower) a regular, repetitive motion. There are two main types of cams. **Plate cams** (or **disk cams**) are curved, often ovoid, plates mounted on a shaft (the **camshaft**). On rotation, the cam pushes the follower in a direction PERPENDICULAR to the shaft. (The follower is returned to its position by gravity or a SPRING.) **Cylindrical cams** consist of parallel raised lips on the surface of the camshaft and angled such that, when a projection of the follower lies in the groove so formed, rotation of the camshaft imparts to it a motion (usually to and fro) parallel to the shaft. Many engines make use of one or more camshafts: notably, in some forms of INTERNAL-COMBUSTION ENGINE camshafts are used to regulate and actuate the cylinder valves.

CAMBIUM, a meristematic tissue that lies between the XYLEM and PHLOEM in the vascular tissue of plants. In woody plants, a complete ring of cambium develops which produces new xylem cells on the inside and phloem on the outside by a process known as secondary thickening. In some trees, under the epidermis a layer of cork cambium develops which then produces a corky bark. (See BARK; PLANT.)

CAMBRIAN, the earliest period of the PALEOZOIC (see GEOLOGY), dated roughly 570–500 million years ago, and immediately preceding the ORDOVICIAN. Cambrian rocks contain the oldest FOSSILS that can be used for dating (see PRECAMBRIAN).

CAMBRIDGE PLATONISTS, an influential group of philosophers centered on the U. of Cambridge in the mid-17th century, founded by Benjamin Whichcote and including Henry More and Ralph Cudworth. Their philosophy was Platonist, their outlook was tolerant and one of their chief aims was the reconciliation of faith with scientific knowledge and rational philosophy. Their influence long survived their eclipse, and largely inspired the religious and political toleration which was characteristic of Restoration England.

CAMERA, device for forming an optical image of a subject and recording it on a photographic film or plate or (in television cameras) on a photoelectric mosaic. The design of modern cameras derives from the ancient camera obscura, represented in recent times by the pinhole camera. This consists of a light-tight box with a small hole in one side and a ground-glass screen for the opposite wall. A faint image of the objects facing the hole is formed on the screen and this can be exposed on a photographic plate substituted for the screen.

Although the image produced in the pinhole camera is distortion-free and perfectly focused for objects at any distance, the sensitive materials used when photography was born in the 1830s required so long exposure times that the earliest experimentalists turned to the already available technology of the LENS as a means of allowing more light to strike the plate. From the start cameras were built with compound lenses to overcome the effects of chromatic aberration (see ABERRATION, OPTICAL) and the subsequent history of camera design has seen constant improvement in lens performance.

Today's simple camera consists of a light-tight box, a fixed achromatic lens, a simple shutter, a view finder and a film support and wind-on mechanism. The lens will focus all subjects more than a few feet distant and

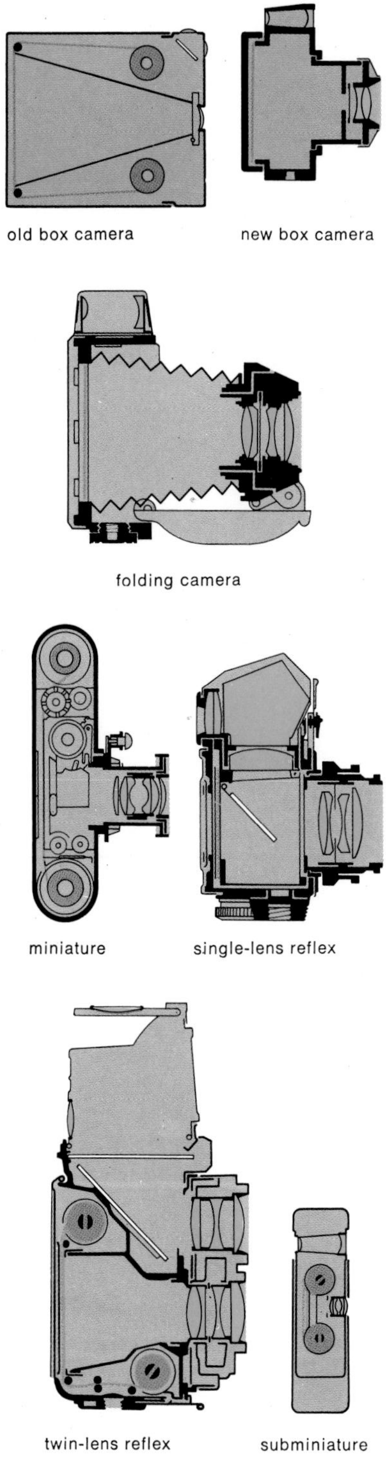

Some of the many kinds of still camera. Lens and viewfinder systems are shown in blue and film in brown. The box camera, in its old and new forms, has always been popular for its simplicity and cheapness, yielding excellent results in ideal conditions. The quest for smaller cameras led to the folding format, using collapsible bellows, but these have now been largely superseded by modern 'miniature' types, using 35mm film. The twin-lens reflex uses matched viewing and exposing lenses set close together to produce the same image in the viewfinder as will appear on the film; this is done even more efficiently by the single-lens reflex, in which a mirror (shown in white), which folds out of the way when the picture is taken, transmits the image through a prism to the viewfinder. Formerly the province of spies, the subminiature camera is now becoming increasingly popular with snapshotters.

the shutter (usually giving an exposure of $\frac{1}{30}$s or $\frac{1}{50}$s) admits sufficient light to expose negative materials on a sunny day. If exposures are to be made for reversal processing (see PHOTOGRAPHY) or of close-by or rapidly moving subjects or in poor light, a more complex camera is required. This may include a movable lens perhaps coupled to a RANGE FINDER (allowing the precise focusing of objects at different distances), a variable diaphragm (aperture) and shutter-speed mechanism (allowing adjustment to meet a wide range of light conditions) perhaps coupled to an exposure meter (LIGHT METER), a flash synchronization unit (allowing use of a flash gun—see FLASH BULB) or a facility for interchangeable lenses (allowing the photographer to alter the width of the camera's field of view). These refinements are realized in a wide range of different types of camera including the miniature camera and the single-lens reflex.

Special types of camera include the POLAROID LAND CAMERA (which produces prints almost instantaneously), stereo cameras (which take pairs of pictures of the same subject from slightly different angles—see PHOTOGRAMMETRY), motion-picture cameras (which make 16 or 24 successive exposures each second on long reels of film) and TELEVISION cameras (See also PHOTOGRAPHY.)

CAMERA LUCIDA AND CAMERA OBSCURA, simple optical devices which assist artists in drawing faithful reproductions of distant scenes, plans and diagrams or microscopic specimens. The principle of the **camera obscura** was known to ARISTOTLE; light admitted through a small hole into a darkened chamber projects a real image of the scene outside on the opposite wall. Later versions have used LENSES and MIRRORS to give an evenly-illuminated horizontal image. The **camera lucida**, invented in 1807 by WOLLASTON, employs a four-sided prism to allow the artist to see a virtual image of an object in the plane of the paper on which he copies the image. It is of particular use in enlarging or reducing artwork and in drawing from the MICROSCOPE.

CAMPHOR, a white crystalline compound distilled from the wood and young shoots of the camphor tree (*Cinnamomum camphora*). Camphor has a strong characteristic odor which repels insects. It is also used medicinally—internally as an anodyne and antispasmodic and externally in linaments. In large doses it is a narcotic poison.

camphor

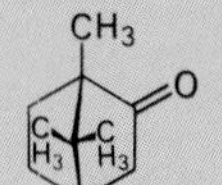

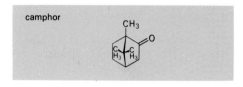

CAMSHAFT, a shaft on which a CAM (or cams) is mounted.

CANADA BALSAM, or Canada turpentine, a sticky exudate from the Balsam fir (*Abies balsamea*). It is a pale yellow oleoresin that has a refractive index similar to glass and is primarily used as a cement for glass in optics.

CANADIAN SHIELD, or **Laurentian Shield**, that area of North America (including the E half of Canada and small portions of the US) which has remained more or less stable since PRECAMBRIAN times. Its surface rocks, which are igneous and metamorphic (see IGNEOUS ROCK; METAMORPHIC ROCK), are amongst the oldest in the world, younger structures having disappeared through EROSION, in some areas by GLACIERS of the PLEISTOCENE.

CANCER (the Crab), a spring constellation in the N Hemisphere, the fourth sign of the ZODIAC. At the time the zodiacal system was adopted, Cancer marked the northernmost limit of the ECLIPTIC. A hazy object near the center of Cancer is a cluster of stars named Praesepe, the Beehive.

CANCER, a group of diseases in which some body cells change their nature, start to divide uncontrollably and may revert to an undifferentiated type. They form a malignant TUMOR which enlarges and may spread to adjacent tissues; in many cases cancer cells enter the BLOOD or LYMPH systems and are carried to distant parts of the body. There they form secondary "colonies" called **metastases**. Such advanced cancer is often rapidly fatal, causing gross emaciation. Cancer may present in very many ways—as a lump, some change in body function, bleeding, ANEMIA or weight loss—occasionally the first symptoms being from a metastasis. Less often tumors produce substances mimicking the action of HORMONES or producing remote effects such as NEURITIS.

Cancers are classified according to the type of tissue in which they originate. The commonest type, **carcinoma**, occurs in glandular tissue, SKIN, or visceral linings. **Sarcoma** occurs in connective tissue, MUSCLE, BONE and CARTILAGE. **Glioma** is a sarcoma of BRAIN neuroglia, unusual in that it does not spread elsewhere. **Lymphoma**, including HODGKIN'S DISEASE, is a tumor of the lymphatic system (see LYMPH); LEUKEMIA can be regarded as a cancer of white blood cells or their precursors. The cause of cancer remains unknown, but substantial evidence points to damage to or alteration in the DNA of CHROMOSOMES. Certain agents are known to predispose to cancer including RADIOACTIVITY, high doses of X RAYS and ULTRAVIOLET RADIATION and certain chemicals, known as **carcinogens**. These include tars, oils, dyes, ASBESTOS and tobacco smoke (see SMOKING). A number of cancers are suspected of being caused by a VIRUS and there appear to be hereditary factors in some cases.

Prevention of cancer is mainly by avoiding known causes, including smoking, excess radiation and industrial carcinogens. People suffering from conditions known to predispose to cancer need regular surveillance. Treatments include surgical excision, RADIATION THERAPY, CHEMOTHERAPY, or some combination of these. The latter two methods destroy cancer cells or slow their growth; the difficulty is to do so without also damaging normal tissue. They have greatly improved the outlook in lymphoma and certain types of leukemia. Treatment can be curative if carried out in the early stages, but if the cancer has metastasized, therapy is less likely to succeed; all that may be possible is the relief of symptoms. Thus, if cure is sought, early recognition is essential.

CANDELA (cd), the photometric base unit in SI UNITS. It is defined as the luminous intensity, in the perpendicular direction, of a surface of 1/600 000 sq m of a black body at the temperature of freezing platinum under a pressure of 101 325 newtons per sq m. (See BLACKBODY RADIATION; PHOTOMETRY.)

CANDLEPOWER, the ability of a light source to radiate as expressed in CANDELAS.

CANIS MAJOR (the Dog), a constellation of the S Hemisphere visible during winter in N skies. It contains SIRIUS, the brightest star in the night sky. Mythologically, Canis Major and CANIS MINOR were ORION'S hunting dogs.

CANIS MINOR (the Little Dog), a constellation on the celestial equator (see CELESTIAL SPHERE) visible in N skies during winter. It contains the binary star PROCYON (see DOUBLE STAR).

CANNABIS. See HEMP; MARIJUANA.

CANNIBALISM, or **anthropophagy,** consumption by humans of human flesh, common throughout the world at various times in the past and still occasionally practiced, though now generally TABOO. Among PRIMITIVE MAN the motive appears to be belief that eating an enemy or a respected elder transfers to the eater the strength, courage or wisdom of the dead.

CANNING, the process of preserving foods in sealed metal containers, developed by the French chef Nicolas Appert in 1809 and first patented in the US by Ezra Daggett in 1815. The fragile glass jars originally used were replaced by tin-coated iron cans after 1810. Today, a production line process is used. The food may reach the cannery a few hours after picking; it is first cleaned, and then prepared by removing inedible matter. After it has been peeled, sliced or diced as necessary, the food is blanched: hot water and steam are used to deactivate enzymes that might later spoil the flavor and color, and to shrink the product to the desired size and weight. The cans are then filled, and heated to drive out dissolved gases in the food and to expand the contents, thus creating a partial vacuum when they are cooled after sealing. Finally, the cans are sterilized, usually by steam under pressure. (See also FOOD PRESERVATION.)

CANNIZZARO, Stanislao (1826–1910), Italian chemist who discovered the Cannizzaro reaction in which BENZALDEHYDE is converted to BENZYL ALCOHOL and BENZOIC ACID (1853), but he is chiefly remembered for the republication of AVOGADRO'S hypothesis at the Karlsruhe conference of 1860. This at last allowed chemists to distinguish EQUIVALENT WEIGHTS from true ATOMIC WEIGHTS.

CANNON, term used loosely of early ARTILLERY,

though some modern devices are still described as cannon. Cannon were present at the Battle of Crécy (1346) but not until the 15th century did artillery power become of note. Early cannon were muzzle-loaded, GUNPOWDER being pushed down the barrel, followed by a pad of material (often cloth), then the stone or metal ball. Flame applied to a touchhole at the rear of the barrel set off the gunpowder. Important developments were increasing maneuverability; the invention of breech-loading; rifling (see RIFLE) of the barrel; and use of explosive AMMUNITION. (See EXPLOSIVES; GUN.)

CANOPIC JARS, covered vessels in which, in ancient Egypt, the embalmed viscera removed during mummification (see MUMMY) were placed for burial. After c1000 BC the embalmed viscera were more generally replaced in the mummy and these jars rarely used.

CANOPUS, Alpha Carinae, the brightest star in the S Hemisphere with an apparent magnitude of -0.72. It is 210 times larger than the sun and 30pc from earth.

CANTILEVER, beam or structural member supported at one end only. The simplest example is a diving board. (See also BRIDGE.)

CANTOR, Georg Ferdinand Ludwig Philip (1845–1918), German mathematician who pioneered the theory of infinite sets (see SET THEORY).

CANVAS, heavy woven cloth (see TEXTILES) usually made of cotton or linen fibers, used for centuries as sailcloth, and now also for tents, bags, shoes, hammocks, outdoor chairs, coverings, etc. It can be waterproofed (see WATERPROOFING). Artists' canvas is made of finer material specially treated to take paint.

CANYON, steep-sided VALLEY formed through EROSION by a river of hard rock lying in horizontal strata (see STRATIGRAPHY). The best-known example is the Grand Canyon, Ariz.

CAOUTCHOUC, or "pure rubber," a vegetable gum which is the main constituent of natural RUBBER.

CAPACITANCE, the ratio of the electric charge (see ELECTRICITY) on a conductor to its POTENTIAL, or, for a CAPACITOR, the ratio of its charge to the potential difference between its plates. Capacitance is measured in farads (F).

CAPACITOR, or **condenser,** an electrical component used to store electric charge (see ELECTRICITY) and to provide REACTANCE in alternating-current circuits. In essence, a capacitor consists of two conducting plates separated by a thin layer of insulator. When the plates are connected to the terminals of a BATTERY, a current flows until the capacitor is "charged," having one plate positive and the other negative. The ability of a capacitor to hold charge, its capacitance C, is the ratio of quantity of electricity on its plates, Q, to the potential difference between the plates, V. The electric energy stored in a capacitor is given by $\frac{1}{2}CV^2$. The capacitance of a capacitor depends on the area of its plates, their separation and the DIELECTRIC constant of the insulator. Small fixed capacitors are commonly made with metal-foil plates and paraffin-paper insulation; to save space the plates and paper are rolled up into a tight cylinder. Variable capacitors used in RADIO tuners consist of movable intermeshing metal vanes separated by an air gap. Electrolytic capacitors, which must be connected with the correct polarity, also find use in flash guns.

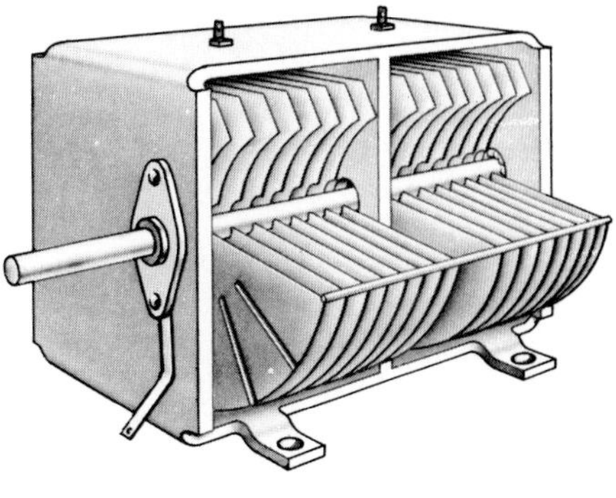

Variable capacitor as used in a radio tuner. The capacitance depends on the effective area of overlap between the moveable metal vanes.

CAPELLA, Alpha Aurigae, the fifth brightest star in the night sky. It is a DOUBLE STAR, 13.8pc distant, each component having apparent magnitude $+0.85$, with possibly two further dim components.

CAPILLARIES, minute BLOOD vessels concerned with supplying OXYGEN and nutrients to and removing waste products from the tissues. In the LUNGS capillaries pick up oxygen from the alveoli and release carbon dioxide. These processes occur by DIFFUSION. The capillaries are supplied with blood by ARTERIES and drained by VEINS.

CAPILLARITY, the name given to various SURFACE-TENSION phenomena in which the surface of a liquid confined in a narrow-bore tube rises above or is depressed below the level it would have if it were unconfined. When the attraction between the molecules of the liquid and those of the tube exceeds the combined effects of gravity and the attractive forces within the liquid, the liquid rises in the tube until EQUILIBRIUM is restored. Capillarity is of immense importance in nature, particularly in the transport of fluids in plants and through the soil.

CAPRICORNUS (the Sea Goat), a fairly inconspicuous constellation of the S Hemisphere, lacking any bright stars, and the tenth sign of the ZODIAC. Lying between AQUARIUS and SAGITTARIUS, Capricôrnus in ancient times lay at the southernmost limit of the ECLIPTIC.

CAPSULE, dry, dehiscent FRUIT, developed from two or more carpels and containing numerous seeds. Seeds are shed by a number of methods including pores at the top of capsule (in the poppy) or detachment of the apex (in the pimpernel).

CARAMEL ($[C_{12}H_{18}O_9]_x$), brown syrupy substance made by heating SUGAR to 180°C with a little water and sometimes sodium carbonate. It is used as a coloring and flavoring in foodstuffs, including a soft candy of the same name.

CARAPACE, the hard outer covering on many INVERTEBRATE and a few VERTEBRATE animals. It is made of bone or the substance **chitin**, which is similar to CELLULOSE. The carapace of insects and other invertebrates forms part of the external skeleton. In vertebrates, such as the turtle, it serves a protective function.

CARAT, a unit of MASS used for weighing precious stones. Since 1913 the internationally accepted carat has been the metric carat (CM) of 200mg. The purity of GOLD is also expressed in carats (usually spelled "karat" in the US). Here one karat is a 24th part; thus, pure gold is 24-karat; "18-karat gold" contains 75% gold and 25% other NOBLE METALS, and so on.

CARBIDES. See CARBON.

CARBOHYDRATES, a large and important class of ALIPHATIC COMPOUNDS, widespread and abundant in nature, where they serve as an immediate energy source; cellulose is the chief structural material for plants. Most carbohydrates have chemical formulas $(CH_2O)_n$, and so were named as hydrates of carbon—which, however, they are not. Systematic names of carbohydrates end in -ose. They are generally divided into four groups, the simplest being the **monosaccharides** or simple SUGARS and the **disaccharides** or double sugars. The **oligosaccharides** (uncommon in nature) consist of three to six monosaccharide molecules linked together. The **polysaccharides** are POLYMERS, usually homogeneous, of monosaccharide units, into which they are broken down again when used for energy. The main plant polysaccharides are CELLULOSE and STARCH; in animals a compound resembling starch, GLYCOGEN, is formed in the muscles and liver. Other polysaccharides include AGAR, ALGIN, CHITIN, DEXTRIN, GUM acacia, INSULIN and PECTIN. Carbohydrates play an important role in food chains (see ECOLOGY): they are formed in plants by PHOTOSYNTHESIS, and are converted by ruminant animals into PROTEIN. They also form one of the major classes of human FOOD (see also NUTRITION). In Europe and the US they provide a third to a half of the calories in the diet, of which starch and the various sugars supply about half each. In less developed countries carbohydrates, especially starch, are even more important.

CARBOLIC ACID, old name for PHENOL, particularly when used as an ANTISEPTIC. The hydroxyl hydrogen atom of phenol is relatively acidic.

CARBOLOY, a common name for the ultrahard BEARING and cutting-tool material, TUNGSTEN carbide (WC).

CARBON (C), nonmetal in Group IVA of the PERIODIC TABLE. It is unique among elements in that a whole branch of chemistry (ORGANIC CHEMISTRY) is devoted to it, because of the vast number of compounds it forms. The simple carbon compounds described below are usually regarded as inorganic.

Carbon occurs in nature both uncombined (COAL) and as CARBONATES, carbon dioxide in the atmosphere, and PETROLEUM. It exhibits ALLOTROPY, occurring in three contrasting forms: DIAMOND, GRAPHITE and "white" carbon, a transparent allotrope discovered in 1969 by subliming graphite. So-called amorphous carbon is actually microcrystalline graphite; it occurs naturally, and is found as COKE, CHARCOAL and **carbon black** (obtained from the incomplete burning of petroleum, and used in pigments and printer's ink, and to reinforce rubber). Amorphous carbon is widely used for ADSORPTION, because of its large surface area. A new synthetic form is carbon fiber, which is very strong and is used to reinforce plastics and to make electrically-conducting fabrics.

Carbon has several ISOTOPES: C^{12} (used as a standard for ATOMIC WEIGHTS) is much the most common, but C^{13} makes up 1.11% of natural carbon. C^{10}, C^{11}, C^{14}, C^{15} and C^{16} are all radioactive. C^{14} has the relatively long half-life of 5730yr, and is continuously formed in the atmosphere by COSMIC RAY bombardment; it is used in RADIOCARBON DATING.

The element (especially as diamond) is rather inert, but all forms will burn in air at a high temperature to give carbon monoxide in a poor supply of oxygen, and carbon dioxide in excess oxygen. Fluorine will attack carbon at room temperature to give carbon tetrafluoride, and strong oxidizing agents will attack graphite. Carbon will combine with many metals at high temperatures, forming carbides. Carbon shows a covalency of four, the bonds pointing toward the vertices of a tetrahedron, unless multiple bonding occurs. AW 12.011.

Carbides, binary compounds of carbon with a

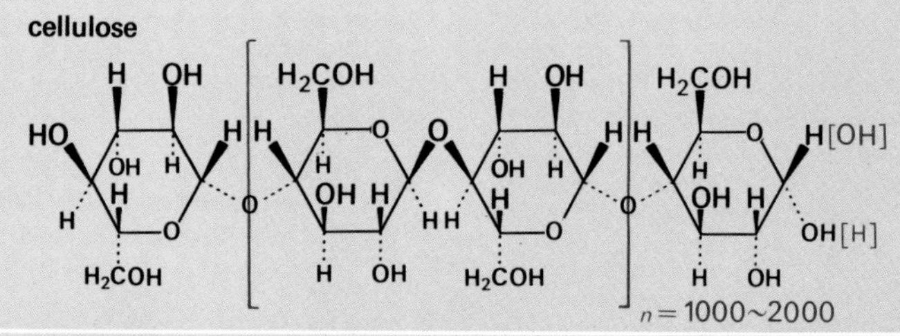

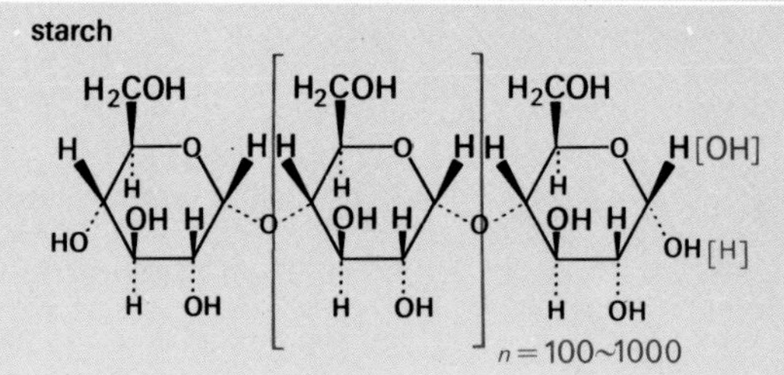

metal, prepared by heating the metal or its oxide with carbon. Ionic carbides are mainly acetylides (C_2^{2-}) which react with water to give ACETYLENE, or methanides (C^{4-}) which give METHANE. There are also metallic interstitial carbides, and the covalent boron carbide (B_4C) and silicon carbide (see CARBORUNDUM). **Carbon Dioxide** (CO_2), colorless, odorless gas. It is nontoxic, but can cause suffocation. The air contains 0.03% carbon dioxide, which is exhaled by animals and absorbed by plants (see RESPIRATION; PHOTOSYNTHESIS; CARBON CYCLE). Carbon dioxide is prepared in the laboratory by reacting a CARBONATE with acid; industrially it is obtained by calcining LIMESTONE, burning coke in excess air, or from FERMENTATION. At atmospheric pressure, it solidifies at −78.5°C to form "dry ice" (used for refrigeration and CLOUD seeding) which sublimes above that temperature; liquid carbon dioxide, formed under pressure, is used in fire extinguishers. Carbon dioxide is also used to make carbonated drinks. When dissolved in water an equilibrium is set up, with CARBONATE, BICARBONATE and HYDROGEN ions formed, and a low concentration of **Carbonic Acid** (H_2CO_3). **Carbon Disulfide** (CS_2), colorless liquid, of nauseous odor due to impurities; highly toxic and flammable. Used as a solvent and in the manufacture of rayon and CELLOPHANE. mp −111°C, bp 46°C, sg 1.261 (22°C). **Carbon Monoxide** (CO), colorless, odorless gas. It is produced by burning carbon or organic compounds in a restricted supply of oxygen, for example, in poorly-ventilated stoves, or the incomplete combustion of gasoline in AUTOMOBILE engines. It is manufactured as a component of WATER GAS. It reacts with the halogens and sulfur, and with many metals, to give carbonyls. Carbon monoxide is an excellent reducing agent at high temperatures, and is used for smelting metal ores (see BLAST FURNACE; IRON). It is also used for the manufacture of METHANOL and other organic compounds. It is a component of manufactured gas, but not of natural gas. Carbon monoxide is toxic because it combines with hemoglobin, the red BLOOD pigment, to form pink carboxyhemoglobin, which is stable, and will not perform the function of transporting oxygen to the tissues. mp −199°C, bp −191°C. **Carbon Tetrachloride** (CCl_4), colorless liquid, nonflammable but toxic, made by chlorinating carbon disulfide. Used as a fire extinguisher, a solvent (especially for dry-cleaning) and in the manufacture of FREON. mp −23°C, bp 77°C. (See also CYANIDES; CYANOGEN.)

CARBONATES, salts of carbonic acid (see CARBON), containing the CO_3^{2-} ion. A solution of carbon dioxide in water reacts with a base to form a carbonate. Carbonates are decomposed by heating to give the metal oxide and carbon dioxide. They also react with acids to give carbon dioxide. Many minerals are carbonates, the most important being CALCITE, DOLOMITE and MAGNESITE. (For sodium carbonate see SODIUM).

CARBON BLACK. See CARBON.

CARBON CYCLE, in biology, a very important cycle by which carbon, obtained from the atmosphere as carbon dioxide, is absorbed by green plants, synthesized into organic compounds and then returned to the atmosphere as carbon dioxide. The organic compounds, particularly CARBOHYDRATES, are synthesized in plants from carbon dioxide and water in the presence of CHLOROPHYLL and light by a process known as PHOTOSYNTHESIS. The carbohydrates are then broken down to carbon dioxide and water either by the plant during RESPIRATION or after death by putrefying BACTERIA and FUNGI. (See also PLANT.)

CARBON CYCLE, or carbon-nitrogen cycle, in physics, the chain of nuclear FUSION reactions, catalyzed by CARBON nuclei (see CATALYSIS), which is the main source of ENERGY in the hotter STARS, though of minor importance in the SUN and cooler stars where proton–proton fusion is the chief reaction. Overall, four PROTONS are converted to one ALPHA PARTICLE, with destruction of some matter and consequent evolution of energy (see RELATIVITY). Most of the gamma-radiation produced is absorbed within the star, and energy is released as heat and light. The carbon cycle was first described by Hans BETHE.

$$
\begin{aligned}
{}_6C^{12} + {}_1H^1 &\rightarrow {}_7N^{13} + \gamma \\
{}_7N^{13} &\rightarrow {}_6C^{13} + e^+ + \nu \\
{}_6C^{13} + {}_1H^1 &\rightarrow {}_7N^{14} + \gamma \\
{}_7N^{14} + {}_1H^1 &\rightarrow {}_8O^{15} + \gamma \\
{}_8O^{15} &\rightarrow {}_7N^{15} + e^+ + \nu \\
{}_7N^{15} + {}_1H^1 &\rightarrow {}_2He^4 + {}_6C^{12}
\end{aligned}
$$

The sequence of reactions in the carbon cycle.

CARBONIFEROUS, collective term used mainly in Europe for the MISSISSIPPIAN and PENNSYLVANIAN.

CARBORANES, compounds derived from polyhedral BORANES by substitution of CARBON for BORON atoms. They are relatively unreactive and thermally stable, and form many derivatives, including highly stable organo-inorganic polymers.

CARBORUNDUM, or silicon carbide (SiC), black, cubic crystalline solid, made by heating COKE with SILICA in an electric furnace. It is almost as hard as DIAMOND (whose structure it resembles), and hence it is used as an abrasive. It is inert, refractory and a good heat conductor, so is used in making high-temperature bricks; at high temperatures it is a SEMICONDUCTOR. subl 2700°C (with decomposition).

CARBOXYLIC ACIDS, a major class of organic compounds, of general formula RCOOH where R is an organic group; those acids where R is a straight-chain alkyl group (see ALKANES) are sometimes known as **fatty acids**, and are named systematically as alkanoic acids from the corresponding alkane. Some carboxylic acids occur free in nature, including FORMIC and ACETIC acids—the two simplest—and CITRIC, LACTIC, MALIC and TARTARIC acids. These, and others including BENZOIC, OXALIC and SALICYLIC acids, are found also as their salts and ESTERS. Many of the fatty acids, including OLEIC, PALMITIC and STEARIC acids, occur in oils and FATS as esters of GLYCEROL (see also SOAPS AND DETERGENTS). Carboxylic acids are made by HYDROLYSIS of esters, ACID ANHYDRIDES or ACID CHLORIDES, or by oxidation of ALDEHYDES or primary ALCOHOLS. They are weak ACIDS, the exact strength depending on the ELECTRONEGATIVITY of the group R. and so are often used with their salts as BUFFERS. Many derivatives of carboxylic acids are important in nature or chemical synthesis: they include acid anhydrides, acid chlorides, AMIDES,

Carboxylic acids

methanoic acid (formic acid) MW 46.03, mp 9 °C, bp 101 °C	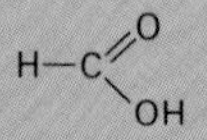
ethanoic acid (acetic acid) MW 60.05, mp 17 °C, bp 118 °C	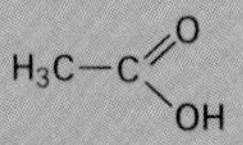
ethanedioic acid (oxalic acid) MW 90.04, subl 157 °C	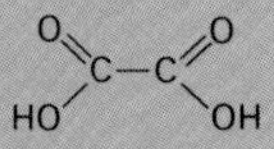
propanoic acid (propionic acid) MW 74.08, mp −21 °C, bp 142 °C	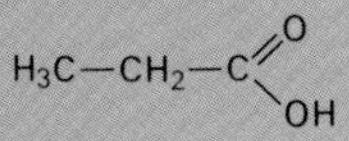
2-hydroxypropanoic acid (lactic acid) MW 90.08, mp (*dl*-) 18 °C, (*d*-, *l*-) 53 °C	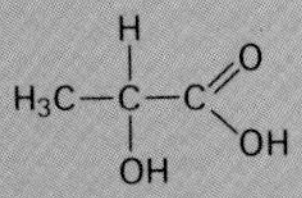
2-oxopropanoic acid (pyruvic acid) MW 88.06, mp 14 °C, bp 165 °C	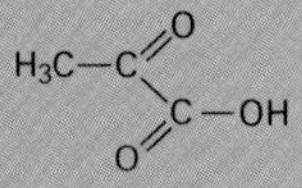
propanedioic acid (malonic acid) MW 104.06, mp 136 °C	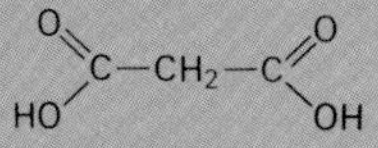
propenoic acid (acrylic acid) MW 72.06, mp 13 °C, bp 142 °C	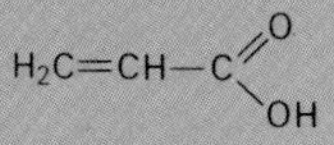
butanoic acid (butyric acid) MW 88.12, mp −4 °C, bp 164 °C	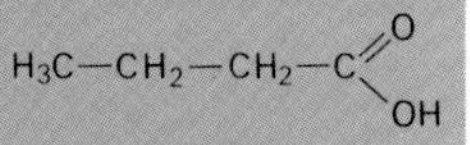
3-oxobutanoic acid (acetoacetic acid) MW 102.09	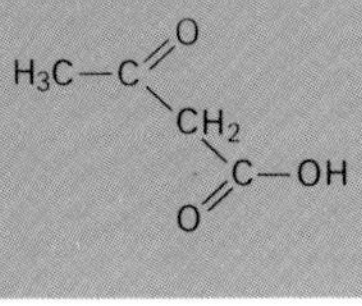
butanedioic acid (succinic acid) MW 118.09, mp 188 °C	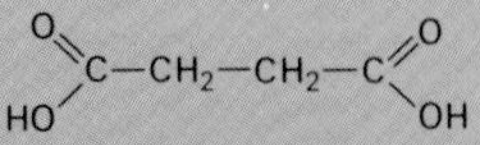
2-hydroxybutanedioic acid (malic acid) MW 134.09, mp (*dl*-) 133 °C, (*l*-) 100 °C	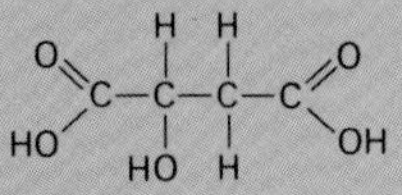
2,3-dihydroxybutanedioic acid (tartaric acid) MW 150.09, mp (*dl*-) 206 °C, (*d*-, *l*-) 170 °C (*meso*-) 140 °C	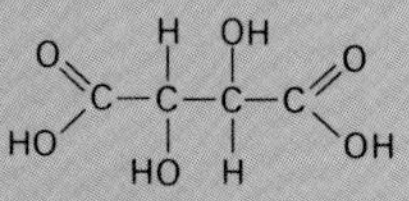
trans-butenedioic acid (fumaric acid) MW 116.07, mp 300 °C	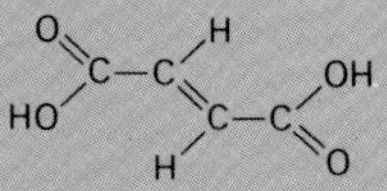
cis-butenedioic acid (maleic acid) MW 116.07, mp 140 °C	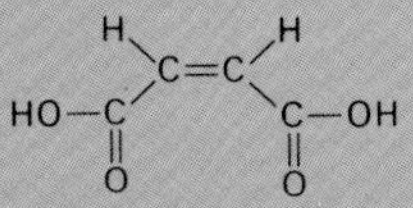
hexadecanoic acid (palmitic acid) MW 256.43, mp 63 °C, bp 390 °C	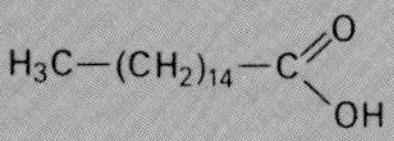
octadecanoic acid (stearic acid) MW 284.50, mp 72 °C	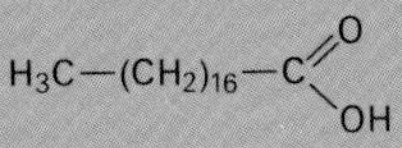
benzoic acid MW 122.13, mp 122 °C, bp 249 °C	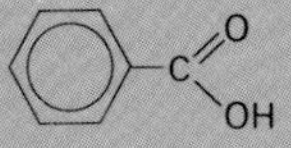
2-hydroxybenzoic acid (salicylic acid) MW 138.12, mp 159 °C	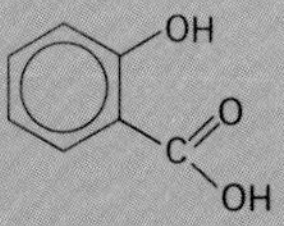
3,4,5-trihydroxybenzoic acid (gallic acid) MW 170.17	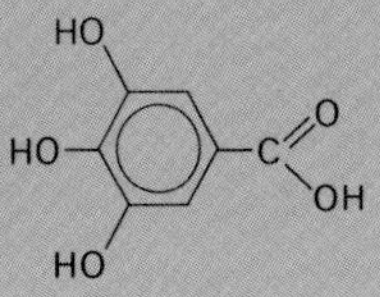

esters, NITRILES and peroxyacids (see PEROXIDES). (See also AMINO ACIDS.)

CARBUNCLE. See BOIL.

CARBURETOR, an important element in most automobile engines, the carburetor mixes air and GASOLINE in the correct ratio for most efficient combustion (usually about 15:1, air:gasoline, by weight). Most simply, a carburetor has a tube constricted at one point into a narrow throat, or VENTURI. The speed of air flowing through the venturi increases and hence its pressure decreases: fuel from a reservoir (the float chamber) is therefore sucked in through a hole, or jet, at this point. The fuel mixture then passes through a throttle valve, which controls the rate at which the mixture enters the engine and hence the engine speed. A CHOKE in the air-intake regulates the air supply and thus the richness of the mixture. In practice, carburetors incorporate various means of ensuring constancy of mixture strength during running. High-performance engines may use more than one carburetor or a FUEL INJECTION system. (See also INTERNAL-COMBUSTION ENGINE.)

CARCINOGENS. See CANCER.

CARCINOMA. See CANCER.

CARDANO, Girolamo (1501–1576), Italian physician and astrologer, chiefly remembered for his contributions to mathematics. In particular, he developed and published a general solution for cubic equations.

CARDIAC, relating to the HEART, as in cardiac surgery; also to the cardia or upper part of STOMACH.

CARDINAL NUMBER, one that describes the number of elements in a SET. For example, in the phrase "my 3 oranges," 3 is a cardinal number. (See also ORDINAL NUMBER; TRANSFINITE CARDINAL NUMBER.)

CARDING, process used in TEXTILE manufacture whereby the fibers are laid out parallel to each other, then gathered into strands (card slivers) about 1in (25mm) thick. This process may be followed by **combing**, where the fibers in the sliver below a certain length are removed, those remaining being set more accurately parallel.

CARIES, decay or softening of hard tissues, usually TEETH, but also used for BONE, especially spinal TUBERCULOSIS. Dental caries is bacterial decay of dentine and enamel (see TEETH), hastened by sugary diet and poor oral hygiene. Fluoride in small quantities protects against caries.

CARNAC, French town near which are extensive alignments of megaliths (see MEGALITHIC MONUMENTS), a DOLMEN and two BARROWS. The monuments are believed to be of late STONE AGE origin.

CARNAP, Rudolf (1891–1970), German-US logician and philosopher of science, a leading figure in the Vienna Circle (see LOGICAL POSITIVISM), who later turned to study problems of linguistic philosophy and the role of probability in inductive reasoning.

CARNELIAN, or **cornelian,** translucent variety of CHALCEDONY, colored red by colloidal HEMATITE; a semiprecious GEM stone widely used in classical times for intaglio signets. (See also SARD.)

CARNIVORA, an order of the MAMMALIA that includes the seals, raccoons, bears, wolves, badgers, hyaenas, lions, tigers and the pandas. All, except the pandas, are meat eaters and have large canine teeth, pointed cheek teeth and claws on all the toes. Most hunt their prey, sometimes in organized packs.

CARNIVORE, any animal that feeds exclusively on other animals. Carnivores are essentially of two types: trappers, that lie in wait for prey, and hunters that actively hunt their prey.

Carnelian: as the raw mineral, and cut and polished as a gemstone.

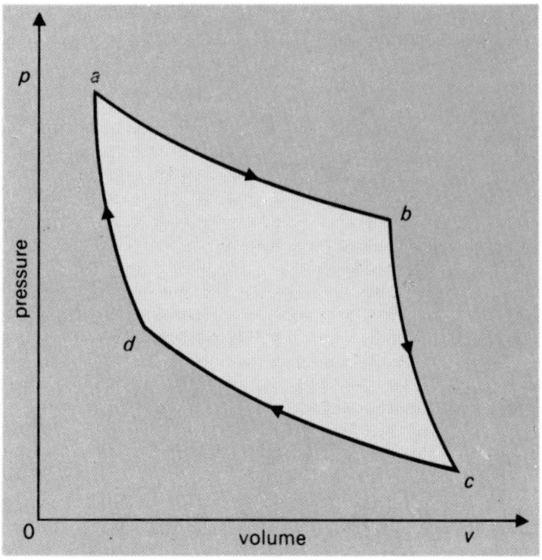

Carnot cycle pressure-volume diagram for an ideal working fluid:
(1) Isothermal expansion: *a*→*b*. Heat energy is absorbed from a high-temperature source at constant temperature. The volume increases while the pressure drops somewhat.
(2) Adiabatic expansion: *b*→*c*. While the working fluid is thermally isolated (so that there is no heat transfer between it and its surroundings), the volume further increases while the temperature drops.
(3) Isothermal compression: *c*→*d*. Heat energy is given up to a cool heat "sink" during compression at constant temperature.
(4) Adiabatic compression: *d*→*a*. The fluid is compressed to its original state without transfer of heat to or from its surroundings.
The net result is that, of the heat abstracted from the hot source, some, represented by the area *abcd*, is converted into mechanical (or electrical) work, the rest being rejected to the sink.

CARNIVOROUS PLANT. See INSECTIVOROUS PLANTS.

CARNOT, Nicholas Léonard Sadi (1796–1832), French physicist who, seeking to improve the EFFICIENCY of the STEAM ENGINE, devised the **Carnot cycle** (1824) on the basis of which Lord KELVIN and R. J. E. CLAUSIUS formulated the second law of THERMODYNAMICS. The Carnot cycle, which postulates a heat engine working at maximum thermal efficiency, demonstrates that the efficiency of such an engine does not depend on its mode of operation but only the TEMPERATURES at which it accepts and discards heat ENERGY.

CARNOTITE, soft, yellow radioactive mineral, potassium uranyl vanadate ($K_2[UO_2]_2[VO_4]_2 . nH_2O$), a major ore of URANIUM and VANADIUM. It is found in Col., Wyo., S.D., Pa., Siberia, Zaire and Australia.

CAROTENOIDS, group of yellow, orange, red and brown pigments found in almost all animals and plants, and responsible for the color of carrots, lobsters, and many flowers and fruits. In leaf CHLOROPLASTS the carotenoid colors are masked by CHLOROPHYLL until this is lost in the fall. The color is due to a long conjugated double-bond system (see

ALKENES; RESONANCE) formed by condensation of ISOPRENE units. There are two main types of carotenoids, the oxygen-containing xanthophylls and the hydrocarbon carotenes. VITAMIN A is formed from certain carotenes.

CARPEL. See FLOWER.

CARREL, Alexis (1873–1944), French surgeon who won the 1912 Nobel Prize for Medicine or Physiology for developing a technique for suturing (sewing together) blood vessels, thus paving the way for organ TRANSPLANTS and blood TRANSFUSION.

CARRIER, person who carries the agents responsible for infectious DISEASES, and is able to infect others, while often remaining quite well.

CARRIER, Willis Haviland (1876–1950), US industrialist and mechanical engineer, pioneer designer of AIR-CONDITIONING equipment. He invented an automatic humidity-control device first used in a New York printing plant in 1902—arguably the first commercial air-conditioning installation.

CARTER, Howard (1873–1938), English Egyptologist, famous for the Valley of the Kings excavations with Lord Carnarvon that led to the discovery of the tomb of Tutankhamen in 1922. Carter spent ten years in careful excavation and exploration of the tomb.

CARTESIAN COORDINATES, the most common system of rectangular coordinates employed in ANALYTIC GEOMETRY. The term is used after DESCARTES.

CARTESIAN PHILOSOPHY. See DESCARTES, RENÉ.

CARTILAGE, tough, flexible connective tissue found in all vertebrates, consisting of cartilage cells (chondrocytes) in a matrix of COLLAGEN fibers and a firm protein gel. The skeleton of the vertebrate embryo is formed wholly of cartilage, but in most species much of this is replaced by BONE during growth. There are three main types of cartilage: hyaline, translucent and glossy, found in the joints, nose, trachea and bronchi; elastic, found in the external ear, Eustachian tube and larynx; and fibrocartilage, which attaches tendons to bone and forms the disks between the vertebrae.

CARTOGRAPHY. See MAP.

CARTRIDGE, a case of metal or paper (sometimes of cloth) containing the charge for a FIREARM (see also AMMUNITION). For small arms, the term generally embraces also the bullet or shot, a blank cartridge being one without bullet or shot. In larger GUNS, cartridge and projectile are usually loaded separately.

CARTWRIGHT, Edmund (1743–1823), British inventor of a mechanical loom (c1787) that was the ancestor of the modern power loom. He also invented a wool-combing machine (c1790). (See also WEAVING.)

CASEHARDENING, in METALLURGY, any process applied to the surface of mild STEEL to increase its wear resistance and surface HARDNESS. Most commonly, the steel is packed around with powdered CHARCOAL and heated for a certain period, so that the surface regions absorb CARBON. The steel is then quenched (rapidly cooled) with water. Other techniques include INDUCTION HARDENING and FLAME HARDENING; and **nitriding**, used for certain suitable steels, where the steel is heated in AMMONIA gas or molten CYANIDE salts so that NITROGEN is diffused into its surface regions.

CASEIN, the chief milk PROTEIN, found there as its calcium salt. It is precipitated from skim MILK with acid or RENNET, washed and dried. Highly nutritious, it is used in the food industry. In alkaline solution casein forms a COLLOID used as a glue, a binder for paint pigments and paper coatings, and to dress leather. Casein is also used to make PLASTICS.

CASHMERE, very fine natural fiber, the soft underhair of the Kashmir goat, bred in India, Iran, China and Mongolia. Cashmere is finer than the best wools, although the name may be applied to some soft wool fabrics.

CASSETTE. See TAPE RECORDER.

CASSINI, Giovanni Domenico, or Cassini, Jean Dominique (1625–1712), Italian–French astronomer who discovered four satellites of SATURN, the Cassini division in the ring of Saturn and estimated the scale of the solar system from the PARALLAX of MARS (1672).

CASSINI OVALS, a plane CURVE produced as follows: Consider two points A and B in the same plane, the distance AB equalling $2x$, $x>0$. Consider then a point P moving such that $PA.PB=y^2$, where y is constant and $y^2>x^2$. The figure traced out by the motion of P is a Cassini Oval. If $y^2<x^2$, the figure is in two parts. In the case where $y=x$, the figure traced out, which looks like a figure eight, is called the **Lemniscate of Bernoulli**.

CASSIOPEIA, in Greek mythology, the mother of Andromeda. In astronomy, a northern circumpolar constellation (see CIRCUMPOLAR STARS) whose five principal stars form a prominent "W."

CASSITERITE, mineral consisting of stannic oxide (SnO_2) with iron impurity; the chief ore of TIN. It is usually brown to black, with submetallic luster; it forms prismatic crystals (tetragonal), but is usually massive granular. sg 6.8–7.1.

CASTING, the production of objects of a desired form by pouring the raw material (e.g., ALLOYS; FIBERGLASS; PLASTICS; STEEL) in liquid form into a suitably shaped mold. Both the mold and the pattern from which it is made may be either permanent or expendable. Permanent-mold techniques include **die casting**, where the molten material is forced under pressure into a DIE; **centrifugal casting**, used primarily for pipes, the molten material being poured into a rapidly rotating mold (see CENTRIFUGE); and **continuous casting**, for bars and slabs, where the material is poured into water-cooled, open-ended molds. Most important of the expendable-mold processes is **sand casting (founding)**: here fine SAND is packed tightly around each half of a permanent pattern, which is removed and the two halves of the mold placed together. The material is poured in through a channel (**sprue**); after setting, the sand is dispersed. In some processes, the mold is baked before use to remove excess water. (See also CAST IRON; METALLURGY.)

CAST IRON, iron ALLOYS containing 1.8–4.5% carbon, used for CASTING; made from PIG IRON in a cupola furnace by melting and purifying it and adding other components. Its properties depend largely on the composition and the ANNEALING process used.

CASTOR, Alpha Geminorum, second brightest star of GEMINI. About 14.4pc from earth, it has at least six components: a binary, each component of which is itself a binary, is orbited at distance by a third binary

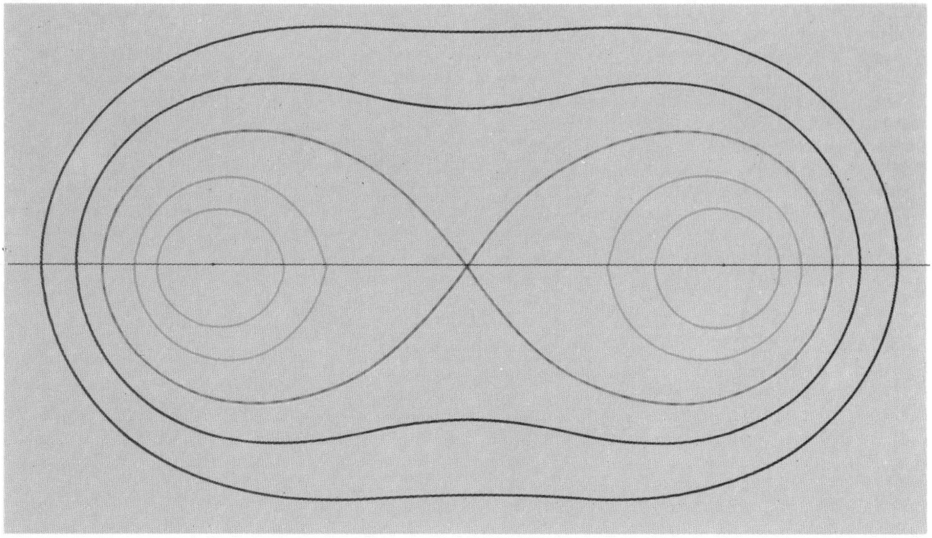

Cassini ovals: five cases including the lemniscate of Bernouilli (red).

(see DOUBLE STAR).

CASTOREUM, a glandular secretion of the Castor beaver. It is soluble in ETHANOL and is used in the blending of perfumes.

CASTOR OIL, a vegetable oil extracted from the purple-streaked seeds of the castor oil plant (*Ricinus communis*). Once widely used as a laxative, castor oil is now mainly used as a lubricant and in the manufacture of oil and varnish.

CASTRATION, removal of the TESTES of a male animal, usually by surgery but also by constricting their blood supply. It is employed for STERILIZATION, particularly for selective BREEDING; to produce docility; to cause the animal to gain more flesh of better quality; and to prevent secondary sexual characteristics from developing. **Spaying**, removal of the OVARIES of a female animal, has similar uses.

CASTRATION COMPLEX, the nexus of fears (see COMPLEX) concerned with possible or threatened loss of the generative organs, especially the penis (see REPRODUCTION). The term is used analogously to describe fears of loss of sexuality or the capacity for erotic pleasure in either males or females. (See OEDIPUS COMPLEX.)

CATALEPSY, rare nervous disorder characterized by episodes of rigid immobility.

CATALYSIS, the changing of the rate of a chemical reaction by the addition of a small amount of a substance which is unchanged at the end of the reaction. Such a substance is called a catalyst, though this term is usually reserved for those which speed up reactions; additives which slow down reactions are called inhibitors. Catalysts are specific for particular reactions. In a reversible reaction, the forward and back reactions are catalyzed equally, and the EQUILIBRIUM position is not altered. Catalysis is either homogeneous (the catalyst and reactants being in the same phase, usually gas or liquid), in which case the catalyst usually forms a reactive intermediate which then breaks down; or heterogeneous, in which ADSORPTION of the reactants occurs on the catalytic surface. Heterogeneous catalysis is often blocked by impurities called poisons. Catalysts are widely used in industry, as in the CONTACT PROCESS, the HYDROGENATION of oils, and the cracking of PETROLEUM. All living organisms are dependent on the complex catalysts called ENZYMES which regulate biochemical reactions.

CATAPLEXY, episodic muscular weakness, often precipitated by emotion, its severity relating to depth of emotion. Associated with NARCOLEPSY.

CATARACT, disease of the EYE lens, regardless of cause: the normally clear lens becomes opaque and light transmission and perception are reduced. Congenital cataracts occur especially in children born to mothers who have had GERMAN MEASLES in early PREGNANCY, and in a number of inherited disorders.

Catalysis. In general, chemical reactions are possible if the reaction products have lower energy than the reactants. But the reaction only proceeds if sufficient energy is available to exceed the "activation energy" ($\Delta H_1^{\ddagger}$) for the reaction. Many catalysts work by providing a reaction pathway which has a lower activation energy ($\Delta H_2^{\ddagger}$) than the uncatalyzed reaction. The overall heat of reaction (ΔH) remains the same.

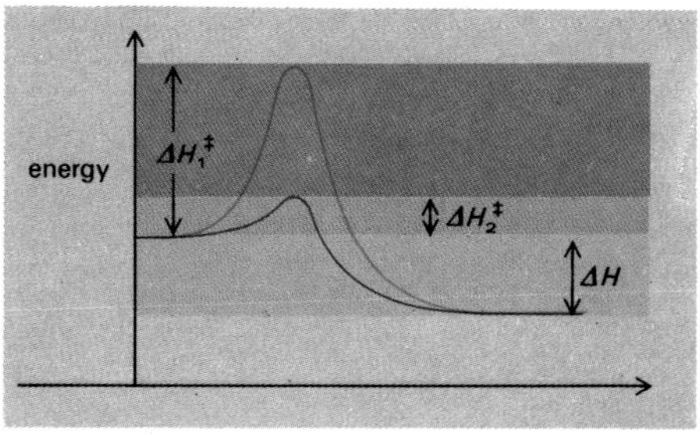

Certain disturbances of METABOLISM or HORMONE production can cause cataracts, especially DIABETES. Eye trauma and INFLAMMATION are other causes in adults. Some degree of cataract formation is common in old age. Once a cataract is formed, vision cannot be improved until the lens is removed surgically. After this, GLASSES are required to correct loss of focusing power. It is among the commonest causes of BLINDNESS in developed countries.

CATARRH, vague term usually referring to excess MUCUS or discharge of PUS from mucous membranes of NOSE or SINUSES but sometimes referring to sputum.

CATASTROPHISM, in geology, the early 19th-century theory that major changes in the geological structure of the earth occurred only during short periods of violent upheaval (catastrophes) which were separated by long periods of comparative stability. The theory fell from prominence after LYELL's enunciation of the rival doctrine of UNIFORMITARIANISM.

CATATONIA, a form of SCHIZOPHRENIA in which the individual oscillates between excitement and stupor. The term is also used for CATALEPSY.

CATECHU, an astringent brown substance containing TANNIN, made from the bark, wood or fruit of plants like Acacias, mainly used as dye for leather and calico.

CATEGORY, a philosophical term given many different meanings by different thinkers. In Aristotelian philosophy, it signifies one of the ultimate classes (e.g., substance, quality, quantity, relation) to which things can be referred. Thus an object might be of the substance, table; the quality, brown, and so on. In the philosophy of KANT, however, the categories are the A PRIORI concepts in terms of which the human mind organizes and interprets its experience of the world. The notions of cause and substance are Kantian categories.

CATERPILLAR, the larva of a moth or a butterfly, with 13 segments, 3 pairs of true legs and up to 5 pairs of soft false legs.

CATGUT, a strong, thin cord used to string musical instruments and rackets, and to sew up wounds in surgery, made from the intestines of herbivorous animals. In surgery, it has the advantage of being eventually absorbed by the body.

CATHARSIS (from Greek *katharsis*, purging), in PSYCHOANALYSIS (where it is also generally termed **abreaction**), the bringing into the open of a previously repressed (see REPRESSION) MEMORY or EMOTION, thus, hopefully, relieving unconscious emotional stress. In medicine, the term is used for the artificial induction of vomiting (see EMETIC). Play, which can be viewed as working off atavistic impulses, is sometimes described as cathartic.

CATHETER, hollow tube passed into body organs for investigation or treatment. **Urinary catheters** are used for relief of BLADDER outflow obstruction and sometimes for loss of nervous control of bladder; they also allow measurement of bladder function and special X-RAY techniques. **Cardiac catheters** are passed through ARTERIES or VEINS into chambers of the HEART to study its functioning and ANATOMY.

CATHODE, a negatively charged ELECTRODE, found particularly in ELECTRON TUBES, CATHODE RAY TUBES, and electrochemical CELLS, and used in combination with its positive counterpart, an ANODE, as a source of ELECTRONS or to produce an electric field.

CATHODE RAYS, ELECTRONS emitted by a CATHODE when heated. First studied by Julius Plücker (1801–1868) in 1858, they can be drawn off in vacuo by the attraction of an ANODE to form a beam which causes fluorescence (see LUMINESCENCE) in appropriately coated screens or X-RAY emission from metal targets.

CATHODE RAY TUBE, the principal component of OSCILLOSCOPES and TELEVISION sets. It consists of an evacuated glass tube containing at one end a heated CATHODE and an ANODE, and widened at the other end to form a flat screen, the inside of which is coated with a fluorescent material. ELECTRONS emitted from the cathode are accelerated toward the anode, and pass through a hole in its center to form a fine beam which causes a bright spot where it strikes the screen. Because of the electric charge carried by the electrons, the beam can be deflected by transverse electric or magnetic fields produced by electrodes or coils between the anode and screen: one such set allows horizontal deflection, and another vertical. The number of electrons reaching the screen can be controlled by the voltage applied to a third electrode, commonly in the form of a wire grid, placed between the cathode and anode so as to divert to itself a proportion of the electrons emitted by the former. It is thus possible to move the spot about the screen and vary its brightness by the application of appropriately timed electrical signals, and sustained images may be produced by causing the spot to traverse the same pattern many times a second. In the oscilloscope, the form of a given electrical signal, or any physical effect capable of conversion into one, is investigated by allowing it to control the vertical deflection while the horizontal deflection is scanned steadily from left to right, while in television sets half-tone pictures can be built up by varying the spot brightness while the spot scans out the entire screen in a series of close horizontal lines.

CATION, a positively-charged ION, normally of a metal, which moves to the CATHODE in ELECTROLYSIS.

CATKIN, type of INFLORESCENCE found on some trees, usually consisting of a spike of unisexual flowers that have no petals or sepals. POLLINATION is by the wind. Typical examples are found on the hazel, birch and willow.

CAT'S EYE, any of several GEM stones which, when cut to form a convex surface, show a thin line of light like a cat's eye (chatoyancy). The commonest type is CHALCEDONY; the rarest is cymophane, a green variety of CHRYSOBERYL, found in Sri Lanka. The effect is due to minute parallel inclusions or cavities.

CAUCASOID, a racial division of man. Caucasoids have straight or curly fine hair, generally mesocephalic (see CEPHALIC INDEX) heads, thin lips, straight faces and well-developed chins. The RACE may have originated in W Asia.

CAUDATA. See URODELA.

CAUSALITY, the philosophical notion that successive events can be related such that the later is dependent on the earlier, the earlier being the cause and the later its effect. According to ARISTOTLE everything has four causes: the material cause, the material substance involved; the formal cause, its shape or structure; the efficient cause, the agency which imposes the shape upon the matter, and the final cause, the end to which it is done (see

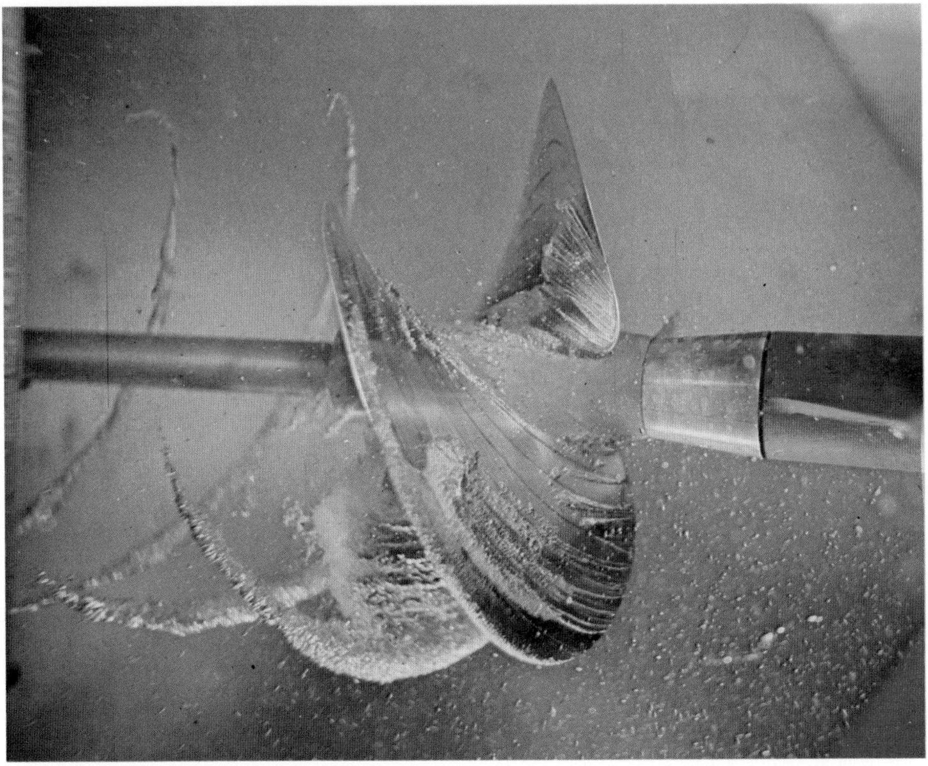

Cavitation caused by the rotation of a model propeller in a water tunnel.

TELEOLOGY). Modern notions of causation approximate most closely to Aristotle's efficient cause, though many "skeptical" philosophers, notably HUME, have doubted whether there is in fact any "necessary connexion" between the events designated cause and effect. Such thinkers instead argue that the conviction that events can be related as cause and effect has merely been learned from experience. KANT, on the other hand, held that causality was one of the *a priori* CATEGORIES necessary for the ordering of experience.

CAUSTIC SODA, or sodium hydroxide. See SODIUM; ALKALI.

CAUTERIZATION, application of heat or caustic substances. Used for minor SKIN or mucous-membrane lesions (especially of the NOSE and cervix uteri) to remove abnormal tissue and encourage normal healing. Before antiseptics, it was used to sterilize wounds.

CAVE, any chamber formed naturally in rock and, usually, open to the surface via a passage. Caves are found most often in LIMESTONE, where rainwater, rendered slightly ACID by dissolved CARBON dioxide from the ATMOSPHERE, drains through joints in the stone, slowly dissolving it. Enlargement is caused by further passage of water and by bits of rock that fall from the roof and are dragged along by the water. Such caves form often in connected series; they may display STALACTITES AND STALAGMITES and their collapse may form a GORGE. Caves are also formed by selective EROSION by the sea of cliff bases. Very occasionally they occur in LAVA, either where lava has solidified over a mass of ice that has later melted, or where the surface of a mass of lava has solidified, molten lava beneath bursting through and flowing on.

CAVENDISH, Henry (1731–1810), English chemist and physicist who showed HYDROGEN (inflammable air) to be a distinct GAS, water to be a compound and not an elementary substance and the composition of the ATMOSPHERE to be constant. He also used a torsion BALANCE to measure the DENSITY of the earth (1798).

CAVIAR, the salted roe of certain STURGEON, a delicacy because of its scarcity. The best caviar comes from the Beluga sturgeon of the Caspian Sea.

CAVITATION, the formation of bubbles of vapor in a liquid, strictly applicable only to formation of bubbles through reduction in PRESSURE without corresponding TEMPERATURE change. Cavitation reduces efficiency in, e.g., TURBINES and PROPELLERS, rapidly erodes (see EROSION) the moving surfaces, and produces unwanted VIBRATIONS.

CAYLEY, Sir George (1773–1857), British inventor who pioneered the science of AERODYNAMICS. He built the first man-carrying GLIDER (1853) and formulated the design principles later used in AIRPLANE construction, although he recognized that in his day there was no propulsion unit which was sufficiently powerful and yet light enough to power an airplane.

CELESTIAL SPHERE, in ancient times, the sphere to which it was believed all the stars were attached. In modern times, an imaginary sphere of indefinite but

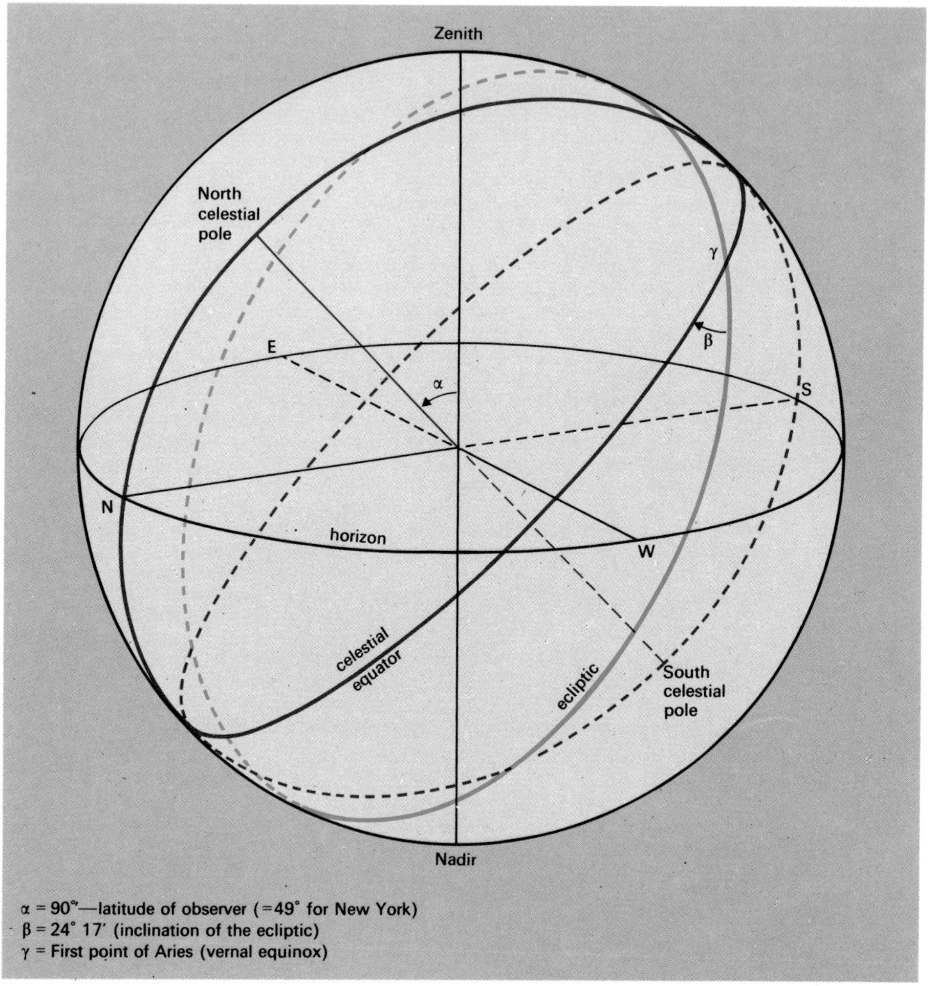

The locations of the principal features on the celestial sphere.

very large radius upon which, for purposes of angular computation, celestial bodies are considered to be situated. The **celestial poles** are defined as those points on the sphere vertically above the terrestrial poles, and the **celestial equator** by the projection of the terrestrial EQUATOR onto the sphere. (See also ECLIPTIC.) Astronomical coordinate systems are based on these great circles (circles whose centers are also the center of the sphere) and, in some cases, on the observer's celestial HORIZON. In the most frequently used, the equatorial system, terrestrial latitude corresponds to declination—a star directly overhead in New York City will have a declination of +41° (S Hemisphere declinations are preceded by a minus sign), New York City having a latitude of 41°N—and terrestrial longitude to right ascension, which is measured eastward from the First Point of ARIES. Right ascension is measured in hours, one hour corresponding to 15° of longitude. (See also SPHERICAL TRIGONOMETRY.)

CELIAC DISEASE, a disease of the small intestine (see GASTROINTESTINAL TRACT), among the commonest causes of food malabsorption. In celiac disease, ALLERGY to part of gluten, a component of wheat, causes severe loss of absorptive surface. In children, failure to thrive and DIARRHEA are common signs, while in adults weight loss, ANEMIA, diarrhea, TETANY and VITAMIN deficiency may bring it to attention. Complete exclusion of dietary gluten leads to full recovery.

CELL, the basic unit of living matter from which all plants and animals are built. A living cell can carry out all the functions necessary for life. BACTERIA, AMOEBA and paramecia are examples of single-celled organisms. In multicellular organisms cells become differentiated to perform specific functions. All cells have certain basic similarities.

Nearly all cells can be divided into three parts: an outer membrane or wall, a **nucleus** and a clear fluid called CYTOPLASM.

Animal cells are surrounded by a plasma membrane. This is living, thin and flexible. It allows substances to diffuse in and out and is also able to select some substances and exclude others. The membrane plays a vital role in deciding what enters a cell. Plant cells are surrounded by a thick, rigid, non-living CELLULOSE cell wall.

Other types of membrane are found in a cell. Around the nucleus is the nuclear membrane, which has in it tiny pores to allow molecules to pass between the cytoplasm and the nucleus. Another type of membrane is the much-folded endoplasmic reticulum which seems to be a continuation of the cell or nuclear membrane. The endoplasmic reticulum is always associated with the RIBOSOMES where PROTEIN SYNTHESIS takes place, controlled by the CHROMOSOMES which are sited in the nucleus and are mainly made of DNA (see NUCLEIC ACIDS).

The components of a typical plant cell.
(1) Plasmalemma; (2) mitochondrion; (3) intercellular space; (4) microbody; (5) ribosomes; (6) endoplasmic reticulum; (7) nuclear envelope; (8) nucleolus; (9) nucleus with DNA; (10) nuclear pore; (11) tonoplast; (12) vacuole; (13) cytoplasm; (14) plasmodesma; (15) DNA; (16) stroma; (17) chloroplast; (18) granum; (19) dictyosome (Golgi body); (20) cell wall; (21) middle lamella; (22) intercellular space.

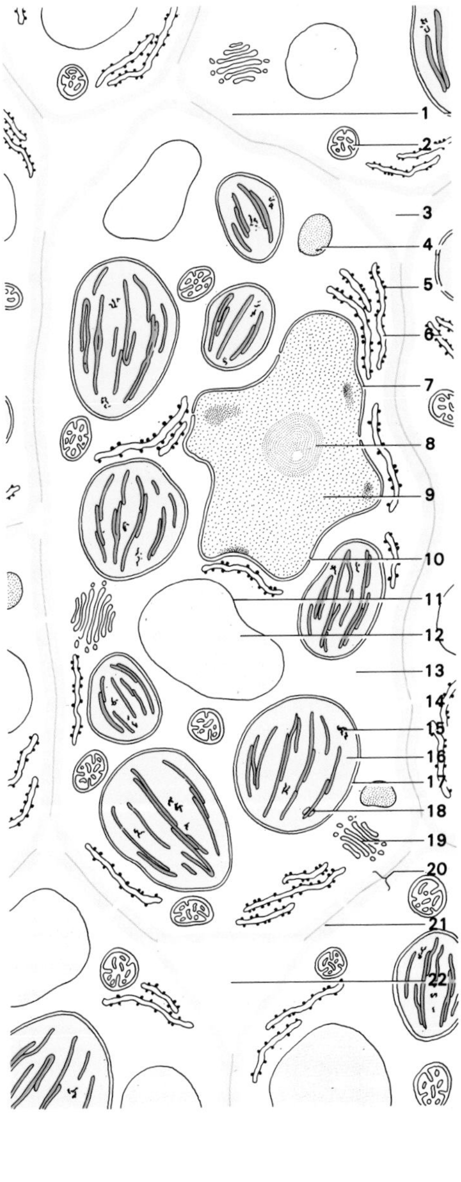

The cytoplasm contains many organelles. Among the most important are the rod-shaped **mitochondria**, containing the enzymes necessary for the release of energy from food by the process of RESPIRATION (see also CITRIC ACID CYCLE). Other organelles whose function is still uncertain are the Golgi bodies, which may be involved in the synthesis of cell wall material; and the lysosomes, which may contain enzymes involved in autolysis and controlled destruction of tissues. The cytoplasm of green plants also contains CHLOROPLASTS, where PHOTOSYNTHESIS occurs.

New cells are formed by a process of division called MITOSIS. Each chromosome duplicates and mitosis involves the transfer of this new set of chromosomes to the new daughter cell. Gamete (reproductive) cells are formed by MEIOSIS which is a division that halves the number of chromosomes; thus a human cell that contains 46 chromosomes will produce gamete cells with 23.

Cells differentiate in a multi-cellular organism to produce cells as different as a nerve cell and a muscle cell. Cells of similar types are grouped into TISSUES.

There are two broad types of cells. Firstly, **prokaryotic cells**, which have the genetic material in the form of loose filaments of DNA not separated from the cytoplasm by a membrane. Secondly, **eukaryotic cells**, which have the genetic material borne on chromosomes made up of DNA and protein that are separated from the cytoplasm by a nuclear membrane. Eukaryotic cells are the unit of basic structure in all organisms except bacteria and blue green algae, which comprise single prokaryotic cells.

CELL, Electrochemical, device for interconverting chemical and electrical ENERGY. For power cells, which transform chemical into electrical energy, see BATTERY. In electrolytic cells the reverse process occurs: by applying an ELECTROMOTIVE FORCE across two electrodes in an electrolyte, a chemical reaction is effected (see ELECTROLYSIS).

CELLOPHANE, transparent, impermeable film of CELLULOSE used in packaging, first developed by J. E. Brandenburger (1911). Wood pulp is soaked in sodium hydroxide, shredded, aged and reacted with CARBON disulfide to form a solution of viscose (sodium cellulose XANTHATE). This is extruded through a slit into an acid bath, where the cellulose is regenerated as a film. It is dried and given a waterproof coating. If the viscose is extruded through a minute hole, rayon is produced (see SYNTHETIC FIBERS).

CELLULOID, the first commercial synthetic PLASTIC, developed by J. W. Hyatt (1869). It is a colloidal dispersion of NITROCELLULOSE and CAMPHOR. It is tough, strong, resistant to water, oils and dilute acids, and thermoplastic. Used in dental plates, combs, billiard balls, lacquers, spectacle frames and (formerly) photographic films and toys, celluloid is highly inflammable, and has been largely replaced by other plastics.

CELLULOSE, the main constituent of the CELL walls of higher plants, many algae and some fungi; cotton is 90% cellulose. Cellulose is a CARBOHYDRATE with a similar structure to starch. In its pure form it is a white solid which absorbs water until completely saturated, but dissolves only in a few solvents, notably strong alkalis and some acids. It can be broken down by heat and by the digestive tracts of some animals, but it passes through the human digestive tract unchanged and is helpful only in stimulating movement of the intestines. Industrially, it is used in manufacturing textile fibers, CELLOPHANE, CELLULOID, and the cellulose PLASTICS, notably NITROCELLULOSE (used also in explosives), cellulose acetate for toys and boxes and cellulose acetate butyrate for typewriter keys.

CELSIUS, Anders (1701–1744), Swedish astronomer, chiefly remembered for his proposal (1742) of a centigrade TEMPERATURE scale which had 100° for the freezing point and 0° for the boiling point of water. The modern centigrade temperature scale (with 0° for the freezing point and 100° for the boiling point of water) is known as the **Celsius scale** in his honor, temperatures being quoted in "degrees Celsius" (°C).

CELSUS, Aulus Cornelius, 1st-century Roman medical writer, renowned in the Renaissance as the "Cicero of medicine" for the fine Latin style of his *De medicina*. His reputation prompted the 16th-century founder of IATROCHEMISTRY to adopt the name PARACELSUS—beyond Celsus—thus boasting his supposed superiority over his Roman predecessor.

CELTS, a prehistoric people whose numerous tribes occupied much of Europe between c2000 and c100 BC, the peak of their power being around 500–100 BC. No European Celtic literature survives, but the later Irish and Welsh sources tell much about Celtic society and way of life. Primarily an agricultural people, though in local areas crafts and iron smelting developed, they grouped together in small settlements. Their social unit, based on kinship, was divided into a warrior nobility and a farming class, from the former being recruited the priests or druids, who ranked highest of all. Celtic art mixes stylized heads with abstract designs of scrolls and spirals (but see LA TÈNE). Remnants of Celtic languages are to be found in the forms of Gaelic, Erse, Manx and Welsh. The Celtic sphere of influence declined during the 1st century BC owing to the simultaneous expansion of the Roman Empire and the incursions of the Germanic races.

CEMENT, common name for Portland cement, the most important modern construction material, notably as a constituent of CONCRETE. In the manufacturing process, limestone is ground into small pieces (about 2cm). To provide the silica (25%) and alumina (10%) content required, various clays and crushed rocks are added, including iron ore (about 1%). This material is ground and finally burned in a rotary kiln at up to 1500°C, thus converting the mixture into clinker pellets. About 5% GYPSUM is then added to slow the hardening process, and the ground mixture is added to sand (for MORTAR) or to sand, gravel and crushed rock (for concrete). When water is added it solidifies gradually, undergoing many complex chemical reactions. The name "Portland" cement arises from a resemblance to stone quarried at Portland, England.

CENOZOIC, or **Cainozoic**, the period of geological time containing the TERTIARY and QUATERNARY.

CENSORSHIP, in psychoanalysis, the term used by FREUD to describe the process in the UNCONSCIOUS whereby repressed (see REPRESSION) EMOTION, ideas, impulses and MEMORIES are prevented from reaching the CONSCIOUS. (See also SUPEREGO.)

CENTAURUS, the Centaur, a constellation in the S Hemisphere. (See ALPHA CENTAURI; PROXIMA CENTAURI.)

CENTER OF GRAVITY, the point about which gravitational FORCES on an object exert no net turning effect, and at which the mass of the object can for many purposes be regarded as concentrated. A freely suspended object hangs with its center of gravity vertically below the point of suspension, and an object will balance, though it may be unstable, if supported at a point vertically below the center of gravity. In free flight, an object spins about its center of gravity, which moves steadily in a straight line; the application of forces causes the center of gravity to accelerate in the direction of the net force, and the rate of spin to change according to the resultant turning effect.

CENTER OF SYMMETRY, a point about which a geometrical figure is symmetrical (see SYMMETRY).

CENTIGRADE DEGREE (C°), a unit of TEMPERATURE difference equal to 1 kelvin (K), originally defined by dividing the interval between the boiling and freezing points of water into 100 equal divisions. It is used as the basis of the Celsius temperature scale (see CELSIUS, ANDERS).

CENTIMETRE (cm), a unit of length equal to 0.01 metre. (See METRE.)

CENTRIFUGAL FORCE. See CENTRIPETAL FORCE.

CENTRIFUGE, a machine for separating mixtures of solid particles and immiscible liquids of different DENSITIES and for extracting liquids from wet solids by rotating them in a container at high speed. The separation occurs because the centrifugal force experienced in a rotating frame increases with particle density. Centrifuges are used in drying clothes and slurries, in chemical ANALYSIS, in separating cream and in atomic ISOTOPE separation. Giant ones are used to accustom pilots and astronauts to large ACCELERATIONS. The **ultracentrifuge**, invented by T. SVEDBERG, uses very high speeds to measure (optically) sedimentation rates of macromolecular solutes and so determine molecular weights.

CENTRIPETAL FORCE, the FORCE applied to a body to maintain it moving in a circular path. To maintain a body of MASS m, traveling with instantaneous VELOCITY v, in a circular path of radius r, a centripetal force F, acting *toward the center* of the circle, given by $F = mv^2/r$ must be applied to it. If a body is resting in a rotating frame, it experiences a **centrifugal force**, apparently acting *away from the center* of rotation, numerically equal to the external centripetal force. Because, from the point of view of an observer external to the rotating frame, the centrifugal force has no real existence, it is often termed a fictitious force.

CENTROID, the point of concurrence of the medians of a TRIANGLE (i.e., those lines drawn from each angle to the midpoint of the facing side). The centroid is the CENTER OF GRAVITY of the triangle.

CEPHALIC INDEX, in ANTHROPOMETRY, an index used originally in attempts to classify RACE, now used

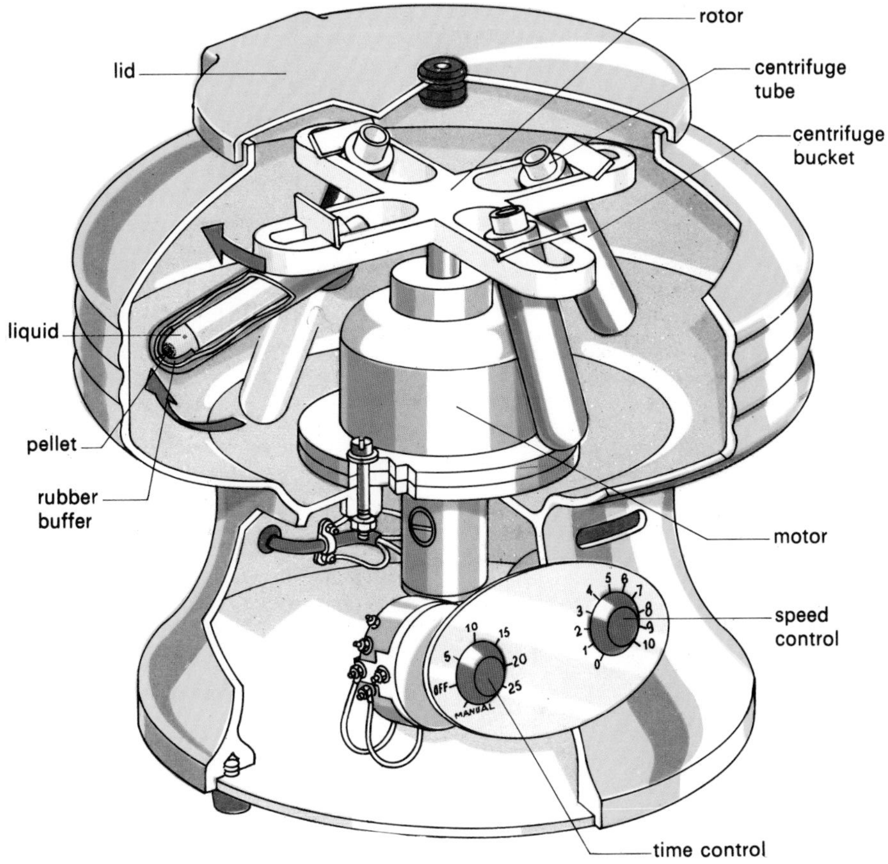

Cutaway of a typical bench centrifuge as used in laboratory work and in teaching. It has a maximum speed of about 3800 revolutions per minute when it produces about 2100*g*. Liquids to be centrifuged are placed in glass or plastic tubes which are held in the small buckets which are hung from the rotor arms on hinges. When the rotor starts, the bucket swings out and any solid material is sedimented as a pellet in the base of the centrifuge tube.

mainly to indicate possible relationships between small groups. The index is given by

$$\frac{\text{maximum head breadth} \times 100}{\text{maximum head length}}$$

Peoples with broad heads (cephalic index over 80) are classed as brachycephalic; those with long heads (less than 76) as dolichocephalic; and those in between as mesocephalic (mesaticephalic).

CEPHALIZATION, the tendency in the evolution of animals for sense organs and associated nervous tissue to be concentrated at one end of the body—the end that "faces" the environment. Cephalization has led to the development of a distinct head in most animals.

CEPHALOCHORDATA, a subphylum of the CHORDATA containing marine fish-like animals that differ from fish by lacking bone or cartilage, paired fins, jaws and teeth. The best known member is *Amphioxus* which is capable of short bursts of swimming activity, but spends most of its life partially buried in sand. It feeds on particles filtered from the water.

CEPHALOPODA, a class of the phylum MOLLUSCA.

CEPHEID VARIABLES, stars whose brightness varies regularly with a period of 1–50 days, possibly, but improbably, due to a fluctuation in size. The length of their cycle is directly proportional to their absolute magnitude, making them useful "mileposts" for computing large astronomical distances. (See VARIABLE STAR.)

CERAMICS, materials produced by treating non-metallic inorganic materials (originally CLAY) at high temperatures. Modern ceramics include such diverse products as porcelain and china, furnace bricks, electric insulators, ferrite magnets (see SPINEL), rocket nosecones and abrasives. In general, ceramics are hard, chemically inert under most conditions, and can withstand high temperatures in industrial applications. Many are refractory metal OXIDES. Primitive ceramics in the form of pottery date from the 5th millennium BC, and improved steadily in quality and design. By the 10th century AD porcelain had been developed in China. (See also CERMETS; CONCRETE; GLASS; POTTERY AND PORCELAIN.)

CEREAL CROPS, annual plants of the grass family including wheat, rice, corn, barley, sorghum, millet, oats and rye. Their grain forms the staple diet for most of the world. Though lacking in calcium and vitamin A, they have more CARBOHYDRATE than any other food, as well as PROTEIN and other VITAMINS. Cereal crops are relatively easy to cultivate and can cope with a wide range of climates. About 1 757 million acres of the world's arable land are sown with cereal crops each year. The US leads in production of corn, oats and sorghum.

CEREBELLUM. See BRAIN.

CEREBRAL PALSY, a diverse group of conditions caused by BRAIN damage around the time of BIRTH and resulting in a variable degree of nonprogressive physical and mental handicap. While abnormalities of MUSCLE control are the most obvious, loss of sensation and some degree of DEAFNESS are common accompaniments. Speech and intellectual development can also be impaired but may be entirely normal. SPASTIC PARALYSIS of both legs with mild arm weakness (diplegia), or of one half of the body (hemiplegia), are common forms. A number of cases have abnormal movements (athetosis) or ATAXIA. Common causes include birth trauma, ANOXIA, prematurity, Rhesus incompatibility and cerebral HEMORRHAGE. PHYSIOTHERAPY and training allow the child to overcome many deficits; deformity must be avoided by ensuring full range of movements at all joints, but surgical correction may be necessary. Sometimes transposition of TENDONS improves the balance of strength around important joints. It is crucial that the child is not deprived of normal sensory and emotional experiences. Improved antenatal care, OBSTETRIC skill and care of premature infants have reduced the incidence.

CEREBROSPINAL FLUID, watery fluid circulating in the chambers (ventricles) of the BRAIN and between layers of the meninges covering the brain and SPINAL CORD. It is a filtrate of BLOOD and is normally clear, containing salts, GLUCOSE and some PROTEIN. It may be sampled and analysed by SPINAL TAP.

CEREBRUM. See BRAIN.

CERENKOV RADIATION, ELECTROMAGNETIC RADIATION emitted when a high-energy particle passes through a dense medium at a velocity greater than the velocity of light in that medium. It was first detected in 1934 by P. A. CHERENKOV and may be seen when radioactive materials are stored under water.

CERES, largest of the ASTEROIDS (470mi in diameter) and the one first discovered (by PIAZZI, 1801). Its orbit was first computed by K. F. GAUSS and found to satisfy BODE's Law. Its "year" is 1 681 days and its maximum apparent magnitude is +7.

CERIUM (Ce), most abundant of the RARE EARTHS; one of the LANTHANUM SERIES. AW 140.1, mp 804°C, bp 3257°C, sg 6.657 (25°C).

CERMETS, or **ceramels**, composite materials made from mixed METALS and CERAMICS. The TRANSITION ELEMENTS are most often used. Powdered and compacted with an oxide, carbide or boride, etc., they are heated to just below their melting point, when bonding occurs. Cermets combine the hardness and strength of metals with a high resistance to corrosion, wear and heat. This makes them invaluable in jet engines, cutting tools, brake linings and nuclear reactors.

CERUSSITE, white mineral form of lead (II) carbonate ($PbCO_3$), a major ore of LEAD, found in Spain, SW Africa, Australia and Col. It is formed by weathering of GALENA. **White lead**, used as a paint pigment, is a basic lead (II) carbonate ($PbCO_3 + Pb(OH)_2$).

CESARIAN SECTION, BIRTH of a child from the WOMB by abdominal operation. The mother is given an anesthetic and an incision is made in the ABDOMEN and lower part of the uterus; the child is delivered and attended to; the PLACENTA is removed and incisions are sewn up. Cesarian section may be necessary if the baby is too large to pass through the PELVIS, if it shows delay or signs of ANOXIA during labor, or in cases where maternal disease does not allow normal labor. It may be performed effectively before labor has started. With modern ANESTHESIA and BLOOD TRANSFUSION, the risks of Cesarian section are not substantially greater than those of normal delivery. It is believed that Julius Caesar was born in this way.

CESIUM (Cs), a very soft, silvery-white ALKALI METAL, found mainly in the mineral pollucite, and made by reduction of cesium chloride. It is highly reactive, similar to potassium, and burns spontaneously in moist air. When exposed to light it emits electrons, and has applications in television cameras. It is used in vacuum tubes to remove oxygen and water; and a recent application is in ion-propulsion rocket engines. AW 132.9, mp 28°C, bp 678°C, sg 1.873 (20°C).

CESTODA, parasitic platyhelminths commonly known as TAPEWORMS.

CETACEA, an order of the MAMMALIA that includes the whales, dolphins and porpoises. Cetaceans are

CGS Units

Quantity	Unit	Symbol	Equivalent in SI units
length	centimetre	cm	0.01 metre
mass	gram	g	0.001 kilogram
time	second	s	1 second
length	angstrom unit	Å	0.1 nanometre
work, energy	erg	erg	10^{-7} joule
force	dyne	dyn	10^{-5} newton
energy	calorie (IT)	$cal_{(IT)}$	4.1868 joule
force	gram-force	gf	9.807×10^{-5} newton

A group of transmission chains and chain drives manufactured by Renold Limited, Manchester, UK.

aquatic and look extremely fish-like but, being mammals, they breathe atmospheric air. The tail operates with an up-and-down stroke making speeds of 25–30 knots possible. The BLUE WHALE is the largest animal of all time.

CGS UNITS, a metric system of units based on the CENTIMETRE (length), GRAM (mass) and SECOND (time), generally used among scientists until superseded by SI UNITS. Several variants are used for electrical and magnetic problems, including electrostatic units (esu or stat-units—see STATAMPERE), electromagnetic units (emu or ab-units—see ABAMPERE) and the Gaussian system. In this last, ab-units are used for quantities arising primarily in an electromagnetic context, stat-units for electrostatic quantities and both the PERMEABILITY and the PERMITTIVITY of free space are set equal to unity. As a result, the electromagnetic constant (*c*) tends to occur in equations in which electrostatic and magnetic quantities are mixed.

CHADWICK, Sir James (1891–1974), English physicist who was awarded the 1935 Nobel physics prize for his discovery of the NEUTRON (1932).

CHAETOGNATHA, a phylum of marine worm-like animals with bodies divided into head, trunk and tail, also known as arrow-worms. The trunk bears lateral fins. Larval chaetognathans are free-swimming; adults are planktonic or tube dwellers.

CHAGAS' DISEASE, PARASITIC DISEASE found only on the American continent, caused by a trypanosome and carried by insects. In the acute form, there is swelling around the eye, FEVER, malaise, enlargement of LYMPH nodes, LIVER and SPLEEN, and EDEMA. Most cases recover fully. The chronic form causes disease of the HEART and GASTROINTESTINAL TRACT.

CHAIN, a series of interlocking links forming a flexible cable. Simple chains are used in, for example, haulage and hoisting. Roller chains are used in power TRANSMISSION (as on a bicycle): here the links of an endless chain mesh with the teeth of toothed wheels called sprockets. EFFICIENCY may be as high as 99%.

CHAIN, Sir Ernst Boris (1906–), German-born UK biochemist who helped develop PENICILLIN for clinical use. For this he shared with FLOREY and FLEMING the 1945 Nobel Prize for Physiology or Medicine.

CHAIN REACTION. See NUCLEAR ENERGY.

CHALCEDONY, mineral consisting of fibrous microcrystalline QUARTZ. Common in gravels, it occurs in several forms, including some semiprecious GEM stones. Notable are BLOODSTONE, CAT'S EYE, ONYX, CHRYSOPRASE, JASPER, CARNELIAN, SARD and AGATE.

CHALCOCITE, dark gray or black mineral, copper (I) sulfide (Cu_2S). A major COPPER ore, it is found (often with BORNITE) in the Ural Mts, Africa, South America, Alaska, Conn. and the southern US.

Chalcedony: as the raw mineral, and cut and polished as a gemstone.

CHALCOPYRITE ($CuFeS_2$), the most important COPPER ore, very similar to PYRITE, with which it is often associated. It is found in the US, Canada, Australia, W Europe and South America.

CHALK, soft, white rock mainly formed of fine-grained, porous LIMESTONE and containing calcareous remains of minute marine animals. There are large deposits in Tex., Kan. and Ark. Chalk is widely used in lime and cement manufacture and as a fertilizer. It is also used in cosmetics, plastics, crayons and oil paints; school chalk is today usually made from chemically-produced calcium carbonate.

CHALLENGER EXPEDITION, a round-the-world oceanographic-survey cruise made by the steam corvette HMS *Challenger* between 1872 and 1876 under the scientific direction of **Sir Charles Wyville Thomson** (1830–1882), the first and most comprehensive voyage of its type. Its results formed the 50-volume *Challenger Report* (1881–95).

CHAMBERLAIN, Owen (1920–), US physicist who shared the 1959 Nobel physics prize with E. Segrè in recognition of their discovery of the antiproton (see ANTIMATTER).

CHAMPOLLION, Jean François (1790–1832), French linguist and historian, the "father of Egyptology." Professor of history at Grenoble U. 1809–16, he was the first to effectively decipher Egyptian HIEROGLYPHICS, a result of his research on the ROSETTA STONE. A chair of Egyptian antiquities was created especially for him at the Collège de France in 1831.

CHANCE. See PROBABILITY.

CHANGE OF LIFE. See MENSTRUATION.

CHAOS, the primordial emptiness that existed before anything came into being in early Greek COSMOGONY. Later this notion was superseded and chaos came to refer to an aboriginal state of confusion. In this sense PARACELSUS applied the term to describe air, hence the modern term, GAS.

CHAPARRAL, a dense thicket of dwarf EVERGREEN trees, usually less than 3m (10ft) high; a locality characterized by such thickets, especially common in the Cal. region.

CHARCOAL, form of amorphous CARBON produced when wood, peat, bones, cellulose or other carbonaceous substances are heated with little or no air present. A highly porous residue of microcrystalline GRAPHITE remains. Charcoal is a fuel and was used in BLAST FURNACES until the advent of COKE. A highly porous form, activated charcoal, is made by heating charcoal in steam; it is used for ADSORPTION in refining processes and in gas masks. Charcoal is also used as a thermal insulator and by artists for drawing.

CHARCOT, Jean Martin (1825–1893), French physician and founder of modern NEUROLOGY, whose many researches advanced knowledge of HYSTERIA, MULTIPLE SCLEROSIS, locomotor ATAXIA, ASTHMA and aging. FREUD was one of his many pupils.

CHARGE, Electric. See ELECTRICITY.

CHARLES, Jacques Alexandre César (1746–1823), French physicist who with Nicholas Robert, made the first ascent in a hydrogen BALLOON (1783). About 1787 he discovered **Charles' Law** which, stated in modern terms, records that for an ideal GAS at constant PRESSURE, its VOLUME is directly proportional to its absolute TEMPERATURE.

CHEESE, nutritious food made from the milk of various animals, with a high protein, calcium and vitamin content. Cheesemaking was already common by 2000 BC. It involves first the curdling of milk by adding an acid or RENNET, so that the fat and protein (mostly CASEIN) coagulate to form the solid curds. After excess liquid whey has been drained off, the curds are compressed and enough moisture is removed to give the cheese the desired degree of hardness. Most cheeses (but not cottage cheese) are then subjected to a period of FERMENTATION, from two weeks to two years, called ripening or curing, during which they are salted and perhaps flavored. The consistency and flavor of the cheese depend on the time, temperature and humidity of storage and on the microorganisms present. Camembert, for instance, is ripened with two molds, *Penicillium candidum* and *P. camemberti*, which make it soft. Process cheese is a blend of several types of cheese melted together.

CHELATE, chemical complex formed from a polydentate LIGAND and a metal ION, thus making a ring. They are more stable than the corresponding unidentate complexes, and are used to sequester metal ions (see HARD WATER), as well as in chemical ANALYSIS and for separating metals. Some biochemical substances, including CHLOROPHYLL and HEMOGLOBIN, are chelates.

CHELONIA, members of the REPTILIA that include turtles, terrapins and tortoises. Chelonians first appeared in the Triassic period, about 200 million years ago, and have changed little since. Their bodies are enclosed in a shell that overlaps a bony CARAPACE above and a plastron below. Some chelonians have a soft or leathery shell: all lack teeth and have a horny beak.

CHEMICAL AND BIOLOGICAL WARFARE, the use of poisons and diseases against an enemy, either to kill or disable personnel or to diminish food supply, natural ground cover, etc. According to legend, Solon defeated a Megaran army c600 BC by poisoning their drinking water. Thucydides records that the Spartans in the 5th century BC used in attack the fumes produced by burning wood, sulfur and pitch. Julius Caesar mentions with disapproval the use of poisons in warfare. GREEK FIRE was in use from about the middle of the 7th century AD. In the US during the French and Indian War, infected blankets were given to the Indians to spread SMALLPOX among them. During the Civil War, John Doughty proposed the use of an artillery shell containing the choking, corrosive gas CHLORINE. Chemical warfare on a large scale was first waged by the Germans in WWI at Yprés (1916), using chlorine against the Allies. Gas warfare on both sides escalated throughout the remainder of WWI; despite the use of the GAS MASK, around 100 000 may have died as a result of chlorine, PHOSGENE and MUSTARD GAS attacks. During WWII the Germans developed nerve gases, which attack the NERVOUS SYSTEM, but these (Sarin, Soman and Tabun) were not used. More deadly nerve gases have since been developed in the US: some may linger for months and kill in seconds. In the Vietnam War, TEAR GASES were used in combat as distinct from their more normal role in riot control. Also in Vietnam, defoliants were sprayed from aircraft on enemy crops and on vegetation to deprive guerrillas of cover (see also NAPALM).

Waging of biological warfare has been rare, mainly because its effects are hard to control. Nevertheless, most developed countries have encouraged military research in this field. Available preparations could, if used, unleash pneumonic PLAGUE, pulmonary ANTHRAX, and BOTULISM, among other fatal diseases; it has been estimated that 1oz (about 30g) of these would, if well distributed, be sufficient to kill the entire population of North America. Less fatal diseases, such as TULAREMIA, and certain HALLUCINOGENIC DRUGS are also available for such use.

Of the international agreements outlawing the use of chemical and biological warfare, the oldest is the 1925 Geneva Protocol; the UK accepted its strictures only in part, reserving the right to retaliate; and the US, although they signed the agreement in 1925, did not ratify until 1975. The Protocol was contravened by Italy against Ethiopia (1936) and by Japan against China (1943). Most countries have signed the Biological Weapons Convention (1972), agreeing to cease production of biological weapons and destroy existing stockpiles.

CHEMILUMINESCENCE, LUMINESCENCE caused by a chemical reaction, usually oxidation, as of phosphorus in air. The molecules are excited to a high ENERGY LEVEL, and emit light as they return to the ground state. This process in living organisms is called BIOLUMINESCENCE.

CHEMISTRY, the science of the nature, composition and properties of material substances, and their transformations and interconvertions. In modern terms, chemistry deals with ELEMENTS and compounds, with the ATOMS and MOLECULES of which they are composed, and with the reactions between them. It is thus basic to natural phenomena and modern technology alike. Chemistry may be divided into five major parts: ORGANIC CHEMISTRY, the study of carbon compounds (which form an idiosyncratic group); INORGANIC CHEMISTRY, dealing with all the elements except carbon, and their compounds; chemical ANALYSIS, the determination of what a sample contains and how much of each constituent is present; BIOCHEMISTRY, the study of the complex organic compounds in biological systems; and PHYSICAL CHEMISTRY, which underlies all the other branches, encompassing the study of the physical properties of substances and the theoretical tools for investigating them. Related sciences include GEOCHEMISTRY and METALLURGY.

Practical chemistry originated with the art of the metallurgists and artisans of the ancient Middle East. Their products included not only refined and alloyed metals but also dyes and glasses, and their methods were and remain shrouded in professional secrecy. Their chemical theory, expressed in terms of the prevailing theology, involved notions such as the opposition of contraries and the mediation of a mediating third. Classical Greek science generally expressed itself in the theoretical rather than the practical, as the conflicting physical theories of THALES and ANAXAGORAS, ANAXIMENES and ARISTOTLE bear witness. An important concept, that matter exists as atoms—tiny individual material particles—emerged about this time (see ATOMISM) though it did not become dominant for another 2000 years. During the HELLENISTIC AGE a new practical chemistry arose in the study of ALCHEMY. These early alchemists sought to apply Aristotelian physical theory to their practical experiments. Alchemy was the dominant guise of chemical science throughout the medieval period. Like the other sciences, it passed through Arab hands after the collapse of the Roman world, though, unlike the case with some other sciences, great practical advances were made during this time with the discovery of alcohol DISTILLATION and methods for preparing NITRIC and SULFURIC ACIDS. Chemical theory, however, remained primitive and practitioners sought to guard their secret recipes by employing obscure and even mystical phraseology. The 16th century saw new clarity brought to the description of metallurgical processes in the writings of Georgius AGRICOLA and the foundation by PARACELSUS of the new practical science of IATROCHEMISTRY with its emphasis on chemical medicines. Jan Baptist van HELMONT, the greatest of his successors, began to use quantitative experiments. In the 17th century mechanist atomism enjoyed a revival with Robert BOYLE leading a campaign to banish obscurantism from chemical description. The 18th century saw the rise and fall of the PHLOGISTON theory of combustion, promoted by STAHL and adopted by all the great chemists of the age: BLACK, SCHEELE and PRIESTLEY (all of whom found their greatest successes in the study of GASES). The phlogiston theory fell before the oxygen theory of LAVOISIER and his associated binomial nomenclature, and the new century (the 19th) saw the proposal of DALTON's atomic theory, AVOGADRO's hypothesis (neglected for 50 years until revived by CANNIZZARO) and the foundation of ELECTROCHEMISTRY which, in the hands of DAVY, rapidly yielded two new elements, SODIUM and POTASSIUM. During the 19th century chemistry gradually assumed its present form, the most notable innovations being the periodic table of MENDELEYEV, the BENZENE ring-structure of KEKULÉ, the systematic chemical THEMODYNAMICS of GIBBS and BUNSEN's chemical SPECTROSCOPY. In the opening years of the present century the new atomic theory revolutionized chemical theory and the interrelation of the elements was deciphered. Since then successive improvements in experimental techniques (e.g., CHROMATOGRAPHY; isotopic labeling; MICROCHEMISTRY) and the introduction of new instruments (infrared, nuclear-magnetic-resonance and mass spectroscopes) have led to continuing advances in chemical theory. These developments have also had a considerable impact on industrial chemistry and biochemistry. Perhaps the most significant recent change in the chemist's outlook has been that his interest has moved away from the nature of chemical substance itself towards questions of molecular structure, the energetics of chemical processes and reaction mechanisms.

CHEMOTHERAPY, the use of chemical substances to treat disease. More specifically, the term refers to the use of nonantibiotic antimicrobials and agents for treating CANCER. The drug must interfere with the growth of bacterial, parasitic or TUMOR cells, without significantly affecting host cells. In antimicrobial chemotherapy, the work of P. ERLICH on aniline dyes and arsenicals (SALVARSAN) and of G. DOMAGK on Prontosil led to the development of sulfonamides (see SULFA DRUGS). Many useful synthetic compounds are now available for BACTERIAL and PARASITIC DISEASE,

The History of Chemistry.

BEFORE 700BC: The anonymous discovery and use of fire, copper, gold and silver, and later of bronzes and iron; also of glazes, dyes and pigments.

700BC–100BC: Greek philosophers contemplate the nature of matter.

Ionian philosophers seek a primal matter:
Thales (water)
Anaximander (*apeiron*)
Anaximenes (air)
Heraclitus (fire)
Anaxagoras (an infinity of kinds of "seeds")

Eleatic philosophers:
thought alone yields truth
Parmenides
Zeno

Atomists:
Empedocles (four kinds of atoms)
Leucippus
Democritus

Socrates
Plato (*Timaeus:* the four elements geometrized)
Aristotle: four elements (also the quintessence) and four qualities.

Lucretius

200BC–400AD: *The high point of Chinese alchemy.*

100BC–1500AD: *Metallurgists and alchemists:*

100BC–300AD: Hellenistic: "Hermes Trismegistus"; "Pseudo-Democritus"

670AD–1050AD: Arab: *Book of Krates*
"Jabir" corpus
al-Razi (Rhazes)
ibn Sina (Avicenna)

Discovery of gunpowder, c900, in China

c1100: distillation of *aqua vitae* (ethanol—in Italy)

13TH CENTURY: preparation of *aqua regia*, nitric and sulfuric acids; Albertus Magnus; Roger Bacon.

14TH CENTURY: "Geber"

16TH CENTURY:
metallurgy: Georg Bauer (Agricola), *De re metallica* (1556)
chemistry: Andreas Libau (Libavius), *Alchemia* (1597)
iatrochemistry: Paracelsus: three principles: mercury, sulfur and salt

17TH CENTURY:
quantitative method: van Helmont ("gas")
practical chemistry: Glauber
theoretical chemistry: Becher

Revival of atomism:
Gassendi
Descartes
Boyle (*The Sceptical Chymist* (1661))

18TH CENTURY:

theoretical chemistry:
Boerhaave
Stahl (phlogiston theory)
Geoffroy (affinities)
Lavoisier (oxygène theory; binomial nomenclature)
Proust

gas chemistry:
Scheele (HF, HCN, O_2—"fire air")
Black (CO_2, thermochemistry)
Cavendish (H_2)
Priestley (NO, NO_2, CO, SO_2, HCl (gas), O_2—"dephlogisticated air")

Electrochemistry:
Galvani
Volta
Davy (K, Na, Cl_2)
Faraday

19TH CENTURY:
Berthollet
Dalton (modern atomic theory)
Gay-Lussac
Avogadro (atoms and molecules)
Prout
Dulong & Petit
Döbereiner (triads)
Newlands (octaves)
Meyer & Mendeleyev (periodic table)

Rayleigh & Ramsey (noble gases)

Organic Chemistry:
Berzelius (coins term)
Liebig & Wöhler (radical theory)
Laurent (unitary theory)
Pasteur (isomerism of tartaric acid)
Gerhardt (new type theory)
Kekulé (teratomic carbon; benzene ring)
Crum Brown (graphic formulae)
van't Hoff (tetrahedral carbon)
Fischer (sugar synthesis)

Physical Chemistry:
Carnot (heat engine)
Clapeyron
Joule
Hess
Helmholtz & Kelvin (conservation of energy)
Clausius (entropy)
Bunsen & Kirchhoff (spectroscopy)
Berthelot
Nernst (heat theorem)
Gibbs (phase rule)
Arrhenius (dilute solutions)
Debye & Hückel (strong electrolytes).

20TH CENTURY:
Major advances in theoretical chemistry dependent on developments in atomic physics.
Studies of chemical bonding (Lewis, Pauling & al.)
Major developments in biochemistry.
Importance of industrial syntheses.
Growth of petrochemical industries.

although ANTIBIOTICS are often preferred for bacteria. Cancer chemotherapy is especially successful in LEUKEMIA and lymphoma; in carcinoma it is usually reserved for disseminated tumor. Nitrogen mustard, ALKALOIDS derived from the periwinkle, certain antibiotics and agents interfering with DNA METABOLISM are used, often in combinations and usually with STEROIDS.

CHEMURGY, the application of science to both the development of new agricultural products (and the realization of the potential of previously uncultivated plants) and the derivation of new uses to which existing agricultural products may be put (see also AGRONOMY). As such, the chemurgist deals not only with techniques of plant and animal BREEDING but also with ways of processing foods and feeds, industrial and agricultural uses of what were previously regarded as agricultural waste by-products, etc.

CHERENKOV (or CERENKOV), Pavel Alekseyevich (1904–), Russian physicist who first observed CERENKOV RADIATION (1934). He shared the 1958 Nobel Prize for Physics with FRANK and TAMM who correctly interpreted this phenomenon in 1937.

CHERNOZEM (Russian: black earth), a group of neutral SOILS with a dark surface layer rich in HUMUS. It occurs in the Russian steppes, Argentina, and in a wide belt from E Kan. to N Alberta. It is the best soil for cereal growing.

CHERT. See FLINT.

CHEST, upper part of the trunk, between the neck and ABDOMEN. The chest wall consists of ribs articulating with the spinal column and sternum, and the related muscles. The DIAPHRAGM separates the chest from the abdomen. The LUNGS fill much of each side of the chest while the HEART, the AORTA and other large vessels lie centrally, the heart slightly to the left; the TRACHEA and ESOPHAGUS pass into the chest from the neck.

CHICKENPOX, or **varicella,** a VIRUS disease due to *Varicella zoster,* affecting mainly children, usually in EPIDEMICS. It is contracted from other cases or from cases of SHINGLES and is contagious. It causes malaise, FEVER and a characteristic vesicular rash—mainly on trunk and face—and cropping occurs. Infrequently it becomes hemorrhagic or LUNG involvement occurs. Chickenpox is rarely serious in the absence of underlying disease but it is important to distinguish it from SMALLPOX.

CHILBLAIN, itchy or painful red swelling of extremities, particularly toes and fingers, in predisposed subjects. A tendency to cold feet and exposure to extremes of temperature appear to be factors in causation. Treatment is symptomatic.

CHILDBIRTH. See BIRTH.

CHILE SALTPETER, an impure form of sodium nitrate (see SODIUM) occurring in large quantities in Chile.

CHILL HARDENING, technique used in CASTING whereby the mold is chilled. This accelerates the cooling of the molten metal poured into it, thereby increasing its surface HARDNESS.

CHINA CLAY. See KAOLIN.

CHINAWARE. See CERAMICS; POTTERY AND PORCELAIN.

CHINOOK, FOEHN wind blowing eastward from the Rocky Mountains, mainly in winter. The term is also applied to the warm, wet WIND coming off the Pacific before it passes over the Rockies.

CHIROPODY. See PODIATRY.

CHIROPRACTIC, a health discipline based on a theory that disease results from misalignment of VERTEBRAE. Manipulation, massage, dietary and general advice are the principal methods used. It was founded by Daniel D. Palmer in Davenport, Ia., in 1895 and has a substantial following in the US.

CHIROPTERA, an order of the MAMMALIA that includes the bats. Unlike other mammals, they can fly, the front limbs being modified to support a membranous wing. All bats are nocturnal in habits and most are found in tropical regions. There are about 800 living species.

CHI-SQUARED TEST, in STATISTICS, a test of the closeness of an observed result to an expected result, and hence how closely a model corresponds to a particular sample of n members of a population. χ^2 is given by

$$\chi^2 = \frac{(n-1)s^2}{\sigma^2},$$

where s^2 is the observed sample VARIANCE and σ is the STANDARD DEVIATION of the model. Values of χ^2 have been tabulated for a range of values of n, and their distributions (chi-square distributions with $(n-1)$ DEGREES OF FREEDOM) derived in each case.

CHITIN. See CARAPACE.

CHLORAL, or trichloroacetaldehyde (CCl_3CHO), colorless, oily liquid made by reacting CHLORINE and ETHANOL or ACETALDEHYDE, used chiefly in the manufacture of DDT. MW 147.4, mp −57.5°C, bp 98°C. **Chloral hydrate** ($CCl_3CH(OH)_2$), colorless crystalline solid made by reacting chloral and water and used as a SEDATIVE, since it is a DEPRESSANT of the central nervous system. It is toxic in excess, and especially when mixed with alcohol to make "Mickey Finns" or "knockout drops." Habitual use produces DRUG ADDICTION and gastritis. mp 57°C.

CHLORINE (Cl), greenish-yellow gas with a pungent odor, a typical member of the HALOGENS, occurring naturally as chlorides (see HALIDES) in seawater and minerals. It is made by electrolysis of SALT solution, and is used in large quantities as a bleach, as a disinfectant for drinking water and swimming pools, and in the manufacture of plastics, solvents and other compounds. Being toxic and corrosive, chlorine and its compound PHOSGENE have been used as poison gases (see CHEMICAL AND BIOLOGICAL WARFARE). Chlorine reacts with most organic compounds, replacing hydrogen atoms (see ALKYL HALIDES) and adding to double and triple bonds. AW 35.5, mp −101°C, bp −35°C. **Chlorides,** the commonest chlorine compounds, are typical HALIDES except for carbon tetrachloride (see CARBON), which is inert (see STEREOCHEMISTRY). Other chlorine compounds include a series of oxides, unstable and highly oxidizing, and a series of oxyanions—HYPOCHLORITE, chlorite, chlorate and perchlorate—with the corresponding OXY-ACIDS, all powerful oxidizing agents. Calcium hypochlorite (see BLEACHING POWDER) and sodium chlorite are used as bleaches; chlorates are used as weedkillers and to make matches and fireworks; perchlorates are used as explosives and rocket fuels. (See also HYDROGEN CHLORIDE.)

CHLORITE, group of green CLAY minerals, related to the MICAS, and consisting of alternate single layers of

The structure of chlorophyll A.

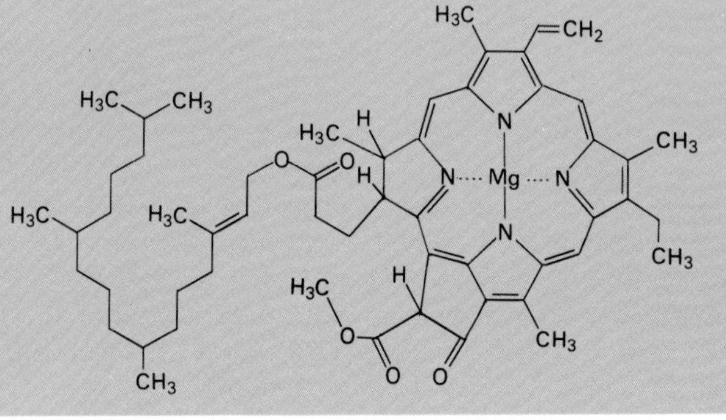

BIOTITE, containing ferrous iron (Fe^{2+}), and brucite [$(Mg,Al)_6(OH)_{12}$]. They are widespread alteration products.

CHLOROFORM, or trichloromethane ($CHCl_3$), dense, colorless, volatile liquid made by chlorination of ETHANOL or ACETONE. One of the first anesthetics (see ANESTHESIA) in modern use (by Sir James SIMPSON, 1847), it is now seldom used except in tropical countries, despite its potency, since it has a narrow safety margin and is highly toxic in excess. It is also used in cough medicines and as an organic solvent; it is nonflammable. MW 119.4, mp −64°C, bp 61°C.

CHLOROMYCETIN, or chloramphenicol, the first broadspectrum ANTIBIOTIC. Owing to risk of aplastic ANEMIA it is restricted to use in TYPHOID FEVER or serious *Hemophilus* infections and for tropical use.

CHLOROPHYLL, various green pigments found in plant CHLOROPLASTS. They absorb light and convert it into chemical energy, thus playing a basic role in PHOTOSYNTHESIS. Chlorophylls are CHELATE compounds in which a magnesium ion is surrounded by a PORPHYRIN system.

CHLOROPHYTA, or green algae and stone-worts. See ALGAE.

CHLOROPLAST, a PLASTID body containing CHLOROPHYLL, found in plant cells. Higher plant cells contain up to 50 chloroplasts, while most algal cells have only one. Variously shaped, they are of the order of 5μm in diameter, and are surrounded by a semipermeable membrane. They contain ordered stacks of membranes within which all PHOTOSYNTHESIS reactions take place.

CHLOROQUINE, antimalarial drug derived from QUININE, acting on the red-cell phase of the parasite. Probably the most commonly used drug for both the prevention and treatment of MALARIA. It is effective against all varieties of malarial parasite.

CHOCOLATE, popular confectionary made from cacao beans. Fermented beans are roasted and the outer husks removed by a process that breaks the kernels into fragments called nibs. Chocolate is made from ground nibs, cocoa butter (the fat released when the nibs are subjected to hydraulic pressure), sugar and sometimes milk. It is a high energy food that contains a small amount of the stimulant CAFFEINE. Chocolate may be molded into bars or used as a beverage and in some liqueurs.

CHOKE, or **inductor**, device used in electric CIRCUITS to oppose changes in the magnitude or direction of current flow (see ELECTRICITY); i.e., it is a coil of high self-inductance (see INDUCTANCE). The term is also used for the VALVE regulating the air supply to the CARBURETOR of an INTERNAL-COMBUSTION ENGINE.

CHOLERA, a BACTERIAL DISEASE causing profuse watery DIARRHEA, due to *Vibrio cholerae*. It is endemic in many parts of the East and EPIDEMICS occur elsewhere. A water-borne infection, it was the subject of a classic epidemiological study by John Snow in 1854. Abdominal pain and diarrhea, which rapidly becomes severe and watery, are main features, with rapidly developing dehydration and SHOCK. Without rapid and adequate fluid replacement, death ensues rapidly; ANTIBIOTICS may shorten the diarrheal phase. It is a disease due to a specific TOXIN; a similar but milder disease occurs due to the El Tor Vibrio. VACCINATION gives limited protection for six months.

CHOLESTEROL ($C_{27}H_{46}O$), STEROL found in nearly all animal tissue, especially in the NERVOUS SYSTEM, where it is a component of MYELIN. Cholesterol is a precursor of BILE salts and of adrenal and sex HORMONES. Large amounts are synthesized in the liver, intestines and skin. Cholesterol in the diet supplements this. Since abnormal deposition of cholesterol in the arteries is associated with ARTERIOSCLEROSIS, some doctors advise avoiding high-cholesterol foods and substituting unsaturated for saturated FATS (the latter increase production and deposition of cholesterol). It is a major constituent of gallstones (see CALCULI).

cholesterol

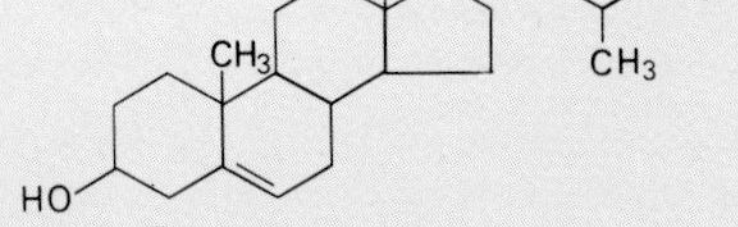

CHOMSKY, Avram Noam (1928–), US linguist whose theory of "transformational grammar" has revolutionized the study of language structure. (See LINGUISTICS.)

CHONDRICHTHYES, or **Elasmobranchii**, or cartilaginous fish, a class of the phylum CHORDATA that includes the sharks, rays, skates and bradyodonts. They are characterized by having cartilaginous (see CARTILAGE) skeletons and scales that are tooth-like. Today, they are marine, but freshwater forms are known from a fossil record that extends back some 300 million years.

CHONDROSTEI, a group of mainly fossil fish of the ACTINOPTERYGII represented today by the STURGEONS and bichirs (POLYPTERUS).

CHONDRULE, small, roughly spherical granules of OLIVINE, PYROXENE, GLASS or other MINERALS found in members of that class of stony meteorites (see METEOR) known as **chondrites**.

CHORD. See CIRCLE.

CHORDATA, a phylum that includes animals called protochordates (HEMICHORDATA, UROCHORDATA, CEPHALOCHORDATA), as well as the better known vertebrate groups AGNATHA and GNATHOSTOMATA. All are characterized by having, at some stage in their lives, a skeletal rod of tissue called a NOTOCHORD.

CHOREA, abnormal, nonrepetitive involuntary movements of the limbs, body and face. It may start with clumsiness, but later uncontrollable and bizarre movements occur. It is a disease of basal ganglia (see BRAIN). **Sydenham's chorea**, or **Saint Vitus' dance**, is a childhood illness associated with STREPTOCOCCUS infection and RHEUMATIC FEVER; recovery is usually full. **Huntingdon's chorea** is a rare hereditary disease, usually coming on in middle age and associated with progressive dementia.

CHROMATIC ABERRATION. See ABERRATION, OPTICAL.

CHROMATOGRAPHY, a versatile technique of chemical separation and ANALYSIS, capable of dealing with many-component mixtures, and large or small amounts. The sample is injected into the moving phase, a gas or liquid stream which flows over the stationary phase, a porous solid or a solid support coated with a liquid. The various components of the sample are adsorbed (see ADSORPTION) by the stationary phase at different rates, and separation occurs. Each component has a characteristic velocity relative to that of the solvent, and so can be identified. In **liquid-solid chromatography** the solid is packed into a tube, the sample is added at the top, and a liquid eluant is allowed to flow through; the different fractions of effluent are collected. A variation of this method is ion-exchange chromatography, in which the solid is an ION-EXCHANGE resin from which the ions in the sample are displaced at various rates by the acid eluant. Other related techniques are paper chromatography (with an adsorbent paper stationary phase) and thin-layer chromatography (using a layer of solid adsorbent on a glass plate). The other main type of chromatography—the most sensitive and reliable—is **gas-liquid chromatography (glc)**, in which a small vaporized sample is injected into a stream of inert eluant gas (usually nitrogen) flowing through a column containing nonvolatile liquid adsorbed on a powdered solid. The components are detected by such means as measuring the change in thermal conductivity of the effluent gas.

CHROMITE, a hard, black SPINEL mineral, an iron (II) chromium (III) mixed OXIDE ($FeCr_2O_4$); the chief ore of CHROMIUM. It is found in southern Africa, USSR, Turkey and the Philippines.

CHROMIUM (Cr), silvery-white, hard metal in Group VIB of the PERIODIC TABLE; a TRANSITION ELEMENT. It is widespread, the most important ore being CHROMITE. This is reduced to a ferrochromium alloy by carbon or silicon; pure chromium is produced by reducing chromium (III) oxide with aluminum. It is used to make hard and corrosion-resistant ALLOYS and for chromium ELECTROPLATING. Chromium is unreactive. It forms compounds in oxidation states +2 and +3 (basic) and +6 (acidic). Chromium (III) oxide is used as a green pigment, and lead chromate (VI) as a yellow pigment. Other compounds are used for tanning leather and as mordants in dyeing. (See also ALUM.) AW 52.0, mp 1890°C, bp 2482°C, sg 7.20 (20°C).

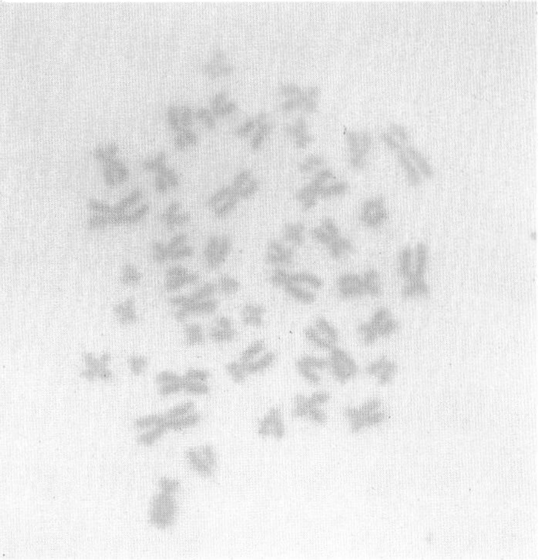

Chromosome spread from a dividing human cell.

CHROMOSOMES, threadlike bodies in cell nuclei, composed of GENES, linearly arranged, which carry genetic information responsible for the inherited characteristics of the organism (see HEREDITY). Chromosomes consist of the NUCLEIC ACID DNA (and sometimes RNA) attached to a protein core. All normal cells contain a certain number of chromosomes characteristic of the species (46 in man), in homologous pairs (diploid). GAMETES, however, are HAPLOID, having only half this number, one of each pair, so that they unite to form a ZYGOTE with the correct number of chromosomes. In man there is one pair of sex chromosomes, females having two X chromosomes, males an X and a Y; thus each EGG cell must have an X chromosome, but each spermatozoon (see SPERM) has either an X or a Y, and determines the sex of the offspring. In cell division, the chromosomes replicate and separate (see MEIOSIS; MITOSIS). Defective or supernumerary chromosomes cause various abnormalities, including MONGOLISM. (See also MUTATION; PROTEIN SYNTHESIS.)

CHROMOSPHERE. See SUN.

CHRONOLOGY, the science of dating involving the accurate placing of events in time and the definition of suitable timescales. In Christian societies, events are dated in years before (BC) or after (AD—*Anno Domini*) the traditional birth date of Christ. In scientific use, dates are often given BP (Before Present). In ARCHAEOLOGY, dating techniques include DENDROCHRONOLOGY and RADIOCARBON DATING. In GEOLOGY, rock strata are related to the geological time scale by examination of the FOSSILS they contain (see also POTASSIUM).

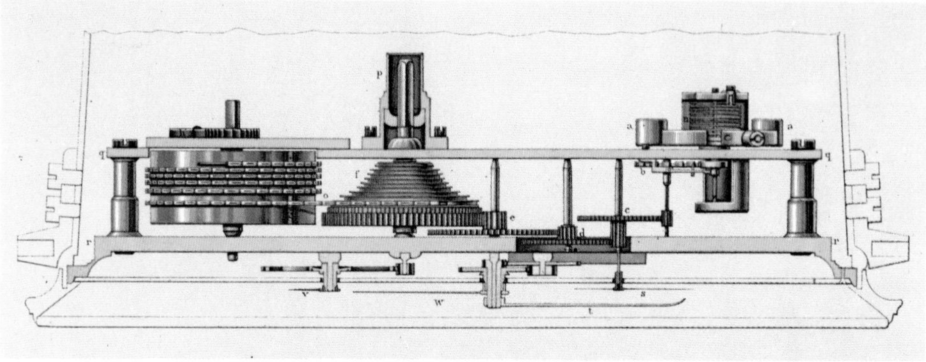

Mechanism of a marine chronometer from the late 19th century: (a) balance; (b) escape wheel; (c) second wheel; (d) third wheel; (e) minute or center wheel; (f) fusee; (g) spring barrel; (h) ratchet; (i) catch-spring mechanism to prevent chronometer stopping during winding; (k) winding stop with spring (s); (m) ratchet spring; (n) hairspring; (o) chain; (p) dust box surrounding winding lug; (q) rear plate; (r) main plate; (s) second hand; (t) minute hand; (v) escapement reserve hand indicating how long has elapsed since rewinding; (w) hour hand. Between the main plate and the engraved face is the "hand work" effecting the 12:1 reduction between the minute and hour hands and controlling the escapement-reserve hand.

CHRONOMETER, an extremely accurate clock, especially one used in connection with celestial NAVIGATION at sea (see also CELESTIAL SPHERE). It differs from the normal clock in that it has a **fusee**, by means of which the power transmission of the mainspring is regulated such that it remains approximately uniform at all times; and a balance made of metals of different coefficients of EXPANSION to minimize the effects of temperature changes. The device is maintained in gimbals to reduce the effects of rolling and pitching. A chronometer's accuracy is checked daily and its error noted; the daily change in error is termed the **daily rate**. Chronometers are always set to GREENWICH MEAN TIME. The first chronometer was invented by John HARRISON (1735). (See also ATOMIC CLOCK; CLOCKS AND WATCHES.)

CHRYSALIS, the PUPA stage in the life-history of many insects, during which the LARVA undergoes METAMORPHOSIS to produce the features characteristic of the adult.

CHRYSOBERYL, mixed OXIDE mineral of BERYLLIUM and ALUMINUM ($BeAl_2O_4$), forming hard crystals in the orthorhombic system, and found mainly in Brazil. ALEXANDRITE and CAT'S EYE are GEM varieties.

CHRYSOPHYTA, or golden algae and diatoms. See ALGAE.

CHRYSOPRASE, a GEM variety of CHALCEDONY, colored apple-green by colloidal nickel silicate. It is found in Silesia and Cal.

CHRYSOTILE, mineral forming the chief variety of ASBESTOS, a fibrous SERPENTINE.

CHURNING, process, known since prehistoric times, for converting cream to BUTTER. Cream is an EMULSION of water in oils: agitation in a butter churn

Chrysoprase as the raw mineral and cut and polished as a gemstone.

makes it an oil-in-water emulsion, and the fatty globules coagulate. The remaining buttermilk may then be skimmed.

CHYTRIDOMYCETES. See FUNGI.

CIDER, an ALCOHOLIC BEVERAGE made from fermented sour apples (see FERMENTATION). In the US this is known as "hard" cider, whereas "sweet" cider is a commercially-prepared nonalcoholic apple juice.

CIERVA, Juan de la (1895–1936), Spanish aviator, the inventor of the AUTOGIRO. He was killed in an airplane crash at Croydon, England.

CILIA, hair-like outgrowths from the surfaces of cells, which provide the means of locomotion for many small animals such as PROTOZOA and worms. In higher animals, cilia provide means for moving fluids over the surfaces of cells, such as in the respiratory tract. Flagella is the name used for longer cilia.

CILIATA. See PROTOZOA.

CINNABAR, red SULFIDE mineral consisting of mercury (II) sulfide (HgS), the chief ore of MERCURY, found in Spain, Italy, Peru and Cal. as massive deposits or hexagonal crystals. It is used as the pigment vermilion. A black, cubic form of mercury (II) sulfide, **metacinnabar**, also occurs.

CIRCADIAN RHYTHM. See BIOLOGICAL CLOCKS.

CIRCLE, a CURVE so drawn that all points on its perimeter are at an equal distance, the **radius**, from a single point, the **center**. A straight line cutting the circle is known as a **secant**, the area within the circle cut off by the secant being termed a **segment**. A secant passing through the center is called a **diameter**. The part of a secant lying with a circle is a **chord**, the part of the perimeter between the two points of intersection of the secant and the circle being termed an **arc**. A line in the same plane touching the circle at a single point is a **tangent** (see TANGENT OF A CURVE). A **central angle** is the angle between two radii, and the area between them is termed a **sector**. The length of the perimeter is the circle's **circumference**. The AREA of a circle is πr^2 (see PI), where r is the radius; and its circumference is $2\pi r$. (See also CYLINDER; MENSURATION; SPHERE; SPHERICAL GEOMETRY.)

CIRCUIT, Electric, assemblage of electrical CONDUCTORS (usually wires) and components through which current from a power source such as a BATTERY or GENERATOR flows (see ELECTRICITY). Components may be connected one after another (in series) or side by side (in parallel). If current may flow between two points their connection is a closed circuit; if not, an open circuit; and if RESISTANCE between them is virtually zero, a short circuit: a switch when off is a closed circuit, when on a short circuit. Short circuits between the terminals of the power source are dangerous (see CIRCUIT BREAKER; FUSE). (See also ELECTRONICS; KIRCHHOFF'S LAWS.)

CIRCUIT BREAKER, device now often used in place of a FUSE to protect electrical equipment from damage when the current exceeds a desired value, as in short-circuiting. The circuit breaker opens the CIRCUIT automatically, usually by means of a coil that separates contacts when the current reaches a certain value (see ELECTROMAGNETISM). One advantage of the circuit breaker is that the contacts may be reset (by hand or automatically) whereas a fuse has to be replaced. Small circuit breakers are used in the home (as in many TELEVISION sets), larger ones in industry.

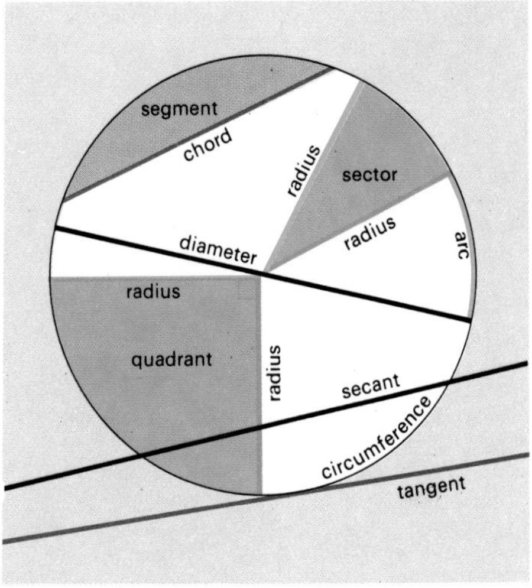

Terms associated with the circle.

CIRCULATION OF THE BLOOD. See BLOOD CIRCULATION.

CIRCUMCENTER, the point of concurrence of the lines drawn perpendicular to and through the midpoints of the sides of a TRIANGLE. This point is equidistant from the triangle's vertices and hence the center of the circle that may be circumscribed about it (see CIRCUMSCRIPTION).

CIRCUMPOLAR STARS, those stars which can be seen every night of the year from any particular latitude, and which appear to circle the celestial pole (see CELESTIAL SPHERE).

CIRCUMSCRIPTION, in plane GEOMETRY, the construction of a CIRCLE such that all vertices of a particular POLYGON lie on its circumference, the polygon then being said to be inscribed (see INSCRIPTION) within the circle. All regular polygons and all TRIANGLES may be circumscribed. In three dimensions circumscription implies the construction of a SPHERE such that on its surface lie all the vertices of a POLYHEDRON.

CIRQUE, corrie or **cwm,** steep-sided hollow formed by glacial erosion, usually occupied by a lake where the GLACIER has retreated, or by névé where the glacier is still present. (See also EROSION.)

CIRRHOSIS, chronic disease of the LIVER, with disorganization of normal structure and replacement by fibrous scars and regenerating nodules. It is the end result of many liver diseases, all of which cause liver-cell death; most common are those associated with ALCOHOLISM and following some cases of hepatitis, while certain poisons and hereditary diseases are rare causes. All liver functions are impaired, but symptoms often do not occur until early liver failure develops with EDEMA, ascites, JAUNDICE, COMA, emaciation, or gastrointestinal-tract HEMORRHAGE; BLOOD clotting is often abnormal and PLASMA proteins are low. The liver damage is not reversible, but if recognized early in the alcoholic, abstention can minimize progression. Treatment consists of measures to protect the liver from excess protein, DIURETICS and the prevention and treatment of hemorrhage.

CITRIC ACID ($C_6H_8O_7$), colorless crystalline solid, a CARBOXYLIC ACID widespread in plant and animal tissue, especially citrus fruits. It is made commercially by FERMENTATION of crude sugar with the fungus *Aspergillus niger*, and is used in the food, pharmaceutical and textile industries, and for cleaning metals. It is vital in cell METABOLISM (see CITRIC ACID CYCLE). MW 192.1, mp 153°C.

CITRIC ACID CYCLE, or **Krebs cycle,** or tricarboxylic acid cycle, a vital cycle of chemical reactions forming the final stage in the oxidation of food in cells (see METABOLISM; RESPIRATION) and thus producing energy for BIOSYNTHESIS. The previous stages of oxidation produce the acetyl derivative of coenzyme A (see ENZYMES). The acetyl group then adds to oxaloacetic acid to produce citric acid, which in a sequence of enzyme-catalyzed oxidation reactions is converted back to oxaloacetic acid. The net result of the cycle is total oxidation of one ACETIC ACID molecule to carbon dioxide and water. The oxidizing agents are pyridine NUCLEOTIDES, which accept hydrogen atoms and are then reoxidized by CYTOCHROMES, the energy produced being stored as ATP (see NUCLEOTIDES). Except in microorganisms, the enzymes needed in the cycle are located in the mitochondria (see CELL).

CIVIL ENGINEERING, that branch of engineering concerned with the design, construction and maintenance of stationary structures such as buildings, bridges, highways, dams, etc. The term "civil engineer" was first used c1750 by John Smeaton, who built the Eddystone Lighthouse. Nowadays, civil engineering incorporates modern technological advances in the structures required by industrial society. The branches of the field include: SURVEYING, concerned with the selection of sites; hydraulic and sanitary engineering, dealing with public WATER SUPPLY and SEWAGE disposal, etc.; transportation engineering, which deals with highways, airports and so on; structural engineering, which is concerned with the actual planning and construction of permanent installations; and environmental, town or city planning, which improves existing urban environments and plans the development of new areas.

The citric acid (Krebs) cycle.

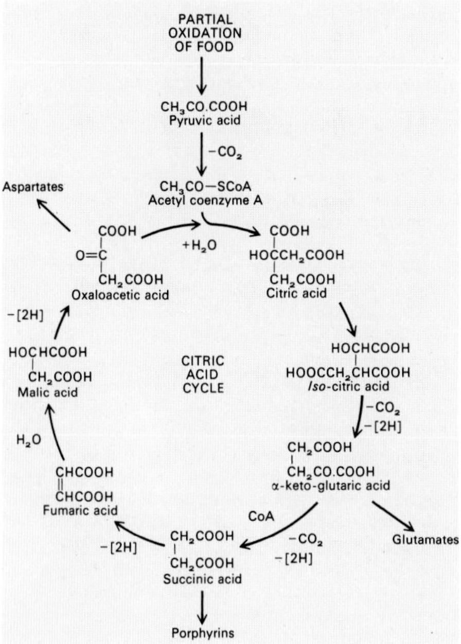

CLAN, a social group whose membership is restricted to those claiming descent from a common ancestor, usually through the male line though sometimes through the female line. Clans are often exogamous (see ENDOGAMY AND EXOGAMY) and usually have a TOTEM or other emblem. The most sophisticated clans currently in existence are those in Scotland.

CLARK, Alvan Graham (1832–1897), US lens-maker and astronomer who first saw the white dwarf companion (see STAR) of Sirius (1862). In 1897 he directed the construction of the world's largest refracting TELESCOPE, the 40in instrument at the Yerkes observatory.

CLAUDE, Albert (1899–), Belgian-US biologist who shared the 1974 Nobel Prize for Physiology or Medicine with G. E. PALADE and C. de DUVE for demonstrating the usefulness of the ELECTRON MICROSCOPE and the CENTRIFUGE in biological studies. He pioneered the study of the internal structure of CELLS.

CLAUSIUS, Rudolf Julius Emanuel (1822–1888), German theoretical physicist who first stated the second law of THERMODYNAMICS (1850) and proposed the term ENTROPY (1865). He also contributed to KINETIC THEORY and the theory of ELECTROLYSIS.

CLAUSTROPHOBIA, an intense feeling of fear or panic experienced by certain people in enclosed places. (See PHOBIA.)

CLAVICLE. See COLLARBONE.

CLAY, any SOIL material with a particle size of less than 2–4μm in diameter, i.e. finer-grained than SILT or SAND; an earthy material which becomes plastic when wet, including mud (which is used in oil drilling). Clays are used as catalysts (see CATALYSIS) in PETROLEUM refining, for making molds for CASTING and, when molded and fired, for CERAMICS, POTTERY AND PORCELAIN, bricks and TILES. They are also used in making CEMENT and RUBBER, and as ION-EXCHANGE agents for softening HARD WATER. (See also FULLER'S EARTH.) Clay rocks, including mudstones and SHALES, are microcrystalline rocks composed mainly of clay-size particles. Their mineralogical composition is highly variable, but they usually contain a high proportion of **clay minerals**, hydrated aluminum and magnesium SILICATES including BENTONITE, CHLORITE, diaspore (hydrated ALUMINUM oxide), illite (hydrated MICA—see also GLAUCONITE), KAOLINITE and MEERSCHAUM.

CLEAR AIR TURBULENCE (CAT). See AIR TURBULENCE.

CLEAVAGE, of a MINERAL, the tendency to split along a definite PLANE parallel to an actual or possible CRYSTAL face: e.g., GALENA, whose crystals are cubic, cleaves along three mutually PERPENDICULAR planes (parallel to 100, 010, 001). Such cleavage is useful in identifying minerals. ROCK cleavage generally takes place between roughly parallel beds whose resistance

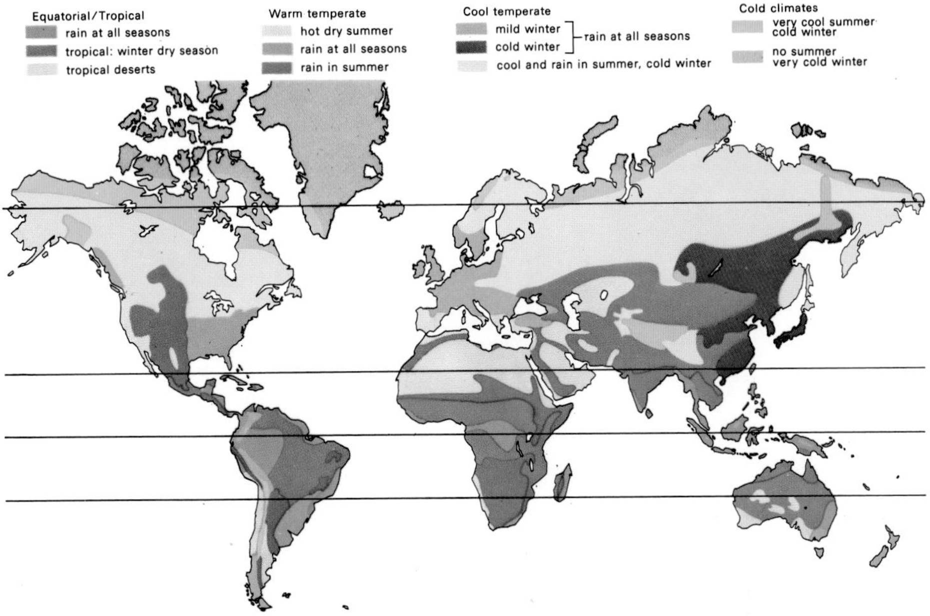

The climates of the world (modified Köppen system).

under deformation to internal SHEARING differs.

CLEFT PALATE, a common developmental deformity of the PALATE in which the two halves do not meet in the midline; it is often associated with HARELIP. It can be familial or follow disease in early PREGNANCY, but may appear spontaneously. It causes a characteristic nasal quality in the cry and voice. Plastic SURGERY can close the defect and allow more normal development of the voice and TEETH.

CLEPSYDRA, or **water clock,** an instrument in which the discharge of water from a storage tank is monitored in order to measure the passing of time. They were used from ancient times until the Renaissance.

CLIMATE, the sum of the weather conditions prevalent in an area over a period of time. Weather conditions include temperature, rainfall, sunshine, wind, humidity and cloudiness. Climates may be classified into groups. The system most used today is that of Vladimir Köppen, with five categories (A, B, C, D, E), broadly defined as follows:

A Equatorial and tropical rainy climates;
B Arid climates;
C Warmer forested (temperate) climates;
D Colder forested (temperate) climates; and
E Treeless polar climates.

These categories correspond to a great extent to zoning by LATITUDE; this is because the closer to the EQUATOR an area is, the more direct the sunlight it receives and the less the amount of ATMOSPHERE through which that sunlight must pass. Other factors are the rotation of the earth on its axis (diurnal differences) and the revolution of the earth about the sun (seasonal differences).

PALEOCLIMATOLOGY, the study of climates of the past, has shown that there have been considerable long-term climatic changes in many areas: this is seen as strong evidence for CONTINENTAL DRIFT (see also PLATE TECTONICS). Other theories include variation in the solar radiation (see SUN) and change in the EARTH's axial tilt. Man's influence has caused localized, short-term climatic changes. (See CLOUDS; METEOROLOGY; RAIN; TROPIC; WEATHER FORECASTING; WIND.)

CLOCKS AND WATCHES, devices to indicate or record the passage of time; essential features of modern life. In prehistory, time could be gauged solely from the positions of celestial bodies; a natural development was the SUNDIAL, initially no more than a vertical post whose SHADOW was cast by the sun directly onto the ground. Other devices depended on the flow of water from a pierced container (see CLEPSYDRA); the rates at which marked candles, knotted ropes and oil in calibrated vessels burned down; and the flow of sand through a constriction from one bulb of an HOURGLASS to the other. Mechanical clocks were probably known in Ancient China, but first appeared in Europe in the 13th century AD. Power was supplied by a weight suspended from a rope, later by a coiled spring; in both cases an escapement being employed to control the energy release. Around 1657–58 HUYGENS applied the PENDULUM principle to clocks; later, around 1675, his hairspring and balance-wheel mechanism made possible the first portable clocks—resulting eventually in watches. Jeweled bearings, which reduced wear at critical points in the mechanism, were introduced during the 18th century, and the first CHRONOMETER was also devised in this century. Electric clocks with synchronous motors are now commonly found in the home and office, while the ATOMIC CLOCK, which can be accurate to within one second in 3 million years, is of great importance in science.

CLOTTING, the formation of semisolid deposits in a

liquid by coagulation, often by the denaturing of previously soluble ALBUMIN. Thus clotted cream is made by slowly heating milk so that the thick cream rises; the curdling of skim milk to make CHEESE is also an example of clotting. Clotting of BLOOD is a complex process set in motion when it comes into contact with tissues outside its ruptured vessel. These contain a factor, **thromboplastin**, which activates a sequence of changes in the PLASMA clotting factors (12 enzymes). Alternatively, many surfaces, such as glass and fabrics, activate a similar sequence of changes. In either case, factor II (prothrombin, formed in the liver), with calcium ions and a platelet factor, is converted to **thrombin**. This converts factor I (fibrinogen) to **fibrin**, a tough, insoluble polymerized protein which forms a network of fibers around the platelets (see BLOOD) that have stuck to the edge of the wound and to each other. The network entangles the blood cells, and contracts, squeezing out the serum and leaving a solid clot. (See also ANTICOAGULANTS; EMBOLISM; HEMOPHILIA; HEMORRHAGE; THROMBOSIS.)

CLOUD CHAMBER, device, invented by C. T. R. WILSON (1911), used to observe the paths of subatomic particles. In simplest form, it comprises a chamber containing saturated VAPOR (see SATURATION) and some liquid, one wall of the chamber (the window) being transparent, another retractable. Sudden retraction of this wall lowers the temperature, and the gas becomes supersaturated (and thus metastable). Passage of SUBATOMIC PARTICLES through the gas leaves charged IONS that serve as seeds for CONDENSATION of the gas into droplets. These fog trails (condensation trails) may be photographed through the window. (See also BUBBLE CHAMBER; SPARK CHAMBER.)

CLOUDS, visible collections of water droplets or ice particles that, because they fall so slowly, may be regarded as suspended in the ATMOSPHERE. Clouds whose lower surfaces touch the ground are usually called FOG. The water droplets are very small, indeed of colloidal size (see COLLOID; AEROSOL); they must coagulate or grow before falling as rain or snow. This process may be assisted by **cloud seeding**; supercooled clouds are seeded with particles of (usually) dry ice (i.e., solid CARBON dioxide) to encourage CONDENSATION of the droplets, ideally causing RAIN or SNOW.

Cloud Formation. Clouds are formed when air containing water vapor cools in the presence of suitable condensation nuclei (e.g., dust particles). This may occur through CONVECTION, when warm, moist air from near the earth's surface penetrates upward into regions of lower PRESSURE; here they expand, thus cooling past the dew point (see CONDENSATION), the temperature at which SATURATION of water in air is reached. The vapor condenses out to form cloud droplets. They may also form when warm air flows over a mountain; the air travels in a vertical wave (see WAVE MOTION), parts of which may be higher than its condensation level,

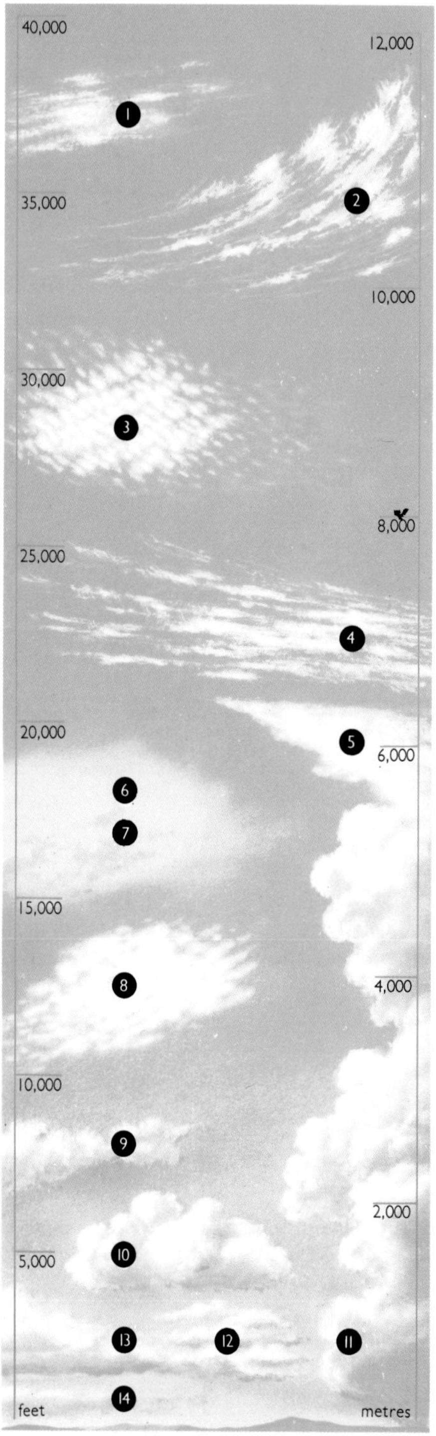

The altitudes at which different types of clouds occur: (1) iridescent; (2) cirrus; (3) cirrocumulus; (4) cirrostratus; (5) anvil head of cumulonimbus; (6) altostratus; (7) fractostratus patches; (8) altocumulus; (9) nimbus; (10) cumulus; (11) cumulonimbus; (12) stratocumulus; (13) nimbostratus; (14) stratus.

resulting in a stationary cloud at the crest of the wave. Other modes of formation occur. (See METEOROLOGY.)

Types of Clouds. There are three main cloud types: Cumulus (heap) clouds, formed by convection, often mountain- or cauliflower-shaped, are found from about 600m (2000ft) up as far as the tropopause, even temporarily into the stratosphere (see ATMOSPHERE). Cirrus (hair) clouds are composed almost entirely of ice crystals. They appear feathery, and are found at altitudes above about 6000m (20000ft). Stratus (layer) clouds are low-lying, found between ground level and about 1500m (5000ft). Other types of cloud include cirrostratus, cirrocumulus, altocumulus, altostratus, cumulonimbus, stratocumulus and nimbostratus.

CLUBFOOT, deformity of the FOOT, with an abnormal relationship of the foot to the ankle; most commonly the foot is turned in and down. Abnormalities of fetal posture and ligamentous or muscle development, including CEREBRAL PALSY and SPINA BIFIDA, may be causative. Correction includes gentle manipulation, PHYSIOTHERAPY, plaster splints and sometimes SURGERY.

CLUB MOSSES, primitive vascular plants of the order Lycopodiales, related to the ferns. They have creeping stems that branch dichotomously and small leaves arranged spirally. These are found mainly in the tropics, but some occur in temperate climates.

CLUBROOT, a plant disease affecting cabbages and other members of the mustard family (Cruciferae). It is caused by the SLIME MOLD *Plasmodiophora brassicae*. Club-like bodies form on the roots. Clubroot is avoided by growing resistant plant strains and adding lime to the soil. (See also PLANT DISEASES.)

CLUSTER. See STAR CLUSTER.

CLUTCH. See TRANSMISSION.

CNIDARIA, a phylum of mainly marine animals once included with the CTENOPHORA in the phylum COELENTERATA. They include the freshwater hydra and the marine jellyfish, Sea anemones, CORALS, Seafirs and Sea fans. Cnidarians have bodies composed of two layers of cells and most species exist in a variety of forms, a phenomenon known as ALTERNATION OF GENERATIONS. The main forms are the polyp, which is cylindrical and sessile, and the hydranth or medusa, which is bell- or saucer-shaped and free swimming.

COAL, hard, black mineral burned as a FUEL. With its by-products COKE and COAL TAR it is vital to many modern industries.

Coal is the compressed remains of tropical and subtropical plants, especially those of the CARBONIFEROUS and PERMIAN periods. Changes in the world climatic pattern explain why coal occurs in all continents, even Antarctica. Coal formation began when plant debris accumulated in swamps, partially decomposing and forming PEAT layers. A rise in sea level or land subsidence buried these layers below marine sediments, whose weight compressed the peat, transforming it under high-temperature conditions to coal; the greater the pressure, the harder the coal.

Coals are analyzed in two main ways: the "ultimate analysis" determines the total percentages of the elements present (carbon, hydrogen, oxygen, sulfur and nitrogen); and the "proximate analysis" gives an empirical estimate of the amounts of moisture, ash, volatile materials and fixed carbon. Coals are classified, or ranked, according to their fixed-carbon content, which increases progressively as they are formed. In ascending rank, the main types are: **lignite**, or brown coal, which weathers quickly, may ignite spontaneously, and has a low calorific value (see FUEL), but is used in Germany and Australia; **subbituminous coal**, mainly used in generating stations; **bituminous coal**, the commonest type, used in generating stations and the home, and often converted into COKE; and **anthracite**, a lustrous coal which burns slowly and well, and is the preferred domestic fuel.

Coal was burned in Glamorgan, Wales, in the 2nd millennium BC, and was known in China and the Roman Empire around the time of Christ. Coal mining was practiced throughout Europe and known to the American Indians by the 13th century AD. The first commercial coal mine in the US was at Richmond, Va., (opened 1745) and anthracite was mined in Pa. by 1790. The Industrial Revolution created a huge and increasing demand for coal. This slackened in the 20th century as coal faced competition from abundant oil and gas, but production is now again increasing. Annual world output is about 3 billion tonnes, 500 million tonnes from the US. World coal reserves are estimated conservatively at about 7 trillion tonnes, enough to meet demand for centuries at present consumption rates. (See also MINING.)

COAL GAS, a mixture of gases produced by the destructive distillation of COAL, consisting chiefly of hydrogen, methane and carbon monoxide. Other products are COKE and COAL TAR. Coal gas is used as a domestic fuel, but has been largely superseded by NATURAL GAS.

COAL TAR, a dense black viscous liquid produced by the destructive distillation of COAL; COKE and COAL GAS are other products. Fractional distillation of coal tar produces a wide variety of industrially important substances. These include ASPHALT (pitch) and CREOSOTE, a wood preservative; also various oils used as fuels, solvents, preservatives, lubricants and disinfectants. Specific chemicals that can be isolated include benzene, toluene, xylene, phenol, pyridine, naphthalene and anthracene—the main source for the pharmaceutical and other chemical industries.

COAST AND GEODETIC SURVEY, US. See NATIONAL OCEANIC AND ATMOSPHERIC ADMINISTRATION.

COBALT (Co), silvery-white, hard, ferromagnetic (see MAGNETISM) metal in Group VIII of the PERIODIC TABLE; a TRANSITION ELEMENT. It occurs in nature largely as sulfides and arsenides, and in nickel and copper ores; major producers are Canada, Zaire and Zambia. An ALLOY of cobalt, aluminum, nickel and iron ("Alnico") is used for magnets; other cobalt alloys, being very hard, are used for cutting tools. Cobalt is used as the matrix for tungsten carbide in drill bits. Chemically it resembles IRON and NICKEL: its characteristic oxidation states are +2 and +3. Cobalt compounds are useful colorants (notably the artists' pigment cobalt blue). Cobalt CATALYSTS facilitate HYDROGENATION and other industrial processes. The RADIOISOTOPE cobalt-60 is used in RADIATION THERAPY. Cobalt is a constituent of the vital VITAMIN B_{12}. AW 58.9, mp 1495°C, bp 2870°C, sg 8.9 (20°C).

COBALT BOMB, device used in CANCER treatment as a source of gamma rays (see RADIOACTIVITY;

RADIATION THERAPY). It uses Co^{60}, a RADIOISOTOPE of cobalt. Because of its long half-life, Co^{60} is used widely as a radioactive tracer; and gives its name to a theoretical nuclear weapon (see ATOMIC BOMB; NUCLEAR WARFARE), also called the cobalt bomb, whose fallout might remain deadly for years.

COBOL (Common Business-Orientated Language), COMPUTER language designed primarily for business use. It has the advantage that it can be easily learned and understood by people without technical backgrounds; and that a program designed for one computer may be run on another with minimal alteration.

COCA, a shrub, *Erythroxylon coca*, whose leaves contain various ALKALOIDS, especially COCAINE. Native to the Andes, it is widely cultivated elsewhere. The leaves have been chewed by South American Indians for centuries to quell hunger and to refresh. Cocaine-free coca extracts are used in making cola drinks (see KOLA).

COCAINE, an ALKALOID from the coca leaf, the first local anesthetic agent and model for those currently used; it is occasionally used for surface ANESTHESIA. It is a drug of abuse, taken for its euphoriant effect by chewing the leaf, as snuff or by intravenous INJECTION. Although physical dependence does not occur, its abuse may lead to acute psychosis. It also mimics the actions of the sympathetic NERVOUS SYSTEM.

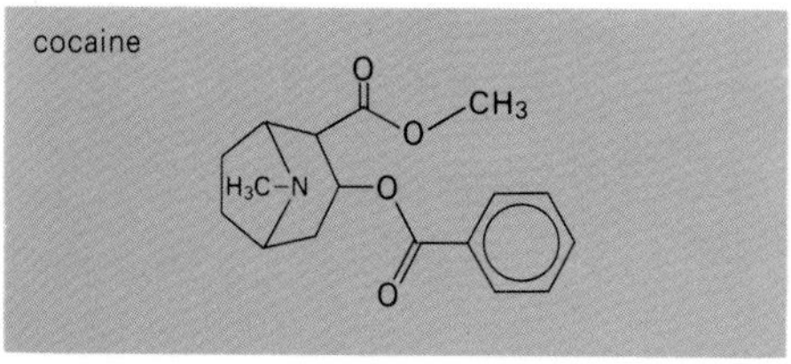

COCCYX, bone at the base of the spine, vestigial in man. If large in rare cases it causes difficulties in childbirth. It may even cause pain, requiring its excision.

COCHINEAL, a red coloring agent obtained from dried bodies of female scale insects. Synthetic ANILINE dyes have largely replaced cochineal commercially.

COCHLEA, part of the inner EAR, concerned with the mechanism of hearing. It is a spiral structure containing fluid and specialized membranes on which receptor nerve cells lie. Sound is conducted to the fluid by the tiny bones of the middle ear.

COCKCROFT, Sir John Douglas (1897–1967), English physicist who first "split the atom." With E. T. S. WALTON, he built a particle ACCELERATOR and in 1932 initiated the first man-made nuclear reaction by bombarding LITHIUM atoms with PROTONS, producing ALPHA PARTICLES. For this work Cockcroft and Walton received the 1951 Nobel Prize for Physics. In 1946 Cockcroft became the first director of the UK's atomic research laboratory at Harwell, and in 1959, the first Master of Churchill College, Cambridge.

COCKERELL, Sir Christopher Sydney (1910–), British engineer who rediscovered the hovercraft principle and stimulated worldwide interest in the possibilities of the AIR-CUSHION VEHICLE.

COCONUT PALM, *Cocus nucifera*, an economically valuable tree found on many tropical coasts. It has a long trunk crowned by a cluster of large fronds. The fruits, coconuts, take one year to develop and a single palm normally produces up to 100 nuts in one year. Each nut is surrounded by a thick fibrous husk and contains a white kernel surrounding the "coconut milk." The kernel is dried to produce **copra**, which is the source of coconut oil, a vegetable oil much used in the US and Europe in detergents, edible oils, margarine, brake fluid etc. The fibers of the husk are used for mats and ropes.

COCOON, the protective capsule enclosing the eggs of many animals; also, the stage in the life history of many insects called the PUPA or CHRYSALIS. In butterflies and moths the cocoon enclosing the chrysalis is formed by secretions of SILK.

CODEINE, a mild NARCOTIC, ANALGESIC and COUGH suppressant related to MORPHINE. It reduces bowel activity, causing CONSTIPATION, and is used to cure DIARRHEA.

COD LIVER OIL, an oil rich in VITAMINS A and D, a convenient way of enriching diets which contain inadequate amounts, particularly used for children. Both vitamins are dangerous in excess.

CODON, the basic "vocabulary" unit in the genetic code which controls PROTEIN SYNTHESIS. Each molecule of messenger-RNA (see NUCLEIC ACIDS) consists of an ordered sequence of NUCLEOTIDES. There are only four different nucleotides in RNA, not enough to code individually for the 20 AMINO ACIDS which make up proteins. So the nucleotides are grouped in threes, giving 64 different combinations called codons. Of these, 61 code for amino acids—several codons usually coding for the same amino acid—and 3 signal for termination of the protein chain. The code appears to be the same for all organisms.

COEFFICIENT, any FACTOR in an algebraic expression. Usually, coefficients are considered to be CONSTANTS by which terms involving one or more VARIABLES must be multiplied for all values of the variable. Thus, in ax^3+bx^2+cx+d the numbers a,b,c are coefficients. The values of coefficients are as important as the values of variables in most POLYNOMIALS and FUNCTIONS (see also EQUATION).

COELACANTH, *Latimeria chalumnae*, a member of the fossil fish group, the CROSSOPTERYGII, that was thought to be extinct until discovered off the east coast of Africa in 1938.

COELENTERATA, a phylum of primitive animals, now divided into two groups, the CNIDARIA and the CTENOPHORA.

COELOM, a body cavity found in most animals—the so-called coelomate animals. The main organs of the body are contained within the coelom.

COFFERDAM, temporary DAM used to divert the stream in early stages of dam-building. The term is used also for a structure similar to an open CAISSON.

COHESION, the tendency of different parts of a substance to hold together. This is due to forces acting between its MOLECULES: a molecule will repel one close to it but attract one that is farther away; somewhere between these there is a position where WORK must be done to either separate the molecules or push them together. This situation results both in cohesion and in ADHESION. Cohesion is strongest in a SOLID, less strong in a LIQUID, and least strong in a GAS.

COHN, Ferdinand Julius (1828–1898), German

botanist renowned as one of the founders of BACTERIOLOGY. He showed that BACTERIA could be classified in fixed species and discovered that some of these formed endospores which could survive adverse physical conditions. He was also the first to recognize the value of KOCH's work on the ANTHRAX bacillus.

COKE, form of amorphous CARBON (also containing ash, volatile residues and sulfur) remaining when bituminous COAL is heated in special furnaces to distill off the volatile constituents. Before the exploitation of NATURAL GAS, much COAL GAS was thus produced. In the US 95% of coke is used in METALLURGY, mostly in BLAST FURNACES. Such coke must be strong (to support the weight of the charge), porous and relatively pure. Some coke is used as a smokeless fuel, and to make WATER GAS.

COLCHICINE ($C_{22}H_{25}NO_6$), poisonous ALKALOID found in the autumn crocus or Meadow saffron (*Colchicum autumnale*). It is used to stimulate genetic changes in plants and animals. mp 156°C.

COLD, Common, or coryza, a mild illness of the NOSE and throat caused by various types of VIRUS. General malaise and RHINITIS, initially watery but later thick and tenacious, are characteristic; sneezing, COUGH, sore throat and headache are also common, but significant FEVER is unusual. Secondary bacterial infection of EARS, SINUSES, PHARYNX or LUNGS may occur, especially in predisposed people. Spread is from person to person. Mild symptomatic relief only is required.

COLD-BLOODED ANIMALS, or poikilotherms, animals, in particular fish, amphibians and reptiles, that cannot maintain a constant body TEMPERATURE and which are therefore greatly affected by climatic changes.

COLD SORE, vesicular SKIN lesion of lips or NOSE caused by *Herpes simplex* VIRUS. Often associated with periods of general ill-health or infections such as the COMMON COLD or PNEUMONIA. The virus, which is often picked up in early life, persists in the skin between attacks. Recurrences may be reduced by special antivirus drugs, applied during an attack.

COLIC, intermittent pain; generally experienced as bouts of severe pain with pain-free intervals. It is due to irritation or obstruction of hollow viscera, in particular the GASTROINTESTINAL TRACT, ureter (see KIDNEY) and GALL BLADDER or bile ducts (see LIVER). Treatment of the cause is supplemented by ANALGESICS and drugs to reduce smooth muscle spasm.

COLITIS, INFLAMMATION of the colon (see GASTROINTESTINAL TRACT). Infection with VIRUSES, BACTERIA or PARASITES may cause it, often with ENTERITIS. Inflammatory colitis can occur without bacterial infection in the chronic diseases, ulcerative colitis and Crohn's disease. Impaired blood supply may also cause colitis. Symptoms include COLIC and DIARRHEA (with slime or blood). Severe colitis can cause serious dehydration or SHOCK. Treatments include ANTIBIOTICS, and, for inflammatory colitis, STEROIDS, ASPIRIN derivatives or occasionally SURGERY.

COLLAGEN, tough, fibrous PROTEIN occurring as a major component of the connective TISSUE of many animals. Animal hide is chiefly collagen, converted by tanning into LEATHER. When collagen is boiled it yields GELATIN and GLUE.

COLLARBONE, or **clavicle**, a bone which forms part of the shoulder girdle in many vertebrates. Man has two S-shaped collarbones running between the breastbone and the shoulder blade. In birds the two collarbones are joined to form the furcula or **wishbone**.

COLLECTIVE UNCONSCIOUS, term used, especially by JUNG, for those parts of the UNCONSCIOUS derived from racial, rather than individual, experience.

COLLODION, solution of pyroxyline in ETHANOL and diethyl ETHER which evaporates to form a fine film impervious to water; used for protection of wounds and BURNS. Inhalation should be avoided.

COLLOID, or **colloidal solution**, a system in which two (or more) substances are uniformly mixed so that one is extremely finely dispersed throughout the other. A colloid may be viewed intuitively as a halfway stage between a SUSPENSION and a SOLUTION, the size of the dispersed particles being larger than simple MOLECULES, smaller than can be viewed through an optical MICROSCOPE (more precisely, they have at least one diameter in the range 1μm–1nm). Typical examples of colloids include FOG and BUTTER. Colloids may be classified in two ways: one by the natures of the particles (dispersed phase) and medium (continuous phase); the other by, as it were, the degree of permanency of the colloid. In the latter case, one may define a **lyophilic colloid** as one that forms spontaneously when the two phases are placed in contact; and a **lyophobic colloid** as one that can be formed only with some difficulty and maintained for a moderate elapse of time only under special conditions. Colloids have interesting properties, perhaps the most notable of which is light DISPERSION: it is due to colloidal particles in the atmosphere that the sky is blue in the daytime and the sunset red. Moreover, the property of ADSORPTION of molecules and IONS at the interface between particles and continuous phase plays a major part in water purification (see WATER SUPPLY). (See also AEROSOL; BROWNIAN MOTION; DIALYSIS; ELECTROPHORESIS; OSMOSIS; TYNDALL; ULTRAMICROSCOPE.)

COLLOTYPE, PRINTING process, akin to lithography, whereby photographs may be reproduced without a HALFTONE screen. A GLASS or ALUMINUM plate is coated in a light-sensitive GELATIN and POTASSIUM bichromate (dichromate, $K_2Cr_2O_7$) solution. On exposure, the gelatin hardens most in areas of brightest light. The plate is soaked in GLYCEROL which is absorbed most in the softest areas; during printing, the plate is kept moist by a glycerol-water mixture, the glycerol-soaked areas absorbing the moisture and repelling the INK. Collotype printing is slow and the plate lacks durability (print run at most about 5000).

COLON. See GASTROINTESTINAL TRACT.

COLOR, the way the brain interprets the wavelength distribution of the LIGHT entering the eye. The phenomenon of color has two aspects: the physical or optical—concerned with the nature of the light—and the physiological or visual—dealing with how the eye sees color.

The light entering the eye is either emitted by or reflected from the objects we see. Hot objects emit light with wavelengths occupying a broad continuous band of the electromagnetic SPECTRUM, the position of the band depending on the temperature of the object—the hotter the object, the shorter the

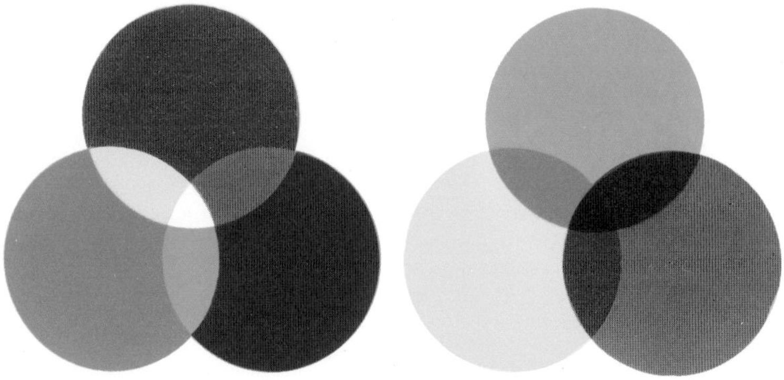

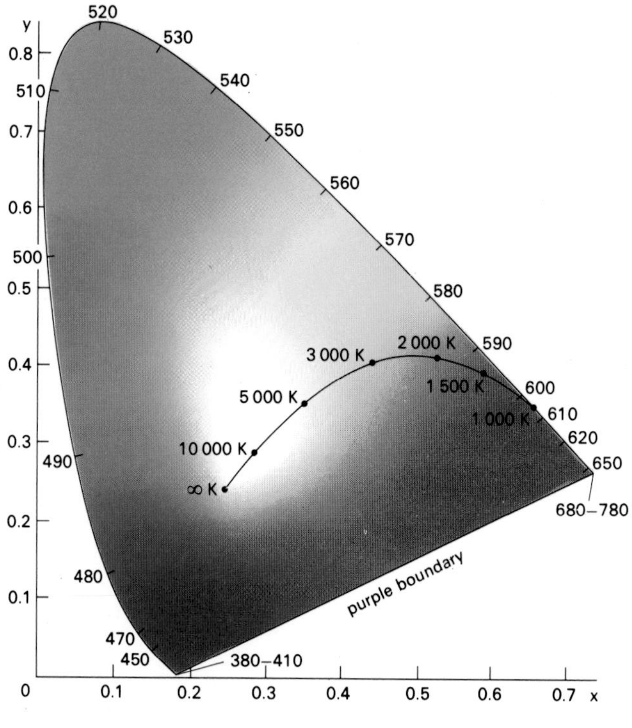
y
0.8
0.7
0.6
0.5
0.4
0.3
0.2
0.1
0
0.1
0.2
0.3
0.4
0.5
0.6
0.7
x
520
530
540
550
560
570
580
590
600
610
620
650
680–780
510
500
490
480
470
450
380–410
3 000 K
2 000 K
1 500 K
1 000 K
5 000 K
10 000 K
∞ K
purple boundary

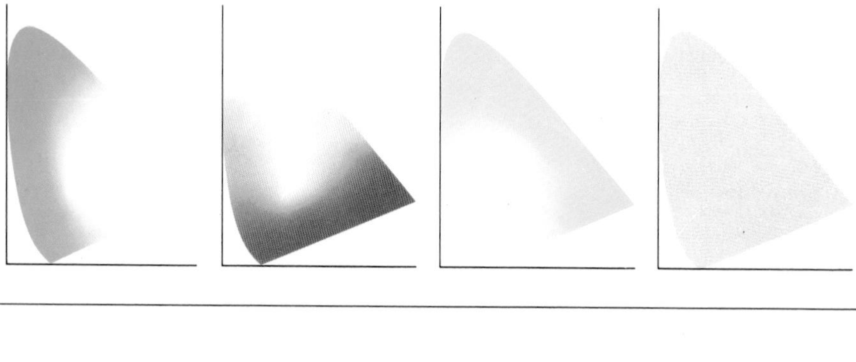

wavelengths emitted (see BLACKBODY RADIATION). (We tend to think of the spectrum of visible light in terms of the band of colors revealed by NEWTON when he split up a beam of sunlight using a PRISM; in these terms, the shorter the wavelength, the bluer the light.) Other objects emitting light do so either at particular wavelengths or in narrow bands of the spectrum (see SPECTROSCOPY). Where these emission bands fall within the visible spectrum, such objects appear colored, otherwise black. Objects reflecting diffuse light appear colored in virtue of combined SCATTERING and ABSORPTION effects, though the nature of the light source is also important.

The EYE can only see colors when the light is relatively bright; the rods used in poor light see only in black and white. The cones used in color VISION are of three kinds, responding to light from the red, green or blue portions of the visible spectrum. The brain adds together the responses of the different sets of cones and produces the sensation of color. The three colors to which the cones of the eye respond are known as the three primary colors of light. By mixing different proportions of these three colors, any other color can be simulated, equal intensities of all three producing white light. This is known as the production of color by addition, the effect being used in color TELEVISION tubes where PHOSPHORS glowing red, green and blue are employed. Color pigments, working by transmission or reflection, produce colors by subtraction, abstracting light from white and displaying only the remainder. Again a suitable combination of a set of three pigments—cyan (blue-green), magenta (blue-red) and yellow (the "complementary" colors of the three primaries)—can simulate most other colors, a dense mixture of all three producing black. This effect is used in color photography but in color printing an additional black pigment is commonly used.

Most colors are not found in the spectrum. These nonspectral colors can be regarded as intermediates between the spectral colors and black and white. Many schemes have been proposed for the classification and standardization of colors. The most widely used is that of Albert Henry Munsell which describes colors in terms of their hue (basic color); saturation (intensity or density), and lightness or brightness (the degree of whiteness or blackness).

Color effects. *Top left:* Additive mixing of light of the three primary colors—red, blue and green. Mixture of two of these gives rise to the secondary colors—cyan, magenta and yellow; mixture of all three gives white light. *Top right:* Color by subtraction from white light by cyan, magenta and yellow pigments. Where the effects of two pure pigments are combined, primary color effects are reproduced; combination of all three yields a black effect. *Middle:* The chromaticity diagram seeks to describe colors in terms of rectangular x and y coordinates. The pure spectral colors are ranged along the upper edge of the diagram, their wavelengths indicated in nanometres (nm). The colors displayed by sources of blackbody radiation at different temperatures (in kelvins) lie along a line that extends into the center of the diagram. The light from such sources can thus be described in terms of its "color temperature". *Bottom:* The standard four-color printing process. The original picture is photographed through color filters and four printing plates are prepared: one for printing with cyan (process blue) ink, one for magenta (process red), one for process yellow, and one for black (for reducing the brightness where necessary). Each of these plates is screened as in the usual halftone process, but the dots produced are carefully oriented so as neither to clash nor give rise to moiré patterns. The combination of the subtractive effects of the four printings yields a full-color effect. Here the four printings which are used to reproduce the chromaticity diagram (*middle*) are reproduced separately, side by side and at a reduced scale. Although the four-color printing process does reproduce color effects reasonably well, it does not allow all colors to appear accurately. Royal blues, for instance, are particularly difficult to reproduce satisfactorily.

COLOR BLINDNESS, inability to discriminate between certain COLORS, an inherited trait. It is a disorder of the RETINA cones in the EYE. The commonest form is red-green color blindness (Daltonism), usually found in men (about 8%), the other types being rare.

COLORIMETRY, the techniques for measuring and describing the phenomena of COLOR. Since the experience of color is largely subjective and similar experiences can occur with quite different sets of physical phenomena, many different instruments (colorimeters) and criteria are employed. Of two basic approaches, one involves the visual comparison of colors (using color comparators), while the other employs the instruments and techniques of PHOTOMETRY and SPECTROSCOPY.

COLOSTRUM, the first thin watery fluid secreted by the BREAST after the BIRTH of a child. Although low in FAT, it is rich in PROTEIN, including ANTIBODIES which are important in giving temporary passive IMMUNITY, notably in cattle and other ungulates.

COLT, Samuel (1814–1862), US inventor and industrialist who devised the REVOLVER, a single barreled pistol with a revolving multiple breech (bullet chamber), in the early 1830s. His factories pioneered mass-production techniques and the use of interchangeable parts.

COLUMBITE, lustrous black OXIDE mineral of iron, manganese and niobium [$(Fe,Mn)Nb_2O_6$], the chief ore of NIOBIUM, found mainly in W Australia, Zaire, Madagascar and S. D. TANTALUM replaces niobium in all proportions up to 100% (tantalite). Columbite crystallizes in the orthorhombic system.

COLUMBIUM, former name of NIOBIUM, especially in the US.

COMA, state of unconsciousness in which a person cannot be roused by sensory stimulation and is unaware of his surroundings. Body functions continue but may be impaired, depending on the cause. These include POISONING, head injury, DIABETES, and BRAIN diseases, including STROKES and CONVULSIONS. Severe malfunction of LUNGS, LIVER or KIDNEYS may lead to coma.

COMA, in optics. See LENS.

COMBINATIONS. See PERMUTATIONS AND COMBINATIONS.

COMBINATORIAL ANALYSIS, the branch of mathematics dealing with subdivisions of sets (see SET THEORY): its primary concerns are PERMUTATIONS AND COMBINATIONS and partitions. A **partition** of a

The comet Humason, discovered in 1961 when it was visible as a bright star in the Southern Hemisphere. The diagonal bars are due to the apparent motion of stars during the exposure.

number n is its expression in the form of a SUM of positive INTEGERS: i.e., setting

$$n=a_1+a_2+\ldots+a_m.$$

For example, the number 5 has 7 partitions: (1+1+1+1+1), (2+1+1+1), (2+2+1), (3+1+1), (3+2), (4+1) and (5). By extension, **combinatorial topology** is the study of complex forms in terms of their being built up from combinations of basic geometric figures. (See TOPOLOGY.)

COMBING. See CARDING.

COMB JELLIES. See CTENOPHORA.

COMBUSTION, or **burning**, the rapid OXIDATION of FUEL in which heat and usually light are produced. In slow combustion (e.g., a glowing charcoal fire) the reaction may be heterogeneous, the solid fuel reacting directly with gaseous oxygen; more commonly, the fuel is first volatilized, and combustion occurs in the gas phase (a flame is such a combustion zone, its luminance being due to excited particles, molecules and ions). In the 17th and 18th centuries combustion was explained by the PHLOGISTON theory, until LAVOISIER showed it to be due to combination with oxygen in the air. In fact the oxidizing agent need not be oxygen: it may be another oxidizing gas such as nitric oxide or fluorine, or oxygen-containing solids or liquids such as nitric acid (used in rocket fuels). If the fuel and oxidant are premixed, as in a BUNSEN BURNER, the combustion is more efficient, and little or no SOOT is produced. Very rapid combustion occurs in an explosion (see EXPLOSIVES), when more heat is liberated than can be dissipated, or when a branched chain reaction occurs (see FREE RADICALS). Each combustion reaction has its own ignition temperature below which it cannot take place, e.g. c400°C for coal. Spontaneous combustion occurs if slow oxidation in large piles of such materials as coal or oily rags raises the temperature to the ignition point. (See also INTERNAL-COMBUSTION ENGINE.)

COMET, a nebulous body which orbits the sun. In general, comets can be seen only when they are comparatively close to the sun, though the time between their first appearance and their final disappearance may be as much as years. As they approach the sun, a few comets develop tails (some comets develop more than one tail) of lengths of the order 1–100Gm, though at least one tail 300Gm in length—more than twice the distance from the earth to the sun—has been recorded. The tails of comets are always pointed away from the sun, so that, as the comet recedes into space, its tail precedes it. For this reason it is generally accepted that comets' tails are caused by the SOLAR WIND.

The head of the comet is known as the nucleus. Nuclei may be as little as 100m or as much as 100km in radius, and are thought to be composed primarily of frozen gases and ice mixed with smaller quantities of meteoritic material. Most of the mass of a comet is contained within the nucleus, though this may be less than 0.000001 that of the earth. Surrounding the nucleus is the bright coma, possibly as much as 100Mm in radius, which is composed of gas and possibly small particles erupting from the nucleus.

Cometary orbits are usually very eccentric ellipses, with some perihelions (see ORBIT) closer to the sun than that of MERCURY, aphelions as much as 100000 AU from the sun. The orbits of some comets take the form of hyperbolas, and it is thought that these have their origins altogether outside the SOLAR SYSTEM, that they are interstellar travelers.

In Greco-Roman times it was generally believed that comets were phenomena restricted to the upper atmosphere of the earth. In the late 15th and 16th centuries it was shown by M. Mästlin and BRAHE that comets were far more distant than the moon. NEWTON interpreted the orbits of the comets as parabolas, deducing that each comet was appearing for the first time. It was not until the late 17th century that HALLEY showed that at least some comets returned periodically.

COMMENSALISM, form of SYMBIOSIS in which one organism—the commensal—benefits from the partnership while the other neither gains nor loses. The commensal often gets shelter, transport and food from its host. (See also EPIPHYTE.)

COMMON SENSE SCHOOL, in philosophy, a group of Scottish thinkers, including Thomas REID and Dugald STEWART, who, reacting against the idealism of BERKELEY and the skepticism of HUME, affirmed that the truths apparent to the common man—the existence of material objects, the reality of CAUSALITY, and so on—were genuine, reliable and not to be questioned.

COMMUNICABLE DISEASES. See DISEASE.

COMMUTATIVE LAW. See ALGEBRA.

COMPASS, device for determining direction parallel to the earth's surface. Most compasses make use of the EARTH's magnetic field (see GEOMAGNETISM); if a bar magnet (see MAGNETISM) is pivoted at its center so that it is free to rotate horizontally, it will seek to align itself with the horizontal component in its locality of the earth's magnetic field. A simple compass consists of a

magnet so arranged and a compass card marked with the four cardinal points and graduated in degrees (see ANGLE). In ship compasses, to compensate for rolling, the card is attached to the magnet and floated or suspended in a liquid, usually alcohol. Aircraft compasses often incorporate a GYROSCOPE to keep the compass horizontal. The two main errors in all magnetic compasses are **variation** (the angle between lines of geographic longitude and the local horizontal component of the earth's magnetic field) and **deviation** (local, artificial magnetic effects, such as nearby electrical equipment). Both vary with the siting of the compass, and may be with more or less difficulty compensated for. (See also GYROCOMPASS; NAVIGATION.) A **radio compass**, used widely in aircraft, is an automatic radio DIRECTION FINDER, calibrated with respect to the station to which it is tuned.

COMPASSES, instrument used in EUCLIDEAN GEOMETRY and technical draftsmanship to measure off distances and to draw CIRCLES or parts thereof. A compass whose two legs are tipped by sharp points is called a **divider**, the term compass (or pair of compasses) being usually reserved for instruments with one leg tipped by a pen or pencil.

COMPETITIVE INHIBITION. See ENZYMES.

COMPILER. See COMPUTER.

COMPLEMENTARITY PRINCIPLE, a philosophical thesis proposed in 1927 by Niels BOHR, seeking to make intelligible the then newly-developed WAVE MECHANICS. Bohr recognized that the hardware used in any subatomic-physics experiment materially affected the results obtained and inferred from this

The head of a Horse fly showing the remarkable color patterns typical of the compound eyes of this family (Tabanidae).

that atomic systems could only be described and understood in terms of a series of complementary partial views.

COMPLEX, a nexus of ideas and feelings from both the CONSCIOUS and UNCONSCIOUS which has an effect, favorable or adverse, on the actions or emotional state of the individual. ADLER used the term in connection with the superiority and INFERIORITY COMPLEXES, FREUD in connection with the CASTRATION COMPLEX and OEDIPUS COMPLEX.

COMPLEX NUMBERS. See IMAGINARY NUMBERS.

COMPOSITION, Chemical, the proportion by weight of each ELEMENT present in a chemical compound. The **law of definite proportions**, discovered by J. L. PROUST, states that pure compounds have a fixed and invariable composition. A few compounds, termed non-stoichiometric, disobey this law: they have lattice vacancies or extra atoms, and the composition varies within a certain range depending on the formation conditions. The **law of multiple proportions**, discovered by DALTON, states that, if two elements A and B form more than one compound, the various weights of B which combine with a given weight of A are in small whole-number ratios. (See also BOND, CHEMICAL; EQUIVALENT WEIGHT.)

COMPOST. See MANURE.

COMPOUND EYE, a type of eye found in the phylum ARTHROPODA, so-called because it consists of a

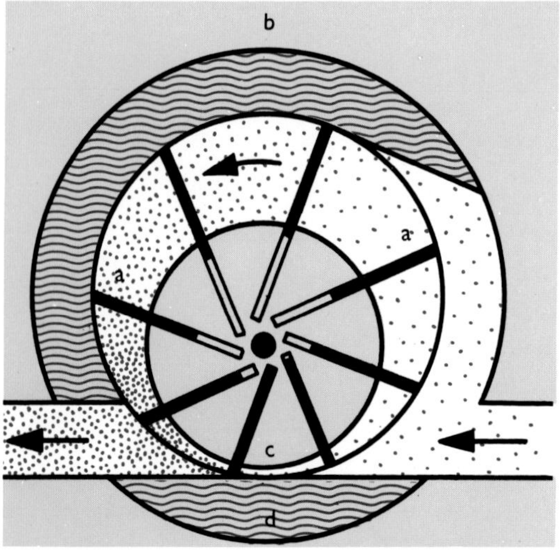

A rotary compressor in which the pressure of a gas is increased by trapping it between vanes which reduce it in volume as the impeller rotates around an axis eccentric to the casing: (a) diffuser; (b) casing; (c) impeller; (d) water coolant.

number of organs called *ommatidia*. Each ommatidium perceives part of the visual field so that the animal's view of the world is composed of a mosaic of images that is more complete the greater the number of ommatidia that are present. Predatory insects such as dragonflies have compound eyes composed of several thousand ommatidia.

COMPRESSION RATIO, in INTERNAL-COMBUSTION ENGINES, the RATIO between the volumes in a cylinder when the piston is at the bottom of its stroke and when it is at the top of its stroke. In AUTOMOBILE engines this ratio is about 8:1, while in truck DIESEL ENGINES it is around 16:1.

COMPRESSOR, device for increasing the pressure of a gas or vapor, important in, for example, GAS TURBINES and pneumatic tools. Gases under very high pressure may liquefy; liquefied gases are of great importance in many fields of everyday life. **Reciprocating compressors** have a cylinder in which runs a piston driven by a crankshaft; their principle of operation is similar to that of a PUMP. Simple **rotary compressors** have a rotating cylinder mounted eccentrically in a cylindrical casing. Vanes are mounted on the cylinder such that they slide in and out of slots as compelled by the walls of the chamber. Gas is introduced at one side of the chamber and drawn round by the vanes; the volume between cylinder and chamber decreases as the cylinder turns, and thus pressurized gas is released at the outlet on the other side of the chamber. A device known as a compressor is used in TELEPHONE systems to reduce the effects of INTERFERENCE.

COMPTON, Arthur Holly (1892–1962), US physicist who discovered the Compton effect (1923), thus providing evidence that X RAYS could act as particles as predicted in QUANTUM THEORY. Compton found that when monochromatic X rays were scattered by light elements, some of the scattered radiation was of longer wavelength, i.e., of lower ENERGY than the incident. Compton showed that this could be explained in terms of the collision between an X-ray PHOTON and an ELECTRON in the target. For this work he shared the 1927 Nobel physics prize with C. T. R. WILSON.

COMPULSION, an irresistible UNCONSCIOUS force which makes an individual perform conscious (see CONSCIOUSNESS) thoughts or actions which he would not normally perform, perhaps even against his will. The force may also come from outside, i.e., from someone whose character dominates the individual (see also BRAINWASHING; OBSESSIONAL NEUROSIS).

COMPUTER, any device which performs calculations. In this light, the ABACUS, CALCULATING MACHINE and SLIDE RULE may all be described as computers; however, the term is usually limited to those electronic devices that are given a program to follow, data to store or to calculate with, and means with which to present results or other (stored) information.

Programming. A computer program consists essentially of a set of instructions which tells the computer which operations to perform, in what order to perform them, and the order in which subsequent data will be presented to it; for ease of use, the computer may already have subprograms built into its memory, so that, on receiving an instruction such as LOG X, it will automatically go through the program necessary to find the LOGARITHM of that piece of data supplied to it as X. Every model of computer has a different machine language or code; that is, the way in which it should ideally be programmed; however, this language is usually difficult and cumbersome for an operator to use. Thus a special program known as a **compiler** is retained by the computer, enabling it to translate computer languages such as ALGOL, COBOL and FORTRAN, which are easily learned and used by operators and programmers, into its own machine code. Programmers also make extensive use of ALGORITHMS to save programming and operating time.

Input. Programs and data are fed into computers using either the medium of punched tape or, more commonly in recent years, that of punched cards. In both cases it is the positions of holes punched in the medium which carry the information. These are read by a card or tape reader which usually consists of a light shining through the holes and activating PHOTOELECTRIC CELLS on the far side. (Mechanical readers exist but are generally regarded as less reliable.) The computer "reads" the resulting electrical pulses.

Storage. Machine languages generally take the form of a binary code (see BINARY NUMBERS), so that the two characters 0 and 1 may be easily represented by + and −. Thus the ideal medium for data storage is magnetic, and may take the form of tapes, disks or drums. Magnetic tapes are used much as they are in a TAPE RECORDER; a magnetic head "writes" on the tape by creating a suitable magnetic flux, and can "read" the spots so created at a later date, retransmitting them in the form of electric pulses. Magnetic disks and drums work on a similar principle; the former are flat disks mounted in groups of up to twenty on a shared shaft, looking rather like a stack of phonograph records; drums are, as the name

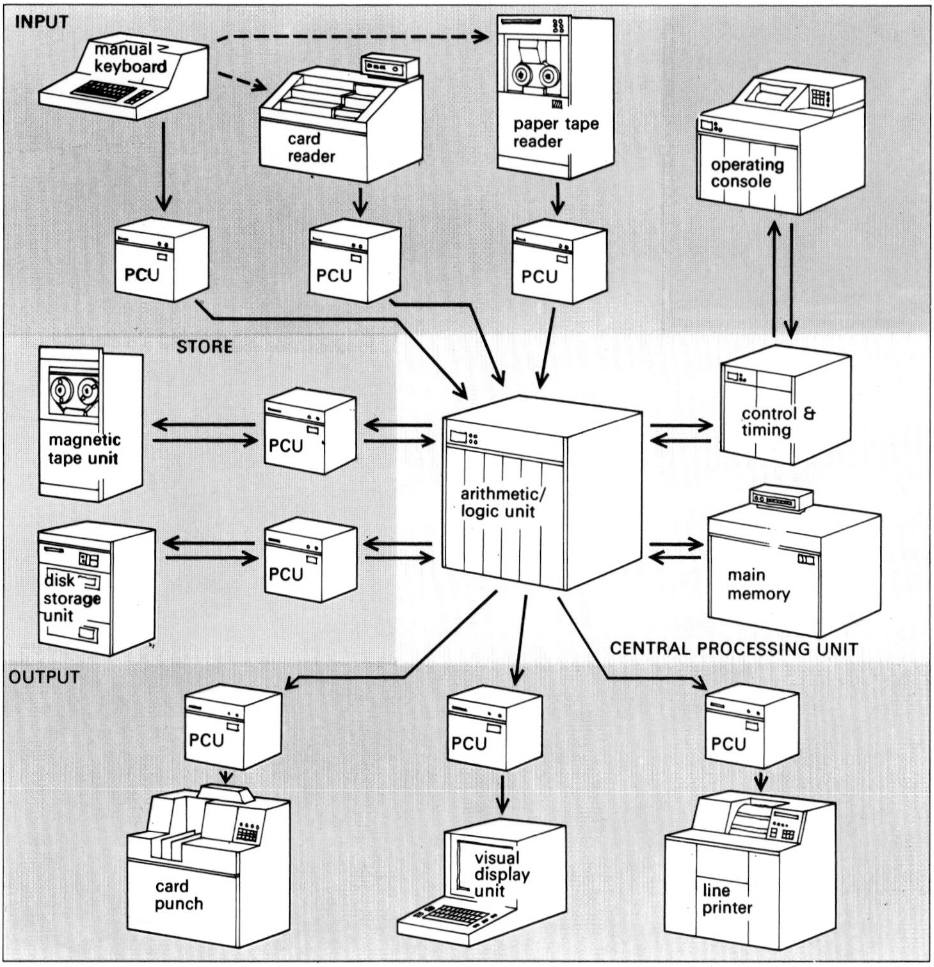

At the heart of a computer is the central processing unit. Under the control of the operator, the control and timing units, and the instructions stored in its main memory, the arithmetic/logic unit draws data from the various input devices and data stores, processes them, and issues the results through the output devices. Each of the "peripheral" units is connected to the central processor through peripheral control units (PCUs) which control the flow of information between the wholly electronic central processor and the peripheral devices which, generally involving mrchanical linkages, can handle information only at somewhat slower rates.

suggests, cylindrical, and are coated with a magnetic medium. Both drums and stacks of disks rotate constantly while the computer is in use, so that the maximum time taken for the read/write head to locate any specific area is that for one revolution. In all cases, each datum must be identified and given a specific "address" in the storage system, so that instructions for its retrieval may be given to the computer and so that the operator may take precautions against erasing it. (See INFORMATION RETRIEVAL.)

The computer's "memory," for longer-term storage, usually takes the form of arrays of cores: tiny (about 0.5mm in diameter) doughnut-shaped magnetic objects. Once again, each core in the array is uniquely identifiable by the computer.

Data processing. All the operations performed by the computer on the information it receives are collectively described as data processing. The main element of data processing is, of course, computation. This is almost exclusively done by addition, and performed using binary arithmetic (see BINARY NUMBERS). More complicated procedures, such as integration (see CALCULUS) or finding ROOTS, are performed algorithmically, suitable subprograms being built into the computer. Again the characters 0 and 1 are represented by + and −, where this may refer to a closed or open switch, a direction of magnetic flux (see MAGNETISM), etc. Moreover, the computer contains logic circuits so that it may evaluate information while performing a calculation. If, for example, it were performing an algorithm to find $\sqrt{2}$ to a specified number of decimal places (see

APPROXIMATION), it has to have a system whereby it can check at the end of each cycle of the algorithm whether or not its result is correct to the accuracy required. These circuits are designed using an application of Boolean algebra (see LOGIC; BOOLE, GEORGE): the three elements are AND, NOT and OR. Information will be passed by the AND and OR elements (switches) if, respectively,

$$a \wedge b \leftrightarrow I,$$
$$a \vee b \leftrightarrow I$$

(for pulses a and b). The NOT switch is also called the **inverter**, and its function is to convert a into a', or vice versa (i.e., + into − or − into +). Combinations of these three switches are capable of handling any logical operation required.

Output. Before being fed out, the information must be converted from machine code back into the programmer's computer language, numerical data being translated from the binary into the DECIMAL SYSTEM. The information is then fed out in the form of paper tape, punched cards or, using an adapted teleprinter, as a printout.

Types of computer. We have been talking almost exclusively about the **digital computer**, since this is the most widely-encountered and certainly the most versatile type. As we have seen, it requires information to be fed into it in "bits." Contrarily, the other main type of computer, the **analog computer**, is designed to deal with continuously varying quantities, such as lengths or voltages; the most everyday example of an analog computer is the slide rule. Electronic analog computers are usually designed for a specific task; as their accuracy is not high, their greatest use is in providing models of situations as bases for experiment. (See also CYBERNETICS; DICTIONARY.)

COMTE, Auguste (1798–1857), French philosopher, the founder of POSITIVISM and a pioneer of scientific sociology. His thinking was essentially evolutionary; he recognized a progression in the development of the sciences: starting from mathematics and progressing through astronomy, physics, chemistry and biology towards the ultimate goal of sociology. He saw this progression reflected in man's mental development. This had proceeded from a theological stage to a metaphysical one. Comte now sought to help inaugurate the final scientific or positivistic era. His social thinking reflected that of Henri de Saint-Simon and in turn his own works, particularly the *Philosophie positive* (1830–42), became widely influential in both France and England.

CONCENTRICITY, in GEOMETRY, the situation in which two or more figures share a common center.

CONCEPTUALISM, a modern term describing a position in scholastic philosophy with respect to the status of universals that was intermediate between the extremes of both NOMINALISM and REALISM. To a conceptualist, UNIVERSALS (general concepts such as chair-ness) indeed exist, but only as concepts common to all men's minds and not as things in the world of particular objects (such as chairs).

CONCHOID, or **Conchoid of Nicomedes**, a shell-shaped CURVE constructed as follows: consider a point O a PERPENDICULAR distance a from a fixed straight line PQ on which there is a moving point A. In each direction along the line AO measure out a distance b,

The conchoid of Nicodemus—the three cases: $a = b$ (blue curve); $a > b$ (white curve); $a < b$ (red curve).

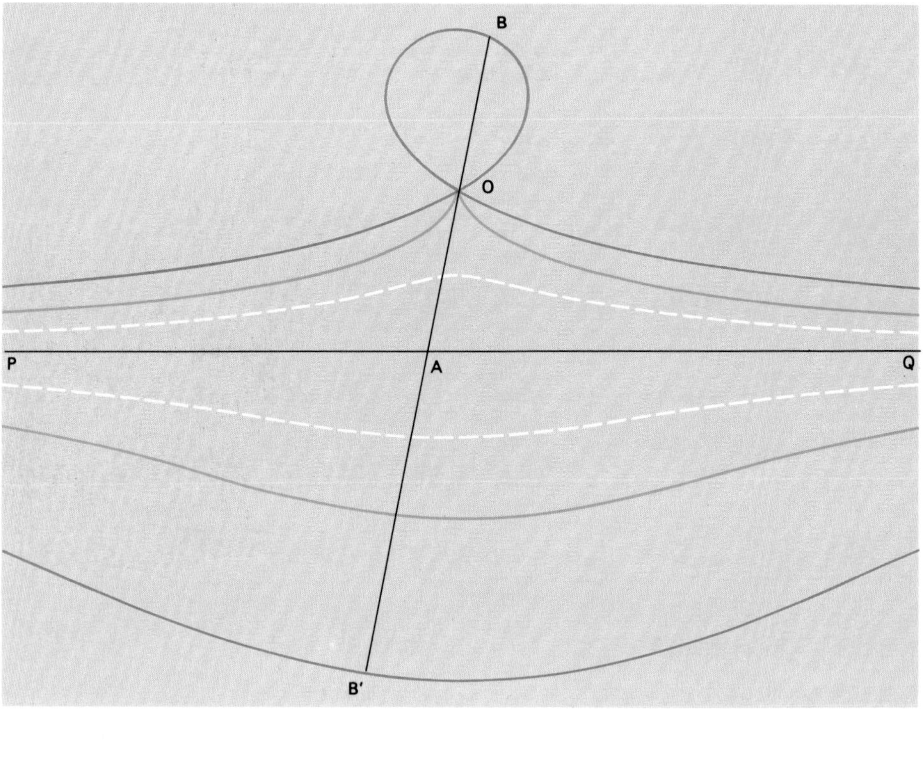

so that $AB' = AB = b$, which distance is fixed. If the line B′ABO is pivoted about O the points B′ and B trace out a conchoid, whose shape depends on the relationship of b to a.

CONCHOID FRACTURE, a form of ROCK and MINERAL fracture producing a curved, ribbed surface resembling the shell of certain mollusks.

CONCRETE, versatile structural building material, made by mixing CEMENT, AGGREGATE and water. Initially moldable, the cement hardens by hydration, forming a matrix which binds the aggregate. Various other ingredients—admixtures—may be added to improve the properties of the concrete; air-entraining agents increase durability. Since concrete is much more able to resist compressive than tensile STRESS, it is often reinforced with a steel bar embedded in it which is able to bear the tension. **Prestressed concrete** is reinforced concrete in which the steel is under tension and the concrete is compressed; it can withstand very much greater stresses. Concrete is used for all building elements and for bridges, dams, canals, highways etc., often as precast units.

CONCRETION, irregularly-shaped nodule of rock embedded in sedimentary rock of a different composition, in which it grew by deposition on a nucleus. Examples include FLINT in chalk.

CONCUSSION, a state of disturbed consciousness following head injury, characterized by AMNESIA for events preceding and following the trauma. Permanent BRAIN damage is only found in cases of repeated concussion, as in boxers who develop the punch-drunk syndrome.

CONDENSATION, passage of substance from gaseous to liquid or solid state: CLOUDS are a result of condensation of water vapor in the ATMOSPHERE (see also RAIN). Warm air can hold more water vapor than cool air; if a body of air is cooled it will reach a temperature (the DEW point) where the water vapor it holds is at SATURATION level. Further decrease in temperature without change in pressure will initiate water condensation. Such condensation is greatly facilitated by the presence of condensation nuclei ("seeds"), small particles (e.g., of smoke) about which condensation may begin. **Condensation trails** behind high-flying jet aircraft result primarily from water vapor produced by the engines increasing the local concentration (see also CLOUD CHAMBER; GAS; VAPOR). Condensation is important in all processes using steam; and in DISTILLATION, where the liquid is collected, and condensed by removal of its LATENT HEAT of vaporization, in an apparatus called a **condenser**. In chemistry a **condensation reaction** is one in which two or more MOLECULES link together with elimination of a relatively small molecule, such as water.

CONDENSER, term sometimes used for CAPACITOR. (See also CONDENSATION.)

CONDILLAC, Étienne Bonnot de (1715–1780), French philosopher, who broke with the teaching of LOCKE to found the doctrine of SENSATIONALISM, holding that all knowledge is derived from the senses.

CONDITIONED REFLEX. See REFLEX.

CONDITIONING, term used to describe two quite different LEARNING processes. In the first, a human or animal response is generated by a stimulus which does not normally generate such a response (see conditioned REFLEX; PAVLOV). In the second, animals (and by extension humans) are trained to perform certain actions to gain rewards or escape punishment (see LEARNING).

CONDORCET, Marie Jean Antoine Nicholas de Caritat, Marquis de (1743–1794), French philosopher, mathematician and revolutionary politician chiefly remembered for his theory that the human race, having risen from barbarism, would continue to progress toward moral, intellectual and physical perfection. His principal mathematical work was in the theory of probability. He played a prominent role in the Revolution, though his moderate opinions led to his outlawry and suicide.

CONDUCTANCE, a measure of the ability of a body to conduct ELECTRICITY. Measured in SIEMENS, it is the RECIPROCAL of RESISTANCE. (See also CONDUCTIVITY.)

CONDUCTION, Heat, passage of heat through a body without large-scale movements of matter within the body (see CONVECTION). Mechanisms involved include the transfer of vibrational ENERGY from one MOLECULE to the next through the substance (dominant in poor conductors), and energy transfer by ELECTRONS (in good electrical conductors) and PHONONS (in crystalline solids). In general, solids, especially metals, are good conductors, liquids and gases poor. (See also RADIATION.)

CONDUCTIVITY, or specific conductance, the CONDUCTANCE of a 1-metre cube of a substance, measured between opposite faces. Measured in siemens per metre, conductivity is the RECIPROCAL of resistivity (see RESISTANCE), and expresses the substance's ability to conduct electricity. The **equivalent conductivity** Λ of an electrolytic solution is the conductivity of a solution divided by its concentration in gram-equivalents (see EQUIVALENT WEIGHT) per cubic metre, and is usually measured with an electrolytic CELL in a WHEATSTONE BRIDGE. The degree of ionic DISSOCIATION (α) was found by ARRHENIUS to be given by

$$\alpha = \frac{\Lambda}{\Lambda_0},$$

where Λ_0 is the value of Λ extrapolated to zero concentration; this equation has since been shown to apply only to weak electrolytes (see ELECTROLYSIS).

CONDUCTORS, Electric, substances (usually metals) whose high CONDUCTIVITY makes them useful for carrying electric current (see ELECTRICITY). They are most often used in the form of WIRES or CABLES. The best conductor is SILVER, but, for reasons of economy, COPPER is most often used. (See also SEMICONDUCTORS; SUPERCONDUCTIVITY.)

CONE, a solid geometrical figure traced by the rotation of a straight line A (the generator) about a fixed straight line B which it intersects, such that each point on A traces out a closed CURVE. A cone has therefore two parts (nappes) which touch each other at the point of intersection, termed the vertex of the cone, of lines A and B; the two parts being skew-symmetrical (see SYMMETRY) about the vertex and of infinite extent. Usually one considers only one of these parts, limited by a PLANE which cuts it. The tracing of the closed curve of rotation on this plane is the directrix and the part of the plane bounded by the directrix is the base of the cone. The lines joining the vertex to each point of the directrix are the cone's elements. The PERPENDICULAR line from the vertex to

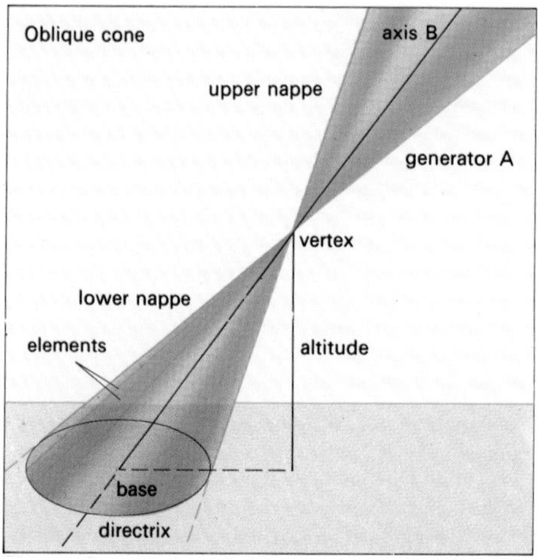

Terms associated with the cone.

this plane is the altitude or height of the cone; the line joining the vertex to the center of the base (if it has a center) is the axis, and in most cases coincides with line B. Should axis and altitude coincide, the cone is a right cone; otherwise it is oblique. A cone whose directrix is a circle is a circular cone, its volume being given by $\pi r^2 h/3$ where r is the radius of the directrix and h is the altitude. (See also CONIC SECTIONS.)

CONFORMATIONAL ANALYSIS, the study of molecular conformations, i.e. the spatial arrangements of the atoms that can be interconverted merely by rotation about single bonds (see BOND, CHEMICAL). In general, different conformations are transient and nonisolable, unlike different configurations (see STEREOCHEMISTRY). But if large substituents, such as bromine or phenyl groups, are attached to adjacent atoms, steric hindrance or interference may cause the eclipsed conformation to be less stable than the staggered. This may affect reaction rates. Conformational effects are most important in ALICYCLIC COMPOUNDS.

CONGLOMERATE, or pudding stone, in geology, a consolidated GRAVEL consisting of rock fragments (rounded by transportation) cemented in a sedimentary matrix. If the fragments are angular it is known as **breccia.** The fragments vary in size from boulders to small pebbles, grading into SANDSTONE. Poorly-sorted conglomerates, having fragments of many different rocks, include **tillite** (of glacial origin) and **graywacke.** (See also AGGLOMERATE.)

CONGREVE, Sir William (1772–1828), English inventor of the Congreve ROCKET (1805), an incendiary rocket used with great effect against Boulogne (1806) and Copenhagen (1807).

CONGRUENCE, in GEOMETRY, the situation in which two geometrical figures are identical except for their positions in space.

CONIC SECTIONS, plane CURVES formed by the intersection of a PLANE with a right circular or right elliptical CONE: the three curves are the ellipse, the parabola and the hyperbola. An **ellipse** occurs when the ANGLE between the axis of the cone and the plane is greater than the angle between the axis and the generator (in special cases a CIRCLE may be produced). It may be defined as the locus of a point P about two fixed foci (singular, focus) F and F′, such that $PF+PF'=c$, where c is a constant greater than the distance FF′. The major axis of an ellipse is its axis of SYMMETRY concurrent with FF′; its minor axis is the axis of symmetry perpendicular to this, their point of intersection being defined as the center of the ellipse. If the length of the minor axis is 2b, b being a constant, then $c^2=(FF')^2+b^2$. The eccentricity of an ellipse is given by the distance FF′ divided by the length of the major axis. A **parabola** occurs when the angle between the axis and the plane equals the angle between the axis and generator (in special cases a straight line may be produced). It may be defined as the locus of a point P such that its distance from a fixed focus F is constantly equal to its PERPENDICULAR distance from a fixed straight line XY. The curve, which has only one axis of symmetry, perpendicular to XY and passing through F, is of infinite extent. The **hyperbola** occurs when the angle between axis and plane is less than that between axis and generator (in special cases a pair of intersecting straight lines may be produced). It may be defined as the loci of two points, P and P′, about two foci, F and F′, such that $PF'-PF=c=P'F-P'F'$, where c is a constant less than the distance FF′. The curve, which is of infinite extent, has a real axis of symmetry passing through F and F′, and an imaginary axis of symmetry passing perpendicularly through the midpoint of FF′. The hyperbola, though of infinite extent in the direction of its real axis, is bounded in the direction perpendicular to this (see ASYMPTOTE).

CONJUGATE, of one ROOT OF AN EQUATION, another number that is a root of the same EQUATION. Thus, if $x^2+2x-3=0$, the numbers 1 and -3 are conjugates. If one root of an equation is a complex number (see IMAGINARY NUMBERS) of the form $a+ib$, then it is a fundamental theorem of algebra that it has a conjugate of the form $a-ib$, also a root of the equation. **Conjugate binomials** (see POLYNOMIAL) are those such as $(a+b)$ and $(a-b)$ which differ only by one sign. Another conjugate of $(a+b)$, though not of $(a-b)$, is $(-a+b)$.

CONJUGATION, primitive mode of REPRODUCTION occurring in single-celled organisms—bacteria, protozoa, and some algae and fungi—in which two similar cells link and then exchange nuclear material containing CHROMOSOMES, or fuse, or one cell may absorb the contents of the other. This first step towards SEX often occurs only after many generations of reproduction by FISSION.

CONJUNCTION occurs when the earth, the sun and a planet are in a straight line (as projected onto the plane of the solar system). A planet on the far side of the sun is in superior conjunction, a planet between earth and sun in inferior conjuction (see also OPPOSITION). Planets may be in conjunction with each other (see ASTROLOGY).

CONJUNCTION, in mathematical LOGIC, the assertion that two statements are both true. For statements a and b, this is written $a \wedge b$, read "a is true and b is true." In terms of Boolean algebra (see BOOLE, GEORGE), $a \leftrightarrow I$ and $b \leftrightarrow I$ implies $a \wedge b \leftrightarrow I$.

CONJUNCTIVITIS, INFLAMMATION of the conjunctiva, or fine skin covering the EYE and inner eyelids. It is a common but usually harmless condition

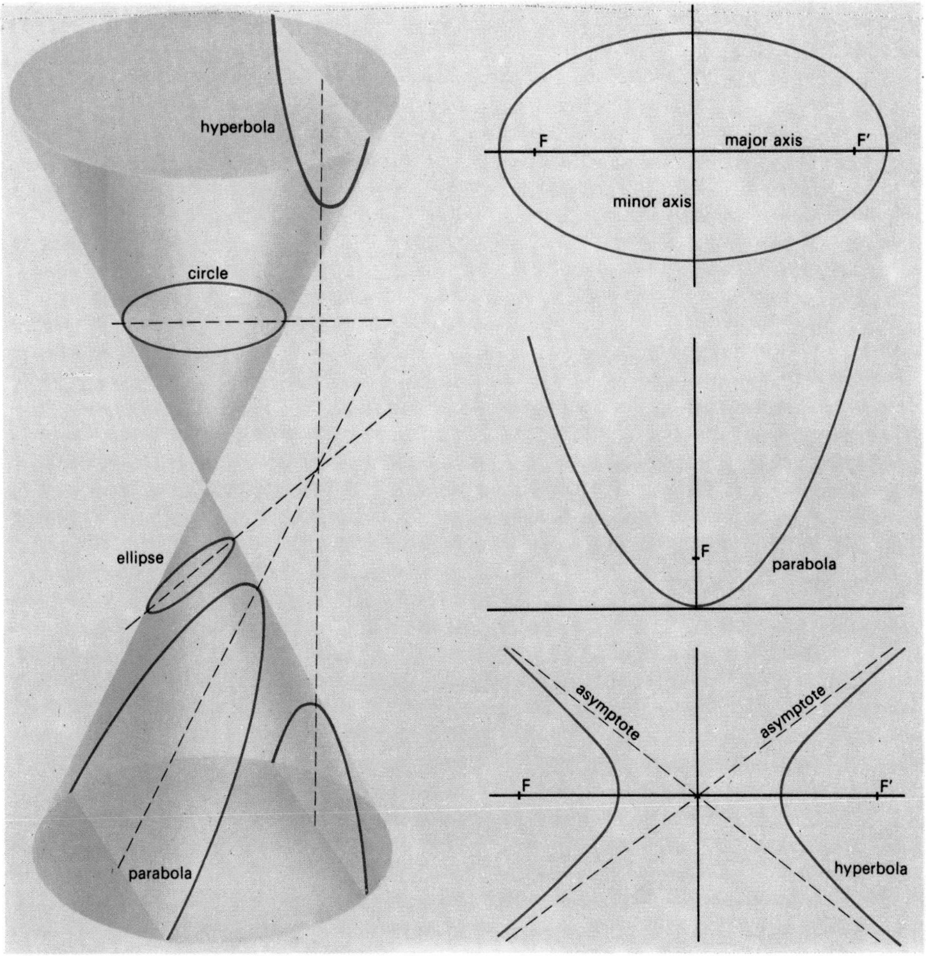

The formation of the conic sections on the surface of a right circular cone and their representation as figures in plane geometry.

caused by ALLERGY (as part of HAY FEVER), foreign bodies, or infection with VIRUSES or BACTERIA. It causes irritation, watering and sticky discharge, but does not affect VISION. Eye drops may help, as can ANTIBIOTICS if bacteria are present.

CONNECTIVE TISSUE. See TISSUE.

CONSCIOUS, in psychoanalysis, the structure of the mind in which logical, conscious thought takes place. Because of confusion over the roles of conscious, CONSCIOUSNESS, UNCONSCIOUS and UNCONSCIOUSNESS (e.g., in dreams), the conscious is now often referred to as the EGO.

CONSCIOUSNESS, in psychology and psychoanalysis, man's capacity for self-awareness as contrasted with its lack in other animals. According to FREUD, it differs from UNCONSCIOUSNESS in that it applies and abides by rules, and in that it recognizes distinctions of space and time.

CONSERVATION, the preservation of the ENVIRONMENT, whether to ensure the long-term availability of natural resources such as FUEL or to retain such intangibles as scenic beauty for future generations.

History. The conservation movement was born in the 19th century as a result of two developments: acceptance of the theory of EVOLUTION and the concept (later proved erroneous) of the BALANCE OF NATURE. It was estimated in that century that over 100 million acres of land in the US had been totally destroyed through EROSION of SOIL caused by the reckless destruction of forests; Congress passed the Forest Reserve Act (1891) and the Carey Land Act (1894) but both were rendered ineffectual by commercial interests. . The first genuinely conservationist president was Theodore Roosevelt, whose Newlands Reclamation Act of 1902 began the struggle for American conservation in earnest. More recently, where officialdom has been dilatory, conservation has been brought to the people by groups such as Friends of the Earth, earning through their efforts a powerful international membership.

The Part of Science. ECOLOGY, the study of the interrelationships of elements of an environment, has

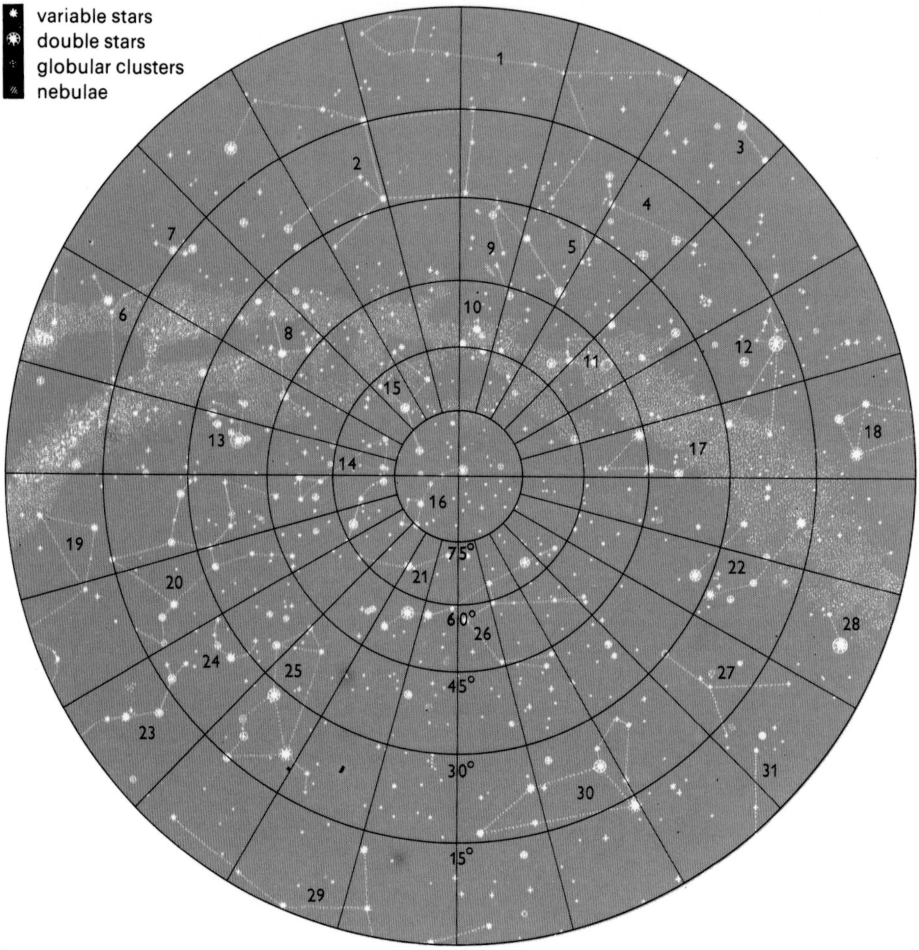

The constellations of the northern sky: (1) Pisces, (2) Pegasus, (3) Cetus, (4) Aries, (5) Triangulum, (6) Aquila, (7) Delphinus, (8) Cygnus, (9) Andromeda, (10) Cassiopeia, (11) Perseus, (12) Taurus, (13) Lyra, (14) Draco, (15) Cepheus, (16) Ursa Minor, (17) Auriga, (18) Orion, (19) Ophiuchus, (20) Hercules, (21) Draco, (22) Gemini, (23) Serpens, (24) Corona Borealis, (25) Boötes, (26) Ursa Major, (27) Cancer, (28) Canis Minor, (29) Virgo, (30) Leo, (31) Hydra.

enabled many scientific disciplines to play a part in conservation. In agriculture, where protection of the soil from erosion is clearly of paramount importance, crop rotation, strip-cropping and other improvements in land use have been made. Important in all fields of human existence and endeavor is the conservation of WATER for IRRIGATION, industrial, drinking and other purposes. Careful use, plus the prevention or amelioration of POLLUTION, especially by industry, are essential. Conservation of raw materials is more complicated, since they cannot be replaced; however, much has been done in the way of good management, and science has developed new processes, artificial substitutes and techniques of RECYCLING. Conservation of wildlife, however, is probably the most dramatically successful of all conservation in this century. Many species, such as the koala and American bison, that were in danger of extinction are now reviving; and most governments are vigilant in areas such as hunting and industrial pollution. Important to all these efforts is the retention of the human population within reasonable limits. In this, CONTRACEPTION has a large part to play, and many governments are now active in their encouragement of it.

CONSISTENCY. See LOGIC.

CONSONANT, SPEECH sound characterized by partial or total obstruction of the breath channel. In English, all the letters of the ALPHABET save the five VOWELS are considered consonants, though *h* is truly an aspirate and *y* may be used as a vowel.

CONSTANT, a mathematical quantity that is not VARIABLE. Constants occur in almost all POLYNOMIALS, in the form of either COEFFICIENTS or ADDENDS. Thus in ax^3+bx^2+cx+d each of a,b,c,d is a constant; should the expression be set to ZERO (see EQUATION) then the values of the constants determine the values of x for

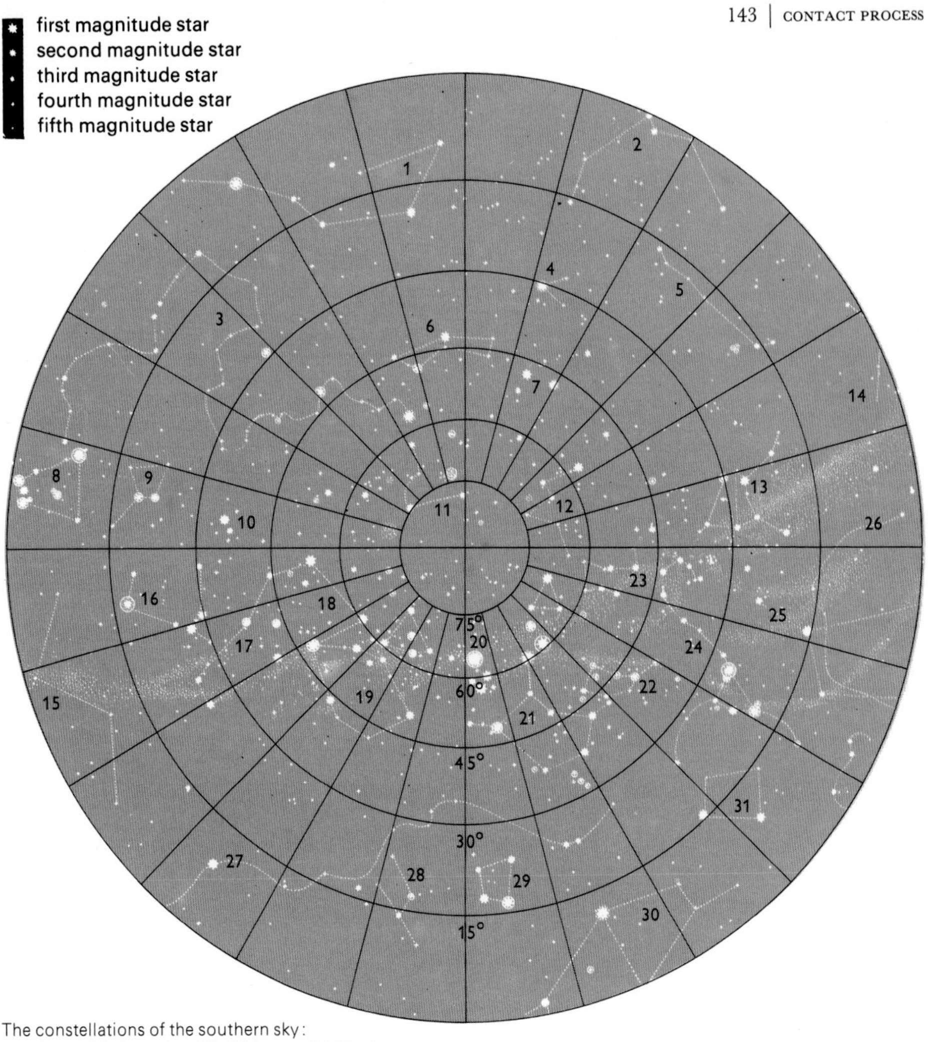

The constellations of the southern sky:
(1) Cetus, (2) Aquarius, (3) Eridanus, (4) Piscis Austrinus, (5) Capricornus, (6) Phoenix, (7) Grus, (8) Orion, (9) Lepus, (10) Columba, (11) Hydrus, (12) Pavo, (13) Sagittarius, (14) Aquila, (15) Monoceros, (16) Canis Major, (17) Puppis, (18) Carina, (19) Vela, (20) Crux, (21) Centaurus, (22) Lupus, (23) Ara, (24) Scorpius, (25) Ophiuchus, (26) Serpens, (27) Hydra, (28) Crater, (29) Corvus, (30) Virgo, (31) Libra.

which the equation holds (see ROOTS OF AN EQUATION); should the expression be a FUNCTION then the constants determine the values of x for which $f(x)$ is defined. Constant addends are usually symbolized by c,C,k or K, though this is by no means universal.

Some constants are naturally occurring. One such is π (see PI), which is of considerable importance in MENSURATION; another is e (see EXPONENTIAL). Physical quantities whose SCALAR magnitude is constant, such as c, the velocity of LIGHT (the electromagnetic constant), or G, the universal gravitational constant, are termed **physical constants**.

CONSTELLATION, a group of stars forming a pattern in the sky, though otherwise unconnected. In ancient times the patterns were interpreted as pictures, usually of mythic characters. The ECLIPTIC passes through twelve constellations, known as the zodiacal constellations (see ZODIAC).

CONSTIPATION, a decrease in the frequency of bowel actions from the norm for an individual; also increased hardness of stool. Often precipitated by inactivity, changed diet or environment, it is sometimes due to GASTROINTESTINAL TRACT disease. Increased dietary fiber, and taking of fecal softeners or intestinal irritants are usual remedies; enema may be required in severe cases.

CONSUMPTION, an obsolete term for TUBERCULOSIS.

CONTACT LENS, a small LENS worn directly on the cornea of the EYE under the eyelid to correct defects of VISION. Generally made of transparent plastic, they sometimes give better results than GLASSES and are certainly less noticeable.

CONTACT PROCESS, major process for manufacturing SULFURIC ACID. SULFUR is burned in air to give sulfur dioxide, which is then oxidized to sulfur trioxide with a vanadium pentoxide or platinum catalyst. Arsenic, which poisons the catalyst, must first be removed. The sulfur trioxide is absorbed in concentrated sulfuric acid to yield oleum, which is diluted with water to the concentration required. Direct reaction of sulfur trioxide with water is too violent.

CONTINENT, one of the seven major divisions of land on earth: Africa, Antarctica, Asia, Australia, Europe, North America, South America. These continents have evolved during the earth's history from a single landmass, PANGAEA (see also CONTINENTAL DRIFT; PLATE TECTONICS; GONDWANALAND; LAURASIA). (See also CONTINENTAL SHELF.)

CONTINENTAL DRIFT, theory first rigorously formulated by WEGENER, later amplified by du TOIT, to explain a number of geological and paleontological phenomena. It suggests that originally the land on earth composed a single, vast CONTINENT, PANGAEA, which broke up. Continental drift is now recognized to be a consequence of the theory of PLATE TECTONICS.

CONTINENTAL SHELF, the portion of a landmass that is submerged in the OCEAN to a depth of less than 200m (650ft), resulting in a rim of shallow water surrounding the landmass. The outer edge of the shelf slopes towards the ocean bottom (see ABYSSAL PLAINS), and is called the continental slope.

CONTINGENCY, in LOGIC, the property whereby a statement may be either true or false, its truth or falsity depending upon the actual state of the world. The statement "the paper in this book is white" is contingent in that although it is in fact true, it is conceivable that it might have been, say, pale green. (See also NECESSITY.)

CONTINUOUS CREATION. See COSMOLOGY.

CONTRACEPTION, the avoidance of conception, and thus of PREGNANCY. Many different methods exist, none of which is absolutely certain. In the **rhythm method**, sexual intercourse is restricted to the days immediately before and after MENSTRUATION, when fertilization is unlikely. **Withdrawal** (*coitus interruptus*) is removal of the PENIS prior to ejaculation, which reduces the sperm released into the vagina. The **condom** is a rubber sheath, fitting over the penis, into which ejaculation occurs; the diaphragm is a complementary device which is inserted into the vagina before intercourse. Both are more effective with **spermicide creams**. **Intrauterine devices** (IUDs) are plastic or copper devices which are inserted into the WOMB and interfere with IMPLANTATION. They are convenient but may lead to infection, or increased blood loss or pain at menstruation. **Oral contraceptives** ("the Pill") are sex HORMONES of the ESTROGEN and PROGESTERONE type which, if taken regularly through the menstrual cycle, inhibit the release of eggs from the ovary. While they are the most reliable form of contraception, they carry a small risk of venous THROMBOSIS, raised blood pressure and possibly other diseases. When the Pill is stopped, periods and ovulation may not return for some time, and this can cause difficulty in assessing fetal maturity if pregnancy follows without an intervening period. While the more effective forms of contraception carry a slightly greater risk, this must be set against the risks of pregnancy and induced ABORTION in the general context of FAMILY PLANNING. Indeed, many risky and costly abortions could be avoided, if more thought were given to contraception.

CONVECTION, passage of heat through a fluid by means of large-scale movements of material within the body of the fluid (see CONDUCTION). If, for example, a liquid is heated from below, parts close to the heat source expand and, because their DENSITY is thus reduced, rise through the liquid; near the top, they cool and begin to sink. This process continues until HEAT is uniformly distributed throughout the liquid. Convection in the ATMOSPHERE is responsible for many climatic effects (see METEOROLOGY). (See also RADIATION.)

CONVERGENCE, of a SEQUENCE, the tending of its terms toward a LIMIT; of a SERIES, the tending of its consecutive partial sums toward a limit. The sequence $\frac{1}{2}, \frac{1}{4}, \frac{1}{8}, \ldots, (\frac{1}{2})^n$ is convergent, since $\lim_{n\to\infty} (\frac{1}{2})^n = 0$. Similarly, the series $\sum_{n=1}^{\infty} (\frac{1}{2})^n$ is convergent, its limit being 1. A sequence or series which does not converge is said to diverge: one such sequence is $1, -1, 1, -1, \ldots, (-1)^n$ though the sequence $1, -\frac{1}{2}, \frac{1}{4}, -\frac{1}{8}, \ldots, (-\frac{1}{2})^n$ is convergent, with limit 0. An example of a divergent series is

$$\sum_{n=1}^{\infty} (-1)^n.$$

CONVERGENT EVOLUTION, the process by which unrelated animals or plants come to resemble one another. Convergence is due to the similar effects of NATURAL SELECTION on the organisms living under the same conditions. An example of convergence is seen in the evolution of the octopus eye; this closely resembles the eye in mammals.

CONVERTIPLANE, a VERTICAL TAKEOFF AND LANDING AIRCRAFT capable of high forward speeds.

CONVULSIONS, or seizures, abnormal involuntary movements, usually rhythmic and associated with disturbance of consciousness; also, a popular term for EPILEPSY.

COOK, James (1728–1779), English navigator and explorer who led three celebrated expeditions to the Pacific Ocean (1768–71; 1772–75; 1776–80), during which he charted the coast of New Zealand (1770), showed that if there were a great southern continent it could not be so large as was commonly supposed, and discovered the Sandwich Islands (1778). He died in an attack by Hawaiian natives.

COORDINATES. See ANALYTIC GEOMETRY; CELESTIAL SPHERE; SPHERICAL COORDINATES.

COPERNICUS, Nicolaus, or Niklas Koppernigk (1473–1543), Polish astronomer who displaced the earth from the center of man's conceptual universe and made it orbit a stationary sun. Belonging to a wealthy German family, he spent several years in Italy mastering all that was known of mathematics, medicine, theology and astronomy before returning to Poland where he eventually settled into the life of lay canon at Frauenburg. His dissatisfaction with the earth-centered (geocentric) cosmology of PTOLEMY was made known to a few friends in the manuscript *Commentariolus* (1514), but it was only on the insistence of Pope Clement VII that he expanded this into the *De revolutionibus orbium coelestium (On the revolutions of the heavenly spheres)* which, when published in 1543, announced the sun-centered (heliocentric) theory to

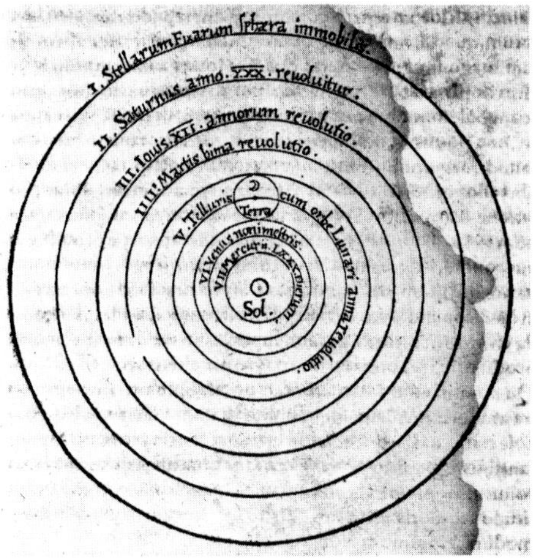

The heliocentric system of Nicolaus Copernicus, after an illustration in *De revolutionibus orbium coelestium* (1543). In the center is the sun, then come the spheres of Mercury, Venus, the earth (with the moon), Mars, Jupiter and Saturn (which orbit the sun in 80 days, 9 months, 1, 2, 12 and 30 years respectively) and finally, on the outside, the sphere of the fixed stars.

the world. Always the theoretician rather than a practical observer, Copernicus' main dissatisfaction with Ptolemy was philosophical. He sought to replace the equant, EPICYCLE and deferent of Ptolemaic theory with pure circular motions, but in adopting a moving-earth theory he was forced to reject the whole of the scholastic physics (without providing an alternative—this had to await the work of GALILEO) and postulate a much greater scale for the universe. Although the heliocentric hypothesis was not immediately accepted by the majority of scientists, its proposal did begin the period of scientific reawakening known as the Copernican Revolution.

COPLANAR, in GEOMETRY, lying in the same PLANE. It is possible to construct a plane through any set of three POINTS, but this is true of sets of four points only in special cases. Any two intersecting or parallel straight LINES are coplanar (see INTERSECTION; PARALLEL LINES).

COPPER (Cu), soft, red metal in Group IB of the PERIODIC TABLE; a TRANSITION ELEMENT. Copper has been used since c6500 BC (see BRONZE AGE). It occurs naturally as the metal in the US, especially Mich., and as the ores CUPRITE, CHALCOPYRITE, ANTLERITE, CHALCOCITE, BORNITE, AZURITE and MALACHITE in the US, Zambia, Zaire and Chile. The metal is produced by roasting the concentrated ores and smelting, and is then refined by electrolysis. Copper is strong, tough, and highly malleable and ductile. It is an excellent conductor of heat and electricity, and most copper produced is used in the electrical industry. It is also a major component of many ALLOYS, including BRASS, BRONZE, GERMAN SILVER, cupronickel (see NICKEL) and beryllium copper (very strong and fatigue-resistant). Many copper alloys are called bronzes, though they need not contain tin: copper+tin+phosphorus is phosphor bronze, and copper+aluminum is aluminum bronze. Copper is a vital trace element: in man it catalyzes the formation of HEMOGLOBIN; in mollusks and crustaceans it is the basic constituent of HEMOCYANIN. Chemically, copper is unreactive, dissolving only in oxidizing acids. It forms cuprous compounds (oxidation state +1), and the more common cupric salts (oxidation state +2), used as fungicides and insecticides, in pigments, as mordants for dyeing, as catalysts, for copper plating, and in electric cells. **Copper (II) Sulfate** ($CuSO_4 . 5H_2O$), or **Blue Vitriol**, blue crystalline solid occurring naturally as chalcanthite; used as above. AW 63.5, mp 1083°C, bp 2567°C, sg 8.96 (20°C).

COPPERAS, or Iron (II) Sulfate. See IRON.

COPPER PYRITES. See CHALCOPYRITE.

CORALS, small marine invertebrates of the class Anthozoa (phylum CNIDARIA) whose limestone skeletons form coral **reefs** and islands in warm seas. Most corals join together in colonies and secrete external LIMESTONE skeletons. Branches and successive layers are formed by budding and by the addition of new members produced sexually which swim freely before attaching themselves and secreting their skeletons. Older members of the colony gradually die, leaving their skeletons behind. Vegetation, such as coraline algae, cements the discarded skeletons, forming coral reefs, of which there are three types: *fringing reefs* along the shore, *barrier reefs* offshore and *atolls*, circular reefs enclosing a lagoon.

CORDILLERA, an extended mountain system, often composed of a number of parallel ranges, associated with a GEOSYNCLINE. In some parts of the world they appear only as chains of islands.

CORDITE, smokeless propellant EXPLOSIVE containing NITROCELLULOSE and NITROGLYCERINE, with petroleum jelly as a stabilizer. Other nitrates were partially substituted for nitroglycerine in WWII to reduce gun-barrel erosion.

CORI, Carl Ferdinand (1896–   ), Czech-born US biochemist who shared the 1947 Nobel Prize for Medicine or Physiology with his wife, **Gerty Theresa Radnitz Cori** (1896–1957), for their joint elucidation of the processes by which GLYCOGEN is broken down and reformed in the body. (B. A. HOUSSAY also shared in the 1947 physiology prize.)

CORIOLIS EFFECT, a FORCE which, like a centrifugal force (see CENTRIPETAL FORCE), apparently acts on moving objects when observed in a frame of reference which is itself rotating. Because of the rotation of the observer, a freely moving object does not appear to move steadily in a straight line as usual, but rather as if, besides an outward centrifugal force, a "Coriolis force" acts on it, perpendicular to its motion, with a strength proportional to its MASS, its VELOCITY and the rate of rotation of the frame of reference. The effect, first described in 1835 by **Gaspard de Coriolis** (1792–1843), accounts for the familiar circulation of air flow around CYCLONES, and numerous other phenomena in METEOROLOGY, oceanography and BALLISTICS.

CORK, protective, waterproof layer of dead cells that have thick walls impregnated with suberin, a waxy material. Cork is found as the outer layer of stems and roots of older woody plants. The cork oak (*Quercus suber*) of S Europe and North Africa produces a profuse amount of cork, which is harvested commercially every 3 to 4 years. (See also CAMBIUM.)

CORM, short, stout, underground stem. It is an organ of vegetative reproduction consisting of a stem base swollen with food material and bearing buds in the remains of the leaves of the previous year's growth. Examples are found in the crocus and gladiolus. (See also BULB; RHIZOMES; TUBER.)

CORNEA, the transparent part of the outer EYE, through which VISION occurs and the iris may be seen; it is responsible for much of the focusing power of the eye. Made of specialized cells and connective tissue, it may be replaced after trauma or infection by a graft from a cadaver.

CORNELIAN. See CARNELIAN.

CORNFORTH, John Warcup (1917–), Australian-born British biochemist who shared the 1975 Nobel Prize for Chemistry with V. PRELOG for his work on the STEREOCHEMISTRY of enzyme-catalyzed reactions.

CORNS AND CALLUSES, localized thickenings of the horny layer of the SKIN, produced by continual pressure or friction. **Calluses** project above the skin and are rarely troublesome; **corns** are smaller, and are forced into the deep, sensitive layers of skin, causing pain or discomfort.

COROLLA. See FLOWER.

CORONA, outer atmosphere of the SUN or other STAR. The term is used also for the halo seen around a celestial body due to DIFFRACTION of its light by water droplets in thin CLOUDS of the earth's ATMOSPHERE; and for a part appended to and within the corolla of some FLOWERS. Around high-voltage terminals there appear a faint glow due to the ionization (see ION) of the local air. The result of this ionization is an electrical discharge known as **corona discharge**, the glow being called a corona.

CORONARY THROMBOSIS, myocardial infarction, or heart attack, one of the commonest causes of serious illness and death in Western countries. The coronary ARTERIES, which supply the HEART with OXYGEN and nutrients, may become diseased with ARTERIOSCLEROSIS which reduces BLOOD flow. Significant narrowing may lead to superimposed THROMBOSIS which causes sudden complete obstruction, and results in death or damage to a substantial area of heart tissue. This may cause sudden death, usually due to abnormal heart rhythm which prevents effective pumping. Severe persistent pain in the center of the CHEST is common, and it may lead to SHOCK or LUNG congestion. Characteristic changes may be seen in the ELECTROCARDIOGRAPH following myocardial damage, and ENZYMES appear in blood from the damaged heart muscle. Treatment consists of rest, ANALGESICS and drugs to correct disordered rhythm or inadequate pumping; certain cases must be carefully observed for development of rhythm disturbance. Recovery may be complete and normal activities resumed. Predisposing factors, including OBESITY, SMOKING, high blood pressure, excess blood FATS (including CHOLESTEROL) and DIABETES must be recognized and treated.

CORPUSCLE. See BLOOD.

CORRELATION, Statistical, the interdependent variation of two or more VARIABLES. Positive correlation between two variables occurs when increase in the value of one implies increase in the value of the other, decrease similarly implying decrease. Negative correlation occurs when increase in the value of one implies decrease in the value of the other. (See also STATISTICS.)

A measure of the correlation between two variables is given by their **correlation coefficient**, r, given by

$$r = \frac{\sum_{i=1}^{n} (a_i - \bar{a})(b_i - \bar{b})}{\sqrt{\sum_{i=1}^{n} (a_i - \bar{a})^2 \sum_{i=1} (b_i - \bar{b})^2}}$$

where a and b are the variables, and $\bar{a}$ and $\bar{b}$ are their mean values (see MEAN, MEDIAN AND MODE). r varies between -1 and $+1$; $r=-1$ indicating perfect negative correlation, $r=0$ the mutual independence of the variables, and $r=+1$ perfect positive correlation.

CORROSION, the insidious destruction of metals and alloys by chemical reaction (mainly OXIDATION) with the environment. The annual cost of corrosion is more than $5 billion in the US alone. In moist air most metals form a surface layer of oxide, which, if it is coherent, may slow down or prevent further corrosion. **Tarnishing** is the formation of such a discolored layer, mainly on copper or silver. (**Rust**—hydrated iron (III) oxide, FeO(OH)—offers little protection, so that iron corrodes rapidly.) Industrial AIR POLLUTION greatly speeds up corrosion, oxidizing and acidic gases (especially sulfur dioxide) being the worst culprits. Corrosion is usually an electrochemical process (see ELECTROCHEMISTRY): small cells are set up in the corroding metal, the potential difference being due to the different metals present or to different concentrations of oxygen or electrolyte; corrosion takes place at the anode. It also occurs preferentially at grain boundaries and where the metal is stressed. Prevention methods include BONDERIZING; a protective layer of paint, varnish or electroplate; or the use of a "sacrificial anode" of zinc or aluminum in electrical contact with the metal, that is preferentially corroded. **Galvanizing**—coating iron objects with zinc—works on the same principle.

CORROSIVE SUBLIMATE, or **mercury (II) chloride.** See MERCURY.

CORTEX, in plant STEMS and ROOTS, the layer of mostly unspecialized packing cells between the EPIDERMIS and the PHLOEM. It is used to store food and other substances including resins, oils and tannins; in stems it may contain CHLOROPLASTS; in roots it transports water and ions inwards. The term is used for the outer layer of the BRAIN, KIDNEYS or ADRENAL GLANDS.

CORTISOL, principal natural STEROID HORMONE.

Cortisol

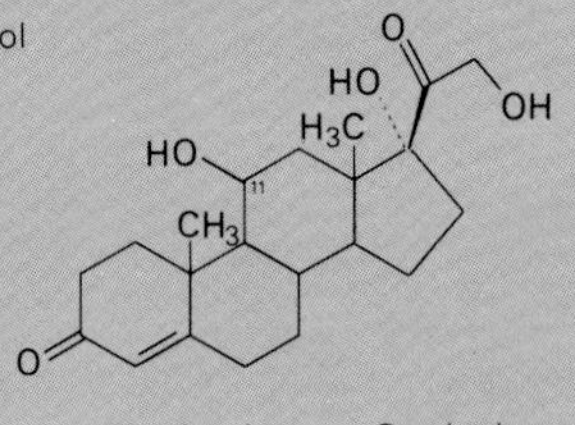

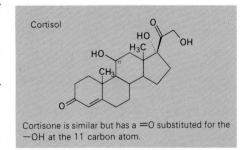

Cortisone is similar but has a =O substituted for the —OH at the 11 carbon atom.

CORTISONE, minor STEROID, converted to CORTISOL.
CORUNDUM, or α-alumina, rhombohedral form of aluminum oxide (see ALUMINUM), occurring worldwide (see also EMERY); transparent varieties include RUBY and SAPPHIRE. Artificial corundum, or β-alumina, is made by calcining BAUXITE. Corundum is the hardest natural substance after diamond (Mohs hardness 9), and is used as an abrasive and for BEARINGS.
COSECANT. See TRIGONOMETRY.
COSHINE (COSH). See HYPERBOLIC FUNCTIONS.
COSINE. See TRIGONOMETRY.
COSINE RULE, in any plane TRIANGLE, with angles P, Q, R, and sides p, q, r,

$$p^2=q^2+r^2-2qr\cos P;$$

i.e., the SQUARE of any one side is equal to the sum of the squares of the other two sides less twice the product of those two sides and the cosine (see TRIGONOMETRY) of the ANGLE they include. In SPHERICAL TRIGONOMETRY the cosine rule as applied to the sides of a spherical triangle is

$$\cos p=\cos q\cos r+\sin q\sin r\cos P$$

and, as applied to angles,

$$\cos P=-\cos Q\cos R+\sin Q\sin R\cos p.$$

(See also SINE RULE.)
COSMIC EGG. See COSMOLOGY; LEMAÎTRE.
COSMIC RAYS, ELECTRONS and the nuclei of HYDROGEN and other ATOMS which isotropically bombard the earth's upper ATMOSPHERE at VELOCITIES close to that of light. These primary cosmic rays interact with molecules of the upper atmosphere to produce what are termed secondary cosmic rays, which are considerably less energetic and extremely shortlived: they are SUBATOMIC PARTICLES that change rapidly into other types of particles. Initially, secondary cosmic rays, which pass frequently and harmlessly through our bodies, were detected by use of the GEIGER COUNTER, though now it is more common to employ a SPARK CHAMBER. It is thought that cosmic rays are produced by SUPERNOVAS, though some may be of extragalactic origin.
COSMOGONY, the science or pseudoscience of the origins of the UNIVERSE.
COSMOLOGY, the study of the structure and evolution of the universe. Ancient and medieval cosmologies were many, varied and imaginative, usually oriented around a stationary, flat earth at the center of the universe, surrounded by crystal spheres carrying the moon, sun, planets and stars, although ARISTARCHUS understood that the earth was spherical and circled the sun. With increasing sophistication of observational techniques and equipment, more realistic views of the universe emerged (see COPERNICUS; BRAHE; KEPLER; GALILEO; NEWTON). Modern cosmological theories, which take into account EINSTEIN's Theory of RELATIVITY and the recession of galaxies shown by the RED SHIFT in their spectra (see SPECTRUM), divide into two main types.
Evolutionary Theories. The most important of these is the **Big Bang Theory** resulting from HUBBLE's observations of the galaxies. This theory proposes that initially the universe existed as a single compact ball of matter, the **cosmic egg**, that exploded to form a mass of gaseous debris which eventually began to condense to form stars. An alternative theory is of an **oscillating universe** that periodically reaches a maximum size, then begins to shrink until it is once more in the form of a cosmic egg, which in turn explodes to create a new universe.
Continuous Creation. The **Steady State Theory** of BONDI, GOLD and HOYLE, first put forward in 1948, proposes that the universe has existed and will exist forever in its current form, the expansion being caused by the continuous creation of new matter so that the average density and appearance of the universe remain the same at all times. This would necessitate a reexamination of the Law of Conservation of ENERGY.

In 1965 A. Penzias and R. Wilson discovered that the universe possesses an inherent radio "background noise," and it was suggested by R. Dicke that this was the relic of the RADIATION produced by the Big Bang. Further researches have indicated the probability that space is filled with uniform BLACKBODY thermal radiation corresponding to a temperature of around 3K. This would support an evolutionary theory and, though it still has adherents, the Steady State Theory has now largely been abandoned. (See also ASTRONOMY; BLACK HOLE; PULSAR; QUASAR.)
COSMOTRON, large proton SYNCHROTRON used to accelerate (see ACCELERATOR) PROTONS to energies in the 1 GeV range (see ELECTRON VOLT).
COTANGENT. See TRIGONOMETRY.
COTTON, a subtropical plant of the genus *Gossypium*, grown for the soft white fibers attached to its seed, which can be woven into cloth. The seeds are planted in the early spring and the plants bloom after four months. The white flowers redden and fall in a few days, leaving the seed pods, which are fully grown in another month or so. These pods then burst, showing the white lint, which is picked either by hand or mechanically. Each fiber is a single cell, with numerous twists along its length, which give it excellent spinning characteristics.

A number of species and their varieties are grown. Cultivated varieties of *Gossypium barbodense* produce fibers 36–64mm (1.4–2.5in) long, *G. hirsutum* (the American upland cotton) fibers 21–32mm (0.8–1.3in) and the Asiatic species *G. herbaceum* and *G. arboreum* fibers 9–19mm (0.4–0.8in). Cotton is produced in more than 60 countries; total world production exceeds 55 million bales annually, each bale weighing 222.6kg (490lb).

Cotton is prone to many pests and diseases, which cause enormous damage to the crop (averaging nearly 300 million dollars in the US every year). The main insect pests are the bollweevil in the US and the pink boll worm in India and Egypt. Destructive fungus diseases that attack the plant include fusarium and verticillium wilt and Texas root rot. Boll rots can cause severe damage to the crop. Mechanization of the cotton processing industry was one of the first stages of the Industrial Revolution. It is still an important industry, although consumption of cotton has not risen since the development of man-made textiles. However, 80% of the yarn from spinning mills is still made into cloth, the remainder being used in industry. The seed is now used for oils and cattle food, while small fibers are made into cellulose. (See also COTTON GIN; PLANT DISEASES; TEXTILES.)
COTTON GIN, device for separating cotton fibers

from the seeds, invented by Eli WHITNEY (1793), which revolutionized the cotton industry in the US South. Whitney's original gin comprised a rotating drum on which were mounted wire spikes that projected through narrow slits in a wire grid. The spikes drew the fibers through these slits, leaving behind the seeds, which are broader. A revolving brush removed the fibers from the drum. In 1794, Hodgen Holmes replaced the wire spikes by a circular SAW; and modern cotton gins still work on this principle.

COTTRELL, Frederick Gardner (1877–1948), US chemist who invented the ELECTROSTATIC PRECIPITATOR used to reduce AIR POLLUTION from factory flues (1910), and founded the nonprofit Research Corporation to plow back the profits of invention into further scientific research (1912).

COTYLEDON, leaf forming part of the embryo in seeds. In MONOCOTYLEDONS there is only one cotyledon and in DICOTYLEDONS there are two. They often contain food reserves used in the early stages of GERMINATION and normally they are brought above the ground, turn green and carry out PHOTOSYNTHESIS, but soon fall off. They bear no resemblance to normal foliage leaves. (See also PLANT; SEED.)

COTYLOSAURIA, fossil members of the REPTILIA that lived during the Permian and Triassic periods, 280–195 million years ago, and which included the ancestors to all later reptiles. They were sluggish creatures with short limbs and long tails. Many were small but some attained a length of 1.5m (5ft).

COUGH, sudden explosive release of air from the LUNGS, which clears respiratory passages of obstruction and excess MUCUS or PUS; it occurs both as a REFLEX and on volition. Air flow may reach high velocity and potentially infectious particles spread a great distance if the mouth is uncovered. Persistent cough always implies disease.

COULOMB (C), the SI UNIT of electric charge, defined as the quantity of ELECTRICITY transported in

Catalytic cracker. The feedstock of long-chain hydrocarbons (1) is mixed with hot catalyst (2) and vaporized. The vapor/powder mixture is carried to the reactor where the cracking reactions occur. Cyclones (3) extract the cracked hydrocarbon vapor and pass it to the fractionating column where it is fractionated, yielding petroleum gases and gasoline (4), light gas oil (5), medium gas oil (6) and heavy gas oil (7). Spent catalyst meanwhile is mixed with steam (8) and carried in a current of hot air (9) to the catalyst regenerator where it is cleaned and recycled (10). Waste gases are drawn off (11) and vented.

one second by a one-ampere current.

COULOMB, Charles Augustin de (1736–1806), French physicist noted for his researches into FRICTION, TORSION, ELECTRICITY and MAGNETISM. Using a torsion BALANCE, he established **Coulomb's Law** of electrostatic forces (1785). This states that the force between two charges is proportional to their magnitudes and inversely proportional to the square of their separation. He also showed that the charge on a charged conductor lies solely on its surface.

COUNTERFACTUAL CONDITIONALS, or contrary-to-fact conditionals, in LOGIC, statements known to be false of which the consequences are nevertheless explored. The epistemic status of these conclusions is hotly debated by philosophers.

COUNTERGLOW or **Gegenschein**. See ZODIACAL LIGHT.

COURNAND, André Fréderic (1895–), French-born US physiologist who shared the 1956 Nobel Prize for Physiology or Medicine with D. W. RICHARDS and W. FORSSMANN for the development of heart catheterization (see CATHETER) and other work which has led to a much fuller understanding of heart and lung diseases.

COURTSHIP. See MATING RITUALS.

COUVADE, a custom in many primitive societies throughout the world in which a prospective father imitates the prospective mother's confinement and labor, perhaps as a ritual stressing of the male reproductive role.

COWPOX, a disease of cattle, caused by a VIRUS related to the smallpox virus. It was by noting the IMMUNITY to SMALLPOX conferred on humans who contracted cowpox by milking infected cattle that JENNER popularized VACCINATION against smallpox.

CRAB NEBULA (M1), a bright NEBULA in the constellation TAURUS, the remnants of the SUPERNOVA of 1054. It is 1 223pc from the earth and is associated with a PULSAR.

CRACKING, process by which heavy HYDROCARBON molecules in PETROLEUM are broken down into lighter molecules and isomerized, by means of high temperatures or CATALYSIS. It yields branched-chain ALKANES and ALKENES of high OCTANE rating for GASOLINE, and also simple gaseous alkenes for chemical synthesis.

CRAMP, the painful contraction of muscle—often in the legs. The cause is usually unknown. It may be brought on by exercise or lack of SALT; it also occurs in muscles with inadequate BLOOD supply. Relief is by forcibly stretching the muscle or by massage.

CRANE, an important device used for hoisting heavy loads, a familiar sight in warehouses, at docks, on building sites and elsewhere. It has a WINCH or winding mechanism operating a rope (usually steel) attached to a PULLEY block on which there is a hook. In many cranes the rope passes over a pulley at the far end of a boom or **jib** from the winch; in others it hangs from a horizontal beam or girder. A **derrick** is a crane where the distance of the end of the jib from the crane's vertical support (e.g., pillar) may be varied: simplest is the ship's derrick, where the jib is pivoted near the base of a braced or guyed mast. The tower derrick is similar, though here the pivoted end of the jib may be moved up and down the pillar as needed. Mobile cranes move on rails or caterpillar tracks: in particular, the **gantry crane** is shaped like a bridge,

Telescopic boom cranes such as this Coles Hydra Truck 45/50T are particularly useful where only a few heavy lifting jobs must be done on a site. The truck is raised off its roadwheels on hydraulic jacks to ensure stability during lifting. When a construction scheme requires the use of a crane over a long period, it is common to install a tower crane.

the two verticals having wheels and the pulley block being suspended from the horizontal. In **traveling bridge cranes** the horizontal runs on wheels on tracks mounted high on facing walls of, say, a workshop. Cranes may also be mounted on trucks, railroad freight cars or ships.

CRANNOG. See LAKE DWELLING.

CREAM OF TARTAR. See TARTARIC ACID.

CREOSOTE, distillation product of COAL TAR or wood tar. Coal-tar creosote is a mixture of HYDROCARBONS, PHENOLS (see also CRESOL) and other compounds, used to preserve wood used outdoors, and often applied under pressure. Wood creosote is an oily mixture of phenols formerly used as an ANTISEPTIC and food preservative.

CRESOL, or methylphenol ($CH_3C_6H_4OH$), a member of the PHENOLS; there are three isomers, synthesized from TOLUENE. Distillation of CREOSOTE or PETROLEUM yields a mixture of cresols and xylenols ("cresylic acid") used as a disinfectant and in the manufacture of resins and tricresyl phosphate (used to make gasoline additives and plasticizers).

CRETACEOUS, final period of the MESOZOIC, about 135 to 65 million years ago, lying after the JURASSIC and before the CENOZOIC. (See GEOLOGY.)

CRETINISM, congenital disease caused by lack of THYROID HORMONE in late fetal life and early infancy, which interferes with normal development, including that of the BRAIN. It may be due to congenital inability to secrete the hormone or, in certain areas of the world, to lack of dietary IODINE (which is needed for hormone formation). The typical appearance, with coarse SKIN, puffy face, large tongue and slow responses, usually enables early diagnosis. It is crucial that replacement therapy with thyroid hormone should be started as early as possible to minimize or prevent the mental retardation that occurs if diagnosis is delayed.

CRICK, Francis Harry Compton (1916–), English biochemist who, with J. D. WATSON, proposed the double-helix model of DNA. For this, one of the most spectacular advances in 20th-century science, they shared the 1962 Nobel Prize for Physiology or Medicine with M. H. F. WILKINS, who had provided them with the X-ray data on which they had based their proposal. Crick's subsequent work has been concerned with deciphering the functions of the individual CODONS in the genetic code.

CRINOIDEA, a class of ECHINODERMATA which includes the sea lilies and feather stars. They have cup-shaped bodies. Sea lilies are sendentary and are attached to the sea bed by a stalk. Feather stars are mobile and have no stalk.

CRITICAL MASS, the MASS of a radioactive material needed to achieve a self-sustaining FISSION process. This requires that the NEUTRONS emitted by a given nuclear fission are more likely to encounter further fissionable nuclei than to escape. It is of the order of several kilograms for URANIUM-235.

CROCODILIA, members of the REPTILIA that include alligators and crocodiles. The crocodilians have short legs which are clumsy on land but long tails that enable them to swim powerfully. Alligators are distinguished from crocodiles by the arrangement of their teeth.

CRO-MAGNON MAN, a race of primitive man named for Cro-Magnon, France, dating from the Upper Paleolithic (see STONE AGE) and usually regarded as AURIGNACIAN, though possibly more recent. Coming later than Neanderthal Man (see PREHISTORIC MAN), Cro-Magnon Man was dolichocephalic (see CEPHALIC INDEX) with a high forehead and a large brain capacity, his face rather short and wide. He was probably around 1.7m tall, powerfully muscled and robust.

CROMPTON, Samuel (1753–1827), English inventor of the SPINNING mule (1779), so-called because it was a cross between ARKWRIGHT's water-frame machine and HARGREAVES' spinning jenny. Although the mule was a great advance on its progenitors, it made little money for its inventor.

CRONSTEDT, Axel Fredrik, Baron (1722–1765), Swedish chemist who discovered the element NICKEL, introduced the use of the blowpipe into chemical ANALYSIS and published one of the first chemical classifications of minerals.

CROOKES, Sir William (1832–1919), English physicist who discovered the element THALLIUM (1861), invented the RADIOMETER (1875) and pioneered the study of CATHODE RAYS. By 1876 he had devised the **Crookes tube**, a glass tube containing two ELECTRODES and pumped out to a very low gas PRESSURE. By applying a high voltage across the electrodes and varying the pressure, he was able to produce and study cathode rays and various glow discharges.

CROP, or craw, in some birds, a pouch-like enlargement of the gullet or of the alimentary canal. In it, food can be stored, or receive partial preparation, before digestion.

CROP DUSTING, the spraying of farm and garden crops with INSECTICIDES and herbicides (see WEEDKILLER). It aims to protect them from pest attack and competition of uneconomic plants. In recent years, concern has been expressed over the use of poisonous chlorinated and phosphate insecticides, which may cause human and animal injury. (See also PESTICIDES; PLANT DISEASES.)

CROSS-EYE. See STRABISMUS.

CROSSING OVER, exchange of genes between the members of a homologous pair of CHROMOSOMES, resulting from breakage and rejoining of the chromosomes, usually during MEIOSIS. It can alter the LINKAGE groups.

CROSSOPTERYGII, a subclass of the OSTEICHTHYES that contains fossil fish which include the ancestors to the AMPHIBIA. The group was thought to be extinct until a member, the COELACANTH, was discovered living off the east coast of Africa. Crossopterygians have paired fins that are supported by fleshy lobes.

CROSS-SECTION, Nuclear, a measure of the strength of the interaction between two atomic or SUBATOMIC PARTICLES. A beam of one kind of particle is directed at a particle of the other kind, and the number of particles scattered or absorbed is measured as an equivalent area (often in BARNS) presented to the beam by the target particle within which all incident particles are affected.

CROUP, a condition common in infancy due to VIRUS infection of LARYNX and TRACHEA and causing characteristic stridor, or spasm of larynx when the child breathes in. Often a mild and short illness, it occasionally causes so much difficulty in breathing that OXYGEN is needed.

CRUCIBLE, a container used for chemical reactions at high temperatures, for CALCINATION, or for melting metals. Crucibles are made of earthenware, porcelain, quartz or graphite; or of metals such as iron, platinum, nickel or silver. Metal crucibles may be damaged by alloy formation if metals are melted in them. Industrial crucibles are generally large and lined with refractory materials.

CRUSTACEA, a class of ARTHROPODA which includes the crabs, shrimps, woodlice, barnacles and water fleas. Most are aquatic, breathing by gills, and have two pairs of antennae.

CRYOGENICS (from Greek *kruos*, frost), the branch of physics dealing with the behavior of matter at very low temperatures, and with the production of those temperatures. Early cryogenics relied heavily on the Joule–Thomson effect (named for James JOULE and William Thomson, later Lord KELVIN) by which temperature falls when a gas is permitted to expand without an external energy source. Using this, James DEWAR liquefied HYDROGEN in 1895 (though not in quantity until 1898), and H. K. Onnes liquefied HELIUM in 1908 at 4.2K (see also ABSOLUTE ZERO). Several cooling processes are used today. Down to about 4K the substance is placed in contact with liquefied gases which are permitted to evaporate, so removing HEAT energy (see LATENT HEAT). The lowest temperature that can be reached thus is around 0.3K. Further temperature decrease may be obtained by paramagnetic cooling (adiabatic demagnetization). Here a paramagnetic material (see PARAMAGNETISM) is placed in contact with the substance and with liquid helium, and subjected to a strong magnetic field (see MAGNETISM), the heat so generated being removed by the helium. Then, away from the helium, the magnetic field is reduced to zero. By this means temperatures of the order of 10^{-2}–10^{-3}K have been achieved (though, because of heat leak, such temperatures are always unstable). A more complex process, nuclear adiabatic demagnetization, has been used to attain temperatures as low as 2×10^{-7}K.

Near absolute zero, substances can display strange properties. Liquid helium II has no VISCOSITY (see SUPERFLUIDITY) and can flow up the sides of its container. Some elements display SUPERCONDUCTIVITY: an electric current started in them will continue indefinitely. (See also CRYOTRON.)

Low temperatures can be used to preserve foods for periods of years. Recently, much publicity has surrounded the idea of freezing people with terminal illnesses, so that at some date in the future, when medical science has advanced sufficiently, they may be revived and cured. Another idea mooted is that of freezing travelers on interstellar spacecraft.

CRYOLITE, a mineral found in Greenland: chemical formula Na_3AlF_6, white monoclinic crystals. Molten cryolite is used to dissolve bauxite ore in the electrolytic production of aluminum (see HALL-HÉROULT PROCESS).

CRYOTRON, miniature switching device used in computers. It makes use of the property of SUPERCONDUCTIVITY exhibited by metals at very low temperatures (see CRYOGENICS). About a single wire of one metal (e.g., TANTALUM) is a coil of another (e.g., NIOBIUM): both are bathed in liquid HELIUM at about 4K. Current passed through the coil creates a magnetic field (see MAGNETISM) which renders the tantalum non-superconductive (niobium can tolerate a greater magnetic field without loss of superconductivity). Such switching can be performed extremely rapidly.

CRYPTIC COLORATION, coloration that aids the concealment of animals. In many, the coloration matches the animal's natural background, but in others it serves to break up or distort the otherwise easily recognizable outline of the animal.

CRYPTOGAM, name given by early botanists to plants such as ALGAE, FUNGI, MOSSES, LIVERWORTS and FERNS, which do not have prominent organs of reproduction when compared to the GYMNOSPERMS and the ANGIOSPERMS.

CRYPTOZOIC, in GEOLOGY, the aeon in which life first appeared (see EVOLUTION) and PRECAMBRIAN rocks were formed. The rocks do not contain FOSSILS that can be used for dating (see CAMBRIAN; CHRONOLOGY); hence the name Cryptozoic (hidden life). By contrast the aeon of visible life, from the end of the Cryptozoic to the present, is called the PHANEROZOIC.

CRYSTALS, homogeneous solid objects having naturally-formed plane faces. The order in their external appearance reflects the regularity of their internal structure, this internal regularity being the

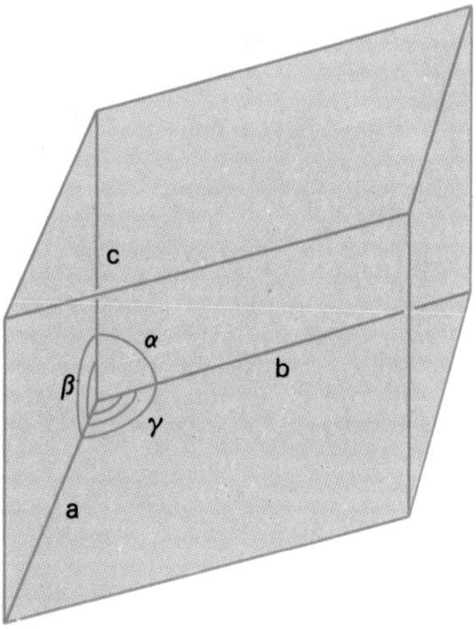

The unit cell of a triclinic crystal, conventionally drawn to show the crystallographic axes *a*, *b*, *c* and the angles α, β, γ between them.

The seven crystal systems

Name	Symmetry axes	Crystallographic axes
triclinic (two classes)	none	$a \neq b \neq c$ $\alpha \neq \beta \neq \gamma$
monoclinic (three classes)	one diad axis	$a \neq b \neq c$ $\gamma \neq \alpha = \beta = 90°$
orthorhombic (three classes)	three diad axes	$a \neq b \neq c$ $\alpha = \beta = \gamma = 90°$
tetragonal (seven classes)	one tetrad axis	$a = b \neq c$ $\alpha = \beta = \gamma = 90°$
cubic (five classes)	four triad axes	$a = b = c$ $\alpha = \beta = \gamma = 90°$
trigonal (five classes)	one triad axis	$a_1 = a_2 = a_3 \neq c$ $90° = \alpha \neq \gamma = 120°$
hexagonal (seven classes)	one hexad axis	$a_1 = a_2 = a_3 \neq c$ $90° = \alpha \neq \gamma = 120°$

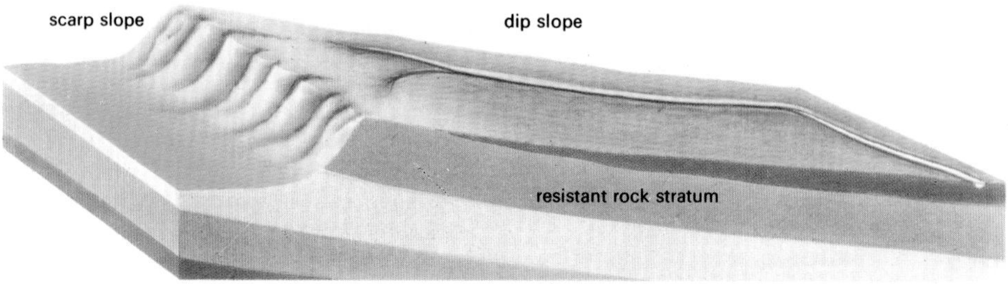

Block diagram of a cuesta showing how the surface topography is related to the underlying stratigraphy.

keynote of the **crystalline state**. Although external regularity is most obvious in natural crystals and those grown in the laboratory, most other inorganic solid substances (with the notable exceptions of PLASTICS and GLASS) also exist in the crystalline state although the crystals of which they are composed are often microscopic in size. True crystals must be distinguished from cut gemstones which, although often internally crystalline, exhibit faces rather chosen according to the whim of the lapidary than developed in the course of any natural growth process. The study of crystals and the crystalline state is the province of **crystallography**. Crystals are classified according to the symmetry elements that they display. (See SYMMETRY; an alternative method of reaching the same classification uses the crystallographic axes used in describing the crystal's faces). This gives the 32 crystal classes, which can be grouped into the seven traditional crystal systems (six if trigonal and hexagonal are counted together). Crystals are allotted to their proper class by considering their external appearance, the symmetry of any etch marks made on their surfaces, and their optical and electrical properties (see DOUBLE REFRACTIONS; PIEZOELECTRICITY). Although such observations do enable the crystallographer to determine the type of "unit cell" which, when repeated in space, gives the overall LATTICE structure of a given crystal, they do not allow him to determine the actual dispositions of the constituent ATOMS or IONS. Such patterns can only be determined using X-RAY DIFFRACTION techniques. When a crystal is composed solely of particles of a single species and the attractive forces between molecules are not directionally localized, as in the crystals of many pure metals, the atoms tend to take up one of two structures which both allow a maximum degree of close-packing. These are known as hexagonal close packed (hcp) and face-centered cubic (fcc) and can both be looked upon as different ways of stacking planes of particles in which each is surrounded by six neighbors. Where there is a degree of directionality in the bonding between particles (as in DIAMOND) or more than one particle species is involved (as in common SALT), the crystals exhibit more complex structures, single substances often adopting more than one structure under different conditions (POLYMORPHISM). Although much crystallography assumes that crystals perfectly exhibit their supposed structure, real crystals, of course, contain minor defects such as grain boundaries and DISLOCATIONS. Many of the most important properties and uses of crystals depend on these defects (see SEMICONDUCTOR).

CTENOPHORA, or **comb jellies**, exclusively marine, gelatinous animals that are characterized by the presence of eight rows of comb plates which bear CILIA used for swimming. Ctenophorans were once classified with the CNIDARIA, but are now considered to represent a separate phylum. Well-known examples are often called sea gooseberries.

CUBE, a regular hexahedron (see POLYHEDRON). In ALGEBRA, the cube (x^3) of a number, x, means $x \times x \times x$: e.g., $10^3 = 10 \times 10 \times 10 = 1000$.

CUBE ROOT. See ROOT.

CUBIT, the name of various ancient units of length, originally the length of the human forearm and typically about 525mm. (See WEIGHTS AND MEASURES.)

CUBOID, a rectangular PARALLELEPIPED.

CUESTA, or (British) escarpment, a ridge or hill with one side gently sloping, the other a steep cliff (SCARP), formed by selective EROSION in areas where the ROCK strata are tilted. (See also HOGBACK.)

CUISENAIRE METHOD, system used in teaching basic mathematics. The students may, by manipulation and comparison of rods of ten different lengths and ten corresponding colors, derive for themselves fundamental laws and relations.

CULTURE, a controlled growth of living cells in an artificial medium, often in a PETRI DISH. The cells may be microorganisms isolated and studied in a pure culture, or made to compete with each other under various environmental and nutritional conditions (see MICROBIOLOGY); or they may be cells from animal or plant tissue cultivated for studies of heredity and function. The culture medium usually contains water, GELATIN or AGAR, salts and various nutrients.

CULTURE. See ANTHROPOLOGY.

CUMULUS. See CLOUDS.

CUNEIFORM (from Latin *cuneus*, wedge, and *forma*, shape), one of the earliest known fully developed writing systems. Each character is formed by a combination of wedge- or nail-shaped strokes. Invented probably by the inhabitants of Sumer before 3000 BC, it was soon adopted by the Akkadians and then by other peoples, such as the Hittites and the Persians. The characters are stylizations of earlier pictographs, and were impressed in clay. Cuneiform was first deciphered in detail by Rawlinson in 1846.

CUPPING, a former medical technique for drawing BLOOD to the surface of the body by producing a partial VACUUM inside a "cupping glass" applied to the skin by burning a few drops of denatured alcohol in it.

CUPRITE, red mineral consisting of copper(I) oxide

(Cu_2O); a major ore of COPPER, found in Europe, Australia, Ariz., SW Africa and South America. It is formed by oxidation of copper sulfide ores.

CURARE, arrow-poison used by South American Indian hunters, extracted from various plants, chiefly of the genera *Strychnos* and *Chondodendron*, killing by respiratory paralysis. Curare is a mixture of ALKALOIDS, the chief being *d*-tubocurarine. By competing with ACETYLCHOLINE it blocks nerve impulse transmission to muscles, producing relaxation and paralysis. It has revolutionized modern surgery by producing complete relaxation without a dangerous degree of ANESTHESIA being required. (See also TETANUS.)

CURIE (Ci), a unit of RADIOACTIVITY now defined as the quantity of a radioactive material in which 3.7×10^{10} disintegrations per second are occurring.

CURIE, Marie (1867–1934), born Marja Sklodowska, Polish-born French physicist who, with her French-born husband **Pierre Curie** (1859–1906), was an early investigator of RADIOACTIVITY, discovering the radioactive elements POLONIUM and RADIUM in the mineral PITCHBLENDE (1898). For this the Curies shared the 1903 Nobel physics prize with A. H. BECQUEREL. After the death of Pierre, Marie went on to investigate the chemistry and medical applications of radium and was awarded the 1911 Nobel Prize for Chemistry in recognition of her isolation of the pure metal. She died of LEUKEMIA, no doubt contracted in the course of her work with radioactive materials. Pierre Curie is also noted for the discovery, with his brother Jacques, of PIEZOELECTRICITY (1880) and for his investigation of the effect of TEMPERATURE on magnetic properties. In particular he discovered the **Curie point**, the temperature above which ferromagnetic materials display only PARAMAGNETISM (1895). The Curies' elder daughter, Irène JOLIOT-CURIE, was also a noted physicist.

CURIUM (Cm), a TRANSURANIUM ELEMENT in the ACTINIDE series. It is prepared by bombarding plutonium-239 with ALPHA PARTICLES or americium-241 with neutrons. Cm^{244} is used as a compact power source for space uses, the heat of nuclear decay being converted to electricity. mp c1340°C, sg 13.51.

CURL of a vector (see VECTOR ANALYSIS), the vector product of DEL with the vector. Thus

$$\text{curl}\,\mathbf{V} = \nabla \times \mathbf{V} = \mathbf{i}\left(\frac{\partial v_z}{\partial y} - \frac{\partial v_y}{\partial z}\right) + \mathbf{j}\left(\frac{\partial v_x}{\partial z} - \frac{\partial v_z}{\partial x}\right) + \mathbf{k}\left(\frac{\partial v_y}{\partial x} - \frac{\partial v_x}{\partial y}\right)$$

(in CARTESIAN COORDINATES).

CURRENT, Electric. See ELECTRICITY.

CURRENT, Ocean. See OCEAN CURRENTS.

CURTISS, Glenn Hammond (1878–1930), US aircraft manufacturer and inventor of the AILERON (1911). In 1908 he won a trophy for the first public flight of more than 1km in the US.

CURVATURE, at any point P on a plane CURVE, the difference between the angle θ made by the tangent (see TANGENT OF A CURVE) at that point with a fixed straight line, and the angle $\theta + \delta\theta$ made by the tangent at the adjacent point P' with the same straight line.

Marie Curie in her laboratory c.1905, with apparatus built largely by herself or Pierre Curie.

The curvature at P may therefore be defined as the rate of change, k, of θ at that point. In terms of DIFFERENTIAL CALCULUS, for the curve $y = f(x)$,

$$k = \frac{\pm d^2y/dx^2}{[1 + (dy/dx)^2]^{3/2}}.$$

The curvature of a curved surface in a specified direction is defined similarly.

CURVE, in mathematics, the map of an infinite set of points ($x_1, x_2, \ldots x_\infty$) into a space S such that each point satisfies a FUNCTION, $f(x)$. In plane GEOMETRY, should $f(x)$ be a polynomial (see ALGEBRA), the curve is described as an algebraic curve; others are termed transcendental curves. Although most curves are of infinite extent in at least one direction, it is generally useful to consider only that section of it lying between two points, $f(p)$ and $f(q)$: a curve on which for every $f(p)$ an $f(q)$ can be chosen which coincides with $f(p)$, $p \neq q$, is a closed curve and may be finitely bounded in all directions (though possibly of infinite extent); curves which do not satisfy these conditions are described as open.

The most commonly encountered algebraic curves are the hyperbola, parabola, ellipse (see CONIC SECTIONS), the CIRCLE, which is a special case of the ellipse, and the straight LINE. The most common transcendental curves are those of the trigonometric functions (see TRIGONOMETRY) sine, cosine and tangent and those of the logarithmic and exponential functions (see LOGARITHM; EXPONENT). In CARTESIAN COORDINATES these are:

(hyperbola)	$y = \pm b\sqrt{x^2 - a^2}/a$,
(parabola)	$y = ax^2 + bx + c$,
(ellipse)	$y = \pm b\sqrt{a^2 - x^2}/a$,
(circle)	$y = \pm\sqrt{a^2 - x^2}$,
(straight line)	$y = ax + b$,
(sine curve)	$y = a.\sin x + b$,
(cosine curve)	$y = a.\cos x + b$,

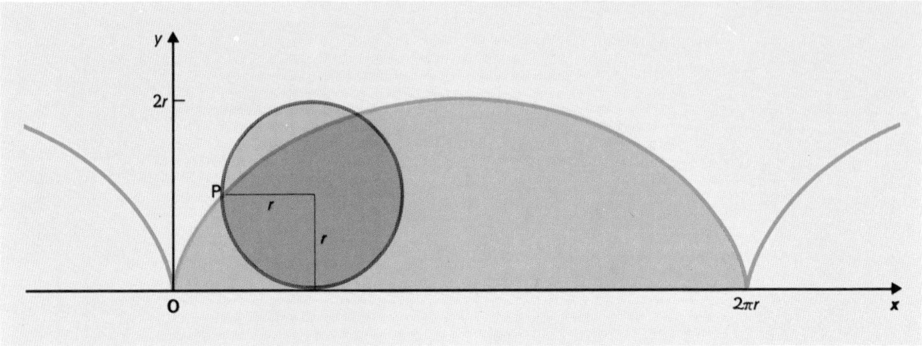

The cycloid (blue) with its generating circle (red).

(tangent curve) $y = a \, . \tan x + b$,

(log curve) $y = a \, . \log_b x + c$,

(exponential curve) $y = a \, . b^x + c$,

where in each case a, b, c are CONSTANTS.

CUSA, Nicholas of. See NICHOLAS OF CUSA.

CUSHING, Harvey Williams (1869–1939), US surgeon who pioneered many modern neurosurgical techniques and investigated the functions of the PITUITARY GLAND. In 1932 he described **Cushing's Syndrome**, a rare disease caused by STEROID imbalance and showing itself in obesity, high blood pressure and other symptoms.

CUSP, a point on a CURVE at which the CURVATURE is infinite.

CUTLERY, general term used for edged cutting instruments such as knives, axes and scissors. PREHISTORIC MAN used cutlery of sharpened stones, SHELLS and FLINTS, as does PRIMITIVE MAN today. Gradually implements of bronze, iron (see BRONZE AGE; IRON AGE) and, later, STEEL became current. Modern table cutlery may be made of SILVER, STAINLESS STEEL, SHEFFIELD PLATE, any of a number of PLASTICS, or other suitable material. (See also FORGING; GRINDING AND POLISHING; METALLURGY.)

CUVIER, Georges Léopold Chrétien Frédéric Dagobert, Baron (1769–1832), French comparative anatomist and the founder of PALEONTOLOGY. By applying his theory of the "correlation of parts" he was able to reconstruct the forms of many fossil creatures, explaining their creation and subsequent extinction according to the doctrine of CATASTROPHISM. A tireless laborer in the service of French Protestant education, Cuvier was perhaps the most renowned and respected French scientist in the early 19th century.

CYANAMIDE PROCESS, process for NITROGEN FIXATION. Calcium carbide (CaC_2) is made by heating CALCIUM carbonate with COKE. Nitrogen is passed through the finely-divided calcium carbide at 1000°C, giving calcium cyanamide ($CaCN_2$), which is then decomposed by steam to give calcium carbonate (for recycling) and AMMONIA.

CYANIDES, compounds containing the CN group. Organic cyanides are called NITRILES. Inorganic cyanides are salts of **hydrocyanic acid** (HCN), a volatile weak ACID; both are highly toxic. Sodium cyanide is made by the Castner process: ammonia is passed through a mixture of carbon and fused sodium. The cyanide ion (CN^-) is a PSEUDOHALOGEN, and forms many complexes. Cyanides are used in the extraction of GOLD and SILVER, ELECTROPLATING and CASEHARDENING.

CYANOGEN (C_2N_2), colorless, toxic gas, a PSEUDOHALOGEN, prepared by oxidizing a CYANIDE. It is reactive, and polymerizes to paracyanogen. mp −27.9°C, bp −21.2°C.

CYANOSIS, the bluish tinge of SKIN and mucous membranes seen when there is too little OXYGEN bound to HEMOGLOBIN in the BLOOD. If generalized it may be due to inadequate oxygen reaching the blood or failure of the BLOOD CIRCULATION; this may be chronic or acute, the latter often needing early treatment or resuscitation. If it is only at the extremities, it indicates slow blood flow due to vasoconstriction.

CYBERNETICS, a branch of INFORMATION THEORY which compares the communication and control systems built into mechanical and other man-made devices with those present in biological organisms. For example, fruitful comparisons may be made between data processing in COMPUTERS and various of the BRAIN functions; the fundamental theories of cybernetics may be applied with equal validity to both.

CYCLADIC CIVILIZATION. See AEGEAN CIVILIZATION.

CYCLAMATES, the sodium or calcium salts of N-cyclohexylsulfamic acid ($C_6H_{11}NHSO_3H$), white crystalline solids widely used as potent SWEETENING AGENTS until they were found to cause CANCER in rats. They were banned in the US in 1969.

CYCLOID, the CURVE traced by a point on the circumference of a rolling CIRCLE. The curve resembles a succession of arches, with CUSPS separated by distances equal to the circumference of the circle. The cycloid is of architectural interest since it forms the strongest known ARCH. (See also EPICYCLOID; HYPOCYCLOID.)

CYCLONE, an atmospheric disturbance in regions of low pressure (see ATMOSPHERE; METEOROLOGY) characterized by a roughly circular ground plan, a center towards which ground WINDS move, and at which there is an upward air movement, usually spiraling. Above the center, in the upper TROPOSPHERE, there is a general outward movement. The direction of spiraling is counterclockwise in the N Hemisphere, clockwise in the S Hemisphere, owing to the CORIOLIS EFFECT. **Anticyclones**, which occur in regions of high pressure, are characterized by an opposite direction of spiral. (See also HURRICANE; TORNADO.)

CYCLONITE. See RDX.

CYCLOTRON, a type of charged particle ACCELERATOR in which the particles travel a spiral path in a strong magnetic field, thus repeatedly traversing the same electric field regions (dees) and achieving energies much greater than those attainable with a linear accelerator.

CYGNUS (the Swan), a large, approximately cruciform constellation of the N Hemisphere containing Deneb (absolute magnitude −7) and Albireo.

CYLINDER, the geometrical solid formed of two congruent (see CONGRUENCE) two-dimensional geometric figures lying in parallel PLANES and the lines (elements) joining each point on the boundary of one to the equivalent point on the boundary of the other. The volume of a cylinder is given by the area of its base multiplied by the vertical distance between the two planes. The most common form of cylinder is the right circular cylinder, whose bases are circles to which the elements are perpendicular (see ANGLE). A cylinder whose elements are not perpendicular to the bases is termed oblique. In ANALYTIC GEOMETRY a cylinder is formed of the lines cutting a plane CURVE, each being parallel to a fixed line not in the plane of the curve, the curve being open or closed.

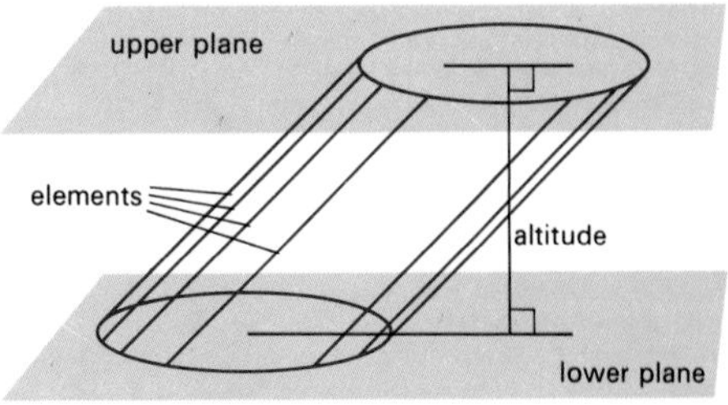

Terms associated with the cylinder.

CYLINDRICAL COORDINATES, coordinate system in which a point is located by its height z above a reference PLANE and the polar coordinates (see ANALYTIC GEOMETRY) of its PROJECTION onto that plane, (r, θ). (r, θ, z) can be related to rectangular CARTESIAN COORDINATES by $x=r\cos\theta$, $y=r\sin\theta$, and $z=z$ (where the original reference plane corresponds to the xy-plane).

CYRILLIC ALPHABET. See ALPHABET.

CYST, a fluid-filled sac, lined by fibrous connective tissue or surface EPITHELIUM. It may form in an enlarged normal cavity (e.g., sebaceous cyst), it may arise in an embryonic remnant (e.g., branchial cyst), or it may occur as part of a disease process. They may present as swellings or may cause pain (e.g., some ovarian cysts). Multiple cysts in KIDNEY and LIVER occur in inherited polycystic diseases; here kidney failure may develop.

CYSTIC FIBROSIS, an inherited disease presenting in infancy or childhood causing abnormal GLAND secretions; chronic LUNG disease with thick sputum and liability to infection is typical, as is malabsorption with pale bulky feces and MALNUTRITION. Sweat contains excessive salt (the basis for a diagnostic test) and heat exhaustion may result. Significant disease of LIVER, SINUSES and salivary glands occurs. Prompt treatment of chest infection with PHYSIOTHERAPY and appropriate ANTIBIOTICS is crucial to minimize lung damage; concentrated PANCREAS extract and special diets encourage normal digestion and growth, and extra salt should be given in hot weather. Although long-term outlook in this disease has recently improved, there is still a substantial mortality before adult life.

CYSTITIS, INFLAMMATION of the BLADDER, usually due to infection. A common condition in women, sometimes precipitated by intercourse. It occasionally leads to pyelonephritis, or upper urinary tract infection. Burning pain and increased frequency of urination are usual symptoms. ANTIBIOTICS are often needed. Recurrent cystitis may suggest an underlying disorder of bladder or its nervous control.

CYTOCHROMES, colored proteins which transfer energy within cells. They contain a CHELATE complex of an iron ion and a PORPHYRIN system, and so include HEMOGLOBIN. Cytochromes undergo a series of OXIDATION AND REDUCTION reactions, transferring electrons from a substrate to each other and finally to an electron-acceptor. In cellular RESPIRATION the substrates are the hydrogen-acceptors in the CITRIC ACID CYCLE, the electron-acceptor is oxygen, and the energy produced is stored as ATP. (See also PHOTOSYNTHESIS.)

CYTOLOGY, the branch of BIOLOGY dealing with the study of CELLS, their structure, function, biochemistry, etc. Techniques used include tissue CULTURE and electron microscopy. (See also HISTOLOGY.)

CYTOPLASM, the PROTOPLASM of which a CELL is made, excluding the nucleus and the boundary membrane.

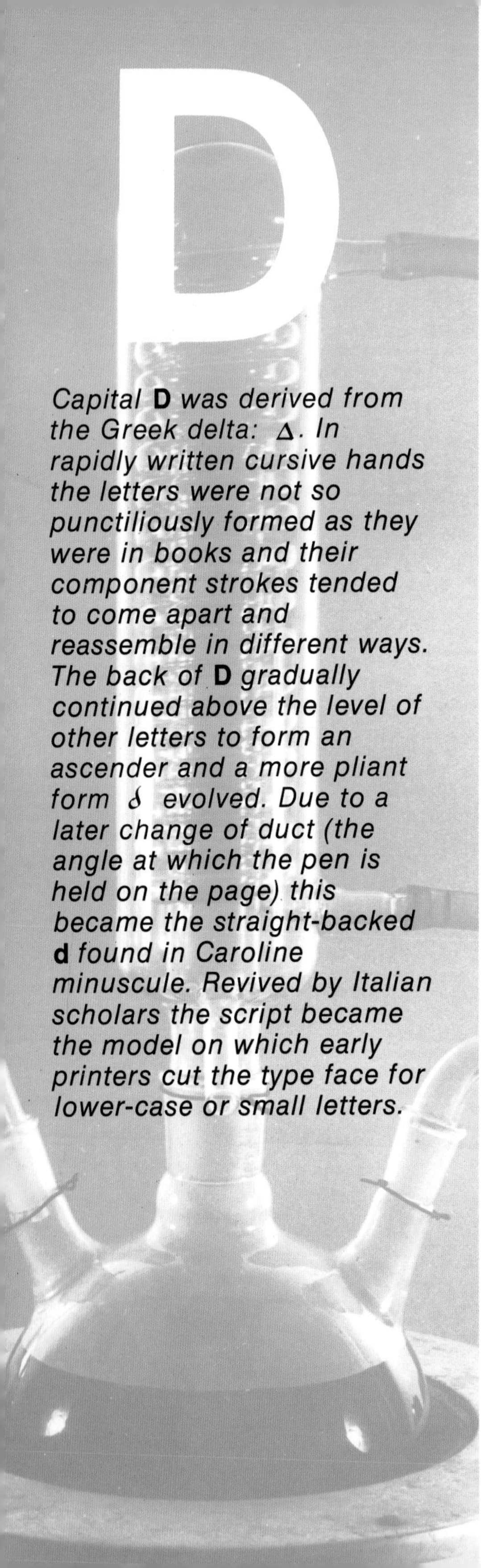

D

*Capital **D** was derived from the Greek delta: Δ. In rapidly written cursive hands the letters were not so punctiliously formed as they were in books and their component strokes tended to come apart and reassemble in different ways. The back of **D** gradually continued above the level of other letters to form an ascender and a more pliant form δ evolved. Due to a later change of duct (the angle at which the pen is held on the page) this became the straight-backed **d** found in Caroline minuscule. Revived by Italian scholars the script became the model on which early printers cut the type face for lower-case or small letters.*

DACRON, or **Terylene,** a polyester SYNTHETIC FIBER.

DAGUERRE, Louis Jacques Mandé (1789–1851), French theatrical designer and former partner of NIÉPCE who in the late 1830s developed the DAGUERREOTYPE process, the first practical means of producing a permanent photographic image.

DAGUERREOTYPE, the first practical photographic process, invented by DAGUERRE in 1837 and widely used in portraiture until the mid-1850s. A brass plate coated with silver was sensitized by exposure to iodine vapor and exposed to light in a CAMERA for several minutes. A weak positive image produced by mercury vapor was fixed with a solution of salt. Hypo soon replaced salt as the fixing agent and after 1840 gold (III) chloride was used to intensify the image. (See PHOTOGRAPHY.)

DAIMLER, Gottlieb Wilhelm (1834–1900), German engineer who devised a high-speed INTERNAL-COMBUSTION ENGINE (1883) and used it in building one of the first AUTOMOBILES about 1886.

DALE, Sir Henry Hallet (1875–1968), British biologist who discovered ACETYLCHOLINE and described its properties together with those of HISTAMINE. In 1936 he shared the Nobel physiology or medicine prize with Otto LOEWI, having identified the chemical that Loewi had found to be secreted by certain nerve endings as acetylcholine and demonstrated its role in transmitting nerve impulses.

D'ALEMBERT, Jean Le Rond. See ALEMBERT, JEAN LE ROND D'.

D'ALEMBERT'S PRINCIPLE, the observation that NEWTON's third law of motion (that to every action there is an equal and opposite reaction) applies not only to systems in static EQUILIBRIUM but also to those in which at least one component is free to move. Thus when a FORCE **F** is applied to an unconstrained body of MASS m endowing it with ACCELERATION **a**, there is an equal inertial reaction $-m\mathbf{a}$.

DALÉN, Nils Gustaf (1869–1937), Swedish engineer who was awarded the 1912 Nobel physics prize in recognition of his invention of the automatic "sun valve." This device, which rapidly went into worldwide use, allowed the gas lights in unmanned buoys and lightships to be controlled by the amount of natural light.

DALTON, John (1766–1844), English Quaker scientist renowned as the originator of the modern chemical atomic theory. First attracted to the problems of GAS chemistry through an interest in meteorology, Dalton discovered his **Law of Partial Pressures** in 1801. This states that the PRESSURE exerted by a mixture of gases equals the sum of the partial pressures of the components and holds only for ideal gases. (The partial pressure of a gas is the pressure it would exert if it alone filled the volume.) Dalton believed that the particles or ATOMS of different ELEMENTS were distinguished from one another by their weights, and, taking his cue from the laws of definite and multiple proportions (see COMPOSITION, CHEMICAL), he compiled and published in 1803 the first table of comparative ATOMIC WEIGHTS. This inaugurated the new quantitative atomic theory. Dalton also gave the first scientific description of COLOR BLINDNESS. The red-green type from which he suffered is still known as **Daltonism**.

DAM, a structure confining and checking the flow of a

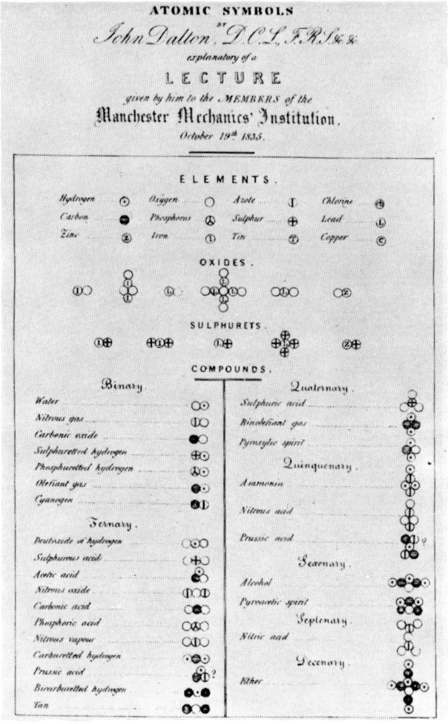

Dalton's table of atomic symbols.

river, stream or estuary to divert its flow, improve navigation, store water for irrigation or city supplies or raise its level for use in power generation. Often a recreation area is made as a by-product. Dams are one of the earliest known man-made structures, records existing from c2900 BC of a 15m-high dam on the Nile. Construction methods were largely empirical until 1866, when the first scientifically designed dam was built in France. Dams are classified by profile and building material, these being determined by availability and site. They must be strong enough to hold back water; withstand ice, silt and uplift pressures, and stresses from temperature changes and EARTHQUAKES. The site must have stable earth or rock that will not unduly compress, squeeze out or let water seep under the dam. Borings, seismic tests, structural models and computer simulations are all design aids. **Masonry** or **concrete dams** are typically used for blocking streams in narrow gorges. The highest are around 300m high. A **gravity dam** holds back water by its own weight and may be solid, sloping downstream with a thick base, or buttressed, sloping upstream and strengthened by buttresses which transfer the dead weight sideways; these require less concrete. **Arch dams**, with one or more ARCHES pointing upstream, are often built across a canyon and transfer some water pressure to its walls. Hoover Dam, built in 1936, is a combination of arch and gravity types. **Embankment** or **earthfill dams** are large barriers of rock, sand, silt or clay for controlling broad streams. As in a gravity dam, their weight deflects the horizontal water thrust downward toward the broad base. The materials may be uniformly mixed or there may be zones of waterproof material such as CONCRETE either on the upstream face or inside the dam. During construction, temporary **cofferdams** are built to keep water away from the site. Automatic **spillways** for excess water, intakes, gates and bypasses for fish or ships all form part of a dam complex.

DAM, Carl Peter Henrik (1895–), Danish biochemist who discovered and investigated the physiological properties of VITAMIN K. For this he shared the 1943 Nobel Prize for Physiology or Medicine with E. A. DOISY who isolated the vitamin and determined its structure.

DAMASCENING, inlaid ornamentation of gold, silver or copper on the surface of metal objects; the process originated in Damascus, Syria before the 12th century. The surface is undercut with a chisel, and wires or chips of metal are hammered in. The term also applies to watered patterns on sword blades, etc., made from Damascus steel.

DAMP, various noxious gases found in mines. **Firedamp** is METHANE, a colorless gas forming a highly explosive mixture with air. Sir Humphry DAVY's safety lamp reduced the danger of such explosions. **Afterdamp** is a mixture of carbon dioxide and nitrogen which results from firedamp explosions. **Chokedamp** (or **blackdamp**) is also mainly carbon dioxide.

The Aswan Dam in cross-section: (1) alluvium bed; (2) clay and sand grouted with cement; (3) clay; (4) crushed stone and sand; (5) rock and sand; (6) drainage wells.

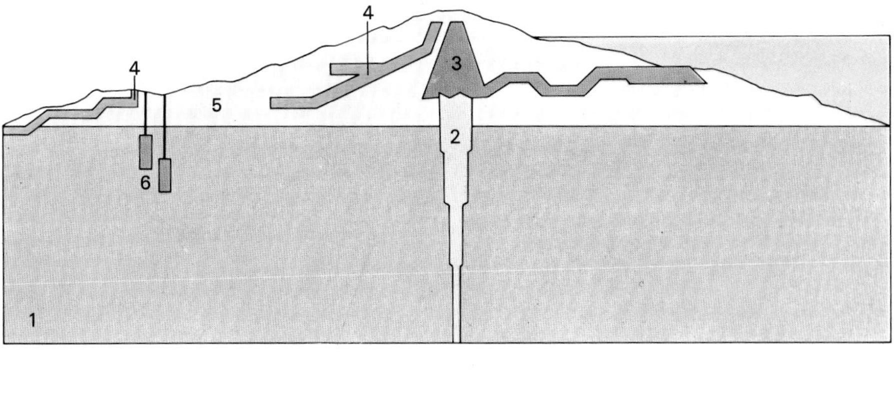

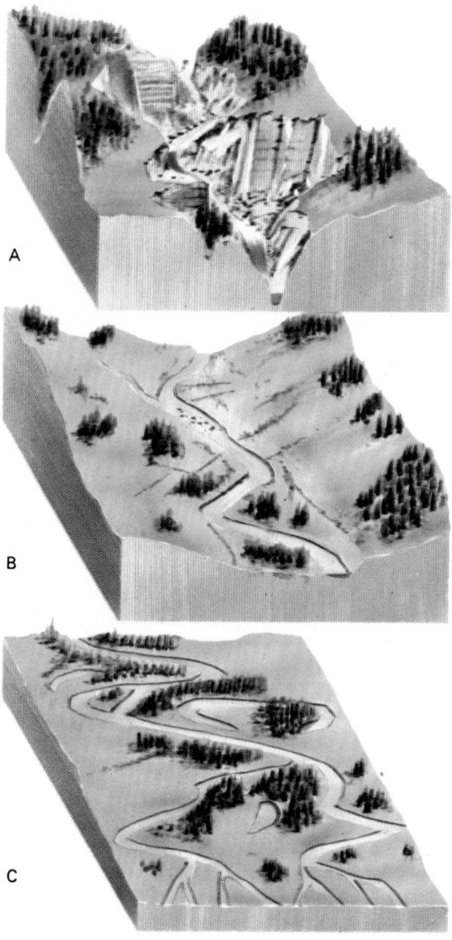

The erosion cycle according to William Morris Davis. (A) Streams cut deep gorges during the "youth" of newly uplifted sections of land. (B) When the landscape reaches "maturity", rivers follow more gently contoured V-sectioned valleys. (C) In the "old age" of a landscape, rivers meander across a gently undulating peneplain.

DANDRUFF, scaling of the SKIN of the scalp, part of the chronic skin condition of seborrheic DERMATITIS; scalp involvement is usually diffuse and itching may occur. The condition is lifelong but usually little more than an inconvenience. Numerous remedies are advertised but few are effective.

DANIELL CELL. See BATTERY.

DARBY, family of English ironmasters. **Abraham Darby** (1677–1717) pioneered IRON smelting using COKE rather than charcoal. His grandson, **Abraham Darby** (1750–1791) built the world's first cast-iron bridge over the Severn R at Ironbridge, Salop.

DARWIN, Charles Robert (1809–1882), English naturalist, who first formulated the theory of EVOLUTION by NATURAL SELECTION. Between 1831 and 1836 the young Darwin sailed round the world as the naturalist on board H.M.S. BEAGLE. In the course of this he made many geological observations favorable to LYELL's uniformitarian geology, devised a theory to account for the structure of coral islands and was impressed by the facts of the geographical distribution of plants and animals. He became convinced that species were not fixed categories as was commonly supposed but were capable of variation, though it was not until he read MALTHUS' *Essay on the Principle of Population* that he discovered a mechanism whereby ecologically favored varieties might form the basis for new distinct species. Darwin published nothing for 20 years until, on learning of A. R. WALLACE's independent discovery of the same theory, he collaborated with the younger man in a short Linnean-Society paper. The next year (1859) the theory was set before a wider public in his *Origin of Species*. The rest of his life was spent in further research in defense of his theory, though he always avoided entering the popular controversies surrounding his work and left it to others to debate the supposed consequences of "Darwinism."

DARWIN, Erasmus (1731–1802), English physician, philosopher, scientist and poet who proposed a theory of EVOLUTION by the inheritance of ACQUIRED CHARACTERISTICS in his *Zoonomia* of 1794–96. Perhaps the foremost physician of his day, he was a leading member of the LUNAR SOCIETY of Birmingham. Charles DARWIN was his grandson.

DATA PROCESSING. See COMPUTER.

DATE LINE, International, an imaginary line on the earth's surface, with local deviations, along longitude 180° from Greenwich. As the earth rotates, each day first begins and ends on the line. A traveler going east over the line sets his calendar back one day, and one going west adds one day.

DATING. See CHRONOLOGY; DENDROCHRONOLOGY; RADIOCARBON DATING.

DAVIS, William Morris (1850–1934), US geographer and leading geomorphologist of the late 19th century. Davis is best remembered for his erosion-cycle concept, characterizing the successive stages in the history of a newly uplifted landmass as its youth, its maturity and its old age (see EROSION).

DAVISSON, Clinton Joseph (1881–1958), US physicist who shared the 1937 Nobel physics prize with G. P. THOMSON for his demonstration of the DIFFRACTION of ELECTRONS. This confirmed DE BROGLIE's view that matter might act in a wavelike manner in appropriate circumstances.

DAVY, Sir Humphry (1778–1829), English chemist who pioneered the study of ELECTROCHEMISTRY. Electrolytic methods yielded him the elements SODIUM, POTASSIUM, MAGNESIUM, CALCIUM, STRONTIUM and BARIUM (1807–08). He also recognized the elemental nature of and named CHLORINE (1810). His early work on nitrous oxide (see NITROGEN) was done at Bristol under T. BEDDOES but most of the rest of his career centered on the ROYAL INSTITUTION where he was assisted by his protégé, M. FARADAY, from 1813. A major practical achievement was the invention of a miner's SAFETY LAMP, known as the **Davy Lamp**, in 1815–16. From his Bristol days, Davy was a friend of S. T. Coleridge.

DAWSON, Sir John William (1820–1899), Canadian geologist and educator who specialized in the geology of Nova Scotia; a determined opponent of

Darwinian theory. His son **George Mercer Dawson** (1849–1901), also a geologist, directed the Canadian Geological Survey from 1895.

DAY, term referring either to a full period of 24 hours (the civil day) or to the (usually shorter and varying) period between sunrise and sunset when a given point on the earth's surface is bathed in light rather than darkness (the natural day). Astronomers distinguish the sidereal day from the solar day and the lunar day depending on whether the reference location on the earth's surface is taken to return to the same position relative to the stars, to the sun or to the moon respectively. The civil day is the mean solar day, some 168 seconds longer than the sidereal day. In most modern states the day is deemed to run from midnight to midnight, though in Jewish tradition the day is taken to begin at sunset.

DDT, dichlorodiphenyltrichloroethane, a synthetic contact INSECTICIDE which kills a wide variety of insects, including mosquitoes, lice and flies, by interfering with their nervous systems. Its use, in quantities as great as 100 000 tonnes yearly, has almost eliminated many insect-borne diseases, including MALARIA, TYPHUS, YELLOW FEVER and PLAGUE. Being chemically stable and physically inert, it persists in the environment for many years. Its concentration in the course of natural food chains (see ECOLOGY) has led to the buildup of dangerous accumulations in some fish and birds. This prompted the US to restrict the use of DDT in 1972. In any case the development of insect strains resistant to DDT was already reducing its effectiveness as an insecticide. Although DDT was first made in 1874, its insecticidal properties were only discovered in 1939 (by P. H. MÜLLER).

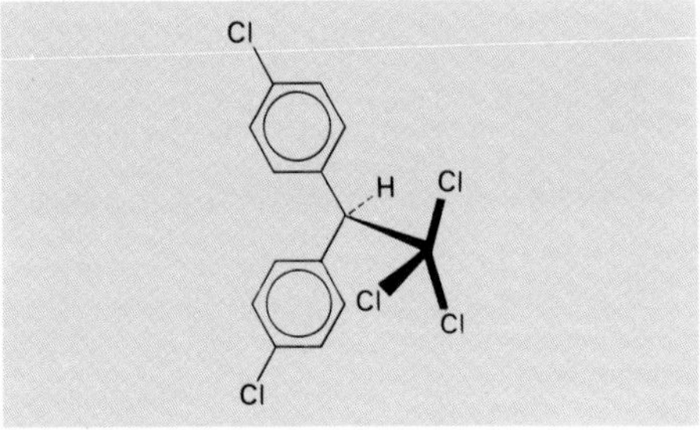

The formula of *pp'*-DDT.

DEAF MUTE. See DUMBNESS.

DEAFNESS, or failure of hearing, may have many causes. Conductive deafness is due to disease of outer or middle EAR, while perceptive deafness is due to disease of inner ear or nerves of hearing. Common physical causes of **conductive deafness** are obstruction with wax or foreign bodies and injury to the tympanic membrane. Middle ear disease is an important cause: in *acute otitis*, common in children, the ears are painful, with deafness, FEVER and discharge; in *secretory otitis* or glue ear, also in children, deafness and discomfort result from poor Eustachian tube drainage; *chronic otitis*, in any age group, leads to a deaf discharging ear, with drum perforation. ANTIBIOTICS in adequate courses are crucial in acute otitis, while glue ear is relieved by tubes or "grommets" passed through the drum to drain the middle ear. In both, the ADENOIDS may need removal to relieve Eustachian obstruction. In chronic otitis, keeping the ears clean and dry is important and antibiotics are used for secondary infection, while SURGERY, including reconstitution of the drum, may be needed to restore hearing. *Otosclerosis* is a common familial disease of middle age in which fusion or ANKYLOSIS of the small bones of the ear cause deafness. Early operation can prevent irreversible changes and improve hearing. **Perceptive deafness** may follow infections in PREGNANCY (e.g., GERMAN MEASLES) or be hereditary. Acute VIRUS infection and trauma to the inner ear (e.g., blast injuries or chronic occupational noise exposure) are important causes. Damage to the ear blood supply or the auditory nerves by drugs, TUMORS or MULTIPLE SCLEROSIS may lead to perceptive deafness, as may the later stages of MÉNIÈRE'S DISEASE. Deafness of old age, or *presbycusis*, is of gradual onset, mainly due to loss of nerve cells. Early recognition of deafness in children is particularly important as it may otherwise impair learning and speech development. HEARING AIDS are valuable in most cases of conductive and some of perceptive deafness. Lip-reading, in which the deaf person understands speech by the interpretation of lip movements, and sign language are useful in severe cases.

DEATH, the complete and irreversible cessation of LIFE in an organism or part of an organism. Death is conventionally accepted as the time when the HEART ceases to beat, there is no breathing and when the BRAIN shows no evidence of function. Ophthalmoscopic examination of the EYE shows that columns of BLOOD in small vessels are interrupted and static. Since it is now possible to resuscitate and maintain heart function and to take over breathing mechanically, it is not uncommon for the brain to have suffered irreversible death but for "life" to be maintained artificially. The concept of "brain death" has been introduced, in which reversible causes have been eliminated, when no spontaneous breathing, no movement and no specific REFLEXES are seen on two occasions. When this state is reached, artificial life support systems can be reasonably discontinued as brain death has already occurred. The ELECTROENCEPHALOGRAPH has been used to diagnose brain death but is now considered unreliable.

After death, ENZYMES are released which begin the process of autolysis or decomposition, which later involves BACTERIA. In the hours following death, changes occur in muscle which cause rigidity or RIGOR MORTIS. Following death, anatomical examination of the body (AUTOPSY) may be performed. Burial, embalming or cremation are usual practices for disposal of the body in Western society.

Death of part of an organism, or necrosis, such as occurs following loss of blood supply, consists of loss of cell organization, autolysis and GANGRENE. The part may separate or be absorbed but if it becomes infected, this is liable to spread to living tissue. Cells may also die as part of the normal turnover of a structure (e.g., SKIN or blood cells), after POISONING or infection (e.g., in the LIVER), from compression (e.g., by TUMOR), or as part of a degenerative disease. They then undergo characteristic involutionary changes.

DE BROGLIE, Louis Victor Pierre Raymond, Prince (1892–), French physicist who was awarded the 1929 Nobel Prize in Physics for his suggestion that acknowledged particles should display wave properties under appropriate conditions in the same way that ELECTROMAGNETIC RADIATION sometimes behaved as if composed of particles.

DEBYE, Peter Joseph Wilhelm (1884–1966), Dutch-born German–US physical chemist chiefly remembered for the Debye-Hückel theory of ionic solutions (1923). He was awarded the 1936 Nobel Prize for Chemistry.

DECAY, Radioactive. See RADIOACTIVITY.

DECIBEL (dB), a unit used to express power ratio, equal to one-tenth of a BEL, widely used by acoustics and telecommunications engineers. It is commonly employed in describing NOISE levels relative to the threshold of hearing. Doubling the noise level adds 3 to the decibel rating.

DECIDUOUS TREES, shrubs and trees that shed their leaves at a certain season, generally in the autumn when the environmental conditions are becoming unsuitable to active growth and just before the tree enters a period of dormancy. (See also EVERGREENS.)

DECIMAL SYSTEM, a number system using the POWERS of ten; our everyday system of numeration. The digits used are 0, 1, 2, 3, 4, 5, 6, 7, 8, 9; the powers of 10 being written $10^0=1$, $10^1=10$, $10^2=100$, $10^3=1000$, etc. To each of these powers is assigned a place value in a particular number; thus $(4\times10^3)+(0\times10^2)+(9\times10^1)+(2\times10^0)$ is written 4092. Similarly, FRACTIONS may be expressed by setting their DENOMINATORS equal to powers of 10—

$$\frac{3}{4}=\frac{75}{100}=\frac{7}{10}+\frac{5}{100}=(7\times10^{-1})+(5\times10^{-2}),$$

which is written as 0.75. Not all numbers can be expressed in terms of the decimal system: one example is the fraction $\frac{1}{3}$ which is written 0.333 3 . . . , the row of dots indicating that the 3 is to be repeated an infinite number of times. Fractions like 0.333 3 . . . are termed **repeating decimals**. APPROXIMATION is often useful when dealing with decimal fractions.

The decimal system probably came about through the number of digits on a pair of hands. In recent years it has been suggested that a more efficient number system would be one using the powers of twelve (see DUODECIMAL SYSTEM).

DECLINATION, the angular distance of a celestial body from the celestial equator (see CELESTIAL SPHERE) along the MERIDIAN through the body. Bodies north of the equator have positive declinations, those south, negative. Together with right ascension, declination defines the position of a body in the sky.

DECLINATION, Magnetic. See EARTH.

DECOMPOSITION, Biological. See PUTREFACTION.

DECOMPOSITION, Chemical, a reaction in which a chemical compound is split up into its elements or simpler compounds. Heat, or light of a suitable wavelength, will decompose many compounds, and some decompose spontaneously. Ionic compounds may be decomposed by ELECTROLYSIS. **Double decomposition** is a reaction of the type

$$AC+BD\rightarrow AD+BC$$

in which radicals are exchanged.

DECOMPRESSION SICKNESS, or **Bends.** See AEROEMBOLISM.

DECONGESTANT DRUGS are used to relieve the stuffy nose and RHINITIS of the COMMON COLD or HAY FEVER. They reduce swelling and secretion of affected mucous membrane and lessen symptoms, but prolonged use leads to chronic changes in mucous membranes.

DEDEKIND, Julius Wilhelm Richard (1831–1916), German mathematician who contributed to the theory of numbers. He proposed the method of "Dedekind cuts" for defining all the REAL NUMBERS in terms of RATIONAL NUMBERS.

DE DUVE, Christian René (1917–), English-born Belgian-US biochemist who shared the 1974 Nobel Prize for Physiology or Medicine with A. CLAUDE and G. PALADE for their pioneer studies of CELL anatomy. De Duve's particular contribution was the discovery of LYSOSOMES.

DEE, John (1527–1608), English mathematician and astrologer to Elizabeth I of England who encouraged the practical application of mathematics and supported the theory of COPERNICUS.

DEEP-SEA ANIMALS. See ABYSSAL FAUNA.

DEERE, John (1804–1886), US inventor who developed and marketed the first steel plows.

DEFENSE MECHANISM, any measure or measures taken by the individual's UNCONSCIOUS to protect his EGO from unpleasant or humilating situations: for example, SUBLIMATION, REPRESSION or PROJECTION. The more embracing term, **defense reaction**, describes the behavior resulting from such measures and also from analogous conscious measures.

DEFERENT. See EPICYCLE.

DEFICIENCY DISEASES. See DISEASE; VITAMINS.

DE FOREST, Lee (1873–1961), US inventor of the TRIODE (1906), an electron tube with three ELECTRODES (CATHODE, ANODE and grid) which could operate as a signal AMPLIFIER as well as a RECTIFIER. The triode was crucial to the development of RADIO.

DEGAUSSING, removing the magnetization (see MAGNETISM) of a permanent magnet or other object by withdrawing it slowly from a strong oscillating magnetic field, so that it is repeatedly magnetized in opposite directions but to progressively smaller degrees.

DEGENERATION, a term sometimes, but erroneously, applied to the evolution of parasitic organisms. Parasitic animals often lack features found in their free-living relatives, but they always possess others that suit them for their specialized way of life. (See PARASITE.)

DEGENERATIVE DISEASES. See DISEASE.

DEGREE, in ALGEBRA, of a POLYNOMIAL such as

$$ax^4+bx^2+cx+d$$

the highest POWER to which the VARIABLE x is raised—in this case, 4. (See also EQUATION.)

DEGREE, in GEOMETRY. See ANGLE.

DEGREES OF FREEDOM, the minimum number of independent VARIABLES in a system whose values must be specified in order to define the system. The number of degrees of freedom equals the number of variables less the number of independent relations between them (constraints). Thus OHM'S LAW provides one relation between three variables, and there are two degrees of freedom: if the resistance of a wire and the current through it are specified, both degrees of freedom have been used, and the voltage across the wire is fixed. In STATISTICS, for a sample of n members, the distribution has $(n-m)$ degrees of freedom where there are m constraints on the distribution. (See also CHI-SQUARED TEST; PHASE EQUILIBRIA; STUDENT'S t-DISTRIBUTION.)

DEHUMIDIFIER, a device for removing water vapor from air, by passing the air over cooling fins to promote CONDENSATION, or by ADSORPTION by such materials as SILICA GEL or alumina. It is a standard part of AIR-CONDITIONING systems in hot climates, where low humidity is desirable.

DEHYDRATION, or drying, the removal of WATER from a substance. To remove the elements of water, as in the dehydration of ALCOHOLS to ETHERS, requires a powerful dehydrating agent such as concentrated SULFURIC ACID. Generally, however, water is present as such, as a HYDRATE or merely absorbed, in which case milder methods suffice, such as equilibration in a desiccator with SILICA GEL or deliquescent compounds (see DELIQUESCENCE). Dry air may be passed over solids in heated drums, causing EVAPORATION. Gases are dried (for AIR CONDITIONING, or before liquefying them) by compression and refrigeration. Foods are preserved by drying; this is done for convenience and compactness. Milk and eggs are dried by spraying into hot air. Modern freeze drying—sublimation of ice from frozen foods under vacuum—retains texture and flavor. In medicine, dehydration of the body occurs through DIARRHEA, VOMITING, CHOLERA or merely lack of water to replace PERSPIRATION.

DEIMOS, the outer moon of MARS, circling the planet in 30.3h at a distance of 23 500km. Its diameter is about 8km.

DE KRUIF, Paul (1890–1971), US bacteriologist and science writer whose *Microbe Hunters* (1926) and later books telling the story of recent advances in medical science achieved large circulations.

DEL (also known as **alted** and **nabla**), the vector operator ∇, expressed in CARTESIAN COORDINATES as

$$\nabla = \mathbf{i}\frac{\partial}{\partial x} + \mathbf{j}\frac{\partial}{\partial y} + \mathbf{k}\frac{\partial}{\partial z}.$$

The product of ∇ and a SCALAR S yields a VECTOR FIELD called grad S (the gradient of S):

$$\text{grad}\, S = \mathbf{i}\frac{\partial S}{\partial x} + \mathbf{j}\frac{\partial S}{\partial y} + \mathbf{k}\frac{\partial S}{\partial z}.$$

The scalar product (see VECTOR ANALYSIS) of ∇ with a vector **V** is the divergence of **V** (see DIV); the vector product of ∇ with **V** is the CURL of **V**.

DELBRÜCK, Max (1906–), German-born US biologist whose discovery of a method for detecting and measuring the rate of mutations in BACTERIA opened up the study of bacterial GENETICS.

DELIQUESCENCE, the absorption of atmospheric moisture by a solid until it dissolves to form a saturated solution. If it merely forms a crystalline HYDRATE it is termed **hygroscopic**. The phenomenon depends on the relative HUMIDITY: sugar, for example, deliquesces above 85% humidity. (See also EFFLORESCENCE.)

DELIRIUM, altered state of consciousness in which a person is restless, excitable, hallucinating and is only partly aware of his surroundings. It is seen in high FEVER, POISONING, drug withdrawal, disorders of METABOLISM and organ failure. SEDATIVES and reassurance are basic measures.

DELIRIUM TREMENS, specific delirium due to

A river delta on the shores of the Arafura Sea.

acute alcohol withdrawal in ALCOHOLISM. It occurs within days of abstinence and is often precipitated by injury, surgery or imprisonment. The sufferer becomes restless, disorientated, extremely anxious and tremulous; FEVER and profuse sweating are usual. Characteristically, hallucinations of insects or animals cause abject terror. Constant reassurance, SEDATIVES, well-lit and quiet surroundings are appropriate measures until the episode is over. Treatment of dehydration and reduction of high fever may be necessary, though fatalities do occur.

DELPHINUS (the Dolphin), a small summer constellation in N skies, four of whose stars form a diamond sometimes known as Job's Coffin.

DELTA, a flat alluvial plain at a rivermouth or at the confluence of two rivers, formed of fertile mud deposited by the slow-moving water. Typically, the stream divides and subdivides until a fan-shaped plain covered by a complex of channels results. The form of a delta depends on the rates of SEDIMENTATION and of EROSION by the sea. (See also ESTUARY.)

DELTA RAYS, short ELECTRON tracks surrounding the track left by a fast charged particle in CLOUD CHAMBERS, BUBBLE CHAMBERS, or photographic emulsions. The primary particle dislodges electrons from the ATOMS of the medium concerned, and the faster of these leave short tracks of their own before being brought to rest.

DEMENTIA, loss of the ability to reason, as distinct from AMNESIA. (See KORSAKOV'S PSYCHOSIS.) *Dementia Praecox* is an obsolete term roughly corresponding to SCHIZOPHRENIA.

DEMOCRITUS OF ABDERA (c460–370 BC), Greek materialist philosopher, one of the earliest exponents of ATOMISM, who maintained that all phenomena were explicable in terms of the nomic motion of atoms in the void.

DEMOGRAPHY, a branch of sociology, the study of the distribution, composition and internal structure of human populations. It draws on many disciplines (e.g., genetics, psychology, economics, geography), its tools being essentially those of STATISTICS: the sample and the census whose results are statistically analyzed. Its prime concerns are birth rate, emigration and immigration. (See also POPULATION.)

DE MOIVRE, Abraham (1667–1754), French-born English mathematician chiefly remembered for **De Moivre's Theorem** which states that for a rational number n,

$$(\cos\theta + i\sin\theta)^n = \cos n\theta + i\sin n\theta.$$

DENATURED ALCOHOL. See ETHANOL.

DENDRITE, a branched, treelike CRYSTAL form, common in ice (especially frost) and certain minerals, and chiefly important in metals, which often consist of dendrites embedded in a matrix of the same or (for alloys) different composition. In anatomy, a dendrite is a tapering receptor branch of a NEURON.

DENDROCHRONOLOGY, the dating of past events by the study of tree-rings. A hollow tube is inserted into the tree trunk and a core from bark to center removed. The ANNUAL RINGS are counted, examined and compared with rings from dead trees so that the chronology may be extended further back in time. Through such studies important corrections have been made to RADIOCARBON DATING.

DENGUE FEVER, or breakbone fever, a VIRUS infection carried by mosquitoes, with FEVER, headache, malaise, prostration and characteristically severe muscle and joint pains. There is also a variable skin rash through the roughly week-long illness. It is a disease of warm climates, and may occur in EPIDEMICS. Symptomatic treatment only is required.

DENIER, a unit used to describe the mass per unit length of silk or nylon yarn. The denier number of a yarn is its weight in grams per 9km length.

DENOMINATOR, the DIVISOR of a common FRACTION.

DENSITY, the ratio of MASS to volume for a given material or object. Substances that are light for their size have a low density, and vice versa. Objects whose density is less than that of water will float in water, while a hot air BALLOON will rise when its average density becomes less than that of air. The term is also applied to properties other than mass: e.g., **charge density** refers to the ratio of electric charge (see ELECTRICITY) to volume.

DENTISTRY, the branch of MEDICINE concerned with the care of TEETH and related structures. Dental CARIES is responsible for most dental discomfort. Here the bacterial dissolution of dentine and enamel leads to cavities, especially in molars and premolars, and these allow accumulation of debris which encourages further bacterial growth; destruction of the tooth will gradually ensue unless treatment restores a protective surface. Each tooth contains sensitive nerve fibers extending into the dentine; exposure of these causes toothache, but the fibers then retract so that the pain often recedes despite continuing caries. The dentist removes all unhealthy tissue under ANESTHESIA and fills the cavity with metal AMALGAM which hardens and protects the tooth, although a severely damaged tooth may require extraction. Traumatic injury to teeth is repaired by a similar process. In some instances a tooth may be reconstructed on a "peg" of the original by using an artificial "crown." Maldeveloped or displaced teeth may need extraction or, during childhood, braces or plates to encourage realignment with growth. Wisdom teeth (rearmost molars), in particular, may need extraction if they erupt out of alignment or if they interfere with the normal bite. Infection of tooth pulp with ABSCESS formation destroys the tooth; PUS can only be drained by extraction. False teeth or dentures, either fitted individually or as a group on a denture plate that sits on the gums, are made to replace lost teeth, to allow effective bite and for cosmetic purposes. Dentistry is also concerned with the prevention of carious decay and periodontal disease by encouragement of oral hygiene, including regular adequate brushing of teeth. Fluoride and protective films are important recent developments in preventive dentistry.

DENTITION. See TEETH.

DEODORANT, a substance used to counteract unpleasant body smells. Deodorants may simply mask a smell, as air fresheners in rooms, or may destroy the bacteria that live in PERSPIRATION and are the cause of the smell (see ANTISEPTICS). However, body odors are most easily prevented by regular washing. Antiperspirants, usually aluminum chlorohydroxide, are ASTRINGENTS.

DEOXYRIBONUCLEIC ACID (DNA). See NUCLEIC ACIDS.

DEPRESSANT, an agent that depresses any organ,

but specifically the functioning of the central NERVOUS SYSTEM. Reduction of the level of consciousness and impaired control of breathing are serious effects, usually seen in overdosage. Many drugs including alcohol, BARBITURATES and SEDATIVES are depressants. The loss of inhibition associated with their use may be due to a differential depressant effect.

DEPRESSION, a common psychiatric disease with pathologically depressed mood and characteristic somatic and sleep disturbance. It is divided into those due to external factors, and those where depression arises without obvious cause, including manic-depressive illness. SHOCK THERAPY, ANTIDEPRESSANTS and psychotherapy are usually successful.

DEPRESSION, a midlatitude CYCLONE.

DEPTH OF FIELD, in PHOTOGRAPHY, the distance between the nearest and farthest planes from the CAMERA for which the image formed in the film plane is reasonably in focus. The available depth of field depends both on the quality of the LENS employed and on the aperture stop. The greater the aperture (the lower the *f*/number), the shorter the depth of field, and vice versa.

DERIVATIVE. See CALCULUS.

DERMATITIS, SKIN conditions in which INFLAMMATION occurs. These include ECZEMA, contact dermatitis (see ALLERGY) and seborrheic dermatitis (see DANDRUFF). Acute dermatitis leads to redness, swelling, blistering and crusting, while chronic forms usually show scaling or thickening of skin. Cool lotions and dressings, and ointments are used in acute cases, whereas tars are often useful in more chronic conditions. Avoidance of allergens in contact or allergic dermatitis is essential.

DERMATOLOGY, subspeciality of MEDICINE concerned with the diagnosis and treatment of SKIN DISEASES: a largely visual speciality, but aided by skin BIOPSY in certain instances. Judicious use of lotions, ointments, creams (including STEROID creams) and tars is the essence of treatment, while the recognition of ALLERGY, infection and skin manifestations of systemic disease are tasks for the dermatologist.

DERRICK. See CRANE.

DESALINATION, or **desalting**, the conversion of salt or brackish water into usable fresh water. DISTILLATION is the most common commercial method; heat from the sun or conventional fuels vaporizes BRINE, the vapor condensing into fresh water on cooling. Reverse OSMOSIS and electrodialysis (see DIALYSIS) both remove salt from water by the use of semipermeable membranes; these processes are more suitable for brackish water. Pure water crystals may also be separated from brine by freezing. The biggest problem holding back the wider adoption of desalination techniques is that of how to meet the high ENERGY costs of all such processes. Only where energy is relatively cheap and water particularly scarce is desalination economic, and even then complex energy conservation procedures must be built into the plant.

DESCARTES, René, or **Renatus Cartesius** (1596–1650), mathematician, physicist and the foremost of French philosophers, who founded a rationalist, à priorist school of philosophy known as **Cartesianism**. After being educated in his native France and spending time in military service (1618–19) and traveling, Descartes spent most of his creative life in Holland (1625–49) before entering the service of Queen Christiana of Sweden shortly before his death. In mathematics Descartes founded the study of ANALYTIC GEOMETRY, introducing the use of CARTESIAN COORDINATES. He found in the deductive logic of mathematical reasoning a paradigm for a new methodology of science, first publishing his conclusions in his *Discourse on Method* (1637). The occult qualities of late scholastic science were to be done away with; only ideas which were clear and distinct were to be employed. To discover what ideas could be used to form a certain basis for a unified A PRIORI science, he introduced the method of universal doubt; he questioned everything. The first certitude he discovered was his famous *cogito ergo sum* (I think therefore I am) and on the basis of this, the existence of other bodies, and of God, he worked out his philosophy. In science, Descartes, denying the possibility of a VACUUM, explained everything in terms of motion in a plenum of particles whose sole property was extension. This yielded his celebrated but ultimately unsuccessful VORTEX theory of the solar system and statements of the principle of INERTIA and the laws of ordinary REFRACTION. In psychology Descartes upheld a strict DUALISM: there were no causal relationships between physical and mental substances. In biology, his views were mechanistic; he regarded animals as but complex machines.

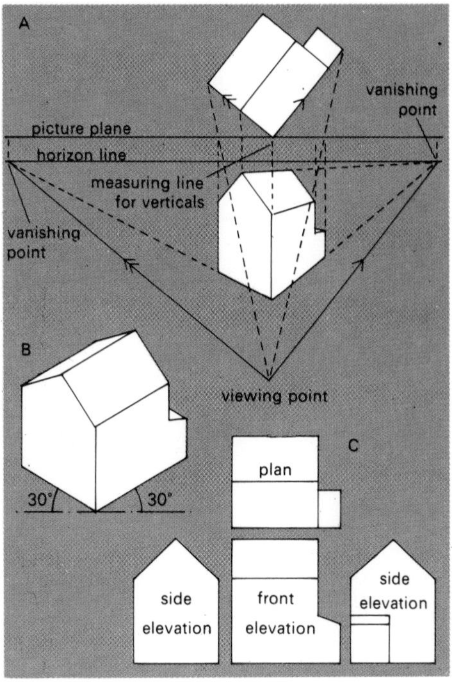

The same object viewed in (A) two-point perspective, (B) isometric projection, and (C) third-angle orthographic projection.

DESCRIPTIVE GEOMETRY, the branch of GEOMETRY concerned with the representation of three-dimensional objects on a PLANE, the basis of architectural and technical drawing. There are two main systems of projection: the PERSPECTIVE view, in which the object is projected from a POINT onto a plane (see PROJECTIVE GEOMETRY); and orthographic projection, where the object is projected from one plane into another. In **orthographic projection**, two planes are considered, at right ANGLES to each other, the vertical plane and the horizontal plane, intersecting in a line termed the ground line. From each point of the object, PERPENDICULARS are dropped to each of the two planes, giving the horizontal and vertical coordinates of that point. The representation of this on a plane surface is achieved by considering the two planes to be hinged along the ground line so that the horizontal plane may be rotated through 90° to coincide with the vertical, thus providing a graphical representation of the object.

DESERTS, areas where life has extreme difficulty in surviving. Deserts cover about one third of the earth's land area. There are two types.

Cold Deserts. In cold deserts, water is unavailable during most of the year as it is trapped in the form of ice. Cold deserts include the Antarctic polar icecap, the barren wastes of Greenland, and much of the TUNDRA. (See also GLACIER.) Eskimos, Lapps and Samoyeds are among the ethnic groups inhabiting such areas in the N Hemisphere. Their animal neighbors include seals and the polar bear.

Hot Deserts. These typically lie between latitudes 20° and 30° N and S, though they exist also farther from the equator in the centers of continental landmasses. They can be described as areas where water precipitation from the ATMOSPHERE is greatly exceeded by surface EVAPORATION and plant TRANSPIRATION. The best known, and largest, is the Sahara. GROUNDWATER exists but is normally far below the surface; here and there it is accessible as SPRINGS or WELLS (see ARTESIAN WELL; OASIS). In recent years, IRRIGATION has enabled reclamation of much desert land. Landscapes generally result from the surface's extreme vulnerability to EROSION (see also SOIL EROSION). Features include arroyos, BUTTES, DUNES, MESAS and WADIS. The influence of man may assist peripheral areas to become susceptible to erosion, and thus temporarily advance the desert's boundaries. (See also DUST BOWL; SANDSTORM.)

Plants may survive by being able to store water, like the cactus; by having tiny leaves to reduce evaporation loss, like the paloverde; or by having extensive ROOT systems to capture maximum moisture, like the mesquite. (See also DRY FARMING.) Animals may be nomadic, or spend the daylight hours underground. Best adapted of all is the camel.

DE SITTER, Willem. See SITTER, WILLEM DE.

DETERGENTS. See SOAPS AND DETERGENTS.

DETERMINANT, a square array of numbers, each a member of a FIELD F, used to represent another number in F. Consider the four numbers a_1, a_2, b_1, b_2. The determinant

$$\begin{vmatrix} a_1 & b_1 \\ a_2 & b_2 \end{vmatrix} = |\mathbf{A}|$$

(see MATRICES) is described as a determinant of order 2 over F and defined to have value

$$+a_1b_2 - a_2b_1.$$

Notice that transposition (i.e., exchange of rows for columns) does not affect the value of $|\mathbf{A}|$:

$$\begin{vmatrix} a_1 & a_2 \\ b_1 & b_2 \end{vmatrix} = +a_1b_2 - b_1a_2 = |\mathbf{A}|$$

(since multiplication in F is commutative: see ALGEBRA). This is true also of higher order determinants, so that every theorem applied to the columns of a determinant may equally be applied to its rows, and vice versa. Notice also that each of the terms in $+a_1b_2 - a_2b_1$ contains exactly one number from each row and one from each column; thus the value of a determinant of order 2 has 2 terms, one of order 3 has 6 terms, and one of order n has $n!$ terms (see FACTORIAL). Other properties of determinants include:

(1) If the elements of any row are multiplied by a factor c, then the new determinant has a value c times that of the original.

(2) Interchanging two rows of a determinant creates a new determinant with a value equal to that of the original multiplied by (-1).

(3) If two rows of a determinant are proportional the determinant has a value 0 (a special case of this is

The world's hot and cold desert regions

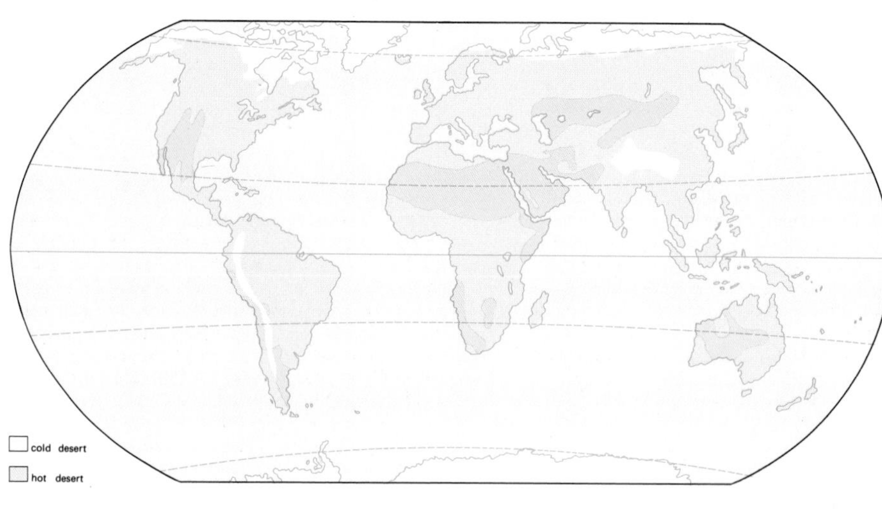

when two rows are identical).
(4) Addition of the elements of one row to the elements of another does not alter the value of the determinant.

The theory of determinants closely parallels that of linear simultaneous EQUATIONS. Consider

$$a_1x+b_1y+c_1=0,$$
$$a_2x+b_2y+c_2=0.$$

These imply that

$$a_2(a_1x+b_1y+c_1)-a_1(a_2x+b_2y+c_2)=0,$$
$$b_2(a_1x+b_1y+c_1)-b_1(a_2x+b_2y+c_2)=0.$$

These may be restated, respectively, as

$$(a_1b_2-a_2b_1)y-(c_1a_2-c_2a_1)=0,$$
$$(a_1b_2-a_2b_1)x-(c_2b_1-c_1b_2)=0.$$

The solutions, assuming that

$$\begin{vmatrix} a_1 & b_1 \\ a_2 & b_2 \end{vmatrix} \neq 0,$$

may therefore be expressed in the form

$$x=\frac{\begin{vmatrix} b_1 & c_1 \\ b_2 & c_2 \end{vmatrix}}{\begin{vmatrix} a_1 & b_1 \\ a_2 & b_2 \end{vmatrix}},\quad y=\frac{\begin{vmatrix} c_1 & a_1 \\ c_2 & a_2 \end{vmatrix}}{\begin{vmatrix} a_1 & b_1 \\ a_2 & b_2 \end{vmatrix}}.$$

DETERMINISM, the philosophical theory that all events are determined (inescapably caused) by preexisting events which, when considered in the context of inviolable physical laws, completely account for the subsequent events. The case for determinism has been variously argued from the inviolability of the laws of nature and from the omniscience and omnipotence of God. Determinism is often taken to be opposed to the principles of FREE WILL and indeterminacy.

DETONATOR, a small explosive device used to initiate the explosion of a main explosive charge in quarrying, mining or tunneling operations and in AMMUNITION. Safer, i.e., less sensitive, EXPLOSIVES need stronger detonators. The typical detonator comprises a thin-walled metal or plastic tube containing a charge of MERCURY fulminate ($Hg(ONC)_2$) or, now more common, LEAD azide (PbN_6), often mixed with other chemicals. The detonator is set off by a flame from a safety fuze or electrically, by passing a current through an attached resistance wire.

DEUTERIUM (D or ${}_1H^2$), or "heavy hydrogen," an ISOTOPE of HYDROGEN discovered in 1931, whose nucleus (the deuteron) has one PROTON and one NEUTRON (see also TRITIUM). It forms 0.014% of the hydrogen in naturally occurring hydrogen compounds, such as water, and is chemically very like ordinary hydrogen, except that it reacts more slowly. It is obtained as HEAVY WATER by the fractional electrolysis of WATER. Deuterium is the major fuel for nuclear FUSION (see HYDROGEN BOMB) and is used in tracer studies. (See also NUCLEAR REACTORS.) AW 2.0, mp −254.6°C, bp −249.7°C.

DEUTEROMYCETES, or fungi imperfecti. See FUNGI.

DEVONIAN, the fourth period of the PALEOZOIC, which lasted from about 400 to 345 million years ago. (See GEOLOGY.)

DE VRIES, Hugo. See VRIES, HUGO DE.

DEW, water droplets produced on clear calm nights by CONDENSATION of water vapor in the air. It is deposited on surfaces freely exposed to the sky which have cooled by RADIATION of heat. Dew forms when the air TEMPERATURE falls to the **dew point**. This, the temperature at which the air becomes saturated with water vapor, rises rapidly with the HUMIDITY. Dew is an important source of moisture for desert plants.

DEWAR, Sir James (1842–1923), British chemist and physicist who proposed various structures (including the "Dewar structures") for BENZENE, invented the VACUUM BOTTLE (Dewar bottle—1892) and pioneered the techniques of low-temperature physics. He developed methods for liquifying gases and discovered the magnetic properties of liquid OXYGEN and OZONE.

DEWEY, John (1859–1952), US philosopher and educator, the founder of the philosophical school known as INSTRUMENTALISM (or experimentalism) and the leading promotor of educational reform in the early years of the 20th century. Profoundly influenced by the PRAGMATISM of William JAMES, Dewey developed a philosophy in which ideas and concepts were validated by their practicality. He taught that "learning by doing" should form the basis of educational practice, though in later life he came to criticize the "progressive" movement in education, which, in abandoning formal tuition altogether, he felt had misused his educational theory.

DEWEY DECIMAL SYSTEM, a system devised by Melvil Dewey (1851–1931) for use in the classification of books in libaries, and based on the DECIMAL SYSTEM of numbers. Dewey divided knowledge into ten main areas, each of these into ten subdivisions, and so on. Thus a book could fall into one of a thousand categories, from 000 to 999. Extensions of this system added further classificatory numbers after the decimal point.

DEXTRIN, a polysaccharide (see CARBOHYDRATES) obtained from STARCH by heating or partial HYDROLYSIS with acids, DIASTASE or the bacterium *Bacillus macerans*. Soluble in water, it is used as an adhesive and a size for paper and textiles.

DIABASE, or **dolerite,** dark IGNEOUS ROCK intermediate in grain size between BASALT and GABBRO. Consisting of plagioclase FELDSPAR and PYROXENE, it is a widespread intrusive rock, quarried for crushed and monumental stone (known as "black granite").

DIABETES, a common systemic disease, affecting between 0.5 and 1% of the population, and characterized by the absent or inadequate secretion of INSULIN, the principal hormone controlling BLOOD sugar. There are many causes, including heredity, VIRUS infection, primary disease of the PANCREAS and OBESITY. Though it may start at any time, two main groups are recognized: juvenile (beginning in childhood, adolescence or early adult life)—due to inability to secrete insulin; and late onset (late middle life or old age)—associated with obesity and with a relative lack of insulin. High blood sugar may lead to coma, often with keto-ACIDOSIS, excessive thirst and high urine output, weight loss, ill-health and liability to infections. The disease may be detected by urine or blood tests and confirmed by a glucose tolerance test. It causes disease of small blood vessels, as well as premature ARTERIOSCLEROSIS, RETINA disease,

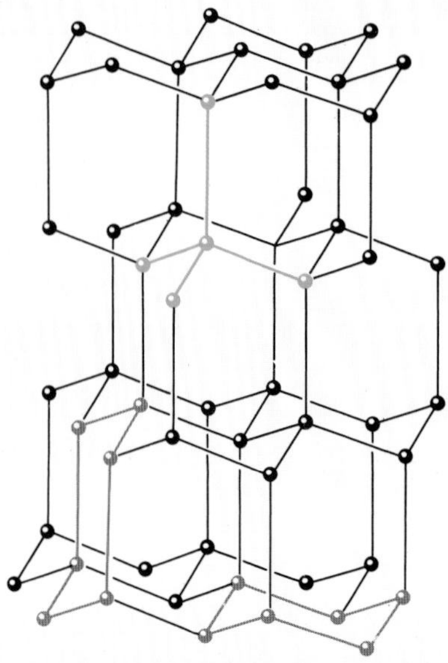

The structure of diamond, the hardest of materials. Each carbon atom is surrounded by neighbors positioned as at the apexes of a tetrahedron (blue), while rings of carbon atoms can be picked out in a configuration similar to that of the "chair" form of the cyclohexane skeleton (red).

CATARACTS, KIDNEY disease and NEURITIS. Poor blood supply, neuritis and infection may lead to chronic leg ULCERS. Once recognized, diabetes needs treatment to stabilize the blood sugar level and keep it within strict limits. Regular medical surveillance and education is essential to minimize complications. Dietary carbohydrate must be controlled and for late onset cases this may be all that is needed; in this group, drugs that increase the body's insulin production are valuable. In juvenile and some late onset cases, insulin itself is needed, given by subcutaneous injection by the patient. Regular dosage, adjusted to usual diet and activity, is used but surgery, PREGNANCY and infection increase insulin requirement. Control can be assessed by a simple urine test. Insulin overdose can occur, with sweating, confusion and COMA, and prompt treatment with sugar is crucial. EYE complications should be recognized early, especially in juvenile onset cases, as early intervention may prevent or delay BLINDNESS.

DIAGONAL, in plane GEOMETRY, a straight line joining two non-consecutive vertices of a POLYGON.

DIAGNOSIS. See DISEASE.

DIALECTIC, in philosophy, variously: a method of forcing a respondent to alter his opinion by leading him into self-contradiction (SOCRATES); the process of getting to know the world of ideal forms (PLATO); sound reasoning from generally accepted opinions rather than from self-evident truths (ARISTOTLE); argument exposing the folly of reasoning that employs the categories of the understanding outside the world of experience (KANT), or a dynamic logic, common to true philosophy and historical process, in which apparent contradictories—theses and antitheses—are reconciled in syntheses (HEGEL).

DIALYSIS, process of selective DIFFUSION of ions and molecules through a semipermeable membrane which retains COLLOID particles and macromolecules. It is accelerated by applying an electric field (see ELECTROPHORESIS). Dialysis is used for DESALINATION and in artificial kidneys (see ARTIFICIAL ORGANS). (See also OSMOSIS.)

DIAMAGNETISM, very weak magnetization (see MAGNETISM) of a material in a direction opposing the magnetizing field, due to ELECTRON orbital distortion. It is a property of all materials, though sometimes masked by stronger PARAMAGNETISM or ferromagnetism.

DIAMETER, the LINE joining any two points on the surface of a geometrical figure and passing through its center. The term is most often used in connection with the CIRCLE and SPHERE, all of whose infinitely many diameters have equal length.

DIAMOND, allotrope of CARBON (see ALLOTROPY), forming colorless cubic crystals. Diamond is the hardest known substance, with a Mohs hardness of 10, which varies slightly with the orientation of the crystal. Thus diamonds can be cut only by other diamonds. They do not conduct electricity, but conduct heat extremely well. Diamond burns when heated in air to 900°C; in an inert atmosphere it reverts to GRAPHITE slowly at 1000°C, rapidly at 1700°C. Diamonds occur naturally in dikes and pipes of KIMBERLITE, notably in South Africa (Orange Free State and Transvaal), Tanzania, and in the US at Murfreesboro, Ark. They are also mined from secondary (alluvial) deposits, especially in Brazil, Zaire, Sierra Leone and India. The diamonds are separated by mechanical panning, and those of GEM quality are cleaved (or sawn), cut and polished. Inferior, or industrial, diamonds are used for cutting, drilling and grinding. Synthetic industrial diamonds are made by subjecting graphite to very high temperatures and pressures, sometimes with fused metals as solvent. sg 3.51.

DIAPHRAGM, thin muscular and fibrous structure dividing the contents of the CHEST from those of the ABDOMEN. It is involved to a varying degree in quiet breathing and also contracts during deep breathing and straining. Through it pass the ESOPHAGUS, AORTA and inferior VENA CAVA, the first by a small opening or hiatus, through which part of the STOMACH may slide into the chest causing a hiatus HERNIA.

DIARRHEA, loose and/or frequent bowel motions. A common effect of FOOD POISONING, GASTROINTESTINAL TRACT infection (e.g., DYSENTERY, CHOLERA) or INFLAMMATION (e.g., COLITIS, ENTERITIS, ABSCESS), drugs and systemic diseases. Benign or malignant TUMORS of the colon and rectum may also cause diarrhea. Slime or blood indicate severe inflammation or tumor.

DIASTASE, a mixture of ENZYMES present in MALT which converts STARCH into MALTOSE. This action forms the basis of the BREWING process.

DIASTROPHISM, the large-scale deformation of the crust of the earth to produce such features as continents, oceans, mountains and rift valleys. (See also EARTH; FAULT; FOLD; PLATE TECTONICS.)

DIATHERMY, the use of electrically generated HEAT, particularly in SURGERY. Two ELECTRODES are connected to the patient: a localized point electrode is used to cause local tissue destruction, while a larger surface electrode, which dissipates the electrical energy over a wider area and thus avoids damage, is used to complete the circuit. Diathermy allows small blood vessels to be occluded and is often used to incise the GASTROINTESTINAL TRACT, when it gives a bacteria-free, nonbleeding edge. It has also been used to remove hairs, but the FOLLICLES may be scarred.

DIATOMS, single-celled fresh and salt water ALGAE, important as food to many small animals. Over millions of years, their skeletons have formed deposits on the sea bed, one in California being over 300m (1 000ft) thick. These deposits are excavated as fuller's or diatomaceous earth and the fine silica grains used in metal polishes and toothpaste. Diatom deposits are sometimes associated with the occurrence of petroleum deposits.

DIAZONIUM COMPOUNDS, salts containing the cation $R{-}\overset{+}{N}{\equiv}N$ where R is an aryl group (see AROMATIC COMPOUNDS), made by reacting a primary aromatic AMINE with nitrous acid (see NITROGEN). Their sensitivity to light is exploited in the **diazo process** or OZALID PROCESS. They couple with phenols and amines to give AZO COMPOUNDS. The aliphatic diazo compounds have the related structure $R_2C{=}\overset{+}{N}{=}\overset{-}{N}$.

DICK, George Frederick (1881–1967) and **Gladys Henry Dick** (1881–1963), US physicians (husband and wife) who discovered the organism responsible for causing SCARLET FEVER (1923) and developed a means (the **Dick test**) for estimating an individual's susceptibility to the disease.

DICOTYLEDONS, flowering PLANTS or ANGIOSPERMS that produce SEEDS with two seed leaves (COTYLEDONS) and thus differ from MONOCOTYLEDONS in which only one is produced. They also have net-veined leaves, a ring of bundles in the stem vascular system and flowering parts in fours or fives or multiples of these.

DICTATING MACHINE, device for recording and subsequently reproducing spoken messages; normally used in offices. Letters or messages are dictated into a MICROPHONE and recorded on magnetic tape, plastic disk or belt. This is played back on a reproducing unit (usually separate from the recording unit) for transcription by a typist.

DICTIONARY, in text processing or analysis by COMPUTER, a section of the computer's memory devoted to common phrases, punctuation, different uses of words spelled similarly, synonyms, etc. Comprehensive dictionaries are not yet practicable, but restricted ones are of great value in some fields of information storage and retrieval (see INFORMATION RETRIEVAL).

DIDEROT, Denis (1713–1784), French ENLIGHTENMENT philosopher, and editor, with Jean le Rond D'ALEMBERT, of the momentous *Encyclopédie* (1751–1772).

DIE, a mold or tool used in the shaping of metals, plastics and other materials. Dies are used in countless areas of METALLURGY, including CASTING, FORGING and EXTRUSION. Rather different are the gripper dies used in the making of NAILS and elsewhere. **Blanking and punching dies** are used together for trimming or punching holes in metal sheet. Both the blanking die and the punch have sharp edges: the workpiece is placed on the blanking die and the punch brought down with sufficient force to shear the metal (see SHEARING). **Deep-drawing** is similar, but the punch, though of the same cross-sectional shape as the die, is smaller, and both have rounded edges. The flat workpiece is placed on the die, into which it is forced by the punch. Cup-shaped and other parts can be produced in this way. WIRE can be produced by **drawing**, being pulled through one or a series of tapered dies to reduce cross-section and improve surface. (See also MATERIALS, STRENGTH OF.)

DIELECTRIC, an electrical insulator in which the application of an electric field (see ELECTRICITY) causes polarization which in turn produces a field opposed to the original field, thus reducing the resultant field strength by a factor known as the **dielectric constant** for the material concerned. Their properties are exploited in CAPACITORS, and to reduce dangerously strong fields. They also have optical applications, since refractive index (see REFRACTION) is the square ROOT of dielectric constant.

Some approximate dielectric constants

air (at room temperature and atmospheric pressure)	1.0005
paraffin wax	2.3
natural rubber	2.4
diamond	5.7
soda glass	7.0
water	80

DIELS, Otto Paul Hermann (1876–1954), German organic chemist who discovered carbon suboxide (C_3O_2) in 1906 and who with Kurt ALDER discovered the DIELS-ALDER REACTION (1928). For this latter work he and Alder shared the 1950 Nobel chemistry prize.

DIELS-ALDER REACTION, or diene synthesis, reaction discovered by Otto DIELS and Kurt ALDER, important in making plastics, insecticides and fungicides. A conjugated diene (see ALKENES; RESONANCE), such as 1, 3-BUTADIENE or ISOPRENE, adds readily to a "dienophile" containing a double or triple bond activated by an adjacent NUCLEOPHILE group.

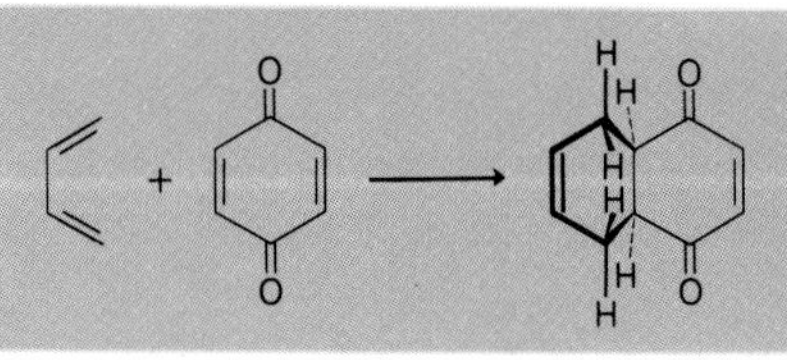

A Diels-Alder reaction in which a diene (butadiene: *left*) adds on to a dienophile (1,4-benzoquinone: *center*), yielding a Diels-Alder adduct (*right*).

DIESEL ENGINE, oil-burning INTERNAL-COMBUSTION ENGINE patented by **Rudolf Diesel** (1858–1913), a German engineer, in 1892 after several years of development work. Air enters a cylinder and is compressed by a piston to a high enough TEMPERA-

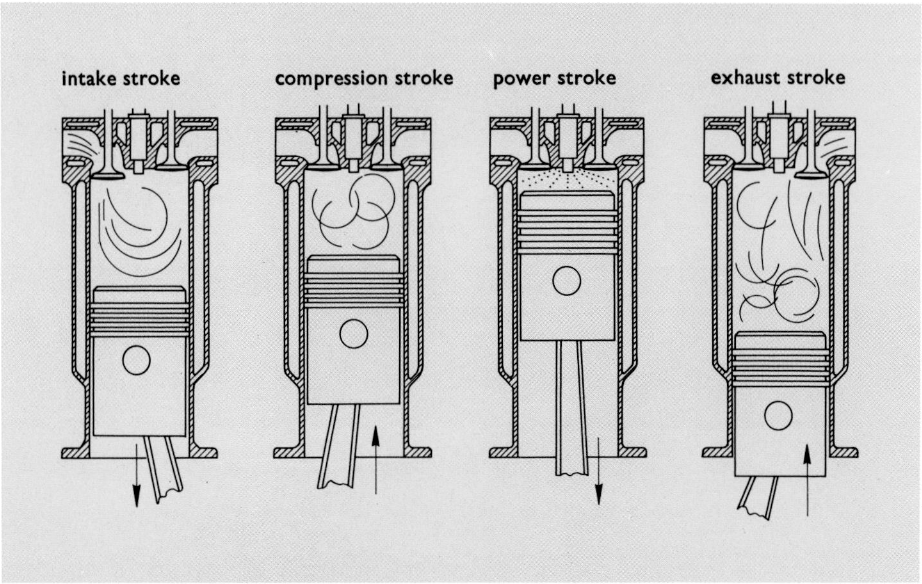

The operating cycle of a four-stroke diesel engine. In the intake stroke, the piston travels downward while the intake valve is open; this draws air into the cylinder. Then, during the compression stroke, the piston returns upward, compressing, and thus heating, the air in the cylinder. Now the fuel is injected; it mixes with the air and ignites, driving the piston violently down in the power stroke. Lastly, the piston again rises, this time with the exhaust valve open, forcing out the waste gases in the exhaust stroke.

TURE and PRESSURE for spontaneous combustion to occur when fuel is sprayed in. This method of operation differs from that of a gasoline engine in which air and fuel are mixed before entering the cylinder, there is less compression and a spark is needed to initiate combustion. In the first (intake) stroke of the cycle of a 4-stroke diesel engine, the piston moves down, drawing in air through a valve. In the second (compression) stroke, the piston returns up, compressing the air and heating it to over 300°C. (The exact value depends on the COMPRESSION RATIO, which may be between 12:1 and 22:1.) Near the end of the stroke, fuel is sprayed into the cylinder at high pressure through a nozzle and ignites in the hot air. In the third (power) stroke, the burning fuel–air mixture increases the pressure in the cylinder, pushing the piston down and driving the crankshaft. Then, in the fourth (exhaust) stroke, the piston moves up again and drives the burnt gases out of the cylinder. There are also 2-stroke diesel engines. These have only compression and power strokes, the exhaust gases being scavenged and new air introduced by a blower while the piston is at the bottom of its stroke. Diesel engines are less smooth-running, heavier and initially more expensive than gasoline engines but make more efficient use of cheaper fuel. They are widely used in ships, heavy vehicles and power installations.

DIET. See DIETING; DIETETIC FOODS; NUTRITION.

DIETETIC FOODS, special foods for conditions in which normal diet leads to disease or ill-health, or where dietary manipulation modifies disease. GLUTEN-free diet for CELIAC DISEASE; polyunsaturated FAT diet for excess BLOOD fats and ARTERIOSCLEROSIS; milk avoidance in lactose intolerance; CARBOHYDRATE restriction in DIABETES; low phenylalanine foods for PHENYLKETONURIA, and low PROTEIN diet for KIDNEY or LIVER failure are important examples. VITAMIN- or PROTEIN-enriched diets may be required for certain conditions, and a high-fat (ketogenic) diet can be used to treat EPILEPSY. Numerous diets exist which have no proven value.

DIETING, a term usually applied to the restriction of food intake in treatment of OBESITY, although the use of DIETETIC FOODS is strictly included. Many special diets have been recommended, but simple CALORIE restriction is the essence of weight reduction. Abstinence from potatoes, most root vegetables, rice, pasta, bread, cakes, biscuits, sugar, candies, cream and alcohol is usually effective; lean meat and green vegetables should be the staple diet. Occasionally more extreme dieting under supervision is needed.

DIFFERENCE EQUATIONS, in the calculus of finite differences, equations which play a role analagous to that played by DIFFERENTIAL EQUATIONS in CALCULUS. The calculus of finite differences deals with discrete quantities; unlike calculus, which deals with continuous quantities. For a FUNCTION $f(x)$ at a particular value x_n, we define $\Delta f(x_n)$ as $f(x_{n+1})-f(x_n)$, where Δ is called the difference operator. From this, we find that $\Delta^2 f(x_n)$—i.e., $\Delta(f(x_{n+1})-f(x_n))$—can be expressed as $f(x_{n+2})-2f(x_{n+1})+f(x_n)$; and so forth for $\Delta^3 f(x_n), \ldots, \Delta^m f(x_n)$. A difference table may be constructed showing values of $\Delta f(x_n)$, $\Delta^2 f(x_n), \ldots, \Delta^m f(x_n), \ldots$, and from this a relationship between the differences may be deduced. Generally, then, a difference equation is any equation which expresses such a relationship; and use may be made of it to find discrete values for $f(x)$ which lie outside the known range. Approximation of a differential equation to a suitable difference equation is often a powerful tool in the solution of the former.

DIFFERENTIAL, in AUTOMOBILES and trucks, an assembly of GEARS which permits the two wheels on a driven axle to rotate at slightly different rates during cornering.

DIFFERENTIAL, in differential CALCULUS, either of dy and dx where $dy/dx = f'(x)$.

DIFFERENTIAL EQUATIONS, EQUATIONS involving derivatives (see CALCULUS). Consider a body accelerating (see ACCELERATION) uniformly at 40 m/s^2. After a time t it has a VELOCITY of $40t$, assuming a stationary start. This velocity may also be expressed as ds/dt, the instantaneous rate of change of distance, s, from the starting point. Thus

$$40t = \frac{ds}{dt}.$$

To find out the distance traveled by the body after a time t we can integrate to find

$$20t^2 + \mathrm{k} = s + \mathrm{c}$$

or

$$s = 20t^2 + (\mathrm{k} - \mathrm{c})$$

where k and c are CONSTANTS. However, we have assumed a stationary start; i.e., that when $t=0$, $s=0$ and hence, by substitution, $(\mathrm{k}-\mathrm{c})=0$. Therefore, to find out how far the body has traveled after a given period of time, we need merely to substitute the value of t into

$$s = 20t^2.$$

This is the solution of a very simple first-order differential equation. In some problems there occur second derivatives of the form d^2y/dx^2, and these involve solution of **second-order differential equations**. Equations involving nth derivatives, d^ny/dx^n, are called nth-order equations, most important of which are the equations of the form

$$f(x) = \mathrm{A}y + \mathrm{B}\frac{dy}{dx} + \mathrm{C}\frac{d^2y}{dx^2} + \cdots + \mathrm{N}\frac{d^ny}{dx^n}$$

where A, B, C, ..., N are constants. This is termed a **linear *n*th-order differential equation**. Differential equations occur in many, if not most, physical problems. (See DIFFERENCE EQUATIONS.)

DIFFERENTIAL GEOMETRY, the branch of GEOMETRY dealing with the basic properties of curves and surfaces, using the techniques of CALCULUS and ANALYTIC GEOMETRY.

DIFFERENTIATION. See CALCULUS.

DIFFRACTION, the property by which a WAVE MOTION (such as ELECTROMAGNETIC RADIATION, SOUND or water waves) deviates from the straight line expected geometrically and thus gives rise to INTERFERENCE effects at the edges of the shadows cast by opaque objects, where the wave-trains that have reached each point by different routes interfere with each other. Opaque objects thus never cast completely sharp shadows, though such effects only become apparent when the dimensions of the obstruction are of the same order as the wavelength of the wave motion concerned. It is diffraction effects which place the ultimate limit on the resolving power of optical instruments, RADIO TELESCOPES and the like. Diffraction is set to work in the diffraction grating. Here, light passed through a series of very accurately

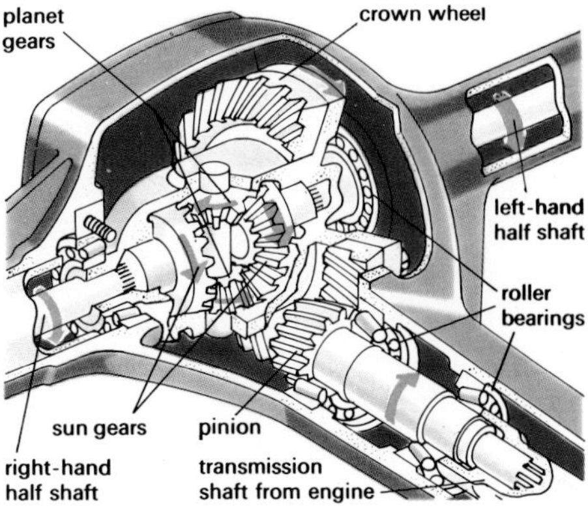

Differential drive as used on trucks. A gear wheel on the transmission shaft drives the crown wheel which is connected directly to the left-hand half-shaft (which drives the left-hand road wheel). Drive is transmitted to the right-hand half-shaft (and wheel) via a "sun" wheel on the left-hand shaft, two "planet" wheels whose axis of rotation rotates with the left-hand shaft, and a further "sun" wheel attached to the right-hand half-shaft. When the truck is going straight ahead both road wheels, and hence both half-shafts, rotate at the same rate: there is no movement in the planet gears. But when the track corners, the road wheel on the outside of the curve together with its half-shaft want to rotate faster than those on the inside. This is made possible by the rotation of the two planet gears in opposite senses, allowing one sun gear to rotate faster than the other. Here the truck is turning right, so the left-hand sun wheel, half-shaft and road wheel are turning faster than their right-hand counterparts.

ruled slits or reflected from a series of narrow parallel mirrors produces a series of spectrums by the interference of the light from the different slits or mirrors. Gratings are ruled with from 70 lines/mm (for infrared work) to 1 800 lines/mm (for ultraviolet work).

DIFFUSION, the gradual mixing of different substances placed in mutual contact due to the random thermal motion of their constituent particles. Most rapid with gases and liquids, it does also occur with solids. Diffusion rates increase with increasing TEMPERATURE; the rates at which gases diffuse through a porous membrane vary as the inverse of the square root of their MOLECULAR WEIGHT. Gaseous diffusion is used to separate fissile URANIUM-235 from nonfissile uranium-238, the gas used being uranium hexafluoride (UF_6).

DIGESTIVE SYSTEM, the mechanism for breaking down or modifying dietary intake into a form that is absorbable and usable by an organism. In unicellular organisms this is by phagocytosis and enzyme breakdown of large molecules; in larger animals it occurs outside cells after liberation of ENZYMES. In higher animals, the digestive system consists structurally of the GASTROINTESTINAL TRACT, the principal absorbing surface which also secretes enzymes, and the related organs: the LIVER and

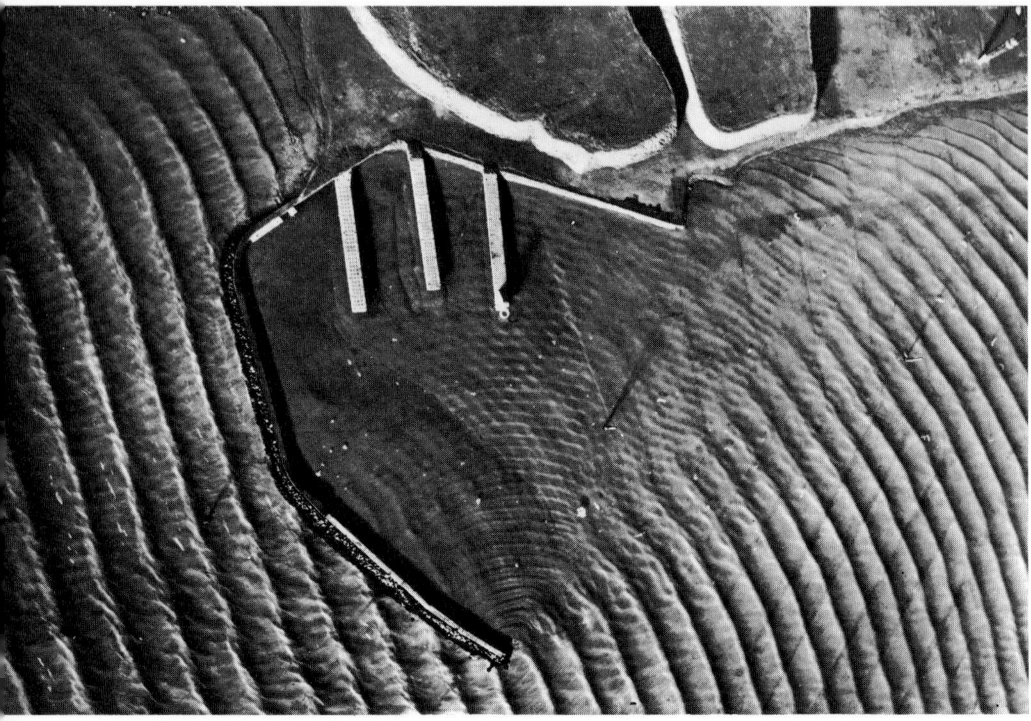

The diffraction of water waves demonstrated in a model of harbor works at the Hydraulics Research Station, Wallingford, UK. The waves approach the breakwater from the bottom left, at right-angles to the line of their crests. Beyond the point of the breakwater the main wave crests bend around into water which might otherwise have been thought to have been in the shadow of the breakwater. The short wave-length waves within the harbor are a separate phenomenon.

PANCREAS, which secrete into the tract via ducts. Different enzymes act best at different pH, and **gastric juice** and BILE respectively regulate the acidity of the STOMACH and alkalinity of the small intestine. PROTEINS are broken down by pepsin in the stomach and by trypsin, chymotrypsin and peptidases in the small intestine. CARBOHYDRATES are broken down by specialized enzymes, mainly in the small intestine. FATS are physically broken down by stomach movement, enzymatically by lipases and emulsified by bile salts. Food is mixed and propelled by PERISTALSIS, while nerves and locally regulated HORMONES, including gastrin and secretin, control both secretion and motility. Absorption of most substances occurs in the small intestine through a specialized, high-surface-area mucous membrane; some molecules pass through unchanged but most in altered form. Absorption may be either by an active transport system involving chemical or physical interaction in the gut wall, or simply by a passive DIFFUSION process. Some VITAMINS and trace metals have specialized transport systems. Most absorbed food passes via the portal system to the liver, where much of it is metabolized and toxic substances removed. Some absorbed fat is passed into the LYMPH. BACTERIA colonize most of the small intestine and are important in certain digestive processes. **Malabsorption** occurs when any part of the digestive system becomes defective. Pancreas and liver disease, obstruction to bile ducts, alteration of bacteria and inflammatory disease of the small intestine are common causes.

DIGITALIS, drug derived from the foxglove and acting on the muscle and conducting systems of the HEART. WITHERING in 1785 described its efficacy in heart failure or dropsy; it increases the force of cardiac contraction. It is also valuable in treatment of some abnormal rhythms; however, overdosage may itself cause abnormal rhythm, nausea or vomiting.

DIHEDRAL ANGLE, the figure formed by the intersection of two PLANES, described by a point on one of the planes, the line of intersection (edge), and a point on the other plane. Should a point P on one of the planes be so positioned that the line through it PERPENDICULAR to the edge, E, intersects E at the same point as a line drawn perpendicular to E through a point P' on the other plane, then the angle PEP' is the **plane angle** of the dihedral angle. All plane angles of a dihedral angle are equal.

DIKE, or **dyke**, a tabular body of IGNEOUS ROCK which, unlike a SILL, cuts across the beds of surrounding rock. Dikes commonly occur in swarms, which may be parallel or radial

DIMENSIONAL ANALYSIS, the branch of applied MATHEMATICS concerned with the analysis of physical problems in terms of DIMENSIONS such as mass, length and time. Its fundamental theorem is that the

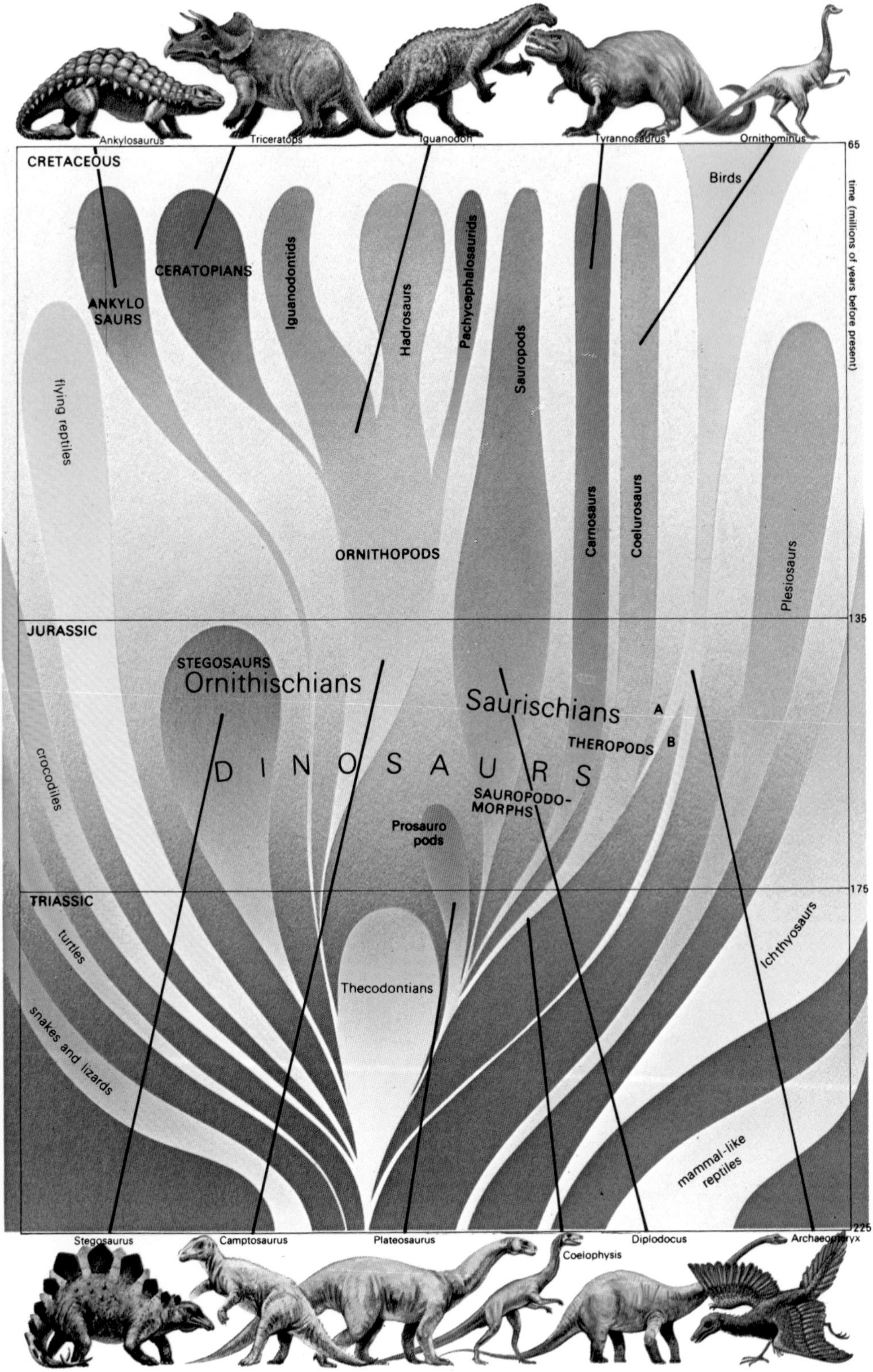

Ankylosaurus
Triceratops
Iguanodon
Tyrannosaurus
Ornithominus
65
CRETACEOUS
Birds
time (millions of years before present)
CERATOPIANS
ANKYLO SAURS
Iguanodontids
Hadrosaurs
Pachycephalosaurids
Sauropods
flying reptiles
Carnosaurs
Coelurosaurs
ORNITHOPODS
Plesiosaurs
135
JURASSIC
STEGOSAURS
Ornithischians
Saurischians
A
THEROPODS
B
D I N O S A U R S
crocodiles
SAUROPODO-MORPHS
Prosauro pods
175
TRIASSIC
turtles
Thecodontians
Ichthyosaurs
snakes and lizards
mammal-like reptiles
225
Stegosaurus
Camptosaurus
Plateosaurus
Coelophysis
Diplodocus
Archaeopteryx

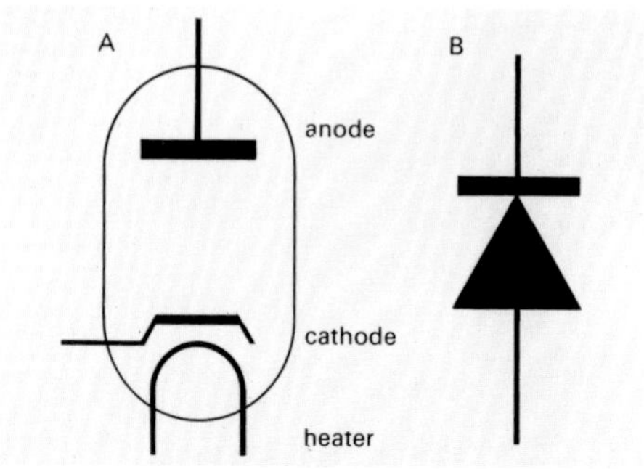

Diode: graphic symbols for (A) electron-tube diode; (B) semiconductor diode.

dimensions of the quantities appearing on opposite sides of an equation are the same.

DIMENSIONS, in GEOMETRY, the three properties of geometrical figures: length, breadth and depth. A POINT is defined as having no dimensions, a straight LINE as having only one, length, a PLANE figure as having two, length and breadth, and a solid figure as having all three. In terms of ANALYTIC GEOMETRY, an object has as many dimensions as it requires AXES to define its spatial position. In RELATIVITY there is considered to be a fourth dimension, TIME. (See also DIMENSIONAL ANALYSIS.)

DIMENSIONS, as applied to a physical quantity, an indication of the role it plays in equations. The dimensions of a mechanical quantity, in terms of mass [M], length [L] and time [T], can be deduced from the units in which it is expressed. Thus VELOCITY, measured in m/s, has dimensions [L]/[T]. Dimensions are purely conventional, having no real physical significance. This is clear in the case of electromagnetic quantities where dimensions vary according to the units system employed. Dimensions nevertheless find use in DIMENSIONAL ANALYSIS.

DINOFLAGELLATES, microscopic single-celled organisms, occurring in vast numbers in fresh and salt water. Some dinoflagellates contain CHLOROPHYLL, others do not, so they may be classed as plants or animals. They are propelled by two whip-like hairs (flagella) and each one is covered by a layer of cellulose; they are important as food for many animals, though some are poisonous.

DINOSAURIA, or **dinosaurs**, extinct members of the REPTILIA that flourished for 120 million years from the TRIASSIC to the CRETACEOUS periods. They ranged in size from small forms no larger than a domestic chicken to giants such as *Diplodocus* which was 27m (90ft) long and weighed about 30 tonnes. Early in their history two distinct dinosaur groups evolved: the SAURISCHIA and the ORNITHISCHIA. At the end of the Cretaceous period, about 80 million years ago, dinosaurs disappeared. The reasons for this sudden extinction are not known and are the subject of much debate and controversy among paleontologists.

DIODE, originally an ELECTRON TUBE having two ELECTRODES (CATHODE and ANODE) used as a RECTIFIER, but now extended to include SEMICONDUCTOR devices performing a similar function (See ELECTRONICS.)

DIOECIOUS PLANTS, those in which the male and female organs are borne in separate flowers which are on separate plants. Examples are willow, hemp and asparagus. (See also MONOECIOUS PLANTS.)

DIOPHANTUS OF ALEXANDRIA, mathematician, probably of the 3rd century AD, whose fame rests on his use of an algebraic notation and his interest in **Diophantine equations**. These are formulations of problems such as that of finding all the right-angled TRIANGLES whose sides are INTEGERS (e.g., 3, 4, 5). Diophantes indeed solved this problem.

DIOPTRE, or reciprocal metre, a unit used to express the focal power of optical lenses. The power of a converging LENS in dioptres is a positive number equal to the reciprocal of its focal length in metres. Diverging lenses have negative powers.

DIPHTHERIA, BACTERIAL DISEASE, now uncommon, causing FEVER, malaise and sore throat, with a characteristic "pseudomembrane" on throat or PHARYNX; also, the LYMPH nodes may enlarge. The LARYNX, if involved, leads to a hoarse voice, breathlessness and stridor; this may progress to respiratory obstruction requiring tracheostomy. The bacteria produce TOXINS which can damage nerves and HEART muscle; cardiac failure and abnormal rhythm, or PARALYSIS of palate, eye movement and peripheral NEURITIS may follow. Early treatment with ANTITOXIN and use of ANTIBIOTICS are important. Protection is given by VACCINATION.

DIPLOCOCCUS, spherical or ovoid nonmotile BACTERIA, so named because they occur in pairs. PNEUMONIA, MENINGITIS and gonorrhea (see VENEREAL DISEASES) are caused by types of diplococcus bacteria.

DIPLODOCUS, quadruped vegetarian DINOSAUR found in the Jurassic strata in the US. It had a long tail and a small head and was up to 27m (90ft) long.

DIPLOID. See MEIOSIS.

DIP NEEDLE, or inclinometer, an instrument for

A dip needle, as illustrated in a French textbook of 1874.

measuring magnetic dip (see EARTH), consisting of a magnetic needle mounted free to pivot in the vertical plane within a graduated circle. For use, the instrument is carefully leveled and oriented into the magnetic meridian.

DIPNOI, or **lungfish**, a subclass of the OSTEICHTHYES that contains a large number of fossil forms and three living freshwater groups, found in Africa, South America and Australia respectively. As their name suggests, the lungfish have lungs, organs not found in other living fish. The African and South American lungfish are unusual in their ability to aestivate in cocoons of mud during droughts.

DIPOLE, in radio; see ANTENNA.

DIPOLE MOMENT. An electric dipole is a pair of equal and opposite electric charges a short distance apart (see ELECTRICITY). All ordinary manifestations of MAGNETISM are the result of magnetic dipoles, whether these arise in the context of permanent magnets or ELECTROMAGNETS. In either case the dipole moment is a VECTOR quantity descriptive of the dipole. Atomic nuclei and asymmetric molecules often exhibit dipole or other multipole properties.

DIPSOMANIA. See ALCOHOLISM.

DIRAC, Paul Adrien Maurice (1902–), English theoretical physicist who shared the 1933 Nobel physics prize with E. SCHRÖDINGER for their contributions to WAVE MECHANICS. Dirac's theory (1928) took account of RELATIVITY and led him to postulate the existence of the positive ELECTRON or positron, later discovered by C. D. ANDERSON. Dirac was also the codiscoverer of FERMI-DIRAC STATISTICS.

DIRECTION FINDER, device used to locate the direction of an incoming RADIO signal. Usually a loop ANTENNA is rotated until maximum reception strength is achieved, giving the line of the transmission. If this is repeated from a different position, the transmitting station may be located. The **radiocompass** used in air and sea NAVIGATION is a direction finder: position is determined by finding the directions of two or more transmitters.

DIRIGIBLE. See AIRSHIP.

DISCRIMINANT, in a POLYNOMIAL equation, a number calculated from the COEFFICIENTS which can tell much about the roots (see ROOTS OF AN EQUATION). For a quadratic EQUATION of the standard form $ax^2+bx+c=0$ the discriminant, D, is b^2-4ac; for a cubic equation ax^3+bx^2+cx+d, $D=18abcd-4b^3d+b^2c^2-4ac^3-27a^2d^2$. For a cubic or quadratic, $D<0$ implies that there are two imaginary roots (see IMAGINARY NUMBERS); $D\geqslant 0$ that the roots are all real (see REAL NUMBERS); $D=0$ that at least two of these are equal.

DISEASE, disturbance of normal bodily function in an organism. MEDICINE and SURGERY are concerned with the recognition or diagnosis of disease and the institution of treatment aimed at its cure. Disease is usually brought to attention by symptoms, in which a person becomes aware of some abnormality of, or change in, bodily function. Pain, HEADACHE, FEVER, COUGH, shortness of breath, DYSPEPSIA, CONSTIPATION, DIARRHEA, loss of BLOOD, lumps, PARALYSIS, numbness and loss of consciousness are common examples. **Diagnosis** is made on the basis of symptoms, signs on physical examination and laboratory and X-RAY investigations; the functional disorder is analyzed and possible causes are examined. Causes of physical disease in man are legion, but certain categories are recognized: trauma, congenital, infectious, inflammatory, vascular, tumor, degenerative, deficiency, poison, metabolic, occupational and iatrogenic diseases.

Trauma to body may cause SKIN lacerations and BONE FRACTURES as well as disorders specific to the organ involved (e.g., CONCUSSION). **Congenital diseases** include hereditary conditions (i.e., those passed on genetically) and diseases beginning in the FETUS, such as those due to drugs or maternal infection in PREGNANCY. **Infectious diseases** include VIRAL DISEASE, BACTERIAL DISEASE and PARASITIC DISEASE, which may be acute or chronic and are usually communicable. Insects, animals and human carriers may be important in their spread and EPIDEMICS may occur. INFLAMMATION is often the result of infection, but **inflammatory disease** can also result from disordered IMMUNITY and other causes. In **vascular diseases**, organs become diseased secondary to disease in their blood supply, such as ARTERIOSCLEROSIS, ANEURYSM, THROMBOSIS and EMBOLISM. **Tumors,** including benign growths, CANCER and LYMPHOMA are diseases in which abnormal growth of a structure occurs and leads to a lump, pressure on or spread to other organs and distant effects such as emaciation, HORMONE production and NEURITIS. In **degenerative disease**, DEATH or premature ageing in parts of an organ or system lead to a gradual impairment of function. **Deficiency diseases** result from inadequate intake of nutrients, VITAMINS, minerals, calcium, iron and trace substances; disorders of their fine control and that of hormones leads to **metabolic disease**. **Poisoning** is the toxic action of chemicals on body systems, some of which may be particularly sensitive to a given poison. An increasingly recognized side-effect of industrialization is the occurrence of **occupational diseases**, in which chemicals, dusts or molds encountered at work cause disease—especially PNEUMOCONIOSIS and other LUNG disease, and certain cancers. **Iatrogenic disease** is disease produced by the intervention of doctors, in an attempt to treat or prevent some other disease. The altered ANATOMY of diseased structures is described as **pathological**. **Psychiatric disease**, including psychoses (schizophrenia and depression) and neuroses, are functional disturbances of the BRAIN, in which structural abnormalities are not recognizable; they may represent subtle disturbances of brain metabolism. **Treatment** of disease by SURGERY or DRUGS is usual, but success is variable; a number of conditions are so benign that symptoms may be suppressed until they have run their natural course.

DISINFECTANTS. See ANTISEPTICS.

DISLOCATION, a CRYSTAL defect in which the normal crystal lattice is distorted (e.g., by the interposition of an extra half plane of atoms—an edge dislocation). The type and number of dislocations in a crystal help to determine its electrical and mechanical properties.

DISPERSION, in optics, the separation of a mixture of LIGHT radiations according to COLOR (i.e., wavelength). This can occur in REFRACTION (when it is responsible for the RAINBOW and the production of a SPECTRUM with a PRISM), in DIFFRACTION (as is applied in the grating spectroscope), or in SCATTERING (giving

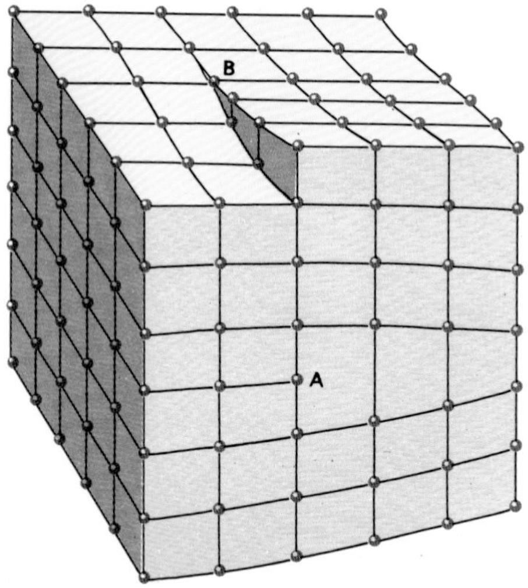

A section of a crystal lattice including an edge dislocation (A) and a screw dislocation (B).

rise, e.g., to the blue color of the sky). The dispersive power of an optical medium is a measure of the extent to which its refractive index varies with wavelength.

DISPLACEMENT ACTIVITY, term used by ethologists for actions performed by an animal in stressful situations that appear to be irrelevant to prevailing conditions. An example is grooming in a situation where fighting would normally be expected. A possible explanation of such activity is that it results from the effects of conflicting stimulae. This diversion of impulses is due either to conflict between opposing drives or the presence of a strong drive in the absence of an appropriate external object for its relief.

DISSOCIATION, in chemistry, the reversible decomposition of a compound, often effected by heat. **Ionic dissociation** is the dissociation of a covalent compound—a weak electrolyte (see ELECTROLYSIS)—into IONS when dissolved in water or other ionizing SOLVENTS. It accounts in part for the phenomena of electrolytic CONDUCTIVITY.

DISSONANCE, in acoustics, the simultaneous sounding of two or more tones which, because of BEATING, seem unpleasant to the human ear. Which musical intervals are considered dissonant is largely a matter of cultural and historical relativism.

DISTANCE MEASURING EQUIPMENT (DME), instrument used in aerial NAVIGATION to determine distance from a RADIO beacon by measuring the time taken for a pulse to reach the beacon and return.

DISTEMPER, term applied to several animal diseases, but particularly referring to a specific VIRAL DISEASE of dogs. It commonly occurs in puppies, with FEVER, poor appetite and discharge from mucous membranes; bronchopneumonia and ENCEPHALITIS may be complications. VACCINATION is protective.

DISTILLATION, process in which substances are vaporized and then condensed by cooling, probably first invented by the ALEXANDRIAN SCHOOL and used in ALCHEMY, the still and the ALEMBIC being employed. It may be used to separate a volatile liquid from nonvolatile solids, as in the production of pure WATER from seawater, or from less volatile liquids, as in the distillation of liquid air to give oxygen, nitrogen and the noble gases. If the boiling points of the components differ greatly, **simple distillation** can be used: on gentle heating, the components distill over in order (the most volatile first) and the pure fractions are collected in different flasks. Mixtures of liquids of similar boiling points require **fractionation** for efficient separation. This technique employs multiple still heads and fractionating columns in which some of the vapor is condensed and returned to the still, equilibrating as it does so with the rising vapor. In effect, the mixture is redistilled several times; the number of theoretical simple distillations, or theoretical plates, represents the separating efficiency of the column. The theory of distillation is an aspect of PHASE EQUILIBRIA studies. For ideal solutions, obeying RAOULT'S law, the vapor always contains a higher proportion than the liquid of the more volatile component; if this is not the case, an AZEOTROPIC MIXTURE may be formed. When two immiscible liquids are distilled, they come over in the proportion of their VAPOR PRESSURES at a temperature below the boiling point of either. This is utilized in **steam distillation**, in which superheated steam is passed into the still and comes over together with the volatile liquid. It is useful when normal distillation would require a temperature high enough to cause decomposition, as is **vacuum distillation**, in which the pressure reduction lowers the boiling points. A further refinement is **molecular distillation**, in which unstable molecules travel directly in high vacuum to the condenser.

DISTILLED LIQUOR, an ALCOHOLIC BEVERAGE of high ETHANOL content produced by DISTILLATION of a fermented mixture (see BREWING; FERMENTATION). Crude distilling of rice beer was practiced in the Far East several centuries BC; in the West distilled mead and wine were made by 500 AD. Whiskey and brandy production flourished in the later Middle Ages, and in the 16th century fractional distillation was introduced. Modern production is in large, continuous stills. As well as ethanol and water, the distillate contains FUSEL OILS and other compounds which give flavor and aroma, further improved by several years' maturation in oak casks. The alcohol content is expressed in the "proof" system: in the US one degree proof equals 0.5% ethanol by volume.

DISTRIBUTIVE LAW. See ALGEBRA.

DISTRIBUTOR, in AUTOMOBILES, a device operated off the CAMSHAFT which ensures that the cylinders of the INTERNAL-COMBUSTION ENGINE are fired in the correct sequence and at the optimum time.

DISULFIRAM. See ANTABUSE.

DIURETICS, drugs that increase urine production by the KIDNEY. Alcohol and CAFFEINE are mild diuretics. Thiazides and other diuretics are commonly used in treatment of HEART failure, EDEMA, high blood pressure, LIVER and KIDNEY disease.

DIV, in VECTOR ANALYSIS, divergence. Div $\mathbf{V}$ is defined as the scalar product of DEL with the vector $\mathbf{V}$: $\nabla.\mathbf{V}$. Hence, in CARTESIAN COORDINATES,

$$\operatorname{div}\mathbf{V}=\nabla.\mathbf{V}=\frac{\partial v_x}{\partial x}+\frac{\partial v_y}{\partial y}+\frac{\partial v_z}{\partial z}.$$

DIVERGENCE, of a SEQUENCE or SERIES. See CONVERGENCE. See also DIV.

DIVIDE, water parting, or **watershed**, a region of high ground that lies between and determines the flow of two unconnected drainage systems. The **Continental Divide** of North America is formed by the Rocky Mountains.

DIVIDEND, a number that is divided (see DIVISION) by another number, the DIVISOR. In a FRACTION such as $\frac{a}{b}$ the dividend, a, is called the numerator.

DIVING, Deep Sea, the descent by divers to the sea bed, usually for protracted periods, for purposes of exploration, salvage, etc. Skin diving is almost as old as man and the Romans had primitive diving suits connected by an air pipe to the surface. This principle was also known in the early 16th century. A breakthrough came when John Lethbridge devised the forerunner of the armored suits used today in deepest waters (1715): it looked much like a barrel with sleeves and a viewport, and was useless for depths of more than a few metres. In 1802 William Forder devised a suit where air was pumped to the diver by bellows. And in 1837 (improving his earlier design of 1819) Augustus Siebe (1788–1872) invented the modern diving suit, a continuous airtight suit to which air is supplied by a pump. The diving suit today has a metal or fiberglass helmet with viewports and inhalation and exhalation valves, joined by an airtight seal to a metal chestpiece, itself joined to a flexible watertight covering of rubber and canvas; and weights, especially weighted boots, for stability and to prevent the diver shooting toward the surface. Air or, more often, an oxygen/helium mixture is conveyed to him via a thick rubber tube. In addition, he has either a telephone wire, or simply a cord which he can tug, for communication with the surface. Nowadays SCUBA diving, where the diver has no suit but carries gas cylinders and an AQUALUNG, is preferred in most cases since it permits greater mobility. In all diving great care must be taken to avoid the bends (see AEROEMBOLISM) through too-rapid ascent to the surface. (See also BATHYSCAPHE; BATHYSPHERE.)

DIVINING ROD, a forked rod (usually of wood) used in DOWSING. The dowser holds the forked end, one prong in each hand, and is allegedly able to detect underground water, metal, etc., by the directions and magnitudes of convulsions of the rod.

DIVISION, the INVERSE operation of MULTIPLICATION, the determination of the number of times one number must be multiplied to equal another number. If DIVIDEND and DIVISOR have like positive or negative sign, then the QUOTIENT is positive; if their signs are different, the quotient is negative:

e.g., $\frac{6}{3}=\frac{-6}{-3}=2$, and $\frac{-6}{3}=\frac{6}{-3}=-2$.

Division involving two POWERS of a number or VARIABLE is performed by subtracting the EXPONENT of the divisor from that of the dividend:

$$x^4/x^2=x^{4-2}=x^2;$$

the division of the COEFFICIENTS being carried out in the normal way:

$$ax^7/bx^9=\frac{}{b}x^{-2}.$$

In the division of INTEGERS, where an integral quotient is required, it may well be that the divisor does not divide exactly into the dividend: thus $\frac{7}{3}=2$ REMAINDER 1 (or, in REAL NUMBERS, $2\frac{1}{3}$).

A tanker raised out of the water in the floating dry dock at Amsterdam, Netherlands.

Division of a polynomial by a binomial can be performed by factoring if the binomial is a FACTOR of the POLYNOMIAL. If not, a more complex procedure is used. To divide a polynomial by a monomial, divide each term separately by the monomial. Thus $ax^4y^2+bxy^3-cx$ divided by $-dxy$ is

$$-\frac{a}{d}x^3y-\frac{b}{d}y^2+\frac{c}{d}(y^{-1}).$$

DIVISOR, a number by which another number, the DIVIDEND, is divided (see DIVISION). In a FRACTION such as $\frac{a}{b}$, the divisor, b, is called the denominator.

DIXON, Joseph (1799–1869), US manufacturer, the inventor of the high-temperature GRAPHITE crucible and other graphite products.

DNA, deoxyribonucleic acid, a NUCLEIC ACID comprising two strands of NUCLEOTIDE wound around each other in a double helix, found in all living things and VIRUSES.

DOBZHANSKY, Theodosius (1900–), Russian-born US biologist, famed for his study of the fruit fly, *Drosophila*, which demonstrated that a wide genetic range could exist in even a comparatively well-defined species. Indeed the greater the "genetic load" of unusual genes in a species, the better equipped it is to survive in changed circumstances. (See EVOLUTION; HEREDITY.)

DOCK, an enclosure of water in a port or harbor in which a ship may be berthed for maintenance or loading. Where access to the hull is required, a **dry dock** may be used. Usually this is a basin of water dug into the shore of a water channel, from which it can be closed off by a gate: the ship is floated in, the gate closed and the water pumped out. **Floating dry docks** are trough-like structures which can, by use of ballast, be partly submerged: with the ship inside, the dock is drained and the ballast discarded. In all cases, the ship must be supported before pumping dry.

The doppler effect: sound waves (compression waves in the air) radiating isotropically (with equal velocities in all directions) from a moving sound source (here a light aircraft), are bunched up ahead of and spaced out behind the source. A nearby stationary observer thus hears a higher-pitched sound than that heard by anyone moving with the source while the source approaches, dropping to a lower-pitched sound as the source passes by and recedes.

DOCTOR, Medical. See MEDICINE.
DODECAHEDRON. See POLYHEDRON.
DOG STAR. See SIRIUS.
DOISY, Edward Adelbert (1893–), US biochemist who first crystallized the female sex HORMONE estrone (1929). He shared the 1943 Nobel Prize for Physiology or Medicine with H. DAM after isolating, determining the structure of, and synthesizing VITAMIN K (1936).
DOLDRUMS, regions of low wind, calms and strong upward air movement around the EQUATOR, produced by the convergence of the SE and NE TRADE WINDS. Sailing ships were often becalmed there.
DOLERITE. British term for DIABASE.
DOLMEN, Neolithic (see STONE AGE) tomb in a roughly table-like form, with one horizontal slab of stone supported by three or more vertical slabs, beneath which is the burial chamber. It is possible that dolmens are manifestations of a megalithic religion that spread from the Aegean.
DOLOMITE, a common mineral, calcium magnesium carbonate ($CaMg(CO_3)_2$), with white or colorless rhombohedral crystals, often found associated with limestone and MARBLE and in magnesium-rich metamorphic rocks. It is used as an ornamental and building stone, and is a source of magnesium.
DOMAGK, Gerhard (1895–1964), German pharmacologist who discovered the antibacterial action of the dye Prontosil Red. This led to the discovery of other SULFA DRUGS. In recognition of this Domagk was offered the 1939 Nobel Prize for Physiology or Medicine, though he was not allowed to accept it at the time.
DOMAIN. See SET THEORY.
DOMAINS, Magnetic. See MAGNETISM.
DOMESTICATION, the process by which some

animals have been adapted to the uses of man. Domestication implies the rearing of animals in captivity or semicaptivity but also usually involves changes in the animals concerned. (See also ARTIFICIAL SELECTION.)

DOMINANCE HIERARCHY, the organization of animals, notably PRIMATES, within groups according to social status. This order is usually established by mutual threatening or warning signals that make fighting for dominance unnecessary. The dominant animal has priority over members of the group in feeding, sex relations, and other activities.

DONATI, Giovanni Battista (1826–1873), Italian astronomer who first studied the spectra of COMETS (1864). He discovered several comets including "Donati's comet" of 1858.

DOPPLER EFFECT, the change observed in the wavelength of a sonic, electromagnetic or other wave (see WAVE MOTION) because of relative motion between the wave source and an observer. As a wave source approaches an observer, each pulse of the wave is closer behind the previous one than it would be were the source at rest relative to the observer. This is perceived as an increase in frequency, the pitch of a sound source seeming higher, the color of a light source bluer. When a sound source achieves the speed of sound, a SONIC BOOM results. As a wave source recedes from an observer, each pulse is emitted farther away from him than it would otherwise be. There is hence a drop in pitch or a reddening in COLOR (see LIGHT; SPECTRUM). The Doppler Effect, named for **Christian Johann Doppler** (1803–1853) who first described it in 1842, is of paramount importance in astronomy. Observations of stellar spectra can determine the rates at which stars are moving towards or away from us, while observed red shifts in the spectra of distant galaxies are generally interpreted as an indication that the universe as a whole is expanding (but see RED SHIFT).

DORADO (the Swordfish or Goldfish), a southern hemisphere constellation containing the Greater Magellanic Cloud (see MAGELLANIC CLOUDS).

DORMANCY, a resting state that occurs in animals and plants, when growth stops and the internal processes, principally RESPIRATION, are slowed down. In animals, dormancy during the winter is termed HIBERNATION, while dormancy in the summer or dry season, such as in the lung fish and earthworm, is termed **aestivation** (or **estivation**). In plants, dormancy during adverse conditions is manifest in the lack of growth of perennial plants such as grasses, the loss of foliage by deciduous trees and the production of underground perennating organs such as BULBS, CORMS and TUBERS.

DOSIMETER, a device worn by persons working in situations where they are exposed to ionizing radiations (see RADIOACTIVITY), which measures the dose of radiation to which they have been exposed.

DOUBLE INTEGRAL, the integral (see CALCULUS) used to treat two FUNCTIONS that are of different VARIABLES. More than two such functions are treated together by use of a multiple integral.

DOUBLE REFRACTION, or **birefringence**, the property of certain CRYSTALS to split a ray of unpolarized light into two rays plane-polarized at right-angles to each other (see POLARIZED LIGHT). One of these, the ordinary ray, is refracted according to the ordinary laws of REFRACTION, but the other, the extraordinary ray, is refracted with a refractive index that depends on the direction from which the original ray was incident. Double refraction in Iceland spar is used in the NICOL PRISM to produce plane-polarized light.

DOUBLE STAR or **binary star**, a pair of stars revolving around a common center of gravity. Less frequently the term "double star" is applied to two stars that merely appear close together in the sky though in reality at quite different distances from the earth (optical pairs) or to two stars whose motions are linked but which do not orbit each other (physical pair). About 50% of all stars are members of either binary or multiple star systems, in which there are more than two components. It is thought that the components of binary and multiple star systems are formed simultaneously. **Visual binaries** are those which can be seen telescopically to be double. There are comparatively few visual binaries, since the distances between components are small relative to interstellar distances, but examples are CAPELLA, PROCYON, SIRIUS and ALPHA CENTAURI. **Spectroscopic binaries**, while unable to be seen telescopically as doubles, can be detected by RED SHIFTS in their spectra, their orbit making each component alternately approach and recede from us. **Eclipsing binaries** are those whose components, due to the orientation of their orbit, periodically mutually eclipse each other as seen from the earth.

DOUGH, thick, elastic mixture of flour, water or milk and other ingredients, often including YEAST, shortening, sugar, salt and eggs, used to make bread, pastry and cakes. Leavened dough (see LEAVEN) undergoes FERMENTATION to make it rise. **Batter** is a thinner, pourable mixture used to make pancakes, scones, biscuits etc. and as a coating for fried foods and fritters.

DOW, Herbert Henry (1866–1930), Canadian-born US chemist and industrialist who developed an electrolytic method for extracting BROMINE from certain natural BRINES. He founded the Dow Chemical Company to exploit this process in 1897.

DOWSING, the detection, usually with a DIVINING ROD, of hidden (usually underground) resources of water, metals, etc. Another frequently-used technique employs a PENDULUM held in the dowser's hand; the direction and magnitude of swing provides information as to the material beneath the ground, the bob itself being made of suitable material. Laboratory tests (see PARAPSYCHOLOGY) have shed little light on dowsing, though some dowsers have remarkable records of success.

DRACO (the Dragon), a large N Hemisphere constellation. Alpha Draconis was the POLESTAR c3000 BC (see PRECESSION). Gamma Draconis is the bright star Eltanin. Draco also contains the planetary NEBULA NGC 6543.

DRAGON'S BLOOD, a reddish resin obtained from a number of plants. Once used in medicine for its astringent properties, it is now mostly used in China to give a red facing to writing paper. Its main commercial source is an Indonesian rattan palm, *Daemonorops draco*.

DRAINAGE, the runoff of water from an area, either naturally or, in agriculture, under artificial control. In nature, drainage generally takes the form of a

A bucket dredger discharges watery mud into the hold of a hopper moored alongside. The hopper will take its cargo out to the dumping ground while the dredger continues work with a second hopper.

pattern of streams, which feed rivers and lakes and flow, usually, to the sea (see HYDROLOGIC CYCLE). An area all of whose rainwater drains into a particular body of water is called a watershed or catchment basin. (See also GROUNDWATER.) Systems of artificial drainage depend on the nature of the SOIL as well as local topography. Two main systems are used: surface and subsurface. Surface systems usually comprise a pattern of ditches; subsurface systems a pattern of conduits and tunnels. These usually lead to a natural stream. (See also IRRIGATION; LAND RECLAMATION.)

DRAKE, Edwin Laurentine (1819–1880), US oil-industry pioneer who drilled the world's first oil well at Oil Creek, Titusville, Pa. in 1859. (See PETROLEUM.)

DRAPER, John William (1811–1882), English-born US chemist who investigated the chemical action of light, first photographed the moon (1840) and obtained a photograph of the solar spectrum (1844). His son, **Henry Draper** (1837–1882), US physician and amateur astronomer, obtained the first photograph of a nonsolar stellar spectrum (1872) and the first photograph of a NEBULA (the Orion) in 1880.

DRAVIDIANS, subgroup of the Hindu race, some 100 000 000 people of (mainly) S India. They are fairly dark-skinned, stocky, have rather more NEGROID features than other Indics, and are commonly dolichocephalic (see CEPHALIC INDEX). (See also RACE.)

DREAMS, fantasies, usually visual, experienced during sleep and in certain other situations. About 25% of an adult's sleeping time is characterized by rapid eye movements (REM) and brain waves that, registered on the ELECTROENCEPHALOGRAPH, resemble those of a person awake (EEG). This REM-EEG state occurs in a number of short periods during sleep, each lasting a number of minutes, the first coming some 90min after sleep starts and the remainder occurring at intervals of roughly 90min. It would appear that it is during these periods that dreams take place, since people woken during a REM-EEG period will report and recall visual dreams in some 80% of cases; people woken at other times report dreams only about 40% of the time, and of far less visual vividness. Observation of similar states in animals suggests that at least all mammals experience dreams. Dreams can also occur, though in a limited way, while falling asleep; the origin and nature of these is not known. **Dream interpretation** seems as old as recorded history. Until the mid-19th century dreams were regarded as supernatural, often prophetic; their possible prophetic nature has been examined in this century by, among others, J. W. Dunne. According to FREUD, dreams have a *latent content* (the fulfilment of an individual's particular UNCONSCIOUS wish) which is converted by *dreamwork* into *manifest content* (the dream as experienced). In these terms, interpretation reverses the dreamwork process. (See also HALLUCINATION; SLEEP.)

DREDGING, the excavation of material from the bottom of harbors and navigation channels, generally with the aim of keeping them open for shipping. In the **bucket dredge** an endless chain of buckets extends down into the bottom mud: the buckets dig into the mud and carry it up to be discharged on board. The **grab dredge** operates just as a CRANE with a grab bucket; the **dipper dredge** much as an EXCAVATOR. In the **hydraulic dredge** the material is drawn up through a pipe by suction.

DRIFT, the material left behind on the retreat of a GLACIER. The unstratified material deposited directly on land is called TILL; fluvio-glacial drift, well stratified, is that transported by melted waters of the glacier. Drift may be composed of particles from fine sand up to huge boulders, and may be up to 100m deep.

DRILLS, tools for cutting or enlarging holes in hard materials. There are two classes: those that have a rotary action, with a cutting edge or edges at the point and, usually, helical fluting along the shank; and those that work by percussive action, where repeated blows drive the drill into the material. **Rotary drills** are commonly used in the home for wood, plastic, masonry and sometimes metal. They are usually hand-turned, though electric motors are increasingly used to power drills in home workshops. In metallurgy the mechanical drilling machine or *drill press* is one of the most important MACHINE TOOLS, operating one or several drills at a time. As great heat is generated, LUBRICATION is very important. Most metallurgical drills are of high-speed STEEL. Dentists' drills rotate at extremely high speeds, their tips (of TUNGSTEN carbide or DIAMOND) being water-cooled: they are powered by an electric motor or by compressed air. Rotary drills are used for deeper oil-well drilling: a cutting bit is rotated at the end of a long, hollow drill pipe, new sections of pipe being added as drilling proceeds. **Percussive drills** are used for rock-boring, for concrete and masonry, and for shallower oil-well drilling. Rock drills are generally powered by compressed air, the tool rotating after each blow to increase cutting speed. The pneumatic drill familiar in city streets is also operated by compressed air. *Ultrasonic drills* are used for brittle materials; a rod, attached to a TRANSDUCER, is placed against the surface, and to it are fed ABRASIVE particles suspended in a cooling fluid. It is these particles that actually perform the cutting. (See also ULTRASONICS.)

DRIVE, in psychology and psychoanalysis, term used for a human or animal INSTINCT.

DROPSY. See EDEMA.

DROSOPHILA, or **fruit fly**, a member of the INSECTA that has been used extensively in the study of GENETICS. It has the advantages of being small and of breeding extremely rapidly.

DROUGHT, temporary, often disastrous climatic condition of extreme dryness when an area's natural water supplies are insufficient for plant, specifically crop, needs. It occurs when loss of water from the SOIL by EVAPORATION or otherwise greatly exceeds the water precipitation from the atmosphere; this may result from high winds, low HUMIDITY and heat. Areas with a well-defined dry season suffer **seasonal drought** (see DRY FARMING). (See also DESERT; DUST BOWL.)

DROWNING, immersion in water causing DEATH by ASPHYXIA, metabolic or blood disturbance, following inhalation of water. On immersion, REFLEX breath-holding occurs but is eventually overcome; if

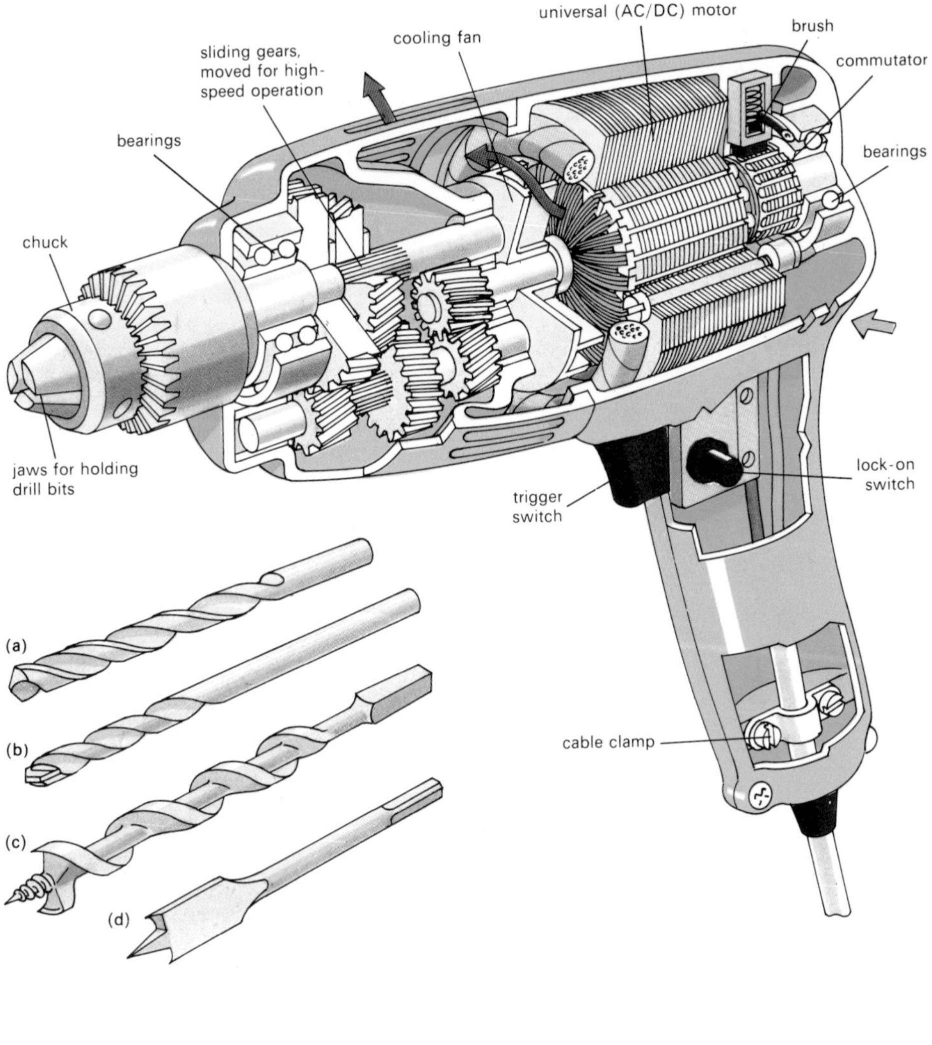

Electric drill as used for light construction work in the home. This versatile instrument employs a universal motor, offering high torque at low speeds and reduced torque at higher speeds, coupled through a two-speed transmission to the chuck assembly. As shown here the gears are set for low-speed running. Electrical safety is ensured through double insulation.
(*Bottom left*) Types of drill bit include: (a) twist drill (for metal, plastic, wood, etc.); (b) masonry bit; (c) auger bit (for wood); (d) high-speed wood bit. The jaws of most domestic electric drills will also accept accessories other than drill bits. Rotary wire brushes, polishing pads, grinding wheels, sanding disks, files and cutting heads are common examples. Alternatively, the whole chuck assembly can be removed and the drill become the power source for a circular saw blade.

immersion continues, water is taken into the LUNGS. Spasm of LARYNX leads to further asphyxia and abnormal HEART rhythm. If death does not follow, water absorbed from the lungs alters the mineral concentration of BLOOD and red blood cells may be damaged. ACIDOSIS, lung EDEMA and distension of STOMACH may occur. Prompt resuscitation at an early stage by clearing the airway, ARTIFICIAL RESPIRATION and, if necessary, cardiac massage and correction of blood abnormalities, may be successful.

DRUG ADDICTION, an uncontrollable craving for a particular DRUG, usually a NARCOTIC, which develops into a physiological or sometimes merely psychological dependence on it. Generally the individual acquires greater tolerance for the drug, and therefore requires larger and larger doses, to the point where he may take doses that would be fatal to the nonaddict. Should his supply be cut off he will suffer **withdrawal symptoms** ("cold turkey") which are psychologically gruelling and often physically debilitating to the point where death may result. Many drugs, such as ALCOHOL and TOBACCO, are not addictive in the strictest sense but more correctly HABIT forming (but ALCOHOLISM does involve addiction). Others, such as the OPIUM derivatives, particularly HEROIN and MORPHINE, are extremely addictive. With others, such as LSD (and most other HALLUCINOGENIC DRUGS), COCAINE, HEMP and the AMPHETAMINES, the situation is unclear: dependence may be purely psychological, but it may be that these drugs interfere with the chemistry of the BRAIN; for example, the hallucinogen MESCALINE is closely related to ADRENALINE. The situation is even less clear with such drugs as MARIJUANA which appear to be neither addictive nor habit forming. An inability to abstain from regular self-dosage with a drug is described as a **drug habit**.

DRUGS, chemical agents that affect biological systems. In general they are taken to treat or prevent disease, but certain drugs, such as the OPIUM NARCOTICS, AMPHETAMINES, BARBITURATES and cannabis, are taken for their psychological effects and are drugs of addiction or abuse (see DRUG ADDICTION). Many drugs are the same as or similar to chemicals occurring naturally in the body and are used either to replace the natural substance (e.g., THYROID HORMONE) when deficient, or to induce effects that occur with abnormal concentrations as with STEROIDS or oral CONTRACEPTIVES. Other agents are known to interfere with a specific mechanism or antagonize a normal process (e.g., ATROPINE, CURARE). Many other drugs are obtained from other biological systems; FUNGI OR BACTERIA (ANTIBIOTICS) or plants (DIGITALIS), and several others are chemical modifications of natural products. In addition, there are a number of entirely synthetic drugs (e.g., barbiturates), some of which are based on active parts of naturally occurring drugs (as with some antimalarials based on QUININE).

In devising drugs for treating common conditions, an especially desirable factor is that the drug should be capable of being taken by mouth; that is, that it should be able to pass into the body unchanged in spite of being exposed to STOMACH acidity and the ENZYMES of the DIGESTIVE SYSTEM. In many cases this is possible but there are some important exceptions, as with INSULIN which has to be given by INJECTION. This method may also be necessary if VOMITING or GASTROINTESTINAL-TRACT disease prevent normal absorption. In most cases, the level of the drug in the BLOOD or tissues determines its effectiveness. Factors affecting this include: the route of administration; the rate of distribution in the body; the degree of binding to PLASMA PROTEINS or FAT; the rates of breakdown (e.g., by the LIVER) and excretion (e.g., by the KIDNEYS); the effect of disease on the organs concerned with excretion, and interactions with other drugs taken at the same time. There is also an individual variation in drug responsiveness which is also apparent with undesired **side-effects**. These arise because drugs acting on one system commonly act on others. Side effects may be nonspecific (nausea, DIARRHEA, malaise or SKIN rashes); allergic (HIVES, ANAPHYLAXIS), or specific to a drug (abnormal HEART rhythm with digitalis). Mild side-effects may be suppressed but others must be watched for and the drug stopped at the first sign of any adverse effect. Drugs may cross the PLACENTA to reach the FETUS during PREGNANCY, interfering with its development and perhaps causing deformity as happens with THALIDOMIDE.

Drugs may be used for symptomatic relief (ANALGESICS, antiemetics) or to control a disease. This can be accomplished by killing the infecting agents; by preventing specific infections; by restoring normal control over MUSCLE (anti-Parkinsonian agents) or mind (ANTIDEPRESSANTS); by replacing a lost function or supplying a deficiency (e.g., VITAMIN B_{12} in pernicious ANEMIA); by suppressing inflammatory responses (steroids, ASPIRIN); by improving the functioning of an organ (digitalis); by protecting a diseased organ by altering the function of a normal one (e.g., DIURETICS for heart failure), or by toxic actions on CANCER cells (cancer CHEMOTHERAPY). The scientific study of drugs is the province of PHARMACOLOGY.

DRUMLINS, elongate hillocks, formed of TILL, found usually in swarms in lowland areas formerly covered by GLACIERS. They usually taper away from a steep slope that faced the oncoming ice.

DRUPE, fleshy FRUIT comprising an outer skin (exocarp), a fleshy pulp (mesocarp) and an inner hard

A drumlin near Soudus, in New York State. The occurence of drumlins in swarms, which may consist of as many as ten thousand individuals, all with the same orientation, gives rise to what is known as "basket-of-eggs topography". Drumlins may be several kilometres long, but rarely exceed 60m in height.

and woody stone (endocarp) enclosing a single seed. Examples are found in the cherry, peach and plum. (See also BERRY.)

DRY CELL. See BATTERY.

DRY CLEANING, process of cleaning fabrics with nonaqueous SOLVENTS and detergents, removing grease-based stains and retaining the shape and texture of the garments. Modern dry cleaning dates from the mid-19th century. GASOLINE was first used, but has been replaced by less inflammable solvents such as Stoddard solvent (a petroleum distillate), or the nonflammable chloroethylenes. Carbon tetrachloride, being highly toxic, is seldom used. The process is analogous to the use of a washing machine which spins and tumble-dries.

DRY FARMING, the raising of crops in semiarid areas without making use of IRRIGATION. The essential principle of dry farming is the encouragement of efficient retention of moisture by the SOIL, coupled with selection of crops that can make best use of that moisture. Among the techniques employed are tilling the land and eradicating weeds; wide spacing of crops; leaving the stubble after harvest to act as a snow-trap over winter; letting clods of earth or dry vegetable matter lie on the land so as to reduce water runoff and SOIL EROSION; contouring of fields; and, occasionally, allowing fields to lie fallow in alternate years. Crops may be of two types: those whose GROWING SEASONS permit them to evade the summer DROUGHT; and those which are able to survive the drought by reduction of their own moisture loss.

DRY ICE, solid CARBON dioxide, used as a refrigerant for transporting perishables. It is made by compressing carbon dioxide gas to about $7MN/m^2$ at $-57°C$, when it liquefies; it is then expanded adiabatically to atmospheric pressure and cools, solid carbon dioxide "snow" separating. This is compressed into blocks. subl $-78.5°C$.

DRY ROT, form of wood decay found in houses, resulting in loss of strength of timbers. The causal agents are the basidiomycete fungi *Merulius lacrymans* (in Britain) and *Poria incrassata* (in North America). Preventative measures include ensuring that conditions ideal for growth of the fungus are avoided, i.e., relative HUMIDITY is kept low, no free moisture is allowed and wood surfaces are covered. Treatment for the disease includes painting the wood surface with CREOSOTE and using FUNGICIDES to kill the fungus.

DUALISM, any religious or philosophical system characterized by a fundamental opposition of two independent or complementary principles. Among religious dualisms are the unending conflict of good and evil spirits envisaged in Zoroastrianism and the opposition of light and darkness in Jewish apocalyptic, Gnosticism and Manichaeism. The Chinese complementary principles of *yin* and *yang* exemplify a cosmological dualism while the mind–body dualism of DESCARTES is the best-known philosophical type. Dualism is often opposed to MONISM and pluralism.

DUBOIS, Marie Eugène François Thomas (1858–1940), Dutch paleontologist who discovered Java Man (*Pithecanthropus erectus*) in 1894. (See PREHISTORIC MAN.)

DUBOS, René Jules (1901–   ), French-born US microbiologist who discovered tyrothricin (1939), the first ANTIBIOTIC to be used clinically.

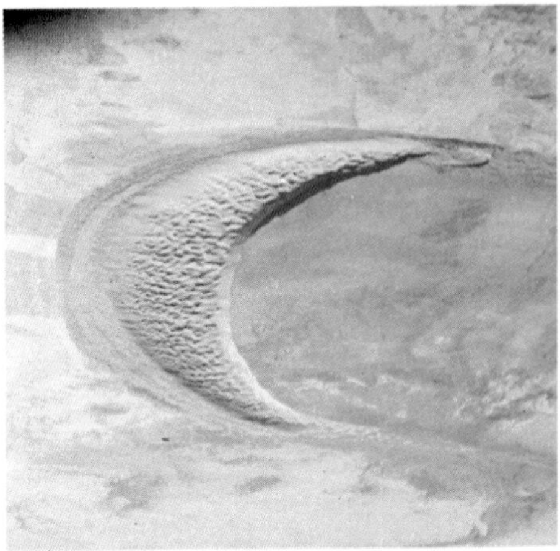

A barchan dune.

DUCTILITY, the property of metals, alloys and some other substances to be drawn out or extruded (see EXTRUSION) without rupture or loss of strength. GOLD is the most ductile metal at normal temperatures. (See MALLEABILITY; MATERIALS, STRENGTH OF.)

DUCTLESS GLANDS. See ENDOCRINE GLANDS.

DULBECCO, Renato (1914–   ), Italian-born Anglo-American physiologist who shared the 1975 Nobel Prize for Physiology or Medicine with D. BALTIMORE and H. TEMIN for their work on cancer-causing VIRUSES.

DULONG, Pierre Louis (1785–1838), French chemist who with A. T. Petit discovered **Dulong and Petit's Law**. This states that the SPECIFIC HEATS of elements are inversely proportional to their ATOMIC WEIGHTS, and is equivalent to the observation that the atomic heats of most elements are about 26.4J/K.mol.

DUMAS, Jean Baptiste André (1800–1884), French organic chemist who discovered that CHLORINE could substitute for HYDROGEN in HYDROCARBONS and was thus led to propose his "law of substitution" (1834) which revolutionized the theory of organic chemistry. From 1868 he was secretary of the Academy of Sciences.

DUMBNESS, inability to speak. Failure of speech development, usually associated with congenital DEAFNESS (deaf-mute) is the most common cause in childhood. APHASIA and hysterical mutism are the usual adult causes. If comprehension is intact, writing and sign language are alternative forms of communication, but in aphasia language is usually globally impaired. (See SPEECH AND SPEECH DISORDERS.)

DUMDUM BULLET, a bullet so devised as to expand on impact, creating greater damage. Most simply, a cross is scored on the bullet's nose. Their military use was outlawed by the Hague Conference, 1899.

DUNE, hillock of SAND built up by a prevailing WIND, found mostly in DESERT areas. They have several forms, commonest being **barchans**, crescent shaped with the horns pointing downwind; and **transverse**, elongate dunes at right angles to the wind direction.

DUNLOP, John Boyd (1840–1921), British inventor who devised the first commercially successful pneumatic TIRE. This was patented in the UK in 1888.

DUNNING, John Ray (1907–1975), US physicist who first measured the ENERGY released in nuclear FISSION (1939) and the next year demonstrated the fission of URANIUM-235. He later helped to develop the gas-diffusion method of ISOTOPE separation used in making the first two ATOMIC BOMBS.

DUODECIMAL SYSTEM, a number system using the POWERS of twelve, which are allotted place values as in the DECIMAL SYSTEM. The number written in decimals 4092 can be expressed as $(2\times 12^3)+(4\times 12^2)+(5\times 12^1)+(0\times 12^0)$ or, in duodecimals, 2450. Fractions are expressed similarly. Two extra symbols are needed for this system to represent the numbers 10 and 11; these are generally accepted as X (dek) and Σ (el) respectively. The advantage of this

Important classes of dyes and pigments

type	example	
azo dyes	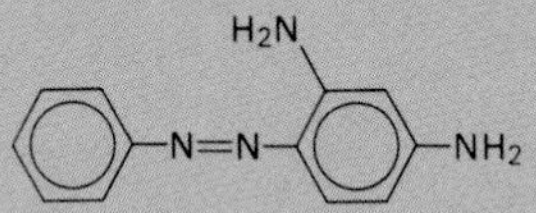	chrysoidine (orange)
anthraquinone dyes	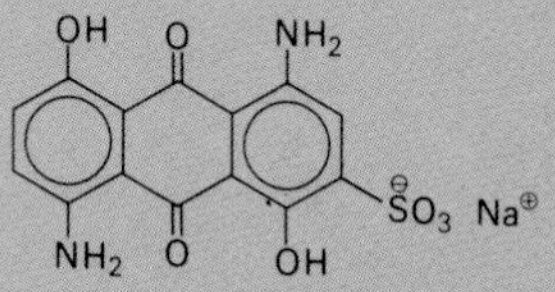	alizarin sapphire (blue)
vat dyes	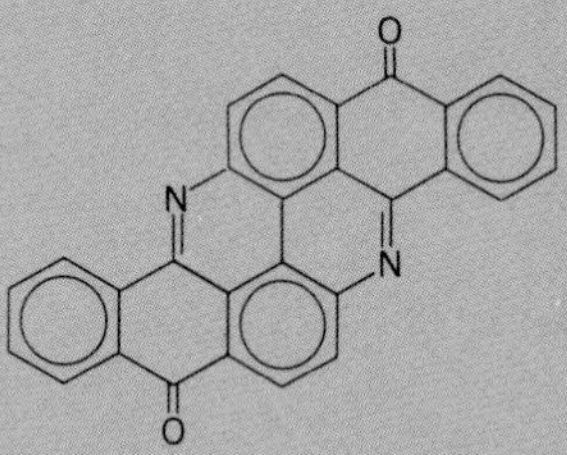	flavanthrone (yellow)
indigoid colors	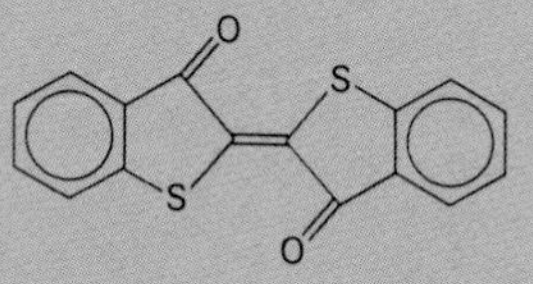	thioindigo (red)
triphenylmethanes	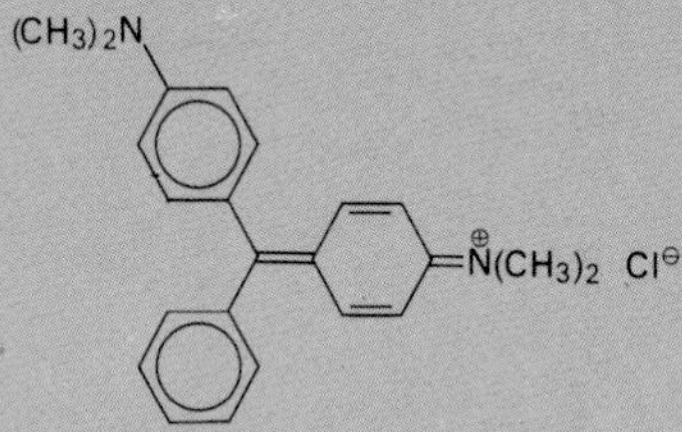	malachite green
phthalocyanines	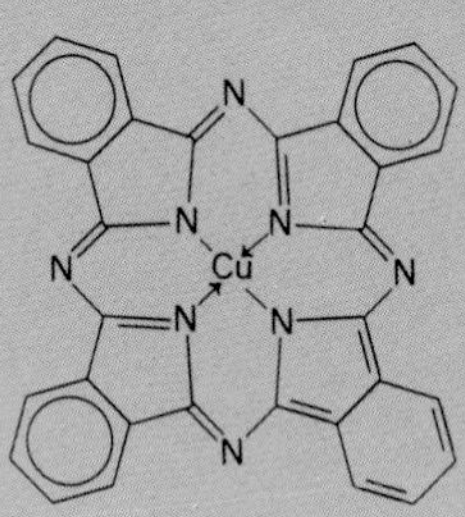	phthalocyanine blue

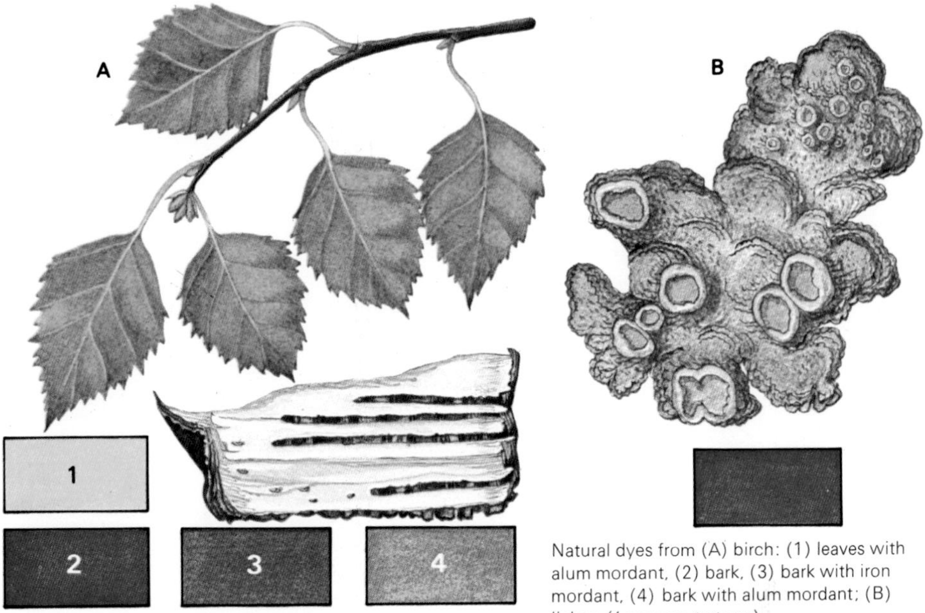

Natural dyes from (A) birch: (1) leaves with alum mordant, (2) bark, (3) bark with iron mordant, (4) bark with alum mordant; (B) lichen (*Lecanora tartarea*).

system can be realized by consideration of the integral (see INTEGERS) FACTORS of 10 and 12: 10 has two (2,5) while 12 has four (2,3,4,6). The most common examples of everyday use of this system are the setting of 12 inches to the foot, 12 months to the year.

DUODENUM, the first part of the small intestine, leading from the STOMACH to the jejunum (see GASTROINTESTINAL TRACT). The BILE and pancreatic ducts end in it and its injury may result in a FISTULA. Peptic ULCERS are common in the duodenum.

DUPLICATING MACHINE, machine to produce copies from an original master. A stencil comprising a porous backing sheet coated on one side with an ink-resisting waxy plastic is clamped round an inked drum: ink seeps through where the plastic has been displaced using a stylus or typewriter. In the **spirit process**, the master sheet bears the impression in a strong dye: the copy paper is moistened with spirit which dissolves some of the dye, so taking the copy. (See also XEROGRAPHY.)

DUPLICATION OF THE CUBE, one of the classical problems of ancient GEOMETRY. Given a CUBE A with edges of length a, the problem is to find the length, b, of the edges of a cube, B, whose VOLUME is twice that of A: i.e., $b^3=2a^3$. This may be expressed as $b^3/a^3=2$, or $b/a=\sqrt[3]{2}$. It is impossible with the tools of classical geometry, the straight edge and compass, to evaluate a cube ROOT (see also IRRATIONAL NUMBERS).

DURALUMIN, aluminum-based ALLOY typically containing 4% copper, 1% magnesium, 0.7% manganese and 0.5% silicon. After heat treatment and aging it is hard and strong as steel, and, being light, is used in aircraft construction.

DURKHEIM, Émile (1858–1917), pioneer French sociologist who advocated the synthesis of empirical research and abstract theory in the social sciences and developed the concepts of "collective consciousness" and the "division of labor."

DURYEA, Charles Edgar (1861–1938), pioneer US AUTOMOBILE manufacturer who with his brother, **J. Frank Duryea** (1870–1967), produced the first successful car in the US.

DUST, fine particles, usually inorganic, which may be easily picked up by the wind and remain suspended in the air for long periods. It may be produced by volcanic action (dust from the Krakatoa explosion of 1883 circled the earth several times, some taking years to settle), by wind EROSION (see DUST BOWL; SANDSTORM), by the breaking up of meteoroids (see METEOR) in the atmosphere, by salt spray from the oceans, and by industrial processes and auto exhausts (see AUTOMOBILE EMISSION CONTROL). POLLEN is an example of an organic dust. Many dusts are serious health risks, especially the radioactive dust present in nuclear FALLOUT. (See also AIR POLLUTION; INTERSTELLAR MATTER.)

DUST BOWL, area of some 400 000km² in the S Great Plains region of the US which, during the 1930s, the Depression years, suffered violent dust storms owing to accelerated SOIL EROSION. Grassland was plowed up in the 1910s and 1920s to plant wheat: a severe drought bared the fields, and high winds blew the topsoil (see SOIL) into huge dunes. (See SANDSTORM.) Despite rehabilitation programs, farmers plowed up grassland again in the 1940s and 1950s, and a repetition of the tragedy was averted only by the action of Congress.

DU VIGNEAUD, Vincent (1901–), US biochemist who was awarded the 1955 Nobel Prize for Chemistry in recognition of his synthesis of the hormone OXYTOCIN (1954). He had earlier worked out the structure of the VITAMIN biotin (1942).

DWARFISM, or small stature. This may be a family characteristic or associated with congenital disease of CARTILAGE or BONE development (e.g., achondroplasia). Failure of growth-HORMONE (see PITUITARY GLAND) or THYROID-hormone production during growth, and excess STEROID, ANDROGEN or

ESTROGEN can cause small stature by altering control of bone development. The condition can also arise from spine or limb deformity (e.g., SCOLIOSIS), MALNUTRITION, RICKETS, chronic infection or visceral disease.

DWARF STAR. See STAR.

DYES AND DYEING. Dyes are colored substances which impart their color to textiles to which they are applied and for which they have a chemical affinity. They differ from PIGMENTS in being used in solution in an aqueous medium. Dyeing was practiced in the Fertile Crescent and China by 3000 BC, using natural dyes obtained from plants and shellfish. These were virtually superseded by synthetic dyes—more varied in color and applicability—after the accidental synthesis of mauve by PERKIN (1856). The raw materials are AROMATIC hydrocarbons obtained from COAL TAR and PETROLEUM. These are modified by introducing chemical groups called chromophores which cause absorption of visible LIGHT (see also COLOR). Other groups, auxochromes, such as amino or hydroxyl, are necessary for substantivity—i.e., affinity for the material to be dyed. This fixing to the fabric fibers is by HYDROGEN BONDING, ADSORPTION, ionic bonding or covalent bonding in the case of "reactive dyes" (see BOND, CHEMICAL). If there is no natural affinity, the dye may be fixed by using a MORDANT before or with dyeing. Vat dyes are made soluble by reduction in the presence of alkali, and after dyeing the original color is re-formed by acidification and oxidation; INDIGO and ANTHRAQUINONE dyes are examples. Dyes are also used as biological stains (see MICROSCOPE), INDICATORS and in PHOTOGRAPHY. (See also AZO COMPOUNDS; FLUORESCEIN; INK.)

DYKE. See DIKE.

DYNAMICS, the branch of mechanics concerned with the actions of forces on bodies, with particular respect to the motions produced. (See MECHANICS.)

DYNAMITE, high EXPLOSIVE invented by Alfred NOBEL, consisting of NITROGLYCERIN absorbed in an inert material such as KIESELGUHR or wood pulp. Unlike nitroglycerin itself, it can be handled safely, not exploding without a DETONATOR. In modern dynamite SODIUM nitrate replaces about half the nitroglycerin. Gelatin dynamite, or **gelignite**, contains also some NITROCELLULOSE.

DYNAMO. See GENERATOR, ELECTRIC.

DYNAMOMETER, any of various devices used to measure the POWER output of a MACHINE.

DYNE (dyn), unit of FORCE in the CGS UNITS system, equal to 0.00001 newtons in SI UNITS.

DYSENTERY, a BACTERIAL or PARASITIC DISEASE causing abdominal pain, DIARRHEA and FEVER. In children, **bacillary dysentery** due to *Shigella* species is a common endemic or EPIDEMIC disease, and is associated with poor hygiene. It is a short-lived illness but may cause dehydration in severe cases. The organism may be carried in feces in the absence of symptoms. ANTIBIOTICS may be used to shorten the attack and reduce carrier rates. **Amebic dysentery** is a chronic disease, usually seen in warm climates, with episodes of diarrhea and CONSTIPATION, accompanied by MUCUS and occasionally BLOOD; constitutional symptoms occur and the disease may resemble noninfective COLITIS. Treatment with emetine, while effective, is accompanied by a high risk of toxicity; metronidazole is a less toxic antiamebic agent introduced recently.

DYSLEXIA, difficulty with reading, often a developmental problem possibly associated with suppressed left-HANDEDNESS, and spatial difficulty; it requires special training. It may be acquired by BIRTH injury, failure of learning, visual disorders or as part of APHASIA (see SPEECH AND SPEECH DISORDERS).

DYSPEPSIA, or indigestion, a vague term usually describing abnormal visceral sensation in upper ABDOMEN or lower CHEST, often of a burning quality. Relationship to meals and posture is important in defining its origin; relief by ANTACIDS or milk is usual. HEARTBURN from esophagitis and pain of peptic (gastric or duodenal) ULCERS are usual causes.

DYSPHASIA. See APHASIA; SPEECH AND SPEECH DISORDERS.

DYSPROSIUM (Dy), one of the LANTHANUM SERIES. AW 162.5, mp 1409°C, bp 2335°C, sg 8.550 (25°C).

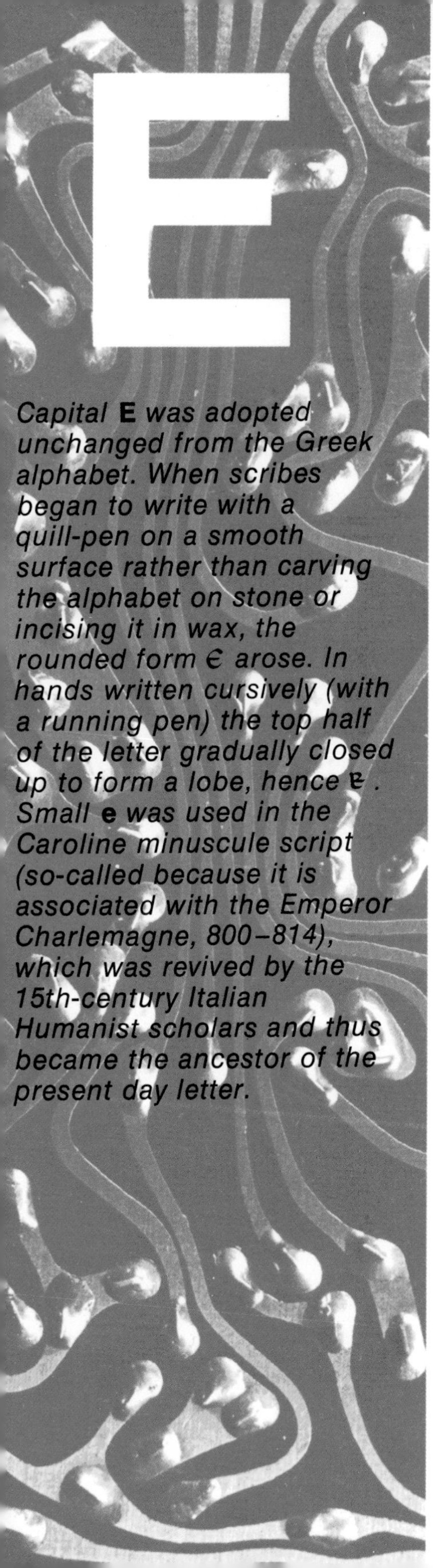

E

Capital **E** *was adopted unchanged from the Greek alphabet. When scribes began to write with a quill-pen on a smooth surface rather than carving the alphabet on stone or incising it in wax, the rounded form* Є *arose. In hands written cursively (with a running pen) the top half of the letter gradually closed up to form a lobe, hence* e *. Small* **e** *was used in the Caroline minuscule script (so-called because it is associated with the Emperor Charlemagne, 800–814), which was revived by the 15th-century Italian Humanist scholars and thus became the ancestor of the present day letter.*

e. See EXPONENTIAL.

EAR, a special sense organ in higher animals, concerned with hearing and balance. It may be divided into the outer ear, extending from the tympanic membrane or ear drum to the pinna, the inner ear embedded in the SKULL bones, consisting of cochlea and labyrinth, and between them the middle ear, containing small bones or ossicles. The cartilaginous pinna varies greatly in shape and mobility in different animals; a canal lined by skin leads from it and ends with the thin tympanic membrane stretched across it. The middle ear is an air-filled space which communicates with the PHARYNX via the **Eustachian tube**. This allows the middle ear to be at the same pressure as the outer and also secretions to drain away. The middle ear is also connected with the MASTOID antrum. Three ossicles (malleus, incus, stapes) form a bony chain articulating between the ear drum and part of the cochlea; tiny muscles are attached to the drum and ossicles and can affect the intensity of SOUND transmission. The inner ear contains both the cochlea, a spiral structure containing fluid and specialized membranes on which hearing receptors are situated, and the labyrinth which consists of three semicircular canals, the utricle and the saccule, all of which contain fluid and receptor cells. Nerve fibers pass from the cochlea and labyrinth to form the eighth cranial nerve.

In **hearing**, sound waves travel into the outer ear, funneled by the pinna, and cause vibration of the ear drum. The drum and ossicular chain, which transmits vibration to the cochlea, effect some amplification. The vibration set up in the cochlear fluid is differentially distributed along the central membrane according to pitch. By a complex mechanism, this membrane movement causes certain groups of receptor cells to be preferentially stimulated, giving rise to auditory nerve impulses, which are conducted via several coding sites to higher centers for perception. These centers can in turn affect the sensitivity of receptors by means of centrifugal fibers. In **balance**, rotation of the head in any of three perpendicular planes causes stimulation of specialized cells in the semicircular canals as fluid moves past them. The utricle and saccule contain small stones which respond to gravitational changes and affect receptor cells in their walls. All these balance receptor cells cause impulses in the vestibular nerve, and this connects to higher centers.

Disease of the ear usually causes DEAFNESS or ringing in the ears. Peripheral disorders of balance include VERTIGO and ATAXIA, which may be accompanied by nausea or VOMITING. MÉNIÈRE'S DISEASE is an episodic disease affecting both systems.

EARTH, the largest of the inner planets of the solar system, the third planet from the sun and, so far as is known, the sole home of life in the solar system. To an astronomer on Mars, several things would be striking about our planet. Most of all, he would notice the relative size of our MOON: there are larger moons in the SOLAR SYSTEM, but none so large compared with its planet—indeed, some astronomers regard the earth as one component of a "double planet," the other being the moon. Our Martian astronomer would also notice that the earth shows phases, just as the moon and Venus do when viewed from earth. And, if he were a

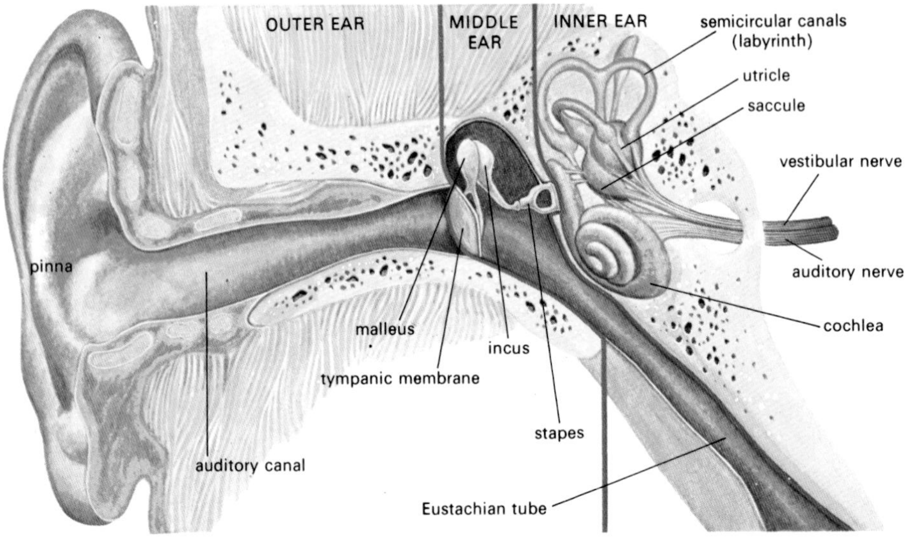

Section through the human ear, showing the principal anatomical divisions.

radio astronomer, he would detect a barrage of radio "noise" from our planet—clear evidence of the presence of intelligent life.

The earth is rather larger than VENUS. It is slightly oblate (flattened at the poles), the equatorial diameter being about 12 756.4km, the polar diameter about 12 713.6km. It rotates on its axis in 23h 56min 4.09s (one **sidereal day**), though this is increasing by roughly 0.00001s annually due to tidal effects (see TIDES); and revolves about the sun in 365d 6h 9min 9.5s (one **sidereal year**: see SIDEREAL TIME). Two other types of year are defined: the **tropical year**, the interval between alternate EQUINOXES (365d 5h 48min 46s); and the **anomalistic year**, the interval between moments of perihelion (see ORBIT), 365d 6h 13min 53s. The earth's equator is angled about 23.5° to the ECLIPTIC, the plane of its orbit. The direction of the earth's axis is slowly changing owing to

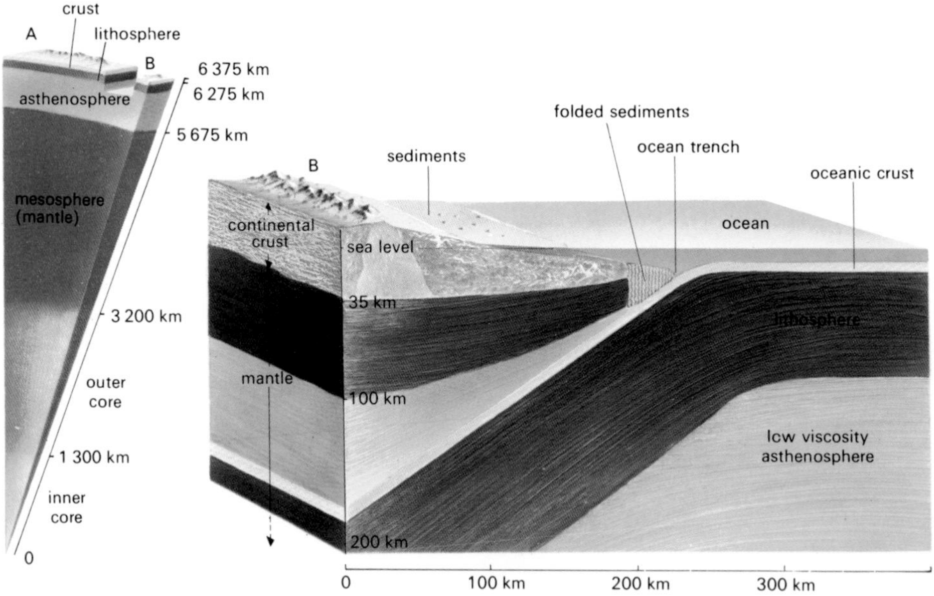

The structure of the earth. (A) A solid section of the earth from the center to the surface. (B) A 200km-deep section through the crust and upper mantle showing a subduction zone where a plate of "continental" crust is advancing over an oceanic plate. A tongue of oceanic crust and lithosphere plunges down into the asthenosphere, remaining identifiable down to depths of about 700km.

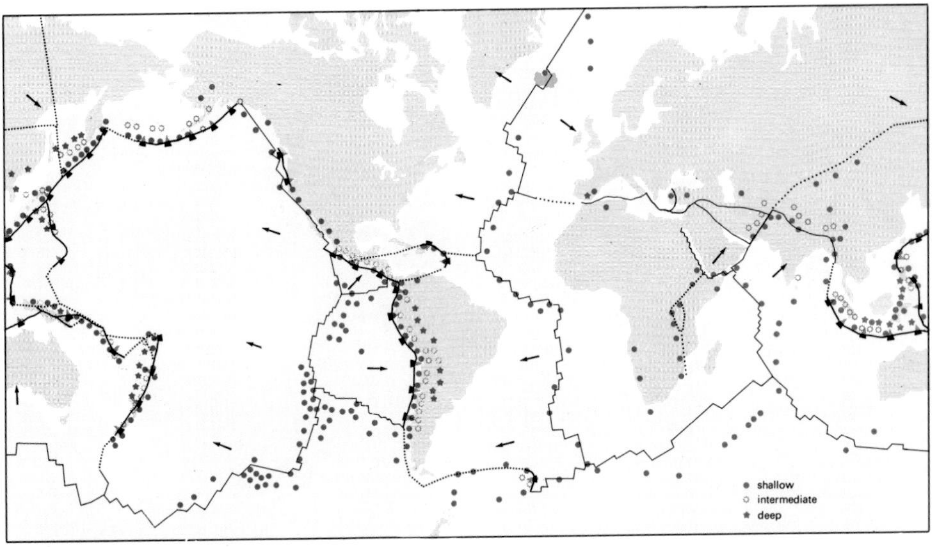

The relation between earthquake foci and plate boundaries.

PRECESSION. The planet has a mass of about 5.98×10^{21} tonnes, a volume of about 1.08×10^{21} m³, and a mean DENSITY of about 5.52 tonnes/m³.

Like other planetary bodies, the earth has a magnetic field (see MAGNETISM). The magnetic poles do not coincide with the axial poles (see NORTH POLE; SOUTH POLE), and moreover they "wander." At or near the earth's surface, **magnetic declination** (or **variation**) is the angle between true N and compass N (lines joining points of equal variation are **isogonic lines**); and **magnetic dip** (or **inclination**) the vertical angle between the MAGNETIC FIELD and the horizontal at a particular point. **Isomagnetic lines** can be drawn between points of equal intensity of the field. There is also evidence to suggest that the direction of the field reverses from time to time. These changes are of primary interest to the paleomagnetist (see PALEOMAGNETISM). The earth is surrounded by radiation belts, probably the result of charged particles from the sun being trapped by the earth's magnetic field (see VAN ALLEN RADIATION BELTS; AURORA).

There are three main zones of the earth: the ATMOSPHERE; the HYDROSPHERE (the world's waters); and the LITHOSPHERE, the solid body of the world. The atmosphere shields us from much of the harmful radiation of the SUN, and protects us from excesses of heat and cold. Water covers much of the earth's surface (over 70%) in both liquid and solid (ice) forms (see GLACIER; OCEANS). There are permanent polar icecaps. The earth's solid body can be divided into three regions: The **core** (diameter about 7 000km), at a temperature of about 3000K, is at least partly liquid, though the central region (the inner core) is probably solid. Probably mainly of NICKEL and IRON, the core's density ranges between about 9.5 and perhaps over 15 tonnes/m³. The **mantle** (outer diameter about 12 686km), probably mainly of OLIVINE, has a density around 5.7 tonnes/m³ toward the core, 3.3 tonnes/m³ toward the **crust**, the outermost layer of the earth and the one to which all human activity is confined. It is some 35km thick (much less beneath the oceans) and composed of three types of ROCKS: IGNEOUS ROCKS, SEDIMENTARY ROCKS and METAMORPHIC ROCKS. FOSSILS in the strata of sedimentary rocks give us a geological time scale (see GEOLOGY). The earth formed about 4550 million years ago; life appeared probably little more than 570 million years ago, and man around 4 million years ago. Life has thus been present for about 12.5% of the earth's history, man for about 0.09%, and civilization for less than 0.000 1%.

It is now known that the earth's configuration of continents and oceans has changed radically through geological time—as it were, the map has changed. Originally, this was attributed to continents drifting, and the process was called CONTINENTAL DRIFT (see also Alfred WEGENER). However, although this term is still used descriptively, the changes are now realized to be a manifestation of the theory of PLATE TECTONICS, and so a result of the processes responsible also for EARTHQUAKES, MOUNTAIN building and many other phenomena.

EARTHENWARE. See POTTERY AND PORCELAIN.

EARTHQUAKE, a fracture or implosion beneath the surface of the earth, and the shock waves that travel away from the point where the fracture has occurred. The immediate area where the fracture takes place is the **focus** or **hypocenter**, the point immediately above it on the earth's surface is the **epicenter**, and the shock waves emanating from the fracture are called seismic waves.

Earthquakes occur to relieve a stress that has built up within the crust or mantle of the EARTH; fracture results when the stress exceeds the strength of the rock. The reasons for the stress build-up are to be found in the theory of PLATE TECTONICS. If a map is drawn of the world's earthquake activity, it can be immediately seen that earthquakes are confined to discrete belts. These belts signify the borders of contiguous plates; shallow earthquakes being generally associated with MID-OCEAN RIDGES where creation of new material occurs, deep ones with regions where one plate is being forced under another.

Seismic waves are of two main types. Body waves travel from the hypocenter, and again are of two types: P (compressional) waves, where the motion of particles of the earth is in the direction of propagation of the wave; and S (shear) waves, where the particle motion is at right angles to this direction. Surface waves travel from the epicenter, and are largely confined to the earth's surface; Love waves are at right angles to the direction of propagation; Rayleigh waves having a more complicated, backward elliptical movement in the direction of propagation.

The experienced intensity of an earthquake depends mainly on the distance from the source. Local intensities are gauged in terms of the Mercalli Intensity Scale, which runs from I (detectable only by SEISMOGRAPH) through to XII ("Catastrophic"). Comparison of intensities in different areas enables the source of an earthquake to be located. The actual magnitude of the event is gauged according to the RICHTER SCALE.

The study of seismic phenomena is known as **seismology**. (See also FAULT; TSUNAMI.)

EASTMAN, George (1854–1932), US inventor and manufacturer who invented the Kodak CAMERA, first marketed 1888. Earlier he perfected processes for manufacturing dry photographic plates (1880) and flexible, transparent film (1884). He took his own life in 1932. (See also PHOTOGRAPHY.)

EBBINGHAUS, Hermann (1850–1909), German psychologist who developed experimental techniques for the study of rote LEARNING and memory. In later life he devised means of intelligence testing and researched into color VISION.

ECCENTRICITY, in Geometry. See CONIC SECTIONS.

ECCLES, Sir John Carew (1903–), Australian physiologist who shared the 1963 Nobel Prize for Physiology or Medicine with Alan HODGKIN and Andrew HUXLEY. Using their findings, he had been able to establish the chemical bases of the electrical changes during transmission of nervous impulses across the SYNAPSES (see also NERVOUS SYSTEM).

ECDYSIS. See MOLTING.

ECHINODERMATA, a phylum of about 5000 marine animals which includes the CRINOIDEA (sea lilies and feather stars), ASTEROIDEA (starfishes), OPHUROIDEA (brittlestars), ECHINOIDEA (sea urchins, heart urchins and sand dollars) and the HOLOTHUROIDEA (sea cucumbers). All echinoderms have a skeleton composed of plates of CALCITE, radial (usually five-fold) symmetry and a water vascular system. Fossil echinoderms include crinoids and blastoids and have a history extending back to the Ordovician period, which began 500 million years ago.

ECHINOIDEA, a class of ECHINODERMATA which includes the sea urchins, heart urchins and sand dollars. They are spherical or heart-shaped and do not possess arms.

ECHIUROIDEA, a phylum of ovoid or bulbous, worm-like marine animals of uncertain affinities, but probably related to the members of the phylum ANNELIDA. They differ from annelids by having a single pair of chaetae and by being unsegmented when adult. Echiuroids inhabit rock crevices or U-shaped burrows excavated in mud or sand.

ECHO, a wave signal reflected back to its point of origin from a distant object, or, in the case of RADIO signals, a signal coming to a receiver from the transmitter by an indirect route. Echoes of the first type can be used to detect and find the position of reflecting objects (echolocation). High-frequency SOUND echolocation is used both by bats for navigation and to detect prey and by man in marine SONAR. RADAR, too, is similar in principle, though this uses UHF radio and MICROWAVE radiation rather than sound energy. The range of a reflecting object can easily be estimated for ordinary sound echoes: since sound travels about 340m/s through the air at sea level, an object will be distant about 170m for each second that passes before an echo returns from it.

During a lunar eclipse the moon passes through the earth's solar shadow. The eclipse is partial while the moon falls only within the region of partial shadow (penumbra—as at 2 and 6) and total while the moon is in full shadow (umbra—as at 4). In the much rarer solar eclipse the moon casts a shadow on the earth (7). Because of the relative distances and dimensions of the earth, sun and moon, the area, if any of the earth's surface experiencing total eclipse is very small.

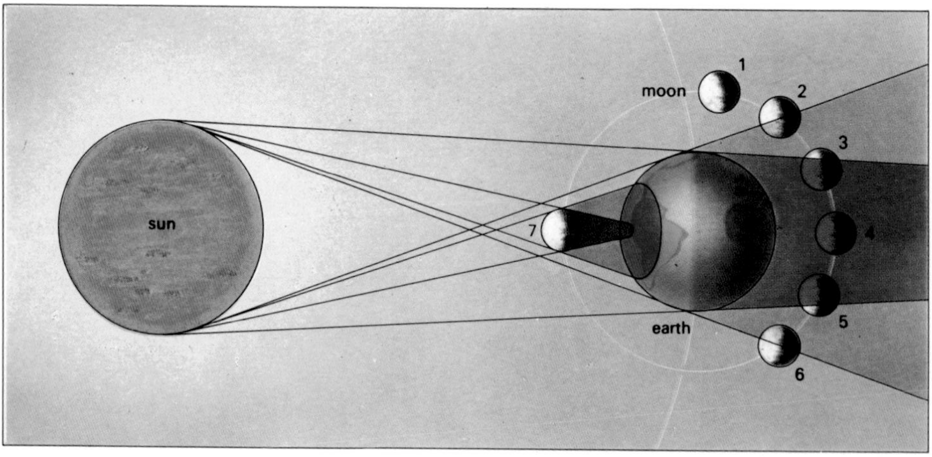

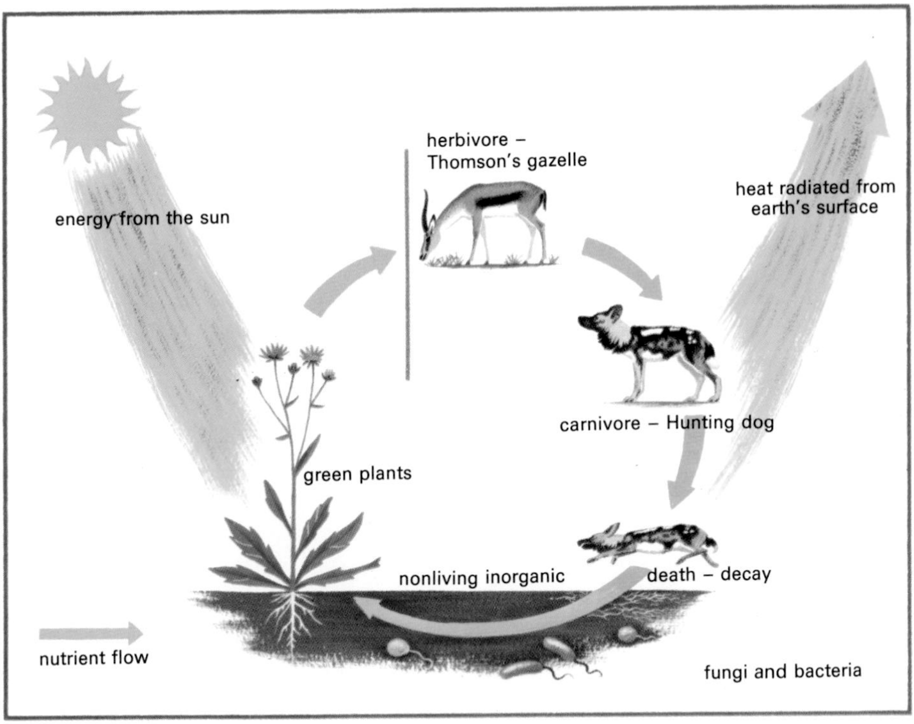

The components of an ecosystem.

ECHO SOUNDER, device for determining the depth of water under a ship's keel from the time taken for SOUND pulses beamed vertically downward from the ship to be reflected from the sea-bed; an application of SONAR.

ECLIPSE, the partial or total obscurement of one celestial body by another; also the passage of the moon through the earth's shadow. The components of a binary star (see DOUBLE STAR) may eclipse each other as seen from the earth, in which case the star is termed an eclipsing binary. The moon frequently eclipses stars or planets, and this is known as OCCULTATION.

A **lunar eclipse** occurs when the moon passes through the umbra of the earth's SHADOW. This happens usually not more than twice a year, since the moon's orbit around the earth is tilted with respect to the ECLIPTIC. The eclipsed moon is blood-red in color due to some of the sun's light being refracted by the earth's atmosphere into the umbra. A partial lunar eclipse occurs when only part of the umbra falls on the moon.

In a **solar eclipse**, the moon passes between the sun and the earth. A total eclipse occurs when the observer is within the umbra of the moon's shadow: the disk of the sun is covered by that of the moon, and the solar corona (see SUN) becomes clearly visible. Total eclipses are particularly important since only during them can astronomers study the solar corona and prominences. The maximum possible duration of a total eclipse is about 7½min. Should the observer be outside the umbra but within the penumbra, or should the earth pass through only the penumbra, a partial eclipse will occur.

An **annular eclipse** is seen when the moon is at its farthest from the earth, its disk being not large enough to totally obscure that of the sun. The moon's disk is seen surrounded by a brilliant ring of light.

ECLIPTIC, the great circle traced out on the CELESTIAL SPHERE by the apparent motion of the sun during the year, corresponding to the motion of the earth around the sun. The ecliptic passes through 12 constellations, known as the constellations of the ZODIAC.

ECOLOGY, the study of plants and animals in relation to their ENVIRONMENT. The whole earth can be considered as a large ecological unit: the term BIOSPHERE is used to describe the atmosphere, earth, surface, oceans and ocean floors within which living organisms exist. However, it is usual to divide the biosphere into a large number of ecological sub-units or **ecosystems**, within each of which the organisms making up the living community are in balance with the environment. Typical examples of ecosystems are a pond, a deciduous forest or a desert. The overall climate and topography within an area are major factors determining the type of ecosystem that develops, but within any ecosystem minor variations give rise to smaller communities within which animals and plants occupy their own particular niches. Within any ecosystem each organism, however large or small, plays a vital role in maintaining the stability of the community.

The most important factor for any organism is its source of energy or food. Thus, within any ecosystem, complex patterns of feeding relationships or **food chains** are built up. Plants are the primary source of food and energy; they derive it through PHOTOSYNTHESIS, utilizing environmental factors such as light, water, carbon dioxide and minerals. Herbivores then obtain their food by eating plants. In their turn, herbivores are preyed upon by carnivores, who may also be the source of food for other carnivores. Animal and plant waste is decomposed by microorganisms (BACTERIA, FUNGI) within the habitat and this returns the raw materials to the environment. The number of links within a food chain are normally three or four, with five, six and seven less frequently. The main reason for the limited length of food chains is that the major part of the energy stored within a plant or animal is wasted at each stage in the chain. Thus if it were possible for a carnivore to occupy, say, position 20 within a food chain, the area of vegetation required to supply the energy needed for the complete chain would be the size of a continent.

The plants within an ecosystem, as well as the major environmental features, help create habitats suitable for other organisms. Thus, in a forest ecosystem, the humid, dimly illuminated environment covered by a thick canopy is suitable for mosses, lichens and ferns and their associated fauna. Within any ecosystem the raw materials nitrogen, carbon, oxygen and hydrogen (in water) are continually being recycled via a number of processes including the NITROGEN CYCLE, CARBON CYCLE and photosynthesis.

Most natural ecosystems are in a state of equilibrium or balance so that few changes occur in the natural flora and fauna. However, when changes occur in the environment, either major climatic changes or minor alterations in the inhabitants, an imbalance results and the ecosystem changes to adapt to the new situation. The sequence of change that leads to a new period of equilibrium is called a succession and may take any length of time from a few years, for the establishment of a new species, to several centuries, for the change from grassland to forest.

Over millions of years, nature has moved toward the overall creation of stable ecosystems. Natural changes, such as adaptations to the slow change of climate, tend to be gradual. However, man often causes much more sudden changes—the introduction of a disease to a hitherto uninfected area, the cutting down of a forest or the polluting of a river. The effects of this type of change upon an ecosystem can be rapid and irreversible. Up to now these changes have not been too serious on a worldwide scale but there is an increasing awareness of what could happen if a worldwide disturbance in the biosphere occurred. The forms of life as they are known today depend entirely upon the sensitive balance within the environment and any change with worldwide effects could have devastating consequences for man and life in general.

ECOSYSTEM. See ECOLOGY.

ECTOPLASM, the outer layer of cellular CYTOPLASM (see also CELL). In PARAPSYCHOLOGY, a substance thought to be emitted by mediums when in trance.

ECZEMA, form of DERMATITIS, usually with redness and scaling. It is often familial, being worst in childhood, and is associated with HAY FEVER and ASTHMA.

EDDINGTON, Sir Arthur Stanley (1882–1944), English astronomer and astrophysicist who pioneered the theoretical study of the interior of STARS and who, through his *Mathematical Theory of Relativity* (1923), did much to introduce the English-speaking world to the theories of EINSTEIN.

EDELMAN, Gerald Maurice (1929–), US biochemist who shared with Rodney PORTER the 1972 Nobel Prize for Physiology or Medicine for his researches into the chemical structures of antibodies (see ANTIBODIES AND ANTIGENS).

EDEMA, the accumulation of excessive watery fluid outside the cells of the body, causing swelling of a part. Some edema is seen locally in INFLAMMATION. The commonest type is gravitational edema (**dropsy**), where fluid swelling is in the most dependent parts, typically the feet. HEART or LIVER failure, MALNUTRITION and nephrotic syndrome of the KIDNEY are common causes, while disease of VEINS or LYMPH vessels in the legs also leads to edema. Serious edema may form in the LUNGS in heart failure and in the BRAIN in some disorders of METABOLISM, trauma, TUMORS and infections. DIURETICS may be needed in treatment.

EDENTATA, an order of the MAMMALIA including only three living families: the anteaters, armadillos and sloths. The anteaters are completely toothless (edentate) but the sloths and armadillos have primitive molars. Extinct edentates include the Giant ground sloth, *Megatherium*.

EDGERTON, Harold Eugene (1903–), US electrical engineer known for his development of rapid-flash STROBOSCOPES and application of them to high-speed PHOTOGRAPHY.

EDISON, Thomas Alva (1847–1931), US inventor, probably the greatest of all time with over 1 000 patents issued to his name. His first successful invention, an improved stock-ticker (1869), earned him the capital to set up as a manufacturer of telegraphic apparatus. He then devised the diplex method of TELEGRAPHY which allowed one wire to carry four messages at once. Moving to a new "invention factory" (the first large-scale industrial-research laboratory) at Menlo Park, N.J., in 1876, he devised the carbon transmitter and a new receiver which made A. G. BELL's TELEPHONE commercially practical. His tin-foil PHONOGRAPH followed in 1877 and in the next year he started to work toward devising a practical incandescent lightbulb. By 1879 he had produced the carbon-filament bulb and electric LIGHTING became a reality, though it was not until 1882 that his first public generating station was supplying power to 85 customers in New York.

Moving his laboratories to West Orange, N.J., in 1887 he set about devising a motion-picture system (ready by 1889) though he failed to exploit its entertainment potential. In all his career he made only one important scientific discovery, the **Edison effect**—the ability of ELECTRICITY to flow from a hot filament in a vacuum lamp to another enclosed wire but not the reverse (1883)—and, because he saw no use for it, he failed to pursue the matter. His success was probably more due to perseverance than any special insight; as he himself said: "Genius is one percent inspiration and ninety-nine percent perspiration."

EFFICIENCY, in THERMODYNAMICS and the theory of MACHINES, the ratio of the useful WORK derived from a machine to the ENERGY put into it. The mechanical efficiency of a machine is always less than 100%, some energy being lost as HEAT in FRICTION. When the machine is a heat engine, its theoretical thermal efficiency can be found from the second law of thermodynamics but actual values are often rather lower. A typical gasoline engine may have a thermal efficiency of only 25%, a STEAM ENGINE 10%.

EFFLORESCENCE, in chemistry, spontaneous loss of water from a crystalline HYDRATE, which crumbles on its surface to an anhydrous powder. SODIUM carbonate and sulfate are common examples. Like its converse, DELIQUESCENCE, it depends on the relative HUMIDITY, occurring if the partial VAPOR PRESSURE of water at the solid exceeds that of the air.

EGG, the female GAMETE, germ cell or OVUM found in all animals and in most plants. Popularly, the term is used to describe those animal eggs that are deposited by the female either before or after fertilization and develop outside the body, such as the eggs of reptiles and birds. The egg is a single cell which develops into the EMBRYO after FERTILIZATION by a single sperm cell or male gamete. In animals, it is formed in a primary sex organ or GONAD called the ovary. In fishes, reptiles and birds there is a food store of yolk enclosed within its outer membrane. In ANGIOSPERMS, the female reproductive organs form part of the FLOWER. The egg cell is found within the ovules, which upon fertilization develop into the embryo and SEED. (See also POLLINATION, REPRODUCTION.)

EGO, the structured part of the individual's psychic makeup, developing, according to FREUD, from the ID through experience. Closely related in concept to the CONSCIOUS, it can be viewed as the objective equivalent of IDENTITY. (See also SUPEREGO.)

EHRLICH, Paul (1854–1915), German bacteriologist and immunologist, the founder of CHEMOTHERAPY and an early pioneer of HEMATOLOGY. His discoveries include: a method of staining (1882), and hence identifying, the TUBERCULOSIS bacillus (see also Robert KOCH); the reasons for immunity in terms of the chemistry of ANTIBODIES AND ANTIGENS, for which he was awarded (with METCHNIKOFF) the 1908 Nobel Prize for Physiology or Medicine; and the use of the drug SALVARSAN to cure syphilis (see VENEREAL DISEASES), the first DRUG to be used in treating the root cause of a disease (1911).

EHRLICH, Paul Ralph (1932–), US biologist and ecologist, author of *The Population Bomb* (1968).

EIDETIC IMAGE, an exceptionally vivid mental image which the individual "sees" either projected onto a suitable background (e.g., a wall) or with eyes closed in a darkened room. The image may be a MEMORY (eidetic memory) or a fantasy, and may be either voluntary or spontaneous. Eidetic imagery is most common among children.

EIFFEL, Alexandre Gustave (1832–1923), French engineer best known for his design and construction of the Eiffel Tower, Paris (1887–89), from which he carried out experiments in AERODYNAMICS. In 1912 he founded the first aerodynamics laboratory.

EIGEN, Manfred (1927–), German physicist awarded, with Ronald NORRISH and George PORTER, the 1967 Nobel Prize for Chemistry for studies of extremely fast chemical reactions.

EIJKMANN, Christiaan (1858–1930), Dutch pathologist. Following a trip to Indonesia (1886) to investigate BERIBERI he was able to show that the disease resulted from a dietary deficiency. This led to the discovery of VITAMINS. For his work he was awarded (with Sir F. G. HOPKINS) the 1929 Nobel Prize for Physiology or Medicine.

EINSTEIN, Albert (1879–1955), German-born Swiss-American theoretical physicist, the author of the theory of RELATIVITY. In 1905 Einstein published several papers of major significance. In one he applied PLANCK'S QUANTUM THEORY to the explanation of photoelectric emission. For this he was awarded the 1921 Nobel Prize for Physics. In a second he demonstrated that it was indeed molecular action which was responsible for BROWNIAN MOTION. In a third he published the special theory of relativity with its postulate of a constant VELOCITY for LIGHT (c) and its consequence, the equivalence of MASS (m) and ENERGY (E), summed up in the famous equation $E=mc^2$. In 1915 he went on to publish the general theory of relativity. This came with various testable predictions, all of which were spectacularly confirmed within a few years. Einstein was on a visit to the US when Hitler came to power in Germany and, being a Jew, decided not to return to his native land. The rest of his life was spent in a fruitless search for a "unified field theory" which could combine QUANTUM MECHANICS with GRAVITATION theory. After 1945 he also worked hard against the proliferation of nuclear weapons, although he had himself, in 1939, alerted President F. D. Roosevelt to the danger that Germany might develop an ATOMIC BOMB, and had thus contributed to the setting up of the Manhattan Project.

EINSTEINIUM (Es), a TRANSURANIUM ELEMENT in the ACTINIDE series, first found in the debris from the first HYDROGEN BOMB, and now prepared by bombardment of lighter actinides.

EINTHOVEN, Willem (1860–1927), Dutch physiologist awarded the 1924 Nobel Prize for Physiology or Medicine for his invention of, and investigation of heart action with, the ELECTROCARDIOGRAPH. In 1903 he devised the **string galvanometer**, a single fine wire placed under tension in a MAGNETIC FIELD. Current passed through the wire causes a deflection which can be measured, for greater accuracy, by microscope. This GALVANOMETER was sensitive enough for him to use it to record the electrical activity of the HEART.

ELASMOBRANCHII. See CHONDRICHTHYES.

ELASTICITY, the ability of a body to resist tension, torsion, shearing or compression and to recover its original shape and size when the stress is removed. All substances are elastic to some extent, but if the stress exceeds a certain value (the elastic limit), which is soon reached for brittle and plastic materials, permanent deformation occurs. Below the elastic limit, bodies obey Hooke's Law (see MATERIALS, STRENGTH OF).

ELASTOMER, any POLYMER suitable for use as a synthetic RUBBER.

ELECTRICAL ENGINEERING, branch of technology dealing with the practical applications of ELECTRICITY and ELECTRONICS, and thus concerned with generation of electric POWER, design and construction of electrical and electronic components, and the use of these components in integrated, functional systems. **Power engineers** are important in modern industry as electric MOTORS are an integral part of most factory work, and electric LIGHTING and AIR CONDITIONING play a vital role in maintaining good working conditions. **Communications engineers** deal with construction of TELEVISIONS, TELEPHONES and other electronic equipment: perhaps their greatest success is the COMPUTER.

ELECTRIC ARC, a high-current electric discharge between two ELECTRODES. The current is carried by the gas PLASMA maintained by the discharge. Sodium and neon lights exemplify large relatively cool arcs, while arc-welding uses a small very hot arc between a slowly consumed electrode and the workpiece. LIGHTNING is a naturally occurring arc. (See also ARC LAMP.)

ELECTRIC-ARC PROCESS, process for NITROGEN FIXATION, now largely uneconomical. Air (see ATMOSPHERE) is blown through an ELECTRIC ARC at 1000°C, and nitric oxide (see NITROGEN) is formed. It is converted into NITRIC ACID as in the OSTWALD PROCESS.

ELECTRIC BELL. See BELL, ELECTRIC.

ELECTRIC CAR, an automobile driven by electric MOTORS and (usually) using storage BATTERIES as the ENERGY source. Although an electrically-powered carriage was built as long ago as 1837, it was only in the 1890s that electric cars became common. After WWI they lost ground to AUTOMOBILES with INTERNAL-COMBUSTION ENGINES although, particularly in Europe, electric traction has remained popular for urban delivery vehicles. With increasing concern being felt at the energy- and pollution-costs of the gasoline automobile, renewed interest is being shown in the electric car in spite of its relatively short range between charges. It is pollution-free, robust and simple to drive and maintain. The only difficulty is its low power-to-weight ratio, largely due to the weight of the lead-acid storage batteries commonly used. Much research is being put into finding alternative, lighter battery systems or powerful-enough FUEL CELLS to make electric cars once again an attractive proposition for urban transportation.

ELECTRIC CURRENT. See ELECTRICITY.

ELECTRIC EYE, popular name for a PHOTOELECTRIC CELL.

ELECTRIC FIELD, what is said to exist where stationary electric charges (see ELECTRICITY) experience a force: the field is defined with the strength and direction of the force on a unit charge. Electric fields are produced by (other) electric charges, and by changing magnetic fields (see ELECTROMAGNETISM).

Electric car developed for city use by the UK's Leyland Cars.

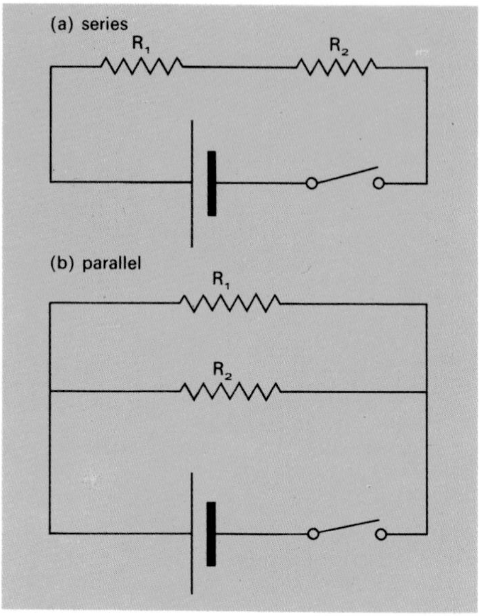

Electricity: series and parallel circuits. When resistances are connected in series (a), the overall resistance (R_s) is the sum of the individual ones:

$$R_s = R_1 + R_2$$

When resistances are connected in parallel (b), the overall resistance (R_p) is given by:

$$\frac{1}{R_p} = \frac{1}{R_1} + \frac{1}{R_2}$$

ELECTRIC FURNACE, any of various industrial furnaces heated electrically. In an **arc furnace**, batches of steel are melted by the heat of ELECTRIC ARCS struck between GRAPHITE ELECTRODES and the charge. In **induction furnaces**, used mainly for remelting batches of steel, a refractory crucible is surrounded by a large water-cooled copper coil. When this is connected to a high-frequency AC supply, eddy currents are induced in the batch which heat it. Often a lower-frequency field is also applied to help stir the charge. **Chamber furnaces** are simply large electric ovens, heated from the walls.

ELECTRIC GENERATOR. See GENERATOR, ELECTRIC.

ELECTRICITY, the phenomena of charged particles at rest and in motion. Electricity provides man with a highly versatile form of ENERGY, electrical devices being used in heating, LIGHTING, machinery, telephony and ELECTRONICS. **Electric charge** is an inherent property of matter, ELECTRONS carrying a negative charge of 1.602×10^{-19} coulomb and atomic nuclei normally carrying a similar positive charge for each electron in the ATOM. When the balance is disturbed (in the case of glass, by rubbing it, for example), a net charge is left on an object, and the study of such isolated charges is called

electrostatics. Like charges repel and unlike charges attract each other with a FORCE proportional to the two charges and inversely proportional to the square of their separation (the inverse square law): electrostatic repulsion, for example, will be familiar in newly combed hair. The force is normally interpreted in terms of an ELECTRIC FIELD produced by one charge with which the other interacts. The fields produced by any number of charges may be superposed independently, and the result conveniently represented graphically by field lines, beginning at positive and ending at negative charges, showing by their direction that of the field, and by their density its strength. Pairs of equal but opposite charges separated by a small distance are called dipoles, the product of charge and separation being called the DIPOLE MOMENT. These experience a TORQUE in an electric field tending to align them with the field, but no net force unless the field is nonuniform. The amount of work done in moving a unit charge from one point to another against the electric field is called the electric potential difference or **voltage** between the points, and is measured in VOLTS (volts (V) = joules/coulomb). The ratio of a charge added to a body to the voltage produced is called the CAPACITANCE of the body; for most practical purposes, the earth provides a reference potential with an infinite capacitance.

In some materials, known as electric CONDUCTORS, there are charges free to move about—for example, valence electrons in metals, or IONS in salt solutions—and in these, the presence of an electric field produces a steady flow of charge in the direction of the field (negative charges moving the opposite way); such a flow constitutes an **electric current**, measured in AMPERES (amperes(A) = coulombs/second). The field implies a voltage between the ends of the conductor, which is normally proportional to the current (OHM'S LAW), the ratio being called the RESISTANCE of the conductor, and measured in OHMS (ohms (Ω) = volts/amps); it normally rises with

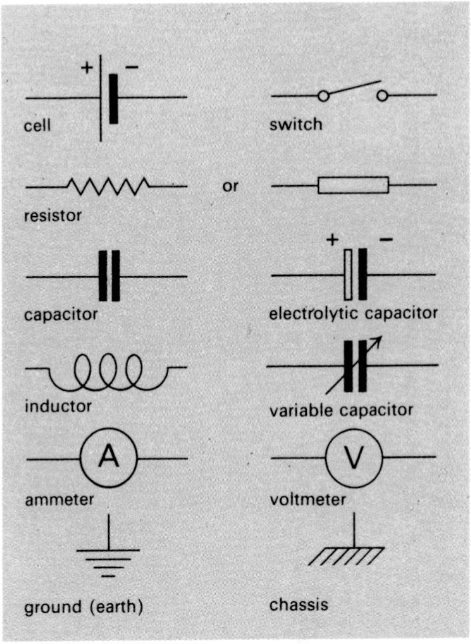

Electricity: some of the commonest graphic symbols used in circuit diagrams.

TEMPERATURE. Materials with high resistance to currents are classed as **insulators**. (See also SEMICONDUCTOR, DIELECTRIC.) The energy acquired by the charges in falling through the field is dissipated as HEAT—and LIGHT, if a sufficient temperature is reached—the total POWER output being the product of current and voltage. Thus, for example, a 1 kW fire supplied at 110 V draws a current of about 9 A, and the hot element has a resistance of about 12 Ω.

Electric sources such as BATTERIES or GENERATORS convert chemical, mechanical, or other energy into electrical energy (see ELECTROMOTIVE FORCE), and will pump charge through conductors much as a water pump circulates water in a radiator heating system (see CIRCUIT, ELECTRIC). Batteries create a constant voltage, and so produce a steady or **direct current** (DC); many generators on the other hand provide a voltage which changes in sign many times a second, and so produce an **alternating current** (AC) in which the charges move to and fro instead of continuously in one direction. This system has advantages in generation, transmission and application, and is now used almost universally for domestic and industrial purposes.

An electric current is found to produce a MAGNETIC FIELD circulating around it, to experience a force in an externally generated magnetic field, and to be itself generated by a changing magnetic field; for more details of these properties on which most electrical machinery depends, see ELECTROMAGNETISM.

Static electricity was known to the Greeks; the inverse square law was hinted at by J. PRIESTLEY in 1767 and later confirmed by H. CAVENDISH and C. A. COULOMB. G. S. OHM formulated his law of conduction in 1826, though its essentials were known before then. The common nature of all the "types of electricity"

For an alternating-current (AC) circuit containing pure resistance, the power (P), voltage (V) and current (I) are related according to $P=VI$. The current and voltage waves are in phase and the frequency of the instantaneous power wave is twice that of the instantaneous voltage and current waves. The effective power (P_{eff}) is given by $V_{max}I_{max}/2$, where the subscript *max* refers to the maximum values; the effective, or root-mean-square voltage (V_{RMS}) by $V_{max}/\sqrt{2}$, and the effective current (I_{RMS}) by $I_{max}/\sqrt{2}$.

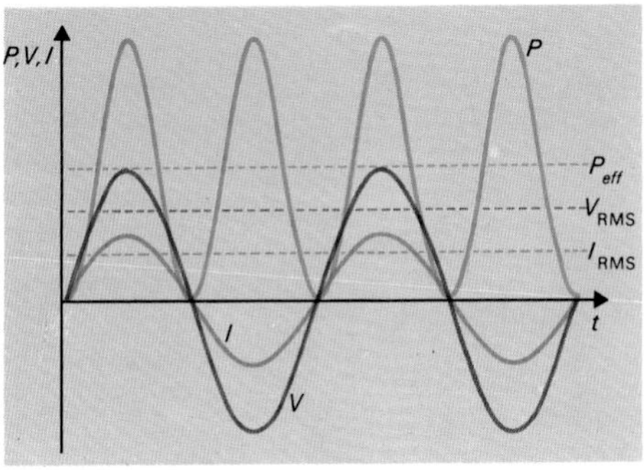

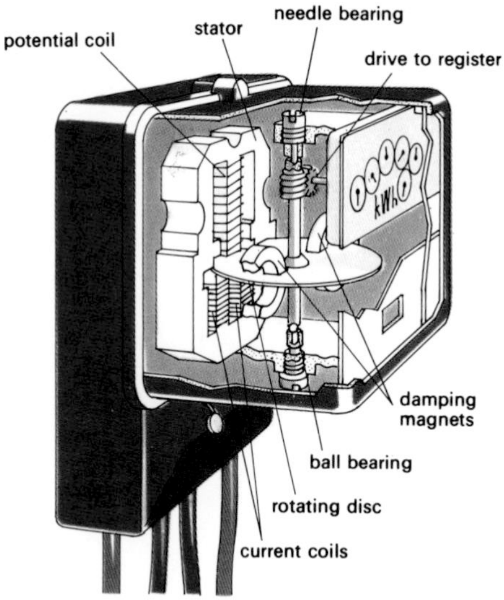

Cutaway of a domestic electric (watt-hour) meter.

then known was demonstrated in 1826 by M. FARADAY, who also originated the concept of electric field lines.

ELECTRIC LIGHT. See LIGHTING.

ELECTRIC METER, a single-phase AC watt-hour meter used to measure the amount of ELECTRICITY used by domestic consumers. It is a sort of simple electric MOTOR with a disk free to rotate in the magnetic field set up by two sets of coils, one in series with (the current coil), and the other in parallel to (the potential coil), the applied load. The rate of rotation is proportional to the POWER being used and so, by counting the rotations mechanically, the total ENERGY consumption can be measured.

ELECTRIC MOTOR. See MOTOR, ELECTRIC.

ELECTRIC POWER. See ELECTRICITY; POWER.

ELECTRIC SHOCK, passage of a small electric current through the body, usually causing unpleasant sensations, powerful MUSCLE spasm and sometimes stunning. Low energy shocks are rarely fatal but may cause superficial burns; high energy shocks cause ELECTROCUTION. Small electric shocks are used in SHOCK THERAPY and in defibrillation of the HEART.

ELECTROCARDIOGRAPH, instrument for recording the electrical activity of the HEART, producing its results in the form of multiple tracing called an electrocardiogram (ECG). These are conventionally recorded with twelve combinations of ELECTRODES on the limbs and CHEST wall. The electrical impulses in the conducting tissue and muscle of the heart pass through the body fluids, while the position of the electrodes determines the way in which the heart is "looked at" in electrical terms. ECGs allow CORONARY THROMBOSIS, abnormal heart rhythm, disorders of the heart muscle and PERICARDIUM to be detected, as well as diseases of the METABOLISM that affect the heart.

ELECTROCHEMICAL SERIES, or electromotive series, a sequence of elements (chiefly metals) listed in order of their standard redox potentials—i.e., the potential developed by an electrode of the element immersed in a molar solution (see MOLECULAR WEIGHT) of one of its salts (see ELECTROCHEMISTRY; OXIDATION AND REDUCTION). Metals high in the series are generally more reactive than those lower down, and displace them from aqueous solutions of their salts. (See also ELECTRONEGATIVITY; IONIZATION POTENTIAL.)

ELECTROCHEMISTRY, branch of PHYSICAL CHEMISTRY dealing with the interconversion of electrical and chemical energy (see CELL, ELECTROCHEMICAL). Many chemical species are electrically-charged IONS (see BOND, CHEMICAL), and a large class of reactions—OXIDATION AND REDUCTION—consists of electron-transfer reactions between ions and other species. If the two half-reactions (oxidation, reduction) are made to occur at different ELECTRODES, the electron-transfer occurs by the passing of a current through an external circuit between them (see BATTERY; FUEL CELL). The ELECTROMOTIVE FORCE driving the current is the sum of the electrode potentials (in volts) of the half-reactions, which represent the free energy (see THERMODYNAMICS) produced by them. Conversely, if an emf is applied across the electrodes of a cell, it causes a chemical reaction if it is greater than the sum of the potentials of the half-reactions (see ELECTROLYSIS). Such potentials depend both on the nature of the reaction and on the concentrations of the reactants. Cells arising through concentration differences are one cause of CORROSION. (See also CONDUCTIVITY; ELECTROMETALLURGY; ELECTROPHORESIS.)

ELECTROCUTION, the usually fatal effect of passing a high-energy electric current through the body. ELECTRICITY passing through the body fluid, which acts as a resistor, causes BURNS at sites of connection and along the electrical pathway. CONVULSIONS and rhythm disturbance in the heart are usual; the latter are the cause of immediate death. ARTIFICIAL RESPIRATION with cardiac massage must be started immediately if resuscitation is to be successful.

ELECTRODE, a component in an electric CIRCUIT at which current is transferred between ordinary metal conductors and a gas or electrolyte. A positive electrode is an ANODE and a negative one a CATHODE.

ELECTRODYNAMICS, the study of the interaction of charged particles with ELECTROMAGNETIC FIELDS, as summarized in the definitions of ELECTRIC and MAGNETIC FIELDS through the FORCES experienced by charged particles and described by the MAXWELL Equations.

ELECTROENCEPHALOGRAPH, instrument for recording the BRAIN's electrical activity using several small electrodes on the scalp. Its results are produced in the form of a multiple tracing called an electroencephalogram (EEG). The "brain waves" recorded have certain normal patterns in the alert and sleeping individual. Localized brain diseases and metabolic disturbances cause abnormal wave forms either in particular areas or as a generalized disturbance. The abnormal brain activity in EPILEPSY, both during CONVULSIONS and when the patient appears normal, usually allows diagnosis. The interpretation of EEGs requires skill and experience.

ELECTROFORMING, process for making an exact replica of an object. A wax mold is made and electroplated with copper (see ELECTROPLATING).

Masters for pressing records and duplicate plates (**"electrotype"**) for letterpress PRINTING are made thus, the copper shells being backed with lead.

ELECTROLUMINESCENCE, the emission of LIGHT from some PHOSPHORS (particularly ZnS) and SEMICONDUCTORS under the influence of an applied ELECTRIC FIELD. Electrophotoluminescent screens (in which the effect is PHOTON activated) are used medically to intensify X-ray pictures. (See also LUMINESCENCE.)

ELECTROLYSIS, production of a chemical reaction by passing a direct current through an electrolyte—i.e., a compound which contains IONS when molten or in solution. (See ELECTROCHEMISTRY.) The CATIONS move toward the CATHODE and the ANIONS toward the ANODE, thus carrying the current. At each electrode the ions are discharged according to FARADAY's laws: (1) the quantity of a substance produced is proportional to the amount of electricity passed; (2) the relative quantities of different substances produced are proportional to their EQUIVALENT WEIGHTS. Hence one gram-equivalent of any substance is produced by the same amount of electricity, known as a **faraday** (96 500 coulombs). Electrolysis is used to extract electropositive metals from their ores (see ELECTROCHEMICAL SERIES), and to refine less electropositive metals; to produce SODIUM hydroxide, CHLORINE, HYDROGEN, OXYGEN and many other substances; and in ELECTROMETALLURGY.

ELECTROLYSIS, Cosmetic, treatment designed to remove cosmetically unwanted HAIR—usually from the face—by a form of DIATHERMY, in which electric current is passed into the hair causing destruction of the hair follicle, and thus preventing regrowth. Permanent scarring may be an undesired end result.

ELECTROMAGNET, a magnet produced (and thus easily controlled) by the electric current in a coil of wire which is usually wound on a frame of highly permeable (see PERMEABILITY) material so as to reinforce and direct the MAGNETIC FIELD appropriately.

ELECTROMAGNETIC FIELD, any combination of ELECTRIC and MAGNETIC FIELDS, the two being closely related physically.

ELECTROMAGNETIC RADIATION, or radiant energy, the form in which ENERGY is transmitted through space or matter using a varying ELECTROMAGNETIC FIELD. Classically, radiant energy is regarded as a WAVE MOTION. In the mid-19th century MAXWELL showed that an oscillating (vibrating) electric charge would be surrounded by varying electric and magnetic fields. Energy would be lost from the oscillating charge in the form of transverse waves in these fields, the waves in the electric field being at right-angles both to those in the magnetic field and to the direction in which the waves are traveling (propagated). Moreover, the VELOCITY of the waves would depend only on the properties of the medium through which they passed; for propagation in a vacuum its value is a fundamental constant of physics—the **electromagnetic constant,** $c =$ 299 792.5km/s. At the beginning of the 20th century PLANCK proposed that certain properties of radiant energy were best explained by regarding it as transporting energy in discrete amounts called quanta. EINSTEIN later proposed the name PHOTON for the electromagnetic quantum. The energy of each photon is proportional to the frequency of the associated radiation (see QUANTUM THEORY).

The different kinds of electromagnetic radiation are classified according to the energy of the photons involved, the range of energies being known as the electromagnetic SPECTRUM. (Looked at in other ways, this spectrum arranges the radiations according to wavelength or frequency.) In order of decreasing energy the principal kinds are GAMMA RAYS, X-RAYS, ULTRAVIOLET RADIATION, LIGHT, INFRARED RADIATION, MICROWAVES and RADIO waves. In general, the higher the energies involved, the better the properties of the radiation are described in terms of particles (photons) rather than waves. Radiant energy is emitted from objects when they are heated (see BLACKBODY RADIATION) or otherwise energetically excited (see LUMINESCENCE; SPECTROSCOPY); man uses it to channel and distribute both energy and information (see INFORMATION THEORY). (See also ABSORPTION.)

ELECTROMAGNETISM, the study of ELECTRIC and MAGNETIC FIELDS, and their interaction with electric charges and currents. The two fields are in fact different manifestations of the same physical field, and are interconverted according to the speed of the observer. Apart from the effects noted under ELECTRICITY and MAGNETISM, the following are found:

1. Moving charges (and hence currents) in magnetic fields experience a FORCE, perpendicular to the field and the current, and proportional to their product. This is the basis of all electric MOTORS, and was first applied for the purpose by M. FARADAY in 1821.

2. A change in the number of magnetic field lines passing through a circuit "induces" an electric field in the circuit, proportional to the rate of the change. This is the basis of most GENERATORS, and was also established by M. Faraday, in 1831.

3. An effect analogous to the above, but with magnetic and electric fields interchanged, and usually much smaller. This was hypothesized by J. C. MAXWELL, who in 1862 deduced from it the possibility of self-sustaining electromagnetic waves traveling at a speed which coincided with that of LIGHT, thereby identifying the nature of visible light, and predicting other waves such as the RADIO waves found experimentally by H. HERTZ shortly afterwards.

ELECTROMETALLURGY, branch of METALLURGY which uses electricity. It includes the use of ELECTRIC FURNACES, and also the use of ELECTROLYSIS for extracting and refining metals, ELECTROPLATING, ANODIZING and ELECTROFORMING.

ELECTROMOTIVE FORCE (emf), loosely, the voltage produced by a BATTERY, GENERATOR or other source of ELECTRICITY, but more precisely, the product of the current it produces in a circuit and the total circuit RESISTANCE, including that of the source itself. The actual voltage across the source is usually somewhat lower.

ELECTRON, a stable SUBATOMIC PARTICLE, with rest MASS 9.1091×10^{-31}kg (roughly 1/1836 the mass of a HYDROGEN atom) and a negative charge of 1.6021×10^{-19}C, the charges of other particles being positive or negative integral multiples of this. Electrons are one of the basic constituents of ordinary MATTER, commonly occupying the ORBITALS surrounding positively charged atomic nuclei. The

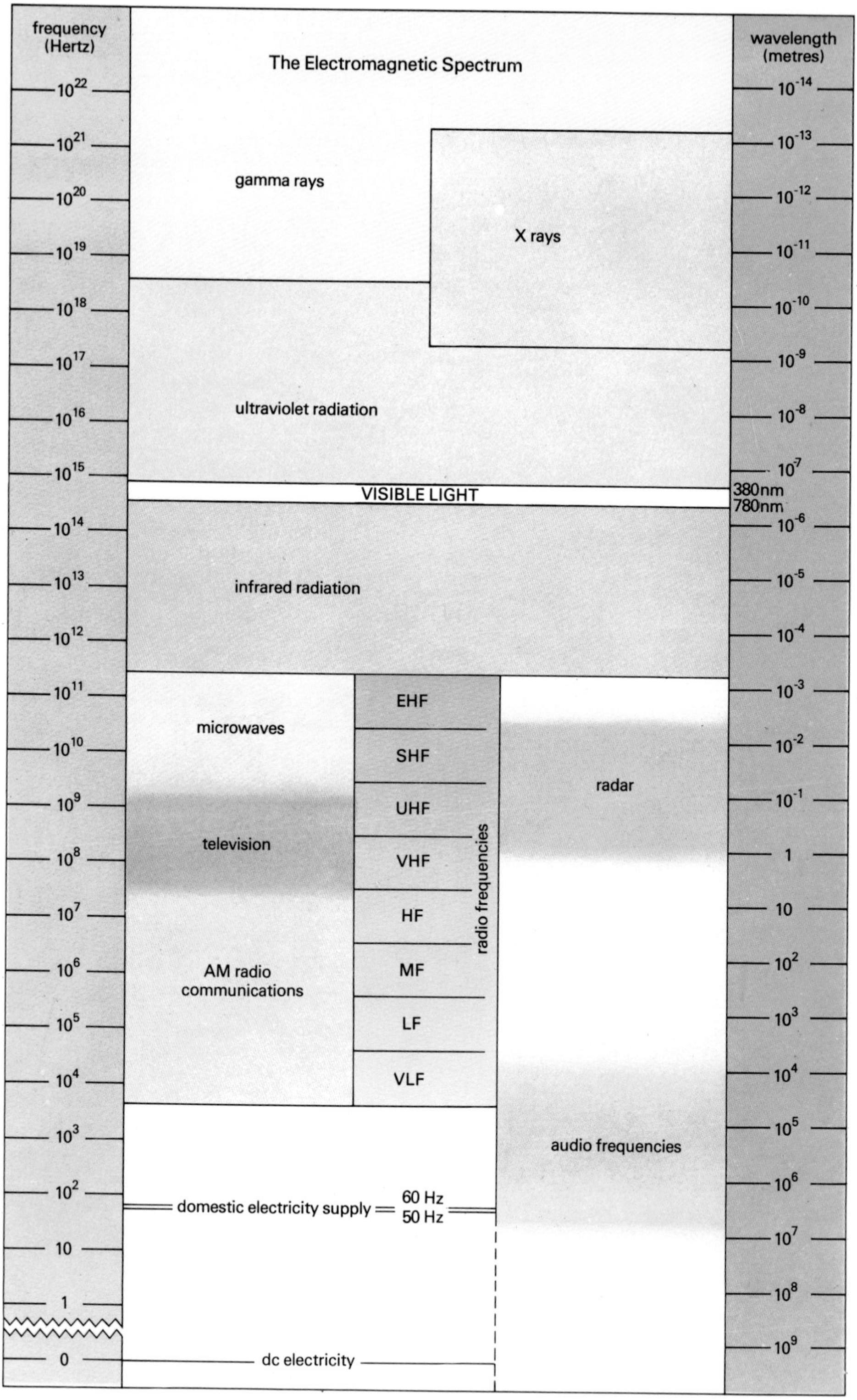
frequency (Hertz)
The Electromagnetic Spectrum
wavelength (metres)
10^{22}
10^{21}
10^{20}
10^{19}
10^{18}
10^{17}
10^{16}
10^{15}
10^{14}
10^{13}
10^{12}
10^{11}
10^{10}
10^{9}
10^{8}
10^{7}
10^{6}
10^{5}
10^{4}
10^{3}
10^{2}
10
1
0
10^{-14}
10^{-13}
10^{-12}
10^{-11}
10^{-10}
10^{-9}
10^{-8}
10^{-7}
380nm
780nm
10^{-6}
10^{-5}
10^{-4}
10^{-3}
10^{-2}
10^{-1}
1
10
10^{2}
10^{3}
10^{4}
10^{5}
10^{6}
10^{7}
10^{8}
10^{9}
gamma rays
X rays
ultraviolet radiation
VISIBLE LIGHT
infrared radiation
microwaves
television
AM radio communications
EHF
SHF
UHF
VHF
HF
MF
LF
VLF
radio frequencies
radar
audio frequencies
domestic electricity supply
60 Hz
50 Hz
dc electricity

chemical properties of ATOMS and MOLECULES are largely determined by the behavior of the electrons in their highest-energy orbitals. Both CATHODE RAYS and BETA RAYS are streams of free electrons passing through a gas or vacuum. The unidirectional motion of electrons in a solid conductor constitutes an electric current. Solid conductors differ from nonconductors in that in the former some electrons are free to move about while in the latter all are permanently associated with particular nuclei. Free electrons in a gas or vacuum can usually be treated as classical particles, though their wave properties become important when they interact with or are associated with atomic nuclei. The anti-electron, with identical mass but an equivalent positive charge, is known as a positron (see ANTIMATTER).

ELECTRONEGATIVITY, the relative power of an atom in a molecule to attract electrons. A concept variously defined and estimated since its proposal by L. PAULING, it depends on the atom's VALENCE state (see BOND, CHEMICAL) and is useful only as a qualitative guide to polarity of bonds and molecules. Electropositive metals are generally high in the ELECTROCHEMICAL SERIES. (See also NUCLEOPHILES.)

ELECTRON GUN, that part of an ELECTRON TUBE which produces, accelerates, focuses and deflects a beam of ELECTRONS. A CATHODE emits electrons which pass through a grid to the steering ANODES.

ELECTRONIC FLASH, a glass tube containing inert gas, used as a light source in photography and having a useful life of over 10000 flashes. A high-voltage pulse applied to electrodes at either end creates a discharge through the gas, giving a daylight-type flash lasting from 1ms to 1μs.

ELECTRONICS, an applied science dealing with the development and behavior of ELECTRON TUBES, SEMICONDUCTORS and other devices in which the motion of electrons is controlled; it covers the behavior of electrons in gases, vacuums, conductors and semiconductors. Its theoretical basis lies in the principles of ELECTROMAGNETISM and solid-state physics discovered in the late 19th and early 20th centuries. Electronics began to grow in the 1920s with the development of RADIO. During WWII, the US and UK concentrated resources on the invention of RADAR and pulse transmission methods and by 1945 they had enormous industrial capacity for producing electronic equipment. The invention of the TRANSISTOR in 1948 as a small, cheap replacement for vacuum tubes led to the rapid development of COMPUTERS, transistor radios, etc. Now, with the widespread use of integrated circuits, electronics plays a vital role in communications (TELEPHONE networks, information storage, etc.) and industry. All electronic circuits contain both active and passive components and transducers (e.g., MICROPHONES) which change ENERGY from one form to another. Sensors of light, temperature, etc., may also be present. **Passive components** are normally conductors and are characterized by their properties of RESISTANCE (R), CAPACITANCE (C) and INDUCTANCE (L). One of these usually predominates, depending on the function required. **Active components** are electron tubes or semiconductors; they contain a source of power and control electron flow. The former may be general-purpose tubes (DIODES, TRIODES, etc., the name depending on the number of ELECTRODES) which

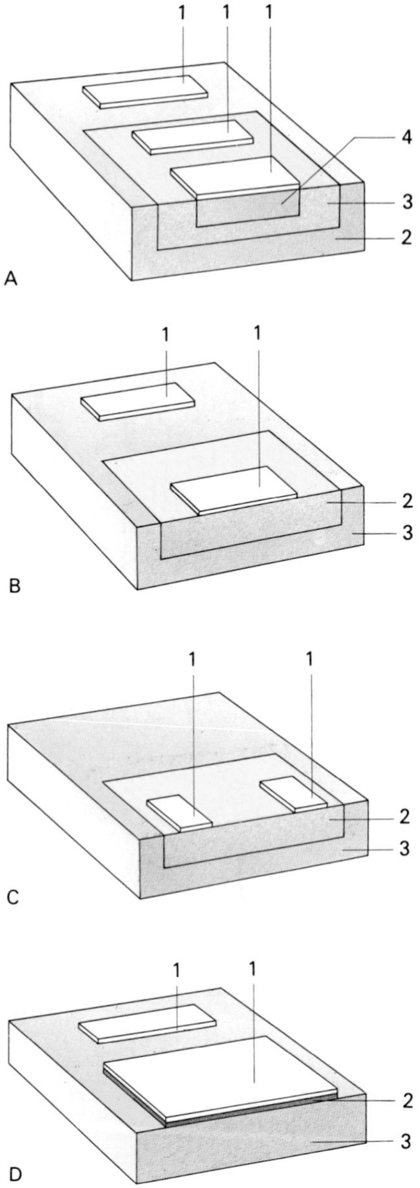

Electronic components in integrated form (about 350× actual size). The leadout wires are attached to the metallized areas (1) in each case. (A) A "diffused" transistor, with collector (2), base (3) and emitter (4). (B) A diode, with p-type anode (2) and n-type cathode (3). (C) A resistor of p-type material (2) in an n-type substrate (3). (D) A capacitor, with a thin layer of silicon dioxide (2) acting as the dielectric; the substrate (3) may be p- or n-type.

rectify, amplify or switch electric signals. Image tubes (in TELEVISION receivers) convert an electric input into a light signal; photoelectric tubes (in television cameras) do the reverse. Semiconductor diodes and transistors, which are basically sandwiches made of two different types of semiconductor, now usually perform the general functions once done by tubes, being smaller, more robust and generating less heat. These few basic components can build up an enormous range of circuits with different functions. Common types include: power supply (converting AC to pulsing DC and then smoothing out the pulsations); switching and timing (the logic circuits in computers are in this category); AMPLIFIERS, which increase the amplitude or power of a signal, and oscillators, used in radio and television transmitters and which generate AC signals. Demands for increased cheapness and reliability of circuits have led to the development of microelectronics. In **printed circuits**, printed connections replace individual wiring on a flat board to which about two components per cm^3 are soldered. INTEGRATED CIRCUITS assemble about 10^5 components and interconnections per cm^3 in a single structure, formed directly by evaporation or other techniques as films about 0.03mm thick on a substrate. In monolithic circuits, components are produced in a tiny chip of semiconductor by selective diffusion.

ELECTRON MICROSCOPE, a microscope using a beam of ELECTRONS rather than LIGHT to study objects too small for conventional MICROSCOPES. First constructed by Max Knoll and Ernst Ruska around 1930, the instrument now consists typically of an

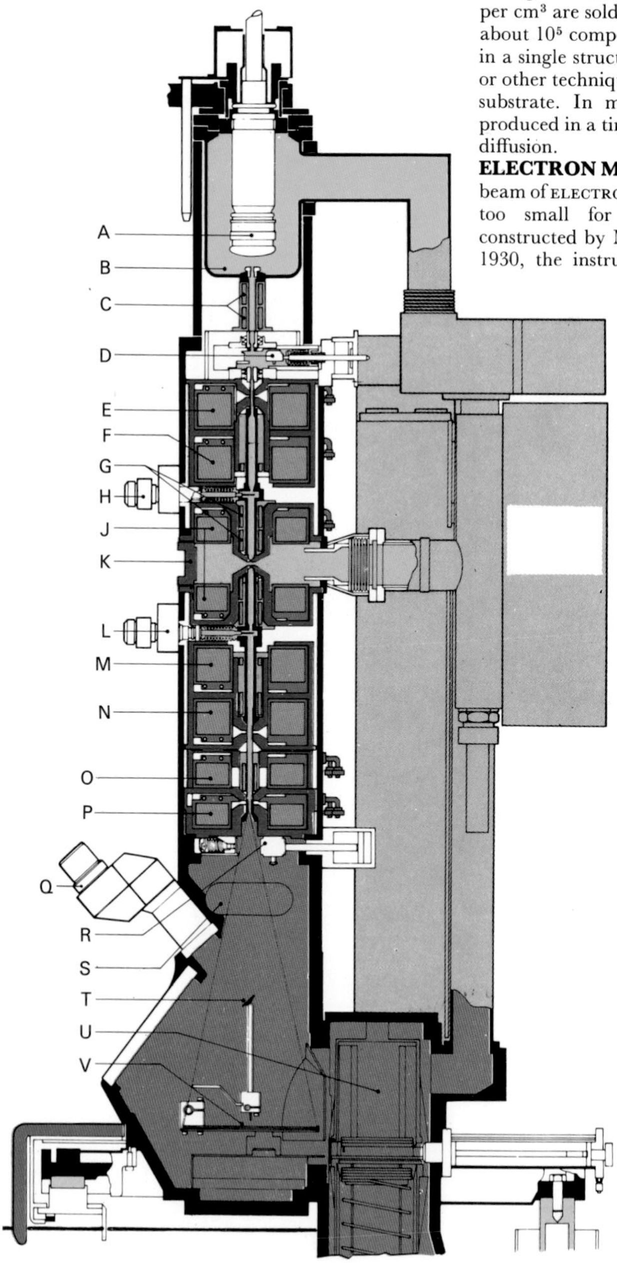

Section through a Philips EM400 electron microscope. (A) Electron gun; (B) anode; (C) centering coils; (D) vacuum lock; (E) first condenser lens; (F) second condenser lens; (G) alignment coils; (H) condenser diaphragm, (J) objective lens; (K) specimen chamber; (L) objective diaphragm; (M) diffraction lens; (N) intermediate lens; (O) first projector lens; (P) second projector lens; (Q) binoculars; (R) vacuum lock; (S) 35mm camera; (T) focusing screen; (U) film magazine; (V) fluorescent screen.

evacuated column of magnetic lenses with a 20–1 000 kV electron gun at the top and a fluorescent screen or photographic plate at the bottom; it can thus be thought of as a kind of CATHODE RAY TUBE. The various lenses allow the operator to see details almost at the atomic level (0.3 nm) at up to a million times magnification (though many specimens deteriorate under the electron bombardment at these limits), and to obtain DIFFRACTION patterns from very small areas. In the scanning electron microscope, the beam is focused to a point and scanned over the specimen area while a synchronized television screen displays the transmitted or scattered electron intensity. Electron microscopes are used for structural, defect and composition studies in a wide range of biological and inorganic materials.

ELECTRON TUBE, or **valve,** an evacuated glass or metal tube which may contain gas at low pressure, through which ELECTRONS flow between two or more ELECTRODES. A heated filament, the cathode, emits electrons which are attracted to the positively-charged anode. By varying the voltage on intermediate (grid) electrodes, the electron flow can be regulated and the tube made to work as an AMPLIFIER. (See also DIODE; CATHODE-RAY TUBE; RECTIFIER; TRIODE.)

ELECTRONVOLT (eV), a unit of ENERGY used in atomic and high-energy physics, defined as the kinetic energy acquired by an ELECTRON in passing through a POTENTIAL difference of 1 volt in a vacuum. The electronvolt is about $1.602\,19 \times 10^{-19}$ joules.

ELECTROPHILES. See NUCLEOPHILES.

ELECTROPHORESIS, the DIFFUSION of charged particles through FLUIDS or GELS under the influence of an ELECTRIC FIELD. Particles with different sizes and charges diffuse at different rates, so that the effect can be used to separate and identify large MOLECULES such as PROTEINS.

ELECTROPLATING, process of depositing a thin metal coating on base-metal objects, to improve their appearance or CORROSION resistance (see also ELECTROFORMING). The object is made the CATHODE of a cell containing a salt of the metal to be deposited, which is made the ANODE; on ELECTROLYSIS the metal dissolves from the anode and deposits on the object. Chromium, nickel, copper, silver and gold are commonly used. In **electropolishing**, the reverse process, the object is made the anode; preferential solution of irregularities yields a high polish. (See also ANODIZING.)

ELECTROSCOPE, gold leaf, an instrument used (mainly historically) for the measurement of electric charge or potential (see ELECTRICITY), based on one or two fine gold foils suspended vertically and free to deflect under electrostatic repulsion when an electric charge was applied. Although never very accurate, the instruments could sometimes detect as little as 10^{-14} coulombs and were also used to measure radiation intensity through the rate of charge leakage caused by ionization of the air.

ELECTROSTATIC COPYING. See XEROGRAPHY.

ELECTROSTATIC GENERATOR, an instrument producing a very high direct voltage, particularly that developed by Robert J. van de Graaff (1901–1967). An insulating belt is driven around a pair of rollers some distance apart, a small electric discharge at one end producing positively charged IONS which are carried by the belt to the other end and lifted off by a small metal comb, the charge accumulating on a polished metal sphere surrounding this end. Such generators can develop up to 20MV with a 15m sphere and are used in various ways as particle ACCELERATORS.

ELECTROSTATIC PRECIPITATOR, a device for removing solid or liquid particles from gases, capable of cleaning dirty ventilating air or removing tar or dust from the smoke of coal-burning plants. An electrical discharge is produced in the gas and the ELECTRONS released collide with the suspended particles carrying them toward the ANODE, where they accumulate for periodic removal.

ELECTROSTATICS, the study of static ELECTRICITY.

ELECTROTYPING. See ELECTROFORMING.

ELECTRUM, pale yellow ALLOY of GOLD with up to 40% SILVER, occurring naturally as gold ore, and used for ornament. The term has been used for AMBER.

ELEMENT, Chemical, simple substance composed of ATOMS of the same atomic number, and so incapable of chemical degradation or resolution. They are generally mixtures of different ISOTOPES. Of the 106 known elements, 88 occur in nature, and the rest have been synthesized (see TRANSURANIUM ELEMENTS). The elements are classified by physical properties as METALS, METALLOIDS and NONMETALS, and by chemical properties and atomic structure according to the PERIODIC TABLE. Most elements exhibit ALLOTROPY, and many are molecular (e.g., oxygen, O_2). The elements have all been built up in STARS from HYDROGEN by complex sequences of nuclear reactions, e.g., the CARBON CYCLE.

ELEMENTARY PARTICLES. See SUBATOMIC PARTICLES.

ELEPHANTIASIS, disease in which there is massive swelling and hypertrophy of the SKIN and subcutaneous tissue of the legs or scrotum, due to the obstructed flow of LYMPH. This may be a congenital DISEASE, due to trauma, CANCER, or infection with FILARIASIS, TUBERCULOSIS and some VENEREAL DISEASES. Recurrent secondary bacterial infections are common and chronic skin ULCERS may form. Elevation, elastic stockings, DIURETICS and treatment of infection are basic to relief, while some cases are helped by SURGERY.

ELEVATOR, in aeronautics, one of the three basic control surfaces of an AIRPLANE, used to control pitch. Elevators are usually pivoted on the stabilizer.

ELEVATOR, or **lift,** a device installed in a building for carrying passengers and freight vertically between levels. Most modern lifts are electrically powered and many are now electronically controlled. The MOTOR is now usually mounted at the top of the elevator well with the elevator car slung below. The weight of the empty car plus about 40% of its designed load is offset by a traveling counterweight at the other end of the suspension cable. Modern elevator installations feature a galaxy of safety devices including automatic door locks, emergency brakes and buffers at the foot of the well.

ELIMINATION, the technique used in the solution of simultaneous equations by which n equations in n VARIABLES are reduced to one solvable equation in one variable, the process being repeated until all the equations are solved for all the variables. The process may be performed by establishing the value of one variable in terms of another and substitution; or by the addition to or subtraction from one equation of another. Thus

if $x-6=y$ (1)

and $2x+3=y$ (2)

we can substitute the value of y in equation (1) into equation (2) to give

$$2x+3=x-6$$

and hence $x=-9$.

Substituting this value into equation (1) gives $y=-15$. (See also EQUATION.)

ELLIPSE. See CONIC SECTIONS.

ELLIPSOID, a three-dimensional geometrical figure whose intersection with any plane is an ellipse (see CONIC SECTIONS). It has three mutually PERPENDICULAR axes of SYMMETRY; taking these as the AXES defining a set of CARTESIAN COORDINATES, the equation of an ellipsoid is $x^2/a^2+y^2/b^2+z^2/c^2=1$, where a, b, c are CONSTANTS. If any one of a, b, c equals any other, the figure is an ellipsoid of revolution or spheroid, sections parallel to the plane defined by two of its axes of symmetry being circular. If $a=b=c$ the figure is a SPHERE.

ELLIS, Henry Havelock (1859–1939), British writer chiefly remembered for his studies of human sexual behavior and psychology. His major work was *Studies in the Psychology of Sex* (1897–1928).

EMBALMING, a process by which a corpse is prevented, at least temporarily, from decomposing. Under some circumstances bodies may be naturally embalmed, but artificial embalming first appeared in ancient Egypt (see MUMMY). Modern embalming began after HARVEY'S discovery of the blood circulation (1628). Embalming fluid is injected into an artery (arterial fluid) while blood is drained from a vein, then a stronger fluid (cavity fluid) is injected into bodily orifices and hollow organs.

EMBOLISM, the presence of substances other than liquid BLOOD in the BLOOD CIRCULATION, causing obstruction to ARTERIES or interfering with the pumping of the HEART. The commonest embolism is from atheromatous plaques (see ARTERIOSCLEROSIS) or THROMBOSIS on a blood vessel or the HEART walls. FAT globules may form emboli from bone MARROW after major bone FRACTURES, and amniotic fluid may cause embolism during childbirth. STROKE or transient cerebral episodes, pulmonary embolism, CORONARY THROMBOSIS and obstruction of limb or organ blood supply with consequent cell DEATH are common results, some of them fatal. Some may be removed surgically, but prevention is preferable.

EMBOSSING, mechanical production of a raised pattern on a surface: suitable materials are plastic, thin metal, paper, leather, fabric, etc. A male DIE is machined such that the required pattern is raised, and a female die such that a mirror image of the pattern is engraved into it. When the two dies are forced together the pattern is embossed on the material between.

EMBRYO, the earliest stage of the life of a FETUS, the development from a fertilized EGG through the differentiation of the major organs. In man, the fertilized egg divides repeatedly, forming a small ball of cells which fixes by IMPLANTATION to the wall of the WOMB; differentiation into PLACENTA and three primitive layers (endoderm, mesoderm and ectoderm) follows. These layers then undergo further division into distinct organ precursors and each of these develops by a process of migration, differentiation and differential growth. The processes roughly correspond to the phylogeny or evolutionary sequence leading to the species. Much of development depends on formation of cavities, either by splitting of layers or by enfolding. The HEART develops early at the front, probably splitting into a simple tube, before being divided into separate chambers; the gut is folded into the body, although for a long time the bulk of it remains outside. The NERVOUS SYSTEM develops as an infolding of ectoderm, which then becomes separated from the surface. Facial development consists of mesodermal migration and modification of the bronchial arches, remnants of the GILLS in phylogeny; primitive limb buds grow out of the developing trunk. The overall control of these processes is not yet understood; however, infection (especially GERMAN MEASLES) in the mother, or the taking of certain DRUGS (e.g., THALIDOMIDE) during PREGNANCY may lead to abnormal development and so to congenital defects, including heart defects (e.g., BLUE BABY), limb deformity, HARELIP and CLEFT PALATE and SPINA BIFIDA. By convention, the embryo becomes a fetus at three months gestation.

Emerald crystals embedded in limestone, and the polished gemstone.

EMBRYOLOGY, the study of the development of EMBRYOS of animals and humans, based on anatomical specimens of embryos at different periods of gestation, obtained from animals or from human ABORTION. The development of organ systems may be deduced and the origins of congenital defects recognized, so that events liable to interfere with development may be avoided. It may reveal the basis for the separate development of identical cells and for control of growth. The ANATOMY of an organism may be better understood and learnt by study of embryology. The principal embryologists of past ages have included ARISTOTLE; William HARVEY and Marcello MALPIGHI in the 17th century, and Karl Ernst von BAER in the 19th century.

EMBRYOPHYTE, any plant of the subkingdom Embryophyta, which is a group characterized by having an embryo and multicellular sex organs. Included in the Embryophyta are MOSSES, LIVERWORTS, FERNS, GYMNOSPERMS and ANGIOSPERMS. (See also PLANT KINGDOM.)

EMERALD, valuable green GEMSTONE, a variety of BERYL. The best emeralds are mined in Colombia, Brazil and the USSR. Since 1935 it has been possible to make synthetic emeralds.

EMERY, an impure CORUNDUM containing MAGNETITE and other minerals, occurring on Naxos island and in Asia Minor. It is used as an abrasive (Mohs hardness 8), and as a non-skid material in floors and stairs.

EMETIC, any agent that causes vomiting. Salt solution or stimulation of the PHARYNX are emetics used from antiquity, while drugs such as ipecacuanha and apomorphine are also effective. They are used when a POISON such as an overdose of tablets has been recently ingested; they should not be used when the poison taken causes damage to the ESOPHAGUS or LUNGS.

EMOTION, in psychology, a term that is only loosely defined. Generally, an emotion is a sensation which causes physiological changes (as in pulse rate, breathing) as well as psychological changes (as disturbance) which result in, usually, compulsive (see COMPULSION) adaptations in the individual's behavior. Some psychologists differentiate types of emotion: one such classification is into primary (e.g., fear), complex (e.g., envy) and sentiment (e.g., love, hate); but such schemata are controversial. The causes of emotion are not fully understood, but the emotional effects of certain drugs suggest that emotions are the result of biochemical change in various parts of the body. (See also IDEA; INSTINCT.) Modern psychoanalysts generally prefer the term **affect** for emotion.

EMPEDOCLES (c490–430 BC), Sicilian Pythagorean philosopher who developed the notion that there were four fundamental elements in matter—earth, air, fire and water. In medicine he taught that blood ebbed and flowed from the heart and that health consisted in a balance of the four HUMORS in the body.

EMPHYSEMA, condition in which the air spaces of the LUNGS become enlarged, due to destruction of their walls. Often associated with chronic BRONCHITIS, it is usually a result of SMOKING but may be a congenital or occupational disease. Subcutaneous emphysema refers to air in the subcutaneous tissues.

EMPIRICISM, in philosophy, the view that knowledge can be derived only from sense experience. Modern empiricism, fundamentally opposed to the RATIONALISM that derives knowledge by deduction from principles known À PRIORI, was developed in the philosophies of LOCKE, BERKELEY and HUME. Other thinkers in the "British empiricist tradition" include J. S. MILL and the Americans J. DEWEY and W. JAMES.

EMULSION, a COLLOID in which both phases are initially liquid.

ENAMEL, vitreous glaze (see CERAMICS; GLASS) fused on metal for decoration and protection. Silica, potassium carbonate, borax and trilead tetroxide (see LEAD) are fused to form a glass (called flux) which is colored by metal oxides; tin (IV) oxide makes it opaque. The enamel is powdered and spread over the cleaned metal object, which is then fired in a furnace until the enamel melts.

ENCEPHALITIS, infection affecting the substance of the BRAIN, usually caused by a VIRUS. It is a rare complication of certain common diseases (e.g., mumps, herpes simplex) and a specific manifestation of less common viruses, often carried by insects. Typically an acute illness with HEADACHE and FEVER, it may lead to evidence of patchy INFLAMMATION of brain tissue, such as personality change, EPILEPSY, localized weakness or rigidity. It may progress to impairment of consciousness and COMA. A particular type, *Encephalitis lethargica*, occurred as an EPIDEMIC early this century leading to a chronic disease resembling PARKINSON'S DISEASE but often with permanent mental changes.

ENCKE, Johann Franz (1791–1865), German astronomer who discovered one of the divisions in the rings of SATURN and, using observations of a transit of Venus, established a good value for the distance of the sun. He also discovered **Encke's Comet**, whose period is only 3.3yr (see COMET).

ENDERS, John Franklin (1897–), US microbiologist who shared the 1954 Nobel Prize for Physiology or Medicine with F. C. ROBBINS and T. H. WELLER for their cultivation of POLIOMYELITIS virus in non-nerve tissues, so opening the gate for the development of polio vaccines.

ENDOCRINE GLANDS, ductless glands in the body which secrete HORMONES directly into the BLOOD stream. They include the PITUITARY GLAND, THYROID and PARATHYROID GLANDS, ADRENAL GLANDS and part of the PANCREAS, TESTES and OVARIES. Each secretes a number of hormones which affect body function, development, mineral balance and METABOLISM. They are under complex control mechanisms including FEEDBACK from their metabolic function and from other hormones. The pituitary gland, which is itself regulated by the HYPOTHALAMUS, has a regulator effect on the thyroid, adrenals and gonads.

ENDOGAMY AND EXOGAMY, social rules requiring a person to marry within (endogamous) or without (exogamous) his or her social, religious or ethnic group. A well-known example of endogamy is to be found among Hindus, where inter-caste MARRIAGES have been traditionally forbidden. In biology, endogamy means INBREEDING, and exogamy outbreeding, or breeding between individuals not closely related.

ENDOSKELETON, any SKELETON that is enclosed within an animal's body. (See also EXOSKELETON).

ENERGY, to the economist, a synonym for fuel; to the scientist, one of the fundamental modes of existence, equivalent to and interconvertible with MATTER. The MASS-energy equivalence is expressed in the Einstein equation, $E = mc^2$, where E is the energy equivalent to the mass m, c being the electromagnetic constant (speed of LIGHT). Since c is so large, a tiny mass is equivalent to a vast amount of energy. However, this energy can only be realized in nuclear reactions and so, although the conversion of mass may provide energy for the STARS, this process does not figure much in physical processes on earth (except in nuclear power installations). The law of the conservation of mass-energy states that the total amount of mass-energy in the UNIVERSE or in an isolated system forming part of the universe cannot change. In an isolated system in which there are no nuclear reactions, this means that the total quantities both of mass and of energy are constant. Energy then is generally conserved.

Energy exists in a number of equivalent forms. The commonest of these is HEAT—the motion of the MOLECULES of matter. Ultimately all other forms of energy tend to convert into thermal motion. Another form of energy is the motion of ELECTRONS, ELECTRICITY. Moving electrons give rise to

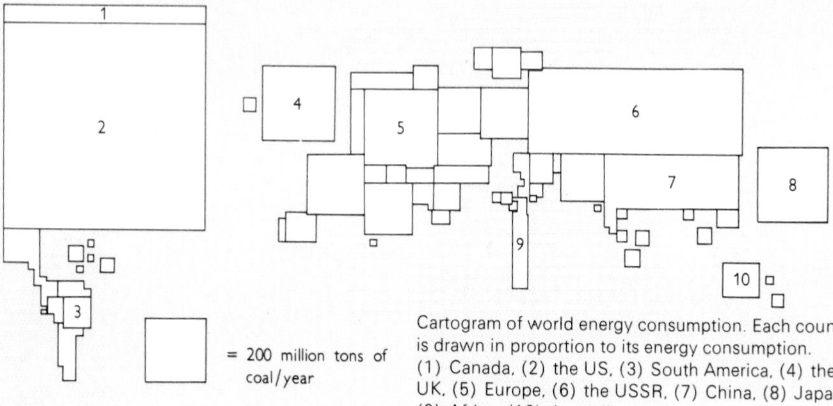

Cartogram of world energy consumption. Each country is drawn in proportion to its energy consumption. (1) Canada, (2) the US, (3) South America, (4) the UK, (5) Europe, (6) the USSR, (7) China, (8) Japan, (9) Africa, (10) Australia.

electromagnetic fields and these too contain energy. A pure form of electromagnetic energy is ELECTROMAGNETIC RADIATION (**radiant energy**) such as light. According to the QUANTUM THEORY, the energy of electromagnetic radiation is "quantized," referable to discrete units called PHOTONS, the energy E carried by a photon of radiation of FREQUENCY ν being given by $E=h\nu$, where h is the PLANCK CONSTANT. When macroscopic bodies move, they too have energy in virtue of their motion; this is their **kinetic energy** and is given by $\frac{1}{2}mv^2$ where m is the mass and v the velocity of motion. To change the velocity of a moving body, or to set it in motion, a FORCE must be applied to it and work must be done. This work is equivalent to the change in the kinetic energy of the body and gave physicists one of their earliest definitions of energy: the ability to do work. When work is done against a restraining force, **potential energy** is stored in the system, ready to be released again. The restraining force may be electromagnetic, torsional, electrostatic, tensional or of any other type. On earth when an object of mass m is raised up to height h, its gravitational potential energy is given by mgh, where g is the acceleration due to gravity. If the object is let go, it falls and it will strike the ground with velocity v, its potential energy having been converted into kinetic energy $\frac{1}{2}mv^2$. SOUND energy is kinetic energy of the vibration of air. Chemical energy is the energy released from a chemical system in the course of a reaction. Although all forms of energy are equivalent, not all interconversion processes go with 100% EFFICIENCY (the energy deficit always appears as heat—see THERMODYNAMICS). The SI UNIT of energy is the joule.

ENERGY LEVEL, a stationary state of a physical system characterized by its having or being able to have a fixed quantity of ENERGY. QUANTUM MECHANICS assumes that physical systems can only exist in a well-defined set of energy levels. The emission of ELECTROMAGNETIC RADIATION is associated with transitions of electronic and molecular systems between energy levels (see SPECTROSCOPY).

ENGINE, a device for converting stored ENERGY into useful WORK. Most engines in use today are heat engines which convert HEAT into work, though the EFFICIENCY of this process, being governed according to the second law of THERMODYNAMICS, is often very low. Heat engines are commonly classified according to the fuel they use (as in gasoline engine); by whether they burn their fuel internally or externally (see INTERNAL-COMBUSTION ENGINE), or by their mode of action (whether they are reciprocating, rotary or reactive). (See DIESEL ENGINE; GAS TURBINE; JET PROPULSION; STEAM ENGINE; TURBINE.)

ENGINEERING, essentially, the managing of

Energy-level diagram for sodium. The ground-state for the sodium atom's valence electron is designated 3*s*. Some of the allowed transitions to other energy levels are shown as red lines; the transitions which principally feature in the sodium emission spectrum are labelled in nanometres. The transition 3*p*→3*s* is responsible for the yellow "sodium D line" of the solar spectrum and the bright yellow color of the sodium flame and of sodium vapor lamps. Electrons raised above the series limit (SL) value are lost from the atom to which they were attached, which is thus ionized.

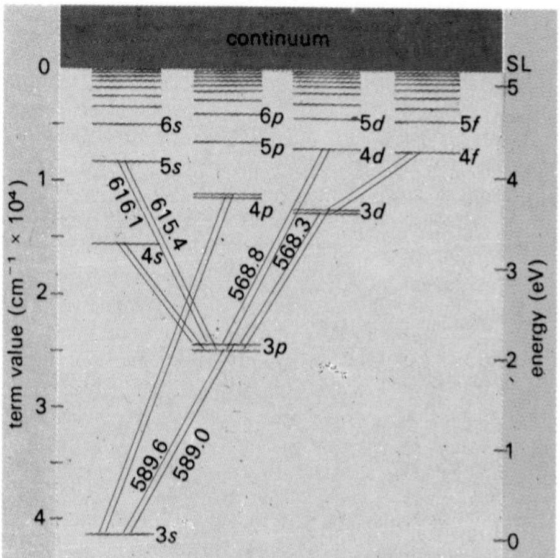

ENGINES. The term also describes the application of the sciences (including mathematics), and the development and uses of technology, to the service of man. Branches include: aeronautical engineering (see AERONAUTICS); chemical engineering, using chemical knowledge and processes in the conversion of raw materials into desired products (see CHEMISTRY); CIVIL ENGINEERING; ELECTRICAL ENGINEERING; HUMAN ENGINEERING; marine engineering, the design and construction of structures and processes for naval purposes; mechanical engineering, the design and use of MACHINES, with its tool, MECHANICAL DRAWING; and nuclear engineering (see NUCLEAR ENERGY).

ENGRAVING, various craft and technological techniques for producing blocks or plates from which to print illustrations, banknotes etc.; also, an individual print made by one of these processes. Line engraving refers to preparing a plate by scratching its smooth surface with a highly-tempered steel tool called a burin or graver. If the desired design is left standing high as is common with woodcuts and linocuts, this is known as a relief process. If the ink is transferred to the paper from lines incised into the plate, the surface of the inked plate having been wiped clean, this is known as intaglio. Drypoint and mezzotint are mechanical engraving processes developed from line engraving; other techniques, including AQUATINT, involve chemical ETCHING processes.

ENLIGHTENMENT, The, also known as The Age of Reason or *Aufklärung*, a term applied to the period of European intellectual history centering on the mid-18th century. The empiricist philosophy of LOCKE and scientific optimism following the success of NEWTON's *Principia* provided men with the confidence to deem reason supreme in all the departments of intellectual enquiry.

ENTERITIS, INFLAMMATION of the small intestine (see GASTROINTESTINAL TRACT) causing abdominal COLIC and DIARRHEA. It may result from VIRUS infection, certain BACTERIAL DISEASES or FOOD POISONING, which are in general self-limited and mild. The noninfective inflammatory condition known as **Crohn's disease** causes a chronic relapsing regional enteritis, which may present with weight loss, ANEMIA, abdominal mass or VITAMIN deficiency, as well as colic and diarrhea. In bacterial enteritis, ANTIBIOTICS may help, while Crohn's disease is sometimes helped by anti-inflammatory drugs or STEROIDS; SURGERY may also be required, but is often hazardous and may lead to FISTULA formation.

ENTOMOLOGY, the study of insects. In a broader sense the term is sometimes erroneously used to describe studies on other arthropod groups. Entomology is important, not only as an academic discipline, but because insects are among the most important pests and transmitters of disease.

ENTOPROCTA, a phylum of colonial animals found in shallow marine water and rarely exceeding a length of 2mm (0.08in). The best-known genus, *Pedicellina*, is composed of a branching stem, or stolon, on which are borne zooids—cup-shaped structures with tentacles arising from the upper rim.

ENTROPY, the name of a quantity in THERMODYNAMICS, statistical mechanics and INFORMATION THEORY variously representing the degree of disorder in a physical system, the extent to which the ENERGY in a system is available for doing WORK, the distribution of the energy of a system between different modes, or the uncertainty in a given item of knowledge. In thermodynamics ABSOLUTE entropies cannot be determined, only changes in entropy. The infinitesimal entropy change δS when a quantity of HEAT δQ is transferred at absolute TEMPERATURE T is defined as $\delta S = \delta Q/T$. One way of stating the second law of thermodynamics is to say that in any change in an isolated system, the entropy (S) increases: $\Delta S \geqslant 0$. This increase in entropy represents the energy that is no longer available for doing work in that system.

ENVELOPE, a CURVE or curves TANGENT to every member of a family of curves. For example, the CIRCLE $x^2+y^2=r^2$ (see ANALYTIC GEOMETRY) is envelope to the family of LINES $x \cos\theta + y \sin\theta = r$.

ENVIRONMENT, the surroundings in which animals and plants live. The study of organisms in relation to their environment is called ECOLOGY. Organisms are affected by many different physical factors in their environment, such as temperature, water, gases, light, pressure and also biotic factors such as food resources, competition with other species, predators and disease.

ENZYMES, PROTEINS that act as catalysts (see CATALYSIS) for the chemical reactions upon which LIFE depends. They are generally specific for either one or a group of related reactions. Enzymes are responsible for the production of all the organic materials present in living CELLS, for providing the mechanisms for energy production and utilization in MUSCLES and in the NERVOUS SYSTEM, and for maintaining the intracellular environment within fine limits. They are frequently organized into subcellular particles which catalyze a whole sequence of chemical events in a manner analogous to a production line. Enzymes are themselves synthesized by other enzymes on templates derived from NUCLEIC ACIDS. An average cell contains about 3000 different enzymes. In order to function

Some notable enzymes

enzyme	source	reaction catalyzed
amylases	saliva, pancreas	hydrolysis of starch to maltose
invertase	yeasts	hydrolysis of sucrose to fructose and maltose
lipases	pancreas	hydrolysis of fats
maltase	intestine, pancreas	hydrolysis of maltose to glucose
pepsin	gastric juice	hydrolysis of proteins to peptides and amino acids
pyruvic kinase	animal and plant tissues	transfer of phosphate group in glycolysis, forming ATP and pyruvic acid from ADP and phospho-enolpyruvic acid
zymase complex	yeasts	fermentation of hexose sugars yielding ethanol and carbon dioxide

correctly, many enzymes require the assistance of metal IONS or accessory substances known as **coenzymes** which are produced from VITAMINS in the diet. The action of vitamins as coenzymes explains some of the harmful effects of a lack of vitamins in the diet. A majority of enzymes function in a neutral aqueous environment although some require different conditions. For instance, those which digest food in the stomach require an ACID environment. Cells also contain special activators and inhibitors which switch particular enzymes on and off as required. In some cases a substance closely related to the substrate (the substance on which the enzyme acts) will compete for the enzyme and prevent the normal action on the substrate; this is termed **competitive inhibition**. Again, the product of a reaction may inhibit the action of the enzyme so that no more product is produced until its level has dropped to a particular threshold, this being known as FEEDBACK control. Enzymes either synthesize or break down chemical compounds or transform them from one type to another. These differing actions form the basis of the classification of enzymes into oxidoreductases, transferases, hydrolases, lyases, isomerases and synthetases. Enzymes normally work inside living cells but some (e.g., digestive enzymes) are capable of working outside the cell. Enzymes are becoming important items of commerce and are used in "biological" washing powders, food processing and brewing.

EOCENE, the second epoch of the TERTIARY, lasting from about 55 million to about 40 million years ago. (See also GEOLOGY.)

EPHEDRINE, a drug related to ADRENALINE. It may act as a central nervous system STIMULANT and is used to dilate the BRONCHI in ASTHMA, to dilate the pupils, and occasionally for its effects on BLADDER function. Recently it has been largely replaced by other drugs.

EPICENTER, the point on the earth's surface directly above the focus of an EARTHQUAKE.

EPICURUS (c341–270 BC), Athenian philosopher, the author of Epicureanism. Reviving the ATOMISM of DEMOCRITUS, he preached a materialist, sensationalist philosophy which emphasized the positive things in life and remained popular for more than 600 years.

EPICYCLE, a circle whose center lies on the circumference of a larger circle. In geocentric cosmologies, such as that of PTOLEMY, the planets were thought to move around the earth in epicycles, the centers of which lay on larger **deferent** circles.

EPICYCLOID, the CURVE traced by a point on the circumference of a CIRCLE that is rolling along the outside of another, fixed, circle. (See also CYCLOID; HYPOCYCLOID.)

EPIDEMIC, the occurrence of a disease in a geographically localized population over a limited period of time; it usually refers to INFECTIOUS DISEASE which spreads from case to case or by carriers. Epidemics arise from importation of infection, after environmental changes favoring infectious organisms or due to altered host susceptibility. A **pandemic** is an epidemic of very large or world-wide proportions. Infectious disease is said to be **endemic** in an area if cases are continually occurring there. Travel through endemic areas may lead to epidemics in nonendemic areas.

EPIDERMIS, the outermost layer of the SKIN.

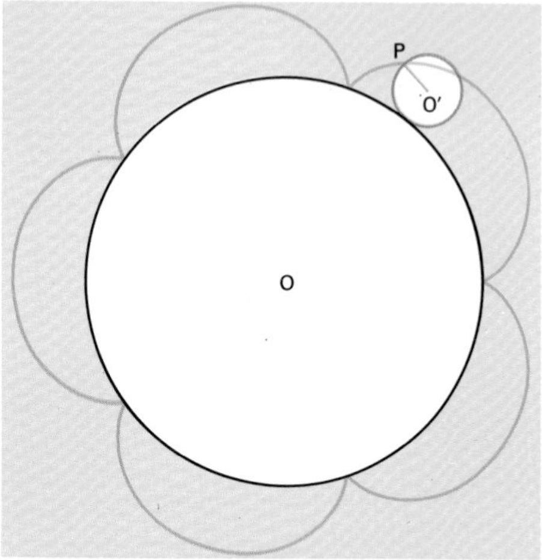

The epicycloid (yellow) traced out by a point on the circumference of an epicycle (orange) as it rolls around the circumference of a deferent circle (black).

EPIGENESIS, the succession of changes by which an embryonic organism passes through stages, more or less distinct from each other, in which new parts and organs appear that were not preformed. The slow acquisition of the characteristic form and function of the individual is called morphogenesis.

EPIGLOTTIS, a structure made of CARTILAGE and covered with mucous membrane, situated in the PHARYNX in front of the glottis or upper windpipe, from which it tends to divert food. It becomes swollen and inflamed in the childhood condition of acute epiglottitis and may cause respiratory obstruction.

EPIGRAPHY, the study of ancient writings inscribed on hard or durable material. (See PALEOGRAPHY.)

EPILEPSY, the "sacred disease" of HIPPOCRATES, a chronic disease of the BRAIN, characterized by susceptibility to CONVULSIONS or other transient disorders of NERVOUS-SYSTEM function and due to abnormal electrical activity within the cerebral cortex. There are many types, of which four are common. **Grand mal** convulsions involve rhythmic jerking and rigidity of the limbs, associated with loss of consciousness, urinary incontinence, transient cessation of breathing and sometimes CYANOSIS, foaming at the mouth and tongue biting. **Petit mal** is largely a disorder of children in which very brief episodes of absence or vacancy occur, when the child is unaware of the surroundings, and is associated with a characteristic ELECTROENCEPHALOGRAPH disturbance. In **focal or Jacksonian epilepsy**, rhythmic movements start in one limb, progress to involve others and may lead to a grand mal convulsion. **Temporal lobe or psychomotor epilepsy** is often characterized by abnormal visceral sensations, unusual smells, visual distortion or memory disorder, and may or may not be followed by unconsciousness. **Status epilepticus** is when attacks of any sort occur repetitively without consciousness being regained in between; it requires emergency treatment.

Epilepsy may be either primary due to an inborn

tendency, often appearing in early life, or it may be symptomatic of brain disorders such as those following trauma or brain SURGERY, ENCEPHALITIS, cerebral ABSCESS, TUMOR, or vascular disease. The ELECTROENCEPHALOGRAPH is the cornerstone of diagnosis in epilepsy, helping to confirm its presence and localize its origin, and suggesting whether there is a structural cause. If epilepsy is secondary, the cause may respond to treatment such as surgery, but all cases require anticonvulsant medication in the long term. Phenytoin (Dilantin), Phenobarbitone (see BARBITURATES), ethosuccimide, carbamazepine, diazepam and related compounds are important anticonvulsants, suitable for different types. DIETARY FOOD (ketogenic diet) may be effective in some cases.

EPINEPHRINE. See ADRENALINE.

EPIPHYSES, the ends of the long bones of the body (e.g., the FEMUR or HUMERUS) which are growth centers during growth and form the articulating surfaces at JOINTS. Each epiphysis remains partly cartilaginous during growth. The CARTILAGE is eventually converted into mature BONE which is then incorporated into the main shaft. Bone-growth disorders such as RICKETS and some HORMONE disorders affect the epiphyses.

EPIPHYTE, or **airplant**, a plant that grows on another but which obtains no nourishment from it. Various LICHENS, MOSSES, FERNS and orchids are epiphytes, particularly on trees. Epiphytes thrive in warm, wet climates. (See also COMMENSALISM; PARASITE.)

EPISTEMOLOGY, the branch of philosophy dealing with the theory of knowledge. Its fundamental questions enquire as to the sources and status of our knowledge. It thus differs from **ontology** which is concerned with the being of things, the nature of things-in-themselves.

EPITHELIOMA, CANCER of the skin EPITHELIUM.

EPITHELIUM, surface tissue covering an organ or structure. Examples include skin and the mucous membranes of the LUNGS, gut and urinary tract. A protective layer specialized for water resistance or absorption, depending on site, it usually shows a high cell-turnover rate.

EPOXY RESIN, class of thermoplastic POLYMERS of the polyether type, formed from epichlorhydrin and a dihydric ALCOHOL or PHENOL (e.g. bisphenol-A), and cross-linked by a curing agent. Inert, strong, adhesive, and good insulators, they are mixed with fillers and plasticizers and used for construction, coating and bonding.

EPSOM SALTS, or epsomite, the mineral MAGNESIUM sulfate heptahydrate, found at Epsom, England, and elsewhere. It has been used as a LAXATIVE.

EQUALITY, in mathematics, two or more expressions which represent the same thing. Thus $3+4=7$ is an equality. For CONSTANT nonzero a, b, c, the equality $a=b$ implies:

$$b=a,$$
$$a+c=b+c,$$
$$ac=bc,$$

and if $b=c$ then $a=c$. (See also EQUIVALENCE.)

EQUATION, a statement of equality. Should this statement involve a VARIABLE it will, unless it is an invalid equation, be true for one or more values of that variable, though those values need not be expressible in terms of REAL NUMBERS: $x^2+2=0$ has two imaginary (see IMAGINARY NUMBERS) roots, $+\sqrt{-2}$ and $-\sqrt{-2}$.

Linear equations are those in which no variable term is raised to a POWER higher than 1. Solution of linear equations in one variable is simple. Consider the equation $x+3=7$. The equation will still be true if we add or subtract equal numbers from each side (see EQUALITY):

$$x+3=7,$$
$$x+3-3=7-3$$

and $$x=4.$$

Linear equations are so called because, if considered as the equation of a CURVE (see ANALYTIC GEOMETRY), they can be plotted as a straight LINE (see also FUNCTION).

Quadratic equations are those in a single variable which appears to the power 2, but not higher. A quadratic equation always has two roots (see ROOTS OF AN EQUATION) though these roots may be equal.

Cubic equations are those in a single variable which appears to the power 3, but not higher. Cubic equations always have three roots, though two or all three of these may be equal.

Degree of an equation. Linear, quadratic and cubic equations are said to be of the 1st, 2nd and 3rd degrees respectively. More generally, the degree of an equation is defined as the SUM of the EXPONENTS of the variables in the highest-power term of the equation. In $ax^5+bx^3y^3+cx^2y^5=0$, the sums of the exponents of each term are, respectively, 5, 6 and 7; hence cx^2y^5 is the highest-power term, and the equation is of the 7th degree.

Radical equations are those in which ROOTS of the variables appear: e.g., $a\sqrt[p]{x}+b\sqrt[q]{x}+c=0$. Radical equations can always be simply converted into equations of the nth order, where $n=1,2,3\ldots$, by raising both sides of the equation to a power, repeating the process where necessary.

Simultaneous equations. A single equation in two or more variables is generally insoluble. However, if there are as many equations as there are variables, it is possible to solve for each variable. Consider

$$2x+xy+3=0 \qquad (1)$$

and $$x+2xy=0. \qquad (2)$$

Multiplying equation (1) by 2 we have

$$4x+2xy+6=0 \qquad (3)$$

and, subtracting equation (2) from this,

$$3x+6=0.$$

Hence $$x=-2.$$

Substituting this value into equation (1) we find the value $y=-\frac{1}{2}$. More complicated simultaneous equations can be solved in the same way (see also ELIMINATION). (See also DIFFERENCE EQUATIONS; DIFFERENTIAL EQUATIONS.)

EQUATION OF A CURVE. See ANALYTIC GEOMETRY.

EQUATOR, an imaginary line drawn about the earth such that all points on it are equidistant from the N and S poles (see NORTH POLE; SOUTH POLE). All points on it have a latitude of 0°. (See also CELESTIAL SPHERE; LATITUDE AND LONGITUDE.)

EQUILATERAL POLYGON, a POLYGON all of whose sides are of equal length. The term is most frequently applied when describing a TRIANGLE.

EQUILIBRIUM, a state in which a mechanical, electrical, thermodynamic or other system will remain if undisturbed. In "stable equilibrium" (as

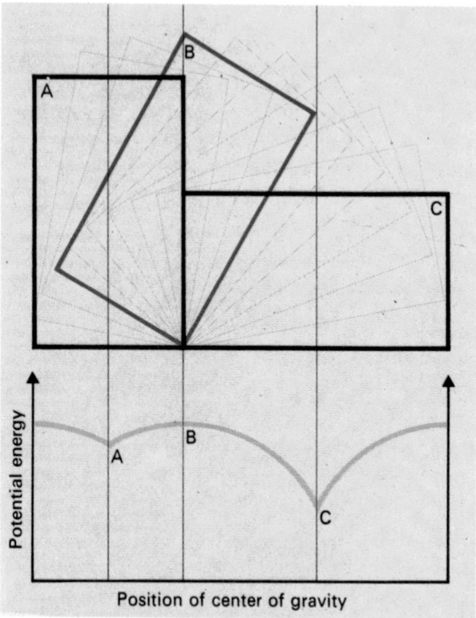

A body is in translational equilibrium when it is motionless and all the physical forces acting on it are mutually balanced. Such conditions are associated with peaks or wells in the body's potential-energy curve. A brick standing on its end (A) is said to be in a position of metastable equilibrium because, although it occupies a well in the potential-energy curve, it is not the deepest well available: by adding to the body's potential energy, it can be pushed through a position of unstable equilibrium (B—a peak in the energy curve), whence it can fall to the position of stable equilibrium (C—the deepest well in the potential-energy curve). A further condition, called neutral equilibrium, is examplified by a sphere resting on a horizontal plane: no horizontal displacement can alter its potential energy.

with a well sprung automobile body) the system returns to its original position if disturbed; the position of stable equilibrium thus determines the rest position of the system. In "unstable equilibrium" (as with a tall pole balanced on one end) it moves farther away. Stable and unstable equilibria correspond to configurations with minimal and maximal ENERGY respectively. Systems in thermodynamic equilibrium, e.g., the contents and air in an unheated room, have the same TEMPERATURE throughout.

EQUILIBRIUM, Chemical, the EQUILIBRIUM condition of a reversible reaction, in which the concentrations of the reactants and products have no tendency to change. In this case, the free energy of the system (see THERMODYNAMICS) is a minimum, and it may be shown that for the general reaction $aA+bB \rightleftharpoons cC+dD$ there is an equilibrium constant K given approximately by

$$K=[C]^c[D]^d/[A]^a[B]^b$$

where [A] is the concentration of A, and so on. Chemical KINETICS yields the same equation, since at equilibrium the rates of the forward and back reactions are equal, and so K is also given by the ratio of the rate constants of these two reactions ($K=k_1/k_{-1}$).

EQUINOXES, (1) the two times each year when day and night are of equal length. The spring or **vernal equinox** occurs in March, the **autumnal equinox** in September. (2) The two intersections of the ECLIPTIC and equator (see CELESTIAL SPHERE). The vernal equinox is in PISCES (see also First Point of ARIES), the autumnal between VIRGO and LEO.

EQUIVALENCE, a relationship * such that if a*b then b*a, and if a*b and a*c then b*c. Moreover, a*a and b*b. One example of equivalence is the relation EQUALITY; another CONGRUENCE. The sign for equivalence is ≡, as in $\frac{1}{4} \equiv 25\%$.

EQUIVALENT WEIGHT, the weight of an element or compound which combines with or displaces the equivalent weight of any other element or compound; for an element, it equals the atomic weight (see ATOM) divided by the VALENCE. This presupposes the **law of equivalent proportions**, which states that the ratio of the weights of two elements A and B which combine with the same weight of an element C, is the same as the ratio of the weights of A and B which combine with each other, or a small integral multiple of it. Since an element may have more than one valence, and a compound may react in more than one way, they may have more than one equivalent weight. The **normality** of a solution is its concentration in gram-equivalent weights per litre; a solution whose normality is 1 is called normal. (See also ELECTROLYSIS; MOLECULAR WEIGHT.)

ERASISTRATUS (3rd century BC), Greek physician of the ALEXANDRIAN SCHOOL who is credited with the foundation of PHYSIOLOGY as a separate discipline.

ERATOSTHENES OF CYRENE (273–192 BC), the Librarian of the ALEXANDRIAN LIBRARY, remembered for his remarkably accurate determination of the circumference of the earth and for his map of the then-known world.

ERBIUM (Er), one of the LANTHANUM SERIES. AW 167.3, mp 1522°C, bp 2510°C, sg 9.066 (25°C).

ERG, the unit of WORK in the CGS system (see CGS UNITS), equal to 10^{-7} joules in SI UNITS.

ERGONOMICS. See HUMAN ENGINEERING.

ERGOT, disease of grasses and sedges caused by fungal species of the genus *Claviceps*. Also, the masses of dormant mycelia (sclerotia) formed in the flower heads of the host plant. Ergots contain toxic ALKALOIDS which if eaten by animals or man, can cause serious poisoning (ergotism or St. Anthony's fire).

ERICSSON, John (1803–1889), Swedish-born US engineer and inventor who developed the screw PROPELLER (patented 1836). He designed and built the *Monitor* (1862), the first warship to bear an armored revolving gun turret. Later he pioneered researches into tapping solar energy for power.

ERIDANUS (the River), large, long CONSTELLATION of the S celestial hemisphere. Of particular interest is Epsilon Eridani: this, at 3.31pc, is the closest star to us to resemble our SUN.

ERIKSON, Eric Homburger (1902–), German-born US psychoanalyst who defined eight stages, each characterized by a specific psychological conflict, in the development of the EGO from infancy to old age. He studied also the IDENTITY, introducing the

Buttes of Jurassic Navajo sandstone overlying Triassic red beds form a spectacular monument to the powers of erosion in Utah, USA.

concept of the identity crisis.

ERLANGER, Joseph (1874–1965), US physiologist who shared with Herbert GASSER the 1964 Nobel Prize for Physiology or Medicine for their discovery that different nerve fibers have different functions, carrying different types of impulses and at different speeds. (See NERVOUS SYSTEM.)

EROS, an ASTEROID measuring roughly 35 × 16 × 8km discovered in 1898 by G. Witt. Eros' eccentric orbit brings it close to earth every seven years, sometimes within 22 million km. Its orbital period is 643 days.

EROSION, the wearing away of the earth's surface by natural agents. Running water constitutes the most effective eroding agent, the process being accelerated by the transportation of particles eroded or weathered farther upstream: it is these that are primarily responsible for further erosion. GROUNDWATER may cause erosion by dissolving certain minerals in the rock (see also KARST). OCEAN WAVES and especially the debris that they carry may substantially erode coastlines. GLACIERS are extremely important eroding agents, eroded material becoming embedded in the ice and acting as further abrasives (see ABRASION). Many common landscape features are the results of glacial erosion (e.g., DRUMLINS, FJORDS). Rocks exposed to the atmosphere undergo **weathering**: mechanical weathering usually results from temperature changes (e.g., in **exfoliation**, the cracking off of thin sheets of rock due to extreme daily temperature variation); chemical weathering results from chemical changes brought about by, for example, substances dissolved in RAIN water. Wind erosion may be important in dry, sandy areas. (See also SOIL EROSION.)

ERYSIPELAS, or St. Anthony's fire, a SKIN infection, usually affecting the face, caused by certain types of STREPTOCOCCUS. It is common in infancy and middle age. ERYTHEMA and swelling spread with a clear margin and cause blistering. It is a short illness with FEVER; if it affects the trunk it may however cause prostration and can prove fatal. PENICILLIN is the ANTIBIOTIC of choice.

ERYTHEMA, redness of the SKIN due to increased CAPILLARY BLOOD flow; it occurs in INFLAMMATION including DERMATITIS, and numerous rashes.

ERYTHROCYTE, or red BLOOD cell, a discoid cell without a nucleus, that contains HEMOGLOBIN and is responsible for OXYGEN transport in the body.

ESAKI, Leo (1925–), Japanese-born US physicist awarded, with I. GIAEVER and B. JOSEPHSON, the 1973 Nobel Prize for Physics for his work on tunneling (see WAVE MECHANICS).

ESCALATOR, a moving stairway to transport passengers or goods from one level to another. Most simply, it comprises steps mounted on an endless belt such that they fold flat at top and bottom to allow passengers more easily to step on and off. (See also ELEVATOR.)

ESCAPE VELOCITY, the velocity that a less massive body must achieve in order to escape from the gravitational attraction of a more massive body;

sometimes known as the parabolic velocity. The earth's escape velocity is 11.2km/s. Less massive planets have smaller escape velocities (Mars: 5.1km/s), more massive planets greater escape velocities (Jupiter: 61.0km/s).

ESCARPMENT. See SCARP.

ESKER, serpentine ridge of glacial DRIFT, up to several kilometres long, formed from deposits at the mouth of a subglacial stream as the GLACIER retreated.

ESOPHAGUS, the thin tube leading from the PHARYNX to the STOMACH. Food passes down it as a bolus by gravity and PERISTALSIS. Its diseases include reflux esophagitis (HEARTBURN), ULCER, stricture and CANCER.

ESP, or **Extrasensory Perception,** the perception other than by the recognized SENSES of an event or object; and, by extension, those powers of the mind (such as **telekinesis**, the moving of distant objects by the exercise of willpower) that cannot be scientifically evaluated. The best known and most researched area of ESP is **telepathy**, the ability of two or more individuals to communicate without sensory contact: though laboratory tests (see PARAPSYCHOLOGY) have been inconclusive, it seems probable that telepathic communication between individuals can exist. Analogous is **empathy**, the communication across distance of EMOTIONS. Another important area of ESP is **precognition**, the prior knowledge of an event: again, despite a mass of circumstantial evidence, laboratory tests have been inconclusive. The term **clairvoyance** is sometimes used for ESP.

ESSENCE, in philosophy, a term referring to the permanent actuality of a thing, the that-by-which it can be recognized, whatever its outward appearance. Different philosophers have used the term with various detailed significations; LOCKE, for instance, distinguished a thing's real essence, the what-it-is-in-itself, from its nominal essence, the what-it-appears-to-be, the name men give it.

ESSENTIAL OILS, volatile oils produced by many plants, flowers and fruits which give them their characteristic odor or flavor. A great number of chemical compounds have been identified from various essential oils. They are used commercially for flavoring food and in perfumes; some have medicinal properties.

Esters

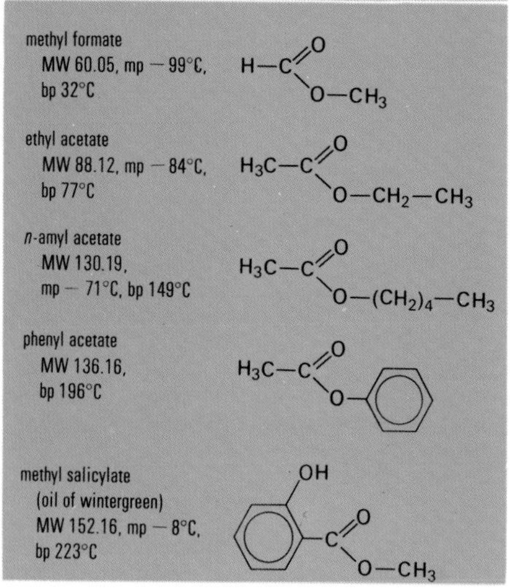

ESTERS, organic compounds formed by CONDENSATION of an ACID (organic or inorganic) with an ALCOHOL, water being eliminated. This reaction, esterification, is acid-catalyzed; its reverse, HYDROLYSIS, is acid- or base-catalyzed; an EQUILIBRIUM is set up in aqueous solution. Many esters occur naturally: those of low molecular weight have fruity odors and are used in flavorings, perfumes and as solvents; those of higher molecular weight are FATS and WAXES.

ESTIVATION. See HIBERNATION.

ESTROGENS, female sex HORMONES concerned with the development of secondary sexual characteristics and maturation of reproductive organs. They are under the control of pituitary-gland GONADOTROPHINS and their amount varies before and after MENSTRUATION and in PREGNANCY. After the menopause, their production decreases. Many pills used for CONTRACEPTION contain estrogen, as do some preparations given to menopausal women. Their administration may lead to venous THROMBOSIS, and some other diseases.

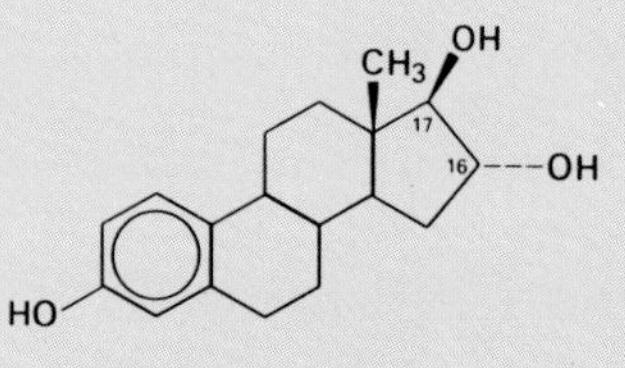

Estriol, one of the principal estrogens. Other important estrogens include 17(β)-estradiol (unsubstituted at the C-16 position) and estrone (unsubstituted at C-16 and having a keto group (=O) at C-17).

ESTUARY, the typically funnel-shaped part of a river near its mouth where fresh- and seawater mix and which is affected by TIDES. At ebb tide both tide and river current assist in the EROSION of the estuary Estuaries may also form by local subsidence of the coast. (See also FIRTH.) Many estuaries provide important harbors.

ETCHING, an ENGRAVING technique in which acid is used to "bite" lines into a metal plate which is then printed, usually intaglio. The plate, usually copper or zinc, is first coated with a resin "ground" through which the design is drawn with a needle. Only the exposed metal is etched away. Different line thicknesses can be obtained by selective stopping out and repeated exposure to the acid. (See also AQUATINT.)

ETHANE (C_2H_6), an ALKANE occurring in natural gas and formed in petroleum CRACKING. It is catalytically dehydrogenated to produce ETHYLENE. MW 30.1, mp −183°C, bp −89°C.

ETHANOL (C_2H_5OH), or ethyl alcohol, also known as grain alcohol, the best-known ALCOHOL; a colorless, inflammable, volatile, toxic liquid, the active constituent of ALCOHOLIC BEVERAGES. Of immense industrial importance, ethanol is used as a solvent, in ANTIFREEZE, as an ANTISEPTIC and in much chemical synthesis. Its production is controlled by law, and it is heavily taxed unless made unfit for drinking by adulteration (denatured alcohol); see METHANOL. Most industrial ethanol is the AZEOTROPIC MIXTURE containing 5% water. It is made by FERMENTATION of sugars or by catalytic hydration of ETHYLENE. MW 46.1 mp −112°C, bp 78°C. (See also DISTILLED LIQUOR.)

ETHER, a hypothetical medium postulated by late-19th-century physicists in order to explain how LIGHT could be propagated as a wave motion through otherwise empty space. Light was thus thought of as a mechanical WAVE MOTION in the ether. The whole theory was discredited following the failure of the MICHELSON–MORLEY EXPERIMENT to detect any motion of the earth relative to the supposed stationary ether.

ETHERS, organic compounds of general formula R-O-R′, where R and R′ are HYDROCARBON radicals. They are prepared by catalytic dehydration of ALCOHOLS, catalytic hydration of ALKENES, or by reacting an ALKYL HALIDE with a sodium alkoxide (see ALCOHOLS). Ethers are chemically fairly inert, though they are split by hydrogen HALIDES, and form explosive PEROXIDES on standing in air. **Diethyl ether** $(C_2H_5)_2O$, the most important, is used as an anesthetic (see ANESTHESIA) and an industrial solvent. It is a volatile, inflammable liquid, partially miscible with water. MW 74.1, mp −116°C, bp 35°C.

Ethers

methoxymethane (dimethyl ether) MW 46.07, mp −139°C, bp −23°C	$H_3C—O—CH_3$
ethoxyethane (ether, diethyl ether) MW 74.12, mp −116°C, bp 35°C	$H_3C—CH_2—O—CH_2—CH_3$
methoxybenzene (anisole) MW 108.15, mp −38°C, bp 155°C	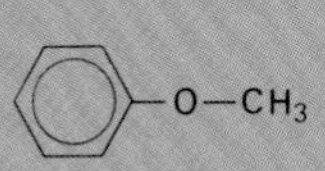
1,2-epoxyethane (ethylene oxide, oxirane) MW 44.05, mp −111°C, bp 14°C	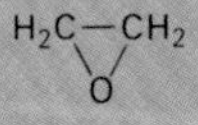
1,4-epoxy-1,3-butadiene (furan, furfuran) MW 68.08, mp −86°C, bp 31°C	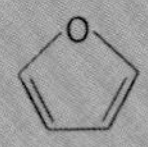

ETHNOGRAPHY, in ANTHROPOLOGY, the descriptive study of different human cultures, as distinct from the more theoretical ETHNOLOGY. The terms ethnology and ethnography are often now used interchangeably, and the latter is gradually falling out of use.

ETHNOLOGY, the science dealing with the differing races of man, where they originated, their distribution about the world, their characteristics and the relationships between them. More generally, the term is used to mean cultural ANTHROPOLOGY.

ETHOLOGY, meaning literally the "study of behavior," is applied particularly to the European school of behavioral scientists. Their concern is the behavior of the animal in the wild, paying special attention to patterns of behavior, or "instinctive" behavior. This contrasts with the interests of most American behaviorists whose work stems largely from comparative psychology and is concentrated on the study of the learning process, studies which are mainly if not entirely laboratory based. The founders of ethology were Konrad LORENZ and Nikolaas TINBERGEN, whose ideas on instinctive behavior supplied the stimulus for renewed study of animals in the wild.

ETHYL COMPOUNDS, organic compounds containing the ethyl group, C_2H_5 (see ALKANES). The most important are: ETHANOL, diethyl ETHER, ethyl ACETATE, LEAD tetraethyl, and the ethyl halides (see ALKYL HALIDES). Ethyl chloride is a refrigerant and local anesthetic.

ETHYLENE (C_2H_4), or ethene, the simplest ALKENE, made by CRACKING of ETHANE and PROPANE. It is a colorless gas, used to ripen fruit and to make ETHANOL, diethyl ETHER, ethyl chloride, POLYETHYLENE and STYRENE. MW 28.05, mp −169°C, bp −104°C. It is oxidized by air over a silver catalyst to **ethylene oxide**, a reactive gas used as a fumigant and to make plastics and emulsifiers, and hydrated to ETHYLENE GLYCOL.

ETHYLENE GLYCOL $(HOCH_2CH_2OH)$, colorless syrupy liquid, an ALCOHOL, made from ETHYLENE. It is used to make polyester POLYMERS, and, mixed with water, as an ANTIFREEZE and deicer. MW 62.1, mp −12°C, bp 198°C.

ETYMOLOGY (from Greek *etymos*, true meaning, and *logos*, word), the history of a word or other linguistic element; and the science, born in the 19th century, concerned with tracing that history, by examining the word's development since its earliest appearance in the language; by locating its transmission into the language from elsewhere; by identifying its cognates in other languages; and by tracing it and its cognates back to a (often hypothetical) common ancestor. **Cognates** (from Latin *co*, together, and *nasci*, to be born) of English words appear in many languages: our "father" is cognate with the German "*Vater*" and French "*père*," all three deriving from the Latin "*pater*." An **etymon** is the earliest known form of a word, though the term is sometimes applied to any early form. (See also LINGUISTICS; PHILOLOGY.)

EUCLID (c300 BC), Alexandrian mathematician whose major work, the *Elements*, is still the basis of much of geometry (see EUCLIDEAN GEOMETRY): its fifth postulate (the Euclidean axiom) cannot be proved, and this lack of proof gave rise to the NON-EUCLIDEAN GEOMETRIES. Other ascribed works include *Phaenomena*, on SPHERICAL GEOMETRY, and *Optics*, treating vision and PERSPECTIVE.

EUCLIDEAN GEOMETRY, the branch of GEOMETRY dealing with the properties of three-dimensional space. It is commonly split up into plane geometry, which is concerned with figures and constructions in two or less dimensions (such as the POLYGON; CIRCLE; ellipse (see CONIC SECTIONS); CURVE; LINE and POINT), and solid or three-dimensional geometry, which deals with three-dimensional figures (such as the POLYHEDRON; SPHERE and ELLIPSOID) and the relative spatial positions of figures of three dimensions or less. It takes its name from EUCLID, whose *Elements*, written c300 BC, summarized all the mathematical knowledge of contemporary ancient Greece into 13 books; those on geometry were taken as the final, authoritative word on the subject for well over a millennium and still form the basis for many school geometry textbooks. (See also ANGLE; AREA; CONGRUENCE; CYLINDER; DUPLICATION OF THE CUBE; GOLDEN SECTION; PLANE; PYRAMID; PYTHAGORAS' THEOREM; TRIANGLE; VOLUME.)

EUDEMUS OF RHODES (4th century BC), Greek mathematician and anatomist, a pupil of ARISTOTLE and a friend of THEOPHRASTUS. He probably edited *Eudemian Ethics*, one of the three works in which Aristotle's ethical statements are preserved.

EUDOXUS OF CNIDUS (c400–c350 BC), Greek mathematician and astronomer who proposed a system of homocentric crystal spheres to explain planetary motions; this system was adopted in ARISTOTLE's cosmology (see ASTRONOMY). He was probably responsible for much of the content of Book V of EUCLID's *Elements*.

EUGENICS, the study and application of scientifically directed selection in order to improve the genetic endowment of human populations. Eugenic control was first suggested by Sir Francis GALTON in the 1880s. People supporting eugenics suggest that those with "good" traits should be encouraged to have children while those with "bad" traits should be discouraged or forbidden from having families. But who is to decide which traits are "good?"

EUGLENA, a genus of the PROTOZOA that some biologists regard as a plant because some species have CHLOROPLASTS. (See also ALGAE.)

EUGLENOPHYTA, or euglenoids. See ALGAE.

EUKARYOTE; eukaryotic cell. See CELL.

EULER, Leonhard (1707–1783), Swiss-born mathematician and physicist, the father of modern ANALYTIC GEOMETRY and important in almost every area of mathematics. He introduced the use of analysis (especially CALCULUS, a field which he also profoundly affected) into the study of MECHANICS; and made major contributions to modern ALGEBRA. **Euler's Relation** links the logarithmic and trigonometric functions: $e^{ix} = \cos x + i \sin x$, or, more generally, $e^{ikx} = \cos kx + i \sin kx$ (see IMAGINARY NUMBERS; TRIGONOMETRY). *e*, the EXPONENTIAL, is often named **Euler's Number** for him; and the VENN DIAGRAM is sometimes called the **Euler Diagram**. He worked also on a theory to explain the motions of the MOON and pioneered the science of HYDRODYNAMICS.

EULER, Ulf Svante von (1905–), Swedish physiologist, President of the Nobel Foundation from 1965, who shared with AXELROD and KATZ the 1970 Nobel Prize for Physiology or Medicine for their independent work on the chemistry of the transmission of nerve impulses (see NERVOUS SYSTEM).

EULER-CHELPIN, Hans Karl August Simon von (1873–1964), German-born Swedish biochemist who shared with Arthur HARDEN the 1929 Nobel Prize for Chemistry for their work on ENZYME action in the FERMENTATION of sugar.

EUMYCOTINA. See FUNGI.

EUROPIUM (Eu), one of the LANTHANUM SERIES. AW 152.0, mp 822°C, sg 5.243 (25°C).

EUSTACHIAN TUBE. See EAR.

EUSTACHIO, Bartolomeo (1524–1574), Italian anatomist. Aiming at first to vindicate GALEN against VESALIUS and others, he brought a new skill and accuracy to dissection. The eustachian tube of the EAR is named for him. (See also ANATOMY.)

EUTHANASIA, the practice of hastening or causing the DEATH of a person suffering from incurable DISEASE. While frequently advocated by various groups, its practical and legal implications are so contentious that it is illegal in most countries.

EUTHERIA, or **Placental Mammals**, a subclass of the MAMMALIA that includes all species except those classified in the MONOTREMATA and METATHERIA (marsupials) and in which the FETUS is nourished in the womb attached to a highly organized PLACENTA until a comparatively late stage in its development. They are included in the orders: INSECTIVORA; RODENTIA; EDENTATA; LAGOMORPHA; CARNIVORA; ARTIODACTYLA; PERISSODACTYLA; CETACEA; HYRACOIDEA; PROBOSCIDIA; SIRENIA; CHIROPTERA, and PRIMATES.

EUTROPHICATION, the increasing concentration of plant nutrients and FERTILIZERS in lakes and estuaries, partly by natural drainage and partly by POLLUTION. It leads to excessive growth of algae and aquatic plants, with oxygen depletion of the deep water, causing various undesirable effects.

EVANS, Sir Arthur John (1851–1941), English archaeologist famous for his discovery of the Minoan Civilization from excavations at Knossos in Crete. He was curator of the Ashmolean Museum, Oxford 1884–1908 and professor of prehistoric archaeology at Oxford from 1909.

EVANS, Oliver (1755–1819), US engineer who constructed the first high-pressure STEAM ENGINE in America (c1802), and possibly the first continuous production-line system.

EVAPORATION, the escape of molecules from the surface of a liquid into the vapor state. Only those molecules with above-average ENERGY are able to overcome the cohesive forces holding the liquid together and escape from the surface. Eventually all the molecules left in the liquid have below-average energy; its temperature is now lower. In an enclosed space, the pressure of the vapor above the surface eventually reaches a maximum, the saturated vapor pressure (SVP). This varies according to the substance concerned and, together with the rate of evaporation, increases with temperature, equalling atmospheric pressure at the liquid's BOILING POINT.

EVAPORITES, sedimentary deposits of salts that have been precipitated from solution owing to the evaporation of a body of water (see EVAPORATION; SOLUTION). Evaporite deposits have the least soluble salts at the bottom (CALCIUM salts), followed by the very soluble halite (common SALT) and MAGNESIUM and POTASSIUM salts. Most important commercially are GYPSUM ($CaSO_4 . 2H_2O$), ANHYDRITE ($CaSO_4$) and halite (NaCl). (See also SALT DOME; SEDIMENTATION.)

EVENT HORIZON, the boundary of a BLACK HOLE beyond which an outside observer can detect nothing.

EVERGREEN, a plant that retains its leaves all the year around, although they are continually being

...ution: the phylogeny of life on earth.

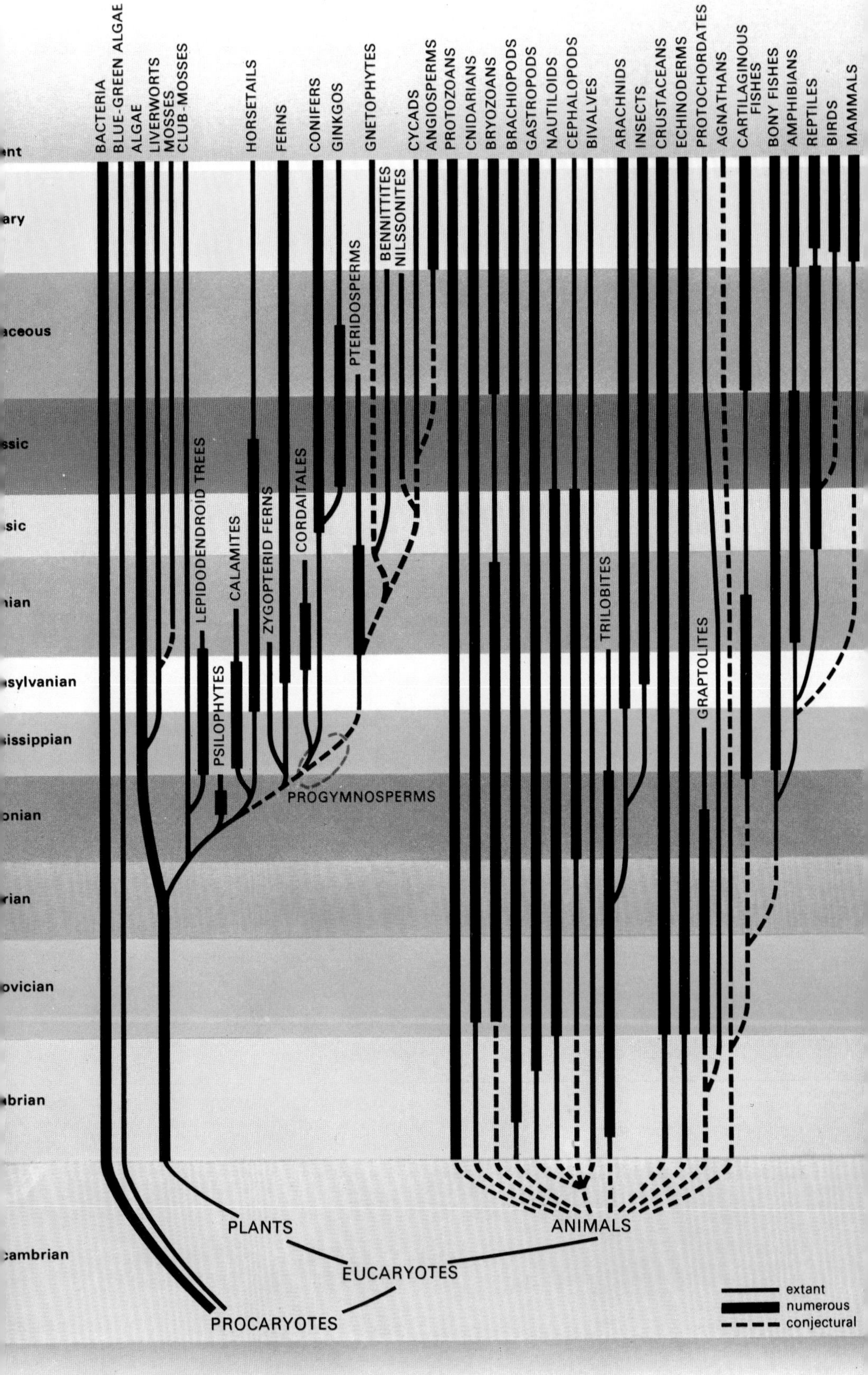

shed and replaced. Examples are the holly, laurel and pine. (See also DECIDUOUS TREES.)

EVOLUTION, the process by which living organisms have changed since the origin of life. The formulation of the theory of evolution by NATURAL SELECTION is credited to Charles DARWIN, whose observations while sailing around the world on H.M.S. BEAGLE, when taken together with elements from MALTHUS' population theory and viewed in the context of LYELL's doctrine of UNIFORMITARIANISM, led him to the concept of natural selection, but the theory also later occurred independently to A. R. WALLACE. Other theories of evolution by the inheritance of ACQUIRED CHARACTERISTICS had earlier been proposed by E. DARWIN and LAMARCK. Darwin defended the mechanism of natural selection on the basis of three observations: that animals and plants produced far more offspring than were required to maintain the size of their population; that the size of any natural population remained more or less stable over long periods, and that the members of any one generation exhibited variation. From the first two he argued that in any generation there was a high mortality rate, and from the third that, under certain circumstances, some of the variants had a greater chance of survival than did others. The surviving variants were, by definition, those most suited to the prevailing environmental conditions. Any change in the environment led to adjustment in the population such that certain new variants were favored and gradually became predominant.

The missing link in Darwin's theory was the mechanism by which heritable variation occurs. Unknown to him, a contemporary, G. MENDEL, had demonstrated the principle of GENETICS and had deduced that the heritable characters were controlled by discrete particles. We now know these particles to be GENES which are carried on the CHROMOSOMES. Mendel's variants were caused by RECOMBINATION and MUTATION of the genes. Natural selection acts to eradicate unfit variants either by mortality of the individual or by ensuring that such individuals do not breed. How then can natural selection lead to the evolution of a new character? The key is that a character that is advantageous to an individual in the normal environment may become disadvantageous if the environment changes. This means that individuals that happen through variation to be well adapted to the new set of circumstances will tend to survive and thus become the norm.

An example of natural selection at work is provided by studies carried out recently on North American sparrows. Large numbers of sparrows were trapped and their various characteristics recorded. In this way the "normal" sparrow was identified. A further collection was made of dead sparrows which had succumbed to the adverse conditions of a particularly severe winter. It was found that the individuals in the second sample were all different in some important respect from the "normal" sparrow. Natural selection could thus be seen to be maintaining a population that was ideally suited to the North American environment.

Today, the evidence for evolution is overwhelming and comes from many branches of biology. For instance, the comparative anatomy of the arm of a man, the foreleg of a horse, the wing of a bat and the flipper of a seal reveals that these superficially different organs have a very similar internal structure, this being taken to indicate a common ancestor. Then, the study of the embryos of mammals and birds reveals that at some stages they are virtually indistinguishable and thus have common ancestors. Again, vestigial organs such as the appendix of man and the wing of the ostrich are of no use to these mammals, but in related species such as herbivores and flying birds they clearly are of vital importance. Evidently these individuals have progressively evolved in different ways from a common ancestor. The hierarchical classification of plants and animals into species, genus, family etc. (see TAXONOMY) is a direct reflection of the natural pattern that would be expected if evolution from common ancestors occurred. Again, the geographical distribution of animals and plants presents many facts of evolutionary significance. For example, the tapir is today centered in two widely separated areas, the E Indies and South America. However, it probably evolved in a single center, migrated across the world and then became extinct in many areas as habitats changed. Indeed FOSSILS of tapirs have been found in Asia, Europe and North America. Fossils in general provide convincing evidence of evolution. Thus, the theory of the evolution of birds indicates descent from now extinct reptiles. The fossil ARCHAEOPTERYX, a flying reptile with some bird-like features, was believed to represent the missing link in this development.

LIFE probably first evolved from the primeval soup some 3000–4000 million years ago when the first organic chemicals were synthesized due to the effects of lightning. Primitive ALGAE capable of synthesizing their own food material have been found in geological formations some 2 000 million years old. Simple forms of animals and fungi then evolved. From that time there has been a slow evolution of multicellular organisms.

EXCAVATORS, machines used for digging ditches, trenches or holes in the ground, as well as for removing heaps of rubble or banks of earth. At the end of a jointed boom (or jib) is a bucket whose angle to the boom may be altered: power is usually supplied by a hydraulic system (see HYDRAULICS). Smaller excavators usually resemble a tractor with the addition of the boom; larger ones comprise a **slewing platform** mounted on caterpillar tracks. The **backhoe** is used usually for digging below the working surface: the bucket is forced into the ground, then drawn back toward the base. In the **face shovel**, used usually to excavate above the working surface, the bucket is forced forward, then lifted toward the base.

EXCLUSION PRINCIPLE, the law accounting for the different chemical properties of the ELEMENTS and numerous other phenomena. Applying to those particles called fermions (see FERMI-DIRAC STATISTICS), particularly ELECTRONS, it is a consequence of the fact that particles of the same kind are indistinguishable, and states that only one such particle can occupy a given quantum state (see QUANTUM MECHANICS) at a time. In a system of such particles, the lack of empty neighboring states often prevents most particles from contributing to the system properties, which thus depend only on the states bordering the filled ones—

A 88bhp Massey-Ferguson tracked excavator. The bucket capacity is 570 litres.

i.e., those at the **Fermi surface**. (See also PAULI, W.)

EXCRETION, the removal of the waste products of METABOLISM either by storing them in insoluble forms or by removing them from the body. Excretory organs are also responsible for maintaining the correct balance of body fluids. In VERTEBRATES the excretory organs are the KIDNEYS: blood flows through these and water and waste products are removed as URINE. Other forms of excretory organs include the Malpighian tubes of insects, arachnids and myriapods, the contractile vacuoles of Protozoa and the nephridia of annelids. In plants, excretion usually takes the form of producing insoluble salts of waste products within the cells.

EXERCISE, or physical exertion, the active use of skeletal muscle in recreation or under environmental stress. In exercise, MUSCLES contract actively, consuming OXYGEN at a high rate, and so require increased BLOOD CIRCULATION; this is effected by increasing the HEART output by raising the PULSE and increasing the blood expelled with each beat. Meanwhile, the CAPILLARIES in active muscles dilate. The raised demand for oxygen and, more especially, the increased production of carbon dioxide in the muscles increase the rate of RESPIRATION. Some energy requirements can be supplied rapidly without oxygen but, if so, the "oxygen debt" must be made good afterward. GASTROINTESTINAL TRACT activity is reduced during exercise. Changes in the autonomic and central NERVOUS SYSTEMS, HORMONES and local regulators are responsible for adaptive changes in exercise. In athletes, exercise increases muscle efficiency and cardiac compensation.

EXOBIOLOGY, or **xenobiology**, the study of life beyond the earth's atmosphere. Drawing on many other sciences (e.g., biochemistry, physics), it is for obvious reasons a discipline dealing primarily in hypotheses (though FOSSIL organic matter has been found in certain meteorites—see METEOR). An important branch deals with the effects on man of nonterrestrial environments.

EXOCRINE GLANDS. See GLANDS.

EXOGAMY. See ENDOGAMY AND EXOGAMY.

EXOSKELETON, any skeletal material that lies on the surface of the animal's body. In this position it not only performs the mechanical functions common to any other SKELETON but, in addition, affords protection. Exoskeletons are particularly well developed in arthropods such as crabs, lobsters and insects.

EXOSPHERE, the outermost zone of the earth's ATMOSPHERE (altitude greater than 500km) where terrestrial GRAVITATION is too weak an effect to prevent the escape of uncharged particles.

EXPANDING UNIVERSE. See COSMOLOGY; UNIVERSE.

EXPANSION, the increase in volume of a body as a result of changing conditions, normally increasing TEMPERATURE or decreasing PRESSURE, the latter being more important for GASES. Contraction is the reverse process. In most solids and liquids, increasing temperature increases the random thermal motion of their atoms, which tend to move apart, i.e., expansion occurs. The amount of expansion is usually expressed as a coefficient of expansion—the fractional change in length or volume per unit temperature change—and is specific for a given material. Water is unusual in that it expands on cooling from 4°C to 0°C. This means that ice floats on water at 0°C and rivers freeze from the surface downward.

EXPERIMENT. See SCIENTIFIC METHOD.

EXPLOSIVES, substances capable of very rapid COMBUSTION (or other exothermic reaction—see THERMOCHEMISTRY) to produce hot gases whose rapid expansion is accompanied by a high-velocity shock wave, shattering nearby objects. The detonation travels 1000 times faster than a flame. The earliest known explosive was GUNPOWDER, invented in China in the 10th century AD, and in the West by Roger BACON (1242). Explosives are classified as **primary explosives**, which explode at once on ignition, and are used as DETONATORS; and **high explosives**, which if ignited at first merely burn, but explode if

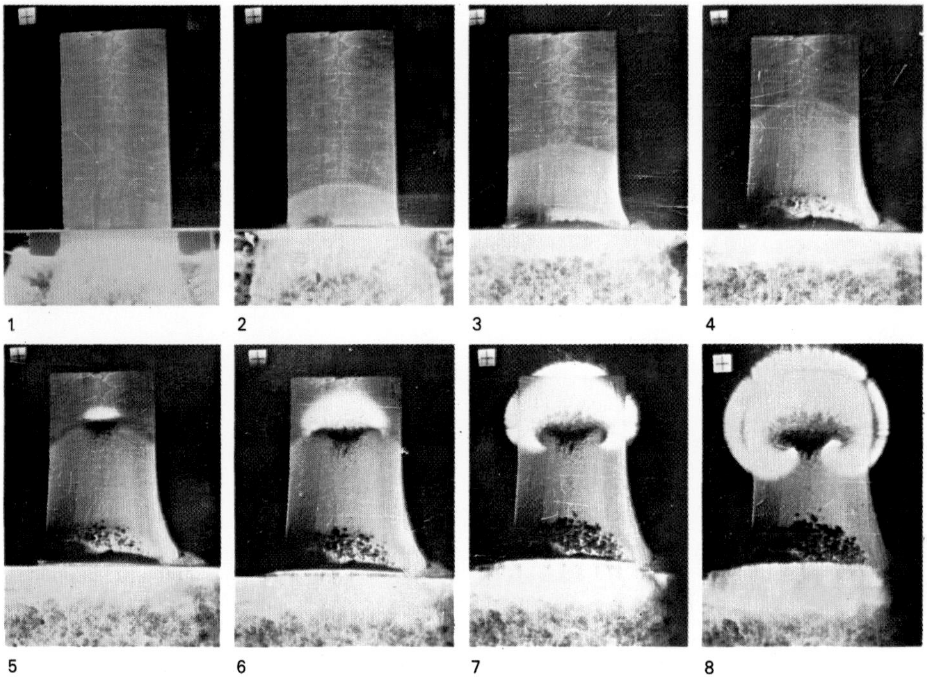

An explosive-driven shock wave (lighter area) travels down a charge of high explosive RDX/TNT (1–4) before initiating a detonation in the charge. The detonation, with its characteristic outburst of light, appears in (5) and then spreads through the charge (6–8). Frames (1–4) succeed each other after intervals of 10 microseconds; 2 microseconds separate successive frames of sequence (4–8). The total time for the whole sequence (1–8) is 38 microseconds.

detonated by a primary explosion. The division is not rigid. Military high explosives are usually mixtures of organic nitrates, TNT, RDX, PICRIC ACID and PETN, which are self-oxidizing. Commercial blasting explosives are less-powerful mixtures of combustible and explosive substances; they include DYNAMITE (containing NITROGLYCERIN, ammonium nitrate and sometimes NITROCELLULOSE), ammonals (ammonium nitrate+aluminum) and Sprengel explosives (an oxidizing agent mixed with a liquid fuel such as nitrobenzene just before use). Obsolete explosives include the dangerous chlorates and perchlorates, and the uneconomical liquid oxygen explosives (LOX). Explosives which do not ignite firedamp (see DAMP) are termed "permissible," and may be used in coal mines. Propellants for guns and rockets are like explosives, but burn fast rather than detonating.

EXPONENT, a number such as x in the expression a^x, a being a number to be used as a FACTOR x times: e.g., $a^3 = a.a.a$. In the expression, a is termed the base. (See also POWER.)

Exponential equation, in ALGEBRA, an equation of the form $a^x = b$ where a and b are numbers. For example, if $3^x = 81$ we can solve for x by restating 81 as $9.9 = 3.3.3.3$, and so $x = 4$. More difficult problems are solved using LOGARITHMS: e.g., if $4^x = 15$, then $x \log_e 4 = \log_e 15$. Therefore $1.3863x = 2.7081$ and $x = 1.9535$ (to 4 decimal places).

Exponential function, in ANALYTIC GEOMETRY, a FUNCTION of the form $f(x) = a^x$ where x is positive and does not equal 1. In terms of differential CALCULUS, $f'(x) = a^x \log_e a$ if $f(x) = a^x$.

Exponential growth. Since the value of a^x increases considerably with increase in x, the term exponential growth is used loosely in STATISTICS to refer to the very rapid growth in number of a quantity over a period of time. More accurately, it refers to an increase which, if plotted against units of time, would approximate closely to an exponential CURVE (the plotting of an exponential function).

Exponential notation, the expression of a number by means of a base and an exponent; e.g., 16 expressed as 4^2. (See also EXPONENTIAL.)

EXPONENTIAL, the base of the natural LOGARITHMS, known also as EULER's number and always symbolized by the letter e. It is defined as the REAL NUMBER such that $\int_1^e x^{-1}dx = 1$, and is an IRRATIONAL NUMBER whose value to six decimal places is 2.718284. (See also EXPONENT.)

EXPONENTIAL SERIES, the SERIES $1/0! + 1/1! + 1/2! + 1/3! + \ldots + 1/n! + \ldots$ (see FACTORIAL). The LIMIT of the series as $n \rightarrow \infty$ is e (see EXPONENTIAL).

EXPOSURE METER. See LIGHT METER.

EXTERIOR ANGLE, the angle between one side of a POLYGON and the extension of the side adjacent to it. The exterior angle at any VERTEX is the supplement of the INTERIOR ANGLE there (see ANGLE). The sum of the exterior angles of a convex polygon is always equal to 360° in plane geometry.

EXTINCT ANIMALS. See PREHISTORIC ANIMALS.

EXTINCTION, the disappearance of any species or

group of organisms. Extinction may be the result of a number of factors including physical changes in the environment, and competition from other species—during recent times notably from man.

EXTRACTION, method of separating a desired substance from a SOLUTION containing one or more other substances. The solution is shaken with a solvent immiscible with it, into which the solute required (but not the others) is largely transferred, according to the distribution law that its concentrations in the two phases are in a constant ratio. The process is repeated if necessary. Continuous extraction is used if the distribution ratio is small. Metal ions are extracted as a CHELATE by adding a ligand to the solvent.

EXTRAPOLATION. See INTERPOLATION AND EXTRAPOLATION.

EXTRASENSORY PERCEPTION. See ESP.

EXTREMUM, a point on a CURVE $f(x)$ at which there is an instantaneous change of sign of the derivative $f'(x)$ (see CALCULUS; FUNCTION). If $f'(x)$ is defined at this point it has value 0. There are two types of extrema: maxima and minima. At a **maximum** $f'(x)$ changes from positive to negative with increasing x; at a **minimum** $f'(x)$ changes from negative to positive. Points of INFLECTION are not extrema.

EXTRUSION, a way of producing metal and plastic components of constant cross-section (e.g., tubes, sheets) by forcing the material through a DIE. In **cold extrusion** a billet of metal is surrounded by liquid lubricant in a suitable chamber; hydrostatic PRESSURE is increased, forcing the metal through orifices at one end of the chamber. In **hot extrusion**, used for plastics, the material is melted throughout before being forced through a die and rapidly cooled.

EYE, the specialized sense organ concerned with VISION. In all species it consists of a lens system linked to a LIGHT receptor system connected to the central

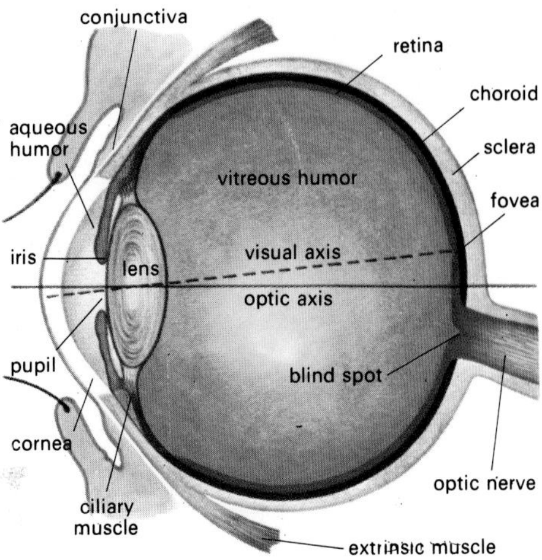

Section through the human eye.

NERVOUS SYSTEM. In man and mammals, the eye is roughly spherical in shape, has a tough fibrous capsule with the transparent CORNEA in front, and is moved by specialized eye muscles. The exposed surface is kept moist with tears from lacrymal glands. Most of the eye contains VITREOUS HUMOR—a substance with the consistency of jelly—which fills the space between the lens and the retina, while in front of the lens there is watery or AQUEOUS HUMOR. The colored iris or aperture surrounds a hole known as the pupil. The focal length of the lens can be varied by specialized ciliary muscles. The RETINA is a layer containing the nerve cells (rods and cones) which receive light, together with the next two sets of cells in the relay pathway for vision. The optic nerve leads back from the retina to the BRAIN. Rods and cones receive light reflected from a pigment layer and contain pigments (e.g., RHODOPSIN) which are bleached by light and thus set off the nerve-cell reaction.

Schematic section through a machine for the hot extrusion of plastic moldings. (A) heating elements· (B) forcing screw; (C) mantle; (D) hard metal cylinder; (E) hot plastic; (F) molding die; (G) product.

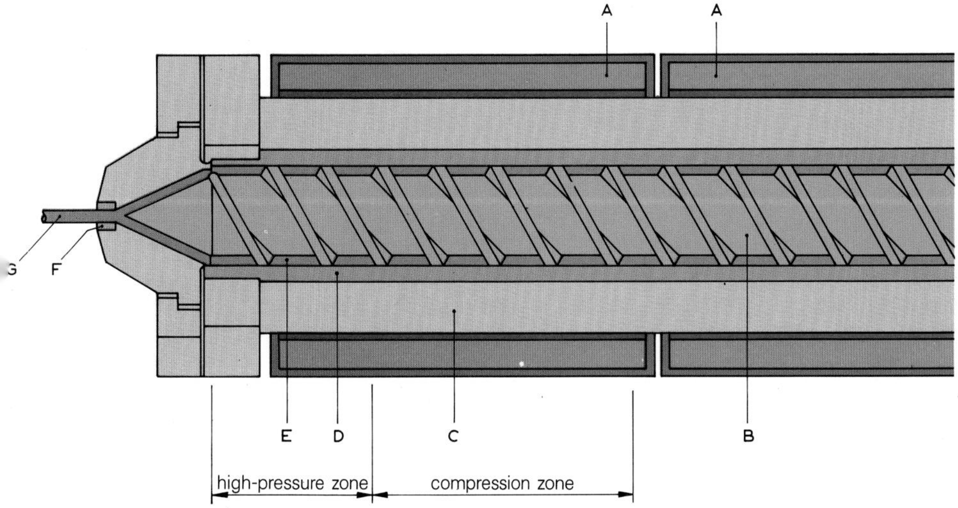

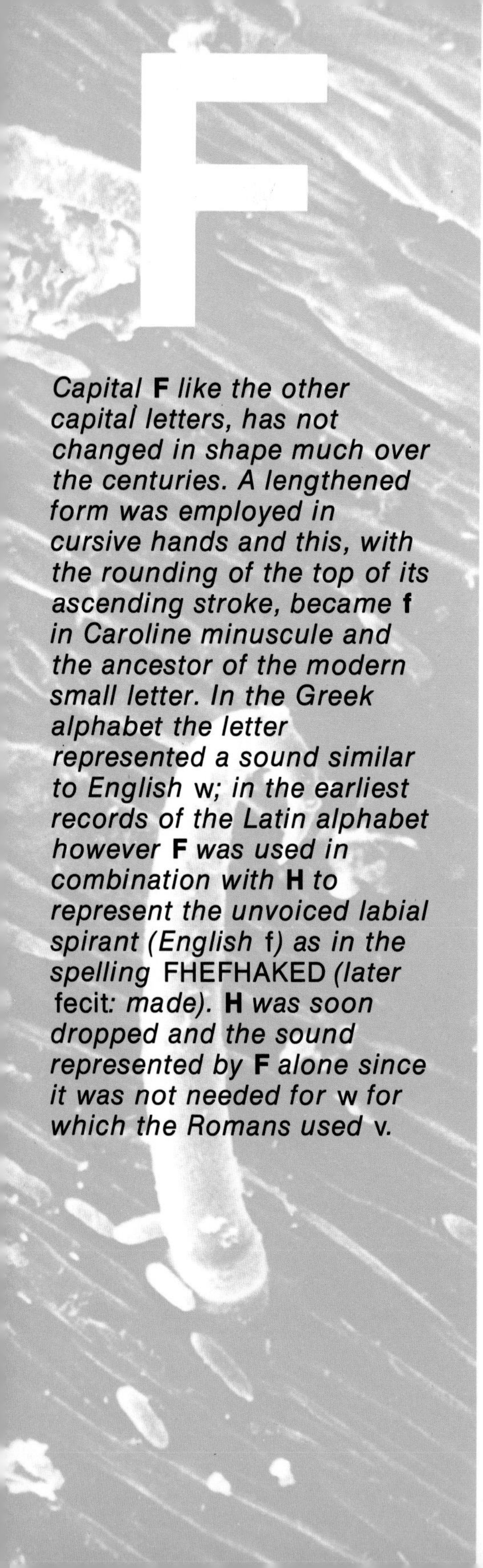

F

Capital **F** *like the other capital letters, has not changed in shape much over the centuries. A lengthened form was employed in cursive hands and this, with the rounding of the top of its ascending stroke, became* **f** *in Caroline minuscule and the ancestor of the modern small letter. In the Greek alphabet the letter represented a sound similar to English* w; *in the earliest records of the Latin alphabet however* **F** *was used in combination with* **H** *to represent the unvoiced labial spirant (English* f*) as in the spelling* FHEFHAKED *(later* fecit: *made).* **H** *was soon dropped and the sound represented by* **F** *alone since it was not needed for* w *for which the Romans used* v.

FABRE, Jean Henri (1823–1915), French entomologist who used direct observations of insects in their natural environments in his pioneering researches into insect instinct and behavior.

FABRICIUS, David (1564–1617), German Protestant minister and astronomer who discovered the first VARIABLE STAR (1596). His son, **Johannes Fabricius** (1587–1615), was one of the first to observe SUNSPOTS and hence report the SUN's rotation.

FABRICIUS AB AQUAPENDENTE, Hieronymus, or Girolamo Fabrizi (1537–1619), Italian physician, the pupil and successor of FALLOPIUS at Padua and teacher of William HARVEY. Fabricius made a detailed study of the valves in VEINS and pioneered the modern study of EMBRYOLOGY.

FACE, front part of the head, bordered by hairline and chin, and consisting of the EYES, NOSE, mouth, EARS, forehead, cheeks and jaw. It is particularly concerned with sensibility and communication: it bears the special sense organs of VISION, hearing, SMELL, TASTE and balance, as well as especially sensitive SKIN. VOICE and facial expression (fine facial movements) are mediated via the face. Neck mobility allows these organs to be directed quickly and easily toward an object without turning the body.

FACSIMILE, in telecommunications, the transmission of graphic images (photographs, diagrams, maps, manuscript or printed matter, etc.) by wire or RADIO, as used by governmental, weather or news agencies and business firms. The copy for transmission is scanned to give an electrical analogue signal in terms of a large number of dot areas and this signal is modulated onto a carrier for transmission. At the receiving station the original image is reconstituted using CATHODE-RAY TUBES, pressure-sensitive paper, XEROGRAPHY, etc.

FACTOR, an INTEGER which may be divided into another integer without REMAINDER. Thus the factors of 12 are 1, 2, 3, 4 and 6, since each of these may be divided exactly into 12. In general it is of use to consider only the factors of a number which are NATURAL NUMBERS. The prime factors of a number are those PRIME NUMBERS which are its factors. The prime factors of 12 are 1, 2 (twice), 3, since $4=2.2$ and $6=3.2$.

In ALGEBRA the factors of a POLYNOMIAL are found by a mixture of guesswork and rules of thumb. This is helped by certain standard rules:

$$x^2-y^2=(x+y)(x-y)$$
$$x^3-y^3=(x-y)(x^2+xy+y^2)$$
$$x^3+y^3=(x+y)(x^2-xy+y^2)$$
$$x^2+2xy+y^2=(x+y)^2.$$

Moreover, to find the factors of a polynomial of the form x^2+bx+c we know that, if the factors are $(x+p)$ and $(x+q)$, $p+q=b$ and $p.q=c$. Hence x^2-3x+2 has factors $(x-2)$ and $(x-1)$, since $(-2)+(-1)=(-3)$ and $(-2)(-1)=2$. (See also FACTORIAL.)

FACTORIAL, a system of notation in which, for a NATURAL NUMBER n, $n!$ (or $\underline{|n}$), read "n factorial," represents the PRODUCT of all the natural numbers up to n. Thus

$12! = 1.2.3.4.5.6.7.8.9.10.11.12 = 479\,001\,600.$

Additionally, $0!$ is defined as 1.

FAHRENHEIT, Gabriel Daniel (1686–1736), German-born Dutch instrument maker who introduced the mercury-in-glass THERMOMETER and discovered the variation of BOILING POINTS with atmospheric PRESSURE, but who is best remembered for his **Fahrenheit temperature scale**. This has 180 divisions (degrees) between the freezing point of water (32°F) and its boiling point (212°F). Although still commonly used in the US, elsewhere the Fahrenheit scale has been superseded by the Celsius scale (see CELSIUS, ANDERS).

FAINTING, or **syncope,** transient loss or diminution of consciousness associated with an abrupt fall in blood pressure. In the upright position, head and BRAIN are dependent on a certain blood pressure to maintain BLOOD CIRCULATION through them; if the pressure falls for any reason, inadequate flow causes consciousness to recede, often with the sense of things becoming more distant. The body goes limp and falls, so that, unless artificially supported, the effect of gravity on brain flow is lost and consciousness is rapidly regained. Fainting may result from sudden emotional shock in susceptible individuals, HEMORRHAGE, ANEMIA or occur with transient rhythm disorders of the HEART.

FAIRY RING, a complete circle or an arc of fruiting bodies of FUNGI, particularly agarics, which commonly appear in fields and lawns. Each colony of fungi is derived from a single spore from which the hyphae grow out in all directions forming an invisible circular colony that only becomes visible when fruiting bodies are formed at its periphery. The name is derived from the old superstition that mushrooms growing in a circle represent the path of dancing fairies.

FAITH HEALING, the treatment of DISEASE by the evocation of faith, usually induced during a public ceremony or meeting; chanting and laying on of hands are common accompaniments. Greatest success is often with disease that tends to remit spontaneously and in HYSTERIA; in some instances, patients are helped to come to terms with disease. Substantiation of cures is rare.

FALLEN ARCHES. See FLATFOOT.

FALLING STAR. See METEOR.

FALL LINE, a line along which a number of nearby rivers have WATERFALLS, marking the progress of the rivers from hard to softer rock. Since this marks the farthest inland point navigable from the sea, and because the falls can supply HYDROELECTRIC POWER, many important industrial centers have sprung up along fall lines.

FALLOPIAN TUBE, narrow tube leading from the surface of each ovary within the female PELVIS to the WOMB. Its abdominal end has fimbria which waft peritoneal fluid and eggs into the tube after ovulation. Fertilization may occur in the tube, and if followed by IMPLANTATION there, the PREGNANCY is ectopic and ABORTION, which may be life-threatening, is inevitable. In STERILIZATION, the tubes are divided.

FALLOPIUS, Gabriel, or **Fallopia** (1523–1562), Italian anatomist, a supporter of VESALIUS. He carried out important work on many anatomical structures; and is best known for his descriptions of the FALLOPIAN TUBES, whose function he discovered.

FALLOUT, Radioactive, deposition of radioactive particles from the ATMOSPHERE on the earth's surface. Three types of fallout follow the atmospheric explosion of a nuclear weapon. Large particles are deposited as intense but short-lived local fallout within about 250km of the explosion; this dust causes radiation burns. Within a week, smaller particles from the TROPOSPHERE are found around the latitude of the explosion. Long-lived RADIOISOTOPES such as strontium-90, carried to the STRATOSPHERE by the explosion, are eventually deposited worldwide.

FALSE-COLOR PHOTOGRAPHY, the processing of special photographic emulsions to display information using unnatural colors, used in terrestrial

A false color infrared photograph of part of the Amsterdam Zoo and the surrounding district. In this picture, taken from a height of 400m, green vegetation appears red. Trees with yellowing leaves and those which have recently been poisoned following local gas leaks appear in a much weaker color than the healthy trees.

infrared PHOTOGRAPHY (e.g., vegetation studies) and astronomy. False-color ultraviolet photographs of the sun can be obtained using the SPECTROHELIOGRAPH.

FAMILY, a social unit comprising a number of persons in most cases linked by birth or MARRIAGE. There are four main types of families: the conjugal or nuclear family, a single set of parents and their children; the extended or consanguine family, which includes also siblings and other relations and generations (e.g., brothers, grandparents, grandchildren, uncles and aunts); the corporate family, a group organized around an important activity such as hunting, sharing of shelter, religion or customs; and the experimental family, a group whose members are generally unrelated to each other genetically, but who choose to live together and perform the traditional roles of the nuclear or consanguine family. The *kibbutzim* of Israel and the commune are examples of experimental families.

The descent within a family is usually either patrilineal, through its male members, or matrilineal, through its female members. Occasionally descent is bilineal, through either male or female lines, or bilateral, through both males and females.

By far the most common forms of families are the **nuclear** and **consanguine**. There are sound reasons for this: psychological security through membership of a close, intimate group; ready sexual and emotional satisfaction between husband and wife; and physical security based on a family's sense of duty and willingness to protect its members. Moreover, it would seem that these types of families are the most efficient insofar as childrearing is concerned, with older generations or siblings acting as mentors during the child's formative years. In the West, the nuclear family has in recent years become generally more democratic, the male's absolute authority being tempered to permit wives and children greater freedom and responsibility.

Centrifugal (A) and axial-flow (B) electric fans as used in ventilation systems. In (B) only the middle set of blades rotate; the outer sets are fixed.

A

B

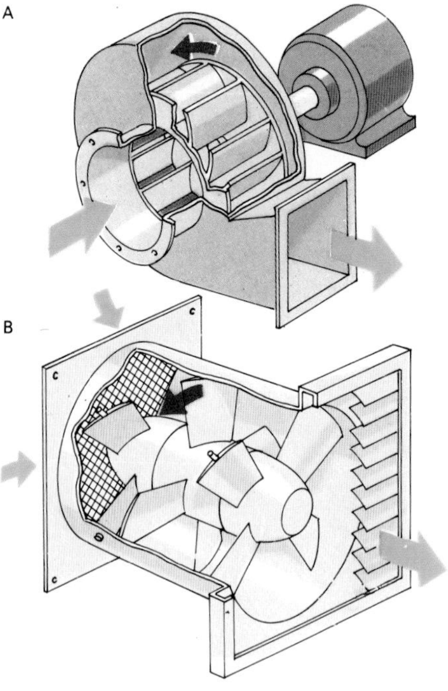

FAMILY, in the classification of living things. See TAXONOMY.

FAMILY PLANNING, the practice of regulation of family size by judicious use of CONTRACEPTION, STERILIZATION and, occasionally, induced ABORTION; increased survival of children and increasing world population have created the need for such an approach. Planning of numbers and timing to accord with economic and social factors are greatly aided by modern methods of contraception, so unwanted PREGNANCY should be a rarity. However, ignorance and neglect have prevented the realization of this ideal. Adoption, ARTIFICIAL INSEMINATION and infertility treatment are used for inability to conceive.

FAN, device to produce a current of air. Originally, fans were hand-held, a small screen mounted on a stick. Later came the folding fan, a number of sticks held together by a pin at one end, and connected by paper or fabric. Modern **electric fans** are of two types: axial-flow, whose principle is roughly that of a PROPELLER; and centrifugal, where air is introduced along the rotation axis and forced outward by the rotation.

FANJET or **Turbofan.** See JET PROPULSION.

FARAD (F), the SI UNIT of CAPACITANCE, defined as the capacitance of a CAPACITOR for which a one coulomb charge raises its potential by one volt. The capacitance of most practical capacitors is measured in micro- or picofarads.

FARADAY, Michael (1791–1867), English chemist and physicist, the pupil and successor of H. DAVY at the ROYAL INSTITUTION, who discovered BENZENE (1824), first demonstrated electromagnetic INDUCTION (see also HENRY, JOSEPH) and invented the dynamo (1831—see GENERATOR, ELECTRIC), and who, with his concept of magnetic lines of force, laid the foundations of classical field theory later built upon by J. Clerk MAXWELL. In the course of many years of researches, he also discovered the laws of ELECTROLYSIS which bear his name and, in showing that the plane of polarization of plane POLARIZED LIGHT was rotated in a strong magnetic field, demonstrated a connection to exist between LIGHT and MAGNETISM.

FARNSWORTH, Philo Taylor (1906–1971), US radio engineer known especially for his early work in the development of TELEVISION.

FARSIGHTEDNESS. See HYPEROPIA.

FASCIOLA, a genus of the TREMATODA, members of which are parasitic in vertebrates including man. Best known is *Fasciola hepatica*, the liver fluke.

FATHOM, a unit of length used to describe depth at sea. One fathom equals six feet (1.8288m).

FATHOMETER, trade name for a recording ECHO SOUNDER.

FATIGUE, or tiredness, a vague term indicating an inability to perform EXERCISE or even normal tasks due to previous exertion or DISEASE. This may consist of general weakness, specific muscular weakness or shortness of breath. Physiological fatigue after exertion may be due to accumulation of waste

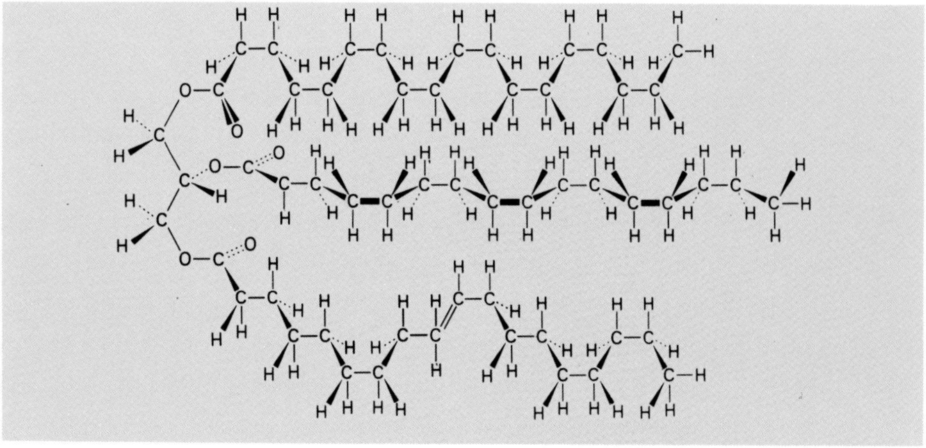

A typical fat (triglyceride). On hydrolysis this one yields glycerol with stearic acid (*top*), palmitic acid (*middle*) and oleic acid (*bottom*).

products of METABOLISM. Virtually any disease may cause fatigue, in particular FEVER, visceral disease, DIABETES, ADDISON'S DISEASE and MUSCULAR DYSTROPHY. MYASTHENIA GRAVIS is characterized by excessive muscle fatigability.

FATIGUE, Metal. See METAL FATIGUE.

FATS, ESTERS of CARBOXYLIC ACIDS with GLYCEROL which are produced by animals and plants and form natural storage material. Fats are insoluble in water and occur naturally as either liquids or solids; those liquid at 20°C are normally termed **oils** and are generally found in plants and fishes. Oils generally contain esters of OLEIC ACID which can be converted to esters of the solid STEARIC ACID by HYDROGENATION in the presence of finely divided nickel. This process is basic to the manufacture of MARGARINE. Fats are the most concentrated sources of energy in the human diet, giving over twice the energy of STARCHES. Diets containing high levels of animal fats have been implicated as causative factors in heart disease, and replacement of animal fat by plant oils (e.g., peanut oil, sunflower oil) has been suggested. Fats particularly of fish and plant origin represent important items of commerce, and world production in 1976 should reach 49 million tonnes. Of major importance are soybean oil (9–10 million tonnes), sunflower, palm, peanut, cottonseed, rapeseed and coconut oil (2.5–4 million tonnes each) and olive and fish oil (over 1 million tonnes each). Major producers include the US (soybean oil), the USSR (sunflower oil and cottonseed oil) and India (peanut oil).

FATTY ACIDS. See CARBOXYLIC ACIDS.

FAULT, a fracture in the earth's crust on either side of which there has been relative movement (see EARTH). Faults seldom occur along a single PLANE: usually a vast number of roughly parallel faults take place in a belt (fault-zone) a few hundred metres across. The side of a fault on which the strata have moved relatively downward is the *downthrow* side; the other the *upthrow* side. The difference in vertical height between the sides is the *throw*, the lateral displacement the *heave*. The angle between the horizontal and the fault plane is the *dip*: where this is roughly 90°, the fault is a *normal fault*. (See also DIASTROPHISM; HORST; RIFT VALLEY.)

FEATHER, the structure which forms the outer covering of birds. No other animal possesses feathers. Feathers function to insulate and waterproof the body, provide flight surfaces and colors that are important as camouflage and in displays. There are two types of feather: pennae which are composed of contour and flight feathers; and plumulae or down feathers.

FECHNER, Gustav Theodor (1801–1887), German physicist and founder of experimental psychology, usually remembered for his reformulation of Ernst Heinrich Weber's (1795–1878) conclusions concerning the increase in a stimulus

Fault terminology: (1) normal fault; (2) reverse fault; (3) displacement; (4) hanging wall; (5) foot wall.

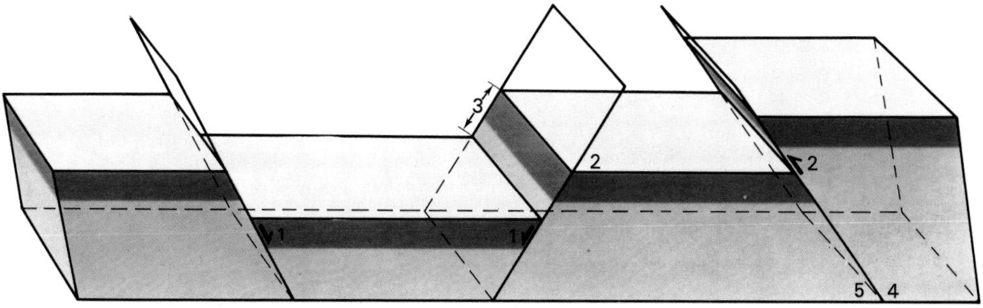

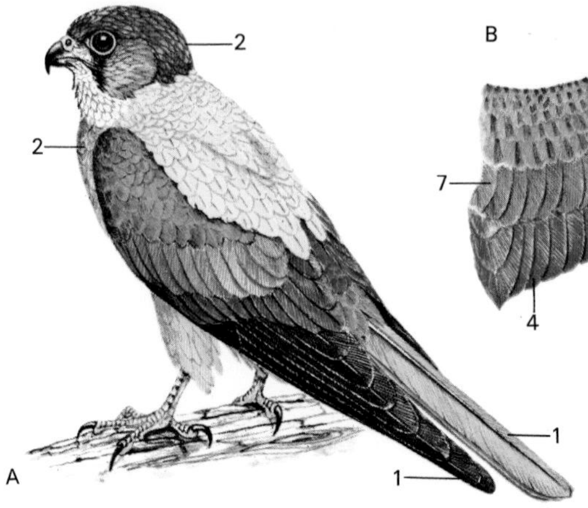

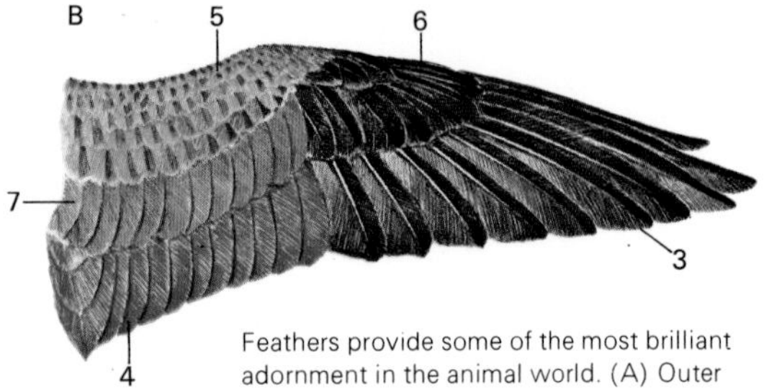

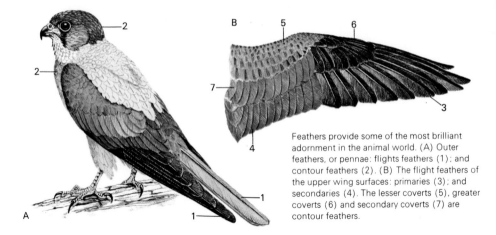

Feathers provide some of the most brilliant adornment in the animal world. (A) Outer feathers, or pennae: flights feathers (1); and contour feathers (2). (B) The flight feathers of the upper wing surfaces: primaries (3); and secondaries (4). The lesser coverts (5), greater coverts (6) and secondary coverts (7) are contour feathers.

needed to make someone aware that there was a difference. **Fechner's law** (or the Weber-Fechner law) states that the perceived intensity of a sensation increases with the logarithm of its stimulus.

FEEDBACK, the use of the output of a system to control its performance. Many examples of feedback systems can be found in the life sciences, particularly in ecology, biochemistry and physiology. Thus the population of a species will grow until it overexploits its food supply. Malnutrition then leads to a reduction in population. In the design of machines, SERVOMECHANISMS and GOVERNORS also exemplify feedback systems. The most important application of the feedback concept in modern technology comes in ELECTRONICS where it is common practice to feed some of the output of an AMPLIFIER back to the input to help reduce NOISE, distortion or instability. Most often used is "negative feedback" in which the effect of the feedback is to reduce the amplifier's output while stabilizing its performance. The howling that can occur when too much sound from a LOUDSPEAKER enters the MICROPHONE of a public address system is an example of positive feedback.

Block diagram of feedback circuit. The output potential $\mathbf{E}_0$ is related to the input signal $\mathbf{E}_s$ as

$$\mathbf{E}_0 = \frac{\mathbf{A}\,\mathbf{E}_s}{1 - \mathbf{A}\beta}$$

where **A** is the amplifier gain and β the fraction of the output signal returned to the input. Thus for an amplifier with gain ×5, and a negative feedback of $\times\frac{1}{10}$, the output potential is 3.3 times the input signal.

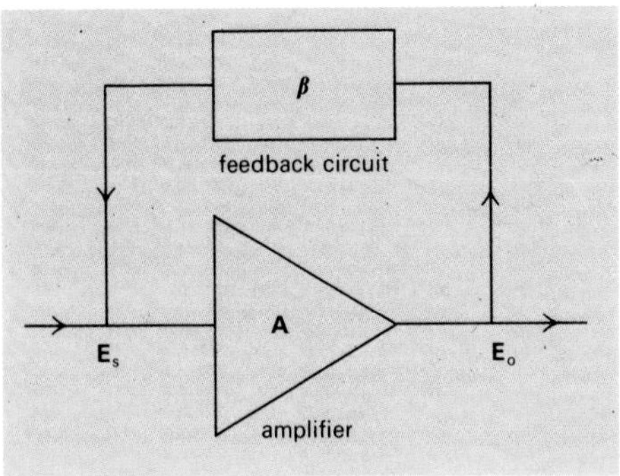

FEHLING'S SOLUTION, reagent composed of COPPER (II) sulfate, Rochelle salt (see TARTARIC ACID) and sodium hydroxide, used in chemical ANALYSIS to detect ALDEHYDES, including some SUGARS. It is reduced to a copper (I) oxide precipitate, from whose amount the quantity of sugar can be determined.

FELDSPAR, widely distributed mineral group. Potash feldspars, principally ORTHOCLASE and MICROCLINE, are potassium aluminum silicates where sodium or barium may partly replace potassium. Plagioclase feldspars form a series derived from sodium aluminum silicate (ALBITE) with calcium replacing sodium in all proportions up to 100% (ANORTHITE). Feldspar is used in making glass and glazes.

FELT, fabric made from wool, hairs and fur. The fibers are matted together by rolling and pressing under heat. Hat felts are of wool or fur. Other types include padding felts and carpet underfelts. Jute roofing felt is not true felt.

FEMUR, in vertebrates the proximal BONE of the hind leg, or thigh bone. In the human skeleton it is the largest and longest bone.

FERENCZI, Sándor (1873–1933), Hungarian psychoanalyst, and an early colleague of FREUD, best known for his experiments in PSYCHOTHERAPY, in course of which he broke away from Freud's classic psychoanalytic theory. (See PSYCHOANALYSIS.)

FERMAT, Pierre de (1601–1665), French mathematician best remembered for **Fermat's Principle,** that the path of light traveling between two points by REFLECTION is that taking least time; and **Fermat's Last Theorem**, that $x^n + y^n - z^n = 0$, where $x, y, z \neq 0$ and $n > 2$, is impossible for integral x, y, z and n.

FERMENTATION, the decomposition of CARBOHYDRATES by microorganisms in the absence of air. Louis PASTEUR first demonstrated that fermentation is a biochemical process, each type being caused by one species (see also BUCHNER, EDUARD). It is an aspect of bacterial and fungal METABOLISM, in which GLUCOSE and other sugars are oxidized by ENZYME catalysis to pyruvic acid (see CITRIC ACID CYCLE). Pyruvic acid is then reduced to LACTIC ACID or degraded to carbon dioxide and ETHANOL. Considerable energy is released in this process: some is stored as the high-energy compound ATP (see NUCLEOTIDES), and the rest is given off as heat. Fermentation by YEAST has been used for centuries in BREWING and making bread and wine; fermentation by lactic acid bacteria is used to make cheese. Special fermentations are used industrially for the manufacture of ACETONE, butanol (see ALCOHOLS), GLYCEROL, CITRIC ACID, glutamic acid (see MONOSODIUM GLUTAMATE) and many other compounds. (See also RESPIRATION.)

FERMI, Enrico (1901–1954), Italian atomic physicist who was awarded the 1938 Nobel Prize for Physics. His first important contribution was his examination of the properties of a hypothetical gas whose particles obeyed Pauli's EXCLUSION PRINCIPLE; the laws he derived can be applied to the ELECTRONS in a metal, and explain many of the properties of metals (see FERMI-DIRAC STATISTICS). Later he showed that NEUTRON bombardment of most elements produced their RADIOISOTOPES.

FERMI-DIRAC STATISTICS, in QUANTUM MECHANICS, the statistical behavior of a system of indistinguishable particles with a number of discrete states, each of which may be occupied at any one time by a single particle only. SUBATOMIC PARTICLES that show this behavior are termed **fermions**. (See also BOSE–EINSTEIN STATISTICS.)

FERMIUM (Fm), a TRANSURANIUM ELEMENT in the ACTINIDE series, found in the debris from the first HYDROGEN BOMB, and now prepared by bombardment of lighter actinides.

FERNS, nonflowering plants of the class Filicineae having creeping or erect RHIZOMES or an erect aerial stem and large conspicuous leaves. Spores are produced on the underside of the leaf within sporangia and germinate to form the GAMETOPHYTE or sexual stage of the life cycle. Ferns are widely distributed throughout the world, but the majority grow in the tropics.

FERREL'S LAW, the proposition that moving air masses tend to be deflected to the right in the northern hemisphere and to the left in the southern, first proposed by US meteorologist, **William Ferrel** (1817–1891). (See also BUYS-BALLOT'S LAW.)

Huge fermentation tanks in the cellar of a large brewery. The albumen foam seen on top of the fluid is formed by escaping carbon dioxide gas.

FERROCENE, or dicyclopentadienyl iron [$(C_5H_5)_2Fe$], an orange crystalline solid; a typical, stable SANDWICH COMPOUND—the first to be prepared (1951). The molecular structure is a pentagonal ANTIPRISM. mp 174°C, subl 100°C.

FERROMAGNETIC MATERIALS. See MAGNETISM.

FERROMAGNETISM. See MAGNETISM.

FERROUS SULFATE, or Iron (II) Sulfate. See IRON.

FERTILIZATION, the union of two GAMETES, or male and female sex cells, to produce a CELL from which a new individual, animal or plant, develops. The sex cells contain half the normal number of CHROMOSOMES, and fertilization therefore produces a cell with the normal number of chromosomes for any particular species. Fertilization may take place outside the organism's body (external fertilization), or inside the female (internal fertilization) as a result of copulation.

FERTILIZERS, materials added to the SOIL to provide elements needed for plant NUTRITION, and so to enable healthy growth of crops with high yield. The elements needed in large quantities are NITROGEN, PHOSPHORUS, POTASSIUM, SULFUR, CALCIUM and MAGNESIUM; the last three are usually adequately supplied in the soil or incidentally in other fertilizers. Small amounts of TRACE ELEMENTS are also needed, and usually supplied in fertilizers. The choice of compounds or materials containing nitrogen, phosphorus and potassium depends mainly on cost. The traditional natural fertilizers—BONE MEAL, GUANO and MANURE—are now too expensive to be much used outside HORTICULTURE. Potassium is supplied as potassium chloride, widely available as SYLVITE. Phosphorus fertilizers are obtained from mineral PHOSPHATES, especially APATITE; some is used as such, but most is converted to ammonium phosphate or superphosphate (see PHOSPHATES). Nitrogen is supplied as AMMONIA (injected under pressure), ammonium salts, NITRATES (ammonium nitrate being most useful) and UREA (see also NITROGEN CYCLE; NITROGEN FIXATION). Fertilizers in excess may harm crops and cause EUTROPHICATION.

FETISH, in abnormal psychology, any object or focus of obsession onto which has been projected an exaggerated power to erotically stimulate. The term also applies to the abnormal attraction itself.

FETISH, in anthropology, an object in which a spirit is thought to reside, distinct from an AMULET whose supernatural power is believed to be externally derived. Unlike idols, fetishes are not intended to be a likeness of the spiritual being.

FETUS, the developing intrauterine form of an animal, loosely used to describe it from the development of the fertilized egg (EMBRYO), but strictly referring in man to the period from three months gestation to BIRTH. During fetal life, organ development is consolidated and specialization extended so that function may be sufficiently mature at birth; some organs start to function before birth in preparation for independent existence. During the fetal period most increase in size occurs, both in the

Cellulose acetate provides a much used man-made fiber. Here cellulose acetate yarn is being made by the "dry spinning" process: a solution of the material is extruded through a multiple-jet assembly. Twenty-eight filaments are produced as the solvent evaporates. These are spun to produce the cellulose acetate yarn.

fetus and in the PLACENTA and WOMB. The fetus lies in a sac of AMNIOTIC FLUID which protects it and allows it to move about. BLOOD CIRCULATION in the fetus is adapted to the placenta as the source of OXYGEN and nutrients and site for waste excretion, but alternative channels are developed so that within moments of birth they may take over. Should the fetus be delivered prematurely, immaturity of the LUNGS may cause respiratory distress, that of the LIVER, JAUNDICE.

FEVER, raising of body TEMPERATURE above normal (37°C or 98.6°F in man), usually caused by DISEASE. Infection, INFLAMMATION, heat stroke and some TUMORS are important causes. Fever is produced by pyrogens, which are derived from cell products, and alter the set level of temperature-regulating centers in the HYPOTHALAMUS. Fever may be continuous, intermittent or remittent, the distinction helping to determine the cause. Anti-inflammatory drugs (e.g., ASPIRIN) reduce fever; STEROIDS mask it.

FEYNMAN, Richard Phillips (1918–   ), US physicist awarded with SCHWINGER and TOMONAGA the 1965 Nobel Prize for Physics for their independent work on quantum electrodynamics (see QUANTUM MECHANICS). With GELL-MANN he has proposed the quark as a fundamental SUBATOMIC PARTICLE.

FIBER, a thin thread of natural or artificial material. **Animal fibers** include wool, from the fluffy coat of the sheep, and silk, the fiber secreted by the silkworm LARVA to form its cocoon. **Vegetable fibers** include COTTON, FLAX, HEMP, JUTE and SISAL: they are mostly composed of LIGNIN, though CELLULOSE is also important. **Mineral fibers** are generally loosely termed ASBESTOS. These fibrous mineral SILICATES are mined in South Africa, Canada and elsewhere. **Man-made fibers** are of two types: regenerated fibers, extracted from natural substances (e.g., rayon is cellulose extracted from wood pulp); and SYNTHETIC FIBERS. Most PAPER is made from wood fiber. (See also COTTON GIN; SPINNING; WEAVING.)

FIBERBOARD, two types of board made by reducing woodchips to fibers which are then compacted with adhesives. **Hardboard**, strong and fairly dense, is used for, e.g., wall paneling. Less dense **insulating board** is used for soundproofing, etc.

FIBERGLASS, GLASS drawn or blown into extremely fine fibers that retain the tensile strength of glass while yet being flexible. The most used form is fused QUARTZ, which when molten can be easily drawn and which is resistant to chemical attack. Most often, the molten glass is forced through tiny orifices in a platinum plate, on the far side of which the fine fibers are united (though not twisted) and wound onto a suitable spindle. Fiberglass mats (**glass wool**) are formed from shorter fibers at random directions bonded together with a thermosetting RESIN: they may be pressed into predetermined shapes. Fiberglass is used in INSULATION, automobile bodies, etc.

FIBIGER, Johannes Andreas Grib (1867–1928), Danish pathologist awarded the 1926 Nobel Prize for Physiology or Medicine for his discovery of a technique of inducing CANCER in rats.

FIBONACCI, Leonardo, or **Leonardo of Pisa** (c1180–c1240), Italian mathematician whose *Liber Abaci* (1202) was probably the first European account of the mathematics of India and Arabia, including some material on ALGEBRA. He also devised the FIBONACCI SEQUENCE.

FIBONACCI SEQUENCE, often misleadingly termed **Fibonacci Series**, an infinite SEQUENCE in which each term $u_n = u_{n-2} + u_{n-1}$. The first few terms are thus 1, 1, 2, 3, 5, 8, 13, ... The term is sometimes applied to other recursive sequences, such as 1, 1+a, 2+a, 3+2a, . . ., as in 1, 4, 5, 9, 14, . . ., where a = 3.

FIBRIN. See CLOTTING.

FIBULA, the smaller of the two BONES in the lower LEG, apposed to the TIBIA, and part of the ankle JOINT.

FIELD, a set (see SET THEORY) F with two operations, $+$ and $\times$, in which: (1) both $+$ and $\times$ are associative and commutative, and the operation $\times$ is distributive over $+$ (see ALGEBRA); (2) there are two identity elements in F, 0 relative to $+$ and 1 relative to $\times$, such that $a+0=a$ and $a\times 1=a$ for any element a of the field; (3) every element a has an inverse $-a$, also a member of the set, such that $a+(-a)=0$; (4) every nonzero element a has an inverse a^{-1}, also a member of the set, such that $a\times a^{-1}=1$. Examples of fields include the RATIONAL NUMBERS and the REAL NUMBERS with, in each case, the operations ADDITION and MULTIPLICATION. A set (with two operations) which satisfies the conditions (1), (2) and (3) but not (4) is an **integral domain**: an example is the set of all INTEGERS under addition and multiplication. (See also GROUP.)

FIELD-EMISSION MICROSCOPE, a lower-resolution relative of the FIELD-ION MICROSCOPE, in which the image is produced by ELECTRONS emitted by the tip itself when negatively charged.
FIELD GLASS. See BINOCULARS.
FIELD-ION MICROSCOPE, an instrument producing very beautiful pictures of the arrangement of individual ATOMS in materials drawn out into, or evaporated on to, a fine tip, typically 40nm in radius. Invented by Erwin Wilhelm Müller (1911–) in 1936, the microscope is lensless, the image being produced on a fluorescent screen by IONS created in a low-pressure gas by the intense ELECTRIC FIELD at the tip when it is positively charged to a few kilovolts
FILAMENT, in flowers, the stalk of the stamen that bears the anther at its apex. (See FLOWER.) Also, the thread-like row of cells found in certain ALGAE.
FILAMENT, a length of tungsten resistance wire which glows white-hot on carrying a suitable electric current, used in incandescent filament lamps (see LIGHTING for early history). Lamp filaments, from 0.015 to 0.045mm in thickness, are coiled once or twice (coiled coil) before being fused into the neck of the lamp. Heater filaments in ELECTRON-TUBE cathodes operate at only red heat.

The surface structure of a platinum-iridium crystal, as viewed in a field-ion microscope. The ions producing the image are helium ions (He^+) and the magnification is about 2 000 000×.

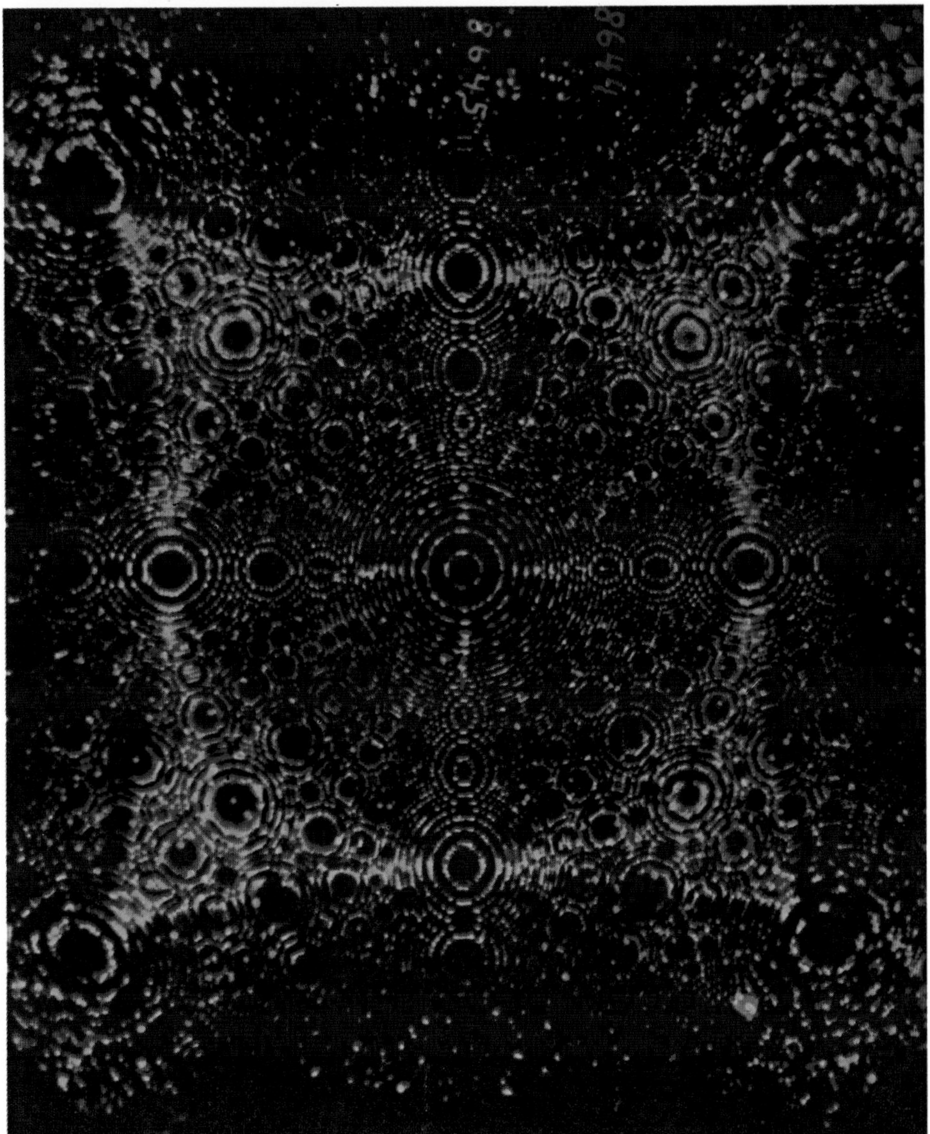

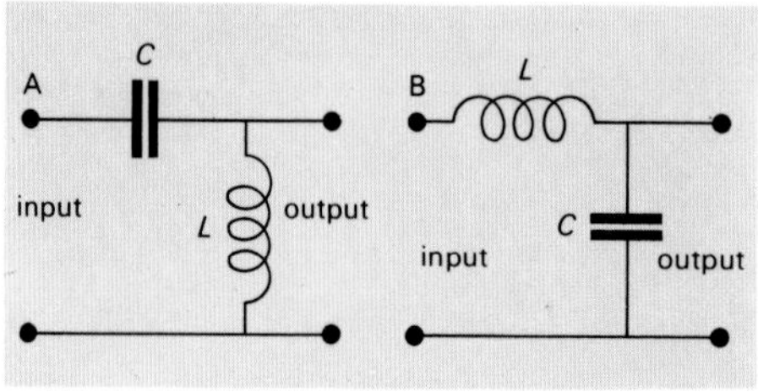

Electric filter circuits: combinations of a capacitor (*C*) and an inductor (*L*) forming a simple high-pass filter (A) and a low-pass filter (B).

FILARIASIS, a group of PARASITIC DISEASES of warm climates, transmitted by MOSQUITOS, causing FEVER, LYMPH node enlargement, ABSCESSES, epididymal inflammation and signs of ALLERGY; ELEPHANTIASIS may result. A specific type, onchocerciasis, or river blindness, leads to SKIN rash, EYE disease, sometimes causing BLINDNESS, and muscle pains or nodules. Some cause LUNG disease and increased BLOOD eosinophils. Diagnosis is by special staining of blood films and skin tests. Treatment and prevention are with diethylcarbamine; mosquito control is needed.
FILICINEAE. See FERNS.
FILM, Photographic. See PHOTOGRAPHY.
FILTER, Electric, an arrangement of electronic components used in a circuit to transmit signals within a given frequency range, rejecting others, used in RADIOS, TELEVISION receivers, TELEPHONE systems, etc. A **low-pass filter** transmits frequencies below a specified cut-off and blocks out higher frequencies; typically it consists of an inductor (see INDUCTANCE) in series with and a CAPACITOR shunted across the load to short out high-frequency current. In a **high-pass filter** the inductor and capacitor are interchanged and only high frequencies are transmitted. A **band-pass filter** blocks all frequencies outside two limits.
FILTRATION, separation of solid particles from a liquid or gas by passing it through a porous mesh on which the solid collects. Commonly used are filter cloths, filter paper, wire mesh, sintered glass, or—where a large volume of water is to be filtered—sand beds. Filters have many varied uses, including use in cigarettes, vacuum cleaners, gasoline and diesel engines, coffee-making, air-conditioning units, water-purification systems, and chemical preparation and analysis. A magnetic filter is used to remove iron or steel particles from oils etc. in machine tools and some engines. (See also ELECTROSTATIC PRECIPITATORS.)
FINGERPRINTS, impressions of the loops and whorls of the papillary ridges of the fingertips, a valuable police tool. The earliest police system was developed by Jean Vucetich (1888, Argentina) and is still in use in the Spanish-speaking world. The system most in use today was developed by Sir Edward Richard Henry from the work of Sir Francis GALTON.

Early firearms adopted various firing mechanisms, of which three of the more significant are illustrated here. The development of the percussion lock and the center-fire cartridge in the 19th century represented an enormous improvement on all these designs and greatly increased the efficiency of firearms.

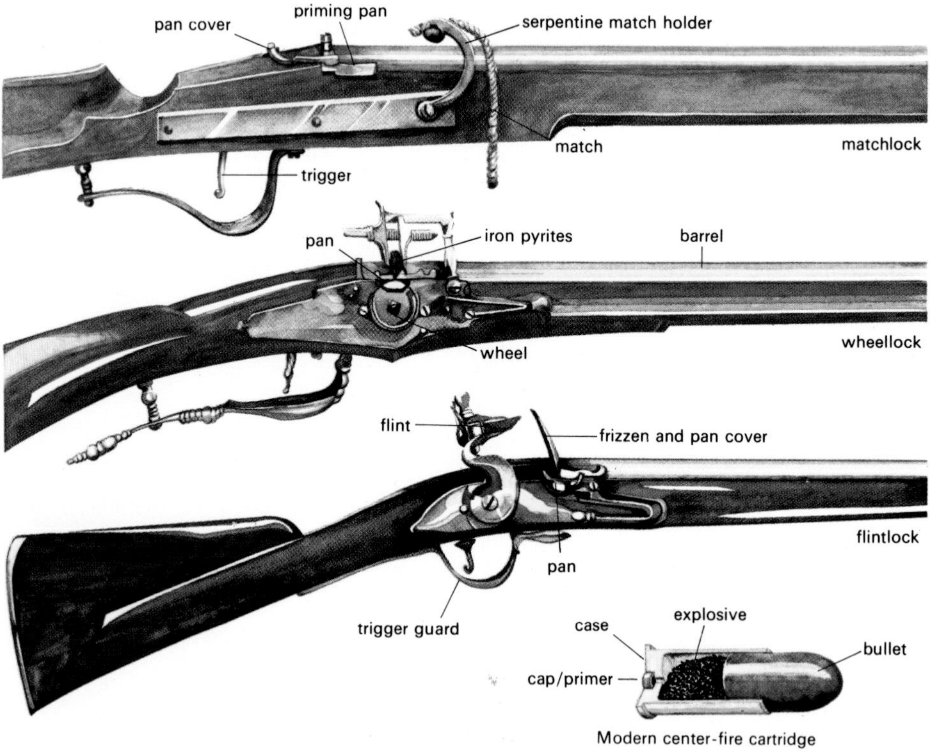

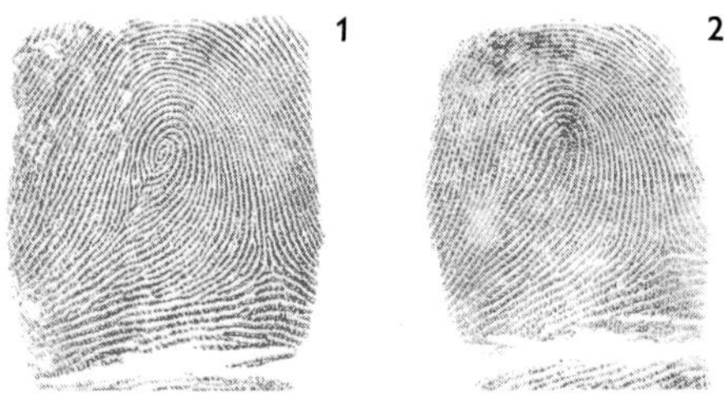

Two of the commonest types of fingerprints: (1) whorl in the center; (2) loop in the center.

Replacing the anthropometric techniques of BERTILLON, it was adopted in the UK in 1901, the US in 1903. In **dactyloscopy** (fingerprinting) the tips are well cleaned, rolled on printer's ink spread on a glass sheet and then onto coated cards.

FINLAY, Carlos Juan (1833–1915), Cuban physician who first proposed (1881) that YELLOW FEVER is transmitted by the MOSQUITO. Despite his considerable research, this was unproved until 1900.

FINSEN, Niels Ryberg (1860–1904), Danish physician awarded the 1903 Nobel Prize for Physiology or Medicine for his discovery and use of the curative properties of certain wavelengths of LIGHT.

FIRE. See COMBUSTION.

FIREARMS, weapons in which missiles are projected by firing explosive charges. They are classified as either ARTILLERY or small arms. The latter seem to have originated in 14th-century Europe in the form of metal tubes, closed at one end, into which GUNPOWDER and the missile were packed, the charge being ignited via a touch hole. The heavy **harquebus** was one of these. The 15th century saw the introduction of the **matchlock** in which a spring-loaded lever mechanism was used to introduce a smoldering match (a hemp cord soaked in SALTPETER) to the powder. This was superseded by the **wheel lock** in the next century. In this a serrated wheel rotated against a flint and ignited the powder with a spark. In the 17th century the **flintlock** was introduced. Here a flint held in a spring-loaded arm, or cock, struck a metal hammer, or frizzen, to produce the spark. The perfecting of the gas-tight breechblock and the modern percussion lock in the early 19th century led to the development of the breech-loading RIFLE and the repeating pistol (see REVOLVER). By the end of the century, MACHINE GUNS were in an advanced state of development. SHOTGUNS, used mainly for sport, fire a cartridge containing numerous small pellets. (See also AIR GUN; AMMUNITION; PISTOL.)

FIREBALL, a particularly bright METEOR, especially a bolide. The term is also applied to LIGHTNING of globular form.

FIREBRICK, bricks made in a variety of shapes from refractory materials and used in constructions, particularly FURNACES, designed for the treatment of molten GLASS and metals. (See METALLURGY; REFRACTORY.)

FIRE CLAY, type of CLAY which, being refractory (mp above 1500°C), is used to make FIREBRICK. It has a high content of ALUMINUM oxide and SILICA.

FIRE CONTROL, the techniques and operations ensuring that a missile, shell or depth charge lands on target. When the target is visible and stationary the problem is relatively simple (see BALLISTICS), and accurate sights, firing tables, etc., are used. In other cases such devices as RADAR, SONAR, RADIO and TELEPHONE communications and COMPUTER control are employed.

FIREDAMP. See DAMP.

FIRE EXTINGUISHER, a portable appliance for putting out small fires. Extinguishers work either by cooling or by depriving the fire of OXYGEN (as typified by the simplest, a bucket of water or bucket of sand), and most do both. The now-obsolete **soda-acid extinguisher** contains a SODIUM bicarbonate solution and a small, stoppered bottle of SULFURIC ACID;

A multipurpose dry powder fire extinguisher, based on a design by Nu-Swift International Ltd., UK. When the striker knob is struck, carbon dioxide gas in the main cylinder escapes through the nozzle, carrying the charge of fire-killing dry powder to the fire. The powder works by physically restricting the oxygen supply to the flame and by chemically and physically combining with the hot combustible materials supporting the flame.

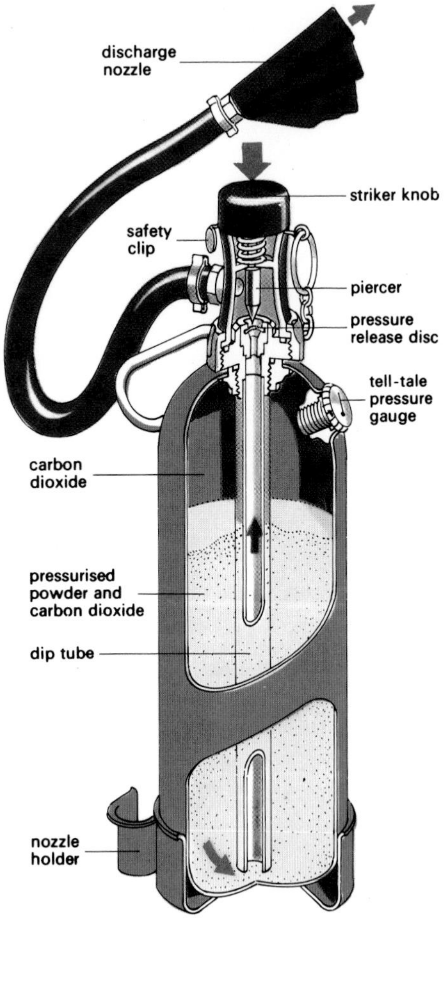

depression of a plunger shatters the bottle, mixing the chemicals so that CARBON dioxide (CO_2) gas is generated, forcing the water out of a nozzle. The safer and more reliable modern **water extinguisher** utilizes a pressure charge of carbon dioxide gas. When the charge is ruptured, the expanding gas immediately expels a powerful jet or spray of water. **Foam extinguishers** employ a foaming agent (usually animal PROTEIN or certain detergents) and an aerating agent: they are effective against oil fires, as they float on the surface. **Carbon dioxide extinguishers** provide a smothering blanket of CO_2; and **dry chemical extinguishers** provide a powder of mainly sodium bicarbonate, from which the fire's heat generates CO_2.

FIREPROOFING, the techniques and materials used in rendering an object, building, etc., resistant to COMBUSTION. One of the commonest for fabrics, etc., is WATER GLASS. Substances such as ASBESTOS may be used to protect structural elements of buildings from excesses of heat.

FIRESTORM, condition where flames from multiple fires combine as a single column. Owing to CONVECTION, violent winds blow in from around, fanning the flames to very high temperatures. In populated areas, firestorms, used as a weapon by the Allies in WWII, caused phenomenal loss of life: in the raid on Dresden (1945) up to 135 000 died.

FIREWORKS, combustible or explosive preparations used for entertainment, probably first devised in ancient China to frighten off devils. Their initial European use was as weaponry (see GREEK FIRE) and not until after about 1500 were they employed for entertainment. Compounds of CARBON, POTASSIUM and SULFUR are the prime constituents in fireworks, colors being produced by metallic salts (e.g., blue, COPPER; yellow, SODIUM; red, LITHIUM or STRONTIUM; green, BARIUM), sparks and crackles by powdered IRON, CARBON or ALUMINUM, or by certain LEAD salts. (See also EXPLOSIVES; GUNPOWDER.)

FIRN. See NÉVÉ.

FIRST AID, treatment that can be given by unqualified people in the event of accident, injury and sudden illness, until more skilled persons arrive or the patient is transferred to hospital. Recognition of the injury or the nature of the illness and its gravity are crucial first measures, along with prevention of further injury to the patient or helpers. Clues such as medical bracelets or cards, tablets, lumps of sugar, alcohol and evidence of external injury should be sought and appropriate action taken. Arrest of breathing should be treated as a priority by clearing the airway of dentures, gum, vomit and other foreign material and the use of ARTIFICIAL RESPIRATION; likewise CARDIAC massage may be needed to restore BLOOD CIRCULATION if major PULSES cannot be felt. In traumatic injury, FRACTURES must be recognized and splinted to reduce pain; the possibility of injury to the spine must be considered before moving the patient to avoid unnecessary damage to the SPINAL CORD. External HEMORRHAGE should be arrested, usually by direct pressure on the bleeding point; tourniquets are rarely needed and may be dangerous. Internal hemorrhage may be suspected if SHOCK develops soon after collapse or trauma without obvious bleeding. BURNS AND SCALDS should be treated by immediately cooling the burnt surface to reduce the continuing injury to SKIN due to retained heat. The use and, if necessary, improvization of simple dressings, bandages, splints and stretchers should be known; simple methods of moving the injured, should this be necessary, must also be understood. Accessory functions such as contacting ambulances or medical help, direction of traffic and different aspects of resuscitation should be delegated by the most experienced person present. The inquisitive should be kept away and a calm atmosphere maintained.

Prevention as a part of first aid includes due care in the home: avoiding highly polished floors and unfixed carpets, obstacles on or near stairs, loose flex, overhanging saucepan handles, unlabeled bottles of poison and DRUG cupboards accessible to children. Attention to fireguards, adequate lighting and suitable education of children are also important. Effective first aid depends on prevention, recognition, organization and, in any positive action, adherence to the principle of "do no harm."

FIRTH, term used primarily in Scotland to describe usually a long narrow ESTUARY, often a FJORD, and sometimes a STRAIT. The best known are those of the rivers Forth, Clyde and Tay; the Solway Firth, into which several rivers drain; the Moray Firth, between the NE tip of Loch Ness and the North Sea; and the Pentland Firth, the strait between NE Scotland and the Orkney Islands.

FISCHER, Emil (1852–1919), German organic chemist and pioneer of BIOCHEMISTRY, awarded the 1902 Nobel Prize for Chemistry for his work on the structures of SUGARS and PURINES, simple members of both of which families he synthesized. He also synthesized PEPTIDES and studied ENZYME action in the breaking down of PROTEINS.

FISCHER, Ernst Otto (1918–), German chemist awarded with G. WILKINSON the 1973 Nobel Prize for Chemistry for their work on ORGANOMETALLIC COMPOUNDS.

FISCHER, Hans (1881–1945), German organic chemist awarded the 1930 Nobel Prize for Chemistry for his elucidation of the structure of, and synthesis of, hemin—a substance closely related to heme (see HEMOGLOBIN). He later worked on the analysis of the CHLOROPHYLLS.

FISHERIES, the commercial harvesting of marine and freshwater animals (and some plants) to provide food for men and animals. The main catch is of fishes, but shellfish and marine mammals including seals and whales are also important. The world harvest now totals about 70 million tonnes per year, and has risen annually by about 6% since WWII. About 75% is caught in the cold and temperate zones of the Northern Hemisphere. The chief fishing nations are Peru, China, the USSR, Norway, Japan and the US; in the next rank are Canada, India, Spain, Great Britain and Iceland. Inland fisheries—in lakes, rivers and ricefields—account for less than 10% of recorded catches. The most important groups of fish caught are herring and its relatives, and cod and its relatives. Modern fishing vessels are equipped with radar, depth sounders and echo sounders to locate fish shoals; increasingly used are factory ships which process the fish and freeze or can them. Modern nets are very strong, being made from synthetic fibers. Trawlers draw a bag-shaped net behind them; drift nets are fastened to a buoy; lining involves trailing many-hooked lines in deep water; and in seining a large net encircles the fish and is gradually closed as it is drawn in. The supply of fish can no longer be regarded as practically inexhaustible: it is depleted by the vast catches of efficient modern fishing and also by POLLUTION. Conservation is therefore important, and there are international agreements against overfishing and to regulate the meshes of nets so that young fish can escape. Fish farming is also being developed. International disputes have often arisen over fishing rights in coastal waters.

FISH. See AGNATHA; CHONDRICHTHYES; OSTEICHTHYES.

FISKE, John (1842–1901), born Edmund Fisk Green, US historian and philosopher who wrote popular accounts of Colonial America and attempted to show that evolutionary theory was compatible with religious ideas.

FISSION, the division of CELLS, or sometimes multicellular organisms, to produce identical offspring. Binary fission results in the production of two equal parts and multiple fission in the production of more than two equal parts. The term is normally applied to the reproduction of multicellular organisms such as members of the phylum PROTOZOA.

FISSION, Nuclear, the splitting of the nucleus of a heavy ATOM into two or more lighter nuclei with the release of a large amount of ENERGY. Fission power is used in NUCLEAR REACTORS and the ATOMIC BOMB.

FISTULA, an abnormal communication between two internal organs, or from an organ to the outside of the body. Infection, inflammatory disease (e.g. Crohn's disease—see ENTERITIS), TUMORS, trauma and SURGERY may lead to fistula. The GASTROINTESTINAL TRACT (particularly DUODENUM), PANCREAS, BLADDER and female genital tract are particularly susceptible.

FITCH, John (1743–1798), US inventor and engineer who built the first viable steamboat (1787), larger vessels being launched in 1788 and 1790. All were paddle-powered; his later attempt to introduce the screw PROPELLER was a commercial failure.

FITZROY, Robert. See BEAGLE, H.M.S.

FIXATION, an ambivalent (see AMBIVALENCE) attachment to an object or habit typical of an earlier stage of psychological development. Specific examples are REGRESSION to infantile behavior during stress, and COMPULSION toward objects reminiscent in some way of the individual's childhood.

FJORD, narrow, deep sea inlet, steep-sided and bounded by mountains, formed by past glacial EROSION of a stream or river valley. Usually there is a shallow rock threshold, probably of terminal MORAINE, at the seaward entrance.

FLAGELLATA, or Mastigophora. See PROTOZOA.

FLAME. See COMBUSTION.

FLAME HARDENING, surface hardening technique in which a steel workpiece is heated using an oxyacetylene flame and then quickly cooled with water.

FLAME TEST, preliminary test in qualitative chemical ANALYSIS. A small amount of the sample is introduced into a nonluminous flame; certain metal ions, on excitation, impart a characteristic color to the flame, e.g. yellow for sodium, red for strontium, green for copper. For **flame photometry** see SPECTROPHOTOMETRY.

FLAMETHROWER, a device—basically a SYRINGE—that propels a petroleum fuel, commonly NAPALM, through a nozzle, igniting it as it emerges, so that a burning stream lands on the enemy. Compressed air provides the driving force. A portable flamethrower has a range of 50m, and a mechanized flamethrower 150m. (See also CHEMICAL AND BIOLOGICAL WARFARE.)

FLAMSTEED, John (1646–1719), the first Astronomer Royal, appointed 1675 with the foundation of the GREENWICH OBSERVATORY. In course

The electronic flash gun produces an intense flash of light when the electric charge stored in a capacitor is discharged through a type of electron tube known as a flash tube. Since the flash tube requires a high-voltage source to power and trigger it, but only low-voltage batteries are conveniently portable, a chopper circuit (C) is used to produce a pulsed voltage which can be stepped up in a transformer (T_1) and rectified by a diode (R) in order to charge the main capacitor (C_1). This is permanently connected across the main electrodes of the gas-filled flash tube (F) but a flash discharge occurs only when there is an ionizing high-voltage discharge between the cathode and a trigger electrode. The trigger discharge occurs when an ignition capacitor (C_2), also charged from T_1, is discharged through the primary winding of the ignition transformer (T_2), the secondary winding of which lies between the cathode and trigger electrode of the flash tube. The ignition capacitor is discharged when the camera switch (S) is closed on pressing the shutter release.

Below right: A toroidal flash tube as used in professional electronic flash units.

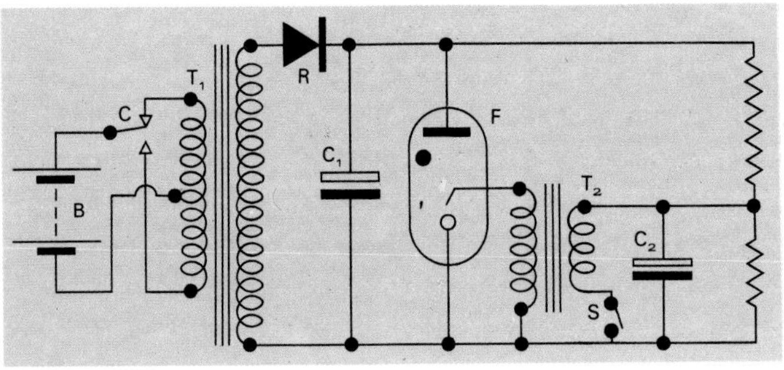

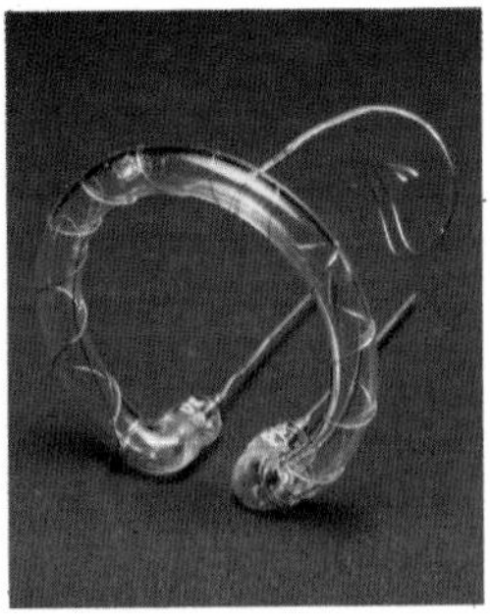

of compiling his major star catalog, *Historia Coelestis Britannica*, he was pestered for his data by NEWTON: Newton prevailed, and a muddled version, edited by HALLEY, was published in 1712. Flamsteed's own edition appeared posthumously (1725).

FLANNEL, in TEXTILES, a soft all-wool fabric with a slightly raised (fluffy) surface used for blankets and clothing. **Flannelette** is a cotton fabric having a similar finish.

FLAPS, AIRPLANE control surfaces fixed at the trailing edges of the wings, lowered to improve control in landing. When lowered, flaps improve the AIRFOIL's lift but at the cost of increased drag. They are consequently raised for normal flying.

FLARE, Solar, a temporary brilliance in the SUN's chromosphere associated with a sunspot or group of sunspots. Large flares may last for as long as an hour or more, small ones only a few minutes. During the flare X RAYS and RADIO waves are emitted from the solar corona, as are slower-moving ALPHA PARTICLES and PROTONS.

FLASHBULB, a light source used in PHOTOGRAPHY to give a brief intense flash of light. A switch in the CAMERA shutter mechanism discharges a CAPACITOR in the flash gun through a tungsten filament in the oxygen-filled bulb, igniting the mass of aluminum, magnesium or zirconium wire present. Bulbs are lacquered to prevent shattering, blue lacquering being used to balance the light for color work. **Flash cubes**, comprising four flash bulbs mounted in one expendable unit, are percussion-ignited by a pin in the camera body. More versatile is the **electronic flash gun** which employs a discharge lamp (see LIGHTING) which can be used repeatedly.

FLASH PHOTOLYSIS, technique for investigating very fast chemical reactions (see KINETICS, CHEMICAL). A very intense flash of light of very short duration (from a LASER or flash lamp) is passed through a reaction mixture, usually gaseous. Instant dissociation occurs, producing FREE RADICALS, whose subsequent fast reactions are followed by automatic spectroscopy. (See also PHOTOCHEMISTRY.)

FLASH POINT, the lowest temperature at which the VAPOR above a volatile LIQUID forms a combustible mixture with the OXYGEN in the air. At the flash point the TEMPERATURE is too low for sustained COMBUSTION to occur, but when a pilot flame is introduced to the mixture a brief flash occurs.

FLATFOOT, deformity of FOOT in which the longitudinal or transverse arches of the feet are flattened or lost; this results in loss of spring and the inefficient use of the feet in walking or running. It may result from muscle weakness or be congenital. Corrective exercises and shoe wedges may relieve the condition.

FLATWORMS. See PLATYHELMINTHES.

FLAX, *Linum usitatissimum*, an important temperate and subtropical plant, grown for fiber and LINSEED OIL. The native flax of Eurasia is a straw-like annual, 600–900mm (2–3ft) high, whose white, blue or pink flowers ripen into seed bolls. The crop is usually harvested after about 14 weeks, when the fiber is separated from the seed. The fibers are then soaked and scraped away from the woody stem and the longer ones are combed out and spun into yarn, which is turned into LINEN. The seeds are squeezed to give oil. The USSR is the chief producer of fiber flax and a leading producer of seed flax. Family: Linaceae.

FLEMING, Sir Alexander (1881–1955), British bacteriologist, discoverer of lysozome (1922) and penicillin (1928). Lysozome is an ENZYME present in many body tissues and lethal to certain bacteria; its discovery prepared the way for that of ANTIBIOTICS. His discovery of PENICILLIN was largely accidental; and it was developed as a therapeutic later, by Howard FLOREY and Ernst CHAIN. All three received the 1945 Nobel Prize for Physiology or Medicine for their work.

FLEXNER, Simon (1863–1946), US medical scientist who did important work in bacteriology and pathology, especially on the nature of POLIOMYELITIS, and contributed toward the development of a serum treatment for spinal MENINGITIS.

FLIGHT, the ability to travel through the air during long periods. The only animals that are capable of sustained flight are the extinct pterodactyls; some insects; a few fish and mammals, and most birds. Very few insects are completely wingless, and most have two pairs of wings that flap together. In flies and beetles one pair is modified. Usually muscles distort the thorax, forcing the wings up and down at rates of up to 1000 beats per second in midges.

The only true flying fish are certain freshwater hatchetfishes and Butterfly fishes. The tropical flying fish in fact glides. In birds, power for flying comes from the breast muscles, which control both the up and down strokes of the wing. Wing shape depends on the kind of flight; narrow for gliding and broad for soaring.

The only flying mammals are the bats, whose wings are composed of skin stretched between the finger and back legs.

FLIGHT, History of. LEONARDO DA VINCI was the first man to attempt the scientific design of flying machines. But in his time no motor was available which was powerful enough to lift a man into the air. Man's first ascents from the ground had to await the late 18th century and the invention of the MONTGOLFIER brothers' hot-air BALLOON and J. CHARLES' hydrogen balloon (1783). The addition of steam engines to the balloon gave the first maneuverable AIRSHIP (1852). Meanwhile G. CAYLEY designed and built flying GLIDERS (1810–1853) and William Henson designed a steam-powered model airplane with twin PROPELLERS (1842). It was not until the advent of the gasoline INTERNAL-COMBUSTION ENGINE, though, that the powered heavier-than-air machine became a practical possibility. The first successful controlled airplane flight was made by the WRIGHT BROTHERS near Kitty Hawk, N.C., on December 17, 1903 and within a few years there were many competing manufacturers and fliers of airplanes. Airplane technology was greatly stimulated by WWI and after 1919 commercial aviation developed rapidly.. Meanwhile, the AUTOGIRO was invented by J. de la CIERVA (1923), to be followed by SIKORSKY's HELICOPTER in 1939. JET PROPULSION was developed in several countries during WWII and by the mid-1950s had come to be used in the majority of military and commercial airplanes. RADAR navigation systems came into general use in this period. The early 1970s saw the introduction of wide-bodied jet airliners (jumbo jets) with vastly increased carrying capacity and the development of the first supersonic jet airliners.

FLINT, or **chert**, sedimentary rock composed of microcrystalline QUARTZ and CHALCEDONY. It is found as nodules in LIMESTONE and CHALK, and as layered beds, and was mainly formed by alteration of marine sediments of silicaceous organisms, and by replacement, preserving many FOSSIL outlines. A hard rock, flint may be chipped to form a sharp cutting edge, and was used by STONE AGE man for their characteristic tools.

FLOODS AND FLOOD CONTROL. River floods are one of mankind's worst enemies. In 1887, when the Hwang Ho overflowed, around 900 000 lost their lives; and, as recently as 1970, 200 000 died in E Pakistan when a cyclone struck the Ganges delta. Clearly the development of ways to control and contain floods must be a preoccupation of man.

Often floods are caused by unusually rapid thawing of the winter snows: the river, unable to hold the increased volume of water, bursts its banks. Heavy rainfall may have a similar effect. Coastal flooding may result from an exceptionally high TIDE combined with onshore winds, or, of course, from a TSUNAMI.

River floods can be forestalled by artificially deepening and broadening river channels or by the construction of suitably positioned DAMS. Artificial levees may also be built (in nature, levees occur as a result of sediment deposited while the river is in flood; they take the form of built-up banks). Vegetation planted on uplands helps to reduce surface runoff (see DRAINAGE).

Flood control can create new problems to replace the old. In Egypt the Aswan Dam has halted the once regular flooding of the Nile, thus robbing farmlands of a rich annual deposit of silt. But flood control made possible the civilization of ancient Mesopotamia and plays a vital rôle in modern water conservation. (See also RIVERS AND LAKES.)

FLOREY, Howard Walter, Baron (1898–1968), Australian-born British pathologist who worked with E. B. CHAIN and others to extract PENICILLIN from *Penicillium notatum* mold for use as a therapeutic drug (1938–44). He shared with Chain and Alexander FLEMING the 1945 Nobel Prize for Physiology or Medicine.

FLOTATION, Froth, a process used to recover valuable minerals from low-grade areas. The pulverized ore is mixed with water and flotation reagents. When air is pumped into the mixture, the mineral particles that preferentially adhere to the air bubbles rise to the surface in a froth which can then be skimmed off.

FLOUR, fine powder ground from the grains or starchy portions of wheat, rye, corn, rice, potatoes, bananas or beans. Plain white flour is produced from wheat; soft wheat produces flour used for cakes and hard wheat, with a higher GLUTEN content, makes flour used for bread. Flour is made from the endosperm, which constitutes about 84% of the grain; the remainder comprises the BRAN, which is the outer layers of the grain, and the germ, which is the embryo. Grain used to be milled by hand between two stones, until the development of wind, water or animal driven mills. In modern mills, the grain is thoroughly cleaned and then tempered by bringing the water content to 15%, which makes the separation of the bran and germ from the endosperm easier. The endosperm is broken up by rollers and the flour graded and bleached. It may then be enriched with VITAMINS. Byproducts are used mainly for cattle food, although wheat germ is an important source of vitamin E.

Froth flotation cells used for the concentration of copper ore.

FLOWER, the part of an ANGIOSPERM that is concerned with REPRODUCTION. There is a great

The success of the flowering plants (angiosperms) is largely due to their sophisticated reproductive system involving the development of flowers, which are often brightly colored and scented. The illustration shows the structure of a typical angiosperm flower with brightly colored petals (1) and sepals (2). The female parts of the flower are the ovules or egg cells (3) contained in the ovary (4), and the style (5), at the tip of which is the stigma (6). Pollen produced by the male parts of the flower, or stamens, must arrive on the stigma if fertilization is to occur. The stamens are composed of anther (7) and filament (8).

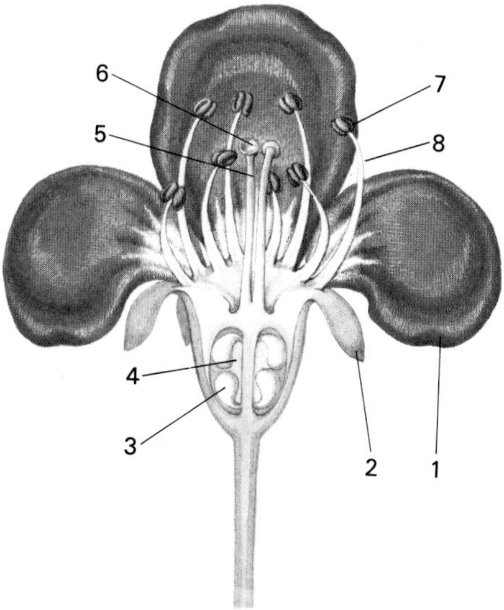

variety of floral structure, but the basic organs and structure are similar. Each flower is borne on a stalk or pedicel, the tip of which is expanded to form a receptacle that bears the floral organs. The **sepals** are the first of these organs and are normally green and leaflike. Above the sepals there is a ring of **petals**, which are normally colored and vary greatly in shape. The ring of sepals is termed the **calyx** and the ring of petals the **corolla**. Collectively the calyx and corolla are called the **perianth**. Above the perianth are the reproductive organs comprising the male organs, the **stamens** (collectively known as the **androecium**) and female organs, the **carpels** (the **gynoecium**.) Each stamen consists of a slender stalk, or FILAMENT, which is capped by the pollen-producing **anther**. Each carpel has a swollen base, the **ovary**, which contains the **ovules** that later form the **seed**. Each carpel is connected by a **style** to an expanded structure called the **stigma**. Together, the style and stigma are sometimes termed the **pistil**.

There are three main variations of flower structure. In hypogynous flowers (e.g. buttercup) the perianth segments and stamens are attached below a superior ovary, while in perigynous flowers (e.g. rose) the receptacle is cup-like enclosing a superior ovary, with the perianth segments and stamens attached to a rim around the receptacle. In epigynous flowers (e.g. dandelion) the inferior ovary is enclosed by the receptacle and the other floral parts are attached to the ovary. In many plants, the flowers are grouped together to form an INFLORESCENCE.

Pollen produced by the stamens is transferred either by insects or the wind to the stigma where POLLINATION takes place. Many of the immense number of variations of flower form are adaptions that aid either insect or wind pollination. (See also PLANT KINGDOM.)

FLOWERING PLANTS. See ANGIOSPERMS.

FLU. See INFLUENZA.

FLUENT. See FLUXIONS.

FLUID, a substance which flows (undergoes a continuous change of shape) when subjected to a tangential or shearing FORCE. LIQUIDS and GASES are fluids, both taking the shape of their container. But while liquids are virtually incompressible and have a fixed volume, gases expand to fill whatever space is available to them.

FLUID COUPLING, or fluid flywheel, a device for transmitting TORQUE between rotating shafts comprising an oil-filled case connected to the driving shaft and with impeller vanes on the inside, and an opposing turbine mounted on the driven. The only coupling between these elements is by means of oil thrown by the impeller against the turbine. Modified forms of this device known as the fluid converter and the converter coupling (torque converter) and including a third stator or reactor element are used in the TRANSMISSIONS of construction machinery and AUTOMOBILES respectively.

FLUID MECHANICS, the study of moving and static FLUIDS, dealing with the FORCES exerted on a fluid to hold it at rest and the relationships with its boundaries that cause it to move. The scope of the subject is wide, ranging from HYDRAULICS, concerning the applications of fluid flow in pipes and channels, to aeronautics, the study of airflow relating to the design of AIRPLANES and ROCKETS. Any fluid process, such as

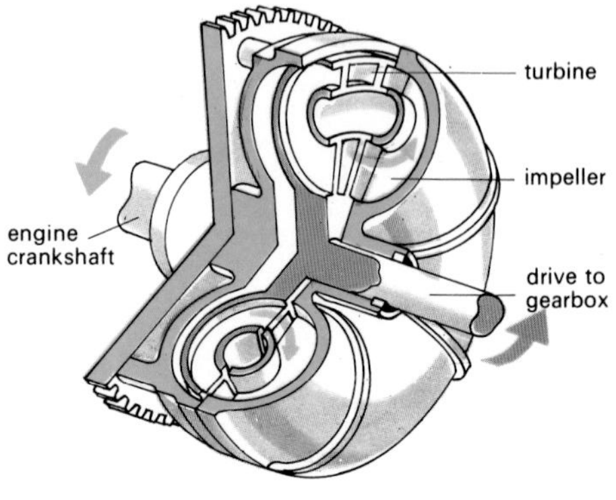

Cutaway of a fluid coupling as used in truck transmissions. The blue-colored portions are driven directly from the engine; the circulation of the fluid is indicated by the green arrows, and the driven elements are colored red.

flow around an obstacle or in a pipe, can be described mathematically by a specific equation that relates the forces acting, the dimensions of the system and its properties such as TEMPERATURE, PRESSURE and DENSITY. Newton's laws of motion and VISCOSITY, the first and second laws of THERMODYNAMICS and the laws of conservation of MASS, ENERGY and MOMENTUM are applied as appropriate. Much use is also made of experimental evidence from models, wind tunnels, etc., to determine the process equation. Many types of flow occur: in laminar flow in a closed pipe, distinct layers of fluid slide over each other, their velocity decreasing to zero at the pipe wall; in turbulent flow the fluid is mixed by eddies and vortices and a statistical treatment is needed. (See also ARCHIMEDES; BERNOULLI; PASCAL'S LAW; REYNOLDS NUMBER.)

FLUKES, members of the PLATYHELMINTHES, sometimes known as flatworms. Flukes are parasitic and are responsible for diseases in a number of animals including man.

FLUORESCEIN, synthetic DYE made by fusing RESORCINOL with phthalic anhydride (see ACID ANHYDRIDES) and zinc chloride catalyst. A red crystalline solid, it dissolves in alkalis to give a deep red solution showing very intense yellow-green FLUORESCENCE, visible at very low concentrations. It is used as a water tracer, and was used in WWII to mark spots on the sea.

FLUORESCENCE. See LUMINESCENCE.

FLUORESCENT LIGHTING. See LIGHTING.

FLUORIDATION, addition of small quantities of fluorides (see FLUORINE) to public water supplies,

The chemical structure of fluorescein (3,6-dihydroxyfluoran).

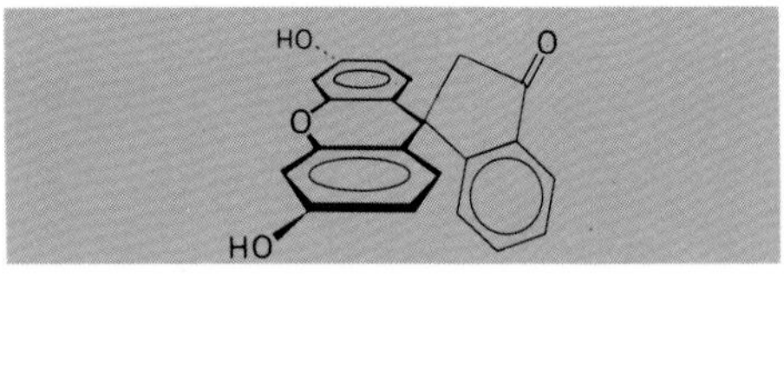

The foehn effect. When warm, moist air (1) crosses a mountain range, it cools as it climbs the windward side and deposits most of its moisture as rain (2). On its descent on the leeward side of the range, the now drier air heats up much more rapidly (3), giving rise to hot, dry weather downwind.

bringing the concentration to 1 ppm, as in some natural water. It greatly reduces the incidence of tooth decay by strengthening the teeth. Despite some opposition, many authorities now fluoridate water. Toothpaste containing fluoride is also valuable.

FLUORINE (F), the lightest of the HALOGENS, occurring naturally as FLUORITE, CRYOLITE and fluorapatite (see APATITE). A pale-yellow toxic gas, fluorine is made by electrolysis of potassium fluoride in liquid HYDROGEN FLUORIDE. It is the most reactive, electronegative and oxidizing of all elements, reacting with almost all other elements to give fluorides (see HALIDES) of the highest possible oxidation state. It displaces other nonmetals from their compounds. Most nonmetal fluorides are highly reactive, but sulfur hexafluoride (used as an electrical insulator) and carbon tetrafluoride are inert (see STEREOCHEMISTRY). Fluorine is used in rocket propulsion, in URANIUM production, and to make FLUOROCARBONS. (See also FLUORIDATION.) AW 19.0, mp −220°C, bp −188°C.

FLUORITE, a common mineral composed of calcium fluoride (see CALCIUM), also called fluorspar. It forms cubic crystals with a wide color range. Fluorite is used as a flux in the iron and steel industries, and as a FLUORINE ore.

FLUOROCARBONS, HYDROCARBONS in which hydrogen atoms are replaced (wholly or in part) by fluorine. Because of the stability of the carbon-fluorine bond, they are inert and heat-resistant. Thus they can be used in artificial joints in the body, and where hydrocarbons would be decomposed by heat, such as in spacecraft heatshields, the coating of nonstick pans or as lubricants. Liquid fluorocarbons are used as refrigerants. (See also TEFLON; FREON.)

FLUOROSCOPE, device used in medical diagnosis and engineering quality control which allows the direct observation of an X-RAY beam which is being passed through an object under examination. It contains a fluorescent screen which converts the X-ray image into visible light (see LUMINESCENCE) and, often, an image intensifier.

FLUTTER AND WOW. See HIGH-FIDELITY.

FLUX, in WELDING and SOLDERING, a material used to clean the surface of the workpieces, particularly to remove OXIDE films; also, in metallurgical processes, a material (such as the LIMESTONE used in IRON smelting) added to a melt to abstract impurities in the form of a SLAG.

FLUXIONS, term used by NEWTON for CALCULUS. In his terminology, later abandoned in favor of that of LEIBNIZ, what we now call a FUNCTION was called a fluent, its derivative a fluxion.

FLYWHEEL, a device used for storing mechanical ENERGY and for smoothing the POWER output of an engine, comprising a heavy shaft-mounted wheel. Since energy is stored as kinetic energy of rotation, it is advantageous for the wheel to have as large as possible a MOMENT of INERTIA, and thus for its MASS to be concentrated near the rim.

FM (Frequency Modulation). See RADIO.

FOAM RUBBER, a material used widely for cushioning and sometimes for thermal INSULATION. It is about 10% rubber, 90% air. Usually LATEX is whipped into a froth, then gelled and vulcanized in a mold. For insulation, hydrogen PEROXIDE is introduced into latex: this liberates oxygen during vulcanization to form unlinked cells.

FOCAL LENGTH. See LENS, OPTICAL.

FOCUS (Geometry). See CURVE.

FOCUS (Optics). See LENS, OPTICAL.

FOCUS OF EARTHQUAKE. See EARTHQUAKE.

FOEHN, or **föhn,** dry, warm wind coming down the leeward slopes of mountains due to air having lost its moisture while ascending the windward slope, then warming on its descent. (See also CHINOOK.)

FOETUS. See FETUS.

FOG, in essence, a CLOUD touching or near to the earth's surface. A fog is a SUSPENSION of tiny water (sometimes ice) particles in the air. Fogs are a result of the air's HUMIDITY being high enough that CONDENSATION occurs around suitable nuclei; and are found most often near coasts and large inland bodies of water. In industrial areas, fog and smoke may mix to give **smog.** Persistent **advection fogs** occur when warm, moist air moves over cold land or water.

FOIL, Metal, metal rolled or beaten into a very thin sheet. Commonest is ALUMINUM foil (having largely replaced TIN foil), used extensively in packaging and in the home in the storage or cooking of food. GOLD foil is used decoratively and in electronics (see also GOLD LEAF); LEAD foil as a radiation shield.

FOLD, a buckling in rock strata. Folds convex upward are called **anticlines**; those convex downward, **synclines**. They may be tiny or up to hundreds of kilometres across. Folds result from horizontal pressures in the EARTH's crust. The upper portions of anticlines have often been eroded away. (See also GEOSYNCLINE; MOUNTAIN.)

FOLIC ACID. See VITAMINS.

FOLLICLE, dry single-seeded FRUIT that splits along one margin only to liberate the seeds, e.g., larkspurs, columbine. (See also LEGUME.) Also, in animals, deep pit surrounding the root of a HAIR.

FOLLICLE-STIMULATING HORMONE, or FSH, a pituitary-gland GONADOTROPHIN, concerned with ovarian follicle and SPERM development in females and males.

FOLSOM CULTURE, prehistoric (c8000 BC) American Indian culture named for Folsom, N. M., where stone arrow- or spearheads were discovered in the 1920s in close association with the bones of an extinct breed of bison. Other tools, such as stone scrapers, bone needles, etc., have been found; and it is thought that the culture was a nomadic, hunting one.

FOOD CHAIN. See ECOLOGY.

FOOD POISONING, disease resulting from ingestion of unwholesome food, usually resulting in COLIC, VOMITING, DIARRHEA and general malaise. While a number of VIRUS, contaminant, irritant and allergic factors may play a part in some cases, three specific types are common: those due to STAPHYLOCOCCUS, Clostridium and SALMONELLA bacteria. Inadequate cooking, allowing cooked food to stand for long periods in warm conditions and contamination of cooked food with bacteria from humans or uncooked food are usual causes. *Staphylococci* may be introduced from a BOIL or from the NOSE of a food handler; they produce a TOXIN if allowed to grow in cooked food. Sudden vomiting and abdominal pain occur 2–6 hours after eating. *Clostridium* poisoning causes colic and diarrhea, 10–12 hours after ingestion of contaminated meat. *Salmonella* enteritis causes colic, diarrhea, vomiting and often fever, starting 12–24 hours after eating; poultry and human carriers are the usual sources. BOTULISM is an often fatal form of food poisoning. In general, food poisoning is mild and self-limited and symptomatic measures only are needed; ANTIBIOTICS rarely help.

FOOD PRESERVATION, a number of techniques used to delay the spoilage of food. There are two main causes of spoilage: one is the PUTREFACTION that follows the death of any plant or animal; the other over-ripening, the result of the action of certain plant ENZYMES. Heating destroys these enzymes and the BACTERIA responsible for putrefaction but, before it cools, the food must be isolated in cans or bottles from air-borne bacteria. Freezing slows the enzyme action and the REPRODUCTION of the bacteria and preserves flavor better. DEHYDRATION, IRRADIATION and preservatives are also used. Traditional means of preservation include smoking, salting and pickling (see VINEGAR). (See also FOOD POISONING; REFRIGERATION.)

FOOL'S GOLD. See PYRITE.

FOOT, weight-bearing structure of animal LEGS, and, in man, of the lower limbs only. It consists of numerous BONES which are connected at the ankle with the bones of the lower leg. Muscles and tendons in the feet are concerned with walking and running and in sustaining the transverse and longitudinal arches. Muscles in the calf act across the ankle JOINT to move the foot. The SKIN of the sole is thickened.

FOOT (ft), an old Imperial and US Customary unit of length equal to one-third of a yard or 0.3048m.

FOOT AND MOUTH DISEASE. See HOOF AND MOUTH DISEASE.

FOOT-CANDLE, in PHOTOMETRY a former unit of illumination. It represents the illumination of a surface receiving an incident luminous flux of 1 lumen per square foot.

FOOT-POUND (weight) or **foot-pound-force** (ft lbf), a unit of WORK, that done when a MASS of one pound is lifted through one foot against gravity, equal to 1.35582 joules.

FORCE, in mechanics, the physical quantity which, when it acts on a body, either causes it to change its state of motion (i.e., imparts to it an ACCELERATION), or tends to deform it (i.e., induces in it an elastic strain—see MATERIALS, STRENGTH OF). Dynamical forces are governed by NEWTON's laws of motion, from the second of which it follows that a given force acting on a body produces in it an acceleration proportional to the force, inversely proportional to the body's MASS and occurring in the direction of the force. Forces are thus VECTOR quantities with direction as well as magnitude. They may be manipulated graphically like other vectors, the sum of two forces being known as their resultant. The SI UNIT of force is the newton, a force of one newton being that which will produce an acceleration of $1 m/sec^2$ in a mass of 1 kilogram.

FORCING, the technique of bringing plants into a flowering or fruiting condition out of season. By artificially changing such environmental conditions as heat, light and moisture, plants can be forced into growth and the production of fruits and flowers at times when they would not do so under natural conditions.

FOREBRAIN, in EMBRYOLOGY, the division of the BRAIN that develops into the cerebral cortex (particularly large in man), the basal ganglia, THALAMUS, HYPOTHALAMUS, olfactory bulbs, RETINA and optic nerves.

FORENSIC MEDICINE, the branch of MEDICINE concerned with legal aspects of DEATH, DISEASE or injury. Forensic medical experts are commonly required to examine corpses found in possibly criminal circumstances. They may be asked to elucidate probable cause and approximate time of death, to investigate the possibility of POISONING, trauma or suicide, to analyse links with possible murder weapons and to help to identify decayed or mutilated bodies.

FORESTRY, management of forests for productive purposes. In the US, a forestry program emerged in the 1890s because of fears of a "timber famine" and following exploitation of the Great Lakes pine forests. Congress authorized the first forest reserves in 1891; creation of the Forest Service in 1905 put forestry on a scientific basis.

The most important aspect of forestry is the production of lumber. Because of worldwide depletion of timber stocks, it has become necessary to view forests as renewable productive resources, and because of the time scale and area involved in the growth of a forest, trees need more careful planning

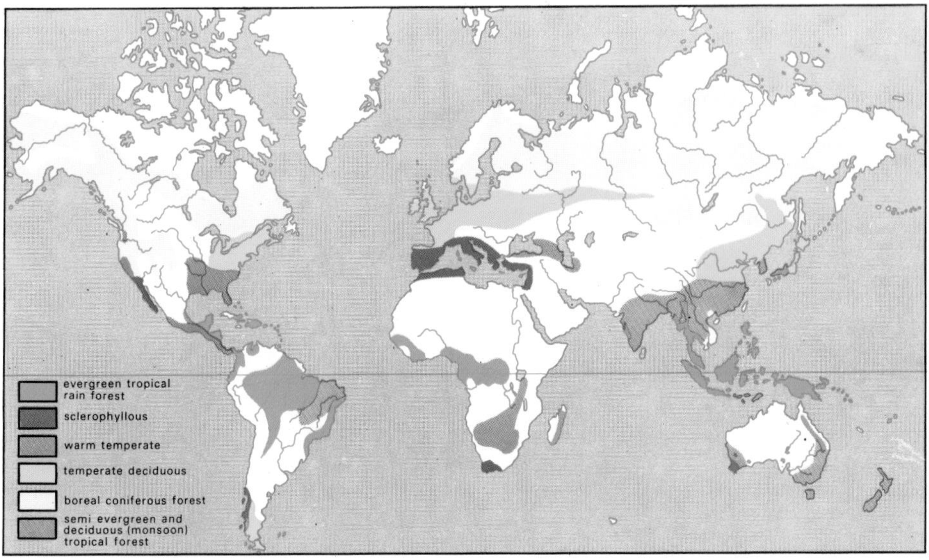

The forest zones of the earth.

than any other crop. Forestry work plans for a continuity of timber production by balancing planting and felling. Other important functions are disease, pest, fire and flood control. The forester must control the density and proportions of the various trees in a forest and ensure that man does not radically disturb a forest's ecological balance.

The science of forestry is well advanced in the US, which is the world's largest timber producer and has more than 25 forestry schools across the country. However, only 20% of the world's forests are being renewed, and timber resources are declining. (See also CONSERVATION; FORESTS.)

FORESTS, extensive tracts of land whose vegetation is composed most notably of trees. They are of considerable importance to man: they have provided fuel and building material since prehistory; their fruits have served as food; and, particularly today, they play a valuable role in countering SOIL EROSION (see also CONSERVATION).

Coniferous Forests have as characteristic trees the conifers—pine, spruce and fir—and are found in temperate regions as far N as the edges of the TUNDRA. They are similarly distributed in the S Hemisphere.

Deciduous Forests, found in temperate zones, are characterized by DECIDUOUS TREES with some conifers. They contribute most of the world's commercially important hardwood; e.g., oak, elm, beech and birch.

Fossil or Petrified Forests, best known of which is that of E Ariz., occur where collections of trees have been petrified, usually by mineralization (see FOSSILS).

Monsoon Forests are found in MONSOON regions, and resemble rain forests except that they are more open. Their trees have adapted to the marked dry season. They are a source of teak, a commercially important hardwood.

Rain or Equatorial Forests (*Selvas*) are found in equatorial regions, where there is no dry season. These dense jungles support the wild RUBBER tree, and are important as sources of mahogany and ebony.

Thorn Forests occur in tropical and subtropical areas where rainfall is insufficient to maintain larger trees. They are characterized by shrubs and small trees; e.g., mesquite, acacia. The thorn forest of Brazil is known as the *caatinga*.

(See also DENDROCHRONOLOGY; TREES; WOOD.)

FORGING, the shaping of metal by hammering or pressing, usually when the workpiece is red hot (about 700–1000K) but sometimes when it is cold. Unlike CASTING, forging does not alter the granular structure of the metal, and hence greater strength is possible in forged than in cast metals. The most basic method of forging is that of the blacksmith, who heats the metal in an open fire (forge) and hammers it into shape against an anvil. Today, metals are forged between two DIES, usually impressed with the desired shape. Techniques include: **drop forging**, where the workpiece is held on the lower, stationary die, the other being held by a massive ram which is allowed to fall; **press forging**, where the dies are pressed together; and **impact forging**, where the dies are rammed horizontally together, the workpiece between. (See also METALLURGY.)

FORMALDEHYDE (HCHO), colorless, acrid, toxic gas; the simplest ALDEHYDE, more reactive than the others. It is made by catalytic air oxidation of METHANOL vapor or of NATURAL GAS. Formaldehyde gas is unstable, and is usually stored as its aqueous solution, **formalin**, used as a disinfectant and preservative for biological specimens; on keeping, formalin deposits a polymer, **paraformaldehyde**, which regenerates formaldehyde on heating. Formaldehyde is condensed with UREA and PHENOLS to make PLASTICS, with AMMONIA to give hexamethylenetetramine (a urinary ANTISEPTIC also used to make RDX), and with ACETALDEHYDE to give pentaerithrytol and hence PETN. It is also used in TANNING and textile manufacture. MW 30.0, mp −92°C, bp −21°C.

FORMIC ACID (HCOOH), colorless, acrid liquid;

A fine fossil of *Acanthonemus*, a member of the Osteichthyes (bony fish) from the Cretaceous period.

the simplest CARBOXYLIC ACID, a stronger ACID than the others. It occurs in the stings of ants and nettles. It is made (via sodium formate) by heating sodium hydroxide with carbon monoxide under pressure, and used in TANNING, as a latex coagulant and to reduce dyes. MW 46.0, mp 8°C, bp 101°C.

FORMULA, Chemical, a symbolic representation of the composition of a MOLECULE. The **empirical formula** shows merely the proportions of the atoms in the molecule, as found by chemical ANALYSIS, e.g., water H_2O, acetic acid CH_2O. (The subscripts indicate the number of each atom if more than one.) The **molecular formula** shows the actual number of atoms in the molecule, e.g., water H_2O, acetic acid $C_2H_4O_2$. The atomic symbols are sometimes grouped to give some idea of the molecular structure, e.g., acetic acid CH_3COOH. This is done unambiguously by the **structural formula**, which shows the chemical BONDS and so distinguishes between ISOMERS. The **space formula** shows the arrangement of the atoms and bonds in three-dimensional space, and so distinguishes between STEREOISOMERS; it may be drawn in PERSPECTIVE or represented conventionally. Loosely-associated compounds, such as LIGAND complexes, are often shown with a dot, e.g., copper (II) sulfate pentahydrate, $CuSO_4.5H_2O$. Special symbols are sometimes used for common groups and ligands, e.g., ethyl Et, phenyl Ph, ethylenediamine en.

FORSSMAN, Werner (1904–), German surgeon awarded, with A. COURNAND and D. W. RICHARDS, the 1956 Nobel Prize for Physiology or Medicine for his discovery of the technique of introducing medicine directly into the heart via a CATHETER.

FORTIN, Jean, French physicist. See BAROMETER.

FORTRAN (*for*mula *tran*slation), one of the most widely used COMPUTER languages. Originally developed for purely scientific work, it is now, as Fortran IV, used also in commerce and elsewhere.

FOSSILS, the remains, traces or impressions of living organisms that inhabited the earth during past ages. Traces may be, for example, footprints, burrows or preserved droppings.

Fossil remains take a number of forms. **Petrification** describes two ways in which the shape of hard parts of the organism may be preserved. In **permineralization**, the pore spaces of the hard parts are infilled by certain minerals (e.g., SILICA, PYRITE, CALCITE) that infiltrate from the local GROUNDWATER. The resulting fossil is thus a mixture of mineral and organic matter. In many other cases, **mineralization** (or **replacement**) occurs, where the hard parts are dissolved away but the form is retained by deposited minerals. Where this has happened very gradually, even microscopic detail may be preserved; but generally only the outward form remains.

Exceptional fossils are those where the organism has been preserved in its entirety: e.g., mammoths in the Siberian PERMAFROST, or insects in AMBER (though dehydrated). Sometimes teeth and shells may be preserved unaltered.

Often the organism is dissolved entirely, so that only a cast or impression remains in the rock. In the process of **carbonization**, the tissues decompose leaving only a thin CARBON film that shows the outline of the organism's form. (See also PALEONTOLOGY; PREHISTORIC ANIMALS; PREHISTORIC MAN; RADIOCARBON DATING.)

FOUCAULT, Jean Bernard Léon (1819–1868), French physicist best known for showing the rotation of the earth with the FOUCAULT PENDULUM, inventing the GYROSCOPE and for the first reasonably accurate determination of the velocity of LIGHT.

FOUCAULT PENDULUM, a PENDULUM comprising an iron ball at the end of a long steel wire which, on being set swinging, maintains its direction of swing while the earth rotates beneath it. When one was demonstrated by J. B. L. FOUCAULT in 1851, it provided the first direct evidence for the rotation of the earth. Foucault pendulums are demonstrated in several major science museums.

FOULARD, a lightweight twill or plain-woven silk fabric, used for scarves and dresses; also, a garment made of foulard. Imitation foulards are made of mercerized cotton or synthetic fibers.

FOUNDING. See CASTING.

FOUR-COLOR PROBLEM, an unsolved topological (see TOPOLOGY) problem concerning the minimum number of colors required to color any map on a PLANE surface such that no two adjoining regions are in the same color. It can be proved that five colors will always be sufficient; and one can easily draw maps requiring more than three colors; but a general proof that four colors will always be sufficient has not yet been given. On non-plane surfaces, more colors may be required: a map on a TORUS, for example, may require up to seven.

FOURIER, Jean Baptiste Joseph, Baron (1768–1830), French mathematician best known for his equations of HEAT propagation and for showing that all periodic oscillations can be reduced to a SERIES of simple, regular WAVE MOTIONS, as represented by Fourier Series, which have the general form

$$\tfrac{1}{2}a_0+\sum_{n=1}^{\infty}(a_n \cos nx + b_n \sin nx).$$

FOURTH DIMENSION. See SPACE-TIME.

FOVEA, the tiny area in the macula (the area of the RETINA used for central VISION) where the retina is thinned of nonreceptor cells and there are no rods. The concentration of cones and lack of additional tissue makes this the most sensitive part of the retina for both acuity and color vision.

FOX TALBOT, William Henry. See TALBOT, WILLIAM HENRY FOX.

FRACTION, a RATIONAL NUMBER that cannot be expressed as an INTEGER. Fractions are normally expressed either as RATIOS between two integers (e.g., $\frac{3}{4}$) or in the DECIMAL SYSTEM (e.g., 0.75). Fractions greater than 1 are usually expressed as an integer plus a fraction less than 1: e.g., $\frac{10}{9} = 1\frac{1}{9}$; $\frac{22}{3} = 7\frac{1}{3}$. Fractions less than -1 are treated analogously.

FRACTIONATION. See DISTILLATION.

FRACTURES, mechanical defects in BONE caused by trauma or underlying DISEASE. Most follow sudden bending, twisting or shearing forces, but prolonged stress (e.g., long marches) may lead to small fractures. Fractures may be *open*, in which bone damage is associated with SKIN damage, with consequent liability to infection; or *closed*, in which the overlying skin is intact. Comminuted fractures are those in which bone is broken into many fragments. *Greenstick* fractures are partial fractures where bone is bent, not broken, and occur in children. Severe pain, deformity, loss of function, abnormal mobility of a bone and HEMORRHAGE, causing swelling and possibly SHOCK, are important features; damage to nerves, ARTERIES and underlying viscera (e.g., LUNG, SPLEEN, LIVER and BRAIN) are serious complications. Principles of treatment are: reduction, or restoring the bone to satisfactory alignment, by manipulation or operation; immobilization, with plaster, splints or internal fixation with metal or bone prostheses, until bony healing has occurred, and rehabilitation, which enables full recovery of function in most cases. Early recognition and appropriate treatment of associated soft tissue injury is crucial. **Pathological fractures** occur when congenital defect, lack of mineral content, TUMORS etc. weaken the structure of bone, allowing fracture with trivial or no apparent injury.

FRANCIUM (Fr), a radioactive ALKALI METAL, resembling CESIUM, which has been obtained only in tracer quantities. Its most stable isotope, Fr^{223}, has a half-life of 21 minutes.

FRANCK, James (1882–1964), German physicist who shared with G. HERTZ the 1925 Nobel Prize for Physics for their experiments showing the internal structure of the atom to be quantized (see QUANTUM THEORY). With E. V. Condon, he was responsible for the **Franck-Condon principle**, which assumes that the nuclei in vibrating molecules do not have time to move during electronic transitions.

FRANK, Ilya Mikhailovich (1908–), Russian physicist who, with TAMM, provided an explanation for CERENKOV RADIATION (1937), first observed by CHERENKOV (1934). For their work all three shared the 1958 Nobel Prize in Physics.

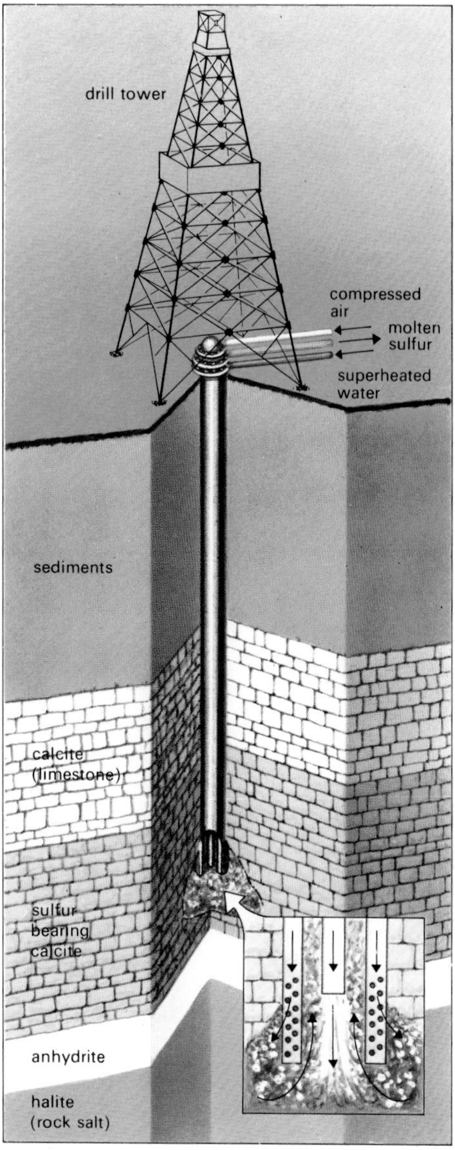

In the Frasch process for the recovery of sulfur, compressed air and superheated water are pumped down the outermost and innermost passages of a triple pipe into the sulfur-bearing stratum. Molten sulfur is forced up the other passage to the surface where it solidifies as it cools.

FRANKLIN, Benjamin (1706–1790), American printer, statesman and scientist who first achieved fame and wealth as publisher of *Poor Richard's Almanack* (1732). His famous kite experiment (1752) demonstrated the electrical nature of lightning. His inventions included the lightning conductor, the efficient Pennsylvanian fireplace (Franklin stove) and bifocal lenses. His most significant theoretical contribution was the single-fluid theory of electricity.

FRASCH PROCESS, process for extracting SULFUR from sulfur-bearing CALCITE deposits, invented by Herman Frasch in 1891. Three concentric pipes are lowered down a bore-hole to the ore. Superheated water at 165°C is pumped down the outer pipe and compressed hot air is blown down the inner pipe. This forces a frothy mixture of molten sulfur and water up the middle pipe. Very pure sulfur is produced.

FRAUNHOFER, Joseph von (1787–1826), German optician who mapped the dark lines (FRAUNHOFER LINES) in the solar spectrum and reinvented the DIFFRACTION grating.

FRAUNHOFER LINES, dark lines that appear in the SPECTRUM of the SUN. They are due to the ABSORPTION of the radiation of particular FREQUENCIES (and thus ENERGIES) by ATOMS in the outer layers of the solar ATMOSPHERE. Analysis of the solar spectrum thus leads to the identification of these atoms. The lines were first accurately mapped by J. von FRAUNHOFER. The more prominent are denoted by letters: A and B being due to terrestrial OXYGEN; C to HYDROGEN; D to SODIUM; E to IRON, and so on.

FRAZER, Sir James George (1854–1941), British social anthropologist. In *The Golden Bough: A Study in Magic and Religion* (1890; enlarged 1907–15) he proposed a parallel evolution of thought in all peoples: from magic through religion to science, each with its distinct notion of cause and effect. Despite the apparent error of his conclusions, his work in surveying primitive customs and beliefs was of great value to cultural anthropology.

FRECKLE, small brown pigmented spot in the SKIN of exposed areas, numerous in some individuals; they are brought out by sunlight and are benign.

FREE ASSOCIATION, the ideas which occur spontaneously in the CONSCIOUSNESS without concentration on the part of the individual. The encouragement in a patient of free ASSOCIATION is one of the primary techniques of PSYCHOANALYSIS.

FREE-PISTON ENGINE, a type of DIESEL ENGINE in which the power output is extracted from a TURBINE driven by the exhaust gases and not via any direct rod and crank drive. The combustion chamber is formed between two opposed pistons in a closed metal tube, the pistons oscillating in and out in phase with each other, regularly exposing the exhaust and air-inlet ports. Fuel is injected at the midpoint. The engine, which has a good power-to-weight ratio and is used in some ships, has the advantage that the gasifier and turbine units need not be sited adjacently.

FREE RADICALS, molecules or atoms which have one unpaired electron (see ORBITAL), and hence an unused VALENCE. Most are very reactive and shortlived, but if the odd electron can be delocalized by RESONANCE (e.g., in triphenylmethyl) they may be stable. Free radicals can be studied by SPECTROSCOPY, chiefly electron-spin resonance. They are produced by heat, irradiation, FLASH PHOTOLYSIS and ELECTROLYSIS, and are important in forming POLYMERS and in explosive chain reactions.

FREE WILL, in philosophy, a faculty that man is alleged to require if he is to be able to make moral choices. Philosophical theories in which man is assumed to have free will formally conflict with those in which his actions are considered to be determined by causes beyond his control. However, the choice between theories of free will and DETERMINISM may admit of other, intermediate alternatives.

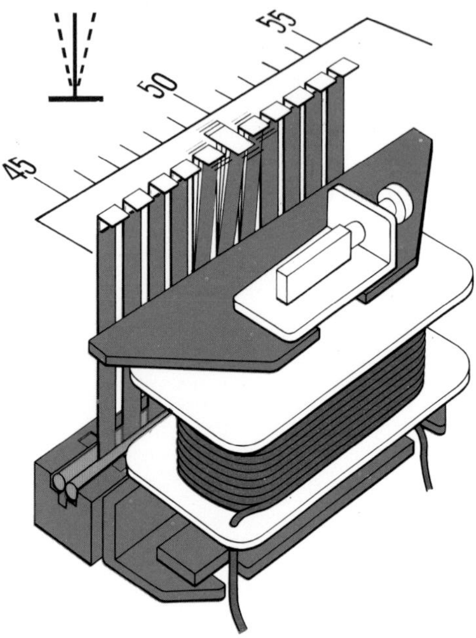

The principle of the vibrating-reed frequency meter as used to monitor commercial electricity supply. A set of tempered steel reeds, tuned to vibrate at a range of frequencies, are arranged in a "comb" such that the electricity supply can excite them. When the electricity supply is connected, the reed whose natural frequency corresponds to the supply frequency vibrates visibly, resonating with the supply, while the others move slightly or remain motionless. The reeds are labelled on an adjoining scale.

FREEZE DRYING, FOOD processing technique in which the produce is deep frozen before being pumped down to a high vacuum so that constituent ice sublimes out as water vapor leaving a high-quality dehydrated product.

FREEZING POINT, the TEMPERATURE at which a liquid begins to solidify—not always well-defined or equal to the melting point (see FUSION). It usually rises with PRESSURE, solids being slightly denser than liquids, though water is a notable exception; it is lowered by solutes in the liquid, the amount providing an accurate means of determining MOLECULAR WEIGHTS. The solid separating from a solution usually has a different composition from the liquid, and repeated freezing can be used to separate substances. Pure substance can often be "supercooled" below their freezing point for limited periods, as the formation of the solid CRYSTAL requires enucleation by rough surfaces in contact with or particles suspended in the liquid.

FREGE, Gottlob (1848–1925), German logician, father of mathematical LOGIC. Inspired by the similar, earlier work of LEIBNIZ, he tried to show that all mathematical truths could be derived logically from a few simple axioms. After RUSSELL's criticism that his system allowed at least one paradox, he wrote little more; but his work influenced later thinkers such as PEANO, Russell and WHITEHEAD. (See also LOGICAL POSITIVISM.)

FREON, trade name for a group of volatile CARBON compounds (derivatives of METHANE and ETHANE) containing fluorine and chlorine or bromine. They are nonflammable and nontoxic, and are used as refrigerants and aerosol propellants.

FREQUENCY, the rate at which a periodic WAVE MOTION executes complete cycles of its variation. Frequencies are measured in hertz (Hz), i.e., in cycles per second. Musical sounds typically have frequencies in the range 30–20000Hz; alternating-current ELECTRICITY supply is at 60Hz in the US, but usually 50Hz in the rest of the world. According to QUANTUM THEORY, the frequency (ν) of ELECTROMAGNETIC RADIATION provides a measure of the ENERGY (E) of its quanta (PHOTONS): $E = h\nu$, where h is the PLANCK CONSTANT. Again, the wavelength (λ), frequency (f) and velocity (v) of a harmonic wave are related by the equation $v = f\lambda$.

FREQUENCY CURVE. See HISTOGRAM.

FREQUENCY MODULATION (FM). See RADIO.

FREQUENCY POLYGON. See HISTOGRAM.

FRESNEL, Augustin Jean (1788–1827), French physicist who evolved the transverse-wave theory of LIGHT through his work on optical INTERFERENCE. He worked also on REFLECTION, REFRACTION, DIFFRACTION and polarization, and developed a compound LENS system still used for many lighthouses.

FREUD, Anna (1895–), Austrian-born British pioneer of child psychoanalysis. Her book *The Ego and Mechanisms of Defense* (1936) is a major contribution to the field. After escaping with her father Sigmund FREUD from Nazi-occupied Austria (1938), she established an influential child-therapy clinic in London.

FREUD, Sigmund (1856–1939), Austrian neurologist and psychiatrist, founder and author of almost all the basic concepts of PSYCHOANALYSIS. He graduated as a medical student from the University of Vienna in 1881; and for some months in 1885 he studied under J. M. CHARCOT. Charcot's interest in HYSTERIA converted Freud to the cause of psychiatry. Dissatisfied with HYPNOSIS and electrotherapy as analytic techniques, he evolved the psychoanalytic method, founded on DREAM analysis and FREE ASSOCIATION. Because of his belief that sexual impulses lay at the heart of NEUROSES, he was for a decade reviled professionally, but by 1905 disciples such as Alfred ADLER and Carl Gustav JUNG were gathering around him; both were later to break away. For some thirty years he worked to establish the truth of his theories, and these years were especially fruitful. Fleeing Nazi anti-Semitism, he left Vienna for London in 1938, and there spent the last year of his life.

FRIAR'S BALSAM, or tincture of benzoin, a resin, taken by inhalation, used to ease COUGH and loosen MUCUS secretion in respiratory diseases.

FRICTION, resistance to motion arising at the boundary between two touching surfaces when it is attempted to slide one over the other. As the FORCE applied to start motion increases from zero, the equal force of "static friction" opposes it, reaching a maximum "limiting friction," just before sliding begins. Once motion has started, the "sliding friction" is less than the limiting. Friction increases with the load pressing the surfaces together, but is nearly independent of the area in contact. For a given pair of surfaces, limiting friction divided by load is a dimensionless constant known as the coefficient of friction. LUBRICATION is used to overcome friction in the BEARINGS of machines.

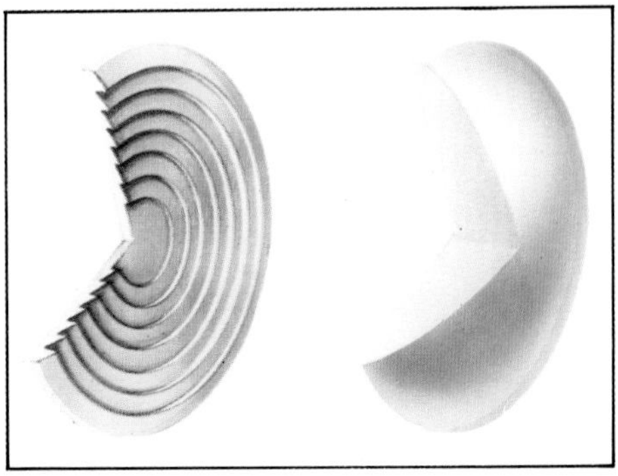

A fresnel lens (*left*) compared with the solid lens (*right*) to which it is optically equivalent. By reproducing the curved surface of the solid lens as a series of rings separated by steps on a thin lens, the fresnel lens effects a great saving in weight, allowing mechanically robust lenses of relatively short focal length to be fitted to searchlights, spotlights and automobile headlights.

FRIEDEL-CRAFTS REACTION, class of substitution reactions, catalyzed by acidic metal halides (e.g., aluminum chloride); an ALKYL HALIDE or an ACID CHLORIDE reacts with an AROMATIC COMPOUND, the alkyl or acyl group replacing a hydrogen atom on the aromatic ring: the carbonium ion formed attacks the ring as an electrophile (see NUCLEOPHILES). (See also ALKYLATION.)

FRISCH, Karl von (1886–), Austrian zoologist best known for his studies of bee behavior, perception and communication, discovering the "Dance of the Bees." With TINBERGEN and LORENZ he was awarded the 1973 Nobel Prize for Physiology or Medicine for his work.

FROEBEL, Friedrich Wilhelm August (1782–1852), German educator noted as the founder of the kindergarten system. He believed in play as a basic form of self-expression, and in the innate nature of mystical understanding. Though much criticized, he has profoundly influenced later educators.

FROMM, Erich (1900–), German-born US psychoanalyst who combines many of the ideas of Freud and Marx in his analysis of human relationships and development in the context of social structures and in his suggested solutions to problems such as alienation.

FRONT, a boundary between two air masses, one cold and dense, the other warm and less dense. Fronts are regions of uncertain weather: cloud and rain, variable HUMIDITY and WIND direction, and low air pressure (see ATMOSPHERE). A **warm front** is where warm air advances, displacing cold; the reverse happens in a **cold front**. An **occluded front** occurs often toward the end of a system of STORMS: a mass of warm air is surrounded and forced upward by cold air.

In the N Hemisphere the **polar front** separates the W-moving cold air of polar regions from the E-moving warm air flowing up from the TROPICS. Warm air flows into "bays" in the front, then cold air in from the rear, resulting in the rotating air system known as a **depression**. (See also METEOROLOGY.)

FRONTAL LOBE, the foremost of the four main lobes of cerebral cortex in the BRAIN of man. It is concerned with personality and behavior, and with the emotional aspects of perception. Its removal by LOBOTOMY leads to a disinhibited type of behavior

FROST, frozen atmospheric moisture formed on objects when the temperature is below 0°C, the freezing point of water (see FREEZING POINT). **Hoarfrost** forms in roughly the same way as DEW but, owing to the low temperature, the water VAPOR sublimes (see SUBLIMATION) from gaseous to solid state to form ice crystals on the surface. The delicate patterns often seen on windows are hoarfrost. **Glazed frost** usually forms when RAIN falls on an object below freezing: it can be seen, for example, on telegraph wires. **Rime** occurs when supercooled (see SUPERCOOLING AND SUPERHEATING) water droplets contact a surface that is also below 0°C; it may result from FOG or drizzle. The first frost of the year signifies the end of the GROWING SEASON. (See also ICE; SNOW.)

FROSTBITE, damage occurring in SKIN and adjacent tissues caused by freezing. (The numbness caused by cold allows considerable damage without pain.) DEATH of tissues follows and they separate off.

The fronts associated with a mid-latitude cyclone (depression). Here a wedge of warm air (red) is enclosed in an indentation into an air mass of cold air (blue). The system is dynamic, the cold air (*left*) advancing into the warm sector along a boundary called a cold front. On a weather chart the surface location of a cold front is indicated by a line with pointed teeth. The warm air is also advancing into the cold air region (*right*) along a boundary called a warm front. The symbol for the surface location of a warm front is a line with solid lobes. The fronts are not just surface features, however, and extend up through the troposphere, each showing a characteristic profile. The cold front usually advances more rapidly than the warm front, the air in the warm sector escaping upward into the lower stratosphere. Where the warm front has already been overtaken by the cold front at the surface, but distinct fronts are still present at higher altitudes, there is said to be an occluded front, and this is indicated on charts by a line with alternate pointed and lobate projections. The air in the warm sector of a mid-latitude cyclone is usually quite moist. As it is forced upward, its temperature falls, the air becomes saturated and precipitation in the form of rain, hail or snow occurs. It is thus that belts of rain (shaded areas) are commonly associated with all types of cyclonic fronts. Because of the more gentle slope of the warm front, the rain associated with this front is commonly more prolonged but less intense than that found along the steeper cold front. In the northern hemisphere, cyclonic systems such as this are usually carried along in the mid-latitude westerly wind belt.

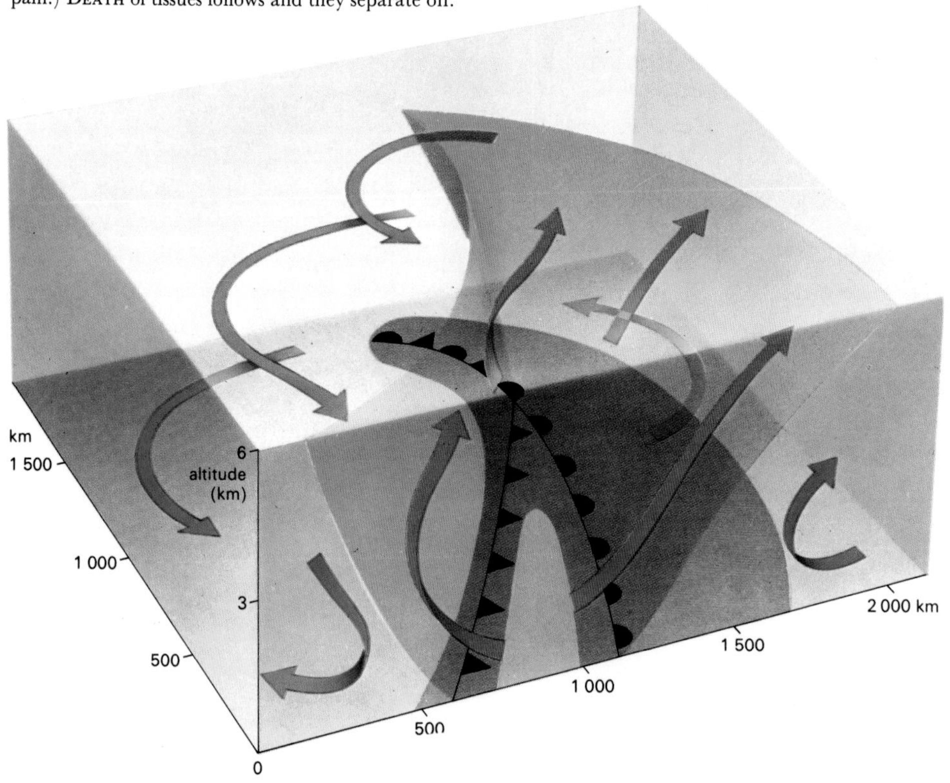

Judicious rewarming, pain relief and measures to maximize skin blood flow may reduce tissue loss.

FROZEN FOODS. See FOOD PRESERVATION; BIRDSEYE, CLARENCE.

FRUCTOSE ($C_6H_{12}O_6$), or "fruit sugar," a naturally occurring simple SUGAR (monosaccharide) found in a wide variety of fruits and in honey. It is the sweetest of the simple sugars.

FRUIT, botanically, the structure that develops from the ovary and accessory parts of a FLOWER after FERTILIZATION. True fruits are formed from the carpels, while in false fruits other parts of the flower are involved, for example in apple the fleshy pulp is derived from the receptacle (see POME). Fruits may be simple (derived from the ovary of one pistil), aggregate (formed by a single flower with several separate pistils, e.g. raspberry) or multiple (formed from the flowers of an INFLORESCENCE, e.g. fig). Simple fruits may be fleshy or dry. Fleshy fruits include the BERRY and the DRUPE. Dry fruit may split open to disperse the seeds (dehiscent), the main types being the LEGUME (or pod), FOLLICLE, CAPSULE and SILIQUE. Some dry fruits do not break open (indehiscent) the main types here being the ACHENE, GRAIN, SAMARA and NUT.

The main function of the fruit is to protect the seeds and disperse them when ripe.

FRUITFLY. See DROSOPHILA.

FRUSTUM, the base of a cone; more rigorously, a part of one nappe of a CONE lying between two PLANES, usually parallel, that intersect the cone.

FUEL, a substance that may be burned (see COMBUSTION) to produce heat, light or power. Traditional fuels include dried dung, animal and vegetable oil, wood, PEAT and COAL, supplemented by the manufactured fuels CHARCOAL, COAL GAS, COKE and WATER GAS. In this century PETROLEUM and NATURAL GAS have come into widespread use. The term "fuel" has also been extended to include chemical and nuclear fuels (see FUEL CELL; NUCLEAR ENERGY), although these are not burned. Specialized high-energy fuels such as HYDRAZINE are used in ROCKET engines. The chief property of a fuel is its **calorific value**—the amount of heat produced by complete combustion of a unit mass or volume of fuel. Also of major importance is the proportion of incombustibles—ASH and moisture—and of sulfur and other compounds liable to cause AIR POLLUTION.

FUEL CELL, a direct-current power source similar to a BATTERY but differing in that a chemical fuel must be supplied while the CELL is in use. Various chemical reactions are utilized in different types of cell. The most common is the hydro-oxygen cell in which HYDROGEN reacts with hydroxyl (HYDROXIDE) ions in the electrolyte to form water at the ANODE, and OXYGEN reacts with water to form hydroxyl ions at the CATHODE, the overall reaction resulting in the formation of water and a flow of ELECTRONS through an external circuit. The cell is divided into three compartments by the ELECTRODES, the two outer ones containing the hydrogen fuel and the oxygen, and the central one the aqueous electrolyte. The electrodes are porous so that the gases can penetrate to meet the electrolyte. Platinum and nickel are typical catalysts. Fuel cells are much more efficient converters of chemical energy than heat ENGINES.

FUEL INJECTION, various systems for delivering a

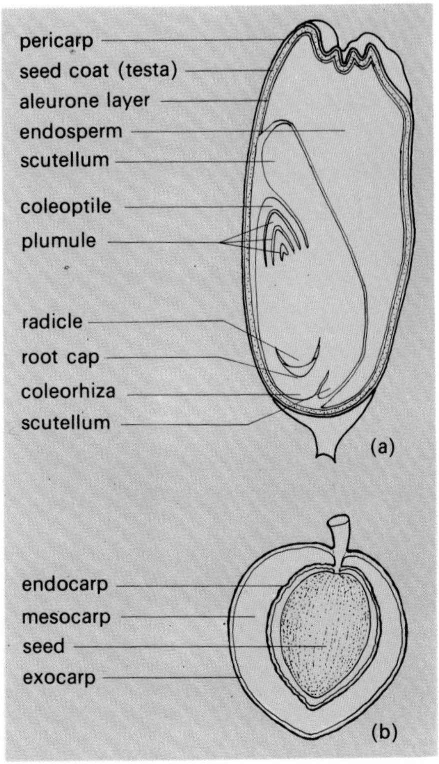

Above: Vertical sections of (a) maize (*Zea mays*) grain and (b) almond (*Prunus*).

Below: Schematic diagram of a hydro-oxygen fuel cell using a hot alkaline electrolyte and biporous sintered nickel electrodes. At the hydrogen electrode (cathode), hydrogen reacts with hydroxyl ion to yield water and electrons: $H_2 + 2OH^- \rightleftharpoons 2H_2O + 2e^-$

At the oxygen electrode oxygen reacts with water, abstracting electrons from the electrode and producing hydroxyl ions: $O_2 + 2H_2O + 4e^- \rightarrow 4OH^-$

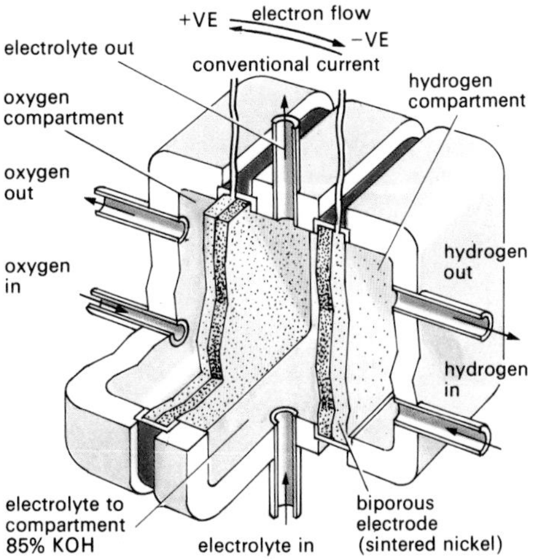

metered quantity of fuel into the cylinder of an INTERNAL-COMBUSTION ENGINE, usually driven by pumps run off the CAMSHAFT and controlled from the throttle. Fuel injection has always been essential in DIESEL ENGINES but it is only in recent years that it has begun to replace the CARBURETOR in gasoline-engined passenger cars. Some recent systems have been electronically controlled.

FUGUE STATE. See AMNESIA.

FULLER, Richard Buckminster (1895–), US inventor, philosopher, author, mathematician and perhaps the 20th century's most original and prolific thinker. He is best known for his concept "Spaceship Earth" and for inventing the GEODESIC DOME.

FULLER'S EARTH, natural CLAY material of variable composition, once used to clean wool and cloth (fulling); now used for decolorizing OIL of all sorts by selective chemical ADSORPTION, and also as a pesticide carrier, and in drilling muds.

FULMINATE, or **mercury (II) cyanate.** See MERCURY.

FULTON, Robert (1765–1815), US inventor who improved both the submarine and the steamboat. His submarine *Nautilus* was launched at Rouen, France (1800), with the aim of using it against British warships: in fact, these repeatedly escaped and the French lost interest. His first steamship was launched on the Seine (1803), and after this success he returned to the US, launching the first commercially successful steamboat (see FITCH, John), the *Clermont*, from New York (1807). He built several other steamboats and the *Demologus*, the first steam warship (launched 1815).

FUMAROLE, found in volcanic regions, a hole in the ground that emits steam, carbon dioxide and other vapors typical of volcanic eruption. (See GEYSER; VOLCANISM; VOLCANO.)

FUNCTION, a rule in which each element of one set (see SET THEORY) is assigned one or more elements, not necessarily unique, of another set. (On occasion, the two sets may share some or all elements.) In CARTESIAN COORDINATES a function may be plotted by setting x along one axis and $f(x)$ (read "function of x") along the other. (See also ANALYTIC GEOMETRY; CALCULUS; VARIABLE.)

A function may be thought of as a rule. If the rule is SQUARE the number, add twice the number and subtract three, this is expressed as $y=x^2+2x-3$, or more usually $f(x)=x^2+2x-3$.

A transcendental function is one that cannot be expressed algebraically (see ALGEBRA; CURVE). For example, sin x (see TRIGONOMETRY) cannot be expressed in algebraic terms and hence, if $f(x)=\sin x$, $f(x)$ is a transcendental function.

FUNCTIONAL GROUP, a group of connected atoms whose presence in a molecule gives rise to characteristic chemical properties and infrared absorptions; e.g., hydroxyl —OH, carboxyl =C=O. The properties of a molecule are roughly the sum of those of its constituent functional groups, though their interactions are also significant.

FUNGAL DISEASES, DISEASES caused by FUNGI which, apart from common SKIN and nail ailments such as ATHLETE'S FOOT, tinea cruris and RINGWORM, develop especially in people with disorders of IMMUNITY or DIABETES and those on certain DRUGS (STEROIDS, immunosuppressives, ANTIBIOTICS). THRUSH is common in the mouth and vagina but rarely causes systemic disease. Specific fungal diseases occur in some areas (e.g., HISTOPLASMOSIS, blastomycosis) while aspergillosis often complicates chronic LUNG disease. In addition, numerous fungi in the environment lead to forms of ALLERGY and lung disease.

FUNGI, a subdivision (Eumycotina) of the PLANT KINGDOM which comprises simple plants that reproduce mostly by means of SPORES and which lack CHLOROPHYLL, hence are either SAPROPHYTES or PARASITES.

The closely related SLIME MOLDS produce naked (no cell walls) amoeboid states, and the YEASTS are single-celled, but the majority of true fungi produce microscopic filaments (hyphae) that group together in an interwoven weft, the mycelium or spawn. REPRODUCTION is sometimes by budding (yeasts) but more normally by the production of asexual and sexual spores. Some fungi produce large fruit bodies, which are the structures commonly associated with fungi. The classification of fungi is complicated and several systems have evolved, mostly based on the types of spore produced. Fungi belong to the division MYCOTA, which also includes the Myxomycetes or slime molds. The true fungi are divided into a number of classes, the main ones being: the Chytridomycetes, which produce motile gametes or zoospores that have a single flagellum; the Oomycetes, which have biflagellate zoospores and produce dissimilar male and female reproductive organs and gametes; Zygomycetes, which do not produce motile zoospores and reproduce sexually by fusion of identical gametes; the Ascomycetes, including yeasts, which reproduce asexually by budding or by the production of spores (conidia) and sexually by the formation of ascospores within sac-like structures (asci) that are often enclosed in a fruiting body or ascocarp; the Basidiomycetes, including bracket fungi and agarics in which the sexual spores are produced or enlarged cells called basidia, that often occur on large fruiting bodies; and the Deuteromycetes, or Fungi Imperfecti, which are only known to reproduce asexually, although sexual forms are often classified in the Ascomycetes and Basidiomycetes. (See also FUNGAL DISEASES; MOLD; PLANT DISEASES; RUST; SMUT.)

FUNGICIDE, substance used to kill FUNGI and so to control FUNGAL DISEASES in man and plants. In medicine some ANTIBIOTICS, SULFUR, CARBOXYLIC ACIDS and potassium iodide are used. In agriculture a wide variety of fungicides is used, both inorganic—BORDEAUX MIXTURE and sulfur—and organic—many different compounds, generally containing sulfur or nitrogen. They are applied to the soil before planting or around seedlings, or are sprayed or dusted onto foliage. (See also PESTICIDE.)

FUNGI IMPERFECTI or Deuteromycetes. See FUNGI.

FUNK, Casimir (1884–1967), Polish-born biochemist. Following the work of C. EIJKMAN and F. G. HOPKINS, he proposed that BERIBERI, RICKETS, SCURVY and PELLAGRA arose from lack of certain trace dietary constituents, which he christened vitamines, or "life-amines." (See also AMINES; VITAMINS.)

FUNNY BONE, part of the ulna at the elbow over which passes the ulnar nerve. If this point is struck sharply, it causes transient unpleasant, electric shock-like tingling and numbness in the ARM and SKIN area served by the ulnar nerve.

FUR, the soft, dense, hairy undercoat of certain mammals. Fur is an excellent heat insulator and protects against the cold of the northern regions where most furbearing animals are found. It is generally interspersed with guard HAIRS, longer and stiffer, that form a protective outer coat and prevent matting. Skins or pelts are cleaned, softened and converted to a leatherlike state by "dressing," a process resembling TANNING. In some cases the guard hairs are sheared or plucked. The pelt is then dyed or bleached and then glazed, chemically or by heat, to give it a lustrous sheen. To make the furs into a garment, they are matched for color and texture, cut to shape, sewn together, and finally dampened and nailed to the pattern to dry smooth and the exact shape wanted. Fur clothing has long been valued for its beauty and

The Culham Conceptual Tokamak Reactor Design Mk II. Research aimed at designing an economic nuclear fusion power plant is being undertaken at laboratories in the USA, USSR, Japan and Western Europe. Seven EEC countries, including France, the UK and West Germany, are banded together in the Euratom fusion power program. Britain's contribution to this is centered on Culham Laboratory near Oxford. Before a working reactor can be built, a whole new nuclear fusion technology must be developed; research problems in both physics and engineering must first be identified and then solved. The Fusion Reactor Studies program at Culham extrapolates the results of the latest available research and uses such data to produce "conceptual" reactor designs, study of which reveals the areas requiring further research effort. The design illustrated here is for a reactor producing about 2 000MW of electricity from the DT reaction. The tritium fuel is bred from lithium contained in a neutron-absorbing blanket surrounding the torus. Heat extraction is by helium gas at high pressure (this gives up its heat to steam which then powers a turbogenerator). The "tokamak" method of plasma confinement is used. Here the plasma is confined within a torus by a magnetic field having two components: a toroidal component induced by external field coils, and a poloidal component induced by the heating current flowing toroidally through the plasma.
(1) toroidal field coils; (2) poloidal field coils; (3) core; (4) blanket module; (5) cooling ducts; (6) duct joints; (7) shield structure and vacuum wall; (8) shield door; (9) shield cooling; (10) shield support; (11) servicing floor; (12) injector, refuel and control access.

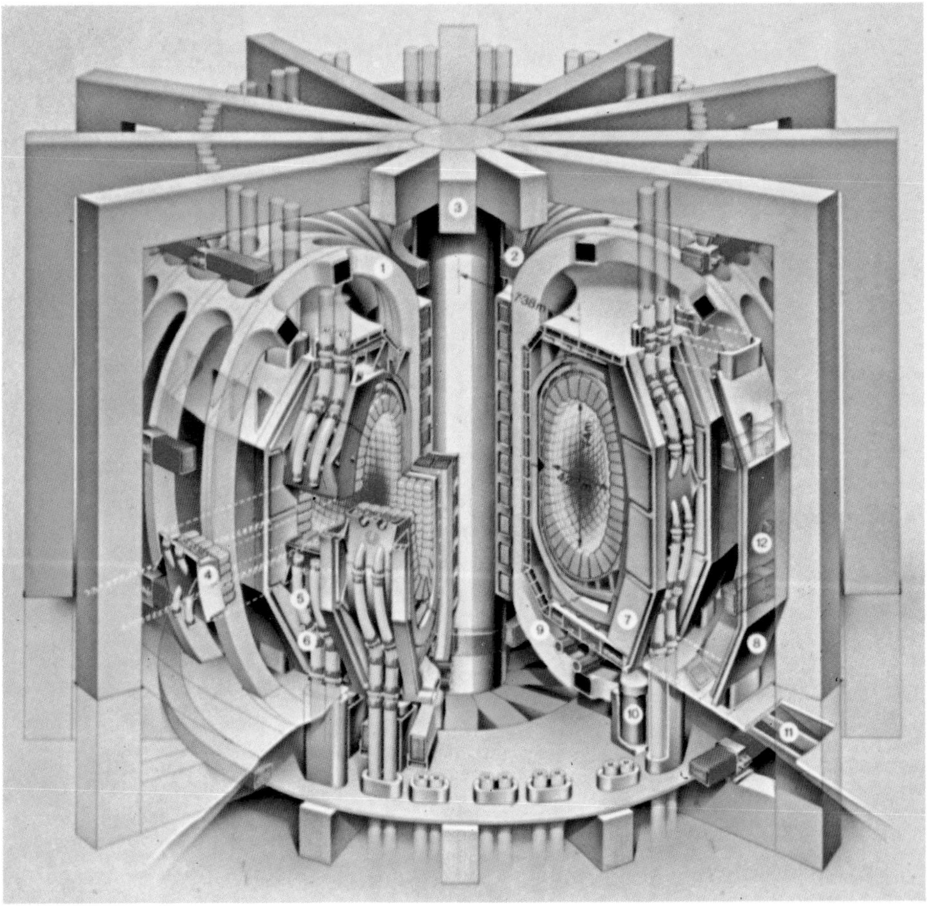

warmth, and was an aristocratic luxury until the discovery of America, in whose exploration and economic development trapping and fur trading played a major role. Demand is still high, threatening some furbearing species with extinction; this has led to fur-farming of suitable animals such as mink and to the development of artificial furs made of synthetic fibers.

FURFURAL ($C_4H_3O.CHO$), colorless liquid, an ALDEHYDE similar to BENZALDEHYDE and a HETEROCYCLIC COMPOUND derived from furan. It is made by digesting corncobs, oat and rice hulls, etc. with acid, and is used as a solvent, a pesticide, and an intermediate in making PLASTICS and other compounds.

FURNACE, a construction in which heat can be generated, controlled and used. The heat may be produced by burning a fuel such as coal, oil or gas; by electricity; by concentrating the heat of the sun; or by atomic energy (see FISSION, NUCLEAR). Simple furnaces are often used in the home to heat water; but much larger ones are used in industry, particularly in the heat treatment of metals (see METALLURGY). These are usually lined with FIREBRICKS (see also REFRACTORY), which may also be water-cooled. (See also BESSEMER PROCESS; BLAST FURNACE; ELECTRIC FURNACE; OPEN-HEARTH PROCESS.)

FUSE, safety device placed in an electric circuit to prevent overloading. It usually comprises a wire of low-melting-point metal mounted in or on an insulated frame. Current passing through the wire heats it (see RESISTANCE); and excessive current heats it to the point where it melts, so breaking the CIRCUIT. In most domestic plugs, the fuse consists of a cylinder of glass, capped at each end by metal, with a wire running between the metal caps. Similar, but larger, cartridge fuses are used in industry. Resettable CIRCUIT BREAKERS offer an alternative to fuses.

FUSEL OIL, a mixture mainly of amyl and butyl ALCOHOLS, produced as a by-product of FERMENTATION. It helps to flavor alcoholic beverages, but is separated by DISTILLATION from industrial ETHANOL, and used for solvents.

FUSION, the process of melting, or passing from the solid state to liquid state. This is accompanied by an absorption of LATENT HEAT and usually occurs at a well-defined TEMPERATURE, the **melting point** which rises slightly with PRESSURE and chemical purity. Many amorphous solids like GLASS have no melting point, and simply reduce their VISCOSITY over a wide temperature range.

FUSION, Nuclear, a nuclear reaction in which the nuclei of light ATOMS combine to produce heavier, more stable nuclei, releasing a large quantity of ENERGY. Fusion reactions are the energy source of the SUN and the HYDROGEN BOMB. If they could be controlled and made self-sustaining man would have a safe and inexhaustible energy source using DEUTERIUM or TRITIUM extracted from seawater. Only small amounts of fuel would be needed and the products are not radioactive. But if they are to fuse, the light, positively charged nuclei must collide with sufficient energy to overcome their electrostatic repulsion. This can be done by using a particle ACCELERATOR, but to get a net energy release the material must be heated to very high temperatures (around 10^9K) when it becomes a PLASMA. However, plasmas are difficult to contain and, as yet, no apparatus has been designed which allows more energy to be extracted than is used in heating and containing the plasma. (See also NUCLEAR ENERGY.)

FUSTIAN, term formerly applied to a fabric with a linen warp and cotton weft but now referring to heavy-piled cotton fabrics including velveteen and corduroy.

FUZE, primary explosive charge set off by electrical, chemical or mechanical means which detonates the main charge of an explosive device. Commonly fuzes operate on the principle of the firecracker fuze. **Proximity fuzes** may be set in a moving projectile, to function when it is a certain distance from target. (See also AMMUNITION.)

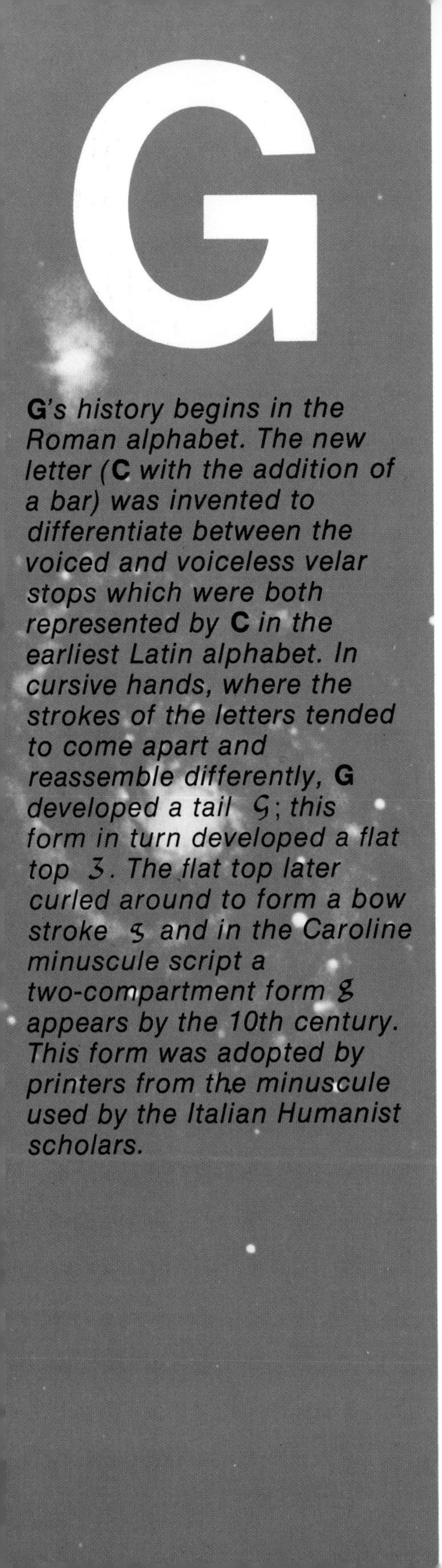

G's history begins in the Roman alphabet. The new letter (C with the addition of a bar) was invented to differentiate between the voiced and voiceless velar stops which were both represented by C in the earliest Latin alphabet. In cursive hands, where the strokes of the letters tended to come apart and reassemble differently, G developed a tail ς; this form in turn developed a flat top ʒ. The flat top later curled around to form a bow stroke ᵹ and in the Caroline minuscule script a two-compartment form g appears by the 10th century. This form was adopted by printers from the minuscule used by the Italian Humanist scholars.

G, the Universal Gravitational Constant. See GRAVITATION.

***g*,** acceleration due to gravity. See ACCELERATION.

GABBRO, a dense IGNEOUS ROCK, resembling BASALT, composed of coarse-grained plagioclase FELDSPAR with PYROXENE and OLIVINE. Often rhythmically banded, it arises from fractional crystallization of MAGMA.

GABOR, Dennis (1900–), Hungarian-born UK physicist who invented HOLOGRAPHY, for which he was awarded the 1971 Nobel Prize for Physics. He had developed the basic technique in the late 1940s, but practical applications had to wait for the invention of the LASER (1960) by C. H. TOWNES.

GADOLINIUM (Gd), element of the LANTHANUM SERIES. AW 157.3, mp 1313°C, bp 3266°C, sg 7.9004 (25°C).

GAJDUSEK, Daniel Carleton (1923–), US physiologist who shared the 1976 Nobel Prize for Physiology or Medicine with S. B. BLUMBERG. Gajdusek identified the rare neurodegenerative disease, kuru, as a "slow virus" infection.

GALACTIC CLUSTERS, clusters of stars lying in or near the galactic plane, each of which contains a few hundred stars. Due to their irregular shape they are also termed **open clusters**. The best known galactic cluster in N skies is the PLEIADES.

GALAXY, the largest individual conglomeration of matter, containing stars, gas, dust and planets. Galaxies start life as immense clouds of gas, out of which stars condense. Initially a galaxy is **irregular** in form; that is, it has neither a specific shape nor any apparent internal structure. It contains large amounts of gas and dust in which new stars are constantly forming. It rotates and over millions of years evolves into a **spiral** form, looking rather like a flying saucer, with a roughly spherical nucleus surrounded by a flattish disk and orbited by GLOBULAR CLUSTERS. In the nucleus there is little gas and dust and a high proportion of older stars; in the spiral arms a great deal of gas and dust and a high proportion of younger stars (our SUN lies in a spiral arm of the MILKY WAY). Over further millions of years the spiral arms "fold" toward the nucleus, the end result being an **elliptical** galaxy containing a large number of older stars and little or no gas and dust. The ultimate form of any galaxy is a sphere, after which it possibly evolves into a BLACK HOLE. The nearest external galaxy to our own is the ANDROMEDA Galaxy. Similarly spiral, though rather larger, it has two satellite galaxies which are elliptical in form. Originally it was thought to be a NEBULA within our own galaxy, but in 1924 HUBBLE showed that it was a galaxy in its own right. Study of the Andromeda Galaxy is important as it enables us better to understand our own, most of which is obscured from us by clouds of gas and dust. Galaxies emit radio waves, and the strongest sources are known as **radio galaxies**. One group of these, spiral with active nuclei, are named **Seyfert galaxies** for the US astronomer Carl Seyfert. (See also PULSAR; QUASAR.) Galaxies tend to form in clusters. The Milky Way and the Andromeda Galaxy are members of a cluster of around 20 galaxies.

GALEN OF PERGAMUM (c130–c200 AD), Greek physician at the court of the Emperor Marcus Aurelius. His writings drew together the best of classical medicine and provided the form in which the

M31, the Great Spiral Galaxy in Andromeda. M32, an elliptical galaxy appears above M31, NGC 205, a further elliptical galaxy to the bottom left.

science was transmitted through the medieval period to the Renaissance. He himself contributed many original and careful observations in anatomy and physiology.

GALENA, gray mineral consisting of lead (II) sulfide (PbS), forming cubic crystals; the main ore of LEAD. Deposits occur in Germany, the US, Britain and Australia.

GALILEO GALILEI (1564–1642), Italian mathematical physicist who discovered the laws of falling bodies and the parabolic motion of projectiles. The first to turn the newly invented TELESCOPE to the heavens, he was among the earliest observers of SUNSPOTS and the phases of Venus. A talented publicist, he helped to popularize the pursuit of science. However, his quarrelsome nature led him into an unfortunate controversy with the Church. His most significant contribution to science was his provision of an alternative to the Aristotelian dynamics. The motion of the earth thus became a conceptual possibility and scientists at last had a genuine criterion for choosing between the Copernican and Tychonic hypotheses in ASTRONOMY.

GALL, Franz Joseph (1758–1828), German-born Viennese physician who was one of the earliest proponents of the theory of cerebral localization (that different areas of the BRAIN control different functions) and who founded the pseudoscience of PHRENOLOGY.

GALL BLADDER, small sac containing BILE, arising from the bile duct which leads from the LIVER to DUODENUM. It lies beneath the liver and serves to concentrate bile. When food, especially fatty food, reaches the STOMACH, local HORMONES cause gall bladder contraction and bile enters the GASTRO-INTESTINAL TRACT. In some people the concentration of bile favors the formation of **gall stones**, usually containing CHOLESTEROL. These stones may cause no symptoms; they may obstruct the gall bladder causing biliary COLIC or INFLAMMATION (cholecystitis), or they may pass into the bile duct and cause biliary obstruction with JAUNDICE or, less often, pancreatitis. Acute episodes are treated with ANALGESICS, antispasmodics and ANTIBIOTICS, but SURGERY is frequently necessary later. Recent advances suggest that in some instances stones may be dissolved by DRUG therapy.

GALLIC ACID, or 3,4,5-trihydroxybenzoic acid, $[(HO)_3C_6H_2COOH]$, an acid occurring in plants, including oak galls, sumach and tea. Obtained from TANNINS by HYDROLYSIS, it is used in INK, dyes and as a mild ANTISEPTIC and ASTRINGENT. On heating it gives **pyrogallol**, a photographic developer.

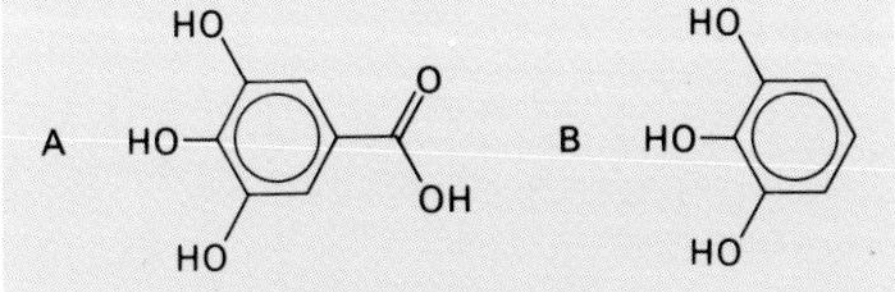

(A) gallic acid; (B) pyrogallol

Above: Galena (PbS) crystals, exhibiting the usual cubic form

Below: Two sketches of the moon's cratered surface made by Galileo. He was the first to use the recently invented telescope to observe the heavens. His discovery of the Medici planets (four moons of Jupiter) and the phases of Venus lent great support to the antagonists of the geocentric theory.

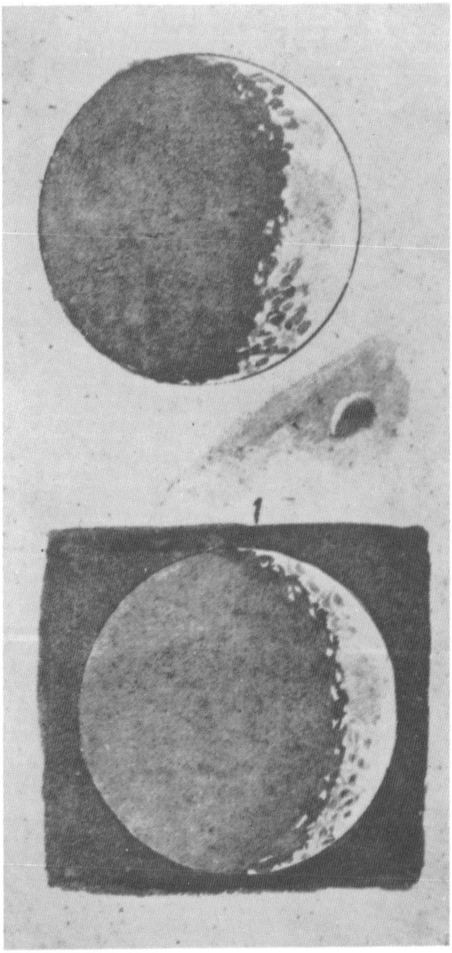

GALLIUM (**Ga**), bluish-white metal in Group IIIA of the PERIODIC TABLE, resembling ALUMINUM; found as a trace element in SPHALERITE, PYRITE, BAUXITE and germanite. Gallium forms trivalent salts and a few monovalent compounds. It contracts on melting, and is liquid over a greater temperature range than any other element. Its few uses include doping SEMICONDUCTORS and producing TRANSISTORS. AW 69.7, mp 29.78°C, bp 2403°C, sg 5.904 (29.6°C), 6.095 (29.8°C).

GALLON (gal), name of various units of volume. The US gallon is 0.003 785m³. The UK gallon, the volume of 10lbf of pure water under specified conditions, is 0.004 546m³ or 1.201 US gallons.

GALLS, growth abnormalities that occur in plants caused by insects, mites, nematodes, fungi and bacteria. By definition, galls are self-limiting growths requiring the continued presence of the inciting organism for full development. They thereby differ from TUMORS, where abnormal growth continues in the absence of the inciting organism. However, in crown gall disease of crucifers (e.g., cabbage), the bacterium *Agrobacterium tumefaciens* is required for the initiation of the gall, but subsequent growth becomes tumorous and does not require the presence of the bacterium.

GALLSTONES. See GALL BLADDER.

GALOIS, Évariste (1811–1832), French mathematician best-known for applying GROUP theory in his investigations of the solubility of EQUATIONS. He received no encouragement from his contemporaries and died after a duel at 20, but his work profoundly influenced many later algebraists.

GALTON, Sir Francis (1822–1911), British scientist, the founder of EUGENICS and biostatistics (the application of statistical methods to animal populations); the coiner of the term "anticyclone" and one of the first to realize their meteorological significance; and the developer of one of the first FINGERPRINT systems for identification.

GALVANI, Luigi (1737–1798), Italian anatomist who discovered "animal electricity" (about 1786). The many varying accounts of this discovery at least agree that it resulted from the chance observation of the twitching of frog legs under electrical influence. A controversy with VOLTA over the nature of animal electricity was cut short by Galvani's death.

GALVANIZING. See CORROSION.

GALVANOMETER, an instrument used for detecting and measuring very small electric currents. Most modern instruments are of the moving-coil type in which a coil of fine wire wrapped around an aluminum former is suspended by conducting ribbons about a soft iron core between the poles of a permanent magnet. When an electric current flows through the coil, a magnetic field is set up which interacts with that of the permanent magnet producing a TORQUE. This turns the coil until it is fully resisted by the suspension, the displacement produced being proportional to the current. The result is read from a scale onto which a light beam is reflected from a mirror carried on the suspension ribbons. If all electromagnetic and mechanical damping can be eliminated from such an instrument, it can also be used as a **ballistic galvanometer** to measure small charges and CAPACITANCES. (See also AMMETER.)

GAMETE, or **germ cell**, a sexual reproductive CELL

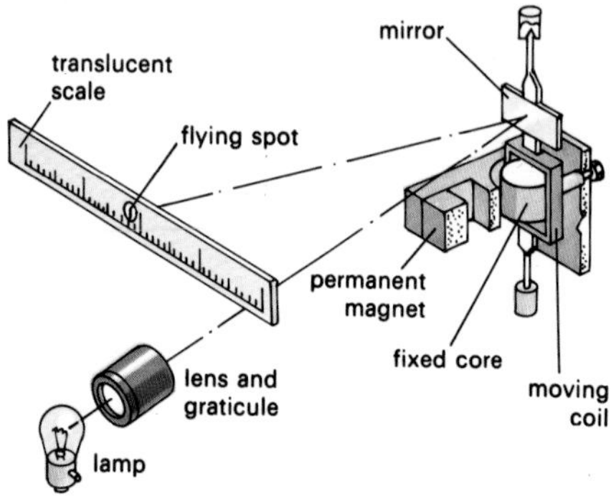

The principle of the flying spot (light beam) galvanometer.

capable of uniting with a gamete of the opposite sex to form a new individual or ZYGOTE; this process is termed FERTILIZATION. Each gamete contains one set of dissimilar chromosomes and is said to be HAPLOID. Thus when gametes unite, the resultant cell contains a DIPLOID or paired set of CHROMOSOMES. The gametes of some primitive organisms are identical cells capable of swimming in water, but in most species only the male gamete (sperm) is mobile while the female gamete (ovary or egg) is a larger static cell. In higher PLANTS the male gametes or pollen are produced by the anthers and the female gametes (ovules) by the ovary. In animals gametes are produced by the GONADS, namely the testes in the male and ovaries in the female.

GAME THEORY, an application of mathematical LOGIC to decision-making in games and, by extension, in commerce, politics and warfare. In singular games (e.g., solitaire) the player's strategy is determined solely by the rules. In dual games (e.g., chess, football) one side's strategy must take into account the possible strategies of the other. Dual games are usually zero-sum: one side's gain exactly equals the other's loss. In practical situations, however, they may be non-zero-sum, as where two conflicting nations negotiate a truce that benefits both. Two major strategies are available to players of dual games: the minimax, in which a player evaluates his probable maximum loss and attempts to minimize it; and the maximin, in which a player evaluates his probable minimum gain and attempts to maximize it. VON NEUMANN showed in his minimax theorem (first stated 1928) that, since, statistically, minimax and maximin strategies negate each other, most dual games are not worth playing in that their outcome is determined solely by the rules.

In plural or *n*-person games (e.g., poker), an individual's gain does not necessarily imply another's loss; and more complex considerations must affect each player's choice of action. Moreover, the outcome may be affected by the formation of coalitions, possibly reducing the *n*-person game to a dual one.

GAMETOPHYTE, phase of the life cycle of a plant representing the haploid generation. (See ALTERNATION OF GENERATIONS.)

GAMMA GLOBULIN, the fraction of BLOOD

PROTEIN containing antibodies (see ANTIBODIES AND ANTIGENS). Several types are recognized. Although they share basic structural features they differ in size, site, behavior and response to different antigens. Absence of all or some gamma globulins causes disorders of IMMUNITY, increasing susceptibility to infection, while the excessive formation of one type is the basis for **myeloma**, a disease characterized by BONE pain, pathological FRACTURES and liability to infection. Gamma globulin is available for replacement therapy, and a type from highly immune subjects is sometimes used to protect against certain diseases (e.g., serum hepatitis, TETANUS). (See also GLOBULINS.)

GAMMA RAYS, high-energy PHOTONS of wavelength shorter than 0.1nm emitted from atomic nuclei during radioactive decay (see RADIOACTIVITY). Usually their emission follows the ejection of an electron (BETA RAY) from the nucleus. The most penetrating of the radioactive emissions, gamma rays find use in engineering quality control—as a source for exposing RADIOGRAPHS—and in RADIATION THERAPY.

GAMOW, George (1904–1968), Russian-born US physicist and popular science writer, best known for his work in nuclear physics, especially related to the evolution of STARS; and for his support of the "big bang" theory of COSMOLOGY. In GENETICS, his work paved the way for the discovery of the role of DNA.

GANGLION, a small collection of nerve cells, sometimes with SYNAPSE formation, common in autonomic or peripheral NERVOUS SYSTEMS. Also, a benign lump, often at the WRIST, found close to TENDONS and containing jelly-like fluid. Traditionally treated by a blow from the family Bible, they are less likely to recur after surgical removal.

GANGRENE, DEATH of tissue following loss of blood supply, often after obstruction of ARTERIES by trauma, THROMBOSIS OR EMBOLISM. **Dry gangrene** is seen when arterial block is followed by slow drying, blackening and finally separation of dead tissue from healthy. Its treatment includes improvement of the blood flow to the healthy tissue and prevention of infection and further obstruction. **Wet gangrene** occurs when the dead tissue is infected with BACTERIA. **Gas gangrene** involves infection with gas-forming organisms (*Clostridium*) and its spread is particularly rapid. ANTIBIOTICS, HYPERBARIC CHAMBERS and early AMPUTATION are often required.

GARNET, common SILICATE mineral group of general formula $M_3^{II}M_2^{III}(SiO_4)_3$, having six end-members It is found in metamorphic and some igneous rocks, often as rhombododecahedral crystals. It is hard, and used as an ABRASIVE. The color is variable; GEM varieties are red, green or transparent. The garnet is popularly the birthstone for January.

GAS, one of the three states (solid, liquid, gas) into which nearly all matter above the atomic level can be classified. Gases are characterized by a low DENSITY and VISCOSITY; a high compressibility; optical transparency; a complete lack of rigidity, and a readiness to fill whatever volume is available to them and to form molecularly homogeneous mixtures with other gases. Air and STEAM are familiar examples. At sufficiently high temperatures, all materials vaporize, though many undergo chemical changes first. Gases, particularly steam and CARBON dioxide, are common products of COMBUSTION, while several available naturally or from PETROLEUM or COAL (e.g., HYDROGEN, METHANE) are used as fuels themselves. The great bulk of the universe is gaseous, in the form of interstellar hydrogen. Gases will often dissolve in liquids, the solubility rising with PRESSURE and falling with TEMPERATURE; a little dissolved carbon dioxide is responsible for the bubbles in soda.

Garnet: as crystals in gneiss and cut and polished as a gemstone.

In contrast to solids and liquids, the MOLECULES of a gas are far apart compared with their own size, and move freely and randomly at a wide range of speeds of the order of 100m/s. For a given temperature and pressure, equal volumes of gas contain the same number of molecules ($2.7 \times 10^{25}\,m^{-3}$ at room temperature and atmospheric pressure). The impacts of the molecules on the walls of the container are responsible for the pressure exerted by gases, which is much larger than is often appreciated: the atmosphere exerts on everything a pressure many times larger than a person's weight and without it we would quite simply boil and burst.

For a given mass of an **ideal gas** (i.e., one in which the molecules are of negligible size and exert no forces on each other), the product of the pressure (P) and the volume (V) is proportional to the absolute temperature (T):

$$PV = RT \text{ (the general gas law).}$$

The constant of proportionality (R) is known as the **universal gas constant** and has the value 8.314 joules/kelvin-mole. The general gas law and most of the other properties of gases can be explained in terms of the KINETIC THEORY without reference to the internal structure of the molecules. Real gases deviate from this ideal behavior at high pressures because of the actual presence of small intermolecular forces.

GAS, Fuel, combustible GAS used as FUEL for domestic or industrial heating, furnaces, engines etc. The main types are NATURAL GAS, COAL GAS, PRODUCER GAS and WATER GAS (blue gas). Town gas, now little used, is a mixture of coal gas and water gas. (See also BOTTLED GAS.)

GAS GUN. See AIR GUN.

GAS MASK, or **respirator**, head-piece to protect the wearer from poisonous fumes or gases. It essentially comprises a filter of CHARCOAL, through which air enters; plastic eyepieces; and a VALVE for exhalation. Various chemicals are added to the charcoal to render specific poisons harmless; e.g., hopcalite (mixed copper and manganese oxides) to

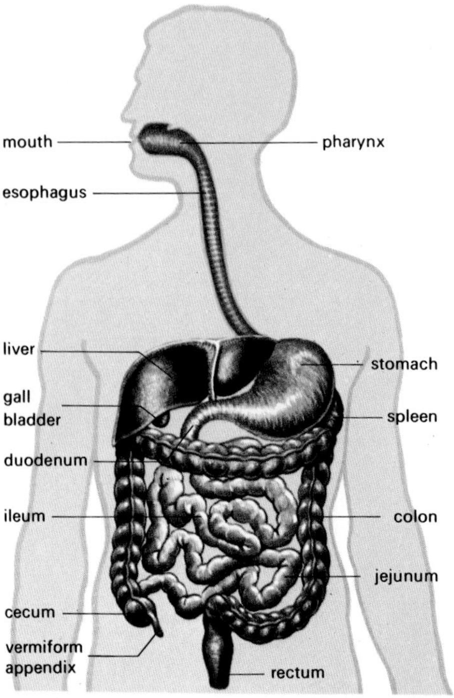

The principal features of the gastrointestinal tract and associated organs.

oxidize CARBON monoxide (CO) to carbon dioxide (CO_2). (See also CHEMICAL AND BIOLOGICAL WARFARE.)

GASOLINE, or petrol, a mixture of volatile HYDROCARBONS having 4 to 12 carbon atoms per molecule, used as a FUEL for INTERNAL-COMBUSTION ENGINES, and as a solvent. Although gasoline can be derived from oil, coal and tar, or synthesized from carbon monoxide and hydrogen, almost all is produced from PETROLEUM by refining, CRACKING and ALKYLATION, the fractions being blended to produce fuels with desired characteristics. Motor gasoline boils between 30°C and 200°C, with more of the low-boiling components in cold weather for easy starting. If, however, the fuel is too volatile, vapor lock can occur—i.e., vapor bubbles form and hinder the flow of fuel. Aviation gasoline contains less of both low- and high-boiling components. The structure of gasoline components is also carefully controlled for maximum power and efficiency, as reflected in the OCTANE rating; this may be further improved by ANTIKNOCK ADDITIVES. Other additives include lead scavengers (ethylene dibromide and dichloride), antioxidants, metal deactivators (which remove metal ions that catalyze oxidation), anti-icing agents, and detergents. Total US gasoline production in 1972 was 272 million tonnes.

GASOLINE ENGINE. See INTERNAL-COMBUSTION ENGINE.

GASSENDI, or **Gassend, Pierre** (1592–1655), French philosopher important for his role in tipping the balance away from the old and toward the new science. A friend and ally of KEPLER and GALILEO, he attacked the prevalent Aristotelianism and supported ATOMISM. He also made a number of important astronomical observations, and named the *aurora borealis* (see AURORA).

GASSER, Herbert Spencer (1888–1963), US physiologist who shared with Joseph ERLANGER the 1964 Nobel Prize for Physiology or Medicine for their investigations of the functions of nerve fibers.

GASTRIC JUICE. See DIGESTIVE SYSTEM.

GASTRIN, group of HORMONES, derived from part of the STOMACH, that stimulate acid secretion by the stomach, PANCREAS secretion, and possibly GALL-BLADDER contraction. Its secretion is stimulated by food in the stomach, and the vagus nerve.

GASTROENTERITIS, group of conditions, usually due to viral or bacterial infection of upper GASTRO-INTESTINAL TRACT, causing DIARRHEA, VOMITING and abdominal COLIC. While these are mostly mild illnesses, in young infants and debilitated or elderly adults, dehydration may develop rapidly and fatalities may result. (See also ENTERITIS; FOOD POISONING.)

GASTROINTESTINAL SERIES, X-RAY examination of the GASTROINTESTINAL TRACT using radio-opaque substances, usually barium salts. In **barium swallow** and **meal**, an emulsion is taken and the ESOPHAGUS, STOMACH and DUODENUM are X-rayed. A follow-through may be performed later to outline the small intestine. For **barium enema**, a suspension is passed into rectum and large intestine. CANCER, ULCERS, diverticulae and forms of ENTERITIS and COLITIS may be revealed.

GASTROINTESTINAL TRACT, or gut, or **alimentary canal**, the anatomical pathway involved in the DIGESTIVE SYSTEM of animals. In man it starts at the PHARYNX, passing into ESOPHAGUS and STOMACH. From this arises the small intestine, consisting of the DUODENUM and the great length of the jejunum and ileum. This leads into the large bowel, consisting of the cecum (from which the vermiform APPENDIX arises), colon and rectum. The parts from the stomach to the latter part of the colon lie suspended on a MESENTERY, through which they receive their blood supply, and lie in loops within the peritoneal cavity of the ABDOMEN. In each part, the shape, muscle layers and epithelium are specialized for their particular functions of secretion and absorption. Movement of food in the tract occurs largely by PERISTALSIS, but is controlled at key points by SPHINCTERS. There are many gastrointestinal tract diseases. In GASTROENTERITIS, ENTERITIS and COLITIS, gut segments become inflamed. Peptic ULCER affects both the duodenum and stomach, while CANCER of the esophagus, stomach, colon and rectum are relatively common. Disease of the small intestine tends to cause malabsorption. Methods of investigating the tract include GASTROINTESTINAL SERIES, and endoscopy, in which viewing tubes are passed in via the mouth or anus to examine the gut epithelium.

GASTROPODA, a class of the phylum MOLLUSCA.

GAS TURBINE, a heat ENGINE in which hot gas, generated by burning a fuel or by heat exchange from a nuclear reactor, drives a TURBINE and so supplies power. Straightforward and reliable, they were developed in the late 1930s, and are now used to power aircraft, ships and locomotives, to generate electricity and to drive compressors in pipelines. The

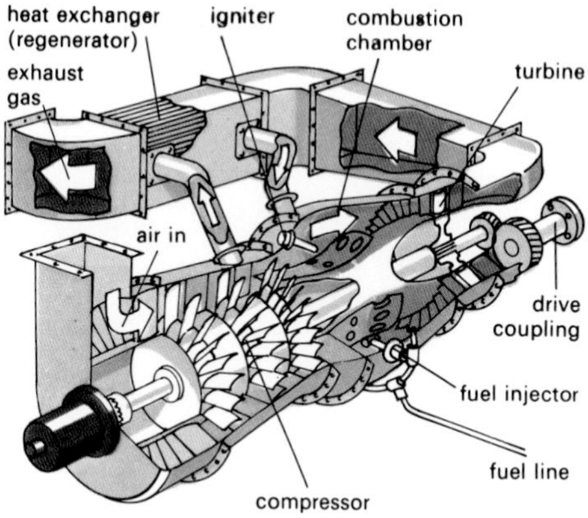

The working cycle of a simple open-cycle gas turbine engine in which waste heat from the turbine exhaust is used to preheat some of the air before combustion.

fuel used may be fuel GAS, gasoline, kerosine or even powdered coal. Some gas turbines are external-combustion engines, the working gas being heated in a heat exchanger and passed round the system in a closed cycle. Most, however, are INTERNAL-COMBUSTION ENGINES working on an open cycle: in the combustors fuel is injected into compressed air and ignited; the hot exhaust gases drive the turbines and are vented to the atmosphere, heat exchangers transferring some of their heat to the air from the compressors. (See also JET PROPULSION.)

GATLING, Richard Jordan (1818–1903), US inventor of the Gatling gun, a multi-barreled MACHINE GUN capable of a high rate of fire (patented 1862). After decades in eclipse, it is now returning to use where extremely rapid fire (up to 7000 shots/min) is required.

GAUSS (Gs), unit of magnetic flux density in CGS UNITS, equalling one maxwell per square centimetre.

GAUSS, Johann Karl Friedrich (1777–1855), German mathematician who discovered the method of LEAST-SQUARES (for reducing experimental errors), made many contributions to the theory of NUMBERS (including the proof that all algebraic equations have at least one root of the form $(a+ib)$ where i is the IMAGINARY OPERATOR and a and b are real numbers), and discovered a NON-EUCLIDEAN GEOMETRY. He won fame when he showed how to rediscover the lost ASTEROID Ceres (1801), and later (1831) turned to the study of MAGNETISM, particularly terrestrial magnetism. He is also remembered for his contributions to STATISTICS and CALCULUS.

GAUSSIAN DISTRIBUTION. See NORMAL DISTRIBUTION.

GAUSSIAN SYSTEM. See CGS UNITS.

GAY-LUSSAC, Joseph Louis (1778–1850), French chemist and physicist best known for **Gay-Lussac's Law** (1808), which states that, when gases combine to give a gaseous product, the ratio of the volumes of the reacting gases to that of the product is a simple, integral one. AVOGADRO's hypothesis is based on this and on DALTON's law of multiple proportions (see COMPOSITION, CHEMICAL). He also showed, independently of CHARLES, that all gases increase in volume by the same fraction for the same increase in temperature, 1/273.2 for 1C°; and (once with BIOT) made two balloon ascents to investigate atmospheric composition and the intensity of the EARTH's magnetic field at altitude. His many contributions to inorganic chemistry include the identification of CYANOGEN.

GEAR, machine part—a toothed wheel—used to transmit rotation from one shaft to another without slip. They are used in pairs, or in threes if both shafts are to rotate in the same sense. The **gear ratio**, the ratio of the number of teeth on the two gears, is equal to the ratio of the TORQUES (neglecting friction), and the inverse of the ratio of the angular velocities. (In the rare case of noncircular gears the torque and angular velocity vary periodically.) Gear teeth are designed to mesh and turn with minimal friction: thus their sides are shaped as involutes of the circular wheel, so that they roll on each other. The commonest type of gear for parallel shafts is the spur gear, with straight teeth parallel to the axis; the helical gear has teeth cut along sections of a HELIX, the double helical gear—the most efficient type—having a herring-bone-like arrangement to avoid axial thrust. Bevel gears, whose tapering teeth are set on a FRUSTUM of a cone, connect intersecting shafts. Skew shafts are connected by a gear and **worm**: the worm is a SCREW, equivalent to a one-toothed gear, so the gear ratio is high.

GEBER, Ibn-Hayyan. See JABIR.

GEDDES, Sir Patrick (1854–1932), Scottish biologist and sociologist who played a formative role in early sociological and urban planning studies.

GEGENSCHEIN, or **Counterglow.** See ZODIACAL LIGHT.

GEIGER COUNTER, or **Geiger-Müller tube,** an instrument for detecting the presence of and measuring radiation such as ALPHA PARTICLES, BETA-, GAMMA- and X-RAYS. It can count individual particles at rates up to about 10000/s and is used widely in medicine and in prospecting for radioactive ores. A fine wire ANODE runs along the axis of a metal cylinder which has sealed insulating ends, contains a mixture of ARGON or NEON and METHANE at low pressure, and acts as the CATHODE, the potential between them being about 1kV. Particles entering through a thin window cause ionization in the gas; ELECTRONS build up around the anode and a momentary drop in the inter-electrode potential occurs which appears as a voltage pulse in an associated counting circuit. The methane quenches the ionization, leaving the counter ready to detect further incoming particles.

GEIKIE, Sir Archibald (1835–1924), Scottish geologist, Director General of the Geological Survey of Great Britain (1882–1901) and President of the Royal Society (1908–12), best known for books such as *A Textbook of Geology* (1882) and *The Ancient Volcanoes of Great Britain* (1897).

GEISSLER TUBE, a forerunner of the modern ELECTRON TUBE, invented in 1858 by Heinrich Geissler (1849–1879). It is a glass tube containing a gas at low pressure which glows with a characteristic COLOR when a high voltage is applied to the metal ELECTRODES at the ends of the tube. Modified forms are used as spectroscopic light sources and in neon or argon signs.

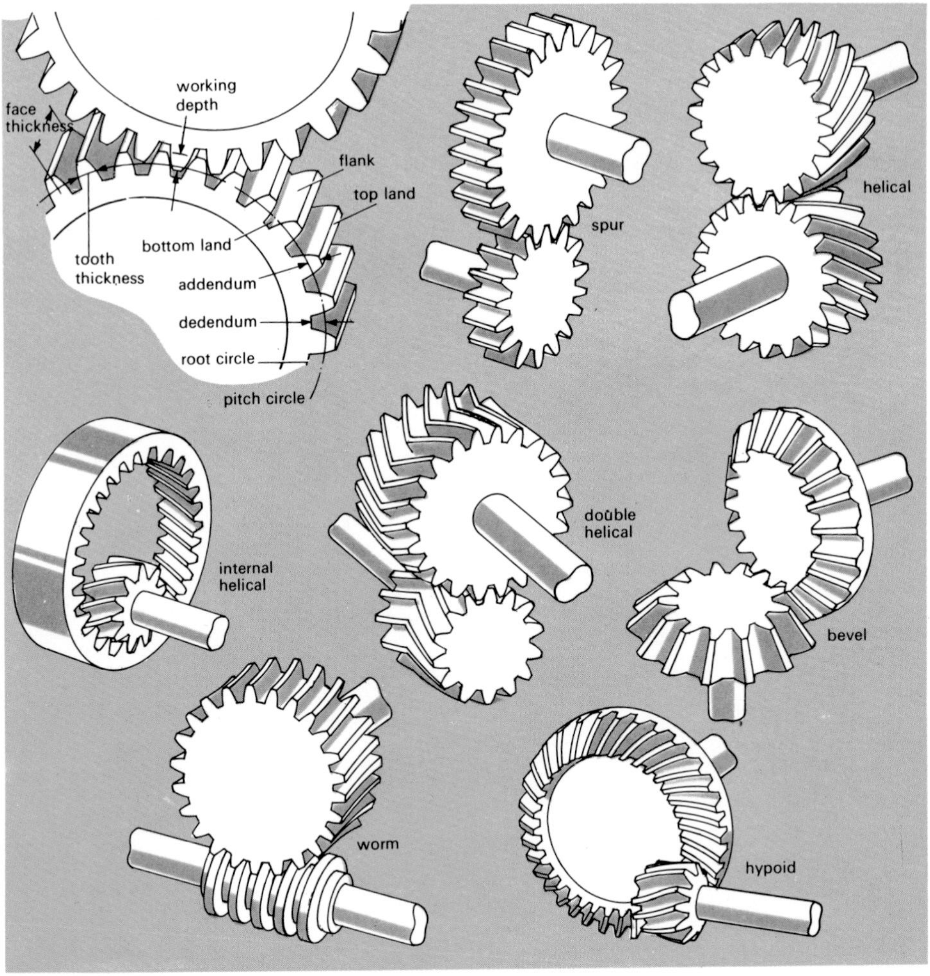

Common gear configurations and nomenclature.

GEL, a COLLOID in which one phase is initially solid, the other initially liquid, but in which both phases are continuous. (See also SOL.) A gel has usually a solid or semisolid form.

GELATIN, yellowish animal PROTEIN, derived from COLLAGEN and obtained by treating animal hides and bones with acid or alkali and boiling them. It dissolves in hot water to form a sol (see COLLOID) which sets to a GEL on cooling. It is used in jellies, soups and other foods, capsules for pharmaceuticals, photographic emulsion, lithography and plastics.

GELIGNITE. See DYNAMITE.

GELL-MANN, Murray (1929–   ), US physicist awarded the 1969 Nobel Prize for Physics for his work on the classification of SUBATOMIC PARTICLES (notably K-mesons and hyperons) and their interactions. With FEYNMAN he has proposed the quark as a basic component of all subatomic particles.

GEMINI (the Twins), a constellation on the ECLIPTIC named after its two brightest stars, Castor and Pollux. The third sign of the ZODIAC, Gemini gives its name to the Geminid METEOR shower.

GEMINI MISSIONS, US space program designed to develop docking and rendezvous procedures, a vital preparation for the Apollo Project (see SPACE EXPLORATION). Gemini 1 was launched April 8, 1964; 3 (the first manned), March 23, 1965; 12 (the last), Nov. 11, 1966. From Gemini IV E.H. White II became the 2nd man ever to float free in space.

GEMS, stones prized for their beauty, and durable enough to be used in jewelry and for ornament. A few—AMBER, CORAL, PEARL and JET—have organic origin, but most are well-crystallized MINERALS. Gems are usually found in IGNEOUS ROCKS (mainly pegmatite dikes) and in contact METAMORPHIC zones. The chief gems have HARDNESS of 8 or more on the Mohs scale, and are relatively resistant to CLEAVAGE and fracture, though some are fragile. They are identified and characterized by their SPECIFIC GRAVITY (which also determines the size of a stone with a given weight in CARATS) and optical properties, especially refractive index (see REFRACTION). Gems of high refractive index show great brilliancy (also dependent on transparency and polish) and prismatic DISPERSION ("fire"). Other attractive optical effects include chatoyancy (see CAT'S EYE), dichroism (see DOUBLE REFRACTION), opalescence and asterism—a star-shaped gleam caused by regular intrusions in the crystal lattice. Since earliest times gems have been engraved in intaglio and cameo. Somewhat later cutting and polishing were developed, the cabochon (rounded) cut being used. Not until the late Middle Ages was faceting developed, now the commonest cutting style, its chief forms being the brilliant cut and the step cut. Some gems are dyed, impregnated, heated or irradiated to improve their color. Synthetic gems are made by flame-fusion or by crystallization from a melt or aqueous solution.

GENE POOL, the total amount of information present at any time in the GENES of the reproductive members of a biological population. The frequency of any particular gene in the gene pool changes owing to NATURAL SELECTION, MUTATION and GENETIC DRIFT. This change forms the basis of evolutionary change.

GENERATOR, Electric, or **dynamo**, a device converting mechanical ENERGY into electrical energy. Traditional forms are based on inducing ELECTRIC FIELDS by changing the MAGNETIC FIELD lines through a circuit (see ELECTROMAGNETISM). All generators can be, and sometimes are, run in reverse as electric MOTORS.

The simplest generator consists of a permanent magnet (the **rotor**) spun inside a coil of wire (the **stator**); the magnetic field is thus reversed twice each revolution, and an AC voltage is generated at the frequency of rotation (see also MAGNETO). In practical designs, the rotor is usually an ELECTROMAGNET driven by a direct current obtained by rectification of a part of the voltage generated, and passed to the rotor through a pair of CARBON **brush**/slip ring contacts. The use of three sets of stator coils 120° apart allows generation of a three-phase supply. (See also ARMATURE.)

Simple DC generators consist of a coil rotating in the field of a permanent magnet: the voltage induced in the coil alternates at the frequency of rotation, but it is collected through a **commutator**—a slip ring broken into two semicircular parts, to each of which one end of the coil is connected, so that the connection between the coil and the brushes is reversed twice each revolution—resulting in a rapidly pulsating direct voltage. A steadier voltage can be achieved through the use of multiple coil/commutator arrangements, and except in very small generators, the permanent magnet is again replaced by an electromagnet driven by part of the generated voltage.

For large-scale generation, the mechanical power is usually derived from fossil-fuel-fired steam TURBINES, or from dam-fed water turbines, and the process is only moderately efficient. The magnetohydrodynamic generator, currently under development, avoids this step, and has no moving parts either. A hot conducting fluid (treated coal gas, or reactor-heated liquid metal) passes through the field of an electromagnet, so that the charges are forced in opposite directions producing a DC voltage. In another device, the electrogasdynamic generator, the voltage is produced by using a high speed gas stream to pump charge from an electric discharge, against the electric field, to a collector. (See also ELECTROSTATIC GENERATOR.)

Generators originated with the discovery of induction by M. FARADAY in 1831; the considerable advantages of electromagnets over permanent magnets were first exploited by E. W. von SIEMENS in 1866.

GENES, the carriers of the genetic information which is passed on from generation to generation by the combination of GAMETES. Genes consist of chain-like molecules of NUCLEIC ACIDS, DNA in most organisms and RNA in some VIRUSES. The genes are normally located on the CHROMOSOMES found in the nucleus of the CELL. The genetic information is coded by the sequences of the four bases present in nucleic acids, with a differing 3-base code for each AMINO ACID so that each gene contains the information for the synthesis of one PROTEIN chain.

GENETIC DRIFT, a process by which genetic information controlling certain features is lost from a population because it is not transmitted to the offspring. It only occurs in small isolated populations. In large populations any specific trait is carried by so many individuals that unless it is unfavorable its loss is highly unlikely. The almost total absence of BLOOD group B in American Indians may be due to genetic drift.

GENETICS, the branch of biology dealing with HEREDITY, which studies the way in which GENES operate and the way in which they are transmitted from parent to offspring. Genetics can be subdivided into a number of more specialized subjects including classical genetics (which deals with the inheritance of parental features in higher animals and plants), cytogenetics (which deals with the cellular basis of genetics), microbial genetics (which deals with inheritance in microorganisms), molecular genetics (which deals with the biochemical basis of inheritance) and human genetics (which deals with inheritance of features of social and medical importance in man). **Genetic counseling** is a branch of human genetics of growing importance. Here couples, particularly those with some form of

The simplest type of electric generator (a simple alternator) comprises a coil (the armature winding) which is rotated in the field of a permanent magnet. The AC current induced in the coil is taken off via slip-ring contacts. A soft-iron armature core helps to maintain a radial field through the area swept by the rotating coil. Replacing the slip rings with a commutator device converts the alternator into a simple DC generator, albeit one generating a fluctuating potential.

inherited defect, are advised on the chances that their children will have similar defects.

GENOTYPE, the total genetic makeup of a particular organism consisting of all the GENES received from both parents. For any individual the genotype determines their strengths and weaknesses during their whole life and is unique and constant for each individual. Duplication of the genotype except in identical twins is statistically impossible except in the simplest organisms. (See also PHENOTYPE.)

GENTIAN VIOLET, or methylrosaniline chloride, green powder used in aqueous solution (violet) as an ANTISEPTIC and medicinal FUNGICIDE, and against intestinal WORMS (see PARASITIC DISEASES).

GENUS. See TAXONOMY.

GEOCHEMISTRY, the study of the CHEMISTRY of the EARTH (and other planets). Chemical characterization of the earth as a whole relates to theories of planetary formation. Classical geochemistry analyzes rocks and MINERALS. The study of PHASE EQUILIBRIA has thrown much light on the postulated processes of ROCK formation. (See also GEOLOGY.)

GEODE, a hollow mineral formation found in certain rocks. Typically, it is almost filled by inward-growing crystal "spikes," usually of QUARTZ. Geodes range between 20mm and 1m across.

GEODESIC DOME, architectural dome-like structure composed of polygonal (usually triangular) faces of lightweight material. It was developed by Buckminster FULLER. A geodesic dome housed the US exhibit at Expo '67 (Montreal).

GEODESY, the branch of geophysics concerned with the determination and explanation of the precise shape and size of the EARTH. The first recorded measurement of the earth's circumference that approximates to the correct value was that of ERATOSTHENES in the 3rd century BC. Modern geodesists use not only the techniques of SURVEYING but also information received from the observations of artificial SATELLITES.

GEOFFROY DE SAINT-HILAIRE, Étienne (1772–1844), French naturalist who argued, against G. CUVIER, for "unity of plan," the notion that all animals contain the same anatomical elements, some of these being developed at the expense of others in particular species. He acknowledged that there had been development in previously existing species.

The different branches of geography (red) and their links with allied sciences (yellow).

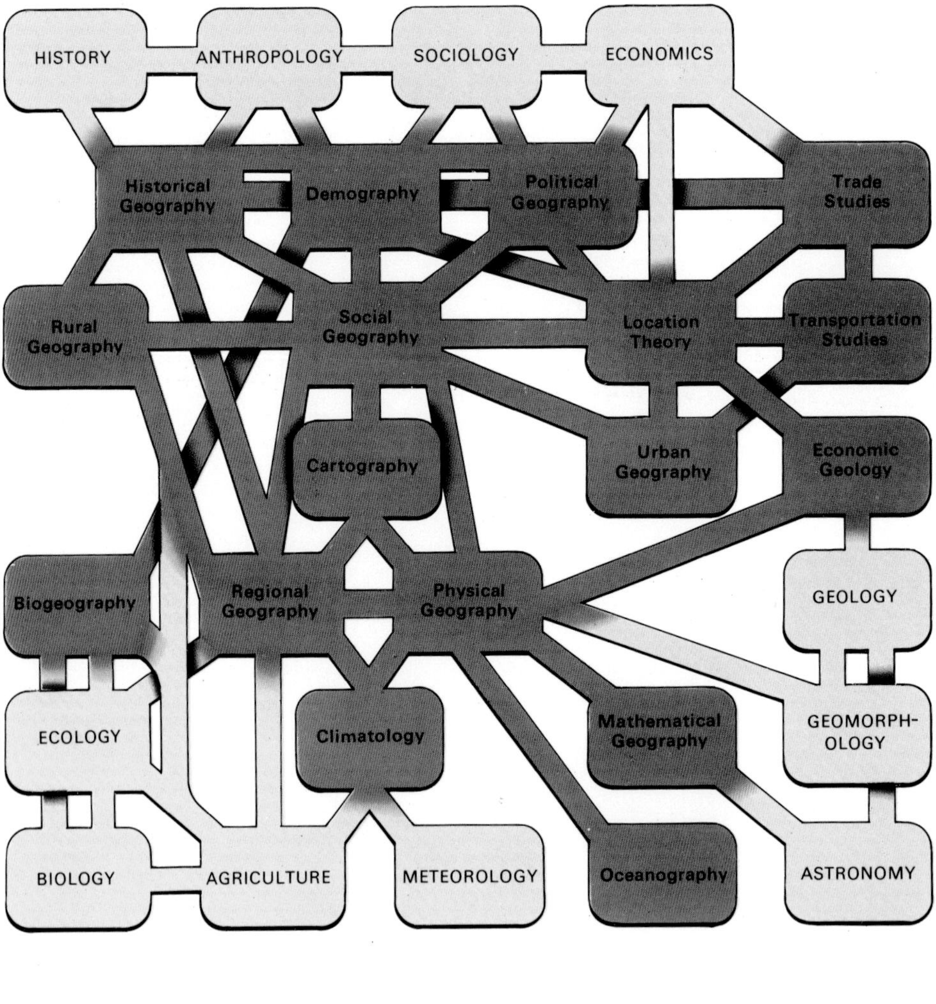

GEOGRAPHY, the group of sciences concerned with the surface of the earth, including the distribution of life upon it, its physical structures, etc. Geography relies on surveying and mapping, and modern cartography (mapmaking) has rapidly adapted to the new needs of geography as it advances and develops. (See MAP; SURVEYING.) **Biogeography** is concerned with the distribution of life, both plant and animal (including man), about our world. It is thus clearly intimately related to BIOLOGY and ECOLOGY. **Economic geography** describes and seeks to explain the patterns of the world's commerce in terms of production, trade and transportation, and consumption. It relates closely to economics. **Mathematical geography** deals with the size, shape and motions of the EARTH, and is thus linked with ASTRONOMY (see also GEODESY). **Physical geography** deals with the physical structures of the earth, also including climatology and OCEANOGRAPHY, and is akin to physical GEOLOGY. **Political geography** is concerned with the world as nationally divided; **regional geography** with the world in terms of regions separated by physical rather than national boundaries. **Historical geography** deals with the geography of the past: PALEOGEOGRAPHY at one level, exploration or past political change or settlement at another. **Applied geography** embraces the applications of all these branches to the solution of socioeconomic problems. Its subdivisions include urban geography and social geography; and it contributes to the science of SOCIOLOGY. (See also ETHNOLOGY; HYDROLOGY; METEROLOGY.) **Development of Geography.** Geography had its origins in the Greek attempts to understand the world in which they found themselves. Once it was realized that the earth was round, the next step was to estimate its size. This was achieved in the 3rd century BC by ERATOSTHENES. The classical achievement in geography, like that in astronomy, was summed up by Claudius PTOLEMY. His world MAP was used for centuries. Geographical knowledge next leapt forward in the age of exploration that opened with the voyages of Dias and Columbus. The 17th century saw continuing discovery and greatly improved methods of survey. The earliest modern geographical treatises, including that of Varenius, also appeared in this era. The 19th century brought with it the works of F. H. A. von HUMBOLDT and Karl Ritter, the former stressing physical and systematic geography, the latter the human and historical aspects of the science. Encompassing so many different studies, geography since the mid-19th century has become a battleground for the strife between different schools of geographers. While some have encouraged a regional approach, others have preferred to develop a landscape-concept. Others have stressed the study of the physical environment while others still have concentrated on political and economic factors. Perhaps the most recent group to come to the fore favors the collection of precise numerical data. With this they try to build mathematical models of geographic phenomena.

GEOID, geodesic model of the earth, the shape the earth would have to have for the pull of GRAVITY to be constant for all points, taken at sea level over the oceans and at corresponding level on land. The result is an oblate spheroid (see ELLIPSOID; OBLATENESS) with irregularities due to differing local densities. (See also EARTH; GEODESY.)

GEOLOGY, the group of sciences concerned with the study of the earth, including its structure, long-term history, composition and origins.

Physical Geology deals with the structure and composition of the EARTH and the forces of change affecting them. The sciences that make up physical geology thus include GEODESY, GEOMORPHOLOGY, GEOPHYSICS and seismology (see EARTHQUAKE). Much of modern physical geology is based on the theory of PLATE TECTONICS.

Historical Geology deals with the earth in past ages, and with the EVOLUTION of life upon it. It embraces such sciences as PALEOCLIMATOLOGY, PALEOMAGNETISM, PALEONTOLOGY and STRATIGRAPHY; and relies heavily on dating (see CHRONOLOGY), events being related to the geological time scale, whose derivation is primarily stratigraphical, to a lesser extent paleontological.

Economic Geology lies between these two, and borrows from both. Concerned with the location and exploitation of the earth's natural resources (see ORE), it is generally taken to include also the disciplines of crystallography, mineralogy and petrology (see CRYSTAL; MINERALS; ROCKS). Its practical manifestations are PROSPECTING and MINING.

Geology of other Planets. Except with the MOON, it is not yet possible to examine the rocks of other planets, but telescope and spectroscopic examinations have revealed much, as have those of unmanned probes. VOLCANISM is known on the moon and MARS (one volcano is some 600km across), and "moonquakes" have been detected.

Development of Geology. Most early geological knowledge came from the experience of mining engineers, some of the earliest geological treatises coming from the pen of Georgius AGRICOLA. The interest of the 16th century in FOSSILS was also reflected in the writings of K. von GESNER. In the 17th century the biblical timescale of about 6000 years from the Creation to the present largely constrained the many speculative "Theories of the Earth" that were issued. The century's most notable geological observations were made by N. STENO. The late 18th century saw the celebrated controversy between A. G. WERNER's "Neptunists" and J. HUTTON's "Plutonists" as to the origin of the rocks. The first decades of the 19th century, however, witnessed the decline of speculative geology as field observations became ever more detailed. William Smith (1769–1839), the "father of stratigraphy," showed how the succession of fossils could be used to index the stratigraphic column, and he and others produced impressive geological maps. C. LYELL's classic *Principles of Geology* (1830–33) restated the Huttonian principle of UNIFORMITARIANISM and provided the groundwork for much of the later development of the science. L. AGASSIZ pointed to the importance of glacial action in the recent history of the earth (1840), while mining engineering continued to contribute to the pool of geologic data. The most significant recent development in the earth sciences has been the acceptance of the theory of PLATE TECTONICS, foreshadowed in A. WEGENER's 1912 theory of CONTINENTAL DRIFT.

GEOMAGNETISM, the magnetic field of the

aeon	era	period	epoch	time since commencement (million years)
Phanerozoic	Cenozoic	Quaternary	Holocene (Recent)	0.01
			Pleistocene	4
		Tertiary	Pliocene	10
			Miocene	25
			Oligocene	40
			Eocene	55
			Paleocene	65
	Mesozoic	Cretaceous		135
		Jurassic		190
		Triassic		225
	Paleozoic	Permian		280
		Pennsylvanian	Carboniferous	315
		Mississippian	Carboniferous	345
		Devonian		400
		Silurian		440
		Ordovician		500
		Cambrian		570
Cryptozoic	Proterozoic	Precambrian		
	Archeozoic (Archean)			
Azoic				4550

The geologic timescale.

EARTH; and the study of it, both as it is in the present and as it was in the past (see PALEOMAGNETISM). (See also GEOPHYSICS; MAGNETISM.)

GEOMETRY, the branch of MATHEMATICS which studies the properties both of space and of the mathematical constructs—lines, curves, surfaces and the like—which can occupy space. Today it divides into ALGEBRAIC GEOMETRY; ANALYTIC GEOMETRY; DESCRIPTIVE GEOMETRY; DIFFERENTIAL GEOMETRY; EUCLIDEAN GEOMETRY; NON-EUCLIDEAN GEOMETRY, and PROJECTIVE GEOMETRY, but many of these divisions have grown up only in the last few hundred years. The name geometry reminds us of its earliest use—for the measurement of land and materials. The Babylonian and Egyptian civilizations thus gained great empirical knowledge of elementary geometric figures, including how to construct a right-angled triangle. The Greek philosophers transformed this practical art into an intellectual pastime through which they sought access to the secrets of nature. About 300 BC EUCLID collected together and added to the Greek rationalization of geometry in his *Elements*. Later Alexandrian geometers began to develop TRIGONOMETRY. The revival of interest in life-like painting in the Renaissance led to the development of projective geometry, though it is to the philosopher–scientist DESCARTES that we owe the invention of the algebraic (coordinate) geometry which allows algebraic FUNCTIONS to be represented geometrically. The next new branch of geometry to be developed followed fast upon the invention of CALCULUS: differential geometry. The greatest upset in the history of geometry came in the 19th century. Men such as J. K. F. GAUSS, N. I. LOBACHEVSKI and BOLYAI János began to question the Euclidean parallel-lines axiom and discovered hyperbolic geometry, the first non-Euclidean geometry. The elliptic non-Euclidean geometry of G. F. B. RIEMANN aided A. EINSTEIN in the development of the theory of general RELATIVITY.

GEOMORPHOLOGY, the surface features of the EARTH; and their study, especially as to their origins and the processes acting on them. (See GEOLOGY.)

GEOPHYSICS, the physics of the EARTH, as such including studies of the ATMOSPHERE, EARTHQUAKES

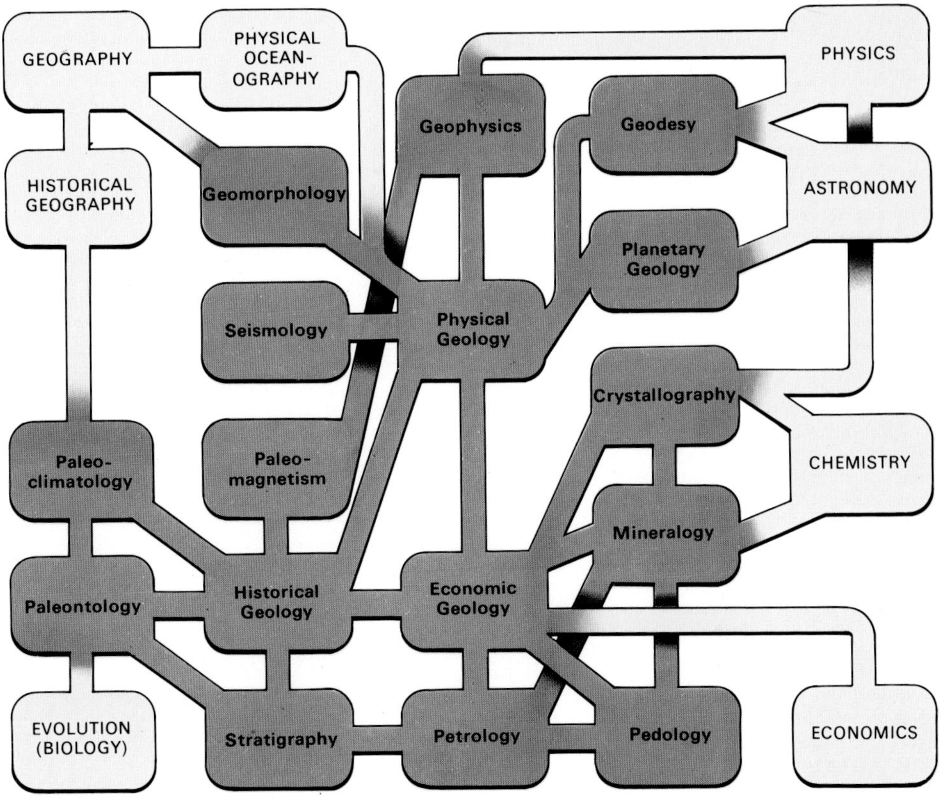

The different branches of geology (red) and their links with allied sciences (yellow).

and VOLCANISM, as well as GEODESY, GEOMAGNETISM, HYDROLOGY and OCEANOGRAPHY.

GEOSYNCLINE, a large basin or syncline (concave FOLD in the rock strata) which has become filled with sediment and whose floor has subsided, so that it contains vast thicknesses of SEDIMENTARY ROCKS. (See also MOUNTAIN; OROGENIES.)

GEOTROPISM. See TROPISMS.

GERIATRICS, the branch of MEDICINE specializing in the care of the elderly. Although concerned with the same DISEASES as the rest of medicine, the different susceptibility of the aged and a tendency for multiple pathology make its scope different. In particular the psychological problems of old age differ markedly from those encountered in the rest of the population and require special management. The social and medical aspects of long-term care involve the coordination of family, voluntary and hospital services; the geriatrician must nevertheless seek to maximize the individuality and freedom available to the geriatric patient.

GERM, microorganism capable of causing disease, including VIRUSES, BACTERIA, PARASITES and PROTOZOA.

GERMANIUM (Ge), silvery-gray metalloid in Group IVA of the PERIODIC TABLE; brittle crystalline solid whose structure resembles that of DIAMOND. It occurs naturally in sulfide ores of SILVER, COPPER and ZINC, and in COAL, and is extracted as a by-product of processing them. Its chemical properties are intermediate between those of SILICON and TIN; it reacts with the halogens, oxidizes in air at 600°C, and is attacked by concentrated oxidizing acids and by fused alkalis. Germanium is a SEMICONDUCTOR, and is used in electronic devices, especially TRANSISTORS; it is also used in alloys and for lenses and windows for INFRARED RADIATION. AW 72.6, mp 937°C, bp 2830°C, sg 5.323 (25°C).

Germanium forms covalent tetra- and divalent compounds. **Germanium (IV) Oxide** (GeO_2) is used in high-refractive-index GLASS. mp 1086°C. **Germanium (IV) Chloride** ($GeCl_4$), colorless liquid intermediate in the extraction of germanium and the preparation of most of its compounds. mp −50°C, bp 84°C. **Germanes**, series of volatile hydrides resembling silanes (see SILICON).

GERMAN MEASLES, or **rubella**, mild VIRUS infection, usually contracted in childhood and causing FEVER, SKIN rash, malaise and LYMPH node enlargement. Its importance lies in the fact that infection of a mother during the first three months of PREGNANCY leads to infection of the EMBRYO via the PLACENTA and is associated with a high incidence of congenital DISEASES including CATARACT, DEAFNESS and defects of the HEART and ESOPHAGUS. VACCINATION of intending mothers who have not had rubella is advisable. If rubella occurs in early pregnancy, ABORTION may be induced to avoid the BIRTH of malformed children.

GERMAN SILVER, or **nickel silver**, an ALLOY composed of copper, nickel and zinc. It resembles silver, and is used for cheap jewelry, cutlery etc., and as the base for silver-plated ware.
GERM CELL. See GAMETE.
GERM-FREE ANIMALS. See GNOTOBIOTICS.
GERMICIDES. See ANTISEPTICS.
GERMINATION, the resumption of growth of a plant embryo contained in the SEED after a period of reduced metabolic activity or dormancy. Conditions required for germination include an adequate water supply, sufficient oxygen and a favorable temperature. Rapid uptake of water followed by increased rate of respiration are often the first signs of germination. During germination, stored food reserves are rapidly used up to provide the energy and raw materials required for the new growth. The embryonic root and shoot which break through the seed coat are termed the radicle and plumule, respectively. There are two general forms of germination: hypogeal and epigeal. In the former, the seed leaves, or COTYLEDONS, remain below the ground, as in the broad bean, while in the latter they are taken above the ground and become the first photosynthetic organs, as in the castor oil seed.
GERM PLASM, a special type of PROTOPLASM present in the reproductive cells or gametes of plants and animals, which A. WEISMANN suggested passed on unchanged from generation to generation. Although it gave rise to the body cells, it remained distinct and unaffected by the offspring.
GESNER, Konrad von (1516–1565), Swiss naturalist whose major work, *Historia Animalium* (4 vols., 1551–58), an encyclopedic study of many varieties of animals, is considered the foundation stone of modern zoology.
GESSO, a paste used to prepare surfaces for painting or gilding, made by mixing CHALK or whiting with GLUE.
GESTALT PSYCHOLOGY, a school of psychology concerned with the tendency of the human (or even PRIMATE) mind to organize PERCEPTIONS into "wholes"; for example, to hear a symphony rather than a large number of separate notes of different tones. Gestalt psychology, whose main proponents were WERTHEIMER, KOFFKA and KÖHLER, maintained that this was due to the mind's ability to complete patterns from the available stimuli. The school emerged as a reaction against such schools as BEHAVIORISM.
GESTATION, the development of young mammals in the mother's uterus from FERTILIZATION to BIRTH. With some exceptions, the gestation period is proportional to the adult size of the animal, thus, for the human young the gestation period is about 270 days, but for the elephant it is closer to two years. (See EMBRYO; FETUS; PREGNANCY.)
GEYSER, a hot spring, found in currently or recently volcanic regions (see VOLCANISM), that intermittently jets steam and superheated water into the air. It consists essentially of a tube leading down to a heat source. GROUNDWATER accumulates in the tube, that near the bottom being kept from boiling by the PRESSURE of the cooler layers above. When the critical temperature is reached, bubbles rise, heating the upper layers which expand and well out of the orifice. This reduces the pressure enough for substantial STEAM formation below, with subsequent eruption. The process then recommences. The famous Old Faithful used to erupt every 66½min, but has recently become less reliable. (See also FUMAROLE; HOT SPRINGS; SPRING.)
GIAEVER, Ivar (1929–), Norwegian-born US physicist awarded, with L. ESAKI and B. JOSEPHSON, the 1973 Nobel Prize for Physics for his work on tunneling (see WAVE MECHANICS).
GIANTISM. See GIGANTISM.
GIANT STAR. See STAR.
GIAUQUE, William Francis (1895–), US chemist who discovered the isotopes of OXYGEN (1929). He has also contributed to the science of CRYOGENICS by inventing and applying the process of adiabatic demagnetization, for which he was awarded the 1949 Nobel Prize for Chemistry.
GIBBERELLINS, a group of plant HORMONES mainly found in the seeds, young leaves and roots of green plants and in certain fungi. They were originally isolated from a Japanese fungus (*Gibberella fujikuroi*) which lengthens the stems of rice plants. **Gibberellic acid**, found in green plants, is involved in the "bolting" of plants like carrots. Various attempts to use gibberellins commercially to increase crop yield have been unsuccessful.

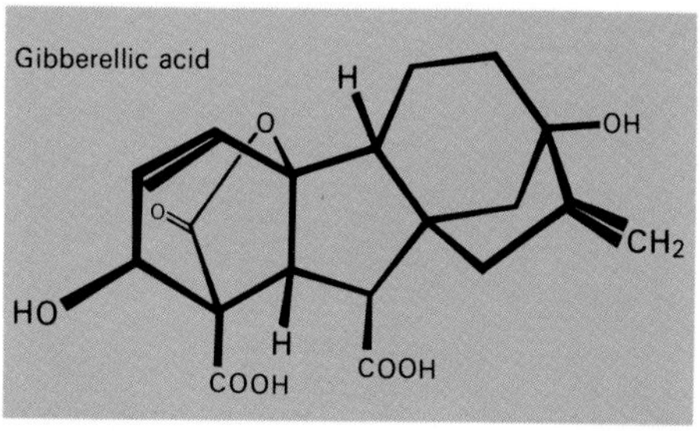

GIBBS, Josiah Willard (1839–1903), US physicist best known for his pioneering work in chemical THERMODYNAMICS. In *On the Equilibrium of Heterogeneous Substances* (2 vols., 1876 and 1878) he states Gibbs' Phase Rule (see PHASE EQUILIBRIA). In the course of his research on the electromagnetic theory of LIGHT, he made fundamental contributions to the art of VECTOR ANALYSIS.
GIFFARD, Henri (1825–1882), French engineer who built the first steam-powered AIRSHIP, flown in Paris in 1852; and invented (1859) an injector for steam boilers, used for many decades.
GIGANTISM, or abnormally large stature starting in childhood, may be caused by a constitutional trait or by HORMONE disorders during growth. The latter are usually excessive secretion of growth hormone or thyroid hormone before the EPIPHYSES have fused.
GILBERT (Gb), unit of magnetomotive force in CGS UNITS, equalling $10/4\pi$ ampere-turns.
GILBERT, William (1544–1603), English scientist, the father of the science of MAGNETISM. Regarding the earth as a giant magnet, he investigated its field in terms of dip and variation (see EARTH), and explored many other magnetic and electrostatic phenomena. The GILBERT is named for him.
GILL (gi), name of various units of volume, usually equalling one-fourth of a pint. The US gill is 0.1183 litres; the Imperial gill is 0.1421 litres.

GILLS, the respiratory organs of many aquatic animals. They take in OXYGEN from the water and give off CARBON dioxide waste. They are thin-walled so that gases pass easily through and usually take the form of thin flat plates or finely divided feathery filaments. The higher invertebrates, crabs and lobsters for instance, have gills protected by an EXOSKELETON and maintain an adequate oxygen supply by pumping water over them. The gills of fish are protected by a bony operculum and movements of the throat provide a water current over them. (See RESPIRATION.)

GILSONITE, natural asphaltic BITUMEN found in veins near the Col./Ut. border, in the Uinta Basin. It is a lustrous black solid, used in paints and coating and insulating materials, and, more recently, converted to coke, gasoline and gas.

GIN, liquor distilled from grain flavored with juniper berries. Sometimes coriander, orange or lemon peel, cardamon and orris roots are added as flavoring agents. It contains 40–47% alcohol (80–94 US Proof). It originated in the Netherlands, apparently from a juniper-berry medicine. (See ALCOHOLIC BEVERAGES.)

GIN. See COTTON GIN.

GINGIVITIS, or gum INFLAMMATION, due to bacterial infection (e.g. VINCENT'S ANGINA) or disease of the TEETH and poor mouth hygiene.

GIORGI SYSTEM. See MKSA UNITS.

GIZZARD, part of the alimentary canal developed by a variety of animals for the mechanical breakdown of food. Situated before the main digestive region of the gut it has very muscular walls. Fragmentation of the food may be by chitinous "teeth" in the inner wall or by stones and grit, swallowed expressly for this purpose.

GLACIER, a large mass of ice that can survive for several years. In most cases, glaciers are heavy enough to flow downhill under their own weight. There are three recognized types of glacier: ice sheets and caps; mountain or valley glaciers; and piedmont glaciers. Glaciers form wherever conditions are such that annual PRECIPITATION of snow, sleet and hail is greater than the amount that can be lost through evaporation or otherwise (see ABLATION). The occurrence of a glacier thus depends much on latitude (see LATITUDE AND LONGITUDE) and also on local topography: there are several glaciers on the EQUATOR. Glaciers account for about 75% of the world's fresh water, and of this the Antarctic ice sheet accounts for about 85%. **Mountain glaciers** usually result from snow accumulated in CIRQUES coalescing to form glaciers; and **piedmont glaciers** occur when such a glacier spreads out of its valley into a contiguous lowland

Features associated with glaciers and glaciated landforms: (1) head of glacier; (2) firn or névé; (3) region of ground moraine deposition; (4) terminal moraine; (5) drumlin; (6) braided stream; (7) kettle; (8) medial moraine; (9) lateral moraine; (10) U-shaped valley; (11) arête; (12) hanging valley; (13) cirque; (14) tarn; and (15) ice fall.

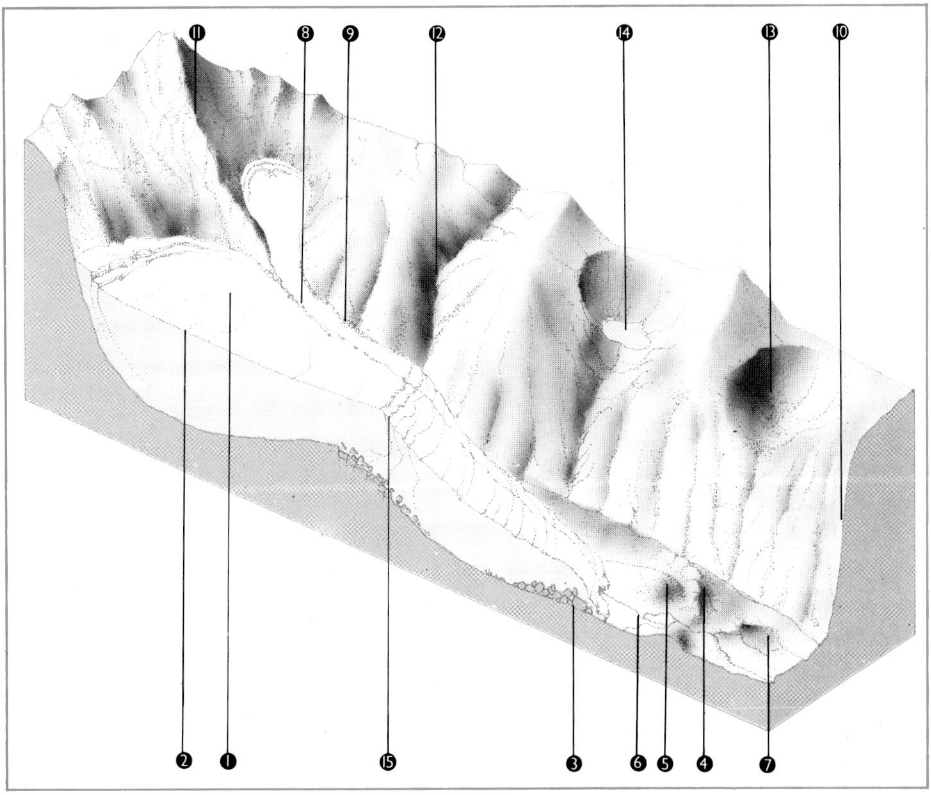

area. (See also DRIFT; DRUMLIN; EROSION; ESKER; FJORD; ICE; ICE AGES; ICEBERG; MORAINE; NÉVÉ; TILL.)

GLANDERS, BACTERIAL DISEASE of horses, rarely transmitted to man. In horses, ABSCESSES are common and are a source of infection, which causes FEVER, ulceration of the NOSE and PHARYNX, and multiple abscesses. PNEUMONIA, LUNG abscess, ARTHRITIS and MENINGITIS occur; without ANTIBIOTICS, death ensues.

GLANDS, structures in animals and plants specialized to secrete essential substances. In plants they may discharge their secretions to the outside of the plant (via glandular hairs), or into special secretory canals. External secretions include NECTAR and insect attractants; internal secretions, pine resin and RUBBER latex. In animals they are divided into ENDOCRINE GLANDS, which secrete HORMONES into the BLOOD stream, and **exocrine glands** which are the remainder, usually secreting materials via ducts into internal organs or onto body surfaces. LYMPH nodes are sometimes termed glands. In man, SKIN contains two types of gland: *sweat glands*, which secrete watery fluid (PERSPIRATION) and *sebaceous glands* which secrete sebum. *Lacrimal glands* secrete TEARS. The cells of mucous membranes or the EPITHELIUM of internal organs secrete MUCUS, which serves to lubricate and protect the surface. *Salivary glands* (parotid, submandibular and sublingual) secrete SALIVA to facilitate swallowing. In the GASTROINTESTINAL TRACT, mucus-secreting glands are numerous, particularly in the STOMACH and colon, where solid food or feces need lubrication. Other stomach glands secrete hydrochloric acid and pepsin as part of the DIGESTIVE SYSTEM. Small-intestinal juices containing ENZYMES are similarly secreted by minute glandular specializations of the epithelium. The part of the PANCREAS secreting enzyme-rich juice into the DUODENUM may be regarded as an exocrine gland. Analysis of gland secretion may be helpful in diseases of digestion, of the EYES and salivary glands and in CYSTIC FIBROSIS.

GLANDULAR FEVER. See MONONUCLEOSIS.

GLASER, Donald Arthur (1926–), US physicist awarded the 1960 Nobel Prize for Physics for his invention of the BUBBLE CHAMBER (1952).

GLASS, material formed by the rapid cooling of certain molten liquid so that they fail to crystallize (see CRYSTAL) but retain an amorphous structure. Glasses are in fact supercooled LIQUIDS which, however, have such high viscosity that they behave like solids for all practical purposes. Some glasses may spontaneously crystallize or devitrify. Few materials form glasses, and almost all that are found naturally or used commercially are based on SILICA and the SILICATES. Natural glass is formed by rapid cooling of MAGMA, producing chiefly OBSIDIAN, or rarely by complete thermal metamorphism (see also TEKTITES). The earliest known manufactured glass was made in Mesopotamia in the 3rd millennium BC. Glass was shaped by molding or core-dipping, until the invention of **glassblowing** by Syrian craftsmen in the 1st century BC. Essentially still used, the process involved gathering molten glass on the end of a pipe, blowing to form a bubble, and shaping the vessel by further blowing, swinging, or rolling it on a surface. They also blew glass inside a shaped mold; this is now the chief process used in mechanized automatic glassblowing. Modern glass products are very diverse, including windows, bottles and other vessels, optical devices, building materials, fiberglass products, etc. Most are made of **soda-lime glass**. Although silica itself can form a glass, it is too viscous and its melting point is too high for most purposes. Adding soda lowers the melting point, but the resultant sodium silicate is water-soluble (see WATER GLASS), so lime is added as a stabilizer, together with other metal oxides as needed for decolorizing etc. The usual proportions are 70% SiO_2, 15% Na_2O, 10% CaO. **Crown glass,** used in optical systems for its low DISPERSION, is of this type, with BARIUM oxide (BaO) often replacing the lime. **Flint glass**, or **crystal**, is a brilliant clear glass with high optical dispersion, used in high-quality glassware and to make lenses and prisms. It was originally made from crushed flints to give pure, colorless silica; later, sand was used, with increasing amounts of lead (II) oxide. For borosilicate glass, used where high thermal stresses must be withstood, see PYREX. The manufacture of the various kinds of glass begins by mixing the raw materials—sand, limestone, sodium nitrate or carbonate, etc.—and melting them in large crucibles in a furnace. The molten glass, having been refined (free from bubbles) by standing, is formed to the shape required and then annealed (see ANNEALING). Some SAFETY GLASS is not annealed, but rapidly cooled to induce superficial compressive stresses which yield greater strength. **Plate glass** is made by passing a continuous sheet of soft glass between rollers, GRINDING AND POLISHING it on both sides, and cutting it up so as to eliminate flaws. A newer method (the float glass process) involves pouring the molten glass onto molten metal, such as tin, and to allow it to cool slowly: the surface touching the metal is perfectly flat and needs no polishing. Special glass products include **foam glass**, made by SINTERING a mixture of glass and an agent that gives off a gas on heating, used for insulation; **photosensitive glass**, which darkens reversibly in bright light; and FIBERGLASS. (See also ENAMEL.)

GLASSES, or spectacles, LENSES worn in front of the EYES to correct defects of VISION or for protection. Converging lenses have been worn to correct farsightedness (HYPEROPIA) since the late 13th century and diverging lenses for shortsightedness (MYOPIA) since the 16th. Glasses with cylindrical lenses are used to correct ASTIGMATISM and those having bifocal lenses (i.e., having two different powers in the upper and lower areas of each lens) or even trifocals (three powers) may be worn for PRESBYOPIA. Most spectacle lenses are worn in a metal or plastic frame which rests on the nose and ears, though in some cases CONTACT LENSES fitting directly onto the eyeball may be suitable. Protective glasses include sunglasses and safety glasses.

GLAUBER, Johann Rudolf (1607–1670), German chemist who prepared a wide variety of organic and inorganic compounds to make use of their (often non-existent) medicinal powers. In preparing hydrochloric acid from sulfuric acid and common salt, he found a residue which he claimed as a cure-all. In fact it was sodium sulfate, a mild laxative, and still often called **Glauber's salt**.

GLAUCOMA, raised fluid pressure in the EYE, leading in chronic cases to a progressive deterioration of VISION. It arises from a variety of causes, often involving block to aqueous humor drainage.

Glaucoma is relieved using drugs or surgically.

GLAUCONITE, a CLAY mineral of the illite type, a hydrated MICA containing considerable iron and magnesium. Formed by alteration of BIOTITE in a shallow, reducing marine environment, it commonly occurs as small green pellets in sedimentary deposits. **Greensand** is a mixture of glauconite with quartz SAND. (See also MARL.)

GLAZE. See POTTERY AND PORCELAIN.

GLIDER, or **sailplane,** nonpowered airplane which, once launched by air or ground towing, or by using a winch, is kept aloft by its light, aerodynamic design and the skill of the pilot in exploiting "thermals" and other rising air currents. Sir George CAYLEY built his first model glider in 1804 and in 1853 he persuaded his coachman to undertake a short glide—the first manned heavier-than-air flight. Otto LILIENTHAL made many successful flights in his hang-gliders (planes in which the pilot hangs underneath and controls the flight by altering his body position, hence moving the craft's CENTER OF GRAVITY) from 1891 until his death in a gliding accident in 1896. Later, the WRIGHT brothers developed gliders in which control was achieved using moving control surfaces, as a prelude to their experiments with powered flight. Gliding as a sport was born in Germany after WWI and is now popular throughout the world. Recent years have seen a particular resurgence of interest in hang gliding.

GLIOMA, TUMOR of glial cells, the supporting cells of the BRAIN. They never metastasize (see CANCER), but produce signs of focal damage to the brain, such as weakness, visual disturbance, personality change or EPILEPSY, and often a characteristic type of HEADACHE. SURGERY and RADIATION THERAPY may be helpful, but glial cell destruction cannot be reversed.

GLOBE, a representation of the earth as a small sphere, mounted on an axis so that it may revolve. The oldest extant globe is from Nuremberg, 1492. Celestial and lunar globes are also made.

GLOBIGERINA, a single-celled organism found floating on the surface of the oceans. It is of the order Foraminiferida of the phylum PROTOZOA. It has a perforated shell, through which it can extend PSEUDOPODIA. After death, the shells sink to the OCEAN floor to form much of the organic OOZES.

GLOBULAR CLUSTERS, apparently ellipsoidal densely packed clusters of up to a million stars orbiting a GALAXY. The MILKY WAY and the ANDROMEDA Galaxy have each around 200 such clusters. They contain high proportions of cool red stars and RR Lyrae VARIABLE STARS. Study of the latter enables the distances of the clusters to be calculated.

GLOBULINS, PROTEINS insoluble in water but soluble in dilute solutions of mineral salts. They are widely distributed in plants and animals, e.g., lactoglobulin in milk and plant globulins in seeds. In man, serum globulins (in BLOOD) are concerned in resistance to disease and in various ALLERGIES.

GLOSSOLALIA, speech in an unknown or fabricated language uttered by individuals under HYPNOSIS, suffering from certain MENTAL ILLNESSES or in trance, or by groups undergoing religious ecstasy.

GLUCAGON, HORMONE produced by specialized cells in the PANCREAS, tending to counteract the effect of INSULIN on blood sugar, but having other actions.

GLUCOSE ($C_6H_{12}O_6$), also dextrose or "grape sugar," a naturally occurring simple SUGAR (monosaccharide) found in honey and sweet fruits. It circulates in the BLOOD of mammals, providing their cells with energy. Other sugars and CARBOHYDRATES are converted to glucose by digestion before they can be utilized.

GLUES, widely used adhesive substances of animal or vegetable origin. Animal glues are made from bones, hides, fish bones, fish oil, or the milk protein CASEIN; vegetable glues from natural GUMS, STARCH (e.g., flour and water) or soybeans. Though in use for millennia, it is not yet fully understood how glues work. Nowadays, synthetic RESINS are replacing glues for many purposes. (See also ADHESIVES.)

GLUTEN, a mixture of two proteins (gliadin and glutenin) found in wheat and other cereal flours. In the rising of BREAD gluten forms an elastic network which traps the carbon dioxide, giving a desirable crumb structure on baking. The proportion of gluten in wheat flour varies from 8% to 15%. The level determines the suitability of the flour for different uses. The high gluten content of hard wheat is right for bread and pasta, while soft wheat (low gluten) is used for biscuits.

GLYCEROL, or **glycerin,** colorless, viscous liquid with a sweet taste; a trihydric ALCOHOL. Its fatty-acid esters constitute natural FATS and OILS, from which glycerol is obtained as a by-product of SOAP manufacture; it is also synthesized from propylene, a petroleum product. It is used to make resins for paints and varnishes, in foods, medicines and cosmetics, as a moistening agent and plasticizer, and to make NITROGLYCERIN.

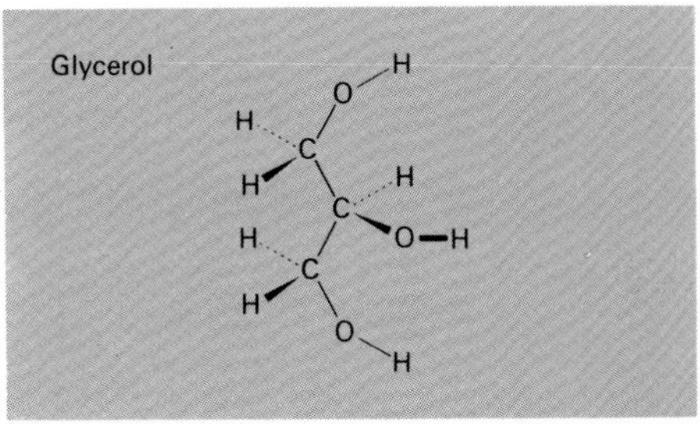

GLYCOGEN, or animal starch, a soluble CARBOHYDRATE consisting of chains of glucose units. It is produced by all vertebrates and stored in muscle and in the liver where it forms a readily available reserve of GLUCOSE.

GLYCOL, or diol, class of dihydric ALCOHOLS, the lower members being viscous, hygroscopic liquids. ETHYLENE GLYCOL is the most important; propylene glycol has similar uses, but being nontoxic is also used in foods, pharmaceuticals and cosmetics.

GLYCOLYSIS, an enzyme-mediated cellular process by which GLUCOSE is broken down to pyruvate. In all, 10 different steps are involved. The process does not require the presence of OXYGEN. In some anaerobic organisms it is the sole source of energy.

GNATHOSTOMATA, a group of the phylum CHORDATA that is distinguished from the AGNATHA by the presence of jaws and which includes the CHONDRICHTHYES, OSTEICHTHYES, AMPHIBIA, REPTILIA, AVES and MAMMALIA.

GNEISS, broad class of coarse-grained METAMORPHIC

ROCKS with a banded, foliated structure and poor CLEAVAGE (see also SCHIST). Their composition is variable, but often approximates to that of GRANITE.

GNOTOBIOTICS, term used to describe laboratory organisms which are either free of all known contaminating organisms (e.g., BACTERIA, FUNGI, YEASTS)—and which are thus "germ-free"—or germ-free organisms specifically contaminated with a known organism. Animals which are not contaminated with a specific organism (but otherwise normal), the so-called Specific Pathogen Free (SPF) animals, are not gnotobiotic. SPF and gnotobiotic animals are widely used in medical research.

GODDARD, Robert Hutchings (1882–1945), US pioneer of rocketry. In 1926 he launched the first liquid-fuel rocket. Some years later, with a Guggenheim Foundation grant, he set up a station in N.M., there developing many of the basic ideas of modern rocketry: among over 200 patents was that for a multistage rocket. He died before his work received US Government recognition.

GÖDEL, Kurt (1906–), Austrian-born US mathematician who in 1931 proposed GÖDEL'S THEOREM, arguably the most significant mathematical achievement of the 20th century.

GÖDEL'S THEOREM, theorem showing the futility of attempting to set up a complete axiomatic formalization of mathematics. GÖDEL proved (1931 onward) that any consistent mathematical system must be incomplete; i.e., that in any system formulae must be constructed that can be neither proved nor disproved within that system. Moreover, no mathematical system can be proved consistent without recourse to axioms beyond that system. Gödel's Theorem has had profound effects on attitudes toward the foundations of MATHEMATICS. (See also LOGIC.)

GOETHITE, brown OXIDE mineral of composition FeO(OH); a major IRON ore of widespread occurrence, formed in bogs or by weathering of other iron minerals. Goethite is very similar to LIMONITE, but is crystalline (in the orthorhombic system). X-RAY DIFFRACTION analysis has shown that much supposed limonite is in fact goethite.

GOITER, enlargement of the THYROID gland in the neck, causing swelling below the LARYNX. It may represent the smooth swelling of an overactive gland in **thyrotoxicosis** or more often the enlargement caused by multiple CYSTS and nodules without functional change. **Endemic goiter** is enlargement associated with IODINE deficiency, occurring in certain areas where the element is lacking in the soil and water. Rarely, goiter is due to CANCER of the thyroid. If there is excessive secretion or pressure on vital structures SURGERY may be needed, although DRUG or RADIATION THERAPY for excess secretion are often adequate.

GOLD (Au), yellow NOBLE METAL in Group IB of the PERIODIC TABLE; a TRANSITION ELEMENT. Gold has been known and valued from earliest times and used for jewelry, ornaments and coinage. It occurs as the metal and as tellurides, usually in veins of QUARTZ and PYRITE; the chief producing countries are South Africa, the USSR, Canada and th US. The metal is extracted with CYANIDE or by forming an AMALGAM, and is refined by electrolysis. The main use of gold is as a currency reserve, a store of value. Like SILVER, it is used for its high electrical conductivity in printed circuits and electrical contacts, and also for filling or repairing teeth. It is very malleable and ductile, and may be beaten into GOLD LEAF or welded in a thin layer to another metal (rolled gold). For most uses pure gold is too soft, and is alloyed with other noble metals, the proportion of gold being measured in CARATS. Gold is not oxidized in air, nor dissolved by alkalis or pure acids, though it dissolves in AQUA REGIA or cyanide solution because of LIGAND complex formation, and reacts with the HALOGENS. It forms trivalent and monovalent salts. Gold (III) chloride is used as a toner in photography. AW 197.0, mp 1063°C, bp 2966°C, sg 19.32 (20°C).

GOLD, Thomas (1920–), Austrian-born US cosmologist who, with Hermann BONDI, proposed (1948) the steady-state model of the universe (see COSMOLOGY).

GOLDEN SECTION, a proportion of interest in classical GEOMETRY. If a straight line AB is cut at a point P so that AP:AB=PB:AP, the division is described as a golden section or divine proportion.

$$\frac{AP}{AB} \simeq 0.618,$$

the SEQUENCE 1/1, 1/2, 2/3, 3/5, 5/8, 8/13, 13/21, ..., the DENOMINATORS of whose terms form a FIBONACCI SERIES, providing successively closer approximations to this ratio.

GOLD LEAF, thin GOLD foil produced by beating gold ribbon placed between vellum and animal skins until the leaf is only 0.1μm thick. It is used for decorative gilding, lettering on leather-bound books, and for coating artificial satellites etc. to reflect infrared radiation.

GOLGI, Camillo (1844–1926), Italian histologist who developed a staining technique (1873) with which he was able to explore the NERVOUS SYSTEM in great detail. He shared with RAMÓN Y CAJAL the 1906 Nobel Prize for Physiology or Medicine.

GONADOTROPHINS, HORMONES secreted by the PITUITARY GLAND and PLACENTA, which stimulate the production of sex hormones by gonads: ESTROGEN and PROGESTERONE in females and ANDROGENS in males. They control the maturation and release of EGGS from the ovaries and the development of SPERM. Gonadotrophin secretion is controlled by releasing hormones from HYPOTHALAMUS.

GONADS, the reproductive organs of animals, which produce GAMETES. The female gonad is the OVARY, producing eggs, and the male gonad the testis (see TESTES), producing spermatozoa. (See also REPRODUCTION.)

GONDWANALAND, hypothetical S Hemisphere supercontinent formed after the split of PANGAEA (see also CONTINENTAL DRIFT; LAURASIA). Stratigraphic and FOSSIL evidence suggest it comprised what are now Antarctica, Australia, India, South America and other, smaller, units.

GONIOMETER, instrument for measuring ANGLES, especially those between CRYSTAL faces. The simplest form is the **contact goniometer**, a protractor whose base is laid against one face, a movable arm being turned until it contacts the adjacent face. The more accurate **reflecting goniometer** mounts the crystal axially on a graduated circle, or more usually two graduated circles, horizontal and vertical, rotatable independently. The crystal is rotated until each face in turn reflects a collimated light beam into a fixed telescope, and so the direction of the normal to each face is determined. (The term may also be used for the DIRECTION FINDER.)

GONORRHEA. See VENEREAL DISEASES.

GOODYEAR, Charles (1800–1860), US inventor of the process of VULCANIZATION (patented 1844). In 1839 he bought the patents of Nathaniel Manley Hayward (1808–1865), who had had some success by treating RUBBER with SULFUR. Working on this, Goodyear accidentally dropped a rubber/sulfur mixture onto a hot stove, so discovering vulcanization.

GOOSEFLESH, fanciful description of SKIN appearance in cold or acute anxiety; contraction of tiny muscles causes the HAIRS to be erected and the papillae to rise, looking like pimples. Cold prompts reflex contraction as a means of increasing skin insulation, while anxiety leads to stimulation of the muscles by the sympathetic NERVOUS SYSTEM.

GORGAS, William Crawford (1854–1920), US Army sanitarian. After Walter REED's commission had proved (1900) Carlos FINLAY's theory that YELLOW FEVER is transmitted by the MOSQUITO, Gorgas conducted in Havana a massive control program; he repeated this in Panama (1904–1913), facilitating the digging of the Panama Canal.

GOUT, a DISEASE of PURINE metabolism characterized by elevation of uric acid in the BLOOD and episodes of ARTHRITIS due to uric acid crystal deposition in SYNOVIAL FLUID and the resulting INFLAMMATION. Deposition of urate in CARTILAGE and subcutaneous tissue (as *tophi*) and in the KIDNEYS and urinary tract (causing stones and renal failure) are other important effects. The arthritis is typically of sudden onset with severe pain, often affecting the great toe first and large JOINTS in general. Treatment with allopurinol prevents recurrences.

GOVERNOR, device that maintains the speed of a machine constant despite load variation, often by controlling the fuel supply. The common flyball governor works by the centrifugal force of two rotating weights, acting against a spring. (See also SERVOMECHANISM.)

GRABEN. See RIFT VALLEY.

GRAD. See DEL; GRADIENT.

GRADIENT, in plane ANALYTIC GEOMETRY, an increase in y corresponding to a unit increase in x. The gradient of a CURVE may be found at any point along it by use of differential CALCULUS. Gradient is a VECTOR quantity. In vector notation, the gradient of a SCALAR field W is the vector field

$$\text{grad } W = \mathbf{i}\frac{\partial W}{\partial x} + \mathbf{j}\frac{\partial W}{\partial y} + \mathbf{k}\frac{\partial W}{\partial z},$$

that is, ∇W (see DEL).

GRAFFITO or **sgraffito**, from Italian, "scratching," in the visual arts a technique (and its results) in which a second covering of color is partially scraped away to reveal a primary covering of color below. In archaeology, the term graffito is used to mean a casual writing on an interior or exterior wall. Graffiti are found in great numbers on ancient Egyptian monuments, the walls of Pompeii, etc., and are of special interest in PALEOGRAPHY as they show the corruptions and transmutations of alphabetical characters. Ancient graffiti, like their modern counterparts, are mainly of a political or obscene nature.

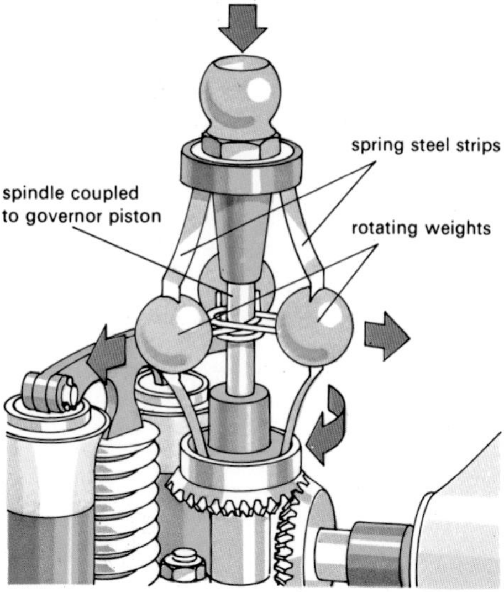

A flyball governor.

GRAFT, Surgical. See PLASTIC SURGERY; TRANSPLANTS.

GRAFTING, the technique of propagating plants by attaching the stem or bud of one plant (called the scion) to the stem or roots of another (the stock or rootstock). Only closely related varieties can be grafted. Roses and fruit trees are often grafted so that good flowering or fruiting varieties have the benefit of strong roots.

GRAHAM, Sylvester (1794–1851), US temperance advocate who recommended the use of coarsely-ground unsifted flour, often now called **graham flour**.

GRAHAM, Thomas (1805–1869), British chemist who formulated **Graham's Law**: the DIFFUSION rate of a gas is proportional to the inverse of the square ROOT of its density. While working further on diffusion and OSMOSIS he discovered the colloidal state, coining the term colloid (see COLLOID).

GRAIN, or caryopsis, a dry one-seeded FRUIT, usually containing a high percentage of starch, produced by, for example, corn, oats, barley, rye and other CEREAL CROPS. Grain crops have a high food value, store well and are a primary food stuff, contributing over half the world's calorie intake. (See also FLOUR.)

GRAIN (gr), the fundamental Anglo-American unit of weight, shared between the avoirdupois, troy and apothecaries' systems. The Imperial grain is defined equal to 0.064 798 91 grams exactly. (See WEIGHTS AND MEASURES.)

GRAM (g), the fundamental unit of mass in the CGS version of the METRIC SYSTEM. It approximates to the mass of a cubic centimetre of water.

GRAMICIDIN, an ANTIBIOTIC contained in the tyrothricin first prepared by DUBOS in 1939. It is rarely used today since less toxic alternatives exist.

GRAMMAR, the structures of language and of its

constituents; and the science concerned with the study of those structures. The grammarian concentrates on three main aspects of language: syntax, the ways that words are put together to form sentences; accidence, or morphology, the ways that words alter to convey different senses, such as past and present or singular and plural (see INFLECTION); and phonology, the ways that sounds are used to convey meaning.

Syntax. In English, the simplest sentence has a noun followed by a verb: "Philip thinks." More complicated is "Philip seldom thinks," where the verb is qualified by an adverb. In both of these, order is important: in "Seldom, Philip thinks" the change in order has brought about a change in meaning. In contrast, sentences of widely different outward form may have the same meaning (for example, using active and passive forms of the verb), and this suggests to many grammarians that superficial structure is not ultimately important, that there is a deep-lying structure of language which can be resolved into a few basic elements whose combinations can be used to produce an infinite number of sentences. Here grammatical studies are probing at the very roots of the human psyche; and ethnographical studies of the syntaxes of different languages, primitive and civilized, have been of primary importance in cultural ANTHROPOLOGY. (See also CHOMSKY; ETHNOGRAPHY.)

Accidence. Most English nouns have different endings for singular and plural: "knight" and "knights." Again, there is a change of ending for the genitive (possessive) case: "knight's" (the obsolete full form is "knightes") and "knights'." Most other cases are dealt with by prepositions: "to the knight"

Sentence analysis is an important exercise in grammar. Each sentence has a subject (the thing, person or idea the sentence is about) and a predicate (what is said about the subject). A clause is a group of words containing a subject and predicate but not forming a complete sentence by itself. A simple sentence has only one clause. A compound sentence has more than one clause of equal force. A complex sentence has one main clause and one or more subordinate clauses. A phrase is a small group of words equivalent to a noun, adjective or adverb.

Box analysis of the structure of a simple sentence

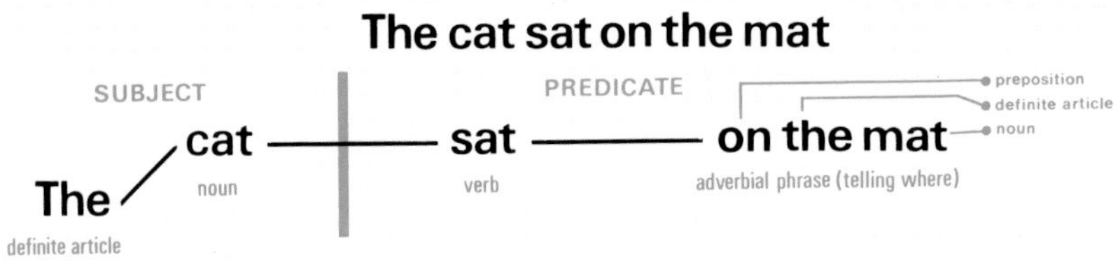

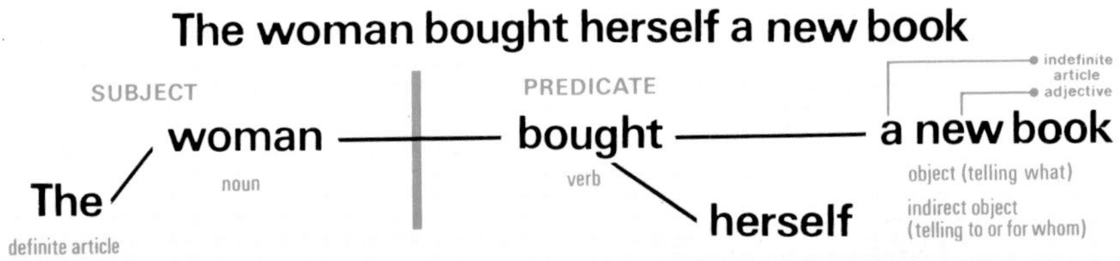

Box analysis of the structure of a complex sentence

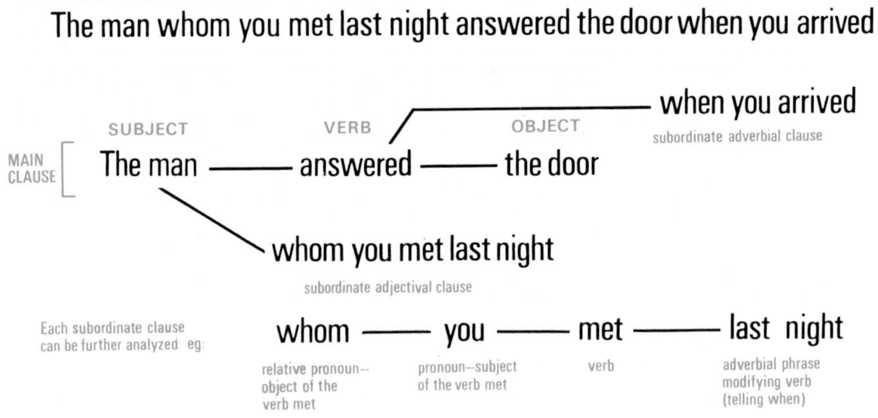

Grammar: the Parts of Speech

Noun: a word used as the name of a person, place, thing or concept (e.g.: Mary, house, cat, song)
Verb: a word that expresses action or a state of being (e.g.: runs, is)
Adjective: a word which qualifies (adds meaning to) a noun or pronoun (e.g.: good, happy)
Pronoun: a word used in place of a noun (e.g.: he, she, it, I, you, we, which, who)
Adverb: a word which modifies a verb or adjective (e.g.: swiftly, very)
Preposition: a word with a noun or pronoun to show its relation with some other word in the sentence (e.g.: on, by, at, from)
Conjunction: a word joining words, phrases or clauses (e.g.: and, but, although)
Interjection: an exclamatory word bearing no formal relation to the rest of the sentence (e.g.: oh! alas! whew!)
Definite article: the
Indefinite article: a or an

(dative); "from the knight" (ablative). Similarly, verb-endings are changed for two tenses only, past and present, the remainder being dealt with by use of the "auxiliary" verbs "to be" and "to have." Most other languages have a profusion of noun-and verb-endings to deal with different cases and tenses, and so have a lesser flexibility than English.

Phonology. Much of our speech depends for meaning on our tone of voice: "Philip is thinking" may have several meanings, depending on the stress placed on each of the words. These stresses are thus an important part of grammar, less so in English than in many other tongues: in the Sino-Tibetan languages, for example, a word may have two utterly different meanings depending upon the tone of voice in which it is said. (See also ETYMOLOGY; LANGUAGE; LINGUISTICS; MORPHEME; PHILOLOGY; PHONEME; PHONETICS; PRONUNCIATION; PUNCTUATION; SEMANTICS.)

GRAM'S STAIN, a stain for BACTERIA which divides them into Gram-positive and Gram-negative groups. Since the cell walls determine not only the staining difference but also behavior and ANTIBIOTIC sensitivity of bacteria, the stain has considerable medical value.

GRAND MAL. See EPILEPSY.

GRANIT, Ragnar Arthur (1900–), Finnish-born Swedish physiologist who shared the 1967 Nobel Prize for Physiology or Medicine with H. K. HARTLINE and G. WALD. Granit demonstrated that individual nerves in the EYE could distinguish light of different colors.

GRANITE, coarse- to medium-grained plutonic IGNEOUS ROCK, composed of FELDSPAR (orthoclase and microcline predominating over plagioclase) and QUARTZ, often containing BIOTITE and AMPHIBOLE. It is the type of the family of **granitic rocks**, plutonic rocks rich in feldspar and quartz, of which the CONTINENTS are principally made. Most granite was formed by crystallization of MAGMA, though some is METAMORPHIC, and some was formed by replacement ("granitization"). It occurs as DIKES and SILLS, large masses, and enormous BATHOLITHS. A hard, weather-resistant rock, usually pink or gray, granite is used for building, paving and road curbs.

GRANULATION TISSUE, the bright red, granular tissue that develops during healing. The tissue consists initially of fine blood vessels and therefore bleeds easily. Later, fibrous tissue is laid down and a SCAR replaces the granulation tissue.

GRAPHITE, allotrope of CARBON (see ALLOTROPY), forming soft, black, metallic crystals, in which the atoms are arranged in layers of hexagons that easily slide over each other. It is found naturally in GNEISS and SCHIST, and synthesized from COKE. Graphite is a good conductor of HEAT and ELECTRICITY. It is used for ELECTRODES, nuclear-reactor moderators and lubricants. subl 3660°C, sg 2.25 (20°C).

GRAPHOLOGY, the study of handwriting, particularly the deduction, from its form, of information about the character of the writer.

GRAPHS, plottings of sets of points whose coordinates are of the form $(x, f(x))$, where $f(x)$ is a FUNCTION of x (see ANALYTIC GEOMETRY). These points may define a CURVE or straight LINE. Graphs are a powerful tool of STATISTICS, since it is often profitable to plot one variable (such as age) along one axis, against another (such as height) plotted along the other (see also NORMAL DISTRIBUTION): the points on statistical graphs need not define a continuous curve (see HISTOGRAM). The AXES on a graph are not always marked off regularly: in some cases it is useful to mark off one or both on a nonlinear scale—e.g., using logarithmic (see LOGARITHM) or exponential (see EXPONENT) scales.

GRAPTOLITES, extinct marine organisms that occur in rocks of Carboniferous to Cambrian age, 280–570 million years old. Some zoologists believe that graptolites are the ancestors of the CHORDATA.

GRASSLAND, the areas of the earth whose predominant type of vegetation consists of grasses,

The structure of graphite. This structure comprises plane layers of atoms, disposed in hexagons spatially and electronically similar to aromatic ring structure, the layers being stacked in an alternating pattern, ABA . . ., 0.34nm apart. The C—C distance within the layers is 0.142nm.

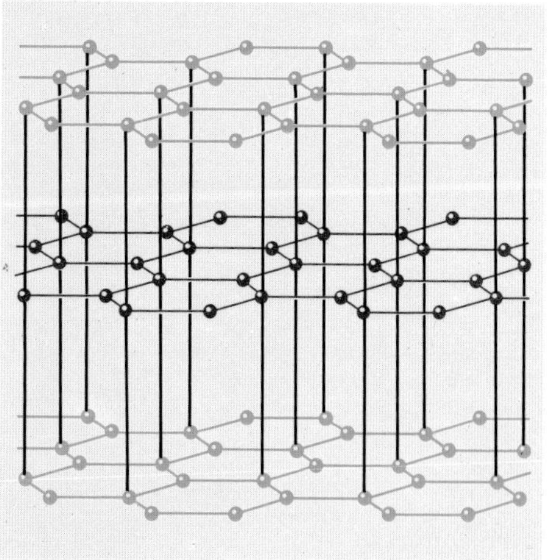

rainfall being generally insufficient to support higher plant forms. There are three main types: SAVANNA, or tropical grassland, has coarse grasses growing 1m to 4m high, occasional clumps of trees and some shrubs; it is found in parts of Africa and South America. PRAIRIE has tall, deep-rooted grasses and is found in Middle and North America, Argentina, the Ukraine, South Africa and N Australia. STEPPES have short grasses and are found mainly in Central Asia. Grasslands are of great economic importance as they provide food for domestic animals and often excellent cropland for cultivation.

GRAVEL, in geology, a collection of rock particles whose diameter ranges from 2mm to 4mm. In general terms, gravel particles may be as large as pebbles. Gravel is used commercially in the making of CONCRETE (see also CONGLOMERATE; SAND).

GRAVES, Robert James (1796–1853), Irish physician remembered for his work on exophthalmic GOITRE (Graves' disease).

GRAVIMETER, an instrument for detecting small variations in the earth's gravitational field, frequently used in mineral and oil prospecting. Variations in the gravitational FORCE on a weight suspended from a SPRING cause it to stretch or be deflected in a way which is then measured.

GRAVIMETRIC ANALYSIS, method of quantitative chemical ANALYSIS in which the substance to be estimated is converted to a substance which can be separated pure and entire, and which is then weighed. Commonly a highly insoluble precipitate is formed, filtered off, washed and dried (see DEHYDRATION). The weight of substance sought is calculated from the weight and composition of the precipitate.

GRAVITATION, one of the fundamental forces of nature, the force of attraction existing between all MATTER. It is much weaker than the nuclear or electromagnetic forces and plays no part in the internal structure of matter. Its importance lies in its long range and in its involving all masses. It plays a vital role in the behavior of the UNIVERSE: the gravitational attraction of the SUN keeps the PLANETS in their orbits, and gravitation holds the matter in a STAR together. NEWTON's **law of universal gravitation** states that the attractive FORCE F between two bodies of MASSES M_1 and M_2 separated by distance d is $F = GM_1M_2/d^2$ where G is the **Universal Gravitational Constant** (6.670×10^{-11} N m^2 kg^{-2}). The force of gravity on the earth is a special case of the attraction between masses and causes bodies to fall toward the center of the earth with a uniform ACCELERATION $g = GM/R^2$ where R and M are the radius and mass of the earth. Assuming, with Newton, that the inertial mass of a body (that which is operative in the laws of motion) is identical with its gravitational mass, application of the second law of motion gives the WEIGHT of a body of mass m, the force with which the earth attracts that body, as mg. Bodies on the earth and moon thus have the same mass but different weights. Again, the gravitational force on a body is proportional to its mass but is independent of the type of material it is. Newton's theory explains most of the observed motions of the planets and the TIDES and is still sufficiently accurate for most applications. The Newtonian analysis of gravitation remained unchallenged until, in the early 20th century, EINSTEIN introduced radically new concepts in his theory of general RELATIVITY. According to this, mass deforms the geometrical properties of the space around it. Einstein reaffirmed Newton's assumption regarding the equivalence of gravitational and inertial mass, proposing that it was impossible to distinguish experimentally between an accelerated coordinate system and a local gravitational field. From this he predicted that LIGHT would be found to be deflected toward massive bodies by their gravitational fields and this effect indeed was observed for starlight passing close to the sun. It was also predicted that accelerated matter should emit gravitational waves with the velocity of light but the existence of these has not as yet been demonstrated.

GRAVITY. See GRAVITATION.

GRAVURE. See PRINTING.

GRAY, Asa (1810–1888), the foremost of 19th-century US botanists. Being a prominent Protestant layman, his advocacy of the Darwinian thesis carried special force. However, he never accepted the materialist interpretation of the evolutionary mechanism and taught that NATURAL SELECTION was indeed consistent with a divine TELEOLOGY.

GRAY MATTER, the parts of BRAIN that are rich in nerve-cell bodies, as opposed to white matter which is mainly nerve fibers, sheathed by MYELIN. The cerebral cortex, basal ganglia, nuclei of the brain stem and the center of the SPINAL CORD are major gray areas.

GRAYWACKE. See CONGLOMERATE.

GREASE. See LUBRICATION.

GREAT BEAR (Ursa Major), a large N Hemisphere constellation containing the seven bright stars known as the **Plow** or **Big Dipper**. Two of these, the **Pointers**, form roughly a straight line with POLARIS and are hence of navigational importance. Five stars of the Plow are, with SIRIUS, members of a widely separated GALACTIC CLUSTER.

GREAT CIRCLE ROUTES, routes of prime importance in air travel as they describe the shortest distances between two points on the earth's surface. Great circles are circles on the surface of a sphere whose centers coincide with the center of the sphere (see SPHERICAL GEOMETRY): the EQUATOR is an example (see also LATITUDE AND LONGITUDE). Great circle navigation is aided by use of MAPS drawn to a **gnomonic projection**, where the center of PERSPECTIVE is the center of the earth. On such maps, great circles appear as straight lines.

GREEK FIRE, liquid mixture of unknown composition that took fire when wet, invented by a Syrian refugee in Constantinople in the 7th century AD and used by the Byzantine Empire and others for the next 800 years. Thrown in grenades or discharged from syringes, it wrought havoc in naval warfare until superseded by gunpowder. It appears to have been a petroleum-based mixture. (See also INCENDIARY BOMB.)

GREEN, Thomas Hill (1836–1882), English idealist philosopher who was the leading critic of the empiricist philosophies of J. S. MILL and H. SPENCER in mid-Victorian Oxford. His influence long survived his death, declining only with the resurgence of the empirical approach in the 20th century.

GREENHOUSE, building used for growing plants in a controlled environment, protecting them from extreme heat and cold. At first, in the 17th century,

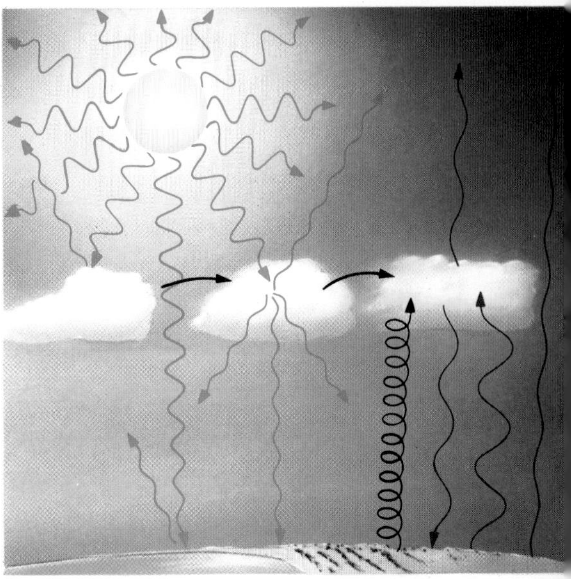

The greenhouse effect can be explained with the aid of Wein's "displacement law". This states that the wavelength at which a black body radiates most intensely varies inversely with its absolute temperature. Thus the radiation originating in the hot sun is of much shorter wavelength than that radiated from the cool earth or its atmosphere. Since the atmosphere, particularly when laden with water vapor, is far more opaque to the long-wave radiation characteristic of the earth than it is to incoming solar radiation, it tends to absorb the former radiation and reradiate it, largely back toward the surface, ensuring that the earth's surface is maintained at a somewhat higher temperature than would be the case were all the energy radiated from the surface lost directly into space. Actually, less than half the short-wave solar radiation arriving at the top of the atmosphere is absorbed at the earth's surface. Much scattered into space by minute particles in the atmosphere or absorbed by atmospheric dust, ozone, carbon dioxide and water vapor. This last, absorbed energy becomes involved in the long-wave-length radiation processes. Some energy is transfered from the surface to the atmosphere by convection and as latent heat of vaporization of water.

they were built of brick or wood, but later glass in a wood or metal frame was used for light, and today transparent plastic is common. Greenhouses are heated artificially and by the sun, whose radiation passes through the glass and is absorbed by the objects inside, which warm the air by convection; some heat is reradiated at longer wavelengths but this is trapped by the glass.

GREENHOUSE EFFECT, a phenomenon whereby the temperature at the earth's surface is some 18C° warmer than would otherwise be the case. Sunlight radiated at visible and near-ultraviolet wavelengths provides most of the earth's energy income. After absorption it is reradiated, but at longer, infrared wavelengths, the earth being much cooler than the sun (see BLACKBODY RADIATION). Although the ATMOSPHERE is transparent to the incoming solar radiation, that reradiated from the earth's surface is strongly absorbed by atmospheric water vapor and carbon dioxide. That absorbed is again reradiated, the majority back toward the surface. A similar effect may account for VENUS' high surface temperature.

GREENSAND. See GLAUCONITE.

GREEN'S THEOREM, in VECTOR ANALYSIS, an extension of the divergence theorem, which latter states

$$\iiint \nabla . \mathbf{V}\, dt = \iint \mathbf{V} . d\mathbf{S}$$

where $\mathbf{V}$ is a VECTOR and $d\mathbf{S}$ an element of a surface. If $\mathbf{V}$ has components (x, y, z), for a volume v and a surface S, Green's theorem states

$$\iiint (x\nabla^2 y - y\nabla^2 x)\, dv = \iint (x\nabla y - y\nabla x) . d\mathbf{S}.$$

(See also CALCULUS; DEL; DIV; SURFACE INTEGRAL.)

GREEN VITRIOL, or Iron (II) Sulfate. See IRON.

GREENWICH MEAN TIME (GMT), local mean time along the Greenwich meridian, used as a TIME standard throughout the world.

GREENWICH OBSERVATORY, Royal, observatory established in 1675 at Greenwich, England, by Charles II to correct the astronomical tables used by sailors and otherwise to advance the art of NAVIGATION. Its many famous directors, the "astronomers royal," have included J. FLAMSTEED (the first), E. HALLEY, and Sir George Airy. The original Greenwich building, now known as Flamsteed House and run as an astronomical museum, was designed by Sir C. WREN. The Observatory is presently sited at Herstmonceux, Sussex, to where it was moved 1948–1957. The Observatory itself is thus no longer sited on the Greenwich meridian, the international zero of longitude.

GREGORIAN CALENDAR. See CALENDAR.

GRENADE, small BOMB (of all sorts) thrown by hand or propelled by a grenade launcher (a specialized gun), used in warfare at short range.

GREW, Nehemiah (1641–1712), English plant anatomist and physician who introduced the term "comparative" anatomy (1676) and was a principal founder of his science. His main work, *The Anatomy of Plants*, appeared in 1682.

GRIGNARD, Francois Auguste Victor (1871–1935), French chemist who shared with SABATIER the 1912 Nobel Prize for Chemistry for his discovery of GRIGNARD REAGENTS, complex compounds used in many organic syntheses.

GRIGNARD REAGENTS, ORGANOMETALLIC COMPOUNDS of great importance in laboratory and industrial chemical synthesis. Made by reacting ALKYL (or aryl) HALIDES with MAGNESIUM in an ETHER solution, they are commonly represented as RMgX where R is an alkyl (see ALKANES) or aryl (see AROMATIC COMPOUNDS) group. They are powerful NUCLEOPHILES, and react with many compounds, introducing the group R.

GRIPPE. See INFLUENZA.

GROUND, Electrical, an electrical connexion between apparatus and the earth or an equivalent conducting body at zero POTENTIAL. Electricity supply systems are grounded to avoid overvoltage and to improve performance. Metal cases, frames etc. of electrical equipment are grounded to minimize risk of ELECTRIC SHOCKS in case a fault should make the exposed part "live."

GROUND-EFFECT MACHINE (GEM). See AIR-CUSHION VEHICLE.

GROUNDWATER, water accumulated beneath the earth's surface in the pores of rocks, spaces, cracks, etc. It may be *meteoric*, rainwater having soaked down from above, or *juvenile*, where water has risen from beneath. Permeable, water-bearing rocks are AQUIFERS; rocks with pores small enough to inhibit the flow of water through them are aquicludes. Build-up of groundwater pressure beneath an aquiclude makes possible construction of an ARTESIAN WELL. The uppermost level of groundwater saturation is the water table. (See also PERMAFROST; SPRING; WELL.)

GROUP, a set of algebraic elements in which there is an operation * such that: (1) for all elements a, b, in the set, * is associative (see ALGEBRA) and $a*b$ is a member of the set; (2) there is an identity element e defined by $a*e=a$ for every element a of the set; (3) every element a has an inverse a^{-1}, also a member of the set, where $a*a^{-1}=e$. If * is commutative, the group is an ABELIAN GROUP. The set of all INTEGERS under the operation ADDITION (+) is such a group; it is not, however, a group under, say, DIVISION, since division is not associative. (See also FIELD; RING; SET THEORY.)

GROUP, in psychology, a collection of individuals that can be regarded as a single unit. The behavior of a group (usually a social unit) or an individual acting in response to his membership of the group is termed **group behavior**; the stimulus that produces such response being **group consciousness**. The study of group behavior and group consciousness is termed **group psychology** or **social psychology**: important factors include the presence of an exterior common enemy, and identification of the individual with not only the group but also another individual within it regarded as leader. An application of this to SOCIOLOGY is **group dynamics**, whose chief proponent was LEWIN. In his field theory he analogized the forces acting on an individual in a group at any given time to a VECTOR FIELD; and extended this to treat the group as a whole. Other key concepts are locomotion (the aims of the group and their achievement of them); cohesiveness (the field of forces binding each member to the group); and communication between members, the nature and extent of which determines the group's structures, hierarchy and cohesiveness. **Group therapy** is a technique of PSYCHOANALYSIS in which several patients are treated by an analyst simultaneously, with the aim that individuals within the group will assist each other in the treatment; recent amateur applications have tended to bring the technique into popular disrepute. The term "group" is also used in GESTALT PSYCHOLOGY to describe a pattern of PERCEPTIONS. Found in primitive societies (see PRIMITIVE MAN) and occasionally more advanced ones is **group marriage**, where a number of individuals of each sex marry in common.

GROWING SEASON, the period of the year in which most active plant growth occurs. In temperate zones it lasts from the final frost of spring through to the first frost of autumn. The length of the growing season in a particular year or in a particular area is of agricultural importance.

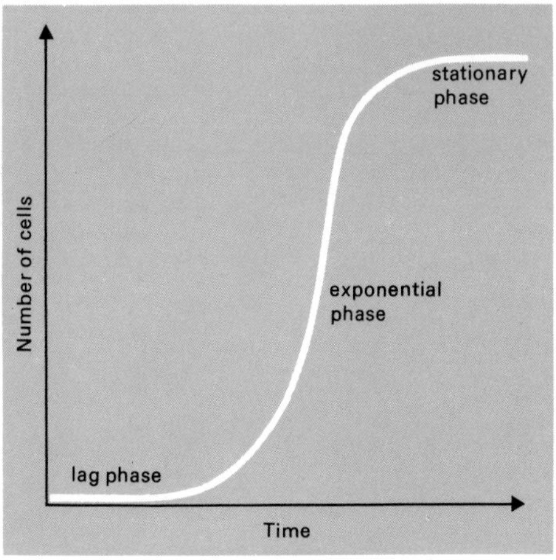

The S-shaped growth curve typical of the growth of the number of cells in a bacterial culture.

GROWTH, the increase in the size of an organism, reflecting either an increase in the number of its CELLS, or one in its protoplasmic material, or both. Cell number and protoplasmic content do not always increase together; cell division can occur without any increase in PROTOPLASM giving a larger number of smaller cells. Alternatively, protoplasm can be synthesized with no cell division so that the cells become larger. Any increase in protoplasm requires the synthesis of cell components such as nuclei, mitochondria, thousands of enzymes, and cell membrane. These require the synthesis of macromolecules such as PROTEINS, NUCLEIC ACIDS and polysaccharides from AMINO ACIDS, SUGARS and fatty acids. These subunits must be synthesized from still simpler substances or obtained from the environment. **Growth curves**, which plot time against growth (such as the number of cells in a bacterial culture, the number of human beings on earth, the size or weight of a plant seedling, an animal or an organ of an animal) all have a characteristic S-shape. This curve is divided into three parts: the lag phase, during which cells prepare for growth; the exponential phase when actual growth occurs, and the stationary phase when growth ceases. The time any particular cell or group of cells remains in any phase depends on their type and the particular condition prevailing. The *lag phase* represents a period of rapid growth of protoplasm so that the cells become larger without any increase in their number. The duration of the lag phase depends on the resynthesis of the enzyme systems required for growth and the availability of the necessary raw materials. Basically each original cell must obtain sufficient components to form two new cells. During the *exponential phase*, each cell gives rise to two cells, the two to four and so on, so that the number of cells after n generations is 2^n. The generation or doubling time for any particular cell is constant throughout the exponential phase. The time for organisms to double their mass ranges from 20 min for some bacteria to 180 days for a human being at birth. If exponential growth were unlimited, one bacterial cell in 24 hours would give rise to some 4 000 tonnes of bacteria. However, the exponential growth usually ceases (giving the *stationary phase*) either because of lack of an essential nutrient or because waste products produced by the cells pollute the environment. Again, in higher animals population growth is often slowed by parasite-carried epidemics.

The S-type growth pattern can be readily seen in unicellular organisms. Although growth in organisms containing different types of cells obeys the same basic rules, the relationships of the different types of cells complicate the pattern. But although all parts of a multicellular organism do not grow at the same rate or stop growing at the same time, the overall growth curve is still S-shaped.

GUANCHES, STONE AGE culture first found by the Conquistadores (15th century) in the W group of the Canary Islands. Thought to have been of Cro-Magnon (see CRO-MAGNON MAN) origin, possibly having come from E or Central Europe in ancient times, they no longer exist as a distinct race.

GUANINE, or 2-amino-6-oxypurine, a PURINE first found in GUANO. It is an important component of several NUCLEOTIDES.

GUANO, a naturally occurring fertilizer composed mainly of the excrement of sea birds; bat and seal guanos are also used. Because of the cheapness of other FERTILIZERS, guano is now little used. (See also MANURE.)

GUERICKE, Otto von (1602–1686), German physicist credited with inventing the vacuum pump. His best-known experiment was with the Magdeburg hemispheres (1654): he evacuated a hollow sphere composed of two halves placed together, and showed that two 8-horse teams were insufficient to separate the halves. (See PUMPS.) He is also credited with inventing the first electric GENERATOR.

GUÉRIN, Camille (1872–1961), French bacteriologist who, with CALMETTE, developed the BCG VACCINE, which is used to counter TUBERCULOSIS.

GUILLAUME, Charles Édouard (1861–1938), Swiss-born French physicist best known for discovering INVAR, an iron-nickel alloy which expands and contracts only very slightly with temperature change. For his work on ferronickels he was awarded the 1920 Nobel Prize for Physics.

GUILLEMIN, Roger (1924–   ), French-born UK physiologist who shared the 1977 Nobel Prize for Physiology or Medicine with R. S. YALLOW and A. V. SCHALLY for his work identifying compounds stimulating peptide HORMONE release from the brain.

GULF STREAM, warm ocean current flowing N, then NE, off the E coast of the US. Its weaker, more diffuse continuation is the E flowing **North Atlantic Drift**, which is responsible for warming the climates of W Europe. The current, often taken to include also the Caribbean Current, is fed by the N Equatorial Current, and can be viewed as the western part of the great clockwise water circulation pattern of the N Atlantic. (See also OCEAN CURRENTS.)

GULLSTRAND, Allvar (1862–1930), Swedish opthalmologist awarded the 1911 Nobel Prize for Physiology or Medicine for his work on the REFRACTION of light in the EYE. Von HELMHOLTZ had shown that the lens's surface curvature altered for focusing: Gullstrand showed that also the internal components of the lens adjust, accounting for about a third of the accommodation.

GUM, sticky substances containing CARBOHYDRATES, exuded from some trees, shrubs, seeds and seaweeds. Gum arabic is produced by the African tree *Acacia senegal*. AGAR is a dried mucilaginous gum extracted from seaweeds.

GUN, weapon able to project a missile by means of an EXPLOSIVE charge (see AMMUNITION; BALLISTICS). Heavy guns, or ARTILLERY, include cannons, howitzers and mortars; lighter guns such as machine guns, pistols, revolvers and rifles count as FIREARMS.

GUNCOTTON, a NITROCELLULOSE with a high nitrate content; an EXPLOSIVE.

GUN METAL, a type of BRONZE, normally 88% copper, 10% tin, 2% zinc. Formerly used for cannons, it is now used for gears, bearings and steam fittings, being wear- and corrosion-resistant.

GUNPOWDER, or **black powder**, a low EXPLOSIVE, the only one known from its discovery in the West in the 13th century until the mid-19th century. It consists of about 75% POTASSIUM (or SODIUM) nitrate, 10% SULFUR and 15% CHARCOAL; it is readily ignited and burns very rapidly. Gunpowder was used in fireworks in 10th-century China, as a propellant for firearms from the 14th century in Europe and for blasting since the late 17th century. It is now used mainly as an igniter, in fuses and in fireworks.

GUTENBERG, Johann (c1400–1468), German printer, usually considered the inventor of PRINTING from separately cast metal types. By 1450 he had a press in Mainz, financed by **Johann Fust** (c1400–c1466) but in 1455 he handed over the press (and his invention) to Fust in repayment of debts. By now the Gutenberg (or Mazarin) Bible was at least well under way: each page has two columns of 42 lines. Gutenberg possibly founded another press some time later.

GUTTA PERCHA, brownish, leathery solid used in the manufacture of golf balls, in dentistry and to insulate marine cables and electrical equipment. It is prepared from the latex obtained from trees native to Malaysia.

GUYOT, submarine table-mountain, found especially in the Pacific. It is thought that guyots originate as volcanic islands associated with MID-OCEAN RIDGES. Wave EROSION reduces the island, and SEA-FLOOR SPREADING causes the guyot to be further submerged. (See also VOLCANISM.)

GYMNOSPERMS, the smaller of the two main classes of seed-bearing plants, the other being the ANGIOSPERMS. Gymnosperms are characterized by having naked seeds usually formed on open scales produced in cones. All are perennial plants and most are EVERGREEN. There are several orders, the main ones being the Cycadales, the cycads or sago palms; the Coniferales, including pine, larch, fir and redwood; the Ginkgoales, the ginkgo; and the Gnetales, tropical shrubs and woody vines.

GYNECOLOGY, branch of MEDICINE and SURGERY, specializing in diseases of women, specifically disorders of female reproductive tract; often linked with OBSTETRICS. CONTRACEPTION, ABORTION, STERILIZATION, infertility and abnormalities of MENSTRUATION are the commonest problems. The early recognition and treatment of CANCER of the WOMB cervix after PAPANICOLAOU smears have become

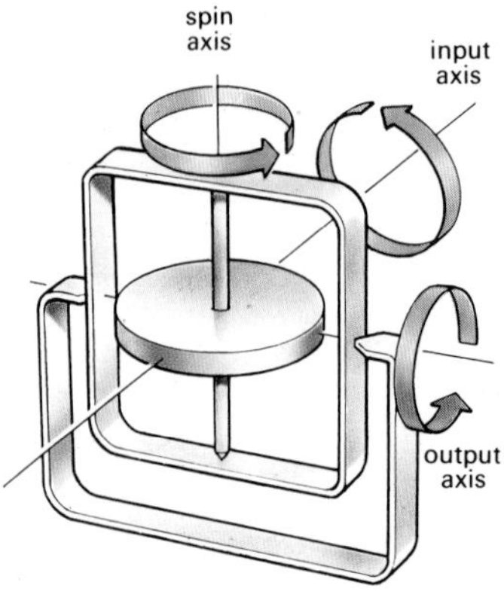

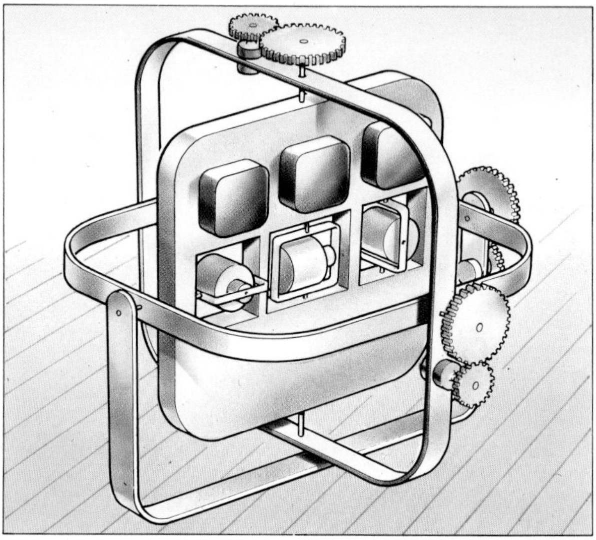

A gyroscope mounted in double gimbals (*left*) demonstrates the phenomenon of gyroscopic intertia. If the gyroscope is spinning in the sense indicated and the indicated torque is applied to the input axis, the gyroscope and the inner gimbal will tend to rotate, also in the sense indicated, about the output axis. This phenomenon is utilized in the gyrostabilized platform (*right*) where three gryroscopes are kept spinning about mutually perpendicular axes. Any torque tending to alter the orientation of the platform relative to its original orientation is "sensed" by at least one of the gyroscopes and a servomechanism system comes into operation restoring the platform to its original orientation whatever the new orientation of the mounting. Instrument packages attached to the platform are thus kept in a constant orientation.

important. Other TUMORS of womb or ovaries, benign or malignant, and disorders of genital tract or closely related BLADDER following PREGNANCY, commonly require gynecological surgery. Dilatation of the cervix and curettage of womb endometrium (D and C) is used frequently for diagnosis and sometimes for treatment of menstrual disorders or postmenopausal bleeding. HYSTERECTOMY or removal of the womb is the commonest major operation of gynecologists.

GYNOECIUM. See FLOWER.

GYPSUM, mineral consisting of CALCIUM sulfate dihydrate. It occurs worldwide as monoclinic crystals of various colors (SELENITE), or as fibrous or massive forms (ALABASTER). Gypsum is used in building and CEMENT.

GYROCOMPASS, a continuously-driven GYROSCOPE which acts as a COMPASS. It is unaffected by magnetic variations and is used for steering large ships. As the earth rotates the gyroscope experiences a TORQUE if it is out of the meridian. The resulting tilting is sensed by a gravity sensing system which itself applies a torque to the gyroscope which returns it to the N–S meridian. The sensitivity of such instruments decreases with latitude away from the equator.

GYROPILOT, an automatic device for keeping a ship or airplane on a given course using signals from a gyroscopic reference. The marine version operates a ship's rudder by displacement signals from the GYROCOMPASS. In an airplane, the device is usually known as an **automatic pilot** and consists of sensors to detect deviations in direction, pitch and roll, and pass signals via a computer to alter the controls as necessary.

GYROSCOPE, a heavy spinning disk mounted so that its axis is free to adopt any orientation. Its special properties depend on the principle of the conservation of angular MOMENTUM. Although the scientific gyroscope was only devised by FOUCAULT in the mid-19th century, the child's traditional spinning top demonstrates the gyroscope principle. The fact that it will stay upright as long as it is spinning fast enough demonstrates the property of **gyroscopic inertia**: the direction of the spin axis resists change. This means that a gyroscope mounted universally, in double gimbals, will maintain the same orientation in space however its support is turned, a property applied in many navigational devices. If a FORCE tends to alter the direction of the spin axis (e.g., the weight of a top tilting sideways), a gyroscope will turn about an axis at right-angles to the force for as long as it is applied; this movement is known as **precession**. Instrument gyroscopes usually consist of a wheel having most of its mass concentrated at its rim to ensure a large moment of inertia and which is kept spinning in frictionless bearings by an electric motor. Once the wheel is set spinning its response to applied TORQUES can be monitored or used in control servomechanisms.

GYROSTABILIZER, a gyroscopic device for stabilizing a ship, airplane or instrument mounting. Originally giant gyroscopes (up to 4m in diameter) were used to counteract roll in ships, but they were found to be too cumbersome. Now fins protruding from the ship's hull are moved hydraulically to oppose roll under the control of signals from small GYROSCOPES that sense roll angle and velocity.

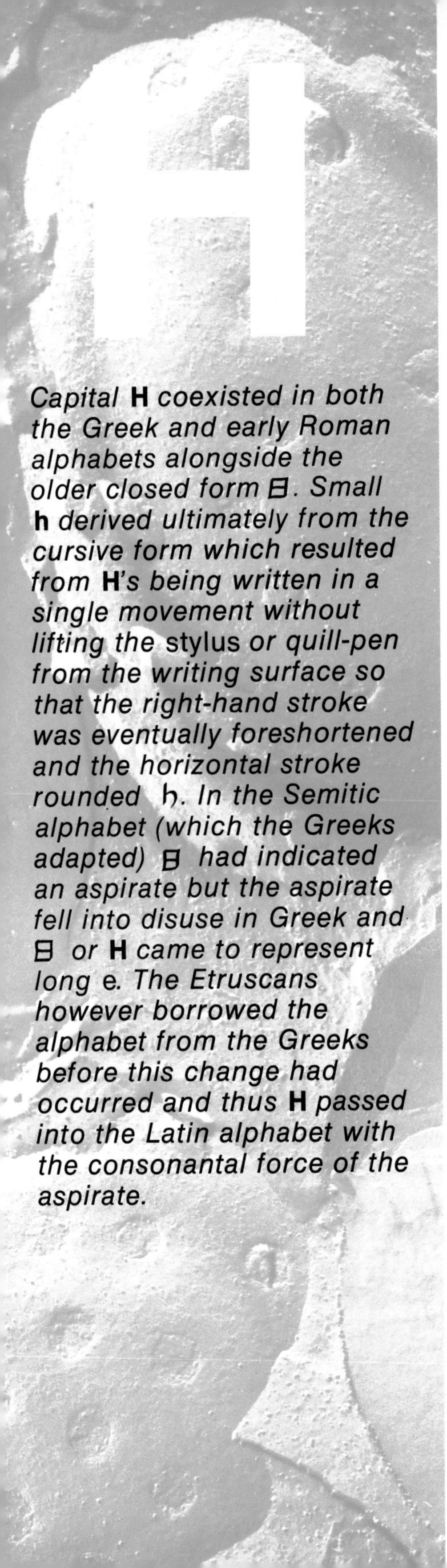

Capital **H** *coexisted in both the Greek and early Roman alphabets alongside the older closed form* 日*. Small* **h** *derived ultimately from the cursive form which resulted from* **H**'s *being written in a single movement without lifting the stylus or quill-pen from the writing surface so that the right-hand stroke was eventually foreshortened and the horizontal stroke rounded* h*. In the Semitic alphabet (which the Greeks adapted)* 日 *had indicated an aspirate but the aspirate fell into disuse in Greek and* 日 *or* **H** *came to represent long e. The Etruscans however borrowed the alphabet from the Greeks before this change had occurred and thus* **H** *passed into the Latin alphabet with the consonantal force of the aspirate.*

HABER, Fritz (1868–1934), German chemist awarded the 1918 Nobel Prize for Chemistry for synthesizing AMMONIA from the elements nitrogen and hydrogen. (See also HABER PROCESS.)

HABER PROCESS, industrial synthesis of AMMONIA invented by HABER and developed by BOSCH. NITROGEN (from the atmosphere) is mixed with HYDROGEN (from natural gas or water gas) and heated to about 500°C under 200–1000atm pressure, with a catalyst of finely divided iron containing aluminum oxide and potassium oxide. The ammonia formed is frozen out, and the unreacted gases recycled. See also NITROGEN FIXATION.

HABIT, in psychology, a REFLEX response to a frequently experienced stimulus; e.g., lighting a cigarette before using the telephone, or singing in the bath. The term is sometimes used loosely for CONSCIOUS reactions to situations. **Habit interference** is the conflict within an individual between two responses to a situation that differs slightly from one to which he has a habitual response. The process of acquiring a habit is **habit formation**. (See also DRUG ADDICTION.)

HABITAT, an area with certain physical characteristics which support a particular community of animals and plants. In general, a habitat can be defined in physical terms, e.g., rocky seashore or sandy desert, but as far as any one animal or plant species is concerned, the habitat cannot be defined without reference to the other animals and plants in the community. (See also ECOLOGY.)

HADRONS, a class of SUBATOMIC PARTICLES including the BARYONS and the mesons. They are influenced by strong interactions (the FORCES binding PROTONS and NEUTRONS within the nucleus), GRAVITATION and, if charged, electromagnetic forces. Protons and neutrons are relatively stable but other hadrons (e.g., π-mesons) produced by collision processes are short-lived.

HAECKEL, Ernst Heinrich (1834–1919), German biologist best remembered for his vociferous support of DARWIN's theory of EVOLUTION, and for his own theory that ontogeny (the development of an individual organism) recapitulates phylogeny (its evolutionary stages), a theory now discarded. (See ONTOGENY AND PHYLOGENY.)

HAFNIUM (Hf), hard TRANSITION ELEMENT in Group IVB of the PERIODIC TABLE, found in ores of ZIRCONIUM, which it closely resembles. It is used in nuclear reactor control rods. AW 178.5, mp 2150°C, bp c4000°C, sg 13.31 (20°C).

HAHN, Otto (1879–1968), German chemist awarded the 1944 Nobel Prize for Chemistry for his work on nuclear FISSION. With Lise MEITNER he discovered the new element PROTACTINIUM (1918); later they bombarded URANIUM with NEUTRONS, treating the uranium with ordinary barium. Meitner showed that the residue was radioactive BARIUM formed by the splitting (fission) of the uranium nucleus.

HAHNEMANN, Christian Friedrich Samuel (1755–1843), German physician, the father of HOMEOPATHY, which has as its basis the fact that the induction by drugs of the symptoms of a disease appears to immunize an individual against that disease.

HAHNIUM (Ha), a TRANSURANIUM ELEMENT in

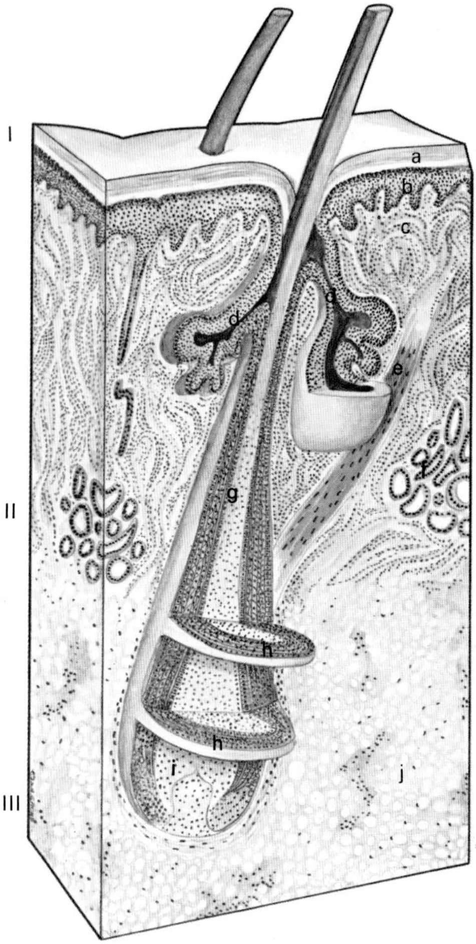

The structure of the hair and skin. (I) Epidermis; (II) dermis; (III) subdermal tissue. (a) Stratum corneum; (b) stratum papillare cutis; (c) connective tissue; (d) sebaceous gland; (e) muscle fiber in the skin; (f) tiny blood vessels; (g) shaft of hair; (h) cross section of hair; (i) hair papilla; (j) fatty tissue.

Group VB of the PERIODIC TABLE, atomic number 105. The priority of its discovery is disputed between the USSR and the US, as in the case of RUTHERFORDIUM. American scientists claimed in 1970 to have synthesized hahnium-260 (half-life 1.6s) by bombarding californium-249 with nitrogen-15 ions.

HAIL, PRECIPITATION of pellets of ice, often associated with thunderstorms. Hailstones have diameters of 2–250mm, 2–5mm being most common. They require a strong updraft, raising them to colder regions, to form. Often this happens several times, the hailstone collecting a new layer of ice each time it rises, until it is too heavy to support and falls to the ground. Larger hailstones may have alternate layers of clear and white ice, due to different rates of freezing. (See also ICE; RAIN; SNOW.)

HAIR, nonliving filamentous structure made of KERATIN and pigment, formed in the skin hair FOLLICLES. Facial and genetic factors determine both coloring and shape (by heat-labile sulfur bridges). In man all skin surfaces except the palms and soles are covered with very fine hair. This assists in TOUCH reception. In the cold, these hairs are erected (see GOOSEFLESH) to create extra insulation. Scalp hair is prominent in man. Pubic and axillary hair develop at PUBERTY in response to sex HORMONES and their patterns differ in the sexes; facial hair is ANDROGEN-dependent. Hair growth is more rapid in the summer. Hormone abnormalities alter hair distribution, while BALDNESS follows hair loss.

HALDANE, John Burdon Sanderson (1892–1964), British geneticist whose work, with that of Sir Ronald Aylmer Fisher (1890–1962) and Sewall WRIGHT, provided a basis for the mathematical study of population GENETICS.

HALDANE, John Scott (1860–1936), British physiologist best known for his researches into industrial (especially mining) diseases caused by poor ventilation. He also contributed to a technique for dealing with the "bends" (see AEROEMBOLISM).

HALE, George Ellery (1868–1938), US astronomer who discovered the magnetic fields of SUNSPOTS, and who invented at the same time as Henri Alexandre Deslandres (1853–1948) the SPECTROHELIOGRAPH (c1892). His name is commemorated in that of the HALE OBSERVATORIES.

HALE OBSERVATORIES, formerly the Mt Wilson and Palomar Observatories, renamed (1970) for G. E. HALE and since 1948 operated jointly by the Carnegie Institution and the California Institute of Technology. At Mt Wilson (Cal.) are two reflecting TELESCOPES and two solar towers; at Palomar Mountain (Cal.) a 200-in reflector, until 1973 the largest in the world, and two Schmidt telescopes.

HALES, Stephen (1677–1761), English plant physiologist and chemist who, in accordance with the Newtonian quantitative paradigm, devised experiments to measure blood pressure in animals, and, realizing the importance of careful weighing and measuring in chemical experiments, applied these principles in his investigations of the life of plants. His *Vegetable Staticks* was published in 1727.

HALF-LIFE, the time taken for the activity of a radioactive sample to decrease to half its original value, half the nuclei originally present having changed spontaneously into a different nuclear type by emission of particles and energy. After two half-lives, the radioactivity will be a quarter of its original value and so on. Depending on the type of nucleus and method of decay, half-lives range from less than a second to over 10^{10} years. The half-life concept can also be applied to other systems undergoing random decay, e.g. certain biological populations.

HALFTONE, reproduction of a photograph or other picture containing a range of continuous tones, by using dots of various sizes but uniform tone. The dots are small enough to blend in the observer's vision to give the effect of the original. The picture is photographed through a screen on which a fine rectangular grid has been scribed (2 to 6 lines/mm); the dots arise by DIFFRACTION. From the screened negative is made a halftone plate used for PRINTING by all processes.

HALIDES, binary compounds of the HALOGENS with

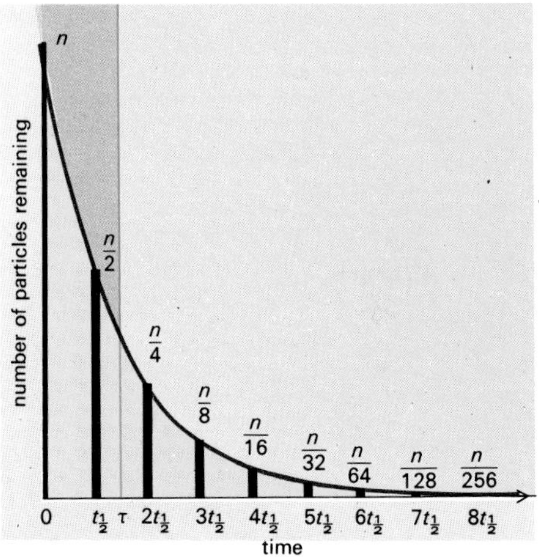

Half-life illustrated in a graph of radioactive decay. After the elapse of each half-life period ($t_{\frac{1}{2}}$), only half the number (n) of undisintegrated nuclei remain that were present at the beginning of the period. The average time that elapses before a nucleus disintegrates (τ), starting from the beginning of the process (t=0) is given by:

$$\tau = \frac{t_{\frac{1}{2}}}{\ln 2} = 1.442\,7t_{\frac{1}{2}}$$

oxidation number −1. Metal halides are mostly ionic salts (X^-), usually very soluble. Nonmetal halides, and a few metal halides such as tin (IV) chloride, are volatile covalent compounds, highly reactive, often violently hydrolyzed by water, and used as halogenating agents. Halide ions form stable LIGAND complexes. Their reducing power increases down the group. Halide minerals include SALT, SYLVITE, FLUORITE, APATITE and CRYOLITE. (See also ALKYL HALIDES; HYDROGEN CHLORIDE; HYDROGEN FLUORIDE.)

HALITE, or **Rock Salt.** See SALT.

HALITOSIS, or bad breath, a condition often caused by excessive bacterial growth in the mouth, e.g., in TONSILLITIS, and associated with poor oral hygiene. Disease of the GASTROINTESTINAL TRACT, including ULCER and APPENDICITIS, may also cause halitosis.

HALL, Charles Martin (1863–1914), US chemist who discovered (c1886), independently of HÉROULT, the electrolytic method of isolating pure aluminum now known as the HALL-HÉROULT PROCESS.

HALL, Granville Stanley (1844–1924), US psychologist and educator best known for founding the *American Journal of Psychology* (1887), the first US psychological journal. He was first president of the American Psychological Institute (1894), a body whose foundation he had assisted.

HALL, James (1811–1898), US geologist and paleontologist, the father of American STRATIGRAPHY. His major work was on the paleontology of the SILURIAN and DEVONIAN of New York State, *The Paleontology of New York* (13 vols., 1847–94).

HALL EFFECT, the POTENTIAL difference that develops across a METAL or SEMICONDUCTOR placed in a transverse magnetic field when an electric current flows in it. This voltage is at right-angles to both the current and magnetic field directions, and arises from the deflection of moving charge carriers (ELECTRONS or holes) by the magnetic field.

HALLER, Albrecht von (1708–1777), Swiss biologist, best known for his work on human anatomy and physiology, and also a poet. A pupil of BOERHAAVE and much influenced by him, he is credited with being the founder of experimental ANATOMY. In physiology he investigated RESPIRATION, the BLOOD CIRCULATION, the NERVOUS SYSTEM and the irritability and sensibility of different types of body tissue; in all cases relying on experiment.

HALLEY, Edmund (1656–1742), English astronomer. In 1677 he made the first full observation of a transit of Mercury; and in 1676–79 prepared a major catalog of the S-hemisphere stars. He persuaded NEWTON to publish the *Principia*, which he financed. In 1720 he succeeded FLAMSTEED as Astronomer Royal. He is best known for his prediction that the comet of 1680 would return in 1758 (see HALLEY'S COMET), based on his conviction that COMETS follow elliptical paths about the sun.

HALLEY'S COMET, the first periodic comet to be identified (by HALLEY, late 17th century) and the brightest of all recurring comets. It has a period of about 76 years. Records of every appearance of the comet since 240 BC, except that of 163 BC, are extant; and it is featured on the Bayeux tapestry. It will next reappear in 1986.

HALL-HÉROULT PROCESS, main ALUMINUM production method. Pure aluminum oxide, extracted from BAUXITE, is dissolved in molten CRYOLITE at 970°C, and electrolyzed (see ELECTROLYSIS) with a current of about 100kA through carbon electrodes. Molten aluminum is formed at the cathode and withdrawn from the bottom of the cell. The process was invented independently in 1886 by Charles HALL in the US and by Paul HÉROULT in France.

HALLSTATT, term referring to the late BRONZE AGE and early IRON AGE in W and Central Europe, from Hallstatt, Austria, where there is a prehistoric cemetery and salt mines which have been in constant operation since 2500 BC. It was characterized by extremely fine, decorated pottery, though the quality deteriorated toward the end of the period.

HALLUCINATION, an experience similar to a normal PERCEPTION but with the difference that

The Hall effect shown schematically.

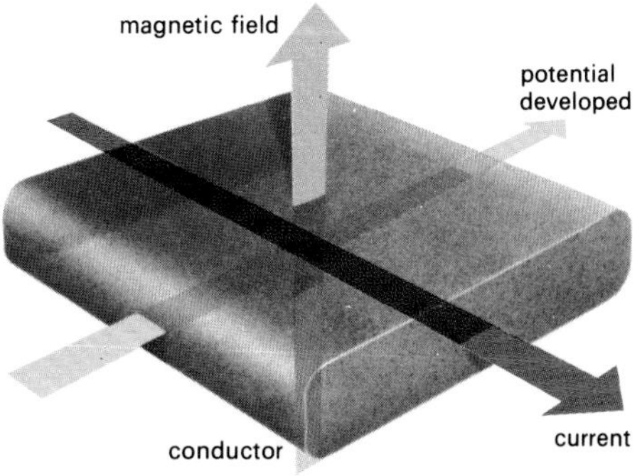

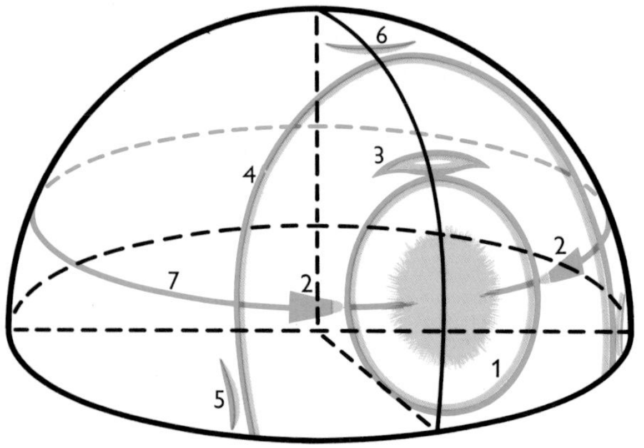

Halo phenomena may take many forms. Besides the common 22° halo (1) we may sometimes see parhelia ("mock suns"), luminous spots about 22° on either side of the sun (2). A tangent arc (3) is often associated with the 22° halo. (4) The large, less luminous halo with an angular radius of 46°; (5) the tangent arc to the 46° halo; and (6) the circumzenithal arc, centered on the zenith and parallel to the horizon. The parhelic circle ("mock sun ring"), which passes through the sun and may extend completely across the sky, is seen at (7).

sensory stimulus is either absent or too minor to explain the experience satisfactorily. Certain abnormal mental conditions (see MENTAL ILLNESS) produce hallucinations, as does taking of HALLUCINOGENIC DRUGS. They may also result from exhaustion or FEVER; or may be experienced while falling asleep (hypnogogic) or waking (hypnopompic), and also by individuals under HYPNOSIS. A **negative hallucination** is lack of perception despite adequate stimulus. **Mass hallucination** is hallucination shared by the members of a GROUP; it may particularly result from mass hypnosis. (See also EIDETIC IMAGE; ILLUSION.)

HALLUCINOGENIC DRUGS, DRUGS which cause hallucinations or illusions, usually visual, together with personality and behavior changes. The last may arise as a result of therapy, but more usually follow deliberate exposure to certain drugs for their psychological effects ("trip"). Lysergic acid diethylamide (LSD), HEROIN, MORPHINE and other OPIUM NARCOTICS, MESCALINE and PSILOCYBIN are commonly hallucinogenic and cannabis sometimes so. The type of hallucination is not predictable and many are unpleasant ("bad trip"). Recurrent hallucinations may follow use of these drugs; another danger is that altered behavior may inadvertently cause death or injury. Although psychosis may be a result of their use, it may be that recourse to drugs represents rather an early symptom of SCHIZOPHRENIA.

HALO, a luminous ring or series of arcs sometimes seen around the sun or moon, the result of REFRACTION or REFLECTION (or both) of their light by crystals of ICE in high, thin clouds. Commonest is the 22° halo, of angular diameter 22° and centered on the sun or moon. (See also CORONA.)

HALOGENS, highly reactive nonmetals in Group VIIA of the PERIODIC TABLE, comprising FLUORINE, CHLORINE, BROMINE, IODINE and ASTATINE; the general symbol X is often used. The elements have molecular formular X_2. They show a regular gradation of physical and chemical properties: with increasing atomic number, they become less volatile, darker in color, less reactive (in particular, less strongly oxidizing), and less electronegative (see ELECTRONEGATIVITY). The typical compounds of the halogens are the HALIDES, with oxidation number −1; compounds with positive oxidation numbers (usually 1, 3, 5 and 7) are also formed, with increasing stability down the group. The halogens react vigorously with almost all other elements and always occur combined in nature. (See also INTERHALOGENS; PSEUDOHALOGENS; ALKYL HALIDES; ACID CHLORIDES.)

HALOPHYTES, plants that are tolerant to saline conditions, found on the seashore and salt marshes, such as glasswort and eel-grasses (*Zostera*).

HAND, structural specialization of the end of the ARM, enabling grip and the fine motor tasks which characterize higher primates, especially man. Numerous small BONES arise from the WRIST in five radiations, which divide into four fingers and, at right-angles to them, the thumb. Manual dexterity reflects the existence of these two planes of operation and the particular mobility of the thumb JOINT. Small MUSCLES in the hand are used in finer tasks, while grip and joint stabilization are controlled by the forearm muscles. The SKIN of the hand is specially sensitive, essential for fine motor tasks and enabling the hand to explore at a distance from the body.

HANDEDNESS refers to the side of the body, and in particular to the hand, that is most used in motor tasks. Most people are right-handed and few are truly ambidextrous (either-handed). In the BRAIN, the

Hardness scales

Mohs' scale	hardness number	modified Mohs' scale
talc	1	talc
gypsum	2	gypsum
calcite	3	calcite
fluorite	4	fluorite
apatite	5	apatite
feldspar	6	orthoclase
quartz	7	vitreous silica
topaz	8	quartz or stellite
corundum	9	topaz
diamond	10	garnet
	11	fused zirconia
	12	fused alumina or tungsten carbide
	13	silicon carbide
	14	boron carbide
	15	diamond

paths for sensory and motor information are crossed, so the right side of the body is controlled by the left cerebral hemisphere and vice versa. The left hemisphere is usually dominant and also contains centers for speech and calculation. The nondominant side deals with aspects of visual and spatial relationships, while other functions are represented on both sides. In some left-handed people, the right hemisphere is dominant. Suppression of left-handedness may lead to speech disorder.

HANSEN'S DISEASE. See LEPROSY.

HAPLOID, term applied to any CELL which contains a single set of unpaired CHROMOSOMES in each nucleus. The number of chromosomes present is termed the haploid number. Haploid cells are produced by MEIOSIS and are typified as GAMETES.

HARDEN, Sir Arthur (1865–1940), British biochemist awarded with EULER-CHELPIN the 1929 Nobel Prize for Chemistry for their work on ENZYME action in the FERMENTATION of sugars. Harden demonstrated the existence of coenzymes, and Euler-Chelpin determined the structure of the first found, cozymase.

HARDENING OF THE ARTERIES. See ARTERIOSCLEROSIS.

HARDNESS, the resistance of a substance to scratching, or to indentation under a blow or steady load. Resistance to scratching is measured on the **Mohs' scale**, named for Friedrich Mohs (1773–1839), who chose 10 MINERALS as reference points, from TALC (hardness 1) to DIAMOND (10). The **modified Mohs' scale** is now usually used, with 5 further mineral reference points. Resistance to indentation is measured by, amongst others, the Brinell, Rockwell and Vickers scales. (See also MATERIALS, STRENGTH OF.)

HARD WATER, water containing CALCIUM and MAGNESIUM ions and hence forming scum with soap and depositing scale in boilers, pipes and kettles. **Temporary hardness** is due to calcium and magnesium BICARBONATES; it is removed by boiling, which precipitates the carbonates. **Permanent hardness** (unaffected by boiling) is due to the sulfates. Hard water may be softened by precipitation of the metal ions using CALCIUM hydroxide and SODIUM carbonate, followed by sodium PHOSPHATE; or by a ZEOLITE or ION-EXCHANGE column, which exchanges the calcium and magnesium ions for sodium ions.

HARDWOOD. See FORESTS; WOOD.

HARELIP, a congenital DISEASE with a cleft defect in the upper lip due to impaired facial development in the EMBRYO, and often associated with CLEFT PALATE. It may be corrected by plastic SURGERY.

HARGREAVES, or **Hargraves, James** (c1720 1778), British inventor of the spinning jenny (c1764),

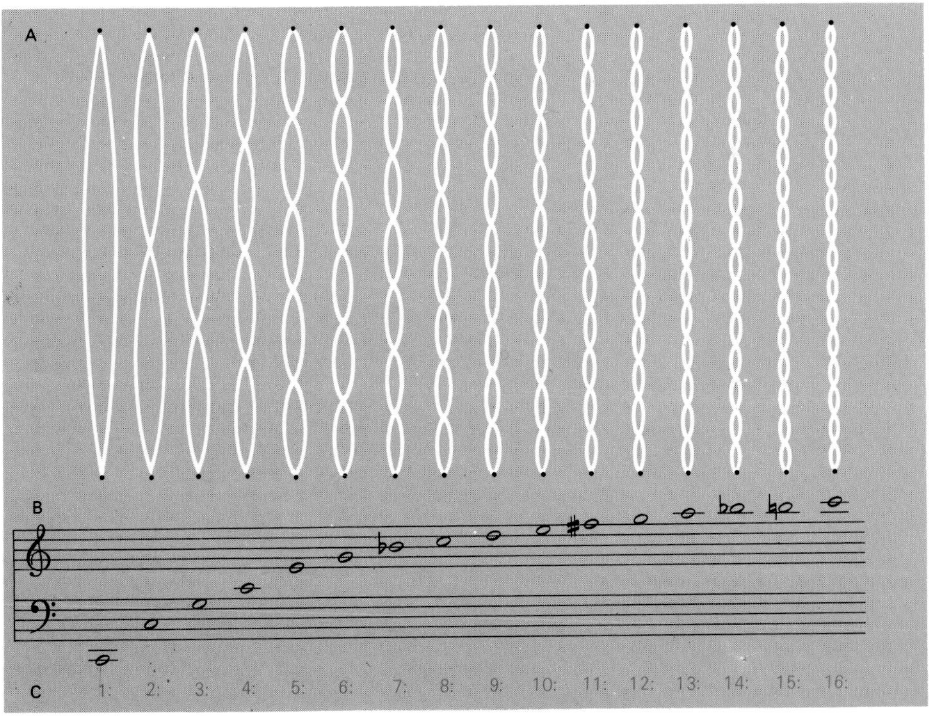

Harmonics: (A) The first 16 modes of harmonic vibration in a stretched string. (B) The musical pitches these would sound if the fundamental (first harmonic) sounded C,,. (C) The ratios between the frequencies of the different modes of vibration. Thus, the frequency of the fundamental is half (1:2) that of the first overtone (second harmonic) and one-third that of the second overtone (third harmonic); that of the second overtone is two-thirds that of the third overtone, and so on. Note that, after the fundamental (the first harmonic), the *n*th overtone is the $(n+1)$th harmonic.

a machine for SPINNING several threads at once. Public uproar forced him to flee his native Blackburn for Nottingham (1768), where he patented the jenny (1770). In 1777 he adopted ARKWRIGHT's more sophisticated machinery.

HARMATTAN, wind blowing W or SW from the S Sahara. In winter it transports vast quantities of dust over the Atlantic Ocean. In summer it interacts with NE-blowing MONSOON winds, often causing TORNADOES.

HARMONIC MOTION. See SIMPLE HARMONIC MOTION.

HARMONIC POINTS, four POINTS A, B, C, D, which divide a LINE such that

$$\frac{AC/BC}{AD/BD} = -1.$$

(The negative is possible only if there are positive and negative directions along the line; e.g., BC = −CB.)

HARMONICS, vibrations at FREQUENCIES which are INTEGER multiples of that of a fundamental vibration: the ascending notes $C_{,}$, C, G, C′, E′, G′ comprise a fundamental with its first five higher harmonics. Apart from their musical consonance, they are important because any periodically repeated signal—a vowel sound, for example—can be produced by superposing the harmonics of the fundamental frequency, each with the appropriate intensity and time lag.

HARPOON, spearlike weapon used in hunting sea creatures such as whales, seals and fish. Most designs have a heavily barbed head and a long line attached to the shaft. Modern whaling harpoons are usually fired from a gun, and often have explosive tips which kill more quickly.

HARRISON, John (1693–1776), British inventor of the marine CHRONOMETER (1735). A prize of £20 000 had been offered for a device to allow the accurate determination of longitude at sea: this he won with his No. 4 Marine Chronometer of 1759.

HARTLEY, David (1705–1757), English physician and a founder of the school of psychology known as ASSOCIATIONISM. In his *Observations on Man* (1749), he taught that sensations were communicated to the brain via vibrations in nerve particles and that the repetition of sensations gave rise to the association of ideas in the mind.

HARTLINE, Haldan Keffer (1903–   ), US physiologist awarded with R. GRANIT and G. WALD the 1967 Nobel Prize for Physiology or Medicine for his work on the functioning of the nerve cells of the retina (see EYE; VISION).

HARVEST MOON, the full moon occurring nearest to the autumnal equinox (around September 23) in the N Hemisphere. For several nights the full moon rises at about the same time (around sunset), and may be bright enough for harvesting to continue into the night. In the S Hemisphere this occurs around the spring equinox. (See EQUINOXES.)

HARVEY, William (1578–1657), British physician who discovered the circulation of the blood. He showed that the HEART acts as a pump and that the blood circulates endlessly about the body; that there are valves in the heart and VEINS so that blood can flow in one direction only; and that the necessary pressure comes only from the lower left-hand side of the heart. His discoveries demolished the theories of GALEN that blood was consumed at the body's periphery and that the left and right sides of the heart were connected by pores. He also made important studies of the development of the EMBRYO.

HASHISH, or **cannabis**, a drug produced from a resin obtained from the HEMP plant (*Cannabis sativa*), particularly from its flowers and fruits. It is a non-addictive drug whose effects range from a feeling of euphoria to fear. Hashish is mainly produced in the Middle East and India, and has been in use for many centuries, although it is still illegal in many countries. (See MARIJUANA.)

HASSEL, Odd (1897–   ), Norwegian chemist awarded with D. BARTON the 1969 Nobel Prize for Chemistry for their work on CONFORMATIONAL ANALYSIS.

HAVERSIAN CANALS. See BONE.

HAWORTH, Sir Walter Norman (1883–1950), British chemist awarded with KARRER the 1937 Nobel Prize for Chemistry for his work on the structures of CARBOHYDRATES and VITAMIN C.

HAY FEVER, common allergic disease causing RHINITIS and CONJUNCTIVITIS on exposure to allergen. The prototype is ALLERGY to grasses, but pollens of many trees, weeds and grasses (e.g., ragweed, Timothy grass) may provoke seasonal hay fever in sensitized individuals. Allergy to FUNGI or to the house-dust mite may lead to perennial rhinitis; animal fur or feathers may also provoke attacks. Susceptibility is often associated with ASTHMA, ECZEMA and ASPIRIN sensitivity in the individual or his family. Treatment consists of allergen avoidance, desensitizing INJECTIONS and cromoglycate, ANTIHISTAMINES or STEROID sprays in difficult cases.

HAYNES, Elwood (1857–1925), US inventor who built one of the first US AUTOMOBILES (1894) and discovered several ALLOYS, including STAINLESS STEEL (patented 1919).

HEADACHE, the common symptom of an ache or pain affecting the head or neck, with many possible causes including FEVER, emotional tension (with spasm of neck MUSCLES) or nasal SINUS infection. **Migraine**, due to abnormal reactivity of blood vessels, is typified by zig-zag or flashing visual sensations or tingling in part of the body, followed by an often one-sided severe throbbing headache. This may be accompanied by nausea, VOMITING and sensitivity to light. There is often a family history. **Meningeal inflammation**, as in MENINGITIS and subarachnoid HEMORRHAGE, may also cause severe headache. The headache of **raised intracranial pressure** is often worse on waking and on coughing and may be a symptom of brain TUMOR, ABSCESS or HYDROCEPHALUS. Headaches are often controlled by simple ANALGESICS, while migraine may need drugs that act on blood vessels (e.g., ERGOT derivatives).

HEARING. See EAR.

HEARING AID, device to amplify SOUND so as to make the best use of remaining HEARING in DEAFNESS. Sounds are picked up by a microphone receiver placed behind the EAR, amplified electronically, and transmitted to the wearer via an earphone (transducer) placed either in the ear or resting behind it. Ear trumpets are still sometimes useful.

HEART, vital organ in the CHEST of animals, concerned with pumping the BLOOD, thus maintaining the BLOOD CIRCULATION. The evolution of the vertebrates shows a development from the simple heart found in fish to the four-chambered heart of

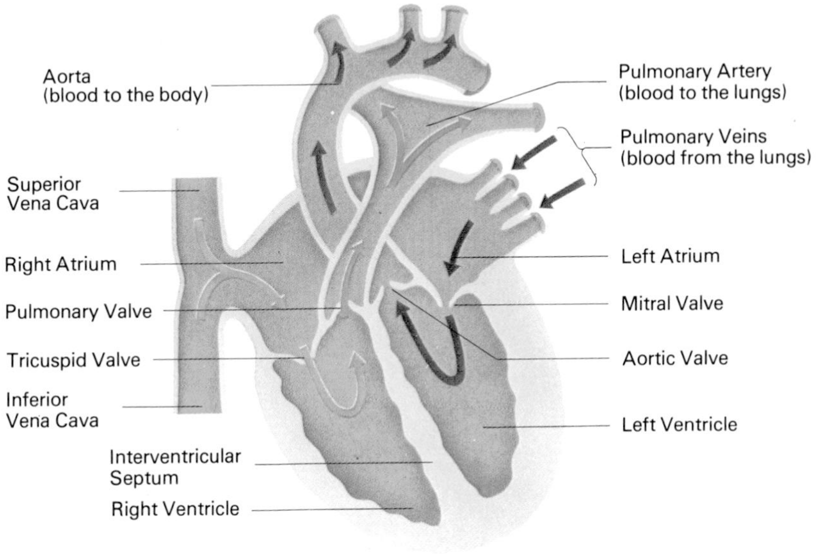

Schematized section through the human heart.

mammals. In man, the circulation may be regarded as a figure-of-eight, with the heart at the cross-over point, but keeping the two systems separate by having two parallel sets of chambers. The pumping in the two sets, right and left, is coordinated, ensuring a balance of flow. Each set consists of an atrium, which receives blood from the LUNGS (left) or body (right), and a ventricle. The atria pump blood into the ventricles, which pump it into the lungs (right) or systemic circulation (left). The bulk of the heart consists of specialized MUSCLE fibers which contract in response to stimulation from a pacemaker region relayed via special conducting tissue. Between each atrium and ventricle are valves, the mitral (left) and tricuspid (right). Similarly, between the ventricles and their outflow tracts are aortic and pulmonary valves. The heart is lined by PERICARDIUM and receives its blood supply from the AORTA via the coronary ARTERIES. The cells in the right atrium have an inbuilt tendency to depolarize and thus to set up an electrical impulse in the conducting tissue. In **heart action** this passes to both atria, which have already filled with blood from the systemic or pulmonary veins. Blood is then pumped by atrial contraction into the ventricles, though much of it passes into the latter before the atria contract. The same electrical impulse is conducted to both ventricles and there sets up a coordinated contraction (*systole*), which leads to the forceful expulsion of blood into the aorta or pulmonary artery and to the closure of the mitral and tricuspid valves. When the contraction ceases (*diastole*), the pressure in the ventricle falls, and the aortic and pulmonary valves close. The force generated by systole is propagated into the major arteries, providing the driving force for the circulation. Heart output may be increased (e.g., in EXERCISE) through several agencies including increased rate (*tachycardia*) and force of contraction (mediated by the sympathetic NERVOUS SYSTEM and ADRENALINE), and the increased return of venous blood (effected by a muscle pumping action on the valved, collapsible VEINS). **Disorders of the heart** include: *Congenital disorders* of the structure of the chambers or valves (e.g., BLUE BABY), and disease following RHEUMATIC FEVER, leading to stenosis or incompetence of the valves, especially the mitral and aortic. These disorders may be improved by DRUG treatment but they frequently require cardiac SURGERY to correct or repair defects or to insert PROSTHETICS (e.g., artificial heart valves). **Coronary thrombosis** causes DEATH or injury to areas of heart muscle. This may lead to defects in pumping and heart failure, rhythm disorder, ANEURYSM or, rarely, cardiac rupture. *Rhythm disturbance* may follow damage to conducting tissue (where abnormal conducting or pacemaker tissue exists), in certain metabolic disorders (thyrotoxicosis—see THYROID GLAND), and in valve disease. *Bradycardia* is very slow heart rate. This may be due to disease but can be normal in fit athletes in whom it indicates increased heart efficiency. Rhythm disorders are often treated with drugs including DIGITALIS, sympathetic-nervous-system stimulants or blockers, ATROPINE, and certain agents used in local ANESTHESIA. *Heart failure*, in which inadequate pumping leads to imbalance between the two parts of the circulation or the failure of both, may be due to coronary thrombosis, cardiac muscle disease or fluid overload. It causes pulmonary EDEMA with shortness of breath on exercise or on lying flat, or peripheral edema. DIURETICS and digitalis are the cornerstone of treatment, relieving edema and increasing pump efficacy. *Infection of abnormal valves* with BACTERIA or FUNGI is a serious disease causing FEVER and other systemic manifestations including EMBOLISM, heart failure and valve destruction. Its prevention, in high-risk patients, and treatment involve careful use of selected ANTIBIOTICS. Valve replacement may also be needed. Investigation of the heart can involve the use of the ELECTROCARDIOGRAPH, chest X-RAY or cardiac CATHETER (to study ANATOMY and flow) and the study of serum ENZYME levels.

HEART ATTACK. See CORONARY THROMBOSIS.

HEARTBURN, or esophagitis, burning sensation of "indigestion" localized centrally in the upper ABDOMEN or lower CHEST. It is frequently worse after large meals or on lying flat, especially with hiatus HERNIA. Acid STOMACH contents irritate the esophageal EPITHELIUM and may lead to ULCER; relief is with ANTACIDS. Heartburn is also loosely applied to other pains in the same situation.

HEART-LUNG MACHINE. See ARTIFICIAL ORGANS.

HEART MURMUR, abnormal sound heard on listening to the CHEST over the HEART with a STETHOSCOPE. Normally there are two major heart sounds due to valve closure, separated by silence. Murmurs arise in the disease of heart valves, with narrowing (stenosis) or leakage (incompetence). Holes between chambers, valve roughening and high flow also cause murmurs.

HEAT, the form of ENERGY that passes from one body to another owing to a TEMPERATURE difference between them; one of the basic functions in THERMODYNAMICS. The energy residing in a hot body is also loosely called heat, but is better termed internal energy, since it takes several different forms. Despite an earlier view by some philosophers that heat was a form of agitation, in the 18th century the CALORIC THEORY OF HEAT predominated, until disproved by the experiments of Sir Humphry Davy and Count RUMFORD (1798) showing that mechanical WORK could be converted to heat. James JOULE confirmed this by many ingenious experiments and found a consistent value for the **mechanical equivalent of heat** (the ratio of work done to heat produced). In the mid-18th century Joseph BLACK first clearly distinguished heat from temperature, a conceptual advance which allowed heat to be measured (see CALORIMETRY) in terms of the temperature rise of a known mass of water, the unit being the CALORIE (or the BRITISH THERMAL UNIT). In SI UNITS heat is measured, as a form of energy, in JOULES. A given mass m of any substance shows a characteristic temperature rise θ when an amount of heat Q is supplied: $Q=ms\theta$ where s is the SPECIFIC HEAT of the substance. If the substance changes its state, however, by melting, freezing, boiling or condensing, LATENT HEAT is absorbed or produced without any temperature change, the internal energy being changed by altering the molecular interrelations, not merely their degree of motion. Heat is commonly produced as required for space HEATING or to power ENGINES, by conversion of chemical energy (burning fuel—see COMBUSTION), electrical energy or nuclear energy. There are three processes by which heat flows from a hotter to a cooler body: CONDUCTION and CONVECTION, in which molecular motion is transferred, and radiation, in which INFRARED RADIATION is emitted and propagates through space. (See also BLACKBODY RADIATION.) Heat transfer may be hindered by means of thermal INSULATION. **Newton's law of cooling** states that the rate of loss of heat by a body in a draft (forced convection) is proportional to its temperature difference from its surroundings. (See also BOLOMETER; PYROMETER; REFRIGERATION; THERMOCHEMISTRY.)

HEAT DEATH, speculative theory of the final state of the UNIVERSE. If the universe is an isolated system (see THERMODYNAMICS) then its ENTROPY must tend to a maximum, at which all energy is degraded to uniform heat, everything is wholly disordered and no change is possible. In the oscillating-universe model (see COSMOLOGY) it is supposed that the entropy inequality is reversed at the universe's greatest extent, so that during contraction entropy tends to a minimum.

HEATING, the supply of heat to buildings to produce a comfortable temperature, usually between 15° and 23°C depending on one's physical activity. For centuries the chief means was an open fire of wood, charcoal or coal in a fireplace with chimney, heating by radiation and by some convection from the fireplace. But most of the heat was lost as the hot exhaust gases rose up the chimney, and the convective iron stove proved more efficient. Other local heaters used today include gas fires and electric heaters. The latter are 100% efficient (converting all the energy input to heat) but expensive to run; they work by passing a current I through a RESISTANCE R, the rate of heat output being IR^2 watts. Bar fires, with the element at the focus of a parabolic mirror, heat by radiation; other types work by convection, including underfloor electric heating. A planned heating system distributes heaters of suitable power to maintain desired temperatures throughout while making good heat loss, which is minimized by INSULATION and draftproofing. **Central heating** systems—thermostatically controlled—are now most common: from a single boiler, fired by gas, fuel oil, coal or (rarely) electricity, hot water or steam is pumped around a system of pipes and "radiators" (really convectors) in each room. Some hot-water systems are not pumped, but use large-bore pipes and gravity flow. In hot-air systems air is heated in a furnace and blown by fans through ducts to diffusers or grilles; the Roman hypocaust was an early example of underfloor hot-air central heating. (See also AIR CONDITIONING; HEAT PUMP.)

HEAT PUMP, device for transferring heat from a cold region to a hotter region by doing WORK (as required by the second law of THERMODYNAMICS). The working fluid or refrigerant is a condensible gas such as ammonia or FREON. A motor-driven compressor compresses the gas adiabatically (raising its temperature) and delivers it to a condenser coil or "radiator" in the space to be heated. As it loses heat it liquefies and passes through an expansion valve into the evaporator, a low-pressure region where the liquid evaporates, taking heat from its relatively cool environment. The gas then returns to the compressor to complete the cycle. Heat pumps are used in domestic refrigerators (see REFRIGERATION), the evaporator being inside the refrigerator, and also in AIR CONDITIONING systems for space HEATING in winter and (by reversing the pump) cooling in summer.

HEAT RASH. See PRICKLY HEAT.

HEAT SHIELD, device to prevent overheating of space capsules on re-entry to the earth's atmosphere. The craft is coated with a layer of ablative material (see ABLATION), often a plastic impregnated with quartz fibers. FRICTION with the air heats and vaporizes the outer regions. Some 80% of the heat is reradiated at the gas/liquid boundary.

HEATSTROKE. See SUNSTROKE.

HEAVES, a lung disease of horses. Symptoms include a wheezy cough, difficult breathing and dilation of the nostrils. Heaves is probably an allergic reaction to poor quality food.

HEAVISIDE, Oliver (1850–1925), British physicist and electrical engineer best known for his work in telegraphy, in course of which he developed operational calculus, a new mathematical system for dealing with changing wave-shapes. In 1902, shortly after KENNELLY, he proposed that a layer of the atmosphere was responsible for reflecting RADIO waves back to earth. This, the E layer of the IONOSPHERE, was found by APPLETON and others (1924), and is often called the **Kennelly-Heaviside Layer**, or **Heaviside Layer**.

HEAVY HYDROGEN. See DEUTERIUM; TRITIUM.

HEAVY WATER, or DEUTERIUM oxide (D_2O), occurs as 0.014% of ordinary WATER, which it closely resembles. It is used as a moderator in nuclear reactors and as a source of deuterium and its compounds. It is toxic in high concentrations. Water containing TRITIUM or heavy isotopes of oxygen (O^{17} and O^{18}) is also called heavy water. mp 3.8°C, bp 101.4°C.

HECTARE (ha), widely used metric unit of area, equal to 10000m² or 100 are. One hectare equals 2.471 acres.

HEGEL, Georg Wilhelm Friedrich (1770–1831), German philosopher of IDEALISM who had an immense influence on 19th and 20th-century thought and history. During his life he was famous for his professorial lectures at the University of Berlin and he wrote on logic, ethics, history, religion and aesthetics. The main feature of Hegel's philosophy was the dialectical method by which an idea (*thesis*) was challenged by its opposite (*antithesis*) and the two ultimately reconciled in a third idea (*synthesis*) which subsumed both. Hegel found this method both in the workings of the mind, as a logical procedure, and in the workings of the history of the world, which to Hegel was the process of the development and realization of the World Spirit (*Weltgeist*). Hegel's chief works were *Phenomenology of the Mind* (1807), and *Philosophy of Right* (1821). His most important follower was Karl Marx.

HEIDELBERG MAN. See PREHISTORIC MAN.

HEISENBERG, Werner Karl (1901–1976), German mathematical physicist generally regarded as the father of QUANTUM MECHANICS, born out of his rejection of any kind of model of the ATOM and use of mathematical MATRICES to elucidate its properties. His famous UNCERTAINTY PRINCIPLE (1927) overturned traditional physics, its implications affecting areas of science far beyond the bounds of atomic physics. (See also SCHRÖDINGER.)

HELICOPTER, exceptionally maneuverable aircraft able to take off and land vertically, hover, and fly in any horizontal direction without necessarily changing the alignment of the aircraft. Lift is provided by one or more rotors mounted above the craft and rotating horizontally about a vertical axis. Change in the speed of rotation or in the pitch (angle of attack) of all the blades at once alters the amount of lift; cyclic change in the pitch of each blade during its rotation alters the direction of thrust. Most helicopters have only a single lift rotor, and thus have also a tail-mounted vertical rotor to prevent the craft from spinning around (see TORQUE); change in the speed of this rotor is used to change the craft's heading.

Helicopter toys were known to the Chinese and in medieval Europe, but, because of problems with stability, it was not until 1939, following the success of the AUTOGIRO (1923), that the first fully successful helicopter flight was achieved by SIKORSKY. (See also VERTICAL TAKEOFF AND LANDING AIRCRAFT.)

US military crane helicopter.

HELIOGRAPH, 19th-century instrument used for mainly military signaling, comprising essentially a mirror and a shutter to cut off the sunlight reflected from it. A further mirror permitted messages to be sent even when the sun was behind the sender. The signals could be interpreted up to 50km away.

HELIOSTAT, an instrument used to observe the sun. A TELESCOPE is used in conjunction with a large flat mirror which rotates so that the image of the sun appears stationary.

HELIUM (He), one of the NOBLE GASES, lighter than all other elements except hydrogen. It is a major constituent of the SUN and other STARS. The main source of helium is natural gas in Tex., Okla. and Kan. ALPHA PARTICLES are helium nuclei. Helium is lighter than air and nonflammable, so is used in balloons and airships. It is also used in breathing mixtures for deep-sea divers, as a pressurizer for the fuel tanks of liquid-fueled rockets, in helium-neon LASERS, and to form an inert atmosphere for welding. Liquid helium He^4 has two forms. Helium I, stable from 2.19K to 4.22K, is a normal liquid, used as a refrigerant (see CRYOGENICS; SUPERCONDUCTIVITY). Below 2.18K it becomes helium II, which is a superfluid with no VISCOSITY, the ability to flow as a film over the side of a vessel in which it is placed, and other strange properties explained by QUANTUM THEORY. He^3 does not form a superfluid. Solid helium can be produced only at pressures above 25atm. AW 4.0, mp 1.1K (25atm), bp 4.22K.

HELIX, the curve traced by a point P moving at a constant angle to the elements of, and on the surface of, a circular CYLINDER or CONE: in the former case it is termed a circular helix, in the latter, a cylindroconical helix. The **Double Helix** is a nickname for DNA (see NUCLEIC ACIDS).

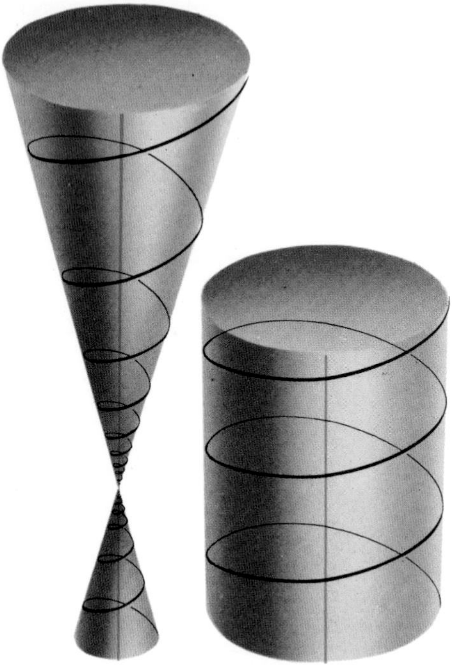

A circular helix (*right*) and cylindroconical helix (*left*). In each case the curve (red) makes a constant angle with the elements (e.g., the blue lines) in the surface on which it is drawn.

HELLADIC CULTURE. See AEGEAN CIVILIZATION.

HELLENISTIC AGE, the period in which Greco-Macedonian culture spread through the lands conquered by Alexander the Great. It is generally accepted to run from Alexander's death (323 BC) to the annexation of the last Hellenistic state, Egypt, by Rome (31 BC) and the death of Cleopatra VII, last of the Ptolemies (30 BC). After Alexander's death, and despite the temporary restraint imposed by Antipater, his empire was split by constant warring between rival generals eager for a share of the territory. Even after the accomplishment of the final divisions (Egypt, Syria and Mesopotamia, Macedonia, the Aetolian and Achaean Leagues in Greece, Rhodes and Pergamum), Greek remained the international language throughout most of the known world and a commercial and cultural unity held sway. The age was marked by cosmopolitanism, sharply contrasting with the parochialism of the earlier Greek era, and by advances in the sciences (see ARCHIMEDES; ARISTARCHUS; ERATOSTHENES; EUCLID; THEOPHRASTUS). The art was powerfully naturalistic if occasionally bathetic. Traditional religious cults weakened and were superseded by others either imported from the east or increasing in influence; such as the cults of Isis, Sarapis, Cybele and Mithras. The Hellenistic age saw the emergence of Stoicism and Epicureanism (see EPICURIS).

HELMHOLTZ, Hermann Ludwig Ferdinand von (1821–1894), German physiologist and physicist. In course of his physiological studies he formulated the law of conservation of ENERGY (1847), one of the first to do so. He was the first to measure the speed of nerve impulses (see NERVOUS SYSTEM), and invented the OPHTHALMOSCOPE (both 1850). He also made important contributions to the study of ELECTRICITY and NON-EUCLIDEAN GEOMETRY.

HELMONT, Jan Baptista van (1580–1644), Flemish chemist and physician, regarded as the father of biochemistry. He was the first to discover that there were airlike substances distinct from air, and first used the name "gas" for them.

HEMATITE, hard, red OXIDE mineral, consisting of iron (III) oxide (α-Fe_2O_3), the chief IRON ore; also used in paints (ochre) and polishes (rouge). Hematite occurs worldwide, mainly in sedimentary rocks, though it is also formed by weathering of other iron minerals. In the US large deposits are found around the Great Lakes. It crystallizes in the rhombohedral system, with the CORUNDUM structure.

HEMATOLOGY, branch of MEDICINE concerned with BLOOD DISEASES (e.g., ANEMIA, LEUKEMIA, CLOTTING disorder).

HEMATOMA. See BRUISE.

HEMICHORDATA, or **acorn worms**, a subphylum of the CHORDATA containing two types of animals; worm-like burrowing forms and sedentary forms. All are marine. The burrowing forms have soft bodies divided into proboscis, collar and trunk. The proboscis and collar are distensible. The sedentary forms are colonial, a number of individuals called zooids being enclosed in a gelatinous "house."

HEMIMORPHITE, colorless or white mineral, formerly called CALAMINE; a hydrated zinc silicate ($Zn_4Si_2O_7[OH]_2.H_2O$) of widespread occurrence, an ore of ZINC formed by alteration of other zinc minerals. It crystallizes in the orthorhombic system and exhibits PIEZOELECTRICITY.

HEMOCYANIN, a blue, copper-containing PROTEIN found in the BLOOD of molluscs and arthropods, especially crustacea. Its function is to carry OXYGEN from the respiratory organs to the tissues.

The heme unit, of which there are four in each hemoglobin molecule, is a planar porphyrin structure, linked to the globin proteins which make up the bulk of the hemoglobin molecule via the amino acid histidine, part of which is shown here above the plane of the porphyrin ring. The histidine coordinates with the iron (II) atom of the heme, which also, in oxyhemoglobin, loosely binds the oxygen.

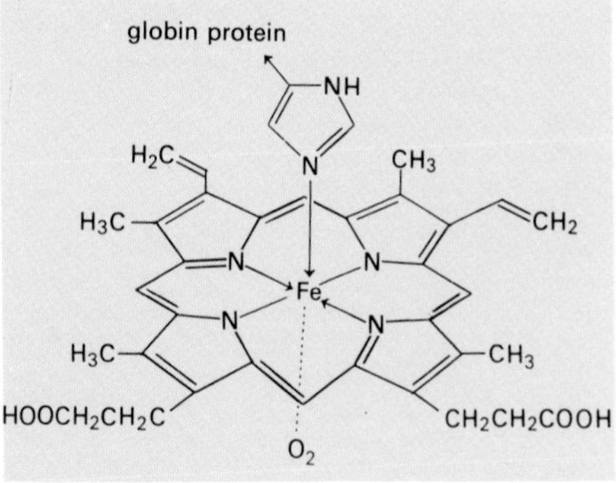

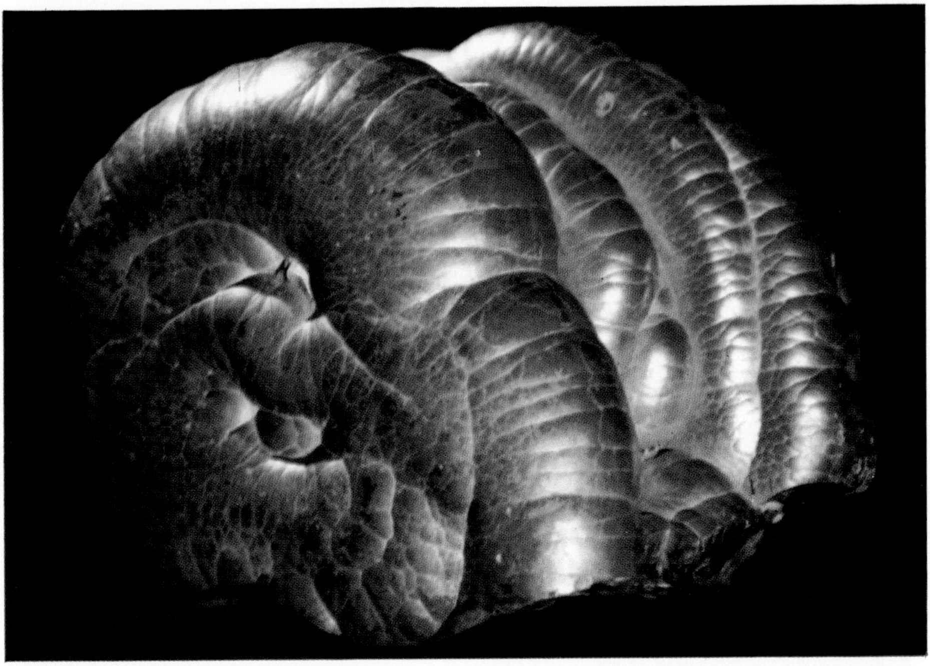

Kidney iron ore is the name given to hematite in mammillated form. Hematite, the most important ore of iron, is widely distributed throughout the world.

HEMOGLOBIN, respiratory pigment found in the BLOOD of many animals including man. It contains heme, an iron-containing molecule, and globin, a large protein, and occurs in red blood cells. The whole molecule has a high affinity for oxygen, being converted to oxyhemoglobin. In the LUNG capillaries, hemoglobin is exposed to a high oxygen concentration and oxygen is taken up. The redder blood then passes via the HEART into the systemic circulation. In the tissues the oxygen concentration is low, so oxygen is released from the ERYTHROCYTES and reduced hemoglobin returns to the lungs. Carbon monoxide has an even higher affinity for hemoglobin than oxygen and thus acts as a poison by displacing oxygen from hemoglobin, causing ANOXIA. Abnormal hemoglobin structures occur in certain races and may cause red-cell destruction and anemia. Lack of hemoglobin, regardless of cause, produces ANEMIA.

HEMOPHILIA, inherited disorder of CLOTTING in males, carried by females who do not suffer from the disease. It consists of inability to form adequate amounts of a clotting factor (VIII) essential for the conversion of soluble fibrinogen in blood to form fibrin. Prolonged bleeding from wounds or tooth extractions, HEMORRHAGE into JOINTS and MUSCLES with severe pain are important symptoms. Bleeding can be stopped by giving PLASMA concentrates rich in factor VIII and, if necessary, BLOOD TRANSFUSION. Similar diseases of both sexes are **Christmas disease** (due to lack of factor IX) and **von Willebrand's disease** (factor VIII deficiency with additional CAPILLARY defect).

HEMORRHAGE, acute loss of BLOOD from any site. Trauma to major ARTERIES, VEINS or the HEART may lead to massive hemorrhage. GASTROINTESTINAL TRACT hemorrhage is usually accompanied by loss of altered blood in vomit or feces and may lead to SHOCK; ULCERS and CANCER of the bowel are important causes. **Antepartum hemorrhage** is blood loss from the WOMB in late PREGNANCY and may rapidly threaten life of both mother and FETUS; **postpartum hemorrhage** is excessive blood loss after BIRTH due to inadequate womb contraction or retained PLACENTA. STROKE due to BRAIN hemorrhage may damage vital structures and cause COMA, while *subarachnoid bleeding* around the brain from ANEURYSM or malformation causes severe HEADACHE. FRACTURES may cause sizeable hemorrhage into soft tissues. Blood loss may be replaced by TRANSFUSION, and any blood clots may need to be removed.

HEMORRHOIDS, or **piles**, enlarged VEINS at the junction of the rectum and anus, which may bleed or come down through the anal canal, usually on defecation, and which are made worse by CONSTIPATION and straining. Sentinal pile is a SKIN tag at the anus. Bleeding from the rectum may be a sign of bowel CANCER and this may need to be ruled out before bleeding is attributed to piles.

HEMP, *Cannabis sativa*, tall herbaceous plant native to Asia, but now widely cultivated for fiber, oil and a narcotic drug called **cannabis**, HASHISH or MARIJUANA. The fibers are used in the manufacture of rope. They are separated from the rest of the plant by a process called RETTING (soaking), during which BACTERIA and FUNGI rot away all but the fibers, which are then combed out. Hemp oil obtained from the seed is used in the manufacture of PAINTS, VARNISHES and SOAPS. (See also DRUGS.)

HENCH, Philip Showalter (1896–1965), US physician who shared with KENDALL and REICHSTEIN the 1950 Nobel Prize for Physiology or Medicine for his use of cortisone (see STEROIDS) to treat rheumatoid ARTHRITIS.

HENEQUEN, *Agave fourcroydes*, a relative of the SISAL, which yields fibers used for making twine. Family: Agavaceae.

HENRY (H), the SI UNIT of inductance, the inductance of a circuit in which current changing at a rate of one ampere per second induces an ELECTROMOTIVE FORCE of one volt.

HENRY, Joseph (1797–1878), US physicist best known for his electromagnetic studies. His discoveries include INDUCTION and self-induction; though in both cases FARADAY published first. He also devised a much improved ELECTROMAGNET by insulating the wire rather than the core; invented one of the first ELECTRIC MOTORS; helped MORSE and WHEATSTONE devise their telegraphs; and found SUNSPOTS to be cooler than the surrounding photosphere. The HENRY is named for him.

HENRY, William (1774–1836), British chemist and physician who formulated **Henry's Law**, that, at a given temperature, the mass of a gas dissolved by a particular solvent is proportional to the pressure on it of the gas.

HEPARIN. See ANTICOAGULANTS.

HEPATICAE. See LIVERWORTS.

HEPATITIS, INFLAMMATION of the LIVER, usually due to VIRUS infection, causing nausea, loss of appetite, FEVER, malaise, JAUNDICE and abdominal pain; liver failure may result. It can occur as part of a systemic disease (e.g. YELLOW FEVER, MONONUCLEOSIS). In two forms infection is restricted to the liver: **infectious hepatitis** is an EPIDEMIC form, transmitted by feces and is of short INCUBATION; it is rarely serious or prolonged. **Serum hepatitis** is transmitted by BLOOD (e.g., used needles and syringes, TRANSFUSION), it develops more slowly but may be more severe, causing death. It is common among drug addicts; carriers may be detected by blood tests and immunization of those at risk may be helpful. Amebiasis and certain DRUGS can also cause hepatitis.

HERACLITUS (c540–c480 BC), Greek philosopher from Ephesus, called "the Weeping Philosopher" for his gloomy views and "the Obscure" for his cryptic style. He is known to us only through other authors. Believing in universal impermanence, and that all things (notably opposites) were interrelated, he considered fire the fundamental element of the universe.

HERB, in botany, any plant with soft aerial stems and leaves that die back at the end of the growing season to leave no persistent parts above ground. In everyday terms, herbs are plants used medicinally and to flavor food.

HERBARIUM, collection of dried and preserved plant specimens systematically arranged and classified. Herbaria are valuable means of plant classification since they are built up over many years of plants from different sources and therefore show the limits of variation within species.

HERBART, Johann Friedrich (1776–1841), German philosopher and educator best remembered for his pedagogical system, now called **Herbartianism**, in which he stressed the importance of ethics (to give social direction) and psychology (to understand the mind of the pupil) acting together.

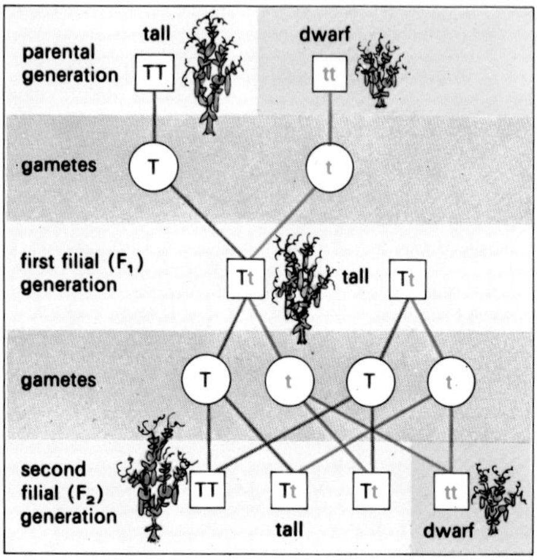

Above: Hereditary relationships for the transmission of a single character. This diagram schematizes Mendel's experiments with tall and short peas. The dominant gene (conferring tallness) is capitalized, the recessive (for dwarfness) is represented by a small letter. Starting from crossing pure tall and pure dwarf strains of peas the second generation shows the mendelian ratio of 3:1, tall: dwarf.

HERBICIDE. See WEEDKILLER.

HERBIVORES, a dietary classification of the Animal Kingdom—including all animals which feed exclusively on plant materials. Preyed on by many carnivorous animals, they form the lower links of food chains.

HERCULES, a large N Hemisphere constellation containing a superb GLOBULAR CLUSTER, M13, of around 500 000 stars, and which is just visible to the naked eye.

HEREDITY, the process whereby progeny resemble their parents in many features but are not, except in some microorganisms, an exact duplicate of their parents. Patterns of heredity for a long time puzzled biologists and it was not until the researches of Gregor MENDEL, an Austrian monk, that any numeric laws of heredity were discovered. Although Mendel's work was published in the mid-1860s, it went ignored by the majority of biologists until the opening of the 20th century.

Mendel showed that hereditary characteristics are passed on in units called GENES. When GAMETES (reproductive cells) are formed by MEIOSIS, the genes controlling any given characteristic "segregate" and become associated with different gametes. Thus, if the height of a pea plant is controlled by the genes *T* (for tallness) and *t* (for shortness) and pollen from a pure-breeding dwarf strain (of genotype *tt*) is used to fertilize ovules of a pure-breeding tall strain (of genotype *TT*), the resulting plants (of the "first filial"—F_1—generation), are of genotype *Tt*. Now the gametes of the F_1 generation contain equal numbers of genes *T* and *t*, both in the pollen and the ovules. The second filial (F_2) generation thus contains 50% of

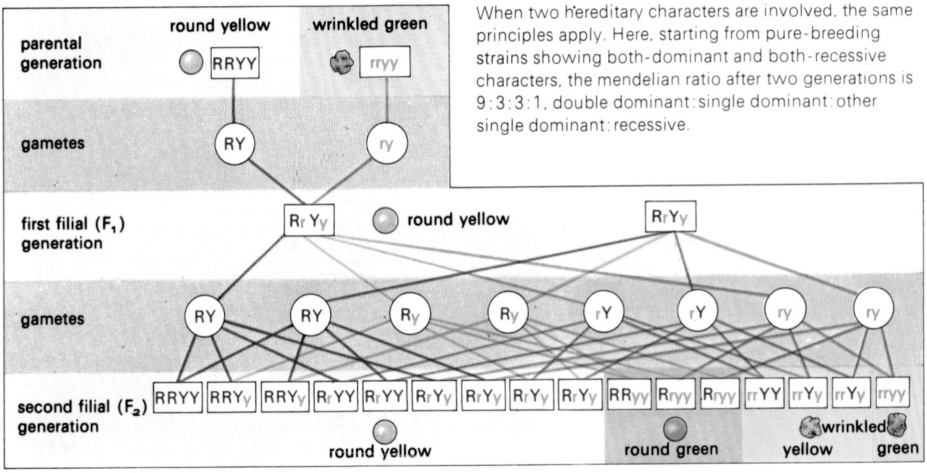

When two hereditary characters are involved, the same principles apply. Here, starting from pure-breeding strains showing both-dominant and both-recessive characters, the mendelian ratio after two generations is 9:3:3:1, double dominant: single dominant: other single dominant: recessive.

the "heterozygote" *Tt*, together with 25% each of the "homozygotes" *TT* and *tt*. In many cases, the heterozygote is indistinguishable from one of the homozygotes. In this case the gene that is expressed in the heterozygous condition is called a *dominant* gene; that which only manifests itself when homozygous is termed *recessive*. In the case of Mendel's peas, since *T* was dominant and *t* recessive, in the F_1 generation (100% *Tt*) all the plants were tall, while in the F_2 generation 75% were of the tall phenotype (i.e., the 25% *TT* and the 50% *Tt*) and 25% (the *tt*) of the short one.

Mendel also showed that when two or more pairs of genes segregate simultaneously the distribution of any one is independent of the distribution of the others. This work was done by crossing peas pure-breeding for round yellow seeds (*RRYY*) with peas pure-breeding for wrinkled green seeds (*rryy*). All the first-cross seeds were round yellow showing that round is dominant over wrinkled and yellow over green. The possible number of GENOTYPES at second cross is 16, but only 4 PHENOTYPES appeared: in the ratio of 9 round yellow seeds, to every 3 round green, 3 wrinkled yellow, and 1 wrinkled green. This "independent segregation" applies only to genes on different CHROMOSOMES; genes on the same chromosome are "linked" and do not segregate independently.

It is now known the genes are normally located on the chromosomes in the nucleus of the CELL. Each chromosome carries many genes which may be transmitted together and are said to be in *coupling*. However, genes are exchanged between chromosome pairs so that RECOMBINATION occurs. Because of the occurrence of recombination the LINKAGE of genes is not complete.

In the vast majority of animals and higher plants sex is determined by a special sex chromosome which in humans is the XY chromosome. Men are XY and women XX so that all ova are X while a sperm is either X or Y. Therefore there should be an equal number of males and females in a population. In practise Y-bearing sperm are more successful in fertilizing ova than X, so that more boys are born than girls.

Genes not only replicate themselves to pass on genetic information and direct the synthesis of PROTEINS within individual cells, they also interact with each other both directly at the chromosomal level and indirectly through gene products. Although a particular characteristic of an organism is probably under the control of a single gene, the characteristic may be modified by a large number of other genes. For example, mice have a gene which can either slightly shorten the tail or result in early death through kidney failure, depending on the presence of other genes. Other genes exist for the sole function of suppressing the effects of another gene. The translocation of genes on chromosomes probably plays an important role in gene interaction.

In most organisms the majority of abnormal or mutant genes are recessive. But in man mutant genes tend either to be dominant or show no dominance. As humans generally avoid marrying close relatives, different combinations of genes are always being formed which give rise to the great variation seen among human beings. A reduction of variability occurs in thoroughbred animals where matings are controlled so as to select for desired constant features.

It has been estimated that throughout EVOLUTION there have been over 500 million different species of plants and animals, therefore there must have been at least 500 million different genes. Genes are composed of DNA (see NUCLEIC ACIDS) which is capable of an enormous number of variations. A sequence of 15 nucleotides composed of four different bases is capable of over 500 million alternatives. It is possible using the four different nucleotides in DNA to construct a code of 64 three-nucleotide sequences capable of indicating all the differing AMINO ACIDS (see CODON).

HERMAPHRODITE, any organism in which the functions of both sexes are combined. Usually, an individual functions in only one sexual role at a time, but in a few species, e.g., earthworms, each of a pair of partners fertilizes the other during copulation. Hermaphrodite plants are usually referred to as being bisexual.

HERMES TRISMEGISTUS (Hermes the Thrice-greatest), Greek name for the Egyptian god Thoth. To him was ascribed the invention of writing and the authorship of the **Hermetic Writings**, books of occult wisdom which date from the 1st–4th centuries AD. They were of great importance in medieval European thought, since they were held to date from before the writings of Moses, and thus provide access to the body of primordial knowledge that had since been lost.

HERNIA, protrusion of abdominal contents through the abdominal wall in the inguinal or femoral part of the groin, or through the DIAPHRAGM (**hiatus hernia**). Hernia may occur through a congenital defect or through an area of MUSCLE weakness. Bowel and omentum are commonly found in hernial sacs and if there is a tight constriction at the neck of the sac (the hernia is "strangulated"), the bowel may be obstructed or suffer GANGRENE. In hiatus hernia, part of the STOMACH lies in the CHEST. Hernia may need SURGERY to reposition the bowel and close the defect, but this is rare in hiatus hernia.

HERODOTUS (c484–425 BC), Greek historian, renowned as "the Father of History" for his work seeking to describe and explain the causes of the Greco-Persian wars of 499–479 BC. This involved him in a monumental survey of the whole of mankind's previous history, collected from the stories he had heard during his extensive travels. He is also famed as a geographer and ethnologist.

HEROIN, OPIUM alkaloid with narcotic ANALGESIC and euphoriant properties, a valuable DRUG in severe pain of short duration (e.g., CORONARY THROMBOSIS) and in terminal malignant disease. It is abused in DRUG ADDICTION, taken intravenously for its psychological effects and later because of physical addiction. SEPTICEMIA and hepatitis may follow unsterile INJECTIONS and early death is common.

HERO OF ALEXANDRIA (c62 AD), or **Heron**, Greek scientist best known for inventing the **aeolipile**, a steam-powered engine that used the principle of jet propulsion, and many other complex steam- and water-powered toys. Other works ascribed to him deal with MENSURATION, optics (containing an early version of FERMAT's Principle) and MECHANICS.

HEROPHILUS (c300 BC), Alexandrian physician regarded as the father of scientific ANATOMY, and one of the first dissectors. He distinguished nerves from tendons and partially recognized their role. His work survives only through GALEN's writings.

HÉROULT, Paul Louis Toussaint (1863–1914), French metallurgist who invented at the same time as C. M. HALL the process now known as the HALL–HÉROULT PROCESS (c1886).

HERPES SIMPLEX. See COLD SORE.

HERPES ZOSTER. See SHINGLES.

HERPETOLOGY (from the Greek *herpeton*, a creeping thing), the study of REPTILIA and AMPHIBIA.

HERSCHEL, family of British astronomers of German origin. **Sir Frederick William Herschel** (1738–1822) pioneered the building and use of reflecting TELESCOPES, discovered URANUS (1781), showed the sun's motion in space (1783), found that some DOUBLE STARS were in relative orbital motion (1793), and studied NEBULAE. His sister **Caroline Lucretia** (1750–1848) assisted him and herself discovered eight COMETS. His son **Sir John Frederick William Herschel** (1792–1871), with BABBAGE and Peacock helped establish Leibnitzian CALCULUS notation in Britain, was the first to use SODIUM thiosulfate (hypo) as a photographic fixer, studied POLARIZED LIGHT and made many contributions to ASTRONOMY, especially that of the S Hemisphere.

HERSHEY, Alfred Day (1908–), US biologist who shared with DELBRÜCK and LURIA the 1969 Nobel Prize for Physiology or Medicine for their various researches on BACTERIOPHAGES.

HERTZ (Hz), SI UNIT of FREQUENCY, equal to 1 cycle per second; much used in RADIO technology.

HERTZ, Gustav Ludwig (1887–1975), German physicist who shared with J. FRANCK the 1925 Nobel Prize for Physics for their experiments showing the internal structure of the atom to be quantized, and so the value of the QUANTUM THEORY.

HERTZ, Heinrich Rudolph (1857–1894), German physicist who first broadcast and received RADIO waves (c1886). He showed also that they could be reflected and refracted (see REFLECTION; REFRACTION) much as light, and that they traveled at the same velocity though their wavelength was much longer (see ELECTROMAGNETIC RADIATION). In doing so he showed that light (and radiant heat) are, like radio waves, of electromagnetic nature.

HERTZSPRUNG, Ejnar (1873–1967), Danish astronomer who showed there was a relation between a STAR's brightness and color: the resulting **Hertzsprung-Russell Diagram** (named also for Henry RUSSELL) is important throughout astronomy and cosmology. He also conceived and defined absolute MAGNITUDE; and his work on CEPHEID VARIABLES has provided a way to measure intergalactic distances.

HERZBERG, Gerhard (1904–), German-born Canadian spectroscopist, awarded the 1971 Nobel Prize for Chemistry for work on the electronic structure and geometry of molecules. In particular he pioneered the study of the spectra of FREE RADICALS.

HESS, Victor Franz (1883–1964), Austrian-born US physicist who shared with C. D. ANDERSON the 1936 Nobel Prize for Physics for his discovery of COSMIC RAYS.

HESS, Walter Rudolf (1881–1973), Swiss physiologist awarded with MONIZ the 1949 Nobel Prize for Physiology or Medicine for his determination of the control exerted by certain parts of the BRAIN over the functioning of internal organs.

HETEROCYCLIC COMPOUNDS, major class of organic compounds in which the atoms are linked to form one or more rings, at least one of which includes one or more atoms other than carbon—commonly nitrogen, oxygen or sulfur. Many such compounds are of great biochemical or industrial importance. As with ALICYCLIC COMPOUNDS, saturated heterocyclic compounds resemble their ALIPHATIC analogues except for the effect of strain in small rings. There is also a large class of heterocyclic AROMATIC COMPOUNDS, highly distinctive in properties.

HETEROTROPH. See AUTOTROPH.

HEURISTICS, an approach to problem-solving in which a formally unjustifiable solution is assumed as an aid in exploring the implications of the problem. In science, even theories which ultimately prove misconceived can be of great heuristic value in research.

HEVESEY, George Charles de (1885–1966), Hungarian-born chemist awarded the 1943 Nobel Prize for Chemistry for his work on radioactive tracers (see RADIOISOTOPES). He was also the co-discoverer of the element HAFNIUM.

HEWISH, Antony (1924–), radio astronomer,

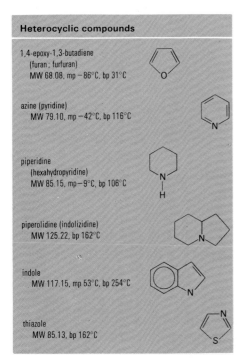

Heterocyclic compounds

1,4-epoxy-1,3-butadiene
(furan; furfuran)
MW 68.08, mp −86°C, bp 31°C

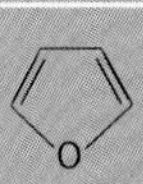

azine (pyridine)
MW 79.10, mp −42°C, bp 116°C

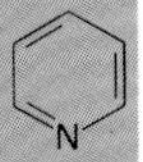

piperidine
(hexahydropyridine)
MW 85.15, mp −9°C, bp 106°C

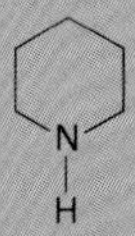

piperolidine (indolizidine)
MW 125.22, bp 162°C

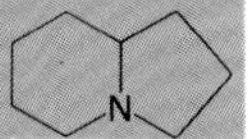

indole
MW 117.15, mp 53°C, bp 254°C

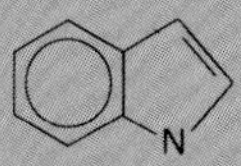

thiazole
MW 85.13, bp 162°C

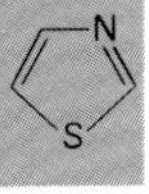

corecipient with RYLE of the 1974 Nobel physics prize, Professor of Radio Astronomy at Cambridge since 1971. He headed the team which discovered the first PULSAR (1967).

HEXAGON. See POLYGON.

HEXAHEDRON. See POLYHEDRON.

HEYERDAHL, Thor (1914–), Norwegian ethnologist famous for his expeditions to prove the feasibility of his theories of cultural diffusion, and for his books. In the **Kon-Tiki**, a primitive balsawood raft, he and his crew sailed from the W coast of South America to Polynesia, demonstrating the possibility that the Polynesians originated in South America (1947). In **Ra**, a facsimile of an ancient Egyptian papyrus reed boat, he and his cosmopolitan crew succeeded at the second attempt in sailing from Morocco to Barbados, showing the possibility that the pre-Columbian cultures of South America were influenced by Egyptian civilization (Ra I, 1969, Ra II, 1970).

HEYMANS, Corneille Jean François (1892–1968), Belgian physiologist awarded the 1938 Nobel Prize for Physiology or Medicine for discovering sensory organs, close by the carotid artery and aorta, that play a part in regulating RESPIRATION.

HEYROVSKY, Jaroslav (1890–1967), Czech physical chemist awarded the 1959 Nobel Prize for Chemistry for inventing the polarograph (see POLAROGRAPHY).

HIBERNATION, a protective mechanism whereby certain animals reduce their activity and apparently sleep throughout winter. At its most developed it is a characteristic of warm-blooded animals but a comparable phenomenon, **diapause**, is found in cold-blooded forms. Diapause is a direct physiological response to cold temperatures: metabolic activity in cold-blooded animals is entirely dictated by external temperature. In hibernating animals, internal preparations, such as laying down a store of fat, begin several weeks before the onset of hibernation. Then, when temperatures drop, the animal goes to sleep. Pulse rate and breathing drop to a minimum. With metabolism reduced, the animal can live on food stored in its body till spring. Winter food supplies would not be sufficient to maintain the animal in a fully-active state. When an animal remains torpid throughout the summer, this is known as **aestivation**.

A comparison between the form of certain symbols as they appeared in the hieroglyphics of various ages, and in the hieratic and the demotic scripts. The form used in the Book of the Dead is known as "linear hieroglyphics"

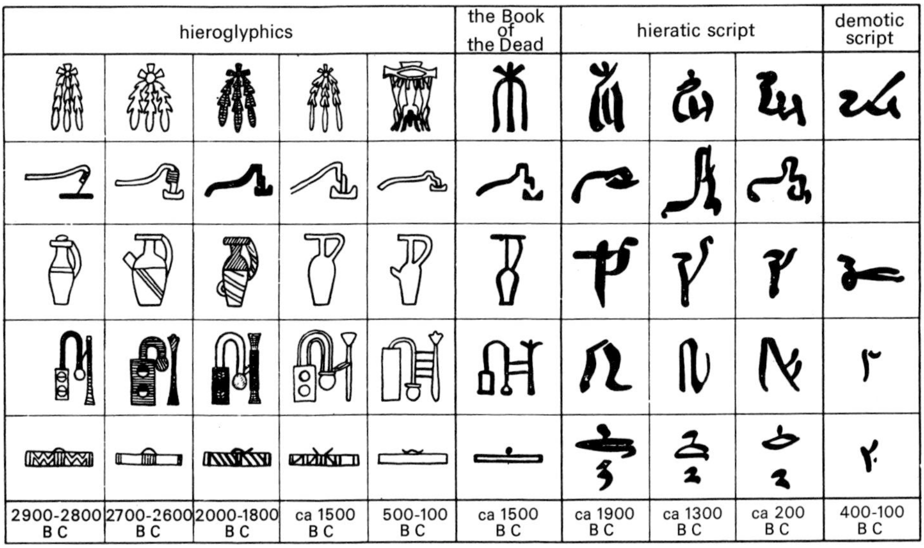
hieroglyphics
the Book of the Dead
hieratic script
demotic script
2900-2800 B C
2700-2600 B C
2000-1800 B C
ca 1500 B C
500-100 B C
ca 1500 B C
ca 1900 B C
ca 1300 B C
ca 200 B C
400-100 B C

HICCUP, brief involuntary contraction of the DIAPHRAGM that may follow dietary or alcoholic excess and rapid eating. It may also be a symptom of UREMIA, mineral disorders and brain-stem disease. Rebreathing into a paper bag or repeated swallowing are effective remedies; chlorpromazine also suppresses hiccups.

HIEROGLYPHICS, system of writing using pictorial characters (hieroglyphs), especially that found on Egyptian monuments. Egyptian hieroglyphics are first found from c3000 BC, their use declining during the 3rd century AD. Initially there were a fairly limited number of hieroglyphs. This was followed by a rapid expansion of the number of characters in order to reduce ambiguity, and by a further expansion around 500 BC. There were two derived cursive scripts, hieratic and demotic. **Hieratic script,** initially used only for sacred texts, coexisted with true hieroglyphics from early on until c100 AD. The less legible, more cursive **demotic script** appeared around 660 BC and disappeared around 450 AD. The writings of other ancient peoples, e.g., the Hittites and Mayas, are also termed hieroglyphics. (See also ROSETTA STONE.)

HIGH BLOOD PRESSURE. See BLOOD CIRCULATION.

HIGHEST COMMON FACTOR (hcf), of two or more INTEGERS, the largest integer which they share as a FACTOR. For example, the hcf of 15, 18 and 27 is 3; similarly, the hcf of 15 and 17 is 1, since 17 is a PRIME NUMBER. In ALGEBRA, the hcf of two or more algebraic expressions may be found by examination of the factors of each: hence the hcf of $9ax^2$ $(=3.3.a.x.x)$ and $3a^2x$ $(=3.a.a.x)$ is $3ax$. (See also LOWEST COMMON MULTIPLE.)

HIGH-FIDELITY, an adjective applicable to systems carrying a signal with very little distortion, such as a good CAMERA or RADIO transmitter, but also a generic noun ("Hi-Fi") for a wide range of domestic equipment for SOUND reproduction. The input signal may arise from a phonograph disc, in which case a high-compliance (flexibility) stylus following a groove produces a piezoelectric (see PIEZOELECTRICITY) or induced (see ELECTROMAGNETISM) voltage; from magnetic tape, on which the signal is recorded in the variations of magnetization of a ferromagnetic (see MAGNETISM) coating, produced by a finely focused ELECTROMAGNET (the recording head) and inducing a voltage in the small playback head coil; or from a radio receiver which detects the slight variations in intensity (AM) or frequency (FM) of a broadcast electromagnetic wave. The resulting voltage is amplified electronically and passed to a LOUDSPEAKER, consisting typically of a paper cone, vibrated by an electromagnet, in an enclosure which attempts to compensate for the uneven response of the cone for different directions and frequencies. The most important measures of the overall faithfulness are the frequency response (the range of frequencies passed with intensities unchanged within a quoted tolerance), the harmonic distortion (the change in the balance of the HARMONICS of a signal—particularly a boost in the high harmonics), the hum and NOISE levels (in the absence of a signal), and the flutter and wow (fluctuations in speed of record or tape decks).

HILBERT, David (1862–1943), German mathematician whose most important contributions were in the field of mathematical LOGIC. With the advent of the NON-EUCLIDEAN GEOMETRIES it had become clear that the axiomatic basis of EUCLID's work needed further examination. This Hilbert did, establishing a logical axiomatic system for geometry.

HILL, Archibald Vivian (1886–), British physiologist who shared with MEYERHOF the 1922 Nobel Prize for Physiology or Medicine for their independent work on the biochemistry of MUSCLE contraction and relaxation.

HINSHELWOOD, Sir Cyril Norman (1897–1967), British physical chemist awarded with SEMYONOV the 1956 Nobel Prize for Chemistry for his work on reaction rates and mechanisms (see KINETICS, CHEMICAL), especially in the reaction of hydrogen and oxygen to form WATER.

HIPPARCHUS (c130 BC), Greek scientist, the father of systematic ASTRONOMY, who compiled the first star catalog and ascribed stars MAGNITUDES, made a good estimate of the distance and size of the moon, probably first discovered PRECESSION, invented many astronomical instruments, worked on plane and SPHERICAL TRIGONOMETRY, and suggested ways of determining LATITUDE AND LONGITUDE.

HIPPOCRATES (c460–c377 BC), Greek physician generally called "the Father of Medicine" and the probable author of at least some of the **Hippocratic Collection**, some 60 or 70 books on all aspects of ancient MEDICINE. The authors probably formed a school centered around Hippocrates during his lifetime and continuing after his death. The **Hippocratic Oath**, traditionally regarded as the most valuable statement of medical ethics and good practice, probably represents the oath sworn by candidates for admission to an ancient medical guild.

HIRUDINEA, a class of the phylum ANNELIDA.

HISTAMINE, AMINE concerned with the production of INFLAMMATION, and particularly of HIVES and the allergic spasm of the BRONCHI in ASTHMA and ANAPHYLAXIS; it enhances STOMACH acid secretion and has several effects on BLOOD CIRCULATION. ANTIHISTAMINES and cromoglycate can interfere with its release; ADRENALINE counteracts its serious effects.

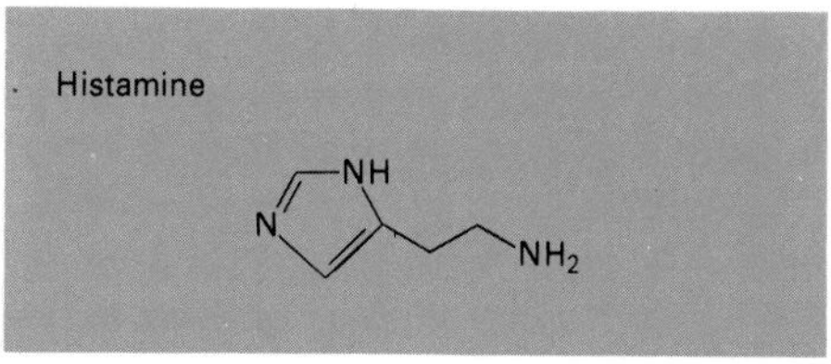

HISTIDINE ($C_3H_3N_2CH_2CH(NH_2)COOH$), a basic AMINO ACID found in many PROTEINS; HEMOGLOBIN contains 8.5% histidine. Man is capable of synthesizing histidine so it need not appear in the diet. It is the metabolic precursor of HISTAMINE.

HISTOGRAM, a graphical way of representing statistical information (see GRAPHS; STATISTICS). The data is classified, and the classes marked off along the x-axis (see AXES); rectangles whose bases are centered on the class midpoints, and whose heights are proportional to the frequencies in the classes, are then constructed. Similar is the **frequency polygon**, constructed by plotting the class midpoints against their respective frequencies and joining the plotted points. If it is meaningful to do so (depending on the shape of the polygon, the certainty felt that the sample is representative of the population, the way that the data has been classified, etc.) a **frequency curve** may be drawn that best "fits" the plotted points.

HISTOLOGY, the study of the microscopic ANATOMY of parts of organisms after DEATH (autopsy) or removal by SURGERY (BIOPSY). Tissue is fixed by agents that denature PROTEINS, preventing autolysis and bacterial degradation; they are stained by dyes that have particular affinity for different structures. Histology facilitates the study both of normal tissue and of diseased organs, or pathological tissue.

A large number of people were asked how many hours a week they spent watching television. The results of the survey can be represented either on a histogram (A); as a frequency polygon (B—with data coordinates indicated), or as a frequency curve (C).

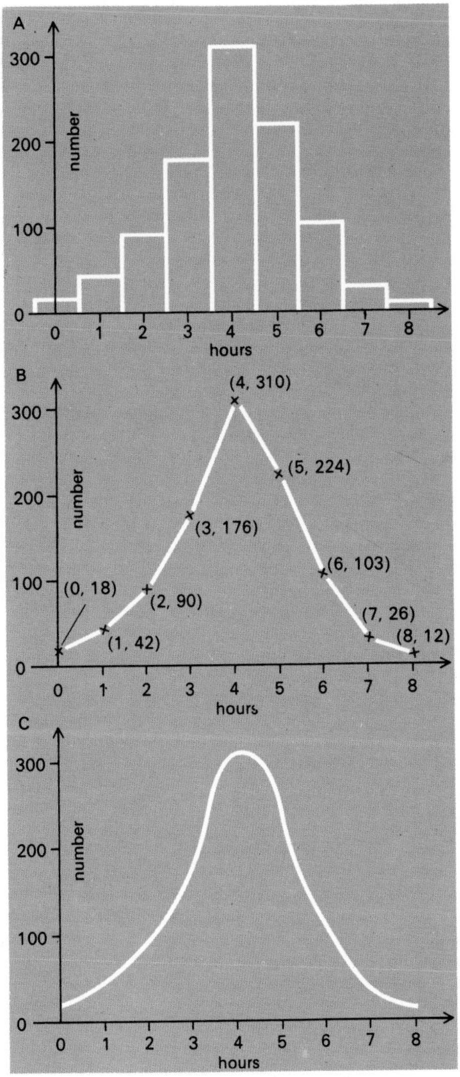

HISTOPLASMOSIS, FUNGAL disease prevalent in parts of North America and Africa, and carried by poultry, birds and bats. It may cause acute respiratory infection, a rapidly developing disseminated form, or a chronic type with FEVER, debility and specific organ involvement (especially of the LUNGS and resembling TUBERCULOSIS). Specific antifungal ANTIBIOTICS are needed for the disseminated disease.

HIVES, or **urticaria,** an itchy SKIN condition characterized by the formation of weals with surrounding ERYTHEMA, and due to HISTAMINE release. It is usually provoked by ALLERGY to food (e.g., shellfish, nuts, fruits), pollens, FUNGI, DRUGS (e.g., PENICILLIN) or parasites (SCABIES, worms). But it may be symptomatic of infection, systemic disease or emotional disorder. **Dermographism** is a condition in which slight skin pressure may produce marked hives, as in the linear marks which appear after writing on the skin.

HOARFROST. See FROST.

HOBBES, Thomas (1588–1679), English political philosopher who sought to apply rational principles to the science of human nature. In both the physical and the moral sciences reasoning was to proceed from cause to effect: certain knowledge could only flow from deductive reasoning upon known principles. Hobbes' view of man was materialistic and pessimistic—men's actions were motivated solely by self-centered desires. This led Hobbes to consider that the existence of a sovereign authority in a state was the only way to guarantee its stability. *Leviathan* (1651), which gave voice to this opinion, was his most celebrated work. Hobbes saw matter in motion to be the only reality: even consciousness and thought were but the outworkings of the motion of atoms in the brain. During and after his lifetime, Hobbes was well known as a materialist and suspected as an atheist, but in the 20th century his fame as an able thinker has overshadowed his former notoriety.

HODGKIN, Alan Lloyd (1914–), British physiologist awarded with A. F. HUXLEY and J. ECCLES the 1963 Nobel Prize for Physiology or Medicine for his work with Huxley on the chemical basis of nerve impulse transmission (see NERVOUS SYSTEM).

HODGKIN, Dorothy Mary Crowfoot (1910–), British chemist awarded the 1964 Nobel Prize for Chemistry for determining the structure of VITAMIN B_{12}.

HODGKIN'S DISEASE, the most important type of LYMPHOMA or malignant proliferation of LYMPH tissue. Usually occurring in young adults, it may present with lymph node enlargement, weight loss, FEVER or malaise; the SPLEEN, LIVER; LUNGS and BRAIN may be involved. Treatment has radically improved the outlook in this disease, with cure obtained in a substantial proportion of cases; it consists of local RADIATION THERAPY or systemic intermittent CHEMOTHERAPY with a combination of agents and STEROIDS.

HODOGRAPH, the CURVE defined by the ends of the VECTORS drawn from any point O representing changes in the VELOCITY of a particle.

HOE, Richard March (1812–1886), US inventor who developed many machines associated with PRINTING and invented the first successful rotary printing press (c1847).

HOFF, Jacobus Henricus van 't. See VAN 'T HOFF, JACOBUS HENRICUS.

HOFMANN, August Wilhelm von (1818–1892), German organic chemist. While he was teaching in London, W. H. PERKIN, his pupil, prepared (1856) the first synthetic dye (see DYES AND DYEING). Hofmann returned to Germany and synthesized many new dyes, and for some 50 years thereafter Germany had the world's largest dye industry. He also discovered the **Hofmann degradation** process (see AMIDES).

HOFSTADTER, Robert (1915–), US physicist who shared with MÖSSBAUER the 1961 Nobel Prize for Physics for his discoveries of the structure of PROTONS and NEUTRONS.

HOGBACK, or hog's-back, a CUESTA both of whose slopes are steep and of approximately equal gradients.

HOHOKAM CULTURE, pre-Columbian North American Indian culture based along the Gila and Salt Rivers, Ariz., from c300 BC to c1400 AD. They built a complex network of irrigation canals, made pottery, and built their houses over shallow pits.

HOIST. See CRANE; PULLEY.

HOLBACH, Paul Henri Dietrich, Baron d' (1723–1789), French encyclopedist and materialist philosopher, best known for *The System of Nature* (1770), published as by "J. B. Mirabeau," which included a scathing attack on religion. He translated many scientific articles for DIDEROT's *Encyclopédie*.

HOLE. See SEMICONDUCTOR.

HOLLAND, John Philip (1840–1914), Irish-born US inventor who built the first fully successful SUBMARINE, the *Holland*, launched in 1898 and bought by the US Navy in 1900.

HOLLEY, Robert William (1922–), US biochemist who shared with NIRENBERG and KHORANA the 1968 Nobel Prize for Physiology or Medicine for his work in establishing for the first time the nucleotide structure of a NUCLEIC ACID.

HOLMIUM (Ho), element of the LANTHANUM SERIES. AW 164.9, mp 1470°C, bp 2720°C, sg 8.795 (25°C).

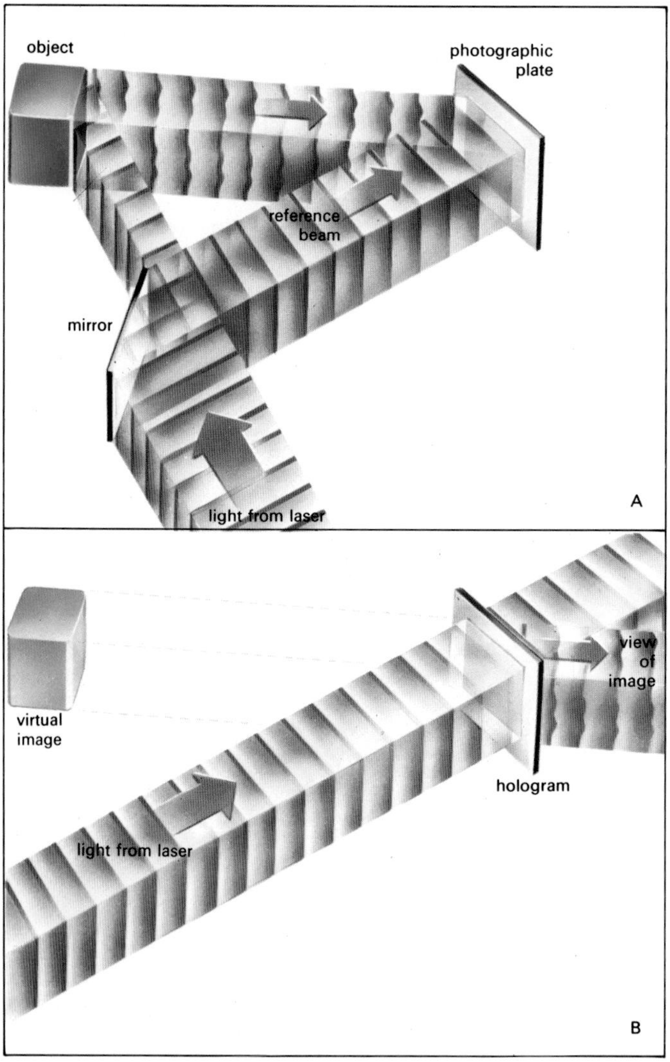

The principle of holography. (A) To form a hologram, an object is illuminated in coherent light (usually from a laser). A photographic plate is then exposed in a combination of the light scattered from the object and some of the original beam of coherent light (the reference beam) reflected in a mirror. (B) To view the hologram, it is illuminated in coherent light similar to the original reference beam. The light scattered by the hologram then affords views of an image of the original object in whatever orientation the hologram is viewed. Holography depends on interference in the plane of the hologram between the light scattered by the object and the coherent reference beam.

HOLOCENE, the later epoch of the QUATERNARY, representing the elapse from the end of the last ICE AGE up to and including the present; i.e., about the last 10 000 years. (See also GEOLOGY.)

HOLOGRAPHY, a system of recording LIGHT or other waves on a photographic plate or other medium in such a way as to allow a three-dimensional reconstruction of the scene giving rise to the waves, in which the observer can actually see round objects by moving his head. The apparently unintelligible plate, or **hologram**, records the INTERFERENCE pattern between waves reflected by the scene and a direct reference wave at an angle to it; it is viewed by illuminating it from behind and looking through rather than at it. The high spatial coherence needed prevented exploitation of the technique, originated in 1948 by D. GABOR, until the advent of LASERS. Color holograms are possible, and three-dimensional TELEVISION may ultimately be feasible.

HOLOSTEI, a group of mainly fossil fish of the ACTINOPTERYGII represented today by the garpikes (*Lepisosteus*) and bowfin (*Amia*).

HOLOTHUROIDEA, a class of ECHINODERMATA which includes the sea cucumbers. They are cylindrical or sausage-shaped with a thick leathery skin. The CALCITE plates of the skeleton are separated and modified as rods, hooks, anchors and wheels.

HOMATROPINE, short-acting ATROPINE-like DRUG used as EYE drops to dilate the pupil for eye examination.

HOMEOMORPHISM. See TOPOLOGY.

Two photographs of the same hologram illuminated by the same reference beam, taken looking into the hologram from different directions.

HOMEOPATHY, system of treatment founded in the early 19th century by C. F. S. HAHNEMANN, based on a theory that DISEASE is cured by DRUGS whose effects mimic it and whose efficacy is increased by the use of extremely small doses, achieved by multiple dilutions.

HOMEOSTASIS, the self-regulating mechanisms whereby biological systems attempt to maintain a stable internal condition in the face of changes in the external environment. It was the 19th-century French physiologist Claude BERNARD who first realized that the internal environment of any free living organism was maintained constant within certain limits. Homeostasis is generally achieved through two types of regulating systems: on-off control and FEEDBACK control. HORMONES often play a vital role in maintaining homeostatic stability.

HOMOGENIZATION, process to delay the separation of fat in milk. Milk, a rather unstable emulsion (see COLLOID), contains fat globules that tend to coalesce. In homogenization the MILK is heated to about 60°C and passed at pressure through small openings. The fatty clusters are broken up by SHEARING as they pass through the holes, by the action of pressure, and by impact with components of the homogenizer.

HOMOLOGOUS SERIES, sequence of chemical compounds which differ one from the next by a simple structural unit, and so can be given a general FORMULA. The members of a series are chemically similar, having the same FUNCTIONAL GROUPS, and their physical properties change regularly with molecular weight. Organic homologous series, such as the ALKANES, are built up by adding the methylene (CH_2) group.

HOMOLOGUE, in biology, a structure or organ that has the same evolutionary origin as an apparently different structure in another species. For instance, there is little apparent similarity between a horse's leg and the flipper of a whale, but both have similar embryonic history. (See EVOLUTION.)

HOMO SAPIENS. See PREHISTORIC MAN; RACE.

HOMOSEXUALITY, mutual sexual attraction by members of the same sex; in women it is termed **lesbianism**. While based on physical attraction, the degree of sexual involvement is variable. It is widely regarded as acceptable social conduct between consenting adults, but the involvement of children is considered undesirable. A number of VENEREAL and other DISEASES can be transmitted by homosexual practice in males.

HONEY, a sweet, sticky confection, formed of partially-digested SUGARS. NECTAR, collected from flowers by foraging worker BEES, is returned to the hive, mixed with digestive "saliva" and often a little pollen, and stored in the cells, of a wax honeycomb to act as a winter food supply for the hive. Combs, with their familiar hexagonal cells, are used for a variety of purposes in the hive, and honeycombs are not always distinct from combs of grubs. Where honey is taken from domestic hives for man's use, the beekeeper must replace the food supply by feeding sugar throughout the winter.

HOOF, a curved horny structure covering the end of the digits in UNGULATES. Since in the horse and related species the foot is effectively reduced to a single toe, the term hoof is applied to the whole foot.

Major hormones in humans

origin	hormone	chemical type	main function
pituitary gland (anterior lobe)	adrenocorticotropic hormone —*ACTH*	polypeptide	controls growth of and secretion from adrenal glands
	thyrotropic hormone (thyroid stimulating hormone—*THS*)	protein	controls growth of and secretion from thyroid gland
	gonadotropic hormones:		
	luteinizing hormone—*LH* (interstitial-cell stimulating hormone—*ICSH*)	protein	promotes secretion of androgens (in males) and estrogens (in females)
	follicle stimulating hormone —*FSH*	protein	initiates development of ovarian follicles (in females); stimulates spermatogenesis (in males)
	lactogenic hormone (prolactin)	polypeptide	stimulates secretion of milk from mammary glands (in females)
	growth hormone (somatropin)	polypeptide	promotes growth, and lactation
pituitary gland (posterior lobe)	oxytocin	polypeptide	uterine stimulation
	vasopressin	polypeptide	vasoconstriction and diuretic action
thyroid gland	thyroxine triiodothyronine	phenol derivatives	acceleration of cell metabolism
	thyrocalcitonin	polypeptide	lowers blood calcium level
parathyroid gland	parathyroid hormone	polypeptide	controls bone resorbtion; increases blood calcium and lowers blood phosphate levels
adrenal gland (medulla)	adrenaline (epinephrine)	phenol derivative	vasodilation; raises blood pressure in stress situations
	noradrenaline (norepinephrine)	phenol derivative	vasoconstriction; normal control of blood pressure
adrenal gland (cortex)	corticosteroids (seven) including:		
	cortisol cortisone	steroids	control carbohydrate metabolism
	aldosterone	steroid	controls water and salt metabolism
pancreas	insulin	polypeptide	reduces blood sugar (glucose) level
	glucagon	polypeptide	increases blood glucose level
duodenum	secretin		stimulates digestive juice production by pancreas
ovary (in females)	estrogens including: estradiol estrone estriol	steroids	control menstrual and reproductive cycles; also produce secondary sexual characteristics
corpus luteum (in females)	progestogens including: progesterone	steroid	promotes development of lining of uterus in menstrual cycle
placenta (in females)	various hormones		various roles in control of pregnancy
testis (in males)	androgens including: androsterone testosterone	steroid	control of primary and secondary sexual characteristics

HOOF AND MOUTH DISEASE, or foot and mouth disease, a VIRUS infection of cattle and pigs, rarely affecting domestic animals and man. Vesicles of the SKIN and mucous membranes, and FEVER are usual. It is highly contagious and EPIDEMICS require the strict limitation of stock movements and the slaughter of affected animals.

HOOKE, Robert (1635–1703), English experimental scientist whose proposal of an inverse-square law of gravitational attraction (1679) prompted NEWTON into composing the *Principia.* From 1655 Hooke was assistant to R. BOYLE but he entered into his most creative period in 1662 when he became the ROYAL SOCIETY OF LONDON's first curator of experiments. He invented the compound MICROSCOPE, the universal joint and many other useful devices. His microscopic researches were published in the beautifully illustrated *Micrographia* (1665), a work which also introduced the term "cell" to biology. He is best remembered for his enunciation in 1678 of **Hooke's Law**. This states that the deformation occurring in an elastic body under stress is proportional to the applied stress (see MATERIALS, STRENGTH OF).

HOOTON, Earnest Albert (1887–1954), US physical anthropologist best remembered for his attempts to relate behavior to physical or racial type, and for books such as *Up From the Ape* (1931) and *The American Criminal* (1939).

HOPEWELL CULTURE, pre-Columbian culture of MOUND builders, flourishing c500 BC–c500 AD and centered in S Ohio. They appear to have had a fairly sophisticated social structure, made decorated pottery, carved stone and were skilled metallurgists.

HOPKINS, Sir Frederick Gowland (1861–1947), British biochemist who shared with EIJKMAN the 1929 Nobel Prize for Medicine or Physiology for his work showing the necessity of certain dietary elements, now identified and known as VITAMINS, for the maintenance of health.

HORIZON, the apparent line where the sky meets the land or sea. At sea, its distance varies in proportion to the square ROOT of the height of the observer's eyes above sea level: if this is, say, 2m the horizon will be about 5.57km distant. The **celestial horizon** is the great circle on the CELESTIAL SPHERE at 90° from the ZENITH (the point immediately above the observer).

HORMONES, substances produced in living organisms to affect GROWTH, differentiation, METABOLISM, digestive function, mineral and fluid balance, and usually acting at a distance from their site of origin. Plant hormones, AUXINS and GIBBERELLINS, are particularly important in growth regulation. In animals and man, hormones are secreted by ENDOCRINE GLANDS, or analogous structures, into the BLOOD stream which carries them to their point of action. The rate of secretion, efficacy on target organs and rate of removal are all affected by numerous factors including FEEDBACK from their metabolic effects, mineral or sugar concentration in the blood, and the action of controlling hormones. The latter usually originate in the PITUITARY GLAND and those controlling the pituitary in the HYPOTHALAMUS. Important hormones include INSULIN, THYROID hormone, ADRENALINE, STEROIDS, PARATHYROID GLAND hormone, GLUCAGON, GONADOTROPHINS, ESTROGEN, PROGESTERONE, ANDROGENS, pituitary growth hormone, VASOPRESSIN, thyroid stimulating hormone, adrenocorticotrophic hormone, GASTRIN and SECRETIN.

HORNBLENDE, group of common rock-forming minerals; dark monoclinic AMPHIBOLES of general composition $Ca_2Na(Mg, Fe, Al)_5Si_6(Si, Al)_2O_{22}(OH)_2$.

HORNEY, Karen (1885–1952), German-born US psychoanalyst best known for her concentration on the importance of environment in character development, so rejecting many of the basic principles of FREUD's classic psychoanalytic theory, especially his stress on the LIBIDO as the root of personality and behavior (see PSYCHOANALYSIS).

HORNS, strictly, keratinous structures (see KERATIN) borne on the forehead of many UNGULATES. They show a variety of forms. Horns are usually permanent structures though the antlers of many deer are cast and regrown annually. Horns appear occasionally to be purely ornamental, but usually they are used for defense or in intra-specific AGGRESSION. In such species horns are borne only by the males.

HORNWORTS, or horned liverworts, small group of BRYOPHYTA belonging to the class Anthocerotae. They have a very simple thallus, the cells of which contain a single chloroplast and a pyrenoid, the latter only occurring elsewhere in ALGAE.

HOROLOGY, the measurement of time, and of the construction of timepieces. See CLOCKS AND WATCHES.

HORSE LATITUDES, two belts, characterized by low winds, about 30°N and S of the equator. Sailing ships bound for America carrying horses were often becalmed here: many horses died from lack of fresh water and were cast overboard. (See also TRADE WINDS.)

HORSEPOWER (hp), unit of power introduced by James WATT, equivalent to 745.70 watt. Brake horsepower (bhp) is power output measured by applying a brake (usually a Prony brake) to the driving shaft. The metric horsepower, or *cheval-vapeur* (CV), originally the power required to raise a 75-kg weight through one metre in one second, is 735.5 watt.

HORSESHOE WORMS. See PHORONIDA.

HORSETAILS, or Sphenopsida, subdivision of the PLANT KINGDOM which reached its evolutionary peak in the CARBONIFEROUS period. Present-day sphenopsids comprise a single genus *Equisetum,* which are green, rush-like weeds found throughout the world except Australia and New Zealand.

HORST, an area that has been thrust upward between two roughly parallel FAULTS.

HORTICULTURE, branch of agriculture concerned with producing fruit, flowers and vegetables. It can be divided into pomology (growing fruit), olericulture (growing vegetables) and floriculture (growing shrubs and ornamental plants). About 3% of US cropland is devoted to horticulture. It was originally practiced on a small scale, but crops such as the potato and tomato are now often grown in vast fields.

HOST, an animal or plant that supports a PARASITE.

HOT SPRINGS, or **thermal springs,** springs supplied by underground water heated usually by vapor from the MAGMA, and most common in recently active volcanic regions (see VOLCANISM). The water contains dissolved minerals that may form terraces around the outlet. (See also GEYSER.)

HOTTENTOTS, people of South Africa similar to the Bushmen and Hamites. Small in stature, they have brown skins, prominent cheekbones, broad noses, coarse hair and pointed chins, and are dolichocephalic (see CEPHALIC INDEX) and commonly steatopygic (see STEATOPYGIA). Originally known to themselves as the Khoikhoin, they were nomadic herdsmen and farmers, but this way of life has largely disappeared.

HOUR (h), unit of time equal to 60 minutes or 3 600 seconds. In one hour, the earth rotates on its axis through 15°. The time of day at any point on earth is expressed as the number of hours and minutes that have elapsed since midnight for the time zone in which the point is situated, the time zones being fixed intervals behind or ahead of Greenwich Mean Time.

HOURGLASS, ancient instrument to measure the passage of time. A quantity of fine, dry sand is contained in a bulb constricted at its center to a narrow neck. The device is turned so that all the sand is in the upper chamber: the time taken for the sand to trickle into the lower chamber depends on the amount of sand and on the diameter of the neck. Small hourglasses are used in the home as eggtimers.

HOUSSAY, Bernardo Alberto (1887–1971), Argentinian physiologist awarded (with C. and G. CORI) the 1947 Nobel Prize for Physiology or Medicine for his discovery that certain HORMONES produced by the PITUITARY GLAND were responsible for regulating the blood's sugar and insulin content.

HOVERCRAFT. See AIR-CUSHION VEHICLE.

HOWE, Elias (1819–1867), US inventor of the first viable SEWING MACHINE (patented 1846). The early machines were sold in Britain, as in the US there was at first no interest. Later Howe fought a protracted legal battle (1849–54) to protect his patent rights from infringement in the US.

HOYLE, Sir Fred (1915–   ), British cosmologist best known for formulating with T. GOLD and H. BONDI the steady state theory (see COSMOLOGY); and for his important contributions to theories of stellar evolution, especially concerning the successive formation of the elements by nuclear FUSION in STARS. He is also well known as a science fiction writer and for popular books such as *Frontiers of Astronomy* (1955).

HRDLIČKA, Aleš (1869–1943), Bohemian-born US physical anthropologist best known for expounding the theory that the AMERINDS are of Asiatic origin, a theory still generally accepted today.

HUBBLE, Edwin Powell (1889–1953), US astronomer who first showed (1923) that certain NEBULAE are in fact GALAXIES outside the MILKY WAY. By examining the RED SHIFTS in their spectra, he showed that they are receding at rates proportional to their distances (see HUBBLE'S CONSTANT).

HUBBLE'S CONSTANT, ratio between the distance of a GALAXY and the rate at which it is receding from us. HUBBLE first calculated this as around 500km/s per Mpc; however, he incorrectly estimated the distances of the galaxies, and the constant has been more recently calculated to be about 75km/s per Mpc.

HUGGINS, Charles Brenton (1901–   ), Canadian-born US surgeon awarded (with F. P. ROUS) the 1966 Nobel Prize for Physiology or Medicine for his discovery that TUMORS of the male PROSTATE GLAND could be controlled by injection of female sex HORMONES, the first demonstration that CANCER might be controlled by chemical agents.

HUMAN BODY, the physical substrate of man, *Homo sapiens*. In terms of ANATOMY, it consists of the head and neck, a trunk divided into the CHEST, ABDOMEN and PELVIS, and four limbs; two ARMS and two LEGS. The head contains (within the bony structure of the SKULL) the BRAIN, which is connected by cranial nerves to the special sense organs for VISION (EYES), hearing and balance (EARS), SMELL (NOSE), and TASTE. On the front of the head is the FACE, specialized for communication (including the special senses, and through which the VOICE emanates—see SPEECH AND SPEECH DISORDERS). The head sits at the top of the **spinal column** of VERTEBRAE, which continue through the neck, thorax and lumbar region to the sacrum and COCCYX. The spinal column is the central structural pillar of the musculoskeletal system, and that onto which the ribs, chest and abdominal walls, and pelvic bones articulate. Within the bony spinal canal is the SPINAL CORD, the downward extension of the brain concerned with relaying information to and from the body and with segmental REFLEX behavior. It is linked with the various parts of the body by the peripheral and autonomic NERVOUS SYSTEMS. The chest, abdomen and pelvis contain many vital organs comprising the various functional systems.

The **internal functions** of the human body include: the BLOOD CIRCULATION, which supplies all organs with OXYGEN and nutrients and removes waste products from them (see AORTA; ARTERIES; CAPILLARIES; HEART; VEINS; VENA CAVA); RESPIRATION, which provides oxygen for the blood and removes carbon dioxide via the LUNGS, BRONCHI and TRACHEA; the DIGESTIVE SYSTEM, which starting at the MOUTH and PHARYNX, leads into the GASTROINTESTINAL TRACT; systems for EXCRETION and METABOLISM, including particularly the LIVER, KIDNEYS and BLADDER; BLOOD, formed in the BONE MARROW and circulating throughout the body; the *lymphoreticular system*, which has a major role in IMMUNITY and blood degradation (see LYMPH; SPLEEN); sites of HORMONE secretion, or ENDOCRINE GLANDS; and the *reproductive system*, essential for the propagation of the species (including the OVARIES, FALLOPIAN TUBES, WOMB, vagina, TESTES and PENIS). The *limbs* are primarily concerned with locomotion and fine movement (see JOINTS; MUSCLES; TENDONS) and with tactile sensibility (see HANDS; TOUCH).

The whole body surface is covered with SKIN, which is a protective layer also concerned with temperature and fluid regulation, specialized for tactile sensation and bearing many HAIRS. The mucous membranes of the nose, mouth, respiratory tract and gastrointestinal tract and of exocrine GLANDS are made of surface EPITHELIUM. The thin inner layer of blood vessels (*endothelium*) is a surface which, when intact prevents the CLOTTING of the contained blood. Other TISSUES found in the body include muscle, both striated (or skeletal) and smooth (or visceral); connective tissue; ADIPOSE TISSUE; BONE; CARTILAGE, and the specialized tissues of the organs discussed above.

The basic unit of the body is the CELL. In the early stages of the development of an individual, this is a multipotential structure containing all the genetic

The anatomy of the human body. The muscles and bones are shown on the left, the principal organs (female) on the right.

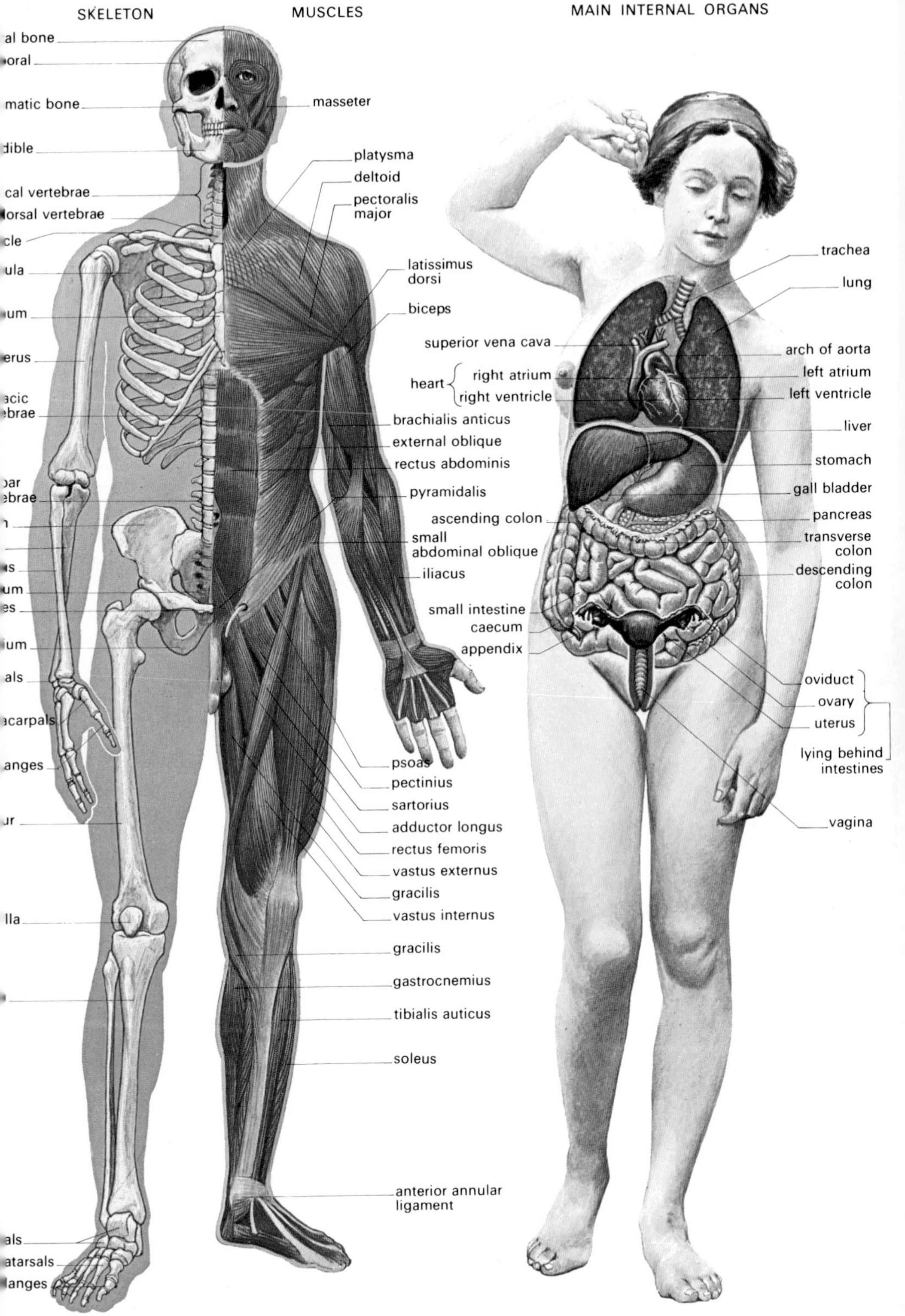
SKELETON
MUSCLES
MAIN INTERNAL ORGANS
al bone
oral
matic bone
dible
cal vertebrae
orsal vertebrae
cle
ula
um
erus
acic
ebrae
bar
ebrae
s
um
es
ium
als
acarpals
anges
ur
lla
als
atarsals
anges
masseter
platysma
deltoid
pectoralis major
latissimus dorsi
biceps
brachialis anticus
external oblique
rectus abdominis
pyramidalis
small abdominal oblique
iliacus
psoas
pectinius
sartorius
adductor longus
rectus femoris
vastus externus
gracilis
vastus internus
gracilis
gastrocnemius
tibialis auticus
soleus
anterior annular ligament
trachea
lung
superior vena cava
arch of aorta
heart
right atrium
right ventricle
left atrium
left ventricle
liver
stomach
gall bladder
ascending colon
pancreas
transverse colon
descending colon
small intestine
caecum
appendix
oviduct
ovary
uterus
lying behind intestines
vagina

information (in GENES on the CHROMOSOMES of its nucleus) required for the development, differentiation, growth and function of any and all the parts of the body. Each cell divides repeatedly and forms a cell line which gradually specializes into a particular aspect of body structure and PHYSIOLOGY, while suppressing its other potentialities. In this way the highly complex and subspecialized body develops, the integration and control of development being perhaps genetically inbuilt or, alternatively, controlled by hormones and the nervous system. The early development of the EMBRYO and FETUS before BIRTH continue in childhood with subtler differentiation and increase of bulk, a further growth spurt and much sexual development occurring at PUBERTY. After early adult life, the optimal function of organs begins to become impaired with the degenerative processes of AGING. Added to these are the DISEASES to which the human body is susceptible; these range from environmental influences such as trauma and infection to specific diseases such as CANCER. Disease and degeneration determine the finite quality of human life, leading to its termination in DEATH.

HUMAN ENGINEERING, or **ergonomics,** research into physical and psychological human characteristics with particular reference to the environments in and the tools with which people work, and the application of the information so received to the design of equipment, factories, etc. In fact, the techniques of human engineering are now applied to a wide range of other problems involving humans and technology; e.g. POLLUTION control.

HUMBOLDT, Friedrich Heinrich Alexander, Baron von (1769–1859), German naturalist. With the botanist **Aimé Jacques Alexandre Bonpland** (1773–1858) he traveled for five years through much of South America (1799–1804), collecting plant, animal and rock specimens and making geomagnetic and meteorologic observations. Humboldt published their data in 30 volumes over the next 23 years. In his most important work, *Kosmos* (1845–62), he sought to show a fundamental unity of all natural phenomena.

HUMBOLDT, Karl Wilhelm, Baron von (1767–1835), German philologist regarded as the father of comparative PHILOLOGY. He maintained both that the nature of language reflects the culture of which it is a product, and that man's perception of the world is governed by the language available to him.

HUME, David (1711–1776), Scottish Enlightenment philosopher, economist and historian, whose *Treatise of Human Nature* (1739–40) is one of the key works in the tradition of British EMPIRICISM. But it was his shorter *Enquiry Concerning Human Understanding* (1748) which prompted KANT to his most radical labors. His influential *Dialogues Concerning Natural Religion* were published posthumously in 1779, long after their composition. In EPISTEMOLOGY Hume argued that men had no *reason* to associate distinct impressions as cause and effect; if they did so, it was only because experience had shown them that this was possible. His SKEPTICISM in this respect has always been controversial. In his own day, Hume's most successful work was possibly his *History of England* (1754–63).

HUMERUS, BONE of the upper ARM, linked to the scapula and clavicle at the SHOULDER and to the radius and ulna at the elbow.

HUMIDIFIER, device to maintain the air inside a building at a desirable humidity. In hot weather moisture is removed from the air by refrigeration; in cold weather it is increased by allowing water to evaporate. (See also AIR CONDITIONING; HUMIDITY.)

HUMIDITY, the amount of water vapor in the air, measured as mass of water per unit volume or mass of air, or as a percentage of the maximum amount the air would support without condensation, or indirectly via the DEW point. Saturation of the air occurs when the water vapor pressure reaches the VAPOR PRESSURE of liquid water at the TEMPERATURE concerned; this rises rapidly with temperature. The physiologically tolerable humidity level falls rapidly with temperature, as humidity inhibits cooling by EVAPORATION of sweat.

HUMORS, in ancient and medieval medicine, the four bodily fluids whose balance was required for the individual's health. They correspond to the four elements (see ARISTOTLE): *blood*:fire; *phlegm*:water; *choler* (or *yellow bile*):air; and *melancholy* (or *black bile*):earth. Excess of blood (hot and dry), for example, made one sanguine; phlegm (cold and wet), phlegmatic; etc. Cure was by enantiopathy (see ALLOPATHY), so that a fever would be treated with cold, and so forth. The idea may have originated with EMPEDOCLES in the 5th century BC, and we still retain something of it in modern words such as "choleric" and "phlegmatic."

HUMUS, the organic component of soil, decomposed plant and animal material (see PUTREFACTION). Dark brown to black, it is intimately mixed with the inorganic SOIL particles. Mainly for its good CARBON, NITROGEN, PHOSPHORUS and SULFUR content, it is of great agricultural importance.

HUNCHBACK, or kyphosis, deformity of the spine causing bent posture with or without twisting (SCOLIOSIS) and abnormal bony prominences. TUBERCULOSIS of the spine may cause sharp angulation, while congenital diseases, ankylosing spondylitis, vertebral collapse and spinal TUMORS cause smooth kyphosis.

HUNDREDWEIGHT (cwt), name of two units of WEIGHT. In the US the short hundredweight of 100lb is used. The Imperial (long) hundredweight is 112lb.

HUNGER. See THIRST AND HUNGER.

HUNTER, John (1728–1797), British anatomist and biologist who made many contributions to SURGERY, ANATOMY and PHYSIOLOGY. He is often regarded as the father of scientific surgery.

HURRICANE, a **tropical cyclone,** usually of great intensity. High-speed winds spiral in toward a low-pressure core of warm, calm air (the **eye**): winds of over 300km/hr have been measured. The direction of spiral is clockwise in the S Hemisphere, counterclockwise in the N (see CORIOLIS EFFECT). Hurricanes form (usually between latitudes 5° and 25°) when there is an existing convergence of air near sea level toward a center. The air ascends, losing moisture as precipitation as it does so. If this happens rapidly enough, the upper air is warmed by the water's LATENT HEAT of vaporization. This reduces the surface pressure, so accelerating air convergence. Hurricanes of the N Pacific are often called **typhoons.** (See also CYCLONE; WIND.)

HUSSERL, Edmund (1859–1938), Czech-born German philosopher who founded PHENOMENOLOGY. Professor at Göttingen and Freiburg universities, he was concerned by what constitutes acts of consciousness and how they relate to experience. He held that consciousness is "intentional" in that it does not exist apart from the objects of awareness.

HUTTON, James (1726–1797), Scottish geologist who proposed, in *Theory of the Earth* (1795), that the earth's natural features result from continual processes, occurring now at the same rate as they have in the past (see UNIFORMITARIANISM). These views were little regarded until LYELL's work some decades later. (See also CATASTROPHISM.)

HUXLEY, distinguished British family. **Thomas Henry Huxley** (1825–1895) is best known for his support of DARWIN's theory of EVOLUTION, without which acceptance of the theory might have been long delayed. Most of his own contributions to paleontology and zoology (especially taxonomy), botany, geology and anthropology were related to this. He also coined the word "agnostic." His son **Leonard Huxley** (1860–1933), a distinguished man of literature, wrote *The Life and Letters of Thomas Henry Huxley* (1900). Of his children, three earned fame. **Sir Julian Sorell Huxley** (1887–1975) is best known as a biologist and ecologist. His early interests were in development and growth, genetics and embryology. Later he made important studies of bird behavior, studied evolution and wrote many popular scientific books. **Aldous Leonard Huxley** (1894–1963) was one of the 20th century's foremost novelists. **Andrew Fielding Huxley** (1917–) shared the 1963 Nobel Prize for Physiology or Medicine with A. L. HODGKIN and Sir J. ECCLES for his work with Hodgkin on the chemical basis of nerve impulse transmission (see NERVOUS SYSTEM).

HUYGENS, Christiaan (1629–1695), Dutch scientist who formulated a wave theory of LIGHT, first applied the PENDULUM to the regulation of CLOCKS, and discovered the surface markings of MARS and that SATURN has rings. In his optical studies he stated **Huygens' Principle**, that all points on a wave front may at any instant be considered as sources of secondary waves that, taken together, represent the wave front at any later instant.

HYBRIDIZATION, the crossing of individuals belonging to two distinct species. Mules, for example, are the result of hybridization between a horse and an ass. Hybrid offspring are often sterile, especially in animals.

HYDRA (the Water Monster), a large S Hemisphere constellation containing the bright star Alphard and a cluster of galaxies over 30Mpc distant.

HYDRATE, a compound (usually ionic) containing a definite proportion of WATER, known as water of crystallization, which may be bound as a LIGAND to the cation, to the anion, or to both. Other hydrates, with more or less water, may be formed under different conditions. Water may be lost from a hydrate spontaneously (EFFLORESCENCE) or by heating (see also DEHYDRATION), the compound becoming anhydrous. Common hydrates include COPPER (II) sulfate, $CuSO_4.5H_2O$, SODIUM carbonate, $Na_2CO_3.10H_2O$, and the ALUMS. (See also DELIQUESCENCE.)

HYDRAULIC BRAKE, a BRAKE system in which the power is transmitted by hydraulic pressure.

HYDRAULICS, application of the properties of liquids (particularly WATER), at rest and in motion, to engineering problems. Since any machine or structure that uses, controls or conserves a liquid makes use of the principles of hydraulics, the scope of this subject is very wide. It includes methods of WATER SUPPLY for consumption, IRRIGATION or navigation and the design of associated DAMS, canals and pipes; HYDROELECTRICITY, the conversion of water power to electric energy using hydraulic TURBINES; the design and construction of ditches, culverts and hydraulic jumps (a means of slowing down the flow of a stream by suddenly increasing its depth) for controlling and discharging FLOOD water, and the treatment and disposal of industrial and human waste. Hydraulics applies the principles of HYDROSTATICS and HYDRODYNAMICS and is hence a branch of FLUID MECHANICS. Any hydraulic process, such as flow of liquid through a turbine, may be described mathematically in terms of four basic equations derived from the conservation of ENERGY, MASS, MOMENTUM and the relationship

A section through a hurricane. Air violently spirals around a calm "eye", strong updrafts giving rise to heavy rain.

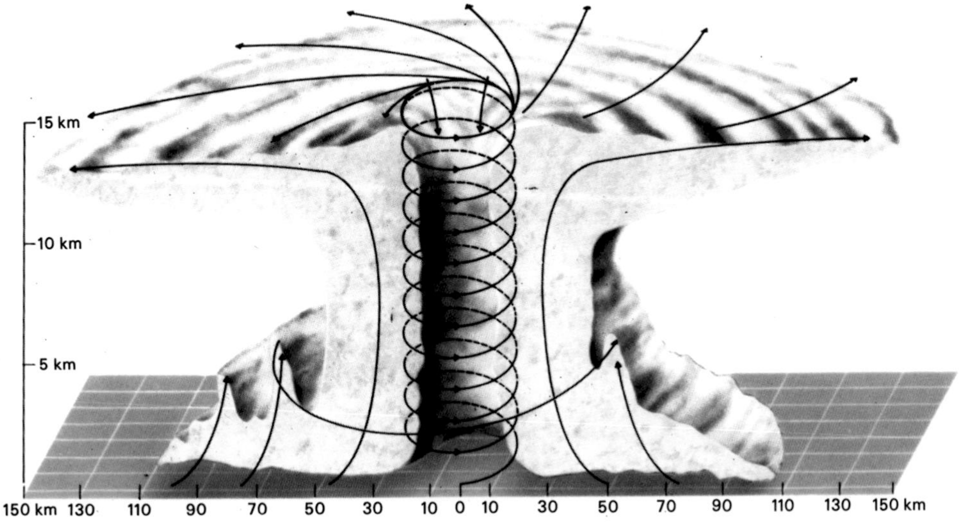

between the specific FORCES and internal mechanics of the problem. In hydraulic machines which transmit energy through liquids and convert it into mechanical power, three principles of liquid behavior that have been known for centuries are applied. TORICELLI's law states that the speed of liquid flow from a hole in the side of a vessel increases with the depth of the hole below the surface of the liquid in it. PASCAL's law states that the PRESSURE (force per unit area) in an enclosed body of liquid is transmitted equally in all directions. (This law is applied directly in a **hydraulic press**, in which a force applied over a small area by a piston is transmitted through the liquid filling the system to another piston with a larger area on which a much larger force will be exerted.) BERNOULLI's law states that at any point in a tube through which liquid is flowing, if no work is done, the sum of energies due to the pressure, motion (kinetic energy) and elevation (potential energy) of the liquid is constant. Thus by increasing the cross-section of the tube and slowing the flow down, kinetic energy is converted to pressure energy. The development of pumps in the 19th century, which converted mechanical to hydraulic energy and produced greater fluid velocities and pressures than had previously been obtainable, meant that hydraulic principles could usefully be applied to operate a wide variety of machines. Self-contained hydraulic units consisting of an engine, a pump, control valves, a motor to convert hydraulic to mechanical energy, and a load were soon developed for use in industry and transportation. Hydraulics is now one of the main technologies for transmitting energy, comparing well with mechanical and electrical systems and having the advantages of being fast and accurate and good at multiplying forces. Hydraulic systems containing water, oil or special fire-resistant fluids are now used in AIRPLANE landing systems, AUTOMOBILE braking systems and many other industrial applications.

HYDRAZINE (N_2H_4), colorless liquid, a covalent HYDRIDE resembling AMMONIA, prepared by oxidizing UREA with hypochlorite in the presence of gelatin. It reacts with ALDEHYDES to form hydrazones ($RCH{=}NNH_2$); it is a weak BASE, forming hydrazonium salts. Hydrazine is a powerful reducing agent, and so is used as a rocket fuel (being oxidized by nitric acid), as well as a corrosion inhibitor in boilers. It is also used to cure rubber, and in the production of plastics, explosives and fungicides.

HYDRIDES, binary compounds of HYDROGEN and another element. They fall into three classes. **Covalent hydrides** are formed by the elements in Groups IB–VIIA of the PERIODIC TABLE, i.e., the nonmetals and some metals. They are mainly volatile, reactive compounds, though those of Groups IB and IIB, and aluminum, are nonvolatile polymers. (See the individual elements, and also BORANES; HYDROCARBONS; AMMONIA; HYDRAZINE; WATER; SULFIDES; HYDROGEN FLUORIDE; HYDROGEN CHLORIDE.) **Ionic hydrides** are formed by the ALKALI METALS and ALKALINE-EARTH METALS. They are crystalline solids containing the ion H^-, which is a very powerful BASE and reducing agent (see OXIDATION AND REDUCTION), and react violently with water to give hydrogen. **Metallic hydrides**, formed by the TRANSITION ELEMENTS (Groups IIIB–VIII), are mostly nonstoichiometric (see COMPOSITION, CHEMICAL) and electrically conducting. They resemble ALLOYS, and some have interstitial structures.

HYDROCARBONS, organic compounds composed of CARBON and HYDROGEN only. Like organic compounds in general, which are derived formally from hydrocarbons by adding FUNCTIONAL GROUPS, they are best divided into ALIPHATIC, ALICYCLIC and AROMATIC hydrocarbons; aliphatic hydrocarbons are further subdivided into ALKANES, ALKENES and ALKYNES. Some hydrocarbons, especially TERPENES, occur in plant oils, and solid, high-molecular-weight hydrocarbons occur as BITUMEN, but by far the largest sources of all sorts of hydrocarbons are PETROLEUM, NATURAL GAS and COAL GAS. They are used as FUELS, for LUBRICATION, and as starting materials for a wide variety of industrial syntheses.

HYDROCEPHALUS, enlargement of the BRAIN ventricles with increased CEREBROSPINAL FLUID (CSF) within the SKULL. In children it causes a characteristic enlargement of the head. Brain tissue is attenuated and damaged by long-standing hydrocephalus. It may be caused by block to CSF drainage in the lower ventricles or brain stem aqueduct (e.g., by TUMOR and malformation, including those seen with SPINA BIFIDA), or by prevention of its reabsorption over the brain surface (e.g., following MENINGITIS). Apart from attention to the cause, treatment may include draining CSF into the atrium of the HEART.

HYDROCHLORIC ACID, solution of HYDROGEN CHLORIDE in water; a strong ACID of major industrial importance.

HYDROCYANIC ACID. See CYANIDES.

HYDRODYNAMICS, the branch of FLUID MECHANICS dealing with the FORCES, ENERGY and PRESSURE of FLUIDS in motion. A mathematical treatment of ideal frictionless and incompressible fluids flowing around given boundaries is coupled with an empirical approach in order to solve practical problems.

HYDROELECTRICITY, or **hydroelectric power**, the generation of ELECTRICITY using water power, is the source of about a third of the world's electricity. Although the power station must usually be sited in the mountains and the electricity transmitted over long distances, the power is still cheap since water, the fuel, is free. Moreover, running costs are low. An exciting modern development is the use in coastal regions of the ebb and flow of the tide as a source of electric power. Hydroelectric power uses a flow of water to turn a TURBINE, which itself drives a GENERATOR.

Convenient heads of water sometimes occur naturally (see WATERFALL), but more often must be created artificially by damming a river (see DAM); an added advantage is that the reservoir that forms behind the dam may be tapped for drinking or irrigation water.

The powerhouse, which contains the turbines and generators, may be at the foot of the dam or some distance away, the water then being transported in tunnels and long pipelines called **penstocks**. The turbines are of two main types: impulse (e.g., the Pelton wheel) and reaction (e.g., the Francis and Kaplan wheels). The Pelton wheel has buckets about its edge, into which jets of water are aimed, so turning the wheel. The Francis wheel has spiral vanes: water enters from the side and is discharged along the axis.

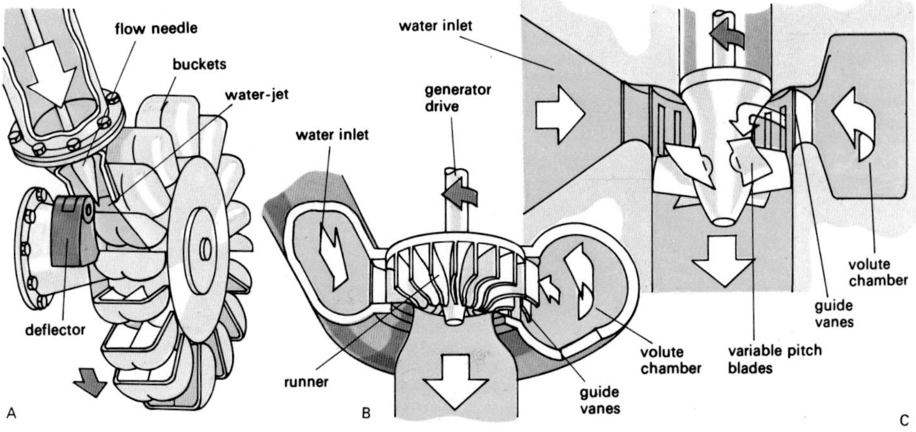

The three principal types of turbine: (A) the Pelton wheel; (B) the Francis turbine; (C) the Kaplan turbine.

The Kaplan wheel is rather like a huge propeller immersed in the water.

HYDROFOIL, a structure which, when moved rapidly through water, generates lift in exactly the same way and for the same reasons as does the AIRFOIL (see also AERODYNAMICS). It is usually mounted beneath a vessel (also called a hydrofoil). Much of a conventional boat's power is spent in overcoming the drag (resistance) of the water; as a hydrofoil vessel builds up speed, it lifts out of the water until only a small portion of it (struts, hydrofoils and PROPELLER) is in contact with the water. Thus drag is reduced to a minimum. Hydrofoils can exceed 125km/h as compared with conventional craft, whose maximum speeds rarely approach 80km/h.

HYDROGEN (H), the simplest and lightest element, a colorless, odorless gas. Hydrogen atoms make up about 90% of the UNIVERSE, and it is believed that all other elements have been produced by fusion of hydrogen (see STAR; FUSION, NUCLEAR). On earth most hydrogen occurs combined with oxygen as WATER and mineral HYDRATES, or with carbon as HYDROCARBONS (see PETROLEUM). Hydrogen is produced in the laboratory by the action of a dilute ACID on zinc or other electropositive metals. Industrially it is made by the catalytic reaction of hydrocarbons with steam, or by the WATER GAS process, or as a by-product of some ELECTROLYSIS reactions. Two-thirds of the hydrogen manufactured is used to make ammonia by the HABER PROCESS. It is also used in HYDROGENATION, PETROLEUM refining, and metal smelting. METHANOL and HYDROGEN CHLORIDE are produced from hydrogen. Being flammable, it has now been largely superseded by helium for filling BALLOONS and AIRSHIPS. Hydrogen is used in oxy-hydrogen WELDING; liquid hydrogen is used as fuel in rocket engines, in BUBBLE CHAMBERS, and as a refrigerant (see CRYOGENICS). Hydrogen is fairly reactive, giving HYDRIDES with most other elements on heating, and a moderate reducing agent. It belongs in no definite group of the PERIODIC TABLE, but has some resemblance to the HALOGENS in forming the ion H^-, and to the ALKALI METALS in forming the ion H^+ (see

Common hydrofoil configurations: (*left*) surface-piercing foils; (*right*) fully submerged foils.

ACIDS); it is always monovalent. (See also HYDROGENATION; HYDROGEN BONDING.) A hydrogen atom consists of one ELECTRON orbiting a nucleus of one PROTON. A hydrogen molecule is two atoms combined (H_2). In parahydrogen both the protons have the same SPIN: in orthohydrogen the protons have opposite spin. They have slightly different properties. At room temperature, hydrogen is 75% orthohydrogen, 25% parahydrogen. DEUTERIUM (H^2) and TRITIUM (H^3) are ISOTOPES of hydrogen. (See also HYDROGEN BOMB.) AW 1.008, mp −259°C, bp −253°C.

HYDROGENATION, a reaction in which hydrogen is added to a compound. Hydrogenation converts unsaturated organic compounds (see BOND, CHEMICAL) into saturated ones. Catalysts (commonly Raney nickel, palladium and platinum) are used. Hydrogenation is used notably to turn vegetable oils into margarine and in PETROLEUM refining.

HYDROGEN BOMB, or thermonuclear bomb, very powerful BOMB whose explosive energy is produced by nuclear FUSION of two DEUTERIUM atoms or of a deuterium and a TRITIUM atom. The extremely high temperatures required to start the fusion reaction are produced by using an ATOMIC BOMB as a fuze. Lithium-6 deuteride (Li^6D) is the explosive; neutrons produced by deuterium fusion react with the Li^6 to produce tritium. The end products are the isotopes of HELIUM He^3 and He^4. In warfare hydrogen bombs have the advantage of being far more powerful than atomic bombs, their power being measured in megatons of TNT, capable of destroying a large city. In defensive and peaceful uses they can be modified so that the radioactivity produced is minimal. Hydrogen bombs were first developed in the US (1949–52) by Edward TELLER and others, and have been tested also by the USSR, Great Britain, China and France.

HYDROGEN BONDING, the formation of a weak bond (see BOND, CHEMICAL) between a HYDROGEN atom (bound to a small electronegative atom, usually fluorine, oxygen, nitrogen or chlorine) and another such electronegative atom, in another or the same molecule. It is an electrostatic effect, but can be well described in molecular-orbital terms, especially when the three-atom system is symmetric (e.g. HF_2^-), resembling the hydrogen-bridge bonding in BORANES. Hydrogen bonding leads to anomalous physical properties due to molecular association: high melting point and boiling point, low vapor pressure, high viscosity, etc. It is important in the hydrogen halides, WATER, ICE, ALCOHOLS, OXY-ACIDS, AMMONIA, AMINES and AMIDES, and hence in vital molecules such as AMINO-ACIDS, PROTEINS and DNA.

HYDROGEN CHLORIDE (HCl), colorless acrid gas, fuming in air; a covalent HYDRIDE prepared by heating SALT with concentrated sulfuric acid or by direct combination of HYDROGEN and CHLORINE. It is unreactive when completely dry. mp −115°C, bp −85°C. **Hydrochloric acid**, a solution of hydrogen chloride in water, is a strong ACID, and reacts with active metals and bases to give chlorides (see HALIDES). It is used to make chlorine compounds and in the extraction and processing of metals. Dilute hydrochloric acid is produced in the stomach (see DIGESTIVE SYSTEM) but in excess causes gastric ULCERS. The concentrated acid is caustic. (See also AQUA REGIA.)

HYDROGEN FLUORIDE (HF), colorless liquid, fuming in air; a covalent HYDRIDE, prepared by distilling FLUORITE with concentrated sulfuric acid. Its physical properties show typical anomalies due to HYDROGEN BONDING. It is a very strong ACID, and an ionizing solvent for many inorganic and organic compounds; it is used to make FLUORINE and its compounds, especially FREON and FLUOROCARBONS. mp −83°C, bp 20°C. **Hydrofluoric acid**, a solution of hydrogen fluoride in water, is (anomalously) a weak acid, but causes very severe burns and is toxic. It dissolves silica to give fluorosilicic acid (H_2SiF_6), and so is used to etch glass.

HYDROGEN PEROXIDE. See PEROXIDES.

HYDROGEN SULFIDE. See SULFIDES.

HYDROGRAPHY, branch of hydrology dealing with bodies of water, such as oceans, lakes and rivers, on the earth's surface; and especially with the charting of their boundaries, currents, underwater contours and shipping hazards, as well as with the composition of their beds. (See also EARTH; HYDROLOGY; OCEANOGRAPHY.)

HYDROLOGIC CYCLE, the circulation of the waters of the earth between land, oceans and atmosphere. Water evaporates from the oceans into the ATMOSPHERE, where it may form CLOUDS (see also EVAPORATION). Much of this water is precipitated as RAIN back into the ocean, but much also falls on land. Of this, some is returned to the atmosphere by the TRANSPIRATION of plants, some joins rivers and is returned to the sea, some joins the GROUNDWATER and eventually reaches a sea, lake or river, and some evaporates back into the atmosphere from the surface of the land or from rivers, streams, lakes, etc. Over 97% of the earth's water is in the oceans; of the remaining fresh water, about 75% is in solid form (see GLACIER). However, although at a particular moment there is very little water in rivers, lakes and the atmosphere, the annual passage of water through

The hydrologic cycle.

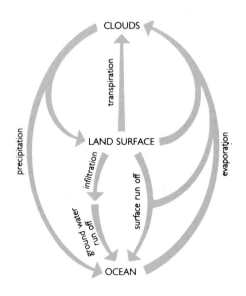

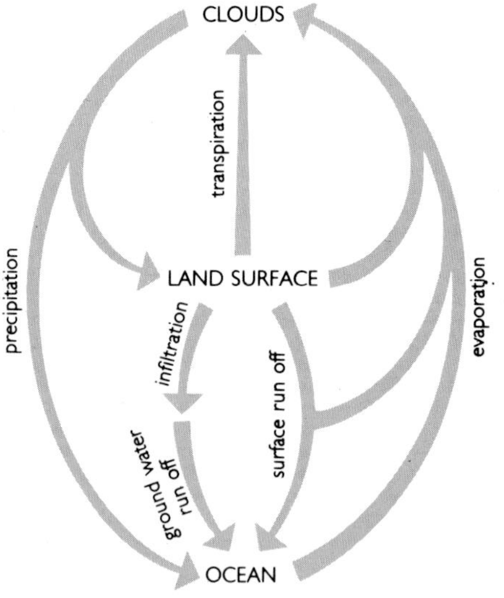

them is quite high. (See also HYDROLOGY; HYDROSPHERE.)

HYDROLOGY, the branch of geophysics concerned with the HYDROSPHERE (all the waters of the EARTH), with particular reference to the HYDROLOGIC CIRCLE. The science was born in the 17th century with the work of Pierre Perrault and Edme MARIOTTE.

HYDROLYSIS, a double decomposition effected by WATER, according to the general equation

$$XY + H_2O \rightarrow XOH + YH.$$

If XY is a salt of a weak ACID or a weak BASE, the hydrolysis is reversible, and affects the pH of the solution (see BUFFER). Reactive organic compounds such as ACID CHLORIDES and ACID ANHYDRIDES are rapidly hydrolyzed by water alone, but others require acids, bases, or ENZYMES as catalysts (as in digestion). Industrial hydrolysis processes include the alkaline saponification of oils and fats to glycerol and SOAP, and the acid hydrolysis of starch to glucose.

HYDROMETER, an instrument to measure the density of a liquid. In essence it consists of a closed glass tube calibrated along its stem and blown into a bulb, which is weighted, at the other. The hydrometer floats upright in the liquid to be tested: the denser the liquid, the more the instrument is buoyed up (according to ARCHIMEDES' Principle). The scale is read at the level of the liquid's surface. Though the instrument measures DENSITY directly, it is usual to calibrate the stem in terms of SPECIFIC GRAVITY, the ratio of the density of the liquid to the density of water at that temperature. Other scales are used for specific purposes.

HYDROPHOBIA. See RABIES.

HYDROPHONE, an adaptation of the MICROPHONE for use underwater. As with the microphone, the device converts SOUND (pressure) waves into electrical impulses, which are passed by cable to an external AMPLIFIER. Its prime use is in SONAR.

HYDROPHYTES, plants that are entirely submerged in water, whose upper leaves are floating, or which are entirely floating.

HYDROPLANE, or hydrofoil. See HYDROFOIL.

HYDROPLANING, or **aquaplaning,** dangerous phenomenon experienced by fast-moving cars on wet roads. A layer of water is built up between the tires and the road: since the water acts as an efficient lubricant (see LUBRICATION), the car goes out of control.

HYDROPONICS, the technique by which plants are grown without soil. It is also known as soilless culture. All the minerals required for plant growth are provided by nutrient solutions in which the roots are immersed. The technique has been highly developed as a tool in botanical research, but commercial exploitation is limited primarily because of the difficulty of aerating the water and providing support for the plants. Gravel culture has overcome these problems to some extent and is used to grow some horticultural crops.

HYDROSPHERE, all the waters of the earth, in whatever form: solid, liquid, gaseous. It thus includes the water of the ATMOSPHERE, water on the EARTH's surface (e.g., oceans, rivers, ice sheets) and GROUNDWATER. (See also HYDROGRAPHY; HYDROLOGY; LITHOSPHERE.)

HYDROSTATICS, the branch of FLUID MECHANICS dealing with FORCES and PRESSURES in stationary FLUIDS, and their effects on bodies immersed in them. The concept of **fluid pressure** (a normal force per unit area acting across any surface in the fluid or at its boundary) enables problems of flotation, buoyancy etc. to be treated.

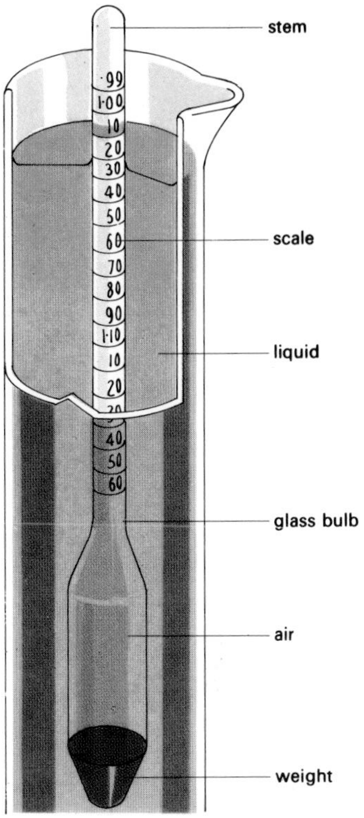

A hydrometer, calibrated with specific gravities.

HYDROTHERAPY, system of treatment by use of water, usually by treading or bathing in special pools, often supplied from mineral springs historically credited with healing properties.

HYDROXIDES, compounds containing the OH group, or the ion OH^-. Hydroxides of metals are generally BASES, and, if soluble, ionize to produce alkaline solutions (see ALKALIS) containing hydroxide ions. Nonmetals form ACID hydroxides, or OXYACIDS, which dissolve to produce hydrogen ions. Some metal hydroxides, such as zinc hydroxide, are amphoteric, that is, both basic and acidic. Hydroxides are formed by hydration of the oxide, or, if insoluble, by precipitation with an alkali. The OH^- ion acts as a LIGAND, forming hydroxo complexes. Organic compounds containing the OH group are ALCOHOLS, PHENOLS and CARBOXYLIC ACIDS.

HYGROMETER, device to measure HUMIDITY (the amount of water vapor the air holds). Usually, hygrometers measure relative humidity, the amount of moisture as a percentage of the SATURATION level at that temperature. The **hair hygrometer**, though of limited accuracy, is common, The length of a hair increases with increase in relative humidity. This length change is amplified by a lever and registered by a needle on a dial. Human hair is most used. The **wet**

and dry bulb hygrometer (psychrometer) has two THERMOMETERS mounted side by side, the bulb of one covered by a damp cloth. Air is moved across the apparatus (e.g., by a fan) and evaporation of water from the cloth draws LATENT HEAT from the bulb. Comparison of the two temperatures, and the use of tables, gives the relative humidity. The **dewpoint hygrometer** comprises a polished container cooled until the DEW point is reached: this temperature gives a measure of relative humidity. The **electric hygrometer** measures changes in the electrical RESISTANCE of a hygroscopic (water-absorbing) strip.

HYOSCINE. See SCOPALAMINE.

HYOSCYOMINE, an ALKALOID poison found in henbane, belladonna and jimson weed. It is a major source of ATROPINE.

HYPATIA (d. 415 AD), probably the first and one of the most famous women philosophers and mathematicians. She probably occupied the chair of Neo-Platonic philosophy at Alexandria. She was murdered by a Christian mob in an Alexandrian riot.

HYPERBARIC CHAMBER, chamber built to withstand and be kept at pressures above atmospheric. The high OXYGEN pressures achieved in them may destroy the anaerobic bacteria (*Clostridia*) responsible for gas GANGRENE; SURGERY may be done in the chamber. It is also used for AEROEMBOLISM in decompression.

HYPERBOLA. See CONIC SECTIONS.

HYPERBOLIC FUNCTIONS, the FUNCTIONS that arise in hyperbolic or LOBACHEVSKIAN GEOMETRY and in the study of complex VARIABLES. They are interrelated similarly to the trigonometric functions (see TRIGONOMETRY).

Hyperbolic Functions

The base definitions for these functions are

$$\cosh x = \tfrac{1}{2}(e^{x}+e^{-x})$$

$$\text{and } \sinh x = \tfrac{1}{2}(e^{x}-e^{-x})$$

(see EXPONENTIAL). From these come

$$\text{sch} = 1/\cosh x;$$

$$\text{csch } x = 1/\sinh x;$$

$$\tanh x = \sinh x/\cosh x,$$

$$\text{and } \coth x = \cosh x/\sinh x.$$

HYPERON. See SUBATOMIC PARTICLES.

HYPEROPIA, or hypermetropia or far- or longsightedness, a defect of VISION in which light entering the EYE from nearby objects comes to a focus behind the retina. The condition may be corrected by use of a converging spectacle LENS.

HYPERTENSION. See BLOOD CIRCULATION.

HYPERTHYROIDISM. See THYROID GLAND.

HYPNOSIS, an artificially induced mental state characterized by an individual's loss of critical powers and his consequent openness to SUGGESTION. It may be induced by an external agency or by the individual himself (**autohypnosis**). Hypnotism has been widely used in medicine (usually to induce ANALGESIA) and especially in PSYCHIATRY and PSYCHOTHERAPY. Here, the particular value of hypnosis is that, while in trance, the individual may be encouraged to recall deeply repressed memories (see MEMORY; REPRESSION) that may be the heart of, for example, a COMPLEX; once such causes have been elucidated, therapy may proceed.

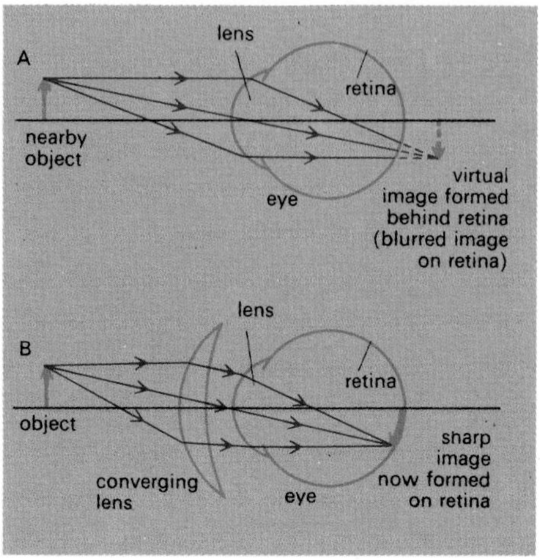

Hyperopia (A), and its correction using a converging lens (B).

Hypnosis seems to be as old as man. However, the first definite information on it comes in the late 18th century with the work of MESMER, who held that disease was the result of imbalance in the patient's "animal magnetism", and hence attempted to cure by use of magnets. In fact, some of his patients *were* cured, presumably by suggestion; and the term **mesmerism** is still sometimes used for hypnotism. Early psychotherapeutic uses include that of CHARCOT and his pupil FREUD; though Freud later rejected hypnosis and used instead his own technique, FREE ASSOCIATION. Little is known of the nature or root cause of hypnosis, and its amateur use is dangerous.

HYPO, or sodium thiosulfate. See SODIUM.

HYPOCHLORITES, salts containing the ClO^- ion derived from **hypochlorous acid**, HOCl, an unstable, weak acid. They are made by absorbing CHLORINE into an ALKALI, and are used for BLEACHING (see also BLEACHING POWDER) and as disinfectants.

HYPOCHONDRIA, or **hypochondriasis,** a PSYCHOSIS involving undue ANXIETY about real or supposed ailments, usually in the belief that these are incurable. The source of hypochondria was once thought to be the hypochondrium, the part of the ABDOMEN containing SPLEEN and LIVER.

HYPOCYCLOID, the CURVE traced by a point on the circumference of a CIRCLE that is rolling along the inside of a larger, fixed circle. (See also CYCLOID; EPICYCLOID.)

HYPODERMIC. See INJECTION; SYRINGE.

HYPOTENUSE, the side opposite the right angle of a right-angled TRIANGLE. (See also PYTHAGORAS' THEOREM.)

HYPOTHALAMUS, central part of the base of the BRAIN, closely related to the PITUITARY GLAND. It contains vital centers for controlling the autonomic NERVOUS SYSTEM, body temperature and water and food intake. It also produces HORMONES for regulating pituitary secretion and two systemic hormones (e.g., VASOPRESSIN).

HYPOTHESIS. See SCIENTIFIC METHOD.

HYPOXEMIA, or **hypoxia.** See ANOXIA.

Hysteresis loop for steel. If an unmagnetized specimen is placed in an increasing magnetic field, it exhibits increasing magnetization (0*a*) until it is saturated (*a*). This magnetism not entirely lost when the magnetizing field is removed (*ab*). The degree of magnetization remaining in the absence of the magnetizing field is known as the residual magnetism (0*b*). A reverse field is required fully to demagnetize the specimen (*bc*). Increasing the reverse field induces a reverse magnetization in the specimen (*cd*) which, again, is not entirely lost when the external field is reduced to zero (*de*). Applying an increasing forward field returns the specimen magnetization to saturation again (*efa*). The area contained within the hysteresis loop is a measure of the energy dissipated as heat in the process (the hysteresis loss). Soft iron is thus used for transformer cores because its hysteresis loop is far narrower than that for steel shown here. It might appear from the diagram that, once magnetized, it is impossible entirely to demagnetize a specimen in the absence of a field. This is not the case: it is possible to demagnetize a specimen by placing it in a repeatedly reversing field of steadily decreasing intensity: the hysteresis loop thus becoming smaller until it vanishes in the origin (0).

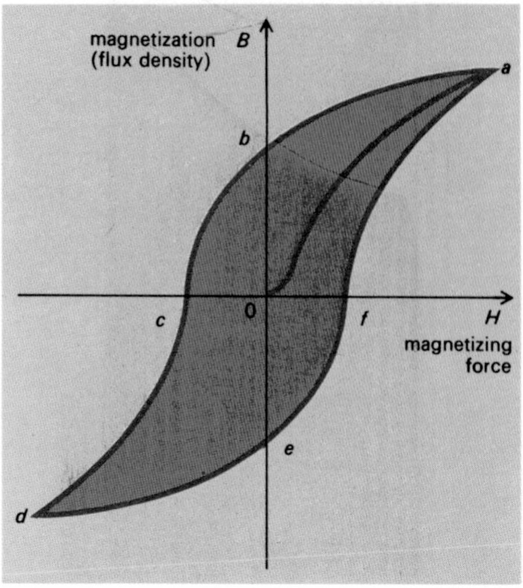

HYRACOIDEA, an order of the MAMMALIA that includes the tree and rock hyraxes, also known as dassies or coneys. They are small and similar in appearance to rodents but in fact are related to the PERISSODACTYLA and PROBOSCIDIA. The front feet have four functional toes and the back feet only three.

HYSTERECTOMY, or surgical removal of the WOMB, with or without the OVARIES and FALLOPIAN TUBES. It may be performed via either the ABDOMEN or the vagina and is most often used for fibroids, benign TUMORS of womb muscle, CANCER of the cervix or body of womb, or for diseases causing heavy MENSTRUATION. If the ovaries are preserved, HORMONE secretion remains intact, though periods cease and infertility is inevitable.

HYSTERESIS, a memory phenomenon in ferromagnetic materials (see MAGNETISM), the magnetization depending not only on the MAGNETIC FIELD applied, but also on how it was applied. In electric MOTORS, GENERATORS, and TRANSFORMERS, the refusal of the magnetization to follow the field directly often results in substantial ENERGY loss (hysteresis loss).

HYSTERIA, psychiatric disorder characterized by exaggerated responses, emotional lability with excess tears and laughter, over-activity and often overbreathing, occasionally leading to TETANY. It is often a manifestation of attention-seeking behavior. **Conversion systems** or mimicry of organic disease are often termed hysterical; the simulation of a particular disorder fulfils some psychological need in response to certain stresses and results in an unconscious gain or release from anxiety.

I

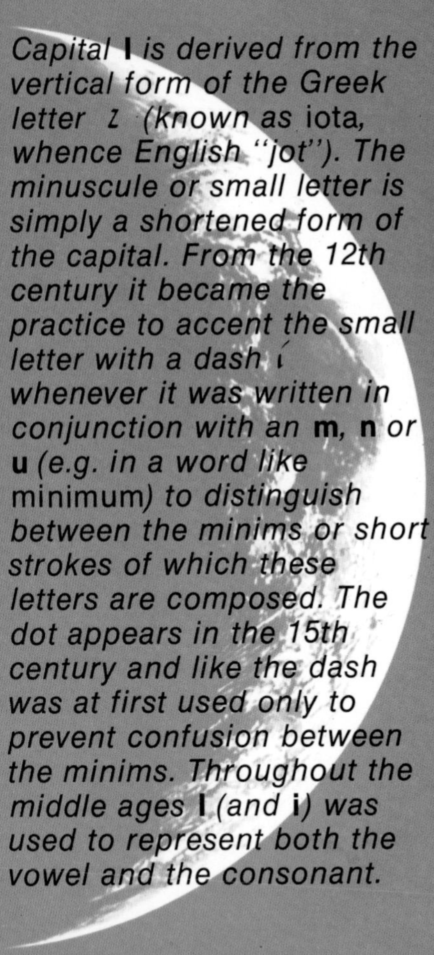

Capital **I** *is derived from the vertical form of the Greek letter* ι *(known as iota, whence English "jot"). The minuscule or small letter is simply a shortened form of the capital. From the 12th century it became the practice to accent the small letter with a dash* ı́ *whenever it was written in conjunction with an* **m, n** *or* **u** *(e.g. in a word like minimum) to distinguish between the minims or short strokes of which these letters are composed. The dot appears in the 15th century and like the dash was at first used only to prevent confusion between the minims. Throughout the middle ages* **I** *(and* **i***) was used to represent both the vowel and the consonant.*

IATROCHEMISTRY, a species of ALCHEMY which sought to find chemical treatments for disease, particularly as promoted in the 16th century by PARACELSUS and his followers. The analytic methods used in iatrochemistry were highly significant in the development of modern CHEMISTRY and the search for new remedies led to the discovery of many new chemical substances.

IBERIANS, a BRONZE AGE people of S and E Spain, culturally influenced by the Carthaginians and Greeks. Their sphere of influence overlapped that of the CELTS, who migrated into N and Central Spain from the 8th–6th centuries BC onward. They had a sophisticated written language and were fine potters.

IBN-SINA. See AVICENNA.

ICARUS, in Greek mythology, the son of DAEDALUS. To escape from King MINOS of Crete, he and his father attached wings to their shoulders with wax. Ignoring his father's warning, Icarus flew too close to the sun; the wax melted and he plunged to his death in the sea.

ICE, frozen WATER: a colorless crystalline solid in which the strong, directional HYDROGEN BONDING produces a structure with much space between the molecules. Thus ice is less dense than water, and floats on it. The expansion of water on freezing may crack pipes and automobile radiators. Since dissolved substances lower the freezing point, ANTIFREEZE is used. For the same reason, seawater freezes at about −2°C (see OCEAN). Ice has a very low coefficient of FRICTION, and some fast-moving sports (ice hockey, ice skating and iceboating) are played on it; however, slippery, ice roads are dangerous. Ice deposited on AIRPLANE wings reduces lift. Ice is used as a refrigerant, and to cool some beverages. (See also FROST; GLACIER; HAIL; ICE AGES; ICEBERG; ICEBREAKER; SNOW.) mp 0°C, sg 0.92 (0°C).

ICE AGES, periods when glacial ice covers large areas of the earth's surface that are not normally covered by ice. Ice ages are characterized by fluctuations of climatic conditions: a cycle of several glacial periods contains interglacial periods, perhaps of a few tens of thousands of years, when the climate may be as temperate as between ice ages. It is not known whether the earth is currently between ice ages or merely passing through an interglacial period.

There seem to have been several ice ages in the PRECAMBRIAN, and certainly a major one immediately prior to the start of the CAMBRIAN. There were a number in the PALEOZOIC, including a major ice age with a complicated cycle running through the MISSISSIPPIAN, PENNSYLVANIAN and early PERMIAN. The ice age that we know most about, however, is that of the QUATERNARY, continuing through most of the PLEISTOCENE and whose last glacial period ended about 10000 years ago, denoting the start of the HOLOCENE. (See GEOLOGY.) At their greatest, the Pleistocene glaciers covered about a third of the earth's surface, or some 45 million km², and may have been up to 3km thick in places. They covered most of Canada, N Europe and N Russia, N parts of what is now the US, and, in the S Hemisphere, Antarctica, parts of South America, and some other areas.

Theories about the cause of ice ages include that the SUN's energy output varies, that the earth moves with respect to its axis, that CONTINENTAL DRIFT may alter global climatic conditions, and that volcanic dust in the ATMOSPHERE could reduce the amount of solar

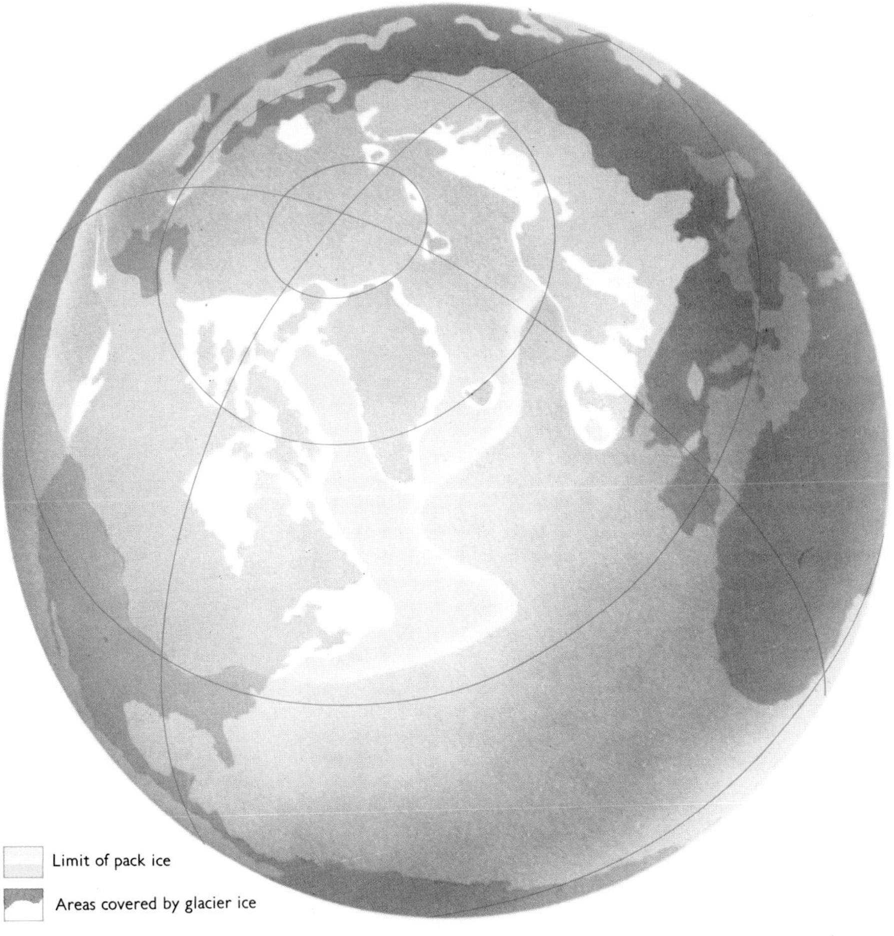

The maximum extent of the Pleistocene ice age.

heat received by the surface. (See EARTH; GLACIER; VOLCANO.)

ICEBERG, a large, floating mass of ice. In the S Hemisphere, the Antarctic ice sheet overflows its land support to form shelves of ice on the sea; huge pieces, as much as 200km across, break off to form icebergs. In the N Hemisphere, icebergs are generally not over 150m across. Most are "calved" from some 20 GLACIERS on Greenland's W coast. Small icebergs (growlers) may calve from larger ones. Some 75% of the height and over 85% of the mass of an iceberg lies below water. Northern icebergs usually float for some months to the Grand Banks, off Newfoundland, there melting in a few days. They endanger shipping, the most famous tragedy being the sinking of the *Titanic* (1912). The International Ice Patrol now keeps a constant watch on the area.

ICEBREAKER, a ship designed to break a channel through pack ice by riding up on the ice, which breaks beneath the weight. Characteristics are: one or more powerful propellers at the stern; a very broad beam; a shallow angle of bow and stern; and armorplating to protect against impact from floating ice and resist pressure if the vessel becomes trapped. Most have an extra propeller at the bow so that they can reverse if trapped in ice.

ICECAP. See GLACIER.

ICE CREAM, popular frozen dairy food whose main constituents are sugars, milk products, water, flavorings and air. Ice cream has a high calorific value, and a very high VITAMIN A content, as well as being protein and calcium-rich. It is also a source of, in smaller quantities, iron, phosphorus, riboflavin and thiamin. Water ices, which contain no milk products, have been known since ancient times in Europe and Asia. Ice cream probably reached the US in the 17th century, and was first commercially manufactured by Jacob Fussel (1851). Today, the US is the world's largest producer and consumer.

ICELAND SPAR. See CALCITE.

ICHTHYOLOGY (from Greek *ichthys*, fish, and *logos*, knowledge), the study of fishes. The word was first used in 1646, respectively 60 and 120 years before the study of birds and insects achieved similar scientific recognition.

ICHTHYOSAURIA, fossil members of the REPTILIA

that lived from the Triassic to the Cretaceous periods, 225–65 million years ago. They were aquatic and appear almost fish-like, with streamlined bodies, limbs modified as fins and a fish-like tail. They gave birth to live young, an ability not found in most living reptiles.

ICHTHYOSTEGA, an extinct member of the AMPHIBIA that lived during the Devonian period, about 350 million years ago, and which was the first vertebrate to colonize land.

ICONOSCOPE, early form of TELEVISION camera tube which converts an optical image into an electrical signal. An ELECTRON beam scans a mosaic of photoemissive particles which become electrically charged depending on the amount of incident LIGHT; a signal plate produces an electric signal corresponding to this charge pattern.

ICOSAHEDRON. See POLYHEDRON.

ID, the formless collection of all parts of the mind present at birth, part of which develops to form the EGO. The id thus contains such parts of the psychic makeup as EMOTIONS and INSTINCTS. (See also PSYCHOANALYSIS; UNCONSCIOUS.)

IDEA, in psychology, a loosely defined term describing any conscious (see CONSCIOUSNESS) mental event that is not stimulated by immediate PERCEPTION (e.g., a MEMORY). Some psychologists hold that idea-forming is present in perception also; e.g., when bread has just been baked, one smells an aroma and recognizes from experience that it is the aroma of new-baked bread, rather than sensing the bread directly. Others, particularly behaviorists (see BEHAVIORISM), hold that ideas do not exist but are merely reflections of other mental processes.

IDEALISM, name adopted by several schools of philosophy, all of which in some way assert the primacy of ideas, either as the sole authentic stuff of reality or as the only medium through which we can have knowledge or experience of the world. Idealisms are commonly contrasted both with the various types of REALISM and with philosophical MATERIALISM. They are often associated with methodological RATIONALISM because they usually seem to owe more to reasoning upon A PRIORI principles than to any appeal to experience. The idealism of PLATO, in which ideas were held to have an external objectivity, is unrepresentative of modern varieties, of which that of BERKELEY is archetypal. KANT and HEGEL were foremost in the German idealist tradition, while T. H. GREEN, F. H. Bradley and J. Royce were representative of more recent English-speaking idealists. Idealism has, however, been in eclipse in the 20th century.

IDENTIFICATION, in PSYCHOLOGY, the process of recognizing a specific MEMORY. In PSYCHOANALYSIS, the fusion of one's IDENTITY with another's either through inability to recognize one's own identity (a state common in infancy), or as a DEFENSE MECHANISM or reaction despite recognition of one's own identity.

IDENTITY, in mathematics. See ALGEBRA.

IDENTITY, or personal identity, in psychology, the individual's sense of being a distinct, continuous entity; roughly corresponding to self-awareness (see CONSCIOUSNESS). (See also EGO.)

IDEOGRAM, a written symbol which directly conveys an idea or represents a thing, rather than representing a spoken word, phrase or letter. **Logograms**, symbols that each represent an entire word, are also often called ideograms. Egyptian HIEROGLYPHICS comprised a writing system partly ideogramic, partly logogrammatic and partly phonetic. (See also WRITING, HISTORY OF.)

IGNEOUS ROCKS, one of the three main types of rocks, those whose origin is related to heat. They crystallize from the MAGMA either at the earth's surface (extrusion) or beneath (intrusion). There are two main classes: **Volcanic rocks** are extruded (see VOLCANISM), typical examples being LAVA and PYROCLASTIC ROCKS. **Plutonic rocks** are intruded

The classification of the plutonic igneous rocks according to their silica and combined sodium oxide and potassium oxide contents. The fine-grained equivalents of these rocks are similarly placed.

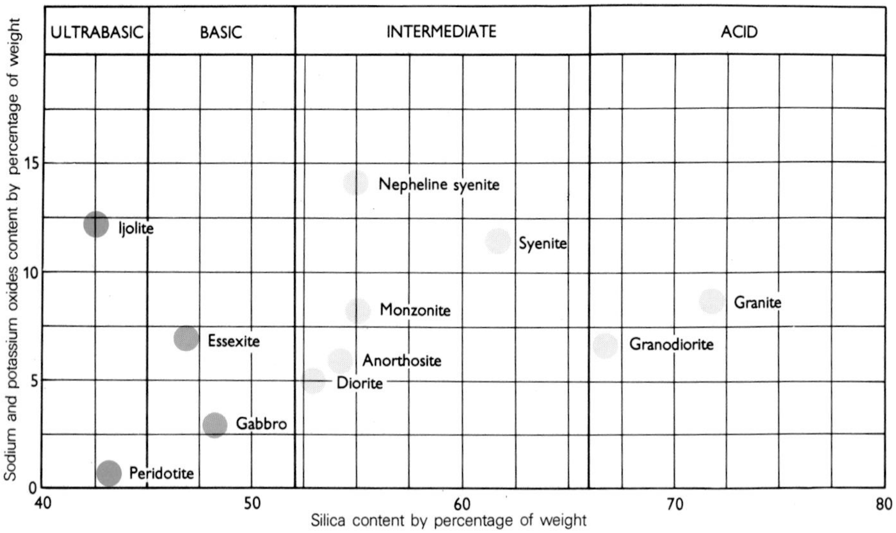

into the rocks of the EARTH's crust at depth, a typical example being GRANITE: those forming near to the surface are sometimes called **hypabyssal rocks**. Types of intrusions include BATHOLITHS, DIKES, SILLS and LACCOLITHS. As plutonic rocks cool more slowly than volcanic, they have a coarser texture, more time being allowed for crystal formation. (See also ROCKS.)

IGNIS FATUUS. See WILL-O'-THE-WISP.

IGNITION SYSTEM, the system in an INTERNAL-COMBUSTION ENGINE for igniting the fuel/air mixture In the DIESEL ENGINE the heat generated by the compression stroke is enough to ignite the fuel when it is sprayed in. In gasoline engines ignition is produced by the SPARK PLUGS whose operation is timed by the DISTRIBUTOR. The high electrical POTENTIAL is provided in automobiles by an INDUCTION COIL working off the BATTERY, and in light engines such as aircraft engines by a MAGNETO. The current in the coil's primary winding charges a CAPACITOR, without which an electric arc would burn out the breaker points when the MAGNETIC FIELD collapses. High-performance engines use TRANSISTORS to switch the current very rapidly.

IGNITRON, gas tube RECTIFIER which controls a wide range of currents, used in resistance WELDING. Current passes as an arc between the single ANODE and an electron-emitting spot formed on the surface of a MERCURY pool by a current pulse through an ignitor ELECTRODE dipping in it.

ILEITIS, INFLAMMATION of the ileum, part of small intestine (see ENTERITIS; GASTROINTESTINAL TRACT).

ILEUM. See GASTROINTESTINAL TRACT.

ILLUSION, an erroneous perception of reality, often the result of misinterpretation by the brain of information received by the SENSES. Most commonly the sense involved is sight: one of the exploitations of optical illusion is the use by artists of PERSPECTIVE. Optical illusions may also have external causes, such as REFRACTION, as in the observation of a stick held in water (see also BROCKEN SPECTER). Examples of auditory illusions include BEATING and the apparent change in pitch of a railroad train's whistle as it passes (see DOPPLER EFFECT). Rather different classes of illusion are HALLUCINATIONS and EIDETIC IMAGES. The unconscious falsification of the MEMORY of a past experience is also termed an illusion.

ILMENITE, hard, black OXIDE mineral, iron (II) titanium (IV) oxide ($FeTiO_3$), the chief ore of TITANIUM. It occurs widely in IGNEOUS ROCKS, notably in the USSR, Norway, Quebec and Wyo. It crystallizes in the rhombohedral system.

IMAGE, Optical, a representation of an object formed in an optical instrument. Although a **virtual image** has no physical existence—light only seems to come from its apparent position—light actually comes to a focus in a **real image** and these can be made visible by using a suitable screen.

IMAGINARY NUMBERS, numbers of the form ai, where a is a REAL NUMBER and i is defined such that $i^2 = -1$. For example, $\sqrt{-16}$ is an imaginary number that can, since $\sqrt{-16} = \sqrt{-1}.\sqrt{16}$, be expressed as $4i$.

Complex numbers are SUMS of real and imaginary numbers, and are usually expressed in the form $a+bi$, *where* a and b are real numbers. They are frequently represented by an ARGAND DIAGRAM on a set of CARTESIAN COORDINATES, real components are plotted along the x-axis, imaginary components along the y-axis. Similarly, they may be represented in polar coordinates (see ANALYTIC GEOMETRY) in the form $r(\cos\theta + i\sin\theta)$.

In the FIELD of complex numbers, algebraic operations are carried out roughly as in the field of real numbers. Thus

$$\begin{aligned}(a+ib)+(c+id) &= (a+c)+i(b+d),\\ (a+ib)-(c+id) &= (a-c)+i(b-d),\\ (a+ib)\ .\ (c+id) &= ac+ida+ibc+i^2bd\\ &= i(da+bc)+ac-bd\\ &= (ac-bd)+i(da+bc),\end{aligned}$$

$$\text{and}\quad \frac{(a+ib)}{(c+id)} = \frac{(a+ib)}{(c+id)}\frac{(c-id)}{(c-id)} = \frac{(ac+bd)+i(bc-ad)}{(c^2+d^2)} = \frac{(ac+bd)}{(c^2+d^2)}+\frac{i(bc-ad)}{(c^2+d^2)}$$

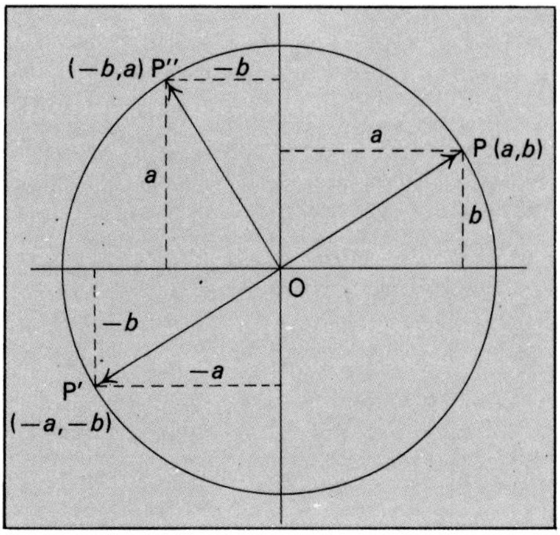

The imaginary operator j rotates the vector **OP** into **OP″** and—on a second application—rotates **OP″** into **OP′**

In matrix notation, $\begin{pmatrix}0 & -1\\ 1 & 0\end{pmatrix}\begin{pmatrix}a\\ b\end{pmatrix} = \begin{pmatrix}-b\\ a\end{pmatrix}$ Thus

$$j^2 = \begin{pmatrix}0 & -1\\ 1 & 0\end{pmatrix}\begin{pmatrix}0 & -1\\ 1 & 0\end{pmatrix} = \begin{pmatrix}-1 & 0\\ 0 & -1\end{pmatrix} = -1$$

IMAGINARY OPERATOR (**i-operator** or **j-operator**). Consider a VECTOR **OP** where O is the origin of a set of CARTESIAN COORDINATES and P is the POINT (a, b)—see diagram. Multiplying **OP** by -1 will produce a VECTOR **OP′**; that is, will rotate **OP** through 180° such that **OP′** $= -$**OP.** Now consider a number j such that multiplying **OP** by j produces **OP″**, a vector PERPENDICULAR to **OP.** Multiplying **OP″** by j will produce **OP′**, and so in a sense $j.j$ (or j^2) $= -1$. This j is therefore imaginary, and is called the imaginary operator. The number i such that $i^2 = -1$ is fundamental in complex numbers (see IMAGINARY NUMBERS). (See also VECTOR ANALYSIS.)

IMAGO, a term referring to the adult insect emerging from a pupa in complete METAMORPHOSIS.

IMIDES, organic compounds of general formula R.CO.NH.CO.R′, obtained by reacting an AMIDE with an ACID CHLORIDE, or AMMONIA with an ACID ANHYDRIDE, and used in synthesis.

IMMERSION FOOT, or **trench foot,** disease of the

FEET after prolonged immersion in water, due to a combination of vasoconstriction and waterlogging. It usually starts with red, cold and numb feet, which on warming develop through EDEMA and blistering to ulceration and sometimes skin GANGRENE.

IMMUNITY, the system of defense in the body which gives protection against foreign materials, specifically infectious microorganisms—BACTERIA, VIRUSES, PARASITES and their products. For many DISEASES, prior exposure to the causative organism in disease itself or by VACCINATION provides acquired resistance to that organism; further infection with it is unlikely or will be less severe. This type of immunity is usually mediated by ANTIBODY AND ANTIGEN reactions and is known as **humoral immunity**. The antigens of microorganisms provoke the formation of the antibody specific to that antigen. Once formed the antibody tends to neutralize (viruses) or to bind to antigen encouraging phagocytosis and destruction (bacteria). In some diseases the development of antibodies is of value in the phase of recovery from the primary infection; once immunity has been thus primed, the easy and rapid availability of antibody protects against further infection. ALLERGY and ANAPHYLAXIS are also largely mediated by humoral immunity. A number of diseases are due to the systemic effects of **immune complexes** (antibody linked to antigen) which may arise in the appropriate response to an infection, or in serum sickness, and these especially affect the KIDNEYS, SKIN and JOINTS. In **autoimmunity** antibodies are produced to antigens of the body's own tissues for reasons that are not always clear; secondary tissue destruction may occur. The second major type of immunity is **cell-mediated immunity** (delayed type hypersensitivity); this system is mediated by lymphocytes and monocytes (including tissue macrophages). It is a reaction only occurring with certain types of infection (TUBERCULOSIS, HISTOPLASMOSIS and FUNGAL DISEASES) and in certain probable auto-immune diseases; it is also important in the immunity of TRANSPLANTS. Lymphocytes are primed by infection with the appropriate organisms or by the autoimmune or graft reaction and produce substances which affect both lymphocytes and monocytes and result in a type of INFLAMMATION with much tissue damage. The understanding of the role of immunity and its disorders in the causation and manifestations of many diseases has seen a substantial advance in recent years. This has led to the development of DRUGS and other agents which are able to interfere with abnormal or destructive immune responses. **Immune deficiency diseases**, although rare, have provided models for the separate parts of the immune system, and have led to methods of replacement of absent components of immunity. **Passive immunity** is the transfer of antibody-rich substances from an immune subject to a nonimmune subject who is susceptible to disease. It is important in infancy, where maternal antibodies protect the child until its own immune responses have matured. In certain diseases such as TETANUS and RABIES, immune serum gives valuable immediate passive protection in nonimmune subjects.

IMPEDANCE, the ratio of the AC voltage applied to an electric circuit to the current it produces. It is a generalization of the concept of electric resistance (see ELECTRICITY) to include cases where the current oscillates ahead of or behind the voltage (i.e., out of PHASE with it) based on the mathematics of complex numbers. The term is also applied to the ratio between the driving force and response of other oscillatory or wave systems.

IMPETIGO, superficial SKIN infection, usually of the FACE, caused by STREPTOCOCCUS or STAPHYLOCOCCUS. It starts with small vesicles which burst and leave a characteristic yellow crust. It is easily spread by fingers from a single vesicle to affect several large areas and may be transmitted to others. It is common in children and requires ANTIBIOTIC creams, and systemic PENICILLIN in some cases.

IMPLANTATION, the earliest stage of EMBRYO development in which the embryo invades the WOMB. After fertilization by SPERM, the EGG divides into a small ball of cells, whose outer layer is specialized to invade the endometrium, which is itself prepared for implantation. The interface between embryo and endometrium develops into the PLACENTA. The term is also used to refer to the placing of DRUGS, PROSTHETICS or grafts in the body in treatment of DISEASE.

IMPLOSION, a bursting inward, as opposed to explosion, a bursting outward. Implosion due to gravitational collapse is an important end-stage in the lives of STARS (see also BLACK HOLES). Domestic lamp bulbs implode on fracture.

IMPRINTING, very rapid LEARNING by a newborn creature. (See ANIMAL BEHAVIOR.)

IMPULSE, the integral (see CALCULUS) of a FORCE over an interval of time. By NEWTON's second law the impulse (a VECTOR quantity) equals the change of MOMENTUM produced by the force. It is a useful concept when a large and variable force acts for a short time, as in an impact.

INBREEDING, the breeding of individual plants or animals that are closely related. Inbreeding tends to bring together recessive GENES with, usually, deleterious effects. This is because recessive genes are often harmless in the heterozygous condition but harmful in the homozygous condition (see GENETICS). For this reason, inbreeding has long been regarded as a practice to be discouraged; in human cultures, consanguinity is frequently forbidden by law or discouraged by custom.

INCANDESCENT LAMP. See LIGHTING.

INCENDIARY BOMB, bomb designed to cause casualty and damage by fire. They may be made of MAGNESIUM, which burns vigorously and with great heat; be packed with THERMITE, which is scattered over a wide area by initial explosion; or contain liquid burning agents such as NAPALM which, burning, cling to people and objects. (See also FIRESTORM.)

INCENSE, aromatic substances such as frankincense, sandalwood and balsam, which produce a fragrant odor when burned. Incense has been widely used in religious ceremonies from earliest times.

INCENTER, the point of concurrence of the lines drawn bisecting the angles of a TRIANGLE. As this point is equidistant from the sides, it is the center of the circle that may be inscribed in the triangle (see INSCRIPTION).

INCH (in), a unit of length in the US customary and Imperial systems. Since 1959 it has been defined exactly equal to 25.4mm.

INCLINED PLANE. See MACHINE.

INCUBATION, a method of keeping microorganisms such as BACTERIA or VIRUSES warm and in an appropriate medium to promote their growth (e.g., in identification of the organisms causing DISEASE); also, the period during which an organism is present in the body before causing disease. INFECTIOUS DISEASE is contracted from a source of infective microorganisms. Once these have entered the body they divide and spread to different parts and it is some time before they cause symptoms due to local or systemic effects. This incubation period may be helpful in diagnosis and in determining length of QUARANTINE periods.

INCUBATOR, a device used for incubating microorganisms (see INCUBATION); also an enclosed cot in which a baby, particularly if premature, is placed to create an ideal protective and controllable environment. Temperature is regulated thermostatically and the possibility of infection is minimized; the air in the incubator may be enriched with controlled amounts of OXYGEN. Nursing is carried out through portholes. The use of incubators has been significant in reducing mortality in premature infants.

INDETERMINACY PRINCIPLE. See UNCERTAINTY PRINCIPLE.

INDETERMINATE EQUATIONS, simultaneous EQUATIONS which for some reason have an infinite number of solutions. Examples are

$$x+y=4$$
$$2x+2y=8,$$

where the two equations are equivalent, and

$$x+y=z$$
$$\tfrac{1}{2}x+6y+3=8z,$$

where there is insufficient data.

INDIAN SUMMER, period of unusually warm, sunny weather often occurring in the late fall in the central and eastern US. It is caused by a warm anticyclone stabilized by a strong temperature inversion, hindering vertical air motions. Thus days are hazy and nights cold.

INDICATOR, substance which indicates when the concentration of a chemical species has passed a certain threshold value, by a change of color, turbidity or fluorescence. They are generally used to find the end-point of a TITRATION. The indicator is a substance existing in two visibly different forms in an EQUILIBRIUM that is the same kind as that of the reaction being followed. Thus, to follow an acid-base titration, a conjugate ACID-base system is used as indicator which changes color over a narrow range of pH corresponding to the end-point. (For this to happen, the equilibrium constant K of the indicator must approximately equal the hydrogen-ion concentration at the end-point.) A **universal indicator** is a mixture of indicators which changes color continuously over a wide pH range, used as a quick guide to acidity. To follow an OXIDATION-reduction reaction, an indicator is used which exists reversibly in an oxidized or reduced state, its oxidation potential being about the same as that of the reaction. For a precipitation or complexing reaction, an indicator is used which itself forms a colored precipitate or complex with excess added reagent. A good indicator must be visible at such low concentrations that it does not interfere with the reaction.

INDICES, the small numbers written as superscripts following a number or term to indicate the POWER to which it is to be raised. (See also EXPONENT.)

INDIGESTION. See DYSPEPSIA.

INDIGO, a blue dye obtained from LEGUMINOUS PLANTS of the genus *Indigofera*. The dye is produced by natural acidation of a solution containing pieces of the plants. Cultivation of indigo plants was once carried out on a large scale in India, but cheap synthetic indigo is now mainly used.

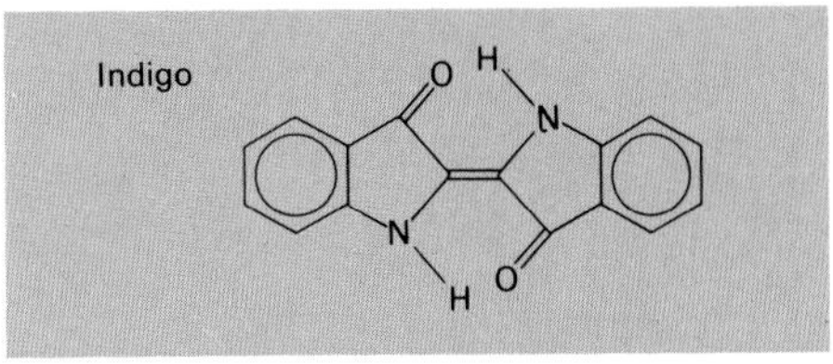

INDIUM (In), rare metal, very soft and silvery-white, in Group IIIA of the PERIODIC TABLE, resembling ALUMINUM. It is prepared from flue dust residues of zinc processing. Indium forms trivalent compounds and some monovalent ones. It is used in solders, low-melting-point alloys, germanium TRANSISTORS, glass-seals, bearing alloys, and (combined with Group VA elements) in SEMICONDUCTORS. AW 114.8, mp 157°C, bp 2080°C, sg 7.31 (20°C).

INDUCTANCE (self-), the ratio of the voltage induced in an electric circuit (see INDUCTION, ELECTROMAGNETIC) to the rate of change of the current in it. It depends on the circuit geometry, being large for coils and small for extended circuits, and is greatly increased by the presence of ferromagnetic materials (see MAGNETISM). Voltages induced by currents in a different circuit are measured in terms of **mutual inductance**. Inductors have an IMPEDANCE to AC currents proportional to the current, and are widely used in electronics. The SI unit of inductance is the HENRY (H).

INDUCTION, in philosophy, the process of reasoning from particular instances to general propositions.

INDUCTION, Electromagnetic, the phenomenon in which an ELECTRIC FIELD is generated in an electric circuit when the number of MAGNETIC FIELD lines passing through the circuit changes, independently discovered by M. FARADAY and J. HENRY. The voltage induced is proportional to the rate of the change of the field, and large voltages can be produced by switching off quite small magnetic fields suddenly. Frequently, the magnetic field is itself generated by an electric current in a coil, in which case the voltage induced is proportional to the rate of change of the current (see INDUCTANCE).

The principle finds numerous applications in electric GENERATORS and MOTORS, TRANSFORMERS, MICROPHONES, and engine ignition systems (see INDUCTION COIL, MAGNETO). In the less familiar technique of **induction heating**, widely used in metal working, an object is heated by currents created in it by the voltage induced by a high-frequency current in a nearby coil; as the coil field will pass through insulators without heating them, the principle can be applied to produce "cold hob" electric stoves.

INDUCTION COIL, a device generating very high voltages, usually for sparking, and particularly in engine IGNITION SYSTEMS. A large secondary and small

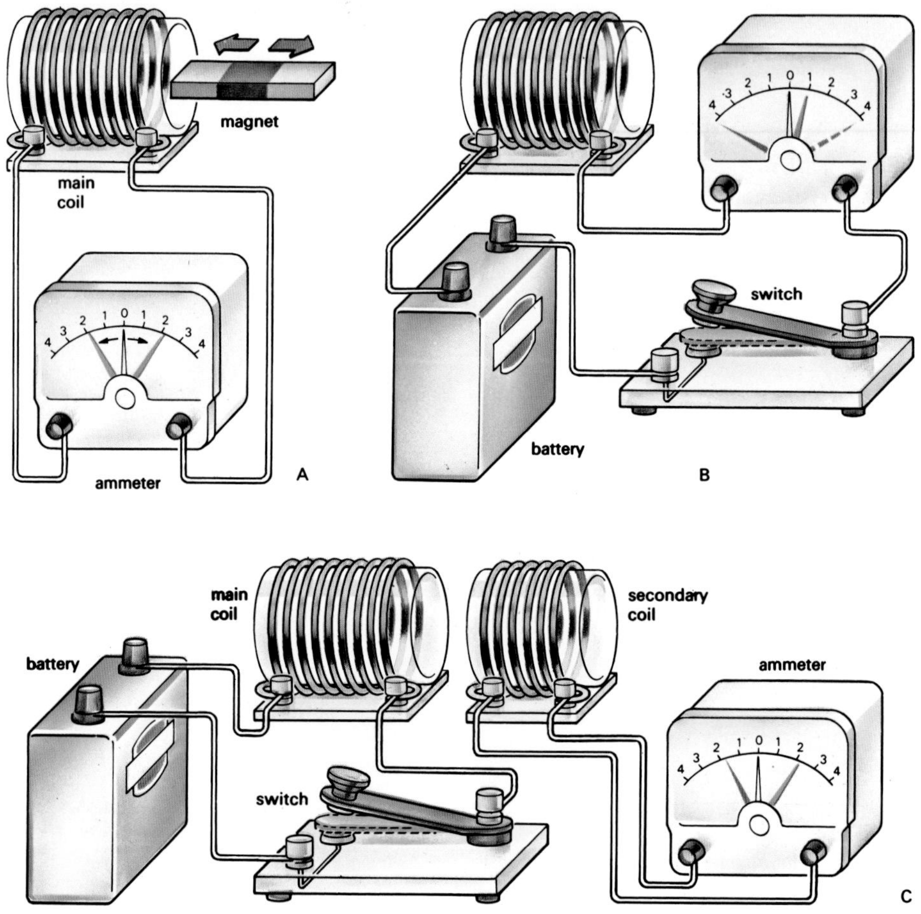

Simple experiments in electromagnetic induction. (A) When a permanent magnet is being moved in and out of a small coil, an "induced" current flows in the coil. But it is only while the magnetic field in the coil is changing that any induced current flows. (B) Self-induction occurs with a circuit contains an inductive component such as a coil. When a circuit is completed through a coil (by closing a switch), the increasing current in the coil gives rise to a changing magnetic field which interacts with the coil itself, setting up an electromotive force which opposes the original current. A sensitive ammeter in a series with such a coil might show a momentary current (red) in the direction opposite to the subsequent steady current (green). On breaking the circuit, a similar forward current surge (broken red) occurs. (C) Mutual induction is exhibited when there is a magnetic linkage between two circuits containing inductive elements. Closing the switch in the primary circuit creates a magnetic field in the secondary circuit, giving rise to a momentary current as in (A). Opening the switch in the primary causes an equivalent current in the reverse direction in the secondary.

primary coil are wound together on a ferromagnetic (see MAGNETISM) core. When a low current in the primary is interrupted by a switch, the rapid change induces (see INDUCTION) a large voltage in the secondary. In a variant producing high-frequency AC voltage pulses, the primary current switches itself on and off through a relay, much as in an electric BELL mechanism.

INDUCTION HARDENING, the use of electromagnetic INDUCTION to heat metals rapidly in order to harden them (see METALLURGY; STEEL). A very high-frequency current is passed through an induction coil surrounding the workpiece.

INDUCTION MOTOR, an alternating-current electric MOTOR in which the current in the moving part is induced (see INDUCTION) rather than supplied via slip-ring contacts. Also called an asynchronous motor, it is the commonest motor in domestic and industrial use.

INDUSTRIAL PSYCHOLOGY, or **occupational psychology,** the study of the mental responses and attitudes of people at work, particularly in their relations with each other, with the organization, and with the devices which they operate. Its purpose is to increase efficiency and provide conditions in which people derive maximum satisfaction from their work. Industrial psychologists use personality tests to determine workers' aptitudes, and contribute to the design of machines and factories. (See also AUTOMATION; HUMAN ENGINEERING; MECHANIZATION AND AUTOMATION; PSYCHOLOGY.)

INDUS VALLEY CIVILIZATION, centered round the Indus R in India and Pakistan, the earliest known urban culture of the Indian subcontinent. Superimposed on earlier stone- and bronze-using (see STONE AGE; BRONZE AGE) cultures dating from c4000 BC, the Indus Valley civilization, centered around Harappa and Mohenjo-Daro, lasted from c2500 to c1750 BC. About 100 of its towns and villages, some with fortified citadels, have been identified.

INEQUALITIES, statements that two quantities are not equal (see EQUALITY), specifically statements as to which is the larger. The relationships are written $>$ (greater than) and $<$ (less than); and these may be coupled with the symbol of equality, as $\geqslant$ (greater than or equal to) and $\leqslant$ (less than or equal to). (See also ALGEBRA.)

INERT GASES, former name for the NOBLE GASES.

INERTIA, property of all MATTER, representing its resistance to any alteration of its state of MOTION. The MASS of a body is a quantitative measure of its inertia; a heavy body has more inertia than a lighter one and needs a greater FORCE to set it in motion. NEWTON's laws of motion and his principle of RELATIVITY depend on the concept of inertia. In EINSTEIN's theory of relativity, the inertial properties of matter are interrelated with its total ENERGY content.

INERTIAL GUIDANCE, an automatic navigational apparatus carried in guided missiles, airplanes, ships and submarines, which depends on the forces of INERTIA for sensing changes in the magnitude and direction of the vehicle's motion. ACCELEROMETERS are mounted on gyrostabilized platforms to isolate them from the vehicle's angular motion; by measuring the FORCES needed to keep a suspended mass stationary with respect to the moving vehicle, they sense changes in its motion and gravitational fields. The orientations of the accelerometers are found from reference directions provided by GYROSCOPES. A COMPUTER calculates VELOCITIES or distances from the instrument signals and can compare these with stored data. The accuracy of the system is improved by a method of FEEDBACK called Schuler tuning.

INFANTILE PARALYSIS. See POLIOMYELITIS.

INFANTILE SEXUALITY, general term embracing those aspects of sexuality (see SEX) exhibited by most children less than about five years old. They do not in general persist into adulthood, though the adult may suffer from a COMPLEX caused by GUILT concerning them. (See also OEDIPUS COMPLEX.)

INFECTIOUS DISEASES, DISEASES caused by any microorganism, but particularly VIRAL and BACTERIAL DISEASES and PARASITIC DISEASES, in which the causative agent may be transferred from one person to another (directly or indirectly). Knowledge of the stages at which a particular disease is liable to infect others and of its route (via SKIN scales, COUGH particles, clothing, urine, feces, SALIVA, or by insects, particularly MOSQUITOS and TICKS) helps physicians to limit the spread of diseases in EPIDEMICS.

INFERIORITY COMPLEX, term used by ADLER and now mainly by psychoanalysts to describe the COMPLEX of fears and EMOTIONS arising out of feelings of inferiority or inadequacy, particularly those concerned with (usually imagined) inferiority of the sexual organs. (See SUPERIORITY COMPLEX.)

INFINITESIMAL, originally, an infinitely small quantity, smaller than any finite quantity. The term is now applied to a part of a quantity whose magnitude is vanishingly small in terms of that of the quantity itself; and hence to that part of a FUNCTION which is

The induction coil: (*left*) schematic diagram; (*right*) circuit diagram. In each case the primary circuit is in blue, the secondary in red. The capacitor prevents sparking between the terminals of make-and-break relay.

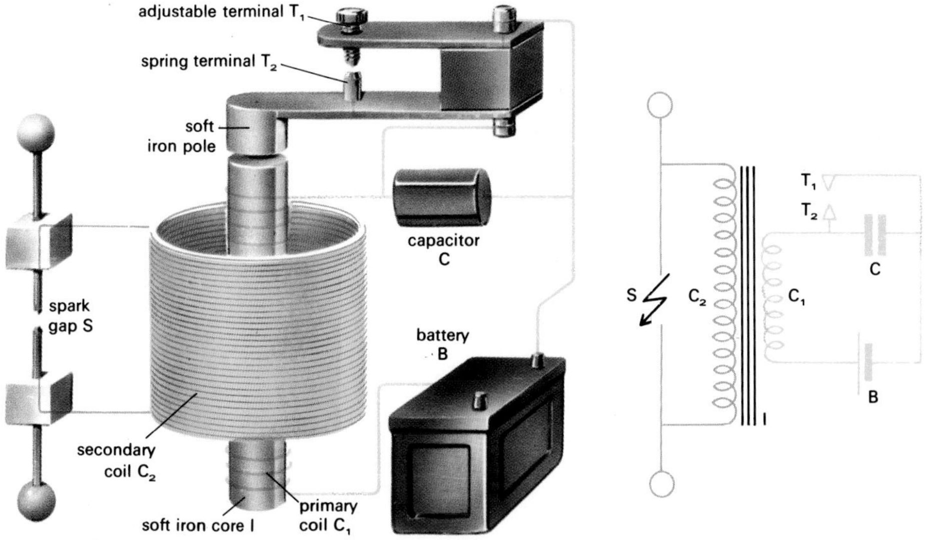

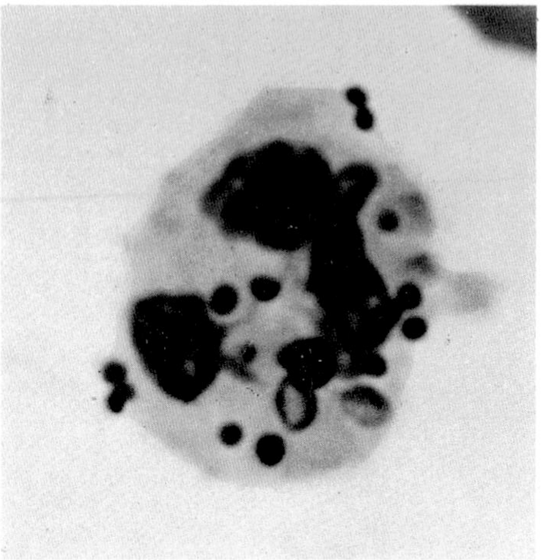

Inflammation: A white blood cell (actually a neutrophil) showing phagocytosis of stained bacteria (magnification × 2 100).

vanishingly small for all values of the VARIABLE, as in $f(x)=x+10^{-10}x$. CALCULUS has sometimes been called **infinitesimal calculus.**

INFINITY (∞)**,** a quantity greater than any finite quantity. In modern mathematics infinity is viewed in two ways. In one, the word infinity has a definite meaning; and with TRANSFINITE CARDINAL NUMBERS, for example, it may have a plurality of meanings. In the other, infinity is seen as a LIMIT: to say that PARALLEL LINES intersect at infinity, for example, means merely that the point of INTERSECTION of two lines may be made to recede indefinitely by making the lines more and more nearly parallel. Similarly, in $f(x)=1/x$, it is meaningful to say that $f(x)$ tends to infinity as x tends to ZERO (see CALCULUS); again, the SEQUENCE $1, 2, 3, \ldots, n$, tends to infinity since, however large n is chosen, there is an $(n+1)$ greater than it. (See also SERIES.)

INFLAMMATION, the complex of reactions established in body TISSUES in response to injury and infection. It is typified by redness, heat, swelling and pain in the affected part. The first change is in the CAPILLARIES, which dilate, causing ERYTHEMA, and become more permeable to cells and PLASMA (leading to EDEMA). White BLOOD cells accumulate on the capillary walls and pass into affected tissues; foreign bodies, dead tissue and bacteria are taken up and destroyed by phagocytosis and ENZYME action. Active substances produced by white cells encourage increased blood flow and white cell migration into the tissues. LYMPH drainage is important in removing edema fluid and tissue debris. ANTIBODY AND ANTIGEN reactions, ALLERGY and other types of IMMUNITY are concerned with the initiation and perpetuation of inflammation. Inflammatory DISEASES comprise VIRAL and BACTERIAL DISEASE, PARASITIC DISEASE and disorders in which the inflammatory response is activated inappropriately (e.g., by autoimmunity) causing tissue damage.

INFLECTION, a change in the tone, pitch or volume of the voice; and in linguistics more importantly a change in the form of a word for grammatical reasons. In modern English, inflections are usually suffixes: "to kick," "kicks" and "kicked." As syntax increases in importance in a language, inflections become less frequent. (See also GRAMMAR; LINGUISTICS.)

INFLECTION, Point of, a point on a CURVE at which the direction of CURVATURE changes. At such a point the *second* DERIVATIVE of the curve's FUNCTION has a value of zero, its value changing from negative to positive, or *vice versa*, between the two adjacent points on each side. (See also EXTREMUM.)

INFLORESCENCE, term applied to the con-

Types of inflorescence.

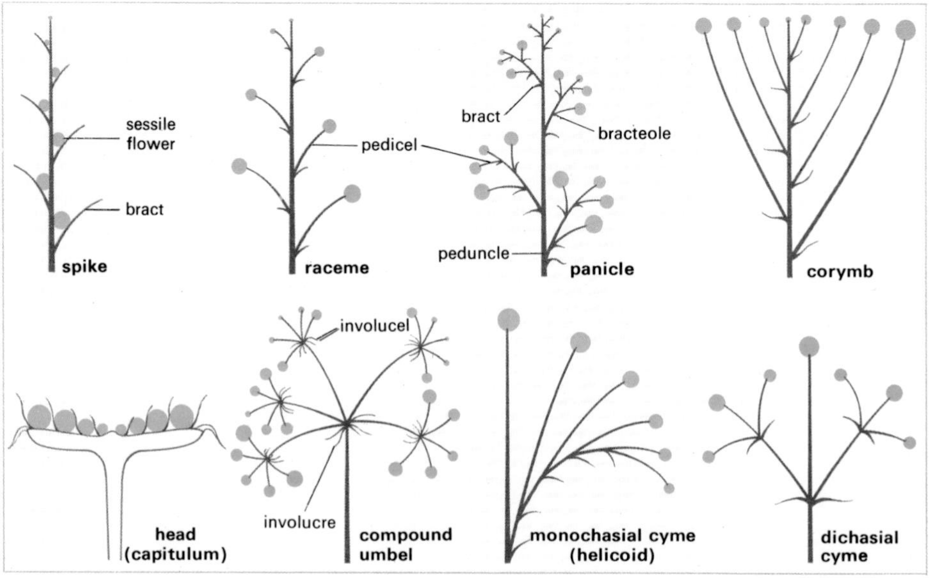

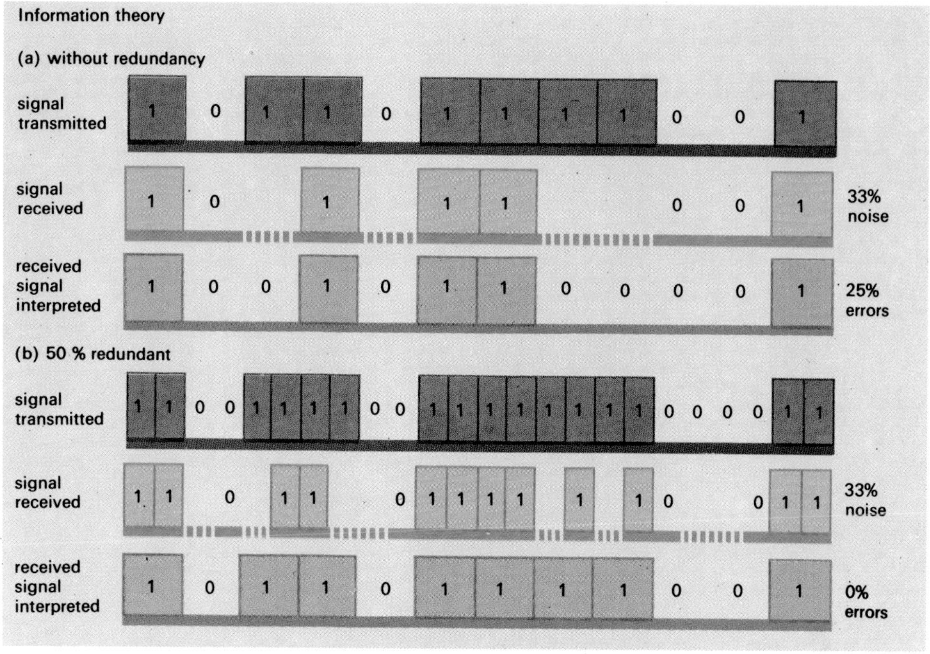

spicuous clusters of FLOWERS that are produced by many ANGIOSPERMS. There are several types of inflorescence, the forms of which vary according to the arrangement of individual flowers. In the type of inflorescence known as a **raceme** the flowers are attached to the main flower axis by short stalks, or pedicels, of equal length, for example the hyacinth, while in the spike there are no pedicels and the flowers are directly attached to the main axis, for example the gladiolus. Plants such as lilac and oats have an inflorescence similar to a raceme, but the pedicels bear more than one flower. This formation is called a panicle. In the corymb, the pedicels are of unequal length so that the inflorescence has a flat-topped appearance, for example hawthorn. In some plants, particularly those of the family Compositae, all the flowers are bunched on a flat disk, this arrangement being known as a head. In the simple umbel the pedicels appear to arise from a central point, while in the compound umbel several simple umbels are borne on a single stalk or ray and each inflorescence comprises a number of rays growing from the tip of the main axis. A simple umbel is produced by the milkweed and most members of the carrot family (Umbelliferae) produce compound umbels.

INFLUENZA, grippe, or **'flu,** a group of VIRAL DISEASES causing mild respiratory symptoms, FEVER, malaise, muscle pains and HEADACHE, and often occurring in rapidly spreading EPIDEMICS. GASTROINTESTINAL TRACT symptoms may also occur. Rarely, it may cause a severe viral PNEUMONIA. A characteristic of influenza viruses is their property of changing their antigenic nature frequently, so that IMMUNITY following a previous attack ceases to be effective. This also limits the usefulness of influenza VACCINATION.

INFORMATION RETRIEVAL, a branch of technology of ever-increasing importance as man attempts

Information theory: how a measure of redundancy in the transmission of a message can improve the probability of its being correctly interpreted on reception. In (A), a simple message in binary digits is transmitted, losing 33% of its information in transmission; on receipt 25% of the message is incorrectly interpreted. (B) By transmitting the message with 50% redundancy, i.e., with each digit repeated, and the same loss in transmission, sufficient information is received for the original message to be correctly reconstructed.

to cope with the "information explosion." To store and have reference to the vast amount of printed matter produced annually is impossible for most libraries. The problem can be solved by microphotography. Pages are photographed at a reduction (typically to about $\frac{1}{20}$) and stored on 35mm or 16mm film (microfilm), on transparent cards measuring about 100 × 150mm (microfiches) or as positive prints on slightly smaller cards (microcards). VIDEOTAPE is also used. Reference may be manual or by machine, usually computer. The information must be classified so that the user may gain rapid access *either* to a particular book or paper *or* to all the relevant material on a particular subject.

In COMPUTERS, information retrieval involves a reverse of those operations used for data storage. The operator inserts a classification which the computer matches with the classification in its memory.

INFORMATION THEORY, or communication theory, a mathematical discipline that aims at maximizing the information that can be conveyed by communications systems, at the same time as minimizing the errors that arise in the course of transmission. The information content of a message is conventionally quantified in terms of "bits" (*bi*nary

Color thermogram of a house revealing the pattern in which it radiates energy in the infrared region of the spectrum. This effectively shows the extent to which different parts of the house lose heat. Each color isotherm represents a temperature change of 1 K; the temperature variation over the brickwork is thus 3K and the windows are some 7K warmer than the coolest brickwork. From this evidence it is clear that the house loses most of its heat through the windows.

digi*ts*). Each BIT represents a simple alternative—in terms of a message, a yes-or-no; in terms of the components in an electrical circuit, that a switch is opened or closed. Mathematically, the bit is usually represented as a 0-or-1. Complex messages can be represented as series of bit alternatives. Five bits of information only are needed to specify any letter of the alphabet, given an appropriate code. Thus able to quantify "information," information theory employs statistical methods to analyze practical communications problems. The errors that arise in the transmission of signals, often termed NOISE, can be minimized by the incorporation of **redundancy**. Here more bits of information than are strictly necessary to encode a message are transmitted, so that if some are altered in transmission, there is still enough information to allow the signal to be correctly interpreted. Clearly, the handling of redundant information costs something in reduced speed of or capacity for transmission, but the reduction in message errors compensates for this loss. Information theoreticians often point to an analogy between the thermodynamic concept of ENTROPY and the degree of misinformation in a signal.

INFRARED RADIATION, ELECTROMAGNETIC RADIATION of wavelength between 780nm and 1mm, strongly radiated by hot objects and also termed heat radiation. Detected using PHOTOELECTRIC CELLS, BOLOMETERS and photographically, it finds many uses—in the home for heating and cooking and in medicine in the treatment of muscle and skin conditions. Infrared absorption SPECTROSCOPY is an important analytical tool in organic chemistry. Military applications (including missile-detection and guidance systems and night-vision apparatus) and infrared PHOTOGRAPHY (often FALSE-COLOR PHOTOGRAPHY) exploit the **infrared window**, the spectral band between 7.5 and 11μm in which the ATMOSPHERE is transparent. This and the high infrared reflectivity of foliage give infrared photographs their striking, often dramatic clarity, even when exposed under misty conditions.

INHIBITION, in biochemistry. See ENZYMES.

INHIBITION, the action of a mental process or function in restraining the expression of another mental process or function; e.g., fear of social condemnation inhibiting fulfilment of sexual desire. Most often the EGO or SUPEREGO inhibits instinctual behavior (see INSTINCT). (See also REPRESSION.)

INJECTION, the administration of a substance, usually a DRUG or vaccine, by SYRINGE and needle which allows the SKIN to be broached and the substance to be delivered intradermally, subcutaneously, intramuscularly, intravenously, intra-arterially or into body organs or cavities. This bypasses the GASTROINTESTINAL TRACT which may be ineffective, slow or unreliable, or may destroy the agent. VACCINATION, desensitization for ALLERGY, drugs for seriously ill or VOMITING patients and INSULIN for diabetics are almost always given by injection.

INJECTION MOLDING, technique used in forming thermoplastic materials. The hot, plastic material is forced into a chilled (usually by water) mold, in which it sets. (See also PLASTICS.)

INK, liquid or paste, containing DYES or PIGMENTS, used for writing or PRINTING. Writing inks date from the mid-3rd millennium BC in China and Egypt. The most common today is blue-black permanent ink, made by dissolving GALLIC ACID, IRON (II) sulfate and TARTARIC ACID in water. Since the blue-black color is produced only when the ink dries, a dye is usually added to color the ink during writing. In ballpoint inks, dyes are dissolved in GLYCOLS and other liquids, and wetting agents are added. Black, waterproof India ink is a suspension of CARBON particles stabilized by GELATIN, glue etc. Carbon black is also the pigment used in black printing ink. Printing inks—diverse in their composition and uses—are viscous pastes made by grinding pigments with varnishes or petroleum solvents, and contain various additives for printability and drying speeds.

INKBLOT TEST, or **Rorschach Test.** See RORSCHACH, H.

INOCULATION, the INJECTION or introduction of microorganisms or their products into living TISSUES or culture mediums. It is used in man to establish antibody formation and IMMUNITY in VACCINATION.

INORGANIC CHEMISTRY, major branch of CHEMISTRY comprising the study of all the elements and their compounds, except carbon compounds containing hydrogen (see ORGANIC CHEMISTRY). The elements are classified according to the PERIODIC TABLE. Classical inorganic chemistry is largely descriptive, synthetic and analytical; modern theoretical inorganic chemistry is hard to distinguish from PHYSICAL CHEMISTRY.

INSANITY, term descriptive of an individual's mental state employed in legal and popular usage. In psychology and psychoanalysis, the term is considered a loose synonym for PSYCHOSIS.

INSCRIPTION, in plane GEOMETRY, the construction of a CIRCLE such that each side of a POLYGON is a tangent to it (see TANGENT OF A CURVE). An inscribed ANGLE is one whose sides are chords of a circle and whose vertex lies on the circle. In three dimensions, inscription implies the construction of a sphere such that every side of a regular POLYHEDRON is a tangent plane to it. (See also CIRCUMSCRIPTION.)

INSECTA, a class of the ARTHROPODA, members of which have bodies divided into three parts, the head, thorax and abdomen. There are about 2 million known species, making the insects the largest class in the Animal Kingdom. Fossil forms are known from the Permian period, 280 million years ago, and the

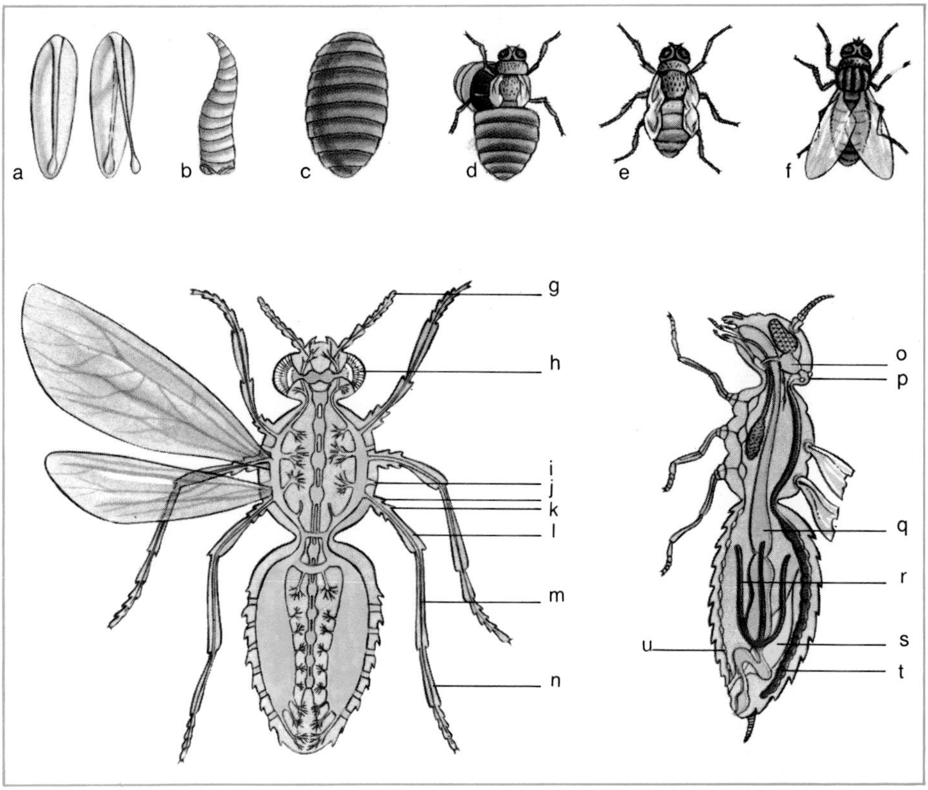

Insect metamorphosis is one of the most remarkable of natural phenomena. The metamorphosis of a fly (*top*) begins with an egg (a), from which the larva hatches (b). The pupa stage (c) appears dormant, but within the puparium important changes are taking place under the influence of the hormones. The immature fly (d) creeps out of the puparium already equipped with legs and eyes but with rudimentary wings (e) and finally reaches the adult stage, called the imago (f). Insects display all the characteristics of their phylum, Arthropoda: a skin covered with a hard layer (cuticula), which also serves as an exoskeleton; a body which is clearly segmented; and jointed appendages. The body is divided into a head, thorax and abdomen. The head has a pair of jointed antennae (g), in which the sense of smell is located, and three pairs of legs and two pairs of wings. Each leg consists of a number of segments, namely, the coxa (j), trochanter (k), femur (l), tibia (m) and tarsus with five segments (n). On the sides of the head are large compound eyes composed of many facets (h), and on the forehead there are generally three simple eyes or ocelli (p). The respiratory organs consist of a system of air tubes (tracheae), the finest branches of which penetrate to all the tissues. The openings (spiracles) of this system (i) are equipped with muscular rings which keep them closed as long as the animal is not active and needs only a small amount of oxygen; thus gas exchange and water loss are minimized. The internal organs of an insect are located primarily in the abdomen. Nearest the back lies the tube-like heart (t), which consists of a number of segments separated by valves. Toward the front, it passes over into the aorta, the only part of the blood-vessel system that has its own wall: everywhere else the blood flows freely among the organs. In the middle is the intestinal canal (q), consisting of the fore-gut, mid-gut and hind-gut. The long, tubular excretory organs (r) arise at the division between mid- and hind-gut. The abdomen also contains the reproductive organs, such as the ovary (s). The nervous system consists of the brain (o) and the abdominal nerve cord (u), along which are strung clusters of nerve cell bodies called ganglia.

The Venus' fly trap (*Dionaea muscipula*) about to close shut on an unsuspecting insect. Trigger hairs touched by a visiting insect cause the trap to close and secrete digestive juices. In this way the plant obtains nitrogen, lacking in the impoverished soils in which it grows.

success of the group can be attributed to the evolution of a variety of feeding methods and the ability to fly. Most insects go through egg, larva, pupa and adult stages during their life histories, each stage being separated by a change of body form called METAMORPHOSIS. Many species are of considerable economic importance as crop pests.

INSECTICIDE, any substance toxic to insects and used to control them in situations where they cause economic damage or endanger the health of man and his domestic animals. There are three main types: **stomach insecticides**, which are ingested by the insect with their food; **contact insecticides**, which penetrate the cuticle, and **fumigant insecticides**, which are inhaled. Stomach insecticides are often used to control chewing insects like CATERPILLARS and sucking insects like APHIDS. They may be applied to the plant prior to attack and remain active in or on the plant for a considerable time. They must be used with considerable caution on food plants or animal forage. Examples include ARSENIC compounds which remain on the leaf, and organic compounds which are absorbed by the plant and transported to all its parts (systemic insecticides). Contact insecticides include the plant products NICOTINE, derris and pyrethrum, which are quickly broken down, and the synthetic compounds such as DDT (and other chlorinated HYDROCARBONS, organophosphates (malathion, parathion) and carbamates. Polychlorinated biphenyls (PCBs) are added to some insecticides to increase their effectiveness and persistance. Highly persistent insecticides may be concentrated in food chains and exert harmful effects on other animals such as birds and fish (see ECOLOGY).

INSECTIVORA, an order of the MAMMALIA that includes the shrews, hedgehogs, moles and tenrecs. Insectivores are the most primitive living mammals. All are small and have pointed snouts, often with prominent whiskers, small brains and sharp teeth.

INSECTIVOROUS PLANTS, or **carnivorous plants**, specialized plants whose leaves are adapted to trap and digest insects, which supplement their food supply. They normally live in boggy habitats or as EPIPHYTES. The insects may be caught in vase-like traps (e.g. pitcher plant), by leaves that spring shut (e.g. venus fly trap), by a trapdoor (e.g. bladderwort) or on sticky leaves (e.g. sundew). The captured insects are broken down by ENZYMES secreted from the plants and the products absorbed.

INSTINCT, a phenomenon whose effects can be observed in animals and man, but whose precise nature is little understood. In general one can say that instinctive behavior comprises those fixed reactions to external stimuli that have not been consciously learned (see ANIMAL BEHAVIOR). In fact, such behavior seems to stem from a complex of hereditary and environmental factors (see ENVIRONMENT; HEREDITY), since animals placed from birth in artificial environments display some, but not all, instinctive reactions characteristic of their species. It has been further suggested that EMBRYOS may have some LEARNING ability; i.e., that some learning before birth is possible. Numbered among the instincts are the sex drive, AGGRESSION, TERRITORIALITY and the food urge; but much debate surrounds such classification. In psychology, "instinct" (sometimes called **drive**) has a similar meaning, with special emphasis on the response as a complex one (see REFLEX). Frustration of, or conflict between, instincts engenders NEUROSES. FREUD suggested the existence of two fundamental instincts: the life instinct, rather akin to the LIBIDO; and its opposite, the death instinct.

INSTRUMENTALISM, or experimentalism, in philosophy, the development of PRAGMATISM promoted by John DEWEY. It is based on the contention that ideas are validated solely by their usefulness in solving problems.

INSTRUMENT LANDING SYSTEM (ILS), radio system used in conditions of poor visibility to guide aircraft towards runways. Two RADIO transmitters are used, each producing two beams which overlap at a slight angle. The **localizer**, at the far end of the runway, transmits in a horizontal plane; the **glide slope**, at the near end of the runway, does so vertically, the overlap sloping up from the runway at an angle of about 3°. Instruments in the pilot's cockpit indicate deviations from the "cone" of the two overlaps, so guiding him into the point where direct observation of the runway is possible.

INSTRUMENTS, Scientific, devices used for measurement and hence for scientific investigation and control (see also MECHANIZATION AND AUTOMATION). They extend the observing faculties of the human senses, providing accuracy and a greater range. They can also detect and measure phenomena such as X rays which man cannot sense. Early instruments, used mainly in the fields of astronomy, navigation and surveying, measured the basic quantities of mass, length, time and direction (see ASTROLABE; BALANCE; CLOCKS AND WATCHES; COMPASS). With the rise of modern science came several instruments including the MICROMETER, MICROSCOPE, TELESCOPE and THERMOMETER. During the Industrial Revolution and after were invented instruments too many to mention; and today in science, industry and even the home there are a host of instruments to measure every conceivable quantity. A few simple instruments, such as the ruler or balance, work by direct comparison, but most are **transducers**, representing the quantity measured by another sensible quantity (usually the position of a pointer on a scale). All instruments require initial CALIBRATION against a known or calculable standard. In general, an instrument interacts with the measured phenomenon, and the resultant change in its state is amplified if necessary, displayed by means of a pointer, pen, light beam, oscilloscope etc., and recorded, usually on chart paper or by photography. Although precision instruments are designed for high accuracy, inevitably errors are introduced: amplification produces NOISE, the slowness of the instrument's response results in lag and damping, and the intrinsic nature of the response may be defective owing to hysteresis or drift; moreover, the observer may misread the scale because of PARALLAX or INTERPOLATION errors. Most fundamental of all, the act of measuring a system may significantly alter the state of the system (see also UNCERTAINTY PRINCIPLE).

INSULATION, Electric, the containment of electric currents or voltage by materials (insulators) that offer a high resistance to current flow, will withstand high voltages without breaking down, and will not deteriorate with age. Resistance to sunlight, rain, flame or abrasion may also be important. The electrical resistance of insulators usually falls with temperature (paper and asbestos being exceptions) and if chemical impurities are present. The mechanical properties desired vary with the application: cables require flexible coatings, such as polyvinyl chloride, while glass or porcelain are used for rigid mountings, such as the insulators used to support power cables. In general, good thermal insulators are also good electrical ones.

INSULATION, Thermal, the reduction of transfer of heat from a hot area to a cold. Thermal insulation is used for three distinct purposes: to keep something hot; to keep something cold; and to maintain something at a roughly steady temperature. HEAT is transferred in three ways, CONDUCTION, CONVECTION and RADIATION. The VACUUM BOTTLE thus uses three different techniques to reduce heat transfer: a vacuum between the walls to combat conduction and convection; silvered walls to minimize the transmission of radiant heat from one wall and maximize its reflection from the other; and supports for the inner bottle made of CORK, a poor thermal conductor. (See also FIREBRICK; POLYSTYRENE; REFRACTORY.)

INSULIN, HORMONE important in METABOLISM, produced by the islets of Langerhans in the PANCREAS, which act as an ENDOCRINE GLAND. Insulin is the only hormone which reduces the level of SUGAR in the BLOOD and is secreted in response to a rise in blood sugar (e.g., after meals, or in conditions of stress); the sugar is converted into GLYCOGEN in the cells of MUSCLE and the LIVER under the influence of insulin. Absence or a relative failure in secretion of insulin occurs in DIABETES, in which blood sugar levels are high and in which sugar overflows into the urine. The isolation of insulin as a pancreatic extract by F. G. BANTING and C. H. BEST in 1921 was a milestone in medical and scientific history. It is a PROTEIN made up of fifty AMINO ACIDS as two peptide chains linked by sulfur bridges. Because it is destroyed in the GASTROINTESTINAL TRACT, it has to be taken by subcutaneous INJECTION in diabetics with severe insulin lack. Its use in diabetics has revolutionized treatment of this disease; the aim in its administration is to be as close to natural secretion patterns as possible. If insufficient insulin is taken, diabetic COMA may result, while in excess hypoglycemia supervenes; both require prompt medical treatment.

INTEGERS, the set (see SET THEORY) of whole numbers, including ZERO and negative whole numbers. (See also NATURAL NUMBERS; RATIONAL NUMBERS; TRANSFINITE CARDINAL NUMBER.)

INTEGRAL DOMAIN. See FIELD.

INTEGRATED CIRCUIT, a single structure in which a large number of individual electronic components are assembled. (See ELECTRONICS.)

INTEGRATION. See CALCULUS.

INTELLIGENCE, the general ability to solve problems. The ability to solve specific problems only may arise as a result of INSTINCT or through experience, neither of which can be regarded as contributing to intelligence; though intelligence affects the ability of the individual to apply experience to problem-solving.

By far the most intelligent animal on this planet is man. For this reason, among others, most investigations of intelligence have been carried out in human beings. Intelligence tests are structured upon seven main bases: numerical ability (the speed and accuracy with which the individual can solve problems of simple arithmetic); verbal fluency; verbal meaning (the ability to understand words); the ability to remember; speed of perception and, most importantly, the ability to reason. Such tests are of considerable use, though their limitations must be recognized: they are of little value in comparisons between ethnic groups, or even between social classes, since in both cases environment plays a very large part in an individual's performance. Moreover, their accuracy in measuring IQs outside the normal limits is suspect. Such qualms have led to a definition of human intelligence as "that which can be measured by intelligence tests" (see also IQ; PSYCHOLOGICAL TESTS).

Throughout the animal kingdom, there is a good correlation between the intelligence of an animal and the size of its brain relative to that of its body. There is an even better one when the surface area of the BRAIN is considered: the higher mammals have a more convoluted cortex (outer layer) than do the lower. After man, the most intelligent animal is the dolphin. Perhaps surprisingly, ANTS show an ability to solve mazes that compares with that of some mammals.

The ways in which animals solve problems are a useful pointer to their intelligence. The two important ways are trial-and-error, which is a LEARNING process dependent upon intelligence, and insight. This latter is displayed only by the higher animals.

The evolution of intelligence is unclear, though obviously it has had a profound effect on the emergence of man as earth's dominant animal. Equally obviously, intelligence is a considerable aid to species survival. Much effort has been used in recent years to examine how much of an individual's intelligence is determined by hereditary factors, how much by environmental factors. Although results have not been conclusive, it would seem that over half the difference in intelligence between people is determined by inheritance, the remainder by early environmental conditions (see also HEREDITY). (See also MEMORY; MENTAL RETARDATION; MIND.)

INTELLIGENCE QUOTIENT. See IQ.

INTELLIGENCE TEST. See PSYCHOLOGICAL TESTS.

INTERCEPT, in mathematics the point at which a CURVE or straight LINE intersects the x, y or z axis of a coordinate system (see ANALYTIC GEOMETRY). The parabola (see CONIC SECTIONS) $y = x^2 - 4$, for example has a y-intercept at -4 and x-intercepts at $+2$ and -2. The term is also used to refer to the portion of a line lying between two other lines or planes that intersect it.

INTERFERENCE, the interaction of two or more similar or related WAVE MOTIONS establishing a new pattern in the AMPLITUDE of the waves. It occurs in all wave phenomena including SOUND, LIGHT and water waves. In most cases the resulting amplitude at a point is found by adding together the amplitudes of the individual interfering waves at that point. Interference patterns can only result if the interfering waves are of related wavelength and exhibit a definite PHASE relationship.

Optical interference. Light from ordinary sources is "incoherent"—there is no definite relationship between the phases of the waves associated with different PHOTONS. Until recently the only way to demonstrate optical interference was to use light from a single source which had been divided and led to the interference zone along paths of differing length, thus ensuring that the interfering beams were coherent at least with each other. In this way Thomas YOUNG in 1801 first demonstrated optical interference, showing, because interference effects cannot be explained on either ray or particle models, that light was indeed to be regarded as a wave phenomenon. Young passed light from a single pinhole source through two parallel slits in an opaque screen and found that interference fringes—alternate bands of light and dark—were formed on another screen placed beyond the slits. The bright bands resulted from the

Newton's rings, a familiar interference phenomenon, can be seen whenever a transparent object or film closely overlies a reflecting surface. Examples include a convex lens on an optical flat, a film of oil floating on water and, all too frequently, a transparency film in a glass slide mounting. Coherent light (light from a single source) reflected from both sides of a thin air gap or oil film interferes with itself, giving rise to a pattern of colored or, for monochromatic light, light and dark fringes.

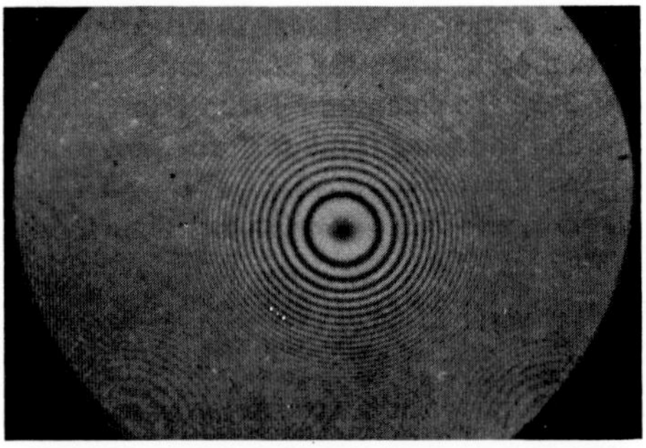

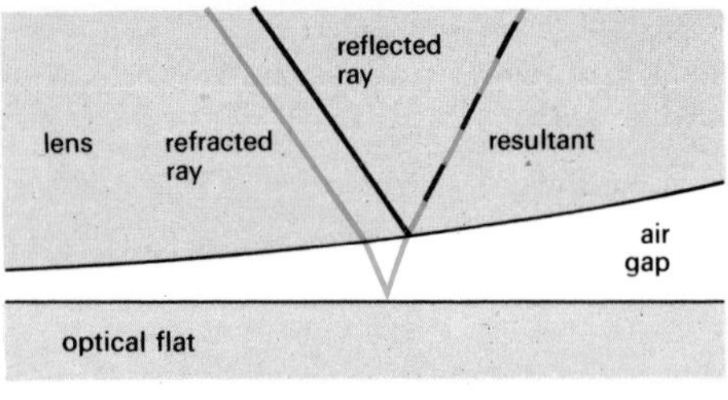

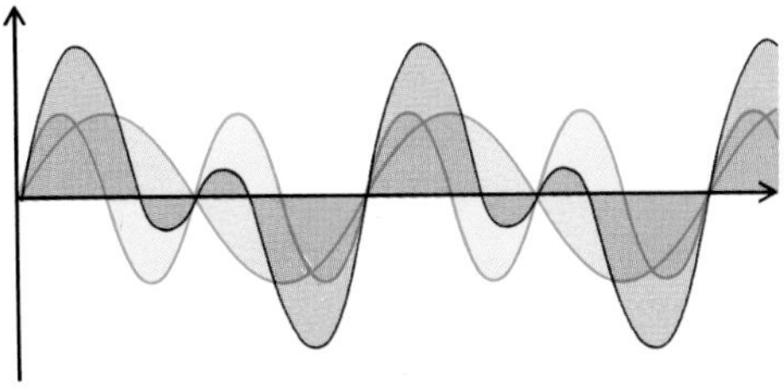

The interference of two related regular wave motions (blue) yields a third regular wave motion (red), the amplitude of which at any point is the algebraic sum of the amplitudes of the component waves.

constructive interference of the two beams, the wave amplitude of each reinforcing the other; the dark bands, destructive interference, the amplitude of one wave effectively canceling the effect of that of the other. Newton's rings, colored fringes seen in thin transparent films, are a similar interference effect. In recent years LASERS (which produce coherent light—radiation having a uniform and controllable phase structure) have enabled physicists to produce optical interference effects much more easily, an important application being HOLOGRAPHY. (See also INTERFEROMETER.)

INTERFERENCE, Radio, unwelcome additions to a desired signal in RADIO and TELEVISION receivers and TELEPHONE circuits, often arising in nearby power circuits or other electrical appliances.

INTERFEROMETER, any instrument employing INTERFERENCE effects used: for measuring the wavelengths of LIGHT, RADIO, SOUND or other wave phenomena; for measuring the refractive index (see REFRACTION) of gases (Rayleigh interferometer); for measuring very small distances using radiation of known wavelength, or, in ACOUSTICS and RADIO ASTRONOMY, for determining the direction of an energy source. In most interferometers the beam of incoming radiation is divided in two, led along paths of different but accurately adjustable lengths and then recombined to give an interference pattern. Perhaps the best known optical instrument is the Michelson interferometer devised in 1881 for the MICHELSON-MORLEY EXPERIMENT. More accurate for wavelength measurements is the Fabry-Perot interferometer in which the radiation is recombined after multiple partial reflections between parallel lightly-silvered glass plates.

INTERFERON, substance produced by living tissues following infection with VIRUSES, BACTERIA etc., which interferes with the growth of any organism. It is responsible for a transient and mild degree of nonspecific IMMUNITY following infection.

INTERHALOGENS, group of binary compounds of the HALOGENS with each other, of general formula XX'_n where X′ is the more electronegative halogen, and $n=1$, 3, 5 or 7. They are volatile, covalent, reactive substances. Bromine (III) fluoride (BrF_3) is a very powerful fluoridating agent, and a good solvent

Cutaway of a four-cylinder four-stroke gasoline engine as used in small automobiles. The four pistons are connected to a crankshaft which supplies power to the wheels via the clutch (*left*). The crankshaft also drives an AC generator and a cooling fan via a belt, and the distributor, oil pump and the inlet and exhaust valves via a chain drive and camshaft (*right*). *Left* to *right*, the cylinders are: just completing the exhaust stroke; drawing in the air/fuel mixture; commencing the compression stroke, and sparking the mixture (ignition).

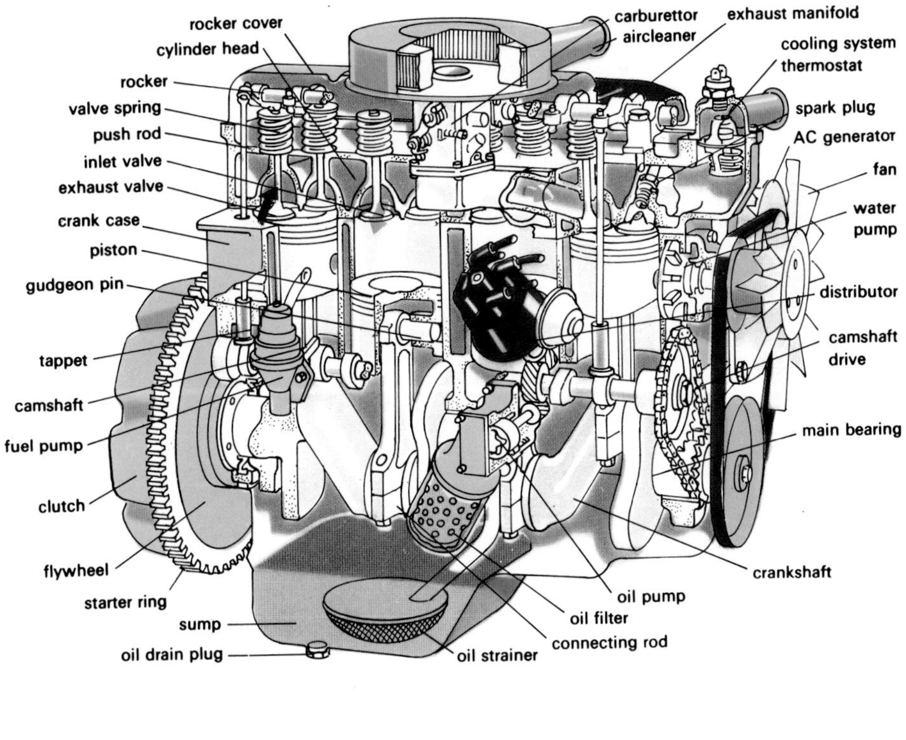

for fluorine compounds. Polyhalide salts, containing ions XX'_{n+1}^{-}, are formed by reacting halogens or interhalogens with HALIDES.

INTERIOR ANGLES, the angles within a POLYGON formed by its adjacent sides.

INTERNAL-COMBUSTION ENGINE, type of ENGINE—the commonest now used—in which the fuel is burned inside the engine and the expansion of the combustion gases is used to provide the power. Because of their potential light weight, efficiency and convenience, internal-combustion engines largely superseded STEAM ENGINES in the early 20th century. They are used industrially and for all kinds of transport, notably to power AUTOMOBILES. There are three classes of internal-combustion engine: RECIPROCATING ENGINES, which include the **gasoline engine**, the DIESEL ENGINE and the FREE-PISTON ENGINE; rotary engines, including the GAS TURBINE, the turbojet (see JET PROPULSION) and the WANKEL ENGINE; and ROCKET engines and non-turbine jet engines, working by reaction. Although originally coal gas and even powdered coal were used as fuel, now almost all fuels used are PETROLEUM products: diesel oil, GASOLINE, BOTTLED GAS and NATURAL GAS. The first working (though not usable) internal-combustion engine was a piston engine made by HUYGENS (1680) that burned gunpowder. In 1794 Robert Street patented a practicable though inefficient engine into which the air had to be pumped by hand. In 1876 N. A. OTTO built the first four-stroke engine, using the principles stated earlier by Alphonse Beau de Roçhas. The cycle is (1) intake of fuel/air mixture; (2) compression of mixture; (3) ignition (see IGNITION SYSTEM) and expansion of burned gases; (4) expulsion of gases as exhaust. Only the third stroke is powered, but the engine is highly efficient, and modern gasoline engines are basically the same. Generally four, six or eight cylinders are linked to provide balanced power. The engine is cooled by water circulating through pipes or by air from a fan. The fuel/air mixture is produced in the CARBURETOR; greater power is given by **supercharging**, by which the proportion of air and the initial pressure of the mixture are increased. The two-stroke engine, giving greater power for a given size, but less efficient in fuel use, does not usually have valves, but an inlet and an exhaust port in the cylinder, blocked and uncovered in turn by the piston. At the end of the powered stroke, the piston drives fresh fuel mixture from the crankcase into the cylinder, pushing out the exhaust gases. The EFFICIENCY of an internal-combustion engine increases with the COMPRESSION RATIO; if this is too high, however, "knocking" occurs due to irregular burning and detonations. It is avoided by using fuel of high OCTANE number, and by using ANTIKNOCK ADDITIVES. (See also AIR POLLUTION.)

INTERNATIONAL DATE LINE. See DATE LINE, INTERNATIONAL.

INTERNATIONAL SYSTEM OF UNITS. See SI UNITS.

INTERPOLATION AND EXTRAPOLATION, techniques used in mathematical ANALYSIS to estimate undetermined values of a dependent VARIABLE, a number of values of which, corresponding to determined values of an independent variable, are known. This is done by finding a FUNCTION $f(x)$ of the independent variable x such that, for any value, x_a, the known corresponding value y_a of the dependent variable y closely approximates to the value $f(x_a)$. In interpolation it is then assumed that, within the range covered by the known values of the variables, $y=f(x)$ for all intermediate values of x and y. In extrapolation, one assumes this relationship to hold outside the range of known values—a rather less justifiable assumption. The simplest (and most commonly used) technique is that of drawing the best straight LINE or CURVE through a set of points on a graph, and assuming that it represents a genuine relationship between the two variables in question (see also LEAST SQUARES). Others include use of partial DIFFERENCE EQUATIONS.

INTERSECTION, in plane GEOMETRY, the crossing of two LINES or CURVES at a point known as the point of intersection. In terms of ANALYTIC GEOMETRY, if two lines have equations $y=f(x)$ and $y=g(x)$ where $f(x)$ and $g(x)$ are FUNCTIONS of x, their points of intersection are given by those values of x for which $f(x)=g(x)$. For example, the line $y=2x$ intersects the curve $y=x^2$ in two points whose coordinates are given by solution of the EQUATION $2x=x^2$; this has two roots (see ROOTS OF AN EQUATION), 0 and 2, and hence the points of intersection are (0,0) and (2,4). Should the equation $f(x)=g(x)$ have roots that are not unique (i.e., two or more are equal), then the curves are tangential (see TANGENT OF A CURVE) at the point or points defined by the equal roots. (See also SET THEORY.)

INTERSTELLAR MATTER, thinly dispersed matter, in the form of gas and dust, between the stars, detectable through its light-absorbing effects. Thicker clouds are seen as NEBULAS. There is in the arms of the MILKY WAY almost as much interstellar as stellar matter. It is thought that STARS form out of interstellar matter.

INTESTINE. See GASTROINTESTINAL TRACT.

INTOXICATION, state in which a person is overtly affected by excess of a DRUG or poison. It is often used to describe the psychological effects of drugs and particularly ALCOHOL, in which behavior may become disinhibited, facile, morose or aggressive and in which judgment is impaired. Late stages of intoxication affecting the BRAIN include stupor and COMA. Ingestion of very large amounts of water causes water intoxication and may lead to coma and death. POISONING with TOXINS and drugs may cause intoxication of other organs (e.g., HEART with DIGITALIS overdosage).

INTRAVENOUS FEEDING, method of supplying nutrients essential for growth or maintenance of body mass when the GASTROINTESTINAL TRACT is unable to provide adequate nutrition in certain DISEASES of the gut and following SURGERY.

INTROVERSION AND EXTRAVERSION, terms coined by JUNG for two opposite character traits. Introverts are shy, introspective, "ingoing"; extraverts sociable, little concerned with their own inner thoughts and feelings, "outgoing." We all display both traits, one or other dominating at different times: Jung suggested that conflict between them was a cause of NEUROSIS. The terms are little used in modern psychology, though introversion is sometimes used to describe the withdrawal characteristic of SCHIZOPHRENIA.

INVAR, an ALLOY composed of 64% iron, 36% nickel and a trace of carbon. Having a very small coefficient of thermal EXPANSION, it is used for pendulums, tuning-forks, measuring devices and other components whose dimensions must be independent of temperature.

INVENTION, the act of devising an original process or device which facilitates or makes possible what was previously more difficult or impossible; also, such a process or device. Inventiveness is one of man's most valuable characteristics. Some of his earliest inventions—the stone ax, painting, wood and ivory carving—are shrouded in the mists of prehistory. But, although invention continued at a steady rate throughout the ancient and medieval periods, most of the inventions that have created the modern world date from 1500 AD at the earliest and the majority belong to the 20th century. If the 19th century was the age of the independent inventor, individually patenting (legally protecting) and marketing his invention, Thomas EDISON pointed the way to a later era in 1876 when he opened his first "invention factory." Today the majority of inventions flow from industrial research laboratories and the costly *development* of a new product is as important as the *research* which produces the basic idea for it: invention has become an industrial activity. The relations of science and invention have often been disputed; on balance it seems fair to admit that benefits have flowed in both directions.

INVERSE, of a number a, the number b such that $a*b=e$, where * is an algebraic operation and e is the identity element relative to the operation * of the set (see SET THEORY) of which a and b are members. For example, 1 is the identity element of the set of REAL NUMBERS relative to multiplication: hence if $a.b=1$, a is the inverse of b, b the inverse of a, relative to the multiplication of real numbers. (Moreover, a and b are RECIPROCALS in this case.)

Inverse operation. If two operations negate each other, they are termed inverse operations. For example, since $a+b-b=a$, ADDITION and SUBTRACTION are inverse operations.

Inverse of a proposition. For a proposition $h \rightarrow c$ (read "h implies c"), the proposition not-$h \rightarrow c$ is described as its inverse.

Inverse function. For a FUNCTION $f(x)$, the function $g(x)$ such that $f(a)=b$ implies that $g(b)=a$ is described as the inverse of $f(x)$. In practice, the inverse function of $f(x)$ is written $f^{-1}(x)$. For example, the inverse of $f(x)=\mathrm{a}x+\mathrm{b}$ is $f^{-1}(x)=(x-\mathrm{b})/\mathrm{a}$, since $f^{-1}(\mathrm{a}x+\mathrm{b})=(\mathrm{a}x+\mathrm{b}-\mathrm{b})/\mathrm{a}=\mathrm{a}x/\mathrm{a}=x$.

Inverse Trigonometric function. For a function of the form $y=\sin x$, the function of the form $x=\sin^{-1}y$ (read as "x is the angle whose sine is y") is described as its inverse. This may also be written as $x=\arcsin y$. (See TRIGONOMETRY.)

INVERSE SQUARE LAW, relationship according to which the intensity of a spherical wave (see WAVE MOTION) varies inversely (see INVERSE VARIATION) with the SQUARE of its distance from the source. The law applies only where the source is small compared with the distance and the medium is unbounded, homogeneous isotropic and nondissipative. Many other natural phenomena, e.g., the gravitational forces between bodies and the electrostatic forces between isolated charges, are also governed by inverse square laws.

INVERSE VARIATION, a relationship between two VARIABLES in which increase in value of one

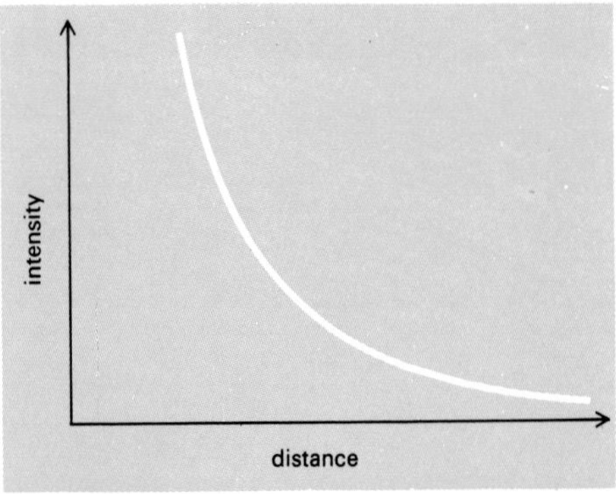

The graph of the intensity of a property against the distance from one center for an inverse-square law relation.

implies decrease in value of the other; e.g.

$$y=\frac{\mathrm{k}}{x}$$

where k is a CONSTANT. y may vary with the RECIPROCAL or a POWER of x, as in the INVERSE SQUARE LAW. The idea is of particular interest in STATISTICS; as in, for example, a case where the consumption of alcohol might vary inversely with the level of taxation on it.

INVERSION, Temperature, a condition of the lower part of the atmosphere in which temperature increases with increase in height above the surface. Normally, temperature decreases upward through most of the ATMOSPHERE, but certain atmospheric disturbances (e.g., a FRONT) can create inversions. The condition occurs also on cold nights. Inversions sometimes aggravate AIR POLLUTION, as the cooler air trapped near the surface cannot rise and so carry away the pollutants.

INVERTEBRATES, animals without backbones, a miscellaneous collection of groups from single-celled PROTOZOA to highly-specialized insects and spiders. Apart from the universal lack of an internal backbone of VERTEBRAE, many of these groups have little in common.

INVERTER. See COMPUTER.

INVISIBLE INK, an INK, used for secret communication, that is colorless when written but that can be developed by heat or chemical reaction. For example, potassium hexacyanoferrate(II) ink can be developed by an iron(III) salt, giving PRUSSIAN BLUE. Fluorescent inks fluoresce in ultraviolet light.

IODINE (I), the least reactive of the HALOGENS, forming black lustrous crystals which readily sublime to pungent violet vapor. Most iodine is produced from calcium iodate ($Ca[IO_3]_2$), found in CHILE SALTPETER. In the US, much is recovered from oil-well brine, which contains sodium iodide (NaI). Chemically it resembles BROMINE closely, but has a greater tendency to covalency and positive oxidation states. It is large enough to form 6-coordinate oxy-anions. Most plants (especially seaweeds) contain traces of iodine; in the higher animals it is a constituent of the thryoxine hormone secreted by the THYROID GLAND. Iodine deficiency can cause GOITER. Iodine and its compounds are used as antiseptics, fungicides and in the production of dyes. The RADIOISOTOPE I^{131} is used as a tracer and to treat goiter. Silver iodide, being light-sensitive, is used in PHOTOGRAPHY. (See also HALIDES.) AW 126.9, mp 113.5°C, bp 184°C, sg 4.93 (20°C).

ION, an ATOM or group of atoms that has become electrically charged by gain or loss of negatively-charged ELECTRONS. In general, ions formed from metals are positive (CATIONS), those from nonmetals negative (ANIONS). Compound ions are usually anions derived from OXYACIDS, e.g., sulfate SO_4^{2-}. CRYSTALS of ionic compounds consist of negative and positive ions arranged alternately in the lattice and held together by electrical attraction (see BOND, CHEMICAL). Many covalent compounds undergo ionic DISSOCIATION in solution. Ions may be formed in gases by radiation or electrical discharge, and occur in the IONOSPHERE (see also ATMOSPHERE). At very high temperatures gases form PLASMA, consisting of ions and free electrons. In solution, many simple ions combine with LIGANDS to give complex ions, including HYDRATES. (See ELECTROLYSIS; ION EXCHANGE; IONIZATION CHAMBER; IONIZATION POTENTIAL; ION PROPULSION; ZWITTERION.)

ION COUNTER. See GEIGER COUNTER; SCINTILLATION COUNTER.

ION EXCHANGE, chemical reaction in which IONS in a solution are replaced by others of like charge. An insoluble solid is used that has an open, netlike molecular structure: a ZEOLITE, or a synthetic organic polymer called an **ion-exchange resin**, whose composition and properties can be tailored for the use required. The solid has attached anionic groups, which are neutralized by small mobile cations in the interstices. It is these cations which are exchanged for others when a solution is passed through. The principle of anion exchange is similar. Ion exchange is used for softening HARD WATER, purifying sugar, and concentrating ores of uranium and the NOBLE METALS. Ion-exchange CHROMATOGRAPHY is used to separate the RARE EARTHS, and in chemical ANALYSIS.

IONIZATION CHAMBER, instrument for measuring the amount of ionization created in it by radiation such as ALPHA PARTICLES, BETA RAYS, GAMMA RAYS or X RAYS. A gas-filled chamber contains two ELECTRODES with a variable POTENTIAL difference between them. The IONS produced move towards the oppositely charged electrode, forming an electric current which is a measure of the amount of incoming radiation.

IONIZATION POTENTIAL, the ENERGY needed to remove an ELECTRON from the ground state of a given type of ATOM to infinity. It increases for removal of successive electrons, which are bound by the atom's positive charge. Ionization potentials can be determined by SPECTROSCOPY.

ION MICROSCOPE. See FIELD-ION MICROSCOPE.

IONOSPHERE, the zone of the earth's ATMOSPHERE extending outward from about 75km above the surface in which most atoms and molecules exist as electrically charged IONS. The high degree of ionization is maintained through the continual ABSORBTION of high-energy solar radiation. Several distinct ionized layers, known as the D, E, F_1, F_2 and G layers, are distinguished. These are somewhat variable, the D layer disappearing and the F_1, F_2 layers merging at night. Since the free ELECTRONS in these layers strongly reflect RADIO waves, the ionosphere is of great importance for long-distance radio communications.

ION PROPULSION, or **ion drive,** drive proposed for spacecraft on interstellar or longer interplanetary trips. The vaporized propellant (liquid CESIUM or MERCURY) is passed through an ionizer, which strips each atom of an ELECTRON. The positive IONS so formed are accelerated rearward by an ELECTRIC FIELD. The resultant thrust is low, but in the near-vacuum of space may be used to build up huge velocities by constant acceleration over a long period of time. The drive has been tested in orbit around the earth.

IPECAC, a drug produced from the dried stems and roots of *Cephalaelis ipecacuanha* and *C. acuminala*, plants native to South America, but also cultivated in Malaysia. The drug is used as an emetic and to treat amebic dysentery, its medicinal properties being due to the ALKALOIDS, emetine, cephaeline and psychtrine.

IQ (*I*ntelligence *Q*uotient), a measure of an individual's INTELLIGENCE. If the intelligences of a large number of people are measured—by any valid means—the result will be very close to a NORMAL DISTRIBUTION. The mean (see MEAN, MEDIAN and MODE) intelligence can be defined as 100; and a person with that intelligence is said to have an IQ of 100. Using this basic, the IQ of an individual can be calculated from

$$IQ = 100 + \frac{16}{\mu}(x - \bar{x}),$$

where $\bar{x}$ is the mean score for individuals of the subject's age, μ is the STANDARD DEVIATION of the distribution of those scores, and x is his own score. (See also PSYCHOLOGICAL TESTS.)

IRIDESCENCE, production of colors of varied hue by INTERFERENCE of light reflected from front and back of thin films (as in soap bubbles) or from faults and boundaries within crystalline solids such as mica or opal. The colors of mother-of-pearl and some insects are due to iridescence.

IRIDIUM (Ir), hard, white metal in the PLATINUM GROUP, the most resistant element to corrosion at room temperature. AW 192.2, mp 2410°C, bp 4527°C, sg 22.4 (20°C).

IRIS, the pigmented diaphragm forming the aperture of the vertebrate EYE.

IRON (Fe), silvery-gray, soft, ferromagnetic (see MAGNETISM) metal in Group VIII of the PERIODIC TABLE; a TRANSITION ELEMENT. Metallic iron is the main constituent of the earth's core (see EARTH), but is rare in the crust; it is found in meteorites (see METEORS). Combined iron is found as HEMATITE, MAGNETITE, LIMONITE, SIDERITE, GOETHITE, TACONITE, CHROMITE and PYRITE. It is extracted by smelting oxide ores in a BLAST FURNACE to produce PIG IRON which may be refined to produce CAST IRON or WROUGHT IRON, or converted to STEEL in the OPEN-HEARTH PROCESS or the BESSEMER PROCESS. Many other iron ALLOYS are used for particular applications. Pure iron is very little used; it is chemically reactive, and oxidizes to RUST in moist air. It has four allotropes (see ALLOTROPY). The stable oxidation states of iron are +2 (ferrous) and +3 (ferric), though +4 and +6 states are known. The ferrous ion (Fe^{2+}) is pale green in aqueous solution; it is a mild reducing agent, and does not readily form LIGAND complexes. **Iron(II) Sulfate** ($FeSO_4.7H_2O$), or **green vitriol**, or **copperas**, green crystalline solid, made by treating iron ore with sulfuric acid, used in tanning, in medicine to treat iron deficiency, and to make ink, fertilizers, pesticides and other iron compounds. mp 64°C. The ferric ion (Fe^{3+}) is yellow in aqueous solution; it resembles the ALUMINUM ion, being acidic and forming stable LIGAND complexes, especially with CYANIDES (see PRUSSIAN BLUE). **Iron(III) Oxide** (Fe_2O_3), red-brown powder used as a pigment and as jewelers' rouge (see ABRASIVES); occurs naturally as HEMATITE. mp 1565°C. (See also ALUM; SANDWICH COMPOUNDS.) In the human body, iron is a constituent of HEMOGLOBIN and the CYTOCHROMES. Iron deficiency causes ANEMIA. AW 55.8, mp 1535°C, bp 2750°C, sg 7.874 (20°C).

IRON AGE, the stage of man's material cultural development, following the STONE AGE and BRONZE AGE, during which iron is generally used for weapons and tools. Though used ornamentally as early as 4000 BC in Egypt and Mesopotamia, iron's difficulty of working precluded its general use until efficient techniques were developed in Armenia, c1500 BC. By c500 BC the use of iron was dominant throughout the known world, and by c300 BC the Chinese were using cast iron. Some cultures, as those in America and Australia, are said never to have had an iron age.

IRRADIATION, exposure of a sample to RADIATION, usually for a definite purpose. Biological and pharmaceutical materials may have their properties altered by exposure to ULTRAVIOLET RADIATION; X RAYS are widely used in medicine and industry. Materials may be irradiated directly with radiation of a given type and ENERGY by placing them in a particle ACCELERATOR or NUCLEAR REACTOR, but it is often more practical to use the radiation from manufactured radioactive ISOTOPES to change their physical and chemical properties as required. Neutrons and GAMMA RAYS are used to sterilize foodstuffs and control the reproduction of insect pests.

IRRATIONAL NUMBERS, those REAL NUMBERS that cannot be expressed as the RATIO of two INTEGERS (see RATIONAL NUMBERS). Common examples include π (see PI), e (see EXPONENTIAL) and $\sqrt{2}$ (see ROOT).

IRRIGATION, artificial application of water to soil to promote plant growth. Irrigation is vital for agricultural land with inadequate rainfall. The practice dates back at least to the canals and reservoirs of ancient Egypt. Today over 320 million acres of farmland throughout the world are irrigated, notably in the US, India, Pakistan, China, Australia, Egypt and the USSR. There are three main irrigation techniques: **surface irrigation**, in which the soil surface is moistened or flooded by water flowing through furrows or tubes; **sprinkler irrigation**, in which water is sprayed on the land from above; and **subirrigation**, in which underground pipes supply water to roots. The amount of water needed for a particular project is called the **duty of water**, expressed as the number of hectares irrigated by water supplied at a rate of $1 m^3/s$.

IRRITABILITY, in biology and psychology, the ability to be affected by external stimuli (see PERCEPTION). (See also ANIMAL BEHAVIOR.)

ISINGLASS, white form of GELATIN, made from dried membranes of swim bladders of fish. It is used as an adhesive, a fabric size, and an additive to clarify wines, beer and vinegar. For the mineral isinglass, see MUSCOVITE.

ISLAND, comparatively small land area entirely surrounded by water, a result of the buildup of the cone of a submarine VOLCANO, EROSION by the sea or GLACIERS of parts of coastal regions, DIASTROPHISM or other process. **Island arcs** are curving chains of islands. They are often associated with EARTHQUAKE activity, and have deep OCEAN trenches on the convex sides. (See also ATOLL; CORAL; PLATE TECTONICS.)

ISLETS OF LANGERHANS. See INSULIN; PANCREAS.

ISOBAR, line drawn on a meteorological map joining points which are, at a given moment in time, experiencing the same air pressure (see ATMOSPHERE).

ISOMERISM, in nuclear physics, the existence of metastable states of an atomic nucleus (see ATOM), having the same atomic number and mass number as the ground state, but higher energy. Nuclear isomers are formed by bombardment or in a radioactive decay chain (see RADIOACTIVITY). They usually have very short HALF-LIVES and decay by emitting GAMMA RAYS.

ISOMERS, chemical compounds having identical

Examples of the principal varieties of chemical isomerism.

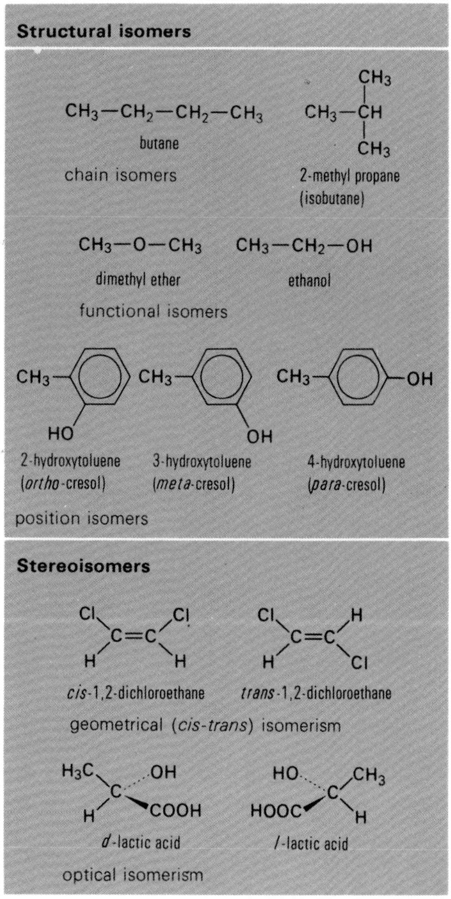

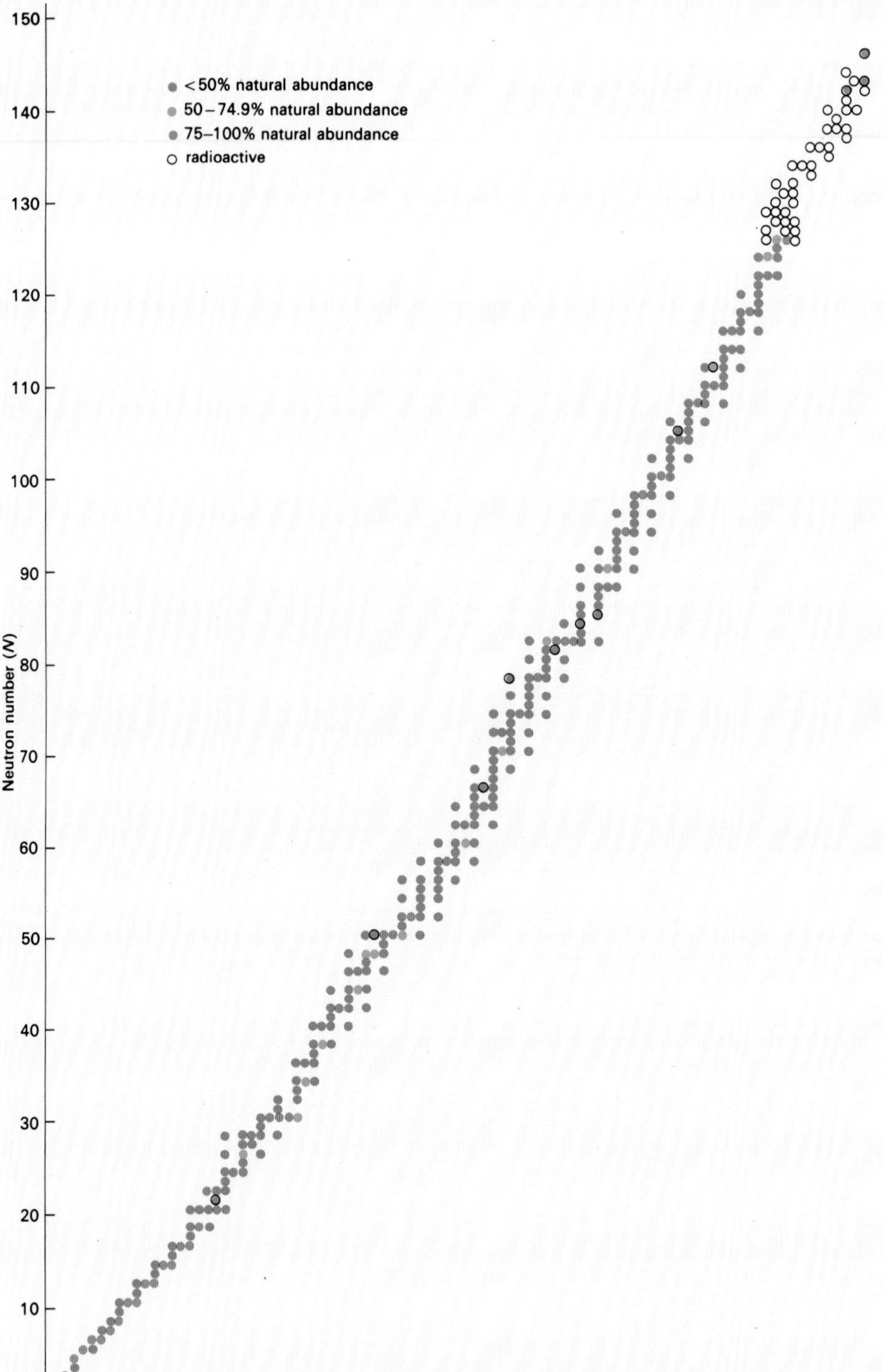

<50% natural abundance
50–74.9% natural abundance
75–100% natural abundance
radioactive
Neutron number (N)
Atomic number (Z)
0
10
20
30
40
50
60
70
80
90
100
110
120
130
140
150

chemical COMPOSITION and molecular FORMULA, but differing in the arrangement of atoms in their molecules, and having different properties. The two chief types are STEREOISOMERS, which have the same structural formula, and **structural isomers**, which have different structural formulas. The latter may be subdivided into positional isomers, which have the same FUNCTIONAL GROUPS occupying different positions on the carbon skeleton; and functional isomers, which have different functional groups. (See also CRACKING; TAUTOMERISM.)

ISOMETRICS, exercises in which MUSCLES are contracted against resistance, but without movement at JOINTS; the muscles remain at the same length but their tension is increased. It is used in some systems of physical training and in PHYSIOTHERAPY.

ISOMORPHISM, the formation by different compounds or MINERALS of CRYSTALS having closely similar external forms and lattice structure. Isomorphous compounds have similar chemical composition—ions of similar size, charge, and electrical polarizability being substituted for each other—and form mixed crystals.

ISOPRENE, or 2-methyl-1,3-butadiene, derivative of 1,3-BUTADIENE, a conjugated diene (see ALKENES; DIELS-ALDER REACTION; RESONANCE). Isoprene is a colorless liquid made by destructive distillation of rubber or from PETROLEUM, and used to make synthetic RUBBER. It is the basic unit of plant products including CAROTENOIDS, STEROLS and TERPENES.

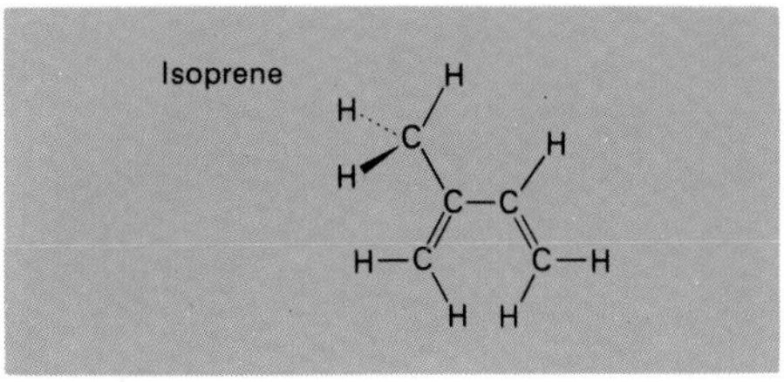

ISOSTASY, the theoretical tendency of the earth's crust to maintain equilibrium as it floats on the MANTLE; assumed to result from flows of the dense plastic SIMA in the lower crust in response to local changes in the pressure on it of the lighter SIAL above. Local differences in the proportion of sima to sial thus maintain an equal weight of crust all around the EARTH.

ISOTHERM, line drawn on a meteorological map joining points that are, at a given moment in time, experiencing the same temperature. (See METEOROLOGY.)

ISOTOPES, ATOMS of a chemical ELEMENT which have the same number of PROTONS in the nucleus, but different numbers of NEUTRONS, i.e., having the same atomic number but different MASS NUMBER. Isotopes of an element have identical chemical and physical properties (except those determined by atomic mass). Most elements have several stable isotopes, being found in nature as mixtures. The natural proportions of the isotopes are expressed in the form of an **abundance ratio**. Because some isotopes have particular properties (e.g., 0.015% of HYDROGEN atoms have two neutrons and combine with oxygen to form HEAVY WATER, used in NUCLEAR REACTORS), mass-dependent methods of separating these out have been devised. These include MASS SPECTROSCOPY, DIFFUSION, DISTILLATION and ELECTROLYSIS. A few elements have natural radioactive isotopes (RADIOISOTOPES) and others of these can be made by exposing stable isotopes to RADIATION in a reactor. These are widely used therapeutically and industrially; their radiation may be employed directly, or the way in which it is scattered or absorbed by objects can be measured. They are useful as tracers of a process, since they may be detected in very small amounts and behave virtually identically to other atoms of the same element. They may also be used to "label" particular atoms in complex molecules, in attempts to work out chemical reaction mechanisms.

ISOTROPY, property exhibited by a medium in which physical properties are independent of direction. Most liquids and materials composed of small randomly oriented crystals are isotropic in all properties, while crystalline materials are, in general, anisotropic.

ISTHMUS, narrow strip of land joining two large landmasses, or a peninsula to the mainland. Best known is the Panama Isthmus. (See also STRAIT.)

The isotopes found in measurable quantities in nature, together with the members of the three naturally occurring radioactive decay series. Neutron number (*N*) is plotted against atomic number (*Z*).

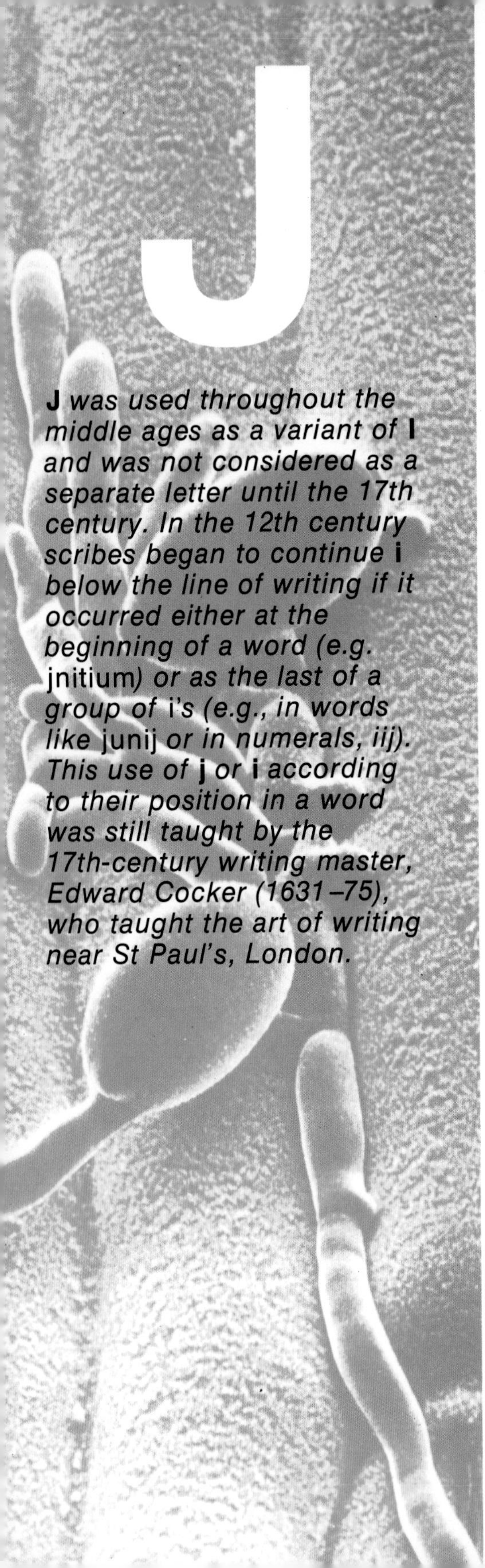

J

J *was used throughout the middle ages as a variant of* **I** *and was not considered as a separate letter until the 17th century. In the 12th century scribes began to continue* **i** *below the line of writing if it occurred either at the beginning of a word (e.g.* jnitium*) or as the last of a group of i's (e.g., in words like* junij *or in numerals,* iij*). This use of* **j** *or* **i** *according to their position in a word was still taught by the 17th-century writing master, Edward Cocker (1631–75), who taught the art of writing near St Paul's, London.*

***j*-OPERATOR.** See IMAGINARY OPERATOR.

JABIR, or Abu-Musa-Jabir-ibn-Haiyan, or (Latin) **Geber**, 9th-century Arab alchemist to whom the authorship of the "Jabir corpus" of writings was formerly ascribed. The early Jabir writings contained many novel contributions to the theory of ALCHEMY.

JACK O'LANTERN. See WILL-O'-THE-WISP.

JACOB, François (1920–), French biologist who shared with MONOD and LWOFF the 1965 Nobel Prize for Physiology or Medicine for his work with Monod on regulatory GENE action in BACTERIA.

JACQUARD, Joseph Marie (1752–1834), French inventor of the **Jacquard loom** (completed 1801), which could weave complex patterns according to instructions coded on punched cards (a technique adopted by BABBAGE for his calculator and still used for COMPUTERS). Modern looms are still based on Jacquard's principles. (See also WEAVING.)

JADE, either of two tough, hard minerals with a compact interlocking grain structure, commonly green but also found as white, mauve, red-brown or yellow; used as a GEM stone to make carved jewelry and ornaments. Jade carving in China dates from the 1st millennium BC, but the finest examples are late 18th century AD. **Nephrite**, the commoner form of jade, is an AMPHIBOLE, a combination of tremolite and actinolite, occurring in China, the USSR, New Zealand and the western US. **Jadeite**, rarer than nephrite and prized for its more intense color and translucence, is a sodium aluminum PYROXENE, found chiefly in upper Burma.

JAKOBSON, Roman (1896–), Russian-born US linguist and philologist best known for his pioneering studies of the Slavic languages.

JAM. See JELLY AND JAM.

JAMES, William (1842–1910), US philosopher and psychologist, the originator of the doctrine of PRAGMATISM. Intermittently dogged by ill-health, his first major contribution was *The Principles of Psychology* (1890). Turning his attention to questions of religion, he published in 1902 his Gifford Lectures, *The Varieties of Religious Experience*, which has remained his best-known work.

JANET, Pierre Marie Félix (1859–1947), French psychologist and neurologist, best known for his studies of HYSTERIA and NEUROSIS, who played an important role in reconciling the theories of psychology and the practice of clinical treatment of mental disease.

JANSSEN, Zacharias (c1600), Dutch spectacle-maker credited with inventing the compound MICROSCOPE (1590).

JASPER, common variety of CHALCEDONY containing admixed HEMATITE or GOETHITE, normally red, brown or yellowish, often with banding and spotting. Good grades are used for semiprecious GEM stones.

JASPERS, Karl Theodor (1883–1969), German philosopher, noted for his steadfast opposition to National Socialism and his acute yet controversial analyses of the state of German society. Early work in psychopathology led him into the Heidelberg philosophical faculty in 1913. He there became one of Germany's foremost exponents of existentialism.

JAUNDICE, yellow color of the SKIN and sclera of the EYE caused by excess bilirubin pigment in the BLOOD. HEMOGLOBIN is broken down to form bilirubin which

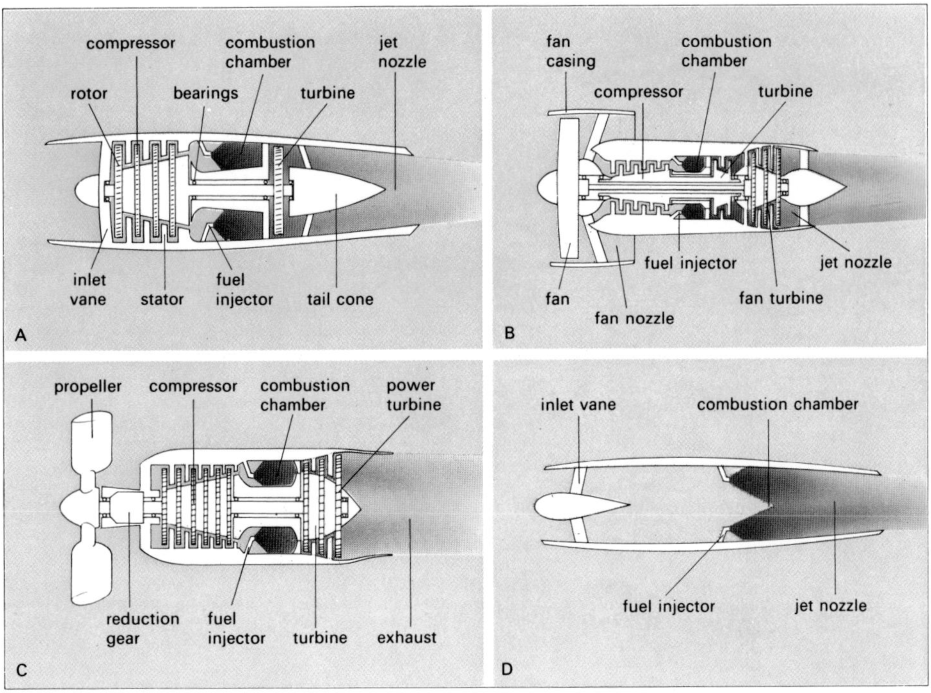

Schematized sections through four important types of jet propulsion unit: (A) turbojet; (B) fanjet (turbofan) (C) turboprop; (D) ramjet.

is excreted by the LIVER in the BILE. If blood is broken down more rapidly than normal (hemolysis), the liver may not be able to remove the abnormal amount of bilirubin fast enough. Jaundice occurs with liver damage (HEPATITIS, late CIRRHOSIS) and when the bile ducts leading from the liver to the DUODENUM are obstructed by stones from the GALL BLADDER or by CANCER of the PANCREAS or bile ducts.

JAVA MAN. See PREHISTORIC MAN.

JEANS, Sir James Hopwood (1877–1946), British physicist and mathematician best known for his contributions to astronomy and for his popular science books. He played a valuable role in proving the invalidity of the NEBULAR HYPOTHESIS, but his own theory of the formation of the SOLAR SYSTEM, that the planets were "drawn out" of the sun by a star passing close by, has now in turn been largely discarded.

JELLY AND JAM, sweet foods made by heating fruit juice (jelly) or crushed or chopped fruit (jam) with sugar. They set because of the PECTIN, present in all fruits: in some cases, extra pectin and CITRIC ACID must be added. **Marmalades** contain peel as well as pulp; **preserves** contain whole fruits.

JENNER, Edward (1749–1823), British pioneer of VACCINATION. He examined in detail the country maxim that dairymaids who had had COWPOX would not contract SMALLPOX: in 1796 he inoculated a small boy with cowpox and found that this rendered the boy immune from smallpox.

JENSEN, Johannes Hans Daniel (1907–), German physicist who shared with M. G. MAYER and E. H. WIGNER the 1963 Nobel Prize for Physics for his suggestion (independent of Mayer's) that the PROTONS and NEUTRONS of the atomic nucleus are arranged in concentric shells.

JESPERSON, Jens Otto Harry (1860–1943), Danish philologist and educator best known for his studies of English GRAMMAR and for devising the international language **Novial** (c1928).

JET, compact, hard variety of lignite COAL, deep black and polishable, mined at Whitby, England, and used as a GEM material.

JET PROPULSION, the propulsion of a vehicle by expelling a fluid jet backward, whose MOMENTUM produces a reaction that imparts an equal forward momentum to the vehicle, according to NEWTON's third law of motion. The squid uses a form of jet propulsion. Jet-propelled boats, using water for the jet, have been built, and air jets have been used to power cars, but by far the chief use is to power AIRPLANES and ROCKETS, since to attain high speeds, jet propulsion is essential. The first jet engine was designed and built by Sir Frank WHITTLE (1937), but the first jet-engine aircraft to fly was German (Aug. 1939). Jet engines are INTERNAL-COMBUSTION ENGINES. The **turbojet** is the commonest form. Air enters the inlet diffuser and is compressed in the air compressor, a multistage device having sets of rapidly rotating fan blades. It then enters the combustion chamber, where the fuel (a kerosine/gasoline mixture) is injected and ignited, and the hot, expanding exhaust gases pass through a TURBINE that drives the compressor and engine accessories. The gases, sometimes heated further in an AFTERBURNER, are expelled through the jet nozzle to provide the thrust. The nozzle converges for subsonic flight, but for supersonic flight one that converges and then diverges is needed. The **fanjet** or

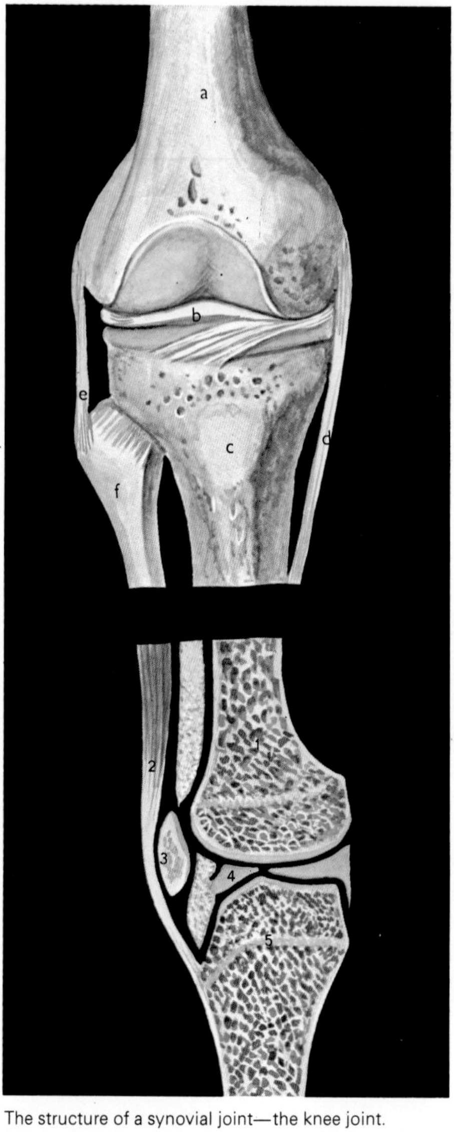

The structure of a synovial joint—the knee joint.
(A) the flexed knee joint from the rear: (a) femur (thigh bone); (b) joint surfaces (cartilage) and cavity; (c) tibia (shin bone); (d) and (e) ligaments; (f) fibula.
(B) Lateral section through the knee: (1) femur; (2) quadriceps muscle with tendon; (3) patella (knee cap); (4) cavity with meniscus; (5) tibia.

turbofan engine uses some of the turbine power to drive a propeller fan in a cowling, for more efficient subsonic propulsion; the **turboprop**, similar in principle, gains its thrust chiefly from the propeller. The **ramjet** is the simplest air-breathing jet engine, having neither compressor nor turbine. When accelerated to supersonic speeds by an auxiliary rocket or turbojet engine, the inlet diffuser "rams" the air and compresses it; after combustion the exhaust gases are expelled directly. Ramjets are used chiefly in guided missiles.

JET STREAM, a narrow band of very fast E-flowing winds, stronger in winter than in summer, found around the level of the tropopause (see ATMOSPHERE). Speeds average about 60km/h in summer, about 125km/h in winter, though over 300km/h has been recorded. (See also METEOROLOGY; WIND.)

JEWELS. See GEMS.

JODRELL BANK EXPERIMENTAL STATION, England, radio astronomy observatory pioneered by Alfred Charles Bernard Lovell (1913–), and including one of the largest steerable RADIO TELESCOPES ("dish" 250ft (76.2m) across) (1957).

JOINT, specialized surface between BONES allowing movement of one on the other. Major joints, especially of limbs, are **synovial joints** which are lined by synovial membrane and CARTILAGE and surrounded by a fibrous capsule; they contain SYNOVIAL FLUID, which lubricates the joint surfaces. Parts of the capsule (e.g., in the ankle) or overlying TENDONS (e.g., in the knee) form LIGAMENTS important in joint stability, though at some joints (e.g., the SHOULDER) resting activity in MUSCLES ensures stability, while in others (e.g., the hip) it is due to the shape of the bony surfaces. **Fibrous** and **cartilaginous joints** between bones are relatively fixed except under special circumstances (e.g., the widening of the symphysis pubis in PREGNANCY). Joint disease causes ARTHRITIS, with pain, limitation of movement and sometimes increase in fluid.

JOLIOT-CURIE, Irène (1897–1956), French physicist, the daughter of Pierre and Marie CURIE. She and her husband, **Jean Frédéric Joliot** (1900–1958), shared the 1935 Nobel Prize for Chemistry for their discovery of artificial RADIOACTIVITY. Both later played a major part in the formation of the French atomic energy commission but, because of their communism, were removed from positions of responsibility there (Frédéric 1950, Iréne 1951). Like her mother, Irène died from LEUKEMIA as a result of prolonged exposure to radioactive materials.

JONES, Alfred Ernest (1897–1958), British psychoanalyst who played a major role in gaining recognition for PSYCHOANALYSIS in Britain and North America.

JORDAN, David Starr (1851–1931), US naturalist and authority on fishes best known for his many books and for his pacifist activities.

JOSEPHSON, Brian David (1940–), British physicist awarded, with I. GIAEVER and L. ESAKI, the 1973 Nobel Prize for Physics for his discovery of the **Josephson Effect**, the passage of ELECTRICITY through an insulator between two superconductors (see SUPERCONDUCTIVITY). Pairs of ELECTRONS form in the superconductors and tunnel (see WAVE MECHANICS) through the insulating layer.

JOULE (J), the SI UNIT of ENERGY, defined as the WORK done when a FORCE of one NEWTON acts through the distance of one metre. Also equivalent to the energy dissipated by one WATT in one second, it equals 10^7 erg in CGS UNITS.

JOULE, James Prescott (1818–1889), British physicist who showed that HEAT energy and mechanical energy are equivalent and hinted at the law of conservation of ENERGY. From 1852 he and Thomson (later Lord KELVIN) performed a series of experiments in THERMODYNAMICS, especially on the Joule–Thomson effect (see CRYOGENICS). The JOULE (unit) is named for him. (See also JOULE'S LAW.)

JOULE'S LAW, law derived by J. P. JOULE that the heat evolved in a given time by passage of electricity through a conductor is proportional to the RESISTANCE of the conductor times the square of the electrical intensity. We now write it $H=I^2R$, where H is the rate of generation of heat in WATTS, I the current in AMPERES and R the resistance in OHMS.

JUGULAR VEINS, a pair of veins on each side of the neck which collect venous blood from the BRAIN (internal jugular vein) and the rest of the head (external jugular vein). Their proximity to the surface makes them liable to trauma with HEMORRHAGE and AEROEMBOLISM.

JULIAN CALENDAR. See CALENDAR.

JUNG, Carl Gustav (1875–1961), Swiss psychiatrist who founded analytical psychology. He studied PSYCHIATRY at Basel University, his postgraduate studies being of PARAPSYCHOLOGY. After working with BLEULER and JANET, he met FREUD (1907), whom he followed for some years. But he disagreed with, particularly, Freud's belief in the purely sexual nature of the LIBIDO, and in 1913 broke away completely. In *Psychological Types* (1921) he expounded his views on INTROVERSION AND EXTRAVERSION. Later he investigated anthropology and the occult to form the idea of ARCHETYPES, the universal symbols present in the COLLECTIVE UNCONSCIOUS. (See also ANIMA AND ANIMUS; PERSONA; PSYCHOANALYSIS; PSYCHOLOGY.)

JUPITER, the largest and most massive planet in the solar system (diameter about 143Mm, mass 317.8 times that of earth), fifth from the sun. Jupiter is larger than all the other planets combined and, with a mean solar distance of 5.20AU and a "year" of 11.86 earth-years, is the greatest contributor to the solar system's angular MOMENTUM. Its atmosphere consists mainly of METHANE, AMMONIA and HYDROGEN. Its disk is marked by prominent cloud-belts paralleling its equator, these being occasionally interrupted by turbulences and particularly the **Great Red Spot**, an elliptical area 40Mm long and 13Mm wide: unlike most other features of Jupiter's disk, which have a lifetime of a few days, it has been observed for about 150 years. Another long-term feature, the **South Tropical Disturbance**, was first observed in 1901 and disappeared in 1939. The nature of these features is not

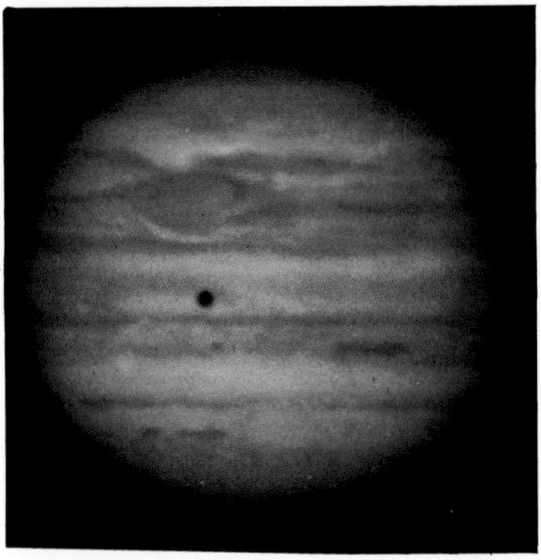

The planet Jupiter. The Great Red Spot is visible just above the equator.

yet known. Jupiter's day is about 9.92h and this high rotational velocity causes a visible flattening of the poles: the equatorial diameter is some 7% greater than the polar diameter. Jupiter has 13 moons, the two largest of which, Callisto and Ganymede, are larger than MERCURY: Io has an atmosphere. Jupiter radiates energy, possibly because of nuclear reactions in its core or a gravitational contraction of the planet.

JURASSIC, the middle period of the MESOZOIC era, lasting from about 190 to 135 million years ago. (See also GEOLOGY.)

JUTE, the fibers produced from the annual herbs *Corchorus capsularis* and *C. olitorius*, which are cultivated in India and Bangladesh. The stems are soaked in water (retted) until the fibers can be separated and then spun into yarn that is used for sacking. Family: Tiliaceae.

JUVENILE HORMONE, or **neotenin,** an insect hormone which maintains the presence of juvenile features in the LARVA. In its absence, adult features appear on molting. It is also involved in normal egg production by female insects. In years to come it may find application in insect control by preventing the emergence of adults.

K

Capital **K** *was taken over unchanged into the Roman alphabet from the Greek; small* **k** *is a slight modification of the capital form with the first stroke written as an ascender and the two limbs foreshortened. Its function in Greek had been to represent the unvoiced velar stop but in the Latin alphabet this function was usurped by* **C. K** *fell into disuse except in initials or formulae such* as Kalendae. *It was reintroduced into the English alphabet in the 12th century when the unvoiced velar had become palatalized before front vowels to distinguish between the palatal and the velar; thus the Anglo-Saxon word* cyng *came to be spelt* kyng *in early medieval English, and later* king.

KALA-AZAR. See LEISHMANIASIS.

KALM, Peter (1716–1779), Swedish botanist best known for his survey of North American natural history. His results were published between 1753 and 1761.

KAMERLINGH-ONNES, Heike (1853–1926), Dutch physicist awarded the 1913 Nobel Prize for Physics for his work on low-temperature physics (see CRYOGENICS). He discovered SUPERCONDUCTIVITY and was the first to liquefy HELIUM (1908).

KANT, Immanuel (1724–1804), German philosopher, one of the world's greatest thinkers. He was born and lived in Königsberg (Kaliningrad). The starting point for Kant's "critical" philosophy was the work of David HUME, who awakened Kant from his "dogmatic slumber" and led him to make his "Copernican revolution in philosophy." This consisted of the radical view found in *Critique of Pure Reason* (1781) that objective reality (the phenomenal world) can be known only because the mind imposes the forms of its own intuitions—time and space—upon it. Things that cannot be perceived in experience (noumena) cannot be known, but as Kant confesses in *Critique of Practical Reason* (1788) their existence must be presumed in order to provide for man's free will. In his third major work, *Critique of Judgment* (1790) he makes aesthetic and teleological judgments serve to mediate between the sensible and intelligible worlds which he divided sharply in the first two *Critiques.*

KAOLIN, or **china clay**, soft, white CLAY composed chiefly of KAOLINITE, and mined in England, France, Saxony, Czechoslovakia, China and the S US. It is used for filling and coating paper, filling rubber and paints, and for making POTTERY AND PORCELAIN.

KAOLINITE, prototypical member of the kaolinite group of CLAY minerals. It consists of hexagonal flakes of composition $Si_4Al_4O_{10}(OH)_8$. It is formed by alteration of other clays or FELDSPAR.

KAPITZA, Peter Leonidovich (1894–), Russian physicist best known for his work on low-temperature physics (see CRYOGENICS), especially his discovery of the SUPERFLUIDITY of HELIUM II.

KAPOK, water-resistant fibers obtained from the seeds of the silk-cotton or kapok tree (*Ceiba pentandra*). The tree is native to tropical America but is naturalized in many parts of the world and extensively cultivated in the Far East, particularly in Indonesia. Kapok is used in the manufacture of mattresses, life-preservers and insulation materials, but its importance has diminished since the advent of man-made fibers. Family: Bombacaceae.

KARAT. See CARAT.

KÁRMÁN, Theodore von (1881–1963), Hungarian-born US aeronautics engineer best known for his mathematical approach to problems in aeronautics (especially in jet engineering) and astronautics.

KARRER, Paul (1889–1971), Russian-born Swiss chemist awarded with W. N. HAWORTH the 1937 Nobel Prize for Chemistry for his work on the CAROTENOIDS and flavins, and on VITAMINS A and B_2.

KARST (from the Karst region of Yugoslavia), a LIMESTONE topography, typically including collapsed caverns, SINK HOLES where streams disappear underground, and areas of bare "limestone pavement."

KARYOTYPE, the characteristic CHROMOSOMES of an individual organism or cell-line arranged in a

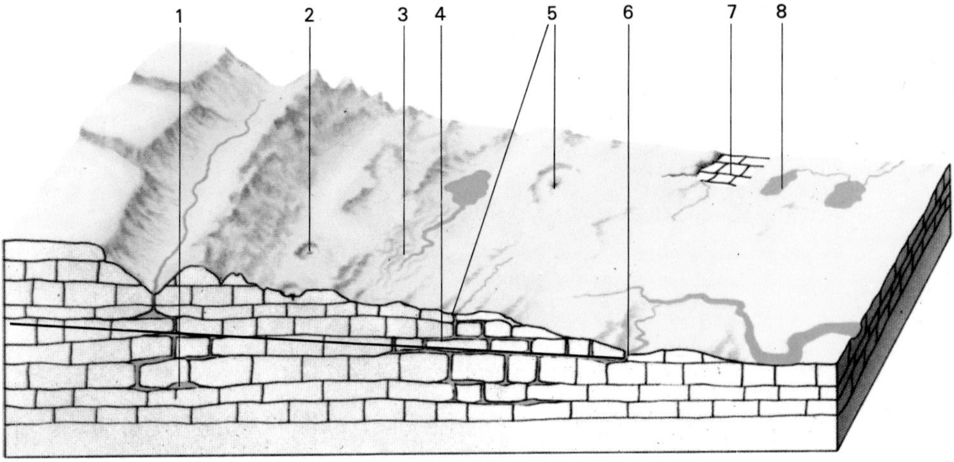

Features of karst landscapes: (1) phreatic zone; (2) sinkhole or doline; (3) polje; (4) vadose region; (5) swallow holes; (6) vauclusian spring; (7) limestone pavement; (8) kamenitza.

systematized form and obtained by microscopic examination of CELLS during MEIOSIS. The chromosomes are numbered by pairs and grouped by their appearance in descending order of size.

KASTLER, Alfred (1902–), French physicist awarded the 1966 Nobel Prize for Physics for his work on the structure of the ATOM, work which led eventually to the development of the LASER.

KATZ, Sir Bernard (1911–), German-born British biophysicist who shared with AXELROD and von EULER the 1970 Nobel Prize for Physiology or Medicine for their independent work on the chemistry of the transmission of nerve impulses.

KAY, John (1704–c1764), British inventor of the flying shuttle (patented 1733), which greatly increased the speed of WEAVING while reducing the number of workers required.

KEKULÉ VON STRADONITZ, Friedrich August (1829–1896), German chemist regarded as the father of modern ORGANIC CHEMISTRY. At the same time as **Archibald Scott Couper** (1831–1892) he recognized the quadrivalency of CARBON and its ability to form long chains. With his later inference of the structure of BENZENE (the "benzene ring"), structural organic chemistry was born

Karyotype of human chromosomes revealing E trisomy—chromosome 18 in group E appears in triplicate—a syndrome giving rise to severe mental and developmental defects in affected babies.

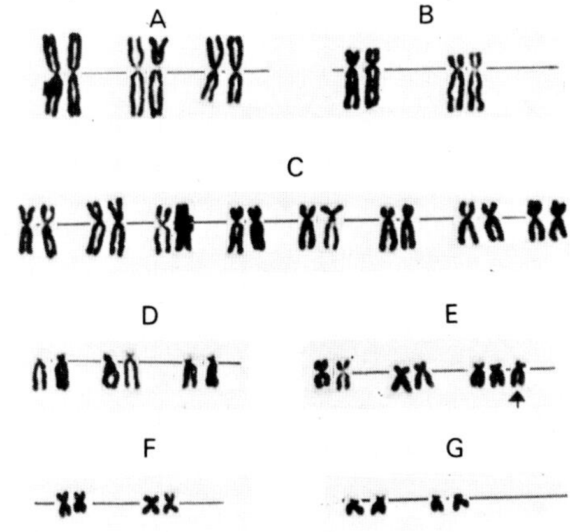

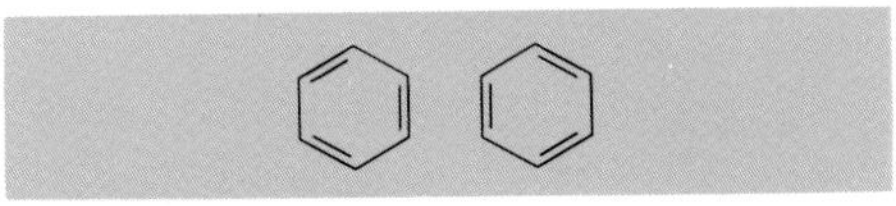

Kekulé structures for benzene.

KELLY, William (1811–1888), US inventor, simultaneously with Sir Henry BESSEMER, of the steelmaking process now known as the BESSEMER PROCESS.

KELVIN (K), the SI UNIT of thermodynamic TEMPERATURE, defined as 1/273.16 of the thermodynamic temperature of the triple point of WATER. It is used both as a unit of temperature difference (when the centigrade degree is defined equal to it) and for expressing ABSOLUTE temperatures (in kelvins above ABSOLUTE ZERO). Temperatures expressed in degrees Celsius equal temperatures expressed in kelvins less 273.15. The older terms "degree Kelvin" and "Kelvin scale" are obsolete.

KELVIN, William Thomson, 1st Baron (1824–1907), British physicist who made important contributions to many branches of physics. In attempting to reconcile CARNOT's theory of heat engines and JOULE's mechanical theory of HEAT he both formulated (independently of CLAUSIUS) the 2nd Law of THERMODYNAMICS and introduced the ABSOLUTE temperature scale, the unit of which is called KELVIN for him. His and FARADAY's work on ELECTROMAGNETISM gave rise to the theory of the electromagnetic field, and his papers, with those of Faraday, strongly influenced J. Clerk MAXWELL's work on the electromagnetic theory of LIGHT (though Kelvin himself rejected Maxwell's over-abstract theory). His work on wire-telegraphic signaling played an essential part in the successful laying of the first ATLANTIC CABLE.

KENDALL, Edward Calvin (1886–1972), US biochemist awarded with HENCH and REICHSTEIN the 1950 Nobel Prize for Physiology or Medicine for his work on the corticoids and isolation of cortisone (see STEROIDS), applied by Hench to the treatment of rheumatoid ARTHRITIS.

KENDREW, John Cowdery (1917–), British biochemist awarded with PERUTZ the 1962 Nobel Prize for Chemistry for his first determining the structure of a globular PROTEIN (myoglobin).

KENNELLY, Arthur Edwin (1861–1939), US electrical engineer who, independently of HEAVISIDE, proposed the existence of that layer of the IONOSPHERE (the E layer) now often called the Kennelly-Heaviside Layer.

KENNY, Sister Elizabeth (1886–1952), Australian nurse best known for developing the treatment of infantile paralysis (see POLIOMYELITIS) by stimulating and reeducating the muscles affected.

KEPLER, Johannes (1571–1630), German astronomer who, using BRAHE's superbly accurate observations of the planets, advanced COPERNICUS' heliocentric model of the SOLAR SYSTEM in showing that the planets followed elliptical paths. His three laws (see KEPLER'S LAWS) were later the template about which NEWTON formulated his theory of GRAVITATION. Kepler also did important work in optics, discovering a fair approximation for the law of REFRACTION.

KEPLER'S LAWS, three laws formulated by Johannes KEPLER to describe the motions of the planets in the solar system. (**1**) Each planet orbits the sun in an ELLIPSE of which the sun is at one focus. (**2**) The line between a planet and the sun sweeps out equal areas in equal times: hence the planet moves faster when closer to the sun than it does when farther away. (**3**) The square of the time taken by a planet to ORBIT the sun is proportional to the cube of its mean distance from the sun.

KERATIN, a fibrous insoluble PROTEIN high in sulfur, found in the skin of vertebrates where it forms the major component of hair, feathers, nails, claws and hooves.

KERNITE, a mineral form of hydrated sodium tetraborate ($Na_2B_4O_7.4H_2O$), found associated with BORAX in Kern Co., Cal. It forms colorless monoclinic crystals, and occurs as veins in clay shale beds. Kernite is a major source of BORON.

KEROSINE, or paraffin oil, a mixture of volatile HYDROCARBONS having 10 to 16 carbon atoms per molecule, used as a FUEL for jet engines (see JET PROPULSION), for heating and lighting and as a solvent and paint thinner. Although it can be derived from oil, coal and tar, most is produced from PETROLEUM by refining and CRACKING. Kerosine boils between 150°C and 300°C.

KETONES, class of organic compounds of general formula RR′CO, containing a carbonyl group, but less reactive than ALDEHYDES, which in some ways they resemble. They are used as solvents and in industrial synthesis. The simplest and most important is ACETONE (see also CAMPHOR). Ketones are formed by dehydrogenation or oxidation of secondary ALCOHOLS, FRIEDEL-CRAFTS acylation of aromatic compounds, and by other methods. They may be reduced by hydrogen or metal HYDRIDES to secondary alcohols, and undergo addition and condensation reactions with NUCLEOPHILES (see also OXIMES). The presence of α-hydrogen yields greater reactivity because of keto-enol TAUTOMERISM.

Ketones

2-propanone (dimethyl ketone, acetone)
MW 58·08, mp −95°C, bp 56°C

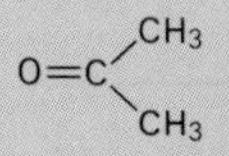

3-pentanone (diethyl ketone, propione)
MW 86·14, mp −40°C, bp 102°C

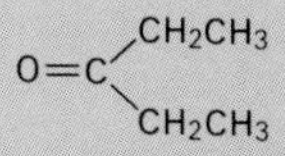

Benzophenone (diphenyl ketone, benzoyl benzene)
MW 182·21, mp 48°C, bp 306°C

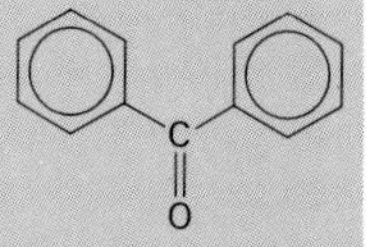

2,3-butandione (dimethyl diketone, biacetyl, dimethylglyoxal, diacetyl)
MW 86·09, mp −2°C, bp 88°C

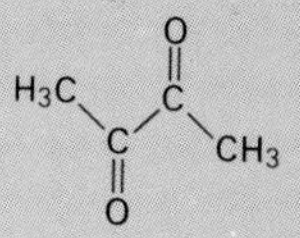

KETOSIS, metabolic state in which breakdown of body FATS leads to the production of KETONE bodies (β-hydroxybutyrate and acetoacetate). These break down to ACETONE, which may be smelled in the breath. Ketosis occurs in diabetic coma where INSULIN lack alters the pattern of fat and glucose METABOLISM. It is also seen in starvation.

KETTERING, Charles Franklin (1876–1958), US inventor of the first electric cash register and the electric self-starter, who made many significant contributions to AUTOMOBILE technology.

KETTLEHOLE, a depression in an area covered by glacial drift, formed where a mass of ice, submerged in the DRIFT, has melted. Kettle lakes are water-filled kettleholes. (See also GLACIER.)

KHORANA, Har Gobind (1922–), Indian-born US biochemist who shared with HOLLEY and NIRENBERG the 1968 Nobel Prize for Physiology or Medicine for his major contributions toward deciphering the genetic code (see GENETICS).

KHWARIZMI, Muhammad ibn-Musa al-, or **al-Khwarizmi,** 9th-century Arab mathematician. The title of his treatise on the solution of many basic mathematical problems contained the word *al-jabr*, from which comes the modern English term, ALGEBRA.

KIDNEYS, two organs concerned with the excretion of waste products in the urine and the balance of salt and water in the body. They lie behind the peritoneal cavity of the ABDOMEN and excrete urine via the ureters, thin tubes passing into the PELVIS to enter the BLADDER. The basic functional unit of the kidney is the *nephron*, consisting of a glomerulus and a system of tubules; these feed into collecting ducts, which drain into the renal pelvis and ureter. BLOOD is filtered in the glomerulus so that low-molecular-weight substances, minerals and water pass into the tubules; here most of the water, sugar and minerals are reabsorbed, leaving behind wastes such as urea in a small volume of salt and water. Tubules and collecting ducts are concerned with the regulation of salt and water reabsorption, which is partly controlled by two HORMONES, VASOPRESSIN and aldosterone). Some substances are actively secreted into the urine by the tubules and the kidney is the route of excretion of many DRUGS. Hormones concerned with ERYTHROCYTE formation and regulation of aldosterone are formed in the kidneys, which also take part in protein METABOLISM. DISEASES affecting the kidney may result in acute NEPHRITIS, including BRIGHT'S DISEASE, the nephrotic syndrome (EDEMA, heavy protein loss in the urine and low plasma albumin) or acute or chronic renal failure. In acute renal failure, nephrons rapidly cease to function, often after prolonged SHOCK, SEPTICEMIA, etc. They may, however, recover. In chronic renal failure, the number of effective nephrons is gradually and irreversibly reduced so that they are unable to excrete all body wastes. Nephron failure causes UREMIA. Disease of the kidneys frequently causes hypertension (see BLOOD CIRCULATION). Advanced renal failure may need treatment with DIETARY FOODS, dialysis and renal TRANSPLANT.

Section through a kidney: (1) capsule; (2) cortex; (3) pyramid; (4) calyx; (5) renal pelvis; (6) ureter; (7) renal artery; (8) renal vein; (9) fatty tissue.

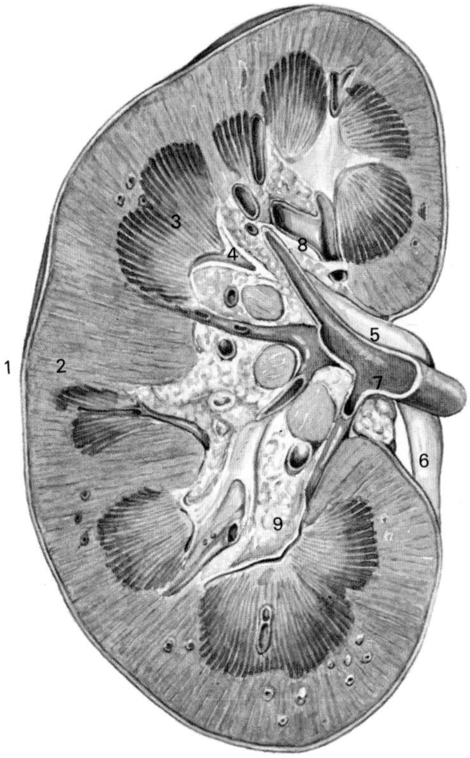

KIESELGUHR, or diatomaceous earth, a fine, porous, chalklike material (amorphous SILICA) formed by the accumulation on ocean floors of the shells of DIATOMS. It is used as an abrasive, a filter and an absorbent, especially in DYNAMITE.

KILN, a FURNACE used for firing CERAMICS. The best known type is the **rotary kiln**, used chiefly for CEMENT: the burners are at the lower end of a slowly rotating cylinder, perhaps 100m long, which is at a slight angle to the horizontal. Material is placed in the upper end and slowly falls to the lower, where it is removed. In **tunnel kilns** the burners are at the center of a tunnel through which slowly pass cars loaded with material. (See also POTTERY AND PORCELAIN.)

KILO- (k), the SI prefix multiplying a unit one-thousandfold. Examples include kilohertz (kHz), kilometre (km), kiloton and kilowatt (kW). (See SI UNITS.)

KILOGRAM (kg), the base unit of MASS in SI UNITS, defined as the mass of a platinum-iridium prototype kept under carefully controlled conditions at the International Bureau of Weights and Measures, near Paris, France.

KILOWATT-HOUR (kWh), the commercial unit of electrical ENERGY, being the energy dissipated by a one-kilowatt device in one hour.

KIMBERLITE, basic IGNEOUS ROCK, often altered and fragmented, which contains DIAMONDS formed *in situ*. It consists of OLIVINE with MICA, SERPENTINE, CALCITE and other minerals. Its chief occurrence is as pipes and DIKES at Kimberley, South Africa.

KINEMATICS, the branch of MECHANICS concerned with describing the motions of objects without consideration of the forces causing those motions. It thus deals with quantities such as distance, time, VELOCITY and ACCELERATION. With KINETICS it makes up DYNAMICS.

KINESIS, an animal's random movement made in response to a stimulus. The animal tends to spend most time in regions where the stimulus is least (or most). **Taxis** is directed movement, away from or towards a directional stimulus.

KINETIC ENERGY. See ENERGY.

KINETICS, the branch of applied MATHEMATICS concerned with the effects of FORCES on the motions of objects (see also KINEMATICS).

KINETICS, Chemical, branch of PHYSICAL CHEMISTRY dealing with reaction rates and mechanisms. In a chemical system, several reactions may be possible according to THERMODYNAMICS, but in practice the fastest reaction predominates, not necessarily the most energetically favored. The reaction rate—the rate at which the concentration of one reactant decreases—is normally proportional to a certain POWER of the concentrations of the reactants, the sum of the exponents being called the *reaction order*. Thus for the reaction $A+B \rightarrow C+D$ it may be found that the rate is given by

$$\frac{d[A]}{dt} = k[A]^2[B]$$

(see CALCULUS; EQUILIBRIUM, CHEMICAL): such a reaction is third order overall (second order in A, first order in B). The *rate constant*, k, depends exponentially (see EXPONENT) on the absolute TEMPERATURE (so that at room temperature most reactions double in rate for a 10K rise in temperature) and on the ACTIVATION ENERGY. CATALYSIS speeds up a reaction by providing an alternative mechanism with lower activation energy. Reaction rates are studied by measuring concentration as a function of time, regular or continuous chemical ANALYSIS being used. (See also FLASH PHOTOLYSIS.)

KINETIC THEORY, widely used statistical theory based on the idea that matter is made up of randomly moving ATOMS or MOLECULES whose kinetic ENERGY increases with TEMPERATURE. It is closely related to statistical mechanics, and predicts macroscopic properties of solids, liquids and gases from motions of individual particles using MECHANICS and PROBABILITY theory. Gases are particularly suited to treatment by kinetic theory, and useful laws connecting their pressure, temperature, density, diffusion and other properties have been deduced with its aid.

KINGDOM. See TAXONOMY.

KING'S EVIL. See SCROFULA.

KININS, plant HORMONES which affect GROWTH and are used to preserve cut flowers; also, in animals, PEPTIDES released by ENZYME action which cause contraction of smooth muscle and elevate blood pressure, thus playing an important role in INFLAMMATION and SHOCK.

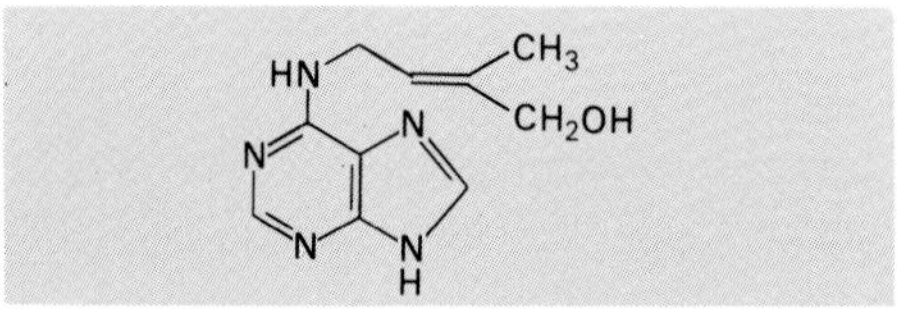

Zeatin, a (cyto)kinin which promotes particularly rapid cell division in plant cultures.

KINSEY, Alfred Charles (1894–1956), US zoologist best known for his statistical studies of human sexual behavior, published as *Sexual Behavior in the Human Male* (1948) and *Sexual Behavior in the Human Female* (1953).

KIRCHHOFF, Gustav Robert (1824–1887), German physicist best known for his work on electrical conduction, showing that current passes through a conductor at the speed of light, and deriving KIRCHHOFF'S LAWS. With BUNSEN he pioneered spectrum analysis (see SPECTROSCOPY), which he applied to the solar spectrum, identifying several elements and explaining the FRAUNHOFER LINES.

KIRCHHOFF'S LAWS, two laws governing electric circuits involving Ohm's-law conductors and sources of electromotive force, stated by G. R. KIRCHHOFF. They assert that the sums of outgoing and incoming currents at any junction in the circuit must be equal, and that the sum of the current-resistance products around any closed path must equal the total electromotive force in it.

KITASATO, Shibasaburo (1852–1931), Japanese bacteriologist who discovered, independently of YERSIN, the PLAGUE bacillus; and with BEHRING discovered that graded injections of toxins could be used for immunization (see ANTITOXINS).

KITCHEN MIDDEN, or **Shell Mound,** refuse heap of usually STONE AGE origin in which, among bones, shells, etc., archaeologists may find potsherds and

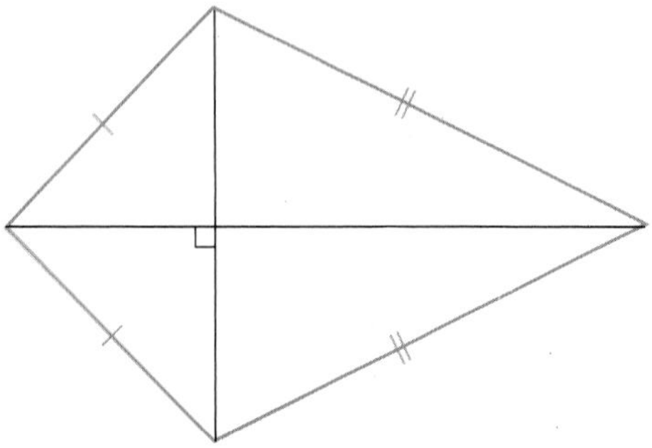

The diagonals of a kite intersect at right angles.

implements of stone, horn and bone. They are 1–3m high, 40–70m wide and up to 400m long.

KITE, recreational aircraft consisting of a light frame covered with thin fabric (e.g. paper) and tethered to a long line. Kites fly in the wind by AERODYNAMIC lift. Originating in the ancient Far East, kite flying has long been a popular sport, and has been used for meteorological observations.

KITE, in GEOMETRY, a QUADRILATERAL with two pairs of equal adjacent sides.

KLAPROTH, Martin Heinrich (1743–1817), German chemist noted for his pioneering work in chemical ANALYSIS. He discovered the elements ZIRCONIUM (1789) and URANIUM (in fact, uranium oxide: 1789), and rediscovered and named TITANIUM (1795).

KLEBS, Edwin (1834–1913), German pathologist and bacteriologist whose work permitted LÖFFLER to isolate the DIPHTHERIA bacillus (1884), often now called the Klebs-Löffler bacillus.

KLEIN, Christian Felix (1849–1925), German geometer whose application of GROUP theory to the unification of mathematics was particularly important in the development of modern GEOMETRY.

KLEIN, Melanie (1882–1960), Austrian-born psychoanalyst whose development of a psychoanalytic therapy for small children radically affected techniques of child psychiatry and theories of child psychology.

KLEIN BOTTLE, a topological space (see TOPOLOGY) of interest in that it has only one side. Consider a hollow CYLINDER made of some flexible material. If one end is bent toward the other, passed through the side of the cylinder and then joined to the other end, the result is a figure which has no "inside." The figure cannot be constructed in Euclidean space (see EUCLIDEAN GEOMETRY); however, if it were correctly

Kirchhoff's laws of electric circuits.

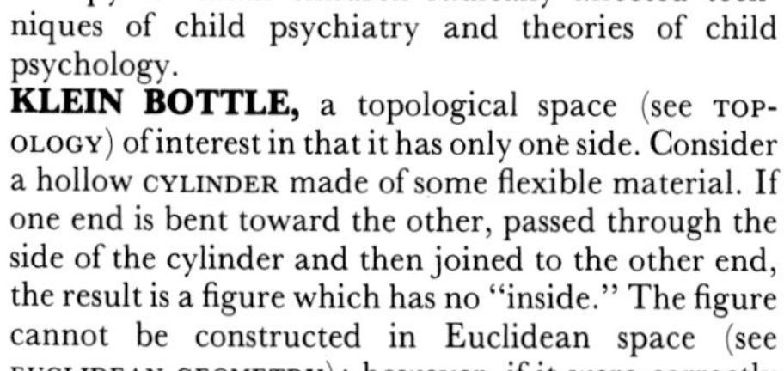

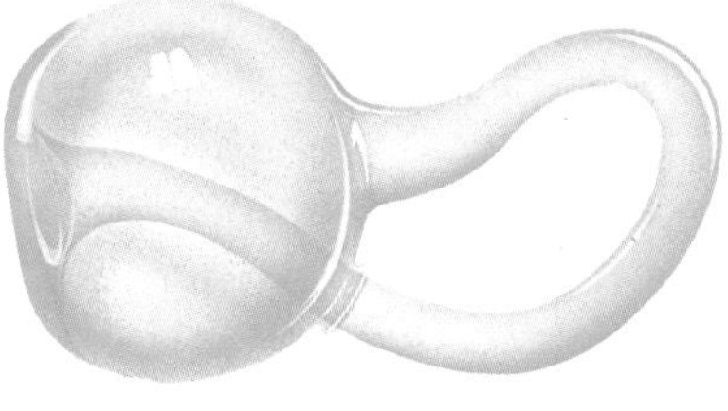

The Klein bottle: a practical vessel which has only one continuous surface.

cut in two, the result would be two MÖBIUS STRIPS.

KLEPTOMANIA, an individual's obsessive (see OBSESSION) urge, usually the result of internal conflict (see COMPLEX; NEUROSIS) and often linked with sexual perversion, to steal objects which, in general, he does not actually wish to possess. Most children pass through a kleptomaniac phase.

KNOT (kn), unit of speed used at sea, defined as one nautical mile per hour. The international knot (1 852m/h) used in the US differs slightly from the UK knot (6080ft/h). The name comes from the old practice of measuring speed at sea by counting the number of knots in a knotted line payed out over the stern in a given time.

KOCH, Robert (1843–1910), German medical scientist regarded as a father of BACTERIOLOGY, awarded the 1905 Nobel Prize for Physiology or Medicine for his work. He isolated the ANTHRAX bacillus and showed it to be the sole cause of the disease; devised important new methods of obtaining pure cultures; and discovered the bacilli responsible for TUBERCULOSIS (1882) and CHOLERA (1883).

KOCHER, Emil Theodor (1841–1917), Swiss surgeon awarded the 1909 Nobel Prize for Physiology or Medicine for his discovery of the relation between the THYROID GLAND and CRETINISM.

KOFFKA, Kurt (1886–1941), German-born US psychologist who, with KÖHLER and WERTHEIMER, was responsible for the birth of GESTALT PSYCHOLOGY.

KÖHLER, Wolfgang (1887–1967), German-born US psychologist who, with KOFFKA and WERTHEIMER, was responsible for the birth of GESTALT PSYCHOLOGY.

KOLA, *Cola acuminata* and *C. nitida*, trees native to tropical W Africa. They produce edible nuts that contain CAFFEINE. The nuts are an important food for local populations. They are also exported for use in manufacture of soft drinks. Family: Sterculiaceae.

KON-TIKI. See HEYERDAHL, THOR.

KORNBERG, Arthur (1918–), US biochemist awarded with OCHOA the 1959 Nobel Prize for Physiology or Medicine for discovering an ENZYME (DNA polymerase) that could produce from a mixture of NUCLEOTIDES exact replicas of DNA molecules. He thus extended Ochoa's related work.

KORSAKOV'S PSYCHOSIS, or **Korsakov Syndrome,** a condition of AMNESIA, unaccompanied by DEMENTIA, observed particularly among alcoholics (see ALCOHOLISM), but also among sufferers from localized BRAIN damage. It is named for the Russian neurologist S. S. Korsakov (1854–1900).

KORZYBSKI, Alfred Habdank Skarbek (1879–1950), Polish-born US scientist who formulated the philosophical linguistic system, General SEMANTICS.

KOSSEL, Albrecht (1853–1927), German biochemist awarded the 1910 Nobel Prize for Physiology or Medicine for his work on PROTEINS and NUCLEIC ACIDS. His main contribution was to show that "nuclein" from cellular sources was not one substance but comprised of a protein and a nonprotein (nucleic acid) component.

KOVALEVSKI, Sonya (1850–1891), Russian mathematician and novelist who made important contributions to the theory of DIFFERENTIAL EQUATIONS. Her brother-in-law, **Alexandr Onufrievich Kovalevski** (1840–1901) did pioneering work in EMBRYOLOGY.

KRAFFT-EBING, Richard, Baron von (1840–1902), German psychologist best known for his work on the psychology of SEX. He also showed there was a relation between syphilis and general PARALYSIS.

KREBS, Sir Hans Adolf (1900–), German-born British biochemist awarded (with LIPMANN) the 1953 Nobel Prize for Physiology or Medicine for his discovery of the CITRIC ACID CYCLE, or "Krebs cycle."

KROEBER, Alfred Louis (1876–1960), US anthropologist who made contributions to many aspects of cultural ANTHROPOLOGY and ARCHAEOLOGY, particularly with reference to the AMERINDS.

KROGH, Schack August Steenberg (1874–1949), Danish physiologist awarded the 1920 Nobel Prize for Physiology or Medicine for his discovery that CAPILLARIES contract and expand so as to vary the amount of BLOOD-oxygen supplied to parts of the body in accordance with their requirements.

KRYPTON (Kr), one of the NOBLE GASES, used to fill high-wattage electric light bulbs, flash lamps and electric-arc lamps. It combines with fluorine in an electric discharge to give krypton (II) fluoride (KrF_2), a highly reactive, colorless crystalline solid, which decomposes slowly at 20°C and is hydrolyzed by water. Other compounds have been claimed. AW 83.8, mp −157°C, bp −152°C.

KUHN, Richard (1900–1967), German chemist awarded the 1938 Nobel Prize for Chemistry for his work on the CAROTENOIDS and VITAMINS.

KURCHATOV, Igor Vasilevich (1903–), Russian nuclear physicist largely responsible for the development of Soviet nuclear armaments and for the first Soviet nuclear power station. The Soviets have named RUTHERFORDIUM *kurchatovium* for him.

KURCHATOVIUM. See RUTHERFORDIUM.

KUSCH, Polykarp (1911–), German physicist who shared with W. E. LAMB the 1955 Nobel Prize for Physics for showing that the ELECTRON had a magnetic moment (see MAGNETISM) greater than that theoretically calculated. This result inspired radical changes in nuclear theory.

KWASHIORKOR, PROTEIN malnutrition simultaneous with the maintenance of relatively adequate calorie intake. In affected children it causes EDEMA, SKIN and HAIR changes, loss of appetite, DIARRHEA, LIVER disturbance and apathy. Its name derives from its occurrence in children rejected from the breast at the birth of the next sibling. Treatment involves rehydration, treatment of infection and a balanced diet with adequate protein.

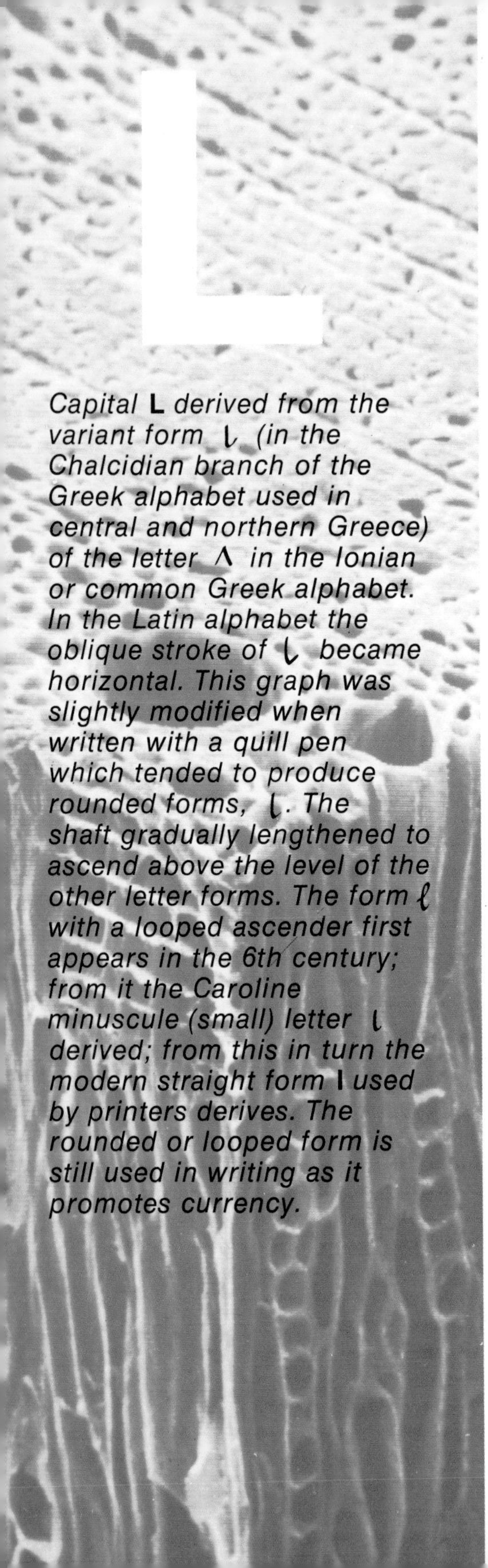

L

*Capital **L** derived from the variant form L (in the Chalcidian branch of the Greek alphabet used in central and northern Greece) of the letter Λ in the Ionian or common Greek alphabet. In the Latin alphabet the oblique stroke of L became horizontal. This graph was slightly modified when written with a quill pen which tended to produce rounded forms, L. The shaft gradually lengthened to ascend above the level of the other letter forms. The form ℓ with a looped ascender first appears in the 6th century; from it the Caroline minuscule (small) letter l derived; from this in turn the modern straight form **l** used by printers derives. The rounded or looped form is still used in writing as it promotes currency.*

LABRADORITE, variety of plagioclase FELDSPAR, consisting of ALBITE and ANORTHITE, commonly occurring in BASALT and GABBRO. Gray to black in color, it often shows red, blue or green iridescence, and hence is used as a GEM stone and for building.

LABYRINTHODONTIA, an extinct order of the AMPHIBIA. They lived between the Carboniferous and Triassic periods, 345–190 million years ago. Some were large, up to 5m (16ft) long, and all had teeth in which the dentine was folded into a pattern of wavy lines seen when the teeth are sectioned.

LAC. See SHELLAC.

LACCOLITH, a dome-like intrusion of igneous rock, usually arching the overlying strata and with an approximately flat floor. **Phacoliths** are similar but lens-shaped, with a concave side facing downward and a convex side facing upward.

LACERTILIA, members of the REPTILIA that include the lizards. Lizards usually have well developed limbs and the 3 000 known species are found in tropical and subtropical regions.

LACTATION, the production of MILK by female mammals. Shortly before the birth of her young, hormonal changes in the mother result in increased development of the mammary glands and teats. Glandular cells in the body of the mammaries secrete milk which is released to the young when the teats are stimulated. Lactation and the feeding of young on milk are characteristic of the MAMMALIA.

LACTIC ACID (CH_3.CHOH.COOH), hydroxypropionic acid, a syrupy odorless liquid produced by the action of lactic-acid bacteria on MILK. It is responsible for the flavor of plain YOGHURT. MUSCLE tissue which is being vigorously exercised is unable to obtain sufficient OXYGEN from the blood to supply the ENERGY it needs. Under these conditions additional energy can be obtained by the reduction of GLUCOSE to lactate. This lactate tends to build up in the muscle and is probably responsible for the CRAMPS which are experienced in vigorous EXERCISE.

LACTOSE ($C_{12}H_{22}O_{11}$), or "milk sugar," a disaccharide SUGAR forming about 4.5% of MILK. It yields GLUCOSE and galactose with the ENZYME lactase.

LAGOMORPHA, an order of the MAMMALIA that includes the hares, rabbits and pikas. Lagomorphs were once thought to be close relatives of the RODENTIA because they too have constantly growing incisor teeth. They are native to many parts of the world and have been introduced to Australia, where they are important crop pests, by man.

LAGOON, stretch of water separated from the sea by a bank or reef, a common phenomenon in CORAL reef areas (see also ATOLL). A lagoon may also be formed when the sea throws up a barrier beach at high tide. **Haffs** are lagoons created by a sandy spit at a river mouth.

LAGRANGE, Joseph Louis (1736–1813), French mathematician who made important contributions to CALCULUS, DIFFERENTIAL EQUATIONS and especially the application of techniques of ANALYSIS to MECHANICS. He worked also on celestial mechanics, in particular explaining the MOON's libration.

LAKE. See RIVERS AND LAKES.

LAKE DWELLING, dwelling built on stilts or piles in the waters of a lake. In parts of Europe can be found STONE AGE and BRONZE AGE lake dwellings, and in some parts of the world they are still built. **Crannogs**, strongholds built on artificial islands, were built in Ireland, Scotland and England from the Late Stone Age until the Middle Ages.

LAMARCK, Jean Baptiste Pierre Antoine de Monet, Chevalier de (1744–1829), French biologist who did pioneering work on taxonomy (especially that of the INVERTEBRATES) which led him to formulate an early theory of EVOLUTION. Where DARWIN was to propose NATURAL SELECTION as a mechanism for evolutionary change, Lamarck felt that organisms could develop new organs in response to their need for them, and that ACQUIRED CHARACTERISTICS could be inherited.

LAMB, Willis Eugene, Jr. (1913–), US physicist who shared with KUSCH the 1955 Nobel Prize for Physics for his examinations of the hydrogen spectrum. He devised new techniques whereby he was able to show that the positions of certain lines differed from the theoretical predictions, thus necessitating a revision of atomic theory.

LAMBERT (L), in PHOTOMETRY, a former unit of LUMINANCE, being that of a uniformly diffusing surface reflecting or emitting 1 lumen per cm^2.

LA METTRIE, Julien Offray de (1709–1751), French physician and philosopher who took the idea of "man as machine" to its extreme. He held that all mental phenomena resulted from organic changes in the NERVOUS SYSTEM.

LAMINATES, components where several laminae (thin sheets) of different substances are bonded together with RESINS. Laminated plastics comprise layers of cloth, paper, plastic, etc., impregnated with synthetic resin, bonded together by heat and pressure. Laminated glass is used in auto and airplane windows and as bulletproof glass; and laminated woods for many purposes (see also VENEER).

World language families

Family	Group	Subgroup	Languages
Indo–European	Germanic	Scandinavian	Icelandic Danish Norwegian Swedish
		Western Germanic	Dutch English German
	Romance (Italic: derived from Latin)	Hispanic	Spanish Portuguese
		French	French Provençal
		Eastern Romance	Italian Romanian
	Celtic	Brythonic	Welsh Breton
		Goidelic	Gaelic Erse Manx
	Balto–Slavic	Baltic	Lettish
		Slavic	Russian Ukrainian Polish Czech Slovak Serbo–Croat Bulgarian
	Greek	—	Greek
	Indo–Iranian	Indic	Sanskrit Hindi Urdu Bengali Gujarati
		Iranian	Persian
	Thracian	—	Armenian Albanian
Malayo–Polynesian	—	—	Javanese
Sino–Tibetan	Chinese	—	Mandarin
Dravidian	—	—	Tamil
Ural–Altaic	Altaic	Turkic	Turkish
	Uralic	Ugric Finnic	Hungarian Finnish
Japanese–Korean	—	—	Japanese Korean
African	Khoisan	—	Bushman
	Niger–Congo	—	Swahili
Hamito–Semitic	Semitic	South Semitic	Arabic
		North Semitic	Hebrew
	Hamitic	Egyptian	Coptic
American	Eskimo–Aleut	—	Eskimo

LAMP. See LIGHTING.

LAMPSHELLS. See BRACHIOPODA.

LANCELET, fish-like animal also called AMPHIOXUS.

LAND, Edwin Herbert (b. 1909), US physicist and inventor of Polaroid, a cheap and adaptable means of polarizing light (1932), and the POLAROID LAND CAMERA (1947). In 1937 he set up the Polaroid Corporation to manufacture scientific instruments and antiglare sunglasses incorporating Polaroid.

LANDAU, Lev Davidovich (1908–1968), Soviet physicist who made important contributions in many fields of modern physics. His work on CRYOGENICS was rewarded by the 1962 Nobel Prize for Physics for his development of the theory of liquid HELIUM and his predictions of the behavior of liquid He^3.

LAND RECLAMATION, the transformation of useless land into productive land, usually for agricultural purposes. The major techniques are irrigation, drainage, fertilization and desalination (see DESALINATION; DRAINAGE; FERTILIZERS; IRRIGATION). The most spectacular examples of land drainage are those where land has been reclaimed from the sea. Best known is that of the Netherlands, where about 8000km² of land have been reclaimed, mainly in the 20th century. (See also CONSERVATION; DRY FARMING; SOIL EROSION.)

LANDSTEINER, Karl (1868–1943), Austrian-born US pathologist awarded the 1930 Nobel Prize for Physiology or Medicine for discovering the major BLOOD groups and developing the ABO system of blood typing.

LANGLEY, Samuel Pierpont (1834–1906), US astronomer, physicist, meteorologist and inventor of the BOLOMETER (1878) and of an early heavier-than-air flying machine. His most important work was investigating the sun's role in bringing about meteorological phenomena.

LANGMUIR, Irving (1881–1957), US physical chemist awarded the 1932 Nobel Prize for Chemistry for his work on thin films on solid and liquid surfaces (particularly oil on water), which gave rise to the new science of surface chemistry.

LANGUAGE, the spoken or written means by which man expresses himself and communicates with others. The word "language" comes from the Latin *lingua*, tongue, demonstrating that speech is the primary form of language and writing the secondary. Language comprises a set of sounds that symbolize the content of the message to be conveyed. It is, on this planet, peculiar to man, constituting as it does a formal system with rules whereby complex messages can be built up out of simple components (see GRAMMAR). Languages are the products of their cultures, arising from the cooperative effort required by societies. There are some 3000 different languages spoken today, added to which are many more regional dialects. Languages may be classified into families, groups and subgroups. To us the most important language family is the Indo-European, to which many Asian and most European languages (including English) belong. Other important families are the Hamito-Semitic, Altaic, Sino-Tibetan, Austro-Asiatic and Dravidian, among others. (See also ETYMOLOGY; LINGUISTICS; PHILOLOGY; PRONUNCIATION; SEMANTICS; SEMIOLOGY; SHORTHAND; SIGN LANGUAGE; WRITING, HISTORY OF.)

LANOLIN, soft, yellow-white unctuous solid, a hydrated grease or wax from sheep's wool. It is a mixture of CHOLESTEROL and its ESTERS of FATTY ACIDS, and is used as a base for ointments and cosmetics.

Lanthanides

element	Z	outermost electrons	atomic weight
lanthanum (La)	57	$5d^1 6s^2$	138.91
cerium (Ce)	58	$4f^2 6s^2$	140.12
praseodymium (Pr)	59	$4f^3 6s^2$	140.91
neodymium (Nd)	60	$4f^4 6s^2$	144.24
promethium (Pm)	61	$4f^5 6s^2$	145*
samarium (Sm)	62	$4f^6 6s^2$	150.4
europium (Eu)	63	$4f^7 6s^2$	151.96
gadolinium (Gd)	64	$4f^7 5d^1 6s^2$	157.25
terbium (Tb)	65	$4f^9 6s^2$	158.93
dysprosium (Dy)	66	$4f^{10} 6s^2$	162.50
holmium (Ho)	67	$4f^{11} 6s^2$	164.93
erbium (Er)	68	$4f^{12} 6s^2$	167.26
thulium (Tm)	69	$4f^{13} 6s^2$	168.93
ytterbium (Yb)	70	$4f^{14} 6s^2$	174.04
lutetium (Lu)	71	$4f^{14} 5d^1 6s^2$	174.97

*radioactive, the best known isotope.

LANTHANIDES, the 14 elements with atomic numbers (see ATOM) 58–71, immediately following LANTHANUM in the PERIODIC TABLE. They comprise CERIUM, PRASEODYMIUM, NEODYMIUM, PROMETHIUM, SAMARIUM, EUROPIUM, GADOLINIUM, TERBIUM, DYSPROSIUM, HOLMIUM, ERBIUM, THULIUM, YTTERBIUM and LUTETIUM. (See also LANTHANUM SERIES.)

LANTHANUM (La), the second most abundant of the RARE EARTHS, and the prototypical member of the LANTHANUM SERIES. AW 138.9, mp 921°C, bp 3457°C, sg 6.145 (25°C).

LANTHANUM SERIES, the 15 elements with atomic numbers (see ATOM) 57–71, comprising LANTHANUM and the LANTHANIDES (see PERIODIC TABLE). Their electronic structures are very similar, differing only in inner ORBITALS; hence their properties are very similar. This also produces a decrease (the *lanthanide contraction*) in ionic radii through the series, so that the third-row TRANSITION ELEMENTS following the lanthanum series have ionic radii almost identical to those of their analogues in the second row, and hence have similar properties. The lanthanum series elements are all reactive metals resembling SCANDIUM, forming trivalent salts and LIGAND complexes. Cerium, praseodymium and terbium also form tetravalent compounds, and europium, ytterbium and samarium form divalent ones. All form divalent ionic hydrides and sulfides. (See also RARE EARTHS.)

LAPIS LAZULI, deep blue METAMORPHIC ROCK, found in crystalline LIMESTONE, and consisting of LAZURITE mixed with other silicates, CALCITE and PYRITE. It chiefly occurs in Afghanistan and Chile, and has long been valued as a GEM stone and as the source of the pigment ultramarine.

LAPLACE, Pierre Simon, Marquis de (1749–1827), French scientist known for his work on celestial mechanics, especially for his NEBULAR HYPOTHESIS; for his many fundamental contributions to mathematics, and for his PROBABILITY studies.

LARVA, a pre-adult stage in the life history of many animals, differing structurally from the adult in more than merely sexual immaturity. The possession of a larva usually enables a species to exploit a different food source from that used by the adult. Again, it may be important in dispersing individuals to new areas, or, in many parasites, as an infective phase. The larva undergoes a change of structure to adult form known as METAMORPHOSIS.

LARYNGITIS, or INFLAMMATION of LARYNX, usually due to either VIRUS or bacterial infection or chronic VOICE abuse, and leading to hoarseness or loss of voice.

LARYNX, specialized part of the respiratory tract used in VOICE production (see SPEECH AND SPEECH DISORDERS). It lies above the TRACHEA in the neck, forming the Adam's apple, and consists of several CARTILAGE components linked by small MUSCLES. Two folds, or *vocal cords*, lie above the trachea and may be pulled across the airway so as to regulate and intermittently occlude air flow. It is the movement and vibration of these that produce voice.

LASER, a device producing an intense beam of parallel LIGHT with a precisely defined wavelength. The name is an acronym for "*l*ight *a*mplification by *s*timulated *e*mission of *r*adiation," and the device is in fact a MASER operating as an oscillator at visible wavelengths.

The light produced by lasers is very different from that produced by conventional sources. In the latter, all the source atoms radiate independently in all directions, whereas in lasers they radiate in step with each other and in the same direction, producing **coherent light.** Such beams spread very little as they travel, and provide very high capacity communication links. They can be focused into small intense spots, and have been used for cutting and WELDING—notably for refixing detached retinas in the human EYE. Lasers also find application in distance measurement by INTERFERENCE methods, in SPECTROSCOPY and in HOLOGRAPHY.

The principles of laser operation are described under MASER. The active material is enclosed between a pair of parallel MIRRORS, one of them half-silvered; light traveling along the axis is reflected to and fro and builds up rapidly by the stimulated emission process, passing out eventually through the half-silvered mirror, while light in other directions is rapidly lost from the laser.

In pulsed operation, one of the end mirrors is concealed by a shutter, allowing a much higher level of pumping than usual; opening the shutter causes a very intense pulse of light to be produced—up to 100MW for 30ns—while other pulsing techniques can achieve 10^{13} W in picosecond pulses.

Among the common laser types are ruby lasers (optically pumped, with the polished crystal ends serving as mirrors), liquid lasers (with RARE EARTH ions or organic dyes in solution), gas lasers (an electric discharge providing the high proportion of excited states), and the very small SEMICONDUCTOR lasers (based on electron-hole recombination).

LATENCY PERIOD, in psychoanalysis, the stage in human development starting around the age of five and ending at PUBERTY, marking the transition to adult from INFANTILE SEXUALITY.

LA TÈNE, late IRON AGE culture of European CELTS, named for the site of the same name at the E end of Lake Neuchâtel, Switzerland. Originating c450 BC, when the Celts came into contact with Greco-Etruscan influences, it died out c50 AD as the Celts became subservient to Rome. La Tène ornaments are decorated with round, S-shaped and spiral patterns.

LATENT HEAT, the quantity of HEAT absorbed or released by a substance in an isothermal change of state, such as FUSION or vaporization. The temperature of a heated lump of ice will increase to 0°C and then remain at this temperature until all the ice has melted to water before again rising. The heat energy absorbed at 0°C overcomes the intermolecular forces in the ordered ice structure and increases the kinetic ENERGY of the water molecules.

LATERAL LINE, a system of sensory cells embedded

A laser beam penetrating a ruby plate 0.5mm thick, photographed using light from the beam itself.

The Rainbow trout *Salmo gairdneri* displays a prominent lateral line.

in pits or canals along the side of the body in fish and tailed amphibians. These organs probably detect the low-frequency vibrations in water which come from the movement of prey and shoaling companions. It is also possible that water movements set up by the animal itself and reflected off distant objects may be detected in a form of echo-location.

LATERITE, residual red CLAY soil, usually soft and porous, consisting mainly of hydrated oxides of iron and aluminum. (Some is used as IRON ore.) It is formed from various iron-containing parent rocks by a process known as *laterization*: secular weathering with powerful leaching out of silica, alkalis and alkaline earths, under oxidizing conditions. A tropical climate with heavy seasonal rainfall and good drainage is required.

LATEX. See RUBBER.

LATHE, MACHINE TOOL used to shape components whose cross-sections are circular, but whose diameter usually varies along their length. Typically, the workpiece is held in a stock or lathe bed, and supported and rotated by components at each end, the *headstock* and *tailstock*. Cutting tools are introduced

The layout of a typical manually operated engine lathe.

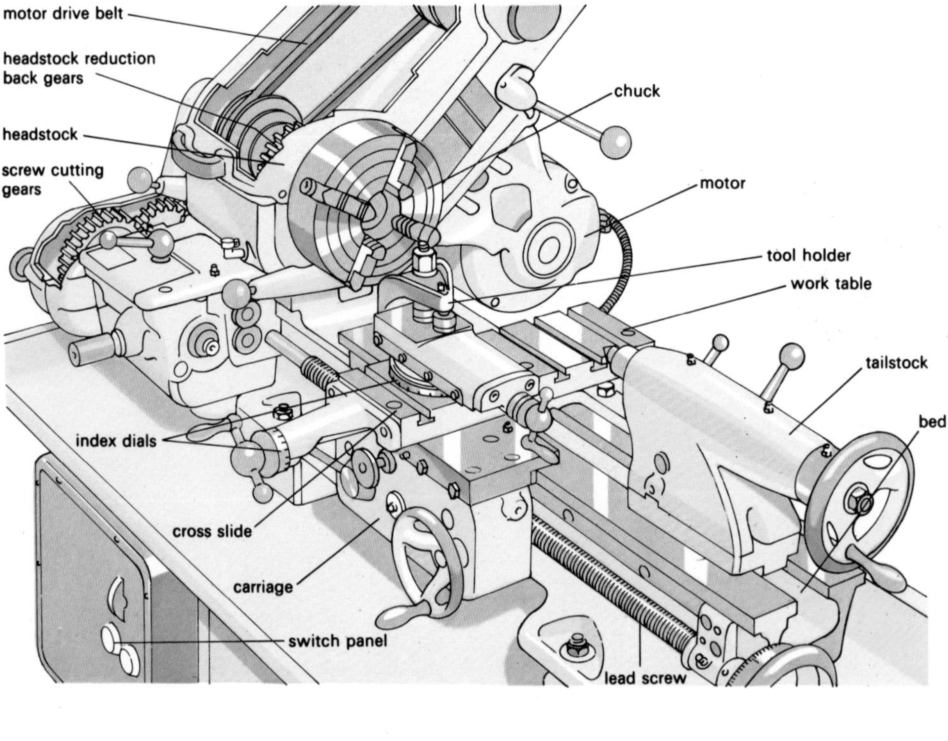

from the side to shape the part. In a screw-cutting lathe, the tool is moved along the piece to produce a thread of given depth and pitch (see BOLTS AND SCREWS). Modern lathes may hold several workpieces and use several cutting tools.

LATIMERIA. See COELACANTH.

LATITUDE AND LONGITUDE, the coordinate system used to locate points on the earth's surface. **Longitude** "lines" are circles passing through the poles whose centers are at the center of the earth; they divide the earth rather like an orange into segments. Longitudes are measured 0°–180°E and W from the line of the GREENWICH OBSERVATORY. Assuming the EARTH to be a sphere, we can think of the **latitude** of a point as the ANGLE between a line from the center of the earth to the point and a line from the center to the equator at the same longitude. Each pole, then, has a latitude of 90°, and so latitude is measured from 0° to 90° N and S of the EQUATOR, latitude "lines" being circles parallel to the equator that get progressively smaller towards the poles. (See CELESTIAL SPHERE.)

LATTICE, in ALGEBRA, a partially ordered set S in which, for any pair of elements a and b, there is in S a least element which is greater than both a and b, and a greatest element which is less than both a and b: for example, the set of REAL NUMBERS forms a lattice. (See also SET THEORY.)

LATTICE, infinite three-dimensional periodic array of points in space, each point being surrounded in an identical way by its neighbors. An assembly of ATOMS placed in the same way at each lattice point makes up a CRYSTAL structure. (See also BRAVAIS LATTICES.)

LATUS RECTUM, a LINE drawn PERPENDICULAR to the major axis of an ellipse (see CONIC SECTIONS), bounded at either end by the ellipse, and passing through one of the foci.

LAUDANUM, or tincture of opium, an extract of OPIUM in alcohol formerly much used in medicine.

LAUE, Max Theodor Felix von (1879–1960), German physicist awarded the 1914 Nobel Prize for Physics for his prediction (and, with others, subsequent experimental confirmation) that X rays can be diffracted by crystals (see X-RAY DIFFRACTION).

LAUGHING GAS, or nitrous oxide. See NITROGEN.

LAURASIA, ancient N-Hemisphere supercontinent formed, with Gondwanaland to the S, after the splitting of Pangaea (see CONTINENTAL DRIFT; GONDWANALAND; PANGAEA). It appears to have comprised present Europe, North America and N Asia.

A latus rectum (red) can be drawn through each focus of an ellipse.

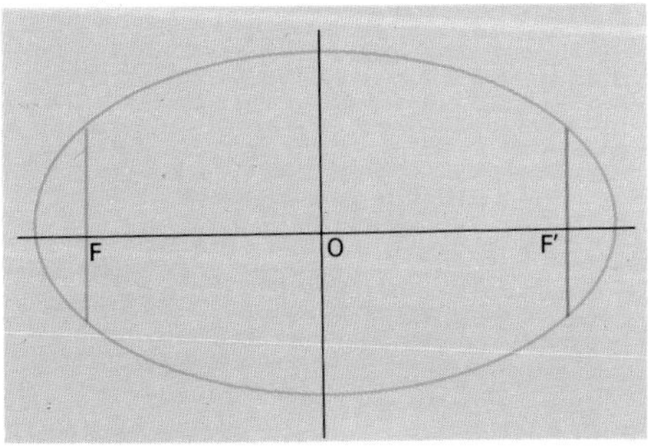

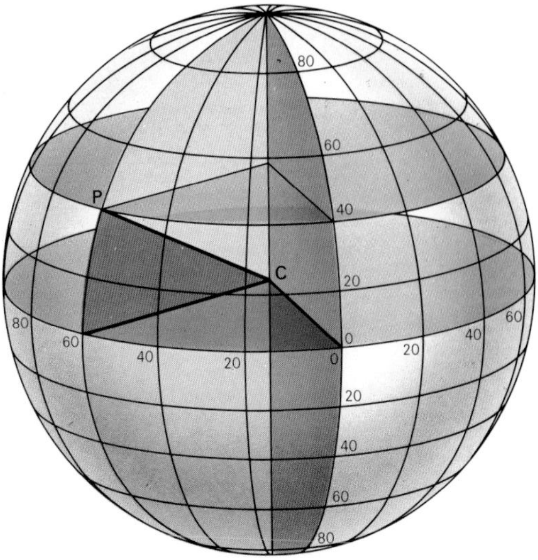

Parallels of latitude are circles on the surface of the earth, parallel to the equator and labeled according to their angular distance from the equator. Meridians of longitude are great circles passing through the poles and labeled according to their angular distance from a standard meridian—that passing through Greenwich, England. The position of a place on the surface of the earth can be specified by citing the parallel of latitude and the meridian of longitude which intercept at that place. Here, the coordinates of P are thus 40°N 60°W, a position in the North Atlantic Ocean some 1200km east of New York.

LAVA, both molten ROCK rising to the earth's surface through VOLCANOES and other fissures, and the same after solidification. Originating in the MAGMA deep below the surface, most lavas (e.g., BASALT) are basic and flow freely for considerable distances. The acidic, SILICA-rich lavas such as RHYOLITE are much stiffer.

Lava of the *pahoehoe* type. As it cools, its surface is wrinkled into rope-like strands by the movement of liquid lava underneath.

Basic lavas solidify in a variety of forms, the commonest being *aa* (Hawaiian, rough) or block lava, forming irregular jagged blocks, and *pahoehoe* (Hawaiian, satiny) or ropy lava, solidifying in ropelike strands. Pillow lava, with rounded surfaces, has solidified under water, and slowly-cooled basalt may form hexagonal columns.

LAVAL, Carl Gustaf Patrik de (1845–1913), Swedish inventor best known for his pioneering work on high-speed steam TURBINES.

LAVERAN, Charles Louis Alphonse (1845–1922), French physician awarded the 1907 Nobel Prize for Physiology or Medicine for his discovery of the MALARIA parasite, a protozoan of the genus *Plasmodium*. (See also PARASITE.)

LAVOISIER, Antoine Laurent (1743–1794), French scientist who was foremost in the establishment of modern CHEMISTRY. He applied gravimetric methods to the process of COMBUSTION, showing that when substances burned, they combined with a component in the air (1772). Learning from J. PRIESTLEY of his "dephlogisticated air" (1774), he recognized that it was with this that substances combined in burning. In 1779 he renamed the gas *oxygène*, because he believed it was a component in all acids. Then, having discovered the nature of the components in water, he commenced his attack on the PHLOGISTON theory, proposing a new chemical nomenclature (1787), and publishing his epoch-making *Elementary Treatise of Chemistry* (1789). In the years before his tragic death on the guillotine, he also investigated the chemistry of RESPIRATION, demonstrating its analogy with combustion.

LAWRENCE, Ernest Orlando (1901–1958), US physicist awarded the 1939 Nobel Prize for Physics for his invention of the CYCLOTRON (1929; the first successful model was built in 1931).

LAWRENCIUM (Lr), a TRANSURANIUM ELEMENT; the final member of the ACTINIDE series, prepared by bombardment of lighter actinides. The most stable isotope, Lr^{256}, has a half-life of only 35s.

LAXATIVE, DRUG or DIETETIC FOOD taken to promote bowel action and to treat CONSTIPATION. They may act as irritants (cascara, senna, phenolphthalein, castor oil), softeners (mineral oil), or bulk agents (bran, methylcellulose and magnesium sulfate—Epsom salts). Laxative abuse may cause GASTROINTENSTINAL TRACT disorders, POTASSIUM deficiency and LUNG disease.

LAYARD, Sir Austen Henry (1817–1894), British archaeologist known for his excavations of Assyrian and Babylonian remains, and especially for his confirmation of the site of Nineveh.

LAZEAR, Jesse William (1866–1900), US physician, a member of W. REED's commission investigating C. FINLAY's theory that YELLOW FEVER is spread by the MOSQUITO. Lazear's death five days after a mosquito bite was a tragic demonstration of the truth of the theory.

LAZURITE ($Na_4Al_3Si_3O_{12}S$), sulfur-bearing feldspathoid SILICATE mineral, forming deep blue granular masses; the chief constituent of LAPIS LAZULI.

LEACHING, the process whereby water, as it percolates through the soil, dissolves out various mineral salts. RAIN water is slightly acidic, because of dissolved CARBON dioxide from the ATMOSPHERE, and thus important in the leaching of SOILS.

Lazurite: the raw mineral and a polished stone.

LEAD (Pb), soft, bluish-gray metal in Group IVA of the PERIODIC TABLE, occurring as GALENA, and also as CERUSSITE and anglesite (lead sulfate). The sulfide ore is converted to the oxide by roasting, then smelted with coke. Lead dissolves in dilute nitric acid, but is otherwise resistant to corrosion, because of a protective surface layer of the oxide, sulfate etc. It is used in roofing, water pipes, coverings for electric cables, RADIATION shields, ammunition, storage BATTERIES, and alloys, including solder (see SOLDERING), PEWTER, BABBITT METAL and type metal. Lead and its compounds are toxic (see LEAD POISONING). AW 207.2, mp 327.5°C, bp 1740°C, sg 11.35 (20°C).

Lead forms two series of salts; the lead(II) compounds are more stable than the lead(IV) compounds. **Lead(II) Oxide** (PbO), or **Litharge**, yellow crystalline solid, made by oxidizing lead; used in lead-acid storage batteries, glass and glazes. mp 888°C. **Lead(IV) Oxide** (PbO_2), brown crystalline solid, a powerful oxidizing agent used in matches, fireworks, and dyes; it decomposes at 290°C. **Trilead Tetroxide** (Pb_3O_4), or **Red Lead**, orange-red powder, made by oxidizing litharge, used in paints, inks, glazes and magnets. **Lead Tetraethyl** ($Pb[C_2H_5]_4$), colorless liquid, made by reacting a lead/sodium alloy with ethyl chloride. It is used as an ANTIKNOCK ADDITIVE to GASOLINE.

LEAD-CHAMBER PROCESS, process for manufacturing SULFURIC ACID, now largely superseded by the CONTACT PROCESS. Sulfur dioxide, produced by burning SULFUR, is mixed with air and nitric oxide (see NITROGEN) and passed in turn through the Glover tower, several lead chambers (where steam is sprayed in), and the Gay-Lussac tower, where concentrated sulfuric acid is sprayed in. The mixture of sulfuric acid and nitrosylsulfuric acid from the Gay-Lussac tower is sprayed back into the Glover tower. The net product is impure 70% sulfuric acid, tapped off from the lead chambers. The nitric oxide is catalytic.

LEAD POISONING, DISEASE caused by excessive LEAD levels in TISSUES and BLOOD. It may be taken in through the industrial use of lead, through AIR POLLUTION due to lead-containing fuels or, in children, through eating old paint. BRAIN disturbance with COMA or CONVULSIONS, peripheral NEURITIS, ANEMIA and abdominal COLIC are important effects. Chelating agents (see CHELATE) are used in treatment but preventitive measures in the community are essential.

LEAF, green outgrowth from the stems of higher plants and the main site of PHOTOSYNTHESIS. The form of leaves varies from species to species but the basic

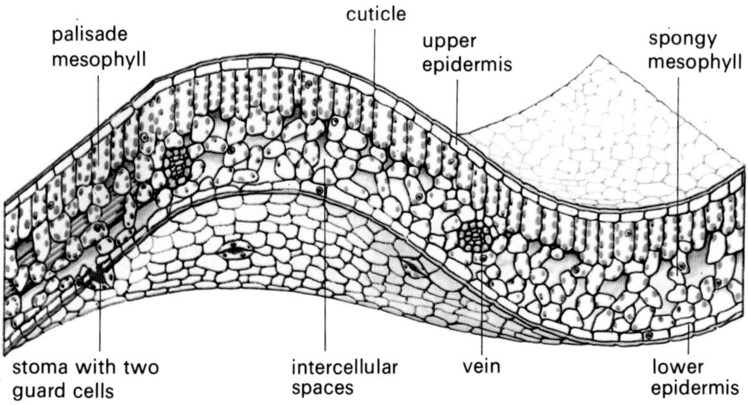

Section through the typical leaf of a flowering plant (Angiosperm). The chloroplasts containing the chlorophyll are represented by green dots—these occur mainly in the palisade mesophyll, though also in the spongy mesophyll.

features are similar. Each leaf consists of a flat blade or lamina, attached to the main stem by a leaf stalk or petiole. Leaf-like stipules may be found at the base of the petiole. The green coloration is produced by CHLOROPHYLL which is sited in the CHLOROPLASTS. Most leaves are covered by a waterproof covering or cuticle. Gaseous exchange takes place through small openings called STOMATA, through which water vapor also passes (see TRANSPIRATION). The blade of the leaf is strengthened by veins which contain the vascular tissue that is responsible for conducting water around the plant and also the substances essential for METABOLISM.

In some plants the leaves are adapted to catch insects (see INSECTIVOROUS PLANTS), while in others they are modified to reduce water loss (see SUCCULENTS; XEROPHYTE). Leaves produced immediately below the FLOWERS are called bracts and in some species, e.g., poinsettia, they are more highly colored than the flowers.

LEAKEY, Louis Seymour Bazett (1903–1972), British archaeologist and anthropologist best known for his findings of human FOSSILS, especially in the region of Olduvai Gorge, Tanzania, and for his (sometimes controversial) views on their significance.

LEAP YEAR. See CALENDAR.

LEARNING. Almost everything that we do derives from learning. If it were not for early school learning you would be able neither to read these words nor to understand them. The concepts of learning and MEMORY are clearly closely related, though learning is usually considered to be the result of practice, and there is usually considered to be a particular stimulus that encourages such practice. The simplest learned response is the conditioned REFLEX. The most powerful learning stimulus is the satisfaction of instinctive drives (see INSTINCT). For example, a dog might learn that if he sits up and "begs" he will be fed by his owner. Here the stimulus is positive, in that the result of his response is a reward, rather than negative, where the correct response earns only escape from punishment: positive stimuli are more effective encouragements to learning than are negative. All animals display the ability to learn, and even some of the most primitive have the ability to become bored with the tests of experimenters (where the reward is not an adequate stimulus). In humans, learning ability depends to a great extent on INTELLIGENCE, though social and environmental factors clearly play a part. (See also CONDITIONING; HABIT; IMPRINTING.)

LEAST SQUARES, technique, widely used in science, for finding the curve which best fits a set of experimental results by minimizing the sum of the squares of the differences between the results and the corresponding points on the curve.

LEATHER, animal hide or skin that has been treated by TANNING to preserve it from decay and to make it strong, supple, water-resistant and attractive in appearance. The skins are typically preserved temporarily by soaking in brine and kept in cold storage. They are washed and soaked in an alkaline solution, and then scraped to remove the hair. Rapidly rotating blades remove residual fat and flesh. The hides are then neutralized and softened by soaking in pancreatic enzymes, and pickled in dilute acid to make them ready for TANNING. The tanned leather is finished by being squeezed to remove excess liquid, lubricated with oil, slowly dried, and impregnated with resins. It is commonly dyed, and a shiny surface is produced by compression. Most leather is made from the hide of sheep, cows, calves, goats, kids and pigs; and for exotic products from the skins of crocodiles, sharks and snakes. Leather is used to make shoes, gloves, coats and other garments, upholstery, bags and luggage, wallets, transmission belts, etc. and to bind books. Chamois leather, for cleaning, is now made from sheepskin. (See also SUEDE).

LEAVEN, substance used to make DOUGH rise during baking by producing gases which expand to make the food light and porous. YEAST produces CARBON dioxide by FERMENTATION; BAKING POWDER and SODIUM bicarbonate produce carbon dioxide by chemical reaction. Air may be introduced by vigorous whipping.

LEBLANC, Nicolas (1742–1806), French chemist who invented the **Leblanc process** for obtaining alkali (SODIUM carbonate) from common SALT. The salt was treated with SULFURIC ACID to give sodium sulfate, which was then heated with CHALK and CHARCOAL to give a "black ash" from which the sodium carbonate could be washed with water. The process was supplanted late in the 19th century by the SOLVAY PROCESS.

LE CHATELIER, Henri Louis (1850–1936), French chemist best known for formulating **Le**

Chatelier's principle (1888), that if a change occurs in one of the conditions of a system initially in equilibrium, the system will adjust, tending to nullify the change and return to equilibrium.

LECLANCHÉ CELL. See BATTERY.

LEDERBERG, Joshua (1925–), US geneticist awarded with G. W. BEADLE and E. L. TATUM the 1958 Nobel Prize for Physiology or Medicine for his work on bacterial genetics. With Tatum, he showed that the offspring of different mutants of *Escherichia coli* had genes recombined from those of the original generation, thus establishing the sexuality of *E. coli*. Later he showed that genetic information could be carried between *Salmonella* by certain bacterial viruses. (See also BACTERIA; GENETICS; VIRUS.)

LEE, Tsung Dao (1926–), Chinese-born US physicist who shared with YANG the 1957 Nobel Prize for Physics for their investigations of examples of the principle of PARITY being violated, which led to significant improvements in our understanding of SUBATOMIC PARTICLES.

LEEUWENHOEK, Anton van (1632–1723), Dutch microscopist who made important observations of CAPILLARIES, red BLOOD corpuscles and SPERM cells, and who is best known for being the first to observe BACTERIA and PROTOZOA (1674–6), which he called "very little animalcules."

LEFT-HANDEDNESS. See HANDEDNESS.

LEG, in man, the lower limb, which is attached to the trunk at the PELVIS by the hip joint. Its major bones are the femur, passing from the hip to the knee, and the tibia and fibula, which pass from the knee to the ankle JOINT, where the leg articulates with the FOOT. The legs are concerned with maintenance of posture, walking, running and jumping. The powerful MUSCLES of the buttock, pelvis and thigh act about the hip and knee joints, while calf muscles act via the ACHILLES TENDON across the ankle on the foot.

LEGENDRE, Adrien Marie (1752–1833), French mathematician best known for his work on elliptic integrals; for first publishing the method of LEAST SQUARES (1806); and for deriving the Legendre functions, which find many applications in physics.

LEGUME, or pod, multiseeded dry fruit that liberates its seeds by splitting along two margins, in for example the bean and pea. (See also FOLLICLE.)

LEGUMINOUS PLANTS, general name for plants of the pea family (Leguminosae) the fruit of which are called LEGUMES (pods). In terms of number of species, this family is second in size only to the Compositae. There are many economically important species including acacia, alfalfa, bean, lentil, pea and soybean. The roots of leguminous plants produce nodules containing nitrogen-fixing bacteria (see NITROGEN FIXATION.)

LEIBNIZ, Gottfried Wilhelm von (1646–1716), German philosopher, historian, jurist, geologist and mathematician, codiscoverer of the CALCULUS and author of the theory of monads. His discovery of the calculus was independent of though later than that of NEWTON, yet it is the Leibnizian form which predominates today. He devised a calculating machine and a symbolic mathematical logic. By theologians he is remembered for his theodicy (analysis of the problem of evil) which, together with his philosophical speculations, helped mold the mind of the Enlightenment.

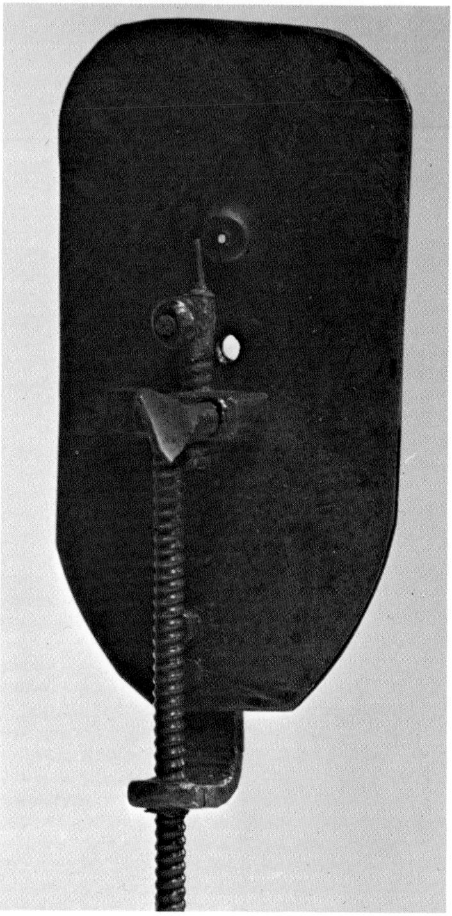

One of the microscopes built and used by Anton van Leeuwenhoek. Only 60mm in height, its optical system comprised a single spherical glass bead held in the small hole in the copper plate, the specimen being mounted on the adjustable vertical peg in front of the hole. Such spherical lenses had very short focal length and their use required considerable skill on the part of the user. This one offered a magnification of ×125.

LEISHMANIASIS, or **Kala-azar,** a chronic tropical DISEASE, particularly of the young, caused by protozoa and carried by sandflies; it causes FEVER, systemic disturbance, ANEMIA, enlargement of the SPLEEN and LIVER and susceptibility to infection. It also causes a chronic SKIN condition (oriental sore) with ulceration and crusting, which may also affect mucous membranes of the mouth, NOSE and PHARYNX. Specific treatment is with antimony compounds.

LELOIR, Luis Federico (1906–), Argentinian biochemist who won the 1970 Nobel Prize for Chemistry for his discovery of the existence and biological significance of the sugar NUCLEOTIDES.

LEMAÎTRE, Georges Édouard (1894–1966), Belgian physicist who first proposed the "big bang" model of the universe, explaining the RED SHIFTS of the galaxies as due to recession (see DOPPLER EFFECT),

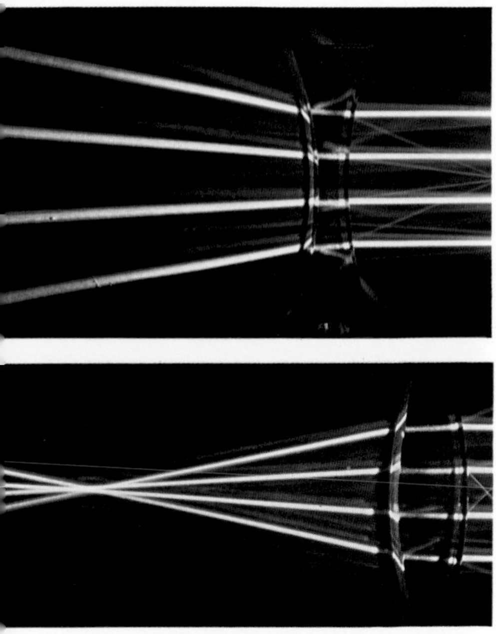

The passage of light rays through a diverging lens (top) and a converging lens (below).

thereby inferring that the universe is expanding. The theory holds that the origins of the universe lie in the explosion of a primeval atom, the "cosmic egg." (See also COSMOLOGY.)

LEMMA, in logic and particularly in mathematics, a subsidiary THEOREM either proved or assumed to be true during the proof of a more major theorem.

LEMNISCATE OF BERNOULLI. See CASSINI OVALS.

LEMONNIER, Pierre Charles (1715–1799), French astronomer best known for his lunar observations and for introducing to France more sophisticated British techniques and instruments.

LENARD, Philipp Eduard Anton (1862–1947), Hungarian-born German physicist awarded the 1905 Nobel Prize for Physics for his investigations of COSMIC RAYS, during which he showed that the ATOM is mainly empty space. He also made pioneering studies of the PHOTOELECTRIC EFFECT, showing that cathode rays are generated thereby.

LENGTH, in GEOMETRY, the first DIMENSION. The length of a straight LINE is given by the difference in the spatial coordinates of its end-points; it is customary to measure this distance along the line. The length of a CURVE is measured along it similarly, coordinates being determined by a moving TRIHEDRON. (For units of length see METRIC UNITS; SI UNITS; WEIGHTS AND MEASURES.)

LENOIR, Jean Joseph Étienne (1822–1900), Belgian-French inventor of the first practical INTERNAL COMBUSTION ENGINE (patented 1860). In 1862 he used it to power an automobile.

LENS, Optical, a piece of transparent material having at least one curved surface and which is used to focus light radiation in CAMERAS, GLASSES, MICROSCOPES, TELESCOPES and other optical instruments. The typical thin lens is formed from a glass disk, though crystalline minerals and molded plastics are also used and, as with spectacle lenses, shapes other than circular are quite common.

The principal axis of a lens is the imaginary perpendicular to its surface at its center. Lenses which are thicker in the middle than at the edges focus a parallel beam of light traveling along the principal axis at the principal focus, a point on the axis on the far side of the lens from the light source. Such lenses are converging lenses. The distance between the principal focus and the center of the lens is known as the focal length of the lens; its focal power is the reciprocal of its focal length and is expressed in dioptres (m^{-1}).

A lens thicker at its edges than in the middle spreads out a parallel beam of light passing through along its principal axis as if it were radiating from a virtual focus one focal length out from the lens center on the same side as the source. Such a lens is a diverging lens.

Lens surfaces may be either inward curving (concave), outward bulging (convex) or flat (plane) and it is the combination of the properties of the two surfaces which determines the focal power of the lens. In general, IMAGES of objects produced using single thin lenses suffer from various defects including spherical and chromatic aberration (see ABERRATION, OPTICAL), coma (in which peripheral images of points are distorted into pear-shaped spots) and astigmatism. The effects of these are minimized by designing compound lenses in which simple lenses of different shapes and refractive indexes (see REFRACTION) are combined. ACHROMATIC LENSES reduce chromatic aberration; aplanatic lenses reduce this and coma, and anastigmatic lenses combat astigmatism. (See also CONTACT LENS; LIGHT.)

LENZ'S LAW, of electromagnetic INDUCTION, states that the ELECTROMOTIVE FORCE (emf) induced in a circuit is such as to oppose the flux change giving rise to it.

LEO (the Lion), a constellation on the ECLIPTIC and fifth sign of the ZODIAC. It contains the bright star REGULUS (apparent magnitude +1.35). Leo gives its name to the annual Leonid METEOR shower.

LEONARDO DA VINCI (1452–1519), Italian Renaissance painter, sculptor, architect, engineer and naturalist. Leonardo's notebooks reveal him a sensitive observer of nature, particularly of anatomy. Although much of his modern scientific acclaim rests on sketches of "inventions" (as of flying machines) which were in his day unrealizable, he indeed possessed considerable insight into the workings of the natural world and was a capable exponent of practical technology.

LEONARDO OF PISA. See FIBONACCI, LEONARDO.

LEPISOSTEUS, or garpike, a North American fish that is a survivor of the fossil group HOLOSTEI.

LEPOSPONDYLA, an extinct order of the AMPHIBIA. They lived during the Carboniferous and Permian periods, 345–225 million years ago, and were small, up to 600mm (2ft) long. Some were snake-like, having no limbs, others displayed bizarre modifications of the skull.

LEPROSY, or **Hansen's disease**, chronic disease caused by a mycobacterium and virtually restricted to tropical zones. It leads to SKIN nodules with loss of

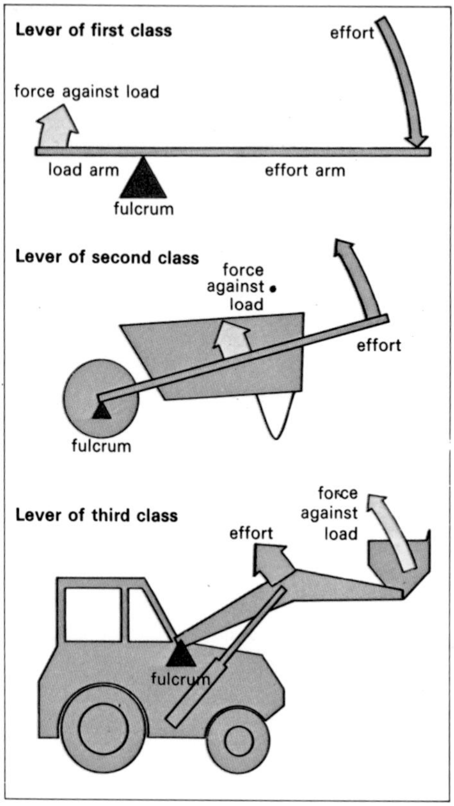

The three classes of lever.

pigmentation, mucous membrane lesions in NOSE and PHARYNX, and NEURITIS with nerve thickening, loss of pain sensation and patchy weakness, often involving FACE and intrinsic HAND muscles. Diagnosis is by demonstrating the organisms in stained scrapings or by skin or nerve BIOPSY. The type of disease caused depends on the number of bacteria encountered and basic resistance to the disease. Treatment is with sulfones (Dapsone).

LEPTON. See SUBATOMIC PARTICLES.

LESBIANISM. See HOMOSEXUALITY.

LEUCIPPUS (5th century BC), Greek philosopher who, according to ARISTOTLE, originated the theory that matter is made up of indivisible, infinitely small atoms (see ATOMISM). His pupil DEMOCRITUS further developed the theory.

LEUKEMIA, malignant proliferation of white blood cells in BLOOD or BONE MARROW. It may be divided into acute and chronic forms for both granulocytes and lymphocytes. In acute forms, primitive cells predominate and progression is rapid with ANEMIA, bruising and infection. Acute lymphocytic leukemia is commonest in young children. Chronic forms present in adult life with mild systemic symptoms, susceptibility to infection and enlarged LYMPH nodes (lymphatic) or SPLEEN and LIVER (granulocytic). Cancer CHEMOTHERAPY and ANTIBIOTICS have greatly improved survival prospects.

LEUKOCYTES. See BLOOD.

LEVEE. See FLOODS AND FLOOD CONTROL.

LEVEL, or **spirit level**, device used to determine whether a surface is level or not. Usually it is a glass tube, curved upward at the center, almost filled with alcohol or ether, leaving a bubble of vapor. This bubble floats to the highest point of the tube which is, when the level is on a perfectly horizontal surface, the center.

LEVER, the simplest MACHINE, a rigid beam pivoted at a *fulcrum* so that an *effort* acting at one point of the beam may be used to shift a *load* acting at another point on the beam. There are three classes of lever: those with the fulcrum between the effort and the load; those with the load between the fulcrum and the effort, and those with the effort between the fulcrum and the load. The part of the beam between the load and the fulcrum is the load arm; that between the effort and the fulcrum, the effort arm. The effort multiplied by the length of the effort arm equals the load multiplied by the length of the load arm: a load of 50kg, 5m from the fulcrum, may be moved by any effort 10m from the fulcrum greater than 25kg (the longer the effort arm, the less effort required). Load divided by effort gives the mechanical advantage; in this case 2. A first-class lever (e.g., a crowbar) has a mechanical advantage greater, less than or equal to 1; a second-class (e.g., a wheelbarrow), always more than 1; a third-class (e.g., the human arm), always less than 1. (See also ARCHIMEDES; MECHANICS; MOMENT.)

LEVERRIER, Urbain Jean Joseph (1811–1877), French astronomer whose calculations enabled **Johann Gottfried Galle** (1812–1910) to discover the planet NEPTUNE (1846). (See also ADAMS, J. C.)

LÉVI-STRAUSS, Claude (1908–), Belgian-born French social anthropologist, best known for his advocacy of *structuralism*, an analytical system whereby different cultural patterns may be related such that the universal logical substructure underlying them may be elicited.

LEWIN, Kurt (1890–1947), Prussian-born US psychologist, an early member of the GESTALT PSYCHOLOGY school, best known for his development of the concept of GROUP dynamics, especially field theory.

LEWIS, Gilbert Newton (1875–1946), US chemist who suggested that covalent bonding consisted of the sharing of valence-electron pairs. His theory of ACIDS and bases involved seeing acids (Lewis acids) as substances which are able to accept electron pairs from bases which are electron-pair donating species (Lewis bases). In 1933, Lewis became the first to prepare HEAVY WATER (D_2O).

LEWIS, Isaac Newton (1859–1931), US inventor of the Lewis MACHINE GUN (patented 1911), which could fire 600 rounds per minute and had the advantage of being light. Because of its low recoil it was much used as an airplane armament.

LEYDEN JAR, the simplest and earliest form of CAPACITOR, a device for storing electric charge. It comprises a glass jar coated inside and outside with unconnected metal foils, and a conducting rod which passes through the jar's insulated stopper to connect with the inner foil. The jar is usually charged from an ELECTROSTATIC GENERATOR. The device is now little used outside the classroom.

LIBBY, Willard Frank (1908–), US chemist awarded the 1960 Nobel Prize for Chemistry for discovering the technique of RADIOCARBON DATING (1947).

LIBIDO, originally and still popularly, the sexuality or general SEX drive of the individual. In PSYCHOANALYSIS, following FREUD, the libido, with its source in the ID, is a type of mental energy (though it may, as in sexuality, generate physiological energy or activity) responsible for all human constructive action.

LIBRA (the Scales), an average size constellation on the ECLIPTIC, the seventh sign of the ZODIAC.

LIBRATION. See MOON.

LICHEN, name given to plants that are in fact an association between FUNGI and ALGAE. The fungus prevents the alga from drying-out, while the alga probably provides assistance to the fungus in mineral absorption. This relationship is a form of SYMBIOSIS. Lichens occur on the bark of trees, rotting wood, rock and soil. They are particularly important because they are primary colonizers of bare rock.

LIEBIG, Baron Justus von (1803–1873), German chemist who with WÖHLER proposed the radical theory of organic structure. This suggested that groups of atoms such as the benzoyl radical (C_6H_5CO–), now known as the benzoyl group (see BENZOIC ACID), remained unchanged in many chemical reactions. He also developed methods for organic quantitative analysis and was one of the first to propose the use of mineral fertilizers for feeding plants. (See also MIRROR.)

LIE DETECTOR, or **polygraph,** device which gives an indication of whether or not an individual is lying. Though much used in criminal investigation, its results are not admissible as legal evidence. Its use is based on the assumption that lying produces emotional, and hence physiological (see EMOTION), reactions in the individual. It usually measures changes in BLOOD pressure, PULSE rate and RESPIRATION; sometimes also muscular movements and PERSPIRATION. Success varies with the individual.

LIFE, the property whereby things live. Despite the vast knowledge that has been gained about life and the forms of life, the term still lacks any generally accepted definition. Indeed, biologists tend to define it in terms that apply only to their own specialisms. Physiologists regard as living any system capable of eating, metabolizing, excreting, breathing, moving, growing, reproducing and able to respond to external stimuli. Metabolically, life is a property of any object which is surrounded by a definite boundary and capable of exchanging materials with its surroundings. Biochemically, life subsists in cellular systems containing both NUCLEIC ACIDS and PROTEINS. For the geneticist, life belongs to systems able to perform complex transformations of organic molecules and to construct from raw materials copies of themselves which are more or less identical, although in the long term capable of EVOLUTION by natural selection. In terms of THERMODYNAMICS, it has been said that life is exhibited by localized regions where net order is increasing (or net ENTROPY decreasing). But the scientist has no monopoly over the use of the term, and for poets, philosophers and artists, it carries another myriad significations.

Life on Earth is manifest in an incredible variety of forms—over 1 million species of animals and 350 000 species of plants. Yet, despite superficial differences, all organisms are closely related. The form and matter of all life on earth is essentially identical, and this implies that all living organisms shared a common ancestor and that life on earth has originated only once.

LIFE CYCLE, the series of stages through which an individual organism passes in its progression through life. It may be simple, as in VERTEBRATES—from the union of the GAMETES in FERTILIZATION to the DEATH of the organism—or rather more complex, as in types exhibiting ALTERNATION OF GENERATIONS.

LIFT, in AERODYNAMICS, the vertical force acting on an aircraft due to the flow of air over its airfoils. See also ELEVATOR.

LIGAMENT, specialized fibrous thickening of a JOINT capsule, providing tensile strength against forces tending to force the joint beyond its normal range. Sudden distraction or twisting forces may cause ligamentous strain or tears (sprain). External joint support encourages its healing.

LIGAND, an ION or molecule linked to a central metal ion by a coordinate bond (see BOND, CHEMICAL) to form a so-called **complex compound**. Almost any ion or molecule that can act as a BASE, having an atom able to donate an electron-pair, may act as a ligand—common examples include NH_3, H_2O, Cl^-, OH^-, SO_4^{2-}, CO, NO^+, H^-, $C_5H_5^-$, CH_3COO^-. The complex formed may be cationic, uncharged or anionic. The *coordination number* of the central ion in the complex is the number of ligand-to-ion bonds; this equals the number of ligands unless they are polydentate—having more than one donating atom—when they may occupy more than one coordination site forming a CHELATE complex. Coordination numbers of 2 to 10 are known, but 6 (octahedral) and 4 (tetrahedral or square planar) are commonest. Many complexes with more than one kind of ligand have STEREOISOMERS. Complexes vary greatly in their lab-

The complex ion formed when the hexadentate ligand EDTA (ethylenediaminetetraacetic acid) chelates a calcium (Ca^{2+}) ion.

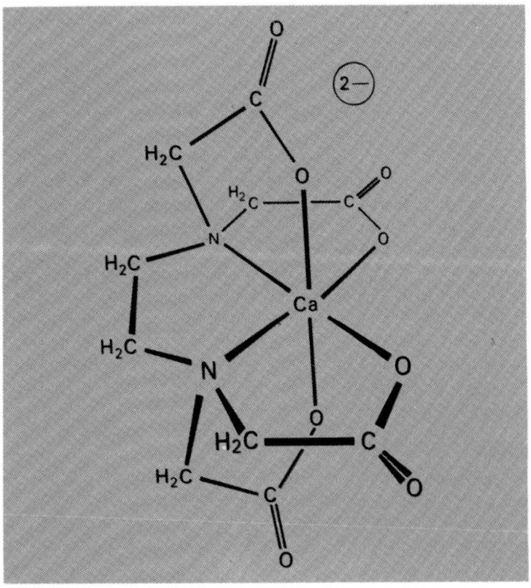

ility, i.e., the rapidity with which the ligands are replaced by others: they are described as labile or inert. The bonding in complexes has been described by several theories: crystal field theory considers the effect that the electrostatic field due to the ligands has on the energies of the central ion *d*-ORBITALS; ligand field theory includes the mixing of ligand and ion orbitals.

LIGHT, ELECTROMAGNETIC RADIATION to which the human EYE is sensitive. Light radiations occupy the small portion of the electromagnetic SPECTRUM lying between wavelengths 400nm and 770nm. The eye recognizes light of different wavelengths as being of different COLORS, the shorter wavelengths forming the blue end of the (visible) spectrum, the longer the red. The term light is also applied to radiations of wavelengths just outside the visible spectrum, those of energies greater than that of visible light being called ultraviolet light, those of lower energies, infrared. (See ULTRAVIOLET RADIATION; INFRARED RADIATION.) White light is a mixture of radiations from all parts of the visible spectrum, typified by the BLACKBODY RADIATION reaching the earth from the sun. Bodies which do not themselves emit light are seen by the light they reflect or transmit. In passing through a body or on reflection from its surface, particular wavelengths may be abstracted from white light, the body consequently displaying the colors which remain. Objects which reflect no visible light at all appear black.

For many years the nature of light aroused controversy among physicists. Although HUYGENS had demonstrated that REFLECTION and REFRACTION could be explained in terms of waves—a disturbance in the medium—NEWTON preferred to think of light as composed of material corpuscles (particles). YOUNG's INTERFERENCE experiments reestablished the wave hypothesis and FRESNEL gave it a rigorous mathematical basis. At the beginning of the 20th century, the nature of light was again debated as PLANCK and EINSTEIN proposed explanations of blackbody radiation and the PHOTOELECTRIC EFFECT respectively which assumed that light carried ENERGY in discrete quanta (see PHOTON). Today physicists explain optical phenomena in terms either of waves (reflection, refraction, DIFFRACTION, interference, polarization—see POLARIZED LIGHT—and SCATTERING) or quanta (blackbody radiation, photoelectric emission and the interaction of light with substantial MATTER) as is most convenient in each case (see also WAVE MOTION; QUANTUM THEORY).

Light from the sun is the principal source of energy on earth, being absorbed by plants in PHOTOSYNTHESIS. Many other chemical reactions involve light (see CHEMILUMINESCENCE; PHOTOCHEMISTRY; PHOTOGRAPHY) though few artificial light sources are chemical in nature (but see FLASHBULB). Most light sources employ radiation emitted from bodies which have become hot or have been otherwise energetically excited (see ENERGY LEVEL; LASER; LIGHTING; LUMINESCENCE). Light can be converted into electricity using the PHOTOELECTRIC CELL. Light used for illumination is the subject of the science of PHOTOMETRY. (See also OPTICS.)

LIGHTING, Artificial, the illumination of sectors of man's physical environment in the absence of natural LIGHT. In the course of EVOLUTION, EYES sensitive to the solar radiation penetrating the earth's atmosphere to the surface developed in many of the planet's animals. Man's eyes are thus sensitive to light of these same wavelengths, so artificial light sources must be designed to produce radiations having an intensity SPECTRUM similar to that of natural sunlight.

Oil lamps, brushwood torches and candles formed man's earliest means of artificial lighting, developments leading towards the KEROSENE lamp of the late 19th century (see also ARGAND BURNER). Gas lighting dates from 1792 when the British engineer William Murdock used coal gas to light his Cornish home. The modern portable camping lamp burns BUTANE gas to heat an incandescent mantle.

In the 20th century the industrialized nations have come to use ELECTRICITY for most lighting purposes because it offers an instant source of bright, clean, fume-free light. One of the earliest electrical lighting sources was the ARC LAMP which utilizes the flame arcing between two pointed carbon electrodes maintained with a moderate voltage between them. Successful incandescent filament lamps date from 1879 when Sir Joseph William Swan and EDISON demonstrated lamps in which a carbon filament enclosed in an evacuated glass bulb was heated electrically until it glowed. After 1913 these gave way to lamps having tungsten filaments, coiled to improve efficiency (from 1918), and filled with an unreactive gas such as nitrogen. In 1937 efficiency was further improved by coiling the coiled filament (coiled-coil lamp). A more recent development is the tungsten-halogen lamp (an early type of which was the quartz-iodine lamp) in which efficiency is improved and life extended by filling the bulb with a HALOGEN with which tungsten evaporating from the filament can combine, preventing deposition of the metal on the envelope (which is sometimes made of quartz). Because discharge lamps (in which a glow discharge is set up in mercury or sodium vapor—glowing blue-green and yellow respectively) do not produce light in all parts of the solar spectrum they find their greatest use in highway rather than domestic lighting. More recent high-pressure sodium lamps, however, offer a fuller light spectrum. Cold-discharge tubes containing neon (glowing red) or argon (glowing blue) are contorted into exotic shapes for use in advertising signs. Fluorescent lamps produce light similar to sunlight by using a PHOSPHOR coating on the inside of the tube to convert ultraviolet light produced in a mercury-vapor discharge. Although they require more complex circuitry than filament lamps, they are much more efficient and last longer. Other light sources such as light-emitting diodes (LEDs—see SEMICONDUCTORS) and electroluminescent panels find use in instrument display panels. (See also FIREWORKS; FLASHBULB; PHOTOMETRY.)

LIGHT METER, a device for measuring LIGHT levels, particularly in PHOTOGRAPHY where they are often coupled directly to the exposure controls of a CAMERA. Most light meters employ either PHOTOVOLTAIC CELLS (e.g., SELENIUM type) or PHOTOCONDUCTIVE DETECTORS (e.g., CADMIUM sulfide—"CdS"—type).

LIGHTNING, a discharge of atmospheric electricity resulting in a flash of light in the sky. Most occur between two parts of a single cloud, some between cloud and ground, and a few between one cloud and

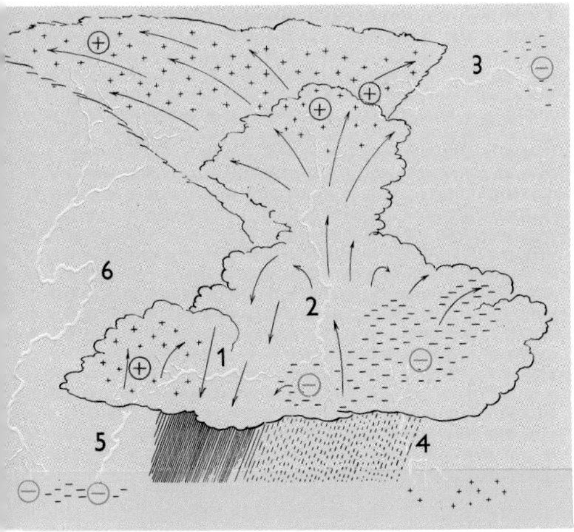

Lightning occurs when electrical charge separation occurs in thunderstorms, creating some regions with overall positive charge (+) and others with predominantly negative charge (−). Six types of lightning are indicated on the diagram, numbered (1), the most frequent, through (6), the least common. Air currents in the cloud are indicated by black arrows.

another. Flashes range from a few km to about 150km in length, and typically have an energy of around 300kWh and an electromotive force around 100MV.

Cloud-to-ground lightning usually appears forked. A relatively faint light moves towards the ground at about 125km/s in steps, often branching or forking. As this first pulse (leader stroke) nears the ground, electrical discharges (streamers) arise from terrestrial objects; where a streamer meets the leader stroke a brilliant, high-current flash (return stroke) travels up along the ionized (see ION) path created by the leader stroke at about 100Mm/s (nearly $\frac{1}{3}$ the speed of light). Several exchanges along this same path may occur. If strong wind moves the ionized path, **ribbon lightning** results.

Sheet lightning occurs when a cloud either is illuminated from within or reflects a flash from outside, in the latter case often being called **heat lightning** (often seen on the horizon at the end of a hot day). **Ball lightning**, a small luminous ball near the ground, often vanishing with an explosion, and **bead lightning**, the appearance of luminous "beads" along the channel of a stroke, are rare.

Lightning results from a buildup of opposed electric charges in, usually, a cumulonimbus CLOUD, negative near the ground and positive on high (see ELECTRICITY). There are several theories which purport to explain this buildup. Understanding lightning might help us probe the very roots of life, for lightning was probably significant in the formation of those organic chemicals that were to be the building blocks of life. (See also SAINT ELMO'S FIRE; THUNDER.)

LIGHTNING ROD, or **lightning conductor**, safety device on buildings, etc., as protection from the destructive effects of LIGHTNING. One end of a strip of conducting metal (e.g., COPPER) is earthed, the other mounted high on the building; any discharge thus passes directly to the ground.

LIGHT YEAR, in astronomy, a unit of distance equal to the distance traveled by light in a vacuum in one sidereal year, equal to 9461Tm (about 6 million million miles). The unit has largely been replaced by the PARSEC (1 ly = 0.3069pc).

LIGNIN, complex CARBOHYDRATE which gives strength and rigidity to the woody tissue of plants and which may account for from 25% to 30% of the WOOD of some trees. In making PAPER the lignin must be separated from the CELLULOSE.

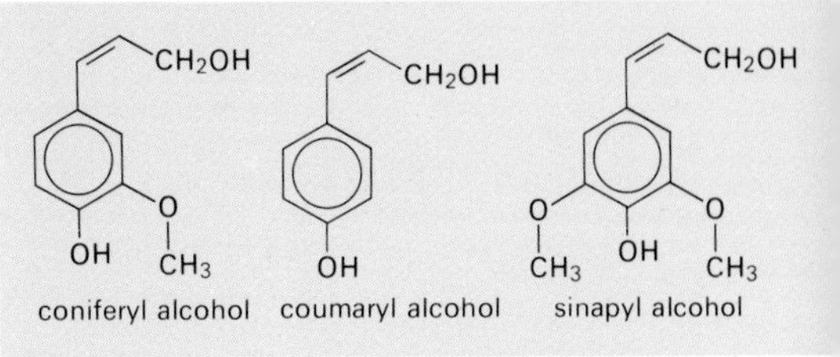

Lignin is a complex polymer formed by condensation of precursors such as coniferyl, coumaryl and sinapyl alcohols.

LIGNITE, or brown coal. See COAL.

LILIENTHAL, Otto (1848–1896), German pioneer of aeronautics, credited with being the first to use curved, rather than flat, wings, as well as first to discover several other principles of AERODYNAMICS. He made over 2000 glider flights, dying from injuries received when one of his gliders crashed. (See also FLIGHT, HISTORY OF.)

LIMAÇON, the GRAPH, in polar coordinates (see ANALYTIC GEOMETRY) of the FUNCTION $r = a\ \cos\theta + b$ where a and b are constants. (See also TRIGONOMETRY.) The light reflected by the inner walls of a cup delineates a limaçon upon the surface of the liquid within.

LIME, or calcium oxide or hydroxide. See CALCIUM.

LIMESTONE, sedimentary rock consisting mainly of

A limaçon.

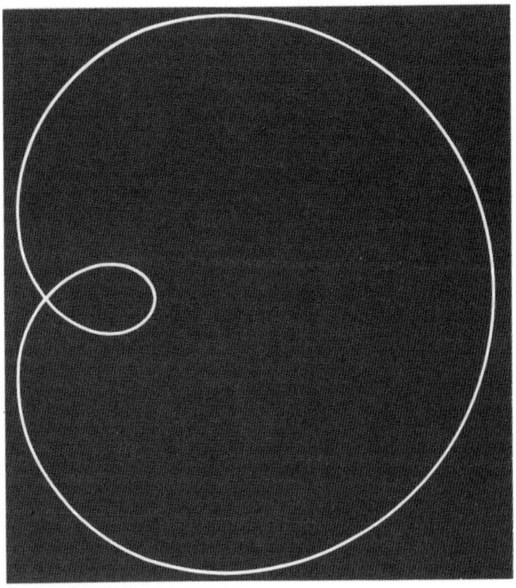

calcium carbonate (see CALCIUM), in the forms of CALCITE and aragonite. Some limestones, such as CHALK, are soft but others are hard enough for use in building. Limestone may be formed inorganically (oolites) by evaporation of seawater or freshwater containing calcium carbonate, or organically from the shells of mollusks or skeletons of coral piled up on sea beds and compressed. In such limestone fossils usually abound.

LIMITS. The limit of a SEQUENCE is a fixed number towards which the terms tend. This number may or may not be a member of the sequence. Similarly, the limit of a FUNCTION $y=f(x)$ is the value which $f(x)$ approaches as x tends to a particular value. The limit may or may not be a value of $f(x)$ (see ASYMPTOTE). The limit point of a point set (or CURVE) is a POINT such that, no matter how small a distance from it is chosen, there is a member of the set closer to it. Limit points may or may not be members of the set.

LIMNOLOGY, a branch of BIOLOGY that deals with the study of freshwater habitats and the PLANTS and ANIMALS within these habitats.

LIMONITE, a dark brown, amorphous OXIDE mineral consisting of hydrated iron(III) oxide ($FeO[OH].nH_2O$). A major IRON ore of widespread occurrence, often with GOETHITE, it is formed by alteration of other iron minerals.

LINE, a CURVE. In general the term is used to denote a straight line; i.e., one whose equation is of the type $y=ax+b$, where a and b are constants (see ANALYTIC GEOMETRY). A straight line has only one DIMENSION, length, may be infinite in extent, and is the shortest distance between any two points. (See also LINEAR RELATIONSHIP.)

LINEAR ACCELERATOR. See ACCELERATORS, PARTICLE.

LINEAR MOTOR, a form of induction motor first built by WHEATSTONE (1845). It comprises a series of ELECTROMAGNETS, current being passed through each along the line in turn, so that a moving component (usually of ALUMINUM) is drawn along the line. (See also INDUCTION MOTOR.)

LINEAR PROGRAMMING, an application of ALGEBRA and CALCULUS to, usually, commercial problems involving a number of unrelated VARIABLES, in order to provide an optimum result. Each of the variables is expressed in the form of a line FUNCTION within specified LIMITS or constraints. The overall most favorable result is then found, usually by COMPUTER. (See also OPERATIONS RESEARCH.)

LINEAR RELATIONSHIP, a relationship between two VARIABLES such that, if one is plotted as a FUNCTION of the other (see ANALYTIC GEOMETRY), the result is a straight LINE. A linear relationship can be said to exist between x and y in the EQUATION $y=ax+b$ (a and b CONSTANTS).

LINE INTEGRAL, the integral (see CALCULUS) along a directed CURVE of any FUNCTION that is continuous and single-valued along that curve.

LINEN, yarn and fabric manufactured from the fibers of the FLAX plant. The stems of the flax plant must first be softened by soaking in water (retting). Next, the fibers are separated from the woody core in a "scutching" mill. The short fibers (tow) are combed out from the long fibers (line) in the "hackling" mills. The tow is finally spun into yarn.

LINGUISTICS, the scientific study of spoken language. Different branches of linguistics deal with the development of languages over time; the structures of languages (see GRAMMAR); the description of different languages at a particular time; the physical and mental factors involved in the production of speech and use of language, and the study of meaning (see SEMANTICS). *Applied linguistcs* makes use of the discoveries of other areas of linguistics to deal with practical problems, in particular language teaching and the deduction of information about a culture, past or present, from the language that it uses (see ANTHROPOLOGY; PHILOLOGY). (See also ETYMOLOGY; LANGUAGE; PHONETICS; SIGN LANGUAGE; WRITING, HISTORY OF.)

LINKAGE, the occurrence together on the same CHROMOSOME of two or more GENES. Linked genes are normally transmitted together from generation to generation. These genes are said to be in *coupling*. Examples of genes linked to the male sex chromosome include red-green COLOR BLINDNESS and HEMOPHILIA.

LINNAEUS, Carolus (1707–1778), later **Carl von Linné**, Swedish botanist and physician, the father of TAXONOMY, who brought system to the naming of living things. His classification of plants was based on their sexual organs (he was the first to use the symbols ♂ and ♀ in their modern sense), an artificiality dropped by later workers; but many of his principles and taxonomic names are still used today.

LINOLEUM, a durable floor-covering material. Linoleum cement (oxidized linseed oil with various RESINS) is mixed with fillers (e.g., wood flour) and coloring elements, the whole having a foundation of, usually, burlap (hessian).

LINOTYPE, technique of letterpress printing in which characters are cast from molten type-metal (see ALLOY) a line at a time. Keying by the operator assembles the matrices (molds) for the characters in the correct order, wedge-shaped space bands being placed between the words: when the line is nearly full, pressure is applied to these so that they force apart the words and justify the line (align the right-hand margin). The slug of type is cast from the completed line. If a single character is wrong, the whole line must be recast. (See also MONOTYPE.)

LINSEED OIL, mixture of fatty acids (see

The graph of $y=ax+b$, the prototypical linear relationship.

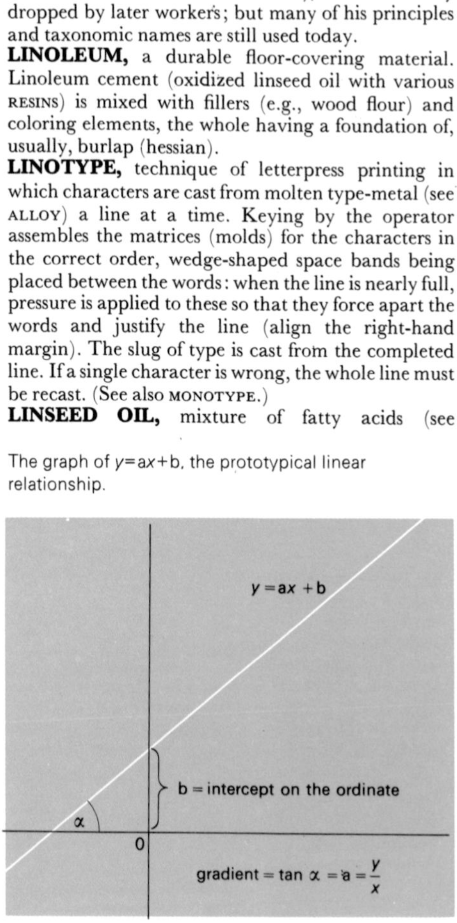

CARBOXYLIC ACIDS) extracted from the seeds of a variety of FLAX. It is used in varnish and oil paints because of the quick-drying durable finish produced. It is also used to manufacture oil cloth, linoleum and printing ink. Leading producers of linseed oil are the US, USSR and Argentina.

LINTON, Ralph (1893–1953), US anthropologist best known for the eclecticism of his studies in cultural ANTHROPOLOGY, as expressed in *The Study of Man* (1936) and *The Tree of Culture* (1955).

LINZ-DONAWITZ PROCESS, or Basic Oxygen Process, major variant of the BESSEMER PROCESS for making STEEL, developed in Linz and Donawitz, Austria (1952) and popular in North America for its low cost. The basic-lined converter has no tuyères; a stream of pure oxygen is blown through a lance onto the surface of the molten pig iron (low in phosphorus) mixed with scrap.

LIPASE, an enzyme which splits fats into CARBOXYLIC ACIDS and GLYCEROL. As the fat is normally water insoluble, lipases act relatively slowly on the surface of a fat globule. They occur widely in oil seeds and are involved in GERMINATION.

LIPIDS, a diverse group of organic compounds found in plants, animals and microorganisms and characterized by their solubility in nonpolar organic solvents such as ETHER, CHLOROFORM and ETHANOL. Lipids include many heterogeneous substances and unlike PROTEINS and CARBOHYDRATES have no characteristic type of building block. They are classified into FATS, phospholipids, WAXES, STEROIDS, TERPENES and other types, according to their products on HYDROLYSIS. Most commonly, these are fatty acids, particularly those having an even number of carbon atoms. Other components found in many lipids include GLYCEROL, choline and derivatives of ISOPRENE. Phospholipids contain PHOSPHORUS, most commonly in the form of phosphoric acid. They are found especially in brain and nervous tissue as cephalins and in egg yolk as lecithin. Phospholipids are good emulsifiers and detergents, hence the use of egg yolk in mayonnaise. Waxes are ESTERS of fatty acids with alcohols other than glycerol and include whale oil and beeswax. Terpenes are a wide range of compounds having usually distinctive fragrances obtained from plants such as pine trees (TURPENTINE), citrus fruits (limonene) and geraniums. They have a wide application in cosmetics, toiletries and medicines.

LIPMANN, Fritz Albert (1899–   ), German-born US biochemist who shared with KREBS the 1953 Nobel Prize for Physiology or Medicine for his discovery in 1947 of coenzyme A (see ENZYMES).

LIPPMANN, Gabriel (1845–1921), French physicist awarded the 1908 Nobel Prize for Physics for inventing (c1891) the first system of color PHOTOGRAPHY. His process required long exposures, and thus is now obsolete. He also invented the coelostat, and predicted PIEZOELECTRICITY.

LIPSCOMB, William Nunn, Jr (1919–   ), US physical chemist who won the 1976 Nobel Prize for Chemistry for his work on structure and bonding in BORANES. He proposed that these might contain three-center (hydrogen-bridge) bonds.

LIQUEFIED PETROLEUM GAS (LPG). See BOTTLED GAS.

LIQUID, one of matter's three states, the others being SOLID and GAS. Liquids take the shape of their container, but have a fixed volume at a particular temperature and are virtually incompressible (see COHESION). Nearly all substances adopt the liquid state under suitable conditions of temperature and pressure. (See also FLUID; KINETIC THEORY; VAPOR.)

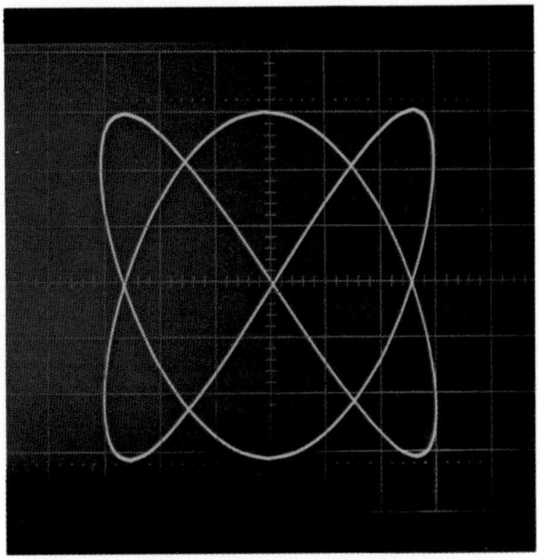

Lissajous' figure displayed on the screen of an oscilloscope.

LISSAJOUS' FIGURES, or **Bowditch curves,** plane CURVES traced by a POINT moving in two SIMPLE HARMONIC MOTIONS that are at right angles to each other. They can most easily be formed by supplying different alternating voltages (see ELECTRICITY) to the *x*- and *y*-deflection plates of an OSCILLOSCOPE. Only if the frequencies are commensurable will a true Lissajous figure (i.e., a closed curve) be formed.

LISTER, Joseph Lister, 1st Baron (1827–1912), British surgeon who pioneered antiseptic SURGERY, perhaps the greatest single advance in modern medicine. PASTEUR had shown that microscopic organisms are responsible for PUTREFACTION, but his STERILIZATION techniques were unsuitable for surgical use. Lister experimented and, by 1865, succeeded by using carbolic acid (see PHENOL).

LITHARGE, or lead(II) oxide. See LEAD.

LITHIUM (Li), a white metallic element somewhat harder and less reactive than the other ALKALI METALS. Physically and chemically, lithium also resembles the ALKALINE EARTH METALS. It is the lightest element which is a solid at room temperature. It is made by ELECTROLYSIS of fused lithium chloride. Lithium metal is used in heat transfer because of its high specific heat; the isotope Li^6 is important in thermonuclear processes. Lithium stearate is an additive to lubricating greases. AW 6.9, mp 180°C, bp 1347°C, sg 0.534 (20°C).

LITHOGRAPHY. See PRINTING.

LITHOSPHERE, the rocks of the earth, as contrasted with the ATMOSPHERE and HYDROSPHERE. Today, use of the term is often restricted to reference to the uppermost 100 km depth of the substance of the earth. (See EARTH.)

LITMUS, mixture of colored compounds, extracted from LICHENS, used as an acid-base INDICATOR.

LITRE (l), a metric unit of volume, originally defined

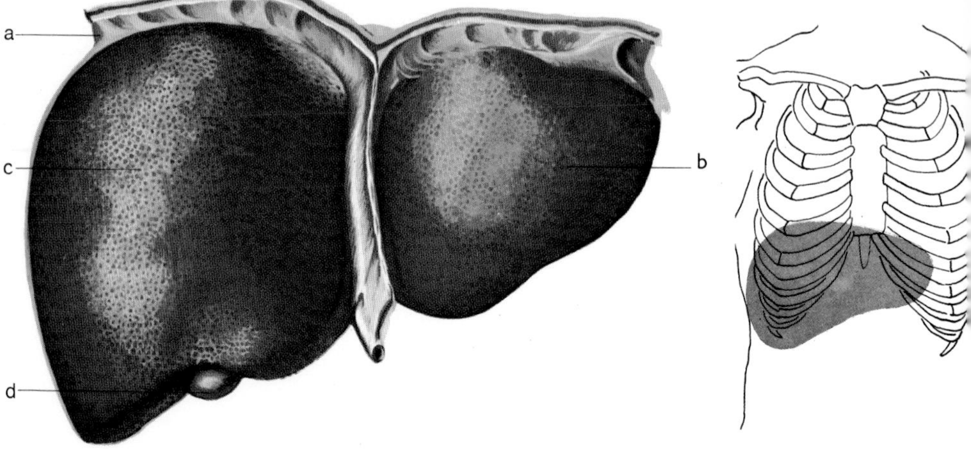

The liver, seen here from the front, lies just below the diaphragm (a) at the top of the abdominal cavity. On the lower surface of the liver, between the left and right lobes (b and c), is the gall bladder (d). This acts as a reservoir for the bile produced by the liver.

as that of 1kg of water at the temperature of its maximum density (= 1.000 028dm³), but redefined in 1964 as exactly equal to one cubic decimetre (= 1dm³). The litre is not recommended for use alongside SI UNITS.

LITTLE DIPPER (Ursa Minor, the Little Bear), N Hemisphere circumpolar constellation containing POLARIS, the N polestar.

LITTORAL FAUNA, term applied to animals which live on the seashore (the littoral or intertidal zone). They comprise a special assemblage which are uniquely adapted to being alternately covered and uncovered by the tide. On any one shore the organisms are "zoned," changing from animals adapted to aquatic conditions in the lower regions, to often closely related semiterrestrial species higher up.

LIVER, the large organ lying on the right of the ABDOMEN beneath the DIAPHRAGM and concerned with many aspects of METABOLISM. It consists of a homogeneous mass of cells arranged round blood vessels and bile ducts. Nutrients absorbed in the GASTROINTESTINAL TRACT pass via the portal VEINS to the liver and many are taken up by it; they are converted into forms (e.g., GLYCOGEN) suitable for storage and release when required. PROTEINS, including ENZYMES, PLASMA proteins and CLOTTING factors, are synthesized from amino acids. The liver converts protein breakdown-products into urea and detoxifies or excretes other substances (including drugs) in the blood. Bilirubin, the HEMOGLOBIN break-down product is excreted in the BILE; this also contains bile salts, made in the liver from CHOLESTEROL and needed for the DIGESTIVE SYSTEM.

Diseases of the liver include CIRRHOSIS and HEPATITIS, while abnormal function is manifested as JAUNDICE, EDEMA, ascites (excessive peritoneal fluid), and a variety of BRAIN and NERVOUS SYSTEM disturbances including DELIRIUM and COMA. Chronic liver disease leads to SKIN abnormalities, a bleeding tendency and alterations in routes of BLOOD CIRCULATION, which may in turn lead to HEMORRHAGE. Hepatitis may be caused by VIRUSES (e.g., infectious and serum hepatitis); their high infectivity has made them a hazard in hospital dialysis units. Many drugs may damage the liver, causing disease similar to hepatitis, and both drugs and severe hepatitis can cause acute liver failure.

LIVERWORTS, primitive prostrate plants growing in moist habitats and occasionally in water. They form the class Hepaticae of the division BRYOPHYTA, which also includes the MOSSES and HORNWORTS. The most familiar types have a flat, lobed thallus attached to the ground by root-like rhizoids. Other types have a thallus bearing leaf-like expansions.

LIZARDS. See LACERTILIA.

LOAM, soil comprised of about 30%–50% SAND particles, 30%–50% SILT particles, and less than 20% CLAY particles. (See SOIL.)

LOBACHEVSKI, Nikolai Ivanovich (1792–1856), Russian mathematician who, independently of BOLYAI, developed the first NON-EUCLIDEAN GEOMETRY, hyperbolic or LOBACHEVSKIAN GEOMETRY, publishing his developments from 1826 onward.

LOBACHEVSKIAN GEOMETRY, or **hyperbolic geometry**, the branch of NON-EUCLIDEAN GEOMETRY based on the hypothesis that for any point P not lying on a line L, there are at least two lines that can be drawn through P parallel to L.

LOBOTOMY, operation in which the FRONTAL LOBES are separated from the rest of the BRAIN, used in the past as treatment for refractory DEPRESSION. It leads to a characteristically disinhibited type of behavior and is now rarely used.

LOCKE, John (1632–1704), English empiricist philosopher whose writings helped initiate the European Enlightenment. His *Essay Concerning Human Understanding* (1690) is one of the highlights of English philosophy. In it he adopted a nominalist view of language, yet believed that our ideas of things, inasmuch as they reflected real essences—the properties of the insensible corpuscles of matter—were founded upon and could be checked against experience. His *A Letter Concerning Toleration* (1689) and *The Reasonableness of Christianity* (1695) were seminal in 18th-century British religious thought.

LOCKJAW. See TETANUS.

LOCKS AND KEYS. The earliest known mechanical lock is from ancient Egypt, c2000 BC. The bolt was hollow, with a number of holes bored in its top; one of the bolt staples held a number of wooden pegs which fell into the holes in the bolt, holding it in place. The key could be fitted into the bolt; it had spikes in the same pattern as the holes, and thus could lift the pegs clear. The ancient Greeks situated their locks on the inside of the door, access being achieved *via* a keyhole to whose shape the key conformed. The Romans improved the Egyptian design by having pegs of different shapes and using springs to drive the pegs home; and invented the **warded lock**, whose key must be slotted to clear wards, obstacles projecting from the back of the lock. Early portable locks, and later padlocks, also used this principle. The modern *lever-tumbler lock* was invented by Robert Barron (1778): levers fit into a slot in the bolt patterned such that each lever must be raised a different distance by the key to free the bolt. Jeremiah Chubb added another lever to jam the lock if the wrong key were tried (1818). The *Bramah lock*, invented by Joseph BRAMAH (1784), has a cylindrical key slotted to push down sprung slides, each of which must be depressed a different distance to clear an obstacle. Most domestic locks are now *Yale locks*, invented by Linus YALE (1861). An inner cylindrical plug has holes into which sprung drivers press pins of different lengths. The key is patterned to raise each pin so that its top is flush with the cylinder, which can then turn. Modern safes have combination locks and time devices so that they can only be opened at certain times.

LOCOMOTION, the means by which animals move from point to point—crawling or running over hard surfaces; burrowing in sand or soil; flying and swimming. Animals which can move are termed *motile*, contrasted with those which cannot, which are *sessile*. There are two anatomical features of the vast majority of animals that make locomotion possible: a skeletal system, and a muscle system. The skeletal system is frequently composed of chitin (see CARAPACE), CARTILAGE or BONE and provides mechanical levers which are operated by MUSCLES. Soft-bodied animals employ a hydrostatic skeleton composed of water-filled cavities that are distorted by muscular walls to produce movement (see also AMOEBA).

Among VERTEBRATES there are many variations of the basic locomotor organ, the limb. In birds and bats it has been modified to form a wing, while in various groups, notably snakes, the limbs are lost and the animal moves by undulations of the body. In aquatic animals, the tail is the most important organ of locomotion, the main function of the fins being steering and stabilization.

LOCOMOTIVE, originally locomotive engine, power unit used to haul railroad trains. The earliest development of the railroad locomotive took place in the UK, where R. TREVITHICK built his first engine c1804. R. STEPHENSON's famous *Rocket* of 1829 proved that locomotive engines were far superior to stationary ones and provided a design that was archetypal for the remainder of the steam era. Locomotives were first built in the US c1830. These pioneered many new design features including the leading truck, a set of wheels preceding the main driving wheels, guiding the locomotives over the usually lightly-constructed American tracks. For most of the rest of the 19th century, locomotives of the "American" type (4-4-0) were standard on US passenger trains, though towards the end of the century, progressively larger types came to be built. Although electric locomotives have been in service in the US since 1895, the high capital cost of converting tracks to electric transmission have prevented their widespread adoption. Since the 1950s, however, most US locomotives have been built with DIESEL ENGINES. Usually the axles are driven by electric motors mounted on the trucks, the main diesel engine driving a generator which supplies power to the motors (diesel-electric transmission). Elsewhere in the world, particularly in Europe, much greater use is made of electric traction, the locomotives usually collecting power from overhead cables via a PANTOGRAPH. Although some GAS-TURBINE locomotives are in service in the US, this and other novel power sources do not seem to be making much headway at present.

LODESTONE. See MAGNETITE.

LODGE, Sir Oliver Joseph (1851–1940), British physicist best known for his work on the propagation of ELECTROMAGNETIC RADIATION, devising an early instrument (the coherer) for detecting it. He also did important work on PARAPSYCHOLOGY.

LOEB, Jacques (1859–1924), German-born US biologist best known for his work on PARTHENOGENESIS, especially his induction of artificial parthenogenesis in sea urchins' and frogs' eggs, thereby highlighting the biochemical nature of FERTILIZATION.

LOESS, fine-grained, wind-deposited SILT found worldwide in deposits up up to 50m thick. Its main components are QUARTZ, FELDSPAR and CALCITE. Extremely porous, it forms highly fertile topsoil, often CHERNOZEM. It is able to stand intact in cliffs.

LOEWI, Otto (1873–1961), German-born US pharmacologist awarded (with Sir Henry DALE) the 1936 Nobel Prize for Physiology or Medicine for his work showing the chemical nature of nerve impulse transmission. (See NERVOUS SYSTEM.)

LÖFFLER, Friedrich August Johannes (1852–1915), German bacteriologist who first isolated the DIPHTHERIA bacillus (1884), which had first been observed by KREBS the previous year. He also discovered the causative organism of GLANDERS; and, with others, showed that HOOF-AND-MOUTH DISEASE is caused by a VIRUS, developing a SERUM against it.

LOG, device to measure the speed of a ship. The oldest type was essentially a float to which was attached a line knotted at regular intervals: the number of knots paid out as the ship sailed for a timed period gave the speed (see KNOT). Nowadays, smaller boats use a device like a propeller whose rate of rotation as it is dragged through the water gives a measure of the speed. Larger ships use a pitometer, a PITOT TUBE with one hole facing forward, one sideways, the water PRESSURE difference indicating the speed.

LOGARITHMS, a method of computation using EXPONENTS. A logarithm is the power (see ALGEBRA) to which one number, the base, must be raised in order to obtain another number. For example, since $10^2 = 100$, $\log_{10}100 = 2$ (read as "log to the base 10 of 100 equals 2"). The most common bases for logarithms are 10 (common logarithms) and the EXPONENTIAL, *e* (natural logarithms).

Since $a^0 = 1$ for any *a*, log 1 = 0 for all bases. In order to multiply two numbers together, one uses the fact that $a^x.a^y = a^{x+y}$, and hence $\log(x.y) = \log x + \log y$. We therefore look up the values of log *x* and log *y* in logarithmic tables, add these values, and then use the tables again to find the number whose logarithm is equal to the result of the addition. Similarly, since $\frac{a^x}{a^y} = a^{x-y}$, $\log\left(\frac{x}{y}\right) = \log x - \log y$; and since $(a^x)^y = a^{xy}$, $\log x^y = y.\log x$. $\text{Log}_x\, x = 1$ since $x^1 = x$. The antilogarithm of a number *x* is the number whose logarithm is *x*; that is, if $\log y = x$, then *y* is the antilogarithm of *x*. A *logarithmic curve* is the plotting of a FUNCTION of the form $f(x) = \log x$.

LOGIC, the branch of PHILOSOPHY concerned with analyzing the rules that govern correct and incorrect reasoning, or inference. It was created by ARISTOTLE, who analyzed terms and propositions and in his *Prior Analytics* set out systematically the various forms of the SYLLOGISM; this work has remained an important part of logic ever since. Aristotle's other great achievement was the use of symbols to expose the form of an argument independently of its content. Thus a typical Aristotelian syllogism might be: all A is B; all B is C; therefore all A is C. This formalization of arguments is fundamental to all logic.

Aristotle's pupil THEOPHRASTUS developed syllogistic logic, and some of the Stoics used symbols to represent not single terms but whole propositions, but apart from this there were no significant developments in later antiquity or the early Middle Ages, although logic (dialectic) was part of the *trivium*. From the 12th century onward there was great revival of interest in logic: Latin translations of Aristotle's logical works (collectively called the *Organon*) were intently studied, and a kind of program emerged, which was based on Aristotle and included much that would nowadays be regarded as GRAMMAR, EPISTEMOLOGY and linguistic analysis. This Scholastic period was a great age of commentaries and compendiums, with much refinement and minute analysis but little original work. Among the most important medieval logicians were William of OCKHAM, Albert of Saxony and Jean BURIDAN. After the Renaissance an anti-Aristotelian reaction set in, and logic was given a new turn by Petrus Ramus and by Francis BACON's prescription that induction (and not deduction) should be the method of the new science. In the work of George BOOLE and Gottlob FREGE the 19th century saw a vast extension in the scope and power of logic. In particular, logic became as bound up with mathematics as it was with philosophy. Logicians became interested in whether particular logical systems were either consistent or complete. (A consistent logic is one in which contradictory propositions cannot be validly derived). The climax of 20th-century logic came in the early 1930s when Kurt GÖDEL demonstrated both the completeness of Frege's first-order logic and that no higher-order logic could be both consistent and complete.

LOGICAL POSITIVISM, the doctrines of the "Vienna Circle," a group of philosophers founded by M. SCHLICK. At the heart of logical positivism was the assertion that apparently factual statements that were not sanctioned by logical or mathematical convention were meaningful only if they could conceivably be empirically verified. Thus only mathematics, logic and science were deemed meaningful; ethics, metaphysics and religion were considered worthless. The influence of logical positivism evaporated after WWII.

LOGOGRAM. See IDEOGRAM.

LOMBROSO, Cesare (1836–1909), Italian physician who pioneered scientific criminology. His view that criminals were throwbacks to earlier evolutionary stages (see ATAVISM) has now been generally discarded. In retrospect his most valuable work is seen to have been his defense of the rehabilitation and more humane treatment of criminals.

LOMONOSOV, Mikhail Vasilievich (1711–1765), Russian scientist and man of letters, best known for his corpuscular theory of matter, in course of developing which he made an early statement of the KINETIC THEORY.

LONG, Crawford Williamson (1815–1878), US physician who first discovered the surgical use of diethyl ETHER as an anesthetic (1842). His discovery followed an observation that students under the influence of ether at a party felt no pain when bruising or otherwise injuring themselves.

LONGITUDE. See LATITUDE AND LONGITUDE.

LORAN (*lo*ng *ra*nge *n*avigation), a NAVIGATION system in which an aircraft pilot may determine his position by comparing the arrival times of pulses from two pairs of RADIO transmitters. Each pair gives him enough information to draw a line of possible positions on a map, the intersection of the two lines marking his true position.

LORENTZ, Hendrik Antoon (1853–1928), Dutch physicist awarded with P. Zeeman the 1902 Nobel Prize for Physics for his prediction of the ZEEMAN effect. Basing his work on J. Clerk MAXWELL's equations, he explained the REFLECTION and REFRACTION of light; and proposed his *electron theory*, that LIGHT occurred through motion of electrons in a stationary electromagnetic ETHER. Thus the wavelength should change under the influence of a powerful magnetic field; and this was experimentally shown by Zeeman (1896).

But the theory was inconsistent with the results of the MICHELSON-MORLEY EXPERIMENT, and so Lorentz introduced the idea of "local time," that the rate of time's passage differed from place to place; and, incorporating this with the proposal of **George Francis Fitzgerald** (1851–1901) that the length of a moving body decreases in the direction of motion (the Fitzgerald contraction), he derived the *Lorentz transformation*, a mathematical statement which describes the changes in length, time and mass of a moving body. His work, with Fitzgerald's, laid the foundations for EINSTEIN's Special Theory of RELATIVITY.

LORENZ, Konrad (1903–), Austrian zoologist and writer, the father of ETHOLOGY, awarded for his work the 1973 Nobel Prize for Physiology or Medicine with FRISCH and TINBERGEN. He is best known for his studies of bird behavior and of human and animal AGGRESSION. His best known books are *King Solomon's Ring* (1952) and *On Aggression* (1966).

LOUDSPEAKER, or **speaker,** device to convert electrical impulses into sound. It commonly comprises a rigid conical diaphragm attached to a coil held such that it may move backward and forward; within the cylinder of the coil is a fixed permanent magnet. Changes in the current supplied to the coil alter its magnetic field so that it, and the cone, vibrate to produce the compression waves that are SOUND. (See also AMPLIFIER; ELECTRICITY; ELECTROMAGNETISM; MAGNETISM.) HIGH FIDELITY sets use two or more loudspeakers of different sizes for more accurate reproduction.

LOUSE, strictly the name given to wingless members of the INSECTA that are parasitic, but more popularly used to describe other types of PARASITE, for example, the Fish louse which is a member of the CRUSTACEA.

LOWELL, Percival (1855–1916), US astronomer and writer who predicted the existence of and initiated the search for PLUTO; but who is best known for his championing the theory (now discarded) that the "canals" of MARS were signs of an irrigation system built by an intelligent race.

LOWEST COMMON MULTIPLE (LCM), in ARITHMETIC, the lowest number of which two or more positive whole numbers are FACTORS.

LSD, lysergic acid diethylamide, a HALLUCINOGENIC DRUG based on ERGOT alkaloids. It may lead to psychotic reaction and bizarre behavior.

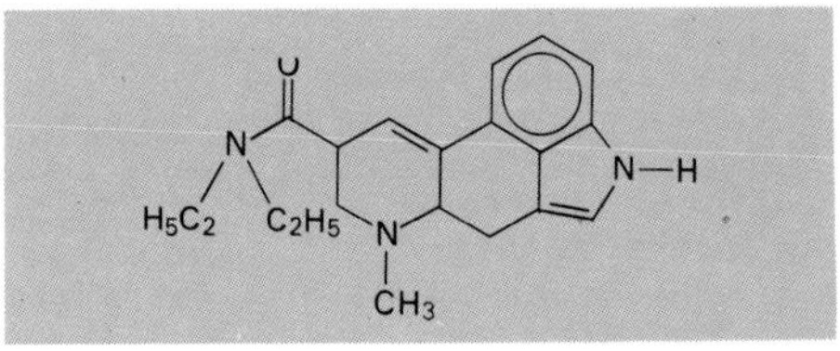

Lysergic acid diethylamide (LSD).

LUBRICATION, the introduction of a thin film of lubricant—usually a semiviscous fluid—between two surfaces moving relative to each other, in order to minimize FRICTION and abrasive wear. In particular, BEARINGS are lubricated in engines and other machinery. Liquid lubricants are most common, usually PETROLEUM fractions, being cheap, easy to introduce, and good at cooling the parts. The VISCOSITY is tailored to the load, being made high enough to maintain the film yet not so high that power is lost. Multigrade oils cover a range of viscosity. The viscosity index represents the constancy of the viscosity over the usual temperature range—a desirable feature. Synthetic oils, including SILICONES, are used for high-temperature and other special applications. **Greases**—normally oils thickened with soaps, fats or waxes—are preferred where the lubricant has to stay in place without being sealed in. Solid lubricants, usually applied with a binder, are soft, layered solids including graphite, molybdenite, talc and boron nitride. TEFLON, with its uniquely low coefficient of friction, is used for self-lubricating bearings. Rarely air or another gas is used as a

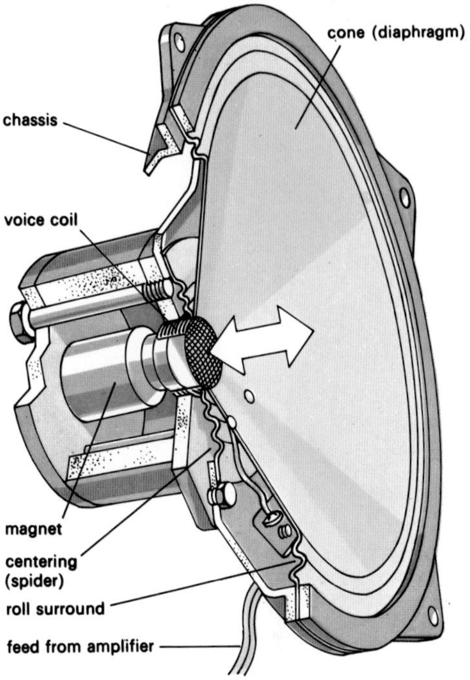

Cutaway of a small moving-coil permanent-magnet loudspeaker.

lubricant. Additives to liquid lubricants include antioxidants, detergents, pour-point depressants (increasing low-temperature fluidity), and polymers to improve the viscosity index.

LUCIFER, an early type of MATCH made by Samuel Jones of London (c1830).

LUCITE (Perspex or Plexiglass), extremely tough, light, very transparent thermoplastic used for auto and aircraft windshields, contact lenses, watchglasses, etc. It is composed of polymethylmethacrylate. (See also PLASTIC.)

LUCRETIUS (c95–55 BC), Roman poet, the author of *De rerum natura*, and the last and greatest classical exponent of ATOMISM. His description of atoms in the void and his vision of the progress of man suffered undeserved neglect on account of his antireligious reputation.

LUMBAGO, popular term for low back pain or lumbar back ache. It may be of various origins including chronic ligamentous strain, SLIPPED DISK (sometimes with SCIATICA), certain types of ARTHRITIS affecting the spine and congenital disease of the spine. Diagnosis and treatment may be difficult.

LUMEN (lm), SI UNIT of luminous flux (see PHOTOMETRY).

LUMIÈRE, Louis (1864–1948), French pioneer of motion PHOTOGRAPHY who, with his brother Auguste (1862–1954), invented an early motion-picture system (patented 1895), the *cinématographe*; and made what is regarded as the first movie (1895).

LUMINANCE, in PHOTOMETRY, the brightness of an extended surface. In SI UNITS luminance is measured in

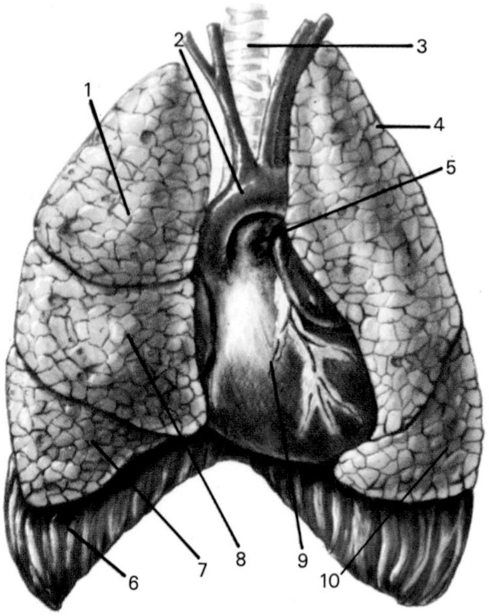

The human lungs: (1) right superior lobe; (2) aorta; (3) trachea; (4) left superior lobe; (5) pulmonary artery; (6) diaphragm; (7) right inferior lobe; (8) middle lobe; (9) heart; (10) left inferior lobe.

CANDELAS per square metre but older units include the APOSTILB, the LAMBERT and the STILB.

LUMINESCENCE, the nonthermal emission of ELECTROMAGNETIC RADIATION, particularly LIGHT, from a PHOSPHOR. Including both fluorescence and phosphorescence (distinguished according to how long emission persists after excitation has ceased, in fluorescence emission ceasing within 10ns but continuing much longer in phosphorescence), particular types of luminescence are named for the mode of excitation. Thus in photoluminescence, X-ray PHOTONS are absorbed by the phosphor and lower-energy radiations emitted; in CHEMILUMINESCENCE the energy source is a chemical reaction; cathodoluminescence is energized by cathode rays (ELECTRONS), and BIOLUMINESCENCE occurs in certain biochemical reactions. (See also ELECTROLUMINESCENCE.)

LUMPY JAW, common name for actinomycosis in cattle, a disease caused by infection with the ACTINOMYCETE, *Actinomyces bovis*. Actinomycosis also affects swine and, occasionally, man.

LUNAR SOCIETY OF BIRMINGHAM, the most illustrious of the British provincial scientific societies of the late 18th century. Its members included the industrialists Matthew BOULTON and Josiah Wedgwood, the physician Erasmus DARWIN and the chemist and theologian, Joseph PRIESTLEY.

LUNGFISH. See DIPNOI.

LUNGS, in vertebrates, the (usually) two largely air-filled organs in the CHEST concerned with RESPIRATION. the absorption of OXYGEN from and release of carbon dioxide into atmospheric air. In man, the right lung has three lobes and the left, two. Their surfaces are separated from the chest wall by two layers of *pleura*, with a little fluid between them; this allows free movement of the lungs and enables the forces of expansion of the chest wall and DIAPHRAGM to fill them with air. Air is drawn into the TRACHEA via mouth or NOSE; the trachea divides into the BRONCHI which divide repeatedly until the terminal airsacs or *alveoli* are reached. In the alveoli, air is brought into close contact with unoxygenated BLOOD in lung CAPILLARIES; the BLOOD CIRCULATION through these comes from the right ventricle and returns to the left atrium of the HEART. Disorders of ventilation or of perfusion with blood leads to abnormalities in blood levels of carbon dioxide and oxygen. Lung DISEASES include ASTHMA, BRONCHITIS, PNEUMONIA, PLEURISY, PNEUMOTHORAX, PNEUMOCONIOSIS, EMBOLISM, CANCER and TUBERCULOSIS; lungs may also be involved in several systemic diseases (e.g., sarcoidosis, LUPUS ERYTHEMATOSUS). Symptoms of lung disease include COUGH, sputum, blood in the sputum, shortness of breath and wheeze. Sudden failure of breathing requires prompt ARTIFICIAL RESPIRATION. Chest X RAY and estimations of blood gas levels and of various lung volumes aid diagnosis.

LUPUS ERYTHEMATOSUS (LE), a rare systemic disorder affecting mainly young women and causing a characteristic SKIN rash with butterfly distribution over the face. It also causes LUNG, KIDNEY and BLOOD disease, largely due to abnormal IMMUNITY directed against substances in cell nuclei. Treatment is with STEROIDS and immunosuppressives.

LUPUS VULGARIS, TUBERCULOSIS of the SKIN.

LURIA, Salvador Edward (1912–), Italian-born US biologist who shared with DELBRÜCK and HERSHEY the 1969 Nobel Prize for Physiology or Medicine for research on BACTERIOPHAGES.

LUTEINIZING HORMONE (LH), a pituitary-gland GONADOTROPHIN which in female mammals (including humans) causes release of EGGS from mature FOLLICLES and promotes the secretion of PROGESTERONE. In males it stimulates testicular ANDROGEN formation. Variations in LH secretion are involved in the onset of PUBERTY and in the menstrual cycle.

LUTEOTROPHIC HORMONE. See PROLACTIN.

LUTETIUM (Lu), the final member of the LANTHANUM SERIES. AW 175.0, mp 1663°C, bp 3315°C, sg 9.840 (25°C).

LUX (lx), SI UNIT of illuminance (see PHOTOMETRY).

LWOFF, André Michael (1902–), French microbiologist awarded with F. JACOB and J. MONOD the 1965 Nobel Prize for Physiology or Medicine for his work on lysogeny, a process of genetic interaction between BACTERIOPHAGES and BACTERIA.

LYCANTHROPY (from Greek *lykanthrōpos*, werewolf), in popular tradition, the assumption by a man of a wolf's form. In psychology, the belief by an individual that he can become a wild animal. The classic study is FREUD's of "The Wolfman."

LYCOPSIDA. See CLUB MOSSES.

LYE, any strong caustic ALKALI, especially POTASSIUM or SODIUM hydroxide.

LYELL, Sir Charles (1797–1875), British geologist and writer whose most important work was the promotion of geological UNIFORMITARIANISM (originally developed by James HUTTON) as an alternative to the CATASTROPHISM of CUVIER and others. The prime expression of these views came in his *Principles of Geology* (1830–33). His other works included the *Elements of Geology* (1838), and *Geological Evidence of the Antiquity of Man* (1863). Here he expressed guarded support for DARWIN's theory of evolution.

LYMPH, fluid which drains from extracellular fluid via lymph vessels and nodes (glands). Important node sites are the neck, axilla, groin, CHEST and ABDOMEN. Fine ducts carry lymph to the nodes, which are filled with lymphocytes and reticulum cells. These act as a filter, particularly for infected debris or PUS and for CANCER cells, which often spread by lymph. The lymphocytes are also concerned with development of IMMUNITY. From nodes, lymph may drain to other nodes or directly into the major thoracic duct which returns it to the BLOOD. Specialized lymph ducts or lacteals carry FAT absorbed in the GASTROINTESTINAL TRACT to the thoracic duct. In addition, there are several areas of lymphoid tissue at the portals of the body as a primary defense against infection (TONSILS, ADENOIDS, Peyer's patches in the gut). Lymph node enlargement may be due to INFLAMMATION following DISEASE in the territory drained (SKIN, PHARYNX), or to development of an ABSCESS in the node (STAPHYLOCOCCUS, TUBERCULOSIS) due to INFECTIOUS DISEASE, secondary spread of cancer and the development of LYMPHOMA or LEUKEMIA. BIOPSY is valuable in diagnosis.

LYMPHOMA, malignant proliferation of LYMPH tissue, usually in the lymph nodes, SPLEEN or GASTROINTESTINAL TRACT. The prototype is HODGKIN'S DISEASE, but a number of other forms occur with varying HISTOLOGY and behavior. Cancer CHEMOTHERAPY and RADIATION THERAPY have much to offer in these disorders.

LYNEN, Feodor (1911–), German biochemist who shared with K. E. BLOCH the 1964 Nobel Prize for Physiology or Medicine for their independent work on the metabolism of CHOLESTEROL and the fatty acids.

LYRA (the Lyre), a medium-sized N Hemisphere constellation containing VEGA and the Ring Nebula (M57), the relic of a stellar explosion 1.66kpc from the earth.

LYSENKO, Trofim Denisovich (1898–1977), Soviet agronomist whose antipathy for GENETICS and position of power under the Stalin regime led to the stifling of any progress in Soviet biological studies for 25 years or more. Refusing on ideological grounds to believe in GENES, he adopted a peculiar form of Lamarckism (see LAMARCK; MICHURIN), and forced other Soviet scientists to support his views. He was removed from power in 1964.

LYSOSOME, a small particle found in the cytoplasm of many living CELLS consisting of granules of ENZYMES surrounded by a lipoprotein membrane. The enzymes are inactive until released into the cytoplasm. This happens when the cells are damaged, the enzymes helping to destroy invading bacteria.

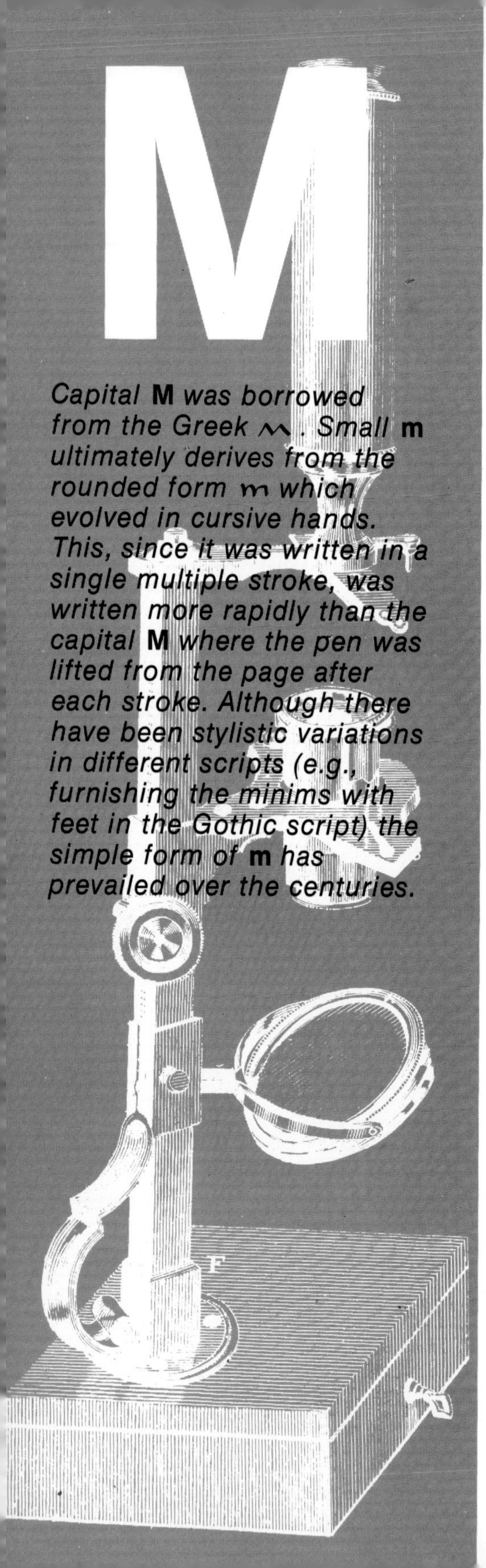

Capital **M** *was borrowed from the Greek* Μ. *Small* **m** *ultimately derives from the rounded form* m *which evolved in cursive hands. This, since it was written in a single multiple stroke, was written more rapidly than the capital* **M** *where the pen was lifted from the page after each stroke. Although there have been stylistic variations in different scripts (e.g., furnishing the minims with feet in the Gothic script) the simple form of* **m** *has prevailed over the centuries.*

MACADAM, road-building system devised by the Scots engineer **John Loudon McAdam** (1756–1836). The soil beneath the road, rather than foundations, takes the load, the road being waterproof and well-drained to keep this soil dry. For modern highways a first layer of larger rocks is laid, then smaller rocks and gravel, the whole being bound with, usually, ASPHALT or TAR.

McCOLLUM, Elmer Verner (1879–1967), US biochemist who is credited with the discovery of fat-soluble VITAMINS A and B (1913), and who contributed to the discovery of vitamins D and E.

McCORMICK, Cyrus Hall (1809–1884), US inventor and industrialist who invented an early mechanical REAPER (patented 1834), the first models appearing under license from 1841 onward.

McDOUGALL, William (1871–1938), British psychologist best known for his fusion of the disciplines of PSYCHOLOGY and ANTHROPOLOGY in order to obtain a better understanding of the roots of social behavior. He also conducted important researches into PARAPSYCHOLOGY.

McDOWELL, Ephraim (1771–1830), US surgeon who performed the first successful ovariotomy (1809) to remove an ovarian TUMOR, as well as a number of other pioneering abdominal operations.

MACE GAS. See TEAR GAS.

MACH, Ernst (1838–1916), Austrian physicist and philosopher whose name is commemorated in MACH NUMBERS. His greatest influence was in philosophy where he rejected from science all concepts which could not be validated by experience.This freed EINSTEIN from the absoluteness of Newtonian space-time (and thus helped him toward his theory of RELATIVITY) and helped inform the LOGICAL POSITIVISM of the Vienna Circle.

MACHINE, a device that performs useful work by transmitting, modifying or transforming motion, forces and energy. There are three basic machines, the inclined plane, the lever, and the wheel and axle: from these, and adaptations of these, are built up all true machines, no matter how complex they may appear. There are two essential properties of all machines: *mechanical advantage*, which is the ratio load/effort, and *efficiency*, the ratio of actual performance to theoretical performance. Mechanical advantage can be less than, equal to or greater than 1; while efficiency, owing to such losses as FRICTION, is always less than 100% (otherwise a PERPETUAL MOTION machine would be possible). (See also EFFICIENCY; ENERGY; FORCE; LEVER; WHEEL; WORK.)

Simple machines derived from the three basic elements include: from the inclined plane, the *wedge* (effort at the top being translated to force at the sides), and the *screw*, (an inclined plane in spiral form); from the lever, the wrench or spanner (the BALANCE also uses the principle of the lever), and from the wheel and axle, the PULLEY, (which can also be viewed as a type of lever). (See also BOLTS AND SCREWS; ENGINE; PUMP.)

MACHINE GUN, a GUN that can fire a number of rounds in rapid succession, a weapon that has changed the face of war in our century. Such guns are known from as early as the 14th century, and LEONARDO DA VINCI produced designs for several. These early guns were little more than a number of single guns arranged so that they could be set off by a single spark. James Puckle patented (1718) the precursor of the modern machine gun: it had a single barrel and a rotating stock holding square bullets: it fired about 9 rounds/min. Flintlock was discarded with the invention of the percussion cap (c1816), and firing reliability much increased. By 1862 GATLING had developed a single-barreled machine gun, used in the Civil War, and later a multi-barreled gun that fired 3000 rounds/min. MAXIM devised the first fully automatic machine gun around 1884; closely followed by John Browning, many of whose designs are in use today.

Three power sources are tapped to operate modern guns: the pressure of the expanding gases in the barrel; the recoil of the bolt and barrel; and the sprung return of a barrel that has recoiled. (See also AMMUNITION; FIREARMS.)

MACHINE TOOLS, nonportable, power-driven tools used industrially for working metal components to tolerances far finer than those obtainable manually. The fundamental processes used are cutting and grinding, individual machines being designed for boring, broaching, drilling (see DRILLS), milling, planing and sawing. Essentially a machine tool consists of a jig to hold both the cutting tool and the workpiece, and a mechanism to allow these to be moved relative to each other in a controlled fashion. A typical example is the LATHE. Auxiliary functions facilitate the cooling and lubrication of the tool and workpiece while work is in progress using a cutting fluid. The rate at which any piece can be worked depends on the material being worked and the composition of the cutting point. High-speed STEEL, TUNGSTEN carbide and CORUNDUM are favored materials for cutting edges. Where several operations have to be performed on a single workpiece, time can be saved by using multiple-function tools such as the turret lathe, particularly if numerically rather than manually controlled. Modern industry would be inconceivable without machine tools. It was only when these began to be developed in the late 18th century that it became possible to manufacture interchangeable parts and thus initiate MASS PRODUCTION.

MACH NUMBER, ratio of the speed of an object or fluid to the local speed of SOUND, which is temperature dependent. Speeds are subsonic or supersonic depending on whether the mach number is less than or greater than one.

MACINTOSH, Charles (1766–1843), Scots chemist who invented the waterproof fabric eventually known as **mackintosh** (1823), a layer of RUBBER dissolved in naphtha sandwiched between two layers of cloth. Quality was much improved with the advent of vulcanized rubber (patented 1844 by GOODYEAR).

MACLEOD, John James Rickard (1876–1935), British-born physiologist who shared with Sir F. G. BANTING the 1923 Nobel Prize for Physiology or Medicine for his role in the isolation of INSULIN. Macleod provided laboratory facilities and a degree of direction for Banting and his collaborator, C. H. BEST, who did not share in the award.

MACLURE, William (1763–1840), British-born US geologist regarded as the father of American geology for his monumental *Observations on the Geology of the United States* (1809), which was accompanied by a geological map.

McMILLAN, Edwin Mattison (1907–), US nuclear physicist who shared with SEABORG the 1951 Nobel Prize for Chemistry for his discovery of the first TRANSURANIUM ELEMENT, number 93, NEPTUNIUM (1940). Independently, he and the Soviet physicist **Vladimir Veksler** (1907–1966) developed the SYNCHROTRON, for which they shared the 1963 Atoms for Peace award.

MACROPHAGE. See RETICULO-ENDOTHELIAL SYSTEM.

MAGDALENIAN, upper-Paleolithic culture named for La Madeleine cave, Dordogne, France, noted for the quality of its painting and bone engraving (see also STONE AGE.)

MAGELLANIC CLOUDS, two irregular GALAXIES that orbit the MILKY WAY, visible in S skies. The Large Magellanic Cloud (Nubecula Major), about 4.5kpc in diameter, has a well marked axis suggesting that it may be an embryonic spiral galaxy. The Small Magellanic Cloud (Nubecula Minor) is about 3kpc across. Both are rich in CEPHEID VARIABLES and about 46kpc from the earth.

MAGENDIE, François (1783–1855), French physiologist, regarded as the father of experimental PHARMACOLOGY. He introduced the medical use of several DRUGS (e.g., MORPHINE); and first proved that in a spinal nerve (see NERVOUS SYSTEM; SPINAL CORD) the ventral (anterior) root has a motor function and the dorsal (posterior) root a sensory function.

MAGIC, Primitive, the prescientific belief that an individual, by use of a ritual or spoken formula, may achieve a result that would otherwise be beyond his, or human, powers. Should the magic fail to work, this is assumed to be due to deviations from the correct formula. FRAZER classified magic under two main heads, imitative and contagious. In **imitative magic** the magician acts upon or produces a likeness of his desired object: rainmakers may light fires, the smoke of which resembles rainclouds; voodoo practitioners stick pins in wax models of their intended victims. In **contagious magic** it is assumed that two objects once close together remain related even after separation: the magician may act upon hair clippings in an attempt to injure their former owner. Magic is crucial to many primitive societies, most tribes having at least an equivalent to a medicine man (see SHAMANISM) who is believed to be able to provide them with extra defense against hostile tribes or evil spirits. (See also AMULET; FETISH.)

MAGIC LANTERN, forerunner of the slide projector. Light shone through a slide (often a simple scene painted on glass) to a LENS, which cast a magnified picture on the screen.

MAGIC SQUARE, square array of numbers such that the sums along each row, column and diagonal are equal; e.g.:

6	7	2
1	5	9
8	3	4

MAGMA, molten material formed in the upper mantle or crust of the EARTH, composed of a mixture of various complex SILICATES in which are dissolved various gaseous materials, including WATER. On cooling magma forms IGNEOUS ROCKS, though any gaseous constituents are usually lost during the solidification. Magma extruded to the surface forms LAVA. The term is loosely applied to other fluid

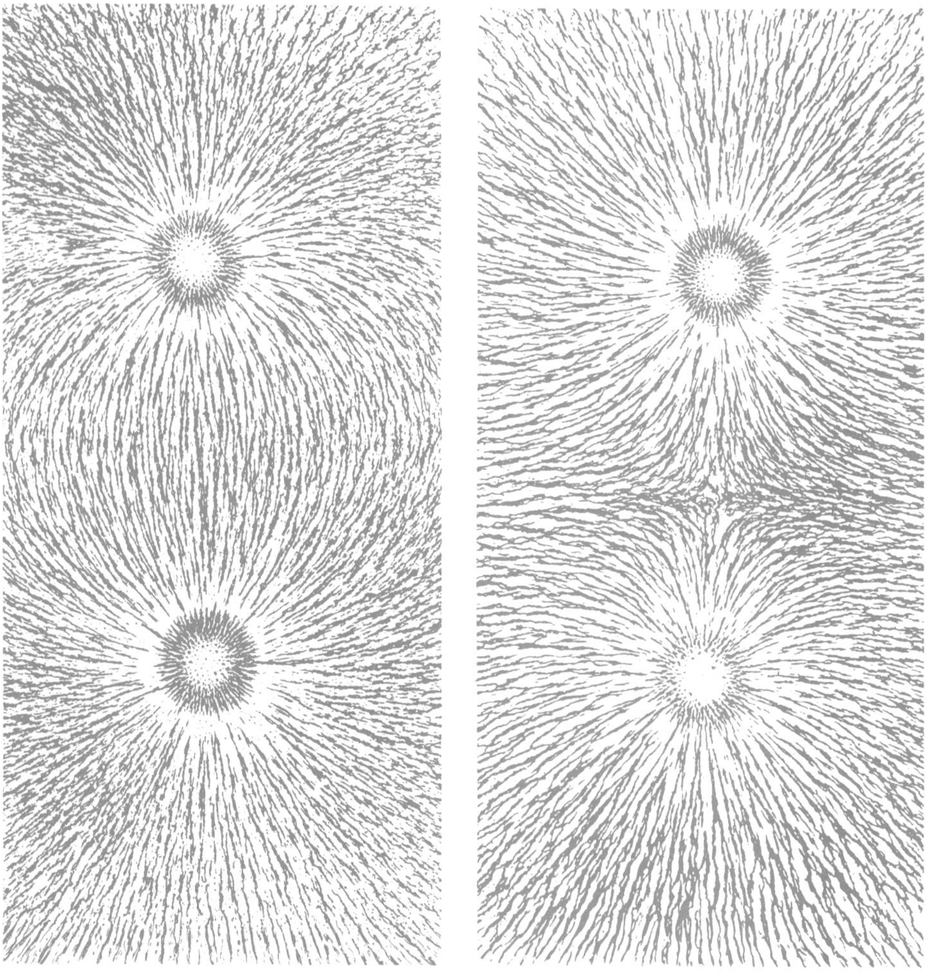

Magnetic "lines of force" revealed by sprinkling iron filings onto a sheet of card below which magnets are concealed. (*Left*) The case of two unlike poles; (*right*) two like poles: here the lines of force "repel" each other.

substances (e.g., molten salt) in the earth's crust. (See also HOT SPRINGS; VOLCANISM.)

MAGNESIA, or magnesium oxide. See MAGNESIUM.

MAGNESITE, mineral form of magnesium carbonate ($MgCO_3$), found in the US, Austria and Manchuria. It forms hexagonal crystals, usually massive and white. Magnesite is mainly used for lining furnaces, since it decomposes to give magnesium oxide (see MAGNESIUM) on heating.

MAGNESIUM (Mg), a reactive, silvery-white ALKALINE-EARTH METAL, the eighth most abundant element. Its chief ores are dolomite, brucite and magnesite. It is also found in many other minerals and in large quantities in the sea. Magnesium is light but strong, and forms useful alloys with aluminum and other metals. Magnesium is manufactured, either by the ELECTROLYSIS of fused magnesium chloride, or by the silicothermic process, in which mixed oxides, formed by the calcination of dolomite, are reduced at high temperature with ferrosilicon, forming magnesium crystals. When heated in air, divided magnesium burns readily with a dazzling white flame: hence its use in FLASHBULBS and flares. It is also used as a powerful reducing agent in the preparation of many other metals. AW 24.3, mp 649°C, bp 1090°C, sg 1.738 (20°C).

Magnesium Hydroxide ($Mg(OH)_2$), occurs naturally as colorless hexagonal plates of brucite. It is formed as a gelatinous precipitate when ALKALI is added to magnesium salts. It is a BASE and loses water when heated to 350°C. (See also MILK OF MAGNESIA.)

Magnesium Oxide (MgO), or **Magnesia**, is a white crystalline solid and also a BASE. Being highly refractory, it is used in furnace linings. When mixed with magnesium chloride solution, magnesia forms a durable cement used as a stucco finish on buildings.

Magnesium Sulfate ($MgSO_4$), or EPSOM SALTS, is a colorless crystalline solid. The aqueous solution is used as a purgative.

MAGNET. See MAGNETISM.

MAGNETIC FIELD, what is said to exist where electric charges (see ELECTRICITY) experience a FORCE proportional to their VELOCITY but at right angles to it, or where magnetic dipoles (see MAGNETISM) experience a torque. The field is defined in the direction of zero torque, with a strength equal to the torque on a unit dipole at right angles to the field. Magnetic fields originate at magnetic dipoles or electric currents.

MAGNETIC LENS, a device using a MAGNETIC FIELD to focus a beam of charged particles. The field is produced between two annular pole pieces around the beam driven by an ELECTROMAGNET. Its imaging properties were first studied by Hans Busch in 1926.

MAGNETIC STORM, an occasional disturbance in the earth's MAGNETIC FIELD, correlated with SUNSPOT activity. A high energy PLASMA ejected from a solar flare sets up large currents in the MAGNETOSPHERE on reaching the earth, causing a rapid rise of around 0.2% in the magnetic field at the surface. The plasma subsequently moves around the earth, often accompanied by auroral displays, while the field drops to about 0.5% below its normal value, recovering over several days.

MAGNETISM, the phenomena associated with "magnetic dipoles," commonly encountered in the properties of the familiar horseshoe (permanent) magnet and applied in a multitude of magnetic devices.

Man first learned of magnetism through the properties of the *lodestone*, a shaped piece of MAGNETITE that had the property of aligning itself in a roughly north-south direction. Eventually he found how to use a lodestone to magnetize a steel bar, thus making an artificial **permanent magnet**. The power of a magnet was discovered to be concentrated in two "poles," one of which always sought the north, and was called a north-seeking pole, or north pole, the other being a south-seeking pole, or south pole. It was early learned that, given two permanent magnets, the unlike poles were attracted to each other and the like poles repelled each other. Furthermore, dividing a magnet in two never resulted in the isolation of an individual pole, but only in the creation of two shorter two-poled magnets. Again, it was found that magnetic poles attracted or repelled each other according to an inverse-square law. The explanation of these properties in terms of magnetic "lines of force" was an early achievement of the science of magnetostatics.

Today, physicists explain magnetism in terms of *magnetic dipoles*. Magnetic dipole moment is an intrinsic property of fundamental particles. ELECTRONS, for example, have a moment of 0.928×10^{-23} A.m^2 parallel or antiparallel to the direction of observation. The forces between magnetic dipoles are identical to those between electric dipoles (see ELECTRICITY). This leads scientists often to regard the dipoles as consisting of two magnetic charges of opposite type, the poles of traditional theory. But unlike electric charges, magnetic poles are believed never to be found in isolation.

In **ferromagnetic materials** such as IRON and COBALT, spontaneous dipole alignment over relatively large regions known as *magnetic domains* occurs. Magnetization in such materials involves a change in the relative size of domains aligned in different directions, and can multiply the effect of the magnetizing field a thousand times. Other materials show much weaker, nonpermanent magnetic properties (see DIAMAGNETISM, PARAMAGNETISM).

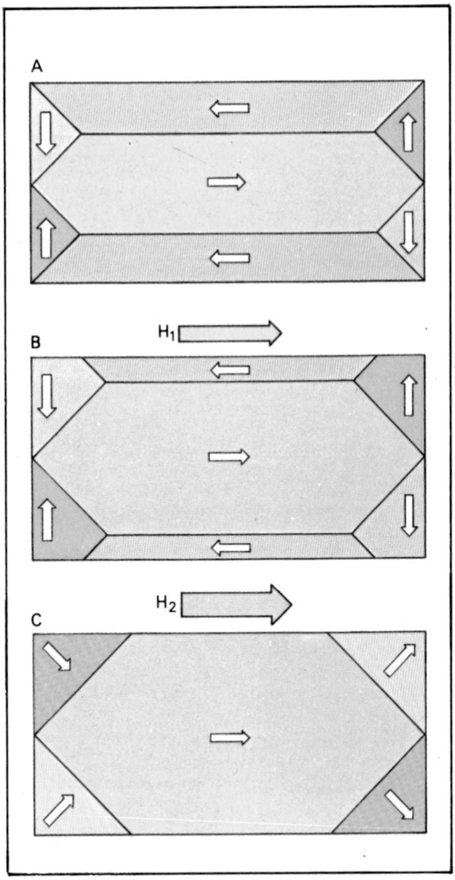

(A) In an unmagnetized rectangular single crystal of iron, the domains are arranged such that the external magnetic field is minimized. (B) In the presence of an external magnetic field (H_1), the domains aligned with the field grow at the expense of those aligned against the field. This process continues until (C) the contrary domains are eliminated and the direction of magnetization in the peripheral domains rotates.

Magnetism is intimately associated with electricity (see ELECTROMAGNETISM). Electric currents generate MAGNETIC FIELDS circulating around themselves— the EARTH's magnetic field is maintained by large currents in its liquid core—and small current loops behave like magnetic dipoles with a moment given by the product of the loop current and area.

MAGNETISM, Terrestrial. See EARTH.

MAGNETITE, hard, black OXIDE mineral of composition Fe_3O_4. As an IRON ore of widespread occurrence in igneous rocks, it is second only in importance to HEMATITE. Magnetite has the inverse SPINEL structure. It is strongly ferromagnetic (see MAGNETISM) and was used in the ancient world as a COMPASS, under the name *lodestone*.

MAGNETO, a simple electrical AC GENERATOR based on a rotating permanent magnet which induces (see ELECTROMAGNETISM) a current in a coil. It is the basis of an IGNITION SYSTEM used in INTERNAL-COMBUSTION ENGINES without BATTERIES, the spark voltage being induced as usual in a large secondary coil when the current in a primary coil is interrupted, but the primary current being itself induced by the rotating magnet rather than being drawn from a battery.

MAGNETOHYDRODYNAMICS (MHD), the DYNAMICS of conducting fluids such as liquid metals or PLASMAS, in ELECTRIC and MAGNETIC FIELDS. It is a macroscopic form of ELECTRODYNAMICS, deriving from fluid dynamics such concepts as magnetic pressure and magnetic viscosity. Its equations often defy exact solution. The most important applications are in magnetohydrodynamic GENERATORS and controlled nuclear FUSION processes. The extremely hot plasma produced by the fusion is contained by strong circulating magnetic fields; various designs are possible, the stability of each being the paramount consideration.

MAGNETOMETER, a device measuring MAGNETIC FIELD strength. Various types exist, exploiting, for instance, the oscillation rate of a small, freely suspended bar magnet or the deflection of a magnet against its suspension or a reference field. Sensitive magnetometers used in space research include the proton precession magnetometer and the helium magnetometer.

MAGNETOSPHERE, term applied to the region of the ATMOSPHERE containing the VAN ALLEN RADIATION BELTS.

MAGNETOSTRICTION, the interaction between the physical dimensions of a ferromagnetic specimen and its magnetization. A long iron rod, for example, contracts slightly in a MAGNETIC FIELD. Magnetization by mechanical strain is exploited in high-frequency vibration detectors.

MAGNIFYING GLASS, or simple MICROSCOPE, a converging LENS used to form an enlarged image of an object. In normal use, the object is held within the focal length of the lens and an enlarged, upright virtual IMAGE is seen through the lens. A magnifying glass can also be used to form a real but inverted image of an object if the object is placed outside the focal length of the lens. A large converging lens can also be used as a **burning glass**, focusing light and heat from the sun.

MAGNITUDE, Stellar, a measure of a star's brightness. The foundations of the system were laid by HIPPARCHUS (c120 BC), who divided stars into six categories, from 1 to 6 in order of decreasing brightness. Later the system was extended to include fainter stars which could be seen only by telescope, and brighter stars, which were assigned negative magnitudes (e.g., Sirius, −1.5). Five magnitudes were defined as a 100-times increase in brightness. These *apparent magnitudes* depend greatly on the distances from us of the stars. *Absolute magnitude* is defined as the apparent magnitude a star would have were it at a distance of 10pc from us: Sirius then has magnitude +1.4. Absolute magnitudes clearly tell us far more than do apparent magnitudes. Stars are also assigned red, infrared, bolometric and photographic magnitude.

MALACHITE, a green mineral consisting of basic copper (II) carbonate ($Cu_2CO_3[OH]_2$). It is of widespread occurrence, usually with AZURITE, and is formed by weathering of other copper minerals. A minor ore of COPPER, it is used for ornamental stone and GEMS.

MALARIA, tropical PARASITIC DISEASE causing malaise and intermittent FEVER and sweating, either on alternate days or every third day; bouts often

Ray diagrams showing different ways in which a magnifying glass can be used: (A) as a simple microscope; (B) to form a real image; (C) as a burning glass.

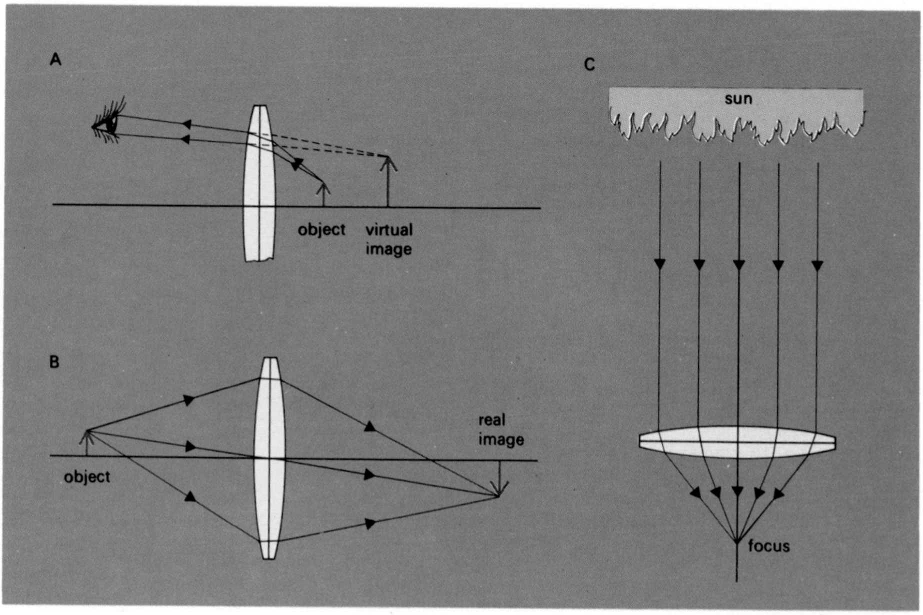

MAGNIFYING GLASS

Malachite: the raw mineral and a polished stone.

reoccur over many years. One form, cerebral malaria, develops rapidly with ENCEPHALITIS, COMA and SHOCK. Malaria is due to infection with *Plasmodium* carried by mosquitos of the genus *Anopheles* from the BLOOD of infected persons. The cyclic fever is due to the parasite's life cycle in the blood and LIVER; diagnosis is by examination of blood. QUININE and its derivatives, especially CHLOROQUINE and primaquine, are used both in prevention and treatment but other chemotherapy (ATABRINE, pyrimethamine) may also be used. Mosquito control, primarily by destroying their breeding places (swamps and pools), provides the best method of combating the disease.

MALEBRANCHE, Nicolas (1638–1715), French philosopher, scientist and Roman Catholic priest. In both philosophy and science he was much influenced by the thought of DESCARTES, in the former field attempting to reconcile Cartesian philosophy with that of St. Augustine, in the latter field researching LIGHT, VISION and the CALCULUS.

MALIC ACID ($HOOC.CH_2.CHOH.COOH$), dibasic acid found in the juice of unripe apples and some other fruits. mp 133°C.

MALIGNANCY. See CANCER.

MALINOWSKI, Bronislaw Kasper (1884–1942), Polish-born British anthropologist, generally accepted as the founder of social ANTHROPOLOGY. In his functional theory all the mores, customs or beliefs of a society perform a vital function in it. From 1927 to 1938 he was a professor at London University; from 1939 until his death, a professor at Yale.

MALLEABILITY, the property of metals and alloys to be deformed by beating, rolling, etc., without breaking. GOLD is the most malleable of metals, and can be beaten into almost any shape. Malleability is not equivalent to DUCTILITY, the ability to be drawn out without breaking: LEAD is malleable but not ductile.

MALNUTRITION, inadequate nutrition, especially in children, which may involve all parts of diet (marasmus), or may be predominantly of PROTEINS (KWASHIORKOR) or VITAMINS (PELLAGRA, BERIBERI, SCURVY). In *marasmus*, essential factors for METABOLISM are derived from the breakdown of body TISSUES; extreme wasting and growth failure result. In adults, starvation is less rapid in onset, as the demands of growth are absent, but similar metabolic changes occur.

MALPIGHI, Marcello (1628–1694), Italian physician and biologist, the father of microscopic ANATOMY, discoverer of the CAPILLARIES (1661), and a pioneer in several fields of medicine and biology.

MALT, the product made from any cereal grain by steeping it in water, germinating and then drying it. This activates dormant ENZYMES such as DIASTASE, which converts the kernel STARCH to MALTOSE. Malt is used as a source of enzymes and flavoring.

MALTHUS, Thomas Robert (1766–1834), English clergyman best known for his *Essay on the Principle of Population* (1798; second, larger edition, 1803). In this he argued that the population of a region would always grow until checked by famine, pestilence or war. Even if agricultural production were improved, the only result would be an increase in population and the lot of the people would be no better. Although this pessimistic view held down the provision of poor relief in England for many decades, it also provided both C. DARWIN and A. R. WALLACE with a vital clue in the formulation of their theory of EVOLUTION by natural selection.

MALTOSE ($C_{12}H_{22}O_{11}.H_2O$), or "malt sugar," disaccharide SUGAR, produced by the action of DIASTASE on STARCH and yielding GLUCOSE with the ENZYME maltase.

MAMMALIA, a class of the CHORDATA, the members of which are distinguished from all other animals by their ability to suckle their young with MILK produced by the mammary glands and by their possession of hair. There are 4000 living species which include man and his major domestic animals. The Mammalia is divided into three subclasses: the MONOTREMATA (monotremes); METATHERIA (marsupials), and EUTHERIA (placental mammals).

MAMMAL-LIKE REPTILES, fossil members of the REPTILIA that lived during the Permian and Triassic periods, 280–190 million years ago, and which include the ancestors of the mammals. Some were large herbivores, but those that gave rise to the mammals were small carnivorous forms.

MAMMARY GLANDS. See BREASTS; LACTATION.

MAN, *Homo sapiens*, the most widespread, numerous, and reputedly the most intelligent (see INTELLIGENCE) of the PRIMATES. For man's evolutionary history see PREHISTORIC MAN; for the varieties of man see RACE, and for his earliest social development see PRIMITIVE MAN.

MANA, a concept found among the more sophisticated Polynesian and Melanesian societies. Mana is the possession of supernatural power by a person or inanimate object, and epitomizes a pre-animistic stage in the evolution of religious thought.

MANGANESE (Mn), hard, grayish metal in Group VIIB of the PERIODIC TABLE; a TRANSITION ELEMENT. It occurs naturally as PYROLUSITE and MANGANITE. Elementary manganese is obtained by the reduction of manganese (IV) oxide with aluminum in a furnace, or by ELECTROLYSIS. When smelted with iron ore, manganese ore gives the alloys SPIEGELEISEN and ferromanganese, widely used in STEEL production. Manganese also forms useful ALLOYS with some nonferrous metals. Manganese is fairly reactive, resembling IRON chemically. Its main oxidation states are +2, +3, +4, +6 and +7. **Manganese (IV) oxide** (MnO_2), a black crystalline solid, is widely used as an oxidizing agent and as a depolarizer in electric dry cells. **Permanganate** (MnO_4^-) is used in nickel refining and tanning, and as a bleach, disinfectant and powerful oxidizing agent. **Mangan-**

ese (II) sulfate ($MnSO_4$) is a component of some fertilizers. AW 54.9, mp 1244°C, bp 1962°C, sg 7.20 (20°C).

MANGANITE, a black mineral consisting of hydrated manganese (III) oxide (MnO[OH]); an ore of MANGANESE, found in W Europe, Mich. and Cal. It occurs as bundles of prismatic crystals, and alters to PYROLUSITE.

MANGE, INFECTIOUS DISEASE of domestic animals, caused by mites in the fur and leading to itching and irritability.

MANHATTAN PROJECT, US project to develop an explosive device working by nuclear FISSION. It was established in Aug. 1942, and research conducted at Chicago, California and Columbia universities, as well as at Los Alamos, N.M., and other centers. By Dec. 1942 a team headed by FERMI initiated the first self-sustaining nuclear chain reaction. On July 16, 1945 the first ATOMIC BOMB was detonated near Alamogordo, N.M., and similar bombs were the following month dropped on Hiroshima (Aug. 6) and Nagasaki (Aug. 9). (See also NUCLEAR WARFARE.)

MANIA, a PSYCHOSIS characterized by high elation, excitement and acceleration of physiological as well as mental processes. **Homicidal mania** is characterized by an uncontrollable urge to kill.

MANIC-DEPRESSIVE PSYCHOSIS, a PSYCHOSIS characterized by alternating periods of deep depression and MANIA. Periods of sanity may intervene.

MANILA HEMP, or **abaca**, a hard fiber obtained from the leaf stalks of several tropical tree species including *Musa textilis*, of the banana family Musaceae. The Philippines provide 95% of the world's output. (See also HEMP.)

MANOMETER, device, usually consisting of a double-legged liquid column in a glass or metal tube, for determining the difference between two fluid pressures. In a simple U-tube manometer, the mercury (or other low-vapor-pressure liquid) rises in the lower pressure side and drops in the other, the difference in heights being measured on a suitably calibrated scale. Sensitivity is increased by inclining the tube or giving the legs different cross-sections.

MANSON, Sir Patrick (1844–1922), British medical scientist known for his pioneering researches in TROPICAL MEDICINE, especially for his naming the MOSQUITO as the transmitter of FILARIASIS and MALARIA. (See also ROSS, SIR RONALD.)

MANTLE, the layer of the earth lying between the crust and the core. (See EARTH.)

MANURE, animal or plant material applied to soil to increase its fertility (see FERTILIZERS). **Animal manure** is composed essentially of animal wastes together with some plant material, such as straw. It supplies NITROGEN, potash (see POTASSIUM) and PHOSPHATES, as well as smaller quantities of COPPER, IRON and other micronutrients (see also NITROGEN CYCLE). **Green manure** consists of growing plants, usually LEGUMINOUS PLANTS, plowed directly into the soil. **Compost** comprises plant and sometimes animal material allowed to rot (see PUTREFACTION) before application to the soil: composts are usually reinforced with nitrogen and phosphorus. (See also CARBON CYCLE; FERTILIZERS; GUANO; HUMUS; SOIL.)

MAP, diagram representing the layout of features on the earth's surface or part of it. Maps have many uses, including routefinding; marine or aerial NAVIGATION (such maps are called *charts*); administrative, political and legal definition, and scientific study. **Cartography**, or mapmaking, is thus an important and an exact art. The techniques of SURVEYING and GEODESY are used to obtain the positional data to be represented. Since the EARTH is roughly spheroidal—the GEOID being taken as the reference level—and since the surface of a sphere cannot be flattened without distortion, no plane map can perfectly represent its original, the distortion becoming worse the larger the area. But spherical maps or *globes* are impractical for large-scale work. Thus plane maps use various **projections**, geometrical algorithms for transforming the spherical coordinates into plane ones. The choice of projection depends on the purpose of the map; one may aim for correct size or correct shape, but not both at once: a suitable compromise is generally reached. Projections fall into three main classes: *Cylindrical projections* are obtained by projection from the earth's axis onto a cylinder touching the equator. MERCATOR's Projection from the center of the earth is a well-known example: its graticule (net of parallels and meridians) takes the form of a rectangular grid with the scale increasing toward the poles, which are infinitely distant; straight lines represent RHUMB LINES. *Conic projections*, best suited to middle latitudes, are obtained by projection onto a cone that caps the earth, touching a given parallel. *Azimuthal projections* are from a single point onto a plane. The *gnomonic* projection, having the point at the center of the earth, represents GREAT CIRCLE ROUTES by straight lines. The *orthographic* projection has the point at infinity, the projective rays being parallel; distortion is great, but the map looks like the globe. These geometric projections are now seldom used as such, but are modified to give correct relative areas, distances or shapes. The *scale* of a map (assuming it to be constant) is the ratio of a distance on the map to the distance that it represents on the earth's surface. It may be expressed directly as the ratio or representative fraction (1:63 360), as a unit ratio (1in to 1mi), or by a graphic graduated scale. Maps use standard symbols and colors to show features, giving the maximum information clearly. Types of map include physical, political, economic, demographic, historical, geological and meteorological maps; there are also star maps (see CELESTIAL SPHERE).

Maps have been drawn from earliest times, but until the Middle Ages most were little more than sketch maps based on impressions and guesswork,

Map projections are attempts to represent the three-dimensional surface of the earth in two dimensions. (A) Mercator's projection—an equatorial cylindrical projection—yields a map (B) on which areas become increasingly distorted toward the poles (each of which is drawn out into a line at an infinite distance from the equator). (C) An azimuthal projection from the north pole provides a useful map (D) of the southern hemisphere. (E) A conic projection, with elements touching the globe at one (as here) or more parallels, is well suited to mapping mid-latitude regions (F). Red circles represent equal areas, showing the type of distortion produced.

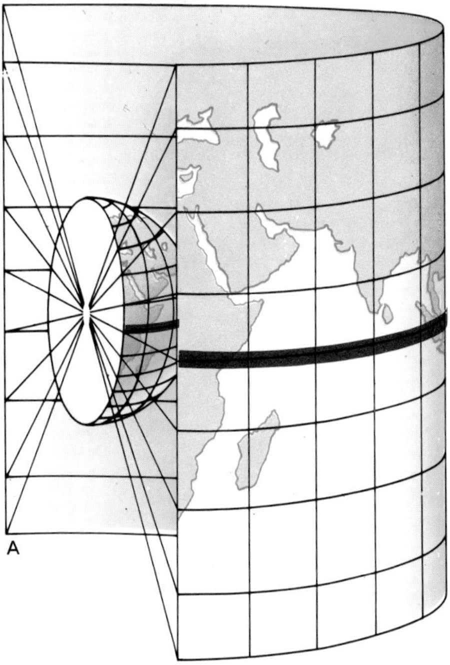

A

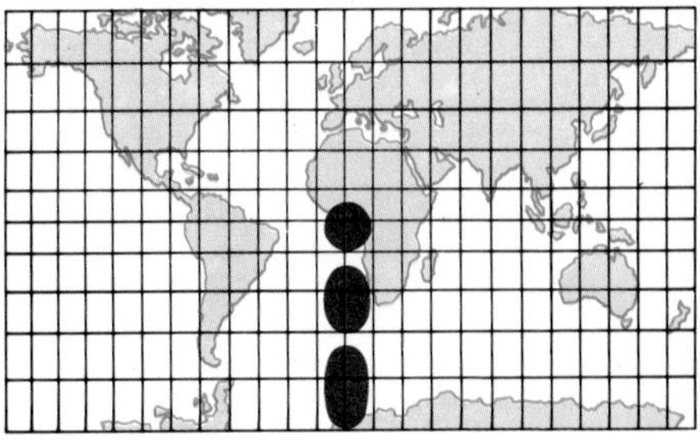

B

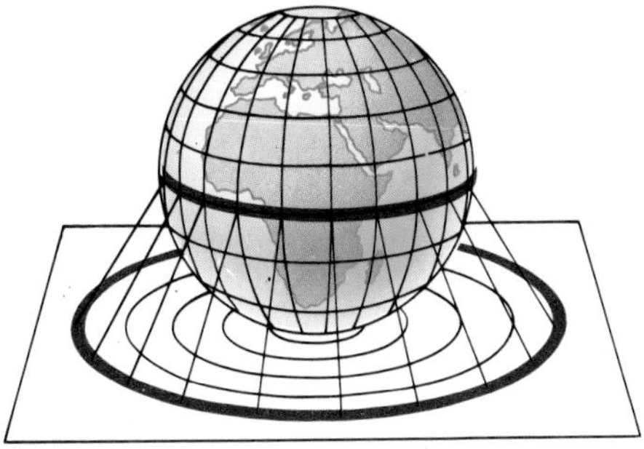

C

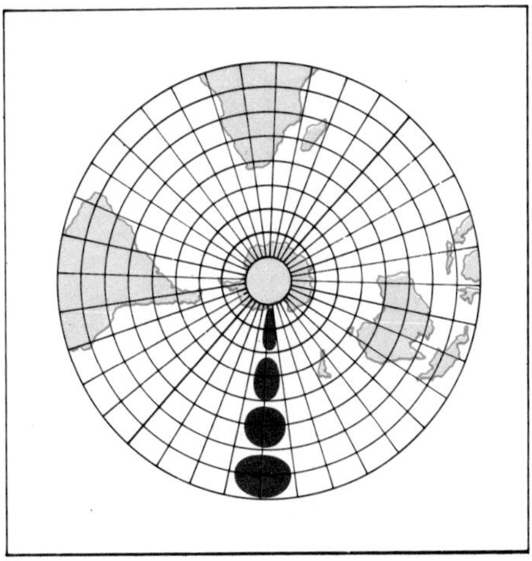

D

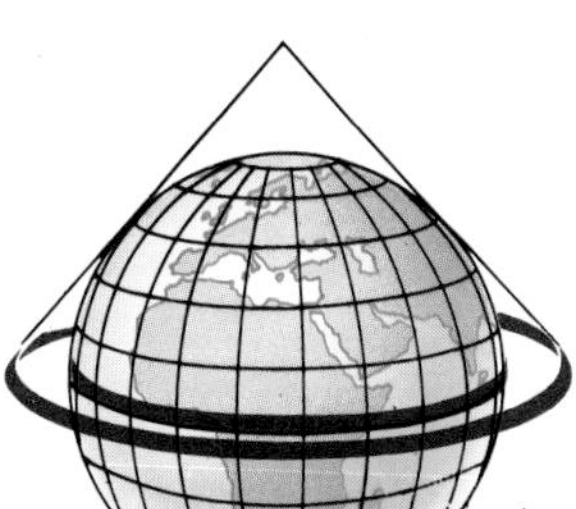

E

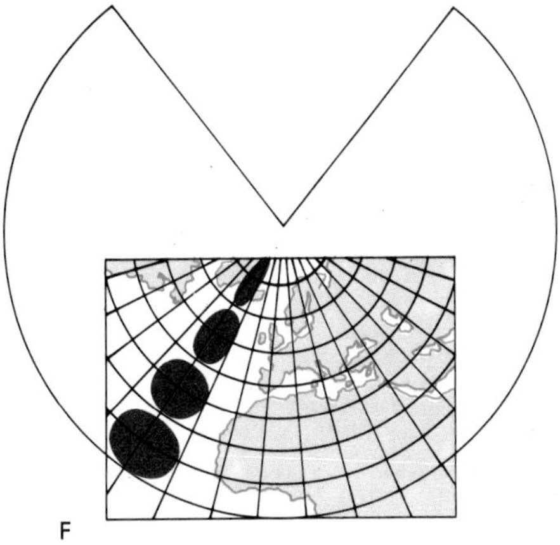

F

except for those of the Greek geographers, notably PTOLEMY of Alexandria. In the 14th century, Mediterranean sea charts were in use which were remarkably accurate, owing to the introduction of the COMPASS and good estimates of distances sailed. The great voyages of discovery, the rediscovery of Ptolemy's map, and accurate surveying in the Low Countries revolutionized cartography in the 16th and 17th centuries, the work of Gerardus MERCATOR and Abraham Ortelius (who produced the first modern atlas) being well-known. Louis XIV promoted a national survey of France, the British Ordnance Survey (1791) followed suit, and in the 19th century most civilized countries produced extensive maps. The US Geological Survey began in 1879. The International Map of the World (IMW), comprising about 1 000 sheets at a scale of 1 : 1 000 000, was started in 1913 but is yet to be completed.

MAPPING, the assignment to each element *a* in a set *A* (see SET THEORY) of at least one element *b* in a set *B* according to a rule, *f*. This is written $f{:}A \rightarrow B$. In a single-valued or one-to-one mapping each element *a* is assigned a single, unique element *b*. In many cases, the term mapping is considered synonymous with FUNCTION.

MARBLE, metamorphic rocks formed by recrystallization of CALCITE or DOLOMITE. The finest Italian marble is pure white calcite and has been prized by sculptors throughout history. Many types of marble contain impurities causing discoloration.

MARCONI, Guglielmo (1874–1937), Italian-born inventor and physicist, awarded (with K. F. BRAUN) the 1909 Nobel Prize for Physics for his achievements. On learning of Hertzian (RADIO) waves in 1894, he set to work to devise a wireless TELEGRAPH. By the following year he could transmit and receive signals at distances of about 2km. He went to the UK to make further developments, and in 1899 succeeded in sending a signal across the English Channel. On Dec. 12, 1901 in St. John's, Newfoundland, he successfully received a signal sent from Poldhu, Cornwall, thus heralding the dawn of transatlantic radio communication.

MARGARINE, a spread high in food value, prepared from vegetable or animal fats together with milk products, preservatives, emulsifiers, butter and salt. It was first developed in the late 1860s by the French chemist Hippolyte Mège-Mouriès, inspired by a competition launched by Napoleon III to find a cheap BUTTER substitute. The fats used were, early on, primarily animal, with whale oil being particularly popular in Europe, but recently vegetable oils (especially soybean and corn) have been used almost exclusively.

MARIJUANA, term applied to any part of the HEMP plant (*Cannabis sativa*) or extract from it. The intoxicating drug obtained from the flowering tops is also called **cannabis** or HASHISH. This drug is usually smoked in cigarettes or pipes, but can also be sniffed or taken as food. It is mainly used for the mild euphoria it produces, although other symptoms include loss of muscular coordination, increased heart beat, drowsiness and hallucination. Its use, the subject of much medical and social debate, is widespread throughout the world.

MARINE BIOLOGY, the study of the flora and fauna in the sea, from the smallest PLANKTON to massive whales. It includes the study of the complex interrelationships between marine organisms that make up the food chains (see ECOLOGY) of the sea. It has become apparent in recent years that if the sea is to remain a major and increasing source of food for man, CONSERVATION measures must be taken, particularly to retain adequate stocks of breeding fish. POLLUTION must also be controlled.

MARINER PROGRAM, US unmanned space probes which have made close "fly-by" observations of VENUS, MARS and MERCURY. Mariner 9 orbited Mars 1971–72, sending back a detailed surface survey. Mariner 10 flew by Venus in 1974, and took the first pictures of Mercury's surface during three fly-bys 1974–75.

MARIOTTE, Edmé (1620–1684), French physicist who independently discovered BOYLE'S LAW (Mariotte's law) and also discovered the blind spot of the EYE (1660).

MARL, natural mixture of clay and calcium carbonate ($CaCO_3$: see CALCIUM; CLAY). If the former predominates, it is a mudstone; if the latter, it is termed calcareous. Greensand marls contain hardly any $CaCO_3$ (see also GLAUCONITE). Marls are used as SOIL conditioners and in making portland CEMENT.

MARRIAGE, a durable union between one or more men and one or more women, sanctioned by society and usually lasting until after the birth and rearing of offspring of their sexual union (see FAMILY). Marriage in the broadest sense is by no means a purely human institution: in the animal world many species practice a union between male and female that may last for one year, or for the time taken for offspring to achieve independence, or, rarely, for life. In man, **monogamy,** the union between one man and one woman, is the most common form of marriage; and it is probably the most common form of union among those animals where long-term union is the normal. **Polygamy** is the marriage of an individual of one sex with a multiplicity of the other sex. The two types of polygamy are *polygyny*, in which one male has two or more wives, and *polyandry*, in which one female has more than one husband.

In most societies, marriage can be sanctioned only under certain circumstances. Generally, the partners must have achieved a certain age, either must not be closely related or of different races (but see ENDOGAMY and EXOGAMY) or colors, and in monogamistic societies must not already be married to someone else. Religious sanction is required in most societies and in many civil authorization is also required.

MARROW, Bone, the material in the center of BONES, in which ERYTHROCYTES, white blood cells and platelets are made. Mature cells only are released unless the marrow is diseased, as in LEUKEMIA, secondary CANCER or in serious infections. Bone marrow aspiration or BIOPSY is often valuable in diagnosis.

MARS, the fourth planet from the sun with a mean solar distance of 228Gm (about 1.52AU) and a "year" of 687 days. During the Martian day of about 24.62h the highest temperature at the equator is about 30°C, the lowest just before dawn being about −100°C. Mars has a mean diameter of 6 750km, with a small degree of polar flattening, and at its closest to earth (see CONJUNCTION) is some 56Gm distant. Its tenuous atmosphere is believed to consist mainly of carbon dioxide, nitrogen and NOBLE GASES, and the distinctive Martian polar caps are thought to be composed of frozen carbon dioxide and ice.

Telescopically, Mars appears as an ocher-red disk marked by extensive dark areas: these latter have in the past been erroneously termed *maria* (seas). Several observers have reported sighting networks of straight lines on the Martian surface—the famous canals—although observations with large telescopes and the photographs sent back by the probes of the MARINER PROGRAM show no signs of these. Mars is spotted with craters, rather as is the MOON. It is not yet known if Mars can support life as we know it. Mars has two moons, PHOBOS and DEIMOS.

MARSH. See SWAMP.

MARSH GAS. See METHANE.

MARSUPIALS. See METATHERIA.

MARTIN, Archer John Porter (1910–), British biochemist awarded with R. L. M. SYNGE the 1952 Nobel Prize for Chemistry for their development of paper CHROMATOGRAPHY, a biochemical tool of great medical importance. He later helped perfect gas chromatography.

MARTIN, Pierre Émile (1824–1915), French engineer who developed the open-hearth smelting process now known as the Siemens-Martin process (1864–65). (See OPEN-HEARTH PROCESS; SIEMENS.)

MASER, a device used as a MICROWAVE oscillator or amplifier, the name being an acronym for "*m*icrowave (or *m*olecular) *a*mplification by *s*timulated *e*mission of *r*adiation." As OSCILLATORS they form the basis of extremely accurate ATOMIC CLOCKS; as AMPLIFIERS they can detect feebler signals than any other kind, and are used to measure signals from outer space.

ATOMS and MOLECULES can exist in various states with different energies; changes from one ENERGY LEVEL to another are accompanied by the emission or absorption of ELECTROMAGNETIC RADIATION of a particular frequency. Maser action is based on the fact that irradiation at the frequency concerned stimulates the process. If more atoms are in the higher energy (excited) state than in the lower state, incident waves cause more emission than absorption, resulting in amplification of the original wave.

The main difficulty is one of maintaining this arrangement of the states, as the EQUILIBRIUM configuration involves more atoms being in the lower than in the excited state. In the AMMONIA gas maser, molecules in the lower state are removed physically through their different response to an ELECTRIC FIELD, while in solid-state masers, often operated at low temperatures, a higher frequency "pumping" wave raises atoms into the excited state from some state not involved in the maser action.

MASOCHISM, mental state in which the individual gains erotic pleasure from experiencing PAIN. In PSYCHOANALYSIS, analogously, masochism describes the unconscious desire to bring humiliation upon oneself, and may again have an erotic basis. In **sadomasochism**, the individual is both sadistic (see SADISM) and masochistic, perhaps inflicting pain on himself.

MASS, a measure of the linear INERTIA of a body, i.e., of the extent to which it resists ACCELERATION when a FORCE is applied to it. Alternatively, mass can be thought of as a measure of the amount of MATTER in a body. The validity of this view seems to receive corroboration when one remembers that bodies of

Noon on the surface of Mars. Orange-red surface materials cover most of the surface, apparently forming a thin veneer over darker bedrock. The surface materials are thought to be limonite: such weathering products form on Earth in the presence of water and an oxidizing atmosphere. The reddish cast of the sky is probably due to scattering and reflection from reddish sediment suspended in the lower atmosphere.

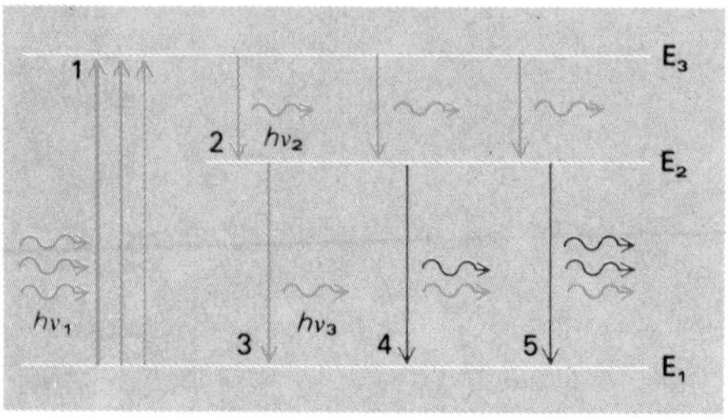

Energy-level diagram illustrating the maser principle for a three-level system. (1) Atoms of the maser material, normally in the ground state (E_1), are "pumped" to an excited state (E_3), in this case by absorbing photons (blue) of suitable energy ($h\nu_1 = E_3 - E_1$), say, from a powerful light source (optical pumping). The atoms return to the ground state by a two-stage process: first (2) falling to level (E_2), spontaneously emitting a photon ($h\nu_2 = E_3 - E_2$), and then (3) decaying to the ground state, spontaneously emitting a further photon ($h\nu_3 = E_2 - E_1$). (4) An $h\nu_3$ photon interacting with another atom in state E_2 can induce it to emit a further $h\nu_3$ photon (red) *which is in phase with* the inducing photon. This process, known as stimulated emission, can continue throughout the maser material (5) as long as there are more atoms in state E_2 than there are in E_1 (otherwise E_1 atoms will absorb more $h\nu_3$ photons than the E_2 atoms emit). This results in the build up of a large quantity of coherent radiation, effectively amplifying the original $h\nu_3$ signal. Three-level systems are also commonly employed in lasers, these using light rather than microwave photons.

equal inertial mass have identical WEIGHTS in a given gravitational field. But the exact equivalence of inertial mass and gravitational mass is only a theoretical assumption, albeit one strongly supported by experimental evidence. According to EINSTEIN's special theory of RELATIVITY, the mass of a body is increased if it gains ENERGY; according to the famous Einstein equation: $\Delta m = \Delta E/c^2$ where Δm is the change in mass due to the energy change ΔE, and c is the electromagnetic constant. It is an important property of nature that in an isolated system mass-energy is conserved. The international standard of mass is the international prototype KILOGRAM.

MASSIF, plateau-like upland area, with abrupt margins and often complex geologic structure. The term is most often applied to the Massif Central, France. (See also MOUNTAIN.)

MASS NUMBER (A), the total number of nucleons (PROTONS and NEUTRONS) in the nucleus of an ATOM, written as a number following its name after a hyphen (e.g., oxygen-16), or as a superscript following its chemical symbol (e.g., O^{16}).

MASS PRODUCTION, the production of large numbers of identical objects, usually by use of mechanization. The root of mass production is the assembly line, essentially a conveyer belt which transports the product so that each worker may perform a single function on it (e.g., add a component). The advantages of mass production are cheapness and speed; the disadvantages are the lack of job satisfaction for the workers and the resultant sociological problems.

MASS SPECTROSCOPY, spectroscopic technique in which electric and magnetic fields are used to deflect moving charged particles according to their mass, employed for chemical ANALYSIS, separation, ISOTOPE determination or finding impurities. The apparatus for obtaining a mass spectrum (i.e., a number of "lines" of distinct charge-to-mass ratio obtained from the beam of charged particles) is known as a mass spectrometer or mass spectrograph, depending on whether the lines are detected electrically or on a photographic plate. In essence, it consists of an ion source, a vacuum chamber, a deflecting field and a collector. By altering the accelerating voltage and deflecting field, particles of a given mass can be focused to pass together through the collecting slit.

MASTECTOMY, removal of a BREAST including the skin and nipple; LYMPH nodes from the armpit and some CHEST wall muscles may also be excised. Mastectomy, often with RADIATION THERAPY, is used for breast CANCER.

MASTIC, resinous exudate obtained from the lentisk pistache (*Pistacia lentiscus*), which is an evergreen shrub native to the Mediterranean region. Mastic was used by the ancient Egyptians for embalming and is now used in varnish, as a theatrical fixative and for temporary teeth fillings.

MASTIGOPHORA. See PROTOZOA.

MASTOID, air spaces lined by mucous membrane lying behind the middle EAR and connected with it; they are situated in the bony protruberance behind the ear. Mastoid infection may follow middle ear infection; block to its drainage by INFLAMMATION and PUS may make eradication difficult. ANTIBIOTICS have reduced its incidence and SURGERY to clear or remove the air spaces is now infrequent.

MATCH, short splint of wood or cardboard having a head that can be ignited by friction, used to kindle fire (see COMBUSTION). Early matches were complex, unreliable and somewhat dangerous (e.g., dipping a match treated with potassium chlorate and sugar into a bottle of concentrated sulfuric acid). Friction matches of the modern type were first produced in 1827, containing antimony (III) sulfide and potassium chlorate. Soon white PHOSPHORUS was introduced for strike-anywhere matches. This, however, caused the disease "phossy jaw" in match-factory workers, and was banned from about 1900, being replaced by phosphorus sesquisulfide (P_4S_3) and potassium chlorate, with iron (III) oxide, ground glass and glue. Safety matches have in the head potassium chlorate, manganese (IV) oxide, sulfur, iron oxide, ground glass and glue. They ignite only when struck on the mixture on the side of the box, which consists of red phosphorus, antimony (III) sulfide and an abrasive. The matchstick is coated with paraffin wax to give a better flame.

MATERIALISM, in philosophy, as opposed to IDEALISM, any view asserting the ontologic primacy of MATTER; in psychology, any theory denying the existence of MIND, seeing mental phenomena to be the mere outworking of purely physico-mechanical processes in the BRAIN; in the philosophy of religion, any synthesis denying the existence of an immortal soul in man. The earliest thoroughgoing materialists were the classical ATOMISTS, in particular DEMOCRITUS and LUCRETIUS. The growth of modern science brought a revival of materialism, which many have argued is a prerequisite for scientific thought, particularly in the field of psychology. Other philosophers, however, have argued against this view, recognizing the arbitrariness of the materialist hypothesis.

MATERIALS, Strength of, a branch of MECHANICS concerned with the behavior of materials when subjected to loads. When force is applied to an object there is a tendency for it to deform: the internal forces resulting from the applied force are called stresses; the deformations are called strains.

Stress. The four main types of stress are: shearing (e.g., the forces set up in a rivet joining two plates that are pulling in opposite directions); bending; tension and compression (which tend to elongate or shorten the member), and torsion (twisting). When we analyze a structure, we are concerned with the stresses that each component is called upon to resist, and its ability to do so without undue strain.

Strain. Materials deform in different ways under load. Basic properties include ELASTICITY, where a material regains its original dimensions when load is removed; plasticity, where the deformation is permanent; brittleness, where deformation is negligible before fracture, and creep, deformation under a constant load over a period of time. (See also DUCTILITY; HARDNESS; MALLEABILITY.) Within limits, elastic materials deform in proportion to the stress; i.e., $\frac{\text{stress}}{\text{strain}} = \text{a constant}$ (Hooke's Law). The value of the constant depends on the material and on the type of stress. For tensile and compressive forces it is called Young's modulus, E (see YOUNG, Thomas); for SHEARING forces, the shear modulus, S; and, for forces affecting the VOLUME of the object, the bulk modulus, B. There comes a point (the elastic limit), however, when further stress results in a permanent deformation. (See also TENSILE STRENGTH.)

MATHEMATICAL LOGIC, or symbolic logic. See BOOLE, GEORGE; LOGIC.

MATHEMATICAL MODELS, physical objects used to represent mathematical abstractions; or, more frequently, mathematical constructions (formulae, FUNCTIONS, GRAPHS, etc.) used to represent physical phenomena. Such models occur throughout applied mathematics and physics, their greatest value being heuristic; i.e., the model may suggest the existence of unsuspected properties in the phenomenon.

MATHEMATICS, commonly abbreviated to Maths or Math, the fundamental, interdisciplinary tool of all science. It can be divided into two main classes, pure and applied mathematics, though there are many cases of overlap between these. Pure mathematics has as its basis the abstract study of quantity, and thus includes the sciences of NUMBER— ARITHMETIC and its broader realization, ALGEBRA—as well as the subjects described collectively as GEOMETRY (e.g., ANALYTIC GEOMETRY, EUCLIDEAN GEOMETRY, MENSURATION, NON-EUCLIDEAN GEOMETRY, TRIGONOMETRY and sometimes TOPOLOGY) and their extensions, the subjects described collectively as ANALYSIS (particularly CALCULUS and some aspects of Analytic Geometry and VECTOR ANALYSIS). In modern mathematics, many of these subjects are treated in terms of SET THEORY.

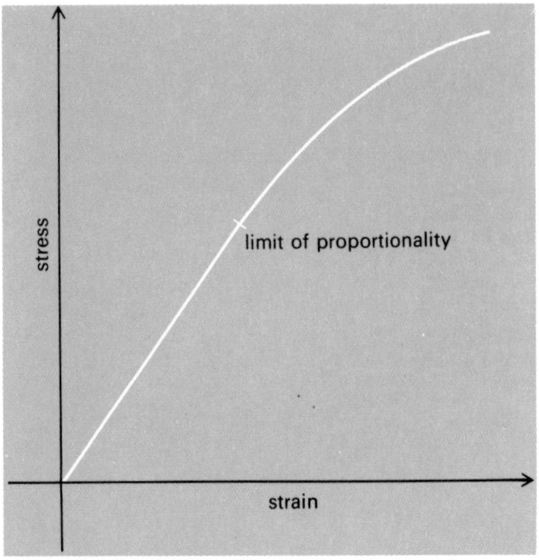

Part of the stress/strain curve for a typical metal. Below the limit of proportionality, Hooke's law applies and the deformation is generally elastic. Above the elastic limit (not the same as the limit of proportionality), deformation is plastic and a permanent set remains when the deforming stress is relaxed. A few alloys, including some steels, also exhibit a yield-point phenomenon in which on reaching the elastic limit a plastic strain wave sweeps through the specimen, instantly reducing the load and leaving a considerable "yield point extention", before renewed increasing stress gives rise to normal plastic deformation.

Applied mathematics deals with the applications of this abstract science. It thus has particulary close associations with PHYSICS and ENGINEERING. Specific subjects that come under its aegis are boolean algebra (an application of Set Theory to LOGIC—see BOOLE, GEORGE); GAME THEORY; INFORMATION THEORY; PROBABILITY; STATISTICS, and VECTOR ANALYSIS.

MATING RITUALS. In most animal species the close approach of one individual to another is an aggressive action. Yet, for reproductive purposes such a male–female approach must be made. Thus complicated behavior patterns have evolved from a combination of submissive gestures and ritualized DISPLACEMENT ACTIVITIES, to appease the sexual partner. In many cases these rituals also involve the development of morphological display characters. Mating rituals may also serve as an isolating mechanism to ensure that mating is with another member of the same species.

MATRICES, arrays of numbers (such as the COEFFICIENTS of a set of simultaneous linear EQUATIONS) of the form

$$\begin{matrix} a_{11} & a_{12} & \cdots & a_{1n} \\ a_{21} & a_{22} & \cdots & a_{2n} \\ \cdot & \cdot & \cdots & \cdot \\ a_{m1} & a_{m2} & \cdots & a_{mn} \end{matrix} = \mathbf{A}.$$

A is described as an $m \times n$ matrix over F, where F is a FIELD, without characteristic, to which all the mn elements of **A** belong. **A** is made up of $m1 \times n$ matrices (row vectors) and $nm \times 1$ matrices (column vectors).

Addition: If two matrices **A** and **B** of F are both

The History of Mathematics

Ancient Babylonia: practical arithmetic and geometry; use of both duodecimal and decimal systems.

Ancient Egypt: arithmetic and cadastral survey; use of decimal system.

Classical Greece:
Thales (idea of pure geometry)
Pythagoras (origin of deductive geometry; special numbers and ratios)
Eudoxus of Cnidus ("method of exhaustion")

Hellenistic Period:
Euclid (systematization of deductive geometry)
Archimedes (applied mathematics)
Apollonius (conics)
Hero (algebraic method)
Diophantus (algebraic notation)

Middle Ages: Khwarizmi (development and name of algebra)

10th–15th Centuries:
Rediscovery of ancient knowledge; introduction of Arabic (Hindu) numerals to West.
Fibonacci (c1180–c1240: rediscovery of Arabic (Greek) mathematics)

16th Century:
Tartaglia (1537), Cardano (1545) (solutions for cubic equations)
Bombelli (1572: complex numbers)
Vieta (1540–1603: algebraic problems solved trigonometrically)
Napier (1550–1617: logarithms); Briggs (1561–1630: logarithm tables)

17th Century:
development of calculus:
Kepler (1571–1630: conics as modified circles)
Wallis, John (1616–1703: mathematics of infinities; limits)
Newton, Isaac (1652–1726), Leibniz (1646–1716) (calculus)
Bernouilli, Jacques & Jean (application of calculus)
Descartes (1596–1650: coordinate geometry)
Desargues (1593–1662: projective geometry)
Pascal (1623–1662), Fermat (1601–1665) (statistical) probability)

18th Century:
Cotes, Roger (1686–1716), De Moivre, Abraham (1667–1754) (complex numbers used in trigonometry)
Maupertuis, Pierre de (1698–1759: principle of least action)
Euler (1707–1783: development of analysis; foundation of topography)
Lagrange (1736–1813: calculus of variations)
Laplace (1749–1827: mathematical physics)

19th Century:
Gauss (1777–1855: number theory; method of least squares)
Bolyai Farkas (1775–1856), Bolyai János (1802–1860), Lobachevski (1792–1856), Riemann, G. F. R. (1826–1866) (non-Euclidean geometries)
Dedekind, J. W. R. (1831–1916: Dedekind cuts; number theory)
Cantor, Georg (1845–1918: theory of infinite sets)
Möbius, A. F. (1790–1868: topology)
Babbage, Charles (1792–1871: calculating engines)
Hamilton, R. W. (1805–1865: quaternions)
Grassmann, Hermann (1809–1877: vector analysis)
Boole, George (1815–1864: Boolean algebra)
Cayley, Arthur (1821–1895: matrices)
Venn, John (1834–1923: graphical solutions in set theory)

20th Century:
Einstein, Albert (1879–1953: theories of relativity)
von Neumann, John (1903–1957: game theory)
Russel, Bertrand (1872–1970: mathematical logic)
Gödel, Kurt (1906–   : Gödel's theorem)

$m \times n$, the result of their ADDITION is defined to be the $m \times n$ matrix of which a typical element is $(a_{ij} + b_{ij})$. Their addition is thus commutative and associative (see ALGEBRA). Moreover, the set of $m \times n$ matrices of F forms an ABELIAN GROUP under addition since it has an identity element (see GROUPS), **O**, the zero matrix whose elements are all ZERO, since $\mathbf{A} + \mathbf{O} = \mathbf{A}$ for all **A**; and since every **A** has an INVERSE $-\mathbf{A}$, whose typical element is $(-a_{ij})$, such that $\mathbf{A} + (-\mathbf{A}) = \mathbf{O}$.

Multiplication: If l is an element of F, the product $l\mathbf{A}$ is defined as the matrix whose typical element is la_{ij}. Multiplication of one matrix by another is possible only if the first has the same number of columns as the second has rows. If **A** is an $m \times n$ matrix and **B** an $n \times p$ matrix then their product **AB** is an $m \times p$ matrix with typical element

$$a_{i1}b_{1j} + a_{i2}b_{2j} + \ . \ . \ . \ + a_{in}b_{nj}.$$

Notice that, in this case, **BA** does not exist unless $p = m$ since otherwise **B** does not have the same number of columns as **A** has rows.

Transposition and Symmetry. The transpose of an $m \times n$ matrix **A** is an $n \times m$ matrix **A′**, obtained by setting the row vectors of **A** as the column vectors of **A′**, the column vectors of **A** as the row vectors of **A′**. Some basic results emerge:

$$(\mathbf{A}')' = \mathbf{A},$$
$$(\mathbf{AB})' = \mathbf{B}'\mathbf{A}',$$
and $$(k\mathbf{A} + l\mathbf{B})' = k\mathbf{A}' + l\mathbf{B}',$$

where k, l, are elements of F. A matrix **A** is termed symmetric when $\mathbf{A} = \mathbf{A}'$ and skew-symmetric when $\mathbf{A} = -\mathbf{A}'$: of course, only matrices where $n = m$ can be symmetric or skew-symmetric. For symmetry the element a_{ij} in **A** must equal a_{ji} for all i, j; for skew-symmetry $a_{ij} + a_{ji} = 0$ for all i, j. (See also DETERMINANT.)

MATTER, material substance, that which has extension in space and time. All material bodies have inherent INERTIA, measured quantitatively by their MASS, and exert gravitational attraction on other such bodies. Matter may also be considered as a specialized form of ENERGY. There are three physical states of matter: solid, liquid and gas. An ideal solid tends to return to its original shape after forces applied to it are removed. Solids are either crystalline or amorphous; most melt and become liquids when heated. Liquids and gases are both FLUIDS: liquids are only slightly compressible but gases are easily compressed. On the molecular scale, the state of matter is a balance between attractive intermolecular forces and the disordering thermal motion of the molecules. When the former predominate, MOLECULES vibrate about fixed positions in a solid crystal LATTICE. At higher temperatures, the random thermal motion of the molecules predominates, giving a featureless gas structure. The short-range intermolecular order of a liquid is an intermediate state between solid and gas.

MAUPERTUIS, Pierre Louis Moreau de (1698–1759), French mathematician and astronomer who showed the EARTH to be flattened at the poles (1738), and who formulated the *principle of least action*, which assumes that, in nature, phenomena such as the motion of bodies occur with maximum economy.

MAUVE, or mauveine, the first synthetic DYE, made by William PERKIN (1856).

MAXIM, US family of inventors, best known for work on FIREARMS and EXPLOSIVES. **Sir Hiram Stevens Maxim** (1840–1916), invented the Maxim MACHINE GUN (c1884) and contributed to the development of CORDITE. He patented hundreds of other inventions. His brother **Hudson Maxim** (1853–1927), a chemist, developed explosives much used in WWII, and notably the high explosive maximite. Sir Hiram's son **Hiram Percy Maxim** (1869–1936), invented the Maxim silencer for firearms and the Maxim MUFFLER for automobiles.

MAXIMA. See EXTREMUM.

MAXWELL, James Clerk (1831–1879), British physicist whose contributions to science have been compared to those of Newton and Einstein. His most important work was in ELECTROMAGNETISM (he pointed to the electromagnetic nature of LIGHT) and THERMODYNAMICS. Most important of all was his derivation of the equations that bear his name, four equations that together describe in terms of the relevant VECTOR quantities the interrelation between electric and magnetic fields in a particular space. (See also ELECTROMAGNETIC RADIATION.)

MAYER, Julius Robert von (1814–1878), German physician and physicist who contributed to the formulation of the law of conservation of ENERGY.

MAYER, Maria Goeppert (1906–1972), German-born US physicist awarded, with J. H. D. JENSEN and E. P. WIGNER, the 1963 Nobel Prize for Physics for her proposal (independent of Jensen's) that the PROTONS and NEUTRONS of the atomic nucleus are arranged in concentric shells.

MAYO, distinguished US family of surgeons. **William Worrall Mayo** (1819–1911), founded St. Mary's Hospital, Rochester, Minn. (1889), which was to become the famous Mayo Clinic. His sons, **William James Mayo** (1861–1939) and **Charles Horace Mayo** (1865–1939), traveled to many countries both to discover new surgical techniques and to attract foreign surgeons to the Clinic; in 1915 they set up the Mayo Foundation for Medical Education and Research. Charles' son, **Charles William Mayo** (1898–1968), was also a distinguished surgeon.

MAZE, a prime tool of animal psychology studies. The animal is introduced at the "start" position and must solve the maze to reach an incentive, usually food. There are two distinct aims in the use of mazes: to examine the animal's learning ability; or to test its intelligence and/or senses. Mazes may be very simple:

Maxwell's equations

$$\operatorname{div} \mathbf{D} = \rho$$
$$\operatorname{div} \mathbf{B} = 0$$
$$\operatorname{curl} \mathbf{H} = \mathbf{i} + \frac{\partial \mathbf{D}}{\partial t}$$
$$\operatorname{curl} \mathbf{E} = -\frac{\partial \mathbf{B}}{\partial t}$$

where **D** is the electric displacement;
ρ is the electric charge density;
B is the magnetic flux density (induction);
H is the magnetic field strength;
i is the current density, and
E is the electric field strength.

some have no more than a single straight passage.

MEAD, George Herbert (1863–1931), US social psychologist and philosopher. Initially influenced by HEGEL, he then moved toward PRAGMATISM. He attempted to explain social psychology in terms of the evolution of the self, and through analyses of spoken language.

MEAD, Margaret (1901–), US cultural anthropologist best known for books such as *Coming of Age in Samoa* (1928), *Growing Up in New Guinea* (1930), *The Mountain Arapesh* (3 vols., 1938–49), and *Male and Female* (1949). A first autobiography, *Blackberry Winter*, appeared in 1972.

MEAN, MEDIAN AND MODE, three terms concerned with different types of averaging processes. An average, in the simplest arithmetical sense, of n terms is the SUM of those terms divided by n. Hence the average of 4, 6 and 9 is $\frac{19}{3}$ or 6.333 . . . . The average value A of a FUNCTION between $x=a$ and $x=b$ is defined as the AREA under the curve of the function (see CALCULUS) divided by $(b-a)$:

$$A = \frac{\int_a^b f(x)\,dx}{b-a}$$

Mean. The simple arithmetical average described above is an arithmetic mean (see also PROGRESSION). The geometric mean of n numbers is defined as the nth ROOT of their PRODUCT. The geometric mean of 4, 6 and 9 is thus $\sqrt[3]{4\times6\times9} = \sqrt[3]{216} = 6$. That of x and y is $\sqrt{xy}$. The geometric mean of a set of positive terms is always less than their arithmetic mean.

Median and Mode. In STATISTICS, ranking all the observations of a sample in increasing order of frequency, the frequency of the middle observation (or, if there is an even number of observations, the arithmetic mean of the two middle observations), is described as the median. The observation with the highest frequency is termed the mode: should a sample contain two (or three) observations of equal frequency greater than that of any of the other observations, it is termed bimodal (or trimodal).

MEASLES, common INFECTIOUS DISEASE, caused by a VIRUS. It involves a characteristic sequence of FEVER, HEADACHE and malaise followed by CONJUNCTIVITIS and RHINITIS, and then the development of a typical rash, with blotchy ERYTHEMA affecting the SKIN of the FACE, trunk and limbs. COUGH may indicate infections in small BRONCHI and this may progress to virus PNEUMONIA. Secondary bacterial infection may lead to middle EAR infection or pneumonia. ENCEPHALITIS is seen in a small but significant number of cases and is a major justification for VACCINATION against this common childhood disease. Recently, an abnormal and delayed IMMUNITY to measles virus has been associated with a number of BRAIN diseases, including MULTIPLE SCLEROSIS.

MEASURE. See WEIGHTS AND MEASURES.

MEASURING INSTRUMENTS. See INSTRUMENTS, SCIENTIFIC.

MEAT, the flesh of any animal, in common use usually restricted to the edible portions of cattle (beef and veal), sheep (lamb) and swine (pork), and less commonly applied to those of the rabbit, horse, goat and deer (venison). Meat consists of skeletal MUSCLE, connective TISSUE, FAT and BONE; the amount of connective tissue determines the toughness of the meat. Meat is an extremely important foodstuff. A daily intake of 100g ($3\frac{1}{2}$oz) provides 45% of daily PROTEIN, 36% of daily iron and important amounts of B VITAMINS, but only 9% of daily energy. Meat protein is particularly valuable as it supplies eight of the AMINO ACIDS which human beings cannot make for themselves. The meat-packing industry employs about 350000 people in the US.

MECHANICAL DRAWING, or engineering drawing, the representation of a component or structure in such a way that it can be formed or assembled by someone else without error or misunderstanding. To this end, various projections are used—isometric and orthographic being the most important (see DESCRIPTIVE GEOMETRY; PROJECTIVE GEOMETRY). A person who executes mechanical drawings is a draftsman. (See also PERSPECTIVE.)

MECHANICS, the branch of applied mathematics dealing with the actions of forces on bodies. There are three branches: kinematics, which deals with relationships between distance, time, velocity and acceleration; dynamics, dealing with the way forces produce motion, and statics, dealing with the forces acting on a motionless body.

Kinematics. In kinematics we deal with distance and time, which are SCALAR quantities, and with VELOCITY and ACCELERATION, which are VECTOR quantities. Velocity is the rate of change of position of a body in a particular direction with respect to time: it thus has both magnitude and direction. Its magnitude is the scalar quantity speed (S), related to distance (s) and time (t) by the equation $S=s/t$; similarly velocity ($\mathbf{v}$) is related to distance and time by $\mathbf{v}=ds/dt$ (see CALCULUS). Acceleration is rate of change of velocity with respect to time: $\mathbf{a}=d\mathbf{v}/dt=d^2s/dt^2$. Thus, if velocity is measured in km/s, acceleration is measured in km/s^2.

Often we have to consider the combination of velocities in different directions. Consider a ship sailing due E at velocity $\mathbf{x}$, and carried N by a current with velocity $\mathbf{y}$. The resultant velocity $\mathbf{z}$ can be found by using a diagram where the arrows represent the velocities in both magnitude and direction or by TRIGONOMETRY. Velocities can be resolved in a similar way (by resolving we mean simply finding its components in different directions).

Dynamics is based on NEWTON's three laws of motion: that a body continues in its state of motion unless compelled by a force to act otherwise; that the rate of change of motion (acceleration) is proportional to the applied force and occurs in the direction of the force, and that every action is opposed by an equal and opposite reaction. The first gives an idea of INERTIA, which is proportional to the MASS and opposes the change of motion: combining the first with the second, we find that $\mathbf{F} \propto m\mathbf{a}$, where $\mathbf{F}$ is the force, m the mass of the body and $\mathbf{a}$ the acceleration produced by the FORCE. In practice we choose units such that $\mathbf{F}=m\mathbf{a}$. (See also MOMENTUM.)

Newton suggested that gravitational attraction existed between all bodies, and proposed a law to describe this: if two bodies, masses m_1 and m_2 are separated by distance d, the force of attraction, $\mathbf{F}$, between them is given by $\mathbf{F}=G(m_1m_2)/d^2$, where G is the universal gravitational constant (see GRAVITATION). Near the surface of the earth we find for a body of mass m that $\mathbf{F} \propto m$ (G and the mass of the earth are constant and the distance from the surface to the center is approximately so). We usually set $\mathbf{F} = m\mathbf{g}$ where $\mathbf{g}$ is another constant, the ACCELERATION due to gravity (usually denoted g although it is a vector).

Statics. We can combine forces much as we do velocities. If forces $\mathbf{P}_1$ and $\mathbf{P}_2$, with resultant $\mathbf{R}$, act at a point, there must be a third force, $\mathbf{P}_3$, equal and opposite to $\mathbf{R}$, for equilibrium. We can combine forces similarly over more complex structures. For a bridge whose weight (w) acts as a downward force $\mathbf{W}$ at the center, the upward reactions $\mathbf{R}_1$ and $\mathbf{R}_2$ at the piers must be such that $\mathbf{R}_1 + \mathbf{R}_2 = -\mathbf{W}$ for it to be in EQUILIBRIUM. Similarly, we examine the members of the bridge to determine the stresses acting on each (see MATERIALS, STRENGTH OF; VECTOR ANALYSIS).

MECHANIZATION AND AUTOMATION, the use of machines wholly or partly to replace human labor. The two words are often used synonymously, but it is of value to distinguish mechanization as requiring human aid, automation as self-controlling.

The most familiar automated device is the domestic THERMOSTAT. This is set to switch off the heating circuit if room temperature exceeds a certain value, to switch it on if the temperature falls below a certain value. Once set, no further human attention is required: a machine is in full control of a machine.

The thermostat is a sensing element; the information it detects is fed back to the production mechanism (the heater), which adjusts accordingly. All automated processes work on this principle. In fact, fully automated processes are still rare: most often the role of sensing element will be taken over by a human being, who will check the accuracy of the machine and adjust it if needed.

The most versatile devices we have are COMPUTERS: very often, the complexity of their physical construction is more than matched by that of the network of subprograms which they contain. Data can be fed in automatically or by human operators and the computer can be programmed to respond in many ways: to present information; adjust and control other machines, or even to take decisions. Computerized automation plays a larger role in our lives than most of us realize: airline and theater agents often book seat reservations with a computer, not a staffed box office; food manufacture is often automatically controlled from raw materials to packaged product; atomic energy is controlled automatically where radiation prohibits the presence of humans, possible leaks or even explosions being forestalled by machine; the justification of the columns of this book has been performed by a fully automated process. In addition, man would not have reached the moon had it not been for computerized automation.

Since mechanization and automation emerged in the "Second Industrial Revolution," they have been associated with all kinds of sociological problems and upheavals. And this is more than ever true today. Long-term benefits to the human race have to be balanced against short-term evils such as unemployment and its attendant human sufferings. (See also LINEAR PROGRAMMING; MASS PRODUCTION; MACHINE TOOLS; SERVOMECHANISM.)

MECHNIKOV, Ilya. See METCHNIKOFF, ÉLIE.

MEDAWAR, Sir Peter Brian (1915–), British zoologist who shared with F. M. BURNET the 1960 Nobel Prize for Physiology or Medicine for their work on immunological tolerance. Inspired by Burnet's ideas, Medawar showed that if fetal mice were injected with cells from eventual donors, skin grafts made onto them later from those donors would "take," thus showing the possibility of acquired tolerance and hence, ultimately, organ TRANSPLANTS.

MEDIAN. See MEAN, MEDIAN AND MODE.

MEDICINE, the art and science of healing. Within the last 150 years or so medicine has become dominated by scientific principles. Prior to this, healing was mainly a matter of tradition and magic. Many of these prescientific attitudes have persisted to the present day.

The earliest evidence of medical practice is seen in Neolithic (see STONE AGE) skulls in which holes have been bored, presumably to let evil spirits out, a practice called trepanning (see TREPHINE). Treatment in primitive cultures was either empirical or magical. Empirical treatment included bloodletting, dieting, primitive surgery and the administration of numerous potions, lotions and herbal remedies (some used in modern medicines). For serious ailments, magical treatment, involving propitiation of the gods, special rituals or the provision of charms, was performed by the medicine man or witch doctor, who was usually both doctor and priest (see SHAMANISM). Exorcism, the casting out of devils, and FAITH HEALING are still practiced in modern societies. ACUPUNCTURE and OSTEOPATHY, both being ancient and empirical, are also practiced today.

The growth of scientific medicine began with the Greek philosophy of nature. The great Greek physician HIPPOCRATES, with whose name is associated the Hippocratic oath which codifies the physician's ideals of humanity and service, has justly been called the father of medicine. Galen of Pergamum, the encyclopedist of classical medicine, clearly distinguished ANATOMY from PHYSIOLOGY. Medieval medicine was basically a corrupted Galenism. The 16th century saw the dawn of modern medicine. Men such as FABRICIUS, VESALIUS and William HARVEY revived the critical, observational approach to medical research. Perhaps the most far-reaching advances since then have been in preventive medicine, anesthesia and drug therapy. Preventive medicine was attempted in medieval times when ships arriving in Europe during the Black Death were "quarantined" for 40 days. More recent major milestones have been Edward JENNER's work on VACCINATION and the "germ theory of disease" proposed by Louis PASTEUR and developed by Robert KOCH. ANAESTHESIA and ASEPSIS (see LISTER, JOSEPH) made possible great advances in SURGERY. Crawford LONG and James SIMPSON were both pioneers of their use. Drug therapy originated with herbal remedies, but perhaps the two most important discoveries in this field both came in the 20th century: that of INSULIN by Frederick BANTING and Charles BEST, and that of PENICILLIN by Alexander FLEMING. (See also ANIBIOTICS; CHEMOTHERAPY; DRUGS; SULFA DRUGS.)

Medical training to high set standards is used to protect society against charlatans and is usually undertaken in universities and hospitals. Since the progress of medical knowledge is very rapid, doctors today undergo continual retraining to keep them up to date. Socialized medicine, under the name MEDICARE, was set up in the US in 1965 and helps to pay the costs of medical care. However, several other

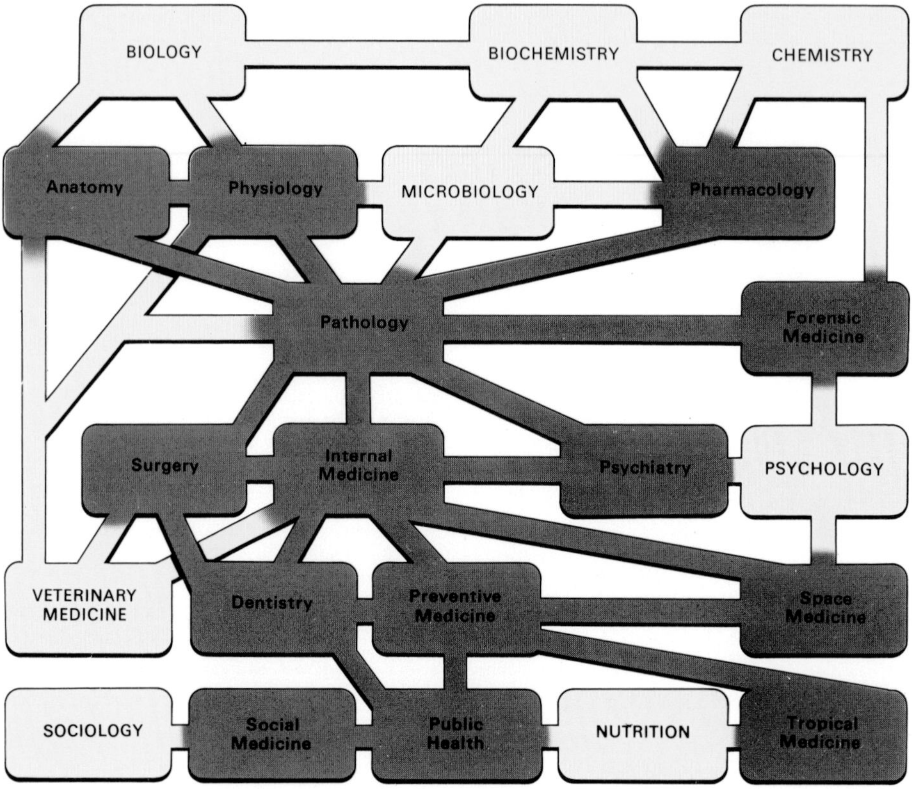

The principal branches of medicine (red) and their links with allied sciences (yellow). The upper part of the diagram represents the medical sciences. These are mediated through pathology—the study of diseased states of organisms—to the clinical and practical specialities appearing in the lower part of the diagram. The specialties of clinical medicine—obstetrics, pediatrics, geriatrics, gynecology, anesthetics, orthopedics, and others dealing with disease in particular organs or anatomical systems—are for the most part omitted, being subdivisions of internal medicine and surgery. Similarly, veterinary, tropical and space medicine are but representatives of a large group of practical subdivisions, identified by their application in particular fields and each drawing on the whole range of the medical sciences. Pseudomedical practices and the varieties of "alternative medicine" are likewise omitted.

countries have more comprehensive programs of socialized medicine.

The success of medicine in preventing disease is largely responsible for today's population explosion. This has stimulated an extensive re-examination of traditional attitudes to medical ethics, particularly in the areas of CONTRACEPTION, ABORTION and EUTHANASIA. (See also DISEASE.)

MEDICINE MAN. See SHAMANISM.

MEERSCHAUM, or **sepiolite**, fibrous CLAY mineral, light and porous, consisting of hydrated magnesium silicate. Chiefly occurring in Turkey, it is used to make tobacco pipes.

MEG-, mega- (M), SI prefix multiplying a unit one-millionfold. Examples include the megahertz (MHz), megametre (Mm) and megohm (MΩ). One megagram is one tonne (t). (See SI UNITS.)

MEGALITHIC MONUMENTS, large, usually undressed stone monuments found principally in Europe but also in many other parts of the world, believed to date usually from the late STONE AGE and early BRONZE AGE. They are of four main types: the **menhir** (from Breton *hir*, long, and *men*, stone) or single standing stone; the **stone circle**, or circles, as exemplified by STONEHENGE; the chamber or room, usually associated with a tomb, the most ancient of which is the DOLMEN; and the **alignment**, or row of stones, such as those at CARNAC.

MEGAPHONE, device to amplify the sound of a voice. It amplifies the sound because of its shape (a hollow cone of metal or plastic), and increases the effective sound power by concentrating the sound in a single direction. Electric megaphones combine a MICROPHONE, AMPLIFIER and LOUDSPEAKER.

MEGATON, unit descriptive of the explosive power of the HYDROGEN BOMB. Each megaton represents the equivalent explosive power of one million tons of TNT.

MEIOSIS, a special mechanism of CELL division which results in the formation of a HAPLOID cell which is normally a GAMETE. Meiosis is similar in all plant and animal cells; there are two divisions of the nucleus in the course of which the CHROMOSOMES divide once so that 1 diploid cell gives 4 haploid daughter cells.

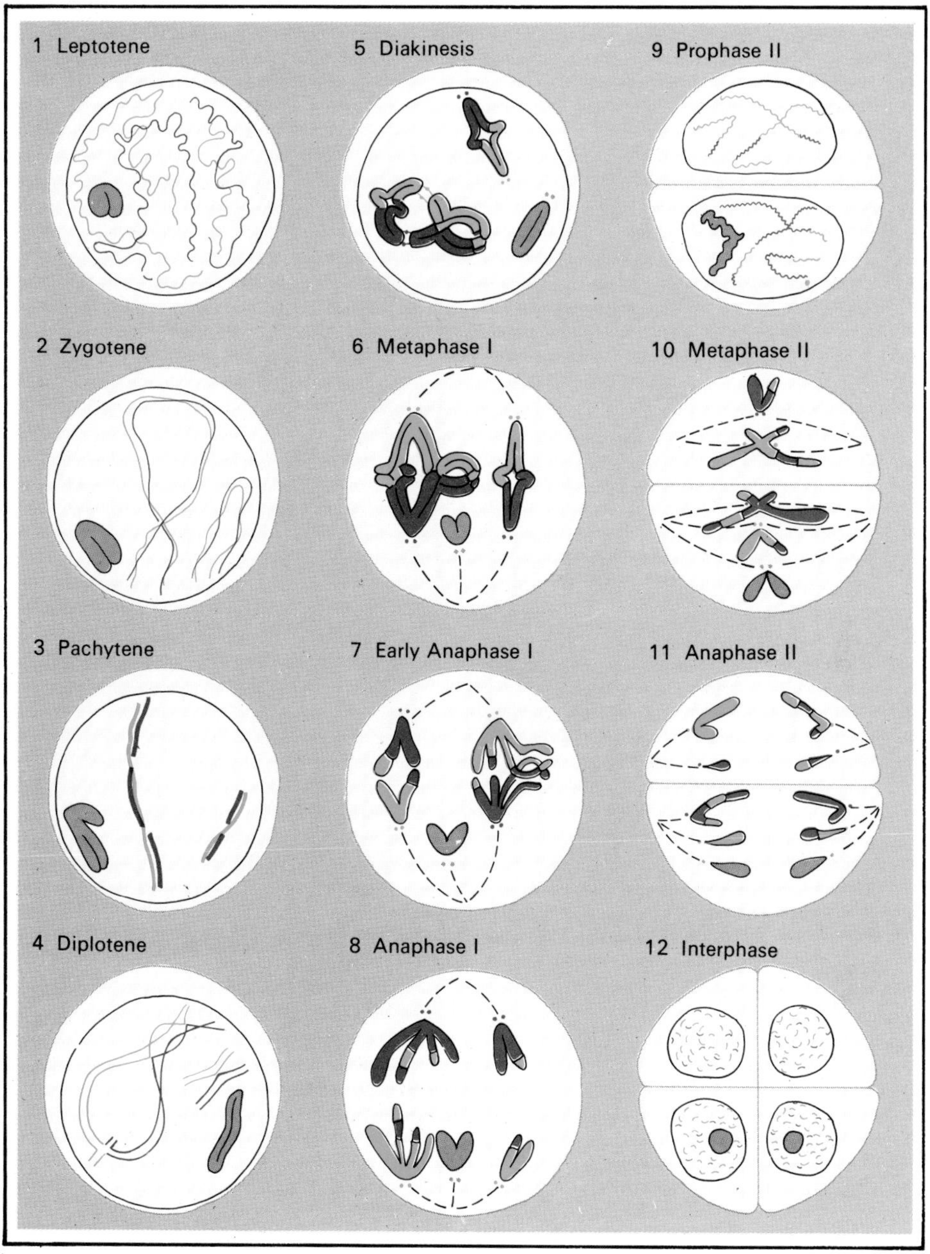

Meiosis is a complex process in diploid cells which results in the formation of the haploid cells necessary to the success of sexual reproduction. It involves two successive divisions. In the prophase of the first division (1–5), homologous chromosomes (this example involves two pairs of homologous chromosomes—the two members in each initial bivalent are colored red and blue—and a single X chromosome) pair (2), contract (3), duplicate and exchange genetic material (4–5). In the subsequent metaphase and anaphase (6–8), this chromosomal material is regrouped so that the second division process (9–12) can proceed as a normal mitosis, albeit one involving only haploid sets of genetic material. It is the opportunity for the exchange of genetic material between different homologous chromosomes during diplotene that gives rise to the evolutionary importance of sexual reproduction. In effect, the genetic variation necessary to the success of evolutionary change can occur only during the meiotic process.

Meiosis, which is often the occasion of genes "crossing over," is at the heart of genetic segregation.

MEITNER, Lise (1878–1968), Austrian physicist who worked with Otto HAHN to discover PROTACTINIUM. Following their experiments bombarding URANIUM with NEUTRONS, Meitner collaborated with her nephew, **Otto Robert Frisch** (1904–), to discover nuclear FISSION and predict the chain reaction.

MELANIN, black pigment which lies in various SKIN layers and is responsible for skin color, including the racial variation. It is concentrated in MOLES and FRECKLES. The distribution in the skin determines skin coloring and is altered by light and certain HORMONES.

MELANISM, an excessive development of the dark pigment, MELANIN, in an animal, as in the panther, a black variety of leopard. Animals without any pigmentation are termed ALBINO.

MELTING POINT, the temperature at which FUSION occurs, so that a solid and the corresponding liquid are at equilibrium (see PHASE EQUILIBRIUM). For pure compounds it has a precise value equal to the FREEZING POINT.

MEMBRANES, layers that form part of the surface of CELLS and which enclose organelles within the cells of all animals and plants. The membranes of the cell wall function to allow some substances into the cell; to exclude others, and actively to transport others into the cell even though the direction of movement may be against existing concentration gradients. Membranes are composed of layers of LIPID or FAT molecules which sandwich a layer of PROTEIN molecules. The protein layer is double but appears as a single layer when viewed under the ELECTRON MICROSCOPE. Thus most membranes appear to be triple-layered, although some appear to be composed of a single layer. Triple-layered membranes are normally 5–10nm thick.

MEMORY, the sum of the mental processes that result in the modification of an individual's behavior in the light of previous experience. There are several different types of memory. In rote memory, one of the least efficient ways of storing information, data is learned by rote and repeated verbatim. Logical memory is far more efficient: only the salient data are stored, and each may be used in its original or in a different context. Mnemonics, which assist rote memory, superimpose what is in effect an artificial logical structure on not necessarily related data. (See also EIDETIC IMAGE.) Testing of the efficiency of memory may be by recall (e.g., remembering a string of unrelated syllables); recognition (as in a multiple-choice test, where the candidate recognizes the correct answer among alternatives); and relearning, in which comparison is made between the time taken by an individual to commit certain data to memory, and the time taken to recommit it to memory after a delay. Though recent studies of certain COMPUTER functions have thrown light on some of the workings of memory (see also CYBERNETICS), little is known of its exact physiological basis. It appears, however, that chemical changes in the brain, particularly in the composition of RNA (see NUCLEIC ACIDS), alter the electrical pathways there. Moreover, it seems that some form of initial learning takes place in the NERVOUS SYSTEM before data are stored permanently in the BRAIN. (See also ELECTROENCEPHALOGRAPH; INTELLIGENCE; LEARNING.)

MENDEL, Gregor Johann (1822–1884), Austrian botanist and Augustinian monk who laid the foundations of the science of GENETICS. He found that self-pollinated dwarf pea plants bred true, but that under the same circumstances only about a third of tall pea plants did so, the remainder producing tall or dwarf pea plants in a ratio about 3:1. Next he cross-bred tall and dwarf plants and found this without exception resulted in a tall plant, but one that did not breed true. Thus, in this plant, both tall and dwarf characteristics were present. He had found a mechanism justifying DARWIN's theory of EVOLUTION by NATURAL SELECTION; but contemporary lack of interest and his later, unsuccessful experiments with the hawkweeds discouraged him from carrying this further. It was not until 1900, when H. de VRIES and others found his published results, that the import ance of his work was realized. (See also HEREDITY; POLLINATION.)

MENDELEVIUM (Md), a TRANSURANIUM ELEMENT in the ACTINIDE series, first made by bombarding einsteinium-253 with ALPHA PARTICLES.

MENDELEYEV, Dmitri Ivanovich (1834–1907), Russian chemist who formulated the Periodic Law, that the properties of elements vary periodically with increasing atomic weight, and so drew up the PERIODIC TABLE (1869). (See also MEYER, J. L.)

MENHIR. See MEGALITHIC MONUMENTS.

MÉNIÈRE'S DISEASE, disorder of the cochlea and labyrinth of the EAR, causing brief acute episodes of VERTIGO, with nausea or VOMITING, ringing in the ears and DEAFNESS. Ultimately permanent deafness ensues and vertigo lessens. It is a disorder of inner-ear fluid and each episode causes some destruction of receptor cells. Drugs can reduce the vertigo.

MENINGITIS, INFLAMMATION of the meninges (see BRAIN) caused by BACTERIA (e.g., meningococcus, pneumococcus, hemophilus) or VIRUSES. **Bacterial meningitis** is of abrupt onset with HEADACHE, vomiting, FEVER, neck stiffness and avoidance of light. Early and appropriate ANTIBIOTIC treatment is essential as permanent damage may occur in some cases, especially in children. **Viral meningitis** is a milder illness with similar signs in a less ill person; symptomatic measures only are required. **Tuberculous meningitis** is an insidious chronic type which responds slowly to antituberculous drugs. Some FUNGI, unusual bacteria and syphilis (see VENEREAL DISEASES) may also cause varieties of meningitis.

MENNINGER, Karl Augustus (1893–), US psychiatrist who, with his brother **William Claire Menninger** (1899–1966) and father **Charles Frederick Menninger** (1862–1953), set up the Menninger Foundation (1941), a nonprofit organization dedicated to the furtherance of psychiatric research.

MENOPAUSE. See MENSTRUATION.

MENSTRUATION, specifically the monthly loss of BLOOD (period), representing shedding of WOMB endometrium, in women of reproductive age; in general, the whole monthly cycle of hormone, structural and functional changes in such women, punctuated by menstrual blood loss. After each period, the womb-lining endometrium starts to proliferate and thicken under the influence of GONADOTROPHINS (FOLLICLE STIMULATING HORMONE) and ESTROGENS. In midcycle a burst of LUTEINIZING HORMONE secretion, initiated by the HYPOTHALAMUS, causes release of an egg from an ovarian follicle (*ovulation*). More PROGESTERONE is then secreted and the endometrium is prepared for IMPLANTATION of a fertilized egg. If the egg is not fertilized, PREGNANCY does not ensue and blood-vessel changes occur leading to the shedding of the endometrium and some blood; these are lost through the vagina for several days, sometimes with pain or COLIC. The cycle then restarts. During the menstrual cycle, changes in the BREASTS, body temperature, fluid balance and mood occur, the manifestations varying from person to person. Cyclic patterns are established at PUBERTY (*menarche*) and end in middle life (age 45–50) at the *menopause*, the "change of life." Disorders of menstruation include heavy, irregular or missed periods; bleeding between periods or after the menopause, and excessively painful periods. They are studied in GYNECOLOGY.

MENSURATION, the branch of GEOMETRY dealing with the measurement of LENGTH, AREA and VOLUME. The base of all such measurements is length, since the areas and volumes of geometric figures can be calculated from suitable length measurements. The area of a rectangle is bh, where b is the length of one side (the base) and h that of the other (the height) (see QUADRILATERAL). It is easy to show that this formula holds also for the parallelogram, if h stands for the altitude (the PERPENDICULAR distance from one side to that facing it) rather than for the height; from this can be found the formula for the area of a TRIANGLE (which can be thought of as half a parallelogram), $\frac{1}{2}bh$; and those for other POLYGONS.

The area and circumference of a CIRCLE, which can be considered as a regular polygon with an infinite number of infinitely small sides, are πr^2 and $2\pi r$ respectively, where π is a constant (see PI) and r is the radius. The area of an ellipse (see CONIC SECTIONS) is given by πab where a is the semi-major and b the semi-minor axis; the circumference of an ellipse cannot be expressed in algebraic terms.

From this information, it is fairly easy to determine formulae for the volumes of regular solids, such as the POLYHEDRON; CONE; CYLINDER; ELLIPSOID; PYRAMID, and SPHERE. The areas of irregular plane shapes can be approximated by considering a large number of extremely small strips, each being almost trapezoidal, formed in them by the construction of a large number of parallel chords; the same principle can be applied to finding the approximate volumes of irregular solids: this process is akin to integral CALCULUS.

MENTAL ILLNESS, or psychiatric DISEASE, disorders characterized by abnormal function of the higher centers of the BRAIN responsible for thought, perception, mood and behavior, in which organic disease has been eliminated as a possible cause. The borderline between disease and the range of normal variability is indistinct and may be determined by cultural factors. Crime may result from mental disease, but modern Western society is careful to eliminate it as far as possible before subjecting a criminal to justice. However, in certain repressive regimes, political or ideological nonconformity can be grounds for admission to mental hospital. Mental disease has been recognized since ancient times and both HIPPOCRATES and GALEN evolved theories as to its origins; but in many cultures, over the centuries, madness has been equated with possession by evil spirits and sufferers were often treated as witches. In the 15th century, PARACELSUS proposed that the moon determined the behavior of mad people (hence "lunacy"), while in the 18th century MESMER favored the role of animal magnetism (from which HYPNOSIS is derived). The first humane *asylum* for the mentally ill was founded in Paris by PINEL (1795). Originally only socially intolerable cases were admitted to such hospitals, but today voluntary admission is more common. The *Viennese school* of psychology, in particular Sigmund FREUD and his pupils, emphasized the importance of past, especially childhood experiences, sexual attitudes and other functional factors. Behavior therapy, PSYCHOANALYSIS and PSYCHOTHERAPY derive from this school. On the other hand, the influence of subtle organic factors (e.g., brain biochemistry) favored by others; this led to using LOBOTOMY, SHOCK THERAPY and DRUGS.

Mental illness may be classified into PSYCHOSIS, NEUROSIS and personality disorder. **Schizophrenia** is a psychosis causing disturbance of thought and perception in which mood is characteristically flat and behavior withdrawn. Features include: auditory hallucinations; delusions of person ("I'm the King of Spain"), of surroundings and other people (e.g., suspicion of conspiracy in PARANOIA); blocking, insertion and broadcasting of thought, and knight's-move thinking, or nonlogical sequence of ideas. Conversation lacks substance and may be in riddles and neologisms; speech or behavior may be imitative, stereotyped, repetitive or negative. Phenothiazine drugs, especially chlorpromazine and long-acting analogues, are particularly valuable in schizophrenia. In **affective psychoses**, disturbance of mood is the primary disorder. Subjects usually exhibit DEPRESSION with loss of drive and inconsolably low mood, either in response to situation (exogenous) or for no apparent reason (endogenous). Loss of appetite, CONSTIPATION and characteristic sleep disturbance also commonly occur. ANTIDEPRESSANTS and SHOCK THERAPY are valuable, but psychotherapy may also be needed. In hypnomania or MANIA, excitability, restlessness, euphoria, ceaseless talk, flight of ideas and loss of social inhibitions occur. Financial, sexual and alcohol excesses may result. Chlorpromazine, haloperidol and lithium are effective. **Neuroses** include ANXIETY, pathological exaggeration of a physiological response. This may coexist with depression but responds to benzodiazepines (Valium) and psychotherapy. Obsessional and compulsive neuroses, manifested by extreme habits, rituals and fixations (which may be recognized as irrational); PHOBIAS, excessive and inappropriate fears of objects or situations (e.g. AGORAPHOBIA), and HYSTERIA are helped by behavior therapy. Psychopathy is a specific disorder of personality characterized by failure to learn from experience. Irresponsibility, inconsiderateness and lack of foresight result, and may lead to crime. Other **personality disorders** are exhibited by a variety of people, often with unstable backgrounds, who seem unable to cope with the realities of everyday adult life; attempted suicide is a common gesture. In sexual disorders with antisocial or perverse sexual fixations, behavior therapy may be of value. (See also ALCOHOLISM; ANOREXIA NERVOSA; DRUG ADDICTION.)

MENTAL RETARDATION, low intellectual

capacity arising, not from MENTAL ILLNESS, but from impairment of the normal development of the BRAIN and NERVOUS SYSTEM. Causes include genetic defect (as in MONGOLISM); infection of the EMBRYO or FETUS; HYDROCEPHALUS or inherited metabolic defects (e.g., CRETINISM), and injury at BIRTH including cerebral HEMORRHAGE and fetal ASPHYXIA. Disease in infancy such as ENCEPHALITIS may cause mental retardation in children with normal previous development. Retardation is initially recognized by slowness to develop normal patterns of social and learning behavior and confirmed through intelligence measurements. It is most important that affected children should receive adequate social contact and education, for their development is truly retarded and not arrested. In particular, special schooling may help them to achieve a degree of learning and social competence.

MENTHOL, or 3-hydroxy-*p*-menthane, a TERPENE alcohol, an ALICYCLIC COMPOUND. It is a white crystalline solid with a pungent odor and mint flavor, having a cooling, soothing effect on the throat and nasal passages, and used in medications, cosmetics, cigarettes and flavorings. Menthol is extracted from peppermint oil or synthesized.

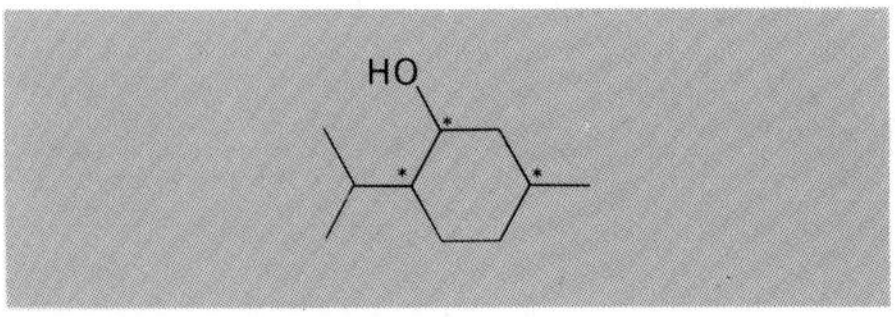

Menthol: three carbon atoms which are centers of asymmetry are starred. Commercially, the most important form is *l*-menthol.

MERBROMIN, or mercurochrome, mercurial salt formerly used for disinfection. It acts by interfering with SULFUR bridges in microorganism PROTEINS.

MERCATOR, Gerardus (1512–1594), Flemish cartographer and calligrapher, best known for Mercator's Projection (see MAP), which he first used in 1569 for a world map. The PROJECTION is from a point at the center of the earth through the surface of the globe onto a cylinder that touches the earth around the equator.

MERCERIZING, technique patented in 1850 by John Mercer (1791–1866) for making textiles take DYE more readily and show more brilliant colors (see TEXTILES). The yarn or cloth (usually under tension to reduce shrinkage) is steeped in a concentrated caustic soda (see SODIUM) solution.

MERCUROCHROME See MERBROMIN.

MERCURY (Hg), or **quicksilver**, silvery-white liquid metal in Group IIB of the PERIODIC TABLE; an anomalous TRANSITION ELEMENT. It occurs as CINNABAR, calomel and rarely as the metal, which has been known from ancient times. It is extracted by roasting cinnabar in air and condensing the mercury vapor. Mercury is fairly inert, tarnishing only slowly in moist air, and soluble in oxidizing acids only; it is readily attacked by the HALOGENS and sulfur. It forms Hg^{2+} and some Hg_2^{2+} compounds, and many important ORGANOMETALLIC COMPOUNDS. Mercury and its compounds are highly toxic. The metal is used to form AMALGAMS; for electrodes, and in barometers, thermometers, diffusion PUMPS, and mercury-vapor lamps (see LIGHTING, ARTIFICIAL). Various mercury compounds are used as pharmaceuticals. AW 200.6, mp −39°C, bp 357°C, sg 13.546 (20°C). **Mercury (II) cyanate** ($Hg[ONC]_2$), or **mercury fulminate**, is a white crystalline solid, sensitive to percussion, and used as a detonator. **Mercury(II) chloride** ($HgCl_2$), or **corrosive sublimate**, is a colorless crystalline solid prepared by direct synthesis. Although highly toxic, it is used in dilute solution as an ANTISEPTIC, and also as a fungicide and a polymerization catalyst. mp 276°C, bp 302°C. **Mercury(I) chloride** (Hg_2Cl_2), or **calomel**, is a white rhombic crystalline solid, found in nature. It is used in ointments and formerly found use as a LAXATIVE. A calomel/mercury cell with potassium chloride electrolyte (the Weston cell) is used to provide a standard ELECTROMOTIVE FORCE. mp 303°C, bp 384°C.

MERCURY, the planet closest to the sun with a mean solar distance of 58Gm. Its highly eccentric ORBIT brings it within 46Gm of the sun at perihelion and takes it 70Gm from the sun at aphelion. Its diameter is about 4870km, its mass about 0.054 that of the earth. It goes around the sun in just under 88 days and rotates on its axis in about 59 days. The successful prediction by Albert EINSTEIN that Mercury's orbit would be found to advance by 43″ per century is usually regarded as a confirmation of the General Theory of RELATIVITY. Night surface temperature is thought to be about 110K, midday equatorial temperature over 600K. The planet's average density (5.2 grams per cubic centimeter) indicates a high proportion of heavy elements in its interior. Mercury has little or no atmosphere and no known moons.

MERCURY POISONING, the cause of acute GASTROINTESTINAL TRACT and KIDNEY disease if mercury(II) salts are ingested. A chronic form, often from vapor inhalation, causes BRAIN changes with tremor, ataxia, irritability and social withdrawal. The mental changes ensuing from the former use of mercury in making felt hats led to the phrase "mad as a hatter." Organic mercury from fish (e.g., tuna) living in contaminated water, or from cereals treated with antifungal agents may cause ataxia, swallowing difficulty, abnormalities of VISION and COMA. Nephrotic syndrome of the kidneys may also be seen. Treatment is with dimercaprol, a chelating agent (see CHELATE).

MERCURY PROGRAM, first US manned space flights 1960–63, using the one-man Mercury capsule. In 1961 Alan Shepard and Virgil Grissom were launched on suborbital flights by the Redstone carrier missile. In 1962 John Glenn made the first US orbital flight, followed by M. Scott Carpenter, Walter Schirra and Leroy Cooper; orbital missions were launched by Atlas carriers. (See also SPACE EXPLORATION.)

MERGENTHALER, Ottmar (1854–1899), German-born US inventor of the first linotype PRINTING machine (1884), first put to commercial use in 1886.

MERIDIAN, on the celestial sphere, the great circle passing through the celestial poles and the observer's ZENITH. It cuts his HORIZON N and S. (See also CELESTIAL SPHERE; TRANSIT.) The term is used also for a line of terrestrial longitude.

MESA (Spanish, table; from Latin *mensa*), steep-sided, flat-topped area formed beneath a horizontal cap of hard rock where the surrounding softer rock has been worn away. Further EROSION of the sides produces a smaller hill, or **butte**.

MESCALINE, HALLUCINOGENIC DRUG, derived from a Mexican cactus, whose use dates back to ancient times when "peyote buttons" were used in religious ceremonies among American Indians. The hallucinations experienced during its use were among the first to be described (by Aldous HUXLEY) and resemble those of LSD.

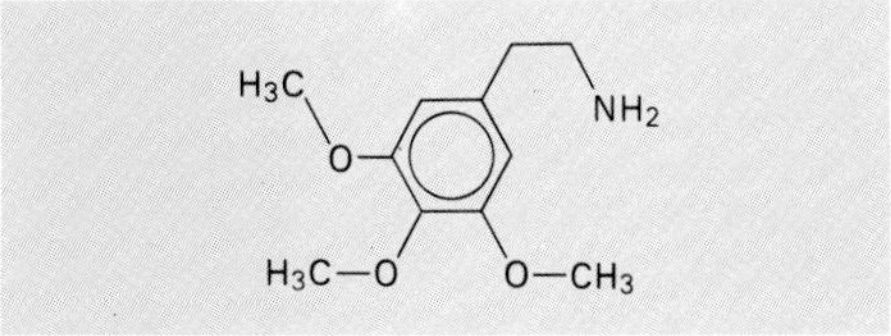

The structure of mescaline.

MESENTERY, the membranous fold in which the GASTROINTESTINAL TRACT is slung from the back wall of the ABDOMEN so that it lies relatively free and mobile in the peritoneal cavity. It consists of a double layer of peritoneum and within it lie the BLOOD vessels and LYMPH nodes and vessels of the gut.

MESMER, Franz Anton (1734–1815), German physician, controversy over whose unusual techniques and theories sparked in CHARCOT and others an interest in the possibilities of using "animal magnetism" (or mesmerism, i.e., HYPNOSIS) for psychotherapy.

MESOLITHIC AGE. See STONE AGE.

MESONS. See SUBATOMIC PARTICLES.

MESOPHYTE, any plant that grows in conditions where the water supply is neither scanty nor abundant. Mesophytes have flat, expanded leaves. (See also HYDROPHYTE; XEROPHYTE.)

MESOSPHERE, the atmospheric zone immediately above the stratosphere, marked by a TEMPERATURE maximum (about 10°C) between altitudes 48km and 53km. (See ATMOSPHERE.)

MESOZOIC, the middle era of the PHANEROZOIC, lasting from 225 until 65 million years ago. It has three periods: the TRIASSIC, JURASSIC and CRETACEOUS. (See GEOLOGY.)

METABOLISM, the sum total of all chemical reactions that occur in a living organism. It can be subdivided into **anabolism** which describes reactions which build up more complex substances from smaller ones, and **catabolism** which describes reactions which break down complex substances into simpler ones. Anabolic reactions require ENERGY while catabolic reactions liberate energy. Metabolic reactions are catalysed by ENZYMES in a highly integrated and finely controlled manner so that there is no overproduction or under utilization of the energy required to maintain life. All energy required to maintain life is ultimately derived from sunlight by PHOTOSYNTHESIS, and most organisms use the products of photosynthesis either directly or indirectly. The energy is stored in most living organisms in a specific chemical compound, adenosine triphosphate (ATP—see NUCLEOTIDES). ATP can transfer its energy to other

Part of the surface of Mercury, viewed from Mariner 10. Both recent (smooth) and older craters are visible.

molecules by a loss of phosphate, later regaining phosphate from catabolic reactions. (See also BASAL METABOLIC RATE.)

METAL, an element with high specific gravity; high opacity and reflectivity to light (giving a characteristic luster when polished); that can be hammered into thin sheets and drawn into wires (i.e., is malleable and ductile), and is a good conductor of heat and electricity, its electrical conductivity decreasing with temperature. Roughly 75% of the chemical elements are metals, but not all of them possess all the typical metallic properties. Most are found as ores and in the pure state are crystalline solids (mercury, liquid at room temperature, being a notable exception), their atoms readily losing electrons to become positive IONS. ALLOYS are easily formed because of the nonspecific nondirectional nature of the metallic bond.

METAL FATIGUE, deterioration and progressive cracking of metal parts caused by repeated and relatively low stresses. Imperfections in crystal grains, which often occur at notches, screw threads, welding defects etc., accumulate after numerous cycles in which a small inelastic strain is applied, often leading to eventual failure. Fatigue failure is guarded against by careful design.

METALLOID, or **semimetal**, an ELEMENT that has properties—physical and chemical—intermediate between those of METALS and those of NONMETALS. The metalloids—BORON, SILICON, GERMANIUM, ARSENIC, ANTIMONY, SELENIUM and TELLURIUM—form a diagonal band in the PERIODIC TABLE. They do not have high ELECTRONEGATIVITY or electropositivity, and form amphoteric OXIDES; they are SEMICONDUCTORS.

METALLURGY, the science and technology of METALS, concerned with their extraction from ores, the methods of refining, purifying and preparing them for use and the study of the structure and physical properties of metals and ALLOYS. A few unreactive metals such as silver and gold are found native (uncombined), but most metals occur naturally as MINERALS (i.e., in chemical combination with nonmetallic elements). Ores are mixtures of minerals from which metal extraction is commercially viable. Over 5000 years man has developed techniques for working ores and forming alloys, but only in the last two centuries have these methods been based on scientific theory. The production of metals from ores is known as process or extraction metallurgy; fabrication metallurgy concerns the conversion of raw metals into alloys, sheets, wires etc., while physical metallurgy covers the structure and properties of metals and alloys, including their mechanical working, heat treatment and testing. Process metallurgy begins with ore dressing, using physical methods such as crushing, grinding and gravity separation to split up the different minerals in an ore. The next stage involves chemical action to separate the metallic component of the mineral from the unwanted nonmetallic part. The actual method used depends on the chemical nature of the mineral compound (e.g., if it is an oxide or sulfide, its solubility in acids etc.) and its physical properties. Hydrometallurgy uses chemical reactions in aqueous solutions to extract metal from ore. ELECTROMETALLURGY uses electricity for firing a furnace or electrolytically reducing a metallic compound to a metal. Pyrometallurgy covers roasting, SMELTING and other high-temperature chemical reactions. It has the advantage of involving fast reactions and giving a molten or gaseous product which can easily be separated out. The extracted metal may need further refining or purifying: electrometallurgy and pyrometallurgy are again used at this stage. Molten metal may then simply be cast by pouring into a mold, giving, e.g., pig iron, or it may be formed into ingots which are then hot or cold worked, as with, e.g., wrought iron. Mechanical working, in the form of rolling, pressing or FORGING, improves the final structure and properties of most metals; it tends to break down and redistribute the impurities formed when a large mass of molten metal solidifies. Simple heat treatment such as ANNEALING also tends to remove some of the inherent brittleness of cast metals. (See also BLAST FURNACE; BRAZING; ROLLING MILLS; SINTERING; STEEL.)

METAMORPHIC ROCKS, one of the three main types of rocks of the earth's crust. They consist of rocks that have undergone change owing to heat, pressure or chemical action. SEDIMENTARY ROCKS undergo *prograde metamorphism*, by which they lose volatiles such as WATER and CARBON dioxide, under conditions of heat and pressure beneath the earth's surface. On exposure to the atmosphere this process may be reversed by weathering (see EROSION). LIMESTONE may be metamorphosed to give MARBLE, SHALE to give SLATE. IGNEOUS ROCKS and previous metamorphic rocks undergo *retrograde metamorphism*, absorbing volatiles, usually from nearby metamorphosing sediments. GRANITE, for example, may be metamorphosed to form a GNEISS.

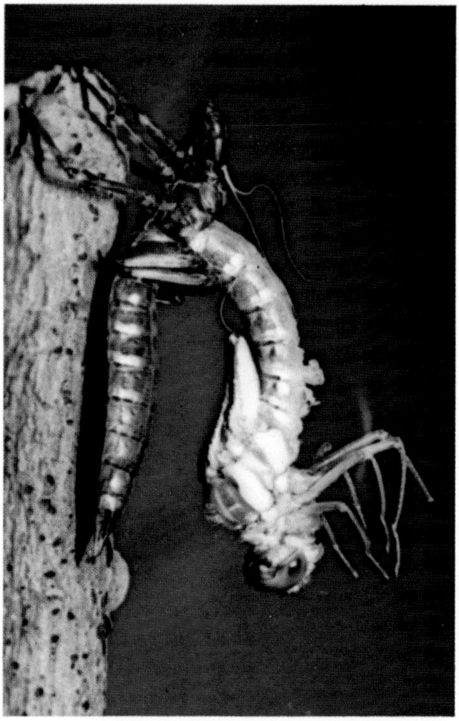

The metamorphosis in which an adult dragonfly (*Aeshna cyanea*) emerges from the nymph stage.

METAMORPHOSIS, in animals (notably frogs, toads and insects), a marked and relatively rapid change in body form. This alteration in appearance is associated with a change in habits. Perhaps the best known example is the change which occurs when a tadpole becomes a frog.

METAPHYSICS, the branch of philosophy concerned with the fundamentals of existence or reality. Although it takes its name merely from the title an early compiler gave a volume of ARISTOTLE's essays which he placed "after the *Physics*" (*meta ta phusika*), many philosophers have assumed that the meta- of metaphysics has the sense of "beyond," thus regarding metaphysics as speculation concerning things that are beyond the scope of science. Positivists have thus denounced the study, not recognizing that their own principles were themselves strictly metaphysical. Indeed all philosophical controversies ultimately reduce to well-worn debates in metaphysics.

METAPHYTA, or **Embryophyta,** taxonomic term proposed to include plants that produce an embryo and multicellular sex organs and exhibit an ALTERNATION OF GENERATIONS. The Metaphyta is divided into two divisions: the BRYOPHYTA, including MOSSES, HORNWORTS and LIVERWORTS and the Tracheophyta (or VASCULAR PLANTS) including the CLUB MOSSES, HORSETAILS, FERNS, GYMNOSPERMS and ANGIOSPERMS. (See also EMBRYOPHYTES; PLANT

KINGDOM.)

METAPSYCHOLOGY, FREUD's term for abstract consideration of psychological phenomena in terms of the *psychic apparatus*, which contains the EGO, SUPER-EGO and ID.

METATHERIA, or **Marsupials**, a subclass of the MAMMALIA that includes the opossums and kangaroos. There are 236 living species, limited to Australasia and the Americas. Metatherians give birth to very immature young that are transferred to the mother's pouch where they continue development.

METCHNIKOFF, Élie (1845–1916), or **Ilya Mechnikov**, Russian biologist who shared with Paul EHRLICH the 1908 Nobel Prize for Physiology or Medicine for his discovery of phagocytes (in man called leukocytes) and their role in defending the body from, for example, bacteria. (See BLOOD.)

METEOR, the visible passage of a meteoroid (a small particle of interplanetary matter) into the earth's atmosphere. Owing to friction it burns up, showing a trail of fire in the night sky. The velocity on entry lies in the range 11–72km/s.

Meteoroids are believed to consist of asteroidal and cometary debris. Although stray meteoroids reach our atmosphere throughout the year, for short periods at certain times of year they arrive in profuse numbers, sharing a common direction and velocity. It was shown in 1866 by SCHIAPARELLI that the annual Perseid meteor shower was caused by meteoroids oribiting the sun in the same orbit as a comet observed some years before; moreover, since their period of orbit is unrelated to that of the earth, the meteoroids must form a fairly uniform "ring" around the sun for the shower to be annual. Other comet-shower relationships have been shown, implying that these streams of meteoroids are cometary debris.

Meteors may be seen by a nighttime observer on average five times per hour: these are known as sporadic meteors or shooting stars. Around twenty times a year, however, a meteor shower occurs and between 20 and 35000 meteors per hour may be observed. These annual showers are generally named for the constellations from which they appear to emanate: e.g., Perseids (PERSEUS), Leonids (LEO). Large meteors are called FIREBALLS, and those that explode are known as bolides.

Meteorites are larger than meteors, and are of special interest in that, should they enter the atmosphere, they at least partially survive the passage to the ground. Many have been examined. They fall into two main categories: "stones," whose composition is not unlike that of the earth's crust; and "irons," which contain about 80%–95% iron, 20%–5% nickel and traces of other elements. Intermediate types exist. Irons display a usually crystalline structure which implies that they were initially liquid, cooling over long periods of time. Sometimes large meteorites shatter on impact, producing large craters like those in Arizona and Siberia.

A small iron meteorite. Minor meteorite falls are not uncommon if the whole earth is taken into account, although comparatively rare in any particular locality. There have been only two major earthfalls of meteoric material this century.

METEOROLOGY, the study of the ATMOSPHERE and its phenomena, weather and climate. Based on atmospheric physics, it is primarily an observational science, whose main application is WEATHER FORECASTING AND CONTROL. The RAIN gauge and WIND vane were known in ancient times, and the other basic instruments—ANEMOMETER, BAROMETER, HYGROMETER and THERMOMETER—had all been invented by 1790. Thus accurate data could be collected; but simultaneous observations over a wide area were impracticable until the development of the telegraph. Since WWI observations of the upper atmosphere have been made, using airplanes, balloons, RADIOSONDE, and since WWII (when meteorology began to flourish) ROCKETS and artificial SATELLITES. RADAR has been much used. Meteorology may be classified by the type of phenomenon observed: CLOUDS, PRECIPITATION and HUMIDITY, WIND and air pressure, air temperature, and STORMS. More basic is the scale of the phenomena: the microscale deals with small, transient phenomena up to about 10km in size and lasting, say, 1h; the mesoscale, those up to 200km across and lasting a few hours; the synoptic scale is that of daily national and continental weather maps, while the macroscale treats of global, seasonal phenomena. The general circulation of the atmosphere is zonal by latitude (see JET STREAM; PREVAILING WESTERLIES; TRADE WINDS). Imposed on this are disturbances—chiefly CYCLONES and anticyclones—due to imbalance of pressure and temperature. An air mass is a large region of air, roughly homogeneous horizontally, which forms by stagnant contact with a land or sea surface and which then moves elsewhere. When two air masses of different properties meet, a FRONT is formed. (See also ISOBAR; ISOTHERM.)

METHANE (CH_4), colorless, odorless gas; the simplest ALKANE. It is produced by decomposing organic matter in sewage and in marshes (hence the name *marsh gas*), and is the "firedamp" of coal mines (see DAMP). It is the chief constituent of NATURAL GAS, occurs in COAL GAS and WATER GAS, and is produced in PETROLEUM refining. Methane is used as a FUEL, for making carbon-black, and for chemical synthesis. MW 16.0, mp −183°C, bp −164°C.

METHANOL (CH_3OH), or methyl alcohol, or wood alcohol, colorless liquid, the simplest ALCOHOL. Formerly made by destructive distillation of wood, it is now almost all made by catalytic reaction of carbon monoxide and hydrogen. It is used to make FORMALDEHYDE and other industrial chemicals, in ANTIFREEZE, rocket fuels and as a solvent. Being highly toxic, it is used to make ETHANOL undrinkable

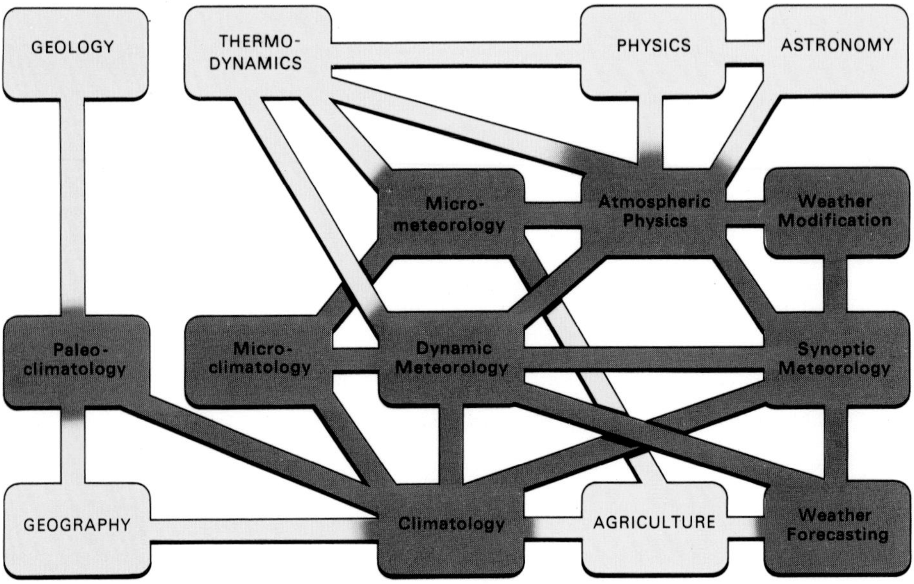

The principal branches of meteorology (red) and their links with allied sciences.

("denatured" or "methylated spirit"). MW 32.0, mp −94°C, bp 65°C.

METHEDRINE, i.e., methamphetamine, one of the AMPHETAMINES.

METHYL COMPOUNDS, organic compounds containing the methyl group, CH_3 (see ALKANES). The most important are: METHANOL, TOLUENE, methyl ethyl ketone (a solvent), and the methyl halides (see ALKYL HALIDES).

METRE (m), the SI base unit of length, defined as the length equal to 1 650 763.73 times the wavelength of radiation corresponding to the transition between the ENERGY LEVELS $2p_{10}$ and $5d_5$ of the krypton-86 atom (see SI UNITS). It was originally intended that the metre represent one ten millionth of the distance from the N Pole to the equator on the MERIDIAN passing through Paris. But the surveyors got their sums wrong and for 162 years (to 1960), the metre was defined as an arbitrary distance marked on a metal bar (from 1889–1960, the "international prototype metre," a bar of platinum-iridium which is still kept under controlled conditions near Paris).

METRIC SYSTEM, a decimal system of WEIGHTS AND MEASURES devised in Revolutionary France in 1791 and based on the METRE, a unit of LENGTH intended to equal a ten-millionth part of the distance from the equator to either geographic pole. The original unit of MASS was the GRAM, the mass of a cubic centimetre of water at 4°C, the TEMPERATURE of its greatest DENSITY. Auxiliary units were to be formed by adding Greek prefixes to the names of the base units for their decimal multiples and Latin prefixes for their decimal subdivisions. The metric system forms the basis of the physical units systems known as CGS UNITS and MKSA UNITS, the present International System of Units (SI UNITS) being a development of the latter. SI also provides the primary standards for the US Customary System of units. This means that exact interconversion can be easily accomplished.

MEYER, Adolf (1866–1950), Swiss-born US psychiatrist best known for his concept of *psychobiology*, the use in psychiatry of both psychological and biological processes together.

Metric system

Quantity	Unit	Symbol	Equivalent in other metric units	Approximate equivalent in US customary units
length	metre	m	—	1.093 613yd
area	are	a	100m²	119.6sq yd
volume	stere	st	1m³	1.308cu yd
capacity	litre	l	0.001m³	1.057qt
mass	gram	g	—	0.035 274oz
mass	quintal	q	100kg	2.205 short cwt
mass	tonne	t	1000kg	1.102 short tons

MEYER, Julius Lothar (1830–1895), German chemist who, independently from MENDELEYEV, drew up the PERIODIC TABLE, publishing his version in 1870. He showed the periodicity of atomic volume.

MEYERHOF, Otto (1884–1951), German physiologist who shared with A. V. HILL the 1922 Nobel Prize for Physiology or Medicine for their independent work on the biochemistry of MUSCLE action.

MHO, or reciprocal OHM, a unit of CONDUCTANCE.

MICA, group of common SILICATE minerals composed of sheets of linked SiO_4 tetrahedra, with aluminum replacing silicon to some extent, and containing cations and hydroxyl groups between the layers. The three main types are BIOTITE, MUSCOVITE and PHLOGOPITE; others include CHLORITE and GLAUCONITE. Micas occur widespread in many igneous, metamorphic and sedimentary rocks, and weather to CLAY minerals. They show perfect basal cleavage, producing thin, flexible flakes which are used as electrical insulators and as the dielectric in CAPACITORS; ground mica is used in paints, inks, wallpaper, rubber and waterproof coatings.

MICHELSON-MORLEY EXPERIMENT, important experiment whose results, by showing that the ETHER does not exist, substantially contributed to EINSTEIN's formulation of RELATIVITY theory. Its genesis was the development by **Albert Abraham Michelson** (1852–1931) of an INTERFEROMETER (1881) whereby a beam of light could be split into two parts sent at right angles to each other and then brought together again. Because of the earth's motion in space, the "drag" of the stationary ether should produce INTERFERENCE effects when the beams are brought together: his early experiments showed no such effects. With E. W. MORLEY he improved the sensitivity of his equipment, and by 1887 was able to show that there was no "drag," and therefore no ether. Michelson, awarded the Nobel Prize for Physics in 1908, was the first US Nobel prizewinner.

The easy insertion of a razor blade into the crystal demonstrates the perfect basal cleavage in biotite, one of the principal types of mica.

MICHURIN, Ivan Vladimirovich (1855–1935), Russian horticulturalist whose theories of heredity, including the inheritance of ACQUIRED CHARACTERISTICS (see also LAMARCK), officially displaced GENETICS in Soviet science from 1948 until LYSENKO's fall from power in 1964.

MICROBIOLOGY, the study of microorganisms, including BACTERIA, VIRUSES, FUNGI and ALGAE. Departments of microbiology include the traditional divisions of ANATOMY, PHYSIOLOGY, GENETICS, TAXONOMY and ECOLOGY, together with various branches of MEDICINE, VETERINARY MEDICINE and plant pathology, since many microorganisms are pathogenic by nature. Microbiologists also play an important role in the food industry, particularly in baking and BREWING. In the pharmaceutical industry, they supervise the production of ANTIBIOTICS.

MICROCHEMISTRY, branch of CHEMISTRY in which very small amounts (1μg to 1mg) are studied. Special techniques and apparatus have been developed for weighing and handling such minute quantities. Tracer methods, especially labeling with radioactive ISOTOPES, are useful, as are instrumental methods of ANALYSIS. Microanalysis is the chief part of microchemistry; another important aspect is the study of rare substances such as the TRANSURANIUM ELEMENTS.

MICROCLINE, common mineral in the FELDSPAR group, occurring in igneous rocks as vitreous crystals of various colors. It has the same chemical composition as ORTHOCLASE, and differs from it only in the arrangement of the silicon and aluminum atoms.

MICROENCAPSULATION, technique for enclosing minute portions of a substance (often a drug or a dye) in tiny capsules from which it is released when these are ruptured or dissolved. Its uses include NCR (no carbon required) copy paper, and the production of slow-release drugs and pesticides.

MICROFICHE; MICROFILM. See INFORMATION RETRIEVAL.

MICROMETER, instrument for measuring accurately dimensions or separations. Its basis is that when a screw is turned once, it advances or retreats a distance equal to its pitch (see BOLTS AND SCREWS). The hairlines in telescope and microscope eyepieces are adjusted by means of a precision micrometer screw to measure separations. The **micrometer caliper** has a G-shaped frame on whose "leg" is a scale; inside the leg runs a screw (of pitch usually 0.5mm) attached to a thimble, calibrated for fractions of a turn, which runs over the scale. An object is placed between the screw's sprindle and an anvil at the far side of the G's opening, and the screw turned until the object is just held. For greater accuracy (of the order of 1μm) a VERNIER SCALE may also be used.

MICRON (μ), unit of length in the CGS system (see CGS UNITS), equal to one millionth of a METRE. In SI UNITS it is replaced by the micrometre (μm).

MICROPHONE, device for converting sound waves into electrical impulses. The **carbon microphone** used in TELEPHONE mouthpieces has a thin diaphragm behind which are packed tiny carbon granules. SOUND waves vibrate the diaphragm, exerting a variable

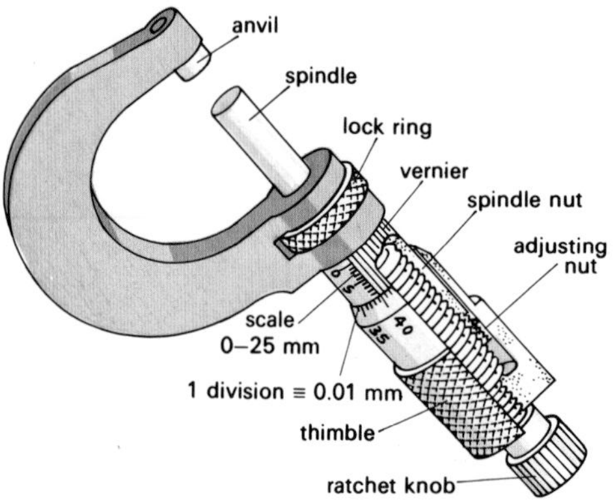

A vernier micrometer allowing small thicknesses to be measured to 1 μm.

pressure on the granules. This varies their RESISTANCE, so producing fluctuations in a DC current (see ELECTRICITY) passing through them. The **crystal microphone** incorporates a piezoelectric crystal in which pressure changes from the diaphragm produce an alternating voltage (see PIEZOELECTRICITY). In the **electrostatic microphone** the diaphragm acts as one plate of a CAPACITOR, vibration producing changes in capacitance. In the **moving-coil microphone** the diaphragm is attached to a coil located between the poles of a permanent magnet: movement induces a varying current in the coil (see INDUCTANCE). The **ribbon microphone** has, rather than a diaphragm, a metal ribbon held in a magnetic field; vibration of the ribbon induces an electric current in it.

MICROSCOPE, an instrument for producing enlarged images of small objects. The simple microscope or MAGNIFYING GLASS, comprising a single converging LENS, was known in ancient times, but the first compound microscope is thought to have been invented by the Dutch spectacle-maker Zacharias JANSSEN around 1590. However, because of the ABERRATION unavoidable in early lens systems, the simple microscope held its own for many years, Anton van LEEUWENHOEK constructing many fine examples using tiny near-spherical lenses. Compound microscopes incorporating ACHROMATIC LENSES became available from the mid 1840s.

In the compound microscope a magnified, inverted image of an object resting on the "stage" is produced by the objective lens (system). This image is viewed through the eyepiece (or ocular) lens (system) which acts as a simple microscope, giving a greatly magnified virtual image. In most biological microscopy the object is viewed by transmitted light, illumination being controlled by mirror, diaphragm and "substage condenser" lenses. The near-transparent objects are often stained to make them visible. As this usually proves fatal to the specimen, phase-contrast microscopy, in which a "phase plate" is used to produce a DIFFRACTION effect, can alternatively be employed. Objects which are just too small to be seen directly can be made visible in dark-field illumination. In this an opaque disk prevents direct illumination and the object is viewed in the light diffracted from the remaining oblique illumination. In mineralogical use objects are frequently viewed by reflected light.

Although there is no limit to the theoretical magnifying power of the optical microscope, magnifications greater than about 2000× can offer no improvement in resolving power (see TELESCOPE) for light of visible wavelengths. The shorter wavelength of ULTRAVIOLET LIGHT allows better resolution and hence higher useful magnification. For yet finer resolution physicists turn to electron beams and electromagnetic focusing (see ELECTRON MICROSCOPE). The FIELD-ION MICROSCOPE, which offers the greatest magnifications, is a quite dissimilar instrument.

MICROTOME, device to prepare thin sections for the microscope. The specimen is embedded in a block of wax, then placed in the microtome. Turning a handle raises and lowers the block against a blade. At the top of its rise the block is advanced by a MICROMETER screw, which controls the thickness of the section cut on the descent.

MICROWAVES, ELECTROMAGNETIC RADIATIONS of wavelength between 1mm and 30cm; used in RADAR,

Two common types of microphone: (A) ribbon microphone; (B) electrostatic microphone.

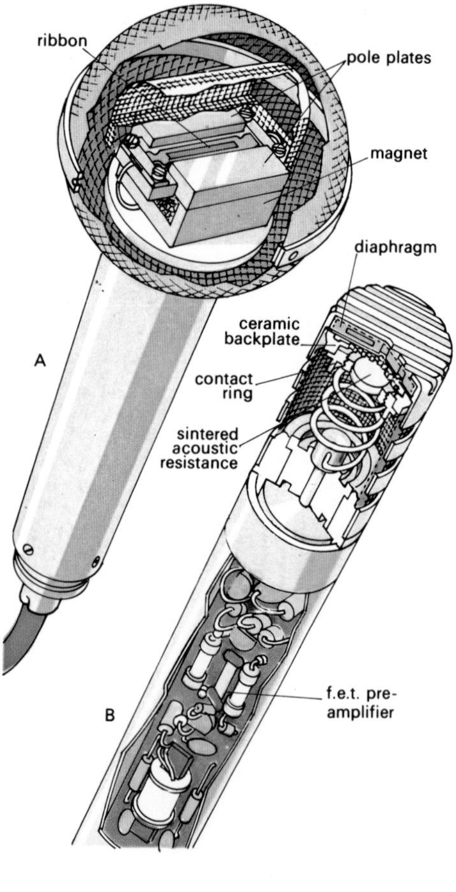

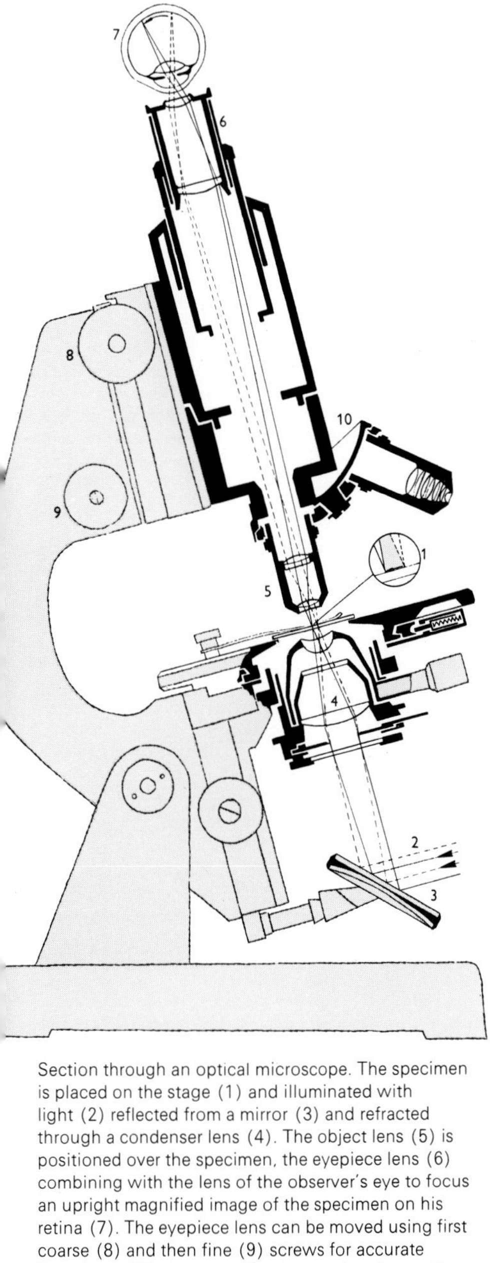

Section through an optical microscope. The specimen is placed on the stage (1) and illuminated with light (2) reflected from a mirror (3) and refracted through a condenser lens (4). The object lens (5) is positioned over the specimen, the eyepiece lens (6) combining with the lens of the observer's eye to focus an upright magnified image of the specimen on his retina (7). The eyepiece lens can be moved using first coarse (8) and then fine (9) screws for accurate focusing, while the magnification can be changed by turning a turret mounting (10) which carries a variety of object lenses.

telecommunications, SPECTROSCOPY and for cooking (microwave ovens). Their dimensions are such that it is easy to build ANTENNAS of great directional sensitivity and high-efficiency WAVEGUIDES for them.

MIDNIGHT SUN, phenomenon observed N of the Arctic Circle and S of the Antarctic Circle. Each summer the sun remains above the horizon for at least one 24-hour period (a corresponding period of darkness occurs in winter), owing to the tilt of the EARTH's equator to the ecliptic.

MID-OCEAN RIDGE. See OCEANS; PLATE TECTONICS.

MIGRAINE. See HEADACHE.

MIGRATION, long-distance mass movements made by animals of many different groups, both vertebrate and invertebrate, often at regular intervals. Generally animals move from a breeding area to a feeding place, returning as the breeding season approaches the following year. This is the pattern of annual movements of migratory birds and fishes. Migrations of this nature may be over great distances, up to 11 000km (7 000mi) in some birds. Navigation is extremely accurate: birds may return to the same nest site year after year; migratory fish return to the exact rivulet of their birth to spawn. In other cases, migrations may follow cycles of food abundance: Gnu in E Africa follow in the wake of the rains grazing on the new grass; caribou in Canada show similar movements. Certain carnivore species may follow these migrations, others capitalize on a temporary abundance as the herds move through their ranges.

MILDEW, general name for the superficial growth of many types of fungi often found on plants and material derived from plants. Powdery mildews are caused by fungi belonging to the Ascomycetes order Erysiphales, the powdery effect being due to the masses of spores. These fungi commonly infest roses, apples, phlox, melons etc. Downy mildews are caused by Phycomycetes. They commonly infest many vegetable crops. Both types of disease can be controlled by use of FUNGICIDES.

MILE, name of many units of length in different parts of the world. The statute mile (st mi) is 1 760 yards (exactly 1 609.344m); the international (US) nautical mile is 1.150 78st mi; the UK nautical mile, 1.151 51st mi. The name derives from the Roman (Latin) *milia passuum*, a thousand paces.

MILK, a white liquid containing water, PROTEIN, FAT, SUGAR, VITAMINS and inorganic salts which is secreted by the mammary glands of female mammals. The secretion of milk (LACTATION) is initiated immediately after birth by the hormone PROLACTIN. The milks produced by different mammals all have the same basic constituents but the proportion of each ingredient differs from species to species and within species. In any species the milk produced is a complete food for the young until weaning. Milk is of high nutritional value and man has used the milk of other animals as a food for at least 5 000 years. Milk for use by man is produced in the largest volume by cows and water buffalo (especially in India); goat milk is also produced in some areas, particularly the Middle East. Milk is an extremely perishable liquid which must be cooled to 10°C within two hours of milking and maintained at that temperature until delivery. The storage life of milk is greatly improved by PASTEURIZATION. Because of the perishable nature of milk large quantities are processed to give a variety of products including BUTTER; CHEESE; cream; evaporated, condensed and dried milk; YOGHURT; milk protein (CASEIN) and LACTOSE.

MILK OF MAGNESIA, emulsion of MAGNESIUM hydroxide, used for its properties both as an ANTACID in ULCER and HEARTBURN, and as a LAXATIVE.

MILKY WAY, our GALAXY. It is a disk-shaped spiral

galaxy containing some 100 billion stars, and has a radius of about 15kpc. Our SOLAR SYSTEM is in one of the spiral arms and is just over 9kpc from the galactic center, which lies in the direction of SAGITTARIUS. The galaxy slowly rotates about a roughly spherical nucleus (diameter about 1.5kpc), though not at uniform speed; the sun circles the galactic center about every 230 million years. The galaxy is surrounded by a spheroidal halo some 50kpc in diameter composed of gas, dust, occasional stars and GLOBULAR CLUSTERS. The Milky Way derives its name from our view of it as a hazy milk-like band of stars encircling the night sky. Irregular dark patches are caused by intervening clouds of gas and dust.

MILL, James (1773–1836) and **John Stuart** (1806–1873), distinguished British economists and philosophers. James Mill rose from humble origins to a senior position in the East India Company. He was an able apologist for the utilitarianism of his friend Jeremy Bentham. A famous account of the education James imposed on his son John Stuart can be found in the latter's *Autobiography* (1873). The younger Mill is noted for his strictly empiricist *A System of Logic* (1843), his *Principles of Political Economy* (1848), and for his essay *On Liberty* (1854). J. S. Mill's circle included F. D. Maurice, Thomas Carlyle and, later, Herbert SPENCER.

MILLIBAR (mbar or mb), CGS UNIT of PRESSURE. See BAR.

MILLIKAN, Robert Andrews (1868–1953), US physicist awarded the 1923 Nobel Prize for Physics for his determination of the charge on a single ELECTRON (the famous oil-drop experiment) and his work on the PHOTOELECTRIC EFFECT. He also studied and named COSMIC RAYS.

MIMICRY, the close resemblance of one organism to another which, because it is unpalatable and conspicuous, is avoided by certain predators. The mimic will thus gain a degree of protection on the strength of the predator's avoidance of the mimicked. Mimicry is well developed among insects. (See also BATES, HENRY W.)

MIND, man's mental organ, with which he thinks, reasons, remembers and wills. The existence of the mind, as separate from the workings of the BRAIN, is denied in MATERIALISM, but affirmed in DUALISMS such as that taught by DESCARTES.

MIND READING, an aspect of telepathy. See ESP.

MINERALS, naturally-occurring substances obtainable by MINING, including COAL, PETROLEUM and NATURAL GAS; more specifically in geology, substances of natural inorganic origin, of more or less definite chemical COMPOSITION, CRYSTAL structure and properties, of which the ROCKS of the earth's crust are composed. (See also GEMS; ORE.) Of the 3000 minerals known, fewer than 100 are common. They may be identified by their color (though this often varies because of impurities), HARDNESS, luster, SPECIFIC GRAVITY, crystal forms and CLEAVAGE; or by chemical ANALYSIS and X-RAY DIFFRACTION. Minerals are generally classified by their ANIONS—in order of increasing complexity: elements, SULFIDES, OXIDES, HALIDES, CARBONATES, NITRATES, SULFATES, PHOSPHATES and SILICATES. Others are classed with those which they resemble chemically and structurally, e.g. arsenates with phosphates. A newer system classifies minerals by their topological structure (see TOPOLOGY).

MINES, concealed explosive devices placed to destroy enemy lives and equipment. They are often triggered by pressure or proximity. but some have a time mechanism and others are detonated by remote control. (See also BOMB; EXPLOSIVES.)

Landmines are of two main types, antivehicular and antipersonnel, differing in little more than sensitivity and in violence of explosion. They are buried just below ground and triggered by pressure or a tripwire. "Bouncing mines" have an auxiliary charge to throw them into the air, so that the explosion kills a maximum number of people.

Naval mines are of three main tyes, floating, moored and bottom. Moored mines, anchored to lie below the surface, may be triggered by contact, but more usually are attached to equipment that detects changes in the local magnetic field or water pressure, or the sound of engines. Bottom mines, usually set off by remote control, are placed on the sea bed in shallow coastal waters.

MINIMA. See EXTREMUM.

MINING, the means for extracting economically important MINERALS and ORES from the earth. Where the desired minerals lie near the surface, the most economic form of mine is the *open pit*. This usually consists of a series of terraces, which are worked back in parallel so that the mineral is always within convenient reach of the excavating machines. *Strip mining* refers to stripping off a layer of overburden to reach a usually thin mineral seam (often COAL). The excavating machines used in open-pit mining are frequently vast. Soft minerals such as KAOLIN can be recovered hydraulically—by directing heavy water jets at the pit face and pumping out the resulting slurry. Where a mineral is found in alluvial (river bed) deposits, bucket or suction dredgers may be used. But where minerals lie far below the surface, various deep mining techniques must be used. Sulfur is mined by pumping superheated water down boreholes into the mineral bed. This melts the sulfur which is then pumped to the surface (see FRASCH PROCESS). Water-soluble minerals such as SALT are often mined in a similar way (*solution mining*). But most often, deep minerals and ores must be won from underground mines. Access to the mineral-bearing strata is obtained via a vertical *shaft* or sloping *incline* driven from the surface, or via a horizontal *adit* driven into the side of a mountain. The geometry of the actual mining area is determined by the type of mineral and the strength of the surrounding material. All underground mines require adequate ventilation and lighting, facilities for pumping out any groundwater or toxic gases seeping into the workings, and means (railroad or conveyor) for removing the ore and waste to the surface. As in open-pit mining, the rock is broken mechanically or with explosives. However, particular care must be exercised when using explosives underground. Several occupational diseases (e.g., PNEUMOCONIOSIS) are associated with mining and extraction metallurgy, particularly where high dust levels and toxic substances are involved. About 900000 persons are employed in the mineral industries in the US.

MINOAN CIVILIZATION. See AEGEAN CIVILIZATION.

MINOAN LINEAR SCRIPTS, two written

languages, samples of which, inscribed on clay tablets, were found in Crete by Sir Arthur EVANS (1900), and named Linear A and B. Linear B was deciphered (1952) by Michael Ventris (1922–1956) and shown to be very early Greek (from c1400 BC). Linear A is yet to be deciphered, though some symbols have been assigned phonetic values. (See also AEGEAN CIVILIZATION.)

MINOT, George Richards (1885–1950), US physician who shared with W. MURPHY and G. WHIPPLE the 1934 Nobel Prize for Physiology or Medicine for his work with Murphy showing daily consumption of large quantities of raw liver to be an effective treatment for pernicious ANEMIA.

MIOCENE, the penultimate epoch of the TERTIARY, which lasted from 25 to 10 million years ago. (See GEOLOGY.)

MIRAGE, optical illusion arising from the REFRACTION of light as it passes through air layers of different densities. In *inferior mirages* distant objects appear to be reflected in water at their bases: this is because light rays traveling initially toward the ground have been bent upward by layers of hot air close to the surface. *In superior mirages* objects seem to float in the air: this occurs where warmer air overlies cooler, bending rays downward.

MIRROR, a smooth reflecting surface in which sharp optical images can be formed. Ancient mirrors were usually made of polished bronze but glass mirrors backed with tin AMALGAM became the rule in the 17th century. Silvered-glass mirrors were first manufactured in 1840, five years after LIEBIG discovered that a silver mirror was formed on a glass surface when an ammoniacal solution of silver nitrate was reduced by an ALDEHYDE (now usually FORMALDEHYDE).

Undistorted but laterally reversed virtual IMAGES can be seen in plane (flat) mirrors. (Such images are "virtual" and not "real" because no light actually passes through the apparent position of the image.) Concave spherical mirrors form real inverted images of objects farther away than half the radius of curvature of the mirror and virtual images of closer objects. Concave mirrors (usually with an unglazed metallic surface) are used in astronomical TELESCOPES because of their freedom from many LENS defects. Parabolic concave mirrors, which focus a parallel beam of light in a single point, also find use as reflectors for solar furnaces and searchlights. Convex spherical mirrors always form distorted virtual images but offer a wider field of view than plane mirrors. Half-silvered glass mirrors are used in many optical instruments and can be used to give a one-way mirror effect between a well-lit and a dimly illuminated room. (See also LIGHT; REFLECTION.)

For reflection at a plane surface, the angle of incidence (*i*) equals the angle of reflection (*r*).

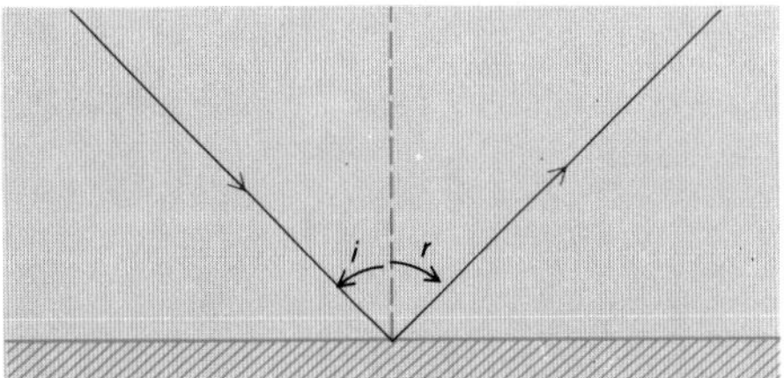

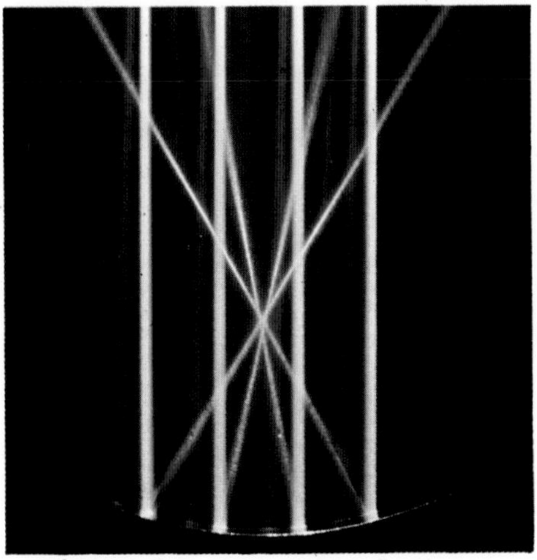

Spherically curved mirrors reflect a parallel beam of light striking them toward a diffuse focus.

MISCARRIAGE, popular term for spontaneous ABORTION.

MISCH METAL, an ALLOY composed of 50% CERIUM, 25% LANTHANUM, 15% NEODYMIUM, 10% other RARE EARTHS, and iron. It is used to make "flints" for cigarette lighters and as a deoxidizer for vacuum tubes.

MISSILE, anything that can be thrown or projected. In modern usage the word most often describes the self-propelled weapons developed during and since WWII, properly called guided missiles.

The first rocket missiles were used by the Chinese in the 13th century, but the ancestors of the modern missile were the German V-1 and V-2 rockets used to bombard London in WWII. The V-2 could reach 5 600km/h and there was no defense against it. Captured V-2s gave Russia and the US the starting point for further ROCKET development culminating in intercontinental missiles capable of delivering nuclear warheads to any spot on the globe—and also of launching mankind into space.

Missiles can be classified according to their range, by the way they approach their target (guided, unguided or ballistic), by their use (surface-to-surface, surface-to-air and so on), or by their target (as with antitank missiles). The *Minuteman*, for example, is an intercontinental ballistic missile (ICBM), with a range of more than 11 000km and a ballistic arc trajectory, like a shell fired from a gun.

Missiles are normally propelled by solid-fueled rockets equipped with an oxidant that allows the fuel to burn outside the atmosphere. Liquid propellants are more volatile and not generally used. Larger missiles, such as the *Minuteman*, comprise several stages, generally mounted on one another, each with its own motor. This arrangement gives the missile increased speed, range and lifting capacity and so is generally also used in spacecraft-launching missiles such as *Atlas* or *Saturn*. A few missiles, such as the *Hound Dog* air-to-surface missile carried by the B-52 bomber, are powered by jets, while Britain's

Bloodhound is driven by a ramjet (see JET PROPULSION).

Guidance systems vary from wire guidance for short-range missiles to elaborate homing systems which eliminate almost all possibility of escape. Some short-range missiles such as *Honest John* need no guidance at all, while those equipped with nuclear warheads cause such widespread destruction that pinpoint accuracy is not required. Wire guidance allows the operator to control the course of the missile through two fine wires paid out behind it. Other more complex systems use a radio beam directed at the target, or even a radar or laser-fed computer tracking and correcting the missile's course. Homing systems involve devices which detect waves emitted by or reflected from the target. Some home on heat in the form of infrared rays; others use radar. Many can be jammed electronically, but some are pre-set with a complete flight plan to guide the missile to a selected target, and to correct any deviation from course. Most long-range missiles have an INERTIAL GUIDANCE system to detect changes in course and velocity.

Missile launchers range from the simple tube of the bazooka to the massive self-propelled mobile launcher/transporter used for *Honest John*. Larger missiles such as ICBMs require so much elaborate support equipment that they must be stored in heavily defended underground silos. Nuclear-powered missile submarines are more difficult to detect and destroy, and the US has over 40 of these, each carrying 16 Polaris missiles. Surface vessels may also use missiles, gyro-stabilized to allow for the movement of the ship. Aircraft-launched missiles usually have a guidance system enabling the aircraft itself to remain a considerable distance from the target. The crippling expense and high risks involved in the uncontrolled development of nuclear missiles and antimissile systems has led to Strategic Arms Limitation Talks between the US and Russia. (See also NUCLEAR WARFARE.)

MISSISSIPPIAN, the antepenultimate period of the PALEOZOIC, lasting from about 345 to 315 million years ago. (See also CARBONIFEROUS; GEOLOGY.)

MISTRAL, cold wind blowing S from the Central Plateau of France to the NW Mediterranean. It occurs mainly in winter and speeds up to about 140km/h have been recorded. It is a hazard to air and surface transport, crops and buildings.

MITOCHONDRIA. See CELL.

MITOSIS, the normal process by which a CELL divides into two. Initially the CHROMOSOMES become visible in the nucleus before longitudinally dividing into a pair of parallel *chromatids*. The chromosomes shorten and thicken and arrange themselves on a spindle across the equator of the cell. The cell then divides so that each daughter contains a full complement of chromosomes.

MKSA UNITS, or **Giorgi System**, a metric system of units based on the METRE (length), KILOGRAM (mass), SECOND (time) and AMPERE (electric current), forming the basis of the now internationally accepted SI UNITS. The system is "rationalized" in that, with the PERMEABILITY of free space set at $4\pi \times 10^{-7}$ henry/metre, equations contain factors reflecting the geometry of the situations they describe: 2π for cylindrical symmetries; 4π for spherical.

MÖBIUS STRIP, a topological space (see TOPOLOGY) formed by joining the two ends of a strip of

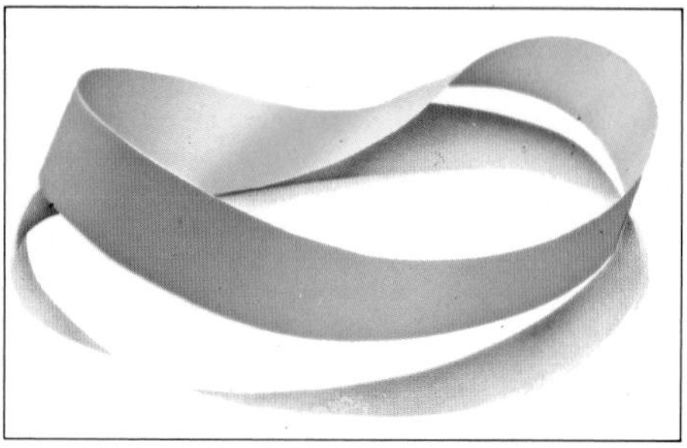

The Möbius strip

paper or other material after having turned one of the ends through an ANGLE of 180°. It is of interest in that it has only one side: if a line is drawn from a point A on the surface parallel to the edges of the strip it will eventually pass through a point A′ directly through the paper from A. This closed curve is known as a nonbounding cycle since it does not bound an area of the surface. (See also KLEIN BOTTLE.)

MODE. See MEAN, MEDIAN AND MODE.

MODULAR ARITHMETIC, an algebraic system of positive INTEGERS in which there is a maximum n such that any number greater than n (the modulus) is expressed as the REMAINDER left after its DIVISION by n.

MODULATION. See RADIO.

MODULUS, a term with several meanings in mathematics. The modulus of a REAL NUMBER is its positive value, shown by placing the number between vertical lines: thus $|-2| = |+2| = 2$. The modulus (or absolute value) of a complex number (see IMAGINARY NUMBERS) z, where $z = x + iy$, is defined as $\sqrt{(x^2+y^2)}$. If z is represented by a point on an ARGAND DIAGRAM, then its modulus, r, is the distance from the origin to that point. Again, the term modulus sometimes refers to a constant of proportionality (see PROPORTION). In LOGARITHMS, if $\log_a x = k \log_b x$, where a and b are different bases, k is known as the modulus. (See also MODULAR ARITHMETIC.)

MOGOLLAN CULTURE, North American Indian culture of c200 BC–c1200 AD in what is now SW N.M. and SE Ariz., believed to have developed from the earlier Cochise culture. As Mogollan pottery, the first in the SW, was from the beginning very fine, it is thought that the art was imported from Mexico.

MOHAIR, the fleece of the Angora goat. The fibers are long, straight and lustrous, and when woven it behaves like WOOL. Highly durable, it is used in many textile fabrics.

MOHL, Hugo von (1805–1872), German botanist who gave the name "PROTOPLASM" to the plastic material he found on the periphery of the CELL.

MOHO, abbreviation for MOHOROVIČIĆ DISCONTINUITY.

MOHOROVIČIĆ DISCONTINUITY, a layer of the earth originally regarded as marking the boundary between crust and mantle (see EARTH), evidenced by a change in the velocity of seismic waves (see EARTHQUAKE). It is now regarded as of little physical significance. The US project Mohole, designed to drill through the "Moho," was abandoned in 1966. More important are the discontinuities between the core and the mantle

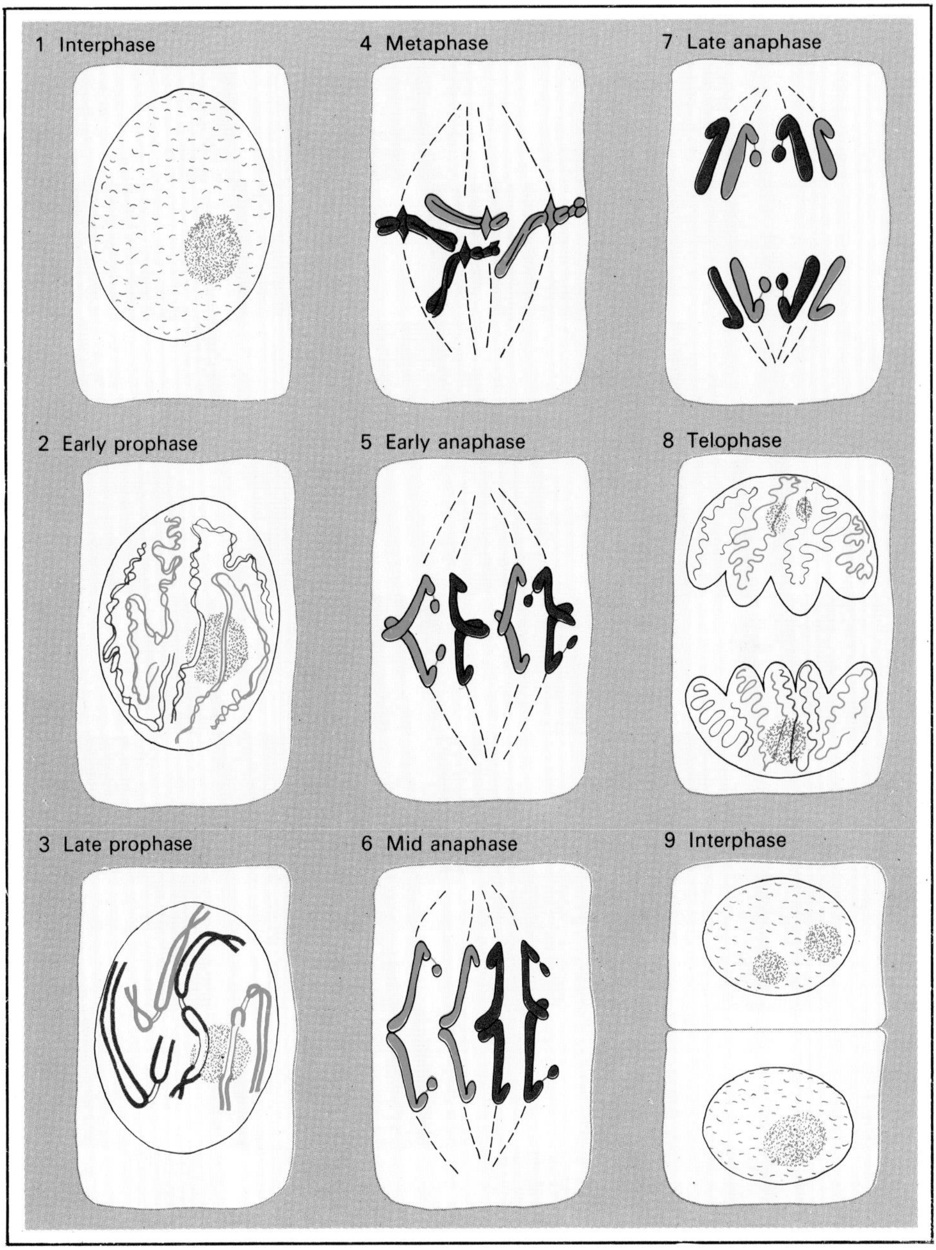

Mitosis is the normal process of cell division in which the number of chromosomes is conserved. Interphase (1) is the resting condition of the cell. During prophase (2–3), the chromosomes appear as long thin threads, each of which consists of two sister chromatids which are wound relationally around one another; these shorten and thicken by internal coiling. During metaphase (4) the chromosomes come into contact with a longitudinally oriented and spindle-shaped system of microtubules which cannot be seen in conventional light-microscope chromosome preparations. Within this spindle the two chromatids of each chromosome establish a bipolar orientation, the two sister centromeres of each chromosome being directed to opposite spindle poles. Next, during anaphase (6–7), sister centromeres move apart, dragging their attached chromatid arms behind them in the spindle. Then, in telophase (8), the two polar groups of single chromatids uncoil while a new nuclear membrane reforms around them. Finally, the two new daughter resting nuclei form (9).

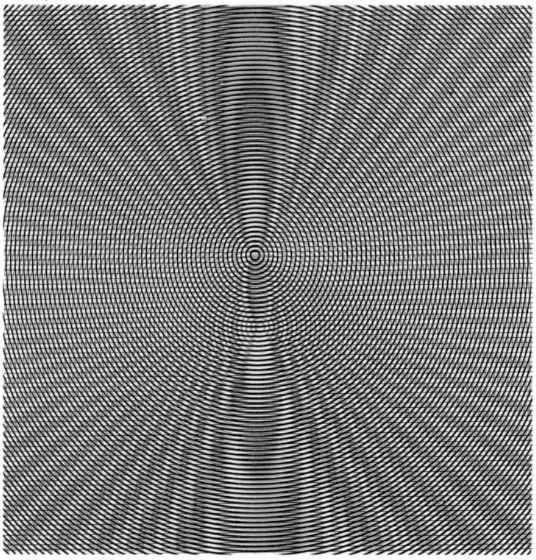

A Moiré pattern seen when two similar regular patterns are almost superimposed.

(Gutenberg or Oldham Discontinuity), with a radius of about 3 500km; and between the inner and outer cores, with a radius of about 1 200km to 1 650km.

MOHS' SCALE. See HARDNESS.

MOIRÉ PATTERN, a family of CURVES formed by the INTERSECTIONS of one family of curves with another over which it has been superimposed. Moiré patterns may be seen by looking through the folds of a gauze or nylon curtain: motion of the curtain or observer will cause dramatic changes in the patterns observed. They are of particular note in color printing, where special techniques are employed to prevent their appearance in HALFTONES; and are used in industry to determine, for example, the degree of flatness of a surface. They are used also as MATHEMATICAL MODELS of physical phenomena, and occasionally in the solution of mathematical problems. Their disturbing optical properties are of interest in psychology.

MOISSAN, Ferdinand Frédéric Henri (1852–1907), French chemist awarded the 1906 Nobel Prize for Chemistry for isolating FLUORINE (1886) and for developing the electric arc furnace (see ELECTRIC FURNACE). He also claimed to have synthesized DIAMONDS (1893).

MOIVRE, Abraham de. See DE MOIVRE, ABRAHAM.

MOLASSES, yellow to dark brown syrup, usually obtained as a byproduct in the production of SUGAR from sugarcane juice. It was originally used for FERMENTATION of industrial ethyl alcohol, but is now mainly used in animal feeds, adhesives, fertilizers and the pharmaceutical industry.

MOLD, general name for a number of filamentous FUNGI that produce powdery or fluffy growths on fabrics, foods and decaying plant or animal remains. Best known is the blue bread mold caused by *Penicillium*, from which the ANTIBIOTIC, PENICILLIN, was first discovered.

MOLE, pigmented spot or nevus in the SKIN, consisting of a localized group of special cells containing MELANIN. Change in a mole, such as increase in size, change of color and bleeding should lead to suspicion of melanoma.

MOLE (mol), the SI base unit of amount of substance, defined as the amount of substance of a system which contains as many elementary entities (of a specified kind) as there are atoms in 0.012kg of carbon-12 (i.e., the AVOGADRO Number). One mole of a compound is its MOLECULAR WEIGHT in grams. The **molarity** of a solution is its concentration in moles per litre; a solution whose molarity is 1 is called molar. (See SI UNITS.)

MOLECULAR BIOLOGY, the study of the structure and function of the MOLECULES which make up living organisms. This includes the study of PROTEINS, ENZYMES, CARBOHYDRATES, FATS and NUCLEIC ACIDS. (See also BIOCHEMISTRY; BIOLOGY, BIOPHYSICS.)

MOLECULAR WEIGHT, the sum of the ATOMIC WEIGHTS of all the atoms in a MOLECULE. It is an integral multiple of the empirical FORMULA weight found by chemical ANALYSIS, and of the EQUIVALENT WEIGHT. Molecular weights may be found directly by MASS SPECTROSCOPY, or deduced from related physical properties including gas DENSITY; effusion; osmotic pressure (see OSMOSIS); and effects on solvents: lowering of VAPOR PRESSURE and freezing point, and raising of boiling point; for large molecules the ultracentrifuge is used. (See also MOLE.)

MOLECULE, entity composed of ATOMS linked by chemical BONDS and acting as a unit; the smallest particle of a chemical compound which retains the COMPOSITION and chemical properties of the compound. The composition of a molecule is represented by its molecular FORMULA. Elements may exist as molecules, e.g., oxygen O_2, phosphorus P_4. FREE RADICALS and IONS are merely types of molecules. Molecules range in size from single atoms to **macromolecules**—chiefly PROTEINS and POLYMERS—with MOLECULAR WEIGHTS of 10000 or more. The chief properties of molecules are their structure (bond lengths and angles)—determined by electron diffraction, X-RAY DIFFRACTION and SPECTROSCOPY—spectra, and DIPOLE MOMENTS. (See also ORBITAL; VAN DER WAALS.)

MOLLUSCA, a phylum of soft-bodied animals which typically have a chalky shell for protection. There are over 80 000 living species, and these are classified into the three main classes: Gastropoda; Bivalvia, and Cephalopoda. **Gastropods**, which include slugs, which lack shells, and snails, which possess shells, have bodies divided into a head, a visceral hump and a muscular foot. The head bears tentacles. During development, the visceral hump twists counterclockwise through 180°, a process known as torsion. **Bivalves**, which include oysters, mussels and clams, are enclosed in shells that form a pair of hinged valves that are drawn together by an elastic ligament. **Cephalopods**, which include squids, octopuses and cuttlefish, typically bear tentacles and their shell is enclosed within their bodies. The nervous system is well developed and the eyes are extraordinarily similar to those found in vertebrate animals such as AMPHIBIA or MAMMALIA. Locomotion is by crawling on the sea bed or by jet propulsion using the sudden ejection of water through a siphon.

MOLTING, the shedding of the skin, fur or feathers by an animal. It may be a seasonal occurrence, as a periodic renewal of fur or plumage in mammals and birds, or it may be associated with GROWTH as in insects or crustaceans. In birds and mammals the molt is primarily to renew worn fur or feathers so that plumage or pelage is kept in good condition for waterproofing, insulation or flight. In addition it may serve to shed breeding plumage in birds, or to change between different summer and winter coats. In invertebrates the rigid external skeleton must be shed and replaced to allow growth within. In larval insects the final molts are involved in the METAMORPHOSIS to adult form.

MOLYBDENITE (MoS_2), soft, gray SULFIDE mineral, the chief ore of MOLYBDENUM, mined mainly in Col. Purified molybdenite has properties very similar to those of GRAPHITE, and is used as a lubricant.

MOLYBDENUM (Mo), silvery-gray metal in Group VIB of the PERIODIC TABLE; a TRANSITION ELEMENT. It is obtained commercially by roasting MOLYBDENITE in air and reducing the oxide formed with carbon in an electric furnace or by the THERMITE process to give ferromolybdenum. Because of its high melting point, it is used to support the filament in electric lamps and for furnace heating elements. It also finds use in corrosion-resistant, high-temperature STEELS and ALLOYS. Molybdenum is unreactive, but forms various covalent compounds. Some are used as industrial CATALYSTS. Molybdenum is a vital trace element in plants and a catalyst in bacterial NITROGEN FIXATION. AW 95.9, mp 2610°C, bp 5560°C, sg 10.2 (20°C).

MOMENT, the product of a quantity and its distance from some specific point connected with it. Statical moments such as the moment of a force (measuring its turning effect on a body by multiplying the force's magnitude by its perpendicular distance from the rotation axis) enter equations of static equilibrium. The **moment of inertia**, I, of a rotating body (the analog of MASS in the dynamics of translation) is the sum of the products of its mass elements, m_i, with the squares of their distances, r_i, from the rotation axis.

$$I=\sum_i m_i r_i^2$$

MOMENTUM, the product of the MASS and linear VELOCITY of a body. Momentum is thus a VECTOR quantity. The linear momentum of a system of interacting particles is the sum of the momenta of its particles, and is constant if no external forces act. The rate of change of momentum with time in the direction of an applied force equals the force (Newton's second law of motion—see MECHANICS). In rotational motion, the analogous concept is **angular momentum**, the product of the moment of inertia and the angular velocity of a body relative to a given rotation axis. If no external forces act on a rotating system, the direction and magnitude of its angular momentum remain constant.

MONAD, in the philosophy of LEIBNIZ, one of the individual substances which together constitute the universe—a sort of metaphysical atom, having properties but not composed of matter.

MONADNOCK, isolated hill formed of erosion-resistant bedrock in an area otherwise well eroded (see EROSION; PENEPLAIN); named for Mt Monadnock, Cheshire Co., N.H.

MONAZITE, PHOSPHATE mineral of widespread occurrence, mined in India, Brazil, the US and South Africa as an ore of the RARE EARTHS and THORIUM. It usually forms small brown prismatic crystals in the monoclinic system.

MOND, Ludwig (1839–1909), German-born British industrial chemist whose discovery (1889) of nickel carbonyl (see TRANSITION ELEMENTS) led him to devise the Mond Process for refining NICKEL, and so to found the Mond Nickel Company.

MONEL METAL, strong, corrosion-resistant ALLOY composed of 68% nickel, 29% copper, 3% iron, manganese, silicon and carbon; used for turbine blades, propellers, etc. Originally made by smelting nickel/copper ore from Sudbury, Ontario.

MONERA, taxonomic term proposed by some authorities to include unicellular, prokaryotic organisms (see CELL) belonging to the PLANT KINGDOM. Included in the Monera is one division, the SCHIZOPHYTA, containing the BACTERIA (Schizomycetes) and BLUE-GREEN ALGAE (Schizophyceae).

MONGOLISM, or Down's syndrome, a relatively common (1 in 600 births) congenital disorder due to a chromosomal abnormality, usually of CHROMOSOME 21. It causes characteristic facial appearance (resembling that of a Mongolian), HAND shape and SKIN patterns; floppiness in the baby; MENTAL RETARDATION, and delayed growth. Congenital diseases of the HEART and GASTROINTESTINAL TRACT are common, as is CATARACT. Mongols also have an increased incidence of LEUKEMIA. As the average age of parenthood advances, mongolism is becoming increasingly common.

MONGOLOID, one of the three racial divisions of man (see RACE). Mongoloids generally have straight black hair, little facial hair, yellow to brown skins and the distinctive epicanthic fold, a fold of skin over the eyes giving them a slanting appearance. The AMERINDS, Eskimos, Polynesians and Patagonians are Mongoloid peoples.

MONISM, any philosophical system asserting the essential unity of things—that all things are material (see MATERIALISM), or mind (see IDEALISM), or of some other essence. Monism is contrasted with the various kinds of DUALISM or pluralism.

MONIZ, Antonio Caetano de Abreu Freire Egas (1874–1955), Portuguese brain surgeon awarded with W. R. HESS the 1949 Nobel Prize for Physiology or Medicine for his development of frontal LOBOTOMY as a treatment for mental disorders. The treatment is now little used (see PSYCHIATRY).

MONOCOTYLEDON, name for flowering plants or ANGIOSPERMS that produce seeds with only one seed leaf (or COTYLEDON). Monocotyledons have parallel-veined leaves and the flowering parts are generally in threes or multiples of threes. (See also DICOTYLEDONS.)

MONOD, Jacques Lucien (1910–1976), French biochemist who, with F. JACOB and A. LWOFF, received the 1965 Nobel Prize for Physiology or Medicine for his work with Jacob on regulatory GENE action in BACTERIA.

MONOECIOUS PLANTS, ones where the male and female organs are borne on the same plant, but in separate flowers. Examples are oak, corn (*Zea mays*) and walnut. (See also DIOECIOUS PLANTS.)

MONOMIAL. See POLYNOMIAL.

MONONUCLEOSIS, Infectious, or **glandular fever**, common VIRUS infection of adolescence causing a variety of symptoms including severe sore throat, HEADACHE, FEVER, malaise and enlargement of LYMPH nodes and SPLEEN. Skin rashes, hepatitis (see

A large boulder on the surface of the moon, photographed during the Apollo 17 mission.

LIVER) with JAUNDICE, pericarditis (see PERICARDIUM) and involvement of the NERVOUS SYSTEM may also be prominent. Atypical lymphocytes in the BLOOD and specific agglutination reactions (see ANTIBODIES AND ANTIGENS) are diagnostic. Severe cases may require STEROIDS and convalescence may be lengthy. It can be transmitted in SALIVA and has thus been nicknamed the "kissing disease."

MONORAIL, railway with only one rail. The cars may hang beneath it, run on top of it (with a guide rail above or a GYROSCOPE to keep them upright), or straddle it. The first was built in London (1824), the cars being horse-drawn. The first successful system was the *Lartigue system*: the cars straddled the rail and had horizontal wheels that ran on guide rails beneath. The *SAFEGE system* has cars hung from a box girder, split at the bottom, inside which the driving and guide wheels run. In the *Alweg system* the cars straddle a broad rail, on either side of which horizontal wheels run to provide stability. Monorails have been proposed as replacements for city subways.

MONOSODIUM GLUTAMATE, white crystalline solid, the acid sodium salt of glutamic acid, an AMINO ACID. Obtained from GLUTEN or soybean protein, it is added to many foods to bring out their flavor.

MONOTREMATA, a subclass of the MAMMALIA that includes the echidnas or spiny anteaters and the platypuses. Monotremes differ from all other mammals in not bearing live young because, like reptiles, they lay eggs. Living members are limited to Australasia.

MONOTYPE, technique of letterpress printing in which each character is individual and fresh-cast out of molten type metal (see ALLOY). Keying by the operator encodes a ribbon with perforations that represent characters and spaces, with special codes punched after sets of characters that almost fill a line to instruct the machine to justify (align the right-hand margin). In response to the perforations, matrices (molds) are drawn from the matrix case, and from these thc characters are cast. (See also LINOTYPE.)

MONSOON, wind system where the prevailing WIND direction reverses in the course of the seasons, occurring where large temperature (hence pressure) differences arise between oceans and large landmasses. Best known is that of SE Asia. In summer, moist winds, with associated HURRICANES, blow from the Indian Ocean into the low-pressure region of NW India that is caused by intense heating of the land. In winter, cold dry winds sweep S from the high-pressure region of S Siberia.

MONTGOLFIER, Joseph Michel (1740–1810) and **Jacques Étienne** (1745–1799), French brothers noted for their invention of the first manned aircraft, the first practical (hot-air) BALLOON, which they flew in 1783. Later in the same year Jacques assisted Jacques CHARLES in the launching of the first gas (hydrogen) balloon.

MONTH, name of several periods of time, mostly defined in terms of the motion of the MOON. The synodic month (lunar month or lunation) is the time between successive full moons; it is 29.531 DAYS. The sidereal month, the time taken by the moon to complete one revolution about the earth relative to the fixed stars, is 27.322 days. The anomalistic month, 27.555 days, is the time between successive passages of the moon through perigee (see ORBIT). The solar month, 30.439 days, is one twelfth of the solar YEAR. Civil or calendar months vary in length throughout the year, lasting from 28 to 31 days (see CALENDAR). In popular usage, the (lunar) month refers to 28 days.

MOON, a SATELLITE, in particular, the earth's largest natural satellite. The moon is so large relative to the earth (it has a diameter two thirds that of MERCURY) that earth and moon are commonly regarded as a double planet. The moon has a diameter of 3476km and a mass 0.0123 that of the earth; its ESCAPE VELOCITY is around 2.4km/s. The orbit of the moon defines the several kinds of MONTH. The distance of the moon from the earth varies between 363Mm and 406Mm (perigee and apogee) with a mean of 384.4Mm. The moon rotates on its axis every 27.322 days, hence keeping the same face constantly toward the earth; however, in accordance with KEPLER's second law, the moon's orbital velocity is not constant and hence there is exhibited the phenomenon known as *libration*: to a particular observer on the earth, marginally different parts of the moon's disk are visible at different times. There is also a very small physical libration due to slight irregularities in its rotational velocity.

The moon is covered with craters, whose sizes range up to 200km diameter. These sometimes are seen in chains up to 1Mm in length. Other features include rilles, trenches a few kilometres wide and a few hundred kilometres long; the *maria* (Latin: seas) or great plains; the bright rays which emerge from the large craters, and the lunar mountains. There are also lunar hot spots, generally associated with those larger craters showing bright rays: these remain cooler than their surrounds during lunar daytime, warmer dur ing the lunar night. It has been shown, both by the samples brought back by the Apollo 11 (1969) and subsequent lunar expeditions (see SPACE EXPLORATION) and measurements of crater circularities carried out in 1968, that the smaller lunar craters are in general of meteoritic (see METEOR) origin, the larger of volcanic origin. It is believed that the earth and the moon formed simultaneously, the greater mass of the earth accounting for its higher proportion of metallic iron; the heat of the young earth's atmosphere, which evaporated silicates, accounting for their higher proportion on the moon.

MOONSTONE, translucent variety of alkali (or plagioclase) FELDSPAR, showing opalescence due to unmixing of sodium and potassium; a GEM stone.

MOORE, Stanford (1913–), US biochemist who shared with C. B. ANFINSEN and W. H. STEIN the 1972 Nobel Prize for Chemistry for his part in determining the structure of the ENZYME ribonuclease (see also NUCLEIC ACIDS).

MORAINE, accumulation of debris carried or dropped by a glacier. *Ground moraine* is DRIFT left in a sheet as the GLACIER retreats. *Terminal moraines* are ridges deposited when the ice is melting prior to the glacial retreat; a series of ridges may mark pauses in the retreat. *Lateral moraines* are formed of debris that falls onto the glacier: when two glaciers merge their lateral moraines may unite to form a *medial moraine*.

MORDANTS, substances which fix DYES to fabrics by precipitating them in the fibers or by forming LIGAND complexes with them. Generally, metal salts and hydroxides (including ALUM), now chiefly CHROMIUM salts and dichromates, are used. They modify the hue and improve fastness.

MORGAGNI, Giovanni Battista (1682–1771), Italian anatomist whose *Of the Seats and Causes of Diseases as Investigated by Anatomy* (1761) established him as the father of morbid anatomy.

MORGAN, Lewis Henry (1818–1881), US ethnologist best known for his studies of kinship systems in his attempts to prove that the AMERINDS had migrated into North America, and to discover their place of origin. His techniques and apparently successful results have earned him regard as a father of the science of cultural ANTHROPOLOGY.

MORGAN, Thomas Hunt (1866–1945), US biologist who, through his experiments with the fruit fly *Drosophila*, established the relation between GENES and CHROMOSOMES and thus the mechanism of HEREDITY. For his work he received the 1933 Nobel Prize for Physiology or Medicine.

MORGANUCODON, a fossil shrew-like animal from rocks of Upper Triassic age, about 200 million years old, that is thought to be the ancestor to all members of the MAMMALIA.

MORLEY, Edward Williams (1838–1923), US chemist who worked on the relative densities of OXYGEN and HYDROGEN, but is best known for his role in the MICHELSON-MORLEY EXPERIMENT.

MORNING SICKNESS. See PREGNANCY.

MORPHEME, any of the smallest meaningful elements of a language. A morpheme may be a whole word (e.g., "talk"), a syllable (e.g., the "ing" of "talking") or merely a letter (e.g., the "s" of "talks"). (See also PHONEME.)

MORPHINE, OPIUM derivative used as a NARCOTIC ANALGESIC and also commonly in DRUG ADDICTION. It depresses RESPIRATION and the COUGH reflex, induces sleep and may cause VOMITING and CONSTIPATION. It is valuable in HEART failure and as a premedication for ANESTHETICS; its properties are particularly valuable in terminal malignant DISEASE (see also HEROIN). Addiction and withdrawal syndromes are common.

MORSE, Samuel Finley Breese (1791–1872), US inventor of an electric TELEGRAPH. His first crude model was designed in 1832, and by 1835 he could demonstrate a working model. With the considerable help of Joseph HENRY (which later he refused to acknowledge) he developed by 1837 electromagnetic relays to extend the range and capabilities of his system. WHEATSTONE's invention had preceded Morse's, so that he was unable to obtain an English patent, and in the US official support did not come until 1843. His famous message, "What hath God wrought!", was the first sent on his Washington-Baltimore line on May 24, 1844. For this he used MORSE CODE, devised in 1838. In early life, Morse was a noted portrait painter.

MORSE CODE, signal system devised (1838) by Samuel MORSE for use in the wire TELEGRAPH, now used in radiotelegraphy and elsewhere. Letters, numbers and punctuation are represented by combinations of dots (brief taps of the transmitting key) and dashes (three times the length of dots).

MORTAR, building material used to bind together stones, blocks or bricks. Early mortar was made from mud and straw. Today a mixture of CEMENT, sand and water is prepared immediately prior to being applied as a paste. When set, it is strong and water-resistant.

MORTON, William Thomas Green (1819–1868), US dentist who pioneered the use of diethyl ETHER as bitter litigation over his refusal to recognize the an anesthetic (1844–46). In later years he engaged in

US morse code		International morse code
.—	A	.—
—...	B	—...
.. .	C	—.—.
—..	D	—..
.	E	.
.—.	F	..—.
——.	G	——.
....	H	....
..	I	..
—.—.	J	.———
—.—	K	—.—
———	L	.—..
——	M	——
—.	N	—.
. .	O	———
.....	P	.——.
..—.	Q	——.—
. ..	R	.—.
...	S	...
—	T	—
..—	U	..—
...—	V	...—
.——	W	.——
.—..	X	—..—
.. ..	Y	—.——
... .	Z	——..
.——.	1	.————
..—..	2	..———
...—.	3	...——
....—	4	....—
———	5	.....
......	6	—....
——..	7	——...
—....	8	———..
—..—	9	————.
———	0	—————

contributions of former colleagues and especially C. W. LONG's prior use of ether in this way.

MOSAIC, general name for VIRUS DISEASES of plants such as tobacco, tomatoes, potatoes, soybeans and peas, which produce a characteristic leaf mottling and stunted growth. A number of viruses cause this type of disease, for example, tobacco mosaic virus. Transmission of the disease may be via aphids or by mechanical contact. Control may be achieved by use of INSECTICIDES and by careful cultivation techniques.

MOSELEY, Henry Gwyn Jeffreys (1887–1915), British physicist who showed that an ELEMENT's properties depend on what he called its atomic number (see ATOM), equivalent to its nuclear charge.

MOSQUITOES, two-winged flies of the family Culicidae, with penetrating, sucking mouthparts. The females of many species feed on vertebrate blood, using their needle-like stylets to puncture a blood capillary, but usually only when about to lay eggs. The males, and the females at other times, feed on sugary liquids such as nectar. Both the larvae and pupae are entirely aquatic, breathing through spiracles at the tip of the abdomen. In all but the Anopheline mosquitoes, the spiracles are at the tip of a tubular siphon, and the larva's body is suspended from this below the surface film. Mosquitoes are involved in the transmission of many diseases in man including YELLOW FEVER, FILARIASIS, and MALARIA.

MÖSSBAUER EFFECT, the recoilless emission of GAMMA RAYS from certain CRYSTALS, discovered by Rudolf Ludwig Mössbauer (1929–   ) in 1957. When gamma rays are emitted from most nuclei, the latter recoil to a variable extent, giving the emitted PHOTONS a broad ENERGY spectrum. Mössbauer found that certain crystals, e.g. Fe^{57}, recoiled as a whole, i.e., their effective recoil was negligible. Gamma rays of closely specified frequency are thus produced and can be used for nuclear clocks and for testing RELATIVITY theory predictions.

MOSSES, large group of plants belonging to the class Musci, of the division BRYOPHYTA. Each moss plant consists of an erect "stem" to which primitive "leaves" are attached. The plants are anchored by root-like rhizoids. Mosses have worldwide distribution and are usually found in woods and other damp habitats. They are often early colonizers of bare soil and play an important role in preventing soil erosion. SPHAGNUM debris is an important constituent of PEAT. (See also ALTERNATION OF GENERATIONS; HORNWORTS; LIVERWORTS.)

MOTHER-OF-PEARL, or **nacre**, the iridescent substance of which PEARLS and the inner coating of bivalved mollusk shells are made. It consists of alternate thin layers of aragonite (CALCIUM carbonate) and conchiolin, a horny substance. Valued for its beauty, it is used in thin sheets for ornament, jewelry and for buttons.

MOTION, change in the aspect of one body relative to another by translation, rotation or revolution, or by combinations of these. (See also MECHANICS; PERPETUAL MOTION; RELATIVITY; VELOCITY.)

MOTION SICKNESS, nausea and VOMITING caused by rhythmic movements of the body, particularly the head, set up in automobile, train, ship or airplane travel. In susceptible people, neither stimulation of the EAR labyrinths nor their action on the vomiting centers in the BRAIN stem are adequately suppressed.

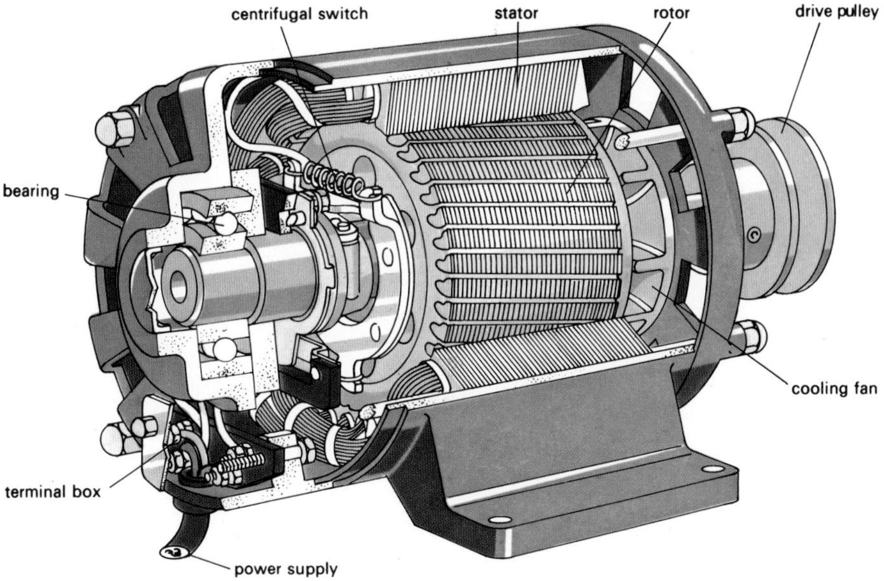

Cutaway of a fractional-horsepower single-phase induction motor, as used in many light-industrial applications. This motor has a squirrel-cage rotor which is not wired into the electrical circuit. Since a single-phase motor with a squirrel-cage rotor cannot start from standstill, a secondary circuit is included in which a capacitor causes the current in a starting winding to lead the applied voltage, thus enabling rotation to begin. When the rotor reaches a certain speed, a centrifugal switch opens and the capacitor circuit is cut out, leaving the motor to run using only its main winding.

Hyoscine and phenothiazines can prevent it if taken before travel.

MOTOR, Electric, a device converting electrical into mechanical energy. Traditional forms are based on the FORCE experienced by a current-carrying wire in a magnetic field (see ELECTROMAGNETISM). Motors can be, and sometimes are, run in reverse as GENERATORS.

Simple direct-current (see ELECTRICITY) motors consist of a magnet or ELECTROMAGNET (the *stator*) and a coil (the *rotor*) which turns when a current is passed through it because of the force between the current and the stator field. So that the force keeps the same sense as the rotor turns, the current to the rotor is supplied via a *commutator*—a slip ring broken into two semicircular parts, to each of which one end of the coil is connected, so that the current direction is reversed twice each revolution.

For use with alternating-current supplies, small DC motors are often still suitable, but **induction motors** are preferred for heavier duty. In the simplest of these, there is no electrical contact with the rotor, which consists of a cylindrical array of copper bars welded to end rings. The stator field, generated by more than one set of coils, is made to rotate at the supply frequency, inducing (see INDUCTION, ELECTROMAGNETIC) currents in the rotor when (under load) it rotates more slowly, these in turn producing a force accelerating the rotor. Greater control of the motor speed and torque can be obtained in "wound rotor" types in which the currents induced in coils wound on the rotor are controlled by external resistances connected via slip-ring contacts.

In applications such as electric clocks, **synchronous motors**, which rotate exactly in step with the supply frequency, are used. In these the rotor is usually a permanent magnet dragged round by the rotating stator field, the induction-motor principle being used to start the motor.

The above designs can all be opened out to form **linear motors** producing a lateral rather than rotational drive. The induction type is the most suitable, a plate analogous to the rotor being driven with respect to a stator generating a laterally moving field. Such motors have a wide range of possible applications, from operating sliding doors to driving trains, being much more robust than rotational drive systems, and offering no resistance to manual operation in the event of power cuts. A form of DC linear motor can be used to pump conducting liquids such as molten metals, the force being generated between a current passed through the liquid and a static magnetic field around it.

MOTT, Sir Neville Francis (1905–), UK physicist who shared the 1977 Nobel Prize for Physics with P. W. ANDERSON and J. H. VAN VLECK for contributions to the understanding of the electronic structure of magnetic and disordered systems.

MOTTELSON, Ben (1926–), US-born Danish physicist who shared the 1975 Nobel Prize for Physics with A. BOHR and L. J. RAINWATER for their work on the physics and structure of the atomic nucleus.

MOULTING. See MOLTING.

MOUNDS, artificial constructions of earth or, on

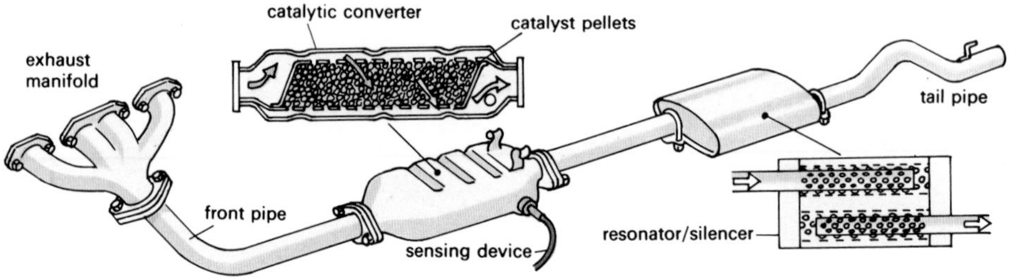

An automobile exhaust system including a two-tube reactive muffler (silencer). Anti-pollution laws sometimes necessitate introducing air into the exhaust manifold to complete the combustion of any partially burned fuel or, as here, completing the combustion process using a catalytic converter.

occasion, piled stones built according to a predetermined plan, found in many areas of the eastern US. The largest known mound is one of the CAHOKIA MOUNDS and the oldest dates from c500 AD. Some mounds have been built in historic times. Dome-shaped burial mounds served the same purpose as BARROWS, while mounds in the form of truncated pyramids were used as bases for temples and other buildings. KITCHEN MIDDENS are sometimes erroneously termed mounds. Less common types of mounds are hill-top forts and mounds in effigy form.

MOUNTAIN, a landmass elevated substantially above its surroundings. The difference between a mountain and a hill is essentially one of size: the exact borderline is not clearly defined. Plateaus, or table-mountains, unlike most other mountains, have a large summit area as compared with that of their base. Most mountains occur in groups, ranges or chains (see also MASSIF). The processes involved in mountain building are termed orogenesis. OROGENIES can largely be explained in terms of the theory of PLATE TECTONICS. Thus the Andes have formed where the Nazca oceanic plate is being subducted beneath (forced under) the South American continental plate, and the Himalayas have arisen at the meeting of two continental plates.

Mountains are traditionally classified as Volcanic, Block or Folded. **Volcanic mountains** occur where LAVA and other debris (e.g., PYROCLASTIC ROCKS) build up a dome around the vent of a VOLCANO. They are found in certain well-defined belts around the world, marking plate margins. **Block mountains** occur where land has been uplifted between FAULTS in a way akin to that leading to the formation of RIFT VALLEYS (see also HORST). **Folded mountains** occur through deformations of the EARTH's crust (see FOLD), especially in geosynclinal areas (see GEOSYNCLINE), where vast quantities of sediments whose weight causes deformation, accumulate (see also SEDIMENTATION). EROSION eventually reduces all mountains to plains. But it may also play a part in the creation of mountains, as where most of an elevated stretch of land has been eroded away, leaving a few resistant outcrops of rock (see MONADNOCK).

MOUNTAIN SICKNESS. See ALTITUDE SICKNESS.

MUCUS, viscid, aqueous solution of glycoproteins secreted by cells of the mucous membranes of the respiratory and gastrointestinal tracts, the salivary and other digestive system glands. It provides a nonliving protective layer which is constantly being renewed, allowing removal of any particulate matter absorbed onto it (in the BRONCHI) and lubrication of food and feces. It contains some GAMMA GLOBULIN, which may have a role in local IMMUNITY.

MUFFLER, or **silencer**, device to reduce the exhaust noise of an INTERNAL COMBUSTION ENGINE. One effect is to cut down the velocity of the exhaust gases and so reduce the pressure waves they create (see SOUND). Some types have a central tube of perforated steel packed around with sound-absorbent fibrous material; in another, passage of the exhaust through different chambers reduces the noise by INTERFERENCE.

MULCH, layer of usually organic material kept on the surface of SOIL in order to reduce surface EVAPORATION of moisture, to protect the soil from wind EROSION, or as a MANURE. Mulches may contain straw, PEAT or scattered topsoil.

MULE, a term now commonly used to describe infertile hybrids between various species. The name is properly restricted to the offspring of a male donkey and a mare. Mules have the shape and size of a horse, and the long ears and small hooves of a donkey. They are favored for their endurance and surefootedness as draft or pack animals.

MULLER, Hermann Joseph (1890–1967), US geneticist awarded the 1946 Nobel Prize for Physiology or Medicine for his work showing that X-RAYS greatly accelerate MUTATION processes.

MÜLLER, Paul Hermann (1899–1965), Swiss chemist awarded the 1948 Nobel Prize for Physiology or Medicine for his discovery of the effective insecticidal properties of DDT (1939), a major contribution to world health and food production.

MULLIKEN, Robert Sanderson (1896–), US chemist and physicist awarded the 1966 Nobel Prize for chemistry for his work on the nature of chemical bonding and hence on the electronic structure of molecules (see BOND, CHEMICAL).

MULTIPLE BIRTH, the delivery of more than one child at the end of PREGNANCY. Twins, the commonest type of multiple BIRTH, are of two distinct varieties. Monozygotic or identical twins originate in a single fertilized egg (zygote) which divides, each half (containing identical genetic material) developing independently into EMBRYO and FETUS, although they may share a common PLACENTA. *Dizygotic* or non-identical twins originate in the release of two eggs at ovulation (see MENSTRUATION), each being fertilized, implanting (see implantation) and developing separately. There is no more relation between their GENES than between those of other siblings. Higher orders of multiple births (triplets, quadruplets, quintuplets, etc.) usually arise from multiple ovulation and are rare unless ovarian follicle stimulants (e.g., GONADOTROPHINS) have been used in the treatment of infertility; here the dosage is critical. Multiple pregnancy may run in families. Prematurity, toxemia, ANEMIA and other complications are more common in multiple pregnancy.

MULTIPLE SCLEROSIS, or **disseminated sclerosis**, a relatively common disease of the BRAIN and SPINAL CORD in which MYELIN is destroyed in plaques of INFLAMMATION. Its cause is unknown although slow VIRUSES, abnormal ALLERGY to viruses and abnormalities of FATS are suspected. It may affect any age group, but particularly young adults. Symptoms and signs indicating disease in widely separate parts of the NERVOUS SYSTEM are typical. They occur episodically, often with intervening recovery or improvement. Blurring of VISION, sometimes with EYE pain; double vision; VERTIGO; abnormal sensations in the limbs; PARALYSIS; ATAXIA, and BLADDER disturbance are often seen, although individually these can occur in other brain diseases. STEROIDS, certain DIETARY FOODS, and DRUGS acting on spasticity in muscles and the bladder are valuable in some cases. The course of the disease is extremely variable, some subjects having but a few mild attacks, while others progress rapidly to permanent disability and dependency.

MULTIPLICATION, a way of combining two numbers to obtain a third; symbolized by ×, ., or merely the juxtaposition of the numbers (where suitable). Where x and y are NATURAL NUMBERS, $x.y$ is commutative and defined by $x+x+\ ...\ +x$, the number x appearing y times (see ADDITION; SUM).

For multiplication of negative INTEGERS, such as $(-x)$ and $(-y)$, $(-x).y=x.(-y)=-(x.y)$; and $(-x).(-y)=x.y$. Multiplication of any number by 0 (see ZERO) is defined to give the PRODUCT 0. FRACTIONS may be multiplied by simple extension of the system. The INVERSE operation of multiplication is DIVISION, since $\frac{x}{y}=x.\frac{1}{y}$. In cases other than with REAL NUMBERS, multiplication must be independently defined (see IMAGINARY NUMBERS; VECTOR ANALYSIS).

MUMMY, a corpse embalmed, particularly in ancient Egypt, in order to ensure its preservation for a protracted period after death. The earliest known attempts artificially to preserve bodies were about 2600 BC, though many bodies from earlier times were naturally preserved through the desiccating effect of the sand in which they were buried. (See EMBALMING.)

MUMPS, common VIRUS infection causing swelling of the parotid salivary GLAND, and occasionally INFLAMMATION of the PANCREAS, an OVARY or a TESTIS. Mild FEVER, HEADACHE and malaise may precede the gland swelling. Rarely a viral MENINGITIS and less often ENCEPHALITIS complicates mumps. Very rarely a bilateral and severe testicular inflammation can cause STERILITY.

MUON. See SUBATOMIC PARTICLES.

MURPHY, John Benjamin (1857–1916), US surgeon who pioneered the use of immediate appendectomy (the surgical removal of the appendix) as a treatment for APPENDICITIS.

MURPHY, William Parry (1892–), US physician who shared the 1934 Nobel Prize for Physiology or Medicine with G. MINOT and G. WHIPPLE for his work with Minot showing that daily consumption of large quantities of raw liver is an effective treatment for pernicious ANEMIA.

MUSCI. See MOSSES.

MUSCLE, the tissue whose contraction produces body movement. In man and other vertebrates there are three types of muscle. **Skeletal or striated muscle** is the type normally associated with the movement of the body. Its action can either be initiated voluntarily, through the central NERVOUS SYSTEM, or it can respond to REFLEX mechanisms. Under the microscope this muscle is seen to be striped or striated. It consists of cylinders of tissue 0.01mm in diameter, showing great variation in length (1–150mm) and containing many nuclei. Each cylinder consists of thousands of filaments, each bathed in cytoplasm (known as sarcoplasm) which is their source of nutrition. Energy for contraction is derived by the OXIDATION of GLUCOSE brought by the BLOOD and stored as granules of GLYCOGEN in the sarcoplasm. The oxidation and breakdown of the glucose takes place in the mitochondria (see CELL), the net result being the formation of adenosine triphosphate (ATP—see NUCLEOTIDES). This molecule provides a "high-energy" bond which enables actin and myosin, two proteins in the muscle filament, to slide into each other, an action which, repeated many times throughout the muscle, results in its contraction. The behavior of a particular fiber is governed by an "all-or-none" law, in that it will either contract completely or not at all. Therefore the extent to which a muscle contracts is dependent solely on the number of individual fibers contracting. If a muscle is starved of oxygen, a process termed GLYCOLYSIS provides the energy. However glycolysis involves LACTIC ACID production with the consequent risk of CRAMP. Skeletal muscle functions by being attached via TENDONS to two parts of the SKELETON which move relative to each other. The larger attachment is known as the muscle's origin. Contraction of the muscle attempts to draw together the two parts of the skeleton. Muscles are arranged in antagonistic groups so that all movements involve the contraction of some muscles at the same time as their antagonists relax. **Smooth or involuntary muscle** is under the control of the autonomic nervous system and we are rarely aware of its action. Smooth muscle fibers are constructed in sheets of cells, each with a single nucleus. They are situated in hollow structures such as the gut, BRONCHI, uterus and BLOOD vessels. Smooth muscle uses the property of "tone" (continual slight tension) to regulate the diameter of tubes such as blood vessels. Being responsive to HORMONES, notably ADRENALINE, it can thus decrease blood supply to nonessential organs during periods of stress. In the gut, the muscle also propels the contents along by contracting along its length in waves (PERISTALSIS). **Cardiac muscle**, found only in the HEART, has the property of never resting throughout life. It combines features of both skeletal and smooth muscle, for it is striped but yet involuntary. The fibers are not discrete but branching and interlinked, thus enabling it to act quickly and in unison when stimulated.

MUSCOVITE, or **isinglass**, the commonest species of MICA, $KAl_2(Si_3Al)O_{10}(OH)_2$, chiefly obtained from pegmatite dikes in India and the US.

MUSCULAR DYSTROPHY, a group of inherited DISEASES in which MUSCLE fibers are abnormal and undergo ATROPHY. Most develop in early life or adolescence. *Duchenne dystrophy* occurs in males although the genes for it are carried by females. It starts in early life and some swelling (pseudohypertrophy) of calf and other muscles may be seen. A similar disease can affect females. Other types, described by muscles mainly affected, include *limb-girdle* and *facio-scapulo-humeral* dystrophies. There are many diverse variants, largely due to structural or biochemical abnormalities in muscle fibers. *Myotonic dystrophy* occurs in older men, causing BALDNESS, CATARACTS, TESTIS atrophy and a characteristic myotonus, in which contraction is involuntarily sustained. Muscular dystrophies usually cause weakness and wasting of muscles, particularly of those close to and in the trunk; a waddling gait and exaggerated curvature of the lower spine are typical. The muscles of RESPIRATION may be affected, with resulting PNEUMONIA and respiratory failure; HEART muscle, too, can also be affected. These two factors in particular may lead to early death in severe cases. Mechanical aids, including if necessary ARTIFICIAL RESPIRATION, may greatly improve well-being, mobility and life-span.

MUSK, a strongly-scented substance used in the manufacture of perfume. The term is strictly applied to that obtained from the musk glands of the male MUSK DEER, but also covers other similar secretions, e.g., civet musk, badger musk.

MUSKEG, found in the far north, particularly in TUNDRA, a BOG or SWAMP almost completely filled with SPHAGNUM moss.

MUSTARD GAS, or dichlorodiethyl sulfide $(CH_2Cl\ CH_2)_2S$, a toxic vesicant gas made from ETHYLENE and sulfur(I) chloride. (See also CHEMICAL AND BIOLOGICAL WARFARE.)

MUTATION, a sudden and relatively permanent change in a GENE or CHROMOSOME set, the raw material for evolutionary change. Chemical or physical agents which cause mutations are known as *mutagens*. Mutations can occur in any type of CELL at any stage in the life of an organism but only changes present in the GAMETES are passed on to the offspring. A mutation may be dominant or recessive, viable or lethal. The majority are changes in individual genes (gene mutations) but in some cases changes in the structure or numbers of chromosomes may be seen. The formation of structural chromosome changes is used to test drugs for mutagenic activity. Mutation normally occurs very rarely but certain mutagens—X-RAYS, GAMMA RAYS, NEUTRONS and MUSTARD GAS—greatly accelerate mutation.

MUTE. See DUMBNESS.

MUTUALISM, a kind of SYMBIOSIS in which the organisms involved depend on each other for their continuing existence. An important example involves the RUMINANTS, which would be unable to digest the material they eat were it not for the cellulase-producing BACTERIA that live within their gut.

MYASTHENIA GRAVIS, a DISEASE of the junctions between the peripheral NERVOUS SYSTEM and the MUSCLES, probably due to abnormal IMMUNITY, and characterized by the fatigability of muscles. It commonly affects EYE muscles, leading to drooping lids and double VISION, but it may involve limb muscles. Weakness of the muscles of RESPIRATION, swallowing and coughing may lead to respiratory failure and aspiration or bacterial PNEUMONIA. Speech is nasal, regurgitation into the nose may occur and the FACE is weak, lending a characteristic snarl to the MOUTH. It is associated with disorders of THYMUS GLAND and THYROID GLAND. Treatment is with cholinesterase inhibitors; STEROIDS and thymus removal may control the causative immune mechanism.

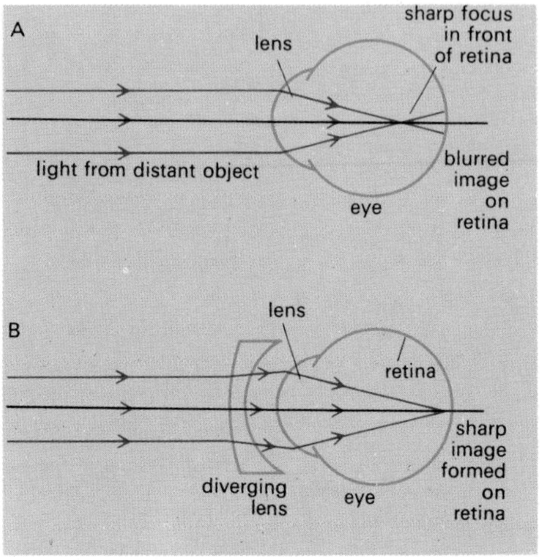

Myopia (A), and its correction using a diverging lens (B).

MYCENAEAN CIVILIZATION. See AEGEAN CIVILIZATION.

MYCOLOGY, the scientific study of FUNGI.

MYCOPLASMAS, minute organisms intermediate in size between BACTERIA and VIRUSES. Structurally they consist of an outer three-layered pliable membrane surrounding proteinaceous cytoplasm that contains both RNA and DNA (see NUCLEIC ACIDS). Mycoplasmas cause diseases of man, animals and plants.

MYCORRHIZA, an association between the roots of vascular plants and a fungus. Mycorrhizas are found in many plant species, typically in forest trees, heath plants and orchids. Some plants grow very poorly in the absence of the fungus. The relationship appears to be symbiotic, that is, of benefit to both plant and fungus. There are two main types of mycorrhiza, the ectotrophic and endotrophic. In the ectotrophic association the fungus forms a sheath around the roots and in the endotrophic, the fungus penetrates the cells within the cortex of the roots. (See SYMBIOSIS.)

MYCOTA, a division of the PLANT KINGDOM that includes the SLIME MOLDS (Myxomycetes) and the true FUNGI.

MYELIN, specialized layering of the membranes of glial cells (in the BRAIN) and the Schwann cells (of the peripheral NERVOUS SYSTEM) which wrap around nerve fibers producing an electrically insulating sheath, interrupted at intervals, which facilitates rapid nerve impulse conduction. Its disorders include MULTIPLE SCLEROSIS, ENCEPHALITIS and NEURITIS.

MYOCARDITIS, a rare INFLAMMATION of the HEART muscle caused by VIRUSES, BACTERIA, some metal poisons and drugs. It is a serious complication of acute RHEUMATIC FEVER. Treatment involves bed rest, but the heart may be permanently damaged.

MYOGLOBIN, molecule related to HEMOGLOBIN found in MUSCLE cells and which serves as a local store for OXYGEN. Its affinity for oxygen encourages the transfer of oxygen to it from the BLOOD.

MYOPIA, or near- or shortsightedness, a defect of VISION in which light entering the EYE from distant objects is brought to a focus in front of the retina. The condition may be corrected by use of a diverging spectacle LENS.

MYRIOPODA, a class of the ARTHROPODA which includes the centipedes and millipedes. All members are terrestrial and have a single pair of antennae, a trunk composed of leg-bearing segments and a long tubular gut. Some zoologists classify the 11 000 species into four separate classes: Chilopoda (centipedes), Diplopoda (millipedes), Pauropoda and Symphyla.

MYRRH, the fragrant resin obtained from small thorny trees of the genus *Commiphora* from the family, Burseraceae. Myrrh has been used for embalming, in medicines and as incense and is now an important constituent of some PERFUMES.

MYTHOLOGY, a collection of traditional tales, usually of a particular people (such as the ancient Greeks, Indians or Norsemen), handed down orally through the generations. Most mythologies are of earlier date than the invention of writing in the cultures from which they spring. There are three main classes of myths: myths proper, which are imaginative and serious attempts to explain natural phenomena and are often concerned with gods and supernatural events occurring in a timeless past; folk tales, including fairy tales, which are narrative stories in historical time of social concerns; and sagas and legends, which recount embellished exploits of heroes of the past who may or may not have existed. Although most of the cultures of primitive men have mythologies, not all people do: one example is the Romans, who appear to have had very little in the way of oral tradition, but who borrowed and adapted Greek mythology. The term mythology is also applied to the study of such tales, from which past cultural exchanges can be inferred and historical archaeological sites can often be identified. The most famous such study was Sir James FRAZER's *The Golden Bough* (1890) which contains an extensive synthesis of many disparate myths and mythologies.

MYXEDEMA, severe hypothyroidism. See THYROID GLAND.

MYXOPHYTA. See SLIME MOLDS.

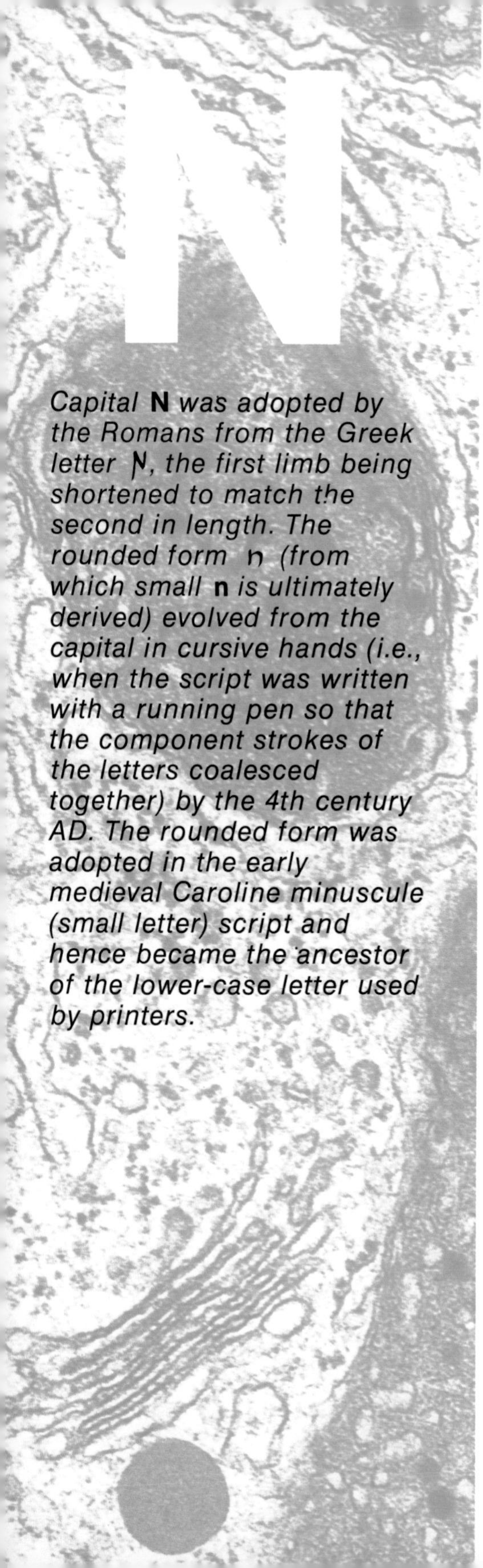

N

*Capital **N** was adopted by the Romans from the Greek letter Ν, the first limb being shortened to match the second in length. The rounded form ɲ (from which small **n** is ultimately derived) evolved from the capital in cursive hands (i.e., when the script was written with a running pen so that the component strokes of the letters coalesced together) by the 4th century AD. The rounded form was adopted in the early medieval Caroline minuscule (small letter) script and hence became the ancestor of the lower-case letter used by printers.*

NABLA. See DEL.

NACRE. See MOTHER OF PEARL.

NADIR, the point on the CELESTIAL SPHERE directly opposite an observer's ZENITH.

NAIL, metal shaft, pointed at one end and usually with a head at the other, that can be hammered into pieces of wood or other materials to fasten them together. In the making of common nails, steel wire is fed discontinuously between a pair of gripper DIES, which hold it while a hammer forms the head. The grippers part and the wire moves forward; nippers then shear the shaft and pliers form the point. Other forms are masonry nails, stamped from a plate, and U-shaped staples.

NAILS, a keratinous covering protecting the tips of the digits in VERTEBRATES. Primitively always present as claws, they are now modified as claws, true nails or, in UNGULATES, as HOOVES. (See also KERATIN.)

NANSEN, Fridtjof (1861–1930), Norwegian explorer, scientist and humanitarian, awarded the 1922 Nobel Peace Prize, best known for his explorations of the Arctic. His most successful attempt at reaching the NORTH POLE was in 1895, when he achieved latitude 86° 14′, the farthest north then reached. He also designed the **Nansen Bottle**, a device for obtaining water samples at depth.

NAPALM, a SOAP consisting of the aluminum salt of a mixture of CARBOXYLIC ACIDS, with aluminum hydroxide in excess. When about 10% is added to GASOLINE it forms a GEL, also called napalm, used in flame throwers and incendiary bombs; it burns hotly and relatively slowly, and sticks to its target. Developed in WWII, it was used in the Vietnam War and caused great havoc. (See also CHEMICAL AND BIOLOGICAL WARFARE.)

NAPHTHA, volatile mixture of liquid HYDROCARBONS boiling in the range 80°C to 180°C, used as a solvent. It is obtained by distilling COAL TAR (yielding aromatic products) or shale oil, or from the refining and cracking of PETROLEUM.

NAPHTHALENE ($C_{10}H_8$), white crystalline solid, an AROMATIC HYDROCARBON consisting of two fused BENZENE rings, and more reactive than benzene. It is produced from COAL TAR, and is used to make phthalic anhydride (see ACID ANHYDRIDES), as an intermediate in DYE manufacture, and for mothballs. MW 128.2, mp 81°C, bp 218°C.

NAPIER, John (1550–1617), Scottish mathematician credited with the invention of LOGARITHMS (before 1614). Natural logarithms (to the base *e* (see EXPONENTIAL)) are often called **Napierian Logarithms** for him. He also developed the modern notation for the DECIMAL SYSTEM.

NARCISSISM, exaggerated self-love, often at a sexual level, characteristic of early sexual development though sometimes retained. It may also develop through IDENTIFICATION with an object of desire as a DEFENSE MECHANISM against possible loss. (See also SEX.)

NARCOLEPSY. See AMPHETAMINES.

NARCOTICS, DRUGS that induce sleep; specifically, the OPIUM-derived ANALGESICS. These affect the higher BRAIN centers causing mild euphoria and sleep (narcosis). They may act as HALLUCINOGENIC DRUGS and are abused in DRUG ADDICTION.

NASMYTH, James (1808–1890), Scottish engineer who invented many MACHINE TOOLS, including the steam hammer (1839).

NATIONAL OCEANIC AND ATMOSPHERIC ADMINISTRATION (NOAA), US government agency set up in 1970 to coordinate scientific research into atmosphere and oceans. Its specific aims are the monitoring and control of POLLUTION and the investigation of potential resources and weather-control techniques. The NOAA is responsible for the work of several formerly independent agencies including the Coast and Geodetic Survey (founded in 1807) and the Weather Bureau (founded in 1870).

NATTA, Giulio (1903–), Italian chemist awarded (with ZIEGLER) the 1963 Nobel Prize for Chemistry for his synthesis of POLYMERS of propene (one of the ALKENES). These have industrially desirable properties such as high melting point and strength.

NATURAL GAS, mixture of gaseous HYDROCARBONS occurring in reservoirs of porous rock (commonly sand or sandstone) capped by impervious strata. It is often associated with PETROLEUM, with which it has a common origin in the decomposition of organic matter in sedimentary deposits. Natural gas consists largely of METHANE and ETHANE, with also propane and butane (separated for BOTTLED GAS), some higher ALKANES (used for GASOLINE), nitrogen, oxygen, carbon dioxide, hydrogen sulfide, and sometimes valuable HELIUM. It is used as an industrial and domestic FUEL, and also to make carbon-black and in chemical synthesis. Natural gas is transported by large pipelines or (as a liquid) in refrigerated tankers. Total world reserves are estimated at $54 \times 10^{12} m^3$, of which the USSR and the US together have about half; there are major fields also in the Netherlands, Algeria and the Middle East.

NATURAL NUMBERS, the SET of all positive INTEGERS.

NATURAL RESOURCES, Conservation of. See CONSERVATION; RECYCLING.

NATURAL SELECTION, mechanism for the process of EVOLUTION discovered by Charles DARWIN in the late 1830s, but not made public until 1858. According to Darwin, evolution occurs when an organism is confronted by a changing environment. A degree of variety is always present in the members of an interbreeding population. Normally, the possession of a variant character by an individual confers no particular advantage on it, and the proportion of individuals in the population with a given variation remains constant. But if it ever arises in a changed environment that a given variation increases the chances of an individual's survival, then individuals possessing that character will be more liable to survive—and breed. The frequency with which the variant character occurs in future generations of the organism will thus increase, and, over a large number of generations, the general form of the population will change. The name "natural selection" derives from the analogy Darwin saw between this selection on the part of "Nature" and the "artificial selection" practiced by animal breeders.

NAUTICAL MILE. See MILE.

NAVAL STORES, products derived from coniferous trees, especially the pine, such as PITCH, rosin (see RESIN), TAR and TURPENTINE. The name comes from the use of these (and other) materials in building and repairing wooden sailing ships.

Liquified-natural-gas tanker *Aquarius*, built by General Dynamics, Quincy Shipbuilding Division, carries 125 000m^3 liquified natural gas in five insulated spherical tanks. Any gas which boils off the liquid cargo helps to fuel the ship's engines.

NAVEL. See UMBILICAL CORD.

NAVIGATION, the art and science of directing a vessel from one place to another. Originally navigation applied only to marine vessels, but now air navigation and, increasingly, space navigation are also important. Although the techniques and applications of navigation have radically changed through time, the basic problems, and hence the principles, have remained much the same.

Marine Navigation. Primitive sailors could not venture out of sight of land without the risk of getting lost. But soon they learned to use sunset and sunrise, the prevailing winds, the POLE STAR and so forth as aids to direction. Early on, the first fathometer, a weighted rope used to measure depth, was developed. Before the 10th century AD the magnetic COMPASS had appeared. But it was not until the 1730s that the inventions of the SEXTANT and CHRONOMETER heralded the dawn of accurate sea navigation. Both LATITUDE AND LONGITUDE could now be determined within reasonable tolerances. (See also ASTROLABE; GREENWICH OBSERVATORY.)

Modern navigation uses electronic aids such as LORAN and the radiocompass; celestial navigation, the determination of position by sightings of celestial bodies, and dead reckoning where, by knowing one's position at a particular past time, the time that has elapsed since, one's direction and speed (see LOG), one can tell one's present position. (See also DIRECTION FINDER; ECHO SOUNDER; MAP; SONAR; SUBMARINE.)

Air Navigation uses many of the principles of marine navigation. In addition, the pilot must work in a third dimension, must know his altitude (see ALTIMETER), and in bad visibility must use aids like the INSTRUMENT LANDING SYSTEM. RADAR is also used.

Space Navigation is a science in its infancy. Like air navigation, it works in three dimensions, but the problems are exacerbated by the motions both of one's source (the earth) and one's destination, as well as by the distances involved. But, prior to developments in new areas, it seems that SPACE EXPLORATION has inaugurated a new era in navigation by the stars. (See also CELESTIAL SPHERE; GYROCOMPASS; GYROPILOT; REMOTE CONTROL.)

NEANDERTHAL MAN. See PREHISTORIC MAN.

NEARSIGHTEDNESS. See MYOPIA.

NEBULA, an interstellar cloud of gas or dust. The term is Latin, meaning "cloud," and was initially used to denote any fuzzy celestial object, including COMETS and external GALAXIES: this practice has now largely been abandoned. There are two main types of nebula. **Diffuse nebulae** are large, formless clouds of gas and dust and may be either bright or dark. *Bright nebulae*, such as the ORION Nebula, appear to shine due to the proximity or more usually presence within them of bright stars, whose light they either reflect (reflection nebula) or absorb and reemit (emission nebula). *Dark nebulae*, such as the Horsehead Nebula, are not close to, or do not contain, any bright stars, and hence appear as dark patches in the sky obscuring the light from stars beyond them. Study of diffuse nebulae is particularly important since it is generally accepted that they are in the process of condensing to form new STARS. **Planetary nebulae** are very much smaller, and are always connected with a star that has gone NOVA some time in the past. They are, in fact, the

Nebulae: The Orion nebula (NGC 1976; M42—*below left*) and the Trifid nebula in Sagittarius (NGC 6514; M20—*below right*) are both bright nebulae. The Horsehead nebula in Orion (NGC 2024—*bottom left*) is a dark nebula; the Ring nebula in Lyra (NGC 6720; M57—*bottom right*) is a planetary nebula. NGC numbers refer to the New General Catalog and M numbers to Messier's catalog of nebulae and star clusters.

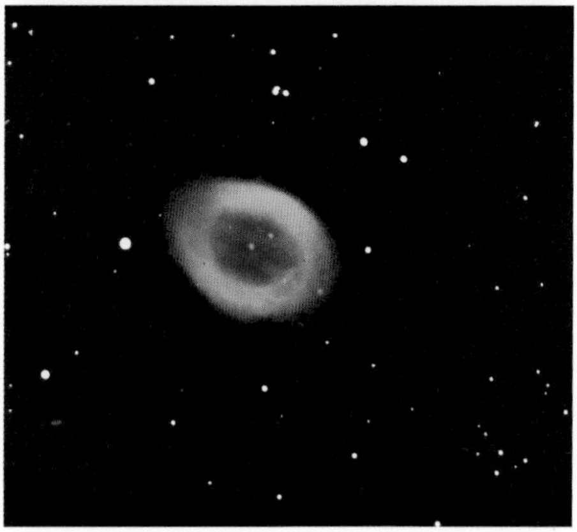

material that has been cast off by the star. They are usually symmetrical, forming an expanding shell around the central star, which is often still visible within. The Ring Nebula is an outstanding example.

NEBULAR HYPOTHESIS, theory accounting for the origin of the solar system put forward by LAPLACE. It suggested that a rotating NEBULA had formed gaseous rings which condensed into the planets and moons, the nebula's nucleus forming the sun.

NECESSITY, as contrasted with CONTINGENCY, in LOGIC, the property whereby a statement must be either true or false, this depending on the correct use of language in framing the statement. The statement "this leaf of paper has two sides" is a necessary truth since if it had either more or less than two sides it would not be a leaf according to the normal usage of the term.

NECESSITY AND SUFFICIENCY, in mathematical ANALYSIS, conditions which describe the validity of a statement. For example, for $4 < a < 10$ it is *necessary* that, say, $3 < a < 11$ since if this were not true then the original statement would not be true; and *sufficient* that $5 < a < 9$, since if this is true it implies the truth of the original statement. More generally, in LOGIC, a condition is necessary for the truth of a statement if the falsehood of the condition implies the falsehood of the statement, and sufficient if the truth of the condition implies the truth of the statement. (See also NECESSITY.)

NECROSIS, the death of body cells due to disease. (See DEATH.)

NECTAR, a sweet viscous secretion containing from 5% to 80% SUGAR, produced by the stems, leaves and flowers of higher plants. By attracting insects it facilitates POLLINATION. Nectars with over 15% sugar are used by honey BEES to make HONEY.

NÉEL, Louis Eugène Félix (1904–), French physicist awarded (with ALFVÉN) the 1970 Nobel Prize for Physics for his work on the magnetic properties of solids. His researches not only permitted manufacture of products used in, e.g., computers, but also explained phenomena such as the recording by certain of the earth's rocks of past geomagnetic fields (see EARTH; PALEOMAGNETISM).

NEGROID, one of the racial divisions of man. The RACE is characterized by woolly hair and yellow, dark brown or black skin. Most negroid peoples originated in Africa, but Melanesians and Negritos are also negroid. (See also ANEMIA.)

The axolotl provides a classic example of neoteny. Although the axolotl (found only in certain lakes around Mexico City) can, and usually does, breed in this, its larval, condition, under certain conditions it can undergo metamorphosis, becoming an adult Mole salamander, *Ambystoma mexicanum*.

NEKTON, the animals of the open sea: squids cuttlefishes; innumerable species of pelagic fishes, and whales. The term is intended to complement plankton, another pelagic assemblage, but of small animals with limited locomotor ability.

NEMATODES. See ASCHELMINTHES.

NEMERTIN(E)A, or **Nemertea**, a phylum of soft-bodied worm-like animals known as ribbonworms or proboscis worms. Nearly 600 species are known; most are marine, but some are freshwater, and a few live on land. Most species grow to a length of 20mm (0.8in) but one, the Bootlace worm *Lineus longissima* attains a length of 55m (180ft).

NEODYMIUM (Nd), one of the LANTHANUM SERIES. AW 144.2, mp 1021°C, bp 3068°C, sg 6.80 (20°C).

NEOLITHIC AGE. See PRIMITIVE MAN; STONE AGE.

NEON (Ne), one of the NOBLE GASES. It is used in discharge tubes, low-wattage glow lamps, SPARK CHAMBERS, and in helium-neon LASERS. Liquid neon is used as a refrigerant in the range 25–40K (see CRYOGENICS), and is added to liquid-hydrogen BUBBLE CHAMBERS to provide a better particle target. AW 20.2, mp −248.7°C, bp −246.0°C. Neon, the earliest of the noble gases to be employed in discharge tubes, glows orange when excited.

NEOPILINA, a member of the MOLLUSCA that was thought to have been extinct by the end of the Devonian period, 345 million years ago, but was discovered alive in 1952.

NEOPLASM. See CANCER, TUMOR.

NEORNITHES, a group of the AVES that includes all living birds and which are therefore classified separately from *Archaeopteryx*, which is included in the ARCHAEORNITHES.

NEOTENY, the retention of larval characters by an organism which has reached sexual maturity. Classically, neoteny is exhibited by certain salamanders (see AXOLOTL), but it also plays a part in the LIFE CYCLE of many invertebrates. Several times, the development of neotenic larvae may have been important in the course of EVOLUTION.

NEPER (Np), named for mathematician John

NAPIER, unit used to express the ratio of quantities such as current or voltage. Thus the ratio N, in nepers, of currents I_1 and I_2 is $N = \ln(I_2/I_1)$. The neper is thus the natural logarithmic analogue of the DECIBEL. Power ratios, say of powers P_1 and P_2, can also be expressed in nepers; here $N = \frac{1}{2}\ln(P_2/P_1)$, and 1 neper = 8.686dB.

NEPHRITE. See JADE.

NEPHRITIS, INFLAMMATION affecting the KIDNEYS. The term **glomerulonephritis** covers a variety of diseases, often involving disordered IMMUNITY, in which renal glomeruli are damaged by immune complex deposition (e.g., BRIGHT'S DISEASE); by direct autoimmune attack (Goodpasture's syndrome), or sometimes as a part of systemic disease (e.g., LUPUS, endocarditis, DIABETES or hypertension). Acute or chronic renal failure or nephrotic syndrome may result. The treatment is immunosuppressive or with STEROIDS. Acute **pyelonephritis** is bacterial infection of the kidney and renal pelvis, following SEPTICEMIA or lower urinary tract infection. Typically, this involves FEVER, loin pain and painful, frequent urination. The treatment requires ANTIBIOTICS. Chronic pyelonephritis includes recurrent kidney infection with permanent scarring and functional impairment.

NEPHRON. See KIDNEYS.

NEPHROSIS, or nephrotic syndrome, EDEMA associated with kidney disease (see NEPHRITIS).

NEPTUNE, the fourth largest planet in the SOLAR SYSTEM and the eighth in position from the sun, with a mean solar distance of 30.07AU. Neptune was first discovered in 1846 by J. G. Galle using computations by LEVERRIER based on the perturbations of URANUS' orbit. The calculation had been performed independently by John Couch ADAMS in England but vacillations on the part of the then Astronomer Royal had precluded a rigorous search for the planet. Neptune has two moons, Triton and Nereid, the former having a circular, retrograde orbit (see RETROGRADE MOTION), the latter having the most eccentric orbit of any moon in the Solar System. Neptune's "year" is 164.8 times that of the earth, its day being 15.8h. Its diameter is about 51Mm and its mass 17.45 times that of the earth. Its structure and constitution are believed to resemble those of JUPITER.

NEPTUNISM, late 18th-century geological theory propagated by the school of A.G. WERNER in which it was claimed that the rocks originally forming the crust of the earth had been precipitated out of aqueous solution.

NEPTUNIUM (Np), the first TRANSURANIUM ELEMENT; one of the ACTINIDES. It is produced in breeder NUCLEAR REACTORS as a by-product of PLUTONIUM production by neutron irradiation of URANIUM (U^{238}). The stablest isotope is Np^{237} (half-life 2.2×10^6yr). Chemically neptunium resembles uranium. (For the **Neptunium Series** see RADIOACTIVITY.) mp 640°C, sg 20.45 (α).

NERNST, Walther Hermann (1864–1941), German physical chemist awarded the 1920 Nobel Prize for Chemistry for his discovery of the Third Law of THERMODYNAMICS.

NERVE. See NERVOUS SYSTEM.

NERVE GAS. See CHEMICAL AND BIOLOGICAL WARFARE.

NERVOUS BREAKDOWN, popular term used to describe various kinds of MENTAL ILLNESS, often associated with fatigue or emotional stress, which drastically impair a person's normal efficiency and disturb his social behavior.

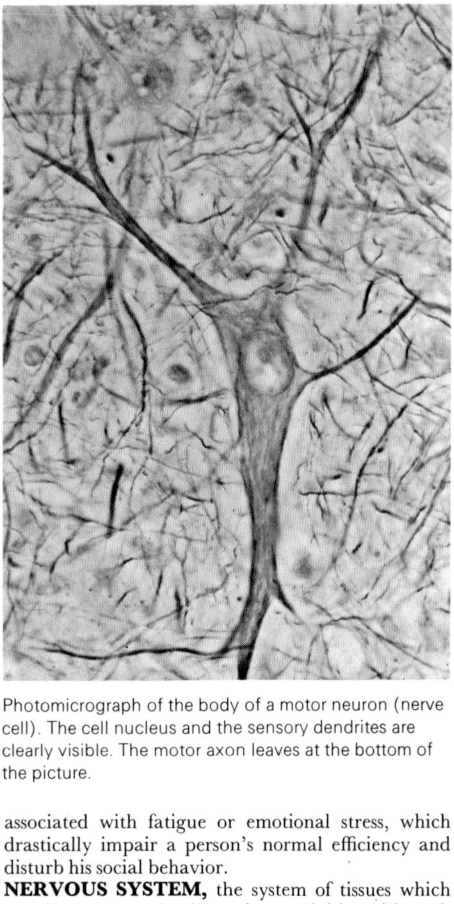

Photomicrograph of the body of a motor neuron (nerve cell). The cell nucleus and the sensory dendrites are clearly visible. The motor axon leaves at the bottom of the picture.

NERVOUS SYSTEM, the system of tissues which coordinates an animal's various activities with each other and with external events by means of nervous impulses conducted rapidly from part to part via nerves. Its responses are generally rapid, whereas those of the endocrine system with which it shares its coordinating and integrating function are generally slow (see GLANDS, HORMONES).

The nervous system can be divided into two parts. The **central nervous system** (CNS), consisting of BRAIN and SPINAL CORD, stores and processes information and sends messages to muscles and glands. The **peripheral nervous system**, consisting of 12 pairs of cranial nerves arising in and near the medulla oblongata of the brain and 31 pairs of spinal nerves arising at intervals from the spinal cord, carries messages to and from the central nervous system.

A third system, the **autonomic nervous system**, normally considered part of the peripheral nervous system, controls involuntary actions such as heartbeat and digestion. It is divisible into two complementary parts: the *sympathetic system* prepares the body for "fight or flight," and the *parasympathetic system* controls the body's vegetative functions. Most internal organs are innervated by both parts.

The nervous system's basic anatomical and functional unit is the highly specialized nerve cell or NEURON, the shape of which varies greatly in different regions. It possesses two kinds of processes: *dendrites* which together with the cell body receive impulses from other neurons, and an *axon* which conducts impulses to other neurons. Axons vary greatly in length (up to a few metres) and speed of conduction (up to about 90m/s).

Sensory or *afferent neurons* carry information to the central nervous system from sensory receptors (such as skin receptors and muscle stretch receptors), whereas *efferent neurons* carry information away from it. Efferent neurons passing to muscle are called *motor neurons*.

Nerves are formed from many axons, both afferent and efferent, surrounded by their associated sheaths which insulate them from each other. Axons surrounded by a fat and protein sheath, called a MYELIN sheath, conduct fastest. Just prior to entering the spinal cord each spinal nerve divides into a *dorsal root* containing afferent axons only and a *ventral root* containing efferent axons only.

Adjacent neurons communicate through specialized contact points or *synapses* which are either excitatory or inhibitory. The elaborate neural circuitry arising from synaptic contact in the central nervous system is responsible for much of behavior, from simple reflex action to complex thought-communication patterns.

The nerve impulse, or action potential, is an electrical signal conducted at speeds far slower than ELECTRICITY. An electrical potential difference of about 70mV, called the resting potential, exists between the inside and outside of the neuron due to the ionic concentration imbalance between inside and outside, and a metabolic pump moving IONS across the cell membrane. If the resting potential is reduced below a certain threshold level, as may occur when impulses are received from other neurons, an impulse is initiated. Impulses are all the same strength ("all-or-none" law), and travel to the end of the axon to the synapse where a chemical transmitter substance (see ACETYLCHOLINE) is released which initiates a new electrical signal in the next neuron. (See also NEURALGIA; NEURITIS; NEUROLOGY.)

NEST, a structure prepared by many animals for the protection of their eggs and young, or for sleeping purposes. In social insects, the nest provides the home of the whole colony, and may have special structures for temperature control and ventilation. The sleeping nests of, for example, the great apes, are commonly no more than crudely woven hammocks of twigs and branches. The sleeping nests of other mammals (which may be used for hibernation) are as complex and woven as any breeding nest. Both these and the nests used by birds and mammals for breeding, must protect the animals within from both weather and predators. Nests can be built of mud, leaves, twigs, down, paper, and kinds of human garbage.

NEURALGIA, pain originating in a nerve and characterized by sudden sharp, often electric shock-like pain or exacerbations of pain. Nerves commonly affected include the digital nerves of toes and intercostal nerves. Neuralgia may be due to INFLAMMATION or trauma.

NEURITIS, or peripheral neuropathy, any disorder of the peripheral NERVOUS SYSTEM which interferes with sensation, the nerve control of MUSCLE, or both. Its causes include DRUGS and heavy metals (e.g., gold); infection or allergic reaction to it (as with LEPROSY or DIPHTHERIA); inflammatory disease (rheumatoid ARTHRITIS); infiltration, systemic and metabolic disease (e.g., DIABETES or PORPHYRIA); VITAMIN deficiency (BERIBERI); organ failure (e.g., of the LIVER or KIDNEY); genetic disorders, and the nonmetastatic effects of distant CANCER. Numbness, tingling, weakness and PARALYSIS result, at first affecting the extremities. Diagnosis involves electrical studies of the nerves and nerve BIOPSY.

NEUROLOGY, branch of MEDICINE concerned with diseases of the BRAIN; SPINAL CORD, and peripheral NERVOUS SYSTEM. These include MULTIPLE SCLEROSIS, EPILEPSY, migraine (HEADACHE), STROKE, PARKINSON'S DISEASE, NEURITIS, ENCEPHALITIS, MENINGITIS, brain TUMORS (GLIOMAS), MUSCULAR DYSTROPHY and MYASTHENIA GRAVIS.

NEURON, or nerve cell, the basic unit of the NERVOUS SYSTEM (including the BRAIN and SPINAL CORD). Each has a long AXON, specialized for transmitting electrical impulses and releasing chemical transmitters that act on MUSCLE or effector cells or other neurons. Branched processes called dendrites integrate the input to neurons.

NEUROSIS, originally any NERVOUS SYSTEM activity; later, any disorder of the nervous system; though in PSYCHOANALYSIS, those mental disorders (e.g., HYSTERIA) unconnected with the nervous system. It is usually seen as based in UNCONSCIOUS conflict, with an unconscious attempt to conform to reality (not escape from it, as in PSYCHOSIS). **Actual neurosis** is based in disorders of current sexual behavior (see SEX); **psychoneurosis** is rooted in the past life; **anxiety neurosis** is characterized by exaggerated ANXIETY. (See OBSESSIONAL NEUROSIS.)

NEUROSURGERY. See SURGERY.

NEUTRALIZATION, a chemical reaction between two compounds of opposite chemical character, giving a relatively inactive product. Common examples include neutralization of an ACID by a BASE

The anatomy of a neuron (nerve cell).

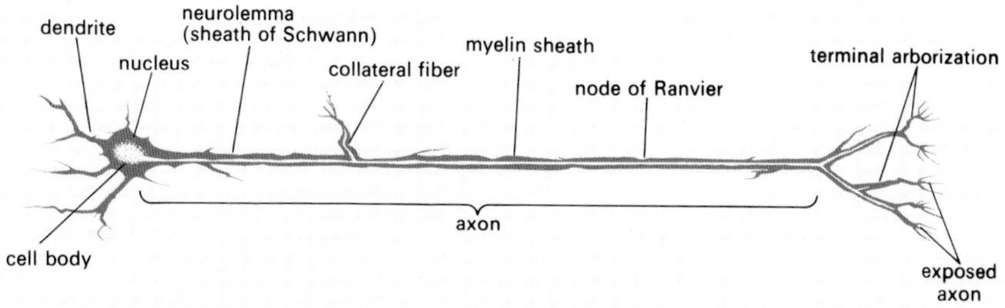

to give a SALT, and of an oxidizing agent by a reducing agent (see OXIDATION AND REDUCTION). (See also EQUIVALENT WEIGHT; TITRATION.)

NEUTRINO. See SUBATOMIC PARTICLES.

NEUTRON (n), uncharged SUBATOMIC PARTICLE with rest mass 1.6748×10^{-27}kg (slightly greater than that of the PROTON) and SPIN $\frac{1}{2}$. A free neutron is slightly unstable, decaying to a proton, an ELECTRON and an antineutrino with HALF-LIFE 680s:

$$n \rightarrow p^{+} + e^{-} + \bar{\nu}$$

But neutrons bound within the nucleus of an ATOM are stable. All nuclei save hydrogen contain neutrons, which contribute to the nuclear cohesive forces and separate the mutually repulsive protons. Free neutrons are produced in many nuclear reactions, including nuclear FISSION, and hence nuclear reactors and particle ACCELERATORS are used as sources. The neutron was discovered in 1932 by Sir James CHADWICK, who bombarded beryllium with ALPHA PARTICLES emitted by a radioisotope. Neutrons are highly penetrating, and are moderated (slowed down) by colliding with the nuclei of light atoms. They induce certain heavy atoms to undergo fission. Shielding requires thick concrete walls. Neutrons are detected by counting the ionizing particles or GAMMA RAYS produced when they react with nuclei. Neutrons have wave properties, and their DIFFRACTION is used to study crystal structures and magnetic properties. (See also CROSS-SECTION, NUCLEAR.)

NEUTRON BOMB, hypothetical "clean" variant of the HYDROGEN BOMB that would produce intense lethal neutron radiation but not much structural damage or radioactive fallout.

NÉVÉ, compacted snow that lasts from year to year, representing one of the earliest stages in the development of a GLACIER. Further compaction, until there is no air left in the SNOW, results in the formation of *firn*.

NEVUS. See BIRTHMARK, FRECKLE, MOLE.

NEWCOMEN, Thomas (1663–1729), British inventor of the first practical STEAM ENGINE (before 1712). His device, employed mainly to pump water from mines, used steam pressure to raise the piston and, after condensation of the steam, atmospheric pressure to force it down again: it was thus called an "atmospheric" steam engine. (See also SAVERY, T.)

NEWTON, Sir Isaac (1642–1726), the most prestigious natural philosopher and mathematician of modern times, the discoverer of the CALCULUS and author of the theory of universal GRAVITATION. Newton went up to Trinity College, Cambridge, in 1661, retiring to Woolsthorp, Lincolnshire, during the Plague of 1665–66, but becoming a fellow in 1667 and succeeding Isaac BARROW in the Lucasian Chair of Mathematics in 1669. He was elected Fellow of the ROYAL SOCIETY in 1672, on the strength of his optical discoveries. In Cambridge, Newton spent much time in alchemical experiments, though, toward the end of the century, he tired of the academical life and accepted a position at the Royal Mint, becoming Master of the Mint in 1699. He resigned his chair and entered Parliament in 1701 and two years later began his presidency of the Royal Society, which he retained until his death. His whole life was one of ceaseless energy—investigating mathematics, optics, chronology, chemistry, theology, mechanics, dynamics and the occult—broken only by a period of mental illness about 1693. His achievements were legion: the method of FLUXIONS and fluents (calculus); the theory of universal gravitation and his derivation of KEPLER'S LAWS; his formulation of the concept of FORCE as expressed in his three laws of motion (see MECHANICS); the corpuscular theory of LIGHT, and the BINOMIAL THEOREM, among many others. These were summed up in his two greatest works: *Philosophiae Naturalis Principia Mathematica* (1687)—the "Principia", which established the mathematical representation of nature as the paradigm of what counted as "science"—and the *Opticks* (1704). Newton's often bitter controversies with his fellow scientists (notably HOOKE and LEIBNIZ) are famous, but his influence is undoubted, even if, in the cases of optical theory and the Newtonian calculus notation, it retarded rather than accelerated the advance of British science.

NEWTON'S RINGS. See INTERFERENCE.

NIACIN. See VITAMINS.

NICCOLITE, coppery-red mineral consisting of nickel arsenide (NiAs); a minor NICKEL ore often found with PYRRHOTITE. It crystallizes in the hexagonal system.

NICHE, the way of life adopted by an animal or plant by which it survives in a particular habitat. An organism requires space and food to live successfully, the description of its niche therefore involves consideration of these aspects of its habitat.

NICHOLAS OF CUSA (1401–1464), German cardinal best known for his advanced cosmological views: he held that the earth rotates on its axis, that space is infinite, and that the sun is a star like other stars (see also ASTRONOMY). He also suggested the use of concave lenses for the shortsighted (see GLASSES; LENS).

NICKEL (Ni), hard, gray-white, ferromagnetic (see MAGNETISM) metal in Group VIII of the PERIODIC TABLE; a TRANSITION ELEMENT. About half the total world output comes from deposits of PYRRHOTITE and PENTLANDITE at Sudbury, Ontario; garnierite in New Caledonia is also important. Roasting the ore gives crude nickel oxide, refined by electrolysis or by the Mond process (see MOND, LUDWIG). Nickel is widely used in ALLOYS, including MONEL METAL, INVAR and GERMAN SILVER. In many countries "silver" coins are made from cupronickel (an alloy of copper and nickel). Nickel-chromium alloys ("nichrome"), resistant to oxidation at high temperatures, are used as heating elements in electric fires, etc. Nickel is used for nickel plating and as a catalyst for HYDROGENATION. Chemically nickel resembles IRON and COBALT, being moderately reactive, and forming compounds in the +2 oxidation state; the +4 state is known in LIGAND complexes. AW 58.7, mp 1453°C, bp 2732°C, sg 8.902 (25°C).

NICKEL SILVER. See GERMAN SILVER.

NICOLLE, Charles Jules Henri (1866–1936), French bacteriologist awarded the 1928 Nobel Prize for Physiology or Medicine for his discovery that the body louse is a carrier and a main transmitter of TYPHUS (1909).

NICOL PRISM, optical device for producing a beam of plane POLARIZED LIGHT. Two pieces of CALCITE crystal are cemented together with CANADA BALSAM. Incident light is split into ordinary and extraordinary linearly polarized rays in the prism. The ordinary ray hits the balsam layer obliquely and is totally

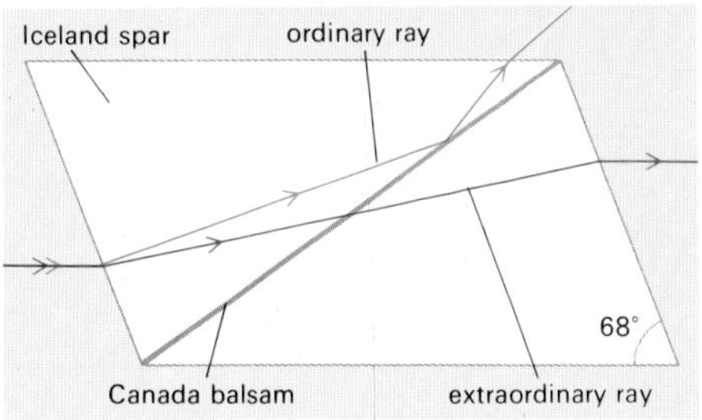

Schematic section through a Nicol prism. The original calcite rhombohedron is split along a diagonal plane, cemented back together again with Canada balsam, and then ground down to the desired shape. Optical polarimeters employ pairs of Nicol prisms.

internally reflected; the other ray emerges plane polarized for a certain range of incidence angles.

NICOTINE, colorless oily liquid, an ALKALOID occurring in tobacco leaves and extracted from tobacco refuse. It is used as an insecticide and to make nicotinic acid (see VITAMINS). Nicotine is one of the most toxic substances known; even the small dose ingested by SMOKING causes blood-vessel constriction, raised blood pressure, nausea, headache and impaired digestion.

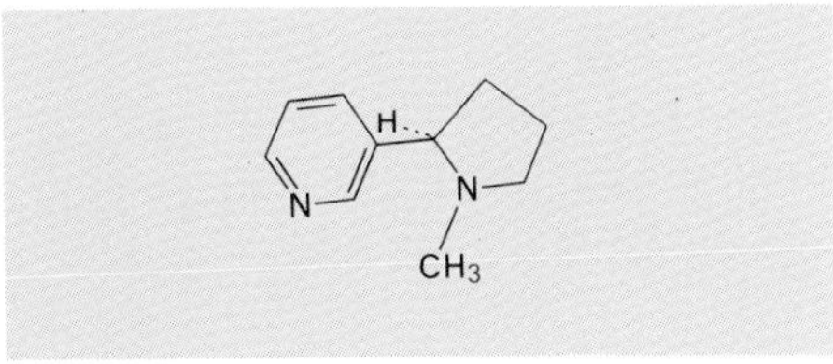

The structure of nicotine.

NICOTINIC ACID. See VITAMINS.

NICTITATING MEMBRANE, or third eyelid, a membrane found in many vertebrates that can be drawn horizontally across the EYE. In birds it lubricates and cleans the eyeball without the necessity of blinking. It is present in a few mammals, and in man is represented by the pink triangle of flesh in the corner of the eye.

NIÉPCE, Joseph Nicéphore (1765–1833), French inventor who in 1826 made the first successful permanent photograph. The image, recorded in asphalt on a pewter plate, required an 8hr exposure in a camera obscura (see CAMERA LUCIDA AND CAMERA OBSCURA). The method used derived from heliography, Niépce's photoengraving process. In 1829 Niépce went into partnership with DAGUERRE.

NIGHT BLINDNESS, or nyctalopia, inability to accommodate in or adapt to darkness. It may be a hereditary defect or an early symptom of VITAMIN A deficiency in adults. It is due to a defect in rod VISION.

NIOBIUM (Nb), or **columbium,** soft silvery-white metal in Group VB of the PERIODIC TABLE; a TRANSITION ELEMENT. It occurs as COLUMBITE and pyrochlore. Niobium is unreactive and corrosion-resistant, and is used in STEELS, high-temperature ALLOYS and superconducting alloys (see SUPERCONDUCTIVITY). Being permeable to neutrons, it is also used in nuclear reactors. At high temperatures it reacts with nonmetals to give pentavalent covalent compounds. AW 92.9, mp 2468°C, bp 4927°C, sg 8.57 (20°C).

NIRENBERG, Marshall Warren (1927–), US biochemist who shared with HOLLEY and KHORANA the 1968 Nobel Prize for Physiology or Medicine for his major contributions toward the decipherment of the genetic code (see GENETICS).

NITRATES, salts of NITRIC ACID, containing the nitrate ion (NO_3^-). Almost all nitrates are soluble in water, and only SODIUM and POTASSIUM nitrate occur significantly in nature, others being made by the action of nitric acid on the metal or its salts. Nitrates are used in medicine, explosives, fertilizers and fireworks. (See also NITROGEN CYCLE; NITROGEN FIXATION.) ESTERS of nitric acid ($RONO_2$) are also called nitrates.

NITRATION, important process in ORGANIC CHEMISTRY, in which a nitro group ($-NO_2$) is introduced into a compound. AROMATIC COMPOUNDS are nitrated with a mixture of concentrated SULFURIC and NITRIC ACIDS, which contains the electrophile (see NUCLEOPHILES) NO_2^+; the production of TNT and nitrobenzene are important cases. "Nitration" is also used for the formation of NITRATE esters including NITROCELLULOSE and NITROGLYCERIN.

NITRIC ACID (HNO_3), strong mineral ACID, a colorless, fuming liquid when pure. Nitric acid is usually made by the OSTWALD PROCESS, the ELECTRIC-ARC PROCESS, or by the action of sulfuric acid on sodium nitrate. It is a powerful oxidizing agent, reacting with most metals to give NITRATES and oxides of NITROGEN; with many organic compounds NITRATION occurs. Nitric acid is used to make nitrates, plastics, explosives and dyes, and as a rocket fuel. mp −42°C, bp 83°C.

NITRILES, or organic CYANIDES, class of organic compounds of general formula RCN, named for the CARBOXYLIC ACID to which they can be hydrolyzed. The simplest is acetonitrile (or methyl cyanide), CH_3CN. Nitriles are prepared by dehydration of AMIDES or by reaction of sodium cyanide with ALKYL HALIDES or aryl sulfonates. They may be catalytically hydrogenated to AMINES, and react with GRIGNARD REAGENTS to yield KETONES. Acrylonitrile is used to make POLYMERS; some other nitriles are used as softening agents for rubber etc.

NITROCELLULOSE, properly called cellulose nitrate, mixture of highly inflammable NITRATE esters of CELLULOSE made by NITRATION of cotton or wood pulp. The degree of nitration depends on the conditions used. Highly nitrated nitrocellulose—**guncotton**—was used as a high EXPLOSIVE and is still used as a propellant. Nitrocellulose with less than 12% nitrogen—**collodion**—is used in making propellants, DYNAMITE, lacquers and CELLULOID.

NITROGEN (N), nonmetal in Group VA of the PERIODIC TABLE; a colorless, odorless gas (N_2) comprising 78% of the ATMOSPHERE, prepared by fractional distillation of liquid air. Combined nitrogen occurs mainly as NITRATES. As a constituent of AMINO ACIDS, it is vital (see also NITROGEN CYCLE). Molecular nitrogen is inert because of the strong triple bond between the two atoms, but it will react with some elements, especially the ALKALINE-EARTH METALS, to give nitrides; with oxygen (see ELECTRIC ARC PROCESS); and with hydrogen (see HABER PROCESS); it also forms N_2 LIGAND complexes with Group VIII transition metals. Activated nitrogen, formed in an electric discharge, consists of nitrogen atoms and is much more reactive. Nitrogen is used in NITROGEN FIXATION and to provide an inert atmosphere; liquid nitrogen is a CRYOGENIC refrigerant. AW 14.0, mp −210°C, bp −196°C.

Nitrogen forms mainly trivalent and pentavalent compounds. **Nitric Oxide** (NO), is a colorless gas formed in the ELECTRIC-ARC PROCESS; it is readily oxidized further to nitrogen dioxide. The NO molecule is unusual in having an odd number of electrons, and so gives the nitrosyl ions NO^+ and NO^-. mp −164°C, bp −152°C. **Nitrites** are salts (or esters) of **Nitrous Acid** (HNO_2) and are mild reducing agents. **Nitrous Oxide** (N_2O), or laughing gas, is a colorless gas with a sweet odor, prepared by heating ammonium nitrate, and used as a weak anesthetic, sometimes producing mild hysteria, and also as an aerosol propellant. mp −91°C, bp −88°C. **Nitrogen Dioxide** (NO_2), a red-brown toxic gas in equilibrium with its colorless dimer (N_2O_4), is a constituent of automobile exhaust and smog. A powerful oxidizing agent, it is used in the manufacture of SULFURIC ACID and in rocket fuels. It is also an intermediate in the manufacture of NITRIC ACID. mp −11°C, bp 21°C. See also AMMONIA; OSTWALD PROCESS; CYANAMIDE PROCESS; HYDRAZINE.

NITROGEN CYCLE, the cycle of chemical changes exchanging NITROGEN between the air and the soil. NITROGEN FIXATION, industrial (producing FERTILIZERS) or by microorganisms, yields combined nitrogen as AMMONIA and NITRATES, which can be absorbed from the soil by plants, which use them to make protein. Animals ingest nitrogen by eating these plants. Excretion products and animal and plant remains return nitrogen to the soil as complex compounds which are converted by fungi and bacteria to ammonium salts, which may then be oxidized to nitrites (see NITROGEN) and nitrates by other bacteria. These are either reused by plants, or converted to nitrogen (by denitrifying bacteria) which returns to the air, thus completing the cycle. (See also ECOLOGY.)

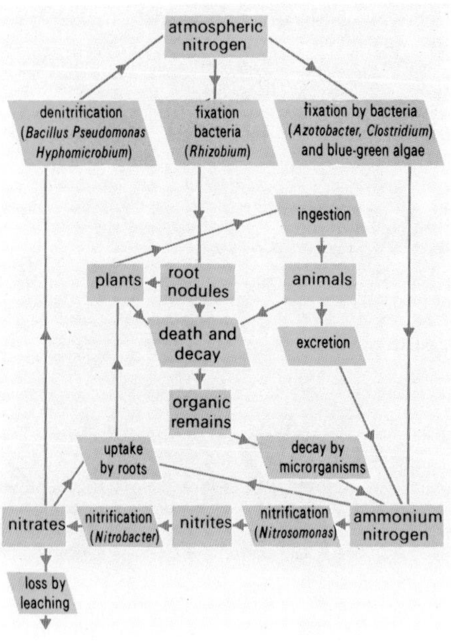

NITROGEN FIXATION, conversion of NITROGEN gas into nitrogen compounds. Nitrogen usually reacts only at high temperatures and pressures. Industrially, it is fixed in the HABER PROCESS, the CYANAMIDE PROCESS and the ELECTRIC-ARC PROCESS.

Some bacteria (*Rhizobium*) in the root nodules of LEGUMINOUS PLANTS, and also some of those living free in the soil (e.g., *Azotobacter*), can fix nitrogen from the air as AMMONIA, NITRATES and nitrites (see NITROGEN). Some fungi and blue-green algae also fix nitrogen. (See also NITROGEN CYCLE.)

NITROGLYCERIN ($C_3H_5(ONO_2)_3$), properly called glyceryl trinitrate, the NITRATE ester of GLYCEROL, made by its NITRATION. Since it causes VASODILATION, it is used to relieve ANGINA PECTORIS. Its major use, however, is as a very powerful high EXPLOSIVE, though its sensitivity to shock renders it unsafe unless used in the form of DYNAMITE or blasting gelatin. It is a colorless, oily liquid. MW 227.1, mp 13°C.

NITROUS OXIDE. See NITROGEN.

NOBEL, Alfred Bernhard (1833–1896), Swedish-born inventor of dynamite and other explosives. About 1863 he set up a factory to manufacture liquid NITROGLYCERIN, but when in 1864 this blew up, killing his younger brother, Nobel set out to find safe handling methods for the substance, so discovering DYNAMITE, patented 1867 (UK) and 1868 (US). Later he invented gelignite (patented 1876) and ballistite (1888). A lifelong pacifist, he wished his explosives to be used solely for peaceful purposes, and was much embittered by their military use. He left most of his fortune for the establishment of the Nobel Foundation and this fund has been used to award Nobel Prizes since 1901.

NOBELIUM (No), a TRANSURANIUM ELEMENT in the ACTINIDE series, prepared by bombardment of lighter actinides. The most stable isotope, No^{255}, has a HALF-LIFE of only 3min.

NOBLE GASES, the elements in Group O of the PERIODIC TABLE, comprising HELIUM, NEON, ARGON, KRYPTON, XENON and RADON. They are colorless, odorless gases, prepared by fractional distillation of

Noble gases

element	Z	atomic weight	oxidation numbers
helium (He)	2	4.00	—
neon (Ne)	10	20.18	—
argon (Ar)	18	39.95	—
krypton (Kr)	36	83.80	+II
xenon (Xe)	54	131.30	+II, IV, VI, VIII
radon (Rn)	86	222*	+II? +IV?

* radioactive, longest-lived isotope.

liquid air (see ATMOSPHERE), except helium and radon. Owing to their stable filled-shell electron configurations, the noble gases are chemically unreactive: only krypton, xenon and radon form isolable compounds. They glow brightly when an electric discharge is passed through them, and so are used in advertising signs: neon tubes glow red, xenon blue, and krypton bluish-white; argon tubes glow pale red at low pressures, blue at high pressures.

NOBLE METALS, the unreactive, corrosion-resistant precious metals comprising the PLATINUM GROUP, SILVER and GOLD, and sometimes including RHENIUM.

NODE, in ANALYTIC GEOMETRY, a point of INTERSECTION of two or more parts of a single CURVE.

NOETIC SCIENCES, field of study embracing researches into ESP and PARAPSYCHOLOGY.

NOGUCHI, Hideyo (1876–1928), Japanese bacteriologist best known for his work on syphilis (see VENEREAL DISEASES) and YELLOW FEVER.

NOISE, unwanted SOUND. As far as man is concerned this is a subjective definition: people vary in their sensitivity to noise; many sounds are agreeable to some and noisy to others. Blasts or explosions can cause sudden damage to the ear and prolonged exposure to impulsive sounds such as a pneumatic drill may cause gradual HEARING impairment. In general, any sound that is annoying, interferes with speech, damages the hearing or reduces concentration or work efficiency may be considered as noise. From the physical view point, sound waves (either in air or vibrations in solid bodies) that mask required signals or cause fatigue and breakdown of equipment or structures are noise and should be minimized. In air, sound is radiated spherically from its source as a compressional wave, being partly reflected, absorbed or transmitted on hitting an obstacle. Noise is usually a nonperiodic sound wave, as opposed to a periodic pure musical tone or a sine-wave combination. It is characterized by its intensity (measured in DECIBELS or NEPERS), frequency and spatial variation; a sound level meter and frequency analyzer measure these properties. Noise may be controlled at source (e.g., by a MUFFLER), between it and the listener (e.g., by sound absorbing material) or at the listener (e.g., by wearing ear plugs).

NOISE, in electronics, any unwanted or interfering current or voltage in an electrical device or system. Its presence in the amplifying circuits of RADIOS, TELEVISION receivers etc. may mask or distort signals. Unpredictable random noise exists in any component with RESISTANCE because of the thermal motion of the current-carrying ELECTRONS, and in electron tubes due to random CATHODE emission. Thermal radiations and variations in the atmosphere also cause random noise. Nonrandom noise arises from spurious oscillations and unintended couplings between components.

NOMINALISM, in philosophy, usually as opposed to REALISM, the view that the names of abstract ideas (e.g., beauty) used in describing things (as in, a *beautiful* table) are merely conventions or conveniences, and should not be taken to imply the actual existence of universals corresponding to those names.

NON-EUCLIDEAN GEOMETRY, those branches of GEOMETRY that challenge EUCLIDEAN GEOMETRY'S tenet that through any POINT A not on a LINE L there

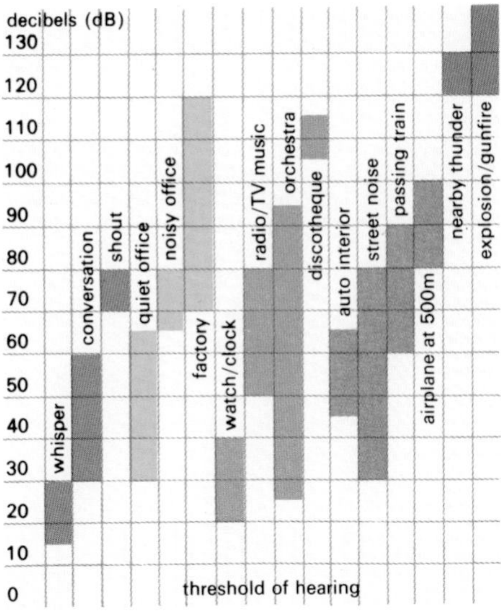

Decibel levels associated with various common sources of noise. Prolonged exposure to high noise levels can lead to impairment of hearing.

passes one and only one line parallel to L, but which accept in general all other Euclidean axioms with at most minor changes. The geometry based on the hypothesis that no lines pass through A parallel to L is RIEMANNIAN GEOMETRY; that based on the hypothesis that there is more than one such line is LOBACHEVSKIAN GEOMETRY. The first mathematician to open the doors for non-Euclidean geometry was GAUSS in the 19th century. He did not publish his work, however, and the fathers of non-Euclidean geometry are usually considered to be BOLYAI János and LOBACHEVSKY.

NONMETAL, a substance—in particular, an ELEMENT—showing none of the properties characteristic of METALS. (See also METALLOID.) The 17 or so nonmetallic elements fill the top right-hand corner of the PERIODIC TABLE. Their atoms are in general relatively small, with nearly-filled electron shells, and have high IONIZATION POTENTIALS. They have high ELECTRONEGATIVITIES, and tend to form covalent BONDS with each other, and to form ANIONS.

NORADRENALINE. See ADRENALINE.

NORMAL DISTRIBUTION, in STATISTICS, an ideal distribution which is symmetrical (see SYMMETRY) about its mean (see MEAN, MEDIAN AND MODE). It can be written in the form

$$f(x) = \frac{1}{\sigma\sqrt{2\pi}} \; e^{-\frac{1}{2}\left(\frac{x-\mu}{\sigma}\right)^2}$$

where $f(x)$ is the relative frequency, σ is the STANDARD DEVIATION, μ is the mean, and π (see PI) and e (see EXPONENTIAL) have their usual values. Note that in a normal distribution the mean, median and mode coincide. Many naturally-occurring distributions approximate to the normal. The typically bell-shaped curve shown when $f(x)$ is plotted against x is called the *normal curve*. The distribution is also named the "Gaussian distribution" for Karl GAUSS.

NORMALITY. See EQUIVALENT WEIGHT.

NORRISH, Ronald George Wreyford (1897–), British chemist awarded, with Manfred EIGEN and George PORTER, the 1967 Nobel Prize for Chemistry for studies of extremely fast chemical reactions, and, in particular, for the development of FLASH PHOTOLYSIS.

NORTH ATLANTIC DRIFT, eastward-flowing continuation of the GULF STREAM, notable for its warming effect on the climates of W Europe.

NORTHERN LIGHTS. See AURORA.

NORTH POLE, the point on the earth's surface some 750km N of Greenland through which passes the earth's axis of rotation. It does not coincide with the earth's N Magnetic Pole, which is over 1 000km away (see EARTH). The Pole lies roughly at the center of the Arctic Ocean, which there is permanently ice-covered, and experiences days and nights each of six months. It was first reached by Robert E. Peary (April 6, 1909). (See also CELESTIAL SPHERE; MAGNETISM; SOUTH POLE.)

NORTHROP, John Howard (1891–), US biochemist who shared with W. M. STANLEY and J. B. SUMNER the 1946 Nobel Prize for Chemistry for his work crystallizing the ENZYMES pepsin (c1930), trypsin (c1932) and chymotrypsin (c1935). He was also the first to isolate a BACTERIOPHAGE (1938).

NORTH STAR. See POLARIS.

NOSE, the midline organ of the FACE, concerned with the perception of SMELL and the preparation of the air stream for RESPIRATION. It is a CARTILAGE extension of the facial bones with two external openings or nostrils. These pass into the nasal cavities, which are separated by a septum and contain turbinates which increase the mucous membrane surface and direct the air flow. The chemoreceptors for smell lie mainly in the roof of the nasal cavities, but fine nerve fibers throughout the nose contribute both to tactile sensation and smell. Nasal MUCUS protects the BRONCHI and LUNGS by removing dust particles, humidifying and warming inspired air. The COMMON COLD, RHINITIS and HAY FEVER are common afflictions of the nose.

NOTOCHORD, the primitive longitudinal skeletal element characterizing the class Chordata, the first stage in the development of a flexible internal skeleton. All chordates possess a notochord at some time during life. Though replaced by cartilage or bone in the adult VERTEBRATE and absent in the adults of other chordate groups, e.g., Tunicates, it is well developed in the embryos or larvae of all these groups, confirming evolutionary relationships within the class.

NOVA, a star which over a short period (usually a few days) increases in brightness by 100 to 1 000 000 times. This is thought to be due to the star undergoing a partial explosion: that is to say, part of the star erupts, throwing out material at a speed greater than the ESCAPE VELOCITY of the star. The initial brightness fades quite rapidly though it is usually some years before the star returns to its previous luminosity, having lost about 0.0001 of its mass. At that time a rapidly expanding planetary NEBULA may be seen to surround the star. Recurrent novae are stars which go nova at irregular periods of a few decades. Dwarf novae are subdwarf stars which go nova every few weeks or months. Novae have been observed in other galaxies besides the MILKY WAY. (See also SUPERNOVA.)

NOVOCAINE, alternative name for PROCAINE, a local anesthetic.

NUCLEAR ENERGY, energy released from an atomic nucleus during a nuclear reaction in which the atomic number (see ATOM), MASS NUMBER or RADIOACTIVITY of the nucleus changes. The term atomic energy, also used for this energy, which is produced in large amounts by NUCLEAR REACTORS and NUCLEAR WEAPONS, is not strictly appropriate, since nuclear reactions do not involve the orbital ELECTRONS of the atom. Nuclear energy arises from the special forces (about a million times stronger than chemical bonds) that hold the PROTONS and NEUTRONS together in the small volume of the atomic nucleus (see NUCLEAR PHYSICS). Lighter nuclei have roughly equal numbers of protons and neutrons, but heavier elements are only stable with a neutron:proton ratio of about 1.5:1. If one could overcome the electrostatic repulsion between protons and assemble them with neutrons to form a stable nucleus, its mass would be less than that of the constituent particles by the *mass defect* Δm, of the nucleus, and the *binding energy*, BE, given by $BE = \Delta m c^2$ (where c is the electromagnetic constant), would be released. Because c is large, a vast amount of energy would be released, even for a very small value of the mass defect. The binding energy (equivalent to the work needed to split up the nucleus into separate protons and neutrons) is always positive—nuclei are always more stable than their separate nucleons (protons or neutrons)—but is greatest for nuclei of medium mass, decreasing slightly for lighter and heavier elements. The low binding energy of very light elements means that energy can be released by combining e.g., two DEUTERIUM nuclei to form a helium nucleus. This combination of two protons and two neutrons is particularly stable (see FUSION, NUCLEAR). For heavy elements the decrease in binding energy indicates that the more positively charged the nucleus becomes, the less stable it is, even though it contains more neutrons than protons. This sets a limit on the number of elements, and also explains why the nuclear-fission process, in which a heavy nucleus splits into two or more medium-mass nuclei with higher total binding energy, releases energy. The first nuclear reaction was performed experimentally in 1919 by RUTHERFORD who exposed NITROGEN to ALPHA PARTICLES (helium nuclei) from the radioactive element RADIUM, producing OXYGEN and HYDROGEN:

$$N^{14} + He^{4} \rightarrow O^{17} + H^{1}$$

But, because nuclei are positively charged and repel each other, it was found difficult to bring them close enough together to react with each other. The discovery of the neutron in 1932 helped overcome this problem. Being uncharged and heavy (on the atomic scale), the neutron has high energy even when moving slowly and is good for initiating nuclear reactions. By 1939 many nuclear reactions had been studied, but none seemed feasible as an energy source. Although energy might be released in a reaction, more energy was expended in producing particles able to initiate the reaction than could be recovered from it. Moreover, only a small fraction of the reagent particles would react as desired and any product particles would have little chance of reacting again. The situation was like trying to set fire to a damp forest with a box of matches! A breakthrough came around

1939 when the violent reaction of the heavy element uranium on bombardment with slow nuetrons (first observed experimentally by Fermi in 1934) was successfully interpreted. It was realized that this was an example of nuclear fission, the slow neutrons delivering enough energy to the small proportion of U^{235} nuclei in natural uranium to split them into two parts. This split does not always occur in the same way, and many radioactive fission products are formed, but each fission is accompanied by the release of much energy and two or three neutrons (these because the lighter nuclei of the fission products have a lower neutron:proton ratio than uranium). These neutrons were the key to the large scale production of nuclear energy; they could make the uranium "burn" by setting up a chain reaction. Even allowing for the loss of some neutrons, sufficient are left to produce other fissions, each producing two or three more neutrons, and so on, leading to an explosive release of energy. The first controlled chain reaction took place in Chicago in 1942, using pure graphite as a moderator to slow down neutrons and natural uranium as fuel. Rods of neutron-absorbing material kept the reaction under control by limiting the number of neutrons available to cause fissions. The possibilities of nuclear energy as a weapon were exploited at once and WWII ended shortly after the United States dropped two ATOMIC BOMBS on Japan. Later, more powerful bombs exploiting nuclear fusion were developed. An increasing quantity of man's energy is produced in NUCLEAR REACTORS from nuclear fission, although the earth's natural supplies of fissionable material are surprisingly limited. Moreover, because the fission products from these reactors are radioactive with long HALF-LIVES, atomic waste disposal is a major environmental problem. At present the waste is stored in concrete vaults lined with stainless steel, though the possibilities of converting waste to an insoluble glass are being explored. Disposal in space, in geologically stable parts of the earth's crust or by chemical conversion to safer materials are ideas for the future. Nuclear fusion seems to offer much better long-term prospects for energy supply, although fusion reactors have not yet progressed beyond the research stage.

NUCLEAR PHYSICS, the study of the physical properties and mathematical treatment of the atomic nucleus and SUBATOMIC PARTICLES. The subject was born when RUTHERFORD postulated the existence of the nucleus in 1911. The nature of the short-range exchange forces which hold together the nucleus, acting between positively charged protons and neutral neutrons is still uncertain. Experimental data from MASS SPECTROSCOPY and scattering experiments have enabled various partially successful theoretical models to be devised. Despite the special techniques required to produce nuclear reactions, the subject has rapidly grown with the technical exploitation of NUCLEAR ENERGY.

NUCLEAR REACTOR, device containing sufficient fissionable material, arranged so that a controlled

Schematized diagram showing the principal designs of commercial nuclear reactors.

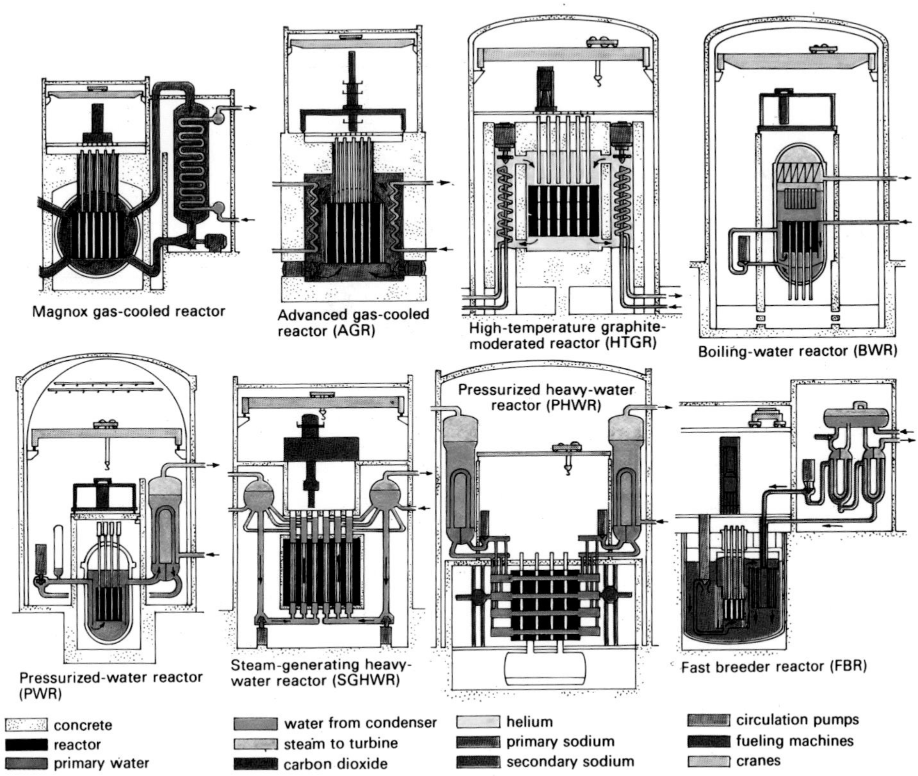

chain reaction may be started up and maintained in it. Many types of reactor exist: all produce NEUTRONS, GAMMA RAYS, radioactive fission products and HEAT, but normally use is made of only one of these. Neutrons may be used in nuclear research or for producing useful RADIOISOTOPES. Gamma rays are dangerous to man and must be shielded against, but have some uses (see IRRADIATION). The fragments produced by fission of a heavy nucleus have a large amount of energy and the heat they produce may be used for carrying out a variety of high-temperature processes or for heating a working fluid (such as steam) to operate a TURBINE and produce ELECTRICITY. This is the function of most commercial reactors, although a number are used to power ships and submarines, since a small amount of nuclear fuel gives these a very long range. In an electricity-generating reactor, the fuel is normally uranium pellets surrounded by a moderator and the cooling fluid heavy water or liquid sodium (which in turn heats the turbine fluid). There is much insulation and radiation shielding. The fuel is expensive, but produces several thousand times the heat of the same weight of coal. After some time it must be replaced (although only partly consumed) because of the build-up of neutron-absorbing fission products. This replacement, and the reprocessing of the radioactive products, needs costly remote handling equipment. New fast breeder reactors with no moderator avoid this problem, since as well as producing fission of U^{235}, they convert nonfissionable U^{238} to plutonium which also undergoes fission chain reactions—they effectively breed fuel! Research is continuing into more efficient reactors as power sources for the future.

NUCLEAR WARFARE, the use of nuclear weapons—the ATOMIC BOMB and the HYDROGEN BOMB—in warfare. The possible appalling consequences of large-scale nuclear warfare have overshadowed world politics since soon after WWII. Following FERMI's discovery of nuclear FISSION, preliminary experiments were done in the US starting in 1939, owing largely to fears of Germany's developing the atomic bomb first. With the US entry into the war, the MANHATTAN PROJECT was started, culminating in the atomic bombs dropped on Hiroshima and Nagasaki. The USSR tested its first such weapon in 1949; from then on an arms race between those two "great powers" escalated until the late 1960s. The US exploded its first hydrogen bomb in 1952: the USSR in 1953. The UK, France and China have also developed nuclear weapons; certain other countries may well have done so without testing them. Development and deployment of nuclear weapons has been dictated by the theory of the nuclear deterrent, the aim being the capability of "assured destruction" of the enemy nation, a stable strategic balance being achieved. Essential to this is the protection of a country's missiles against destruction in a "first strike," i.e., before any can be used in retaliation; thus the guided missiles are sited underground and in nuclear-powered SUBMARINES. The problem of radioactive FALLOUT pollution from nuclear tests led to the partial Nuclear Test-Ban Treaty, signed by the UK, US and USSR in 1963: this bans tests in space, in the atmosphere and underwater, but permits underground explosions that do not release fission products beyond national frontiers. Enforcement of the treaty requires explosions to be detected; this is done by detecting seismic or acoustic disturbances, radiation, or radioactive debris. Further disarmament negotiations led to the Treaty on the Non-Proliferation of Nuclear Weapons, ratified 1970 by 43 nations. There have also been attempts to limit delivery systems, especially intercontinental ballistic missiles (ICBMs), and defensive systems. Strategic arms limitation talks (SALT) began in 1969 and have achieved some limitation. Very stringent precautions are taken to prevent accidental or irresponsible use of nuclear weapons. Several persons validly authorized are required, and there is an electronic failsafe system. Military strategy in recent years has been directed toward limited nuclear warfare, using tactical nuclear weapons in the battle zone, and toward preventing middle-scale nuclear wars from escalating to a total nuclear holocaust.

NUCLEAR WEAPONS, those depending for their destructive capability on the energies released in nuclear fission (the ATOMIC BOMB) or in nuclear fusion (the HYDROGEN BOMB). (See also NUCLEAR ENERGY.)

NUCLEIC ACIDS, the vital chemical constituents of living things; a class of complex threadlike molecules comprising two main types: the deoxyribonucleic acids (DNA) and the ribonucleic acids (RNA). DNA is found almost exclusively in the nucleus of the living CELL, where it forms the chief material of the CHROMOSOMES. It is the DNA molecule's ability to duplicate itself (replicate) that makes cell reproduction possible; and it is DNA, by directing PROTEIN SYNTHESIS, that controls HEREDITY in all organisms other than certain VIRUSES which contain only RNA. RNA performs several important tasks connected with protein synthesis, and is found throughout the cell.

In both DNA and RNA the backbone of the molecule is a chain of alternate phosphate and sugar groups. To each sugar group is bonded one or other of four nitrogenous side groups, which are either purines or pyrimidines. Each unit consisting of a side group, a sugar and a phosphate is called a NUCLEOTIDE. DNA differs chemically from RNA in that its sugar group has one less oxygen atom (hence the prefix "deoxy-") and one of its side groups, thymine, is replaced in RNA by uracil. DNA molecules are usually very much longer than RNA and may contain a million or so phosphate-sugar links.

It is the sequence in which the side groups are arranged along the DNA molecule that constitutes stored genetic information and so makes the difference between one inherited characteristic and another. This information, in the form of coded instructions for the synthesis of particular protein molecules, is carried outside the cell nucleus by molecules of "messenger RNA," each incorporating a side-group sequence determined by DNA. Floating freely outside the nucleus are AMINO ACIDS, the "building blocks" of proteins, and molecules of another, smaller kind of RNA, "transfer RNA." Each of these RNA molecules is able to capture an amino acid molecule of a particular type and locate it in its proper place in a sequence dictated by messenger RNA.

The DNA molecule has not one but two sugar-phosphate chains twisted around each other to form a

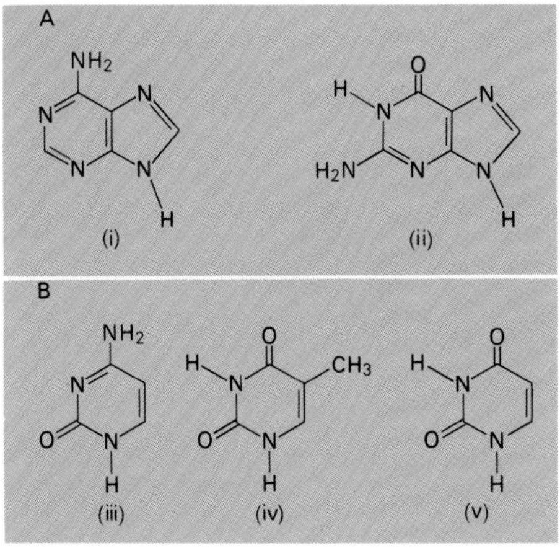

The structural formulae of nucleotide bases commonly found in DNA and RNA. (A) The purines: (i) adenine; (ii) guanine. (B) The pyrimidines: (iii) cytosine; (iv) thymine—found in DNA only—and (v) uracil—found in RNA only.

double helix. Linking the chains rather like the rungs of a ladder are the side groups, each interlocking with its appropriate opposite number, for a particular side group can be partnered by a side group of only one other kind. The molecule replicates by splitting down the middle, whereupon the side groups of each half bond with the appropriate side groups of free phosphate-sugar units to form a pair of identical DNA molecules. The elucidation of DNA structure, one of the greatest advances of 20th-century biology, is chiefly associated with the work of the Nobel prizewinners James WATSON, Francis CRICK and Maurice WILKINS.

NUCLEOPHILES, chemical species that donate ELECTRONS, or a share in electrons, to another molecule in a substitution reaction. They are thus in general BASES, especially ANIONS, which are attracted to partially positively-charged atoms in a molecule; examples are CYANIDE, HYDROXIDE, HALIDE and AMMONIA. **Electrophiles** are the exact opposite: species which receive electrons, or a share in electrons, from another molecule; in general ACIDS, especially CATIONS, attracted to partially negatively-charged atoms in a molecule; examples are HALOGENS, H^+ and NO_2^+.

NUCLEOTIDES, organic chemicals of central importance in the life chemistry of all plants and animals. Some nucleotides provide the basic molecular units for the synthesis of various more complex molecules, notably the NUCLEIC ACIDS—DNA and RNA; others— preeminently adenosine triphosphate (ATP)—provide a means of storing and releasing the ENERGY needed to drive biochemical processes.

The nucleotide molecule is a three-part structure, comprising a phosphate group linked to a 5-carbon sugar group (pentose) linked in turn to a nitrogenous side group (base). The five commonest bases are the purines adenine and guanine, and the pyrimidines cytosine, thymine and uracil. Adenine, guanine, cytosine and thymine serve in the DNA molecule as the four key "letters" of the genetic code. The nucleotide's pentose is either ribose or deoxyribose, the latter differing from the former only in having one less oxygen atom. The base-pentose component of a nucleotide is called a nucleoside.

The phosphate group consists of one or a combination of two or three phosphate units, and is accordingly termed a mono-, di- or triphosphate. The nucleotide adenosine triphosphate (ATP) is a nucleoside consisting of adenine and ribose bonded to a triphosphate group. The importance of ATP as an energy store depends on the third phosphate unit. When a third unit is added to adenosine diphosphate (ADP) to form ATP, an energy-rich chemical bond is formed; and it is this energy, released when ATP is converted back to ADP, that the CELL utilizes. (The human body daily builds up and breaks down approximately its own weight in ATP.) By releasing its energy, ATP activates or accelerates the action of ENZYMES, the catalysts of biochemical reactions, and so belongs to a class of substances called coenzymes. Most coenzymes are nucleotides.

NUCLEUS, Atomic. See ATOM; SUBATOMIC PARTICLES.

NUCLEUS, Cell. See CELL.

NULL SET. See SET THEORY.

Adenosine triphosphate (ATP) is the best known of the energy-storing nucleotides.

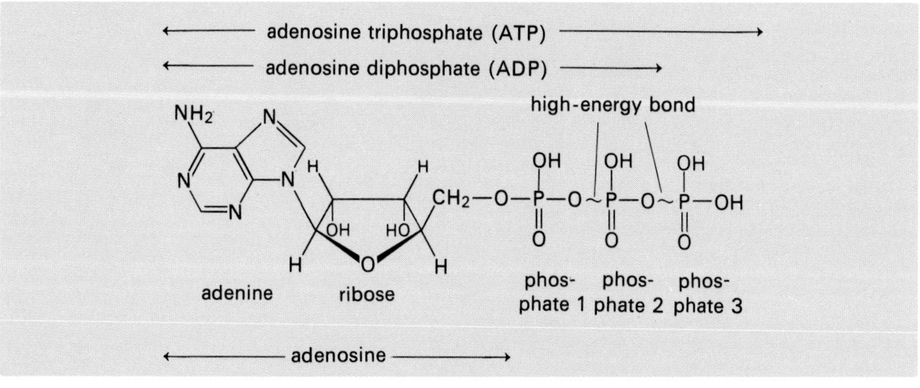

NUMBER, an expression of quantity. In everyday terms, numbers are usually used with UNITS: e.g., "three metres" (or 3m); "6.5893 kilograms" (or 6.5893kg).

For ways of expressing numbers see CUBE; FRACTION; RECIPROCAL; ROOTS; SQUARE; SUM. For systems of expressing numbers see BINARY NUMBER SYSTEM; DECIMAL SYSTEM; DUODECIMAL SYSTEM. Types and systems of numbers include IMAGINARY NUMBERS; INTEGERS; IRRATIONAL NUMBERS; NATURAL NUMBERS; RATIONAL NUMBERS; REAL NUMBERS; TRANSCENDENTAL NUMBERS. (See also ALGEBRA; CARDINAL NUMBER; ORDINAL NUMBER; TRANSFINITE CARDINAL NUMBER and PERFECT NUMBER; PRIME NUMBER; ZERO.)

NUMERATOR, the DIVIDEND of a common FRACTION.

NUT, dry, indehiscent FRUIT with one seed, similar to an ACHENE, but having a hard outer shell or pericarp. Nuts are produced by, for example, hazel, oak and chestnut.

NUTATION, Astronomical, irregularities in the PRECESSION of the equinoxes owing to variations in the torque produced by the gravitational attractions of the sun and moon on the earth.

NUTATION, Mechanical, a "bobbing" superimposed on the PRECESSION of a rigid spinning body such as a GYROSCOPE. With increasing spin rate there is an increase in the frequency and decrease in the magnitude of the nutation.

NUTRITION, the processes by which living organisms take in and utilize nutrients—the substances or foodstuffs required for GROWTH and the maintenance of LIFE. Vital substances that cannot be synthesized within the CELL and must be present in the food are termed "essential nutrients." Organisms such as green plants can derive ENERGY from sunlight and synthesize their nutritional requirements from simple inorganic chemicals present in the soil and air (see PLANT; PHOTOSYNTHESIS). Animals, on the other hand, depend largely on previously synthesized organic materials obtainable only by eating plants or other animals (see ANIMAL; DIGESTIVE SYSTEM; METABOLISM; ECOLOGY).

Human nutrition involves five main groups of nutrients: PROTEINS, FATS, CARBOHYDRATES, VITAMINS and minerals. Proteins, fats and carbohydrates are the body's sources of energy, and are required in relatively large amounts. They yield this energy by OXIDATION in the body cells, and nutritionists measure it in heat units called food CALORIES (properly called kilocalories, each equalling 1000 gram calories). Carbohydrates (food STARCHES and SUGARS) normally form the most important energy source, contributing nearly half the calories in a well-balanced diet. Cereal products and potatoes are rich in starch; SUCROSE (table sugar) and LACTOSE (present in milk) are two common sugars. Fats, which provide about 40% of the calorie requirement, include butter, edible oils and shortening, and are present in such foods as eggs, fish, meat and nuts. Fats consist largely of fatty acids (see carboxylic acids), which divide into two main classes: saturated and unsaturated. Certain fatty acids are essential nutrients; but if there is too much saturated fatty acid in the diet, an excess of CHOLESTEROL may accumulate in the blood. Proteins supply the remaining energy needs, but their real importance lies in the fact that the body tissues, which are largely composed of protein, need certain essential AMINO ACIDS, found in protein foods, for growth and renewal. Protein-rich foods include meat, fish, eggs, cereals, peas and beans. Too little protein in the diet results in malnutritional diseases such as KWASHIORKOR.

Minerals (inorganic elements) and vitamins (certain complex organic molecules) provide no energy, but have numerous indispensable functions. Some minerals are components of body structures. Calcium and phosphorus, for example, are essential to BONES and TEETH. Iron in the BLOOD is vital for the transport of oxygen to the tissues: an iron deficiency results in ANEMIA. Milk and milk products are good sources of calcium and phosphorus; liver, red meat and egg yolk, of iron. Other important minerals, normally well supplied in the Western diet, include chlorine, iodine, magnesium, potassium, sodium and sulfur. Vitamins, which are present in small quantities in most foods, are intimately associated with the action of ENZYMES in the body cells, and particular vitamin deficiencies accordingly impair certain of the body's synthetic or metabolic processes. A chronic lack of vitamin A, for example, leads to a hardening and drying of the skin and can result in irreversible damage to the conjunctiva and cornea of the eye. BERIBERI is caused by a vitamin B_1 deficiency, SCURVY by a vitamin C deficiency, RICKETS by a lack of vitamin D.

Despite the fact that nutritionists now understand the basic requirements of a healthy diet, the difficulty of applying their knowledge world-wide is immense. About two-thirds of the world's population remains severely undernourished and subject to deficiency diseases. Even in the richer countries malnutrition occurs, but here it is likely to be due to an ill-chosen rather than impoverished diet. In the US, the Food and Nutrition Board of the United States Academy of Sciences National Research Council publishes a table of recommended daily nutrient allowances. (See also DIETETIC FOODS; DIETING; OBESITY.)

NYCTALOPIA. See NIGHT BLINDNESS.

NYCTINASTY, periodic response of plant organs to the alternation of day and night, such as the opening and closing of flowers. It is caused by changes in LIGHT intensity and TEMPERATURE. (See BIOLOGICAL CLOCKS.)

NYLON, group of POLYMERS containing AMIDE groups recurring in the chain. The commonest nylon is made by condensation of adipic acid and hexamethylene diamine. Nylon is chemically inert, heat-resistant, tough and very strong, and is extruded and drawn to make SYNTHETIC FIBERS, or cast and molded into bearings, gears, zippers etc.

NYMPH, a pre-adult stage in certain insects, strictly, used to contrast with LARVA. Nymphs typically resemble the adult in structure but are sexually immature and lack wings. Their METAMORPHOSIS is gradual and adult characters are developed progressively with each molt.

NYSTAGMUS, oscillation of the EYES, usually with a relatively slow drift in one direction and a correcting flick in the other. Looking out of a moving vehicle at passing objects induces nystagmus. It can be caused by failure of VISION fixation, a hereditary defect; weakness of the eye muscles, or diseases of the EAR labyrinths, BRAIN stem or cerebellum.

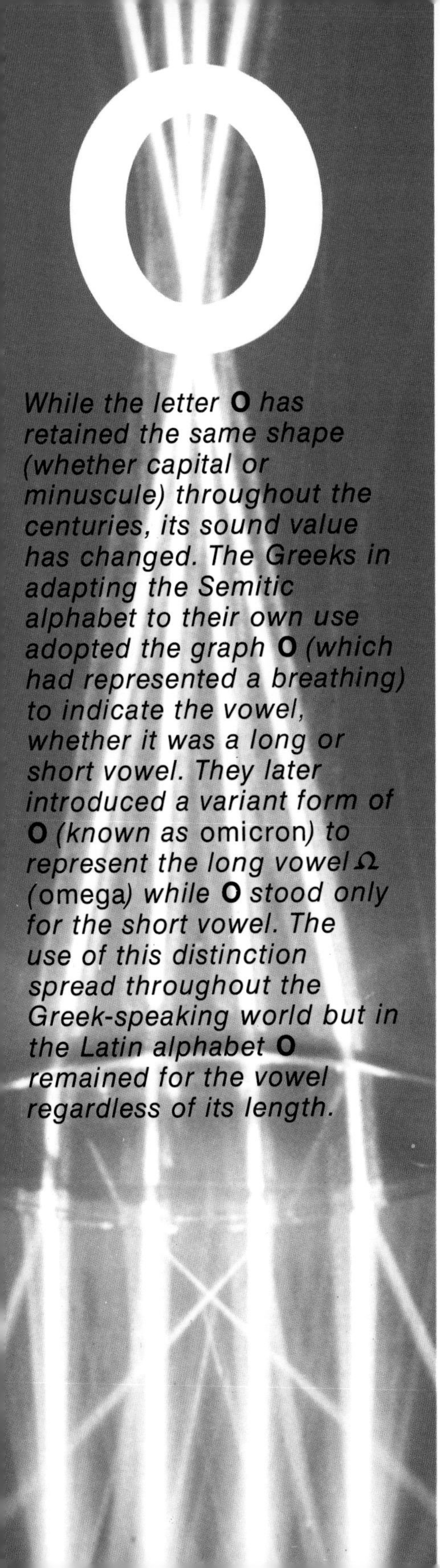

While the letter **O** *has retained the same shape (whether capital or minuscule) throughout the centuries, its sound value has changed. The Greeks in adapting the Semitic alphabet to their own use adopted the graph* **O** *(which had represented a breathing) to indicate the vowel, whether it was a long or short vowel. They later introduced a variant form of* **O** *(known as omicron) to represent the long vowel Ω (omega) while* **O** *stood only for the short vowel. The use of this distinction spread throughout the Greek-speaking world but in the Latin alphabet* **O** *remained for the vowel regardless of its length.*

OAKUM, fibers of HEMP or FLAX used to make the seams of wooden ships watertight, a process known as "caulking." Oakum is made either from short flax fibers produced during LINEN manufacture or from old ropes picked apart and tarred.

OASIS, a fertile area in the midst of a desert. The water source is generally a SPRING; though in the Sahara many oases have sprung up around WELLS. They may be hundreds of square kilometres in area, or merely a few trees clumped together. (See also DESERT; IRRIGATION.)

OBESITY, the condition of a subject's having excessive weight for his height, build and age. It is common in Western society, overfeeding in infancy being a possible cause. Excess ADIPOSE TISSUE is found in subcutaneous tissue and the ABDOMEN. Obesity predisposes to or is associated with numerous DISEASES including ARTERIOSCLEROSIS and high blood pressure; here premature DEATH is usual. Strict diet is essential for cure.

OBLATENESS, of a spheroid (see ELLIPSOID), the situation in which two of its axes of symmetry have an equal length greater than that of the third. The earth, in common with the other planets of the SOLAR SYSTEM, is oblate, its polar diameter being some 45km greater than that of its equator. A *prolate spheroid* is one with two axes of symmetry equal in length and shorter than the third.

OBSERVATORY, place from which a variety of astronomical observations are made. Ancient observatories such as STONEHENGE were used to predict SOLSTICES and EQUINOXES. With Tycho BRAHE (1546–1601) and the advent shortly after his death of the TELESCOPE, the modern observatory was born. Apart from the telescope, modern instruments used by observatories include the spectroscope (see SPECTROSCOPY), the transit instrument and the meridian circle (used to measure the right ascension and declination of stars: see CELESTIAL SPHERE), the coelostat and the coronagraph (for observing the sun), the PHOTOELECTRIC CELL (for measuring stellar brightnesses) and, in RADIO ASTRONOMY, the RADIO TELESCOPE.

OBSESSIONAL NEUROSIS, a NEUROSIS characterized by obsessions (inability to rid the CONSCIOUSNESS of certain ideas despite the individual's desire to do so and recognition of their abnormality) and COMPULSIONS.

OBSIDIAN, volcanic GLASS formed by rapid cooling of LAVA, usually with the composition of GRANITE. Commonly jet-black, but sometimes red, brown or variegated, it may be used as a GEM stone. There is a well-known occurrence at Yellowstone Park, Wyo. STONE AGE man used obsidian for implements.

OBSTETRICS, the care of women during PREGNANCY, delivery and the puerperium, a branch of MEDICINE and SURGERY usually linked with GYNECOLOGY. Antenatal care and the avoidance or control of risk factors for both mother and baby—ANEMIA, toxemia, high blood pressure, DIABETES, VENEREAL DISEASE, frequent MISCARRIAGE, etc.—have greatly contributed to the reduction of maternal and fetal deaths. The monitoring and control of labor and BIRTH, with early recognition of complications; induction of labor and the prevention of post-partum HEMORRHAGE with OXYTOCIN; safe forceps delivery and CESARIAN SECTION, and improved ANESTHETICS are

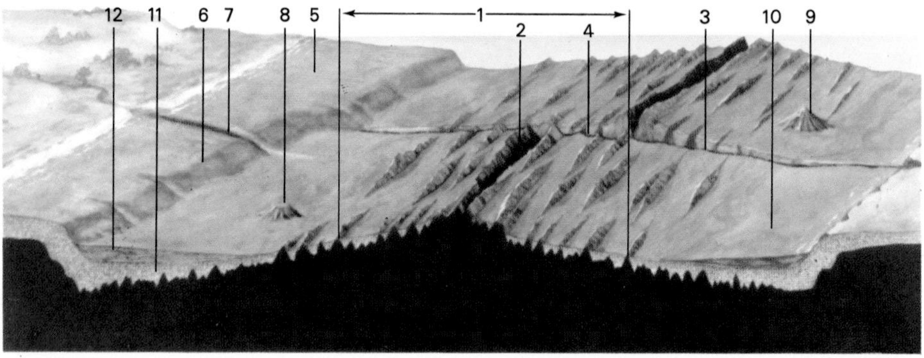

The morphological features of an ocean basin lying between passive continental plates (not to scale). Occupying a central position is a midocean ridge (1) on whose slopes are abyssal hills and along whose crest runs a median valley (2). At right angles to the line of the ridge is a fracture zone, or transform fault (3), motion along which has resulted in an offset in the ridge (4). Also shown are the continental shelf (5), the continental rise (6), a submarine canyon (7), a guyot (8), a seamount (9) and the abyssal plain (10). Near to the ridge, which is a center of sea-floor spreading, basaltic rocks are found: closer to the continents, pelagic sediments (11) are overlain by turbidites (12).

important factors in obstetric safety. ASEPSIS has made PUERPERAL FEVER a rarity.

OBTUSE, of an ANGLE, one between 90° and 180°.

OCCAM'S RAZOR. See OCKHAM, WILLIAM OF.

OCCLUSION, or occluded front. See FRONT.

OCCULTATION, the ECLIPSE of one celestial body by another. The term is usually applied to eclipses of stars by planets and particularly eclipses of planets or stars by the moon.

OCCUPATIONAL THERAPY, the ancillary speciality to MEDICINE concerned with practical measures to circumvent or overcome disability due to DISEASE. It includes the design or modification of everyday items such as cutlery, dressing aids, bath and lavatory aids, and wheelchairs. Assessment and education in domestic skills and industrial retraining are also important. Diversional activities are arranged for long-stay patients.

OCEAN CURRENTS, large-scale permanent or semipermanent movements of water at or beneath the surface of the OCEANS. Currents may be divided into those caused by winds and those caused by differences in DENSITY of seawater. In the former case, FRICTION between the prevailing wind and the water surface causes horizontal motion, and this motion is both modified by and in part transferred to deeper layers by further friction. Density variations may result from temperature differences, differing salinities, etc. The direction of flow of all currents is affected by the CORIOLIS EFFECT. Best known, perhaps, are the GULF STREAM and Humboldt current. (See also TIDES; WHIRLPOOL.)

OCEANOGRAPHY, the study of all aspects of and phenomena associated with seas and OCEANS.

OCEANS. The oceans cover some 71% of the earth's surface and comprise about 97% of the water of the planet (see HYDROSPHERE). They provide man with food, chemicals, minerals and transportation; and, by acting as a reservoir of solar heat energy, they ameliorate the effects of seasonal and diurnal temperature extremes for much of the world. With the atmosphere, they largely determine the world's climate (see also HYDROLOGIC CYCLE).

Oceanography is the study of all aspects of, and phenomena associated with, the oceans and seas. Most modern maps of the seafloor are compiled by use of ECHO SOUNDERS (see also SONAR), the vessel's position at sea being accurately determined by RADAR or otherwise. Water sampling, in order to determine, for example, salinity and oxygen content, is also important. Sea-floor sampling, to determine the composition of the sea-floor, is carried out by use of dredges, grabs, etc. (see DREDGING), and especially by use of hollow DRILLS which bring up cores of rock. OCEAN CURRENTS can be studied by use of buoys, drift bottles, etc., and often simply by accurate determinations of the different positions of a ship allowed to drift. Further information about the sea bottom can be obtained by direct observation (see BATHYSCAPHE; BATHYSPHERE) or by study of the deflections of seismic waves (see EARTHQUAKE).

Oceanographers generally regard the world's oceans as a single, large ocean. Geographically, however, it is useful to divide this into smaller units: the Atlantic, Pacific, Indian, Arctic and Antarctic (or Southern) Oceans (though the Arctic is often considered as part of the Atlantic, the Antarctic as parts of the Atlantic, Pacific and Indian). Of these, the Pacific is by far the largest and, on average, the deepest. However, the Atlantic has by far the longest coastline: its many bays and inlets, ideal for natural harbors, have profoundly affected W civilization's history.

Ocean trenches are long, narrow depressions of V-shaped cross-section running, typically, roughly parallel to continental coastal MOUNTAIN ranges or volcanic island arcs (see VOLCANISM). **Midocean ridges** are submarine mountain belts: the first to be discovered was that running roughly N-S in the Atlantic. They are important sites of earthquakes and volcanic activity (see PLATE TECTONICS; SEA-FLOOR SPREADING).

(See also ABYSSAL PLAINS; FISHERIES; HYDROGRAPHY; HYDROLOGY; ICEBERG; MARINE

BIOLOGY; OCEAN WAVES; OOZES; SUBMARINE CANYON; TIDES; TSUNAMI.)

OCEAN WAVES, undulations of the ocean surface, generally the result of the action of wind on the water surface. At sea, there is no overall translational movement of the water particles: they move up and forward with the crest, down and backward with the trough, describing a vertical circle. Near the shore, FRICTION with the bottom causes increased wave height, and the wave breaks against the land. Waves thus cause much coastal EROSION. (See also TSUNAMI.)

OCHER, reddish-yellow IRON ore (oxides and silicates), ground and roasted for use as a PIGMENT; also, CLAY colored yellow by such iron ore.

OCHOA, Severo (1905–), Spanish-born US biochemist who shared with KORNBERG the 1959 Nobel Prize for Physiology or Medicine for his first synthesis of a NUCLEIC ACID (or RNA).

OCKHAM (or Occam), William of (c1280–1349), English scholar who formulated the principle now known as **Occam's Razor:** "Entities must not unnecessarily be multiplied." This principle, interpreted roughly as "the simplest theory that fits the facts corresponds most closely to reality," has many applications throughout science.

OCTAGON. See POLYGON.

OCTAHEDRON. See POLYHEDRON.

OCTANE (C_8H_{18}), liquid ALKANE with 18 ISOMERS, constituents of GASOLINE. *Normal* octane occurs in PETROLEUM; the branched isomers, which have high antiknock values, are made by ALKYLATION. The **octane number** of a gasoline is the percentage of *iso*-octane (2,2,4-trimethylpentane) in the mixture of *iso*-octane with *n*-heptane which, in standard tests, knocks to the same degree as the gasoline. ANTIKNOCK ADDITIVES can raise the octane number above 100.

OCULIST, one who practices OPHTHALMOLOGY.

ODOMETER. See SPEEDOMETER.

OEDIPUS COMPLEX, COMPLEX typical of INFANTILE SEXUALITY, sometimes retained in the adult, comprising mainly UNCONSCIOUS desires to exclude the parent of one's own SEX and possess the other. In boys, mother FIXATION and consequent father rivalry may lead to a CASTRATION COMPLEX.

OERSTED (Oe), the unit of MAGNETIC FIELD strength in CGS electromagnetic units (see ABAMPERE). It is defined as the field at the center of a single-turn circular coil of radius 1cm and carrying a current of $1/(2\pi)$ abamperes.

OERSTED, Hans Christian (1777–1851), Danish physicist whose discovery that a magnetized needle can be deflected by an electric current passing through a wire (1820) gave birth to the science of ELECTROMAGNETISM.

OFFSET LITHOGRAPHY. See PRINTING.

OHM (Ω), the SI UNIT of electric RESISTANCE. A conductor has a resistance of one ohm when a potential difference of one VOLT across it gives rise to a current of one AMPERE.

OHM, Georg Simon (1789–1854), Bavarian-born German physicist who formulated OHM'S LAW. He also contributed to ACOUSTICS, recognizing the ability of the human ear to resolve mixed SOUND into its component pure (sinusoidal-wave) tones.

OHMMETER, instrument for providing a rapid, if approximate, value for the RESISTANCE of part of an electric circuit. It consists of an AMMETER (reverse calibrated in OHMS) in series with a fixed resistor, a BATTERY and a variable resistor. This last is adjusted to zero the meter with the terminals of the instrument shorted. The test resistance is then introduced between these terminals. The reduction in the current so caused is a measure of the resistance.

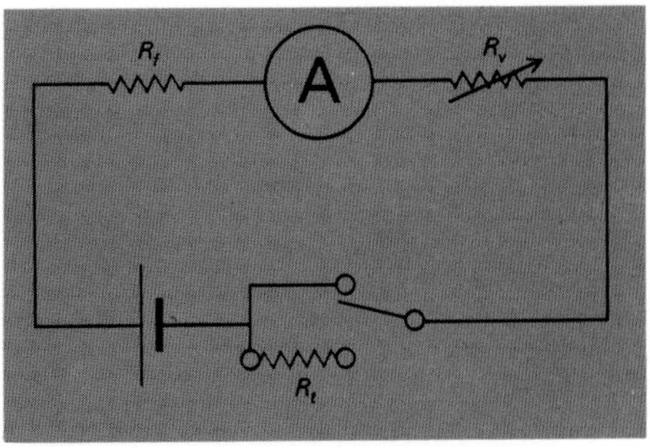

Simplified circuit diagram for an ohmmeter. A represents the ammeter, R_f the fixed resistor, R_v the variable resistor and R_t the resistor under test.

OHM'S LAW, the statement due to G. S. OHM in 1827 that the electric POTENTIAL difference across a conductor is proportional to the current flowing through it, the constant of proportionality being known as the RESISTANCE of the conductor. It holds well for most materials and objects, including solutions, provided that the passage of the current does not heat the conductor, but ELECTRON TUBES and SEMICONDUCTOR devices show a much more complicated behavior.

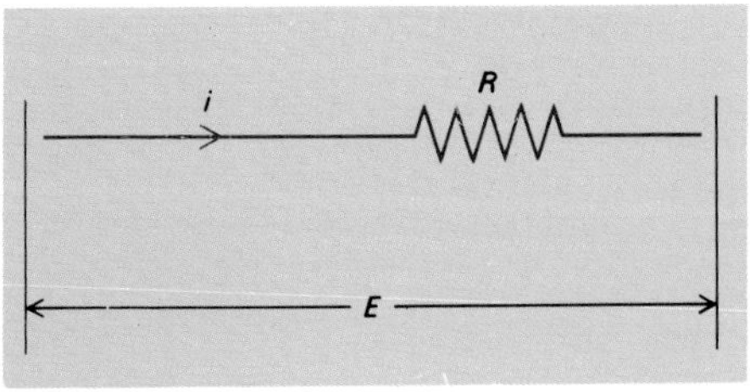

Ohm's law for direct-current circuits states that the current (i) in a conductor is proportional to the potential difference (E) along it, the constant of proportionality being the resistance of the conductor (R). Thus $E=iR$.

OIL, any substance that is insoluble in water, soluble in ETHER and greasy to the touch. There are three main groups: mineral oils (see PETROLEUM); fixed vegetable and animal oils (see FATS; LIPIDS), and volatile vegetable oils (see ESSENTIAL OILS). Oils are classified as fixed or volatile according to the ease with which they vaporize when heated. Mineral oils include GASOLINE and many other fuel oils, heating oils and lubricants. Fixed vegetable oils are usually divided into three subgroups depending on the physical change that occurs when they absorb oxygen: oils such as linseed and tung, which form a hard film, are known as "drying oils"; "semidrying oils," such as cottonseed or soybean oil, thicken considerably but do not harden; "nondrying oils," such as castor and olive oil, thicken only slightly. Fixed animal oils include the "marine oils," such as cod-liver and whale oil. Fixed animal and vegetable fats, such as butterfat and palm oil are often also classified as oils. Examples of volatile vegetable oils, which usually have a very distinct odor and flavor, include such oils as bitter almond, peppermint and TURPENTINE. When dissolved in alcohol, they are called "essences."

OIL REFINING. See PETROLEUM.

OILS, Natural. See FATS.

OIL SANDS, loose sand or sandstone containing viscous oil. Depending on the proportion of oil to sand, they may occur as ASPHALT lakes, such as those in Trinidad, or BITUMINOUS SANDS, such as the Athabaska tar sands (Alberta, Canada). Despite extraction problems, they are a potentially important OIL source.

OIL SHALE, a fine-grained, dark-colored sedimentary rock from which oil suitable for refining can be extracted. The rock contains an organic substance called kerogen, which may be distilled to yield OIL (see also DISTILLATION). It is nowadays an important oil source. (See also SEDIMENTARY ROCKS; SHALE.)

OIL WELL. See PETROLEUM.

OLBERS, Heinrich Wilhelm Matthäus (1758–1840), German astronomer who discovered the ASTEROIDS Pallas (1802) and Vesta (1807), rediscovered Ceres (1802), and found five comets, one of which bears his name. He also formulated a method of calculating cometary orbits (1779); proposed that the pressure of light is responsible for COMETS' tails always pointing away from the sun (1811); and stated **Olbers' Paradox,** that, in an infinite, isotropic universe, the night sky should be uniformly illuminated, whereas in fact it is not. This he explained by suggesting the existence of clouds of INTERSTELLAR MATTER, later discovered; but the "paradox" was not fully resolved until HUBBLE showed that the UNIVERSE is expanding.

OLD AGE, strictly, a chronological division of human life; however, the term also suggests the manifestations and diseases associated with AGING. The SKIN becomes wrinkled and thins, largely due to the effect of ULTRAVIOLET RADIATION on COLLAGEN; HAIR production may be disordered, causing BALDNESS or graying. BONE alters (osteoporosis) with thinning of texture and susceptibility to FRACTURE, while the disks between the VERTEBRAE shrink with resulting loss of height. Wear and tear of the JOINTS frequently leads to osteoarthritis. ARTERIOSCLEROSIS, which starts in early life, becomes established in the elderly, resulting in STROKE, CORONARY THROMBOSIS and limb GANGRENE. Degenerative disease due to cell loss is especially important in the BRAIN. IMMUNITY may be less effective, leading to more frequent and serious infections (PNEUMONIA), and may account for the increased incidence of CANCER with age. VITAMIN deficiency diseases due to inadequate diet are common. Special social and psychiatric problems arise from the social isolation, decreased mobility and poor health associated with old age. (See also GERIATRICS.) *Progeria* is a rare condition of children causing premature aging.

OLEFINS. See ALKENES.

OLEIC ACID ($C_{17}H_{33}COOH$), an unsaturated CARBOXYLIC ACID used in lubricating oils and varnishes. Triolein, its ester with GLYCEROL is present in many natural oils and fats, helping to keep them liquid at room temperature. mp 16°C, bp 286°C.

The mineral olivine and a peridot gemstone.

OLEOMARGARINE, early name for MARGARINE.

OLEUM. See SULFURIC ACID.

OLFACTORY SYSTEM. See SMELL.

OLIGOCENE, the third epoch of the TERTIARY, of duration about 40–25 million years ago. (See also GEOLOGY.)

OLIGOCHAETA, a class of the phylum ANNELIDA.

OLIVE OIL, edible OIL obtained from the fruit of the olive tree. Oil is extracted at a number of stages during processing. The fruit is first pulped, and then pressed at least twice. Further extraction is achieved by using solvents such as hot water and carbon disulfide. Olive oil is used as salad oil, frying oil and in canning foods such as sardines. The lowest grades are used for making soap. Olive oil contains between 67% and 83% OLEIC ACID.

OLIVINE, group of SILICATE minerals, orthosilicates of magnesium and iron $(Mg,Fe)_2SiO_4$. It forms olive-green crystals in the orthorhombic system, and occurs commonly in IGNEOUS ROCKS, chiefly BASALT, GABBRO and PERIDOTITE. The transparent variety **peridot** is used as a GEM stone.

OMEGA-MINUS. See SUBATOMIC PARTICLES.

OMNIVORE, any animal that feeds on both plants and animals. (See also CARNIVORE; HERBIVORES.)

ONCHOCHERCIASIS. See FILARIASIS.

ONSAGER, Lars (1903–1976), Norwegian-born US chemist awarded the 1968 Nobel Prize for Chemistry for his fundamental work on irreversible chemical and thermodynamic processes.

ONTOGENY AND PHYLOGENY, terms respectively descriptive of the developmental (particularly embryonic) history of an individual organism and of the evolutionary history of its race. E. H. HAECKEL's "biogenetic law" proposed that "ontogeny recapitulates (repeats) phylogeny."

ONTOLOGY. See EPISTEMOLOGY.

ONYCHOPHORA, a class of ARTHROPODA which includes a small group of terrestrial species found mainly in the Southern Hemisphere. Onychophorans were once believed to provide a link between annelids and arthropods but this theory is now disputed.

ONYX, variety of CHALCEDONY with variegated bands, straight rather than curved as in AGATE. Sardonyx has white and brown bands; carnelian onyx white and red. It is used as a GEM stone, especially for cameos and intaglios.

OOLITH, a more or less spherical particle of rock which has developed by the accretion of material

Opal: as a raw mineral and as prepared gemstones. The white variety is more highly prized than the colored.

about an initial nucleus. The accretion may be concentric (so that in cross-section circular bands of material may be seen) or radial, and combinations are known. Larger ooliths are termed *pisoliths*. Concentration of ooliths can form, for example, oolitic limestones, often called *oolites*.

OOMYCETES. See FUNGI.

OOZES, sediments on the deep-sea plains, consisting of PLANKTON remains, wind-borne volcanic dust, etc. Beneath about 5000m no organic matter is present (*red clays*); between 5 000 and 3 900m Radiolaria predominate in the *siliceous oozes* (see also PROTOZOA); and between 3 900 and 2 000m Foraminifera (especially GLOBIGERINA) are most common in the *calcareous oozes*.

OPAL, cryptocrystalline variety of porous hydrated SILICA, deposited from aqueous solution in all kinds of rocks, and also formed by replacement of other minerals. Opals are variously colored; the best GEM varieties are translucent, with milky or pearly opalescence and iridescence due to light scattering and interference from internal cracks and cavities. Common opal is used as an abrasive, filler and insulator.

OPEN CLUSTERS. See GALACTIC CLUSTERS.

OPEN-HEARTH PROCESS, technique which was, until recently, responsible for most of the world's STEEL production. It derives its alternative name, the Siemens-Martin Process, from the work of Sir William SIEMENS and Pierre Émile MARTIN in the 1850s and 1860s. The FURNACE has two ducts leading each to a chamber of brick checkerwork. In use, the hot exhaust fumes are passed out through one duct, heating the brick to high temperature, while air intake is through the other. Periodically the streams are reversed, so that the incoming air is preheated by the hot brick before passing the burners, thus greatly increasing the flame temperature. Some furnaces are liquid-fueled, but if the fuel is gaseous it may be fed in with the air. In the *basic process*, the charge is of IRON ore, scrap steel and LIMESTONE. The impurities in the ore combine with the limestone to form a basic SLAG. In the less important *acid process*, REFRACTORIES of SILICA result in an acid slag.

A polished piece of onyx showing the typical banding, and a small onyx gemstone.

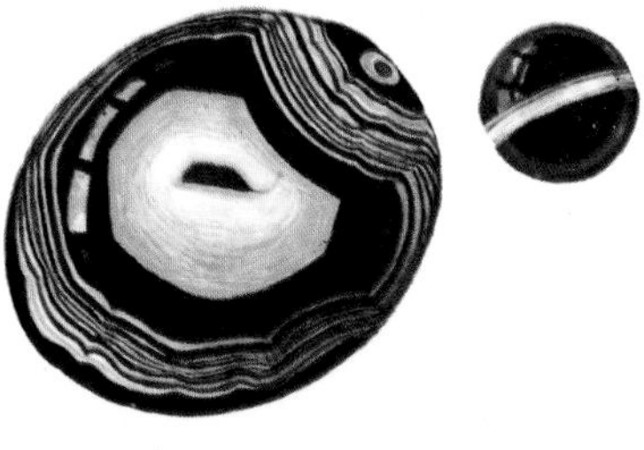

OPERA GLASSES. See BINOCULARS.

OPERATION. See SURGERY.

OPERATIONS RESEARCH, improvement of the efficiency of commercial, military and governmental organizations by techniques of numerical analysis. Its aims are to increase the result/effort ratio by either decreasing the effort required or increasing the result (or output) without increase in effort. (See also HUMAN ENGINEERING; LINEAR PROGRAMMING.)

OPHIDIA, members of the REPTILIA that include the snakes. Snakes have elongated bodies and lack limbs although traces of them may be found internally in some species. The skull is modified to enable large prey to be swallowed whole. Prey is often killed or stunned by venom which is responsible for the fear accorded to many species by man.

OPHTHALMIA, INFLAMMATION of the EYE. This may be CONJUNCTIVITIS, as in **neonatal ophthalmia** (often gonococcal—see VENEREAL DISEASE) or uveitis (or penophthalmitis) as in **sympathetic ophthalmia**. In this, an inflammatory reaction in both eyes follows injury to one. Infection may need ANTIBIOTICS, while sympathetic ophthalmia may benefit from STEROIDS.

OPHTHALMOLOGY, the branch of MEDICINE and SURGERY concerned with diseases of VISION and the EYE. In infancy, congenital BLINDNESS and STRABISMUS, and in adults, glaucoma, uveitis, CATARACT, retinal detachment and vascular diseases are common, as are ocular manifestations of systemic diseases— hypertension and DIABETES. Disorders of eye movement, lids and TEAR production; color vision; infection, and injury are also seen. Surgery to the lens, CORNEA (including corneal grafting), eye muscles and lids may be used, and cryosurgery (freezing) or coagulation employed in retinal disease.

OPHTHALMOSCOPE, instrument for examining the RETINA and structures of the inner EYE. A powerful light and lens system, combined with the CORNEA and lens of the eye allows the retina and eye blood vessels to be seen at high magnification. It is a valuable aid to diagnosis in OPHTHALMOLOGY and internal MEDICINE.

OPHUROIDEA, a class of ECHINODERMATA which includes the brittlestars. They have a central disk-shaped body and (usually) five slender arms that are quite distinct from the body.

OPIUM, NARCOTIC extract from the immature fruits of the opium poppy, *Papaver somniferum*, which is native to Greece and Asia Minor. The milky juice is refined to a powder which has a sharp, bitter taste. Drugs, some drugs of abuse (see DRUG ADDICTION), obtained from opium include the narcotic ANALGESICS, HEROIN, MORPHINE and CODEIN. (Synthetic analogues of these include methadone and pethidine.) Older opium preparations, now rarely used, include LAUDANUM and PAREGORIC. The extraction of opium outside the pharmaceutical industry is strictly controlled in the West.

OPPENHEIMER, Julius Robert (1904–1967), US physicist whose influence as an educator is still felt today and who headed the MANHATTAN PROJECT, which developed the ATOMIC BOMB. His main aim was the peaceful use of nuclear power (he fought against the construction of the HYDROGEN BOMB but was overruled by Truman in 1949); but, because of his left-wing friendships, was unable to pursue his researches in this direction after being labeled a security risk (1954). He also worked out much of the theory of BLACK HOLES.

OPPOSITION, in astronomy, the situation in which the earth lies directly between another planet (or the moon) and the sun.

OPTICAL ACTIVITY, the property, possessed by certain substances, of rotating the plane of polarization of plane-POLARIZED LIGHT passing through them. This optical rotation is measured by POLARIMETRY. Optical activity is shown by asymmetric CRYSTALS which have two mirror-image forms—the rotation being to the left or right respectively—and by compounds with asymmetric molecules showing optical STEREOISOMERISM.

OPTICAL ILLUSION. See ILLUSION.

OPTICIAN, one who practices OPTOMETRY.

OPTICS, the science of light and vision. Physical optics deal with the nature of LIGHT (see also COLOR; DIFFRACTION; INTERFERENCE; POLARIZED LIGHT; SPECTROSCOPY). Geometrical optics consider the behavior of light in optical instruments (see ABERRATION, OPTICAL; CAMERA; DISPERSION; LENS, OPTICAL; MICROSCOPE; MIRROR; PRISM; REFLECTION; REFRACTION; SPECTRUM; TELESCOPE). Physiological optics are concerned with vision (see EYE).

OPTOMETRY, measurement of the acuity of VISION and the degree of lens correction required to restore "normal vision" in subjects with refractive errors (MYOPIA, HYPEROPIA, ASTIGMATISM). Its principal instrument is a chart of letters which subtend specific angles to the EYE at a given distance; temporary lenses being used to correct each eye. (See GLASSES.)

ORBIT, the path followed by one celestial body revolving under the influence of gravity (see GRAVITATION) about another. In the SOLAR SYSTEM, the planets orbit the sun, and the moons the planets, in elliptical paths, although Triton's orbit of NEPTUNE is as far as can be determined perfectly circular. The point in the planetary, asteroidal or cometary orbit closest to the sun is called its *perihelion*; the farthest point is termed *aphelion*. In the case of a moon or artificial satellite orbiting a planet or other moon, the corresponding terms are *perigee* and *apogee*. (See also APSIDES, LINE OF; KEPLER'S LAWS.) Celestial objects of similar masses may orbit each other, particularly DOUBLE STARS.

ORBITAL, in chemistry, the mathematical wave function (see QUANTUM MECHANICS) that describes the motion of an ELECTRON around the nucleus of an ATOM or several nuclei in a molecule. The orbital represents the probability distribution of the electron in space; in effect, for each point, the likelihood of finding the electron there. Orbitals are defined and characterized by three quantum numbers, representing the energy level (and hence the size), the angular momentum (and hence the shape), and the orientation. An orbital can be occupied by one or two electrons (of opposite spin), according to FERMI-DIRAC STATISTICS. The precise energy of each orbital depends on the local electromagnetic field, and is found by SPECTROSCOPY. In the formation of a covalent BOND, molecular orbitals are formed by linear combination of the outer atomic orbitals. (See also AROMATIC COMPOUNDS; RESONANCE.)

ORDER. See TAXONOMY.

ORDINAL NUMBER, one that describes the order of an element in a SET. For example, in the phrase "my second and third candies," 2 and 3 are ordinal numbers. (See also CARDINAL NUMBER.)

ORDINATE, the perpendicular distance from the x-axis of a point in a system of CARTESIAN COORDINATES (see also ANALYTIC GEOMETRY); the y-coordinate of the point. (See also ABSCISSA.)

ORDOVICIAN, the second period of the PALEOZOIC, which lasted from about 500 to 440 million years ago and immediately followed the CAMBRIAN. (See GEOLOGY.)

ORE, aggregate of minerals and rocks from which it is commercially worthwhile to extract minerals (usually metals). An ore has three parts: the country rock in which the deposit is found; the gangue, the unwanted ROCKS and minerals of the deposit; and the desired MINERAL itself. Ore deposits may be, e.g., VEINS; infillings of breccia (consolidated TALUS); sedimentary formations, as of the EVAPORITES; certain DIKES; or, especially with the SULFIDES, hydrothermal replacement deposits (where hot or superheated water has dissolved existing rocks and deposited in their place minerals held in SOLUTION). MINING techniques depend greatly on the form and position of the deposit (see PLACER MINING; STRIP MINING).

ORGAN, in biology, a functionally adapted part of an organism. Organs, such as the vermiform APPENDIX in man, which persist in a species although no longer of any use to it, are termed *vestigial organs*.

ORGANIC CHEMISTRY, major branch of CHEMISTRY comprising the study of CARBON compounds containing hydrogen (simple carbon compounds such as carbon dioxide being usually deemed inorganic). This apparently specialized field is in fact wide and varied, because of carbon's almost unique ability to form linked chains of atoms to any length and complexity; far more organic compounds are known than inorganic. Organic compounds form the basic stuff of living tissue (see also BIOCHEMISTRY), and until the mid-19th century, when organic syntheses were achieved, a "vital force" was thought necessary to make them. The 19th-century development of quantitative ANALYSIS by J. LIEBIG and J. B. A. DUMAS, and of structural theory by S. CANNIZZARO and F. A. KEKULÉ, laid the basis for modern organic chemistry. Organic compounds are classified as ALIPHATIC, ALICYCLIC, AROMATIC and HETEROCYCLIC COMPOUNDS, according to the structure of the skeleton of the molecule, and are further subdivided in terms of the FUNCTIONAL GROUPS present.

ORGAN OF CORTI, the apparatus in the cochlea of the inner EAR, which, in hearing, converts pressure waves in the cochlear fluid into auditory nerve impulses.

ORGANOMETALLIC COMPOUNDS, class of compounds containing bonds from carbon atoms to metal (or metalloid) atoms, and thus at the crossroads of INORGANIC and ORGANIC CHEMISTRY. In the last 30

Organometallic compounds

ethyl sodium MW 52·05	$NaCH_2CH_3$
sodium anilide MW 115·11	$NaNHC_6H_5$
diphenyl magnesium MW 178·53	$Mg(C_6H_5)_2$
Grignard reagents	RMgX
tetraethyl lead MW 323·44, mp −137°C	$Pb(CH_2CH_3)_4$
ferrocene (dicyclopentadienyliron) MW 186·04, mp 173°C	$Fe(C_5H_5)_2$

years the subject has expanded enormously, with the development of many medical, industrial and synthetic uses. There are three main types of organometallic compounds: (1) the alkyl derivatives of Group IA and IIA metals of the PERIODIC TABLE (including GRIGNARD REAGENTS), which have ionic BONDS, and are powerful BASES and reducing agents (see OXIDATION AND REDUCTION); (2) derivatives of other Main Group metals, which are volatile, covalently-bonded compounds; and (3) TRANSITION ELEMENT derivatives (including SANDWICH COMPOUNDS) with special *d*-ORBITAL bonding. Organometallic compounds are prepared by reacting a metal with an ALKYL HALIDE or a reactive HYDROCARBON, or by substitution.

ORIGIN OF SPECIES, short title of book (published 1859) by Charles DARWIN in which he set forth his theory of EVOLUTION by NATURAL SELECTION.

ORION (the Hunter), a large constellation on the celestial equator visible during winter in N skies, containing RIGEL, BETELGEUSE and the Orion NEBULA.

ORLON, SYNTHETIC FIBER made by copolymerization of acrylonitrile with VINYL compounds.

ORMOLU, golden-colored BRASS used for furniture mountings, chiefly 18th century. It was cast, chased and often gilded by brushing with gold AMALGAM and firing.

ORNITHISCHIA, or bird-hipped dinosaurs, a group of DINOSAURS having bird-like pelvic girdles, with four prongs to each side. All were herbivorous. Four-legged types include the stegosaurs, with triangular bony plates along the back, and the armadillo-like ankylosaurs. The two-legged duck-billed dinosaurs were well equipped for swimming.

ORNITHOLOGY, the scientific study of birds. The observation of birds in their natural environment has a long history and is now so popular as to be the most widespread of zoological hobbies.

ORNITHOPTER, flying machine whose wings flap like those of a bird. Models date back as far as 400 BC, and LEONARDO DA VINCI made many designs and models. Though working models have been made, and toy ornithopters mass-produced, no larger, load-carrying design has yet been successful.

OROGENIES, periods of mountain building (orogenesis) in a particular area. They occur usually in geosynclinal regions (see GEOSYNCLINE). **Epeirogeny**, elevation or depression of large land masses, plays comparatively little part in MOUNTAIN building. (See PLATE TECTONICS.)

ORPIMENT, soft, lemon-yellow SULFIDE mineral, arsenic(III) sulfide (As_2S_3), used as a pigment. It crystallizes in the monoclinic system, but is usually massive. It forms by deposition from GEYSERS or by alteration of REALGAR.

ORRERY, a mechanical model to show the motions of planets and/or their satellites, named for the 4th Earl of Orrery, patron of the probable inventor (c1710), George Graham (1673–1751). Time scales may be quite accurate, but size and distance proportions are of course wildly inaccurate.

ORTHOCLASE, common mineral in igneous rocks, consisting of potassium aluminum silicate ($KAlSi_3O_8$); vitreous crystals of various colors. It is one of the three end-members (pure compounds) of the FELDSPAR group. See also MICROCLINE.

ORTHODONTICS, the treatment of malformed and displaced teeth. (See DENTISTRY.)

ORTHOGENESIS, evolutionary change that appears to be directed over a long period of time. Orthogenesis is no longer regarded as an evolutionary mechanism but as the result of consistent selection for the same character in an animal.

ORTHOGRAPHIC PROJECTION. See DESCRIP-

The principal orogenies (mountain-building episodes) of Europe and North America. The positions of the names indicate the times of greatest orogenic activity, although the episodes often extended for long periods before and after their maxima and showed considerable geographic variation in intensity. Orogenic episodes are usually associated with nearby collisions between adjacent crustal plates.

Period	Europe	North America
Tertiary	Alpine	Pasadenian
Cretaceous		Laramide
[illegible]		
Triassic		
Permian		
Pennsylvanian	Hercynian (Variscan)	Appalachian
Mississippian		
Devonian		Acadian
Silurian		
Ordovician	Caledonian	Taconic
Cambrian		

TIVE GEOMETRY.

ORTHOPEDICS, speciality within SURGERY, dealing with BONE and soft-tissue disease, damage and deformity. Its name derives from 17th-century treatments designed to produce "straight children." Until the advent of anesthetics, ASEPSIS and X-RAYS, its methods were restricted to AMPUTATION and manipulation for dislocation, etc. Treatment of congenital deformity; FRACTURES and TUMORS of bone; OSTEOMYELITIS; ARTHRITIS, and JOINT dislocation are common in modern orthopedics. Methods range from the use of splints, PHYSIOTHERAPY and manipulation, to surgical correction of deformity, fixing of fractures and refashioning or replacement of joints. Suture or transposition of TENDONS, MUSCLES or nerves are performed.

OSCILLATING UNIVERSE. See COSMOLOGY.

OSCILLATOR, a device converting direct to alternating current (see ELECTRICITY), used, for example, in generating RADIO waves. Most types are based on an electronic AMPLIFIER, a small portion of the output being returned via a FEEDBACK circuit to the input, so as to make the oscillation self-sustaining. The feedback signal must have the same PHASE as the input: by varying the components of the feedback circuit, the frequency for which this occurs can be varied, so that the oscillator is easily "tuned." "Crystal" oscillators incorporate a piezoelectric crystal (see PIEZOELECTRICITY) in the tuning circuit for stability; in "heterodyne" oscillators, the output is the beat frequency between two higher frequencies.

OSCILLOSCOPE, a device using a CATHODE RAY TUBE to produce line GRAPHS of rapidly varying electrical signals. Since nearly every physical effect can be converted into an electrical signal, the oscilloscope is very widely used. Typically, the signal controls the vertical deflection of the beam while the horizontal deflection increases steadily, producing a graph of the signal as a function of time. For periodic (repeating) signals, synchronization of the horizontal scan with the signal is achieved by allowing the attainment by the signal of some preset value to "trigger" a new scan after one is finished. Most models allow two signals to be displayed as functions of each other; dual-beam instruments can display two as a function of time. Oscilloscopes usually operate from DC to high frequencies, and will display signals as low as a few millivolts.

OSLER, Sir William (1849–1919), Canadian-born physician and educator best known for his work on platelets (see BLOOD) and for the informality of his educational techniques.

OSMIUM (Os), silvery-gray hard metal in the PLATINUM GROUP. It is slowly oxidized in air. AW 190.2, mp 3045°C, bp c5000°C, sg 22.48 (20°C). **Osmium (VIII) oxide** (OsO_4), a toxic oxidizing agent, is used to stain tissues for microscope slides.

OSMOSIS, the diffusion of a solvent through a SEMIPERMEABLE MEMBRANE that separates two solutions of different concentration, the movement being from the more dilute to the more concentrated solution, owing to the thermodynamic tendency to equalize the concentrations. The liquid flow may be opposed by applying pressure to the more

The electron tube used in the cathode-ray oscilloscope. Brightness is controlled by varying the grid current and focusing by varying the potential difference between the anodes. In use it is common to apply a time base to the X-plates to show the variation of a parameter with time. The electron beam represents a negative electric current which is drained from the phosphor screen via the graphite tube lining.

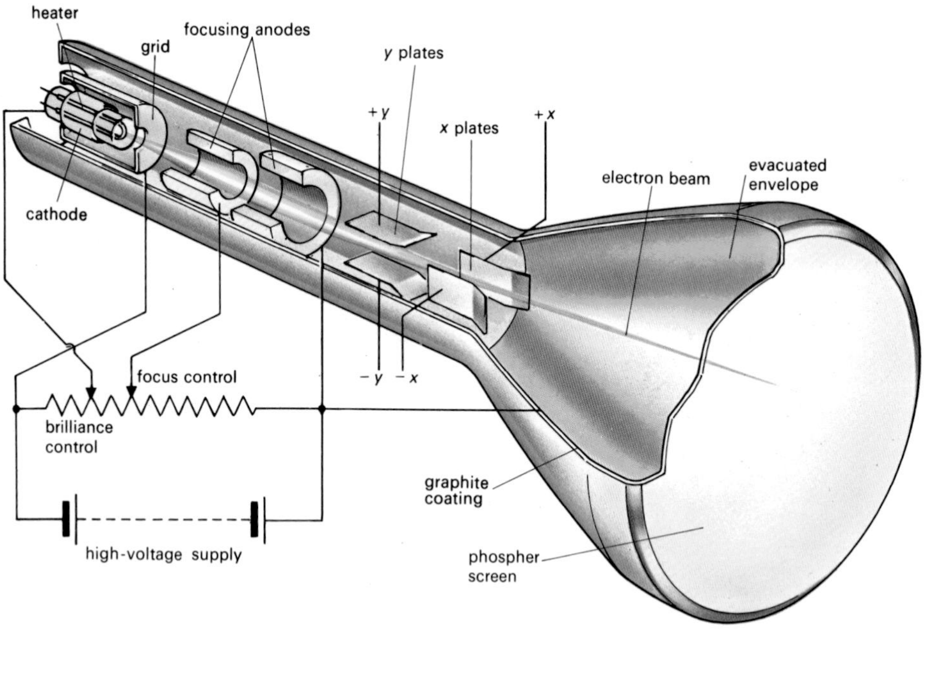

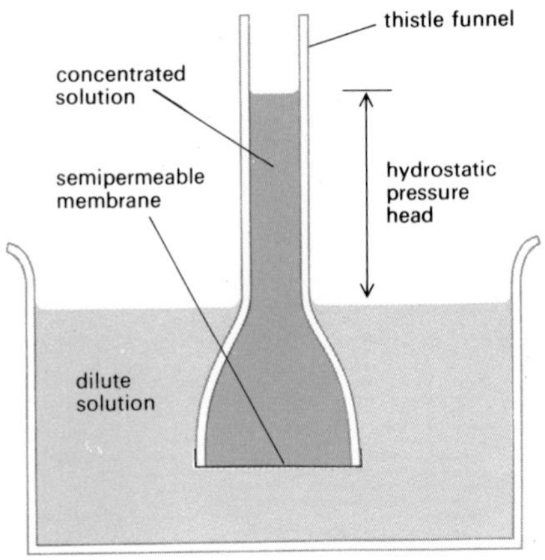

The principle of osmosis. The osmotic pressure difference across a semipermeable membrane separating two solutions of different concentrations causes transport of the solvent through the membrane tending to equalize the concentrations of the two solutions. The process continues until an equilibrium state is achieved, either equal concentrations, or, in this case, balance between the osmotic pressure difference and the hydrostatic pressure excess of the solution in the inverted thistle funnel. The hydrostatic pressure difference thus provides a measure of the osmotic pressure difference.

concentrated solution: the pressure required to reduce the flow to zero from a pure solvent to a given solution is known as the *osmotic pressure of the solution.* Osmosis was studied by Thomas GRAHAM, who coined the term (1858); in 1886 VAN'T HOFF showed that, for dilute solutions (obeying Henry's Law), the osmotic pressure varies with temperature and concentration as if the solute were a GAS occupying the volume of the solution. This enables MOLECULAR WEIGHTS to be calculated from osmotic pressure measurements, and degrees of ionic DISSOCIATION to be estimated. Osmosis is important in DIALYSIS and in water transport in living tissue.

OSSICLES. See EAR.

OSSIFICATION. See BONE.

OSTEICHTHYES, or bony fish, a class of the CHORDATA that includes the ACANTHODII, CROSSOPTERYGII, DIPNOI and ACTINOPTERYGII.

OSTEOLOGY, the study of the structure, function and diseases of BONE.

OSTEOMYELITIS, BACTERIAL infection of BONE, usually caused by STAPHYLOCOCCUS, STREPTOCOCCUS and SALMONELLA carried to the bone by the BLOOD, or gaining access through open FRACTURES. It commonly affects children, causing FEVER and local pain. If untreated or partially treated, it may become chronic with bone destruction and a discharging SINUS. ANTIBIOTICS and surgical drainage are frequently necessary.

OSTEOPATHY, system of treatment based on theory that DISEASE arises from the mechanical and structural disorder of the body skeleton. Prevention and treatment are practiced by manipulation, often of the spine. While it may have a role in treatment of chronic musculo-skeletal pain, its methods may be hazardous, especially to the SPINAL CORD. Furthermore, serious disease may be overlooked. Osteopathy is best regarded as an adjunct to, rather than a replacement for, orthodox medicine.

OSTWALD, Friedrich Wilhelm (1853–1932), Latvian-born German physical chemist regarded as a father of physical chemistry, and awarded the 1909 Nobel Prize for Chemistry for his work on CATALYSIS. He also developed the OSTWALD PROCESS.

OSTWALD PROCESS, process for manufacturing NITRIC ACID, developed by OSTWALD (1902). AMMONIA is oxidized by air to nitric oxide, at 900°C and 1–8atm using a platinum/rhodium catalyst. The nitric oxide is further oxidized to nitrogen dioxide, which is dissolved in water to give 60% nitric acid.

OTIS, Elisha Graves (1811–1861), US inventor of the safety ELEVATOR (1852), first installed for passenger use in 1856, New York City.

OTOSCLEROSIS. See DEAFNESS.

OTOSCOPE, instrument for examining the outer EAR and eardrum. It has a light, a lens and a conical earpiece.

OTTO, Nikolaus August (1832–1891), German engineer who built the first four-stroke INTERNAL-COMBUSTION ENGINE (1876), which rapidly replaced the STEAM ENGINE in many applications and facilitated the development of the AUTOMOBILE.

OUNCE (oz), unit of WEIGHT in the apothecaries', avoirdupois and troy systems:
1oz ap = 1oz troy = 1.0971oz avoirdupois = 31.103g. (See APOTHECARIES' WEIGHT; TROY WEIGHT; WEIGHTS AND MEASURES.)

OVARY, the female reproductive organ. In plants it contains the ovules (see FLOWER); in humans, the FOLLICLES in which the eggs (*ova*) develop (see ESTROGEN; FERTILIZATION; GAMETE; PROGESTERONE; REPRODUCTION).

OVERWEIGHT. See DIETING; OBESITY.

OVIPAROUS ANIMALS, animals that reproduce by laying EGGS.

OVULATION. See MENSTRUATION.

OVULE. See FLOWER.

OVUM. See EGG.

OXALIC ACID $(COOH)_2$, white crystalline solid, a toxic, dibasic CARBOXYLIC ACID occurring in many plants. It is made by heating sodium formate or by fusing sawdust with sodium hydroxide. Oxalic acid is a mild reducing agent used as a standard for VOLUMETRIC ANALYSIS, in the leather and dye industries, to remove ink and rust stains, and to dissolve radiator scale. MW 90.0, mp 189°C.

OXBOW LAKE, C-shaped lake formed when a river meanders almost in a full circle, the water cuts across the narrow neck, and deposits of SILT at the entrances to the original meander eventually cut it off from the main body of the river. (See also RIVERS AND LAKES.) (Illustration page 420.)

OXIDATION AND REDUCTION, or **redox reactions,** large class of chemical reactions, including many familiar processes such as COMBUSTION, CORROSION and RESPIRATION. Oxidation was originally defined simply as the combination of an element or compound with oxygen, or the removal of hydrogen from a compound; and reduction as

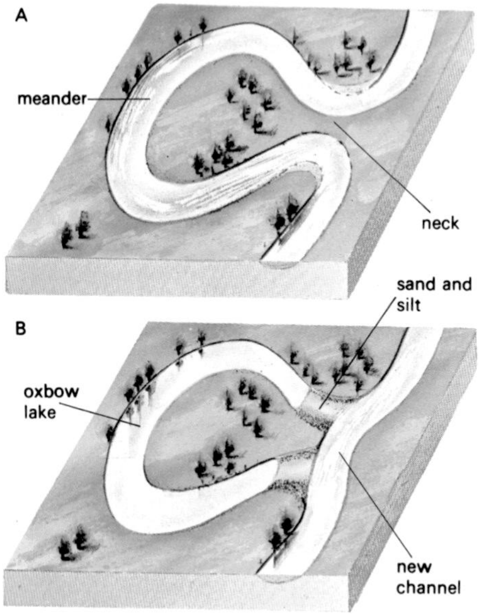

Oxbow (cut off) lakes are formed when the neck of a looping meander is broken through (A), usually during times of flood. The entrances to the bypassed meander are soon blocked by silt (B).

combination with hydrogen or removal of oxygen. In the modern theory this has been generalized: oxidation is defined as loss of electrons, and reduction as gain of electrons. The two always go together: there is an oxidizing agent which is reduced, and a reducing agent which is oxidized. Thus, in the reaction

$$Fe^{3+} + I^- \rightarrow Fe^{2+} + \tfrac{1}{2}I_2$$

the iron (III) ion gains an electron and is reduced to iron (II), and the iodide ion loses an electron and is oxidized to iodine. The strength of a redox reagent, expressing its tendency to react, is measured by the electrode potential of the half-reaction (see ELECTROCHEMISTRY), and so redox reagents may be ranked in an extended ELECTROCHEMICAL SERIES. In a covalent compound or complex ion, each atom is assigned an *oxidation number* (O.N.), which is the charge it would have if all the BONDS were ionic—the electrons in a covalent bond between two atoms are assigned to the atom with the higher ELECTRONEGATIVITY. Thus the sulfur atom in sulfur dioxide has an O.N. of +4, and each of the two oxygen atoms has an O.N. of −2. When sulfur is oxidized by oxygen, $S + O_2 \rightarrow SO_2$, its O.N. increases from 0 (for elements, by definition) to +4, corresponding to a virtual loss of electrons. (See also INDICATOR.)

OXIDES, binary compounds of OXYGEN with the other elements (see also PEROXIDES). All the elements form oxides except helium, neon, argon and krypton. Metal oxides are typically ionic crystalline solids (containing the O^{2-} ion), and are generally BASES, though the less electropositive metals form amphoteric oxides with acidic and basic properties (e.g., ALUMINUM oxide). Nonmetal oxides are covalent and typically volatile, though a few are macromolecular refractory solids (e.g., SILICON dioxide); most are acidic (see ACIDS), some are neutral (e.g., CARBON monoxide), and WATER is amphoteric. Oxides may be prepared by direct synthesis, or by heating hydroxides, nitrates or carbonates. Many oxide minerals are known; simple oxides are binary metal oxides, complex oxides contain several cations, and SPINELS are intermediate between the two. Some metal oxides (e.g., TITANIUM (II) oxide) are nonstoichiometric (see COMPOSITION, CHEMICAL).

OXIMES, derivatives of ALDEHYDES or KETONES (termed aldoximes and ketoximes respectively) formed by condensation with hydroxylamine (NH_2OH), and containing the group C=N—OH. Being easily isolated and with characteristic melting points, they are used for identification. Ketoximes undergo the BECKMANN rearrangement.

OXYACETYLENE. See ACETYLENE; WELDING.

OXY-ACIDS, those ACIDS that contain an acidic hydroxyl (see HYDROXIDES) group, i.e., whose acidic hydrogen is bound to oxygen. They include CARBOXYLIC ACIDS and PHENOLS, but are typically (e.g., sulfuric acid) the hydration products of acidic (nonmetal) OXIDES. Acids, such as hydrochloric acid, whose acidic hydrogen is bound to an element other than oxygen, are termed *hydracids*.

OXYGEN (O), gaseous nonmetal in Group VIA of the PERIODIC TABLE, comprising 21% by volume of the ATMOSPHERE and about 50% by weight of the earth's crust. It was first prepared by SCHEELE and PRIESTLEY, and named *oxygine* by LAVOISIER. Gaseous oxygen is colorless, odorless and tasteless; liquid oxygen is pale blue. Oxygen has two allotropes (see ALLOTROPY):

Oximes

formaldehyde oxime (formaldoxime) MW 45·04, mp 3°C	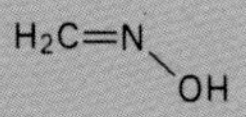
acetaldehyde oxime (acetaldoxime) MW 59·07, mp 47°C, bp 115°C	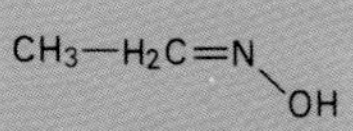
benzophenone oxime (diphenyl ketoxime) MW 197·24, mp 144°C	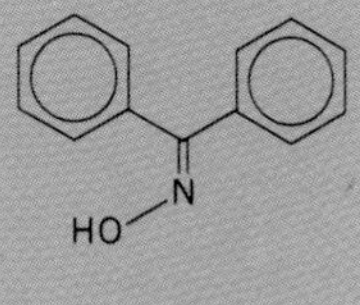
benzaldehyde oxime (*anti*) (*anti*-benzaldoxime) MW 121·14, mp 130°C	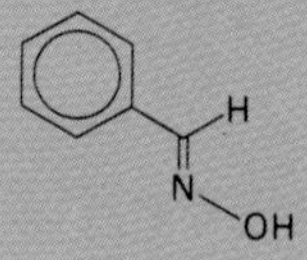
benzaldehyde oxime (*syn*) (*syn*-benzaldoxime) MW 121·14, mp 37°C, bp 200°C	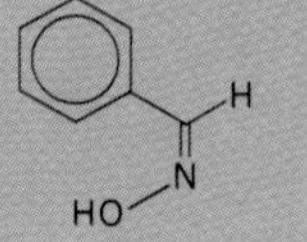

OZONE (O_3), which is metastable; and normal oxygen (O_2), which shows PARAMAGNETISM because its diatomic molecule has two electrons with unpaired spins. Oxygen is prepared in the laboratory by heating mercuric oxide or potassium chlorate (with manganese dioxide catalyst). It is produced industrially by fractional distillation of liquid air. Oxygen is very reactive, yielding OXIDES with almost all other elements, and in some cases PEROXIDES. Almost all life depends on chemical reactions with oxygen to produce energy. Animals receive oxygen from the air, as do fish from the water (see RESPIRATION): it is circulated through the body in the blood stream. The amount of oxygen in the air, however, remains constant because of PHOTOSYNTHESIS in plants and the decomposition by the sun's ultraviolet rays of water vapor in the upper atmosphere.

Oxygen is used in vast quantities in metallurgy: smelting and refining, especially of iron and steel. Oxygen and ACETYLENE are used in oxyacetylene torches for cutting and WELDING metals. Liquid oxygen is used in rocket fuels. Oxygen has many medical applications (see OXYGEN TENT; ANESTHESIA) and is used in mixtures breathed by divers and high-altitude fliers. It is also widely used in chemical synthesis. AW 16.0, mp −218°C, bp—183°C.

OXYGEN TENT, enclosed space, often made of plastic, in which a patient may be nursed in an atmosphere enriched with OXYGEN. It is mainly used for small children with acute respiratory DISEASES, or in adults when the use of a face mask is impractical.

OXYHEMOGLOBIN. See HEMOGLOBIN.

OXYTOCIN, HORMONE secreted by the posterior PITUITARY GLAND (see also HYPOTHALAMUS). It causes WOMB contraction and is used in OBSTETRICS. It is also concerned with milk secretion by the BREAST.

OZALID PROCESS, also known as the diazo or whiteprint process, a copying method in which paper coated with DIAZONIUM COMPOUNDS is exposed to ULTRAVIOLET LIGHT through a transparent original. Only the diazo in the shadows cast by the original survives to be developed with ammonia vapor to give a positive print.

OZONE (O_3), triatomic allotrope of OXYGEN (see ALLOTROPY); blue gas with a pungent odor. It is a very powerful oxdizing agent, and yields ozonides with ALKENES. It decomposes rapidly above 100°C. The upper ATMOSPHERE contains a layer of ozone, formed when ULTRAVIOLET RADIATION acts on oxygen; this layer protects the earth from the sun's ultraviolet rays. Ozone is made by subjecting oxygen to a high-voltage electric discharge. It is used for killing germs, bleaching, removing unpleasant odors from food, sterilizing water and in the production of azelaic acid. mp −193°C, bp—112°C.

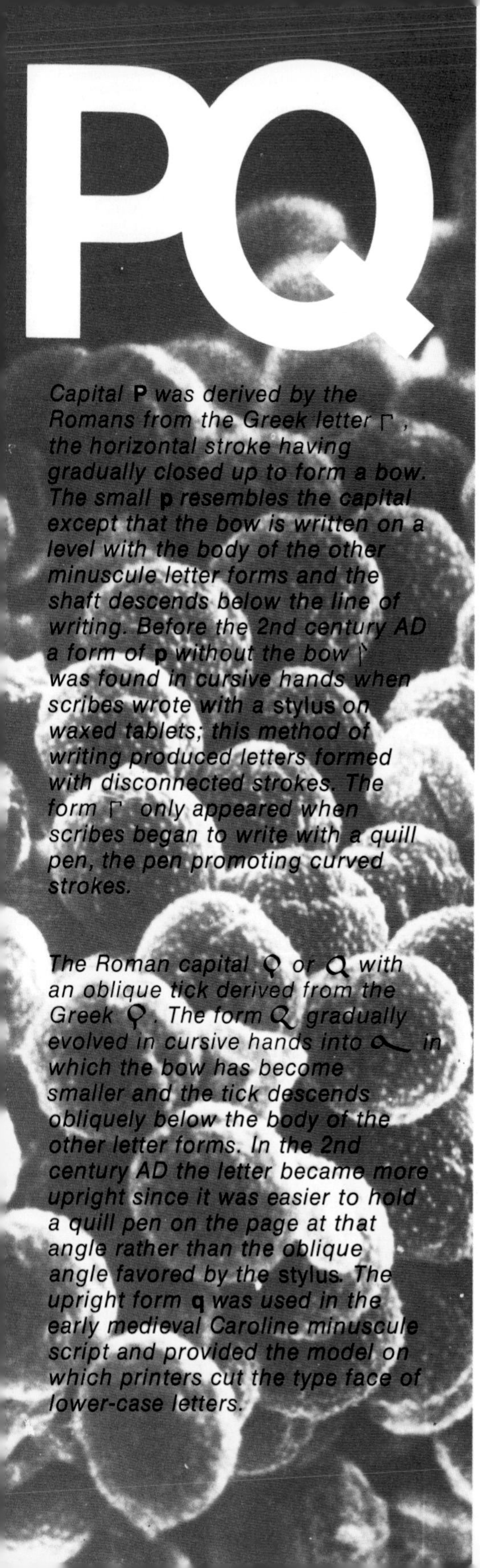

PQ

*Capital **P** was derived by the Romans from the Greek letter Γ, the horizontal stroke having gradually closed up to form a bow. The small **p** resembles the capital except that the bow is written on a level with the body of the other minuscule letter forms and the shaft descends below the line of writing. Before the 2nd century AD a form of **p** without the bow ꟾ was found in cursive hands when scribes wrote with a stylus on waxed tablets; this method of writing produced letters formed with disconnected strokes. The form Γ only appeared when scribes began to write with a quill pen, the pen promoting curved strokes.*

*The Roman capital Ϙ or Q with an oblique tick derived from the Greek Ϙ. The form Q gradually evolved in cursive hands into ɑ in which the bow has become smaller and the tick descends obliquely below the body of the other letter forms. In the 2nd century AD the letter became more upright since it was easier to hold a quill pen on the page at that angle rather than the oblique angle favored by the stylus. The upright form **q** was used in the early medieval Caroline minuscule script and provided the model on which printers cut the type face of lower-case letters.*

PACEMAKER. See PROSTHETICS.

PAHOEHOE, ropy lava. (See LAVA.)

PAIN, the detection by the nervous system of harmful stimuli. The function of pain is to warn the individual of imminent danger: even the most minor tissue damage will cause pain, so that avoiding action can be taken at a very early stage. The level at which pain can only just be felt is the *pain threshold*. This threshold level varies slightly between individuals, and can be raised by, for example, HYPNOSIS, anesthetics, ANALGESICS and the drinking of alcohol. In some psychological illnesses, especially the NEUROSES, it is lowered. The receptors of pain are unencapsulated nerve endings (see NERVOUS SYSTEM), distributed variably about the body: the back of the knee has about 230 per cm^2, the tip of the nose about 40. Deep pain, from the internal organs, may be felt as surface pain or in a different part of the body. This phenomenon, *referred pain*, is probably due to the closeness of the nerve tracts entering the SPINAL CORD. In psychoanalysis, "pain" refers to the distress felt when tension caused through frustration of INSTINCT goes unrelieved.

PAINT, a fluid applied to a surface in thin layers, forming a colored, solid coating for decoration, visual representation and protection (see also VARNISH). Paint consists of a PIGMENT dispersed in a "vehicle" or binder which adheres to the substrate and forms the solid film, and usually a solvent or thinner to control the consistency. Natural binders used, now or formerly, include GLUE, natural RESINS and OILS which dry by OXIDATION—linseed oil used to be the basis of the paint industry. These have been largely displaced by synthetic resins, latex and oils (to which drying agents are added). The solvents used are hydrocarbons or oils, except for the large class of water-thinned paints in which the binder forms an EMULSION or is dissolved in the water. Many specialized paints have been developed, e.g., to resist heat or corrosion. After applying a primer, the paint is brushed, rolled or sprayed on; dip coating and electrostatic attraction are more recent methods. (See also WHITEWASH.)

PALADE, Georg Emil (1912–   ), Romanian-born US cell biologist awarded with A. CLAUDE and C. de DUVE the 1974 Nobel Prize for Physiology or Medicine for their researches into intercellular structures. Palade discovered the nature of what are now called RIBOSOMES.

PALATE, structure dividing the mouth from the NOSE and bounded by the upper gums and TEETH; it is made of BONE and covered by mucous membrane. At the back, it is a soft mobile connective-tissue structure which can close off the naso-PHARYNX during swallowing and speech.

PALEOBOTANY. See PALEONTOLOGY.

PALEOCENE, the first epoch of the TERTIARY period, which extended between about 65 and 55 million years ago. (See also GEOLOGY.)

PALEOCLIMATOLOGY, the determination of climatic conditions of the geological past by study of the FOSSIL (*paleoecology*) and sedimentary (*sedimentology*) evidence.

PALEOGEOGRAPHY, the construction from geologic, paleontologic and other evidence of maps of parts or all of the earth's surface at specific times in the earth's past. Paleogeography has proved of

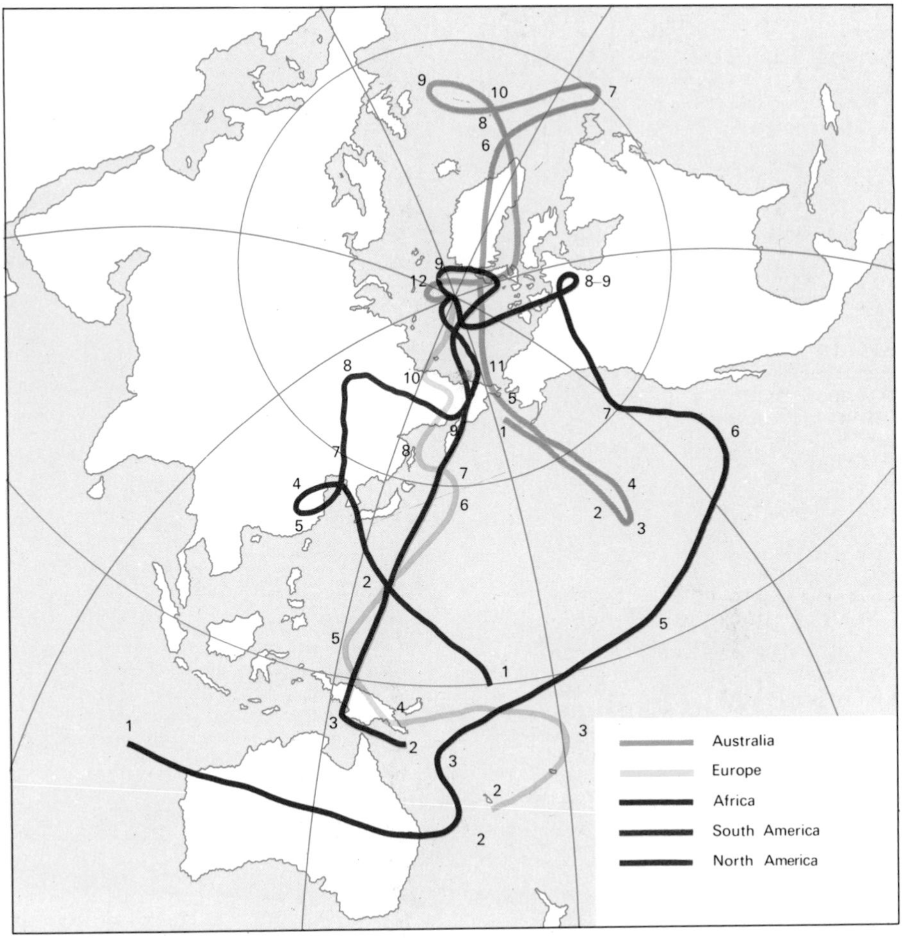

Polar wandering curves for the various continents through geological time based on paleomagnetic data. The numbers refer to: 1, Precambrian; 2, Cambrian; 3, Ordovician; 4, Silurian; 5, Devonian; 6, Carboniferous; 7, Permian; 8, Triassic; 9, Jurassic; 10, Cretaceous; 11, Tertiary; 12, Quaternary. Since these curves, which describe the apparent positions of the north magnetic pole as determined from different continents, are not only not colinear but also show remarkable differences of form, the only apparent explanation of polar wandering would seem to be that the continents have moved relative to each other.

considerable importance in CONTINENTAL DRIFT studies.

PALEOGRAPHY, the study of handwritten material from ancient and medieval times, excluding that on metal or stone (see EPIGRAPHY), for purposes of interpretation and the dating of events, and to trace the evolution of the written ALPHABET.

PALEOLITHIC AGE. See PRIMITIVE MAN; STONE AGE.

PALEOMAGNETISM, the study of past changes in the EARTH's magnetic field by examination of rocks containing certain iron-bearing minerals (e.g., HEMATITE, MAGNETITE). Reversals of the field and movements of the magnetic poles can be charted and information on CONTINENTAL DRIFT may be obtained. (See also SEA-FLOOR SPREADING.)

PALEONTOLOGY, or **paleobiology**, study of the remains of living organisms of past eras. The two branches are *paleobotany* and *paleozoology*, dealing with plants and animals respectively. Such studies are essential to STRATIGRAPHY, and provide important evidence for EVOLUTION and CONTINENTAL DRIFT theories. (See also FOSSILS; PALEOCLIMATOLOGY; RADIOCARBON DATING.)

PALEOZOIC, the earliest era of the PHANEROZOIC, comprising two sub-eras: the **Lower Paleozoic**, 570–400 million years ago, containing the CAMBRIAN, ORDOVICIAN and SILURIAN periods; and the **Upper Paleozoic**, 400–225 million years ago, containing the DEVONIAN, MISSISSIPPIAN, PENNSYLVANIAN and PERMIAN periods. (See GEOLOGY.)

PALEY, William (1743–1805), English theologian and utilitarian philosopher whose *Principles of Moral and Political Philosophy* (1785); *A View of the Evidences of Christianity* (1794), and *Natural Theology* (1802) featured largely in early 19th-century liberal education on both shores of the Atlantic.

PALLADIUM (Pd), white, soft, ductile metal in the PLATINUM GROUP. In addition to the general uses of these metals, palladium is used in dental and other ALLOYS. It absorbs 900 times its volume of hydrogen, forming a metallic HYDRIDE, and, being permeable to hydrogen at high temperatures, it is used in hydrogen purifiers. AW 106.4, m 1552°C, bp 2927°C, sg 11.97 (0°C).

PALLAS, Peter Simon (1741–1811), German naturalist best known for his *Travels through Various Provinces of the Russian Empire* (3 vols., 1771–76), an account of his 6-year expedition collecting extant and fossil plant and animal specimens.

PALMER, Daniel David (1845–1913), Canadian-born US founder of CHIROPRACTIC (1895).

PALMITIC ACID ($C_{15}H_{31}COOH$), CARBOXYLIC ACID whose glyceryl ESTER is a major component of FATS. Palmitates are used in SOAPS.

PALOMAR OBSERVATORY. See HALE OBSERVATORIES.

PALYNOLOGY, the study of fossil pollen and spores. These are relatively resistant to decay and can be used both to index the dates of recent strata and reconstruct the flora and climate prevailing when they were laid down.

PANCREAS, organ consisting partly of exocrine GLAND tissue, secreting into the DUODENUM, and partly of ENDOCRINE GLAND tissue (the *islets of Langerhans*), whose principal HORMONES include INSULIN and GLUCAGON. The pancreas lies on the back wall of the upper ABDOMEN, much of it within the duodenal loop. Powerful digestive-system ENZYMES (pepsin, trypsin, lipase, amylase) are secreted into the gut; this secretion is in part controlled by intestinal hormones (SECRETIN) and in part by nerve REFLEXES. Insulin and glucagon have important roles in glucose and fat METABOLISM (see DIABETES); other pancreatic hormones affect GASTROINTESTINAL-TRACT secretion and activity. Acute INFLAMMATION of the pancreas due to VIRUS disease, ALCOHOLISM or duct obstruction by gallstones, may lead to severe abdominal pain with SHOCK and prostration caused by the release of digestive enzymes into the abdomen. Chronic pancreatitis leads to functional impairment and malabsorption. CANCER of the pancreas may cause JAUNDICE by obstructing the BILE duct.

PANDEMIC. See EPIDEMIC.

PANGAEA, primeval supercontinent which, under plate tectonic action, split up to form Laurasia in the N and Gondwanaland in the S hemisphere (see GONDWANALAND; LAURASIA; PLATE TECTONICS). In turn these, too, split up to form our modern continents. (See also CONTINENTAL DRIFT.)

PANTOGRAPH, instrument for enlarging or reducing a geometric figure or motion. It consists of four hinged bars forming a parallelogram, one vertex being fixed. It is used in technical drawing and mapmaking. A spring-loaded pantograph linkage is used on electric locomotives to collect current from an overhead wire.

PANTOTHENIC ACID. See VITAMINS.

PAPANICOLAOU, George Nicholas (1883–1962), Greek-born US anatomist largely responsible for the development of cytologic PATHOLOGY (see also CYTOLOGY). He devised the PAP SMEAR TEST, used to

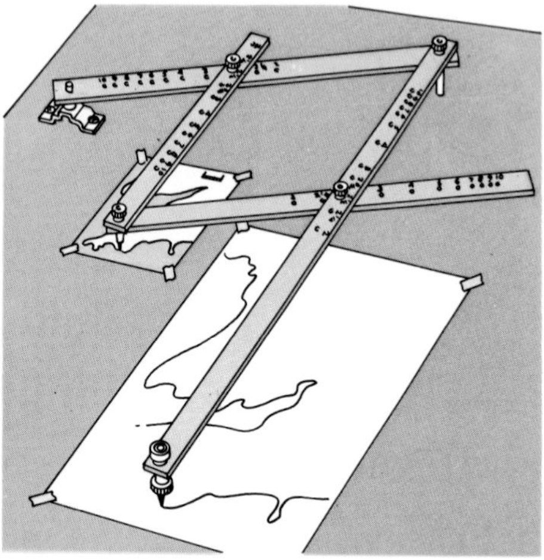

A draftsman's pantograph, here being used to reproduce a map outline at 2½ times the size of the original.

reveal early signs of CANCER.

PAPER, felted or matted sheets of CELLULOSE fibers, formed on a wire screen from a water suspension, and used for writing and printing on. Rags and cloth—still used for special high-grade papers—were the raw materials used until generally replaced by wood pulp processes developed in the mid-19th century. Logs are now pulped by three methods. Mechanical pulping normally uses a revolving grindstone. In full chemical pulping, wood chips are cooked under pressure in a solution that dissolves all but the cellulose: the kraft process uses alkaline sodium sulfide solution; the sulfite process uses various bisulfites with excess sulfur dioxide. Semichemical pulping employs mild chemical softening followed by mechanical grinding. The pulp is bleached, washed and refined—i.e., the fibers are crushed, frayed and cut by mechanical beaters. This increases their surface area and bonding power. At this stage various substances are added: fillers (mainly clay and chalk) to make the paper opaque, sizes (rosin and alum) for water resistance, and dyes and pigments as necessary. A dilute aqueous slurry of the pulp is fed to the paper machine, flowing onto a moving belt or cylindrical drum of fine wire mesh, most of the water being drained off by gravity and suction. The newly-formed continuous sheet is pressed between rollers, dried by evaporation, and subjected to CALENDERING. Some paper is coated to give a special surface.

PAP SMEAR TEST, or Papanicolaou test, CANCER screening test in which cells scraped from the cervix of the WOMB are examined for abnormality under the microscope using the method of G. N. PAPANICOLAOU.

PAPYRUS, or paper reed, *Cyperus papyrus*, a stout, reed-like sedge, used in ancient civilizations as a writing material. It was also used for making sails, baskets and clothing, and the pith as food. Family: Cyperaceae.

PARABOLA. See CONIC SECTIONS.

PARACELSUS, Philippus Aureolus (1493–1541), Swiss alchemist and physician who channeled the arts

A 25100-class dual-voltage mixed-traffic locomotive of the French National Railways (SNCF) with two single-armed pantographs raised ready for starting on a 1.5kV DC section of track. Immediately after moving off, when the danger is past of arcing on the high initial current surge, the forward pantograph is lowered for normal running. This precaution is not necessary on 25kV AC sections.

of ALCHEMY toward the preparation of medical remedies (see IATROCHEMISTRY). Born Theophrastus Bombast von Hohenheim, he adopted the name Paracelsus boasting that he was superior to CELSUS.

PARACHUTE, collapsible umbrella-like structure used to retard movement through the air. It was invented in the late-18th century, being used for descent from balloons, and made successively from canvas, silk and nylon. When opened—either manually by pulling a ripcord or by a line attached to the aircraft—the canopy fills with air, trapping a large air mass which, because of the parachute's movement, is at a higher pressure than that outside, producing a large retarding force. The canopy consists of numerous strong panels sewn together. Parachutes are used for safe descent of paratroops and others, for dropping airplanes or missiles, and returning space capsules. Sport parachuting, or skydiving, has become popular.

PARADOX, commonly a literary or rhetorical device whereby a supposedly true statement is couched for effect in apparently contradictory terms: e.g., "The last shall be first." A "theoretical paradox" is the conclusion of an apparently convincing argument that is apparently inconsistent with a generally accepted body of theory.

PARAFFINS. See ALKANES.

PARALLAX, the difference in observed direction of an object due to a difference in position of the observer. Parallax in nearby objects may be observed by closing each eye in turn so that the more distant object appears to move relative to the closer. The brain normally assembles these two images to produce a stereoscopic effect (see BINOCULAR VISION). Should the length and direction of the line between the two points of observation be known, parallax may be used to calculate the distance of the object. In astronomy, the parallax of a star is defined as half the greatest parallactic displacement when viewed from earth at different times of the year (see PARSEC).

PARALLELEPIPED, a hexahedron (see POLYHEDRON) whose opposite faces are parallel, each side being a parallelogram (see QUADRILATERAL). Should the sides be rectangular, the figure is a cuboid; should they be square, the figure is a CUBE.

PARALLEL LINES, LINES which would have to be extended until they were of infinite length in order to intersect.

PARALLELOGRAM. See QUADRILATERAL.

PARALYSIS, temporary or permanent loss of MUSCLE power or control. It may consist of inability to move a limb or part of a limb or individual muscles, paralysis of the muscles of breathing, swallowing and VOICE production being especially serious. Paralysis may be due to disease of the BRAIN (e.g., STROKE; TUMOR); SPINAL CORD (POLIOMYELITIS); nerve roots (SLIPPED DISK); peripheral NERVOUS SYSTEM (NEURITIS); neuromuscular junction (MYASTHENIA GRAVIS), or muscle (MUSCULAR DYSTROPHY). Disturbance of blood POTASSIUM levels can also lead to paralysis.

PARAMAGNETISM, weak magnetization of a material in the same direction as an applied MAGNETIC

Cytinus hypocistis is a flowering plant of the family Rafflesiaceae which parasitizes the roots of *Cistus* species. The specimen shown here has been dug up to reveal the roots of the host from which the fleshy red flowers of the parasite are emerging, the flowers being the only part to grow above ground level.

FIELD. Normally stronger than DIAMAGNETISM, the effect varies inversely with TEMPERATURE, and involves the partial alignment of intrinsic or orbital ELECTRON dipoles.

PARAMETER, a VARIABLE whose value determines a distinct set (see SET THEORY) of cases in a mathematical statement. Parametric equations are those expressed in terms of parameters. For example,

$$f(x) = ax^2 + b$$

is the general equation (see ANALYTIC GEOMETRY) of a set of points corresponding to a parabola (see CONIC SECTIONS), expressed in terms of the parameters a and b. Substitutions of specific values of these defines a particular set. In STATISTICS, the parameters of a distribution are those elements of it which can be used to loosely define it. They may be location parameters, such as the MEAN, MEDIAN AND MODE, or parameters of dispersion, such as the STANDARD DEVIATION.

PARANOIA, a PSYCHOSIS characterized by delusions of persecution (hence the popular term, **persecution mania**) and grandeur, often accompanied by HALLUCINATIONS. The delusions may form a self-consistent system which replaces reality. (See also MENTAL ILLNESS; SCHIZOPHRENIA.)

PARANTHROPUS. See PREHISTORIC MAN.

PARAPLEGIA, PARALYSIS involving the lower part of the body, particularly the legs. Injury to the SPINAL CHORD is often the cause.

PARAPSYCHOLOGY, or **Psychic Research**, a field of study concerned with scientific evaluation of two distinct types of phenomena: those collectively termed ESP, and those concerned with life after death, reincarnation, etc., particularly including claims to communication with souls of the dead (spiritism or, incorrectly, spiritualism). Tests of the former have generally been inconclusive, of the latter almost exclusively negative. But in both cases many "believers" hold that such phenomena, being beyond the bounds of science, cannot be subjected to laboratory evaluation. In spiritism, the prime site of the alleged communication is the séance, in which one individual (the medium) goes into a trance before communicating with the souls of the dead, often through a spirit guide (a spirit associated particularly with the medium). The astonishing disparity between different accounts of the spirit world has led to the whole field being treated with skepticism.

PARASITE, an organism that is for some part of its life-history physiologically dependent on another, the host, from which it obtains nutrition and which may form its total environment. Nearly all the major groups of animals and plants from viruses to vertebrates and BACTERIA and angiosperms, have some parasitic members. The most important parasites, besides the viruses which are a wholly parasitic group, occur in the bacteria, Protozoa, Flatworms and Roundworms. Study of the parasitic worms, the platyhelminths, nematodes and acanthocephalans, is termed helminthology. Blood-sucking arthropods, such as mosquitoes, Tsetse flies and ticks, are also important because they transmit PARASITIC DISEASES and serve as vectors or transport-hosts for other parasites.

PARASITIC DISEASES, infestation or infection by PARASITES, usually referring to nonbacterial and nonviral agents (i.e., to PROTOZOA and helminths). MALARIA, LEISHMANIASIS, trypanosomiasis (see TRYPANOSOMES), CHAGAS' DISEASE, FILARIASIS, SCHISTOSOMIASIS, toxoplasmosis, amebiasis and TAPE WORM are common examples. Manifestations may depend on the life cycle of the parasite; animal or insect vectors are usual. CHEMOTHERAPY is often effective in treatment.

PARATHYROID GLANDS, a set of four small ENDOCRINE GLANDS lying behind the THYROID which regulate CALCIUM metabolism. Parathyroid hormone releases calcium from BONE and alters the intestinal absorption and KIDNEY excretion of calcium and phosphorus. Disease or loss of parathyroid glands may lead to TETANY, CATARACT or mental changes, and may be associated with disorders of IMMUNITY. Parathyroid overactivity or TUMORS cause raised blood calcium leading to bone disease, kidney disease (including stones) and mental abnormalities.

PARATYPHOID FEVER, BACTERIAL DISEASE similar to TYPHOID FEVER and caused by a related organism, causing FEVER, DIARRHEA, rash, SPLEEN and LYMPH node enlargement. Spread is by cases or carriers and contaminated food; ANTIBIOTICS may be helpful in treatment.

PARAZOA, a phylum containing the sponges, the simplest multicellular animals, consisting of a hollow body composed of cells that are not organized into organs or well developed tissue systems. There are some 2 500 species. They live in seas, lakes and rivers, and vary in size from less than 12mm (0.5in) to 1.8m (6ft). They occur as solitary individuals or are grouped together in colonies.

PARCHMENT, the skin of sheep, ewes or lambs, cleaned, polished, stretched and dried to make a material which can be written on, and also used to make drums and for bookbinding. Invented in the 2nd century BC as a substitute for PAPYRUS, it was widely used for manuscripts until superseded by paper in the 15th century, except for legal documents. **Vellum** is fine-quality parchment made from lamb, kid or calf skin. Both terms are now applied to high-quality paper. Vegetable parchment is paper immersed briefly in sulfuric acid and so made strong and parchment-like.

PARÉ, Ambroise (c1510–1590), French surgeon whose many achievements (e.g., adopting ligatures or liniments in place of CAUTERIZATION; introducing the use of artificial limbs and organs) have earned him regard as a father of modern SURGERY.

PAREGORIC, tincture of OPIUM, a NARCOTIC ANALGESIC.

PARESIS, muscular weakness of a part, usually used in distinction from PARALYSIS, which implies complete loss of muscle power. It may be due to disease of the BRAIN, SPINAL CORD, peripheral NERVOUS SYSTEM or MUSCLES.

PARIS GREEN, or **emerald green,** toxic, bright blue-green PIGMENT, copper acetoarsenate(III), $Cu_2(CH_3COO)AsO_3$; now used as a pesticide and wood preservative.

PARITY, physical property of a wave function (see WAVE MECHANICS) in QUANTUM MECHANICS specifying the function's behavior when its spatial coordinates are simultaneously reflected through the origin. If the parity is even, the wave function is unchanged by changing the sign of its coordinates; if it is odd, the wave function's sign changes. Parity has no significance in classical physics, but essentially arises from the symmetry of space. Strong nuclear and electromagnetic interactions conserve parity because they are governed by physical laws which do not distinguish between a right- or left-handed coordinate system (as in nature, where an object and its mirror image are equally realizable). Parity is not conserved in weak nuclear interactions.

PARKINSON'S DISEASE, a common disorder in the elderly, causing a characteristic mask-like facial appearance, shuffling gait, slowness to move, muscular rigidity and tremor at rest; mental ability is preserved except in those cases following ENCEPHALITIS lethargica. It is a disorder of the basal ganglia of the BRAIN and may be substantially helped by DRUGS (e.g., L-Dopa) that affect impulse transmission in these sites.

PARMENIDES (flourished c475 BC), Greek philosopher of Elea in southern Italy; foremost of the Eleatics. His philosophy, anchored on the proposition "What is *is*," denied the reality of multiplicity and change. His uncompromising attempt to deduce the properties of the Real—a single eternal solid, all-embracing yet undifferentiated—marks the beginning of the Western tradition of philosophical reasoning. (See also PRE-SOCRATICS.)

PAROTID GLANDS. See SALIVA.

PARROT FEVER. See PSITTACOSIS.

PARSEC (pc), in astronomy, the distance at which 1 ASTRONOMICAL UNIT would subtend an angle of 1 second. Originally defined as the distance of a star with a PARALLAX of 1″ viewed from earth and sun, the parsec was introduced to replace the LIGHT YEAR. $1pc = 3.258ly = 206\,265AU$.

PARTHENOGENESIS, sometimes popularly termed virgin birth, the development into a new individual of an ovum that has not been fertilized by a sperm. It is usually regarded as an aberrant form of sexual reproduction.

PARTICLES, Elementary. See SUBATOMIC PARTICLES.

PASCAL, Blaise (1623–1662), French mathematician, physicist and religious philosopher. Though not the first to study what is now called PASCAL'S TRIANGLE, he was first to use it in PROBABILITY studies, the mathematical treatment of which he and FERMAT evolved together, though in different ways. His studies of the CYCLOID inspired others to formulate the CALCULUS. His experiments (performed by his brother-in-law) observing the heights of the column of a BAROMETER at different altitudes on the mountain Puy-de-Dôme (1646) confirmed that the atmospheric air had weight. He also pioneered HYDRODYNAMICS and HYDROSTATICS, in doing so discovering PASCAL'S LAW, the basis of HYDRAULICS. His religious thought was dominated by his association with the Jansenist convent of Port Royal, and is expressed in his *Provincial Letters* (1656–57) and his posthumously published *Pensées* (1670 onward).

PASCAL (Pa), the SI UNIT of PRESSURE, being that due to a FORCE of one NEWTON acting per square metre.

PASCAL'S LAW, in HYDROSTATICS, states that the pressure in an enclosed body of fluid arising from forces applied to its boundaries is transmitted equally in all directions with unchanged intensity. This pressure acts at right angles to the surface of the fluid container.

PASCAL'S TRIANGLE, a tabular arrangement of the COEFFICIENTS of the terms in the expansion of a BINOMIAL raised to the POWERS 0 to n (see BINOMIAL THEOREM).

For $(a+b)^n$ where $n = 0, 1, 2, 3, \ldots$, the coefficients are

$(a+b)^0$	1
$(a+b)^1$	1 1
$(a+b)^2$	1 2 1
$(a+b)^3$	1 3 3 1
$(a+b)^4$	1 4 6 4 1
$(a+b)^5$	1 5 10 10 5 1
$(a+b)^6$	1 6 15 20 15 6 1

etc. The first and last figures in each row are 1; also, each figure is the SUM of the two to its right and left in the row above it. Moreover, the sum of each row is equal to the corresponding power of 2—$1+3+3+1 = 8 = 2^3$, etc.—and each row read across is numerically equal to the corresponding power of 11 ($1\,331 = 11^3$).

PASCHEN, Friedrich (1865–1947), German physicist and a pioneer of SPECTROSCOPY, best known for his experimental work on infrared spectra and for explaining the "Paschen-Back Effect"—which concerns the splitting of spectral lines in an intense MAGNETIC FIELD.

PASTEUR, Louis (1822–1895), French microbiologist and chemist. In his early pioneering studies in STEREOCHEMISTRY he discovered optical ISOMERISM. His attentions then centered around FERMENTATION, in which he demonstrated the role of microorganisms. He developed PASTEURIZATION as a way of stopping wine and beer from souring, and experimentally disproved the theory of SPONTANEOUS GENERATION. His "germ theory" of DISEASE proposed that diseases are spread by living germs (i.e., BACTERIA); and his consequent popularization of the STERILIZATION of medical equipment saved many lives. While studying ANTHRAX in cattle and sheep he developed a form of VACCINATION rather different from that of JENNER: he found inocculation with dead anthrax germs gave future IMMUNITY from the disease. Treating RABIES similarly, he concluded that it was caused by a germ too small to be seen—i.e., a VIRUS. The Pasteur Institute was founded in 1888 to lead the fight against rabies.

PASTEURIZATION, a process for partially sterilizing MILK originally invented by L. PASTEUR for improving the storage qualities of wine and beer. Originally the milk was held at 63°C for 30min in a vat. But today the usual method is a continuous process whereby the milk is held at 72°–85°C for 16s. Disease-producing BACTERIA, particularly those causing TUBERCULOSIS, are thus destroyed with a minimum effect on the flavor of the product. Since the process also destroys a majority of the harmless bacteria which sour milk, its keeping properties are also improved.

PATCH TEST, test used in investigation of skin and systemic ALLERGY, especially contact DERMATITIS. Patches of known or likely sensitizing substances are placed on the SKIN for a short period. Local ERYTHEMA or HIVES indicate allergy.

PATENT, a grant of certain specified rights by the government of a particular country, usually to a person whose claim to be the true and first inventor of a new invention (or the discoverer of a new process) is upheld. Criteria of the "novelty" of an invention are defined in law. The term derives from "letters patent"—the "open letters" by which a sovereign traditionally confers a special privilege or right on a subject. An inventor (or his assignee) who files an application for and is granted a patent is exclusively entitled to make, use or sell his invention for a limited period—17 years from the granting date, in the US and Canada. By granting the inventor a temporary monopoly, patent law aims to stimulate inventive activity and the rapid exploitation of new inventions for the public benefit.

PATHOLOGY, study of the ANATOMY of DISEASE. Morbid anatomy, the dissection of bodies after DEATH with a view to discovering the cause of disease and the nature of its manifestations is complemented and extended by HISTOLOGY. In addition to AUTOPSY, BIOPSIES and surgical specimens are examined; these provide information that may guide treatment. It has been said that pathology is to MEDICINE what anatomy is to PHYSIOLOGY.

PATINA, the attractive thin film of CORROSION products formed on metals by weathering, especially the green tarnish on copper or BRONZE.

PAULI, Wolfgang (1900–1958), Austrian-born physicist awarded the 1945 Nobel Prize for Physics for his discovery of the Pauli EXCLUSION PRINCIPLE, that no two fermions (see FERMI-DIRAC STATISTICS) in a system may have the same four quantum numbers. In terms of the ATOM, this means that at most two electrons may occupy the same ORBITAL (the two having opposite SPIN).

PAULING, Linus Carl (1901–), US chemist and pacifist awarded the 1954 Nobel Prize for Chemistry for his work on the chemical BOND (see also ELECTRONEGATIVITY) and the 1962 Nobel Peace Prize for his support of unilateral nuclear disarmament.

PAVLOV, Ivan Petrovich (1849–1936), Russian physiologist best known for his work on the conditioned REFLEX. Regularly, over long periods, he rang a bell just before feeding dogs, and found that

A cluster of cultured pearls formed in an artificially stimulated oyster.

eventually they salivated on hearing the bell, even when there was no food forthcoming. He also studied the physiology of the DIGESTIVE SYSTEM, and for this received the 1904 Nobel Prize for Physiology or Medicine.

PAWL. See RATCHET.

PEANO, Giuseppe (1858–1932), Italian mathematician whose ideas in mathematical LOGIC profoundly influenced those of A. N. WHITEHEAD and B. RUSSELL. He also devised the international language *Interlingua*, still occasionally used.

PEARLS, white spherical gems produced by bivalve mollusks, particularly by Pearl oysters, *Pinctada*. In response to an irritation by foreign matter within the shell, the mantle secretes calcium carbonate in the form of nacre (MOTHER OF PEARL) around the irritant body. Over several years, this encrustation forms the pearl. Cultured pearls may be obtained by "seeding" the oyster with an artificial irritant such as a small bead. Pearls are variable and may be black or pink as well as the usual white. Another bivalve group producing marketable pearls is the freshwater Pearl mussel, *Margaritifera margaritifera*.

PEARSON, Karl (1857–1936), British mathematician best known for his pioneering work on STATISTICS (e.g., devising the CHI-SQUARED TEST) and for his *The Grammar of Science* (1892), an important contribution to the philosophy of mathematics. He was also an early worker in the field of EUGENICS.

PEAT, partly decayed plant material found in layers, usually in marshy areas. It is composed mainly of the peat mosses SPHAGNUM and Hypnum, but also of sedges, trees, etc. Under the right geological conditions, peat forms COAL. It is used as a MULCH and burned for domestic heating.

PECK ORDER, the term given to a DOMINANCE

HIERARCHY in birds. The top bird can peck all others; the second can peck all but the top bird, and so on down to the bottom bird who is pecked by all but can peck none. Frenzied pecking soon decides the rank of any new bird introduced to the group.

PECTIN, a polysaccharide found in plant tissue, capable of forming thick GELS with strong, acid SUGAR solutions and extensively used in the food industry to set JELLIES AND JAMS. Commercial quantities are obtained from citrus and apple wastes after removal of the juice.

PEDIATRICS, branch of MEDICINE concerned with care of children. This starts with newborn, especially premature, babies in whom intensive care is required to protect the baby from and adapt it to the environment outside the WOMB. An important aspect is the recognition and treatment of congenital DISEASES in which structural or functional defects occur due to inherited disease (e.g., MONGOLISM) or disease acquired during development of EMBRYO or FETUS (e.g., SPINA BIFIDA). Otherwise, INFECTIOUS DISEASE, failure to grow or develop normally, MENTAL RETARDATION, DIABETES, ASTHMA and EPILEPSY form the bulk of pediatric practice.

PEDICEL. See FLOWER.

PEGASUS (the Winged Horse), a large N Hemisphere constellation noticeable for its Great Square, formed by four bright stars. Pegasus contains a GLOBULAR CLUSTER (M15).

PEGMATITE, very coarse-grained IGNEOUS or METAMORPHIC ROCK formed by slow crystallization from a melt containing volatiles. They usually have the composition of GRANITE, but often contain unusual MINERALS, including GEMS, and rare elements.

PEIRCE, Charles Sanders (1839–1914), US philosopher, best known as a pioneer of PRAGMATISM. He also made important contributions to LOGIC and the philosophy of science, being also a noteworthy mathematician and experimental scientist.

PEKING MAN. See PREHISTORIC MAN.

PELLAGRA, VITAMIN deficiency DISEASE (due to lack of niacin), often found in maize- or millet-dependent populations. A DERMATITIS, initially resembling sunburn, but followed by thickening, scaling and pigmentation, is characteristic; internal EPITHELIUM is affected (sore tongue, DIARRHEA). Confusion, DELIRIUM, hallucination and ultimately dementia may ensue. Niacin replacement is essential and food enrichment is an important preventative measure.

PELTIER EFFECT, the heating or cooling effect at a junction between two dissimilar METALS when an electric current is driven through a circuit containing the junction. The effect is named for **Jean Charles Athanase Peltier** (1785–1845), the French scientist who discovered it in 1834. (See also THERMOCOUPLE.)

PELVIS, lowest part of the trunk in animals, bounded by the pelvic BONES and in continuity with the ABDOMEN. The principal contents are the BLADDER and lower GASTROINTESTINAL TRACT (rectum) and reproductive organs, particularly in females—the WOMB, OVARIES, FALLOPIAN TUBES and vagina. The pelvic floor is a powerful muscular layer which supports the pelvic and abdominal contents and is important in urinary and fecal continence. The pelvic bones articulate with the LEGS at the hip JOINTS.

PEN, an instrument for writing with INK. The earliest pens were the Chinese brush and the Egyptian reed pen for use on PAPYRUS. Quill pens, usually made from goose feathers, were used until the mid-19th century, when steel pens largely replaced them. The modern fountain pen was invented in 1884; its nib is supplied with ink from a reservoir in the barrel by capillary action. The BALLPOINT PEN is now very popular.

PENCIL, instrument for writing or drawing, usually consisting of a wooden rod with a core of mixed powdered GRAPHITE and CLAY. The mixture is extruded as a soft paste and placed in the grooves of two half-pencils, which are glued together and dried. The term "lead pencil" comes from an early view that graphite is a form of lead. Pencils vary in hardness, the hardest (10H) containing the most clay, the softest and blackest (8B), the least. HB and F are intermediate grades.

PENDULUM, a rigid body mounted on a fixed horizontal axis that is free to rotate under the influence of gravity. Many types of pendulum exist (e.g., Kater's and the FOUCAULT PENDULUM), the most common consisting of a large weight (the bob) supported at the end of a light string or bar. An idealized simple pendulum, with a string of negligible weight and length, l, the weight of its bob concentrated at a point and a small swing amplitude, executes SIMPLE HARMONIC MOTION. The time, T, for a complete swing (to and fro) is given by $T = 2\pi\sqrt{l/g}$, depending only on the string length and the local value of the gravitational ACCELERATION, g. Actual physical or compound pendulums approximate this behavior if they have a small angle of swing. They are used for measuring absolute values of g or its variation with geographical position, and as control elements in CLOCKS.

PENEPLAIN, the state that William Morris DAVIS (1850–1934) proposed would result after millions of years of constant EROSION in an area: a featureless, perfectly horizontal plain at sea-level.

PENICILLIN, substance produced by a class of FUNGI which interferes with cell wall production by BACTERIA and which was one of the first, and remains

Penicillin: (A) the common penicillin nucleus (6-aminopenicillamic acid; R is an acyl side chain); (B) G or benzyl penicillin; (C) cloxacillin.

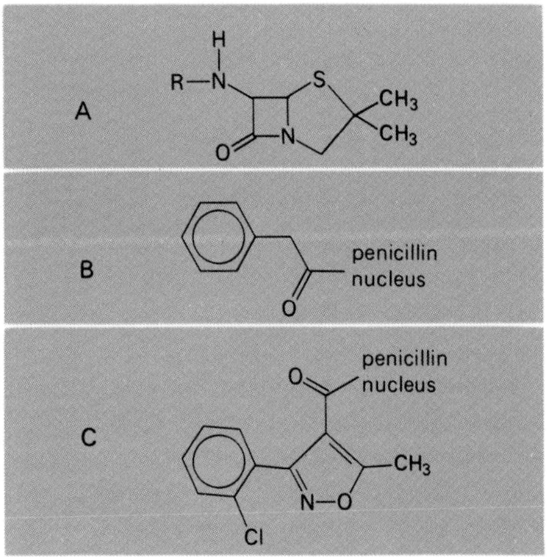

among the most useful, ANTIBIOTICS. The property was noted by A. FLEMING in 1928 and production of penicillin for medical use was started by E. B. CHAIN and H. W. FLOREY in 1940. Since then numerous penicillin derivatives have been manufactured, extending the range of activity, overcoming resistance in some organisms and allowing some to be taken by mouth. STAPHYLCOCCUS, STREPTOCOCCUS and the bacteria causing the VENEREAL DISEASES of gonorrhea and syphilis are among the bacteria sensitive to natural penicillin, while bacilli negative to GRAM's STAIN, which cause urinary-tract infection, SEPTICEMIA, etc., are destroyed by semisynthetic penicillins.

PENIS, male reproductive organ for introducing sperm and semen into the female vagina and WOMB; its urethra also carries URINE from the BLADDER. The penis is made of connective tissue and specialized blood vessels which become engorged with BLOOD in sexual arousal and which cause the penis to become stiff and erect; this facilitates the intromission of semen in sexual intercourse. A protective fold, the foreskin, covers the tip and is often removed for ethnic or medical reasons in circumcision.

PENNSYLVANIAN, the penultimate period of the PALEOZOIC, stretching between about 315 and 280 million years ago. (See CARBONIFEROUS; GEOLOGY.)

PENTAGON. See POLYGON.

PENTLANDITE, bronze SULFIDE mineral with metallic luster, of composition $(Fe,Ni)_9S_8$; the chief ore of NICKEL. Usually found with PYRRHOTITE, its major occurrence is at Sudbury, Ontario.

PENTOTHAL SODIUM, or **thiopentone,** a BARBITURATE drug injected into a vein to produce brief general ANESTHESIA, also used in PSYCHIATRY as a relaxant to remove inhibitions (a so-called "truth drug").

PENUMBRA. See SHADOW.

PEPSIN, an ENZYME which breaks down PROTEINS in the DIGESTIVE SYSTEM.

PEPTIDE, a compound containing two or more AMINO ACIDS linked through the amino group ($-NH_2$) of one acid and the carboxyl group (-COOH) of the other. The linkage -NH-CO- is termed a peptide bond. Peptides containing two amino acids are called dipeptides; with three, tripeptides, and so on; those with many acids are polypeptides.

PERCENT (%—from Latin *per centum*, by one hundred), the expression of a RATIO or FRACTION by setting $a:b=p:100$, where $a:b$ is the original ratio, and p its expression as a percentage.

PERCEPTION, the recognition or identification of something. External perception relies on the SENSES, internal perception, which is introverted, relying on the CONSCIOUSNESS. Some psychologists hold that perception need not be CONSCIOUS: in particular, subliminal perception involves reaction of the UNCONSCIOUS to external stimuli and its subsequent influencing of the conscious (see also SUGGESTION).

PERENNIAL, any plant that continues to grow for more than two years. Trees and shrubs are examples of the perennials that have woody stems that thicken with age. The herbaceous perennials such as the peony and daffodil have stems that die down each winter and regrow in the spring from underground perennating organs, such as TUBERS and BULBS. (See ANNUAL; BIENNIAL.)

PERFECT NUMBER, a NATURAL NUMBER equal to the SUM of its FACTORS. Two such numbers are 6 (divisible by 1, 2, 3, and $1+2+3=6$) and 28 (1, 2, 7, 4, 14). Only 23 perfect numbers are known.

PERFUME, a blend of substances made from plant oils and synthetic materials which produce a pleasant odor. Perfumes were used in ancient times as INCENSE in religious rites, in medicines and later for adornment. Today they are utilized in cosmetics, toilet waters, soaps and detergents, and polishes. A main source of perfumes are ESSENTIAL OILS extracted from different parts of plants, e.g., the flowers of the rose, the leaves of lavender, cinnamon from bark and pine from wood. They are extracted by steam distillation; by using volatile solvents; by coating petals with fat, or by pressing. Animal products, such as AMBERGRIS from the sperm whale, are used as fixatives to preserve fragrance. The development of synthetic perfumes began in the 19th century. There are now a number of synthetic chemicals with flower-like fragrance, for example citronellol for rose and benzyl acetate for jasmine.

PERIANTH. See FLOWER.

PERICARDIUM, two thin connective-tissue layers covering the HEART surface. It may become inflamed due to VIRUS or bacterial infection and in UREMIA.

PERIDOT. See OLIVINE.

PERIDOTITE, dark IGNEOUS ROCK consisting mainly of OLIVINE with some PYROXENE and HORNBLENDE but little FELDSPAR; it alters to SERPENTINE. Some varieties bear chromium ore, platinum or diamonds (see KIMBERLITE).

PERIGEE. See ORBIT.

PERIHELION. See ORBIT.

PERIODIC TABLE, a table of the ELEMENTS in order of atomic number (see ATOM), arranged in rows and columns to illustrate periodic similarities and trends in physical and chemical properties. Such classification of the elements began in the early 19th century, when Johann Wolfgang Döbereiner (1780–1849) discovered certain "triads" of similar elements (e.g. calcium, strontium, barium) whose atomic weights were in arithmetic progression. By the 1860s many more elements were known, and their atomic weights determined, and it was noted by John Alexander Reina Newlands (1837–1898) that similar elements recur at intervals of eight—his "law of octaves"—in a sequence in order of atomic weight. In 1869 MENDELEYEV published the first fairly complete periodic table, based on his discovery that the properties of the elements vary periodically with atomic weight. There were gaps in the table corresponding to elements then unknown, whose properties Mendeleyev predicted with remarkable accuracy. Modern understanding of atomic structure has shown that the numbers and arrangement of the electrons in the atom are responsible for the periodicity of properties; hence the atomic number, rather than the atomic weight, is the basis of ordering. Each row, or period, of the table corresponds to the filling of an electron "shell"; hence the numbers of elements in the periods is 2, 8, 8, 18, 18, 32, 32. (There are n^2 ORBITALS in the nth shell). The elements are arranged in vertical columns or groups containing those of similar atomic structure and properties, with regular gradation of properties down each group. The longer groups, with members in the first three (short)

Periodic table of the elements. The nonmetals occupy the top right-hand corner, where electronegativity is greatest. Melting and boiling points, too, tend to be lower toward the top right. The position of hydrogen is arbitrary; it does not properly fit into any group. In terms of atomic structure, the table is divided into four blocks, *s, p, d* and *f*, according to which kind of orbital is being filled. Helium, although a noble gass, can be considered an *s*-block element in that it contains only *s*-orbitals. The names here offered for elements 104 and 105 are not as yet officially confirmed. Element 104 is also known as kurchatovium, 105 as flerovium. There is considerable interest in the possible existence of yet heavier elements, particularly those of mass numbers 110, 114 and 126. It is believed that certain isotopes of these elements may be relatively stable.

period \ group	Ia	IIa	IIIb	IVb	Vb	VIb	VIIb	VIII	VIII	VIII	Ib	IIb	IIIa	IVa	Va	VIa	VIIa	0
1	1 H hydrogen																	2 He helium
2	3 Li lithium	4 Be beryllium											5 B boron	6 C carbon	7 N nitrogen	8 O oxygen	9 F fluorine	10 Ne neon
3	11 Na sodium	12 Mg magnesium											13 Al aluminum	14 Si silicon	15 P phosphorus	16 S sulfur	17 Cl chlorine	18 Ar argon
4	19 K potassium	20 Ca calcium	21 Sc scandium	22 Ti titanium	23 V vanadium	24 Cr chromium	25 Mn manganese	26 Fe iron	27 Co cobalt	28 Ni nickel	29 Cu copper	30 Zn zinc	31 Ga gallium	32 Ge germanium	33 As arsenic	34 Se selenium	35 Br bromine	36 Kr krypton
5	37 Rb rubidium	38 Sr strontium	39 Y yttrium	40 Zr zirconium	41 Nb niobium	42 Mo molybdenum	43 Tc technetium	44 Ru ruthenium	45 Rh rhodium	46 Pd palladium	47 Ag silver	48 Cd cadmium	49 In indium	50 Sn tin	51 Sb antimony	52 Te tellurium	53 I iodine	54 Xe xenon
6	55 Cs caesium	56 Ba barium	57 La lanthanum	72 Hf hafnium	73 Ta tantalum	74 W tungsten	75 Re rhenium	76 Os osmium	77 Ir indium	78 Pt platinum	79 Au gold	80 Hg mercury	81 Tl thallium	82 Pb lead	83 Bi bismuth	84 Po polonium	85 At astatine	86 Rn radon
7	87 Fr francium	88 Ra radium	89 Ac actinium	104 *Rf* *rutherfordium*	105 *Ha* *hahnium*	106												

58 Ce cerium	59 Pr praseodymium	60 Nd neodymium	61 Pm promethium	62 Sm samarium	63 Eu europium	64 Gd gadolinium	65 Tb terbium	66 Dy dysprosium	67 Ho holmium	68 Er erbium	69 Tm thulium	70 Yb ytterbium	71 Lu lutetium
90 Th thorium	91 Pa protactinium	92 U uranium	93 Np neptunium	94 Pu plutonium	95 Am americium	96 Cm curium	97 Bk berkelium	98 Cf californium	99 Es einsteinium	100 Fm fermium	101 Md mendelevium	102 No nobelium	103 Lr lawrencium

metals

metalloids

nonmetals

all isotopes radioactive

periods, are known as the Main Groups, usually numbered IA to VIIA, and 0 for the NOBLE GASES. The remaining groups, the TRANSITION ELEMENTS, are numbered IIIB to VIII (a triple group), IB and IIB. The characteristic VALENCE of each group is equal to its number $\mathcal{N}$, or to $(8-\mathcal{N})$ for some nonmetals. Two series of 14 elements each, the LANTHANIDES and ACTINIDES, form a hyper-transition block in which the inner *f* ORBITALS are being filled; their members have similar properties, and they are usually counted in Group IIIB. (See also TRANSURANIUM ELEMENTS.)

PERIODONTICS, branch of DENTISTRY concerned with the structures which fix the teeth in the jaw.

PERIPATETIC SCHOOL, in philosophy, the name given to the school of philosophy founded by ARISTOTLE and THEOPHRASTUS. The term derives from the covered arcade (*peripatos*) at the Lyceum where Aristotle taught in Athens.

PERIPATUS, An Australasian animal of the class ONYCHOPHORA.

PERISCOPE, optical instrument that permits an observer to view his surroundings along a displaced axis, and hence from a concealed, protected or submerged position. The simplest periscope, used in tanks, has two parallel reflecting surfaces (prisms or mirrors). An auxiliary telescopic gunsight may be added. Submarine periscopes have a series of lenses within the tube to widen the field of view, crosswires and a range-finder, and can rotate and retract.

PERISSODACTYLA, an order of the MAMMALIA that includes the rhinoceroses, tapirs and horses. All are odd-toed, the weight of the body being supported by the middle toe of each foot.

PERISTALSIS, the coordinated movements of hollow visceral organs, especially the GASTROINTESTINAL TRACT, which cause forward propulsion and mixing of the contents. It is effected by autonomic NERVOUS SYSTEM plexuses acting on visceral MUSCLE layers.

PERITONEUM, two thin layers of connective tissue lining the outer surface of the abdominal organs and the inner walls of the ABDOMEN. A small amount of fluid lies between them in an extensive potential space, allowing free movement of the organs over each other.

PERITONITIS, INFLAMMATION of PERITONEUM, usually caused by BACTERIAL INFECTION or chemical irritation of peritoneum when internal organs become diseased (as with APPENDICITIS) or when GASTROINTESTINAL TRACT contents escape (as with a perforated peptic ULCER). Characteristic pain, sometimes with SHOCK, FEVER, and temporary cessation of bowel activity (ileus), are common. Urgent treatment of the cause is required, often with SURGERY; ANTIBIOTICS may also be needed.

PERKIN, Sir William Henry (1838–1907), English chemist who, in 1856, while studying under von HOFMANN, discovered mauve, the first synthetic dye (see DYES AND DYEING). He manufactured this and other dyes until 1874, then devoted his remaining years to research. In 1868 he synthesized coumarin, the first synthetic PERFUME.

PERMAFROST, permanently frozen ground, typical of the treeless plains of Siberia (see TUNDRA), though common throughout polar regions to depths of as much as 600m.

PERMALLOY, an ALLOY of iron and nickel, often with 5% molybdenum; it has a very high magnetic PERMEABILITY and is used in TRANSFORMERS.

PERMEABILITY (μ), the ratio of the electromagnetic INDUCTION in a material to the MAGNETIC-FIELD producing it. Materials showing DIAMAGNETISM and PARAMAGNETISM have permeabilities just below and above the free space value ($\mu_0 = 4\pi \times 10^{-7}$ H/m); ferromagnets have a permeability a thousand times greater.

PERMIAN, the last period of the PALEOZOIC, stretching between about 280 and 225 million years ago. (See also GEOLOGY.)

PERMITTIVITY (ε), a constant of proportionality between an electric charge and the ELECTRIC FIELD emanating from it. The factor by which it exceeds the free space value ($\varepsilon_0 = 8.85 \times 10^{-12}$F/m) in a given material is known as the *relative permittivity*, or DIELECTRIC constant for the material.

PERMUTATIONS AND COMBINATIONS, respectively, the different orders that can be given to the elements of a SET; and the different selections of elements that may be taken from the set, every selection being of the same size, no element being a member of more than one selection, and order within each selection being immaterial. For a set of n elements there are $n(n-1)(n-2) \ldots 2.1$ ($=n!$—see FACTORIAL) permutations, taking the elements singly; and, taking the elements in *ordered* subsets each containing k elements, there are $n(n-1)(n-2) \ldots (n-(k-1))$ $(=\frac{n!}{(n-k)!})$ permutations. For the same set, taking k elements in each *unordered* subset, there are $\frac{n!}{k!\,(n-k)!}$ combinations.

PEROXIDES, compounds of OXYGEN containing the peroxy group (-O-O-). ALKALI METAL and ALKALINE-EARTH METAL peroxides, containing the peroxide ion O_2^{2-}, are formed by heating the metals or their OXIDES in excess air. Covalent peroxides include peracetic acid and peroxymonosulfuric acid ("Caro's acid"). They are powerful oxidizing agents, used in bleaching. (See also SUPEROXIDES.) **Hydrogen Peroxide** (H_2O_2) is a colorless liquid, usually produced as aqueous solutions by electrolytic or organic oxidation processes; a powerful oxidizing agent which readily decomposes into water and oxygen on heating or with various catalysts. It is used in bleaching, organic synthesis, medicine and in rocket fuels. mp −0.4°C, bp 150°C.

PERPENDICULAR, a LINE drawn through a fixed POINT P and cutting a fixed line L at right angles (see ANGLE). Should it cut L at L's center, it is termed a perpendicular bisector.

PERPETUAL MOTION, an age-old goal of inventors: a machine which would work forever without external interference, or at least with 100% efficiency. No such machine has worked or can work, though many are plausible on paper. Perpetual motion machines of the *first kind* are those whose efficiency exceeds 100%—they do work without energy being supplied. They are disallowed by the First Law of THERMODYNAMICS. Those of the *second kind* are machines that take heat from a reservoir (such as the ocean) and convert it wholly into work. Although energy is conserved, they are disallowed by the Second Law of Thermodynamics. Those of the *third kind* are machines that do no work, but merely continue in motion forever. They are approachable but not actually achievable, because some energy is always dissipated as heat by friction etc. An example, however, of what is in a sense perpetual motion of the third kind is electric current flowing in a superconducting ring (see SUPERCONDUCTIVITY), which continues undiminished indefinitely.

PERRIN, Jean Baptiste (1870–1942), French physical chemist awarded the 1926 Nobel Prize for Physics for his studies of BROWNIAN MOTION in which, by examining colloidal particles, he was able to arrive at a good value of the AVOGADRO number.

PERSECUTION MANIA. See PARANOIA.

PERSEUS, large N Hemisphere constellation containing the eclipsing binary (see DOUBLE STAR) ALGOL, two GALACTIC CLUSTERS, one of which is a double cluster, and the bright star Mirfak. It gives its name to the Perseid METEOR shower.

PERSONA, term used by JUNG to describe the individual's projection of himself; i.e., the role that he plays to conform with others' expectations of his personality. In particular, a man with a strong anima (see ANIMA AND ANIMUS) may project an especially strong male persona.

PERSPECTIVE, in DESCRIPTIVE GEOMETRY, the representation of a three-dimensional object on a PLANE surface by projection (see PROJECTIVE GEOMETRY) of the object onto the plane from a POINT.

PERSPEX. See LUCITE.

PERSPIRATION, or **sweat**, watery fluid secreted by the SKIN as a means of reducing body temperature. Sweating is common in hot climates, after EXERCISE and in the resolution of FEVER, where the secretion and subsequent evaporation of sweat allow the skin and thus the body to be cooled. Humid atmospheres and high secretion rates delay the evaporation, leaving perspiration on the surface. Excessive fluid loss in sweat, and of salt in the abnormal sweat of CYSTIC FIBROSIS, may lead to SUNSTROKE. Most sweating is regulated by the HYPOTHALAMUS and autonomic NERVOUS SYSTEM. But there is also a separate system of sweat glands, especially on the palms, which secretes at times of stress. *Hyperidrosis* is a condition of abnormally profuse sweating.

PERTURBATION, in the elliptical orbit of one celestial body around another, an irregularity caused by the gravitational attraction of a third.

PERTUSSIS. See WHOOPING COUGH.

PERUTZ, Max Ferdinand (1914–), Austrian-born British biochemist who shared with KENDREW the 1962 Nobel Prize for Chemistry for their research into the structure of HEMOGLOBIN and other globular PROTEINS.

PESTICIDE, any substance used to kill plants or animals responsible for economic damage to crops, either growing or under storage, or ornamental plants, or which prejudice the well-being of man and domestic or conserved wild animals. Pesticides are subdivided into INSECTICIDES (which kill insects); miticides (which kill mites); herbicides (which kill plants—see WEEDKILLER); FUNGICIDES (which kill fungi), and rodenticides (which kill rats and mice). Substances used in the treatment of infectious BACTERIAL DISEASES are not generally regarded as pesticides. The efficient control of pests is of enormous economic importance for man, particularly as farming becomes more intensive. A major question with all pesticides is the possibility of unfortunate environmental side effects (see ECOLOGY; POLLUTION).

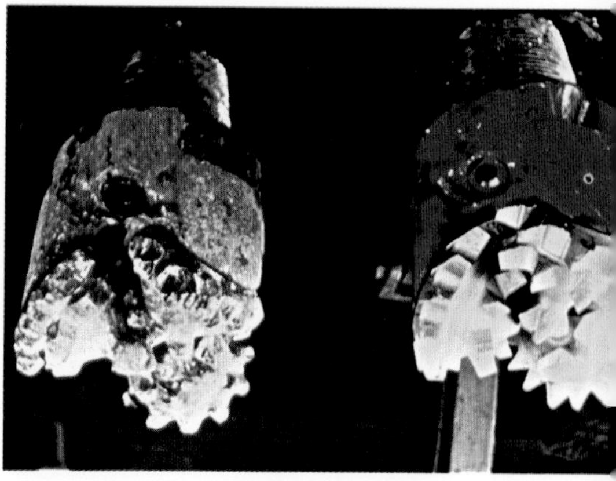

Tricone drilling bits as used for drilling petroleum exploration wells. The cones bite into the rock as the bit is rotated at the bottom of a "string" of drilling pipes. "Rollers" or rock bits such as these are used in drilling through medium hard and very hard rock formations. The bit on the right is new; that on the left is worn, having reached the end of its working life.

PETAL. See FLOWER.

PETIT, Alexis Thérèse (1791–1820), French physicist who worked with P. L. DULONG to discover Dulong and Petit's Law.

PETIT MAL. See EPILEPSY.

PETN, or **pentaerithrytol tetranitrate,** $C(CH_2ONO_2)_4$, colorless crystalline solid made by NITRATION of pentaerithrytol. It is a high EXPLOSIVE used in detonators and grenades.

PETRI DISH, shallow glass dish with a loose lid, used for growing CULTURES, named for Julius Petri (1852–1921), an assistant to R. KOCH.

PETRIE, Sir William Matthew Flinders (1853–1942), British archaeologist who devised a system of sequence dating. A relative CHRONOLOGY could thus be established between sites and dates attributed to the superimposed layers of a site.

PETRIFICATION. See FOSSILS.

PETROCHEMICALS, chemicals made from PETROLEUM and NATURAL GAS, i.e., all organic chemicals, plus the inorganic substances carbon black, sulfur, ammonia and hydrogen peroxide. Many petrochemicals are still made also from other raw materials, but the petrochemical industry has grown rapidly since about 1920. Polymers, detergents, solvents and nitrogen fertilizers are major products.

PETROLATUM, or **Petroleum Jelly.** See VASELINE.

PETROLEUM, naturally-occurring mixture of HYDROCARBONS, usually liquid "crude oil," but sometimes taken to include NATURAL GAS. (See also ASPHALT; BITUMEN.) Petroleum is believed to be formed from organic debris, chiefly of plankton and simple plants, which has been rapidly buried in fine-grained sediment under marine conditions unfavorable to oxidation. After some biodegradation, increasing temperature and pressure cause CRACKING, and oil is produced. As the source rock is compacted, oil and water are forced out, and slowly migrate to

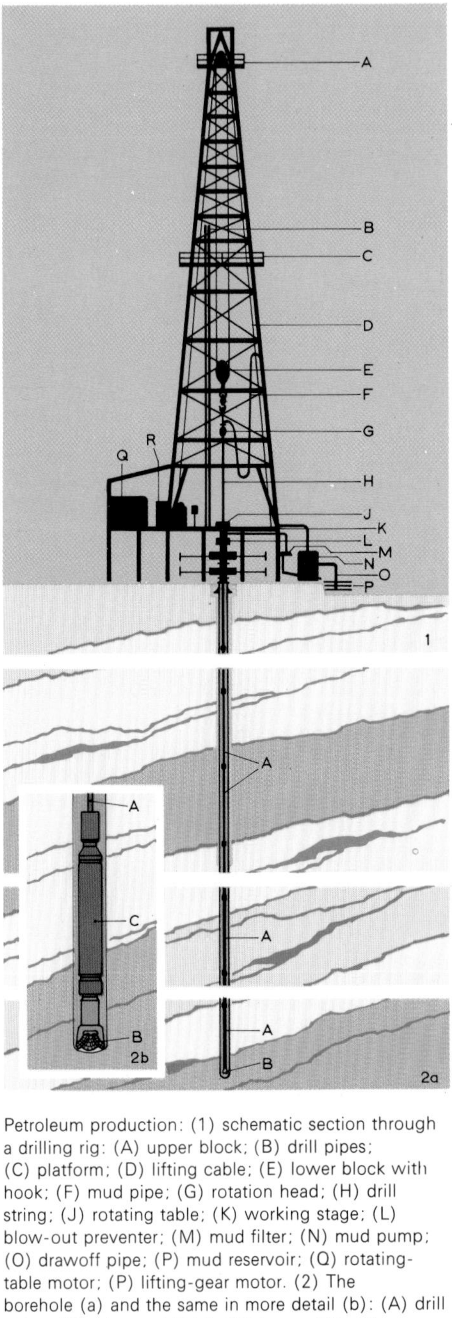

Petroleum production: (1) schematic section through a drilling rig: (A) upper block; (B) drill pipes; (C) platform; (D) lifting cable; (E) lower block with hook; (F) mud pipe; (G) rotation head; (H) drill string; (J) rotating table; (K) working stage; (L) blow-out preventer; (M) mud filter; (N) mud pump; (O) drawoff pipe; (P) mud reservoir; (Q) rotating-table motor; (P) lifting-gear motor. (2) The borehole (a) and the same in more detail (b): (A) drill string; (B) tricone drill bit; (C) heavy drill collar. Each time the drill bit needs to be replaced, the whole drill string has to be removed from the bore hole.

porous reservoir rocks, chiefly sandstone or limestone. Finally, secondary migration occurs within the reservoir as the oil coagulates to form a pool, generally capped by impervious strata, and often associated with natural gas. Some oil seeps to the earth's surface: this was used by the early Mesopotamian civilizations. The first oil well was drilled in western Pa. in 1859. The industry thus begun has grown so fast that it now supplies about half the world's energy, as well as the raw materials for PETROCHEMICALS. Modern technology has made possible oil-well drilling to a depth of 5km, and deep-sea wells in 150m of water. Rotary drilling is used, with pressurized mud to carry the rock to the surface and to prevent escape of oil. When the well is completed, the oil rises to the surface, usually under its own pressure, though pumping may be required. The chief world oil-producing regions are the Persian Gulf, the US (mainly Tex., La., Okla. and Cal.), the USSR, N and W Africa, and Venezuela. After removing salt and water, the petroleum is refined by fractional DISTILLATION producing the fractions GASOLINE, KEROSINE, diesel oil, fuel oil, lubricating oil, and ASPHALT. Undesirable compounds may be removed by solvent extraction, treatment with sulfuric acid, etc., and less valuable components converted into more valuable ones by CRACKING, reforming, ALKYLATION and polymerization. The chemical composition of crude petroleum is chiefly ALKANES, saturated ALICYCLIC COMPOUNDS, and AROMATIC COMPOUNDS, with some sulfur compounds, oxygen compounds (carboxylic acids and phenols), nitrogen and salt. (See also OIL SHALE.)

PETROLOGY, branch of geology concerned with the history, composition, occurrence, properties and classification of rocks. (See GEOLOGY; ROCKS.)

PETTY, Sir William (1623–1687), British physician, statistician, anatomist, musician and economist best known for his *A Treatise of Taxes and Contributions* (1662).

PEWTER, class of ALLOYS consisting chiefly of TIN, now hardened with copper and antimony, and usually containing lead. Roman pewter was high in lead and darkened with age. Pewter has been used for bowls, drinking vessels and candlesticks.

pH, measure of the acidity (see ACID) of an aqueous solution, defined as $pH = \log_{10}\{H^+\}$, where $\{H^+\}$, the thermodynamic activity of the hydrogen ions in the solution, approximates to their concentration in MOLES/litre for dilute solutions. Pure water has a pH of 7 (i.e., contains 10^{-7} mol/l H^+); acidic solutions have pH less than 7, basic ones greater than 7. (See also BUFFER.)

PHACOLITH. See LACCOLITH.

PHAEOPHYTA, or brown algae. See ALGAE.

PHAGE. See BACTERIOPHAGE.

PHAGOCYTE, any CELL, typically a BLOOD leukocyte, able to engulf and thus eliminate foreign bodies.

PHANEROZOIC, the aeon of visible life, the period of time represented by rock strata in which FOSSILS appear, running from about 570 million years ago through to the present and containing the PALEOZOIC, MESOZOIC and CENOZOIC eras. (See also CRYPTOZOIC; GEOLOGY; PRECAMBRIAN.)

PHARMACOLOGY, the study of DRUGS, their chemistry, mode of action, routes of absorption, excretion and METABOLISM, drug interactions, toxicity and side-effects. New drugs, based on older drugs, traditional remedies, chance observations etc., are tested for safety and efficacy, and manufactured by

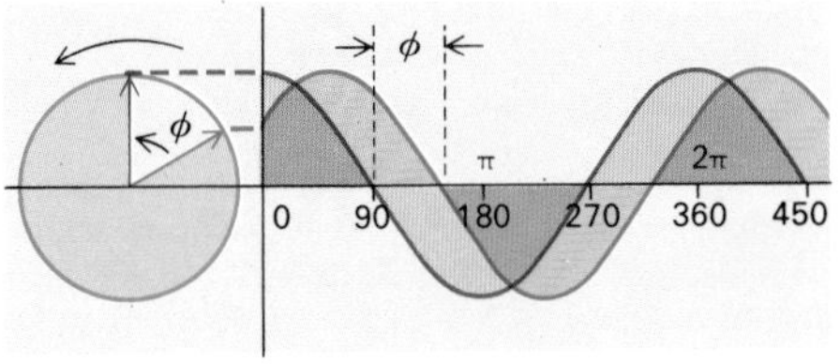

Two wave motions separated by a phase difference ϕ, with their generating vectors. In this case the red wave leads the green wave by ϕ.

the pharmaceutical industry. The dispensing of drugs is PHARMACY. Drug prescription is the cornerstone of the medical treatment of DISEASE.

PHARMACOPEIA, a text containing all available DRUGS and pharmacological preparations, providing a vital source for accurate prescribing in MEDICINE. It lists drugs; their properties and formulation; routes and doses of administration; mode of action, METABOLISM and excretion; known interaction with other drugs; contraindications and precautions in particular DISEASES; toxicity, and side-effects.

PHARMACY, the preparation or dispensing of DRUGS and pharmacological substances used in MEDICINE; also, the place where this is practiced. Most drugs are now formulated by drug companies and the pharmacist need only measure them out and instruct the patient in their use. In the past, however, the pharmacist mixed numerous basic substances to produce a variety of medicines, tonics, etc.

PHARYNX, the back of the throat where the mouth (oropharynx) and NOSE (nasopharynx) pass back into the ESOPHAGUS. It contains specialized MUSCLE for swallowing. The food and air channels are kept functionally separate so that swallowing does not interfere with breathing and speech.

PHASE, the proportion of a cycle already executed by an oscillating system, expressed as an angle (360° or 2π radians corresponding to a full cycle). Thus if an AC voltage is at its maximum value while the current is passing through zero, there is said to be a 90° ($\pi/2$) phase difference between them. In mathematics, the phase of a complex number is the angle between the real axis and a line from the number to the origin (see IMAGINARY NUMBERS).

PHASE EQUILIBRIA, in THERMODYNAMICS, EQUILIBRIA between substances in different phases (solid, liquid or gas). Phase diagrams, basic to engineering, metallurgy and mineralogy, are empirical graphs showing what phases exist at different pressures and temperatures. The *phase rule*, deduced by J. W. GIBBS, states that, for a closed system, $F=C-P+2$, where P is the number of phases, C the number of independent chemical components and F the number of degrees of freedom, i.e., the number of variables (pressure, temperature, composition) whose values must be specified to define the system.

PHENACETIN, mild ANALGESIC commonly used in mixed analgesic preparations. It causes KIDNEY disease if ingested in quantity for any period; its active component **paracetamol** is now used as a safer drug.

PHENOBARBITAL, or **phenobarbitone**. See BARBITURATES.

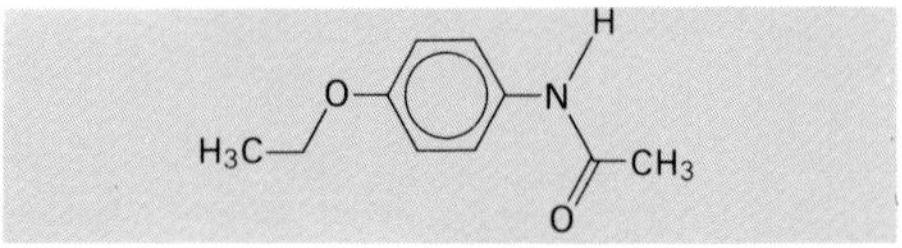

The structural formula of paracetamol (*N*(4-ethoxyphenyl)-ethanamide). Paracetamol is the active component in phenacetin.

PHENOL (C_6H_5OH), or **carbolic acid**, the simplest of the PHENOLS, a white, hygroscopic crystalline solid, isolable from COAL TAR, but made by acid hydrolysis of cumene hydroperoxide, or by fusion of sodium benzenesulfonate (see SULFONIC ACIDS) with sodium hydroxide. Formerly used as an ANTISEPTIC, phenol is

Phenols

phenol (hydroxybenzene, benzenol, carbolic acid) MW 94·11, mp 43°C, bp 182°C	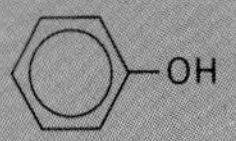
2-methylphenol (*ortho*-cresol 2-hydroxytoluene) MW 108·15, mp 31°C, bp 191°C	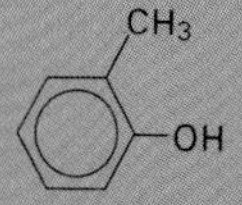
3-methylphenol (*meta*-cresol 3-hydroxytoluene) MW 108·15, mp 12°C, bp 202°C	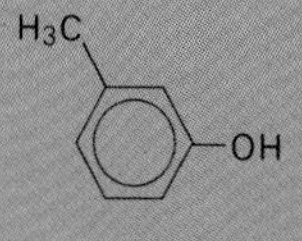
4-methylphenol (*para*-cresol 4-hydroxytoluene) MW 108·15, mp 35°C, bp 202°C	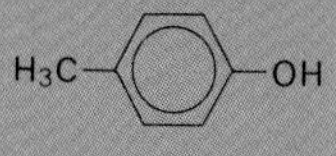
1,2-dihydroxybenzene (catechol) MW 110·11, mp 105°C, bp 245°C	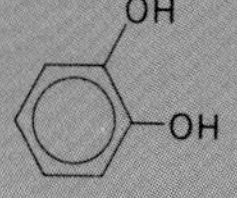
1,3-dihydroxybenzene (resorcinol) MW 110·11, mp 111°C, bp 277°C	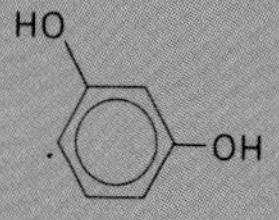
1,4-dihydroxybenzene (hydroquinone, quinol) MW 110·11, mp 173°C, bp 286°C	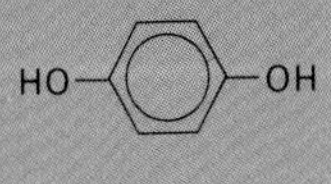
1-hydroxynaphthalene (α-naphthol) MW 144·19, mp 96°C, bp 288°C	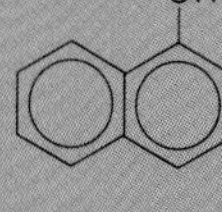
2-hydroxynaphthalene (β-naphthol) MW 144·19, mp 123°C, bp 295°C	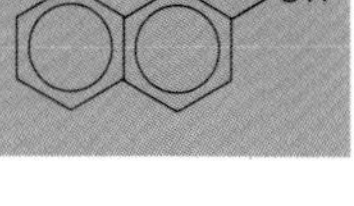

now used to make BAKELITE and many other resins, plastics, dyes, detergents, drugs etc. MW 94.1, mp 43°C, bp 182°C.

PHENOLS, class of AROMATIC COMPOUNDS in which a HYDROXIDE group is directly bonded to an aromatic ring system. They are very weak ACIDS, and, like ALCOHOLS, form ETHERS and ESTERS. They are very liable to undergo electrophilic substitution (see NUCLEOPHILES), and hence condense with formaldehyde to form resins. The main phenols are PHENOL itself, CRESOL, RESORCINOL, pyrogallol (see GALLIC ACID) and PICRIC ACID.

PHENOMENOLOGY, a school of philosophy based on a method of approach due to Edmund HUSSERL. Unlike the naturalist, who describes objects without reference to the subjectivity of the observer, the phenomenologist attempts to describe the "invariant essences" of objects as objects "intended" by consciousness. As a first step toward achieving this, he performs the "phenomenological reduction," which involves as far as possible a suspension of all preconceptions about experience.

PHENOTYPE, the appearance of, and characteristics actually present in an organism, as contrasted with its GENOTYPE (its genetic make-up). Heterozygotes and homozygotes with a dominant GENE have the same phenotype but differing genotypes. Organisms may also have an identical genotype but differing phenotype due to environmental influences.

PHENYLKETONURIA (PKU), inherited DISEASE in which phenylalanine METABOLISM is disordered due to lack of an ENZYME. It rapidly causes MENTAL RETARDATION, as well as irritability and vomiting, unless DIETARY FOODS low in phenylalanine are given from soon after birth and indefinitely. Screening of the newborn by urine tests (with confirmation by blood tests) facilitates prompt treatment.

PHILOLOGY, the study of literature and the language employed in it. The term is used also for those branches of LINGUISTICS concerned with the evolution of languages, especially those dealing with the interrelationships between different languages (comparative philology).

PHILOSOPHER'S STONE. See ALCHEMY.

PHILOSOPHY (from *philosophos*, lover of wisdom), term applied to any body of doctrine or opinion as to the nature and ultimate significance of human experience considered as a whole. It is perhaps more properly applied to the critical evaluation of all claims to knowledge—including its own *and* anything that is presupposed about its own nature and task. In this latter respect, it is widely argued, philosophy differs fundamentally from all other disciplines. What philosophy "is" (what methods the philosopher should employ, what criteria he should appeal to, and what goals he should set himself) is as perennial a question for the philosopher as any other. Traditionally, philosophers have concerned themselves with four main topic areas: LOGIC, the study of the formal structure of valid arguments; METAPHYSICS, usually identified with ontology—the study of the nature of "Being" or ultimate reality; EPISTEMOLOGY, or theory of knowledge, sometimes treated as a branch of metaphysics; and axiology, or theory of value—including aesthetics, the philosophy of taste (especially as applied to the arts), ethics, or moral philosophy, and political philosophy or political science. In modern times as traditional philosophy has yielded up the subject matters of the natural sciences, of other descriptive studies such as PSYCHOLOGY and SOCIOLOGY, and of such formal studies as logic and mathematics, all once numbered among its legitimate concerns, philosophers have become increasingly conscious of their critical role. Most now tend to interest themselves in special philosophies, e.g., philosophy *of* logic, philosophy *of* science (see SCIENTIFIC METHOD) and philosophy *of* religion. The first attempts to answer distinctively philosophical questions were made from about 600 BC by certain Greek philosophers known collectively as the PRESOCRATICS; their intellectual heirs were SOCRATES, PLATO and

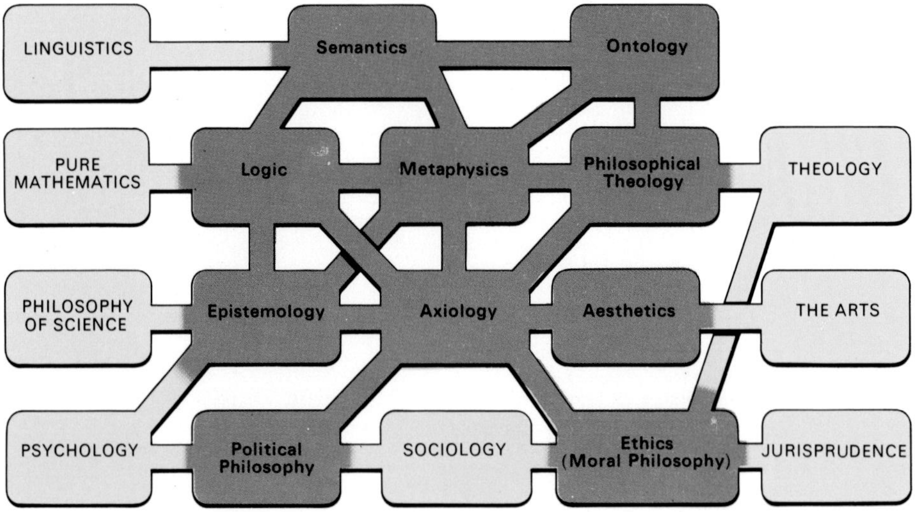

The branches of philosophy (red) and their links with allied sciences (yellow).

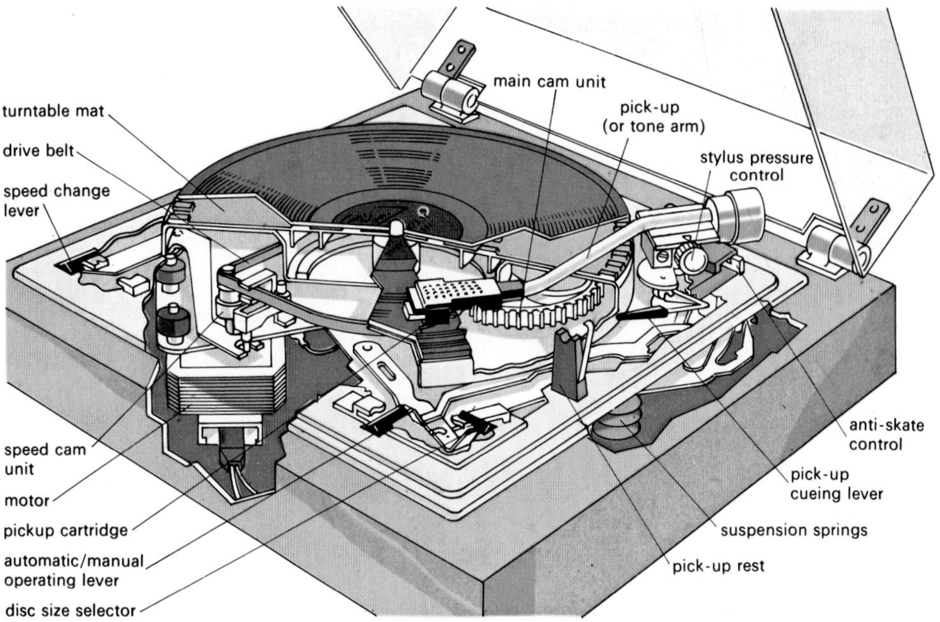

Cutaway of a phonograph (record deck) with belt drive and anti-skate pickup arm.

ARISTOTLE, the three towering figures in ancient philosophy. Later ancient philosophies include epicureanism, stoicism and neoplatonism. Foremost among medieval philosophers were St. Augustine and St. Thomas Aquinas, both leading churchmen. (See also NOMINALISM; REALISM.) Modern philosophy begins with René DESCARTES and a parallel development of RATIONALISM and EMPIRICISM culminating in the philosophy of Immanuel KANT. The IDEALISM of G. F. W. HEGEL and the POSITIVISM of Auguste COMTE were major forces in 19th-century philosophy. The dialectical materialism of Karl Marx had its roots in both. (See also MATERIALISM.) The philosophical orientations of most 20th-century philosophers are developments of marxism, neokantianism, LOGICAL POSITIVISM, PRAGMATISM, PHENOMENOLOGY or existentialism.

PHLEBITIS, INFLAMMATION of the VEINS, usually causing THROMBOSIS (thrombophlebitis) and obstruction to BLOOD flow. It is common in the superficial veins of the legs, especially VARICOSE VEINS, and visceral veins close to inflamed organs or ABSCESSES. Phlebitis may complicate intravenous INJECTIONS of DRUGS or indwelling cannulae for intravenous fluids. Pain, swelling and ERYTHEMA over the vein are typical with it becoming a thick tender cord. Occasionally, phlebitis indicates systemic DISEASE (e.g., CANCER).

PHLOEM, or **bast,** a vascular tissue responsible for the transport of dissolved food substances through the roots, stems and leaves of higher PLANTS. Phloem mainly consists of elongated living sieve tubes, which have perforated end plates.

PHLOGISTON, the elementary principle postulated by G. H. STAHL to be lost from substances when they burn. The phlogiston concept provided 18th-century CHEMISTRY with its unifying principle. The phlogiston theory of COMBUSTION found general acceptance until displaced by its inverse—LAVOISIER's oxygen theory.

PHLOGOPITE, a range of magnesium-rich varieties of MICA, grading into BIOTITE.

PHOBIA, a NEUROSIS characterized by exaggerated ANXIETY on confrontation with a specific object or situation; or the anxiety itself. Phobia is sometimes linked with OBSESSIONAL NEUROSIS, sometimes with HYSTERIA; in each case the object of phobia is usually merely symbolic. Classic phobias are AGORAPHOBIA and CLAUSTROPHOBIA.

PHOBOS, the inner moon of MARS, diameter about 16km, orbiting in 7.65h at a distance of 9 370km. It has the lowest ALBEDO in the SOLAR SYSTEM.

PHON, in ACOUSTICS, a unit of loudness. Loudness in phons is given by the number of DECIBELS above the reference level, 20 μPa, of a pure 1kHz-frequency SOUND which is judged by listeners to be of equal loudness with the original.

PHONEME, any of the smallest units of spoken language serving to differentiate between utterances: e.g., the "p" and "t" of "pin" and "tin." (See also MORPHEME.)

PHONETICS, the systematic examination of the sounds made in speech, concerned not only with the classification of these sounds but also with physical and physiological aspects of their production and transmission, and with their reception and interpretation by the listener. **Phonology**, the study of phonetic patterns in languages, is of importance in comparative LINGUISTICS. **Phonemics** is the study of PHONEMES.

PHONOGRAPH, or **record player**, instrument for reproducing sound recorded mechanically as modulations in a spiral groove (see SOUND RECORDING). It was invented by Thomas EDISON (1877), whose first machine had a revolving grooved cylinder covered with tinfoil. Sound waves caused a diaphragm to vibrate and a stylus on the diaphragm made indentations in the foil. These could then be made to vibrate another stylus attached to a reproducing diaphragm. Wax discs and cylinders soon replaced tinfoil, then, when by etching or electroplating metal master discs could be made, copies were mass-produced in rubber, wax or plastic. The main parts of a phonograph are the turntable to rotate the disc at constant angular velocity; the stylus, which tracks the groove and vibrates with its modulations; the pickup or transducer that converts these movements piezoelectrically or electromagnetically into electrical signals; the AMPLIFIER, and the LOUDSPEAKER. (For high-quality reproduction, see HIGH-FIDELITY.)

PHONON, in SOLID STATE PHYSICS, the particle (quantum) counterpart of the SOUND wave or LATTICE vibration, considered to play an important role in the CONDUCTION of HEAT in electrical insulators.

PHORONID(E)A, or horseshoe worms, a phylum of marine, tube-dwelling, worm-like animals generally less than 200mm (8in) long. Each individual lives with its tube buried in sand or attached to a rock or shell in shallow temperate or tropical seas. Most species are colonial.

PHOSGENE, or **carbonyl chloride** ($COCl_2$), colorless, reactive gas, hydrolyzed by water, made by catalytic combination of CARBON monoxide and CHLORINE, and used to make RESINS and DYES. Highly toxic, it was a poison gas in WWI.

PHOSPHATES, derivatives of phosphoric acid (see PHOSPHORUS): either phosphate ESTERS, or salts containing the various phosphate ions. Like SILICATES, these are numerous and complex, the simplest being orthophosphate, PO_4^{3-}. Of many phosphate minerals, the most important is APATITE. This is treated with sulfuric acid or phosphoric acid to give calcium dihydrogenphosphate ($Ca[H_2PO_4]_2$), known as **superphosphate**—the major phosphate FERTILIZER. The alkaline trisodium phosphate (TSP) (Na_3PO_4) is used as a cleansing agent and water softener. Phosphates are used in making GLASS, SOAPS AND DETERGENTS.

PHOSPHOR, a substance exhibiting LUMINESCENCE, i.e., emitting LIGHT (or other ELECTROMAGNETIC RADIATION) on nonthermal stimulation. Important phosphors include those used in TELEVISION picture tubes (where stimulation is by ELECTRONS) and those coated on the inside wall of fluorescent lamp tubes to convert ULTRAVIOLET RADIATION into visible light.

PHOSPHORESCENCE, or **afterglow.** See LUMINESCENCE.

PHOSPHORUS (P), reactive nonmetal in Group VA of the PERIODIC TABLE, occurring naturally as APATITE. This is heated with silica and coke, and elementary phosphorus is produced. Phosphorus has three main allotropes (see ALLOTROPY): white phosphorus, a yellow waxy solid composed of P_4 molecules, spontaneously flammable in air, soluble in carbon disulfide, and very toxic; red phosphorus, a dark-red powder, formed by heating white phosphorus, less reactive, and insoluble in carbon disulfide; and black phosphorus, a flaky solid, resembling GRAPHITE, consisting of corrugated layers of atoms. Phosphorus burns in air to give the trioxide and the pentoxide, and also reacts with the halogens, sulfur and some metals. It is used in making matches, ammunition, pesticides, steels, phosphor bronze, phosphoric acid and phosphate fertilizers. Phosphorus is of great biological importance. AW 31.0, mp (wh) 44°C, bp (wh) 280°C, sg (wh) 1.82, (red) 2.20, (bl) 2.69. Phosphorus forms phosphorous (trivalent) and phosphoric (pentavalent) compounds. **Phosphine** (PH_3), is a colorless, flammable gas, highly toxic, and with an odor of garlic. It is a weak BASE, resembling AMMONIA, and forms phosphonium salts (PH_4^+). **Phosphoric acid** (H_3PO_4), is a colorless crystalline solid, forming a syrupy aqueous solution. It is used to flavor food, in dyeing, to clean metals, and to make PHOSPHATES. **Phosphorus Pentoxide** (P_4O_{10}), is a white powder made by burning phosphorus in excess air. It is very deliquescent (forming phosphoric acid), and is used as a dehydrating agent.

PHOTOCHEMISTRY, branch of PHYSICAL CHEMISTRY dealing with chemical reactions that produce LIGHT (see CHEMILUMINESCENCE; COMBUSTION), or that are initiated by light (visible or ultraviolet). Important examples include PHOTOSYNTHESIS, PHOTOGRAPHY and bleaching by sunlight. One PHOTON of light of suitable wavelength may be absorbed by a molecule, raising it to an electronically excited state. Re-emission may occur by fluorescence or phosphorescence (see LUMINESCENCE), the energy may be transferred to another molecule, or a reaction may occur, commonly DISSOCIATION to form FREE RADICALS. The *quantum yield,* or efficiency, of the reaction is the number of molecules of reactant used (or product formed) per photon absorbed; this may be very large for chain reactions. (See also FLASH PHOTOLYSIS; LASER; RADIATION CHEMISTRY.)

PHOTOCONDUCTIVE DETECTOR, an electrical component whose CONDUCTIVITY increases as more LIGHT falls on it. Used in light detectors, light-sensitive switches, light meters, and in the Vidicon television camera tube, most employ photoconductive SEMICONDUCTORS such as lead telluride or cadmium sulfide.

PHOTOCOPYING. See OZALID PROCESS; XEROGRAPHY.

PHOTOELECTRIC CELL, a device with electrical properties which vary according to the LIGHT falling on it. There are three types: PHOTOVOLTAIC CELLS; PHOTOCONDUCTIVE DETECTORS and phototubes (see PHOTOELECTRIC EFFECT).

PHOTOELECTRIC EFFECT, properly **photoemissive effect**, the emission of ELECTRONS from a surface when struck by ELECTROMAGNETIC RADIATION such as LIGHT. In 1905 EINSTEIN laid one of the twin foundations of QUANTUM THEORY by explaining photoemission in terms of the action of individual PHOTONS. The Einstein photoelectric law reflects the fact that no electrons are photoemitted

The Einstein photoelectric law

$$E_k = h\nu - \omega$$

where: E_k is the maximum kinetic energy of emitted electrons,
h is Planck's constant,
ν is the frequency of the radiation,
ω is the surface work function for photoemission.

In photographic developing and printing, the exposed film is removed from the camera (1) and, in the darkroom, is wound on to a reel (2) that fits into a light-proof tank. Steps 3 through 11 may be carried out with the lights on. The tank is filled with developer (3) and agitated during development (4). After the proper time has elapsed, the developer is poured out (5) and the tank is filled with the stop bath (6) to halt development. The stop bath in turn is poured out (7), and the fixer is poured in (8) and removed after a specified time (9). The negative is then washed (10) and dried (11). To make enlargements, light is shone through the negative in the enlarger (12a) onto sensitive paper. Alternatively the paper can be exposed directly through the negative with both held together in a frame (12b—"contact" printing). Using red light, the photographer develops the prints (13), washes them (14)—sometimes after using a stop bath—and fixes them (15). The prints are then washed again (16) and dried by hand (17a) or in a dryer (17b)

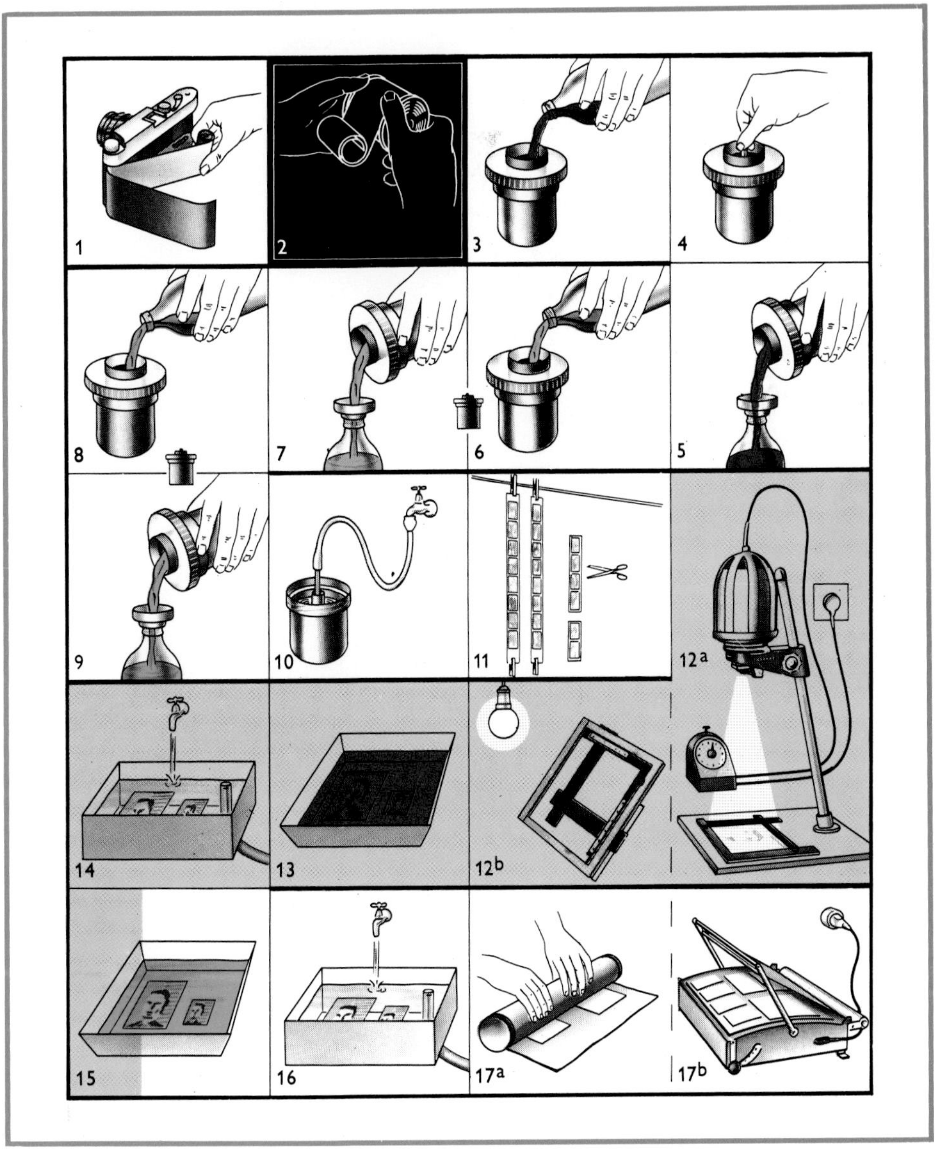

unless the energy of the incident photons exceeds a certain threshold value known as the surface work function for photoemission from the material. The effect is used in phototubes (ELECTRON TUBES having a photoemissive cathode), often employed as "electric eye" switches. Special types are used in image intensifiers and in the Image Orthicon TELEVISION camera.

PHOTOGRAMMETRY, the use of photographs in map-making. Series of overlapping air photographs are generally used. If exposed in stereo pairs at a known altitude, such photographs can be used to make detailed, accurate relief maps.

PHOTOGRAPHY, the use of light-sensitive materials to produce permanent visible images (photographs). The most familiar photographic processes depend on the light-sensitivity of the SILVER halides. A photographic emulsion is a preparation of tiny crystals of these salts suspended in a thin layer of gelatin coated on a glass, film or paper support. On brief exposure to light in a CAMERA or other apparatus, a latent image in activated silver salt is formed wherever light has fallen on the emulsion. This image is made visible in development, when the activated silver halide crystals (but not the unexposed ones) are reduced to metallic silver (black) using a weak organic reducing agent (the developer). The silver image is then made permanent by fixing, in the course of which it becomes possible to examine the image in the light for the first time. Fixing agents (fixers) work by dissolving out the silver halide crystals which were not activated on exposure. The image made in this way is densest in silver where the original subject was brightest and lightest where the original was darkest; it is thus a "negative" image. To produce a positive image, the negative (which is usually made on a film or glass (plate) support) is itself made the original in the above process, the result being a positive "print" usually on a paper carrier. An alternative method of producing a positive image is to bleach away the developed image on the original film or plate before fixing, and reexpose the unactivated halide in diffuse light. This forms a second latent image which on development produces a positive image of the original subject (reversal processing).

The history of photography from the earliest work of NIÉPCE, DAGUERRE and Fox TALBOT to the present has seen successive refinements in materials, techniques and equipment. Photography became a popular hobby after EASTMAN first marketed roll film in 1889. The silver halides themselves are sensitive to light only from the blue end of the SPECTRUM so that in the earliest photographs other colors appear dark. The color-sensitivity of emulsions was improved from the 1870s onward as small quantities of sensitizing dyes were incorporated. "Orthochromatic" plates became available after 1884 and "panchromatic" from 1906.

New sensitizing dyes also opened up the way to infrared and color photography. Modern "tripack" color films have three layers of emulsion, one each sensitive to blue, green and red light from the subject. Positive color transparencies are made using a reversal processing method in which the superposed, positive, silver images are replaced with yellow, magenta and cyan dyes respectively.

Motion-picture photography dates from 1890, when EDISON built a device to expose Eastman's roll film, and motion pictures rapidly became an important art form. Not all modern photographic methods employ the silver-halide process; XEROGRAPHY and the BLUEPRINT and OZALID processes work differently. FALSE-COLOR PHOTOGRAPHY and the diffusion process used in the POLAROID LAND CAMERA are both developments of the silver-halide process.

PHOTOMETRY, the science of the measurement of LIGHT, particularly as it affects illumination engineering. Because the brightness experienced when light strikes the human EYE depends not only on the POWER conveyed by the radiation but also on the wavelength of the light (the visual sensation for a given power reaching a maximum at 555nm), a special arbitrary set of units is used in photometric calculations. In SI UNITS, the photometric base quantity is luminous intensity which measures the intensity of light radiated from a small source. The base unit of luminous intensity is the CANDELA (cd). The luminous flux (the photometric equivalent of the power radiating) from a point source is measured in lumens where 1 lumen (lm) is the flux radiating from a 1 cd source through a solid angle of a steradian. The illuminance falling on a surface (formerly known as its illumination) is measured in luxes where 1 lux (lx) is the level of illuminance occurring when a luminous flux of 1 lm falls on each m^2 of the surface. Up to the 1970s considerable confusion reigned among scientists regarding the concepts and terminology best to be used in photometry and many alternative units—APOSTILBS, BLONDELS, FOOT-CANDLES and LAMBERTS—are still commonly encountered. (See also LUMINANCE.)

PHOTON, the quantum of electromagnetic energy (see QUANTUM THEORY), often thought of as the particle associated with LIGHT or other ELECTROMAGNETIC RADIATION. Its ENERGY is given by $h\nu$ where h is the PLANCK CONSTANT and ν the frequency of the radiation.

PHOTOSPHERE, a 125–190km-thick layer of gas on the sun, visible to us as the sun's apparent surface, emitting most of the sun's light. Its TEMPERATURE is estimated at 6000K.

PHOTOSYNTHESIS, the process by which green plants convert the ENERGY of sunlight into chemical energy which is then stored as CARBOHYDRATE. Overall, the process may be written as:

$$6CO_2 + 6H_2O \xrightarrow{light} C_6H_{12}O_6 + 6O_2$$

Although in detail photosynthesis is a complex sequence of reactions, two principal stages can be identified. In the "light reaction," CHLOROPHYLL (the key chemical in the whole process) is activated by absorbing a quantum of LIGHT, initiating a sequence of reactions in which the energy-rich compounds ATP (adenosine triphosphate—see NUCLEOTIDES) and TPNH (the reduced form of triphosphopyridine nucleotide—TPN) are made, water being decomposed to give free oxygen in the process. In the second stage, the "dark reaction," the ATP and TPNH provide the energy for the assimilation of carbon dioxide gas, yielding a variety of SUGARS from which other sugars and carbohydrates, including STARCH, can be built up.

PHOTOTROPISM. See TROPISMS.

PHOTOVOLTAIC CELL, a device for converting LIGHT radiation into ELECTRICITY, used in LIGHT METERS and for providing spacecraft power supplies. The photovoltage is usually developed in a layer of SEMICONDUCTOR (e.g., SELENIUM) sandwiched between a transparent electrode and one providing support.

PHRENOLOGY, study of the shape and detailed contours of the SKULL as indicators of personality, intelligence and individual characteristics. The

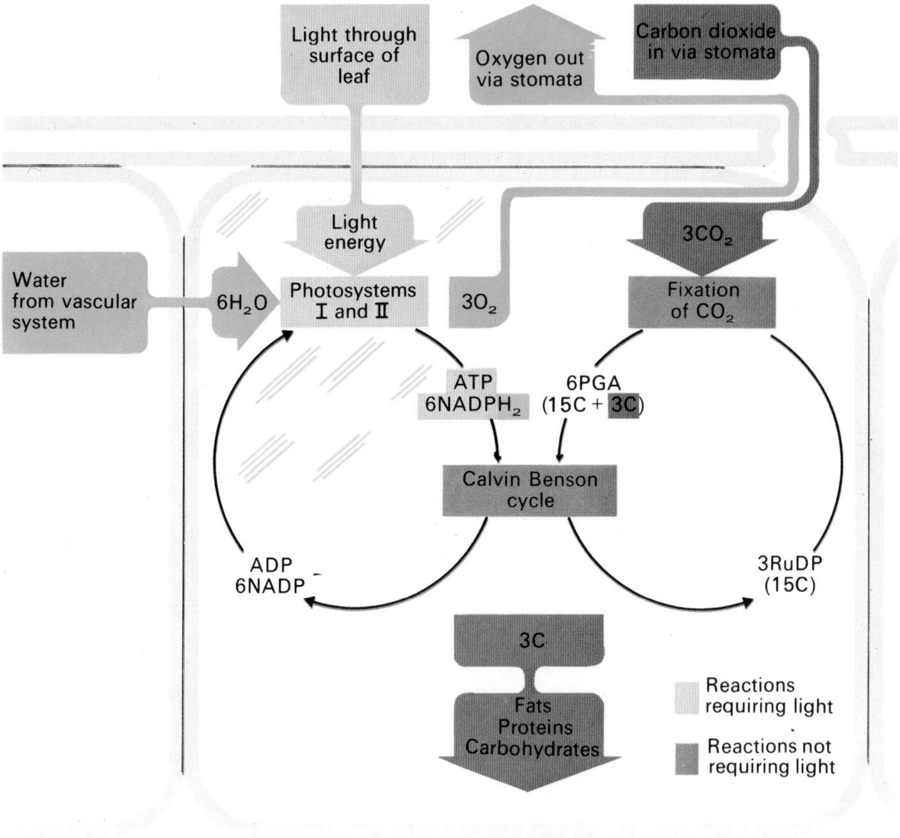

Schematic diagram of photosynthesis in a leaf cell. Energy, in the form of potential chemical energy bound in adenosine triphosphate (ATP) and reduced nicotinamide adenine dinucleotide ($NADPH_2$), is produced by the splitting of water in the presence of light, oxygen being a byproduct. Carbon from carbon dioxide in the atmosphere is fixed (without the need for light), with the chemical ribulose diphosphate (RuDP) to form phosphoglyceric acid (PGA). PGA, ATP and NADP then enter the Calvin-Benson cycle (another process not requiring light) where fats, proteins, carbohydrates and other compounds required for "living" processes of the cell are produced.

method, developed by F. J. GALL and promoted in the UK and US by George Combe (1788–1858), had many 19th-century followers and led to the more enlightened treatment of offenders and the mentally ill.

PHYLOGENY. See ONTOGENY AND PHYLOGENY.

PHYLUM. See TAXONOMY.

PHYSICAL CHEMISTRY, major branch of CHEMISTRY, in which the theories and methods of PHYSICS are applied to chemical systems. Physical chemistry underlies all the other branches of chemistry and includes theoretical chemistry. Its main divisions are the study of molecular structure; COLLOIDS; CRYSTALS; ELECTROCHEMISTRY; chemical EQUILIBRIUM; GAS laws; chemical KINETICS; MOLECULAR WEIGHT determination; PHOTOCHEMISTRY; SOLUTION; SPECTROSCOPY, and chemical THERMODYNAMICS.

PHYSICAL THERAPY. See PHYSIOTHERAPY.

PHYSICIAN. See MEDICINE.

PHYSICS, originally, the knowledge of natural things (=natural science); now, the science dealing with the interaction of MATTER and ENERGY (but usually taken to exclude CHEMISTRY). Until the "scientific revolution" of the Renaissance, physics was a branch of PHILOSOPHY dealing with the natures of things. The physics of the heavens, for instance, was quite separate from (and often conflicted with) the descriptions of mathematical and positional ASTRONOMY. But from the time of GALILEO, and particularly through the efforts of HUYGENS and NEWTON, physics became identified with the mathematical description of nature; occult qualities were banished from physical science. Firm on its Newtonian foundation, classical physics gathered more and more phenomena under its wing until, by the late 19th century, comparatively few phenomena seemed to defy explanation. But the interpretation of these effects (notably BLACKBODY RADIATION and the PHOTOELECTRIC EFFECT) in terms of new concepts due

The History of Physics

THE ANCIENT WORLD:	Greek phiiosophers from Thales to Plato contemplate the nature of matter. Greek physics summed up in the writings of Aristotle. In the Hellenistic period Archimedes works in practical mechanics	Ptolemy writes on optics	
MIDDLE AGES:	mechanics: Buridan (impetus theory) Nicholas of Cusa		
16TH CENTURY:			Gilbert (magnetism)
17TH CENTURY:	Galileo (dynamics; acceleration) Descartes (vortex theory) Huygens (circular motion) Newton (universal gravitation; inertia)	optics: (Descartes, Huygens) } (refraction) (Newton) (theory of color)	Torricelli (barometer) Boyle ("spring" of air) Pascal (weight of atmosphere)
18TH CENTURY:	thermodynamics: Black (caloric theory of heat)		electricity: Franklin (nature of lightning)
19TH CENTURY:	Rumford (heat from friction) Joule (heat and work) Carnot (heat engine theory) Helmholtz (conservation of energy) Kelvin (thermodynamic temperatures) Clausius (entropy) Gibbs (phase rule) "modern physics": Thomson, J. J. (cathode rays) Becquerel and Curies (radioactivity) Planck (quantum theory)	Young, Fresnel } (wave theory) of light) Maxwell (electromagnetic radiation) Michelson and Morley (ether experiment) Roentgen (X-rays)	Volta (current electricity) Oersted (magnetic effect of current) Ampère (forces between currents) Ohm (resistance law) Faraday (electromagnetic induction) Hertz (radio waves)
20TH CENTURY:	Einstein (relativity) Millikan (electronic charge) Bohr (theory of nuclear atom) Rutherford (nuclear transmutation) Compton (X-ray scattering) De Broglie (quantum mechanics) Schrödinger and Dirac (wave mechanics) Heisenberg (uncertainty principle) Chadwick (neutron) Gell-Mann (quark theory)		

to PLANCK and EINSTEIN involved the thoroughgoing reformulation of the fundamental principles of physical science (see QUANTUM THEORY; RELATIVITY). Physics today is divided into many specialisms, themselves subdivided manyfold. The principal of these are ACOUSTICS; ELECTRICITY and MAGNETISM; MECHANICS; NUCLEAR PHYSICS; OPTICS; QUANTUM THEORY; RELATIVITY, and THERMODYNAMICS.

PHYSIOLOGY, the study of function in living organisms. Based on knowledge of ANATOMY, physiology seeks to demonstrate the manner in which organs perform their tasks, and in which the body is organized and maintained in a state of HOMEOSTASIS. Normal responses to various stresses on the whole or on parts of an organism are studied. Important branches of physiology deal with RESPIRATION, BLOOD CIRCULATION, the NERVOUS SYSTEM, the DIGESTIVE SYSTEM, the KIDNEYS, the fluid and electrolyte balance, the ENDOCRINE GLANDS and METABOLISM. Methods of study include experimentation on anesthetized animals and on human volunteers. Knowledge and understanding of physiology is basic to MEDICINE and provides the physician with a perspective in which to view the body's disordered function in DISEASE.

PHYSIOTHERAPY, system of physical treatment for disease or disability. Active and passive muscle movement; electrical stimulation; balancing exercises; HEAT, ULTRAVIOLET or shortwave RADIATION, and manual vibration of the CHEST wall with postural drainage, are some of the techniques used. Rehabilitation after FRACTURE, SURGERY, STROKE or other neurological disease, and the treatment of LUNG infections (PNEUMONIA, BRONCHITIS), are among the aims.

PHYTOPLANKTON. See PLANKTON.

PI (Greek π), the ratio between the circumference of a CIRCLE and its diameter. π is an IRRATIONAL NUMBER whose value to five decimal places is 3.141 59. Approximate values of π have been known to several ancient civilizations, such as Babylonia, where the accepted value was 3.0.

PIAGET, Jean (1896–), Swiss psychologist whose theories of the mental development of children, though now often criticized, have been of paramount importance. His many books include *The Psychology of Intelligence* (1947).

PIAZZI, Giuseppe (1746–1826), Italian astronomer who discovered Ceres, the first ASTEROID (1801). Through illness he lost it again, and it was rediscovered the following year by OLBERS.

PICCARD, name of the Swiss twin brothers **Auguste** (1884–1962), a physicist, and **Jean Félix** (1884–1963), a chemist. Both made famous high-altitude BALLOON ascents in order to study COSMIC RAYS with a minimum of atmospheric interference, Auguste in 1931 and 1932, and Jean in 1936. In 1948 Auguste successfully conducted an unpiloted trial dive of the BATHYSCAPHE, a deep-sea diving device built to his own design; the first piloted dive—in a new bathyscaphe—followed in 1953.

PICKERING, name of two US astronomers, **Edward Charles Pickering** (1846–1919) and his brother **William Henry Pickering** (1858–1938). Edward made important contributions to stellar PHOTOMETRY and was the inventor of the meridian photometer. William, in 1898, discovered Phoebe, the ninth moon of the planet SATURN.

PICKLE, food that has been preserved in VINEGAR or BRINE to prevent the development of putrefying BACTERIA. Spices are usually added for flavor. Cucumbers, onions, beets, tomatoes and cauliflowers are used to make popular pickles. Pigs' feet and corned beef are also sometimes pickled. (See FOOD PRESERVATION.)

PICRIC ACID, or 2,4,6-trinitrophenol, yellow crystalline solid, made by NITRATION of PHENOL or its derivatives. A moderately strong ACID, it has been used as a DYE, as an ANTISEPTIC and ASTRINGENT for treating burns, and as a high EXPLOSIVE. MW 229.1, mp 122°C.

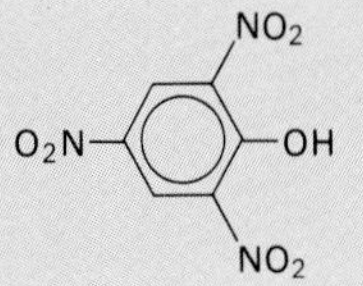

Picric acid (2,4,6-trinitrophenol).

PICTOGRAPHY, WRITING system using pictures and drawings as vehicles of communication. A pictograph used to represent an idea is an IDEOGRAM; one that represents a word, a logogram.

PICTUREPHONE, system of video TELEPHONE, introduced in 1971. The scanned TELEVISION picture, requiring a wide BANDWIDTH, is relayed along auxiliary telephone wires.

PIEZOELECTRICITY, a reversible relationship between mechanical stress and electrostatic POTENTIAL exhibited by certain CRYSTALS with no center of symmetry, discovered in 1880 during investigations of *pyroelectric* crystals (these are also asymmetric and get oppositely charged faces when heated). When pressure is applied to a piezoelectric crystal such as QUARTZ, positive and negative electric charges appear on opposite crystal faces. Replacing the pressure by tension changes the sign of the charges. If, instead, an electric potential is applied across the crystal, its length changes; this effect is linear. A piezoelectric crystal placed in an alternating electric circuit will alternately expand and contract. Resonance occurs in the circuit when its FREQUENCY matches the natural vibration frequency of the crystal, this effect being applied in frequency controllers. This useful way of coupling electrical and mechanical effects is used in MICROPHONES, PHONOGRAPH pickups and ULTRASONIC generators.

PIG IRON, crude CAST IRON produced in a BLAST FURNACE and cast into ingots or "pigs." It is used to make WROUGHT IRON and STEEL. (See also IRON.)

PIGMENTS, Natural, chemical substances imparting colors to animals and plants. In animals the most important examples include MELANIN (black), RHODOPSIN (purple) and the respiratory pigments, HEMOGLOBIN (red) and HEMOCYANIN (blue). (See also MIMICRY; PROTECTIVE COLORATION.) In plants, the CHLOROPHYLLS (green) are important as the key chemicals in PHOTOSYNTHESIS. Other plant pigments include the carotenes and xanthophylls (red-yellow), the anthocyanins (red-blue) and the anthoxanthins (yellow-orange). In nature, whiteness results from the absence of pigment (see ALBINO) and is comparatively uncommon.

Testing a weld on an oil pipeline in the Shetland Isles. This pipeline is part of the 148km long Brent system pipeline bringing oil from the Shell/Esso Brent oilfield to an oil storage terminal at Sullom Voe. It is paralleled in this overland section by British Petroleum's Ninian field pipeline.

PILE, a heavy beam or column made of wood, steel or concrete, sunk into the ground to support a load. When they rest on bedrock they are known as end-bearing piles; when supported by the friction of the soil, friction piles. Some concrete piles are cast in place, but most are driven in by piledrivers, large hammers worked by gravity or by hydraulic or pneumatic power. Vibratory piledrivers are an efficient recent innovation.

PILE, Atomic. See NUCLEAR REACTOR.

PILES. See HEMORRHOIDS.

PILTDOWN MAN, *Eoanthropus dawsoni*. In 1908–15 were found under Piltdown Common, Sussex, UK, a skull with ape-like jaw but large, human cranium and teeth worn down in a way unlike those of any extant ape, surrounded by FOSSIL animals that indicated an early PLEISTOCENE date. Piltdown Man was held by many as an ancestor of *Homo Sapiens* until 1953, when the fraud was exposed: the skull was human but relatively recent; the even more recent jaw that of an orangutan; the teeth had been filed down by hand; and the fossil animals were not of British origin. The remains had been artificially stained to increase confusion.

PIN, peg used for fastening. In engineering the term is applied to a metal peg of any size used to join parts. In ordinary usage, a pin is a headed piece of wire, sharp at one end, used from earliest times to secure cloth to be sewn, clothing or the hair. Safety pins have a spring shield for the point.

PINEAL BODY, or Pineal gland, a gland-like structure situated over the BRAIN stem and which appears to be a vestigial remnant of a functioning ENDOCRINE GLAND in other animals. It has no known function in man, although DESCARTES thought it to be the seat of the soul. It has a role in pigmentation in some species; calcium deposition in the pineal makes it a useful marker of midline in skull X-rays.

PINEL, Philippe (1745–1826), French pioneer of the scientific study of MENTAL ILLNESS and the humane treatment of mental patients, whose remarkably modern ideas have earned him regard as a father of psychiatry.

PINKEYE, common name for CONJUNCTIVITIS.

PINT, name of various units of dry or liquid measure. See WEIGHTS AND MEASURES.

PIONEER PROBES, US space probe series started in 1958. Pioneers 1–3 studied the VAN ALLEN RADIATION BELTS. Pioneers 5–8 were launched into solar orbit to study interplanetary space and the sun itself. Pioneers 10 and 11 were Jupiter "fly-by" probes.

PIPES AND PIPELINES, tubes for conveying fluids—liquids, gases or slurries. Pipes vary in diameter considerably, according to the flow rate required and the pressure gradient: oil pipelines may be up to 1.2m in diameter. Materials used include steel, cast iron, other metals, reinforced concrete, fired clay, plastic, bitumenized-fiber cylinders, and wood. They are often coated inside and out with bitumen or concrete to prevent corrosion. Concrete, plastic and steel pipes can now be made and laid in one continuous process, but most pipes still need to be joined by means of welding, screw joints, clamped flange joints, couplings, or bell-and-spigot joints caulked with lead or cement. Pipelines, consisting of long lengths of pipe with valves and pumps at regular intervals (about 100km for oil pipelines), are used chiefly for transporting water, sewage, chemicals, foodstuffs, crude oil and natural gas.

PISCES (the Fishes), a large, faint constellation on the ECLIPTIC, the 12th sign of the ZODIAC. The vernal EQUINOX now lies in Pisces.

PISTIL. See FLOWER.

PISTOL, small FIREARM that can be conveniently held and operated in one hand. It developed in parallel with the shoulder weapon from the 14th century, first becoming really practical in the early 16th century with the invention of the wheel-lock firing mechanism, soon superseded by the flintlock. Modern rapid-fire pistols are usually either REVOLVERS or automatics. Automatic pistols, such as the Colt .45 Automatic, contain a magazine of cartridges in the butt and are automatically reloaded and cocked by the energy of recoil when a round is fired (see AMMUNITION).

PISTON, solid cylindrical piece that moves up and down inside a hollow cylinder in an ENGINE or PUMP, being driven by, or driving, fluid under pressure. The piston is fitted with rings to fit the cylinder snugly. In the INTERNAL-COMBUSTION ENGINE (but not the FREE-PISTON ENGINE) it bears a connecting rod to transmit the power to the crankshaft.

PITCH, black solid BITUMEN; the residue from distilling COAL TAR, wood tar or PETROLEUM, sometimes occurring naturally. It is used in roadmaking, for waterproofing and for caulking seams.

PITCH, Musical, refers to the FREQUENCY of the

vibrations constituting a SOUND. The frequency associated with a given pitch name (e.g., Middle C) has varied considerably over the years. The present international standard sets Concert A at 440Hz.

PITCHBLENDE, or **Uraninite**, brown, black or greenish mineral, the most important source of URANIUM, RADIUM and POLONIUM. The composition varies between UO_2 and $UO_{2 \cdot 6}$; thorium, radium, polonium, lead and helium are also present. Principal deposits are in Zaire, Bohemia, at Great Bear Lake, Canada, and in the Mountain States.

PITHECANTHROPUS ERECTUS. See PREHISTORIC MAN.

PITOT TUBE, device invented by **Henri Pitot** (1695–1771) in 1732 and widely used in fluid dynamics for measuring FLUID velocities. One open end of a cylindrical tube points directly into the flowing stream, and the other end is connected to a pressure-measuring device. This compares the pitot-tube pressure with the static stream pressure, the difference being a measure of the fluid velocity.

PITUITARY GLAND, major ENDOCRINE GLAND situated just below the BRAIN, under the control of the adjacent HYPOTHALAMUS and in its turn controlling other endocrine glands. The posterior pituitary is a direct extension of certain cells in the hypothalamus and secretes VASOPRESSIN and OXYTOCIN into the BLOOD stream. The anterior pituitary develops separately and consists of several cell types which secrete different HORMONES, including growth hormone, FOLLICLE STIMULATING HORMONE, LUTEINIZING HORMONE, PROLACTIN, thyrotrophic hormone (which stimulates thyroid gland) and adrenocorticotrophic hormone (ACTH). Growth hormone is concerned with skeletal growth and development as well as regulation of blood sugar (anti-INSULIN activity). The anterior pituitary hormones are controlled by releasing hormones secreted by the hypothalamus into local blood vessels; the higher centers of the brain and environmental influences act by this route. FEEDBACK from the organs controlled occurs at both the hypothalamic and pituitary levels. Pituitary TUMORS or loss of blood supply may cause loss of function, while some tumors may be functional and produce syndromes such as GIGANTISM or acromegaly (due to growth hormone imbalance). Pituitary tumors may also affect VISION by compressing the nearby optic nerves. Sophisticated tests of pituitary function are now available.

Automobile piston and connecting rod.

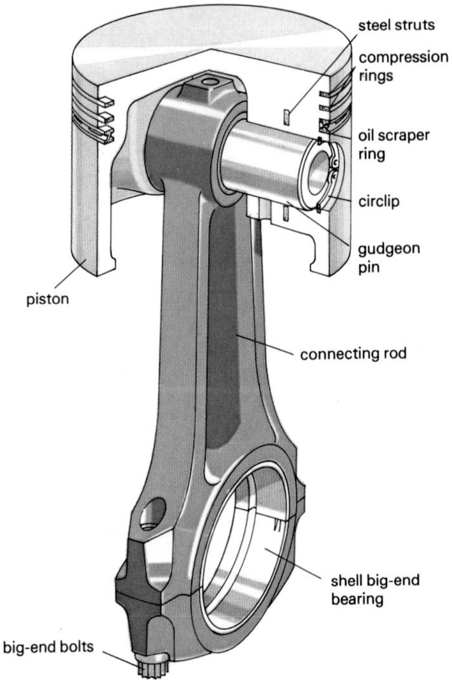

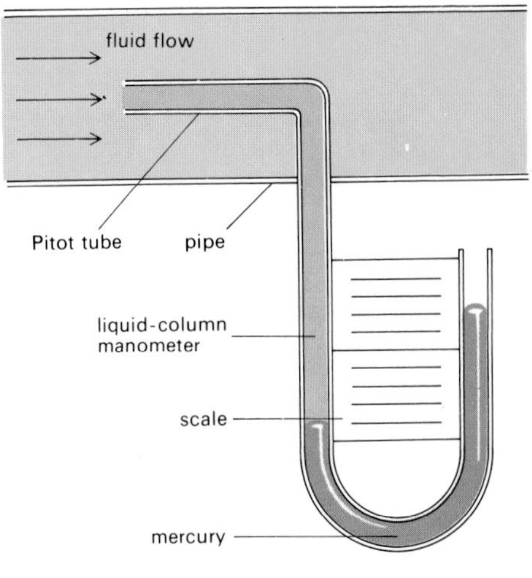

Application of a simple Pitot tube to the measurement of fluid-flow velocity in a pipe. The impact pressure at the mouth of the tube is here measured using a simple liquid-column manometer. The pitot-tube pressure is read off as the difference between the mercury levels in the two branches of the manometer.

PKU. See PHENYLKETONURIA.

PLACEBO, a tablet, syrup or other form of medication which is inactive and is prescribed in lieu of active preparations, e.g., in experimental studies of DRUG effectiveness.

PLACENTA, in PLACENTAL MAMMALS including MAN, specialized structure derived from the WOMB lining and part of the EMBRYO after IMPLANTATION; it separates and yet ensures a close and extensive contact between the maternal (uterine) and fetal (umbilical) BLOOD CIRCULATIONS. This allows nutrients and OXYGEN to pass from the mother to the FETUS, and waste products to pass in the reverse direction. The placenta thus enables the embryo and fetus to live as a PARASITE, dependent on the maternal organs. GONADOTROPHINS are produced by the placenta which prepares the maternal body for delivery and the BREASTS for LACTATION. The placenta is delivered after the child at BIRTH (the afterbirth) by separation of the blood vessel layers; placental disorders may cause ante- or post-partum HEMORRHAGE or fetal immaturity.

PLACENTAL MAMMALS. See EUTHERIA.

PLACER MINING, the extraction of minerals such as gold, platinum and diamonds from ORE that has accumulated through the processes of weathering or EROSION. The earliest and best known form of placer mining is gold panning.

PLAGUE, a highly infectious disease due to a bacterium carried by rodent fleas. It causes greatly enlarged LYMPH nodes (buboes, hence bubonic plague), SEPTICEMIA with FEVER, prostration and COMA; plague PNEUMONIA is particularly severe. If untreated, DEATH is common and EPIDEMICS occur in areas of overcrowding and poverty. It still occurs on a small rural scale in the Far East; massive epidemics such as the **Black Death**, which perhaps halved the population of Europe in the mid-14th century, are rare. Rat and flea control, disinfection and ANTIBIOTICS are the mainstay of current prevention and treatment.

PLANCK, Max Karl Ernst Ludwig (1858–1947), German physicist whose QUANTUM THEORY, with the Theory of RELATIVITY, ushered physics into the modern era. Initially influenced by CLAUSIUS, he made fundamental researches in THERMODYNAMICS before turning to investigate BLACKBODY RADIATION. To describe the electromagnetic radiation emitted from a BLACK BODY he evolved the **Planck Radiation Formula**, which implied that ENERGY, like MATTER, is not infinitely subdivisible—that it can exist only as quanta (see PLANCK CONSTANT). Planck himself was unconvinced of this, even after EINSTEIN had applied the theory to the PHOTOELECTRIC EFFECT and BOHR in his model of the ATOM; but for his achievement he received the 1918 Nobel Prize for Physics.

PLANCK CONSTANT, h ($=6.6256\times10^{-34}$J s), a quantity fundamental to quantum physics, named for Max PLANCK, who in 1900 solved a long-standing problem in radiation physics with the hypothesis that the energy of a system vibrating with frequency ν had to be a whole-number multiple of $h\nu$. The Planck constant also governs the accuracy with which different properties can be measured simultaneously (see UNCERTAINTY PRINCIPLE) and the wavelength of the wave associated with a particle (see QUANTUM MECHANICS).

PLANE, a surface having two DIMENSIONS only, length and breadth, any two POINTS of which can be joined by a straight LINE composed entirely of points also in the plane. A plane may be determined by two intersecting or PARALLEL lines, by a line and a point that does not lie on the line, or by three points that do not lie in a straight line. The intersection of two planes is a straight line; the intersection of a plane and a line in a different plane is a point. An infinite number of planes may pass through a single point or line. A plane is parallel to another plane if all PERPENDICULARS drawn between them are of equal length.

PLANE OF SYMMETRY, a PLANE cutting a geometrical figure such that the parts of it lying on either side are symmetrical. (See also AXIS OF SYMMETRY; SYMMETRY.)

PLANET, in the SOLAR SYSTEM, one of the nine major celestial bodies orbiting the sun; by extension, a similar body circling any other star. In 1963 it was discovered that BARNARD's Star has at least one companion about 1.5 times the size of Jupiter, implying that planets are by no means unique to the sun.

PLANETARIUM, an instrument designed to represent the relative positions and motions of celestial objects. Originally a mechanical model of the SOLAR SYSTEM (see ORRERY), the planetarium of today is an intricate optical device that projects disks and points of light representing sun, moon, planets and stars on to the interior of a fixed hemispherical dome. The various cyclic motions of these bodies as seen from a given latitude on earth can be simulated. Of great assistance to students of ASTRONOMY and celestial NAVIGATION, planetariums also attract large public audiences. The first modern planetarium, built in 1923 by the firm of Carl ZEISS, is still in use at the Deutsches Museum, Munich, West Germany.

PLANETESIMAL HYPOTHESIS, a discarded theory proposed by T. C. Chamberlin and F. R. Moulton to explain the formation of PLANETS. It states that a passing star drew matter out of the sun, some of which condensed to form small solid particles (planetesimals) which in turn coalesced to form planets.

PLANKTON, microscopic animals and plants that live in the sea. They drift under the influence of OCEAN CURRENTS and are vitally important links in the marine food chain (see ECOLOGY). A major part of plankton comprises minute plants (phytoplankton), which are mainly ALGAE, but include DINOFLAGELLATES and DIATOMS. Phytoplankton may be so numerous as to color the water and cause it to have a "bloom." They are eaten by animals (zooplankton), which comprise the eggs, larvae and adults of a vast array of animal types, from Protozoa to jellyfish. Zooplankton is an important food for large animals such as whales and countless fishes such as herring. Phytoplankton is confined to the upper layers of the sea where light can reach, but zooplankton has been found at great depths. (See also OCEANS.)

PLANT, a living organism belonging to the PLANT KINGDOM. Green plants are unique in being able to synthesize their own organic molecules from carbon dioxide and water using light energy by the process known as PHOTOSYNTHESIS. Mineral nutrients are absorbed from the environment. Plants are the primary source of food for all other living organisms (see ECOLOGY). The possession of CHLOROPHYLL, the green photosynthetic pigment, is probably the most important distinction between plants and animals, but there are several other differences. Plants are stationary, have no nervous system and the cell wall contains large amounts of CELLULOSE. But there are exceptions. Some plants, such as ALGAE and BACTERIA, can move about, and others, including FUNGI, bacteria and some PARASITES do not contain chlorophyll and cannot synthesize their own organic molecules, but absorb them from their environment. Some INSECTIVOROUS PLANTS obtain their food by trapping insects.

Although the more primitive plants vary considerably in their overall structure, the higher plants (GYMNOSPERMS and ANGIOSPERMS) are much the same in their basic anatomy and morphology. In a typical angiosperm, four main regions can be recognized: ROOT, STEM, LEAF and FLOWER. Each region has one or more basic functions.

When examined under the microscope, a piece of plant tissue can be seen to consist of thousands of tiny CELLS, generally packed tightly together. The cells are not all alike and each one is adapted to do a certain job. All are derived, however, from a basic pattern. This basic plant cell tends to be rectangular and it has a tough wall of cellulose which gives it its shape, but the living boundary of the cell is the delicate cell membrane just inside the wall. Inside the membrane is the PROTOPLASM, which contains the nucleus, the CHLOROPLASTS and many other microscopic structures. In the center of the protoplasm there is a large sap-filled vacuole, which maintains the cell's shape and plays an important part in the working of the whole plant.

Both sexual and asexual REPRODUCTION are widespread throughout the plant kingdom. Many plants are capable of both forms and in some cases the life cycle of the plant may involve the two different forms (see ALTERNATION OF GENERATIONS). (See also BOTANY; FERTILIZATION; FRUIT; GERMINATION; GROWTH; OSMOSIS; PLANT DISEASES; POLLINATION; TRANSPIRATION.)

PLANT DISEASES cause serious losses to crop production; they may kill plants completely, but more often they simply reduce the yield. Most plant diseases are caused by microorganisms which infect the tissues, the most important being FUNGI, including MILDEW, RUSTS and SMUTS. Control methods are based on FUNGICIDES. VIRUSES are the next most damaging group of plant pathogens. Most of them are carried by aphids and other sap-sucking insects and control is largely a matter of controlling these insect carriers. BACTERIA are less important, their main role being in secondary infection, causing the tissues to rot. Deficiency diseases are caused by a lack of available minerals in the soil. Insect pests, such as the boll weevil on cotton, can also cause serious crop damage.

PLANT KINGDOM, the second great group of living organisms. The plant and ANIMAL KINGDOMS together embrace all living things except VIRUSES, and only overlap in the most primitive organisms. The plant kingdom is extremely diverse (over 400 000 species are now known), and they are found in almost every conceivable habitat. They range in size from microscopic BACTERIA to 100m (330ft) sequoias. The plant kingdom can be arranged into an orderly hierarchical pattern of classification (see TAXONOMY) containing divisions, classes, orders, families, genera and species. Indeed, several systems have been evolved to do this. In the classical Eichler system there are four divisions: the THALLOPHYTA, including bacteria, SLIME MOULDS, ALGAE and FUNGI; the BRYOPHYTA, including LIVERWORTS, HORNWORTS and MOSSES; the PTERIDOPHYTA, including FERNS, CLUB MOSSES and HORSETAILS; and the SPERMATOPHYTA, including GYMNOSPERMS and ANGIOSPERMS, the latter being divided into DICOTYLEDONS and MONOCOTYLEDONS. However, this system has been replaced recently by a more natural arrangement of 11 divisions: Schizophyta, bacteria and blue-green algae; Euglenophyta, euglenoids; Chlorophyta, green algae; Xanthophyta, yellow-green algae; Chrysophyta, golden algae and DIATOMS; Phaeophyta, brown algae; Rhodophyta, red algae; Pyrrophyta, dinoflagellates and cryptomonads; Mycota, slime molds and fungi; Bryophyta, liverworts and mosses; and Tracheophyta, the vascular plants, including horsetails, ferns, gymnosperms and angiosperms. Under this system some authorities break the plant kingdom into three kingdoms: the Monera, including the division Schizophyta; the Metaphyta, including the Bryophyta and Tracheophyta; and the Protista, which includes all the other divisions.

PLASMA, almost completely ionized GAS, containing equal numbers of free ELECTRONS and positive IONS. Plasmas such as those forming stellar atmospheres (see STAR) or regions in an electron discharge tube are highly conducting but electrically neutral and many phenomena occur in them that are not seen in ordinary gases. The TEMPERATURE of a plasma is theoretically high enough to support a controlled nuclear FUSION reaction. Because of this, plasmas are being widely studied particularly in MAGNETOHYDRODYNAMICS research. Plasmas are formed by heating low-pressure gases until the ATOMS have sufficient energy to ionize each other. Unless the plasma can be successfully contained by electric or magnetic fields, rapid cooling and recombination occurs; indeed the high temperatures needed for thermonuclear reactions cannot as yet be maintained in the laboratory for sufficiently long.

PLASMA, the part of the BLOOD remaining when all CELLS have been removed, and which includes CLOTTING factors. It may be used in resuscitation from SHOCK.

PLASMODIUM, genus of PROTOZOA responsible for MALARIA. Four main types are recognized: *P. falciparum*; *P. vivax*; *P. ovale*, and *P. malariae*, which cause variants of malaria and are endemic in different areas. *P. falciparum* causes cerebral malaria.

PLASTER OF PARIS, or calcium sulfate hemihydrate. See CALCIUM.

PLASTIC EXPLOSIVE, putty-like EXPLOSIVE made by mixing RDX with oil. Convenient and weather-resistant, it is used for demolition.

PLASTICS, materials that can be molded (at least in production) into desired shapes. A few natural plastics are known, e.g., BITUMEN, RESINS and RUBBER, but almost all are man-made, mainly from PETROCHEMICALS, and are available with a vast range of useful properties: hardness, elasticity, transparency, toughness, low density, insulating ability, inertness and corrosion resistance, etc. They are invariably high POLYMERS with carbon skeletons, each molecule being made up of thousands or even millions of atoms. Plastics fall into two classes: thermoplastic and thermosetting. **Thermoplastics** soften or melt reversibly on heating; they include celluloid and other cellulose plastics, LUCITE, NYLON, POLYETHYLENE, STYRENE polymers, VINYL polymers, polyformaldehyde and polycarbonates. **Thermosetting** plastics, although moldable when produced as simple polymers, are converted by heat and pressure, and sometimes by an admixed hardener, to a cross-linked, infusible form. These include bakelite and other phenol resins, EPOXY RESINS, polyesters, SILICONES, urea-formaldehyde and melamine-formaldehyde resins, and some polyurethanes. Most plastics are mixed with stabilizers, fillers, dyes or pigments and plasticizers if needed. There are several fabrication processes: making films by calendering (squeezing between rollers), casting or extrusion, and making objects by compression molding, injection molding (melting and forcing into a cooled mold) and casting. (See also LAMINATES; SYNTHETIC FIBERS.)

PLASTIC SURGERY, the branch of SURGERY devoted to reconstruction or repair of deformity, surgical defect or the results of injury. Using bone, cartilage, tendon, and skin from other parts of the body, or artificial substitutes, function and appearance may in many cases be restored. In skin grafting, the most common procedure, a piece of skin is cut, usually from the thigh, and stitched to the damaged area. Bone and cartilage (usually from the ribs or hips), or sometimes plastic, are used in cosmetic remodeling and facial reconstruction after injury. Congenital defects such as HARELIP and CLEFT PALATE can be treated in infancy. "Face lifting," the cosmetic removal of excess fat and tightening of the skin, is a delicate and often unsuccessful operation, carrying the added risk of infection.

PLASTIDS, variously shaped bodies found in the cytoplasm of plant CELLS, containing CHLOROPHYLL (chloroplasts), other PIGMENTS (chromoplasts) or unpigmented (leucoplasts).

PLATELET. See BLOOD.

PLATE TECTONICS, Theory of, fundamental theory of modern geology, arising from studies of CONTINENTAL DRIFT, EARTHQUAKE and VOLCANO distributions, and sea-floor spreading, which phenomena it largely explains. The earth's crust is viewed as consisting of a number of semirigid plates in relative motion. Where plates meet, one edge is subducted beneath (forced under) the other: in midocean, this results in OCEAN trenches, deep seismic activity and arcs of volcanic ISLANDS; at continental margins, similar subduction of the oceanic plate results also in OROGENIES. Where lighter continental blocks are forced together, neither edge is subducted and more complex orogeny occurs. Belts of shallow earthquakes define the midocean ridges where new material is emerging from below. (See SEA-FLOOR SPREADING.)

PLATINUM (Pt), soft, silvery-white metal in the PLATINUM GROUP. In addition to the general uses of these metals, platinum is used as a catalyst for the CONTACT PROCESS and (alloyed with rhodium) for the OSTWALD PROCESS. AW 195.1, mp 1772°C, bp 4010°C, sg 21.45 (20°C).

PLATINUM GROUP, the six NOBLE METALS in Group VIII of the PERIODIC TABLE, i.e., RUTHENIUM, RHODIUM, PALLADIUM, OSMIUM, IRIDIUM and PLATINUM (see also TRANSITION ELEMENTS). They are found together in PYROXENE deposits in South Africa and in the copper and NICKEL ores of Canada and the USSR. All are highly inert and corrosion-resistant, though palladium, osmium and platinum dissolve in AQUA REGIA; the others can be dissolved by fused oxidizing alkalis. Palladium dissolves slowly in oxidizing acids. Ruthenium and osmium show chief oxidation states +3, 4, 6 and 8; the other metals seldom exceed +4. All six metals form numerous HALIDES and complex halogen ions, and many other LIGAND complexes, including carbonyls resembling those of iron, cobalt and NICKEL. The platinum group metals are used, usually as ALLOYS with each other, for jewelry, the tips of pen nibs, electrical contacts, THERMOCOUPLES, crucibles, surgical instruments, standard WEIGHTS AND MEASURES, and (finely divided) as catalysts (see CATALYSIS).

PLATO, Greek philosopher (c427–347 BC). A pupil of SOCRATES, c385 BC he founded the Academy, where ARISTOTLE studied. His early dialogues present a portrait of Socrates as destructive arguer, but in the great middle dialogues he develops his own doctrines such as the Theory of Forms (*Republic*), the immortality of the soul (*Phaedo*), knowledge as recollection of the Forms by the soul (*Meno*), virtue as knowledge (*Protagoras*), and attacks hedonism and the idea that "might is right" (*Gorgias*). The *Symposium*

Plate tectonics: the major plates of the earth's crust and the directions in which they are moving. Dotted lines show probable plate margins, arrowheads destructive plate margins with underthrusting in the direction of the arrowheads.

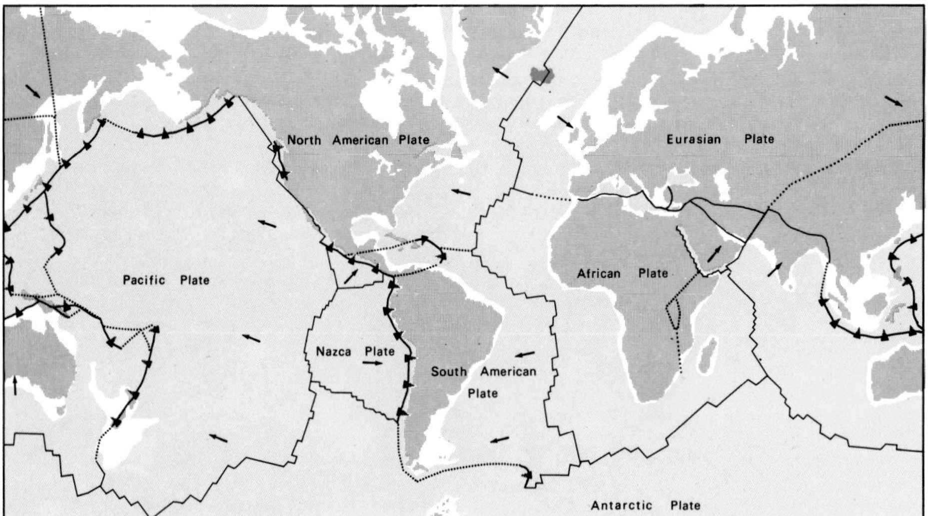

and *Phaedrus* sublimate love into a beatific vision of the Forms of the Good and the Beautiful. The late dialogues (*Sophist, Theaetetus, Politicus, Philebus, Parmenides*) are difficult and technical; the *Timaeus* contains cosmological speculation. In the *Republic* Plato posits abstract Forms as the supreme reality. The highest function of the human soul is to achieve the vision of the Form of the Good. Drawing an analogy between the soul and the state, he presents his famous ideal state ruled by philosophers, who correspond to the rational part of the soul. In the late *Laws* Plato develops in detail his ideas of the state. His idealist philosophy, his insistence on order and harmony, his moral fervor and asceticism and his literary genius have made Plato a dominant figure in Western thought.

PLATYHELMINTHES, or flatworms, a phylum containing free-living TURBELLARIA, parasitic flukes (TREMATODA) and the TAPEWORMS (Cestoda).

PLAY, a distinctive type of behavior of both adult and juvenile animals, of unknown function and involving the incomplete, ritualized expression of normal adult behavior patterns. Movements are extravagant and exaggerated. Play occurs particularly in carnivores, primates and certain birds.

PLAYA, found in undrained areas in arid regions, a level tract formed of deposits from a temporary lake that has formed owing to flooding or heavy rainfall, and then evaporated. (See also ALKALI FLATS; EVAPORITES.)

PLEASURE PRINCIPLE, a concept of FREUD, the avoidance of PAIN or unpleasantness, the sole influence on the mind before the EGO has developed. In later stages, it is modified by the **reality principle**, which recognizes physical and social constraints.

PLEIADES, a GALACTIC CLUSTER in the constellation TAURUS. Seven of the stars can be seen by the naked eye, and these are named after the seven daughters of Atlas. The Pleiades are about 153pc from the sun and are surrounded by a bright NEBULA.

PLEISTOCENE, the earlier epoch of the QUATERNARY, stretching from between about 4 million through 10 000 years ago. (See also GEOLOGY; HOLOCENE.)

PLESIOSAURIA, fossil members of the REPTILIA that lived during the Jurassic and Cretaceous periods, 190–65 million years ago. They were aquatic, with long necks and limbs modified as paddles. Some believe that plesiosaurs still survive as such as the Loch Ness monster but more conventional opinion regards them as having been long extinct.

PLEURISY, INFLAMMATION of the pleura, the two thin connective tissue layers covering the outer LUNG surface and the inner CHEST wall. It causes a characteristic chest pain, which may be localized and is made worse by deep breathing and coughing. It may be caused by infection (e.g., PNEUMONIA, TUBERCULOSIS) or TUMORS and inflammatory disease.

PLEXIGLAS. See LUCITE.

PLIOCENE, the final period of the TERTIARY, immediately preceding the QUATERNARY, lasting from about 10 to 4 million years ago. (See also GEOLOGY.)

PLUTO, the ninth planet of the SOLAR SYSTEM, orbiting the sun at a mean distance of 39.53AU in 248.4 years. Pluto was discovered in 1930 following observations of PERTURBATIONS in NEPTUNE's orbit. Because of its great distance from us, little is known of Pluto's composition, atmosphere, mass (probably less than 0.1 that of the earth) or diameter (probably 5 000–6 000km). Its orbit is very eccentric: indeed, for a period after 1987 it will be closer to the sun than is Neptune.

PLUTONISM, or **Vulcanism**, the geological theory, often associated with the followers of J. HUTTON, that the rocks of the earth were originally volcanic in origin. In the early 19th century, plutonism rivalled NEPTUNISM for acceptance as the fundamental geological principle.

PLUTONIUM (Pu), the most important TRANSURANIUM ELEMENT, used as fuel for NUCLEAR REACTORS and for the ATOMIC BOMB. It is one of the ACTINIDES and chemically resembles URANIUM. Pu^{239} is produced in BREEDER REACTORS by neutron irradiation of uranium (U^{238}); like U^{235}, it undergoes nuclear FISSION, and was used for the Nagasaki bomb in WWII. mp 640°C, bp 3235°C, sg 19.84 (α; 25°C).

PLYWOOD, strong, light wood composite made of layers of VENEER glued with their grain alternately at right angles. Thick plywood may have a central core of sawn lumber. It is made of an odd number of layers, and is termed 3-ply, 5-ply, etc. Being strong in both directions, and almost free from warping and splitting. it is used for construction of all kinds.

PNEUMOCONIOSIS, restrictive disease of the LUNGS caused by deposition of dusts in the lung substance, inhaled during years of exposure, often in extractive industries. SILICOSIS, anthracosis and asbestosis are the principal kinds, although aluminum, iron, tin and cotton fiber also cause pneumoconiosis. Characteristic X-RAY changes are seen in the lungs.

PNEUMONIA, INFLAMMATION and consolidation of LUNG tissue. It is usually caused by bacteria (pneumococcus, STAPHYLOCOCCUS, GRAM'S STAIN negative bacilli), but rarely results from pure VIRUS infection (INFLUENZA, MEASLES); other varieties occur if food, secretions or chemicals are aspirated or inhaled. The inflammatory response causes lung tissue to be filled with exudate and PUS, which may center on the bronchi (**bronchopneumonia**) or be restricted to a single lobe (**lobar pneumonia**). Cough with yellow or green sputum (sometimes containing BLOOD); FEVER; malaise, and breathlessness are common. The involvement of the pleural surfaces causes PLEURISY. ANTIBIOTICS and PHYSIOTHERAPY are essential in treatment.

PNEUMOTHORAX, presence of air in the pleural space between the LUNG and the CHEST wall. This may result from trauma, rupture of lung bullae in EMPHYSEMA or in ASTHMA, TUBERCULOSIS, PNEUMOCONIOSIS, CANCER etc., or, in tall thin athletic males, it may occur without obvious cause. Drainage of the air through a tube inserted in the chest wall allows lung re-expansion.

PODIATRY, or **chiropody**, care of the FEET, concerned with the nails, CORNS AND CALLUSES, bunions and toe deformities. Care of the SKIN of the feet is especially important in the elderly and in diabetics.

PODZOL or **podsol**, SOIL found in moist, cool climates under coniferous FORESTS, TUNDRA, etc. Podzol is unsuitable for agricultural purposes, having little HUMUS.

POGONOPHORA, a phylum of about 100 species of marine worm-like animals that lack both mouth and gut. Pogonophorans are tube dwellers, some growing to a length of 300mm (1ft). The phylum was only recognized in 1950.

POINCARÉ, Jules Henri (1854–1912), French mathematician, cosmologist and scientific philosopher, best known for his many contributions to pure and applied MATHEMATICS and celestial mechanics.

POINT, in GEOMETRY, entity defined as having none of the DIMENSIONS length, breadth or depth. A point may also be defined as the INTERSECTION of two straight LINES or of a straight line and a PLANE.

POINT SET. See SET THEORY.

POISON GAS. See CHEMICAL AND BIOLOGICAL WARFARE.

POISONING, the taking, via ingestion or other routes, of substances which are liable to produce illness or DEATH. Poisoning may be accidental, homicidal, suicidal or as a suicidal gesture. DRUGS and medications are often involved, either taken by children in ignorance of their nature from accessible places, or by adults in suicide or attempted suicide. Easily available drugs such as ASPIRIN, paracetamol and mild SEDATIVES are often taken, though in serious suicidal attempts, BARBITURATES and ANTIDEPRESSANTS are more common. Chemicals, such as disinfectants and weedkillers, cosmetics and paints are frequently swallowed as drinks by children, while poisonous berries may appear attractive. Poisoning by domestic gas or carbon monoxide has been used for suicide and homicide. Heavy metals (see LEAD POISONING, MERCURY POISONING, ARSENIC), INSECTICIDES and CYANIDES are common industrial poisons as well as being a risk in the community. Poisons may act by damaging body structures (e.g., weedkillers); preventing OXYGEN uptake by HEMOGLOBIN (carbon monoxide); acting on the NERVOUS SYSTEM (heavy metals); interfering with essential ENZYMES (cyanides, insecticides); with HEART action (antidepressants), or with the control of RESPIRATION (barbiturates). In some cases, antidotes are available which, if used early, can minimize poisoning, but in most cases, life is supported until the poison is eliminated.

POLAR COORDINATES. See ANALYTIC GEOMETRY.

POLARIMETRY, measurement of OPTICAL ACTIVITY by means of a polarimeter or polariscope, an instrument having two NICOL PRISMS, one fixed (the polarizer) and one rotatable (the analyzer), with the sample between them (see POLARIZED LIGHT). It is used in chemical ANALYSIS (notably for measuring sugar concentrations) and to study molecular configurations. The polariscope is also used to study strain (see MATERIALS, STRENGTH OF) in materials showing DOUBLE REFRACTION.

POLARIS (Alpha Ursae Minoris), a CEPHEID VARIABLE star in the LITTLE DIPPER. Because of its close proximity to the N celestial pole (see CELESTIAL SPHERE), Polaris is also known as the Polestar or North Star, and has been used in navigation for centuries: owing to PRECESSION, Polaris is moving away from the N celestial pole.

POLARIZED LIGHT, LIGHT in which the orientation of the wave vibrations displays a definite pattern. In ordinary unpolarized light the wave vibrations (which occur at right-angles to the direction in which the radiation is propagated) are distributed randomly about the axis of propagation. In *plane-polarized light* (produced in reflection from a DIELECTRIC such as glass or by transmission through a NICOL PRISM or polarizing filter), the vibrations all occur in a single plane. Polaroid filters work by subtracting the components of light orientated in a particular plane; two filters in sequence with their transmission planes crossed transmit no light. In *elliptically polarized light* (produced when plane-polarized light is reflected from a polished metallic surface) and *circularly polarized light* (produced on transmission through certain CRYSTALS exhibiting double refraction), the electric vector of the radiation at any point describes an ellipse or a circle. Much of the light around us—that of the blue sky, or reflected from lakes, walls and highways—is partially polarized. Polarizing sunglasses reduce glare by eliminating the light polarized by reflection from horizontal surfaces. Polariscopes employing two polarizing filters have proved to be valuable tools in organic chemistry (see POLARIMETRY).

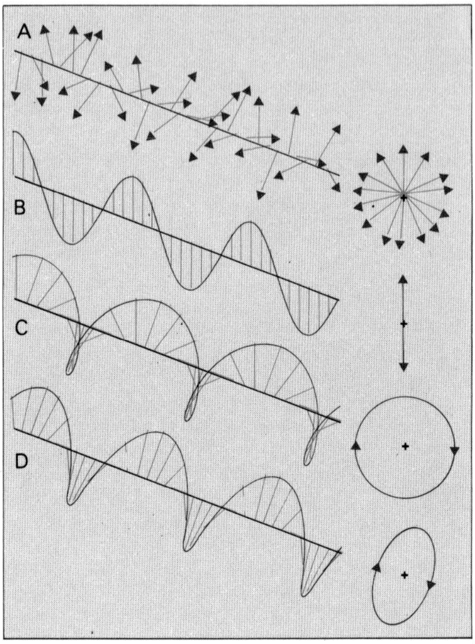

Polarized light: schematized "perspective" (*left*) and "end on" (*right*) views of the instantaneous positions of the electric vector (blue) and the path traced out by the tip of the electric vector (red) of electromagnetic radiation when it is (A) unpolarized; (B) plane polarized; (C) circularly polarized, and (D) elliptically polarized.

POLAROGRAPHY, chemical ANALYSIS, particularly for metals or organic groups, by measuring the saturation current and threshold voltage (see ELECTRICITY) for ELECTROLYSIS of a solution of the substance.

POLAROID LAND CAMERA, photographic CAMERA announced by LAND in 1947 which produces a finished print only seconds after exposure. The optical system is similar to that of other cameras but the film pack used contains positive paper and developing reagent as well as negative film. On advancing the film after exposure, it is pressed against the paper (which is not light-sensitive) and the contents of a reagent pod are spread between them. A positive print formed by a diffusion process on the paper can be stripped off seconds later.

POLESTAR. See POLARIS.

POLIOMYELITIS, or **infantile paralysis**, VIRAL DISEASE causing muscle PARALYSIS as a result of direct damage to motor nerve cells in the SPINAL CORD. The virus usually enters by the mouth or GASTROINTESTINAL TRACT and causes a mild feverish illness, after which PARESIS or paralysis begins, often affecting mainly those muscles that have been most used in preceding days. Treatment is with bed rest and avoidance or treatment of complications: contracture; bed sores; venous THROMBOSIS; secondary infection; MYOCARDITIS; respiratory failure, and swallowing difficulties. Current polio vaccine is a live attenuated strain taken by mouth which colonizes the gut and induces IMMUNITY. Poliomyelitis VACCINATION has been one of the most successful developments in preventive medicine.

The SX-70 Polaroid Land Camera broke away from conventional camera optics to offer a new compact instant-picture camera design. The same object lens system is used both for viewfinding (*above*) and film exposure (*below*), the light paths being switched by swinging upward a plate (red) carrying mirrors on each face. The mirror on the upper side of the plate is, in fact, an eccentrically centered Fresnel mirror on which an image is formed which is viewed through the viewfinder. This facilitates compact design with a minimum of viewfinder distortion.

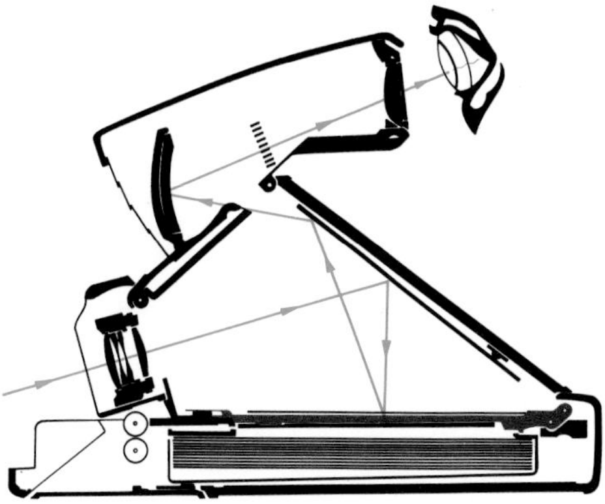

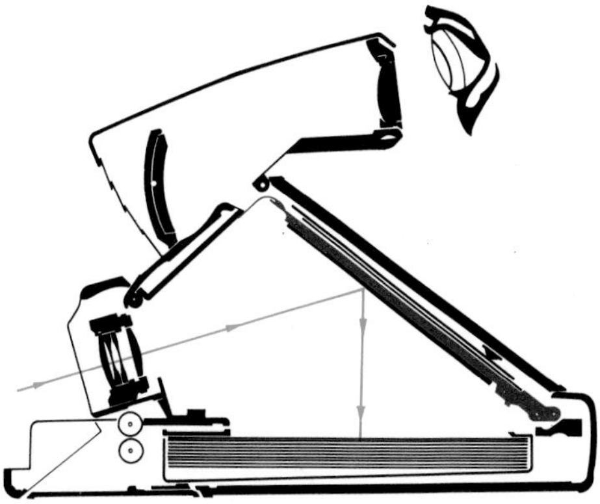

POLLEN. See POLLINATION.

POLLEN COUNT, an estimate of atmospheric pollen each day during spring and summer, which gives a guide to the intensity of the stimulus to ASTHMA and HAY FEVER in subjects with pollen ALLERGY. It is affected by wind, rain and humidity.

POLLINATION, in plants, the transfer of the male GAMETES (*pollen*) from the anthers of a FLOWER to the stigma of the same or another flower, where subsequent growth of the pollen leads to the fertilization of the female gametes (or EGGS) contained in the ovules and the production of SEEDS and FRUIT. Wind-pollinated plants, such as grasses, produce inconspicuous flowers with large feathery stamens and stigmas and usually large quantities of pollen. Insect-pollinated flowers have large conspicuous and colorful flowers, produce NECTAR and have small stigmas. (See PLANT; REPRODUCTION.)

POLLUTION, the contamination of one substance by another so that the former is unfit for an intended use; or, more broadly, the addition to any natural environmental resource on which life or the quality of life depends or any substance or form of energy at a rate resulting in abnormal concentrations of what is then termed the "pollutant." Air (see AIR POLLUTION), water and soil are the natural resources chiefly affected. Some forms of pollution, such as urban sewage and garbage or inshore petroleum spillage, pose an immediate and obvious environmental threat; other forms, such as those involving potentially toxic substances found in industrial wastes and agricultural PESTICIDES, present a more insidious hazard: they may enter biological food chains and, by affecting the metabolism of organisms, create an ecological imbalance (see ECOLOGY). Populations of organisms thriving abnormally at the expense of other populations may themselves be regarded as pollutants. Forms of energy pollution include: NOISE, e.g., factory, airport and traffic noise; THERMAL POLLUTION, e.g., the excessive heating of lakes and rivers by industrial effluents; light pollution, e.g., the glare of city lights when it interferes with astronomical observations, and radiation from radioactive wastes (see RADIOACTIVITY; FALL-OUT). The need to control environmental pollution in all its aspects is now widely recognized. (See also RECYCLING.)

POLLUX. See GEMINI.

POLONIUM (Po), soft, gray metal in Group VIA of the PERIODIC TABLE, occurring in PITCHBLENDE; usually produced by neutron bombardment of bismuth. All its isotopes are highly radioactive; the commonest, Po^{210}, emits ALPHA PARTICLES (half-life 138.4 days), and is used in NEUTRON sources. AW 210, mp 254°C, bp 962°C, sg 9.32 (α).

POLYCHAETA, a class of the phylum ANNELIDA.

POLYCYTHEMIA, excessive number of ERYTHROCYTES in the BLOOD, which leads to plethora,

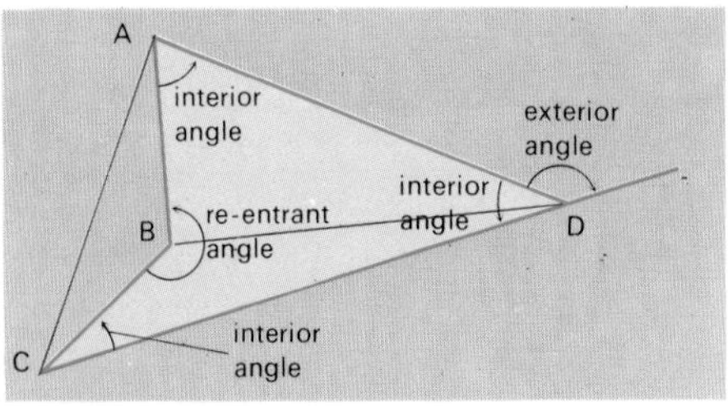

The polygon ABCD is an irregular quadrilateral with acute angles at A, C and D and a re-entrant angle at B. AC and BD are its diagonals.

itching and a tendency to THROMBOSIS. It may be primary or secondary to prolonged hypoxia (with LUNG disease) or certain TUMORS.

POLYETHYLENE, white, translucent RESIN, a POLYMER of ETHYLENE made catalytically at high pressure. Tough, elastic and inert, it is used to make plastic film, molded items, and SYNTHETIC FIBERS.

POLYGON, a closed PLANE figure bounded by three or more straight lines. Polygons with three sides are called TRIANGLES; with four, QUADRILATERALS; with five, pentagons; with six, hexagons; with seven, heptagons; with eight, octagons; with twelve, dodecagons. Polygons may be either convex or concave (except triangles, which are always convex): convex polygons have INTERIOR ANGLES that are all acute or obtuse; in concave polygons one or more of these angles is reflex (see ANGLE). A polygon with equal angles and sides equal in length is called a regular polygon. A *spherical polygon* is a closed figure on the surface of a sphere bounded by arcs of great circles (see also SPHERICAL GEOMETRY). The sum of the interior angles of a plane polygon is given by: $s=(n\times 180°)-360°$, where s is the sum in degrees and n the number of sides of the polygon.

The five regular polyhedrons.

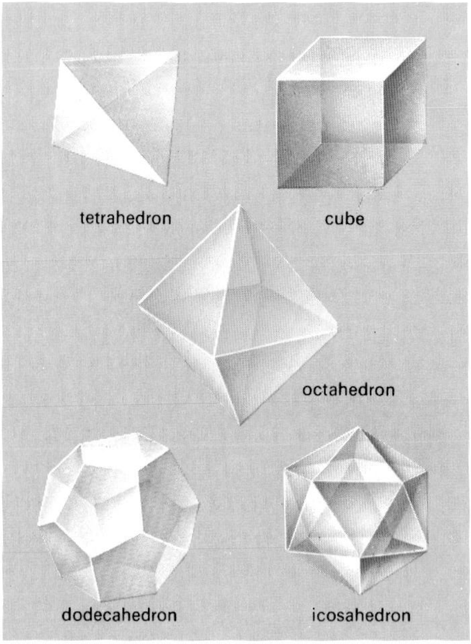

Simple polymers

polyethylene (polythene) from ethylene (ethene)	$[-CH_2-CH_2-]_n$
polypropylene from propylene (propene)	$\left[\begin{array}{c}-CH_2-CH-\\ \quad\quad\quad | \\ \quad\quad\quad CH_3\end{array}\right]_n$
polystyrene from styrene (ethenylbenzene)	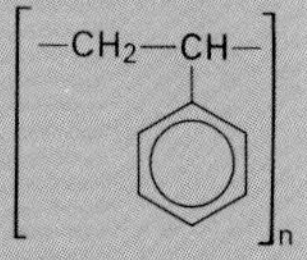
polyvinylchloride (PVC) from vinyl chloride (chloroethene)	$\left[\begin{array}{c}-CH_2-CH-\\ \quad\quad\quad | \\ \quad\quad\quad Cl\end{array}\right]_n$

POLYGRAPH. See LIE DETECTOR.

POLYHEDRON, a three-dimensional figure bounded by four or more PLANE sides. There are only five types of convex polyhedron that can be regular (i.e., have faces that are equal regular POLYGONS, each face being at equal angles to those adjacent to it): these are the tetrahedron, the octahedron and the isocahedron, with 4, 8 and 20 faces respectively, each face being an equilateral TRIANGLE; the hexahedron, with 6 square faces (see CUBE); and the dodecahedron, with 12 pentagonal faces. Regular polyhedrons may be circumscribed about or inscribed in a SPHERE (see CIRCUMSCRIPTION; INSCRIPTION; PYRAMID).

POLYMER, substance composed of very large MOLECULES (macromolecules) built up by repeated linking of small molecules (monomers). Many natural polymers exist, including PROTEINS, NUCLEIC ACIDS, polysaccharides (see CARBOHYDRATES), RESINS, RUBBER, and many minerals (e.g., quartz). The ability to make synthetic polymers to order lies at the heart of modern technology (see PLASTICS; SYNTHETIC FIBERS). Polymerization, which requires that each monomer has two or more FUNCTIONAL GROUPS capable of linkage, takes place by two processes: CONDENSATION with elimination of small molecules, or simple addition. CATALYSIS is usually required, or the use of an initiator to start a chain reaction of FREE RADICALS. If more than one kind of monomer is used, the result is a copolymer with the units arranged at random in the chain. Under special conditions it is possible to form stereoregular polymers, with the groups regularly oriented in space; these have useful properties. Linear polymers may form crystals in which the chains are folded sinuously, or they may form an amorphous tangle. Stretching may orient and extend the chains, giving increased tensile strength useful in synthetic fibers. Some cross-linking between the chains produces elasticity; a high degree of cross-linking yields a hard, infusible product (a thermosetting PLASTIC).

POLYMORPHISM, in zoology the existence of more than two forms or types of individual within the same species of animal. An example is seen in some social insects such as ants and bees in which many different types of worker are structurally adapted for different tasks within the colony.

POLYMORPHISM, in chemistry, the existence of certain chemical compounds in more than one crystalline form (see CRYSTAL). Usually the various forms are stable under different conditions. In some cases one is always stable, the others being metastable; thus CALCIUM carbonate has a stable hexagonal form CALCITE and a metastable orthorhombic form aragonite. (See also ALLOTROPY.)

POLYNOMIAL, an algebraic expression containing more than one term. $ax^5+bx^4+cx^3+dx^2+ex+f$, where a, b, c, d, e, f are CONSTANTS, is a polynomial of 6 terms in x (since $f=fx^0$). A polynomial with two terms is a binomial, with three a trinomial. Algebraic expressions with only one term (e.g., x^a) are called monomials.

POLYP, benign TUMOR of EPITHELIUM extending above the surface, usually on a stalk. Polyps may cause nasal obstruction and some (as in the GASTROINTESTINAL TRACT) may have a tendency to become a CANCER.

POLYP, a column-shaped form of certain CNIDARIA typified in the CORALS and hydras.

POLYPTERUS, or **bichir**, a fish found in rivers of Africa that is a survivor of the fossil group CHONDROSTEI. *Polypterus* has lungs and can briefly live out of water; its exact phylogeny is disputed.

POLYSACCHARIDES. See CARBOHYDRATES.

POLYSTYRENE, polymer of STYRENE ($C_6H_5CH=CH_2$) used as a rigid molded plastic and (for upholstery and thermal insulation) as a foam.

POLYWATER, or **"anomalous water,"** a liquid formerly supposed to be a polymeric form of WATER. First reported in 1962, it is made by condensing water in very fine glass or silica capillary tubes, and has unusual properties (mp −40°C, bp c500°C, sg 1.4). It is now thought to contain substances dissolved from the glass.

POLYZOA, or **bryozoa,** or moss-animals, a phylum of colonial animals found in freshwater and more especially in the sea. Up to 2 million individuals, called zooids, may be linked to form lacy patterns that occur on the surface of seaweeds. There are nearly 4000 living species, and many more fossil forms.

POME, a false fruit, the fleshy part of which is derived from the receptacle of the FLOWER, and not from the ovary. E.g., in the apple, only the core represents the ovary. (See FRUIT.)

POPPER, Sir Karl Raimund (1902–), Austrian-born British philosopher, best known for his theory of falsification in the philosophy of science. Popper contends that scientific theories are never more than provisionally adopted and remain acceptable only as long as scientists are devising new experiments to test (falsify) them.

POPULATION. Population growth is a serious threat to modern society. Some 4 billions inhabit the earth today, compared with some 1.5 billions in 1900: there may be over 6 billions by 2000 AD. Some countries have taken steps (e.g., encouraging BIRTH CONTROL) to counter this trend, and some are even experiencing a population decrease. Four factors affect national populations: births, immigration; deaths, emigration. Ideally, these balance and population stays steady (zero population growth—"ZPG").

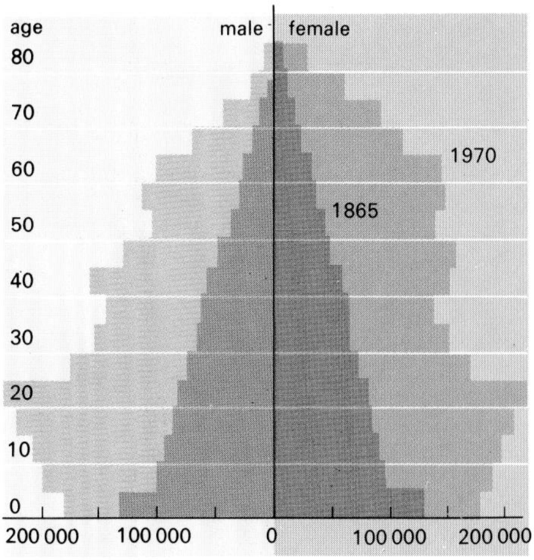

Age-sex pyramids for the population of Finland in 1865 and 1970. The 1865 pyramid is of the "progressive" form typical throughout the world before the impact of modern medicine was felt in decreasing mortality in younger age groups. Progressive pyramids are still typical of developing nations today. The 1970 pyramid approximates to the "regressive" bell shape typical of the developed world today, with declining fertility and low mortality below age 60. The concave section above age 50 represents the effects of World War II, the plateau below age 30, the subsequent increase in the birth rate.

PORCELAIN. See POTTERY AND PORCELAIN.

PORPHYRIA, metabolic disease due to disordered HEMOGLOBIN synthesis. It runs in families and may cause episodic abdominal pain, skin changes, NEURITIS and mental changes. Certain DRUGS can precipitate acute attacks. Porphyria may have been the cause of the "madness" of George III of England.

PORPHYRINS, water-soluble nitrogen-containing pigments, consisting of four substituted pyrrole nuclei joined in a ring structure, occurring widely in nature.

A naturally occurring isomer of protoporphyrin demonstrating the typical porphyrin ring structure.

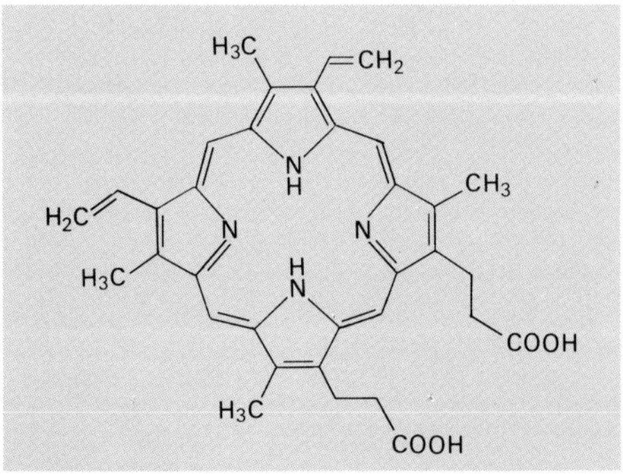

Combinations of porphyrins with metal ions include CHLOROPHYLL (containing MAGNESIUM) and heme (containing IRON); combinations of heme with proteins give HEMOGLOBIN and the CYTOCHROMES. The BILE pigments bilirubin and biliverdin are also related.

PORPHYRY, an IGNEOUS ROCK having many large crystals (phenocrysts) set in a very fine-grained matrix, occurring in DIKES and SILLS. More generally, rocks are said to have porphyritic texture if they contain some phenocrysts in a finer-grained matrix (e.g., porphyritic granite).

PORTER, Sir George (1920–), British chemist awarded the 1967 Nobel Prize for Chemistry with Manfred EIGEN and Ronald NORRISH for their studies of extremely fast chemical reactions, and, in particular, for their development of FLASH PHOTOLYSIS.

PORTER, Rodney Robert (1917–), British biochemist who shared with G. M. EDELMAN the 1972 Nobel Prize for Physiology or Medicine for his work on the molecular structure of ANTIBODIES.

POSITIVISM, philosophical theory of knowledge associated with the 19th-century French philosopher Auguste COMTE. It holds that the observable, or "positive," data of sense experience constitute the sole basis for assertions about matters of fact; only the truths of logic and mathematics are additionally admitted. The speculative claims of theology and metaphysics, regarded as the primitive antecedents of "positive" or scientific thought, are discounted. (See also LOGICAL POSITIVISM.)

POSITRON, the antiparticle corresponding to the ELECTRON. (See ANTIMATTER.)

POSTULATE. See AXIOM.

POTASH, or potassium carbonate. See POTASSIUM.

POTASSIUM (K), a soft, silvery-white, highly reactive ALKALI METAL. It is the seventh most abundant element, and is extensively found as SYLVITE, carnallite and other mixed salts; it is isolated by ELECTROLYSIS of fused potassium hydroxide. Potassium is chemically very like sodium, but even more reactive. It has one natural radioactive isotope, K^{40}, which has a half-life of 1.28 billion yr. K^{40} decays into Ar^{40}, an isotope of argon; the relative amounts of each are used to date ancient rocks. Potassium salts are essential to plant life (hence their use as fertilizers), and are important in animals for the transmission of impulses through the nervous system. AW 39.1, mp 64°C, bp 774°C, sg 0.862 (20°C).

Potassium carbonate (K_2CO_3), or **Potash**, is a hygroscopic colorless crystalline solid, made from potassium hydroxide and carbon dioxide, an ALKALI used for making glass.

Potassium chloride (KCl), is a colorless crystalline solid, found as SYLVITE. Used in fertilizers and as the raw material for other potassium compounds.

Potassium nitrate (KNO_3), or **Saltpeter**, is a colorless crystalline solid, soluble in water, which decomposes to give off oxygen when heated to 400°C. It is made from sodium nitrate and potassium chloride by fractional crystallization, and is used in GUNPOWDER, matches, fireworks, some rocket fuels, and as a fertilizer.

POTENTIAL, Electric, the work done against ELECTRIC FIELDS in bringing a unit charge to a given point from some arbitrary reference point (usually earthed), measured in VOLTS (i.e., joules per coulomb). Charges will tend to flow from points at one potential to those at a lower potential, and potential difference, or **voltage**, thus plays the role of a driving force for electric current. In inductive circuits, the work done in bringing up the charge depends on the route taken, and potential ceases to be a useful concept.

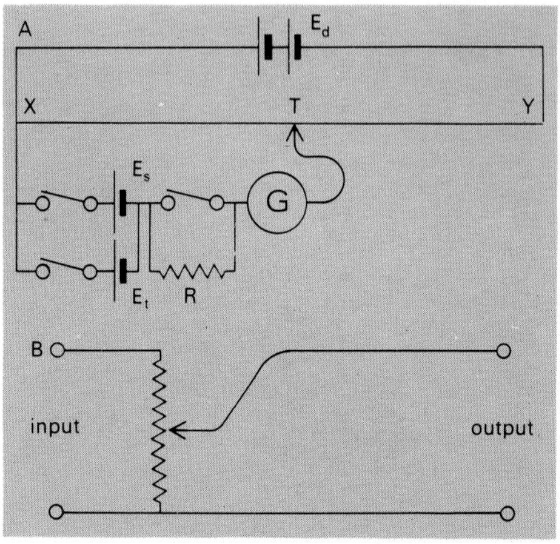

(A) The potentiometer as a means of measuring the electromotive force (emf) developed by a small cell (battery). A potential difference is set up along a resistance wire XY by a drive cell (E_d). The cell under test (E_t) is switched into the circuit and the length of potentiometer wire required to zero the current through the galvanometer (G) is determined, first under the protection of the series resistance (R), and then without it. This process is repeated for a standard cell (E_s) of accurately known emf. The ratio of the lengths of galvanometer wire tapped off then equals the ratio of the emfs of the cells when no current is flowing, that is, while none of the cell's power is expended in overcoming its internal resistance.
(B) The potentiometer (potential divider) as a voltage control device.

POTENTIOMETER, a device for accurate measurement of electric POTENTIAL by comparison with a standard cell potential: numerous DC and AC variants exist, mostly depending on OHM'S LAW. Typically, a potential drop is established in a long wire by a BATTERY, and a sliding contact used to tap a variable proportion of this drop, the lengths needed to balance the standard and unknown potentials being noted in turn. The ratio of these lengths is the ratio of the potentials. The same arrangement is also used to vary an applied voltage, for example, in the thin-carbon-film "potentiometers" used as volume controls in transistor RADIOS.

POTTERY AND PORCELAIN, CERAMIC articles, especially vessels, made of CLAY (generally KAOLIN) and hardened by firing. The simplest and oldest type of pottery, **earthenware** (nonvitreous), is soft, porous and opaque, usually glazed and used for common tableware. TERRA COTTA is a primitive unglazed kind. Earthenware is fired to about 1000°C. **Stoneware**, the first vitreous ware (of low porosity), was developed in China from the 5th to the 7th centuries AD. Fired to about 1200°C, it is a hard, strong, nonabsorbent ware, opaque and cream to brown in color. From stoneware evolved **porcelain** during the Sung dynasty (960–1279). This is a hard, nonporous vitreous ware, white and translucent. Made from flint, kaolin and feldspar, it is fired to about 1350°C.

In the manufacture of pottery the clay is made plastic by blending with water. The article is then shaped: traditionally by hand, by building up layers of strips (coiled pottery), by "throwing" on the potter's wheel or by molding; industrially by high-pressure molding or by a rotating template. The clay is fired in a kiln, slowly at first, then at higher temperatures to oxidize and consolidate it. The **glaze** (if desired) is then applied by spraying or dipping, and the article refired. Glazes are mixtures of fusible minerals and pigments, similar to those used for ENAMEL, powdered and mixed with water.

POUND (lb), the name of various units of weight (see WEIGHTS AND MEASURES). The pound avoirdupois is defined as being exactly 0.453 592 37kg. The "metric pound" commonly used in continental Europe is 500g.

POWELL, Cecil Frank (1903–1969), British physicist and pacifist awarded the 1950 Nobel Prize for Physics for his development of a direct means of photographing the tracks of SUBATOMIC PARTICLES and subsequent discovery of the π-meson.

POWELL, John Wesley (1834–1902), US geologist and ethnologist best known for his geological and topographical surveys, and for his anthropological studies of the AMERINDS.

POWER, the rate at which WORK is performed, or ENERGY dissipated. Power is thus measured in units of work (energy) per unit time, the SI UNIT being the watt (= the joule/second) and other units including the horsepower (= 745.70W) and the *cheval-vapeur* (= 735.5W). Frequently in engineering (and particularly in transportation) contexts, what matters is the power that a given machine can deliver or utilize—the rate at which it can handle energy—and not the absolute energies involved. A high-power machine is one which can convert or deliver energy quickly. While mechanical power may be derived as a product of a FORCE and a VELOCITY (linear or angular), the electrical power utilized in a circuit is a product of the potential drop and the current flowing in it (volts × amperes = watts). Where the electrical supply is alternating, the root-mean-square (rms) value of the voltage must be used.

POWER, the PRODUCT of a number used several times as a factor. Thus 2^4 ("two to the power four") is 16. By extension, 2^{-4} is $1/2^4 \equiv 1/16$; $2^{\frac{1}{4}}$ is $\sqrt[4]{2}$, and by definition $2^0 \equiv 1$.

PRAGMATISM, a philosophical theory of knowledge whose criterion of truth is relative to events and not, as in traditional philosophy, absolute and independent of human experience. A theory is pragmatically true if it "works"—if it has an intended or predicted effect. All human undertakings are viewed as attempts to solve problems in the world of action; if theories are not trial solutions capable of being tested, they are pointless. The philosophy of pragmatism was developed in reaction to late 19th-century IDEALISM mainly by the US philosophers C. S. PEIRCE, William JAMES and John DEWEY. (See also INSTRUMENTALISM.)

PRAIRIES, the rolling GRASSLANDS of North America. There are three types: tallgrass, midgrass (or mixed-grass) and shortgrass, which is found in the driest areas. Typical prairie animals are the coyote, badger, prairie dog and jackrabbit, and the now largely vanished bison and wolf. (See also STEPPES.)

PRASEODYMIUM (Pr), one of the LANTHANUM SERIES. AW 140.9, mp 931°C, bp 3512°C, sg 6.48 (20°C).

PRECAMBRIAN, the whole of geological time from the formation of the planet earth to the start of the PHANEROZOIC (the aeon characterized by the appearance of FOSSILS in rock strata), and thus lasting from about 4 550 to 570 million years ago. (See also GEOLOGY.)

PRECESSION, the gyration of the rotational axis of a spinning body, such as a GYROSCOPE, describing a right circular CONE whose vertex lies at the center of the spinning body. Precession is caused by the action of a TORQUE on the body. **Precession of the equinoxes** occurs because the earth is not spherical, but bulges at the EQUATOR, which is at an angle of 23.5° to the ECLIPTIC. Because of the gravitational attraction of the sun, the earth is subject to a torque which attempts to pull the equatorial bulge into the same plane as the ecliptic, therefore causing the planet's poles, and hence the intersections of the equator and ecliptic (the equinoxes), to precess in a period of about 26 000 years. The moon (see NUTATION) and planets similarly affect the direction of the earth's rotational axis.

PRECIPITATION, in meteorology, all water particles that fall from CLOUDS to the ground; including RAIN and drizzle, SNOW, SLEET and HAIL. Precipitation is important in the HYDROLOGIC CYCLE.

PREGL, Fritz (1869–1930), Austrian chemist awarded the 1923 Nobel Prize for Chemistry for his pioneering work on the microanalysis of organic compounds (see ANALYSIS, CHEMICAL; MICROCHEMISTRY).

Pottery manufacture: slip-casting.

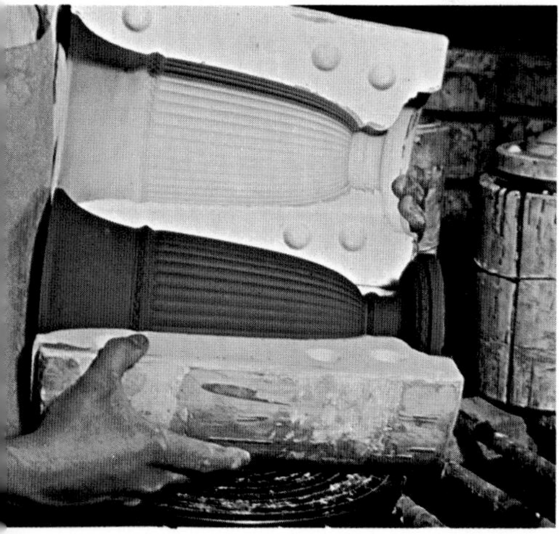

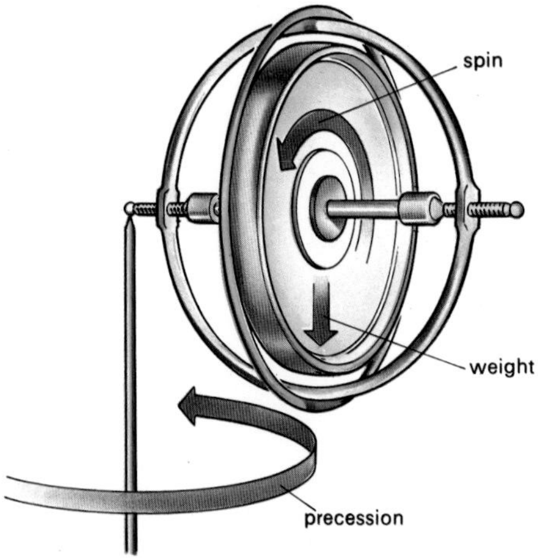

A gyroscope of moment of intertia I and spinning with angular velocity S about a horizontal axis with one end mounted on a vertical mounting will precess in a horizontal circle about the mounting with angular velocity $\omega = T/IS$ where T is the torque acting on the gyroscope about the mounting due to its weight.

PREGNANCY, in humans the nine-month period from the fertilization and IMPLANTATION of an EGG, the development of EMBRYO and FETUS through the BIRTH of a child. Interruption of MENSTRUATION and change in the structure and shape of the BREASTS are early signs; morning sickness, which may be mild or incapacitating is a common symptom. Later an increase in abdominal size is seen and other abdominal organs are pushed up by the enlarging WOMB. LIGAMENTS and JOINTS become more flexible in preparation for delivery. MULTIPLE PREGNANCY, hydatidiform mole, spontaneous ABORTION, antepartum HEMORRHAGE, toxemia and premature labor are common disorders of pregnancy. The time following birth is known as the puerperium.

PREHISTORIC ANIMALS, those members of the ANIMAL KINGDOM such as DINOSAURS, which flourished, reached their prime and then became extinct before the present day. Many of them gave rise to more successful groups which eventually replaced them. A large number are preserved in the FOSSIL record and from such material their evolutionary position and relationships can be worked out. This in turn may provide clues as to the evolutionary history of modern forms.

PREHISTORIC MAN. Study of early man is hampered by the difficulty of distinguishing between what was to become *Homo sapiens* and the ancestors of our modern apes. The actual point of separation of the two strains is so long ago, probably before either bore any resemblance to their modern descendants, that it is unlikely ever to be discovered. The earliest known form of man is **Ramapithecus**, though there is still debate as to whether he should be classed as of the Hominidae (family of man) or of the Pongidae (anthropoid-ape family). Only small fragments of Ramapithecus fossil skeletons exist, the earliest of these dating from some 14 million years ago, the latest from some 10 million years ago. The next earliest human fossil dates from about 3 million years ago and is tentatively designated Ethiopian Man: once again only fragments remain. In 1972 Richard Leakey discovered a skull, known as **Skull 1470**, which dates from 2.6 million years ago and which shows strong resemblances to modern man. It is now thought that men of Skull 1470 type evolved over a period of more than 2 million years into our recent ancestors, Neanderthal Man and Cro-Magnon Man, although there is as yet no fossil evidence to show this evolution. Rather later than Skull 1470, and assumed to be examples of parallel evolution, are **Australopithecus**, containing such subdivisions as Paranthropus and Zinjanthropus, and **Homo habilis**, who may well be merely a variant of Australopithecus. Australopithecines had small brain capacities (300–600cm^3) as compared with those of Skull 1470 (800 cm^3) or modern man (c1300cm^3) and had distinctly ape-like features. They existed from 2 to 1 million years ago. Descended possibly from the australopithecines, possibly from Skull 1470 men, are the various forms of **Pithecanthropus erectus**, including Peking Man and Java Man: the earliest fossil examples date from possibly as much as 1.9 million years ago, the most recent from as little as 200 000 years ago. Their brain capacity was of the order of 1000cm^3. The exact relation of Heidelberg Man, dated c400 000 BC, to Pithecanthropus Erectus and to modern man is not known. **Homo sapiens** remains first appear from about 400 000 years ago onward. The oldest known example is that from Vértesszőllős, with a brain capacity of 1400cm^3. In more recent times, there were two distinct types of prehistoric man: Neanderthal and CRO-MAGNON MAN. **Neanderthal Man**, who was either absorbed or annihilated by invading Cro-Magnon men, did not have a fully erect posture but, despite this and his remarkably simian appearance, had an average brain capacity (1400cm^3) greater than that of modern man. It is thought that Neanderthal Man, who effectively became extinct during the STONE AGE, was probably an offshoot from the mainstream of human evolution. With his disappearance and the emergence of Cro-Magnon Man, the stage was set for the final evolution into the races of modern man. (See PRIMITIVE MAN.)

PRELOG, Vladimir (1906–   ), Yugoslav-born Swiss chemist who shared with CORNFORTH the 1975 Nobel Prize for Chemistry for his work on ENZYMES.

PRESBYOPIA, a defect of VISION coming on with advancing age in which the LENS of the EYE hardens, causing loss of the ability to accommodate (focus) nearby and often distant objects. The condition is corrected by supplying two pairs of GLASSES (one for close work, the other for distant vision), though these may be combined in bifocal lenses.

PRE-SOCRATIC PHILOSOPHY, term applied to the thought of the early Greek philosophers (c600–400 BC) whose work came before the influence of SOCRATES. Their works survive mostly in obscure fragments, but their fame and importance lie in their being the first to attempt rational explanations of the universe. They are grouped into the Ionian school in Asia Minor (THALES, ANAXIMANDER, ANAXIMENES, Xenophanes, HERACLITUS, ANAXAGORAS, and the Atomists, LEUCIPPUS and DEMOCRITUS) and PYTHAGORAS and the Eleatics (PARMENIDES, ZENO, EMPE-

DOCLES) in S Italy and Sicily. Protagoras and the Sophists are usually also included.

PRESSURE, the FORCE per unit area acting on a surface. The SI UNIT of pressure is the PASCAL (Pa = newton/(metre)^2) but several other pressure units, including the atmosphere (101.325kPa), the bar (100kPa) and the millimetre of mercury (mmHg = 133.322Pa), are in common use. In the universe, the pressure varies from roughly zero in interstellar space to an atmospheric pressure of roughly 100kPa at the surface of the earth and much higher pressures within massive bodies and in STARS. According to the KINETIC THEORY of matter, the pressure in a closed container of GAS arises from the bombardment of the container walls by gas molecules: it is proportional to the temperature and inversely proportional to the volume of the gas. Pressure is a stress characterized by its uniformity in all directions and usually produces a decrease in volume. The value of the pressure affects most physical, chemical and biological processes. Consequently many different types of pressure gauges have been developed (see MANOMETER).

PRESSURE COOKER, small AUTOCLAVE used for domestic cooking. Water boils inside the vessel at a pressure greater than atmospheric (commonly 2atm), at which pressure its BOILING POINT is considerably higher than 100°C, so that the food cooks much more rapidly than if boiled normally. The pressure, regulated by a weighted or spring-loaded valve, is due to the steam produced by the boiling water.

PREVAILING WESTERLIES, the predominant WINDS which blow between latitudes 30° and 60° both N and S of the equator. In the N Hemisphere they blow from the SW; in the S Hemisphere, from the NW.

PRICKLY HEAT, or heat rash, an uncomfortable itching sensation due to excessive sweating, mainly seen in Europeans visiting the tropics.

PRIESTLEY, Joseph (1733–1804), British theologian and chemist. Encouraged and supported by Benjamin FRANKLIN, he wrote *The History and Present State of Electricity* (1767). His most important discovery was OXYGEN (1774; named later by LAVOISIER), whose properties he investigated. However, he never abandoned the PHLOGISTON theory of COMBUSTION. He later discovered many other gases—AMMONIA, CARBON monoxide, hydrogen SULFIDE—and found that green plants require sunlight and give off oxygen. He coined the name RUBBER. His association in the 1780s with the LUNAR SOCIETY brought him into contact with scientists such as James WATT and Erasmus DARWIN. His theological writings and activity were important in leading the English Presbyterians into Unitarianism; indeed he is regarded as a principal architect of the Unitarian Church. Hostile opinion over this and his support of the French Revolution led to his emigration to the US (1794).

PRIGOGINE, Ilya (1917–), Russian-born Belgian thermodynamicist who was awarded the 1977 Nobel Prize for Chemistry for contributions to understanding the THERMODYNAMICS of systems which are not in equilibrium states.

PRIMATES, an order of the MAMMALIA to which MAN belongs, other members being the ANTHROPOID APES, monkeys, tarsiers, bush babies, pottos and lemurs. In many ways, primates are primitive; for example, the collar bone and five digits on each hand are retained. In others they are specialized: the hands and feet are modified for grasping branches, the claws have become flat nails and the eyes provide stereoscopic and color vision. Most importantly, the brain is extremely large, especially those regions concerned with the abstract association of sensory signals.

PRIME NUMBER, a NATURAL NUMBER which cannot be expressed as the PRODUCT of other natural numbers, e.g., 1, 2, 3, 5, 7, 11, 13, 17 and 19.

PRIMITIVE MAN, term for societies whose culture has reached a level little, if any, higher than the STONE AGE. Although technologically limited and economically unsophisticated, primitive societies may have extremely complex social structures with extensive rules directing behavior such as MARRIAGE, kinship and religion (see TABOOS). Most contemporary primitive societies are Neolithic; that is, they practice agriculture, make pots, weave textiles and work stone to make tools. A few, however, are of Paleolithic type, such as the AUSTRALIAN ABORIGINES and the recently extinguished TASMANIANS. PREHISTORIC MAN probably first formed primitive societies about 250000 years ago.

PRINTED CIRCUIT. See ELECTRONICS.

PRINTING, the reproduction of words and pictures in ink on paper or other suitable media. Despite the advent of INFORMATION RETRIEVAL systems, the dissemination and storage of knowledge are still based primarily on the printed word. Modern printing begins with the work of Johann GUTENBERG, who probably invented movable type and type metal in the 15th century. Individual characters could be used several times. Little changed for 400 years until the invention of machines that could cast type as it was required (see LINOTYPE; MONOTYPE). Letterpress and lithography are today the two most used printing techniques. **Letterpress** uses raised type that is a mirror image of the printed impression. The type is inked and the paper pressed to it. A number of typeset pages (usually 8, 12, 16, 24 or 32) are tightly locked in a metal form such that, when a sheet of paper has been printed on both sides, it may be folded and trimmed to give a *signature* of up to 64 pages. The arrangement of the pages of type is the *imposition*. Most newspapers use **rotary letterpress**: the forms are not flat but curved backward, so that two may be clamped around a cylinder. Paper is fed between this cylinder and another, the impression cylinder. This technique is especially swift when the paper is fed in as a continuous sheet (a *web*). **Lithography** depends on the mutual repulsion of water and oil or grease. In the fine arts, a design is drawn with a grease crayon on the surface of a flat, porous stone, which is then wetted. The water is repelled by the greasy areas; but ink is repelled by the damp and adheres to the greasy regions. Modern mechanized processes use the same principle. Commonest is **photo-offset**, where the copy to be printed is photographed and the image transferred to a plate such that the part to be printed is oleophilic (oil-loving), the rest hydrophilic (water-loving). The plate is clamped around a cylinder and inked. The impression is made on an intermediate "blanket cylinder," which prints onto the paper. **Gravure** is another major printing technique. The plate is covered with a pattern of recessed cells in which the ink is held, greater depth of cell increasing

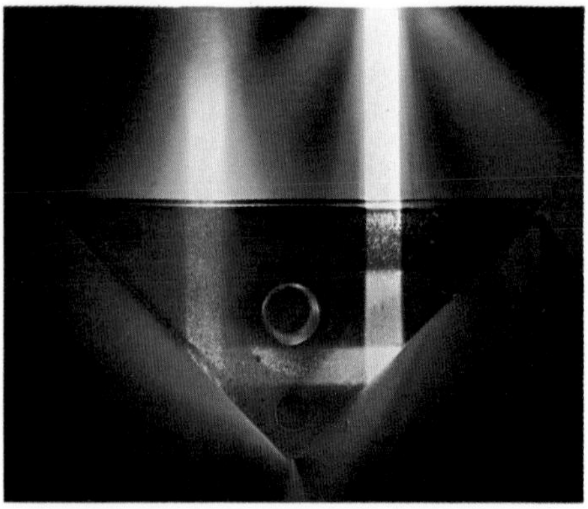

An optical prism can be used to disperse light, as a mirror, to invert an image or, as here (the Porro prism), to reverse the direction of light radiation.

printed intensity. Gravure is good for color and the plates long-lasting, but high initial plate-making costs render it suitable only for long runs. Little-used for books, it is much used in packaging as it also prints well on media other than paper. **Illustrations**, in letterpress, are reproduced using line or HALFTONE blocks (see also ETCHING). In photo-offset black and white illustrations are printed much as is text; and gravure is inherently suitable for printing tones. For COLOR, the illustration is photographed for each of the colors magenta, cyan, yellow and black, and separate plates or blocks made: the four images are superimposed in the printing to give a full-color effect. (See also INK; PAPER; PHOTOGRAPHY.)

PRISM, in GEOMETRY, a solid figure having two faces (the bases) which are parallel equal polygons and several others (the lateral faces) which are parallelograms. Prismatic pieces of transparent materials are much used in optical instruments. In spectroscopes (see SPECTROSCOPY) and devices for producing monochromatic LIGHT, prisms are used to produce DISPERSION effects, just as Newton first used a triangular prism to reveal that sunlight could be split up to give a SPECTRUM of colors. In BINOCULARS and single-lens reflex cameras reflecting prisms (employing total internal reflection—see REFRACTION) are used in preference to ordinary MIRRORS. The NICOL PRISM is used to produce POLARIZED LIGHT.

PROBABILITY, the statistical RATIO between the number n of particular outcomes and the number N of possible outcomes: n/N; where all of N are equally likely. For example, when throwing a die there is 1 way in which a six can turn up and 5 ways in which a "not six" can turn up. Thus $n=1$ and $N=5+1=6$, and the ratio $n/N=1/6$. If two dice are thrown there are 6×6 $(=36)$ possible pairs of numbers that can turn up: the chance of throwing two sixes is 1/36. This does not mean that if a six has just been thrown there is only a 1/36 chance of throwing another: the two events are independent; the probability of their occurring *together* is 1/36.

Consider throwing a die six times with the aim of getting each of the six numbers exactly once. If you want to do this in order (a permutation), say from 1 to 6, working out the probability of your doing so is easy: you have 1/6 chance of throwing a one, 1/6 chance of throwing a two, and so on; so that the probability of a favorable result overall is $(1/6)^6=1/46656$. If you are not concerned with the order (combination) the situation is different: the probability of a favorable result on the first throw is 1 (any number is favorable), on the second 5/6, and so on, so the overall probability is

$$6/6\times5/6\times4/6\times3/6\times2/6\times1/6$$

or $6!/6^6$ (see FACTORIAL)—about 1/65. Clearly one is more likely to succeed with a desired combination than with a permutation. (See PERMUTATIONS AND COMBINATIONS.)

Probability theory is plainly intimately linked with STATISTICS. More advanced probability theory has contributed vital understandings in many fields of physics, as in THERMODYNAMICS, behavior of particles in a COLLOID (see BROWNIAN MOTION) or molecules in a GAS, and atomic physics (see ATOM; ELECTRON; BOSE–EINSTEIN STATISTICS).

PROBOSCIDIA, an order of the MAMMALIA that includes the elephants. The group originated in Africa and today only two species survive, the Indian and African elephants. In both, the characteristic feature is the trunk, an extension of the nose and upper lip.

PROCAINE, prototype local anesthetic agent related to COCAINE; used now mainly for surface ANESTHESIA. Its derivative procainamide is used in suppression of HEART arrhythmias. The hydrochloride of procaine is known as novocaine.

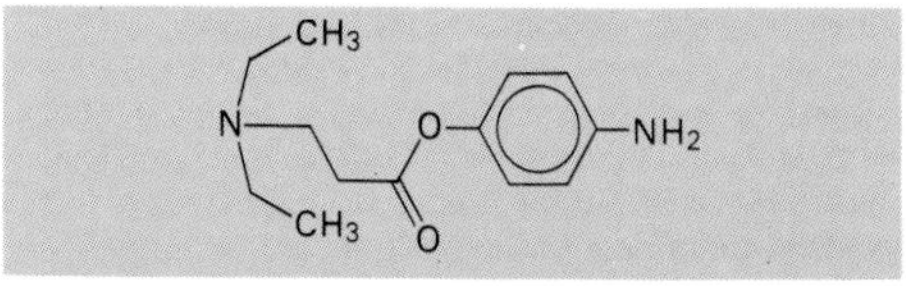

The structure of procaine. The synthetic molecule is structurally much simpler than cocaine.

PROCESS PHILOSOPHY, a radical alternative to positivistic philosophies (see POSITIVISM), in which the physicists' "spatial" notion of TIME is rejected and interest centers in the developmental process. First propounded by Henri BERGSON, process philosophy influenced William JAMES, George Santayana and Alfred North WHITEHEAD.

PROCYON, Alpha Canis Minoris, a visual DOUBLE STAR, the brightest star in CANIS MINOR. It has absolute magnitude $+2.7$ and is 3.53pc distant.

PRODUCER GAS, fuel GAS made by partial combustion of coal or coke in mixed air and steam. It contains carbon monoxide, hydrogen, methane, and incombustible nitrogen and carbon dioxide. It has a low calorific value, 5 MJ/m^3 (see FUEL), but being easy to make it is used in large furnaces.

PRODUCT, the result of MULTIPLICATION. Thus in $a.b=c$, the product is c.

PROGESTERONE, female sex HORMONE produced by the corpus luteum under the influence of LUTEINIZING HORMONE. It prepares the WOMB lining for IMPLANTATION and other body organs for the changes of PREGNANCY. It is used in some oral contraceptives to suppress ovulation or implantation.

PROGRAMMED LEARNING, teaching method

whereby matter to be learned is arranged in a coherent sequence of small clear steps (programmed) and presented in such a way that the student is able to instruct, test and, if necessary, correct himself at each step. The learning program is usually embodied in a book or booklet or adapted for use in conjunction with a teaching machine. It enables the student to learn at his own pace, with a minimum of wasted effort. There are two basic kinds of program. The "linear program," based on the work of the Harvard psychologist B. F. SKINNER, obliges the student to compare his own response at each step with the correct response. The "intrinsic (or branching) program," originally developed for instructing US Air Force technicians, offers a limited choice of responses at each step. The correct response is immediately reinforced; an incorrect response obliges the student to follow a corrective subprogram leading back to the point at which the error occurred.

PROGRAMMING. See COMPUTER.

PROGRESSION, an ordered set of numbers; with the exception of the arithmetic, geometric and harmonic progressions, such sets are more usually termed SEQUENCES.

Arithmetic progression. A progression of the form $a, a+d, a+2d, \ldots, a+(n-1)d$, where a is the first term (x_1) and d is the common difference, $a+(n-1)d$ being the nth term (x_n). The sum to n terms, S_n, of an arithmetic progression is given by

$$S_n = n\left(a + \frac{n-1}{2}d\right)$$

The arithmetic mean of two terms, x_s and x_{s+2}, is given by $(x_s + x_{s+2})/2 = x_{s+1}$.

Geometric progression. A progression of the form $a, ar, ar^2, ar^3, \ldots, ar^{n-1}$, where a is the first term and r the common ratio, ar^{n-1} being the nth term. The sum to n terms of a geometric progression is given by

$$S_n = a\frac{1-r^n}{1-r},$$

and the geometric mean of two terms, x_s and x_{s+2}, by $\sqrt{x_s . x_{s+2}} = x_{s+1}$. The geometric mean of any two different positive numbers is always less than their arithmetic mean.

Harmonic progression. A progression of the form

$$\frac{1}{a}, \frac{1}{a+d}, \frac{1}{a+2d}, \ldots, \frac{1}{a+(n-1)d},$$

the terms being the RECIPROCALS of those in an arithmetic progression. There is no simple expression for S_n in this case. The harmonic mean of two terms, x_s and x_{s+2}, is given by

$$\frac{2x_s x_{s+2}}{x_s + x_{s+2}} = x_{s+1}.$$

(See also MEAN, MEDIAN AND MODE.)

PROJECTILES. See BALLISTICS.

PROJECTION, in PSYCHOLOGY and PSYCHIATRY, the treatment by an individual of mental activity as reality (see DREAMS; EIDETIC IMAGE; HALLUCINATION; ILLUSION). In PSYCHOANALYSIS, the term describes the interpretation of situations or the actions of others in such a way as to justify one's self-opinion or beliefs, as in PARANOIA and paranoid SCHIZOPHRENIA.

PROJECTION, of a POINT P onto a LINE L, the INTERSECTION of the line and the line drawn PERPENDICULAR to it passing through the point.

PROJECTION TEST, test whereby an individual's personality may be gauged by his completion of unfinished sentences, his interpretation of "pictures" from inkblots, etc.

PROJECTIVE GEOMETRY, that branch of GEOMETRY based on PERSPECTIVE. Consider the three-dimensional figure defined by a PLANE figure A and a point P not in that plane; this is a projection of A from P. Should the projection be cut by another plane, a plane figure A′ is formed: the relationship between A and A′ is termed a perspective transformation. If A′ is projected from a new point P′ onto a third plane, the figure A″, the result of two perspective transformations, is formed. The result of a sequence of perspective transformations is termed a *projective transformation*. Projective geometry may be viewed as the study of those of the figure's properties that are unchanged by projective transformation.

PROKARYOTE; PROKARYOTIC CELL. See CELL.

PROKHOROV, Aleksandr Mikhailovich (1916–   ), Soviet physicist awarded with N. G. BASOV and C. H. TOWNES the 1964 Nobel Prize for Physics for work with Basov leading to development of the MASER.

PROLACTIN, or **luteotrophic hormone**, HORMONE secreted by the PITUITARY GLAND concerned with LACTATION after PREGNANCY.

PROLATENESS. See OBLATENESS.

PROMETHIUM (Pm), radioactive element (see RADIOACTIVITY) in the LANTHANUM SERIES. It has no naturally occurring isotopes, and is formed in nuclear reactors. It is used in miniature nuclear-powered batteries. AW 147, mp 1080°C.

PROMINENCE, Solar. See SUN.

PRONUNCIATION, the ways in which words, syllables and letters are spoken. Pronunciation varies from language to language, as well as within a language, different accents deriving from differing geographical locations or educational backgrounds. Studies of pronunciation in the past are assisted by both rhyming verse and the transliteration of words borrowed from other languages. The concept of a "correct" pronunciation is now largely obsolete. (See also PHONETICS.)

PROOF SPIRIT, term describing the proportion of alcohol (ETHANOL) in distilled liquor. In the US the proof value is twice the percentage of ethanol by volume. Thus 90 proof represents 45% ethanol. In the UK, proof spirit is somewhat stronger, containing 57.1% ethanol by volume.

PROPANE (C_3H_8), colorless gas, an ALKANE found in NATURAL GAS and light PETROLEUM. Mixed with butane, it is sold as BOTTLED GAS. It is also used to make ETHYLENE by CRACKING, and is oxidized to ACETALDEHYDE. MW 44.1, mp −190°C, bp −42°C.

PROPELLER, a mechanical device designed to impart forward motion usually to a ship or AIRPLANE, operating on the screw principle. It generally consists of two or more inclined blades (AIRFOILS) radiating from a hub, and the amount of THRUST it produces is proportional to the product of the mass of the fluid it acts on and the rate at which it accelerates the fluid. The inclination, or "pitch," of the propeller blades determines the theoretical distance moved forward with each revolution. A "variable-pitch propeller" can be adjusted while in motion, to maximize its efficiency under different operating conditions; it may also be possible to reverse the propeller's pitch, or to "feather" it—i.e., minimize its resistance when not rotating. John FITCH, in 1796, developed the first marine screw propeller; John ERICSSON perfected the first bladed propeller, in 1837. (See also CAVITATION.)

PROPER MOTION, the rate of motion at right angles to our line of sight of a star, measured in seconds of arc per year. (See RADIAL VELOCITY.)

PROPORTION, a statement that two RATIOS are equivalent, written $a:b=c:d$, as in the statement $2:3=6:9$. Two FUNCTIONS are proportional if, for all x, $f(x)=kg(x)$, where k is the CONSTANT of proportionality. If the value of k is not known, the statement may be written $f(x) \propto g(x)$.

PROSPECTING, the hunt for MINERALS economically worth exploiting. The simplest technique is direct observation of local surface features characteristically associated with specific mineral deposits. This is often done by prospectors on the ground, but increasingly aerial photography is employed (see PHOTOGRAMMETRY). Other techniques include examining the seismic waves caused by explosions—these supply information about the structures through which they have passed; testing local magnetic fields to detect magnetic metals or the metallic gangues associated with nonmagnetic minerals; and, especially for metallic sulfides, testing electric CONDUCTIVITY.

PROSTATE GLAND, male reproductive GLAND which surrounds the urethra at the base of the BLADDER and which secretes semen. This carries sperm made in the TESTES to the PENIS. Benign enlargement of prostate in old age is very common and may cause retention of the urine. CANCER of the prostate is also common in the elderly but responds to HORMONE treatment. Both conditions benefit from surgical removal.

The European nightjar *Caprimulgus europaeus* provides a fine example of cryptic, or protective, coloration. Like other nightjars it nests on the ground; it matches this bark litter superbly.

PROSTHETICS, mechanical or electrical devices inserted into or onto the body to replace or supplement the function of defective or diseased organs. **Artificial limbs** designed for persons with AMPUTATIONS were among the first prosthetics; but metal or plastic JOINT replacements or BONE fixations for subjects with severe ARTHRITIS, FRACTURE or deformity are now also available. Replacement TEETH for those lost by CARIES or trauma are included in **prosthodontics** (see also DENTISTRY). The valves of the HEART may fail as a result of rheumatic or congenital heart disease or bacterial endocarditis, and may need replacement with mechanical valves (usually of ball-and-wire or flap types) sutured in place of the diseased valves under cardiorespiratory bypass. If the **pacemaker** of the heart fails, an electrical substitute can be implanted to stimulate the heart muscle at a set rate.

PROSTHODONTICS. See PROSTHETICS.

PROTACTINIUM (Pa), rare metal in the ACTINIDE series, found in URANIUM ores. More than 12 isotopes are known, all radioactive; Pa^{231} is the longest lived (half-life 34000yr). Protactinium is usually pentavalent, resembling NIOBIUM and TANTALUM. AW 231, mp<1600°C, sg 15.37 (calc.).

PROTECTIVE COLORATION. Many animals have adapted their coloration as a means of defense against predators. Except where selection favors bright coloration for breeding or territorial display, most higher animals are colored in such a way that they blend in with their backgrounds—by pure coloration, by disruption of outline with bold lines or patches, or by a combination of the two. The most highly developed camouflage is found in ground-nesting birds, for example, nightjars, or insects, such as walking sticks or leaf insects. Associated with this coloration must be special behavior patterns enabling the animal to seek out the correct background for its camouflage and to "freeze" against it. Certain animals can change the body texture and coloration to match different backgrounds: octopuses, chameleons, and some flatfishes. An alternative strategy adopted by some animals, particularly insects, is the use of shock-coloration. When approached by a predator these insects flick open dowdy wings to expose bright colors, often in the form of staring "eyes," to scare the predator.

PROTEIN, a high-molecular-weight compound which yields AMINO ACIDS on HYDROLYSIS. Although hundreds of different amino acids are possible, only 20 are found in appreciable quantities in proteins, and these are all α-amino acids. Proteins are found throughout all living organisms. Muscle, the major structural material in animals, is mainly protein; the 20% of blood which is not water is mainly protein. ENZYMES may contain other components, but basically they too are protein. Approximately 700 different proteins are known. Of these 200–300 have been studied and over 150 obtained in crystalline form. Some proteins, such as those found in the hides

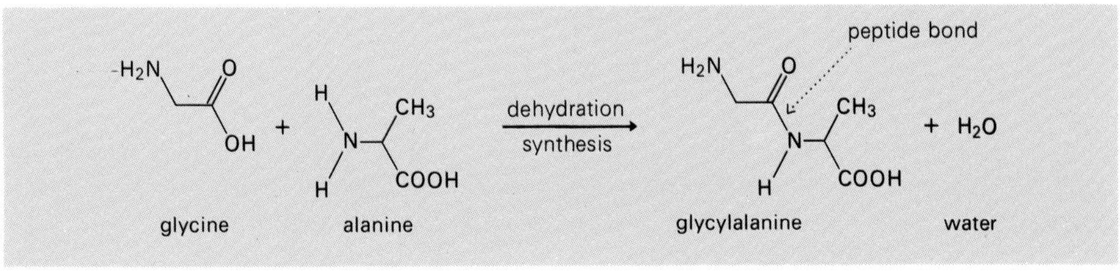

The peptide bond is one of the hallmarks of proteins. Here, a simple reaction between two amino acids yields a dipeptide containing one peptide bond.

of cattle and which can be converted to LEATHER, are very stable, while others are so delicate that even exposure to air will destroy their capability as enzymes. The most important and strongest bond in a protein is the PEPTIDE bond joining the amino acids in a chain. Other bonds hold the different chains together: HYDROGEN BONDING, together with strong disulfide bonds and secondary peptide links are important here. The three dimensional structure of proteins helps to determine their properties; X-RAY studies have shown that the amino acid chain is sometimes coiled in a spiral or helix. Although proteins are very large molecules (with molecular weights ranging from 12 000 to over 1 million), many of them are partly ionized and hence are soluble in water. Such differences in size, solubility and electrical charge are exploited in methods of separating and purifying proteins. The separation of proteins in an electrical field (ELECTROPHORESIS) is widely applied to human serum in the diagnosis of certain diseases.

PROTEIN SYNTHESIS. All the PROTEIN in any living organism is undergoing a continual process of breakdown and resynthesis. The white rat replaces half its protein in 17 days; in man this requires 80 days. However not all proteins are broken down and resynthesized at the same rate: e.g., half the human blood-serum proteins are replaced in 10 days; liver protein requires 20–25 days, while replacement of bone protein is very slow. Protein synthesis is very rapid; within minutes of injecting an animal with a radiolabelled AMINO ACID, radiolabelled protein can be isolated. Protein synthesis takes place within the CELLS of an organism. The first step is the activation of an amino acid by reaction with ATP (see NUCLEOTIDES). This activated amino acid is then bound to a specific soluble form of RNA (see NUCLEIC ACIDS) known as transfer RNA. There are at least 20 different types of transfer RNA, one for each amino acid. The soluble RNA-amino acid complex then travels to the RIBOSOME where the amino acid is added to other amino acids to form a polypeptide (see PEPTIDE). The order in which the specific transfer RNAs bring each amino acid to the ribosome is determined by messenger RNA. The transfer RNA then returns to the CYTOPLASM to collect another molecule of its specific amino acid. Messenger RNA contains the code for a particular protein and is only used a few times before being destroyed. Messenger RNA from one organism can be used with transfer RNA and ribosome from another to synthesize a protein; this may explain how VIRUSES take over the synthetic systems of a cell.

PROTEROZOIC, the portion of the PRECAMBRIAN running from about 2 390 to 570 million years ago, divided into three eras: **Aphebian**, 2390–1640; **Helikian**, 1640–880; and **Hadrynian**, 880–570. (See also ARCHEOZOIC; GEOLOGY.)

PROTISTA, taxonomic term proposed by some authorities to include unicellular, colonial and filamentous forms of eukaryote (see CELL) organisms from both the ANIMAL KINGDOM and PLANT KINGDOM. Included in this group are ALGAE and FUNGI.

PROTON, stable elementary particle found in the nucleus of all ATOMS. It has a positive charge, equal in magnitude to that of the ELECTRON, and rest mass of 1.67252×10^{-27}kg (slightly less than the NEUTRON mass but 1 836.1 times the electron mass). As the HYDROGEN ion, the proton is chemically important, particularly in aqueous solutions (see ACID), and is widely used in physics as a projectile for bombarding atoms and nuclei.

PROTOPLASM, the substance including and contained within the plasma membrane of animal CELLS but in plants forming only the cell's contents. It is usually differentiated into the nucleus and the cytoplasm. The latter is usually a transparent viscous fluid containing a number of specialized structures; it is the medium in which the main chemical reactions of the cell take place. The nucleus contains the cell's genetic material.

PROTOZOA, phylum containing animals that consist of a single cell. All are microscopic and nearly 50 000 species have been described. They occur in great numbers in the sea, in freshwater, in the soil and in the air. Many more are parasitic and cause serious diseases such as MALARIA, SLEEPING SICKNESS and amoebic DYSENTERY. They are therefore of enormous economic importance. The protozoans are divided into four classes: Mastigophora (Flagellata) in which movement is achieved by one or more whip-like flagella; Sarcodina, which use PSEUDOPODIA, outpushings of the cell, for movement; Sporozoa, which are all parasitic, and Ciliata, which move by beating numerous CILIA.

PROTRACTOR, an instrument for measuring ANGLES. Usually semicircular, it is marked off in degrees along the semicircular edge.

PROUST, Joseph Louis (1754–1826), French chemist who established the law of definite proportions, or Proust's Law (see COMPOSITION, CHEMICAL).

PROXIMA CENTAURI, the closest star to the sun (1.33pc distant), a red dwarf star orbiting ALPHA CENTAURI.

PRUNING, practice of cutting off parts of cultivated plants to encourage growth in the rest of the plant. It is commonly used on rose, fruit and ornamental trees and shrubs to regulate growth and improve the quality of flowers and fruits. The timing and degree of pruning is usually critical.

PRUSSIAN BLUE, or potassium ferric ferrocyanide ($KFe[Fe(CN)_6]$), deep-blue paint pigment made by reacting potassium ferrocyanide with any ferric salt.

PRUSSIC ACID, or hydrocyanic acid. See CYANIDES.

PSEUDOHALOGENS, class of monovalent inorganic radicals which chemically resemble the HALIDES and HALOGENS. They include CYANIDE (CN^-) and CYANOGEN ($[CN]_2$), cyanate (OCN^-), thiocyanate (SCN^-), and azide (N_3^-).

PSEUDOPODIUM, the "false limb" by which certain PROTOZOA are able to move. The internal pressure of the CELL forces the elastic cell membrane out in a bulge; the cell contents flow behind into the new position. Pseudopodia can be formed at any position on the cell surface.

PSILOCYBIN, HALLUCINOGENIC DRUG derived from a Mexican fungus (*Psilocybe mexicanus*) and related to LSD.

PSITTACOSIS, or **Parrot fever,** LUNG disease with FEVER, cough and breathlessness caused by a bedsonia, an organism intermediate between BACTERIA and VIRUSES. It is carried by parrots, pigeons, domestic fowl and related birds. TETRACYCLINES provide effective treatment, but any infected birds must be destroyed.

PSORIASIS, common SKIN condition characterized by patches of red, thickened and scaling skin. It often affects the elbows, knees and scalp but may be found anywhere. Several forms are recognized and the manifestations may vary in each individual with time. Coal tar preparations are valuable in treatment but STEROID creams and cytotoxic CHEMOTHERAPY may be needed. There is also an associated ARTHRITIS.

PSYCHE, in psychology, the MIND.

PSYCHEDELIC DRUGS. See HALLUCINOGENIC DRUGS.

PSYCHIATRY, the branch of medicine concerned with the study and treatment of MENTAL ILLNESS. It has two major branches: one is PSYCHOTHERAPY, the application of psychological techniques to the treatment of mental illnesses where a physiological origin is either unknown or does not exist (see also PSYCHOANALYSIS); the other, medical therapy, where attack is made either on the organic source of the disease or, at least, on its physical or behavioral symptoms. (Psychotherapy and medical therapy are often used in tandem.) As a rule of thumb, the former deals with NEUROSES, the latter with PSYCHOSES. (See also PSYCHOLOGY.) DRUGS are perhaps the most widely used tools of psychiatry. Many emotional and other disturbances can be simply treated by the use of mild SEDATIVES or TRANQUILLIZERS. A major area of success for drug therapy is ALCOHOLISM. Other major areas of success are in DRUG ADDICTION and the amelioration of the effects of EPILEPSY. (See also PSYCHOPHARMACOLOGY.) Drastic therapies include shock treatment and BRAIN surgery. The principal shock treatments are insulin shock and electroshock (electroconvulsive therapy—ECT). INSULIN, used primarily in cases of SCHIZOPHRENIA, may be given in increasingly large doses until shock level is achieved. In electroshock treatment, an electric current is passed through the brain, producing convulsions and, often, unconsciousness: it is used in cases of MANIC-DEPRESSIVE PSYCHOSIS. Both techniques are unpredictable in result. LOBOTOMY, a surgical operation which severs certain of the neural pathways, is now rarely used.

PSYCHICAL RESEARCH. See PARAPSYCHOLOGY.

PSYCHOANALYSIS, a system of psychology having as its base the theories of Sigmund FREUD; also, the psychotherapeutic technique based on that system. The distinct forms of psychoanalysis developed by JUNG and ADLER are more correctly termed respectively analytical psychology and individual psychology. Freud's initial interest was in the origins of the NEUROSES. On developing the technique of FREE ASSOCIATION to replace that of HYPNOSIS in his therapy, he observed that certain patients could in some cases associate freely only with difficulty. He decided that this was due to the memories of certain experiences being held back from the CONSCIOUS mind (see REPRESSION) and noted that the most sensitive areas were in connection with sexual experiences. He thus developed the concept of the UNCONSCIOUS (later to be called the ID), and suggested (for a while) that ANXIETY was the result of repression of the LIBIDO. He also defined "resistance" by the conscious to acceptance of ideas and impulses from the unconscious, and TRANSFERENCE, the idea that relationships with people or objects in the past affect the individual's relationships with people or objects in the present. (Other important psychoanalytic ideas include CENSORSHIP; DEFENSE MECHANISM; EGO; INHIBITION; PLEASURE PRINCIPLE; SUPEREGO. See also DREAMS; GROUP INSTINCT; LYCANTHROPY; METAPSYCHOLOGY; PROJECTION; SEX.)

PSYCHOLOGICAL TESTS, experiments devised to elicit information about the psychological characteristics of individuals. Such characteristics may relate to the INTELLIGENCE, vocation, personality or aptitudes of the individual. Tests must be both consistent and accurate so, if possible, a large sample is used. (See also GRAPHOLOGY; IQ; PROJECTION TEST; PSYCHOLOGY; STANFORD-BINET TEST.)

PSYCHOLOGY, originally the branch of philosophy dealing with the mind, then the science of mind, and now, considered in its more general context, the science of behavior, whether human or animal. It is intimately related with ANTHROPOLOGY (the science of man) and Somatology (the science of body). (See also ANIMAL BEHAVIOR.) Clearly, psychology is closely connected with, on one side, MEDICINE and, on the other, SOCIOLOGY. There are a number of closely interrelated branches of human psychology. **Experimental psychology** embraces all psychological investigations controlled by the psychologist. His experiments may center on the individual or GROUP, in which latter case STATISTICS will play a large part in the research. In particular, in **clinical psychology**, information is gained through the treatment of those suffering from MENTAL ILLNESS. **Social psychologists** use statistical and other methods to investigate the effect of the group on the behavior of the individual (see also CYBERNETICS; INDUSTRIAL PSYCHOLOGY). In **applied psychology**, the discoveries and theories of psychology are put to practical use. **Comparative psychology** deals with the different behavioral organizations of animals (including man). In this century, the most important

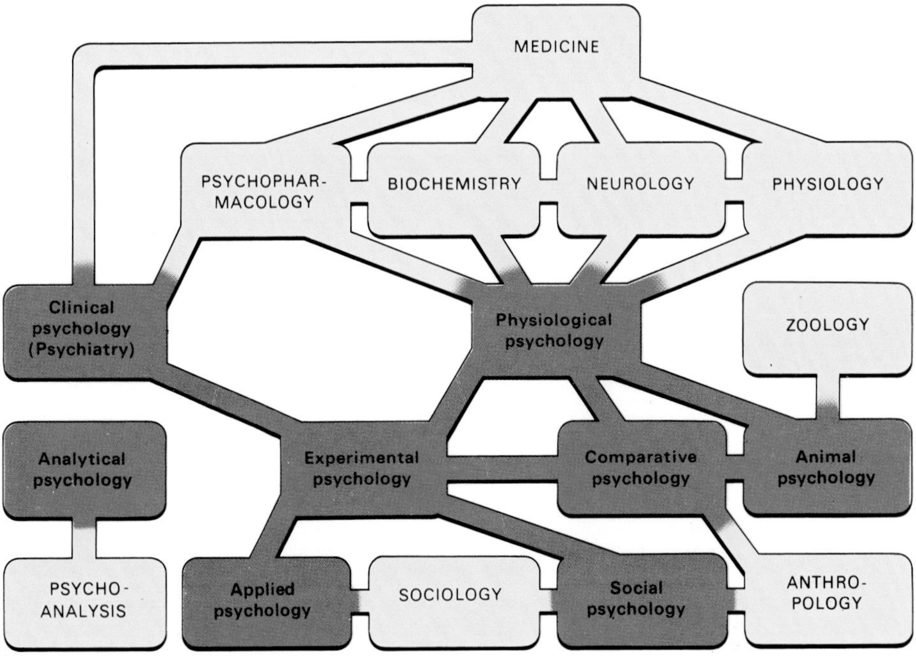

The different branches of psychology (red) and their links with allied sciences (yellow).

offspring of psychology are PSYCHIATRY and PSYCHOANALYSIS. Both are concerned with the treatment of mental illness, but from radically different viewpoints (see also PSYCHOTHERAPY). The former, in particular, is helped by the discoveries of **physiological psychology**, which attempts to understand the NEUROLOGY and PHYSIOLOGY of behavior. Two rather different branches of psychology sprang originally from FREUD's psychoanalysis. They are **analytical psychology**, founded by Carl Gustav JUNG, and **individual psychology**, founded by Alfred ADLER. See also Alfred BINET; Francis GALTON; William JAMES; Wilhelm WUNDT; and BEHAVIORISM; GESTALT PSYCHOLOGY; IDEA; INTELLIGENCE; PARAPSYCHOLOGY; PSYCHOLOGICAL TESTS; PSYCHOPHARMACOLOGY.)

PSYCHOPATH, person emotionally disturbed to a degree approaching MANIA, but who suffers no specific MENTAL ILLNESS; or, loosely, person with psychopathic (see PSYCHOPATHY) symptoms.

PSYCHOPATHY, any specific MENTAL ILLNESS characterized by abnormal, antisocial behavior.

PSYCHOPHARMACOLOGY, the study of the effects of DRUGS on the mind, and particularly the development of drugs for treating MENTAL ILLNESS.

PSYCHOSIS, in contrast with NEUROSIS, any MENTAL ILLNESS, whether of neurological (see NEUROLOGY) or purely psychological origins, which renders the individual incapable of distinguishing reality from unreality or fantasy. If the loss of mental capacity is progressive, the illness is termed a deteriorative psychosis.

PSYCHOSOMATIC ILLNESS, any illness in which some mental activity, usually ANXIETY or the INHIBITION of the EMOTIONS (see also REPRESSION), causes physiological malfunction. There is debate as to which disorders are psychosomatic, but among the most likely candidates are gastric ULCERS, ulcerative COLITIS and certain types of ASTHMA.

PSYCHOTHERAPY, the application of the theories and discoveries of PSYCHOLOGY to the treatment of MENTAL ILLNESS. Psychotherapy does not usually involve physical techniques, such as the use of drugs or surgery (see PSYCHIATRY). The term is sometimes used misleadingly to distinguish other forms of therapy from PSYCHOANALYSIS.

PTERIDOPHYTA, a traditional but artificial division of the PLANT KINGDOM which includes FERNS, HORSETAILS and CLUBMOSSES. Although all these plants contain vascular tissues and have a dominant SPOROPHYTE generation differentiated into ROOTS, STEMS and LEAVES, it is now clear that they are not closely related. (See ALTERNATION OF GENERATIONS; SPERMATOPHYTA.)

PTERODACTYLUS, a fossil member of the PTEROSAURIA that lived during the Jurassic period, 190–135 million years ago. Pterodactylus was small, some specimens being the size of a sparrow.

PTEROPSIDA, a subdivision of the PLANT KINGDOM which includes the FERNS, GYMNOSPERMS and ANGIOSPERMS (flowering plants).

PTEROSAURIA, fossil members of the REPTILIA that lived during the Jurassic and Cretaceous periods, 190–65 million years ago. Pterosaurs, or pterodactyls, were able to fly, and had greatly elongated fourth fingers that supported membranous wings.

PTOLEMY, or **Claudius Ptolemaeus** (2nd century AD), Alexandrian astronomer, mathematician and geographer. Most important is his book on ASTRONOMY, now called *Almagest* ("the greatest"), a synthesis of Greek astronomical knowledge,

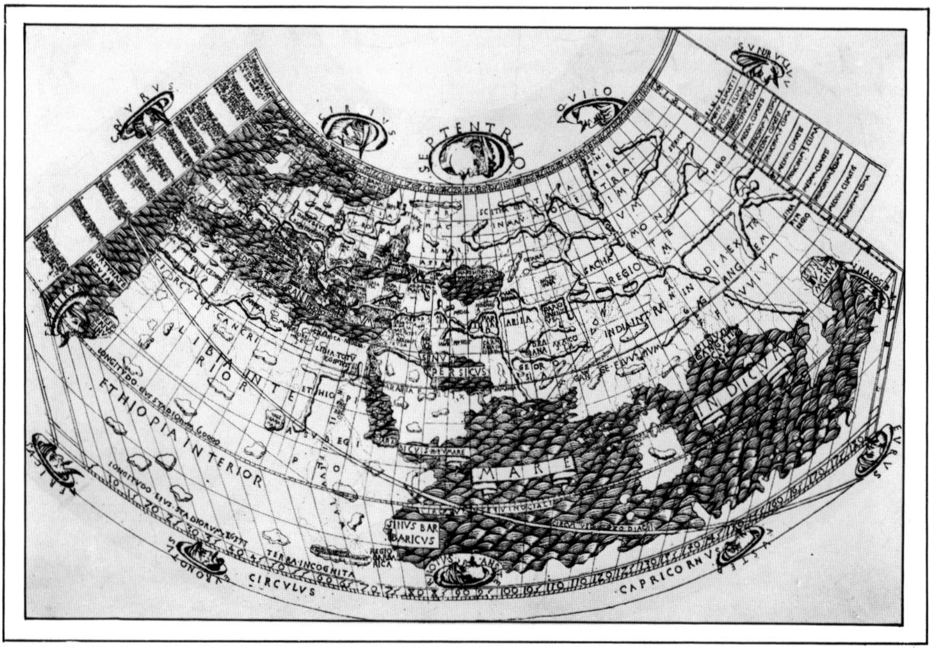

Map of the world according to Claudius Ptolemy as printed at Bologna in 1447. The projection is a hybrid between a conical and a cylindrical.

especially that of HIPPARCHUS: his geocentric cosmology dominated Western scientific thought until the Copernican Revolution of the 16th century (see COPERNICUS, N.). His *Geography* confirmed Columbus' belief in the westward route to Asia. In his *Optics* he attempts to solve the astronomical problem of ATMOSPHERIC REFRACTION. (See also EPICYCLE.)

PTOMAINE POISONING, old name for FOOD POISONING.

PTYALIN, an ENZYME secreted by the salivary GLANDS of the DIGESTIVE SYSTEM, which has a minor role in STARCH breakdown.

PUBERTY, the time during the GROWTH of a person at which sexual development occurs, commonly associated with a growth spurt. Female puberty involves several stages—the acquisition of BREAST buds; of sexual hair, and the onset of MENSTRUATION—which may each begin at different times. Male puberty involves sexual-hair development; VOICE change, and growth of the TESTES and PENIS. Precocious puberty is when the pubertal features develop abnormally early (before 9 years in females). The average age at puberty has fallen in recent years.

PUERPERAL FEVER, disease occuring in puerperal women, usually a few days after the BIRTH of the child and caused by infection of the WOMB, often with STREPTOCOCCUS. It causes FEVER, abdominal pain and discharge of PUS from the womb. The introduction of ASEPSIS in OBSTETRICS by I. P. SEMMELWEISS greatly reduced its incidence. Today, ANTIBIOTICS are required if it develops.

PULLEY, grooved wheel mounted on a block and with a cord or belt passing over it. A pulley is a simple MACHINE applying the equilibrium of TORQUES to obtain a mechanical advantage. Thus, the block and tackle is a combination of ropes and pulleys used for hoisting heavy weights. A belt and pulley combination can transmit motion from one part of a machine to another. Variable speed can be obtained from a single-speed driving shaft by the use of stepped or cone-shaped pulleys with diameters that give the correct speed ratios and belt tensions. To help prevent excessive belt wear and slipping, the rim surface of a pulley is adapted to the material of the belt used.

PULP, or wood pulp. See PAPER.

PULSAR, short for pulsating radio star, a celestial radio source emitting brief extremely regular pulses of ELECTROMAGNETIC RADIATION (with one exception, entirely radio-frequency). Each pulse lasts a few hundredths of a second and the period between pulses is of the order of one second or less. Though the pulse frequency varies from pulsar to pulsar, for each pulsar the period is as regular as man can measure. The first pulsar was discovered in 1967 by HEWISH and S. J. Bell. The fastest pulsar yet observed has a period of 0.033s, emitting pulses of the same frequency in the X-ray and visible regions of the spectrum. It is likely that there are some 10 000 pulsars in the MILKY WAY, though less than 50 have as yet been discovered. It is believed that pulsars are the neutron STAR remnants of SUPERNOVAE, rapidly spinning and radiating through loss of rotational energy.

PULSE, the palpable impulse conducted in the ARTERIES representing the transmitted beat of the HEART. A normal pulse rate is between 68 and 80, but athletes may have slower pulses. FEVER, heart disease, ANOXIA and ANXIETY increase the rate. The pulse character may suggest specific conditions, loss of pulse possibly indicating arterial block or cessation of the heart.

PULSES, the edible seeds produced by LEGUMINOUS PLANTS, such as peas and beans, and a general name for plants yielding such seeds.

PUMICE, porous, frothy volcanic glass, usually silica-rich; formed by the sudden release of vapors as LAVA cools under low pressures. It is used as an ABRASIVE, an AGGREGATE and a railroad ballast.

PUMP, device for taking in and forcing out a fluid, thus giving it kinetic or potential ENERGY. The HEART is a pump for circulating blood around the body. Pumps are commonly used domestically and industrially to transport fluids, to raise liquids, to compress gases or to evacuate sealed containers. Their chief use is to force fluids along pipelines (see PIPES AND PIPELINES). The earliest pumps were waterwheels, endless chains of buckets, and the ARCHIMEDES screw. Piston pumps, known in classical times, were developed in the 16th and 17th centuries, the suction types (working by atmospheric pressure) being usual, though unable to raise water more than about 10.4m (34ft). The STEAM ENGINE was developed to power pumps for pumping out mines. Piston pumps—the simplest of which is the SYRINGE—are reciprocating **volume-displacement pumps**, as are diaphragm pumps, with a pulsating diaphragm instead of the piston. One-way inlet and outlet valves are fitted in the cylinder. Rotary volume-displacement pumps have rotating gear wheels or wheels with lobes or vanes. **Kinetic pumps**, or FANS, work by imparting momentum to the fluid by means of rotating curved vanes in a housing: centrifugal pumps expel the fluid radially outward, and propeller pumps axially forward. **Air compressors** use the TURBINE principle (see also JET PROPULSION). **Air pumps** use compressed air to raise liquids from the bottom of wells, displacing one fluid by another. If the fluid must not come into direct contact with the pump, as in a nuclear reactor, **electromagnetic pumps** are used: an electric current and a magnetic field at right angles induce the conducting fluid to flow at right angles to both (see MOTOR, ELECTRIC); or the principle of the linear INDUCTION MOTOR may be used. To achieve a very high vacuum, the **diffusion pump** is used, in which atoms of condensing mercury vapor entrain the remaining gas molecules.

PUNCTUATION, those marks, distinct from letters and accents, which clarify the meaning of written language. Punctuation does not determine pronunciation, though it often reflects it. It is an important element of GRAMMAR.

PUPA, an immature stage in the development of those insects which have a LARVA completely different in structure from the adult, and in which "complete" METAMORPHOSIS occurs. The pupa is a resting stage in which the larval structure is reorganized to form the adult: all but the nervous system changes. Feeding and locomotion are meanwhile suspended.

PUPIN, Michael Idvorsky (1858–1935), Hungarian-born US inventor who made many contributions to TELEPHONE science, including a technique whereby longer-distance communication can be sustained.

PURCELL, Edward Mills (1912–), US physicist who shared with F. BLOCH the 1952 Nobel Prize for Physics for his independent work on nuclear magnetic moment, discovering nuclear magnetic resonance (NMR) in solids (see SPECTROSCOPY).

PURINE ($C_5H_4N_4$), the parent compound of a class of organic bases of major biochemical importance. The purines adenine and guanine are present in NUCLEIC ACIDS. A combination of a purine and a 5-carbon sugar is termed a nucleoside, which when phosphorylated gives a NUCLEOTIDE. Other important purine derivatives include CAFFEINE and theobromine. The end product of purine metabolism is URIC ACID which is excreted in the urine.

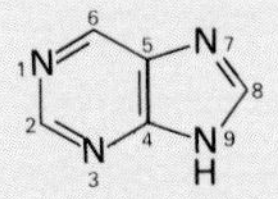

The structure of purine. Many of the most important purine derivatives are substituted on the 2 and 6 carbon atoms.

PURKINJE, Johannes Evangelista (1787–1869), Bohemian-born Czech physiologist and pioneer of HISTOLOGY, best known for his observations of nerve cells (see NERVOUS SYSTEM) and discovery of the Purkinje Effect, that at different overall light intensities the eye is more sensitive to different colors (see VISION).

PUS, off-white or yellow liquid consisting of inflammatory exudate, the debris of white BLOOD cells and BACTERIA resulting from localized INFLAMMATION, especially ABSCESSES. Pus contained in cavities is relatively inaccessible to ANTIBIOTICS and may require drainage by SURGERY. Pus suggests but does not prove the presence of bacterial infection.

PUTREFACTION, the natural decomposition of dead organic matter, in particular the anaerobic decomposition of its PROTEIN by BACTERIA and FUNGI. This process produces foul-smelling substances such as AMMONIA, hydrogen SULFIDE and organic SULFUR compounds. The amino-acid nitrogen of the protein is recycled by incorporation in the bacteria and fungi.

PYGMY, term used to denote those peoples whose adult males are on average less than 1.5m tall. Some Kalahari Desert Bushmen are of pygmy size, but the most notable pygmys are the Mbuti, or Bambuti, of the Ituri Forest, Zaire, who, through their different blood type, skin color and other characteristics, are regarded as distinct from the surrounding peoples and were probably the original inhabitants of the region. A Stone Age people, they are nomadic hunters, living in groups of 50 to 100. Asian pygmies are generally termed **Negritos**. Peoples rather larger than pygmies are described as pygmoid.

PYORRHEA, or flow of PUS, usually used to refer to the pus related to poor oral hygiene and exuding from the margins of the gums and TEETH; it causes loosening of the teeth and HALITOSIS.

PYRAMID, a POLYHEDRON whose base is a POLYGON and whose sides are TRIANGLES having a common VERTEX. A pyramid whose base is triangular is termed a tetrahedron (or triangular pyramid); one whose base is a regular polygon is termed regular; one with a square base, square; one with a rectangular base, rectangular.

PYREX, a borosilicate GLASS used in chemical and industrial apparatus and in ovenware. It is inert, a good insulator and heat-resistant (having a low coefficient of expansion and a high softening temperature).

PYRIDINE (C_5H_5N), a colorless heterocyclic aromatic base with an unpleasant smell, found in bone oil and coal tar. It is used as a solvent and to denature grain alcohol (ETHANOL). Important derivatives are **pyridoxine** (VITAMIN B_6) and niacin.

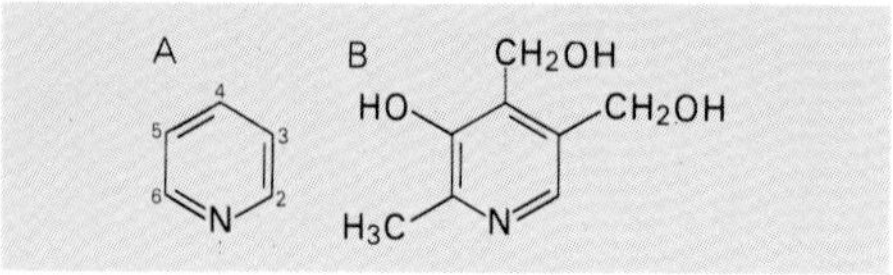

(A) Pyridine (azine), and (B) pyridoxine, one of the three naturally occurring forms of vitamin B_6.

PYRIMIDINE ($C_4H_4N_2$), HETEROCYCLIC COMPOUND with a 6-membered aromatic ring containing nitrogen atoms in the 1 and 3 positions. The nucleus occurs in many important PURINE compounds.

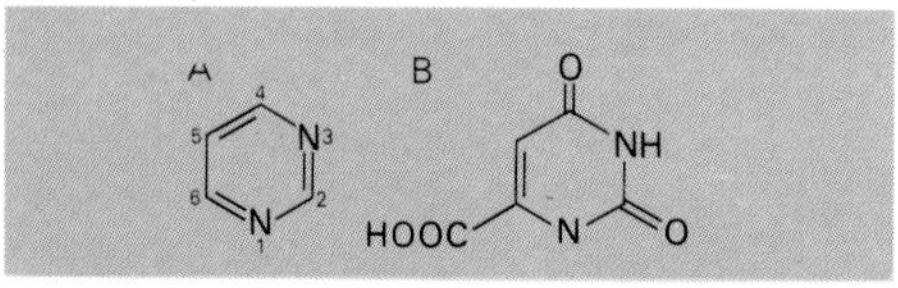

(A) Pyrimidine (1,3-diazine), and (B) orotic acid.

PYRITE, or iron pyrites (FeS_2, iron (II) disulfide), a hard, yellow SULFIDE mineral known as **fool's gold** from its resemblance to gold. Of worldwide occurrence, it is a major ore of SULFUR. It crystallizes in the isometric system, usually as cubes. It alters to GOETHITE and LIMONITE.

PYROCLASTIC ROCKS, rocks made up of particles thrown into the air by volcanic eruptions. (See also LAVA; VOLCANISM; VOLCANO.)

PYROELECTRICITY. See PIEZOELECTRICITY.

PYROLUSITE, soft, gray-black OXIDE mineral composed of manganese (IV) oxide (MnO_2); of widespread occurrence, it is the chief ore of MANGANESE. It is a secondary mineral of aqueous origin crystallizing in the tetragonal system.

The mineral pyrite.

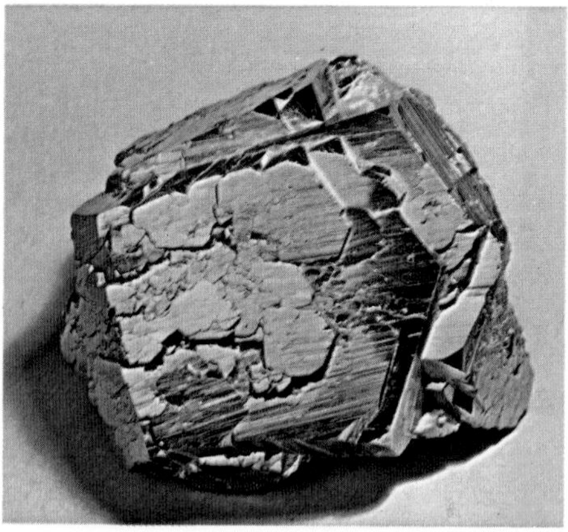

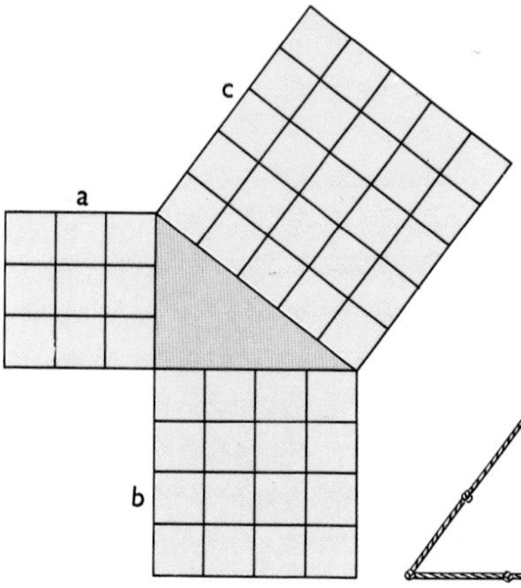

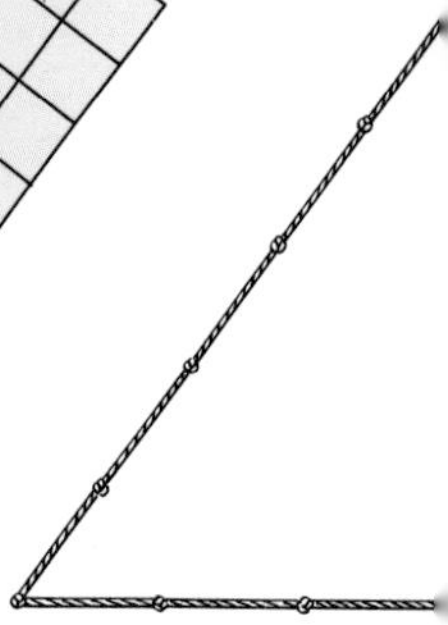

Pythagoras' theorem (*left*) demonstrated in the case of the 3,4,5 triangle (one having sides in the ratio 3:4:5). It can be seen by inspection that in this case $a^2+b^2=c^2$. Other right-angled triangles, the ratios of whose sides can be expressed using only small integers, are the 5,12,13 and 8,15, 17 triangles. *Right*: The ancient method of laying out a right angle using a knotted rope was known and used long before the time of Pythagoras.

PYROLYSIS, chemical DECOMPOSITION of a substance by heat. (See also CALCINATION; COMBUSTION.)

PYROMETER, a temperature measuring device used for high temperatures. Platinum resistance thermometers and pyrometers operating on the principle of the THERMOCOUPLE have the disadvantage that they must be in contact with the hot body, but optical and radiation pyrometers can be used at a distance. Optical pyrometers estimate temperature from the light intensity in a narrow band of the visible spectrum by optical comparison of a glowing filament with an image of the hot body. Radiation pyrometers focus the body's heat radiation on a responsive thermal element such as a thermocouple.

PYROXENES, major group of SILICATE minerals occurring in IGNEOUS ROCKS. They have a chain structure, with prismatic CLEAVAGE close to 90°, and are related to the AMPHIBOLES. They are monoclinic or orthorhombic. Jadeite (see JADE) and SPODUMENE are commercially important.

PYRRHOTITE, bronze-brown iron SULFIDE mineral with the NICCOLITE structure, occurring in basic igneous rocks in Scandinavia and Ontario. It is nonstoichiometric (see COMPOSITION, CHEMICAL) with composition $Fe_{1-x}S$. Improved smelting techniques now make it possible to extract the iron.

PYRROPHYTA, or dinoflagellates and cryptomonads. See ALGAE.

PYTHAGORAS (c570–c500 BC), Greek philosopher who founded the Pythagorean school. Attributed to the school are: the proof of PYTHAGORAS' THEOREM; the suggestion that the earth travels around the sun, the sun in turn around a central fire; observation of the ratios between the lengths of vibrating strings that sound in mutual harmony, and ascription of such ratios to the distances of the planets, which sounded the "harmony of the spheres," and the proposition that all phenomena may be reduced to numerical relations.

PYTHAGORAS' THEOREM, or **Pythagorean Theorem**, the statement that, for any right-angled TRIANGLE, the SQUARE on the HYPOTENUSE is equal to the sum of the squares on the other two sides. The earliest known formal statement of the theorem is in the *Elements* of EUCLID, but it seems that the basis of it was known long before this time and, indeed, long before the time of PYTHAGORAS himself. (See also EUCLIDEAN GEOMETRY.)

PYTHEAS (flourished c300 BC), Greek navigator, the first of his countrymen to explore the Atlantic coast of Europe and visit the British Isles. According to the Greek historian Polybius, he reported the existence of an inhabited island called Thule, six days' sail to the north of Britain—possibly Norway or Iceland.

Q FEVER, or **query fever**, INFECTIOUS DISEASE due to *Coxiella*, an organism intermediate between BACTERIA and VIRUSES, causing FEVER, HEADACHE and often dry cough and chest pain. It is transmitted by ticks from various farm animals and is common among farm workers and veterinarians. Its course is benign but TETRACYCLINES may be used in treatment.

QUADRANGLE, in GEOMETRY, a QUADRILATERAL.

QUADRANT, in plane CARTESIAN COORDINATES, one of the four divisions of the PLANE made by the AXES. In SPHERICAL GEOMETRY, a quadrant is a spherical distance of $\pi/2$. A quadrant of a circle is a sector with a central ANGLE of $\pi/2$.

QUADRANT, a simple astronomical and navigational instrument used in early times to measure the altitudes of the sun and stars. It consisted typically of a pair of sights, a calibrated quadrant (quarter) of a circle, and a plumb line. (See also SEXTANT.)

QUADRAPHONIC SOUND. See HIGH FIDELITY.

QUADRILATERAL, in geometry, a PLANE four-sided POLYGON. Quadrilaterals with two pairs of sides parallel are called parallelograms; with one pair of sides parallel, trapezoids; with no two sides parallel, trapeziums (the word trapezium is often used as a synonym of trapezoid). Parallelograms whose sides are all of equal length are termed rhombuses. Each side of a parallelogram is equal in length to the side parallel to it; and each INTERIOR ANGLE is equal to the interior angle diametrically opposite it. A parallelogram whose interior angles are each 90° is a rectangle: a special case of this is the square, all of whose sides are equal. The sum of the interior angles of a quadrilateral is always 360°.

QUADRUPLETS. See MULTIPLE BIRTH.

QUANTUM MECHANICS, fundamental theory of small-scale physical phenomena (such as the motions of ELECTRONS and nuclei within ATOMS), developed during the 20th century when it became clear that the existing laws of classical mechanics and electromagnetic theory were not successfully applicable to such systems. Because quantum mechanics treats physical events that we cannot directly perceive, it has many concepts unknown in everyday experience. DE BROGLIE struck out from the old QUANTUM THEORY when he suggested that particles have a wavelike nature, with a wavelength $\lambda = h/p$ (h being the Planck constant and p the particle momentum). This wavelike nature is significant only for very small particles such as electrons. These ideas were developed by SCHRÖDINGER and others into the branch of quantum mechanics known as WAVE MECHANICS. HEISENBERG worked along parallel lines with a theory incorporating only observable quantities such as ENERGY, using matrix algebra techniques. The UNCERTAINTY PRINCIPLE is fundamental to quantum mechanics, as is Pauli's EXCLUSION PRINCIPLE. DIRAC incorporated relativistic ideas into quantum mechanics.

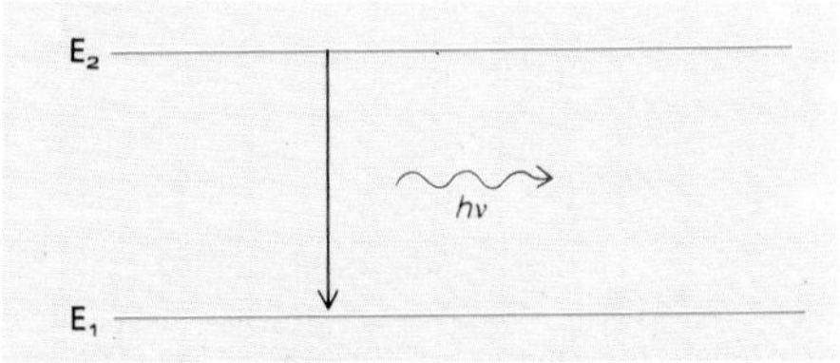

Fundamental to quantum theory is the notion that energy is quantized: that material systems can exist only in certain discrete and defined energy states and that transitions between these states can only occur by the gain or loss of particular quantities of energy. Here a system in energy state E_2 changes to energy state E_1, emitting electromagnetic radiation in the process. The frequency (ν) of this radiation is determined by the energy change ($E_2 - E_1$) involved in the transition according to the equation:

$$E_2 - E_1 = h\nu$$

where h is the Planck constant.

QUANTUM THEORY, theory developed at the beginning of the 20th century to account for certain phenomena that could not be explained by classical PHYSICS. PLANCK described the previously unexplained distribution of radiation from a BLACK BODY by assuming that ELECTROMAGNETIC RADIATION exists in discrete bundles known as quanta, each with an ENERGY $E = h\nu$ (ν being the radiation frequency and h a universal constant—the PLANCK CONSTANT). EINSTEIN also used the idea of quanta to explain the PHOTOELECTRIC EFFECT, establishing that electromagnetic radiation has a dual nature, behaving sometimes as a WAVE MOTION and sometimes as a stream of particle-like quanta. Measurements of other physical quantities, such as the frequencies of lines in atomic spectra and the energy losses of electrons on colliding with atoms, showed that these quantities could not have a continuous range of values, discrete values only being possible. With RUTHERFORD's discovery in 1911 that ATOMS consist of a small positively charged nucleus surrounded by ELECTRONS, attempts were made to understand this atomic structure in the light of quantum ideas, since classically the electrons would radiate energy continuously and collapse into the nucleus. BOHR postulated that an atom only exists in certain stationary (i.e., nonradiating) states with definite energies and that quanta of radiation are emitted or absorbed in transitions between these states; he successfully calculated the stationary states of hydrogen. Some further progress was made along these lines by Bohr and others, but it became clear that the quantum theory was fundamentally weak in being unable to calculate intensities of spectral lines. The new QUANTUM MECHANICS was developed c1925 to take its place.

QUARANTINE, period during which a person or animal must be kept under observation in isolation from the community after having been in contact with an INFECTIOUS DISEASE. The duration of quarantine depends on the disease(s) concerned and their maximum length of INCUBATION. The term derives from the period of 40 days that ships from the Levant had to wait before their crews could disembark at medieval European ports, from fear of their carrying PLAGUE.

QUARK. See SUBATOMIC PARTICLES.

QUART, name of various units of liquid and dry measure. See WEIGHTS AND MEASURES.

QUARTZ, rhombohedral form of SILICA, usually forming hexagonal prisms, colorless when pure ("rock crystal"). A common mineral, it is the chief constituent of SAND, SANDSTONE, QUARTZITE and FLINT, and also occurs as the GEMS: CHALCEDONY; AGATE; JASPER, and ONYX. Quartz is piezoelectric (see PIEZOELECTRICITY) and is used to make oscillators for clocks, radio and radar; and also to make windows for optical instruments. Crude quartz is used to make glass, glazes and abrasives, and as a flux.

QUARTZITE, common rock consisting of QUARTZ, formed by metamorphism of SANDSTONE. It is tough, resisting weathering, and is used as road metal.

QUASAR, or quasi-stellar object, a telescopically star-like celestial object whose SPECTRUM shows an abnormally large RED SHIFT. Quasars may be extremely distant objects, receding from us at high velocities, although some recent research has cast doubt upon this, since the spectra of quasars do not seem to have been affected by the interpolation of intergalactic gas. Quasars show variability in light and radio emission (although the first quasars were discovered by RADIO ASTRONOMY, not all are radio sources), which might indicate that they are comparatively small objects less than 0.3pc across, comparatively close to us (larger—and more distant—objects being unlikely to vary in this way). There are about 200 quasars in each square degree of the sky.

QUATERNARY, the period of the CENOZOIC whose beginning is marked by the advent of man. It has lasted about 4 million years, up to and including the present. (See also TERTIARY; GEOLOGY.)

QUATERNION, a type of complex number (see IMAGINARY NUMBERS) developed by W. R. Hamilton (1805–1865) to operate in three dimensions as ordinary complex numbers do in two.

QUICKSAND, sand saturated with water to form a sand-water SUSPENSION possessing the characteristics of a liquid. Quicksands may form at rivermouths or on sandflats, and are dangerous as they appear identical to adjacent SAND. In fact, the DENSITY of the suspension is less than that of the human body so that, if a person does not struggle, he may escape being engulfed.

QUICKSILVER. See MERCURY.

QUIMBY, Phineas Parkhurst (1802–1866), US pioneer of mental healing, an early user of SUGGESTION as a therapy. A strong influence on Mary Baker Eddy, he is held as a father of the New Thought movement.

QUINE, Willard Van Orman (1908–   ), US philosopher and logician, best known for his rejection of such longstanding philosophical claims as that analytic ("self-evident") statements are fundamentally distinguishable from synthetic (observational) statements, and that the concept of synonymy (sameness of meaning) can be exemplified.

QUININE, substance derived from cinchona bark from South America, for long used in treating a variety of ailments. It was preeminent in early treatment of MALARIA until the 1930s when ATABRINE was introduced; after this more suitable quinine derivatives such as CHLOROQUINE were synthesized. Quinine is also a mild ANALGESIC and may prevent

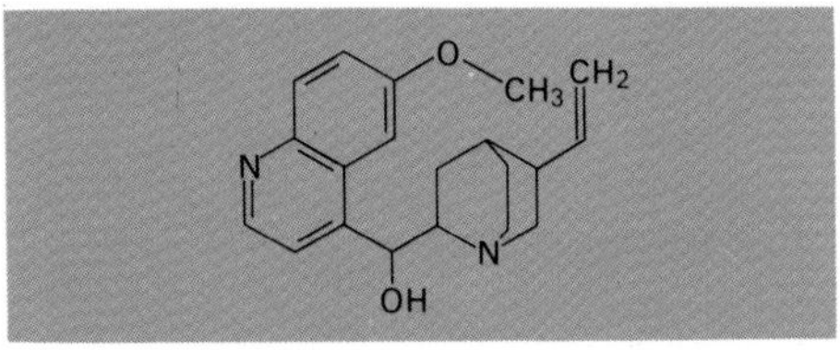

The structure of quinine.

CRAMPS and suppress HEART rhythm disorders. Now rarely used, its side effects include VOMITING, DEAFNESS, VERTIGO and VISION disturbance.

QUINSY, acute complication of TONSILLITIS in which ABSCESS formation causes spasm of the adjacent jaw muscles, FEVER and severe pain. Incision and drainage of the PUS produce rapid relief, though ANTIBIOTICS are helpful and the TONSILS should be excised later.

QUINTAL, Metric (q), unit of MASS equal to 100kg.

QUINTUPLETS. See MULTIPLE BIRTH.

QUOTIENT, in the DIVISION of one positive INTEGER by another, the largest number of times (k) the DIVISOR (a) must be multiplied so that $0 \leqslant (b-ka) < a$, where b is the DIVIDEND and $(b-ka)$ the REMAINDER.

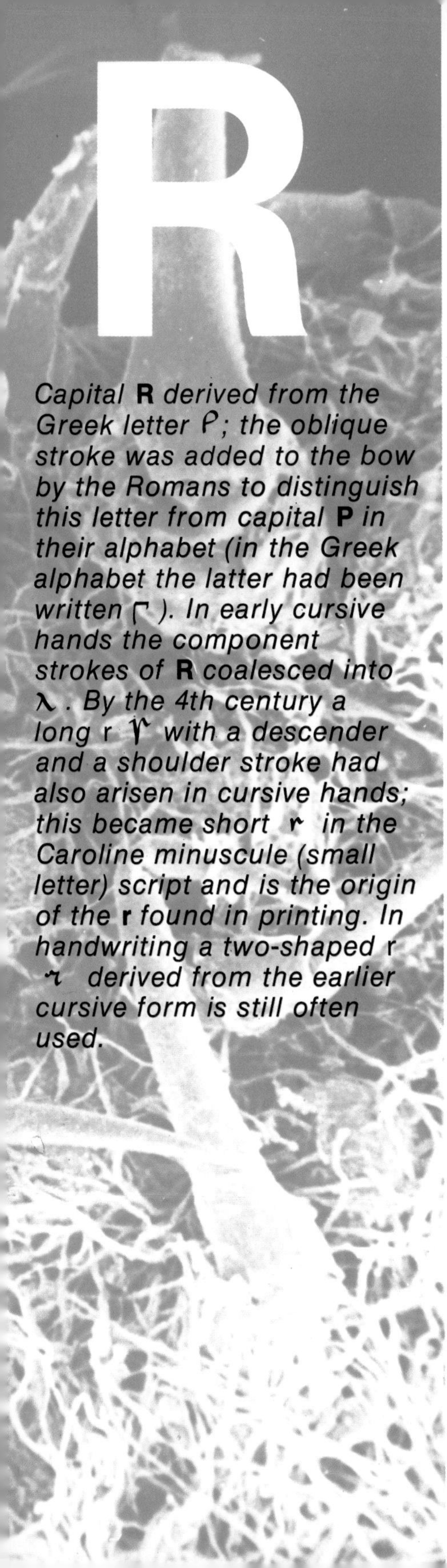

Capital **R** *derived from the Greek letter* Ρ*; the oblique stroke was added to the bow by the Romans to distinguish this letter from capital* **P** *in their alphabet (in the Greek alphabet the latter had been written* Γ *). In early cursive hands the component strokes of* **R** *coalesced into* λ *. By the 4th century a long r* ɼ *with a descender and a shoulder stroke had also arisen in cursive hands; this became short* r *in the Caroline minuscule (small letter) script and is the origin of the* **r** *found in printing. In handwriting a two-shaped r* ꝛ *derived from the earlier cursive form is still often used.*

RABBIT FEVER. See TULAREMIA.

RABI, Isidor Isaac (1898–   ), Austrian-born US physicist whose discovery of new ways of measuring the magnetic properties of ATOMS and MOLECULES both paved the way for the development of the MASER and the ATOMIC CLOCK and earned him the 1944 Nobel Prize for Physics.

RABIES, or **Hydrophobia,** fatal VIRUS disease resulting from the bite of an infected animal, usually a dog. HEADACHE, FEVER, and an overwhelming fear, especially of water, are early symptoms following an INCUBATION period of 3–6 weeks; PARALYSIS, spasm of muscles of swallowing, respiratory paralysis, DELIRIUM, CONVULSIONS and COMA due to an ENCEPHALITIS follow. Wound cleansing, antirabies vaccine and hyperimmune serum must be instituted early in confirmed cases to prevent the onset of these symptoms. Fluid replacement and respiratory support may help, but survival is rare if symptoms appear. Infected animals must be destroyed.

RACE, within a SPECIES, a subgroup most of whose members have sufficiently different physical characteristics from those exhibited by most members of another subgroup for it to be considered as a distinct entity. In particular the term is used with respect to the human species, *Homo sapiens*, the three most commonly distinguished races being CAUCASOID, MONGOLOID and NEGROID (see also AUSTRALOID). However, in practice it is impossible to make unambiguous distinctions between races: a classification by color would yield a quite different result to one by blood-group (see ANTHROPOMETRY). According to DARWIN's theories of EVOLUTION, races arise when different groups encounter different environmental situations. Over generations, their physical characteristics evolve until each group as a whole is physically quite different from its parent stock. Should the isolation of the group continue long enough, and the environment be different enough, the divergent race will eventually become a distinct species, unable to mate with the species from which it originally sprang. This has obviously not happened in the case of man, whose races may interbreed successfully and, in many cases, advantageously. It is not known when man became racially differentiated, but certainly it was at a very early stage in his evolution (see PREHISTORIC MAN): nowadays, with the rise of efficient transportation, racial convergence in man is accelerating and seems likely to result in complete racial fusion, in which individuals will nevertheless vary considerably.

RACEME. See INFLORESCENCE.

RAD (rad, or rd where it might be confused with rad(ian)), a unit used for expressing absorbed dose of ionizing RADIATIONS. It represents the dosage absorbed when 1kg of matter absorbs 0.01 joules of energy.

RADAR (*ra*dio *d*etection *a*nd *r*anging), system that detects long-range objects and determines their positions by measuring the time taken for RADIO waves to travel to the objects, be reflected and return. Radar is used for NAVIGATION, air control, fire control, storm detection, in radar astronomy and for catching speeding drivers. It developed out of experiments in the 1920s measuring the distance of the IONOSPHERE by radio pulses. R. A. WATSON-WATT showed that the technique could be applied to detecting aircraft, and

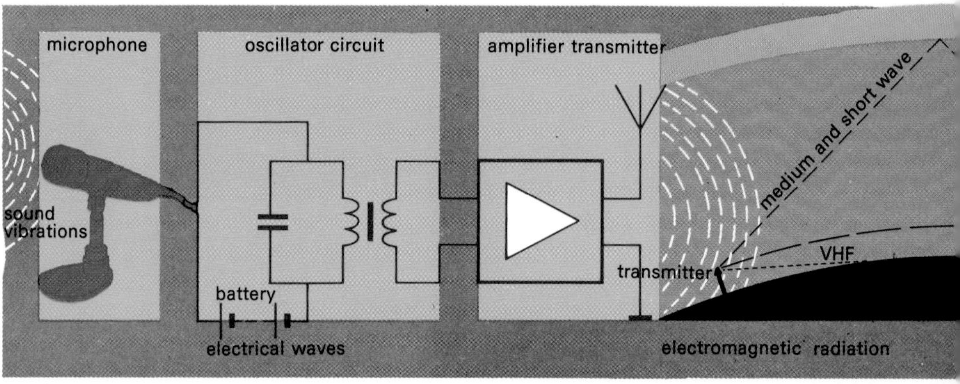

from 1935 Britain installed a series of radar stations which were a major factor in winning the Battle of Britain in WWII. From 1940 the UK and the US collaborated to develop radar. There are two main types of radar: **continuous-wave radar**, which transmits continuously, the frequency being varied sinusoidally, and detects the signals received by their instantaneously different frequency; and the more common **pulsed radar**. This latter has a highly directional antenna which scans the area systematically or tracks an object. A cavity magnetron or klystron emits pulses, typically 400 per second, 1μs across, and at a frequency of 3GHz. A duplexer switches the antenna automatically from transmitter to receiver and back as appropriate. The receiver converts the echo pulses to an intermediate frequency of about 30MHz, and they are then amplified, converted to a video signal, and displayed on a CATHODE-RAY TUBE. A synchronizer measures the time-lag between transmission and reception, and this is represented by the position of the pulse on the screen. Various display modes are used: commonest is the plan-position indicator (PPI), showing horizontal position in polar coordinates. (See also LORAN.)

RADCLIFFE-BROWN, Sir Alfred Reginald (1881–1955), British anthropologist whose comparative studies of social structures provided a firm basis for the development of social ANTHROPOLOGY.

RADIAL SYMMETRY, symmetry about a single polar axis. It is a condition rare in nature though not infrequent among the unicellular PROTOZOA. Less than perfect radial symmetry is more common, five-fold symmetry being characteristic of echinoderms.

RADIAL VELOCITY, the component of the motion of a celestial body in the direction of the line of sight; that is, toward or away from the observer.

RADIAN. See ANGLE.

RADIATION, the emission and propagation through space of ELECTROMAGNETIC RADIATION or SUBATOMIC PARTICLES. Exposure to X-RAYS and GAMMA RAYS is measured in RÖNTGEN units; absorbed dose of any high-energy radiation in RADS.

RADIATION BELTS. See VAN ALLEN RADIATION BELTS.

RADIATION CHEMISTRY, study of the chemical effects produced by interaction of RADIATION with matter. These include the effects of light (see PHOTOCHEMISTRY) but are in general more complex; IONS are often formed. It is important in infrared and X-ray PHOTOGRAPHY and the study of MUTATIONS and of the origin of LIFE.

RADIATION SICKNESS, malaise, nausea, loss of appetite and VOMITING occurring several hours after exposure to ionizing RADIATION in large doses. This occurs as an industrial or war hazard, or more commonly following RADIATION THERAPY for CANCER, LYMPHOMA or LEUKEMIA. Large doses of radiation may cause BONE MARROW depression with ANEMIA, AGRANULOCYTOSIS and bleeding, or gastrointestinal disturbance with distension and bloody DIARRHEA. Skin ERYTHEMA and ulceration, LUNG fibrosis, NEPHRITIS and premature ARTERIOSCLEROSIS may follow radiation and there is a risk of malignancy developing.

RADIATION THERAPY, use of ionizing RADIATION, as rays from an outside source or from radium or other radioactive metal implants, in treatment of malignant DISEASE—CANCER, LYMPHOMA and LEUKEMIA. The principle is that rapidly dividing TUMOR cells are more sensitive to the destructive effects of radiation on NUCLEIC ACIDS and are therefore damaged by doses that are relatively harmless to normal tissues. Certain types of malignancy indeed respond to radiation therapy but RADIATION SICKNESS may also occur.

RADICAL, term formerly meaning FUNCTIONAL GROUP, now meaning FREE RADICAL.

RADIO, the communication of information between distant points using radio waves, ELECTROMAGNETIC RADIATION of wavelength between 1mm and 100km. Radio waves are also described in terms of their FREQUENCY—measured in HERTZ (Hz) and found by dividing the velocity of the waves (about 300Mm/s) by their wavelength. Radio communications systems link transmitting stations with receiving stations. In a transmitting station a piezoelectric OSCILLATOR is used to generate a steady radio-frequency (RF) "carrier" wave. This is amplified and "modulated" with a signal carrying the information (see INFORMATION THEORY) to be communicated. The simplest method of modulation is to pulse (switch on and off) the carrier with a signal in, say, MORSE CODE, but speech and music, entering the modulator as an audio-frequency (AF) signal from tape or a MICROPHONE, is made to interact with the carrier so that the shape of the audio wave determines either the amplitude of the carrier wave (amplitude modulation—AM) or its frequency within a small band on either side of the

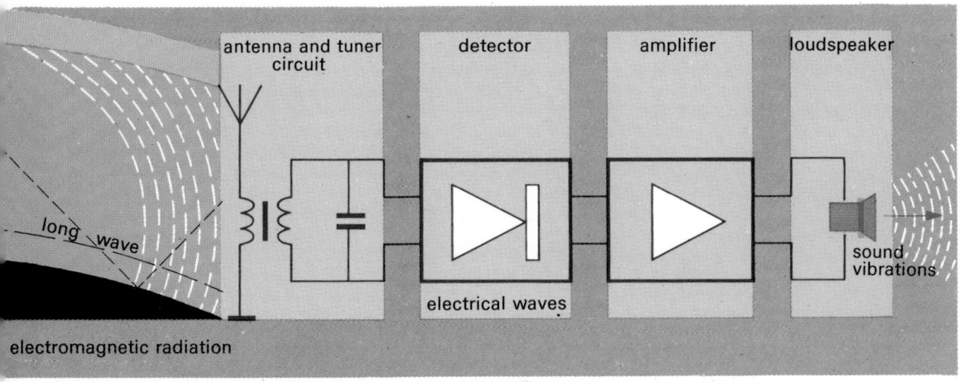

In radio transmission, sound waves are converted into electrical impulses by the microphone. The electrical signal is used to modulate a carrier wave generated in an oscillator circuit. The modulated carrier is then amplified and fed to the transmitter which radiates radio waves which can be picked up by receiving sets. The distance the waves travel is determined by their frequency, though unusual atmospheric conditions may greatly alter the normal range. Long waves follow the curvature of the earth, while VHF waves are limited to line-of-sight transmission; medium- and short-wave signals bounce off the ionosphere and can be received at very great distances, especially at night. Radio signals from the transmitter are picked up by a receiving antenna, creating a weak electrical current. To receive broadcasts from a particular station, a tuner matches the receiver to the transmitting frequency of that station. The detector rectifies the incoming signal, passing on only the audio signal, separated from its carrier wave. The amplifier strengthens the signal, and finally the loudspeaker converts the electrical waves back into sound.

original carrier frequency (frequency modulation—FM). The modulated RF signal is then amplified (see AMPLIFIER) to a high power and radiated from an ANTENNA. At the receiving station, another antenna picks up a minute fraction of the energy radiated from the transmitter together with some background NOISE. This RF signal is amplified and the original audio signal is recovered (demodulation or detection). Detection and amplification often involve many stages including FEEDBACK and intermediate frequency (IF) circuits. A radio receiver must of course be able to discriminate between all the different signals acting at any one time on its antenna. This is accomplished with a tuning circuit which allows only the desired frequency to pass to the detector (see also ELECTRONICS). In point-to-point radio communications most stations can both transmit and receive messages but in **radio broadcasting** a central transmitter broadcasts program sequences to a multitude of individual receivers. Programs are often produced centrally and distributed to a "network" of local broadcasting stations by wire or MICROWAVE link. Because there are potentially so many users of radio communications—aircraft, ships, police and amateur "hams" as well as broadcasting services—the use of the RF portion of the electromagnetic SPECTRUM is strictly controlled to prevent unwanted INTERFERENCE between signals having adjacent carrier frequencies. The International Telecommunication Union (ITU) and national agencies such as the US Federal Communications Commission (FCC) divide the RF spectrum into bands which it allocates to the various users. Public broadcasting in the US uses MF frequencies between 535kHz and 1605kHz (AM) and VHF bands between 88MHz and 108MHz (FM). VHF reception, though limited to line-of-sight transmissions, offers much higher fidelity of transmission (see HIGH-FIDELITY) and much greater freedom from interference. International broadcasting and local transmissions in other countries frequently use other frequencies in the LF, MF and HF (short wave) bands. (See the electromagnetic spectrum table at ELECTROMAGNETIC RADIATION.)

The Development of Radio. The existence of radio waves was first predicted by James Clerk MAXWELL in the 1860s but it was not until 1887 that Heinrich HERTZ succeeded in producing them experimentally. "Wireless" telegraphy was first demonstrated by Sir Oliver LODGE in 1894 and MARCONI made the first trans-Atlantic transmission in 1901. Voice transmission was first achieved in 1900 but transmitter and amplifier powers were restricted before the advent of Lee DE FOREST's triode ELECTRON TUBE in 1906. Only the development of the TRANSISTOR after 1948 has had as great an impact on radio technology. Commercial broadcasting began in the US in 1920.

RADIOACTIVITY, the spontaneous disintegration of certain unstable nuclei, accompanied by the emission of ALPHA PARTICLES (weakly penetrating HELIUM nuclei), BETA RAYS (more penetrating streams of ELECTRONS) or GAMMA RAYS (ELECTROMAGNETIC RADIATION capable of penetrating up to 100mm of LEAD). In 1896, BECQUEREL noticed the spontaneous emission of ENERGY from URANIUM compounds (particularly PITCHBLENDE). The intensity of the effect depended on the amount of uranium present, suggesting that it involved individual atoms. The CURIES discovered further radioactive substances such as THORIUM and RADIUM, and about 40 natural radioactive substances are now known. Their rates of decay are unaffected by chemical changes, pressure, temperature or electromagnetic fields, and each nuclide (nucleus of a particular ISOTOPE) has a

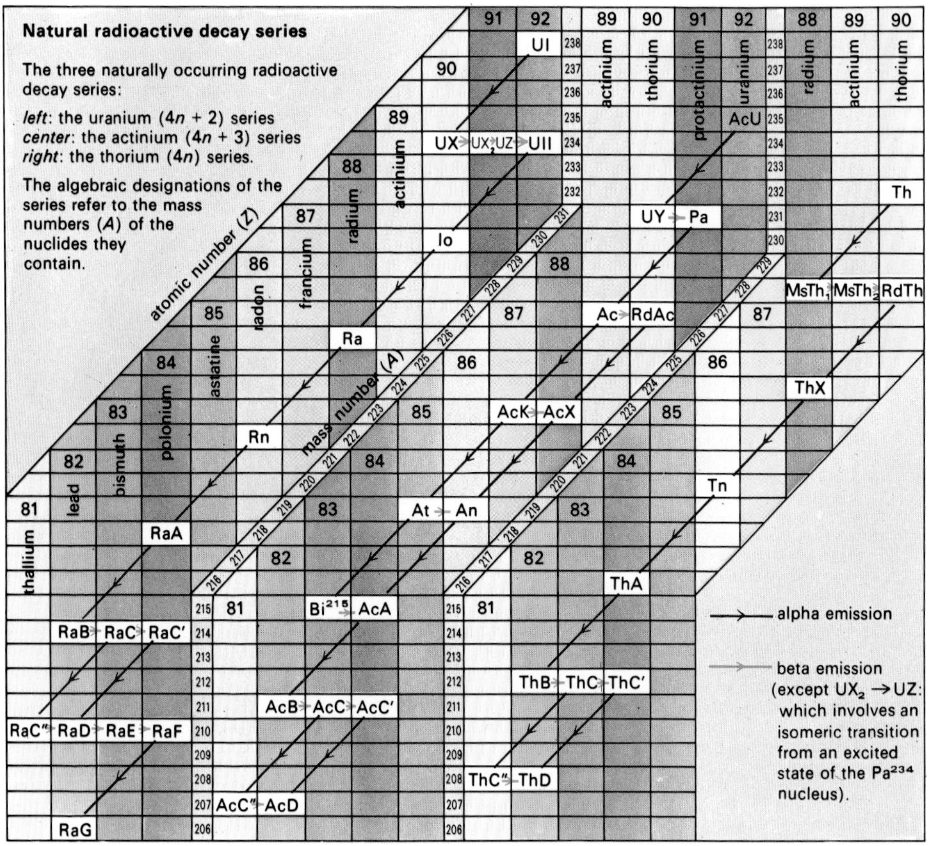

characteristic decay constant or HALF-LIFE. RUTHERFORD and SODDY suggested in 1902 that a radioactive nuclide decays to a further radioactive nuclide, a series of transformations taking place which ends with the formation of a stable "daughter" nucleus. It is now known that for radioactive elements of high ATOMIC WEIGHT, three decay series (the thorium, actinium and uranium series) exist. As well as the natural radioactive elements, a large number of induced radioactive nuclides have been formed by nuclear reactions taking place in ACCELERATORS or NUCLEAR REACTORS (see also IRRADIATION; RADIOISOTOPE). Some of these are members of the three natural radioactive series. Various types of radioactivity are known, but beta emission is the most common, normally caused by the decay of a NEUTRON, giving a PROTON, an electron and an antineutrino (see SUBATOMIC PARTICLES). This results in a unit change of atomic number (see ATOM) and no change in MASS NUMBER. Heavier nuclides often decay to a daughter nucleus with atomic number two less and mass number four less, emitting an alpha particle. If an excited daughter nucleus is formed, gamma-ray emission may accompany both alpha and beta decay. Because the ionizing radiations emitted by radioactive materials are physiologically harmful, special precautions must be taken in handling them.

RADIO ASTRONOMY, the study of the ELECTROMAGNETIC RADIATION emitted or reflected by celestial objects in the approximate wavelength range 1mm–30m, usually by use of a RADIO TELESCOPE. The science was initiated accidentally in 1932 by Karl Jansky who found an INTERFERENCE in a telephone system he was testing: the source proved to be the MILKY WAY. In 1937 an American, Grote Reber, built a 9.5m radio telescope in his back yard and scanned the sky at a wavelength around 2m. After WWII the science began in earnest. Investigation of the sky revealed that clouds of hydrogen gas in the Milky Way were radio sources, and mapping of these confirmed our galaxy's spiral form (see GALAXY).

The sky is very different for the radio astronomer than for the astronomer. Bright stars are not radio objects (our sun is one solely because it is so close), while many radio objects are optically undetectable. Radio objects include QUASARS, PULSARS, supernova remnants (e.g., the CRAB NEBULA) and other galaxies. The work of Martin RYLE in the 1960s and 1970s has enabled radio galaxies that are possibly at the farthest extremities of the universe to be mapped. The universe also has an inherent radio "background noise" (see COSMOLOGY).

RADIOCARBON DATING, a technique of dating organic material up to 70000 years old. The radioactive carbon ISOTOPE C^{14} is produced naturally by the impact of COSMIC RAYS on NITROGEN atoms in the atmosphere, and, in the form of $C^{14}O_2$, enters the ecological CARBON CYCLE. Assuming that the amount of C^{14} produced in this way is constant, one may deduce that the proportion of C^{14} atoms present in living material is uniform throughout time. Knowing that C^{14} decays into N^{14} with a HALF-LIFE of 5730 ± 40 years, examination of the amount of C^{14} remaining in organic material provides a reasonably accurate method of dating. Recent advances in DENDROCHRONOLOGY have shown that production of C^{14} in the atmosphere is not constant, and corrections have accordingly been made to the radiocarbon system.

RADIOCHEMISTRY, the use of RADIOISOTOPES in chemistry, especially in studies involving chemical ANALYSIS, where radioisotopes provide a powerful and sensitive tool. Tracer techniques, in which a particular atom in a molecule is "labeled" by replacement with a radioisotope, are used to study reaction rates and mechanisms (see KINETICS, CHEMICAL). (See also RADIATION CHEMISTRY.)

RADIOCOMPASS. See COMPASS; DIRECTION FINDER.

RADIO CONTROL. See REMOTE CONTROL.

RADIOGRAPH, a photograph exposed with X-RAYS or GAMMA RAYS. Special plates having thick emulsions are used to increase sensitivity.

RADIOISOTOPE, radioactive ISOTOPE of an element. A few elements, such as RADIUM or URANIUM, have naturally occurring radioisotopes, but because of their usefulness in science and industry, a large number of radioisotopes are produced artificially. This is done by IRRADIATION of stable isotopes with PHOTONS, or with particles such as NEUTRONS in an ACCELERATOR or NUCLEAR REACTOR. Radioisotopes with a wide range of HALF-LIVES and activities are available by these means. Because radioisotopes behave chemically and biologically in a very similar way to stable isotopes, and their radiation can easily be monitored even in very small amounts, they are used to "label" particular atoms or groups in studying chemical reaction mechanisms and to "trace" the course of particular components in various physiological processes. The radiation emitted by radioisotopes may also be utilized directly for treating diseased areas of the body (see RADIATION THERAPY), sterilizing foodstuffs or controlling insect pests.

RADIOLOGY, the use of RADIOACTIVITY, GAMMA RAYS and X-RAYS in MEDICINE, particularly in diagnosis but also in treatment. (See also RADIATION THERAPY.)

RADIOMETER, instrument for measuring the intensity of radiant ENERGY. The term is usually applied to a simple vane type instrument (Crookes' radiometer) consisting of an evacuated glass bulb containing a pivot supporting vertical metal vanes blackened on one side. Incident radiation is more strongly absorbed by the blackened side of the vanes and the forces exerted on them by residual gas molecules initiate rotation proportional to the radiation intensity.

RADIOSONDE, meteorological instrument package attached to a small BALLOON capable of reaching the earth's upper ATMOSPHERE. The instruments measure the TEMPERATURE, PRESSURE and HUMIDITY of the atmosphere at various altitudes, the data being relayed back to earth via a RADIO transmitter. Radiosondes provide a cheap and reliable method of getting information for WEATHER FORECASTING.

RADIO TELESCOPE, the basic instrument of RADIO ASTRONOMY. The receiving part of the equipment consists of a large parabola, the big dish, which operates on the same principle as the parabolic mirror of a reflecting TELESCOPE. The signals that it receives are then amplified and examined. In practice, it is possible to build radio telescopes effectively far larger than a single big dish could physically exist by using several connected dishes; this is known as an array.

RADIOTHERAPY. See RADIATION THERAPY.

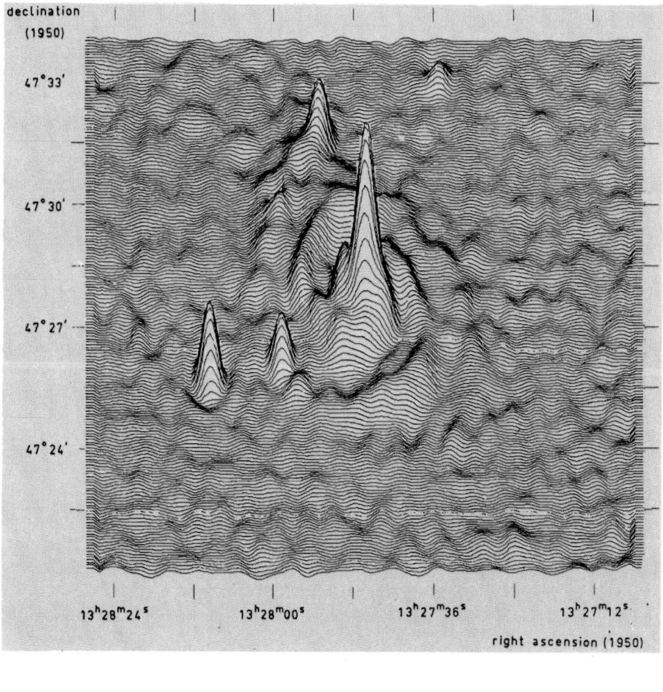

The degree of resolution achievable in radio astronomy, even using radio interferometers, is far below that possible in optical astronomy. Here is a computer-plotted "hill and dale" representation of the Whirlpool Galaxy (M51; NGC 5194) in *Canes Venatici* as it appears in 1415 MHz radiation.

The 250ft (76.2m) diameter paraboloid dish radio telescope at Jodrell Bank, Cheshire, UK, opened in 1957, is still the largest steerable-dish radio telescope in the world. It is operated by the University of Manchester.

RADIUM (Ra), radioactive ALKALINE-EARTH METAL similar to BARIUM, isolated from PITCHBLENDE by Marie CURIE in 1898. It has white salts which turn black as the radium decays, and which emit a blue glow due to ionization of the air by radiation. It has four natural ISOTOPES, the commonest being Ra^{226} with HALF-LIFE 1622 years. Radium is used in industrial and medical radiography. AW 226.0, mp 700°C, bp 1140°C, sg 5.

RADIUS, the distance from the point of INTERSECTION of the axes of symmetry (see AXIS OF SYMMETRY) of a closed CURVE to a point on the curve. The term is usually applied to CIRCLES, ellipses (see CONIC SECTIONS) and SPHERES. All radii of a circle or sphere are equal; and it is generally profitable to consider only the longest and shortest radii (semi-major and semi-minor axes) of an ellipse. (See also CYLINDER; ELLIPSOID; MENSURATION.)

RADON (Rn), a radioactive NOBLE GAS formed in the radioactive decay of RADIUM, ACTINIUM or THORIUM. Found in some radioactive minerals, Rn^{222} has a HALF-LIFE of 3.8 days, making it suitable for use in RADIATION THERAPY. Other natural and synthetic ISOTOPES have shorter half-lives. Radon (II) fluoride (RnF_2) is the only radon compound known. AW 222, mp −71°C, bp −62°C.

RA EXPEDITIONS. See HEYERDAHL, THOR.

RAILROAD, land transportation system in which cars with flanged steel wheels run on tracks of two parallel steel rails. From their beginning railroads provided reliable, economical transport for freight and passengers; they promoted the Industrial Revolution and have been vital to continued economic growth ever since, especially in developing countries. Railroads are intrinsically economical because the rolling friction of wheel on rail is very low, so that a LOCOMOTIVE of only 750W (1hp) per gross tonne pulled is needed—10% of that required for road transport. However, fixed costs of maintenance etc., are high, so high traffic volume is needed. This, together with rising competition and overmanning, has led to the closure of many minor lines in the US and Europe, though elsewhere many new lines are still being built. Maintenance, signalling and many other functions are now highly automated.

Railroads developed out of the small mining tracks or tramways built in the UK and Europe from the mid-16th century. They used gravity or horse power, and the cars generally ran on flanged rails or plateways. These were hard to switch, however, and the system of flanged wheels on plain rails eventually predominated. The first public freight railroad was the Surrey Iron Railway (1801). The modern era of mechanized traction began with TREVITHICK's steam locomotive "New Castle" (1804) (see also STEAM ENGINE). Early locomotives ran on toothed racks to prevent slipping, but in 1813 this was found to be unnecessary. The first public railroad to use locomotives and to carry passengers was the Stockton and Darlington Railway (1825). The boom began when the Liverpool and Manchester Railway opened in 1830 using George STEPHENSON's "Rocket," a much superior and more reliable locomotive. Railroads spread rapidly in Britain, Europe and the US. The first US railroad was the Baltimore and Ohio (1830). The rails were laid on wooden (later also concrete) crossties or sleepers, and were joined by fishplates to allow for thermal expansion. Continuous welded rails are now generally used. Track gauges were at first very varied, but the "standard gauge" of 4ft 8½in (1.435m) soon predominated. Railroads must be built with shallow curves and gentle gradients, using bridges, embankments, cuttings and tunnels as necessary. (See also SUBWAY.)

RAIN, water drops falling through the atmosphere; the chief form of PRECIPITATION. Raindrops range in size up to 4mm diameter; if they are smaller than 0.5mm the rain is called **drizzle**. The quantity of rainfall (independent of the drop size) is measured by a **rain gauge**, an open-top vessel which collects the rain, calibrated in millimetres or inches and so giving a reading independent of the area on which the rain falls. Light rain is less than 25mm/h, moderate rain 25 to 75mm/h, and heavy rain more than 75mm/h. Rain may result from the melting of SNOW or HAIL as it falls, but is commonly formed by direct condensation. When a parcel of warm air rises, it expands approximately adiabatically, cooling about 1K/100m. Thus its relative HUMIDITY rises until when it reaches saturation the water vapor begins to condense as droplets, forming CLOUDS. These droplets may coalesce into raindrops, chiefly through turbulence and nucleation by ice particles or by cloud seeding (see also WEATHER FORECASTING AND CONTROL). Moist air may be lifted by CONVECTION, producing **convective rainfall**; by forced ascent of air as it crosses a mountain range, producing **orographic rainfall**; and by the forces within CYCLONES, producing **cyclonic rainfall**. (See also GROUNDWATER; HYDROLOGIC CYCLE; METEOROLOGY; MONSOON.)

The origins of the primary and secondary rainbows. The concentration of light in the rainbow arcs results from there existing certain preferred angles of deviation for light rays internally reflected within spherical water drops. These angles are about 139° for one internal reflection and about 129° for two internal reflections. Dispersion gives rise to color-fringe effects about the directions of mean deviation and interference effects cause certain other color fringes. The band between the primary and secondary rainbows is noticeably darker than the rest of the sky (Alexander's dark band).

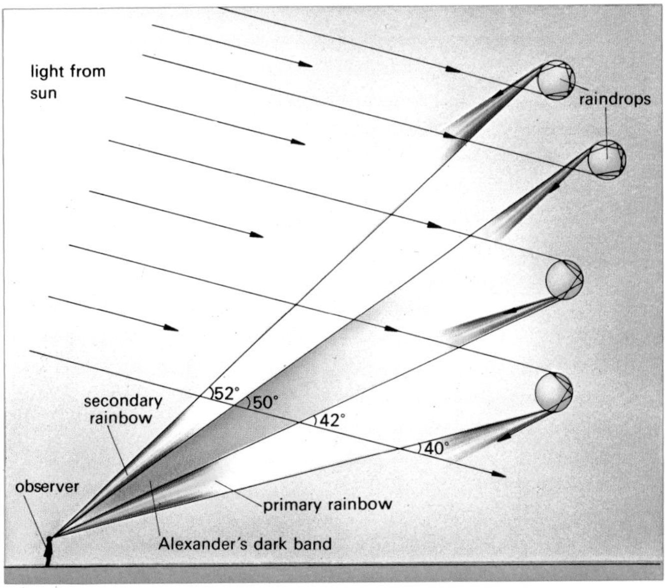

RAINBOW, arch of concentric spectrally colored rings seen in the sky by an observer looking at rain, mist or spray with his back to the sun. The colors are produced by sunlight's being refracted and totally internally reflected (see REFRACTION) by spherical droplets of water. The primary rainbow, with red on the outside and violet inside, results from one total internal reflection. Sometimes a dimmer secondary rainbow with reversed colors is seen, arising from a second total internal reflection.

RAIN FOREST. See FOREST.

RAIN SHADOW, area of low rainfall on the lee side of a mountain barrier, which shelters it from prevailing rainbearing winds. Rainfall on the corresponding windward side of the barrier is extremely high. (See also CLIMATE; RAIN.)

RAINWATER, Leo James (1917–), US physicist who shared the 1975 Nobel Prize for Physics with Aage BOHR and Ben MOTTELSON for their work on the physics and structure of the atomic nucleus.

RAM (the constellation). See ARIES.

RAMAN, Sir Chandrasekhara Venkata (1888–1970), Indian physicist awarded the 1930 Nobel Prize for Physics for his discovery of the **Raman Effect**: when a medium is exposed to a beam of ELECTROMAGNETIC RADIATION, light scattered at right angles to the beam has a range of frequencies characteristic to the medium. This is the basis for Raman SPECTROSCOPY.

RAMAPITHECUS. See PREHISTORIC MAN.

RAMJET. See JET PROPULSION.

RAMÓN Y CAJAL, Santiago (1852–1934), Spanish neurohistologist who shared with GOLGI the 1906 Nobel Prize for Physiology or Medicine for his work showing the NEURON to be the fundamental "building block" of all nervous structures.

RAMSAY, Sir William (1852–1916), British chemist awarded the 1904 Nobel Prize for Chemistry for his discovery, prompted by a suggestion from RAYLEIGH (1892), of all the NOBLE GASES, including (with SODDY) HELIUM, although it had been earlier detected in the solar spectrum (1868).

RANGE FINDER, instrument for remote measurement of distance. RADAR and SONAR provide nonoptical range finders. Optical range finders—used in CAMERAS for correct focusing, and in military applications—are of two types. **Coincidence range finders** measure the angles formed from each end of a base line to the object viewed, and calculate the distance by TRIGONOMETRY. The images from each optical path are made to coincide. **Stereoscopic range finders** use stereobinoculars which are adjusted until the stereo image formed by reticles in the eyepieces appears to be at the same distance as the object.

RANK, Otto (1884–1939), Austrian-born US psychoanalyst best known for his suggestion that the psychological TRAUMA of birth is the basis of later anxiety NEUROSIS; and for applying PSYCHOANALYSIS to artistic creativity.

RANKINE SCALE, scale expressing absolute TEMPERATURES in Fahrenheit degrees, devised by Scottish engineer William Rankine (1820–1872).

RAOULT, François Marie (1830–1901), French physical chemist best known for his work on the theory of SOLUTIONS. **Raoult's Law**, in its most general form, states that the VAPOR PRESSURE above an ideal solution is given by the sum of the PRODUCTS of the vapor pressure of each component and its mole fraction (the number of MOLES of the component divided by the total number of moles of all the components). (See also DISTILLATION.)

RARE EARTHS, the elements SCANDIUM, YTTRIUM and the LANTHANUM SERIES, in Group IIIB of the PERIODIC TABLE, occurring widespread in nature as MONAZITE and other ores. They are separated by CHROMATOGRAPHY and ION-EXCHANGE resins. Rare earths are used in ALLOYS, including MISCH METAL; and their compounds (mixed or separately) are used as ABRASIVES, for making glasses and ceramics, as "getters," as catalysts (see CATALYSIS) in the petroleum industry, and to make PHOSPHORS, LASERS and MICROWAVE devices.

RARE GASES, former name for the NOBLE GASES.

RATCHET, a simple mechanical device consisting of a pivoted bar, or "pawl," resting on a toothed wheel or shaft designed so that the wheel can rotate in one direction only. If a second pawl is used an intermittent rotary motion can be transmitted.

RATIO, a numerical relationship between two quantities of the same kind. Ratios may be expressed in the form $a:b$; as FRACTIONS, a/b; or as percentages, $\frac{100a}{b}\%$ (see PERCENT). Ratios are usually reduced to their lowest terms: e.g., $15:3 \equiv 5:1$. (See also PROPORTION.)

RATIONALISM, a philosophical approach based on the view that reality has a logical structure accessible to deductive reasoning and proof, and holding, as against EMPIRICISM, that reason unsupported by sense experience is a source of "synthetic knowledge"—knowledge, primarily of certain fundamental concepts and principles in logic and mathematics, which, it is argued, cannot be denied without contradiction and yet cannot be dismissed as merely analytic (see LOGIC; UNIVERSALS; REALISM). Major rationalists in modern philosophy include DESCARTES, Spinoza and LEIBNIZ.

RATIONAL NUMBERS, those numbers that can be expressed as the RATIO of two INTEGERS. They include all positive and negative integers as well as, clearly, any number that can be expressed as a FRACTION. The set (see SET THEORY) of all rational numbers is a very dense one, in that it can be proved that between any two rational numbers there is a third; but, despite this, there are infinitely many numbers that are not members of this set (see IRRATIONAL NUMBERS; REAL NUMBERS).

RATITES, the running or flightless birds. Some biologists consider the ratites, which include the ostrich, emu, moa, cassowary and kiwi, to be a homogeneous group, but others believe that flightless birds have evolved from different ancestors.

RATTAN, long thin many-jointed stems of a number of Malaysian climbing palms, especially those of the genera *Calamus* and *Daemonorops*. Rattan canes are used for walking sticks, light construction work and wickerwork.

RAUWOLFIA SERPENTINA, tropical shrub from which **reserpine**, a drug used in hypertension and some mental illnesses, is extracted. The drug affects the HEART and NERVOUS SYSTEMS by reducing the supply of noradrenaline (see ADRENALINE).

RAY, John (1627–1705), British biologist and natural theologian who, with **Francis Willughby** (1635–1672), made important contributions to TAXONOMY, especially in *A General History of Plants* (3 vols., 1686–1704).

RAYLEIGH, John William Strutt, Third Baron (1842–1919), British physicist awarded the 1904 Nobel Prize for Physics for his measurements of the DENSITY of the atmosphere and its component gases, work that led to his isolation of ARGON (see also RAMSAY, SIR WILLIAM). He worked in many other fields of physics, and is commemorated in the terms **Rayleigh scattering** (which describes the way that ELECTROMAGNETIC RADIATION is scattered by spherical particles of radius less than 10% of the wavelength of the radiation—see SCATTERING) and **Rayleigh waves** (see EARTHQUAKES).

RAYNAUD'S DISEASE, a condition in which the fingers (or toes) suddenly become white and numb, often on exposure to mild cold, and become in turn blue and then red and painful. It is caused by digital artery spasm. Raynaud's disease usually occurs in otherwise fit young women; Raynaud's syndrome is the same symptom as a manifestation of an underlying disease (e.g., LUPUS ERYTHEMATOSUS).

RAYON, name for various textile fibers made from regenerated CELLULOSE. (See also SYNTHETIC FIBERS.)

RDX, or cyclonite, or cyclotrimethylenetrinitramine, a high EXPLOSIVE made by NITRATION of the condensation product of FORMALDEHYDE and AMMONIA. First introduced in WWII, it is used in blasting caps and PLASTIC EXPLOSIVE.

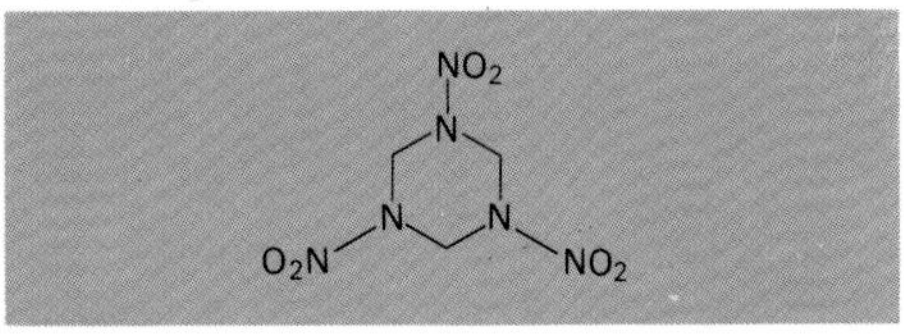

RDX (hexahydro-1,3,5-trinitro-1,3,5-triazine).

REACTANCE, the ratio of an AC voltage applied to a single component of an electric circuit (particularly an inductor or CAPACITOR) to the current produced, the maximum values of each being taken irrespective of their relative PHASE. (See also IMPEDANCE.)

READING, the process of assimilating language in the written form. Initial language development in children is largely as speech (see SPEECH AND SPEECH DISORDERS) and has a primarily auditory or phonetic component; the recognition of letters, words and sentences when written represents a transition from the auditory to the visual mode. The dependence of reading on previous linguistic development with spoken speech is seen in the impaired reading ability of deaf children. Normal reading depends on normal VISION and the ability to recognize the patterns of letter and word order and grammatical variations. In reading, vision is linked with the system controlling EYE movement, so that the page is scanned in an orderly fashion. Reading is represented in essentially the same areas of brain as are concerned with speech, and disorders of the two often occur together (e.g., dysphasia). In DYSLEXIA, pattern recognition is impaired and a specific defect of reading and language development results. The ability to read and write, and thus to record events, ideas, etc., represented one of the most substantial advances in human civilization after the acquisition of speech itself.

REALGAR, soft, bright-red SULFIDE mineral, arsenic disulfide (As_2S_2), an ore of ARSENIC associated with ORPIMENT and found in central Europe, Nev. and Ut. It forms monoclinic CRYSTALS.

REALISM, in philosophy, is a term with two main technical uses. Philosophers who believe, as PLATO did, that UNIVERSALS exist in their own right, and so independently of perceived objects, are traditionally labeled "realists." Realism in this sense is opposed to NOMINALISM. On the other hand, realism also describes the view that perceived objects exist independently of our perceptions of them. Realism in this sense is opposed to the extreme EMPIRICISM of such as BERKELEY and HUME.

REALITY PRINCIPLE. See PLEASURE PRINCIPLE.

REAL NUMBERS, those numbers that can be represented directly by lengths of LINES. The set (see SET THEORY) of real numbers therefore includes ZERO and both positive and negative RATIONAL NUMBERS and IRRATIONAL NUMBERS. It does not include the IMAGINARY NUMBERS.

REAPER, a grain-cutting machine. The first simple reapers were invented in the UK and the US in the early 19th century. They increased in popularity as they were developed to include mechanical gathering and binding.

RÉAUMUR, René Antoine Ferchault de (1683–1757), French scientist whose most important work was in ENTOMOLOGY, but who is best remembered for devising the now little used **Réaumur temperature scale**, in which 0°R = 0°C and 80°R = 100°C.

RECAPITULATION, in embryology. See ONTOGENY AND PHYLOGENY.

RECEPTACLE. See FLOWER.

RECIPROCAL, of a number a, that number b such that $a.b = 1$. With REAL NUMBERS, the reciprocal of a is $1/a$ for all a except ZERO ($1/0$ has no meaning). With IMAGINARY NUMBERS this is true also: the reciprocal of i is $\frac{1}{i}$, which can be expressed as $\frac{i}{i^2}$ and hence as $-i$. (See also INVERSE.) **Reciprocal trigonometric functions,** the functions cosec, sec and cot of an ANGLE (see also TRIGONOMETRY). They are defined as cosec A = 1/sin A; sec A = 1/cos A; cot A = 1/tan A.

RECIPROCATING ENGINE, an ENGINE in which a PISTON oscillates in a cylinder, being driven by the pressure of the working fluid.

RECOMBINATION, the presence in offspring of GENE combinations not found in either parent. Such new combinations may be formed by the crossing over of CHROMOSOMES in MEIOSIS, and, being thus present in either GAMETE, unite randomly at FERTILIZATION.

RECORD PLAYER. See PHONOGRAPH.

RECTANGLE. See QUADRILATERAL.

RECTIFIER, a device such as an ELECTRON TUBE or SEMICONDUCTOR junction which converts alternating electric current (AC—see ELECTRICITY) to direct current (DC) by allowing more current to flow through it in one direction than another. A **half-wave rectifier** transmits only one polarity of the alternating current, producing a pulsating direct current; two such devices are combined in full-wave rectification, giving a continuous pulse train which may be smoothed by a filter.

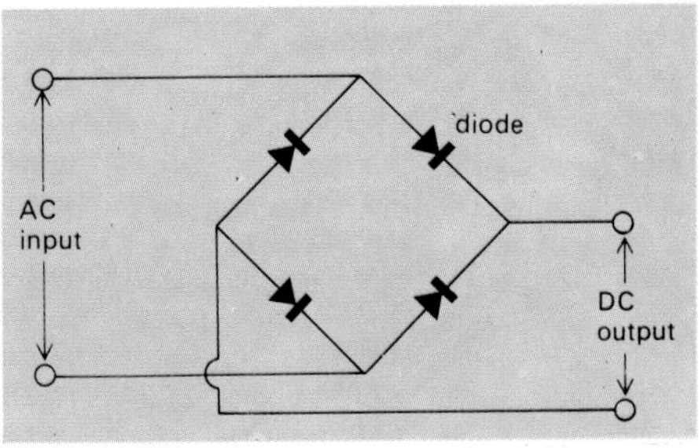

A simple full-wave bridge rectifier circuit using four diodes. Actual circuits usually include components which smooth the DC output, making the rectifier approximate to a steady DC source.

RECYCLING, the recovery and reuse of any waste material. Of obvious economic importance where reusable materials are available more cheaply than fresh supplies of the same materials, the recycling principle is finding ever wider application in the conservation of the world's natural resources and in solving the problems of environmental POLLUTION. The recycling of the wastes of a manufacturing process in the same process—e.g., the resmelting and recasting of metallic turnings and offcuts—is commonplace in industry. So also is the immediate use of wastes or by-products of one industrial process in another—e.g., the manufacture of cattle food from the grain-mash residues found in breweries and distilleries. These are often termed forms of "internal recycling," as opposed to "external recycling": the recovery and reprocessing for reuse of "discarded" materials, such as waste paper, scrap metal and used glass bottles. The burning of garbage to produce electricity and the extraction of pure water from sewage are other common examples of recycling.

RED CORPUSCLES, or erythrocytes. See BLOOD.

REDFIELD, Robert (1897–1958), US cultural anthropologist best known for his comparative studies of primitive and highly civilized cultures, and for his active support of racial integration.

REDI, Francesco (1627–1697 or 1698), Italian biological scientist who demonstrated that maggots develop in decaying meat not through SPONTANEOUS GENERATION but from eggs laid there by flies.

RED LEAD. See LEAD.

RED SHIFT, an increase in wavelength of the light from an object, usually caused by its rapid recession (see DOPPLER EFFECT). The spectra of distant GALAXIES show marked red shifts and this is usually, though far from always, interpreted as implying that they are rapidly receding from us. (See also COSMOLOGY.)

RED TIDE, ocean-coloration effect caused by certain DINOFLAGELLATES which release an alkaloid, poisonous to fish, into the water.

REDUCTION. See OXIDATION AND REDUCTION.

REED, Walter (1851–1902), US Army pathologist and bacteriologist who, in 1900, demonstrated the role of the mosquito *Aëdes aegypti* as a carrier of YELLOW FEVER, so enabling the disease to be controlled.

REEF, line or ridge of rocks just below the surface of the sea. See CORALS.

REFINING, the purification of crude substances, especially metals (see METALLURGY), ORES, PETROLEUM and SUCROSE. Methods used, include DISTILLATION, ELECTROLYSIS and FLOTATION.

REFLECTION, the bouncing back of energy waves (e.g., LIGHT radiation, SOUND or WATER waves) from a surface. If the surface is smooth, "regular" reflection takes place, the incident and reflected wave paths lying in the same plane as, and at opposed equal angles to, the normal (a line perpendicular to the surface) at the point of reflection. Rough surfaces

1

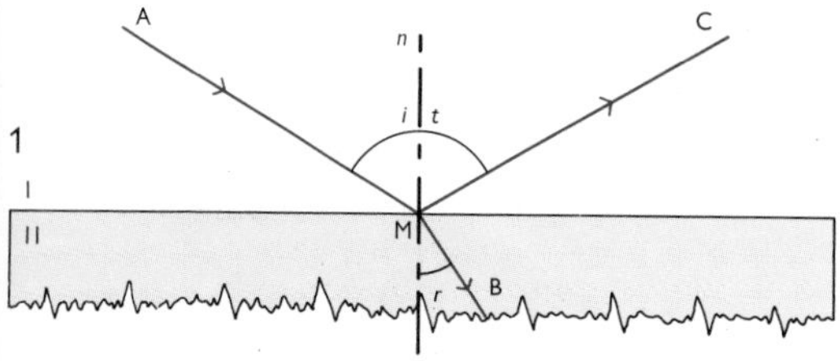

Reflection and refraction of light: a ray coming from A and striking a surface at M is reflected along the line MC, and angle i (the angle of incidence) equals angle t (the angle of reflection). AM, MC and n (the normal) all lie in the same plane. In refraction, waves or rays are bent in passing from one substance to another. If a ray from A in medium I (air) passes at M into medium II (water), it will be refracted so as to follow the path MB. The angle of refractions is then r. By Snell's Law, $\sin i = \mu \sin r$ where μ is the refractive index of medium II.

reflect waves irregularly, so an optically rough surface appears matt or dull while an optically smooth surface looks shiny. Reflected sound waves are known as ECHOES. (See also MIRROR; PRISM; REFRACTION.)

REFLEX, MUSCLE contraction or secretion resulting from nerve stimulation by a pathway from a stimulus via the NERVOUS SYSTEM to the effector organ without the interference of volition. Basic primitive reflexes are stylized responses to stress of protective value to an infant. Stretch or tendon reflexes (e.g., knee-jerk) are muscle contractions in response to sudden stretching of their TENDONS. **Conditioned reflexes** are more complex responses described by PAVLOV that follow any stimulus which has been repeatedly linked with a stimulus of normal functional significance.

REFLEX, of an ANGLE, one between 180° and 360°.

REFRACTION, the change in direction of energy waves on passing from one medium to another in which they have a different velocity. In the case of LIGHT radiation, refraction is associated with a change in the optical density of the medium. On passing into a

A beam of light directed at one face of an equiangular perspex prism is partly reflected, and partly refracted into the prism. Most of the refracted beam undergoes a further refraction on striking the second face, though some is internally reflected and emerges from the third face.

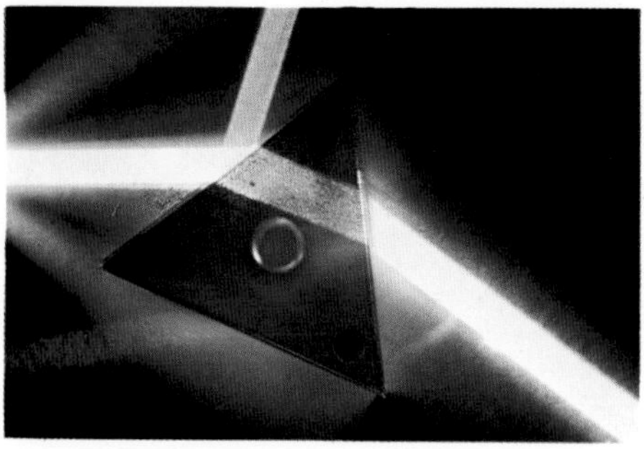

denser medium the wave path is bent toward the normal (the line perpendicular to the surface at the point of incidence), the whole wave path and the normal lying in the same plane. The ratio of the sine of the angle of incidence (that between the incident wave path and the normal) to that of the angle of refraction (that between the normal and the refracted wave path) is a constant for a given interface (Snell's law). When measured for light passing from a vacuum into a denser medium, this ratio is known as the refractive index of the medium. Refractive index varies with wavelength (see DISPERSION). On passing into a less dense medium, light radiation is bent away from the normal but if the angle of incidence is so great that its sine equals or exceeds the index for refraction from the denser to the less dense medium, there is no refraction and total (internal) REFLECTION (applied in the reflecting PRISM) results. Refraction finds its principal application in the design of LENSES. (See also DOUBLE REFRACTION.)

REFRACTOMETER, an instrument measuring the refractive index (see REFRACTION) of liquids for a particular color of light: the index is often very sensitive to impurities besides allowing easy determination of the proportions present in two-component mixtures such as water/ethanol.

REFRACTORIES, substances able to resist high temperatures without melting, decomposing or reacting, and hence used for thermal INSULATION and to line FURNACES. Often made into FIREBRICKS, refractories are composed of various substances (mostly oxides), including the acidic FIRECLAY, SILICA and ZIRCON; the basic CHROMITE, DOLOMITE and MAGNESITE; and the neutral CARBORUNDUM, CORUNDUM and GRAPHITE. (See also CERAMICS.)

REFRIGERATION, removal of HEAT from an enclosure in order to lower its TEMPERATURE. It is used for freezing water or food, for FOOD PRESERVATION, for AIR CONDITIONING and for low-temperature chemical processes and CRYOGENICS studies and applications. The ancient Egyptians and Indians used the evaporation of water from porous vessels; and the Chinese, Greeks and Romans used natural ice, a method which became a major industry in the 19th-century US. Modern refrigerators are insulated cabinets containing the cooling elements of a HEAT PUMP. The pump may use mechanical compression of refrigerants such as AMMONIA or FREON, or may accomplish compression by absorbing the refrigerant in a secondary fluid such as water and pumping the solution through a heat exchanger to a generator where it is heated to drive off the refrigerant at high pressure. Other cycles, similar in principle, using steam or air, are also used. Refrigeration based on the PELTIER EFFECT is being developed but is not yet economically competitive.

REGELATION, the refreezing of ice that has melted under PRESSURE alone, once the pressure is released. Ice skating depends on regelation, but as ice melts readily under pressure only while its temperature is near its FREEZING POINT, skating may not be feasible in very cold weather.

REGENERATION, the regrowing of a lost or damaged part of an organism. In PLANTS this includes the production of, e.g., dormant buds and adventitious organs. All ANIMALS possess some power to regenerate, but its extent varies from that in sponges, in which all the cells in a piece of the body can be almost completely separated and will yet come together again to build up new but smaller sponges, to that in the higher animals, in which regeneration is limited to the healing of wounds.

REGIOMONTANUS (1436–1476), born **Johann Müller,** Prussian mathematician and astronomer whose *Five Books on all Types of Triangles* (1533) laid the foundations for modern TRIGONOMETRY.

REGNAULT, Henri Victor (1810–1878), German-born French chemist best known for work on the physical properties of gases (e.g., showing that BOYLE'S LAW works only for ideal gases), and for inventing an air THERMOMETER and a HYGROMETER.

REGRESSION, any return to an earlier stage of mental development or mode of behavior. In particular the term describes a (usually unsuccessful) DEFENSE MECHANISM whereby the individual avoids some ANXIETY situation by regressing to an earlier stage of LIBIDO development.

REGULUS, Alpha Leonis, the brightest star in LEO. Its apparent magnitude is +1.35 and its distance from the sun 25.75pc. It is a visual triple star.

REICH, Wilhelm (1897–1957), Austrian-born US psychoanalyst best known for his controversial theory that there exists a primal life-giving force, *orgone* energy, in living beings and in the atmosphere. He designed and sold orgone boxes, supposed to concentrate orgone energy in the person within. He died in prison after violating an injunction against selling these boxes.

REICHSTEIN, Tadeus (1897–), Polish-born Swiss chemist awarded (with E. C. KENDALL and P. S. HENCH) the 1950 Nobel Prize for Physiology or Medicine for his work, independent of Kendall's, on the corticoids and isolation of what is now known as cortisone (see STEROIDS).

REID, Thomas (1710–1796), Scottish philosopher who, through his investigations of and rejection of HUME'S skepticism, is regarded as a founder of the COMMON SENSE SCHOOL of philosophy.

RELAPSING FEVER, INFECTIOUS DISEASE caused by a spirochetal bacterium carried by lice or ticks on rodents and causing episodic FEVER; it occurs in epidemics in areas of poverty and overpopulation. Rash, bleeding and respiratory symptoms are common and the central NERVOUS SYSTEM may be affected. TETRACYCLINES are usually effective in treatment.

RELATIVITY, a frequently referred to but less often understood theory of the nature of space, time and matter. EINSTEIN'S "special theory" of relativity (1905) is based on the premise that different observers moving at a constant speed with respect to each other find the laws of physics to be identical, and, in particular, find the speed of LIGHT waves to be the same (the "principle of relativity"). Among its consequences are that events occurring simultaneously according to one observer may happen at different times according to an observer moving past the first (although the order of two causally related events is never reversed); that a moving object is shortened in the direction of its motion; that time runs more slowly for a moving object; that the velocity of a projectile emitted from a moving body is less than the sum of the relative ejection velocity and the velocity of the body; that a body has a greater MASS when moving than when at rest, and that no body can travel as fast as, or faster than, the speed of light (2.998×10^8m/s—at this speed, a body would have zero length and infinite mass, while time would stand still on it).

These effects are too small to be noticed at normal velocities; they have nevertheless found ample experimental verification, and are commonplace considerations in many physical calculations. The relationship between the position and time of a given event according to different observers is known (for H. A. LORENTZ) as the Lorentz transformation. In this, time mixes on a similar footing with the three spatial dimensions, and it is in this sense that time has been called the "fourth dimension." The greater mass of a moving body implies a relationship between kinetic ENERGY and mass; Einstein made the bold additional hypothesis that *all* energy was equivalent to mass, according to the famous equation $E = mc^2$. The conversion of mass to energy is now the basis of NUCLEAR REACTORS, and is indeed the source of the energy of the sun itself.

Einstein's "general theory" (1916) is of importance chiefly to cosmologists. It asserts the equivalence of the effects of ACCELERATION and gravitational fields (see GRAVITATION), and that gravitational fields cause space to become "curved," so that light no longer travels in straight lines, while the wavelength of light falls as the light falls through a gravitational field. The direct verification of these last two predictions, among others, has helped deeply to entrench the theory of relativity in the language of physics.

REMAINDER, in the DIVISION of an INTEGER *a* by another integer *b*, the least difference between *a* and a number less than it which is a multiple of *b*. Thus, dividing 7 by 2, the remainder is 1.

REMINGTON, Eliphalet (1793–1861), US FIREARMS developer and manufacturer. Under his eldest son, **Philo Remington** (1816–1889), the Remington companies for a time manufactured the Remington SEWING MACHINE (1870–82) and the Remington TYPEWRITER (1873–86).

REMOTE CONTROL, control of a mechanized system from afar. Signals are sent from a control center by electrical circuits or radio to trigger or guide an operation elsewhere. There is usually FEEDBACK to the control center. (See also MECHANIZATION AND AUTOMATION.)

REMSEN, Ira (1846–1927), US chemist who, with his student **Constantin Fahlberg** (1850–1910), discovered SACCHARIN (1879).

RENNET, a commercial preparation for curdling or junketing MILK. It is prepared from the ENZYME rennin. This solidifies milk in the stomach of young animals, increasing its retention time and thus improving the animal's digestive efficiency.

REPLICATION. See NUCLEIC ACIDS.

REPRESSION, the DEFENSE MECHANISM whereby an impulse or idea is restricted by the EGO or SUPEREGO to the UNCONSCIOUS (primary repression), or in which derivatives of it are similarly restricted (secondary repression). FREUD considered primary repression essential to ego-development. (See also INHIBITION.)

REPRODUCTION, the process by which an organism produces offspring, an ability that is a unique characteristic of ANIMALS and PLANTS. There are two kinds of reproduction: asexual and sexual. In **asexual reproduction**, parts of an organism split off to form new individuals, a process found in some animals but which is more common in plants: e.g., the FISSION of single-celled plants; the budding of YEASTS; the fragmentation of filamentous ALGAE; SPORE production in BACTERIA, algae and FUNGI, and the production of vegetative organs in flowering plants (bulbs, rhizomes and tubers). In **sexual reproduction**, special (haploid) CELLS containing half the normal number of CHROMOSOMES, called gametes, are produced: in animals, sperm by males in the TESTES and ova by females in the OVARY; in plants, pollen by males in the stamens and ovules by females in the ovary. The joining of gametes (FERTILIZATION) produces a (diploid) cell with the normal number of chromosomes, the zygote, which grows to produce an individual with GENES inherited from both parents (see also HEREDITY). Fertilization may take place inside the female (internal fertilization) or outside (external fertilization). Internal fertilization demands that sperm be introduced into the female—insemination by copulation—and is advantageous because the young spend the most vulnerable early stages of their life-histories protected inside the mother.

At the molecular level, the most important aspect of reproduction is the ability of the chromosome to duplicate itself (see NUCLEIC ACIDS). The production of haploid cells is made possible by a process called MEIOSIS and is necessary to prevent doubling of the chromosome number with each generation in sexually reproducing individuals. The advantage of sexual reproduction is that the bringing together of genes derived from two individuals produces variation in each generation enabling populations to change and thus adapt themselves to changing environmental conditions (see also EVOLUTION; NATURAL SELECTION).

REPTILIA, a class of the CHORDATA that includes the CROCODILIA, SQUAMATA (snakes and lizards) and CHELONIA (turtles and tortoises) as well as a number of extinct groups (see COTYLOSAURIA, DINOSAURIA, ICHTHYOSAURIA, MAMMAL-LIKE REPTILES, PLESIOSAURIA, PTEROSAURIA and RHYNCHOCEPHALIA). Living reptiles have scaly skin and typically lay large yolky eggs. Being cold-blooded, they are most numerous in the tropics. Fossil forms are more numerous than living forms and occur in rocks of the Permian to Cretaceous periods, 280–65 million years ago, the so-called "age of reptiles."

RESEARCH, the use of appropriate methods in attempting to discover new knowledge or to develop new applications of existing knowledge or to explore relationships between ideas or events. Scientific discoveries, technological achievements and scholarly publications are all the fruits of research. Every discipline develops research methods and tools appropriate to its subject matter; but whether undertaken by scholar, technologist or scientist, research always involves three basic steps: the formulation of a problem; the collection and analysis of relevant information, and a concerted attempt to discover a solution or otherwise resolve the problem in a manner dictated by the available evidence. Quite different kinds of initial problem may be formulated. In the field of science and technology, for example, fundamental (or properly scientific) research aims at enlarging man's understanding of observable phenomena; the search is for general explanatory principles. Unlike applied (or technological) research, fundamental research is not explicitly directed toward the solution of a practical problem, although its results may, and usually do, suggest new technological possibilities. Knowledge of atomic structure is a goal of fundamental research; possible applications of this knowledge—nuclear power plants and weapons—demand technological research and development. In practice, however, the distinction is less clear-cut: accidental scientific discoveries are often made by research workers pursuing a technological goal. (See also SCIENTIFIC METHOD.)

RESERPINE. See RAUWOLFIA SERPENTINA.

RESIN, a high-molecular-weight substance characterized by its gummy or tacky consistency at certain temperatures. Naturally occurring resins include congo copal and BITUMEN (found as fossils), SHELLAC (from insects) and **rosin** (from pine trees). Synthetic resins include the wide variety of plastic materials available today and any distinction between PLASTICS and resins is at best arbitrary. The first partially synthetic resins were produced in 1862 using NITROCELLULOSE, vegetable OILS and CAMPHOR, and included Xylonite and later, in 1869, CELLULOID. The first totally synthetic resin was BAKELITE, which was produced by L. H. BAEKELAND in 1910 from PHENOL and FORMALDEHYDE. The work in the 1920s of H. STAUDINGER on the polymeric nature of natural RUBBER and STYRENE resin, which laid the theoretical basis for POLYMER science, was a major factor in stimulating the extremely rapid development of a wide range of synthetic plastics and resins.

RESISTANCE, the ratio of the voltage applied to a conductor to the current flowing through it (see ELECTRICITY; OHM'S LAW), measured in OHMS. It is characteristic of the material of which the conductor is made (the resistance presented by a unit cube of a material being called its **resistivity**) and of the physical dimensions of the conductor, increasing as the conductor becomes longer and/or thinner. Resistance rises with TEMPERATURE in METALS, but falls in SEMICONDUCTORS and SOLUTIONS. Its accurate measurement is performed by the WHEATSTONE BRIDGE method.

RESONANCE, the large response of an oscillatory mechanical, acoustical or electrical system driven near its natural FREQUENCY. The ENERGY dissipation (against FRICTION, etc.) of all practical systems is termed *damping*: the amount of damping controls both the size of the resonant response and the sharpness of the resonance as a function of frequency.

RESONANCE, in chemistry, theory of molecular structure in which the actual state of the bonding in a molecule is expressed as a "resonance hybrid" between two or more valence-bond structures (see

A carboxylate ion can be thought of as a resonance hybrid between valence bond structures (A) and (B). Alternatively, the hybrid structure can be represented as in (C).

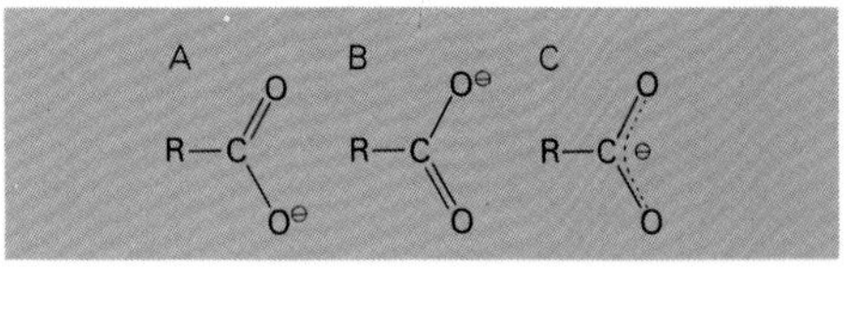

BOND, CHEMICAL), and is intermediate between them, but of lower energy. There is no actual oscillation, and the model is equivalent to molecular ORBITAL theory. First proposed for BENZENE, resonance stabilizes AROMATIC COMPOUNDS and conjugated double-bond systems such as 1,3-BUTADIENE.

RESORCINOL (m-$C_6H_4(OH)_2$), one of the dihydric PHENOLS, a colorless crystalline solid made by sulfonation (see SULFONIC ACIDS) of benzene followed by fusion with sodium hydroxide. It is used to make adhesives, formaldehyde RESINS, DYES, EXPLOSIVES, photographic developers, and as an ANTISEPTIC. (See also FLUORESCEIN.) MW 110.1, mp 111°C, bp 281°C.

RESPIRATION, term applied to several activities and processes occurring in all ANIMALS and PLANTS: e.g., the breathing movements associated with the LUNGS, the uptake of OXYGEN and the release of CARBON dioxide, and the biochemical pathways by which the ENERGY locked in food materials is transferred to energy-rich organic molecules for utilization in the multitude of energy-requiring processes which occur in an organism. Breathing movements, if any, and the exchange of oxygen and carbon dioxide, may be called "external respiration," while the energy-releasing processes which utilize the oxygen and produce carbon dioxide are termed "internal respiration" or "tissue respiration." In man, external respiration is the process whereby air is breathed from the environment into the lungs to provide oxygen for internal respiration. Air, which contains about 20% oxygen, is drawn into the lungs via the NOSE or mouth, the PHARYNX, TRACHEA and BRONCHI. This is achieved by muscular contraction of the intercostal muscles in the CHEST wall and of the DIAPHRAGM; their coordinated movement, controlled by a respiratory center in the BRAIN stem, causes expansion of the chest, and thus of the lung tissue, so that air is drawn in (inspiration). Expiration is usually a passive process of relaxation of the chest wall and diaphragm, allowing the release of the air, which is by now depleted of oxygen and enriched with carbon dioxide. Exchange of gases with the BLOOD circulating in the pulmonary capillaries occurs across the lung alveoli and follows simple diffusion gradients. Disorders of respiration include lung disease (e.g., EMPHYSEMA, PNEUMONIA and PNEUMOCONIOSIS); muscle and nerve disease (e.g., brain-stem STROKE, POLIOMYELITIS, MYASTHENIA GRAVIS and MUSCULAR DYSTROPHY); skeletal deformity; ASPHYXIA, and disorders secondary to metabolic and HEART disease. In man, tissue respiration involves the combination of oxygen with GLUCOSE or other nutrients to form high-energy compounds. This reaction also produces carbon dioxide and water.

RESPIRATION, Artificial. See ARTIFICIAL RESPIRATION.

RETICULO-ENDOTHELIAL SYSTEM, generic name for those CELLS in the body that take up dyes and other foreign material from the BLOOD stream and other body fluids; they are also known as **macrophages**. Blood monocytes are functionally part of the system as are macrophages in the LYMPH nodes, SPLEEN, BONE MARROW, LIVER (Kupffer cells) and LUNG alveoli. When foreign material (e.g., BACTERIA) is introduced into the blood stream, macrophages rapidly take it up and destroy it with intracellular ENZYMES. This constitutes a primary defence system and may play a role in establishing IMMUNITY. Similarly, particulate matter in the lungs or liver is cleared by local macrophages.

RETINA, part of the EYE responsible for conversion of LIGHT into nerve impulses; it contains nerve cells including the rod and cone receptors for light/dark and color VISION respectively.

RETINOL, or vitamin A. See VITAMINS.

RETROGRADE MOTION, the apparent backward (i.e. westward) motion of a PLANET due to the earth's own motion (see ORBIT); also, the motion of any SOLAR SYSTEM body rotating or orbiting in the opposite (clockwise as viewed from the N celestial pole) direction to the majority.

RETROLENTAL FIBROPLASIA, a form of BLINDNESS occurring in premature babies exposed to high OXYGEN concentrations during the treatment of respiratory distress syndrome. Moderation in the use of oxygen and prevention of prematurity have reduced its incidence.

RETTING, the decomposition of the organic tissues of plants by microorganisms, usually while immersed in water. Resistant components such as CELLULOSE fibers are left behind. Retting is used in the processing of FLAX, HEMP and JUTE.

REVERBERATION. See ACOUSTICS.

REVOLVER, a small hand firearm, or PISTOL, incorporating an automatic loading mechanism in the form of a revolving cylinder. The cylinder contains usually five or six chambers into which cartridges are inserted, and activation of the trigger mechanism, in addition to firing a bullet, automatically aligns a fresh chamber with the breech of the barrel. The first practical revolver design was patented in 1836 by Samuel COLT.

REYNOLDS, Osborne (1842–1912), British physicist best known for his important contributions in FLUID MECHANICS, in particular his derivation (1883–84) of the REYNOLDS NUMBER.

REYNOLDS NUMBER, important dimensionless parameter in FLUID MECHANICS, given by $R = \rho ul/\mu$, where ρ is the density, u the velocity, l the length and μ the viscosity of the fluid. The value of R allows for the effects of fluid viscosity on motion and determines whether a given fluid flow is steady or turbulent. It is useful for evaluating the behavior of scale models.

RHAZES, or abu Bakr Muhammad ibn Zakariyya **al-Razi** (c860–925), Persian physician and practical chemist, the author of the *Book of Secret of Secrets* and possibly the earliest to distinguish MEASLES from SMALLPOX.

RHENIUM (Re), very hard, silvery-white TRANSITION ELEMENT in Group VIIB of the PERIODIC TABLE. It is very rare, and is obtained as a by-product of MOLYBDENUM extraction. Analogous to MANGANESE, it forms compounds of all oxidation states between 0 and +7, those of the higher states being volatile and stable. Its uses are similar to those of the PLATINUM GROUP metals. AW 186.2, mp 3180°C, bp 5627°C, sg 20.5 (20°C).

RHEOLOGY, branch of physics concerned with the structure and behavior of flowing and deformed materials, such as the way the shape and size of a body alters with time when subjected to mechanical forces. Properties such as FRICTION, stickiness and roughness are treated, and the study finds application in many fields from engineering to plant physiology.

RHEOSTAT, a variable resistor used to control the current drawn by an electric MOTOR, to dim LIGHTING, etc. It may consist of a resistive wire, wound in a helix, with a sliding contact varying the effective length, or of a series of fixed resistors connected between a row of button contacts. Or, for heavy loads, ELECTRODES dipped in SOLUTIONS can be used, the RESISTANCE being controlled by the immersion depth and separation of the electrodes.
RHESUS FACTOR, or **Rh factor.** See BLOOD.
RHEUMATIC FEVER, feverish illness, following infection with STREPTOCOCCUS and caused by abnormal IMMUNITY to the bacteria, leading to systemic disease. SKIN rash, subcutaneous nodules and a migrating ARTHRITIS are commonly seen. Involvement of the HEART may lead to palpitations, chest pain, cardiac failure, MYOCARDITIS and INFLAMMATION of the PERICARDIUM; murmurs may be heard and the ELECTROCARDIOGRAPH may show conduction abnormality. Sydenham's CHOREA may also be seen, with awkwardness, clumsiness and involuntary movements. Late effects include chronic valve disease of the heart leading to stenosis or incompetence, particularly of the mitral or aortic valves. Such valve disease presents in young to middle age and may require surgical correction. Treatment of acute rheumatic fever includes bed rest, ASPIRIN and STEROIDS. PENICILLIN treatment of streptococcal disease may prevent recurrence. Patients with valve damage require ANTIBIOTICS during operations, especially dental and urinary tract SURGERY, to prevent bacterial endocarditis.
RHEUMATISM, imprecise term describing various disorders of the JOINTS, including RHEUMATIC FEVER and rheumatoid ARTHRITIS.
RHINE, Joseph Banks (1895–), US parapsychologist whose pioneering laboratory studies of ESP have demonstrated the possible occurrence of telepathy (see PARAPSYCHOLOGY).
RHINITIS, INFLAMMATION of the mucous membranes of the NOSE causing runny nasal discharge, and seen in the common COLD, INFLUENZA and HAY FEVER. Irritation in the nose and sneezing are common.
RHIZOME, or **rootstock,** the swollen horizontal underground stem of certain PLANTS that acts as an organ of perennation and vegetative propagation. They last for several years and new shoots appear each spring from the axils of scale leaves.
RHODIUM (Rh), moderately hard metal, the whitest of the PLATINUM GROUP. In addition to the general uses of these metals, rhodium is used for MIRROR surfaces, and a platinum-rhodium alloy is used as a catalyst in the OSTWALD PROCESS. AW 102.9, mp 1966°C, bp 3727°C, sg 12.4 (20°C).
RHODOPHYTA, or **red algae.** See ALGAE.
RHODOPSIN, or visual purple, photosensitive pigment, derived from VITAMIN A, found in the RETINAS of many vertebrates, including man. Its bleaching by incident light is the basis of light-and-dark distinction in VISION.
RHOMBUS. See QUADRILATERAL.
RHUMB LINE, or loxodrome, in NAVIGATION, line of constant compass direction, crossing all lines of LONGITUDE at the same angle, and represented by a straight line on the MERCATOR projection. It spirals toward the poles, and is not as short a distance between two points as the GREAT CIRCLE ROUTE.

RHYNCHOCEPHALIA, members of the REPTILIA that lived during the Triassic period, 200 million years ago. Today, the only surviving species is the Tuatara (*Sphenodon punctatus*) found on islands off the coast of New Zealand.
RHYOLITE, light-colored volcanic IGNEOUS ROCK, of the same composition as GRANITE, and very common and widespread. Often banded from LAVA flow as it solidified, it is fine-grained and usually porphyritic (see PORPHYRY). (See also OBSIDIAN; PUMICE.)
RHYTHMS, Biological. See BIOLOGICAL CLOCKS.
RIB. See SKELETON.
RIBBONWORMS. See NEMERTINA.
RIBOFLAVIN, or vitamin B_2. See VITAMINS.
RIBONUCLEIC ACID (RNA). See NUCLEIC ACIDS.
RIBOSOMES, tiny granules, of diameter about 10nm, found in CELL cytoplasm. They are composed of PROTEIN and a special form of ribonucleic acid (see NUCLEIC ACIDS) known as ribosomal RNA. The ribosome is the site of PROTEIN SYNTHESIS.
RICE PAPER, the edible pith from the stem of the rice paper tree (*Fatsia papyrifera*). Rice paper is used as a decoration on foods and in the Orient as artists' paper and for making artificial flowers. Family: Araliaceae.
RICHARDS, Dickinson Woodruff (1895–1973), US physiologist awarded with COURNAND and FORSSMANN the 1956 Nobel Prize for Physiology or Medicine for his work with Cournand using Forssman's CATHETER technique to probe the heart, pulmonary artery and lungs.
RICHARDS, Theodore William (1868–1928), US chemist awarded the 1914 Nobel Prize for Chemistry for his determination of the ATOMIC WEIGHTS of some 60 ELEMENTS. In particular, his accurate work showed the existence of ISOTOPES, predicted earlier by Frederick SODDY.
RICHARDSON, Sir Owen Willans (1879–1959), British physicist awarded the 1928 Nobel Prize for Physics for his pioneering work on THERMIONIC EMISSION. The **Richardson equation** relates the rate of thermionic emission to the absolute TEMPERATURE of the heated metal.
RICHET, Charles Robert (1850–1935), French physiologist awarded the 1913 Nobel Prize for Physiology or Medicine for his studies of ANAPHYLAXIS, which term he also coined.
RICHTER, Burton (1931–), US physicist who shared the 1976 Nobel physics prize with S. TING for his independent discovery of the massive psi(3095)

Rhodopsin comprises a species-specific protein, opsin, conjugated with the 11-*cis*- isomer of retinal, the aldehyde of vitamin A. On bleaching in light, the retinal is released as the all-*trans*- isomer.

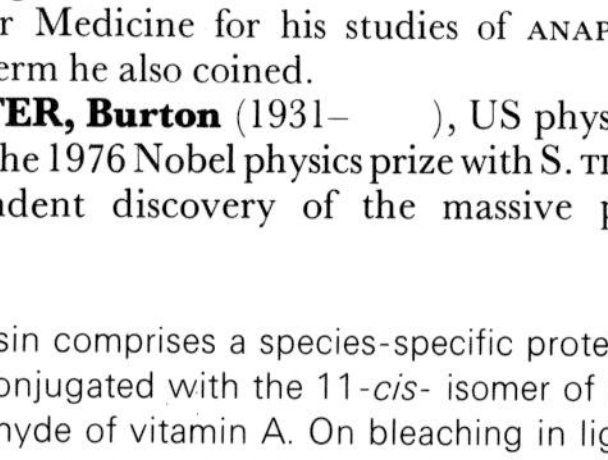

11-*cis*-retinal

Earthquake intensities

intensity	characteristic effects	corresponding magnitude on the Richter scale if the intensity is recorded at the epicenter
instrumental	detected only by seismographs	<3
feeble	noticed by sensitive people only	3–3.4
slight	similar to vibrations caused by heavy trucks	3.5–4
moderate	loose objects disturbed	4–4.4
rather strong	wakes sleeping persons	4.5–4.8
strong	trees sway; damage by overturning and falling of loose objects	4.9–5.4
very strong	walls crack	5.5–6
destructive	chimneys fall; some buildings collapse	6.1–6.5
ruinous	most houses collapse	6.6–7
disastrous	ground badly cracked; landslides	7.1–7.3
very disastrous	few buildings remain; railroads and pipelines destroyed	7.4–8.1
catastrophic	total destruction	>8.1
	(greatest recorded intensity Iran, 1972:	9.5)

The Richter scale, being of empirical numerical origin, does not directly reflect any subjective scale of earthquake intensities. Here an attempt is made to correlate a subjective scale with corresponding Richter magnitudes, imagining the observer to be near the epicenter of the earthquake.

meson, the first particle identified as containing a charmed quark.

RICHTER SCALE, scale devised by C. F. Richter (1900–), used to measure the magnitudes of EARTHQUAKES in terms of the amplitude and frequency of the surface waves. The scale runs from 0 to 10, about one quake a year registering over 8.

RICKETS, VITAMIN D deficiency disease in children causing disordered BONE growth at the EPIPHYSES, with growth retardation, defective mineralization of bone, epiphyseal irregularity on X-RAY, and pliability and tendency to FRACTURE of bones. It is common among the malnourished, especially in cool climates where vitamin D formation in the SKIN is minimal. Treatment is by vitamin D replacement.

RICKETTSIA, organisms partway between BACTERIA and VIRUSES that are obligatory intracellular organisms but have a more complex structure than viruses. They are responsible for a number of diseases (often borne by ticks or lice) including TYPHUS, SCRUB TYPHUS and ROCKY MOUNTAIN SPOTTED FEVER; related organisms cause Q FEVER and PSITTACOSIS. They are sensitive to TETRACYCLINES and cause characteristic serological reactions cross-specific to Proteus bacteria.

RIEMANN, Georg Friedrich Bernhard (1826–1866), German mathematician, the best known among whose contributions to diverse fields of mathematics is the initiation of studies of NON-EUCLIDEAN GEOMETRY. Elliptic geometry is often named RIEMANNIAN GEOMETRY for him.

RIEMANNIAN GEOMETRY, or elliptic geometry, the branch of NON-EUCLIDEAN GEOMETRY based on the hypothesis that for any point P not lying on a line L, no line can be drawn through P parallel to L.

RIFLE, strictly any FIREARM with a "rifled" bore — i.e., with shallow helical grooves cut inside the barrel. These grooves, by causing the bullet to spin, steady it and increase its accuracy, velocity and range. The term "rifle" is more narrowly applied to the long-barreled hand weapon fired from the shoulder. Rifles are generally classified by caliber (see AMMUNITION) or decimal fractions or by mode of action. "Single-shot" rifles are manually reloaded after each discharge; "repeaters" are reloaded from a magazine by means of a hand-operated mechanism that ejects the spent cartridge case and drives a fresh cartridge into the breech. In semiautomatic rifles, these operations are powered by gas produced as the weapon is fired. Today, many rifles have an optional fully automatic action, a single squeeze of the trigger emptying the magazine in seconds.

RIFT VALLEY, or **graben,** a valley formed by the relative downthrow of land between two roughly parallel FAULTS. The best known are the Great Rift Valley of E Africa, and the Rheingraben.

RIGEL, Beta Orionis, the brightest star in ORION. A quadruple star consisting of a visual binary and a spectroscopic binary (see DOUBLE STAR), it is 276pc distant.

RIGHT ASCENSION. See CELESTIAL SPHERE.

RIGHT-HANDEDNESS. See HANDEDNESS.

RIGOR MORTIS, stiffness of the body MUSCLES occurring some hours after DEATH and caused by biochemical alterations in muscle. The body is set in the position held at the onset of the changes.

RIME. See FROST.

RINDERPEST, acute VIRUS disease of, particularly,

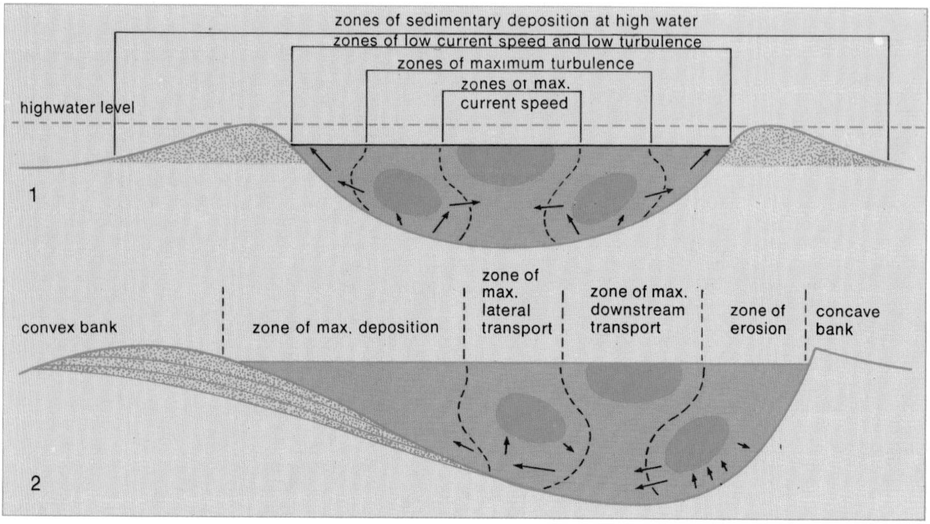

Processes of change at work within a river. The material of the river-bed is being continually worn away, and transported and deposited elsewhere; how much and how far depends on the river's speed and turbulence. The movement of eroded material is indicated by arrows. (1) Straight river with a symmetrical bed; the eroded material is deposited on the banks when the water level is high. (2) Winding river with an asymmetrical bed; material is worn off the concave bank, usually the outer side of a curve, and deposited on the convex.

cattle, common in N Africa and S Asia. Although there have been outbreaks in other parts of the world, North America has hitherto remained unaffected by this usually fatal disease.

RING, a set (see SET THEORY) with the following properties: it has two BINARY OPERATIONS, ADDITION and MULTIPLICATION ($+$ and $\times$); it is an ABELIAN GROUP under addition; multiplication is associative, and distributive over addition (see ALGEBRA); and if a and b are members of the set, $a \times b$ has a unique value c, also a member of the set. If there is a member n of the set such that $a \times n = a$ for every member a of the set, the ring is a "ring with unity." A ring in which multiplication is commutative is a "commutative ring."

RINGWORM, common FUNGUS disease of the SKIN of man and animals which may also affect the HAIR or nails. Ringshaped raised lesions occur, often with central sparing; temporary BALDNESS is seen on hairy skin, together with the disintegration of the nails. ATHLETE'S FOOT is ringworm of the toes, while *tinea cruris* is a variety affecting the groin. Various fungi may be responsible, including *Trichophyton* and *Microspora*. Treatments include topical ointments (e.g., benzyl benzoate) or systemic antifungal ANTIBIOTICS such as Griseofulvin.

RITES OF PASSAGE, ceremonies within a community to mark the achievement by an individual of a new stage in his life cycle (e.g., birth, puberty, marriage) and his consequent change of role in the community.

RITTENHOUSE, David (1732–1796), American astronomer and mathematician, who invented the DIFFRACTION grating, built two famous ORRERIES, discovered the atmosphere of VENUS (1768) independently of LOMONOSOV (1761), and built what was probably the first American TELESCOPE.

RITTER, Johann Wilhelm (1776–1810), Silesian-born German physical chemist, a pioneer of ELECTROCHEMISTRY, who first positively identified ULTRAVIOLET RADIATION (1801).

RIVERS, William Halse Rivers (1864–1922), British anthropologist and psychologist who initiated the study of experimental PSYCHOLOGY at the University of Cambridge (c1893).

RIVERS AND LAKES, bodies of inland water. Rivers flow in natural channels to the sea, lakes or, as tributaries, into other rivers. They are a fundamental component of the HYDROLOGIC CYCLE (see DRAINAGE). Lakes are land-locked stretches of water fed by rivers; though the term may be applied also to temporary widenings of a river's course or to almost-enclosed bays and LAGOONS. In many parts of the world rivers and lakes may exist only during certain seasons, drying up partially or entirely during DROUGHT (see also PLAYA; WADI).

Rivers. The main sources of rivers are SPRINGS, lakes and GLACIERS. Near the source a river flows swiftly, the rocks and other abrasive particles that it carries eroding a steep-sided V-shaped VALLEY (see EROSION). Variations in the hardness of the rocks over which it runs may result in WATERFALLS. In the middle part of its course the gradients become less steep, and lateral (sideways) erosion becomes more important than downcutting. The valley is broader, the flow less swift, and meandering more common. Toward the rivermouth, the flow becomes more sluggish and meandering prominent: the river may form OXBOW LAKES. Sediment may be deposited at the mouth to form a DELTA (see also ESTUARY). (See also CANYON; HYDROELECTRICITY.)

Lakes. Most lakes are the result of glacial erosion during the ICE AGES. Glaciers hollowed out deep basins, often depositing MORAINE to form natural dams. Most lakes have an outflowing stream: where there is great water loss through EVAPORATION there is no such stream and the lake water is extremely saline (see also EVAPORITES), as in the Dead Sea. Lakes are comparatively temporary features on the landscape as they are constantly being infilled by silt. (See also DIVIDE; FIRTH; FJORD; GROUNDWATER.)

RIVETING, the joining of machine or structural parts, usually plates, by rivets. These are headed bolts, usually steel, which are passed through the plates, a second head then being formed on the plain end by pressure, hammering or an explosive charge. Large rivets are heated for satisfactory closing. Although riveting can be automated, it is slowly being displaced by arc WELDING.

RNA, ribonucleic acid. See NUCLEIC ACIDS.

ROBBINS, Frederick Chapman (1916–), US virologist who shared the 1954 Nobel Prize for Physiology or Medicine with J. F. ENDERS and T. H. WELLER for their cultivation of the POLIOMYELITIS virus in nonnerve tissues.

ROBINSON, Sir Robert (1886–), British organic chemist awarded the 1947 Nobel Prize for Chemistry for his pioneering studies of the molecular structures of the ALKALOIDS and other vegetable-derived substances.

ROBOT (from Czech *robota*, work), or automaton, an automatic machine that does work, simulating and replacing human activity; known as an android if humanoid in form (which most are not). Robots have evolved out of simpler automatic devices, and many are now capable of decision-making, self-programming, and carrying out complex operations. Some have sensory devices. They are increasingly being used in industry and scientific research for tasks such as handling hot or radioactive materials. Science fiction from Čapek to Asimov and beyond has featured robots. (See also MECHANIZATION AND AUTOMATION; REMOTE CONTROL.)

ROCHELLE SALT. See TARTARIC ACID.

ROCK CRYSTAL. See QUARTZ.

ROCKET, form of JET-PROPULSION engine in which the substances (fuel and oxidizer) needed to produce the propellant gas jet are carried internally. Working by reaction, and being independent of atmospheric oxygen, rockets are used to power interplanetary space vehicles (see SPACE EXPLORATION). In addition to their chief use to power MISSILES, rockets are also used for supersonic and assisted-takeoff airplane propulsion, and sounding rockets are used for scientific investigation of the upper atmosphere. The first rockets—of the firework type, cardboard tubes containing GUNPOWDER—were made in 13th-century China, and the idea quickly spread to the West. Their military use was limited, guns being superior, until they were developed by William CONGREVE. Later Congreve rockets mounted the guide stick along the central axis; and William Hale eliminated it altogether, placing curved vanes in the exhaust stream, thus stabilizing the rocket's motion by causing it to rotate on its axis. The 20th century saw the introduction of new fuels and oxidants, e.g., a mixture of NITROCELLULOSE and NITROGLYCERIN for solid-fuel

Schematic diagram of a liquid-fuel rocket motor. The liquid fuel is pumped (1) from the fuel tank and circulated through the cooling jacket (2) of the exhaust cone (3). Here it vaporizes and is led off via a turbine (4) to the combustion chamber (5). The turbine drives both the fuel pump and the pump (6) in the oxidant line.

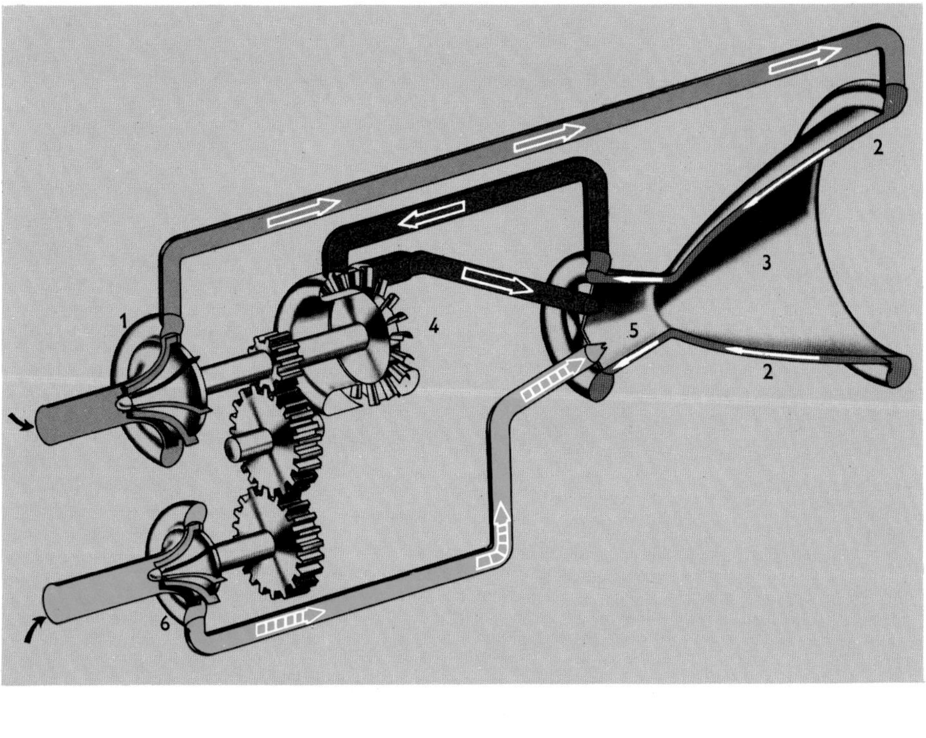

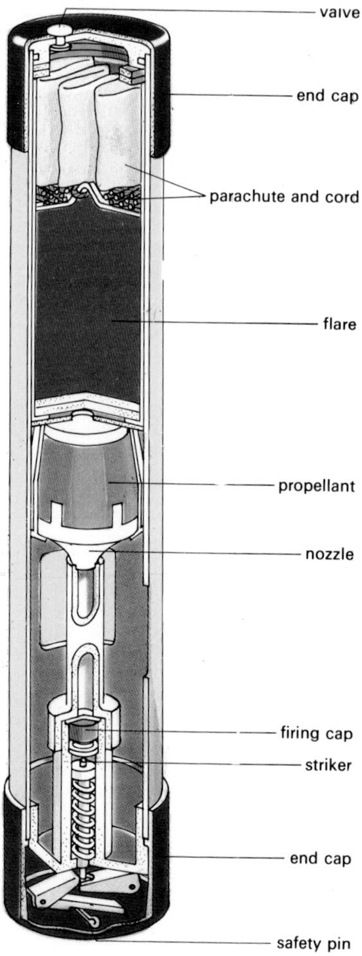

A hand-held distress signal rocket as carried on board all types of sea-going vessel. For use, the end caps and the safety pin are removed, the firing lever is operated and the rocket flies to a height of at least 300m. The flare is ejected and drifts below its parachute for at least 40s. The discharger is made of plastic, the rocket body of aluminum.

rockets, or ETHANOL and liquid oxygen for the more efficient liquid-fuel rockets. The first liquid-fuel rocket was made by R. H. GODDARD, who also invented the multistage rocket. In WWII Germany, and afterward in the US, Wernher von BRAUN made vast improvements in rocket design. Other propulsion methods, including the use of nuclear furnaces, electrically-accelerated PLASMAS and ION PROPULSION, are being developed.

ROCKS, the solid materials making up the earth's crust. They may be consolidated (e.g., sandstone) or unconsolidated (e.g., sand). The study of rocks is **petrology**. Strictly, the term applies only to those materials which, unlike MINERALS, are not homogeneous and have no definite chemical composition. The primary constituents of rocks are OXYGEN and SILICON, combined with each other to form SILICA (see QUARTZ) and with each other and further elements (e.g., aluminum, iron, calcium, potassium, sodium and magnesium) to form SILICATES. Together, silica and silicates make up about 95% of the earth's rocks. There are three main classes of rocks, igneous, sedimentary and metamorphic. IGNEOUS ROCKS form from the MAGMA, a molten, subsurface complex of silicates. They are the primary source of all the earth's rocks. SEDIMENTARY ROCKS are consolidated accumulations of fragmented inorganic and organic material. They are of three types: classic, formed of weathered (see EROSION) particles of other rocks (e.g., SANDSTONE); organic deposits (e.g., COAL, some LIMESTONES); and chemical precipitates (e.g., the EVAPORITES). (See also FOSSILS; STRATIGRAPHY.) METAMORPHIC ROCKS have undergone change within the earth under heat, pressure or chemical action. Sedimentary, igneous and even previously metamorphosed rocks may change structure or composition in this way. (See also EARTH; GEOLOGY.)

ROCK SALT, or **Halite**. See SALT.

ROCKY MOUNTAIN SPOTTED FEVER, tick-borne rickettsial disease (see RICKETTSIA) seen in much of the US, especially the Rocky mountain region. It causes FEVER, HEADACHE and a characteristic rash starting on the palms and soles, later spreading elsewhere. TETRACYCLINES are effective, though untreated cases may be fatal.

RODENTIA, an order of the MAMMALIA that includes the beavers, squirrels, porcupines, guinea pigs, rats and mice. The largest order of mammals, there are almost 1 500 living species. All rodents have a single pair of incisor teeth that grow throughout life and wear to form sharp chisel edges used for gnawing. They are predominantly eaters of seeds and tough vegetable matter and some species are important pests.

RODS AND CONES. See VISION.

ROEBUCK, John (1718–1794), British inventor of the LEAD CHAMBER PROCESS for making sulfuric acid (1746), and patron of James WATT.

ROEMER, Ole or **Olaus**. See RØMER, OLE OR OLAUS.

ROENTGEN (or **Röntgen**), **Wilhelm Conrad** (1845–1923), German physicist, recipient in 1901 of the first Nobel Prize for Physics for his discovery of X RAYS. This discovery was made in 1895 when by chance he noticed that a PHOSPHOR screen nearby a vacuum tube through which he was passing an electric current fluoresced brightly, even when shielded by opaque cardboard. (See also RÖNTGEN (unit).)

ROGERS, Carl Ransom (1902–), US psychotherapist who instituted the idea of the patient determining the extent and nature of his course of therapy, the therapist following the patient's lead.

RÓHEIM, Géza (1891–1953), Hungarian-born US anthropologist best known for his application of the ideas of PSYCHOANALYSIS in ETHNOLOGY studies.

ROLLING MILLS, equipment—essentially sets of power-driven heavy steel rollers—used for flattening metal ingots and producing sheets, bars, rails etc. STEEL is usually rolled hot; nonferrous metals cold.

ROMAN NUMERALS, letters of the Roman ALPHABET used to represent numbers, the letters I, V, X, L, C, D and M standing for 1, 5, 10, 50, 100, 500 and 1000, respectively. All other numbers are represented by combinations of these letters according to certain rules of addition and subtraction; thus, for example, VIII is 8, XL is 40, MCD is 1400 and MCDXLVIII is 1448.

RØMER, Ole or **Olaus** (1644–1710), Danish astronomer who first showed that light has a finite velocity. He noticed that JUPITER eclipsed its moons at times differing from those predicted and correctly concluded this was due to the finite nature of light's velocity, which he calculated as 227 000km/s (a modern value is about 299 800km/s).

RÖNTGEN (R), a unit of radiation exposure named for W. C. ROENTGEN. Its value in SI UNITS is such that exposure to 1 röntgen of X RAYS or GAMMA RAYS causes 0.000258 coulombs of ionization per kg of air.

ROOT, in ALGEBRA, one of the equal FACTORS of a given number: if $x^n = c$ then x is the nth root of c, written $x = \sqrt[n]{c}$ (see EXPONENT; POWER). The 2nd and 3rd roots of a number are specially named the square and cube roots respectively. Of particular interest is i, the imaginary square root of -1 (see IMAGINARY NUMBERS).

ROOTS, that part of a PLANT which absorbs water and nutrients from the soil and anchors the plant to the ground. Water and nutrients enter a root through minute root hairs sited at the tip of each root. Roots need oxygen to function and plants growing in swamps have special adaptations to supply it, like the "knees" of bald cypress trees and the aerial roots of mangrove. There are two main types of root systems: the taproot system, where there is a strong main root from which smaller secondary and tertiary roots branch out; and the fibrous root system where a mass of equal-sized roots are produced. In plants such as the sugar beet, the taproot may become swollen with stored food material. Adventitious roots anchor the stems of climbing plants, such as ivy. Epiphytic plants such as orchids have roots that absorb moisture from the air (see EPIPHYTE). The roots of parasitic plants such as mistletoe and dodder absorb food from other plants.

ROOTS OF AN EQUATION, the values of the VARIABLE for which an EQUATION in it holds. For example, the roots of a quadratic equation $f(x) = ax^2+bx+c$ may be found by application of the formula (where $f(k) = 0$):

$$k = \frac{-b \pm \sqrt{b^2-4ac}}{2a}$$

where b^2-4ac is the DISCRIMINANT.

ROOTSTOCK. See RHIZOME.

ROPE, or cordage, a thick, strong cord made from twisted lengths of natural FIBER. It can be made from MANILA HEMP, HENEQUEN, SISAL, true HEMP, coir (coconut palm fiber), FLAX, JUTE and COTTON. The last three are generally used for lighter ropes such as cords and twines. SYNTHETIC FIBERS, particularly NYLON and polyesters, are used for lighter and more durable rope. Other ropes are made from wire, for example, for suspension cables in bridge building. Rope-making resembles SPINNING.

RORSCHACH, Hermann (1884–1922), Swiss psychoanalyst who devised the **Rorschach Test** (c1920), in which the subject looks at a series of ten symmetrical inkblots, and describes what he sees there. It is intended that, from his description, details of his personality can be deduced.

ROSETTA STONE, an inscribed basalt slab, discovered in 1799, which provided the key to the decipherment of Egyptian HIEROGLYPHICS. About 1.2m long and 0.75m wide, it is inscribed with identical texts in Greek, Egyptian demotic and Egyptian hieroglyphs. Decipherment was begun by Thomas YOUNG (c1818) and completed by Jean-François CHAMPOLLION (c1821–22). Found near Rosetta, Egypt, the stone is now in the British Museum.

ROSIN. See RESIN; TURPENTINE.

ROSS, Sir Ronald (1857–1932), British physician awarded the 1902 Nobel Prize for Physiology or Medicine for his investigations, prompted by Sir Patrick MANSON, of the *Anopheles* MOSQUITO in relation to the transmission of MALARIA, in course of which he isolated malarial cysts in the mosquito's intestinal tract, later correctly identifying these with cysts he found in the bloodstreams of diseased birds (1897–98).

ROT, name given to the symptoms produced by a number of FUNGI and BACTERIA that infect plants. (See also PLANT DISEASES.)

ROT, rotor or **rotation of a vector,** equivalent to CURL. (See also VECTOR ANALYSIS.)

ROTIFERS. See ASCHELMINTHES.

ROUS, Francis Peyton (1879–1970), US physician who shared (with C. B. HUGGINS) in the 1966 Nobel Prize for Physiology or Medicine for his discovery (c1910) of a VIRUS which causes TUMORS in chickens.

ROUX, Pierre Paul Émile (1853–1933), French bacteriologist noted for his work with PASTEUR toward a successful ANTHRAX treatment, with METCHNIKOV on syphilis, and with YERSIN on DIPHTHERIA. Using Roux' and Yersin's results, von BEHRING was able to develop the diphtheria ANTITOXIN.

ROWLAND, Henry Augustus (1848–1901), US physicist and engineer who developed the concave DIFFRACTION grating, in which the lines are ruled directly onto a concave spherical surface, thus eliminating the need for additional MIRRORS and LENSES. (See also SPECTROSCOPY.)

ROYAL INSTITUTION, an English scientific society, founded in 1799 by Benjamin Thompson (see Count RUMFORD) to encourage scientific study and the spread of scientific knowledge. It has associations with many eminent men of science, including Humphry DAVY and Michael FARADAY.

ROYAL JELLY. The social organization of a beehive consists of a queen—a sexually mature female—attended by drones—males—and supported by armies of worker bees—females whose sexual development has been suppressed. Both queen and workers come from the same larval stock. The sexual development of the few grubs chosen to replace the queen depends upon their being fed a "royal jelly" produced by special glands of the worker "nursemaids."

ROYAL SOCIETY OF LONDON FOR THE IMPROVEMENT OF NATURAL KNOWLEDGE, the premier English scientific society. Probably the most famous scientific society in the world, it also has a claim to be the oldest surviving. It had its origins in weekly meetings of scientists in London in the 1640s and was granted a royal charter by Charles II in 1660. Past presidents include Samuel Pepys, Sir Isaac NEWTON and Lord RUTHERFORD.

RUBBER, an elastic substance; that is, one which

quickly restores itself to its original size after it has been stretched or compressed. Natural rubber is obtained from many plants, and commercially from *Hevea brasiliensis*, a tree native to South America and cultivated also in SE Asia and W Africa. A slanting cut is made in the bark, and the milky fluid latex, occurring in the inner bark, is tapped off. The **latex**—an aqueous COLLOID of rubber and other particles—is coagulated with dilute acid, and the rubber creped or sheeted and smoked. Natural rubber is a chain POLYMER of ISOPRENE, known as caoutchouc when pure; its elasticity is due to the chains being randomly coiled, but tending to straighten when the rubber is stretched. Known to have been used by the Aztecs since the 6th century AD, and first known in Europe in the 16th century, it was a mere curiosity until the pioneer work of Thomas Hancock (1786–1865) and Charles MACINTOSH. Synthetic rubbers have been produced since WWI, and the industry has developed greatly during and since WWII. They are long-chain polymers, elastomers; the main types are: copolymers of butadiene/styrene, butadiene/nitriles and ethylene/propylene; polymers of chloroprene (neoprene rubber), butadiene, isobutylene and SILICONES; and polyurethanes, polysulfide rubbers and chlorosulfonated polyethylenes. Some latex (natural or synthetic) is used as an adhesive and for making rubber coatings, rubber thread and foam rubber. Most, however, is coagulated, and the rubber is treated by VULCANIZATION and the addition of reinforcing and inert fillers and antioxidants, before being used in tires, shoes, rainwear, belts, hoses, insulation and many other applications.

RUBELLA. See GERMAN MEASLES.

RUBIDIUM (Rb), a soft, silvery-white, highly reactive ALKALI METAL. It is fairly abundant, but is found only as a minor constituent of POTASSIUM and CESIUM minerals. It is more reactive than potassium, and reacts violently with water and ice. Metallic rubidium, prepared by ELECTROLYSIS of the chloride or reduction of the carbonate, is used in electron tubes, and its salts in making special glasses and ceramics. AW 85.5, mp 39°C, bp 688°C, sg 1.532 (20°C).

RUBY, deep-red GEM stone, a variety of CORUNDUM colored by a minute proportion of chromium ions. It is found significantly only in upper Burma, Thailand and Sri Lanka, and is more precious by far than diamond. The name has been used for other red stones, chiefly varieties of garnet and spinel. Rubies have been synthesized by the Verneuil flame-fusion process (1902). They are used to make ruby LASERS.

RUM, alcoholic liquor, usually produced by distilling fermented MOLASSES. It acquires a brown color from the wooden casks in which it is stored and from added CARAMEL or burnt sugar. It is made mainly in the West Indies. (See also ALCOHOLIC BEVERAGES.)

RUMFORD, Sir Benjamin Thompson, Count (1753–1814), American-born adventurer and scientist best known for his recognition of the relation between WORK and HEAT (inspired by observation of heat generated by FRICTION during the boring of cannon), which laid the foundations for JOULE's later work. He played a primary role in the founding of the ROYAL INSTITUTION (1799), to which he also introduced Humphry DAVY.

RUMINANTS, animals that regurgitate and rechew their food once having swallowed it. They feed by filling one compartment of a three- or four-chambered stomach with unmasticated food, bringing it back up to the mouth again to be fully chewed and finally swallowed. It is an adaptation in many herbivores to increase the time available for the digestion of relatively indigestible vegetable matter.

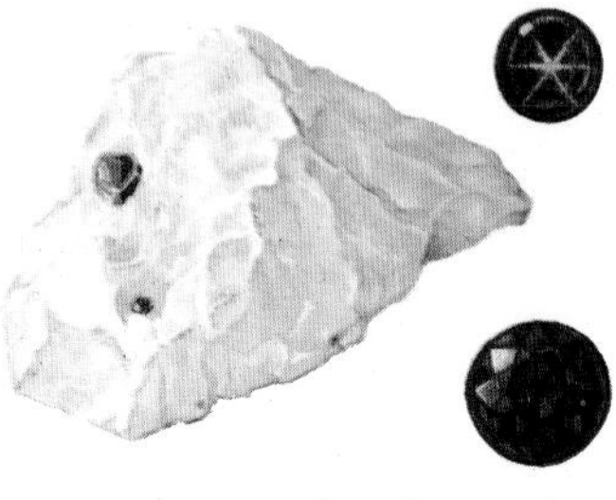

Ruby: crystals bedded in limestone (*left*), with gemstones polished (*above*) and cut and polished (*below*).

RUNES, characters of a pre-Christian writing system used by the Teutonic tribes of N Europe from as early as the 3rd century BC to as late as the 10th century AD, and sometimes after. The three distinct types are Early, Anglo-Saxon and Scandinavian. The Runic alphabet is sometimes known as **futhork** for its first six characters. (See also WRITING, HISTORY OF.)

RUPTURE, common name for HERNIA.

RUSH, Benjamin (1746–1813), US physician, abolitionist and reformer. His greatest contribution to medical science was his conviction that insanity is a disease (see MENTAL ILLNESS): his *Medical Enquiries and Observations upon the Diseases of the Mind* (1812) was the first US book on PSYCHIATRY.

RUSSELL, Bertrand Arthur William, 3rd Earl Russell (1872–1970), British philosopher, mathematician and man of letters. Initially subscribing to IDEALISM, he broke away in 1898 eventually to become an empiricist (see EMPIRICISM). His most important work was to relate LOGIC and MATHEMATICS. After having written to FREGE pointing out a paradox in Frege's attempt to reduce all mathematics to logical principles, Russell endeavored to perform this task himself. His results appeared in *The Principles of Mathematics* (1903) and, in collaboration with A. N. WHITEHEAD, *Principia Mathematica* (3 vols., 1910–13). Russell was a vehement pacifist for much of his life, his views twice earning him prison sentences (1918, 1961): during the former he wrote his *Introduction to Mathematical Philosophy* (1919). His other works include *Marriage and Morals* (1929), *Education and the Social Order* (1932), *An Inquiry into Meaning and Truth* (1940), *History of Western Philosophy* (1945) and popularizations such as *The ABC of Relativity* (1925), as well as his *Autobiography* (3 vols., 1967–69). He received the 1950 Nobel Prize for Literature; and founded the Bertrand Russell Peace Foundation in 1963.

RUSSELL, Henry Norris (1877–1957), US astronomer who, independently of HERTESPRUNG, showed the relation between a STAR's brightness and color: the resulting **Hertzsprung-Russell diagram** is important throughout astronomy and cosmology.

RUST, a large number of FUNGI which cause many

PLANT DISEASES. They form red or orange spots, their spore-bearing organs, on the leaves of infected plants. Spores are carried by the wind to infect new plants. Some rusts are heteroecious: they alternate between two different host plants. The most important rust fungus is probably *Puccinia graminis* which causes black stem rust of wheat. (See also SMUT.)

RUST. See CORROSION.

RUTHENIUM (Ru), hard metal in the PLATINUM GROUP. It is added to platinum or palladium to form hard ALLOYS. Ruthenium is a catalyst (see CATALYSIS) used in organic chemistry. AW 101.1, mp 2250°C bp 3900°C, sg 12.30(20°C).

RUTHERFORD, Sir Ernest, 1st Baron Rutherford of Nelson (1871–1937), New Zealand born British physicist. His early work was with J. J. THOMSON on MAGNETISM and thus on RADIO waves. Following ROENTGEN's discovery of X-RAYS (1895), they studied the CONDUCTIVITY of air bombarded by these rays and the rate at which the IONS produced recombined. This led him to similar studies of the "rays" emitted by URANIUM (see RADIOACTIVITY). He found these were of two types, which he named alpha and beta (see ALPHA PARTICLES; BETA RAYS). As a result of their work on RADIUM, ACTINIUM and particularly THORIUM, he and Frederick SODDY were able in 1903 to put forward their theory of radioactivity. This suggested that the atoms of certain substances spontaneously emit alpha and beta rays, being thereby transformed into atoms of different, but still radioactive, ELEMENTS of lesser ATOMIC WEIGHT. He later showed alpha rays to be positively charged particles, in fact, HELIUM atoms stripped of two ELECTRONS. He was awarded the 1908 Nobel Prize for Chemistry. In 1911 he proposed his nuclear theory of the ATOM, on which BOHR based his celebrated theory two years later. In 1919 he announced the first artificial disintegration of an atom, NITROGEN being converted into OXYGEN and HYDROGEN by collision with an alpha particle. He was President of the ROYAL SOCIETY from 1925 to 1930. His work was commemorated (1969) by the naming of RUTHERFORDIUM.

RUTHERFORDIUM (Rf), a TRANSURANIUM ELEMENT in Group IVB of the PERIODIC TABLE; atomic number 104. Soviet scientists claimed to have synthesized this element, which they called **kurchatovium**, in 1964, but their results are unconfirmed. American scientists claimed synthesis of rutherfordium in 1969 by bombarding californium-249 with carbon ions.

RUTILE, red-to-black OXIDE mineral consisting of impure titanium (IV) oxide (TiO_2), a TITANIUM ore of widespread occurrence. It is used to color porcelain, and synthetic rutile is used for GEMS.

RUZICKA, Leopold (1887–1976), Croatian-born Swiss chemist who shared with BUTENANDT the 1939 Nobel Prize for Chemistry for his work on the TERPENES and for his demonstration that the ring compounds muskone and civetone had respectively 16 and 17 carbon atoms in the ring (see ALICYCLIC COMPOUNDS): previously it had been thought that rings with more than 8 carbon atoms would be unstable.

RYLE, Gilbert (1900–1976), English philosopher, a major figure at Oxford in the tradition of "ordinary language" philosophy, which views philosophical problems as conceptual confusions resulting from an unwary use of language. In his best-known work, *The Concept of Mind* (1949), he seeks to expose the legacy of such confusions bequeathed by the DUALISM of DESCARTES.

RYLE, Sir Martin (1918–), radio astronomer, corecipient with HEWISH of the 1974 Nobel physics prize. He was the first Professor of RADIO ASTRONOMY at Cambridge (1959), knighted in 1966, and became British Astronomer Royal in 1972.

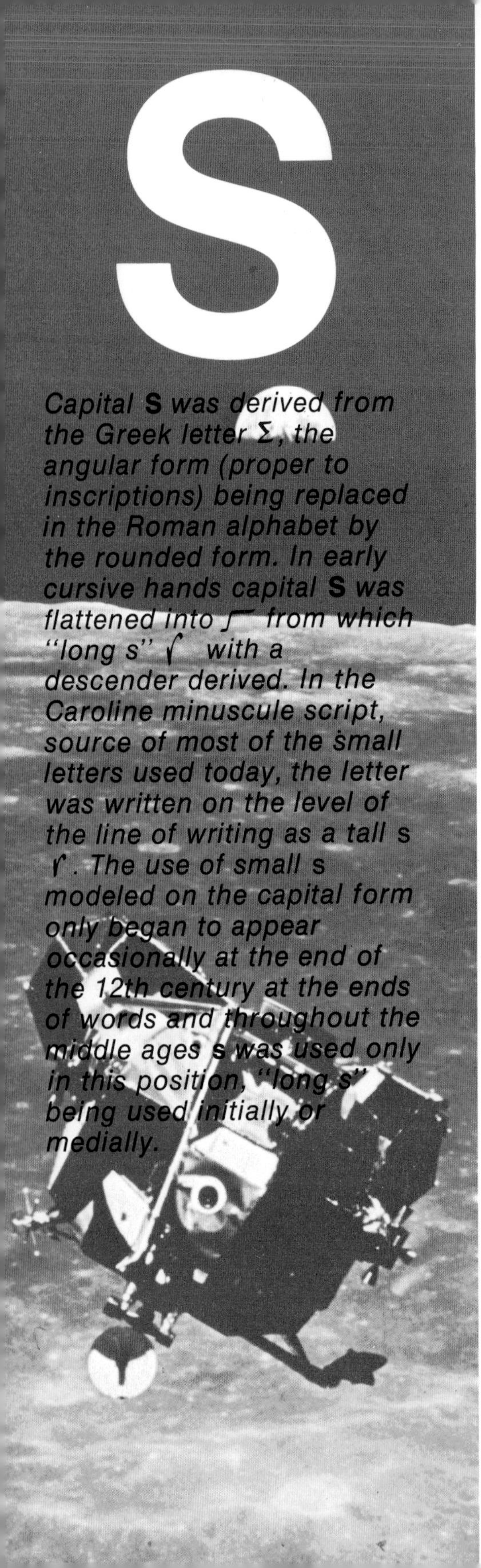
S
Capital S was derived from the Greek letter Σ, the angular form (proper to inscriptions) being replaced in the Roman alphabet by the rounded form. In early cursive hands capital S was flattened into ʃ from which "long s" ſ with a descender derived. In the Caroline minuscule script, source of most of the small letters used today, the letter was written on the level of the line of writing as a tall s ſ. The use of small s modeled on the capital form only began to appear occasionally at the end of the 12th century at the ends of words and throughout the middle ages s was used only in this position, "long s" being used initially or medially.

SABATIER, Paul (1854–1941), French chemist who shared with GRIGNARD the 1912 Nobel Prize for Chemistry for his work on catalyst action in organic syntheses (see CATALYSIS), especially his discovery that finely divided nickel accelerates HYDROGENATION.

SABIN, Albert Bruce (1906–   ), US virologist best known for developing the oral POLIOMYELITIS vaccine, *Sabin* (1955). (See also SALK, J. E.).

SABLE, the FUR of the sable, *Martes zibellina*, a ground-living mustelid of northern Asia.

SACCHARIDES. See CARBOHYDRATES; SUGARS.

SACCHARIN, or *o*-sulfobenzoic imide, a SWEETENING AGENT, 550 times sweeter than sucrose, normally used as its soluble sodium salt. Not absorbed by the body, it is used by diabetics and in low-calorie DIETETIC FOODS. Some concern has been expressed over the continuing widespread use of saccharin because of fears that it may be carcinogenic.

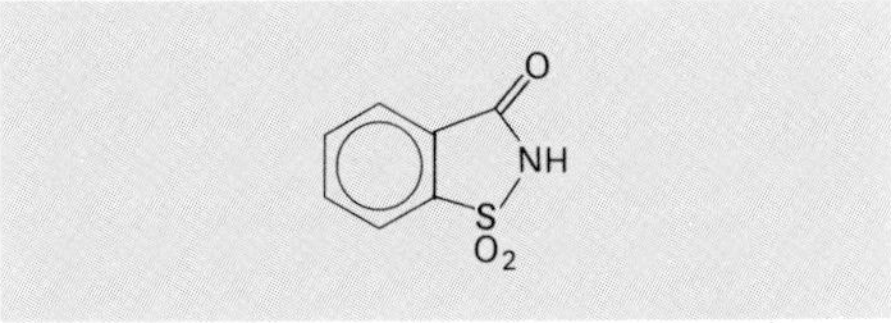

The structure of saccharin (2-sulfobenzoic acid imide).

SACHS, Julius von (1832–1897), German botanist regarded as the father of experimental plant physiology. Among his many contributions are his discovery of what are now called CHLOROPLASTS; the elucidation of the details of the GERMINATION process, and his studies of plant TROPISMS.

SACROILIAC JOINT, the JOINT between the sacrum or lower part of the vertebral column and the iliac bones of the PELVIS. Little movement occurs about the joint but it may be affected by certain types of ARTHRITIS, such as ankylosing spondylitis.

SADISM, the derivation of erotic pleasure from inflicting PAIN on others; possibly a retention of INFANTILE SEXUALITY. (See also MASOCHISM.)

SADOMASOCHISM. See MASOCHISM.

SAFETY GLASS, reinforced GLASS used chiefly in automobile windscreens, aircraft, and where bullet resistance is needed. Some safety glass is glass toughened by being heated almost to softening and then cooled; some has wire mesh embedded to guard against shattering; most is a LAMINATE with a thin layer of PLASTIC (polyvinyl butyral) between two glass layers, so that, if broken, the glass fragments adhere to the plastic.

SAFETY LAMP, lamp used to detect explosive "firedamp" (METHANE) in mines, invented in 1815 by the British chemist Sir Humphry DAVY to provide a safe form of lighting underground. A double layer of wire gauze surrounding the flame dissipated its heat, so preventing a methane atmosphere from reaching its ignition temperature, and yet allowed any methane present to cause a noticeable change in the flame's appearance. A safety lamp was also invented independently by George STEPHENSON.

SAFETY VALVE, a VALVE, sealed by a compressed spring or a weight, that opens to allow fluid above a preset pressure to escape. It is then held open until the pressure has fallen by a predetermined amount. They are used on all pressurized vessels (BOILERS, etc.) to prevent explosion.

SAFFRON, dye extracted from the stigmas and stamens of the Saffron crocus, *Crocus sativus*. Today, it is primarily used for coloring and flavoring foodstuffs. It is the most expensive of all spices.

SAGITTARIUS (the Archer), a constellation on the ECLIPTIC lying in the direction of the galactic center (see MILKY WAY). Sagittarius is the ninth sign of the ZODIAC.

SAINT ANTHONY'S FIRE, name once given to ergotism (see ERGOT); also, the skin infection, ERYSIPELAS.

SAINT ELMO'S FIRE, the glowing electrical discharge seen at the tips of tall, pointed objects—church spires, ships' masts, airplane wings, etc.—in stormy weather. The negative electric charge on the storm clouds induces a positive charge on the tall structure. The impressive display is named (corruptly) for St. Erasmus, patron of sailors.

SAINT VITUS' DANCE, or Sydenham's chorea. See CHOREA.

SAKHAROV, Andrei Dimitrievich (1921–), Soviet physicist who played a prominent part in the development of the first Soviet HYDROGEN BOMB. He subsequently advocated worldwide nuclear disarmament (being awarded the 1975 Nobel Peace Prize) and became a leading Soviet dissident.

SAL AMMONIAC, or ammonium chloride. See AMMONIA.

SALICYLIC ACID (o-$C_6H_4(OH)COOH$), white crystalline solid, made from PHENOL and CARBON dioxide; used in medicine against CALLUSES and WARTS, and to make ASPIRIN and DYES. Its sodium salt is an ANALGESIC and is used for rheumatism. MW 138.1, mp 159°C. **Methyl salicylate**, an ESTER, occurs in oil of WINTERGREEN, and is used as a liniment and a flavoring.

SALIVA, the watery secretion of the slivary GLANDS which lubricates the mouth and food boluses. It contains MUCUS, some gamma globulins and PTYALIN and is secreted in response to food in the mouth or by

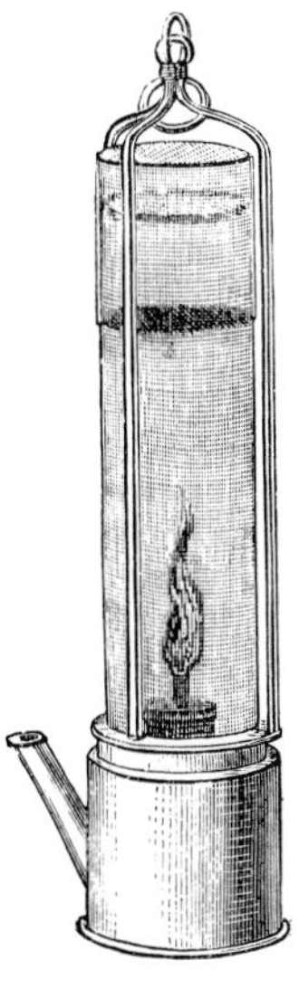

Sir Humphry Davy's miners' safety lamp, invented in 1815. In subsequent years, many inventors sought to improve on this basic design.

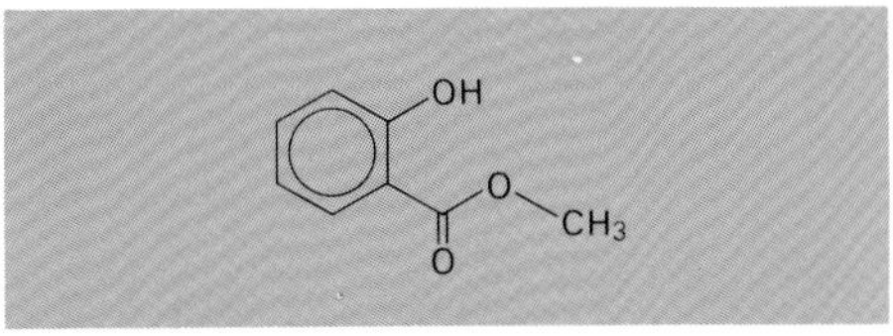

Methyl salicylate, the principal agent in oil of wintergreen. Chemically this ester is closely related to asprin.

conditioned REFLEXES such as the smell or sight of food. Secretion is partly under the control of the parasympathetic autonomic NERVOUS SYSTEM. The various salivary glands—parotid, submandibular and sublingual—secrete slightly different types of saliva, varying in mucus and ENZYME content.

SALK, Jonas Edward (1914–), US virologist best known for developing the first POLIOMYELITIS vaccine, *Salk* (1952–54). (See also SABIN, A. B.)

SALMONELLA, bacteria, some species of which cause FOOD POISONING or ENTERITIS; specific types cause TYPHOID and PARATYPHOID FEVER.

SALT, common name for **Sodium Chloride** (NaCl), found in seawater and also as the common mineral, rock salt or halite. Pure salt forms white cubic CRYSTALS. Some salt is obtained by solar evaporation from salt pans, shallow depressions periodically flooded with seawater; but most is obtained from underground mines. The most familiar use of salt is to flavor food. (Magnesium carbonate is added to table salt to keep it dry.) It is, however, used in much larger quantities to preserve hides in leather-making, in soap manufacture, as a food preservative and in keeping highways ice-free in winter. Rock salt is the main industrial source of chlorine and caustic soda. mp 801°C, bp 1413°C.

SALT, Chemical, an electrovalent compound (see BOND, CHEMICAL) formed by neutralization of an ACID and a BASE. The vast majority of MINERALS are salts, the best known being common SALT, sodium chloride. Salts are generally ionic solids which are good electrolytes (see ELECTROLYSIS); those of weak acids or bases undergo partial HYDROLYSIS in water. Salts may be classified as normal (fully neutralized), acid (containing some acidic hydrogen, e.g., BICARBONATES), or basic (containing hydroxide ions). They may alternatively be classified as simple salts, double salts (two simple salts combined by regular substitution in the crystal lattice) including ALUMS, and complex salts (containing complex ions: see LIGAND).

SALT DOME, a mass of EVAPORITE minerals which has pierced the strata above it and domed the strata near the surface. They often form natural traps for PETROLEUM, and occasionally penetrate to the surface to form *salt glaciers*.

SALT FLAT, dried-up bed of an enclosed stretch of water that has evaporated, leaving the salts that it held in solution as a crust on the ground. Best known are the Lake Bonneville flats, near Salt Lake City, Ut. (See also EVAPORITES.)

SALT LICK, any naturally occurring deposit of common SALT frequented by animals, who lick the salt. Most frequently, the deposit is associated with a salt spring.

SALTPETER, or potassium nitrate. See POTASSIUM.

SALVARSAN, or **arsphenamine,** organic arsenical agent introduced by Paul EHRLICH for the CHEMOTHERAPY of syphilis (see VENEREAL DISEASES) and protozoal infection (e.g., TRYPANOSOMES). It has been superceded by PENICILLIN in syphilis and safer protozoal drugs.

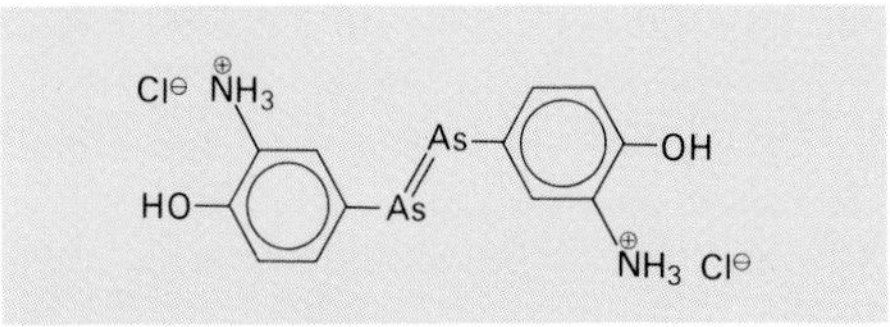

The structure of salvarsan (3,3′-diamino-4,4′-dihydroxyarsenobenzene dihydrochloride).

SALVIA, an ASTRINGENT derived from certain plants (*Salvia*), formerly used for sore throat and ULCERS.

SAL VOLATILE, or ammonium carbonate, $(NH_4)_2CO_3$, colorless crystalline solid made by combining aqueous AMMONIA and CARBON dioxide, the main ingredient of SMELLING SALTS.

SAMARA, winged ACHENE produced by, for example, the maple and sycamore. (See FRUIT).

SAMARIUM (Sm), one of the LANTHANUM SERIES. AW 150.4, mp 1 077°C, bp 1 791°C, sg 7.52 (α).

SAMPLING. See STATISTICS.

SAND, in geology, collection of rock particles with diameters in the range 0.125–2.0mm. It can be graded according to particle size: fine (0.125–0.25mm); medium (0.25–0.5mm); coarse (0.5–1.0mm); and very coarse (1–2mm). Sands result from EROSION by GLACIERS, winds, or ocean or other moving water. Their chief constituents are usually QUARTZ and FELDSPAR. (See also BEACH; DESERT; DUNE; SANDSTONE.)

SANDBLASTING. See ABRASIVE.

SANDSTONE, a SEDIMENTARY ROCK consisting of consolidated SAND, generally cemented by a matrix of CLAY minerals, CALCITE or HEMATITE. The sand grains are chiefly QUARTZ and FELDSPAR. The chief varieties are QUARTZITE, rich in silica; arkose, feldspar-rich; graywacke, coarse-grained and of varied composition; and subgraywacke, with more rounded grains and less feldspar than graywacke. Sandstone grades into CONGLOMERATE and SHALE. Sandstone beds may bear NATURAL GAS or PETROLEUM, and are commonly AQUIFERS. Sandstone is quarried for building, and crushed for use as AGGREGATE.

SANDSTORM, or dust storm, windstorms in which clouds of SAND or DUST are driven across the land. Because of the high wind-velocity, sandstorms are powerful factors in SOIL EROSION. (See also DUST BOWL; EROSION.)

SANDWICH COMPOUNDS, ORGANOMETALLIC COMPOUNDS in which the metal atom is sandwiched between two aromatic-ring LIGANDS—the whole of the ring electron system (see AROMATIC COMPOUNDS) interacting with the metal ORBITALS, giving great stability. FERROCENE and dibenzene-chromium are examples.

SANGER, Frederick (1918–), British biochemist awarded the 1958 Nobel Prize for Chemistry

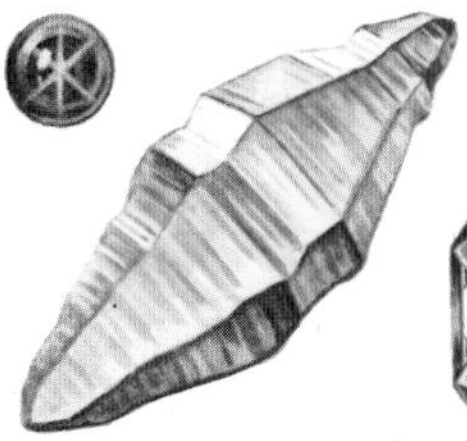

Sapphire: the raw crystal, a polished stone and a cut and polished gemstone.

for his work on the PROTEINS, particularly for first determining the complete structure of a protein, that of bovine INSULIN (1955).

SAPIR, Edward (1884–1939), US anthropologist, poet and linguist whose most important work was on the relation between language and the culture of which it is a product, suggesting that one's perception of the world is dominated by the language with which one may express it.

SAPONIFICATION. See SOAPS AND DETERGENTS.

SAPONINS, substances found in plants which form stable foams with water. They normally occur combined with a SUGAR in glycosides. The component STEROLS or triterpinoids are termed **sapogenins.** Many are of considerable value in medicine in spite of being highly poisonous. The HEART stimulant, DIGITALIS, is a saponin from the Purple foxglove. The sapogenin, diosegnin, from Mexican yams (*Dioscorea*) is a cheap precursor to the STEROID drugs.

SAPPHIRE, all GEM varieties of CORUNDUM except those which, being red, are called RUBY; blue sapphires are best-known, but most other colors are found. The best sapphires come from Kashmir, Burma, Thailand, Sri Lanka and Australia. Synthetic stones, made by flame-fusion, are used for jewel bearings, phonograph styluses, etc.

SAPROPHYTES, plants that absorb food from dead or decaying organic matter instead of carrying out PHOTOSYNTHESIS. The group includes most FUNGI, e.g., the YEASTS that ferment sugar and the MOLDS that decay fruit.

SARCODINA. See PROTOZOA.

SARCOMA, a form of TUMOR derived from connective TISSUE, usually of mesodermal origin in EMBRYOLOGY. It is often distinguished from CANCER as its behavior and natural history may differ, although

Ferrocene (dicyclopentadienyl iron), one of the best known of the sandwich compounds. The bonds represented in red are of hybrid character.

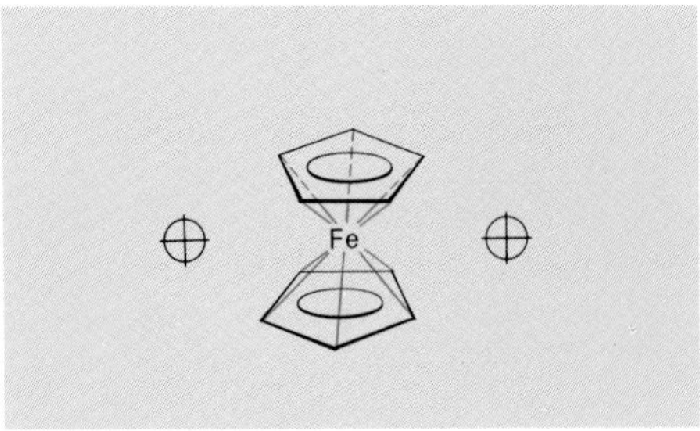

it is still a malignant tumor. It commonly arises from BONE (osteosarcoma), fibrous tissue (fibrosarcoma) or CARTILAGE (chondrosarcoma). Excision is required, though RADIATION THERAPY may be helpful.

SARD, a semiprecious stone, a brown variety of CHALCEDONY, closely related to CARNELIAN.

SARDONYX. See ONYX.

SARGOSSO SEA, oval area of the N Atlantic, of special interest as the spawning ground of American eels, many of whose offspring drift across the Atlantic to form the European eel population. Bounded E by the Canaries Current, S by the N Equatorial Current, W and N by the GULF STREAM, it contains large masses of *Sargassum* weed.

SARSAPARILLA, substance found in the roots of certain vines of the genus *Smilax*, native to Middle and South America. It is used for flavoring carbonated drinks such as root beer, and to disguise the taste of some medicines.

SARSENS, large blocks of SANDSTONE found on the chalk downs of S England, used to construct STONEHENGE and other ancient monuments.

SATELLITE, in astronomy, a celestial object which revolves with or around a larger celestial object. In our SOLAR SYSTEM this includes PLANETS, COMETS, ASTEROIDS and meteoroids (see METEOR), as well as the moons of the planets; although the term is usually restricted to this last sense. Of the 32 known moons, the largest is Callisto (JUPITER IV), the smallest PHOBOS. The MOON is the largest known satellite relative to its parent planet; indeed, the earth-moon system is often considered a double planet.

SATELLITES, Artificial, man-made objects placed in orbit as SATELLITES. First seriously proposed in the 1920s, they were impracticable until large enough ROCKETS were developed. The first artificial satellite, Sputnik 1, was launched by the USSR in Oct. 1957, and was soon followed by a host of others, mainly from the USSR and the US, but also from the UK, France, Canada, West Germany, Italy, Japan and China. They have many scientific, technological and military uses. Astronomical observations (notably X-RAY ASTRONOMY) can be made unobscured by the atmosphere. Studies can be made of the RADIATION and electromagnetic and gravitational fields in which the EARTH is bathed, and of the upper ATMOSPHERE. Experiments have been made on the functioning of animals and plants in space (with zero gravity and increased radiation). Artificial satellites are also used for reconnaissance, surveying, meteorological observation, as navigation aids (position references and signal relays), and in communications for relaying television and radio signals. Manned satellites, especially the historic Soyuz and Mercury series, have paved the way for **space stations**, which have provided opportunities for diverse research and for developing docking techniques; the USSR Salyut and US Skylab projects are notable. The basic requirements for satellite launching are determined by celestial mechanics. Launching at various velocities between that required for zero altitude and the escape velocity produces an elliptical orbit lying on a conic surface determined by the latitude and time of launch. To reach any other orbit requires considerable extra energy expenditure. Artificial satellites require: a power supply—SOLAR CELLS, BATTERIES, FUEL CELLS or nuclear devices; scientific INSTRUMENTS; a communications system to return encoded data to earth; and instruments and auxiliary rockets to monitor and correct the satellite's position. Most have COMPUTERS for control and data processing, thus reducing remote control to the minimum.

SATURATION, term applied in many different fields to a state in which further increase in a variable above a critical value produces no increase in a resultant effect. A saturated SOLUTION is one which will dissolve no more solute, an EQUILIBRIUM having been reached; raising the temperature usually allows more to dissolve: cooling a saturated solution may produce **supersaturation**, a metastable state, in which sudden crystallization depositing the excess solute occurs if a seed crystal is added. In organic chemistry, a saturated molecule has no double or triple bonds and so does not undergo addition reactions.

SATURN, the second largest planet in the SOLAR SYSTEM and the sixth from the sun. Until the discovery of URANUS (1781) Saturn was the outermost planet known. It orbits the sun in 29.46 years at a mean distance of 9.54AU. Saturn does not rotate uniformly: its period of rotation at the equator is 10.23h, rather longer toward the poles. This rapid rotation causes a noticeable equatorial bulge: the equatorial diameter is 120.9Mm, the polar diameter 108.1Mm. Saturn has the lowest density of any planet in the Solar System, less than that of water, and may contain over 60% hydrogen by mass. Its total mass is about 95 times that of the earth. Saturn has ten moons; the largest, Titan, about the same size as MERCURY, is known to have an atmosphere. The most striking feature of Saturn is its ring system: three or more rings thought to be composed of countless tiny particles of ice. The rings are about 16km thick and the outermost has an external diameter of about 280Gm.

SAURISCHIA, or lizard-hipped dinosaurs, group of the DINOSAURIA having pelvic girdles typical of lizards, with three prongs to each side. They included the two-legged carnivorous theropods, such as *Tyrannosaurus* and *Allosaurus*, with enormous skulls and large teeth; and the four-legged herbivorous sauropods, such as *Brontosaurus* and *Diplodocus*, with very small heads and long necks and tails.

SAUSSURE, Ferdinand de (1857–1913), Swiss linguist whose contributions to structural linguistics (e.g., the idea that the structure of a language may be studied both as it changes with time and as it is in the present) have had a formative influence on 20th-century studies of GRAMMAR.

SAVANNA, tropical GRASSLANDS of South America and particularly Africa, lying between equatorial FORESTS and dry DESERTS.

SAVERY, Thomas (c1650–1715), British inventor of an early form of STEAM ENGINE (patented 1698), used for pumping water. His patent covered NEWCOMEN's later invention (c1712), and for this reason the two entered partnership for the development of Newcomen's engine.

SAW, cutting tool consisting of a flat blade or circular disk, having on its edge a row of sharp teeth of various designs, usually set alternately. Excepting jagged stone knives, the first saws (copper and bronze) were used in Egypt c4000 BC, but only with the use of steel did they become efficient. Hand saws include the crosscut saw for cutting wood to length, the backsaw

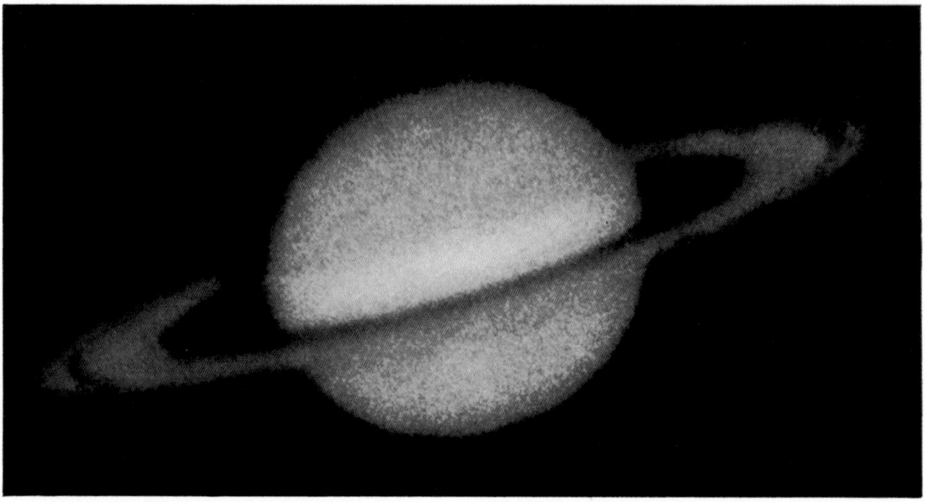

The planet Saturn photographed using the 156cm telescope at Catalina Observatory, Tuscon, Arizona.

for joints, the coping saw for shaping, and the hacksaw for cutting metal. Power saws include circular saws, band saws (with a flexible endless steel band running over pulleys) and chain saws.

SCABIES, infectious SKIN disease caused by a mite which burrows under the skin, often of hands or feet; it causes an intensely itchy skin condition which is partly due to ALLERGY to the mite. Rate of infection has a cyclical pattern. Treatment is with special ointments and should include contacts.

SCALAR, a quantity that can be described fully in terms of its magnitude, as contrasted with a VECTOR, which has both magnitude and direction.

SCALDS. See BURNS AND SCALDS.

SCALE, a flat, rigid or flexible plate forming a protective covering on some animals, notably fish. In plants, the term refers to a small, non-green leaf that protects a young bud.

SCANDIUM (Sc), silvery-white RARE-EARTH metal in Group IIIB of the PERIODIC TABLE; a TRANSITION ELEMENT. It is widely distributed in low concentrations, and is extracted from thortveitite or as a by-product of URANIUM extraction. Scandium forms trivalent ionic compounds and stable LIGAND complexes. The refractory scandium oxide (Sc_2O_3) is used in ceramics and as a catalyst; dilute scandium sulfate solution is used to improve germination of plant seeds. AW 45.0, mp 1541°C, bp 2831°C, sg 2.989 (25°C).

SCAR, area of fibrous tissue which forms a bridge between areas of normal tissue as the end result of wound healing. The fibrous tissue lacks the normal properties of the healed tissue (e.g., it does not tan). The size of a scar depends on the closeness of the wound edges during healing; excess stretching forces and infection widen scars.

SCARLET FEVER, or scarletina, INFECTIOUS DISEASE caused by certain strains of *Streptococcus*. It is common in children and causes sore throat with TONSILLITIS, a characteristic SKIN rash and mild systemic symptoms. PENICILLIN and symptomatic treatment is required. Scarlet fever occurs in EPIDEMICS; a few are followed by RHEUMATIC FEVER or NEPHRITIS.

SCARP, or **escarpment**, steep slope or inland cliff, most often that of a CUESTA. The term is sometimes applied to similar slopes resulting from EROSION or faulting (see FAULT).

SCATTERING, the deflection of moving particles and energy waves (such as ELECTRONS, PHOTONS or SOUND waves) through collisions with other particles. RAYLEIGH scattering of sunlight gives rise both to the blue color of the ATMOSPHERE when the sun is high in the sky and to the reds and yellows of the setting sun because the blue light is scattered more strongly than the red.

SCEPTICISM. See SKEPTICISM.

SCHALLY, Andrew Victor (1926–), Polish-born US physiologist who shared the 1977 Nobel Prize for Physiology or Medicine with R. S. YALLOW and R. GUILLEMIN for his work identifying compounds stimulating peptide HORMONE release from the brain.

SCHEELE, Karl (or **Carl**) **Wilhelm** (1742–1786), Swedish chemist who discovered OXYGEN (c1773), perhaps a year before Joseph PRIESTLEY's similar discovery. He also discovered CHLORINE (1774).

SCHEELITE, calcium tungstate ($CaWO_4$), a major ore of TUNGSTEN. It is of widespread occurrence, and forms tetragonal CRYSTALS of various colors.

SCHIAPARELLI, Giovanni Virginio (1835–1910), Italian astronomer who discovered the ASTEROID Hesperia (1861) and showed that METEOR showers represent the remnants of COMETS. He is best known for terming the surface markings of MARS *canali* (channels). This was wrongly translated as "canals," implying Martian builders: the resulting controversy lasted for nearly a century.

SCHICK, Béla (1877–1967), Hungarian-born US pediatrician who developed the Schick test to determine immunity to DIPHTHERIA (1913).

SCHIST, common group of METAMORPHIC ROCKS which have acquired a high degree of schistosity, i.e., the tendency to split into layers along perfect CLEAVAGE planes. Their major constituents are flaky or platy minerals, especially MICA, TALC, AMPHIBOLES and CHLORITE.

SCHISTOSOMIASIS, or **bilharzia**, a PARASITIC DISEASE caused by *Schistosoma* species of FLUKES. Infection is usually acquired by bathing in infected water, the different species of parasite causing different manifestations. Infection of the BLADDER causes constriction, calcification and secondary infection, and can predispose to bladder CANCER. Another form leads to GASTROINTESTINAL TRACT disease with LIVER involvement. ANTIMONY compounds are often effective in treatment.

SCHIZOMYCETES, a class of the PLANT KINGDOM that includes the BACTERIA. (See also MONERA; SCHIZOPHYTA.)

SCHIZOPHRENIA, formerly called **dementia praecox**, a number of PSYCHOSES characterized by confusion of IDENTITY, HALLUCINATIONS, AUTISM, delusion and illogical thought. The three main types of schizophrenia are CATATONIA; **paranoid schizophrenia**, which is similar to PARANOIA except that the intellect deteriorates, and **hebephrenia**, which is characterized by withdrawal from reality, bizarre or foolish behavior, delusions, hallucinations and self-neglect.

SCHIZOPHYCEAE, a class of the PLANT KINGDOM that includes the BLUE-GREEN ALGAE. (See also MONERA; SCHIZOPHYTA.)

SCHIZOPHYTA, a division of the PLANT KINGDOM that includes the BACTERIA and BLUE-GREEN ALGAE. (See also MONERA.)

SCHLICK, Moritz (1882–1936), German philosopher regarded as the founder of the Vienna Circle, an influential school of LOGICAL POSITIVISM.

SCHLIEMANN, Heinrich (1822–1890), German archaeologist, best known for his discoveries of Troy (1871–90) and Mycenae (1876–78). (See AEGEAN CIVILIZATION.)

SCHMIDT, Bernhard Voldemar (1879–1935), Estonian-born German optician best known for developing the Schmidt TELESCOPE, today one of the most-used tools of ASTROPHOTOGRAPHY. Its special advantage is that it avoids LENS coma.

SCHRIEFFER, John Robert (1931–   ), US physicist awarded with Leon COOPER and John BARDEEN the 1972 Nobel Prize for Physics for their work on SUPERCONDUCTIVITY.

SCHRÖDINGER, Erwin (1887–1961), Austrian-born Irish physicist and philosopher of science who shared with DIRAC the 1933 Nobel Prize for Physics for his elucidation of the **Schrödinger wave equation**, which is of fundamental importance in studies of QUANTUM MECHANICS (1926). It was later shown that his WAVE MECHANICS were equivalent to the matrix mechanics of HEISENBERG.

SCHWANN, Theodor (1810–1882), German biologist who proposed the CELL as the basic unit of animal, as well as of plant, structure (1839), thus laying the foundations of HISTOLOGY. He also discovered the ENZYME pepsin (1836).

SCHWARTZSCHILD RADIUS, the radius of the SPHERE into which a given body of known MASS must be compressed in order to become a BLACK HOLE.

SCHWINGER, Julian Seymour (1918–   ), US physicist who shared with FEYNMANN and TOMONAGA the 1965 Nobel Prize for Physics for his independent work in formulating the theory of quantum electrodynamics.

SCIATICA, a characteristic pain in the distribution of the sciatic nerve in the LEG caused by compression or irritation of the nerve. The pain may resemble an electric shock and be associated with numbness and tingling in the skin area served by the nerve. One of the commonest causes is a SLIPPED DISK in the lower lumbar spine.

SCIENTIFIC METHOD. Science (from Latin *scientia*, knowledge) is too diverse an undertaking to be constrained to follow any single method. Yet from the time of Lord BACON, well into the 20th century, the myth has persisted that true science follows a particular method—Bacon's celebrated "inductive method." This allegedly involved collecting a vast number of individual facts about a phenomenon, and then working out what general statements fitted those facts. After the 17th century nobody attempted to follow that program. In the 19th century, philosophers of science came to recognize the possible existence of the "hypothetico-deductive method." According to this model, the scientist studied the phenomena, dreamed up a hypothetical explanation, deduced some additional consequences of his explanation, and then devised experiments to see if these consequences were reflected in nature. If they were, he considered his theory (hypothesis) confirmed. But K. POPPER pointed to the logical fallacy in this last step—the theory had not been confirmed, but merely not falsified; it could, however, be worked with provisionally, so long as new tests did not discredit it. Philosophers of science now recognize that they cannot justly generalize about the psychology of scientific discovery; their role must be confined to the criticism of theories once they have been devised. Historians of science, meanwhile, have pointed to the importance in scientific discovery of "external factors" such as the contemporary intellectual context and the structures of the institutions of science. Once distinct terms—"theory," "model," "hypothesis," "explanation," "description," and "law"—are all now seen to represent different ways of looking at the same thing—the units in what constitutes scientific knowledge at any given time. Indeed there is still no general understanding of how scientists become dissatisfied with a once deeply-entrenched theory and come to replace it with what, for the moment, seems a better version.

SCINTILLATION COUNTER, instrument for detecting ionizing radiations. A brief localized light flash is produced in a PHOSPHOR when ionizing radiation such as X-rays or protons is incident on it. In early counters, the flashes were counted directly using a microscope, but now they are converted into electric impulses by a photomultiplier and counted electronically.

SCOLIOSIS, a curvature of the spine to one side, with twisting. It occurs as a congenital defect or may be secondary to spinal diseases including neurofibromatosis. Severe scoliosis, often associated with kyphosis, causes HUNCHBACK deformity, loss of height and may restrict CARDIAC or LUNG function.

SCOPOLAMINE, or **hyoscine**, anticholinergic drug related to ATROPINE and used widely in premedication for ANESTHESIA. It tends to be a central-nervous-system DEPRESSANT but otherwise resembles atropine in reducing secretions, GASTROINTESTINAL TRACT activity and vagus effects on the HEART (causing increased PULSE rate) and in dilating the pupils of the EYES. It may cause confusion in the elderly. Other common uses include treatment of MOTION SICKNESS, use as a mild SEDATIVE and in PARKINSON'S DISEASE.

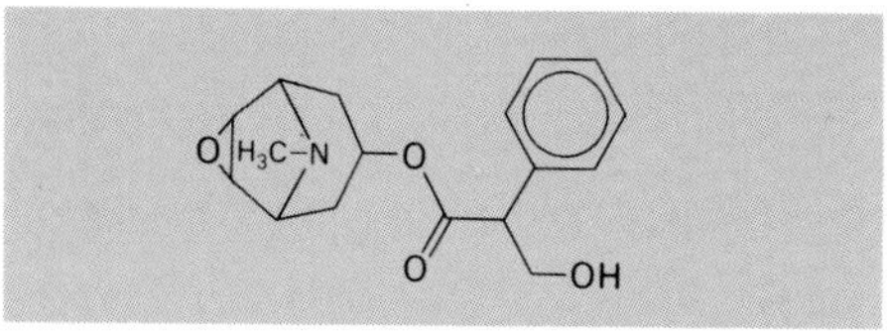

The structure of scopolamine (hyoscine).

SCORIA, a vesicular form of LAVA, the vesicles having been formed by gases escaping from the lava while it was still hot.

SCORPIO (the Scorpion), a medium-sized constellation on the ECLIPTIC; the eighth sign of the ZODIAC. Scorpio contains the bright star ANTARES.

SCREE. See TALUS.

SCREW. See ARCHIMEDES; BOLTS AND SCREWS; MACHINE; PROPELLER.

SCROFULA, TUBERCULOSIS of the LYMPH nodes of the neck, usually acquired by drinking MILK infected with bovine or atypical mycobacteria, and involving enlargement of the nodes with formation of a cold ABSCESS. The eradication of tuberculosis in cattle has substantially reduced the incidence. Treatment includes antituberculous CHEMOTHERAPY. It used to be called the **King's Evil** as the royal touch was believed to be curative.

SCRUB TYPHUS, or **Tsutsugamushi disease**, a disease caused by RICKETTSIA carried by mites, and leading to ulceration at the site of INOCULATION, followed by FEVER, headache, lymph node enlargement and generalized rash. COUGH and chest X-RAY abnormalities are common. ENCEPHALITIS and MYOCARDITIS may occur with fatal outcome. It occurs mainly in the Far East and Australia and is generally seen in people who work on scrubland. TETRACYCLINES eradicate the infection.

SCUBA. See AQUALUNG.

SCURVY, or VITAMIN C deficiency, involving disease of the SKIN and mucous membranes, poor healing and ANEMIA; in infancy BONE growth is also impaired. It may develop over a few months of low dietary vitamin C, beginning with malaise and weakness. Skin bleeding around HAIR follicles is characteristic, as are swollen, bleeding gums. Treatment and prevention consist of adequate dietary vitamin C.

SEA. See OCEANS.

SEABORG, Glenn Theodore (1912–), US physicist who shared the 1951 Nobel Prize for Physics with E. W. MCMILLAN for his work in discovering several ACTINIDES (see TRANSURANIUM ELEMENTS): in 1944 AMERICIUM and CURIUM, and in 1949 BERKELIUM and CALIFORNIUM. Later discoveries were EINSTEINIUM (1952), FERMIUM (1953), MENDELEVIUM (1955) and NOBELIUM (1957).

SEA-FLOOR SPREADING, key phenomenon supporting the theory of PLATE TECTONICS. Along midocean ridges (see OCEANS) material emerges from the EARTH's mantle to form new oceanic crust. This material, primarily BASALT, spreads out to either side of the ridges at a rate of the order of 10–50mm/yr. New laid down basalt is able to "fossilize" the prevailing geomagnetism (see PALEOMAGNETISM): the main evidence for sea-floor spreading comes from the symmetric pattern of alternately magnetized strips of basalt on either side of the ridges.

SEASICKNESS. See MOTION SICKNESS.

SEASONS, divisions of the year, characterized by cyclical changes in the predominant weather pattern. In the temperate zones there are four seasons: spring, summer, autumn (fall) and winter. These result from the constant inclination of the earth's polar axis ($66\frac{1}{2}°$ from the ECLIPTIC) as the earth orbits the sun: during summer in the N Hemisphere the N Pole is tilted toward the sun, in winter—when the solar radiation strikes the hemisphere more obliquely—away from the sun. The summer and winter SOLSTICES (about June 21 and Dec. 22), popularly known as midsummer and midwinter, strictly speaking mark the beginnings of summer and winter, respectively. Thus spring begins on the day of the vernal EQUINOX (about Mar. 21) and autumn at the autumnal equinox (about Sept. 23).

SEBACEOUS GLANDS, small GLANDS in the SKIN which secrete *sebum*, a fatty substance that acts as a protective and water repellant layer on skin and allows the epidermis to retain its suppleness. Sebum secretion is fairly constant but varies from individual to individual. Obstructed sebaceous glands become BLACKHEADS which are the basis for ACNE.

SEBORRHEA. See DANDRUFF.

SECANT. See CIRCLE; TRIGONOMETRY.

SECOND (s), in SI UNITS, the base unit of TIME, defined as the duration of 9 192 631 770 periods of the radiation corresponding to the transition between the two hyperfine levels of the ground state of the CESIUM-133 atom.

SECRETIN, a HORMONE of the GASTROINTESTINAL TRACT secreted by cells in the duodenum EPITHELIUM in response to the presence of food and increasing the secretion of pancreas ENZYMES and BILE.

SECRETION. See GLANDS.

SEDATIVES, DRUGS that reduce ANXIETY and induce relaxation without causing SLEEP; many are also hypnotics, drugs that in adequate doses may induce sleep. BARBITURATES were among the earlier drugs used in sedation, but they have fallen into disfavor because of addiction, side-effects, dangers of overdosage and the availability of safer alternatives. Benzodiazepines (e.g., Valium, Librium) are now the most often used and have proved safe and effective.

SEDIMENTARY ROCKS, one of the three main ROCK types of the earth's crust. They consist of weathered (see EROSION) particles of igneous, metamorphic or even sedimentary rock transported, usually by water, and deposited in distinct strata. They may also be of organic origin, as in COAL, or of volcanic origin, as are PYROCLASTIC ROCKS. Most common are SHALE, SANDSTONE and LIMESTONE. Sedimentary rocks frequently contain FOSSILS, as well as most of the earth's MINERAL resources.

SEDIMENTATION, the processes whereby particles of solid material are transported and deposited elsewhere. The particles are the product of weathering (see EROSION); the transporting agent may be wind, water, GLACIERS or an AVALANCHE or landslide. Products of sedimentation thus include

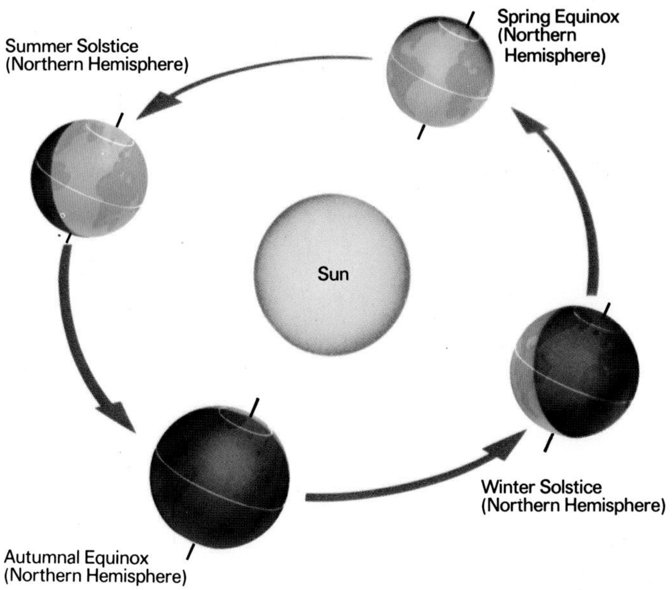

How the inclination of the earth's axis from the normal to the plane of the ecliptic gives rise to the phenomenon of the seasons.

DRIFT, OOZES, TALUS and SEDIMENTARY ROCK.

SEEBECK, Thomas Johann (1770–1831), German physicist who discovered but could not explain the Seebeck effect (1821). This was the first thermoelectric effect to be discovered (see THERMOCOUPLE).

SEED, the mature reproductive body of ANGIOSPERMS and GYMNOSPERMS. It also represents a resting stage which enables the PLANTS to survive through unfavorable conditions. The GERMINATION period varies widely from plant to plant. Seeds develop from the fertilized ovule. Each seed is covered with a tough coat called a testa and it contains a young plant or embryo. In most seeds three main regions of embryo can be recognized. A radicle, which gives rise to the root, a plumule which forms the shoot and one or two seed leaves or COTYLEDONS which may or may not be taken above ground during germination. Plants that produce one seed leaf are called MONOCOTYLEDONS and those that produce two, DICOTYLEDONS. The seed also contains enough stored food (often in the cotyledons) to support embryo growth during and after germination. It is this stored food which is of value to man. Flowering plants produce their seeds inside a FRUIT, but the seeds of conifers lie naked on the scales of the cone. Distribution of seeds is usually by wind, animals or water and the form of seeds is often adapted to a specific means of dispersal. (See also POLLINATION; REPRODUCTION.)

SEGMENT. See CIRCLE.

SEGMENTATION, the serial repetition of structures or organs along the long axis of the animal body. Segmentation is characteristic of many groups, notably the ANNELIDA.

SEGRÈ, Emilio Gino (1950–), Italian-born US nuclear physicist who shared with O. CHAMBERLAIN the 1959 Nobel Prize for Physics for their discovery (1955) of the antiproton (see ANTIMATTER). Earlier he had discovered TECHNETIUM, the first artificially produced ELEMENT (1937).

SEGREGATION, the process whereby a pair of GENES on a particular CHROMOSOME separate singly into different GAMETES during MEIOSIS. On fertilization either the original pair is reconstituted or a new pair is formed in the offspring. Segregation depends on the relative stability of genes.

SEICHE, a standing wave system (see WAVE MOTION) occurring in a lake or bay, set up by a disturbance such as wind, ocean swell, an earth tremor or a sudden change in air pressure. The oscillation period depends on the size and shape of the basin.

SEISMOGRAPH, instrument used to record seismic waves caused by EARTHQUAKES, nuclear explosions,

A simplified seismograph. A horizontal bar (1) is pivoted at one end (2) and suspended from a spring (4). A damping mechanism is incorporated (3). As the ground moves, the mass (5) ensures that the bar remains approximately stationary: thus movements are recorded by the pen (6) which draws a seismogram (7) on a chart moving between rotating drums (8).

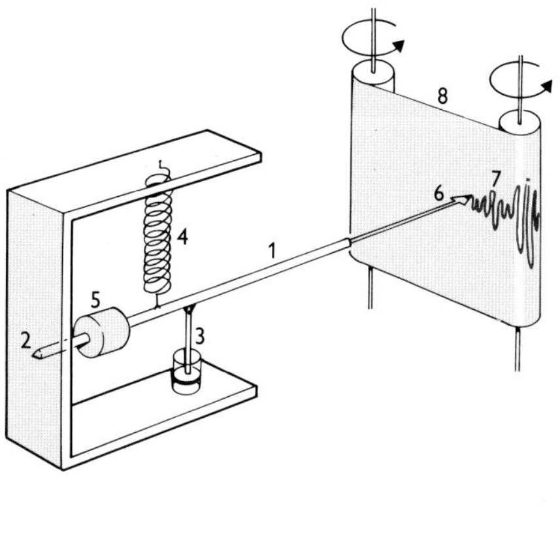

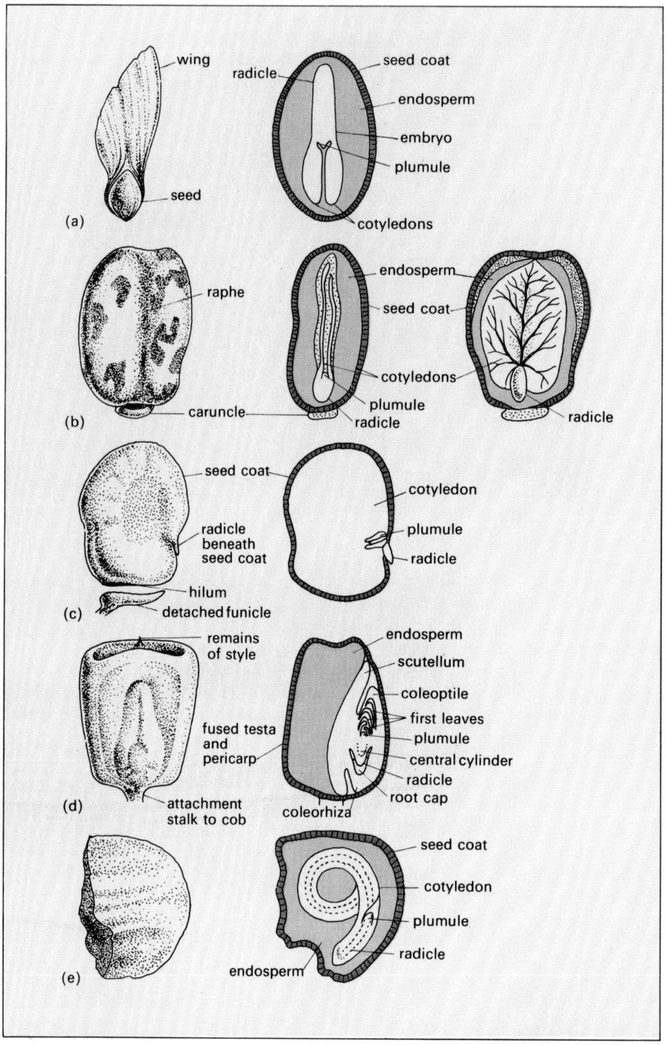

Representative types of seed: side and sectioned views: (a) Winged endospermic seed of pine (*Pinus*); (b) endospermic castor oil seed (*Ricinus communis*); (c) non-endospermic broad bean seed (*Vicia faba*); (d) endospermic "seed" of corn (maize; *Zea mays*) (botanically this is really a fruit but the fruit wall is thin and fused to the seed coat); (e) endospermic seed of the onion (*Allium cepa*).

etc.: the record it produces is a **seismogram**. The simplest seismograph has a horizontal bar, pivoted at one end and with a recording pen at the other. The bar, supported by a spring, bears a heavy weight. As the ground moves, the bar remains roughly stationary owing to the INERTIA of the weight, while the rest of the equipment moves. The pen traces the vibrations on a moving belt of paper. Seismographs are used in PROSPECTING.

SEISMOLOGY, the study of EARTHQUAKE phenomena.

SELACHII, a subclass of the class CHONDRICHTHYES that contains the sharks, rays and skates.

SELECTION. See NATURAL SELECTION; SEXUAL SELECTION.

SELECTIVE PRESSURE, a measure of the forces acting on a population of organisms that stimulate evolutionary change. Selective pressure is proportional to the amount of genetic change produced. (See also EVOLUTION, NATURAL SELECTION.)

SELENITE, well-developed crystals of GYPSUM.

SELENIUM (Se), metalloid in Group VIA of the PERIODIC TABLE; it occurs as rare selenides with heavy-metal SULFIDES, and is obtained as a by-product of copper refining or the LEAD-CHAMBER PROCESS. Selenium has three allotropes (see ALLOTROPY), the most stable being the gray, metallic form. Its chemistry is analogous to that of SULFUR. It is used to make PHOTOELECTRIC CELLS, SOLAR CELLS, RECTIFIERS, in XEROGRAPHY and as a SEMICONDUCTOR; also to make ruby glass and to vulcanize rubber. AW 79.0, mp 217°C (gray), bp 685°C (gray), sg 4.79 (gray).

SELYE, Hans (1907–), Austrian-born Canadian physician best known for his work on the physiological effects of environmental stress, which he suggested might cause certain diseases (see PSYCHOSOMATIC ILLNESS).

SEMANTICS, semasiology or **semology**, the study of meaning, concerned both with the relationship of words and symbols to the ideas or objects that they represent, and with tracing the histories of meanings and changes that have taken place in them. Semantics is thus a branch both of LINGUISTICS and of LOGIC. **General semantics**, propounded primarily by Alfred KORZYBSKI, holds that habits of thought have lagged behind the language and logic of science: it attacks such Aristotelian logical proposals as that nothing can be both not-*x* and *x*, maintaining that these are simplifications no longer valid.

SEMAPHORE, system of visual signalling using movable arms, flags or lights to represent letters and numbers. The first such system was introduced by Claude Chappe (1763–1805): it used towers 8 to 16km apart. Semaphore is still used for signalling between ships and on some railroads.

SEMEN, fluid secreted by the TESTES containing SPERM. (See also REPRODUCTION.)

SEMICIRCULAR CANALS. See EAR.

SEMICONDUCTOR, a material whose electrical CONDUCTIVITY is intermediate between that of an insulator and conductor at room TEMPERATURE and increases with rising temperature and impurity concentration. Typical **intrinsic semiconductors** are single crystals of GERMANIUM or SILICON. At low temperatures their valence electron ENERGY LEVELS are filled and no ELECTRONS are free to conduct ELECTRICITY, but with increasing temperature, some electrons gain enough ENERGY to jump into the empty conduction band, leaving a **hole** behind in the valence band. Thus there are equal numbers of moving electrons and holes available for carrying electric current. Practical **extrinsic semiconductors** are made by adding a chosen concentration of a particular type of impurity atom to an intrinsic semiconductor (a process known as doping). If the impurity atom has more valence electrons than the semiconductor atom, it is known as a donor and provides spare conduction electrons, creating an **n-type semiconductor**. If the impurity atom has fewer valence electrons, it captures them from the other atoms and is known as an acceptor, leaving behind holes which act as moving positive charge carriers and enhance the conductivity of the **p-type semiconductor** that is formed. An n-and p-type semiconductor junction acts as a RECTIFIER; when it is forward biased, holes cross the junction to the negative end and electrons to the positive end, and current flows through it. If the voltage connections are reversed, the carriers will not cross the junction and no current flows. Semiconductor devices, such as the TRANSISTOR, based on the p-n junction have revolutionized ELECTRONICS since the late 1940s.

SEMIMETAL. See METALLOID.

SEMIOLOGY, or **semiotics**, the study of signs (including LANGUAGE), their uses, and the way in which they are used. Its branches are pragmatics (dealing with the relation between the signs and those using them), syntactics (the relation between different words and symbols) and SEMANTICS.

SEMIPERMEABLE MEMBRANE, a MEMBRANE that allows certain substances, usually small molecules or ions, to pass through its pores, but which keeps back others. Examples are CELL membranes, PARCHMENT, cellophane, and copper(II) hexacyanoferrate(II). (See DIALYSIS; OSMOSIS.)

SEMMELWEISS, Ignaz Philipp (1818–1865), Hungarian obstetrician who, through his discovery that PUERPERAL FEVER was transmitted by failure of obstetricians to thoroughly clean their hands between performing autopsies of mothers who had died of the disease and making examinations of living mothers, first practised ASEPSIS.

SEMYONOV, Nikolai Nikolaevich (1896–), Russian physical chemist who shared with HINSHELWOOD the 1956 Nobel Prize for Chemistry for his work on chemical KINETICS, especially concerning chemical chain and branched-chain reactions.

SENILITY, the state of OLD AGE, usually referring to the general mental and physical deterioration, often (but not always) seen in the elderly. Failure of recent MEMORY, dwelling on the past, episodic confusion and difficulty in absorbing new information are common. The degenerative changes in SKIN, BONE and connective TISSUE lead to the altered physical appearance characteristic of the elderly.

SENSATIONALSIM, philosophical theory which in its most extreme forms (see, for example, CONDILLAC) holds that knowledge is composed wholly of sensations, the mind being regarded as a passive *tabula rasa* (clean slate).

SENSES, the media through which stimuli in the environment of an organism act on it (external senses); also, the internal senses which report on the internal state of the organism (through THIRST AND HUNGER; PAIN, etc.). The organs of sense, the eye, ear, skin etc., all contain specialized cells and nerve endings which communicate with centers in the NERVOUS SYSTEM. Sense organs may be stimulated by pressure (in TOUCH, hearing and balance—see EAR), chemical stimulation (SMELL; TASTE), or electromagnetic radiation (VISION; heat sensors).

SENSORY DEPRIVATION, condition of perceptual isolation—which may result in HALLUCINATIONS, thought or emotional disorders or spatiotemporal disorientation—experienced by people confined in highly unstimulating environments, a subject of much recent research.

SEPAL, in plants, one of the outer rings of green leaf-like organs (together called the calyx) which surrounds the petals of the FLOWER. In some plants the calyx is petal-like and colored.

SEPIA, the "ink" of squids and octopuses which can be expelled into the sea to distract potential predators while the animal escapes. The word is taken up in the generic name of the European cuttlefish. Sepia has been a popular drawing medium for 400 years.

SEPIOLITE. See MEERSCHAUM.

SEPSIS, the destructive invasion of TISSUES by BACTERIA. (See also ASEPSIS; SEPTICEMIA.)

SEPTICEMIA, circulation of infective BACTERIA and the white BLOOD cells responding to them in the blood. Bacteria may transiently enter the blood normally but these are removed by the RETICULOENDOTHELIAL SYSTEM. If this system fails and bacteria continue to circulate, their products and those of the white cells initiate a series of reactions that lead to SHOCK, with warm extremities, FEVER or hypothermia. Septic EMBOLISM may occur causing widespread ABSCESSES. GRAM'S STAIN-negative bacteria (usually from urinary or GASTROINTESTINAL TRACT) and STAPHYLOCOCCUS cause severe septicemia. Treatment includes ANTIBIOTICS and resuscitative measures for shock.

SEPTIC TANK, large tank used for SEWAGE disposal

from single residences not linked to the public sewer. The liquid part of the sewage drains off to a cesspool or distribution network in sandy soil. The sludge collects at the bottom of the tank and is largely decomposed by bacteria; it is pumped out every few years.

SEQUENCE, an ordered set of numbers (e.g., 2, 4, 6, 8, ... $2n$, ...) linked by a common mathematical formula. Most sequences are infinite, that is, they have an infinite number of terms, but there are some finite sequences, which end after a finite number of terms: finite sequences are generally of little importance. An infinite sequence may, however, be bounded. The terms of the sequence of typical term $1/n$ all lie between 0 and 1. As the successive terms tend toward 0, this is said to be the LIMIT of the sequence. A recursive series is one in which the value of each term depends on the value of the preceding terms.

SERE. See SUCCESSION.

SERIES, the sum of the terms of a SEQUENCE. Most series are "infinite"—containing an infinite number of terms. Those which tend to a LIMIT are termed "converging," others being "diverging."

SEROTONIN, aromatic amine found in serum and other tissues that acts on BLOOD vessels; it is also involved in PERISTALSIS and in the central NERVOUS SYSTEM as a transmitter at SYNAPSES. It may be involved in ANAPHYLAXIS in some species; its action in increasing CAPILLARY permeability indicates its role in the reactions producing INFLAMMATION.

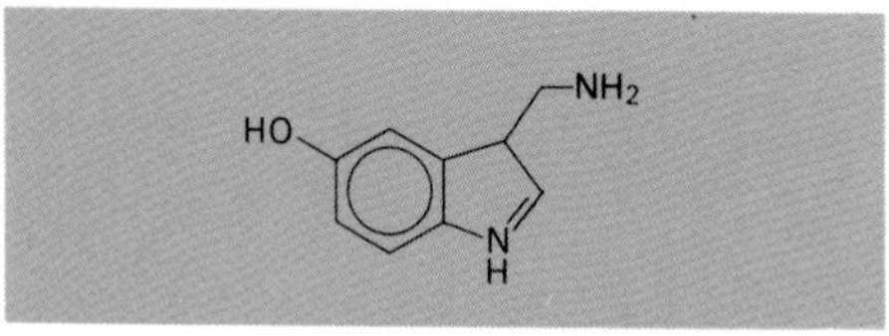

The structure of serotonin (5-hydroxytryptamine).

SERPENTINE, common magnesium SILICATE mineral, $Mg_3Si_2O_5(OH)_4$, whose structure resembles that of KAOLINITE. There are three varieties: CHRYSOTILE, antigorite and lizardite. Gray, green or yellow with an attractive texture and easily polished, serpentine is used as an ornamental stone.

SERUM, the clear yellowish fluid that separates from BLOOD, LYMPH and other body fluids when they clot. It contains water, PROTEINS, fat, minerals, HORMONES and UREA. **Serum therapy** involves injecting (horse or human) serum containing ANTIBODIES (GLOBULINS), which can destroy particular pathogens. Occasionally injected serum gives rise to an allergic reaction known as serum sickness; a second injection of the same serum may induce ANAPHYLAXIS.

SERVETUS, Michael (1511–1553), or **Miguel Serveto**, Spanish biologist and theologian. In *Christianity Restored* (1553) he mentioned in passing his discovery of the pulmonary circulation (see BLOOD CIRCULATION). For heretical views expressed in this book he was denounced by the Calvinists to the Catholic Inquisition: escaping, he foolishly visited Geneva where he was seized by the Protestants, tried for heresy and burned alive.

SERVOMECHANISM, an automatic control device (see MECHANIZATION AND AUTOMATION) which controls the position, velocity or acceleration of a high-power output device by means of a command signal from a low-power reference device. By FEEDBACK, the error between the actual output state and the state commanded is measured, amplified and made to drive a servomotor which corrects the output. The drive may be electrical, hydraulic or pneumatic.

SESSILE ANIMALS, animals that spend part of their lives attached to the ground or sea bed. All sessile animals have, at some stage of their life-histories, a mobile phase facilitating dispersal.

SETSQUARE, in classical GEOMETRY, an instrument used to draw right ANGLES and, with a straight edge particularly, PARALLEL LINES.

SET THEORY. A set is a collection of objects or quantities, symbolized by a capital letter. Thus

$$S = \{2, 4, 6, 8\}$$

means that S is the set of even numbers less than 10. A member of a set is called an element: symbolically, $2 \in S$ means that 2 is an element of S; $3 \notin S$ means that 3 is not an element of S. An ordered set is equivalent to a SEQUENCE. A set may be infinite, finite or empty: the set of all even numbers is infinite, that of all those less than 10 is finite, and that of all those less than 1 is empty. This empty or **null set** is symbolized ϕ or $\{\ \}$; and should not be confused with $\{0\}$, which is a set with one member, ZERO. If there is a one-to-one correspondence between the elements of two sets, then they are termed equivalent, and if the sets have identical elements they are equal:

$$S_1 = \{a, b, c, d\}$$
$$S_2 = \{e, f, g, h\}$$
$$S_3 = \{d, c, b, a\}$$

shows three equivalent sets; moreover $S_1 = S_3$. Two equivalent sets are written $S_1 \leftrightarrow S_2$.

If some elements of one set are also elements of another, then those elements are called the intersection of the two sets, symbolized $S_1 \cap S_2$. The set of all elements that are members of at least one of the two sets is their union, written $S_1 \cup S_2$. A set whose members are all members of another set is termed a subset. Thus, if,

$$S_1 = \{a, b, c, d, e\}$$
$$S_2 = \{b, d, f, g\},$$
$$\text{then } S_1 \cap S_2 = \{b, d\}$$
$$S_1 \cup S_2 = \{a, b, c, d, e, f, g\}.$$

Moreover, S_1 and S_2 are subsets of $S_1 \cup S_2$, written $S_1 \subset S_1 \cup S_2$ and $S_2 \subset S_1 \cup S_2$. The set of all elements of all the sets in a particular discussion is the universal set, or **domain**, symbolized by a capital U. The domain may contain elements in addition to all those under discussion.

Set theory is of importance throughout mathematics. In ANALYTIC GEOMETRY, for example, a CURVE may be considered as a set of points, or **point set**. For two FUNCTIONS $f(x)$ and $g(x)$, represented by the sets S_f and S_g, $S_f \cap S_g$ gives those points at which the curves intersect (see INTERSECTION). Sets can be represented pictorially by use of VENN DIAGRAMS. (See also FIELD; GROUP.)

SEWAGE, the liquid and semisolid wastes from dwellings and offices, industrial wastes, and surface and storm waters. Sewage systems collect the sewage, transport and treat it, then discharge it into rivers, lakes or the sea. Vaulted sewers were developed by the Romans but from the Middle Ages and until the mid-19th century sewage flowed through the open gutters of cities, constituting a major health hazard. Then sewage was discharged into storm-water drains which were developed into sewers. But the dumping of large amounts of untreated sewage into rivers led to serious water POLLUTION, and modern treatment methods arose, at least for major cities. An early solution (still sometimes practiced) was sewage farming, raw sewage being used as FERTILIZER. Chemically-aided precipitation was also tried, but neither proved adequate. Noting that natural watercourses can purify a moderate amount of sewage, sanitary engineers imitated natural conditions by allowing atmospheric oxidation of the organic matter, first by passing it intermittently through a shallow tank filled with large stones (the "trickling filter"), and later much more successfully by the **activated-sludge process**, in which compressed air is passed through a sewage tank, the sludge being decomposed by the many microorganisms that it contains. A by-product is sludge gas, chiefly methane, burned as fuel to help power the treatment plant. Sedimentation is carried out before and after decomposition; the filtered solids are buried, incinerated or dried for fertilizer. The sewer system is designed for fast flow (about 1m/s) to carry the solids; the sewers are provided with manholes, drainage inlets, regulators and, finally, outfalls. Dwellings not connected to the sewers have their own SEPTIC TANK.

SEWING MACHINE, machine for sewing cloth, leather or books: a major industrial and domestic labor-saving device. There are two main types: chain-stitch machines, using a needle and only one thread, with a hook that pulls each looped stitch through the next; and lock-stitch machines, using two threads, one through the needle eye and the other, which interlocks with the first in the material, from a bobbin/shuttle system (to-and-fro or rotary). Chain-stitch machines—the first to be invented, by Barthélemy Thimmonier (1793–1859)—are now used chiefly for sacks or bags. The lock-stitch machines now in general use are based on that invented by Elias HOWE (1846). Zigzag machines differ from ordinary straight-stitch machines in having variously-shaped cams that move the needle from side to side. Almost all US machines are electrically powered, but foot-treadle machines are common elsewhere.

SEX, the totality of the differences between the male and female partners engaged in sexual REPRODUCTION. Examples of sex are found among all levels of life save the VIRUSES. In the higher orders, fertilization is brought about by the fusion of two GAMETES, the male SPERM conveying genetic information to the female EGG, or ovum (see HEREDITY). Many INVERTEBRATES, most PLANTS and some fishes are HERMAPHRODITE; that is, individuals may possess functioning male *and* female organs. This is not the case with birds and mammals, though these on occasion display **intersexuality**, where an individual may possess a confusion of male and female characteristics. **Sexual behavior** is an important

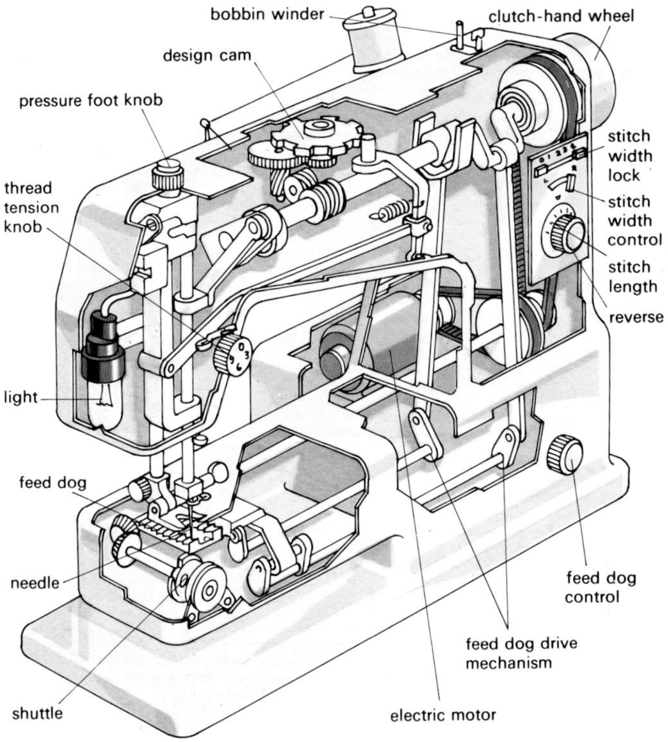

Cutaway of an electrically powered domestic sewing machine using the lock-stitch system. Power is supplied by belt to the main upper shaft which in turn drives the needle bar up and down via a crank. Power is also supplied from the main shaft to the bobbin and rotating hook under the working platform and via cams to the feed dogs which move the material being sewn across the working platform under the presser foot. Zigzag stitches are made by swinging the needle bar to and fro at right angles to the path of the material. Variously shaped cams can be loaded into the top of the machine to control the needle bar, rotary hook and feed dogs for automatic embroidery and fancy stitches.

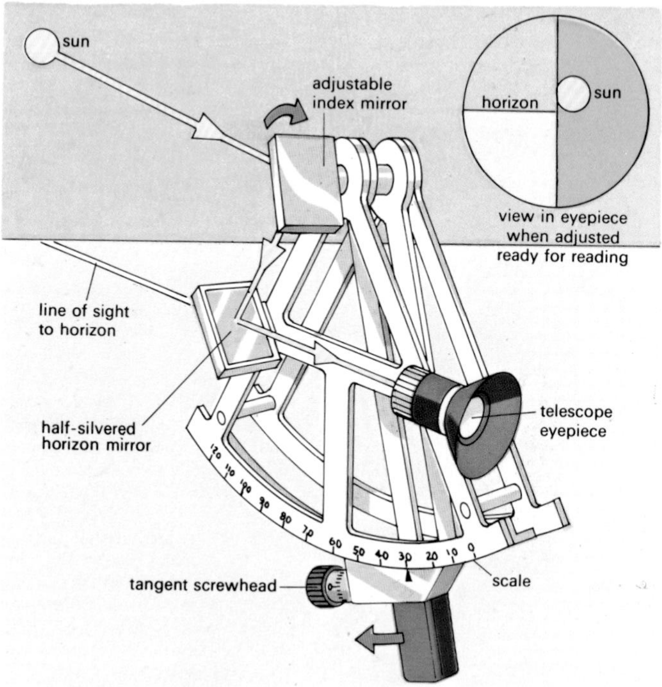

A simple sextant as used for determining the altitude of the sun or a star above the horizon. The horizon is viewed through the telescope while the image of the sun viewed in the horizon and index mirrors is made to coincide with the horizon by rotating the index mirror. The altitude of the sun can then be read off the scale. Practical instruments incorporate a variety of filters to protect the eyes of the user.

facet of animal behaviour (see BREEDING BEHAVIOR; MATING RITUALS): it may also be at the root of AGGRESSION and TERRITORIALITY. To the psychologist, "sex" and "sexual behavior" are used in connection with human drives linked to reproduction, and similarly fantasies, sensations, etc. To the psychoanalyst, sexual behavior has its roots in INFANTILE SEXUALITY as well as INSTINCT; and the term also covers a wide range of behavior derived from or analogous to sexuality and sexual drives. (See also HOMOSEXUALITY.)

SEX CHROMOSOMES. See X AND Y CHROMOSOMES.

SEX HORMONES. See ANDROGENS; ESTROGENS; GLANDS; HORMONES.

SEXTANT, instrument for NAVIGATION, invented in 1730 and superseding the ASTROLABE. A fixed telescope is pointed at the HORIZON, and a radial arm is moved against an arc graduated in degrees until a mirror which it bears reflects an image of a known star or the sun down the telescope to coincide with the image of the horizon. The angular elevation of the star, with the exact time (see CHRONOMETER), gives the LATITUDE. The **air sextant** is a similar instrument, usually periscopic, designed for use in aircraft, and has an artificial horizon, generally a bubble level.

SEXUAL SELECTION, the process whereby mates are chosen in sexually reproducing animal species. Competition for mates is thought to have been responsible for the EVOLUTION of secondary sexual characteristics.

SHADOW, a nonilluminated region or area, shielded from receiving radiation from a light source by an extended object. A point source gives a sharply defined shadow, but an extended source gives rise to a region of full shadow (the umbra), surrounded by one of partial shadow (the penumbra).

SHALE, fine-grained SEDIMENTARY ROCK formed by cementation of SILT particles usually also containing fragments of other minerals. Shales are rich in FOSSILS; and are laminated (they split readily into layers, or laminae). Their metamorphism (see METAMORPHIC ROCKS) produces SLATE. (See also OIL SHALE.)

SHAMANISM, a primitive religious system centered around a shaman, or **medicine man**, who in trance state is believed to be possessed by spirits that speak and act through him. The shaman (from the language of the Tungus of Siberia) is expected to cure the sick, protect the tribe, foretell the future, etc. (See also MAGIC, PRIMITIVE.)

SHAPLEY, Harlow (1885–1972), US astronomer who suggested that CEPHEID VARIABLES are not eclipsing binaries (see DOUBLE STAR) but pulsating stars. He was also the first to deduce the structure and approximate size of the MILKY WAY galaxy, and the position of the sun within it.

SHARPS RIFLE, the first RIFLE with a really satisfactory breech-loading system, invented in Philadelphia in 1848 by Christian Sharps (1811–1874). The breechblock was raised and lowered in a vertical mortise by the action of a lever that also served as a trigger guard.

SHEARING, type of deformation in which parallel planes in an object tend to slide over one another. Shear forces are always present in beams and whenever an object is subjected to bending stresses.

SHEFFIELD PLATE, articles made from SILVER plated on copper by a method of FUSION involving heat and pressure, discovered about 1743 by a Sheffield cutler, Thomas Boulsover. This method was widely used before the advent of ELECTROPLATING.

SHELL, any calcareous external covering secreted by

an invertebrate, enclosing and protecting the body. The term is used particularly for the shells of mollusks, but also refers to those of foraminiferans, and may be used loosely to describe the CARAPACE or chitinous EXOSKELETON of crustaceans and insects.

SHELLAC, brown, flaky RESIN secreted by the **lac insect,** *Laccifer lacca*, a scale insect. Naturally thermoplastic (see PLASTICS), it is used with fillers to make molded articles, and as an ingredient in paints, lacquers, polishes etc. A solution of shellac in ETHANOL is used as varnish.

SHELTER BELT, or **windbreak**, natural or specially planted barrier of vegetation arranged to protect crops and agricultural land from erosion by the wind. (See SOIL EROSION.)

SHERRINGTON, Sir Charles Scott (1857–1952), British neurophysiologist who shared with E. D. ADRIAN the 1932 Nobel Prize for Physiology or Medicine for studies of the NERVOUS SYSTEM which form the basis of our modern understanding of its action.

SHINGLES, or **herpes zoster**, a VIRUS disorder characterized by development of pain, a vesicular rash and later scarring, often with persistent pain, over the SKIN of part of face or trunk. The virus seems to settle in or near nerve cells following CHICKENPOX, which is caused by the same virus, and then becomes activated, perhaps years later and sometimes by disease. It then leads to the acute skin eruption which is in the distribution of the nerve involved.

SHIVERING, fine contractions of MUSCLES, causing slight repetitive movements, employed for increasing heat production by the body, thus raising body temperature in conditions of cold or when DISEASE induces FEVER. Uncontrollable shivering with gross movements of the whole body is a rigor only seen in some fevers.

SHM. See SIMPLE HARMONIC MOTION.

SHOCK specifically refers to the development of low blood pressure, inadequate to sustain BLOOD CIRCULATION, usually causing cold, clammy, gray SKIN and extremities, faintness and mental confusion and decreased urine production. It is caused by acute blood loss; burns with PLASMA loss; acute HEART failure; massive pulmonary EMBOLISM, and SEPTICEMIA. If untreated, death ensues. Early replacement of plasma or BLOOD and administration of DRUGS to improve blood circulation are necessary to prevent permanent BRAIN damage and acute KIDNEY failure.

SHOCK, Electric. See ELECTRIC SHOCK.

SHOCKLEY, William Bradford (1910–), US physicist who shared with BARDEEN and BRATTAIN the 1956 Nobel Prize for Physics for their joint development of the TRANSISTOR.

SHOCK THERAPY, or electroconvulsive therapy (ECT), is a form of treatment used in MENTAL ILLNESS, particularly DEPRESSION, in which carefully regulated electric shocks are given to the BRAINS of anesthetized patients. (Muscular relaxants are used to prevent injury through forceful MUSCLE contractions.) The mode of action is unknown but rapid resolution of severe depression may be achieved.

SHOE, protective covering for the foot. The various types include the boot, whose upper extends above the ankle; the clog, a simple wooden-soled shoe; the moccasin, a hunting shoe whose sole extends around and over the foot; the sandal, an open shoe whose sole is secured to the foot by straps, and the slipper, a soft indoor shoe. Shoes have been made from earliest times, the type depending mainly on the climate; clogs, sandals and moccasins predominated until the early Middle Ages, since when boots and typical shoes in widely varying styles have been most popular. LEATHER has always been the main material used, shaped on a *last* of wood or metal, and hand-sewn, the sole being nailed to the upper. From the mid-19th century the SEWING MACHINE was adapted for sewing shoes, and nailing and gluing were also mechanized, allowing mass-production. Other materials have to some extent displaced leather—natural and synthetic RUBBER for the sole and heel, and various PLASTICS and synthetic fibers for the upper.

SHOLES, Christopher Latham (1819–1890), US inventor (with some help from others) of the TYPEWRITER (patented 1868). He sold his patent rights to the REMINGTON Arms Company in 1873.

SHOOTING STAR. See METEOR.

SHORTHAND, or **stenography**, any writing system permitting the rapid transcription of speech. The three most used today are Isaac Pitman Shorthand, the first to be commercially developed (c1837), Gregg Shorthand, developed c1888 by John Robert Gregg, both of which are phonetic, using symbols to represent recurring sounds; and Speedwriting, which uses abbreviations. Shorthand is much used by secretaries, journalists, court reporters, etc. (See also STENOTYPE.)

SHORTSIGHTEDNESS. See MYOPIA.

SHOTGUN, smooth-bore FIREARM fired from the shoulder, designed to discharge a quantity of small lead pellets ("shot") in a diverging pattern, which increases the chances of hitting small fast-moving targets such as game-birds. Although repeating shotguns have been available since 1860, the double-barreled model has retained its popularity. One of the barrels usually has its bore "choked"—i.e., slightly tapered toward the muzzle to limit the pellet spread and increase the lethal range, which is usually about 50 yards.

SHOULDER, JOINT between the upper ARM (humerus) and the upper trunk (scapula, COLLAR BONE and rib cage). It is an open ball-and-socket joint which is only stable by virtue of the numerous powerful MUSCLES around it; this leads to increased maneuverability.

SHRAPNEL, originally a projectile containing lead bullets, an explosive charge and a time fuze, invented by a British artillery officer, Henry Shrapnel (1761–1842). The term now refers to the fragmenting case of a high-explosive shell (see AMMUNITION).

SHRUB, any small woody plant shorter than a TREE and with side shoots well developed.

SIAL (*si*lica-*al*uminum), collective term for the rocks, lighter and more rigid than the SIMA and composed to a great extent of SILICA and ALUMINUM, that form the upper portion of the EARTH's crust. (See also ISOSTASY.)

SIAMESE TWINS, twins (see MULTIPLE BIRTH) which are physically joined at some part of their anatomy due to a defect in early separation. A variable depth of fusion is seen, most commonly at the head or trunk. SURGERY may be used to separate the twins if no vital organs are shared.

SIDEREAL TIME, time referred to the rotation of the earth with respect to the fixed stars. The sidereal

DAY is about four minutes shorter than the solar day since the earth moves each day about 1/365 of its orbit about the sun. Sidereal time is used in astronomy when determining the locations of celestial bodies.

SIDERITE, brown or gray-green mineral consisting of iron(II) carbonate ($FeCO_3$), often with some magnesium, calcium and manganese. Of widespread occurrence in sedimentary or hydrothermal rocks, it is a major IRON ore. It has the CALCITE structure.

SIEGBAHN, Karl Manne Georg (1886–), Swedish physicist who was awarded the 1924 Nobel physics prize for his pioneer work in X-ray SPECTROSCOPY. He devised a way of measuring X-RAY wavelengths with great accuracy and developed an account of X-rays consistent with the BOHR theory of the ATOM.

SIEMENS, German family of technologists and industrialists. **Ernst Werner von Siemens** (1816–1892) invented, among other things, an ELECTROPLATING process (patented 1842), a differential GOVERNOR (c1844), and a regenerative STEAM ENGINE, the principle of which was developed by his brothers **Friedrich** (1826–1904) and then **Karl Wilhelm** (1823–1883), later **Sir (Charles) William Siemens**, to form the basis of the OPEN-HEARTH PROCESS. Ernst and Sir William both made many important contributions to TELEGRAPH science, culminating in the laying from the *Faraday*, a ship designed by William, of the ATLANTIC CABLE of 1874 by the company he owned.

SIEMENS (S), the SI UNIT of electric CONDUCTANCE, equal in value to the reciprocal OHM (or MHO).

SIEMENS-MARTIN PROCESS. See OPEN-HEARTH PROCESS.

SIGN LANGUAGE, any system of communication using gesture (usually of the hand and arm) rather than speech. The most comprehensive sign language in modern use is that employed by the deaf and dumb, but sophisticated sign languages are also used by many primitive peoples to communicate with other tribes.

SIKORSKY, Igor Ivanovich (1889–1972), Russian-born US aircraft designer best known for his invention of the first successful HELICOPTER (flown in 1939). He also designed several AIRPLANES, including the first to have more than one engine (1913).

SILAGE, or ensilage, winter cattle fodder made from grass, corn, legumes etc., by limited fermentation. The material is harvested when green, chopped and then stored in either a pit or a tower where air access can be carefully controlled. LACTIC ACID is formed from the CARBOHYDRATES while the loss of other nutrients is minimal.

SILENCER. See MUFFLER.

SILICA, or silicon dioxide (SiO_2), a very common mineral, having three crystalline forms, the most common being QUARTZ. Silica is refractory and inert, though it dissolves in hydrogen fluoride and reacts with bases to form SILICATES. It is used to make GLASS, CERAMICS, CONCRETE and CARBORUNDUM. (See also SILICON; KIESELGUHR.)

SILICA GEL, amorphous form of SILICA made by acidifying a SILICATE and dehydrating the silicic acid formed. It is widely used as an adsorbent (see ADSORPTION) and drying agent (see DEHYDRATION). (See also GEL.)

SILICATES, salts of silicic acids. Discrete silicate anions include orthosilicates (SiO_4^{4-}), metasilicates (SiO_3^{2-}), and groups of SiO_4 units linked by Si—O—Si bonds; such condensation also produces infinite anions in chains, layers or three-dimensional arrays. Silicates (including aluminosilicates) are the most important class of minerals, forming 90% of the earth's crust. (See also WATER GLASS.)

SILICON (Si), nonmetal in Group IVA of the PERIODIC TABLE; the second most abundant element (after oxygen), occurring as SILICA and SILICATES. It is made by reducing silica with coke at high temperatures. Silicon forms an amorphous brown powder, or gray semiconducting crystals, metallic in appearance. It oxidizes on heating, and reacts with the halogens, hydrogen fluoride, and alkalis. It is used in alloys, and to make TRANSISTORS and SEMICONDUCTORS. AW 28.1, mp 1410°C, bp 2355°C, sg 2.42 (20°C). Silicon is tetravalent in almost all its compounds, which resemble those of CARBON, except that it does not form multiple bonds, and that chains of silicon atoms are relatively unstable. **Silanes** are series of volatile silicon hydrides, analogous to ALKANES, spontaneously flammable in air and hydrolyzed by water.

Silicon tetrachloride is a colorless fuming liquid, made by reacting chlorine with a mixture of silica and carbon, the starting material for preparing organosilicon compounds, including SILICONES. mp −70°C, bp 58°C. (For silicon carbide, see CARBORUNDUM; silicon dioxide, SILICA).

SILICONES, POLYMERS with alternate atoms of SILICON and oxygen, and organic groups attached to the silicon. They are resistant to water and oxidation, and are stable to heat. Liquid silicones are used for waterproofing, as polishes and anti-foam agents. Silicone greases are high- and low-temperature lubricants, and resins are used as electrical insulators. Silicone rubbers remain flexible at low temperatures.

SILICOSIS, a form of PNEUMOCONIOSIS, or fibrotic LUNG disease, in which long-standing inhalation of fine SILICA dusts in mining causes a progressive reduction in the functional capacity of the lungs. The normally thin-walled alveoli and small bronchioles become thickened with fibrous tissue and the lungs lose their elasticity. Characteristic X-RAY appearances and changes in lung function occur.

SILIQUE, dry, dehiscent FRUIT, similar to a CAPSULE, but formed from two carpels. When ripe, the carpels split apart leaving a thin septum between. They are formed by members of the cabbage family (Cruciferae).

SILK, natural FIBER produced by certain insects and spiders to make cocoons and webs, a glandular secretion extruded from the spinneret and hardened into a filament on exposure to air. Commercial textile silk comes from the various SILKWORMS. The cocooned pupae are killed by steam or hot air, and the cocoons are placed in hot water to soften the gum (sericin) that binds the silk. The filaments from several cocoons are then unwound together to form a single strand of "raw silk," which is reeled. Several strands are twisted together, or "thrown," to form yarn. At this stage, or after weaving, the sericin is washed away. The thickness of the yarn is measured in DENIER. About 70% of all raw silk is now produced in Japan.

SILK-SCREEN PRINTING, method of PRINTING derived from the stencil process (see also DUPLICATING MACHINE). A stencil is attached to a silk screen or fine wire mesh, or formed on it by a photographic process or by drawing the design in tusche (a greasy ink), sealing the screen with glue and washing out the tusche and its covering glue with an organic solvent. The framed screen is placed on the surface to be printed, and viscous ink is pressed through by a rubber squeegee. Each color requires a different screen. The process, which may be mechanized, is used for printing labels, posters, fabrics, and on bottles and other curved surfaces. Since 1938 it has been used by painters, who call it serigraphy.

SILKWORM, the caterpillar of a moth, *Bombyx mori*, which, like many other caterpillars, spins itself a cocoon of silk in which it pupates. The cocoon of *B. mori* is, however, especially thick and may be composed of a single thread commonly 900m (2950ft) long. This is unraveled to provide commercial SILK. Originally a native of China, *B. mori* has been introduced to many countries. The caterpillar, which takes about a month to develop, feeds on the leaves of the mulberry tree.

SILL, tabular body of IGNEOUS ROCK, under 1cm to over 100m thick and perhaps hundreds of kilometres wide, lying parallel to the beds of surrounding rocks.

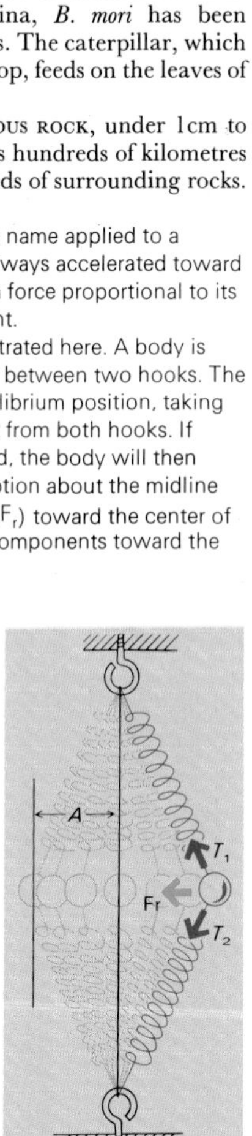

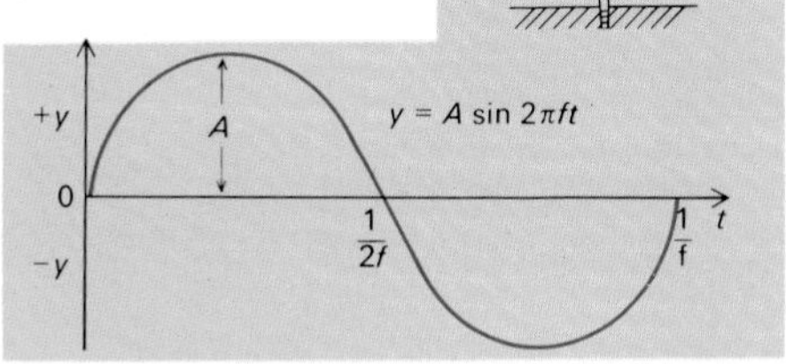

Simple harmonic motion is the name applied to a motion in which a particle is always accelerated toward the midpoint of its motion by a force proportional to its displacement from the midpoint.
A hypothetical example is illustrated here. A body is suspended on springs midway between two hooks. The body is displaced from its equilibrium position, taking care that it remains equidistant from both hooks. If gravitational effects are ignored, the body will then describe a simple harmonic motion about the midline when it is released. The force (F_r) toward the center of motion is the resultant of the components toward the midline of the tensions (T_1,T_2) in the springs. Undamped, the motion will continue indefinitely, potential energy in the springs being converted to kinetic energy in the motion of the body as the body approaches the midline, and the reverse occurring as the body approaches its maximum amplitude (*A*) again. The graphical expression of the motion, displacement (*y*) against time (*t*), is shown below It is the graph of $y=A \sin 2\pi ft$ where *f* is the frequency of the motion.

It has been suggested that sills are an extreme form of LACCOLITH or phacolith.

SILLIMAN, Benjamin (1779–1864), US chemist and geologist who founded *The American Journal of Science* (1819). The mineral **Sillimanite** (a form of aluminum SILICATE, Al_2SiO_5) is named for him.

SILT, soil composed of particles whose diameters range from 1/256 to 1/16mm. SOIL containing over 80% silt particles and less than 12% CLAY particles is often termed silt. In particular, **loess**, accumulations of wind-blown dust, has particles of silt-size in the range 1/32 to 1/16mm.

SILURIAN, the third period of the PALEOZOIC, which lasted between about 440 and 400 million years ago. (See also GEOLOGY.)

SILVER (Ag), soft, white NOBLE METAL in Group IB of the PERIODIC TABLE, a TRANSITION ELEMENT. Silver has been known and valued from earliest times and used for jewelry, ornaments and coinage since the 4th millennium BC. It occurs as the metal, notably in Norway; in COPPER, LEAD and ZINC sulfide ores; and in ARGENTITE and other silver ores. It is concentrated by various processes including cupellation and extraction with CYANIDE (see also GOLD), and is refined by electrolysis. Silver has the highest thermal and electrical conductivity of all metals, and is used for printed circuits and electrical contacts. Other modern uses include dental ALLOYS and AMALGAM, high-output storage batteries, and for monetary reserves. Although the most reactive of the noble metals, silver is not oxidized in air, nor dissolved by alkalis or nonoxidizing acids; it dissolves in nitric and concentrated sulfuric acid. Silver tarnishes by reaction with sulfur or hydrogen sulfide to form a dark silver-sulfide layer. Silver salts are normally monovalent. Ag^+ is readily reduced by mild reducing agents, depositing a silver MIRROR from solution. AW 107.9, mp 960.8°C, bp 2212°C, sg 10.5 (20°C). **Silver Halides** (AgX) are crystalline salts used in PHOTOGRAPHY. The chloride is white, the bromide pale yellow and the iodide yellow. On exposure to light, a crystal of silver halide becomes activated, and is preferentially reduced to silver by a mild reducing agent (the developer). **Silver Nitrate** ($AgNO_3$) is a transparent crystalline solid, used as an ANTISEPTIC and ASTRINGENT, especially for removing WARTS.

SIMA (*si*lica-*ma*gnesium), collective term for the rocks, denser and more plastic than the SIAL and composed to a great extent of SILICA and MAGNESIUM, that form the lower portion of the earth's crust. (See also ISOSTASY.)

SIMOOM, or **Simoon**, hot, dry wind or whirlwind occurring in the deserts of Arabia and N Africa. It usually carries much sand and greatly reduces visibility.

SIMPLE HARMONIC MOTION (SHM), a form of WAVE MOTION in which a moving particle traces a path symmetric (see SYMMETRY) about a midpoint or equilibrium position, through which it passes at regular intervals of time. The force responsible for the motion is always directed towards the midpoint, its magnitude proportional to the displacement of the particle. If the displacement of the particle is plotted as a FUNCTION of time the result is a sinusoidal CURVE,

$$y=A \sin 2\pi ft,$$

where A is the AMPLITUDE, f the FREQUENCY, y the displacement and t the time elapse from a particular zero-point. SHM derives its name from the fact that the vibrations produced by musical instruments (e.g., a string of a violin, the legs of a tuning fork), and hence the SOUND waves they propagate, approximate to it. In fact these, as all other vibrations and wave motions, may be treated as compounded of a number of SHMs.

SIMPSON, Sir James Young (1811–1870), Scottish obstetrician who pioneered the use of CHLOROFORM as an anesthetic, especially for mothers during childbirth (see ANESTHESIA; BIRTH).

SINE. See TRIGONOMETRY.

SINE RULE, in any plane TRIANGLE with angles A,B,C and sides a,b,c,

$$\frac{a}{\sin A} = \frac{b}{\sin B} = \frac{c}{\sin C}$$

that is to say, the ratio between the length of a side and the sine of the ANGLE opposite it is equal for all three sides of the triangle. In SPHERICAL TRIGONOMETRY, the sine rule is

$$\frac{\sin a}{\sin A} = \frac{\sin b}{\sin B} = \frac{\sin c}{\sin C}$$

for a spherical triangle with angles A, B, C and sides a, b, c. (See also COSINE RULE.)

SINGER, Isaac Merrit (1811–1875), US inventor of the first viable domestic SEWING MACHINE (patented 1851). Despite a legal battle with the earlier inventor Elias HOWE, the Singer sewing machine soon became the most popular in the world.

SINGULARITY, in mathematics a point at which a curve, function or property behaves unusually, e.g., by having a node, a cusp or an isolated point. In discussion of BLACK HOLES, singularity refers to the point or ring within which the gravitational field is of infinite strength.

SINH. See HYPERBOLIC FUNCTIONS.

SINK HOLE, doline or **swallow-hole**, well or funnel-shaped hole, typical of KARST landscapes, formed when GROUNDWATER dissolves an underground cavity in LIMESTONE, followed by slump of the surface.

SINTERING, the bonding together of compacted powder particles at temperatures below the melting point. The driving force is the decrease in surface energy that occurs as the particles merge and their total surface area lessens. The smaller the powder particles, the faster is the sintering. It is used to consolidate ORES, in powder METALLURGY, and in making CERAMICS and CERMETS.

SINUS, large air space connected with the NOSE which may become infected and obstructed after upper respiratory infection and cause facial pain and fever (sinusitis). There are four major nasal sinuses: the maxillary, frontal, ethmoid and sphenoid. *Also*, a blind-ended channel which may discharge PUS or other material onto the skin or other surface. These may be embryological remnants or arise from a foreign body or deep chronic infection (e.g., OSTEOMYELITIS). *Also*, a large venous channel, as in the LIVER and in the large vessels draining BLOOD from the BRAIN.

SIPHON, device, usually consisting of a bent tube with two legs of unequal length, which utilizes atmospheric pressure to transfer liquid over the edge of one container into another at a lower level. The flowing action depends on the difference in the pressures acting on the two liquid surfaces and stops when these coincide.

SIPUNCULA, a phylum of sausage-shaped marine worms that attain a length of 25–300mm (1–12in). They are related to the ANNELIDA but differ from members of that phylum by having an unsegmented body and lacking the bristles called chaetae.

SIRENIA, an order of the MAMMALIA, the sea cows, which is now largely extinct and which contains the dugong and manatees. Sea cows are aquatic and similar in habits to seals. They are found in the Atlantic and Indo-Pacific Oceans and are related to the elephants.

SIRIUS, Alpha Canis Majoris, the Dog Star, the brightest STAR in the night sky. 2.7pc distant, it is 20 times more luminous than the sun and has absolute magnitude +1.4. A DOUBLE STAR, its major component is twice the size of the sun; its minor component (the Pup), the first white dwarf star to be discovered, has a diameter only 50% greater than that of the earth, but is extremely dense, its mass being just less than that of the sun.

SIROCCO, in S Europe, warm, humid WIND from the S or SE, originating over the Sahara Desert and gaining humidity from the Mediterranean.

SISAL, or sisal hemp, *Agave sisalana*, a tropical American plant cultivated for its leaf fibers. It has a short stem and its sword-shaped leaves are cropped at intervals and their fibers extracted to make coarse ROPES and twine. In the plantations, flowering is prevented by removing the leaves so that each plant lives about 20 years. Family: Amaryllidaceae.

SITTER, Willem de (1872–1934), Dutch astronomer who helped to get EINSTEIN's theory of RELATIVITY widely known and who proposed a modification to it allowing for a gradual expansion of the universe.

SI UNITS, the internationally adopted abbreviation for the *Système International d'Unités* (International System of Units), a modification of the system known as rationalized MKSA UNITS adopted by the 11th General Conference of Weights and Measures (CGPM) in 1960 and subsequently amended. SI Units are the legal standard in many countries and find almost universal use among scientists.

SKELETON, in VERTEBRATES, the framework of BONES that supports and protects the soft TISSUES and ORGANS of the body. (See also ENDOSKELETON; EXOSKELETON.) It acts as an attachment for the MUSCLES, especially those producing movement, and protects vital organs such as the BRAIN, HEART and LUNGS. It is also a store of calcium, magnesium, sodium, phosphorus and PROTEINS, while its bone marrow is the site of red BLOOD-corpuscle formation. In the adult human body, there are about 206 bones, to which more than 600 muscles are attached. The skeleton consists of two parts: the axial skeleton (the skull, backbone and rib-cage), and the appendicular skeleton (the limbs). The function of the **axial skeleton** is mainly protective. The SKULL consists of 29 bones, 8 being fused together to form the cranium, protecting the brain. The *vertebral column*, or backbone, consists of 33 small bones (or VERTEBRAE): the upper 25 are joined by LIGAMENTS and thick cartilaginous disks and the lower 9 are fused together. It supports the upper body and protects the SPINAL CORD which runs through it. The *rib-cage* consists of 12 pairs of ribs forming a protective cage around the heart and lungs and assists in breathing (see RESPIRATION). The **appendicular skeleton** is primarily concerned with LOCOMOTION and consists of the ARMS and pectoral girdle, and the LEGS and pelvic girdle. The limbs articulate with their girdles in ball and socket JOINTS which permit the shoulder and hip great freedom of movement but are prone to dislocation. In contrast the elbows and knees are hinge joints permitting movement in one plane only, but which are very strong. (*See illustration page* 291.)

SKEPTICISM, philosophical attitude of doubting all claims to knowledge, chiefly on the ground that the adequacy of any proposed criterion is itself questionable. Examples of thoroughgoing skeptics, wary of dogmatism in whatever guise, were Pyrrho of Elis ("Pyrrhonism" and "skepticism" are virtual synonyms) and HUME. Other thinkers, among them Augustine, Erasmus, Montaigne, PASCAL, Bayle and Kierkegaard, sought to defend faith and religion by directing skeptical arguments against the EPISTEMOLOGICAL claims of RATIONALISM and EMPIRICISM. PRAGMATISM and KANT's critical philosophy represent two influential attempts to resolve skeptical dilemmas.

SKIN, the TISSUE which forms a sensitive, elastic, protective and waterproof covering of the HUMAN BODY, together with its specializations (e.g., NAILS, HAIR). In the adult human, it weighs 2.75kg, covers an area of 1.7m^2 and varies in thickness from 1mm (in the eyelids) to 3mm (in the palms and soles). It consists of two layers: the outer, epidermis, and the inner, dermis, or true skin. The outermost part of the **epidermis**, the *stratum corneum*, contains a tough protein called KERATIN. Consequently it provides protection against mechanical trauma, a barrier against microorganisms, and waterproofing. The epidermis also contains cells which produce the MELANIN responsible for skin pigmentation and which provides protection against the sun's ultraviolet rays. The unique pattern of skin folding on the soles and palms provides a gripping surface, and is the basis of identification by FINGERPRINTS. The **dermis** is usually thicker than the epidermis and contains BLOOD vessels, nerves and sensory receptors, sweat glands, SEBACEOUS GLANDS, hair folicles, fat cells and fibers. Temperature regulation of the body is aided by the evaporative cooling of sweat (see PERSPIRATION); regulation of the skin blood flow, and the erection of hairs which trap an insulating layer of air next to the skin (see GOOSEFLESH). The rich nerve supply of the dermis is responsible for the reception of touch, pressure, pain and temperature stimuli. Leading into the hair follicles are sebaceous glands which produce the antibacterial sebum, a fluid which keeps the hairs oiled and the skin moist. The action of sunlight on the skin initiates the formation of VITAMIN D which helps prevent RICKETS.

SKINNER, Burrhus Frederic (1904–), US psychologist and author whose staunch advocacy of BEHAVIORISM has done much to gain it acceptance in 20th-century PSYCHOLOGY. His best known books are *Science and Human Behavior* (1953) and *Beyond Freedom and Dignity* (1971).

SKULL, the bony structure of the head and face situated at the top of the vertebral column. It forms a thick bony protection for the BRAIN with small apertures for blood vessels, nerves, the SPINAL CORD etc., and the thinner framework of facial structure.

SKYSCRAPER, a very tall building. From the mid-19th century the price of land in big cities made it worthwhile to build upward rather than outward, and this became practicable with the development of safe electric ELEVATORS. The first skyscraper was the 40m (130ft) high Equitable Life Assurance Society Building, New York (1870). A major design breakthrough was the use of a load-bearing skeletal iron frame, first used in the 10-story Home Insurance Company Building, Chicago (1885).

SLAG, waste formed as an upper, molten layer in the smelting of ores and refining of metals. It consists of impurities, oxides and ash, with a limestone FLUX, and serves to remove unwanted substances and to protect the metal from oxidation. Solidified slag is used as AGGREGATE, for road making, and as a phosphate FERTILIZER. (See also BLAST FURNACE; OPEN-HEARTH PROCESS.)

SLATE, a dark gray, low-grade metamorphic rock. Because of the comparatively low temperatures and pressures under which it was formed, slate still retains the texture and cleavage properties of the SHALE from which it is derived. For this reason, it is widely used as a roofing material.

SLEEP, a state of relative unconsciousness and inactivity. The need for sleep recurs periodically in all animals. If deprived of sleep humans initially experience HALLUCINATIONS, acute ANXIETY, and become highly suggestible and eventually, COMA and sometimes DEATH result. During sleep, the body is relaxed and most bodily activity is reduced. Cortical, or higher, brain activity, as measured by the ELECTROENCEPHALOGRAPH; blood pressure; body TEMPERATURE; rate of heart beat and breathing are decreased. However, certain activities, such as gastric and alimentary activity, are increased. Sleep tends to occur in daily cycles which exhibit up to 5 or 6 periods of orthodox sleep—characterized by its deepness—alternating with periods of paradoxical, or rapid-eye-movement (REM), sleep—characterized by its restlessness and jerky movements of the eyes. Paradoxical sleep occurs only when we are dreaming and occupies about 20% of total sleeping time. Sleepwalking (SOMNAMBULISM) occurs only during orthodox sleep when we are not dreaming. Sleep-talking occurs mostly in orthodox sleep. Many theories have been proposed to explain sleep but none is completely satisfactory. Separate sleeping and waking centers in the HYPOTHALAMUS cooperate with other parts of the BRAIN in controlling sleep. Sleep as a whole, and particularly paradoxical sleep when dreaming occurs, is essential to health and life. Consequently the key to why animals sleep may reside in a need to DREAM. **Sleep learning** experiments have so far proved ineffective. A rested brain and concentration are probably the most effective basis for LEARNING.

SLEEPING PILLS. Drugs which induce SLEEP are properly termed hypnotics (see SEDATIVES). (See also ANESTHESIA; NARCOTICS.)

SLEEPING SICKNESS, INFECTIOUS DISEASE caused by TRYPANOSOMES occurring in Africa and carried by TSETSE FLIES. It initially causes FEVER, headache, often a sense of oppression and a rash; later the characteristic somnolence follows and the disease enters a chronic, often fatal stage. Treatment is most effective

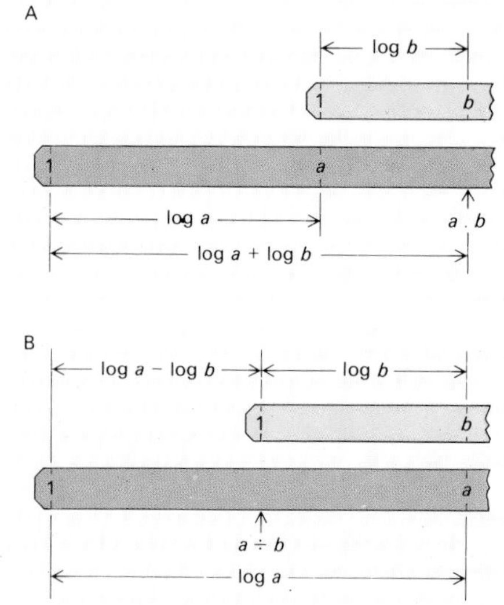

The slide rule works by the addition and subtraction of the logarithms of numbers. (A) To multiply *a* by *b*, the 1 mark on the sliding logarithmic scale (blue) is set opposite the *a* mark on the fixed logarithmic scale (red). The product *a.b* is then read off the fixed scale opposite the *b* mark on the sliding scale. (B) To divide *a* by *b*, the *b* mark on the sliding scale is set opposite the *a* mark on the fixed scale. The quotient *a*÷*b* is then read off the fixed scale opposite the 1 mark on the sliding scale. A standard 250mm long slide rule will give answers correct to about 3 significant figures.

if started before the late stage of BRAIN involvement and uses arsenical compounds.

SLEEPWALKING. See SOMNAMBULISM.

SLEET, PRECIPITATION consisting of small ICE pellets (diameter 5mm or less) formed by the freezing of raindrops or of partially melted snowflakes. A mixture of rain and snow is often termed sleet. (See also HAIL).

SLIDE RULE, an instrument based on LOGARITHMS and used for rapid, though approximate, calculation. Two scales are calibrated identically so that, on each, the distance from the "1" point to any point on the scale is proportional to the logarithm of the number represented by that point. Since log $(a.b) = \log a + \log b$, the multiplication $a.b$ can be performed by setting the "1" point on scale (1) against *a* on scale (2), then reading off the number of scale (2) opposite *b* on scale (1). Division is performed by reversing the procedure. In practice, slide rules have several different scales for different kinds of calculation, and a runner (cursor) to permit more accurate readings.

SLIME MOLDS, organisms belonging to the class Myxophyta, regarded as FUNGI, but which at certain stages of their life cycle are free-living masses of naked PROTOPLASM that move by amoeboid movement (see AMOEBA). They are found in damp, dark woods and ingest solid food particles as they ooze over decaying leaves and wood. Slime molds reproduce by spores like other primitive plants, but lacking CHLOROPHYLL, they do not carry out PHOTOSYNTHESIS.

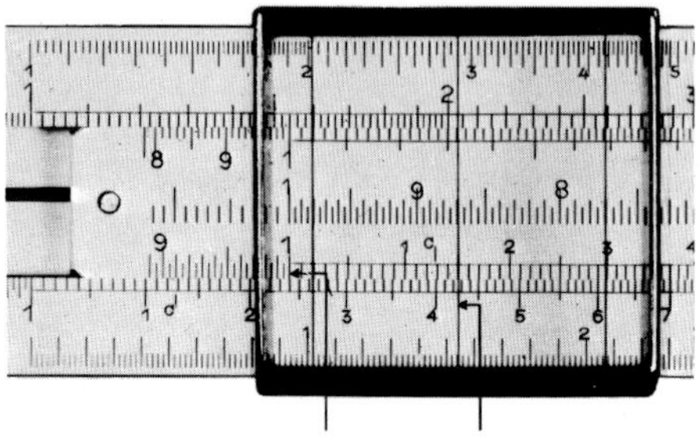

Close up of part of a slide rule showing the cursor.

SLING, weapon for propelling missiles such as rocks, much used in ancient and medieval times. The simple sling was a leather strip, forming a pouch for the missile, with a cord at each end. The sling was whirled above the head, and one cord was loosed to dispatch the missile. The modern **slingshot**—a forked stick with an elastic band for shooting pebbles—is a descendant.

SLIPPED DISK, a common condition in which the intervertebral disks of the spinal column degenerate with extrusion of the central soft portion through the outer fibrous ring. The protruding material may cause back pain, or may press upon the spinal cord or on nerves as they leave the SPINAL CORD (causing SCIATICA). Prolonged bed rest is an effective treatment in many cases, but traction, manipulation or surgery may also be required particularly if there is PARALYSIS or nerve involvement.

SLOANE, Sir Hans, 1st Baronet (1660–1753), British physician and natural historian who served as President of the ROYAL SOCIETY (1727–41) after NEWTON. His collection of books and specimens, left to the nation (for a fee of £20 000 to be paid to his family) formed the basis of the British Museum (founded 1759).

SMALLPOX, INFECTIOUS DISEASE, now restricted to a few areas, causing FEVER, headache and general malaise, followed by a rash. The rash characteristically affects face and limbs more than trunk and lesions start simultaneously. From a maculo-papular appearance, the rash passes into a pustular or vesicular stage and ends with scab formation; the lesions are deep and cause scarring. Major and minor forms of smallpox exist, with high fatality rate in major, often with extensive skin HEMORRHAGE. Transmission is from infected cases by secretions and the SKIN lesions; these are infectious for the duration of the rash. Immunization against smallpox was the earliest form practiced, initially through self-inoculation with the minor form. Later JENNER introduced VACCINATION with the related cowpox VIRUS (vaccinia is now used). QUARANTINE regulations and contact tracing are important in control of isolated outbreaks. It is important to confirm that apparent cases of CHICKENPOX are not indeed of smallpox.

SMELL, SENSE for detecting and recognizing substances at a distance and for assessing the quality of food. One of the earliest senses to develop in EVOLUTION, it may have been based on the chemotaxis of lower forms. Recognition of environmental odors is of vital importance in recognizing edible substances, detecting other animals or objects of danger, and in sexual behavior and attraction. In recent years, particular odors called **pheromones** that have specific physiological functions in insect and mammal behavior have been recognized. Smell reception in insects is localized to the antennae and detection is by specialist (pheromone) receptors and generalist (other odor) receptors. In man and mammals, the NOSE is the organ of smell. Respiratory air is drawn into the nostrils and passes across a specialized receptor surface—the olfactory epithelium. Receptor cells detect the tiny concentrations of odors in the air stream and stimulate nerve impulses that pass to olfactory centers in the BRAIN for coding and perception. It is not possible to classify odors in the same way as the primary colors in VISION and it is probable that pattern recognition is more important. Certain animals depend mainly on the sense of smell, while man is predominantly a visual animal. But with training, he can achieve sensitive detection and discrimination of odors.

SMELLING SALTS, scented solution of SAL VOLATILE in alcohol or ammonia-water, used to revive the weak or faint.

SMELTING, in METALLURGY, process of extracting a metal from its ORE by heating the ore in a BLAST FURNACE or reverberatory furnace (one in which a shallow hearth is heated by radiation from a low roof heated by flames from the burning fuel). A reducing agent (see OXIDATION AND REDUCTION), usually COKE, is used, and a FLUX is added to remove impurities. Sulfide ores are generally roasted to convert them to oxides before smelting.

SMITH, William "Strata" (1769–1839), the "father of stratigraphy." He established that similar sedimentary rock strata in different places may be dated by identifying the fossils each level contains, and made the first geological map of England and Wales (1815).

SMITHSON, James (1765–1829), earlier known as James Lewis and Louis Macie, British chemist and mineralogist who left £100000 (then about $500000) for the foundation of the SMITHSONIAN INSTITUTION.

SMITHSONIAN INSTITUTION, US institution of scientific and artistic culture, located in Washington, D.C., and sponsored by the US Government. Founded with money left by James SMITHSON, it was established by Congress in 1846. It is governed by a board of regents comprising the US Vice-President and Chief Justice, three Senators, three Representatives and six private citizens appointed by Congress. Although it undertakes considerable scientific research, it is best known as the largest US collection of museums, the "nation's attic:" these include the United States National Museum, the National Air and Space Museum, the National Gallery of Art, the Freer Gallery of Art, the National Portrait Gallery and the National Collection of Fine Arts.

SMITHSONITE, white, yellow or green mineral, formerly called CALAMINE, consisting of zinc carbonate ($ZnCO_3$). Smithsonite is of widespread occurrence, an ore of ZINC formed by alteration of other zinc minerals; it has the CALCITE structure. It was named for James SMITHSON.

SMOG. See FOG.

SMOKING, the habit of inhaling or taking into the mouth the smoke of dried tobacco or other leaves from a pipe or wrapped cylinder; it has been practised for many years in various communities, often using leaves of plant with hallucinogenic or other euphoriant properties. The modern habit of tobacco smoking derived from America and spread to Europe in the 16th century. Mass production of cigarettes began in the 19th century. Since the rise in cigarette consumption, epidemiology has demonstrated an unequivocal association with LUNG CANCER, chronic BRONCHITIS and EMPHYSEMA and with ARTERIOSCLEROSIS, leading to CORONARY THROMBOSIS and STROKE. Smoking appears to play a part in other forms of cancer and in other diseases such as peptic ULCER. It is not yet clear what part of smoke is responsible for disease. It is now known that nonsmokers may be affected by environmental smoke. A minor degree of physical and a large degree of psychological addiction occur.

SMUT, parasitic FUNGI, so named for the masses of sooty spores formed on the surface of the host plant. Smuts require only one host plant to complete their life cycle, unlike RUSTS. *Ustilago maydis* is an important parasite of corn and *U. tritici* is the loose smut of wheat. (See also PLANT DISEASES.)

SNAKE BITE. A very small proportion of the world's snakes produce poisonous venom, and most of these live in the tropics. The venom may lead to HEMORRHAGE, PARALYSIS and central NERVOUS SYSTEM disorders as well as local symptoms of pain, EDEMA and ulceration. Treatment aims to minimize venom absorption, neutralize venom with antiserum, counteract the specific effects and support life until venom is eliminated. Antiserum should be used only for definite bites by identified snakes.

SNAKES. See OPHIDIA.

SNEEZE, explosive expiration through the NOSE and MOUTH stimulated by irritation or INFLAMMATION in the nasal EPITHELIUM. It is a REFLEX attempt to remove the source of irritation.

SNELL'S LAW. See REFRACTION.

SNORING, stertorous respiration of certain persons during sleep, the noise being caused by vibration of the soft PALATE. It is predisposed to by the shape of the PHARYNX and by the sleeping position.

SNORKEL, breathing tube used by skin divers swimming near the surface, connecting the diver's mouth with the atmosphere. Also, a similar tube used by submarines in WWII to supply air to their engines without completely surfacing.

SNOW, PRECIPITATION consisting of flakes or clumps of ICE crystals. The crystals are plane hexagonal, showing an infinite variety of beautiful branched forms; needles, columns and irregular forms are also found. Snow forms by direct vapor-to-ice condensation from humid air below 0°C. On reaching the ground, snow crystals lose their structure and become granular. Fresh snow is very light (sg about 0.1), and is a good insulator, protecting underlying plants from severe cold. In time, pressure, sublimation and melting and refreezing lead to compaction into NÉVÉ. The slow melting of mountain snow is important in natural irrigation.

SNOW BLINDNESS, temporary loss of VISION with severe pain, tears and EDEMA due to excessive ultra-

One of the beautiful ice crystals making up snow.

violet light reflected from snow. Permanent damage is rare but protective Polaroid glasses should be used.

SNOWFLAKE CURVE, a closed CURVE of infinite length derived by trisecting the sides of an equilateral TRIANGLE and constructing an equilateral triangle whose sides are one-third the length of those of the original on the center section of each side, indefinitely repeating the process for each side of the resultant polygon.

SNOWMOBILE, or motor sled, motorized vehicle with two skis in front and propelled by an endless track, used for traveling over deep snow. First developed in the 1920s to replace dogsleds, they have become popular for recreation and racing since lightweight models were introduced (1959).

SNOWSHOE, light, broad footwear (about 1m × 0.4m) consisting of a wooden frame, laced with leather, strapped to the shod foot. Spreading the wearer's weight over a large area, they enable him to walk on deep, soft snow. Snowshoe racing is a popular sport, speeds of about 1km in 3min being attainable.

SNUFF, powdered TOBACCO for inhaling through the nose. The practice of taking snuff crossed to Europe from America in the early 16th century.

SOAPS AND DETERGENTS, substances which, when dissolved in water, are cleansing agents. Soap has been known since 600 BC; it was used as a medicine until its use for washing was discovered in the 2nd century AD. Until about 1500 it was made by boiling animal fat with wood ashes (which contain the alkali potassium carbonate). Then caustic soda (see SODIUM), a more effective ALKALI, was used; vegetable FATS and oils were also introduced. **Saponification**, the chemical reaction in soap-making, is an alkaline HYDROLYSIS of the fat (an ESTER) to yield GLYCEROL and the sodium salt of a long-chain CARBOXYLIC ACID. The potassium salt is used for soft soap. In the modern process, the hydrolysis is effected by superheated water with a zinc catalyst, and the free acid produced is then neutralized. Synthetic detergents, introduced in WWI, generally consist of the sodium salts of various long-chain SULFONIC ACIDS, derived from oils and PETROLEUM products. The principle of soaps and detergents is the same: the hydrophobic long-chain hydrocarbon part of the molecule attaches itself to the grease and dirt particles, and the hydrophilic acid group makes the particles soluble in water, so that by agitation they are loosed from the fabric and dispersed. Detergents do not (unlike soaps) form scum in HARD WATER. Their persistence in rivers, however, causes pollution problems, and biodegradable detergents have been developed. Household detergents may contain several additives: bleaches, brighteners, and ENZYMES to digest protein stains (egg, blood, etc.).

SOAPSTONE, or **steatite**, METAMORPHIC ROCK consisting of compacted TALC with SERPENTINE and carbonates, formed by alteration of PERIDOTITE. Soft and soapy to the touch, soapstone has been used from prehistoric times for carvings and vessels. When fired, it becomes hard and is used for insulators.

SOCIALIZATION, the indoctrination of an individual by a society resulting in his obeying the written and unwritten rules of conduct of that society. Socialization of an individual starts during childhood, the norms and basic tenets of parents and teachers generally being adopted, and continues throughout the individual's life.

SOCIALIZED MEDICINE, broad term comprising the various systems of free health care supported by the state from tax revenue. Specific groups of individuals may benefit as in the US Veterans Administration or Medicare, or health care may be entirely state supported as in the UK National Health Service.

SOCIAL PSYCHOLOGY, term now used chiefly as a synonym for GROUP psychology.

SOCIAL SCIENCES, group of studies concerned with man in relation to his cultural, social and physical environment; one of the three main divisions of human knowledge, the other two being the natural sciences and the humanities. Although social scientists usually attempt to model their disciplines on the natural sciences, aspiring to achieve a similar level of consensus, their efforts in this direction continue to be frustrated by the crudeness of their conceptual tools in relation to the complexity of their subject matter and the limited scope afforded for controlled experiments. The social sciences are usually considered to include: ANTHROPOLOGY; ARCHAEOLOGY; criminology; DEMOGRAPHY; economics; education; political science; PSYCHOLOGY; and SOCIOLOGY.

SOCIOLOGY, systematic study that seeks to describe and explain collective human behavior—as manifested in cultures, societies, communities and subgroups—by exploring the institutional relationships that hold between individuals and so sustain this behavior. Sociology shares its subject matter with ANTHROPOLOGY, which traditionally focuses on small, relatively isolated societies, and social PSYCHOLOGY, where the emphasis is on the study of subgroup behavior. The main emphasis in contemporary sociology is on the study of social structures and institutions and on the causes and effects of social change. This gives sociology a special relevance to issues in economics and political science. Not yet a mature discipline, sociology still veers between the tradition of speculative enquiry out of which it arose and the attempt to model its investigations on those of the physical sciences. Mainly because of the complexity of its subject matter and the political implications of social change, questions as to its proper aims and methods remain far from settled. There can be little doubt, however, that pressing sociological problems do exist at all levels in all societies and that sociological concepts (e.g., "internalization"—the processes by which the values and norms of a particular society are learned by its members (see SOCIALIZATION)—and "institutionalization"—the processes by which norms are incorporated in a culture as binding rules of behavior) do often illuminate many of the issues involved. The two great pioneers of modern sociology were Émile DURKHEIM and Max Weber. The anthropologists Bronislaw MALINOWSKI and Alfred RADCLIFFE-BROWN also exerted a profound influence on the development of the subject. Important US sociologists include the pioneers William Sumner and George MEAD, and Talcott Parsons, probably the most influential of postwar sociologists.

SOCIOMETRY, usually refers to techniques of measurement employed mainly by psychologists and sociologists in attempting to determine the relative strengths of interpersonal preferences and the relative status of individuals within groups. The term is sometimes applied to any attempt to quantify interpersonal relationships.

SOCRATES (c469–399 BC), Greek philosopher and mentor of PLATO. He wrote nothing, but much of his life and thought is vividly recorded in the Dialogues of Plato. The exact extent of Plato's indebtedness to Socrates is uncertain—e.g., it is still disputed whether the doctrine of the Forms (see REALISM; UNIVERSALS) is Socratic or Platonic; but Socrates made at least two fundamental contributions to Western philosophy: by shifting the focus of Greek philosophy from COSMOLOGY to ethics; and by developing the "Socratic method" of enquiry. He argued that the good life is the life illuminated by reason and strove to clarify the ideas of his interlocutors by leading them to detect the inconsistencies in their beliefs. His passion for self-consistency was evident even in his death: ultimately condemned for "impiety," he decided to accept the lawful sentence—and so remain true to his principles—rather than make good an easy escape.

SODA, or sodium carbonate. See SODIUM.

SODDY, Frederick (1877–1956), British chemist awarded the 1921 Nobel Prize for Chemistry for his work with RUTHERFORD on radioactive decay and particularly for his formulation (1913) of the theory of ISOTOPES. He also worked with Sir William RAMSAY to discover HELIUM.

SODIUM (Na), a soft, reactive, silvery-white ALKALI METAL. It is the sixth most common element, occurring naturally in common salt and many other important minerals such as cryolite and Chile saltpeter. It is very electropositive, and is produced by ELECTROLYSIS of fused sodium chloride (Downs process). Sodium rapidly oxidizes in air and reacts vigorously with water to give off hydrogen, so it is usually stored under kerosine. Most sodium compounds are highly ionic and soluble in water, their properties being mainly those of the anion. Sodium forms some organic compounds such as alkyls. It is used in making sodium cyanide, sodium hydride and the ANTIKNOCK ADDITIVE tetraethyl lead. Its high heat capacity and conductivity make molten sodium a useful coolant in some nuclear reactors. AW 23.0, mp 98°C, bp 883°C, sg 0.971 (20°C). **Sodium Bicarbonate** ($NaHCO_3$) is a white crystalline solid, made from sodium carbonate and carbon dioxide. It gives off carbon dioxide when heated to 270°C or when reacted with acids, and is used in BAKING POWDER, fire extinguishers and as an ANTACID. **Sodium Borates** are sodium salts of BORIC ACID, differing in their degree of condensation and hydration; BORAX is the most important. They are white crystalline solids, becoming glassy when heated, and used in the manufacture of detergents, water-softeners, fluxes, glass and ceramic glazes. **Sodium Carbonate** (Na_2CO_3), or (Washing) **Soda,** is a white crystalline solid, made by the SOLVAY PROCESS. It is used in making glass, other sodium compounds, soap and paper. The alkaline solution is used in disinfectants and water softeners. mp 851°C. **Sodium Hydroxide** (NaOH), or **Caustic Soda,** is a white deliquescent solid, usually obtained as pellets. It is a strong ALKALI, and absorbs carbon dioxide from the air. It is made by ELECTROLYSIS of sodium chloride solution or by adding calcium hydroxide to sodium carbonate solution. Caustic soda is used in the production of cellulose, plastics, soap, dyestuffs, paper and in oil refining. mp 318°C, bp 1390°C. **Sodium Nitrate** ($NaNO_3$), or **Soda Niter,** is a colorless crystalline solid, occurring naturally in CHILE SALTPETER. Its properties are similar to those of potassium nitrate (see POTASSIUM), but as it is hygroscopic it is unsuitable for gunpowder. mp 307°C. **Sodium Thiosulfate** ($Na_2S_2O_3$), or **Hypo,** is a colorless crystalline solid. It is a mild reducing agent, used to estimate iodine, and as a photographic fixer, dissolving the silver halides which have remained unaffected by light. (For sodium chloride, see SALT.)

SODIUM PENTOTHAL. See PENTOTHAL SODIUM.

SOFT DRINKS, nonalcoholic beverages generally containing fruit acids, SWEETENING AGENTS, and natural or artificial flavorings and colorings. In the early 19th century, carbonated water ("soda water") was developed in imitation of effervescent spa water or mineral water; this was the antecedent of carbonated soft drinks, made by absorption of CARBON dioxide under pressure. The dissolved gas gives a pleasant, slightly acid taste, and acts as a preservative. Still drinks, without carbon dioxide, are frozen or subjected to PASTEURIZATION.

SOFTWARE, term used in the COMPUTER industry to refer to all the non-hardware elements of a computer system, principally the programs.

SOFTWOOD, lumber produced from conifers, accounting for about 80% of world production. The class includes woods that are physically both soft and hard. (See also FORESTRY; WOOD.)

SOIL, the uppermost surface layer of the earth, in which plants grow and on which, directly or indirectly, all life on earth depends. Soil consists, in the upper layers, of organic material mixed with inorganic matter resultant from weathering (see EROSION; HUMUS). Soil depth, where soil exists, may reach to many metres. Between the soil and the bedrock is a layer called the subsoil. Mature soil may be described in terms of four **soil horizons**: A, the

The world's largest solar furnace, built at Odeillo in the Pyrenees in 1969. It consists of 63 mirrors on mountings which may be rotated to reflect solar radiation on to an enormous parabolic reflector made up of 9 000 smaller mirrors. This concentrates the solar radiation on a focal point at which temperatures of over 3 800°C can be produced.

uppermost layer, containing organic matter, though most of the soluble chemicals have been leached (washed out); B, strongly leached and with little or no organic matter (A and B together are often called the topsoil); C, the subsoil, a layer of weathered and shattered rock; and D, the bedrock. Three main types of soil are commonly distinguished: **pedalfers**, associated with temperate, humid climates, have a leached A-horizon but contain IRON and ALUMINUM salts with clay in the B-horizon; **pedocals**, associated with low-rainfall regions, contain soluble substances such as CALCIUM carbonate (soluble in rainwater, which contains CARBON dioxide) and other salts, and **laterites**, tropical red or yellow soils, heavily leached and rich in iron and aluminum. Soils may also be classified in terms of texture (see CLAY; SILT; SAND). LOAMS, with roughly equal proportions of sand, silt and clay, together with humus, are among the richest

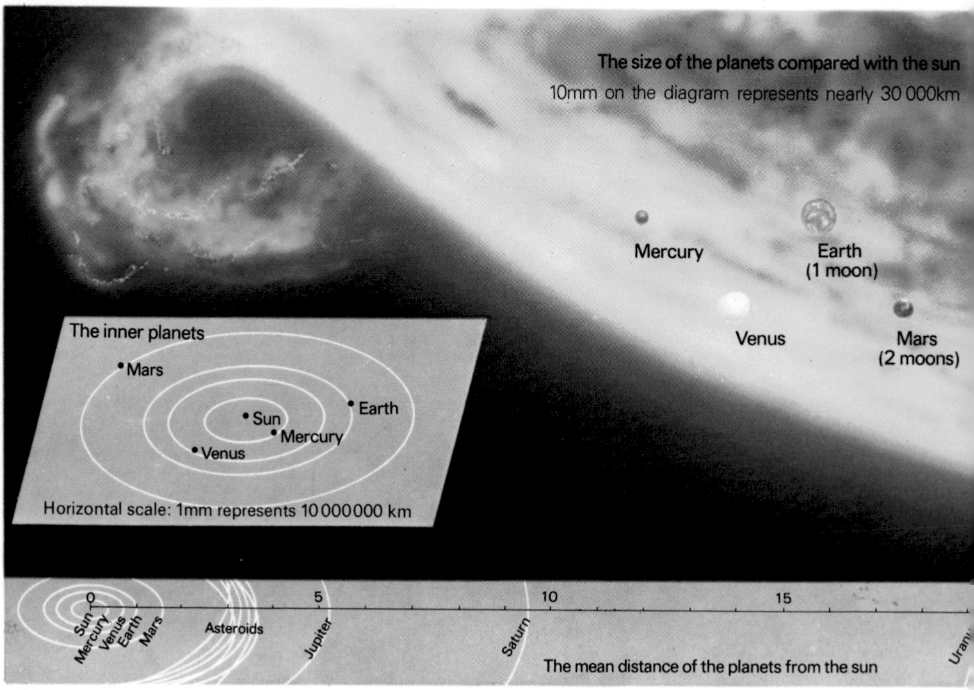

Black

agricultural soils. (See also PERMAFROST; PODZOL.)

SOIL EROSION, the wearing away of soil, a primary cause of concern in agriculture. There are two types: **Geological erosion** denotes those naturally occurring EROSION processes that constantly affect the earth's surface features; it is usually a fairly slow process and naturally compensated for. **Accelerated erosion** describes erosion hastened by the intervention of man. **Sheet erosion** occurs usually on plowed fields. A fine sheet of rich topsoil (see SOIL) is removed by the action of RAIN water. Repetition over the years may render the soil unfit for cultivation. In **rill erosion**, heavy rains may run off the land in streamlets: sufficient water moving swiftly enough cuts shallow trenches that may be plowed over and forgotten until, after years, the soil is found poor. In **gully erosion**, deep trenches are cut by repeated or heavy flow of water. WIND erosion is of importance in exposed, arid areas. (See also CONSERVATION; LAND RECLAMATION.)

SOL, a COLLOID, usually in liquid form, in which the dispersed phase is initially a solid, the continuous phase initially a liquid. (See also GEL.)

SOLAR CELL, device for converting the ENERGY of the sun's radiation into electrical energy. The commonest form is a large array of SEMICONDUCTOR *p-n* junction devices in series and parallel. By the PHOTOELECTRIC EFFECT each junction produces a small voltage when illuminated. Solar cells are chiefly used to power artificial SATELLITES. Their low efficiency (about 12%) makes them uncompetitive on earth except for mobile or isolated devices.

SOLAR ENERGY, the ENERGY given off by the SUN as ELECTROMAGNETIC RADIATION. In one year the sun emits about 5.4×10^{33}J of energy, of which half of one-billionth (2.7×10^{24}J) reaches the earth. Of this, most is reflected away, only 35% being absorbed. The power reaching the ground is at most 1.2kW/m², and on average 0.8kW/m². Solar energy is naturally converted into WIND power and into the energy of the HYDROLOGIC CYCLE, increasingly exploited as hydroelectric power. Plants convert solar energy to chemical energy by PHOTOSYNTHESIS, normally at only 0.1% efficiency; the cultivation of ALGAE in ponds can be up to 0.6% efficient, and is being developed to provide food and fuel. Solar heat energy may be used directly in several ways. Solar evaporation is used to convert brine to SALT and distilled water. Flat-plate collectors—matt black absorbing plates with attached tubes through which a fluid flows to collect the heat—are beginning to be used for domestic water heating, space heating, and to run air-conditioning systems. Focusing collectors, using a parabolic mirror, are used in solar furnaces, which can give high power absorption at high temperatures. They are used for cooking, for high-temperature research, to power heat engines for generating electricity, and to produce electricity more directly by the SEEBECK effect. Solar energy may be directly converted to electrical energy by SOLAR CELLS.

SOLAR PLEXUS, the GANGLION of nerve cells and fibers situated at the back of the ABDOMEN which subserve autonomic NERVOUS SYSTEM function for much of the GASTROINTESTINAL TRACT. A sharp blow on the abdomen over the plexus causes visceral pain and "winding."

SOLAR SYSTEM, the sun and all the celestial objects that move in ORBITS around it, including the nine known planets (MERCURY; VENUS; EARTH; MARS; JUPITER; SATURN; URANUS; NEPTUNE; PLUTO), their 32 known moons, the ASTEROIDS, COMETS, meteoroids (see METEOR) and a large quantity of gas and dust. The

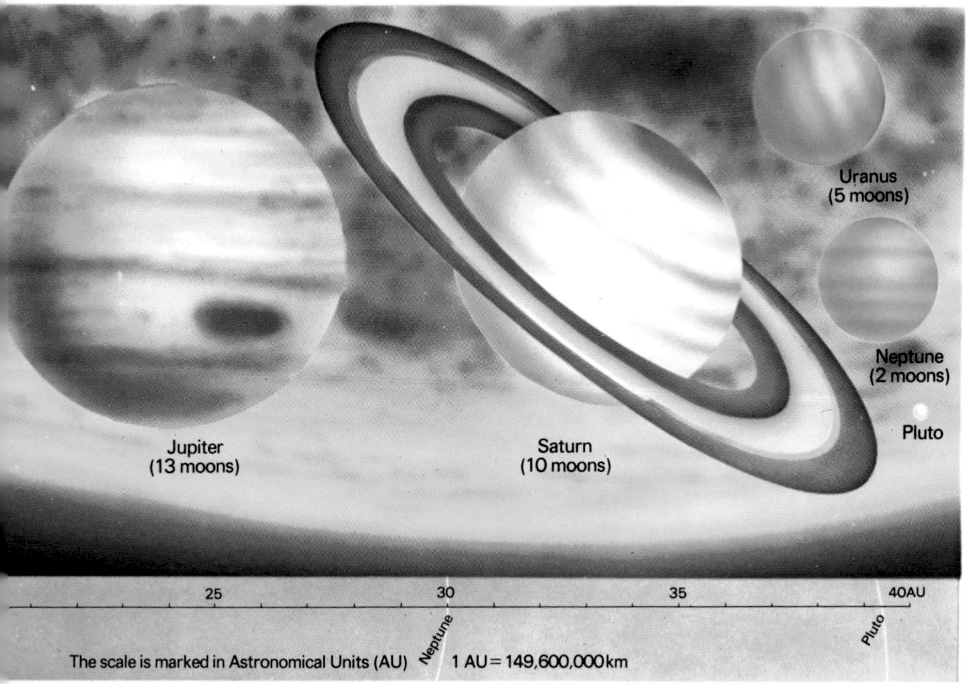

planets all move in their orbits in the same direction and, with the exceptions of Venus and Uranus, also rotate on their axes in this direction: this is known as direct motion. Most of the moons of the planets have direct orbits, with the exception of four of Jupiter's minor moons, the outermost moon of Saturn and the inner moon of Neptune, whose orbits are retrograde (see RETROGRADE MOTION). Most of the planets move in elliptical, near circular orbits, and roughly in the same plane. The origins of the solar system are not known, though various theories have been proposed (see NEBULAR HYPOTHESIS; PLANETESIMAL HYPOTHESIS). It would not appear to be unique among the stars (see PLANET).

SOLAR WIND, the electrically charged material thrown out by the sun at an average speed of 400km/s. The "quiet" component is a continuous stream to which is added an "active" component produced by bursts of activity on the sun's surface. The solar wind affects the magnetic fields of the earth and Jupiter, and causes the tails of COMETS.

SOLDERING, joining metal objects using a low-melting-point ALLOY, **solder**, as the ADHESIVE. Soft solder, commonly used in electronics to join wires and other components, is an alloy of mainly lead and tin. The parts to be joined are cleaned, and heated by applying a hot soldering iron (usually having a copper bit). A FLUX is used to dissolve oxides, protect the surfaces, and enable the solder to flow freely. The solder melts when applied, solidifying again to form a strong joint when the iron is withdrawn. Solder is often supplied as wire with a core of noncorrosive rosin flux. Soldering at higher temperatures is termed BRAZING.

SOLENOID, device, used in CIRCUIT BREAKERS, for producing a short lateral movement of a sliding iron core. This is attracted by the MAGNETIC FIELD produced when an electric current flows in one of the coils surrounding either end of the slider. *Also*, an elongated coil used to produce a region of uniform magnetic field.

SOLID, one of the three physical states of matter, characterized by the property of cohesion: solids retain their shape unless deformed by external forces. True solids have a definite melting point and are crystalline, their molecules being held together in a regular pattern by stronger intermolecular forces than exist in liquids or gases. Amorphous solids are not crystalline, melt over a wide temperature range and are effectively supercooled liquids. GLASS is a familiar amorphous solid.

SOLID ANGLE. See ANGLE.

SOLID STATE PHYSICS, branch of physics concerned with the nature and properties of solid materials, many of which arise from the association and regular arrangement of atoms or molecules in crystalline solids. The term is applied particularly to studies of SEMICONDUCTORS and solid-state electronic devices.

SOLIPSISM, extreme form of subjective IDEALISM based on an argument to the effect that since I can apprehend nothing that is not part of *my* experience, there can be no legitimate grounds for affirming the existence of an external world that is independent of my experience of it.

SOLSTICES, the two times each year when the sun is on the points of the ECLIPTIC farthest from the equator (see CELESTIAL SPHERE). At the summer solstice in late June the sun is directly overhead at noon on the TROPIC of Cancer; at winter solstice, in late December, it is overhead at noon on the Tropic of Capricorn.

SOLUTION, a homogeneous molecular mixture of two or more substances, commonly of a solid and a liquid, though solid/solid solutions also exist. The liquid component is usually termed the SOLVENT, the other component, which is dissolved in it, the *solute.* The **solubility** of a solute in a given solvent at a particular temperature is usually stated as the mass which will dissolve in 100g of the solvent to give a saturated solution (see SATURATION). Solubility generally increases with temperature. For slightly soluble ionic compounds, the **solubility product**—the product of the individual ionic solubilities—is a constant at a given temperature. Most substances are solvated when dissolved: that is, their molecules become surrounded by solvent molecules acting as LIGANDS. Ionic crystals dissolve to give individual solvated ions, and some good solvents of high dielectric constant (such as water) cause certain covalent compounds to ionize, wholly or partly (see also ACID). Analogous to an ideal gas, the hypothetical **ideal solution** is one which is formed from its components without change in total volume or internal energy: it obeys RAOULT's law and its corollaries, so that the addition of solute produces a lowering of the freezing point, elevation of the boiling point and increase in osmotic pressure (see OSMOSIS), all proportional to the number of MOLES added. (See also DISSOCIATION; ELECTROLYSIS; EQUIVALENT WEIGHT.)

SOLVAY PROCESS, for the manufacture of sodium carbonate (see SODIUM). Salt, ammonia, carbon dioxide and water react to give precipitated sodium bicarbonate, which on heating gives sodium carbonate and carbon dioxide for RECYCLING.

SOLVENT, a liquid capable of dissolving a substance to form a SOLUTION. Generally "like dissolves like"; thus a nonpolar covalent solid such as naphthalene dissolves well in a hydrocarbon solvent. Overall, best solvents are those with polar molecules and high DIELECTRIC constant: WATER is the most effective known.

SOMATOTYPES, in ANTHROPOMETRY, descriptions of physique, sometimes supposedly also descriptive of temperament. The individual is classified by three digits representing the extent of his endomorphy (plumpness), mesomorphy (muscularity) and ectomorphy (slenderness), respectively.

SOMNAMBULISM, or **sleepwalking**, state in which the body is able to walk and perform other automatic tasks while consciousness is diminished. Often seen in anxious children, it is said to be unwise to awaken them as intense fear may be felt.

SONAR, *so*und *n*avigation *a*nd *r*anging, technique used at sea for detecting and determining the position of underwater objects (e.g. submarines; shoals of fish) and for finding the depth of water under a ship's keel (see ECHO SOUNDER). Sonar works on the principle of echolocation: high-frequency SOUND pulses are beamed from the ship and the direction of and time taken for any returning ECHOES are measured to give the direction and range of the reflecting objects.

SONIC BOOM, loud noise generated in the form of a shockwave cone when an airplane traveling faster than the speed of sound overtakes the pressure waves it produces. Because of sonic boom damage, supersonic planes are confined to closely defined flight paths.

SOOT, form of carbon black (see CARBON) formed by incomplete combustion of organic material.

SOUND, mechanical disturbance, such as a change of pressure, particle displacement or stress, propagated in an elastic medium (e.g. air or water), that can be detected by an instrument or by an observer who hears the auditory sensation it produces. Sound is a measurable physical phenomenon and an important stimulus to man. It forms a major means of communication in the form of spoken language, and both natural and manmade sounds (of traffic or machinery) contribute largely to our environment. The EAR is very sensitive and will tolerate a large range of sound energies, but enigmas remain as to exactly how it produces the sensation of hearing. The Greeks appreciated that sound was connected with air motion and that the PITCH of a musical sound produced by a vibrating source depended on the vibration FREQUENCY. Attempts to measure the velocity of sound in air date from the 17th century. Sound is carried as a longitudinal compressional wave in an elastic medium: part of the medium next to a sound source is compressed, but its elasticity makes it expand again, compressing the region next to it and so on. The velocity of such waves depends on the medium and the temperature, but is always much less than that of light. Sound waves are characterized by their wavelength and frequency. Humans cannot hear sounds of frequencies below 16Hz and above 20kHz, such sounds being known as infrasonic and ULTRASONIC respectively. The sound produced by a TUNING FORK has a definite frequency, but most sounds are a combination of frequencies. The amount of motion in a sound wave determines its loudness or softness and the intensity falls off with the square of distance from the source. Sound waves may be reflected from surfaces (as in an ECHO), refracted or diffracted, the last property enabling us to hear around corners. The intensity of a sound is commonly expressed in DECIBELS above an arbitrary reference level; its loudness is measured in PHONS.

SOUND BARRIER, term referring to the extra forces acting on an airplane when it goes SUPERSONIC.

SOUND RECORDING, the conversion of SOUND waves into a form in which they can be stored, the original sound being reproducible by use of playback equipment. The first sound recording was made by Thomas EDISON in 1877 (see PHONOGRAPH). In modern electronic recording of all kinds, the sound is first converted by one or more MICROPHONES into electrical signals. In the case of **mechanical recordings** (discs, or records), these signals—temporarily recorded on magnetic tape—are made to vibrate a stylus that cuts a spiral groove in a rotating disc covered with lacquer. The master disc is copied by ELECTROFORMING to produce stamper dies used to press the plastic copies. (See also HIGH-FIDELITY.) In **magnetic recording**, the microphone signals activate an ELECTROMAGNET which imposes a pattern of magnetization on moving magnetic wire, discs or tape with a ferromagnetic coating (see also MAGNETISM; TAPE RECORDER). **Optical recording**, used for many motion-picture sound tracks, converts the microphone signals into a photographic exposure

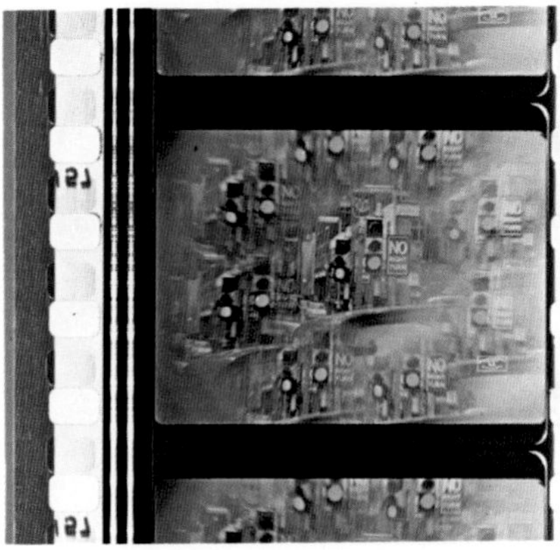

Section of 35mm motion-picture film with a double bilateral variable-area optical soundtrack. Sound for motion pictures is usually recorded magnetically and converted to optical format only for the final combined picture-and-sound negative which is used to make the release prints.

on film using a light beam and a variable shutter. The sound is played back by shining a light beam through the track onto a PHOTOELECTRIC CELL. As with the playback equipment for the other recording methods, this reproduces electrical signals which are amplified and fed to a LOUDSPEAKER.

SOUTHERN CROSS, Crux, a small, bright constellation near the S celestial pole. The four bright stars forming the cross are Acrux (Alpha), Mimosa (Beta), Gacrux (Gamma) and Delta Crucis.

SOUTHERN LIGHTS. See AURORA.

SOUTH POLE, the point in Antarctica through which passes the earth's axis of rotation. It does not coincide with the earth's S Magnetic Pole (see EARTH). It was first reached by Roald Amundsen (Dec. 14, 1911). (See also CELESTIAL SPHERE; MAGNETISM; NORTH POLE.)

SPACE, in MATHEMATICS, a bounded or unbounded extent. In GEOMETRY this extent may be in one, two or three DIMENSIONS, its nature being viewed differently in different geometries. According to EUCLIDEAN GEOMETRY space is uniform and infinite, so that we may talk of a LINE of infinite extent or a POLYGON of infinite AREA. In RIEMANNIAN GEOMETRY, however, all lines are of less than a certain, finite extent; and in LOBACHEVSKIAN GEOMETRY, there is a similar maximum of area. (See also VECTOR SPACE.)

SPACE EXPLORATION. At 10.56pm EDT on July 20, 1969, Neil Armstrong became the first man to set foot on the MOON. This was the climax of an intensive US space program sparked off by the successful launch of the Russian artificial SATELLITE Sputnik 1 in 1957, and accelerated by Yuri Gagarin's flight in Vostok 1, the first manned spacecraft, in 1961. Later that year Alan Shepard piloted the first American manned spacecraft, and President Kennedy set the goal of landing a man on the moon and returning him safely within the decade. On Feb. 20, 1962 John Glenn orbited the earth three times in the first MERCURY craft to be boosted by an Atlas rocket, but it was Valery Bykovsky who set the one-man endurance record with a 5-day mission in June 1963. The next Russian mission involved two craft, Vostoks 5 and 6, and made Valentina Tereshkova the first woman cosmonaut. Alexei Leonov completed the first spacewalk in Oct. 1964: but then it was the turn of the GEMINI MISSIONS to break all records. Both countries lost men, on the ground and in space: among them V. I. Grissom, E. H. White and R. B. Chaffee—in a fire on board Apollo during ground tests—and the crew of Soyuz 11, killed during reentry in 1971, though earlier Soyuz had docked successfully with the first space station and set new records. Unmanned probes such as Orbiter, Ranger and Surveyor were meanwhile searching out Apollo landing sites, while Russian Luna and Lunokhod craft were also studying the moon. In 1968 Apollo 7 carried out an 11-day earth-orbit flight, and at Christmas Apollo 8 made 10 lunar orbits. The lunar landing craft was tested on the Apollo 9 and Apollo 10 missions, leaving the way clear for the triumphant success of Apollo 11. Apollo 12 was equally successful, landing only 600 yards from the lunar probe Surveyor 3, but the Apollo 13 mission was a near disaster: an explosion damaged the craft on its way to the moon, and re-entry was achieved only with great difficulty. Apollo 14 had no such problems in visiting Fra Mauro, and on the Apollo 15 mission a Lunar Roving Vehicle allowed collection of a very wide range of samples. Apollo 16 brought back over 200lbs of moon rock, and in Dec. 1972 Apollo 17 made the last lunar landing. In 1973 the Skylab missions returned attention to the study of world resources from space, and movements toward cooperation in this field were demonstrated by the joint Apollo-Soyuz mission in 1975. Exploration of the planets has been carried out by unmanned probes: the MARINER series to Mars, Venus and Mercury; the PIONEER missions to the outer planets, and a number of Russian contributions, such as the Venera soft-landing missions to Venus, the Zond bypass probe and the Mars soft-landing craft. Results from the two American VIKING probes that soft-landed on Mars in 1976 did not reveal the existence of life there.

SPACE MEDICINE, the specialized branch of MEDICINE concerned with the special physical and psychological problems arising from space flight. In particular, the effects of prolonged weightlessness and isolation are studied, simulated space flight forming the basis for much of this work.

SPACE-TIME, a way of describing the geometry of the physical universe arising from EINSTEIN's special theory of RELATIVITY. Space and time are considered as a single 4-dimensional continuum rather than as a 3-dimensional space with a separate infinite 1-dimensional time. Time thus becomes the "fourth dimension." Events in space-time are analogous to points in space and invariant space-time intervals to distances in space.

SPALLANZANI, Lazzaro (1729–1799), Italian biologist who attacked the contemporary belief in the SPONTANEOUS GENERATION of life by demonstrating that organisms which usually appeared in vegetable infusions failed to do so if the infusions were boiled and kept from contact with the air.

SPARK, momentary electric discharge in a gas.

SPARK CHAMBER, detector for visual investigation of SUBATOMIC PARTICLE paths in high-energy NUCLEAR PHYSICS. A large chamber filled with a NOBLE GAS contains a series of thin parallel metal plates 10–20mm apart. The entry of a particle into the chamber triggers a high voltage pulse (around 10kV) which is applied to the plates. An ELECTRON avalanche quickly builds up around the ionization produced along the particle's path. A bright and well-defined spark appears, which may be photographed, and cuts off the voltage pulse.

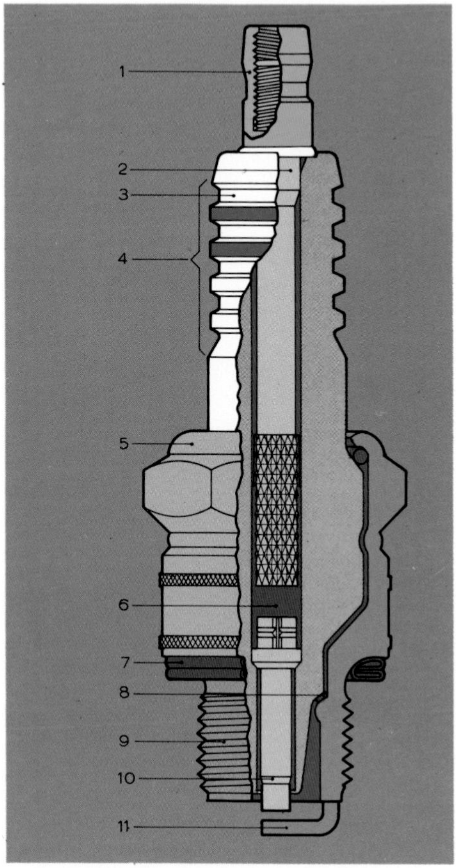

Cutaway of an automobile spark plug: (1) terminal; (2) central conductor; (3) insulator (capable of resisting 30 000V at 900°C); (4) antiflashover ribbing; (5) body with hexagon nut; (6) conducting link between central conductor and center electrode; (7) gasket; (8) gas sealing washer; (9) screw thread; (10) center electrode; (11) ground (earth) electrode.

SPARK PLUG, an IGNITION SYSTEM fitted in the cylinder head of an INTERNAL-COMBUSTION ENGINE. It consists essentially of two electrodes separated by an air gap. When the fuel/air mixture is fully compressed, the DISTRIBUTOR connects a voltage of 20kV across the electrodes, producing a spark that ignites the mixture. The width of the gap between the two electrodes must be correctly set.

SPASTIC PARALYSIS, form of PARALYSIS due to DISEASE of BRAIN (e.g., STROKE) or SPINAL CORD (e.g., MULTIPLE SCLEROSIS), in which the involved MUSCLES are in a state of constantly increased tone (or resting contraction). Spasticity is a segmental motor phenomenon where muscle contraction occurs without voluntary control.

SPEAKER. See LOUDSPEAKER.

SPECIATION, the process by which new species originate. Speciation either involves change in an organism, or divergence of a population that results in the formation of two new species; both are the result of EVOLUTION.

SPECIES. See TAXONOMY.

SPECIFIC GRAVITY (sg), or **relative density**, ratio of the density of a substance to that of a reference material at a specified temperature, usually water at 4°C. If the sg of an inert substance is less than unity (1), it will float in water at 4°C. The sg of liquids is measured with a HYDROMETER.

SPECIFIC HEAT, the HEAT required to raise the temperature of 1kg of a substance through one KELVIN; expressed in J/K.kg, and measured by CALORIMETRY. The concept was introduced by Joseph BLACK; DULONG and Petit showed that the specific heat of elements is approximately inversely proportional to their ATOMIC WEIGHTS, which could thus be roughly determined.

SPECTACLES. See GLASSES.

SPECTROHELIOGRAPH, a device used to obtain spectroheliograms, composite photographs of the SUN in LIGHT of a single wavelength. An optical IMAGE of the sun is scanned by a slit admitting light to a PRISM which forms a SPECTRUM. A second slit is used to select a particular wavelength and expose a photographic plate passing behind it at the same rate as the first slit is scanning the image.

SPECTROPHOTOMETRY, the measurement of the intensity of different colors (wavelengths) present in a beam of light, normally using a PRISM or DIFFRACTION grating, and a PHOTOELECTRIC CELL. It is used for COLOR comparison and in chemical ANALYSIS.

SPECTROSCOPY, the production, measurement and analysis of SPECTRA; an essential tool of astronomers, chemists and physicists. All spectra arise from transitions between discrete energy states of matter, as a result of which PHOTONS of corresponding energy (and hence characteristic FREQUENCY or wavelength) are absorbed or emitted. From the energy levels thus determined, atomic and molecular structure may be studied. Moreover, by using the observed spectra as "fingerprints," spectroscopy may be a sensitive method of chemical ANALYSIS. Most of the different kinds of spectroscopy, corresponding to the various regions of ELECTROMAGNETIC RADIATION, relate to particular kinds of energy-level transitions. **Gamma-ray spectra** arise from nuclear energy-level transitions; **X-ray spectra** from inner-electron transitions in atoms; **ultraviolet** and **visible spectra** from outer (bonding) electron transitions in molecules (or atoms); **infrared spectra** from molecular vibrations; and **microwave spectra** from molecular rotations. There are several more specialized kinds of spectroscopy. **Raman spectroscopy**, based on the effect discovered by C. V. RAMAN, scans the scattered light from an intense monochromatic beam. Some of the scattered light is at lower (and higher) frequencies than the incident light, corresponding to vibration/rotation transitions. The technique thus supplements infrared spectroscopy. **Mössbauer spectroscopy**, based on the MÖSSBAUER EFFECT, gives information on the electronic or chemical environments of nuclei; as does **nuclear magnetic resonance spectroscopy** (nmr), based on transitions between nuclear SPIN states in a strong magnetic field. **Electron spin resonance spectroscopy** (esr) is similarly based on electron spin transitions when there is an unpaired electron in an ORBITAL, and so is used to study FREE RADICALS. The intrument used is a **spectroscope**, called a *spectrograph* if the spectrum is recorded photographically all at once, or a *spectrometer* if it is scanned by wavelength and calibrated from the instrument.

SPECTRUM, the array of colors produced on passing LIGHT through a PRISM; also, by extension, the range of a phenomenon displayed in terms of one of its properties. ELECTROMAGNETIC RADIATION arranged according to wavelength thus forms the electromagnetic spectrum, of which that of visible light is only a minute part. Similarly the mass spectrum of a particular collection of ions displays their relative numbers as a function of their masses. (See SPECTROSCOPY; MASS SPECTROSCOPY.)

SPEECH AND SPEECH DISORDERS. Speech

Infrared absorption spectrogram of polystyrene. The absorbtion bands correspond to modes of intramolecular vibration. Infrared spectroscopy is a standard tool of organic chemical analysis and quality control.

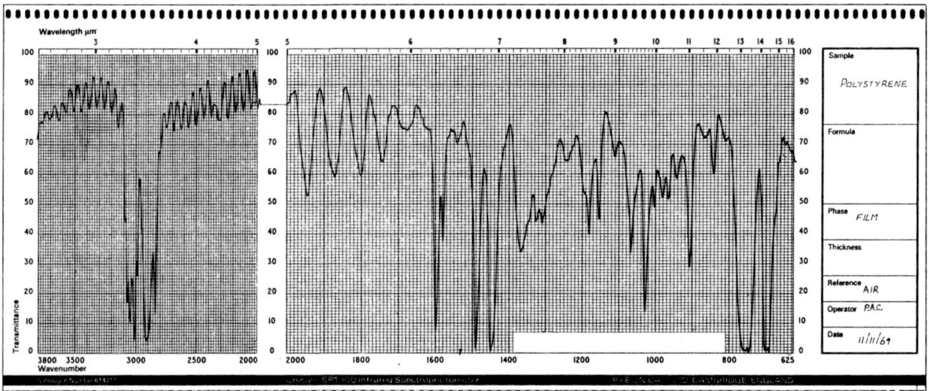

may be subdivided into conception, or formulation, and production, or phonation and articulation, of speech (see VOICE). Speech development in children starts with associating sounds with persons and objects, comprehension usually predating vocalization by some months. Nouns are developed first, often with one or two syllables only; later acquisition of verbs, adjectives, etc. allows the construction of phrases and sentences. A phase of babbling speech, where the child toys with sounds resembling speech, is probably essential for development. READING is closely related to speech development, involving the association of auditory and visual symbols. Speech involves coordination of many aspects of BRAIN function (HEARING, VISION, etc.) but three areas particularly concerned with aspects of speech are located in the dominant hemisphere of right-handed persons and in either hemisphere of left-handed people (see HANDEDNESS). DISEASE of these parts of the brain leads to characteristic forms of dysphasia or APHASIA, ALEXIA, etc. Developmental DYSLEXIA is a childhood defect of visual pattern recognition. Stammering or stuttering, with repetition and hesitation over certain syllables, is a common disorder, in some cases representing frustrated left-handedness. Dysarthria is disordered voice production and is due to disease of the neuromuscular control of voice. In speech therapy, attempts are made to overcome or circumvent speech difficulties, this being particularly important in children (see also DEAFNESS).

SPEED. See VELOCITY; also AMPHETAMINES.

SPEEDOMETER, instrument for indicating the speed of a motor vehicle. The common type works by magnetic INDUCTION. A circular permanent magnet is rotated by a flexible cable geared to the transmission. The rotating magnetic field induces a magnetic field in an aluminum cup, so tending to turn it in the same direction as the magnet. This TORQUE, proportional to the speed of rotation, is opposed by a spiral spring. The angle through which the cup turns against the spring measures the speed. The speedometer is usually coupled with an **odometer**, a counting device geared to the magnet, which registers the distance traveled.

Cutaway of an automobile speedometer with odometer. An alternative to the pointer-and-dial type of readout employs a helical stripe on a cylindrical barrel, rotating behind a calibrated window.

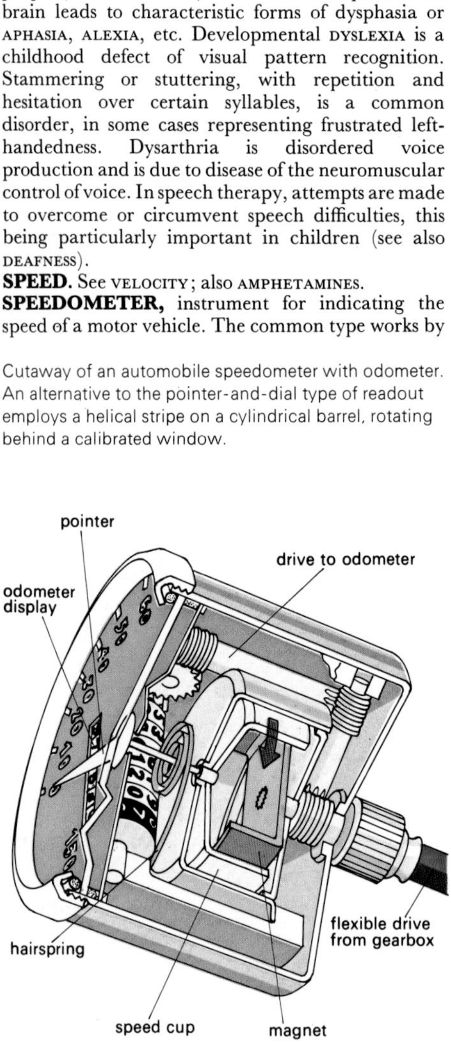

SPEMANN, Hans (1869–1941), German embryologist awarded the 1935 Nobel Prize for Physiology or Medicine for his researches into the development of the EMBRYO, showing that specific CELLS adopted specific functions not through any predetermination of form but because of the action of local chemical "organizers" (in fact, HORMONES).

SPENCER, Herbert (1820–1903), English philosopher, social theorist and early evolutionist. In his multivolume *System of Synthetic Philosophy* (1862–96), he expounded a world view based on a close study of physical, biological and social phenomena, arguing that species evolve by a process of differentiation from the simple to the complex. His political individualism deeply influenced US social thinking. (See also SURVIVAL OF THE FITTEST.)

SPERM, the male GAMETE or sex cell in animals. Sperm are usually motile, having a single flagellum (see REPRODUCTION).

SPERMATOPHYTA, traditional but artificial division of the PLANT KINGDOM which includes the GYMNOSPERMS and ANGIOSPERMS. Although these two groups are related, it is now clear that other groups such as the FERNS (set apart in the PTERIDOPHYTA) are also closely allied.

SPERRY, Elmer Ambrose (1860–1930), US inventor of the GYROCOMPASS (first installed in a ship, 1911) and of a high-intensity arc searchlight (1918).

SPHAGNUM, or **peat moss**, a large genus of MOSSES all of which have remarkable water absorptive powers. Sphagnum is used for packing bulbs and flowers, as compost and on some occasions for surgical dressings. Decomposed Sphagnum is an important consituent of PEAT.

SPHALERITE, or **Blende**, the low-temperature (β) form of zinc sulfide (ZnS); the chief ore of ZINC, occurring worldwide with GALENA. It forms lustrous crystals of the isometric system, white when pure, but usually brown to black with iron impurity. (See also WURTZITE.)

SPHENODON, or tuatara, a member of the otherwise fossil reptile group RHYNCHOCEPHALIA that survives on islands off New Zealand.

SPHENOPSIDA. See HORSETAILS.

SPHERE, the surface produced by the rotation of a CIRCLE through 180° about one of its diameters. The intersection of a sphere and any plane is circular; should the plane pass through the center, the intersection is a great circle (see also SPHERICAL GEOMETRY). The surface AREA of a sphere is $4\pi r^2$, where r is the radius; its VOLUME $4\pi r^3/3$. If mutually perpendicular (see ANGLE) x-, y- and z-AXES are constructed such that they intersect at the center, the sphere's equation is $x^2+y^2+z^2=r^2$. (See also ELLIPSOID.)

SPHERICAL COORDINATES, a system in which a POINT P in SPACE is located by its position relative to three mutually perpendicular AXES (the x-, y- and z-axes) in terms of (1) its distance r from the origin, O, (2) the ANGLE (θ) between the x-axis and the PROJECTION of OP onto the x-y PLANE, and (3) the angle (ϕ) between OP and the z-axis. The coordinates of P are thus expressed in the form (r,θ,ϕ). (See also ANALYTIC GEOMETRY.)

(1) Spherical coordinates system, defining point P in terms of distance r and angles θ and ϕ. (2) The spherical triangle ABC has "sides" a, b, c which are the angles BOC, AOC and AOB respectively and "angles" A, B, C where A is the angle between the tangents to the arcs AC and AB at A, B that between the tangents to the arcs AB and BC at B, and C that between the tangents to the arcs AC and BC at C respectively. Calculations follow the cosine rule:

$$\cos C = \frac{\cos c - \cos a \cos b}{\sin a \sin b}$$

the sine rule:

$$\frac{\sin a}{\sin A} = \frac{\sin b}{\sin B} = \frac{\sin c}{\sin C}$$

and also:

$$\tan \frac{A}{2} = \sqrt{\frac{\sin (s-b) \sin (s-a)}{\sin s \sin (s-a)}}$$

where $s=\frac{1}{2}(a+b+c)$.

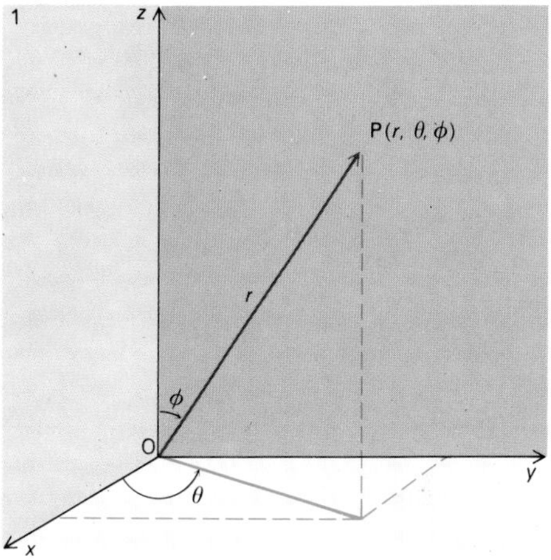

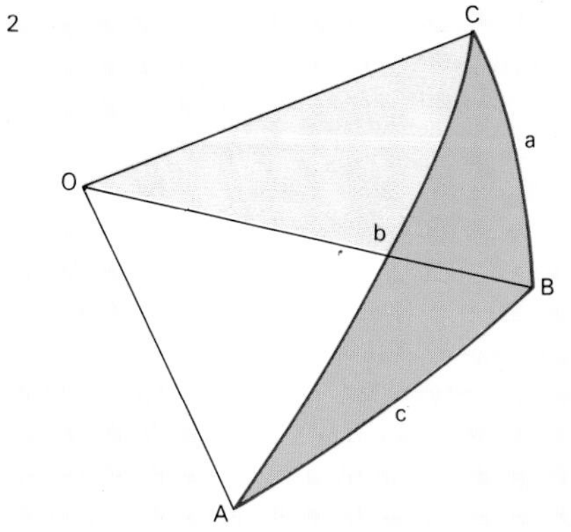

SPHERICAL GEOMETRY, the branch of GEOMETRY dealing with figures drawn on the surface of a SPHERE; sometimes considered as a special case of RIEMANNIAN GEOMETRY. A CIRCLE whose center coincides with that of the sphere is a great circle, other circles on the sphere's surface being small circles: since a great circle may be drawn through any two POINTS on the sphere's surface, one deals primarily with great circles only. The lengths of arcs of great circles are always given in terms of the radius of the sphere and the ANGLE subtended by the arc at the center; i.e., in the form $r\theta$, where r is the radius and θ the angle. It is usually convenient to consider the sphere as of unit radius, thus expressing the length of an arc as an angle. Problems concerning **spherical triangles** are solved using spherical trigonometry.

SPHERICAL TRIGONOMETRY, the branch of SPHERICAL GEOMETRY dealing with the ratios between the sides and ANGLES of spherical TRIANGLES. The sum of the angles of a spherical triangle is always between 180° and 540°, and the amount by which this sum exceeds 180° for a particular triangle is termed the excess; the area of the triangle is $\pi r^2 E/180°$, where r is the radius of the sphere and E the triangle's excess. The relations between the sides and angles of spherical triangles are governed by the COSINE RULE and the SINE RULE. **Spherical Astronomy**, of importance in positional ASTRONOMY and SPACE EXPLORATION, is the application of spherical trigonometry to determinations of stellar positions on the CELESTIAL SPHERE.

SPHEROID. See ELLIPSOID.

SPHINCTER, muscle, group of muscles or aggregation of smooth (visceral) muscle fibers, that can temporarily prevent movement of the contents of hollow viscera under autonomic or voluntary control.

SPHYGMOMANOMETER, instrument for measuring blood pressure by determining the pressure (with a MANOMETER) in a cuff attached around a limb needed to prevent blood flow. A stethoscope is used to hear the changing sounds over the ARTERY as the cuff deflates.

SPIEGELEISEN, an ALLOY based on PIG IRON, containing 5% carbon and 15%–30% MANGANESE. It is added during STEEL manufacture as a reducing agent and to supply manganese.

SPILLWAY. See DAM.

SPIN, intrinsic angular MOMENTUM of a nucleus or SUBATOMIC PARTICLE arising from its rotation about an axis within itself. Every particle has a definite spin, s, given by $nh/4\pi$, where n is an INTEGER and h is the PLANCK CONSTANT.

SPINA BIFIDA, congenital defect of the SPINAL CORD and spinal canal leading to a variable degree of leg PARALYSIS and loss of urine and feces SPHINCTER control; it may be associated with other malformation—particularly HYDROCEPHALUS. It is an embryological disorder due to failure of fusion of the neural tube. SURGERY has made it possible to treat mild cases by closure of the defect and ORTHOPEDIC procedures can be applied to balance muscle power. Also, any hydrocephalus must be shunted.

SPINAL COLUMN. See VERTEBRAE.

SPINAL CORD, the part of the central NERVOUS SYSTEM outside the SKULL. It joins the BRAIN at the base of the skull, forming the *medulla oblongata*, and extends downward in a bony canal enclosed in the VERTEBRAE. Between the bone and cord are three sheaths of connective TISSUE called the *meninges*. A section of the cord shows a central core of *gray matter* (containing the cell bodies of nerve fibers running either to the muscles or within the cord itself), completely surrounded by *white matter* (composed solely of nerve fibers). There is a central canal containing CEREBROSPINAL FLUID, which opens into the cavities of the brain.

SPINAL TAP, or **lumbar puncture**, procedure to remove CEREBROSPINAL FLUID (CSF) from the lumbar spinal canal using a fine needle. It is used in diagnosis of MENINGITIS, ENCEPHALITIS, MULTIPLE SCLEROSIS and TUMORS. In NEUROLOGY, it may be used in treatment, by reducing CSF pressure or allowing insertion of DRUGS.

SPINELS, group of OXIDE minerals of general formula $M_2^{III}M^{II}O_4$, formed in high-temperature IGNEOUS or METAMORPHIC ROCKS. The chief members are spinel itself, aluminum magnesium oxide, with some GEM varieties; CHROMITE; and MAGNETITE. Free substitution occurs. Synthetic spinels are used as refractories and in solid-state components.

SPINNING, the ancient craft of twisting together FIBERS from a mass to form strong, continuous thread suitable for weaving. The earliest method was merely to roll the fibers between hand and thigh. Later two sticks were used: the distaff to hold the bundle of fibers, and a spindle to twist and wind the yarn. Mechanization began with the spinning wheel, invented in India and spreading to Europe by the 14th century. The wheel turned the spindle by means of a belt drive. In the 15th century the flyer was invented: a device on the spindle shaft that winds the yarn automatically on a spool. Improved WEAVING methods in the Industrial Revolution caused increased demand which provoked several inventions. The spinning jenny, invented by James HARGREAVES (c1767), spun as many as 16 threads at once, the spindles all being driven by the same wheel. Richard ARKWRIGHT's "water frame" (1769), so called from being water-powered, had rollers and produced strong thread. Then Samuel CROMPTON produced a hybrid of the two—his "mule"—which had a movable carriage, and was the forerunner of the modern machine. The other modern spinning machine is the ring-spinning frame (1828) in which the strands, drawn out by rollers, are twisted by a "traveler" that revolves on a ring around the bobbin on which they are wound.

SPIRAL, in plane ANALYTIC GEOMETRY, a CURVE traced by a POINT which either approaches or recedes from an origin while moving around it. The most frequently encountered is the Archimedean Spiral, whose equation in polar coordinates is $r = a\theta$ (see ARCHIMEDES). Others include the logarithmic spiral $r = e^{a\theta}$ (see EXPONENT; EXPONENTIAL) and the hyperbolic spiral $r = a/\theta$. Three-dimensional spirals such as the HELIX are traced by points moving about an axis.

SPIRIT LEVEL. See LEVEL.

SPIROCHETE, spiral BACTERIA, species of which are responsible for RELAPSING FEVER, YAWS and syphilis (see VENEREAL DISEASES).

SPIROGYRA, filamentous green ALGAE consisting of long chains of CELLS. Free-floating, they often appear as a scum on still water.

SPLEEN, spongy vascular lymphoid organ (see LYMPH) between the STOMACH and DIAPHRAGM on the left side of the ABDOMEN. A center for the RETICULOENDOTHELIAL SYSTEM, it also eliminates worn-out red BLOOD cells, recycling their iron. Most of its functions are duplicated by other organs. The spleen was classically the source of black bile, or melancholy (see HUMORS).

SPLENIC FEVER, name for ANTHRAX in animals.

SPOCK, Benjamin McLane (1903–), known worldwide as "Dr. Spock," US pediatrician and pacifist best known for his (*Common Sense Book of*) *Baby and Child Care* (1946), which advocated a more liberal attitude on the part of parents, and *Bringing up Children in a Difficult Time* (1974).

SPODUMENE, a lithium-bearing PYROXENE, $LiAlSi_2O_6$; a major LITHIUM ore found in PEGMATITES, forming clear, monoclinic crystals, often large and when colored used as GEM stones.

SPONTANEOUS COMBUSTION, COMBUSTION occurring without external ignition, caused by slow OXIDATION or FERMENTATION which (if heat cannot readily escape) raises the temperature to burning point. It may occur when hay or small coal is stored.

SPONTANEOUS GENERATION, or **abiogenesis**, theory, dating from the writings of ARISTOTLE, that living creatures can arise from nonliving matter. The idea remained current even after it had become clear that higher orders of life could not be created in this way; and it was only with the work of REDI, showing that maggots did not appear in decaying meat to which flies had been denied access, and PASTEUR, who proved that the equivalent was true of microorganisms (i.e., BACTERIA), that the theory was finally discarded.

SPORE, minute single or multicelled body produced during the process of reproduction of many plants, particularly BACTERIA, ALGAE and FUNGI and in some PROTOZOA. The structure of spores varies greatly and depends upon the means of dissemination from the parent. Some, e.g., the zoospores of algae, are motile.

SPOROPHYTE, phase of the life-cycle of a plant representing the diploid generation. (See ALTERNATION OF GENERATIONS.)

SPOROZOA. See PROTOZOA.

SPRAGUE, Frank Julian (1857–1934), US inventor of high-speed electric ELEVATORS and electric RAILROAD systems, including that now used in the New York SUBWAY.

SPRAIN. See LIGAMENT.

SPRING, a naturally occurring flow of water from the ground. This may be, for example, an outflow from an underground stream; but most often a spring occurs where an AQUIFER saturated with GROUNDWATER intersects with the earth's surface. Such an aquifer, if confined above and below by aquicludes, may travel for hundreds of kilometres underground before emerging to the surface, there, perhaps, in desert areas giving rise to OASES. Spring water is generally fairly clean, since it has been filtered through the permeable rocks; but all spring water contains some dissolved MINERALS. (See also GEYSER; HOT SPRINGS; WELL.)

SPRING, mechanical device that exhibits ELASTICITY according to HOOKE's Law. Most springs are made of steel, brass or bronze. The commonest type is the **helical spring**, a helical coil of stiff wire, loose-wound if to be compressed, tight-wound if to be extended under tension. They have many uses, including closing valves, spring BALANCES and ACCELEROMETERS. The **spiral spring** is a wire or strip coiled in one plane, responding to TORQUE applied at its inner end, and used to store energy, notably in CLOCKS AND WATCHES. The **leaf spring**, used in vehicle suspension systems, consists of several steel strips of different lengths clamped on top of each other at one end. When deformed, springs store potential ENERGY, and exert a restoring FORCE. Hydraulic and air springs work by compression of a fluid in a cylinder.

SQUALL, a sudden increase in WIND speed of 8m/s or more, raising the wind speed to at least 11m/s and lasting for one minute or longer. Commonly associated with thunderstorms and heavy rain, squalls may do great damage. A **squall line** is a line of thunderstorms often hundreds of kilometres long with squalls along its advancing edge.

SQUAMATA, members of the REPTILIA that include the LACERTILIA (lizards) and the OPHIDIA (snakes).

SQUARE, a regular QUADRILATERAL. In ALGEBRA, the square (x^2) of a number, x, is defined as $x \times x$: e.g., $10^2 = 10 \times 10 = 100$.

SQUARE ROOT. See ROOTS.

SQUINT. See STRABISMUS.

STABILIZER. See AIRPLANE; GYROSTABILIZER.

STAHL, Georg Ernst (1660–1734), Bavarian-born German physician and chemist who developed the PHLOGISTON theory to explain combustion.

STAINLESS STEEL, corrosion-resistant STEEL containing more than 10% chromium, little carbon, and often nickel and other metals. Made in the ELECTRIC FURNACE, there are four main types: ferritic, martensitic, austenitic and precipitation-hardening. Stainless steel is used for cutlery and many industrial components.

STALACTITES AND STALAGMITES, rocky structures found growing downward from the roof (stalactites) and upward from the floor (stalagmites) of CAVES formed in LIMESTONE. Rainwater percolates through the rocks above the cave and, as it contains atmospheric CARBON dioxide, can dissolve calcium carbonate en route. On reaching the cave, the water drips from the roof to the floor; as a drop hangs, some water evaporates, leaving a little calcium carbonate as CALCITE on the roof. Repetition forms a stalactite; and evaporation of the fallen water on the floor forms a stalagmite. On occasion, the rising stalagmite and descending stalactite fuse to form a pillar.

Over 68% of a population exhibiting a normal distribution for some property lie within one standard deviation (σ) of the mean (μ). Over 95% lie within μ±2σ and over 99% within μ±3σ. The total population in this example is unity (1).

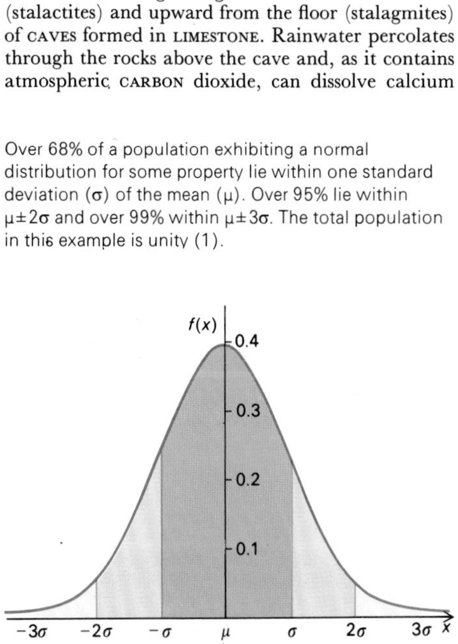

STAMEN. See FLOWER.

STAMMER. See SPEECH AND SPEECH DISORDERS.

STANDARD DEVIATION, in STATISTICS, the positive square ROOT of the sample variance s^2, defined as the arithmetic mean (see MEAN, MEDIAN AND MODE) of the SQUARES of the deviations of the members of a sample from the arithmetic mean of the sample:

$$s^2 = \frac{1}{n}\sum_{i=1}^{n}(x_i - \mu)^2,$$

where n is the number of observations, μ the mean, and x_i the ith value. Here s is the *sample* standard deviation: the *population* standard deviation σ is given by the square root of the population VARIANCE. Standard deviations provide a measure of the dispersion of a distribution about its mean: for example, in a NORMAL DISTRIBUTION, about 68.3% of the population lies within one standard deviation of the mean. (See also CHI-SQUARED TEST.)

STANDARD TIME, the practice of defining the hour of the day throughout a specified geographical area (time zone) as being so many hours behind or ahead of an international standard, GREENWICH MEAN TIME (GMT). The US and Canada are covered by five such zones.

STANFORD-BINET TEST, an adaptation of the Binet-Simon test for INTELLIGENCE, introduced by TERMAN (1916, revised 1937), and used primarily to determine the IQs of children. (See also BINET.)

STANLEY, Wendell Meredith (1904–1971), US biochemist who shared with J. NORTHROP and J. SUMNER the 1946 Nobel Prize for Chemistry for his first crystallization of a VIRUS.

STAPHYLOCOCCUS, BACTERIUM responsible for numerous SKIN, soft tissue and BONE infections, less often causing SEPTICEMIA, a cavitating PNEUMONIA, bacterial endocarditis and enterocolitis. BOILS, CARBUNCLES, IMPETIGO and OSTEOMYELITIS are commonly due to Staphylococci. Treatment usually requires drainage of PUS from ABSCESSES, and ANTIBIOTICS.

STAR, a large incandescent ball of gases held together by its own gravity. The SUN is a fairly normal star in its composition, parameters and color. The lifespan of a star depends upon its mass and luminosity: a very luminous star may have a life of only one million years, the sun a life of ten billion years, the faintest main sequence stars a life of ten thousand billion years. Stars are divided into two categories, Populations I and II. The stars in Population I are slower moving, generally to be found in the spiral arms of GALAXIES, and believed to be younger. Population II stars are generally brighter, faster moving and mainly to be found in the spheroidal halo of stars around a galaxy and in the GLOBULAR CLUSTERS. Many stars are DOUBLE STARS. It is believed that stars originate as condensations out of INTERSTELLAR MATTER. In certain circumstances a protostar will form, slowly contracting under its own gravity, part of the energy from this contraction being

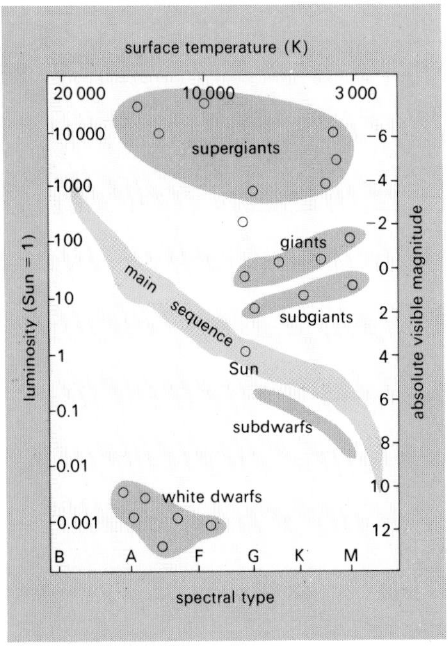

The Hertzsprung-Russel diagram is a plot of stars' luminosities against their spectral types. (These parameters can also be expressed in terms of absolute magnitudes and approximate surface temperatures.) The majority of stars, including the sun, belong to the main sequence. Diagrams of this type are useful for describing the processes of stellar evolution.

radiated, the remainder heating up the core: this stage may last several million years. At last the core becomes hot enough for thermonuclear reactions (see FUSION, NUCLEAR) to be sustained, and stops contracting. Eventually the star as a whole ceases contracting and radiates entirely by the thermonuclear conversion of hydrogen into helium: it is then said to be on the main sequence. When all the hydrogen in the core has been converted into helium, the now purely helium core begins to contract while the outer layers continue to "burn" hydrogen: this contraction heats up the core and forces the outer layers outward, so that the star as a whole expands for some 100–200 million years until it becomes a red **giant star**. Although the outer layers are comparatively cool, the core has become far hotter than before, and thermonuclear conversions of helium into carbon begin. The star contracts once more (though some expand still further to become **supergiants**) and ends its life as a white **dwarf star**. It is thought that more massive stars become **neutron stars**, whose matter is so dense that its PROTONS and ELECTRONS are packed together to form NEUTRONS; were the sun to become a neutron star, it would have a radius of less than 20km. Finally, when the star can no longer radiate through thermonuclear or gravitational means, it ceases to shine. Some stars may at this stage undergo ultimate gravitational collapse to form BLACK HOLES. (See also CARBON CYCLE; CEPHEID VARIABLES; CONSTELLATION; COSMOLOGY; GALACTIC CLUSTER; MAGELLANIC CLOUDS; MILKY WAY; NEBULA; NOVA; PULSAR; QUASAR; SOLAR SYSTEM; STAR CLUSTER; SUPERNOVA; UNIVERSE; VARIABLE STAR.)

STARCH, a CARBOHYDRATE consisting of chains of GLUCOSE arranged in one of two forms to give the polysaccharides amylose and amylopectin. Amylose consists of an unbranched chain of 200–500 glucose units, whereas amylopectin consists of chains of 20 glucose units joined by cross links to give a highly branched structure. Most natural starches are mixtures of amylose and amylopectin; e.g., potato and cereal starches are 20%–30% amylase and 70%–80% amylopectin. Starch is found in plants, occurring in grains scattered throughout the CYTOPLASM. The grains from any particular plant have a characteristic microscopic appearance and an expert can tell the source of a starch by its appearance under the microscope. Starches in the form of rice, potatoes and wheat or other cereal products supply about 70% of the world's food.

STAR CLUSTER, a cluster of stars sharing a common origin. There are two distinct types of star cluster: see GALACTIC CLUSTER; GLOBULAR CLUSTER.

STARK, Johannes (1874–1957), Bavarian-born German physicist awarded the 1919 Nobel Prize for Physics for discovering the **Stark effect** (1913), the splitting of degenerate spectral lines through the application of a powerful ELECTRIC FIELD. The explanation of this was an early triumph of QUANTUM THEORY. (See also ZEEMAN, P.)

STARLING, Ernest Henry, British physiologist. See BAYLISS, SIR WILLIAM MADDOCK.

STATAMPERE, the unit of current in the CGS electrostatic system of units (esu—see CGS UNITS). Like other **stat-units**, the statampere is used in computations in which the PERMITTIVITY of free space is set dimensionless (see DIMENSIONS) and equal to UNITY.

Stat-units

quantity	CGS esu system	SI equivalent
charge	1 statcoulomb	3.336×10^{-10} coulomb
current	1 statampere	3.336×10^{-10} ampere
potential	1 statvolt	2.998×10^{2} volt
resistance	1 statohm	8.987×10^{11} ohm
capacitance	1 statfarad	1.113×10^{-12} farad
inductance	1 stathenry	8.987×10^{11} henry

STATIC, an accumulation of electric charge (see ELECTRICITY) responsible, e.g., for the attractive and repulsive properties produced in many plastics and fabrics by rubbing. It leaks away gradually through warm damp air, but otherwise may cause small sparks (and consequent RADIO interference) or violent discharges such as LIGHTNING.

STATICS, branch of MECHANICS dealing with systems in EQUILIBRIUM, i.e., in which all FORCES are balanced and there is no motion.

STATISTICS, the area of mathematics concerned with the manipulation of numerical information. The science has two branches: descriptive statistics, dealing with the classification and presentation of data, and inferential or analytical statistics, which studies ways of collecting data, its analysis and interpretation. Sampling is fundamental to statistics. Since it is usually impractical to treat of every element in a population (the group under consideration), a representative (often random) sample is instead examined, its properties being ascribed to the whole group. The data is analyzed in series of PARAMETERS such as the STANDARD DEVIATION and the MEAN of the data distribution. The distribution may be presented as a HISTOGRAM, a frequency polygon, or a frequency curve. Ideally, a statistician aims to devise a MATHEMATICAL MODEL of the distribution, especially if it approximates to normality (see NORMAL DISTRIBUTION); there are many tests which he can use to determine whether or not his model "fits" (see CHI-SQUARED TEST; STUDENT'S *t*-DISTRIBUTION). Statistics is used throughout science, wherever there is an element of PROBABILITY involved, and also in industry, politics, market analysis and traffic control. (See also DEGREES OF FREEDOM; STOCHASTIC PROCESS; VARIANCE.)

STAT-UNITS. See STATAMPERE.

STAUDINGER, Hermann (1881–1965), German chemist awarded the 1953 Nobel Prize for Chemistry for showing that the long chains of MOLECULES known as POLYMERS are really giant molecules held together by normal chemical BONDS.

STEADY STATE THEORY. See COSMOLOGY.

STEAM, the vapor formed from WATER at or above the boiling point (100°C). It is colorless, but appears white when it contains droplets of condensed water. Steam is used to make WATER GAS; it is a valuable industrial heat carrier because the LATENT HEAT of water is very high (40.65kJ/mol) and steam has a high SPECIFIC HEAT: it is thus used in central heating and in PRESSURE COOKERS. Because steam occupies about 1 700 times the volume of the water producing it, when water is heated in a boiler, pressure is built up, which is used to drive TURBINES and STEAM ENGINES. Steam occurs naturally in springs, geysers and volcanoes.

STEAM ENGINE, the first important heat ENGINE, supplying the power that made the Industrial Revolution possible, and the principal power source for industry and transport (notably railroad locomotives and steamships) until largely superseded in the 20th century by steam TURBINES and the various INTERNAL-COMBUSTION ENGINES. The steam engine is an external-combustion engine, the steam being raised in a BOILER heated by a furnace; it is also a RECIPROCATING ENGINE. There are two main types: condensing, in which the pressure drop is caused by cooling the steam and so condensing it back to water; and noncondensing, in which the steam is exhausted to the atmosphere. The first major precursor of the steam engine was Thomas SAVERY'S steam pump (1698), worked by the partial vacuum created by condensing steam in closed chambers. It had no moving parts, however, and the first working reciprocating engine was that of Thomas NEWCOMEN (1712): steam was admitted to the cylinder as the piston moved up, and was condensed by a water spray inside the cylinder, whereupon the air pressure outside forced the piston down again. James WATT radically improved Newcomen's engine (1769) by condensing the steam outside the cylinder (thus no longer having to reheat the cylinder at each stroke) and by using the steam pressure to force the piston up. He later found that, if steam were admitted for only part of the stroke, its expansion would do a good deal of extra work. (The principles involved were later studied by CARNOT and became the basis of THERMODYNAMICS.) Watt also invented the double-action principle—both strokes being powered, by applying the steam alternately to each end of the piston—the flyball GOVERNOR, and the crank and "sun-and-planet" devices for converting the piston's linear motion to rotary motion. The compound engine (1781) makes more efficient use of the steam by using the exhaust steam from one cylinder to drive the piston of a second cylinder. Later developments included the use of high-pressure steam by Richard TREVITHICK and Oliver EVANS.

STEARIC ACID ($CH_3(CH_2)_{16}COOH$), white solid CARBOXYLIC ACID obtained from its ESTER, glyceryl tristearate (tristearin), which is found in many natural FATS and oils. Sodium stearate is a principal constituent of many SOAPS.

STEATITE. See SOAPSTONE.

STEATOPYGIA, excessive accumulation of fat in the buttocks, usually of females, common among the Hottentots and some other African peoples.

STEEL, an ALLOY of IRON and up to 1.7% carbon, with small amounts of manganese, phosphorus, sulfur and silicon. These are termed carbon steels; those with other metals are termed alloy steels; low-alloy steels if they have less than 5% of the alloying metal, high-alloy steels if more than 5%. Carbon steels are far stronger than iron, and their properties can be tailored to their uses by adjusting composition and treatment. Alloy steels—including STAINLESS STEEL—are used for their special properties. Steel was first mass-produced in the mid-19th century, and steel production is now one of the chief world industries, being basic to all industrial economies. The US, the USSR and Japan are the major producers. Steel's innumerable uses include automobile manufacture, shipbuilding, skyscraper frames, reinforced concrete and machinery of all kinds. All steelmaking processes remove the impurities in the raw materials—PIG IRON, scrap steel and reduced iron ore—by oxidizing them with an air or oxygen blast. Thus most of the carbon, silicon, manganese, phosphorus and sulfur are converted to their oxides and, together with added FLUX and other waste matter present, form the SLAG. The main processes are the BESSEMER PROCESS, the LINZ-DONAWITZ PROCESS and the similar electric-arc process used for highest-quality steel, and the OPEN-HEARTH PROCESS. Most modern processes use a basic slag and a basic refractory furnace lining: acidic processes are incapable of removing phosphorus. When the impurities have been removed, desired elements are added in calculated proportions. The molten steel is cast as ingots which are shaped while still red-hot in ROLLING MILLS, or it may be cast as a continuous bar ("strand casting"). The properties of medium-carbon (0.25% to 0.45% C) and high-carbon (up to 1.7% C) steels may be greatly improved by heat treatment: ANNEALING, CASEHARDENING and TEMPERING. Steel metallurgy is somewhat complex: unhardened steel may contain combinations of three phases—austenite, ferrite and cementite—differing in structure and carbon content; hardened steel contains martensite, which may be thought of as ferrite supersaturated with carbon.

STEFAN-BOLTZMANN LAW. See BLACKBODY RADIATION.

STEGOSAURUS, a DINOSAUR that grew to a length of about 5.5m (18ft) and was characterized by large bony plates that grew from its back.

STEIN, William Howard (1911–), US biochemist who shared with C. B. ANFINSEN and S. MOORE the 1972 Nobel Prize for Chemistry for his part in determining the structure of the ENZYME ribonuclease (see also NUCLEIC ACIDS).

STEINMETZ, Charles Proteus (1865–1923), formerly **Karl August Rudolf Steinmetz**, German-born US electrical engineer who first worked out the theory of HYSTERESIS (c1892), but who is best remembered for working out the theory of alternating current (1893 onward), so making it possible for AC to be used rather than DC in most applications.

STEM, the part of the PLANT which supports the LEAVES, FLOWERS and FRUITS. It may be short or creeping in low-growing plants or tall as in TREES, or even underground (see RHIZOME). Stems also conduct nutrients, other substances and water between the various organs of the plant. Green stems carry out PHOTOSYNTHESIS.

STENCIL. See DUPLICATING MACHINE; SILK-SCREEN PRINTING.

STENO, Nicolaus (1638–1686), or **Niels Stensen**, Danish geologist, anatomist and bishop. In 1669 he published the results of his geological studies: he recognized that many rocks are sedimentary (see SEDIMENTARY ROCKS); that FOSSILS are the remains of once-living creatures and that they can be used for DATING purposes, and established many of the tenets of modern crystallography.

STENOGRAPHY. See SHORTHAND.

STENOTYPE, system of machine SHORTHAND that uses a keyboard machine like a typewriter except that several keys may be depressed at once. Letter groups phonetically represent words. The machine, silent in operation, is capable of 250 words/min.

STEPHENSON, British family of inventors and railroad engineers. **George Stephenson** (1781–1848) first worked on stationary STEAM ENGINES, reconstructing and modifying one by NEWCOMEN (c1812). His first LOCOMOTIVE, the *Blucher*, took to the rails in 1814: it traveled at 4mph (about 6.5km/h) hauling coal for the Killingworth colliery, and incorporated an important development, flanged wheels. About this time, independently of DAVY, he invented a SAFETY LAMP: this earned him £1 000 (then about $5 000), which helped finance further locomotive experiments. In 1821 he was appointed to survey and engineer a line from Darlington to Stockton: in 1825 his *Locomotion* carried 450 people along the line at a rate of 15mph (about 25km/h), and the modern RAILROAD was born. This was followed in 1829 by the success of the *Rocket*, which ran the 40mi (65km) of his new Manchester–Liverpool line at speeds up to 30mph (about 48km/h), the first main-line passenger rail journey. His only son **Robert Stephenson** (1803–1859) helped his father on both of these lines, and with the *Rocket*, but is best known as a BRIDGE builder, notably for the tubular bridges over the Menai Straits, North Wales (1850), and the St. Lawrence at Montreal (1859).

STEPPES, extensive level GRASSLANDS of Europe and Asia (equivalent to the North American prairies and South American pampas). They extend from SW Siberia to the lower reaches of the Danube R.

STERADIAN. See ANGLE.

STEREOCHEMISTRY, the study of the arrangement in space of atoms in molecules, and of the properties which depend on such arrangements. The two chief branches are the study of STEREOISOMERS and stereospecific reactions (which involve only one isomer); and CONFORMATIONAL ANALYSIS, including the study of steric effects on reaction rates and mechanisms.

STEREOISOMERS, ISOMERS having the same molecular structure, but differing in the spatial arrangement of their atoms. There are two main types. **Optical isomers** are asymmetric molecules—usually having an asymmetric carbon atom with four different groups bonded to it—which hence have two mirror-image forms (enantiomers) and show OPTICAL ACTIVITY. Absolute spatial configurations have now been found for many isomers, and may be represented by projection formulae. Resolution, i.e. separation of the two enantiomers, is achieved by combining them with a single optical isomer, thus producing a pair of diastereoisomers which, not being mirror-images, have different properties and are separable. Inversion of configuration often occurs in substitution reactions. **Geometrical isomers** contain groups which are differently oriented with respect to a double bond or ring where rotation is impossible; they have different properties.

STEREOPHONIC SOUND. See HIGH FIDELITY.

STEREOSCOPE, optical instrument that simulates BINOCULAR VISION by presenting slightly different pictures to the two eyes so that an apparently three-dimensional image is produced. The simplest stereoscope, invented in the 1830s, used a system of mirrors and prisms (later, converging lenses) to view the pictures. In the color separation method the left image is printed or projected in red and seen through a red filter, and likewise for the right image in blue. A similar method uses images projected by POLARIZED LIGHT and viewed through polarizing filters, the polarization axes being at right angles. The pictures are produced by a stereoscopic camera with two lenses a small distance apart. The stereoscope is useful in making relief maps by aerial photographic survey.

STEREOTYPE, in printing, a duplicate plate made by casting molten alloy in a papier mâché or plastic mold made from the original type. It has been largely superseded by electrotype (see ELECTROFORMING) except for newspaper printing.

STERILITY, the condition in which an organism is unable to produce offspring. Its many possible causes include failure by either sex to produce GAMETES; the production of abnormal gametes; the inability of the male to introduce sperm into the female, or inability of the embryo to develop normally. Sterility develops naturally in old age.

STERILIZATION, surgical procedure in which the FALLOPIAN TUBES are cut and tied to prevent eggs reaching the WOMB, thus providing permanent CONTRACEPTION. The procedure is essentially irreversible and should only be performed when a woman has completed her family. It may be done by a small abdominal operation, at CESARIAN SECTION or through an instrument, the laparoscope. (See also

VASECTOMY.) *Also,* the treatment of medical equipment to ensure that it is not contaminated by BACTERIA and other microorganisms. Metal and linen objects are often sterilized by heat (in AUTOCLAVES). Chemical disinfection is also used and plastic equipment is exposed to GAMMA RAYS.

STERN, Otto (1888–1969), German-born US physicist awarded the 1943 Nobel Prize for Physics for his development of the molecular-beam method of studying the magnetic properties of ATOMS, and especially for measuring the magnetic moment of the PROTON.

STEROIDS, HORMONES produced in the body from CHOLESTEROL, mainly by the ADRENAL GLANDS, and related to ESTROGENS and ANDROGENS. All have chemical structures based on that of the STEROLS. Cortisol is the main glucocorticoid (steroids that regulate GLUCOSE metabolism) and aldosterone the main mineralocorticoid (regulating SALT, POTASSIUM and WATER balance). Increased amounts of cortisol are secreted during times of stress, e.g., SHOCK, SURGERY and severe infection. Steroids, mainly of the glucocorticoid type, are also given in doses above normal hormone levels to obtain other effects, e.g., the suppression of INFLAMMATION, ALLERGY and IMMUNITY. Diseases that respond to this include ASTHMA, MULTIPLE SCLEROSIS, some forms of NEPHRITIS, inflammatory GASTROINTESTINAL TRACT disease and cerebral EDEMA; SKIN and EYE conditions may be treated with local steroids. High-dose systemic steroids may have adverse effects if used for long periods; they may cause ACNE, osteoporosis, hypertension, fluid retention, altered facial appearance and growth retardation in children.

STEROLS, naturally occurring secondary ALCOHOLS with a fused ring structure of three six-membered carbon rings and one five-membered carbon ring, all of which are hydrogenated and contain in total one or more double bonds. Sterols are generally colorless crystalline nonsaponifiable compounds. Important sterols include CHOLESTEROL, the major sterol found in most animals, and sitosterol, found in plants.

STETHOSCOPE, instrument devised by René T. H. Laënnec (1781–1826) for listening to sounds within the body, especially those from the HEART, LUNGS, ABDOMEN and blood vessels.

STEVENS, US family of inventors and engineers. **John Stevens** (1749–1838) made many contributions to steamboat development, including the first with a screw PROPELLER (1802) and the first seagoing steamboat (*Phoenix,* 1809). He also built (1825) the first US steam locomotive. His son **Robert Livingston Stevens** (1787–1856) assisted his father, and invented the inverted-T rail still used in modern RAILROADS (1830) as well as the technique of fastening them to wooden sleepers. **Edwin Augustus Stevens** (1795–1868), another son, also made contributions to railroad technology.

STEVINUS, Simon (1548–1620), or **Simon Stevin,** Dutch mathematician and engineer who made many contributions to HYDROSTATICS; disproved, before GALILEO, ARISTOTLE's theory that heavy bodies fall more swiftly than light ones; introduced the DECIMAL SYSTEM into popular use; and first used the triangle of forces in MECHANICS.

STEWART, Dugald (1753–1828), Scottish philosopher, a major member of the COMMON SENSE SCHOOL and a principal disciple of Thomas REID.

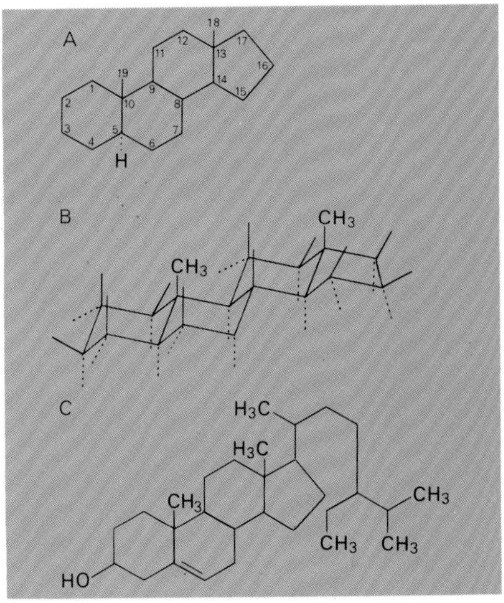

The steroid (cyclopentanophenanthrene) nucleus: (A) structure and labeling of the carbon atoms; (B) conformational structure: hydrogen atoms are to be found at the end of each unlabeled spur bond. These bonds point either above the plane of the molecule (solid lines) or below (broken lines), while the hydrogen atoms are either equatorial (roughly in the plane of the carbon skeleton) or axial (lying above or below the skeletal plane); (C) Beta-sitosterol, a typical plant steroid. Side chains and substituent atoms (in this case the hydroxyl group on carbon atom 3) are either alpha- or beta-, depending on whether their bonding points either below or above the plane of the carbon skeleton.

STIBNITE, soft, gray mineral, antimony(III) sulfide (Sb_2S_3), the chief ore of ANTIMONY, used in the manufacture of fireworks and matches; found in China, Czechoslovakia, Ida., Nev., and Cal.

STIGMA. See FLOWER.

STILB (sb), in PHOTOMETRY a unit of LUMINANCE, equal to 1 CANDELA per square centimetre.

STIMULANT, DRUG that stimulates an organ. NERVOUS SYSTEM stimulants range from ALCOHOL (an apparent stimulant only) and HALLUCINOGENIC DRUGS, to drugs liable to induce CONVULSIONS. CARDIAC stimulants include DIGITALIS and ADRENALINE and are used in cardiac failure and resuscitation respectively. Bowel stimulants have a LAXATIVE effect. WOMB stimulants (OXYTOCIN and ergometrine) are used in OBSTETRICS to induce labor and prevent postpartum HEMORRHAGE.

STOCHASTIC PROCESS, any process governed by the laws of PROBABILITY: for example, the BROWNIAN MOTION of the submicroscopic particles in a colloidal solution (see COLLOID). Most stochastics involve time: in the case of the particles, the state of the system at a time t is a random variable, $x(t)$. In mathematics, particularly STATISTICS, a stochastic is a family of random VARIABLES.

STOMACH, the large distensible hopper of the

DIGESTIVE SYSTEM. It receives food boluses from the ESOPHAGUS and mixes them with hydrochloric acid and the stomach ENZYMES; fats are partially emulsified. After some time, the pyloric SPHINCTER relaxes and food enters the DUODENUM and the rest of the GASTROINTESTINAL TRACT. Diseases of the stomach include ULCER, CANCER and pyloric stenosis, causing pain, anorexia or VOMITING; these often require SURGERY.

STOMATA, tiny pores in the surface of leaves which are automatically opened and closed by a pair of guard cells. They allow exchange of gases and water vapor between the PLANT and the atmosphere.

STONE, former UK unit of weight (=14 pounds—see WEIGHTS AND MEASURES). ROCKS and minerals used in building are often called stone.

STONE AGE, the stage in man's cultural development preceding the BRONZE AGE and the IRON AGE (see also PRIMITIVE MAN). It is characterized by man's use of exclusively stone tools and weapons, though some made of bone, wood, etc., may occur. It is split up into three periods: the **Paleolithic**, or Old Stone Age, began with the emergence of man-like creatures, the earliest stone tools being some 2.5 million years old and associated with the australopithecines (see PREHISTORIC MAN). Paleolithic tools, if worked at all, are made of chipped stone. The **Mesolithic**, or Middle Stone Age, was confined exclusively to NW Europe. Here, between c8000 and c3000 BC, various peoples enjoyed a culture showing similarities with both Paleolithic and Neolithic. In Europe, the **Neolithic**, or New Stone Age, began about 8000 BC, and was signaled by the development of agriculture, with consequent increase in stability of the population and hence elaboration of social structure. The tools of this period are of polished stone. Apart from farming, men also worked mines. The Neolithic merged slowly into the Early Bronze Age.

STONEHENGE, the ruins of a MEGALITHIC MONUMENT, dating from the STONE AGE and early BRONZE AGE, on Salisbury Plain, S England. Its most noticeable features are concentric rings of stones surrounding a horseshoe of upright stones, and a solitary vertical stone, the Heel Stone, some 100m to the NE. Stonehenge was built between c1900 BC and c1400 BC in three distinct phases. It appears to have been both a religious center and an observatory from which predictions of astronomical events could be made.

STONEWARE. See POTTERY AND PORCELAIN.

STORM, a transient but often violent atmospheric disturbance with high WINDS, accompanied by SQUALLS, PRECIPITATION, and often thunder and lightning (see THUNDERSTORMS). Varying in type with latitude and season, storms are associated with CYCLONES. (See also HURRICANE; TORNADO.)

STRABISMUS, cross-eye, or **squint**, a disorder of the EYES in which the alignment of the two ocular axes is not parallel, impairing binocular VISION; the eyes may diverge or converge. It is often congenital and may require SURGERY if orthoptics fail. Acquired squints are usually due to nerve or muscle disease and cause double vision.

STRAIN. See MATERIALS, STRENGTH OF.

STRAIT, a narrow strip of sea joining two large areas of sea, possibly the result of marine EROSION of an ISTHMUS. (See also FIRTH.)

STRANGENESS NUMBER, nonzero integral quantum number assigned to certain "strange" SUBATOMIC PARTICLES (K mesons and hyperons) because of their unusually long lifetimes. (Other particles have zero strangeness number.) Strangeness is conserved in strong nuclear interactions.

STRASSMANN, Fritz (1902–), German physicist who worked on uranium FISSION with Otto HAHN after Lise MEITNER had fled Germany (1938). All three shared the 1966 Fermi Award.

STRATIGRAPHY, the branch of GEOLOGY concerned with the chronological sequence and the correlation of rock strata in different districts. (See also PALEONTOLOGY; ROCKS; SEDIMENTARY ROCKS.)

STRATOSPHERE, the atmospheric zone immediately above the tropopause, including the OZONE layer. (See ATMOSPHERE.)

STRAW, dried stalks of grasses, principally CEREAL CROPS. It has many farm uses: for litter or bedding; as livestock food (less nutritious than hay), and as an ingredient of farmyard MANURE and composts for garden plants. Straw is used in the making of a thick low-quality cardboard (strawboard), straw hats and baskets, and in the thatching of houses.

STREAMLINING, the design of the shape of a body so as to minimize drag as it travels through a fluid; essential to the efficiency of aircraft, ships and submarines. At subsonic speeds turbulent flow is minimized by using a shape rounded in front, tapering to a point behind (see AERODYNAMICS; AIRFOIL; FLUID MECHANICS). At SUPERSONIC speeds a different shape is needed, thin and pointed at both ends, to minimize the shock waves.

STREPTOCOCCUS, BACTERIUM responsible for many common infections including sore throat, TONSILLITIS, SCARLET FEVER, IMPETIGO, cellulitis, ERYSIPELAS and PUERPERAL FEVER; a related organism is a common cause of PNEUMONIA and one type may cause endocarditis on damaged HEART valves. PENICILLIN is the ANTIBIOTIC of choice. RHEUMATIC FEVER and BRIGHT'S DISEASE are late immune responses to streptococcus.

STREPTOMYCIN, early ANTIBIOTIC (discovered by WAKSMAN) with wide spectrum antibacterial activity. For years it was the major drug for TUBERCULOSIS. Toxicity to HEARING and balance led to the use of the related gentamicin and kanamycin, and of rifampicin for tuberculosis.

STRESS. See MATERIALS, STRENGTH OF.

STRIP MINING, technique used where ORE deposits lie close enough to the surface to be uncovered merely by removal of the overlying material; most used for COAL. (See also MINING.)

STROBOSCOPE, instrument that produces regular brief flashes of intense light, used to study periodic motion, to test machinery and in high-speed photography. When the flash frequency exactly equals that of the rotation or vibration, the object is illuminated in the same position during each cycle, and appears stationary. A gas discharge lamp is used, with flash duration about 1 μs and frequency from 2 to 3000Hz.

STROKE, or cerebrovascular accident, the sudden loss of some aspect of BRAIN function due to lack of BLOOD supply to a given area; control of limbs on one side of the body, APHASIA or dysphasia, loss of part of the visual field or disorders of higher function are common. Stroke may result from EMBOLISM, ARTERIOSCLEROSIS and THROMBOSIS, or HEMORRHAGE (then termed apoplexy). Areas with permanent loss of blood supply do not recover but other areas may take over their function.

STRONTIUM (Sr), reactive, silvery-white ALKALINE-EARTH METAL, occurring as strontianite ($SrCO_3$) and celestite ($SrSO_4$), found mainly in Scotland, Ark. and Ariz. Strontium is made by ELECTROLYSIS of the chloride or reduction of the oxide with aluminum. It resembles calcium physically and chemically. The radioactive isotope Sr^{90} is produced in nuclear FALLOUT, and is used in nuclear electric-power generators. Strontium compounds are used in fireworks (imparting a crimson color), and to refine sugar. AW 87.6, mp 769°C, bp 1 384°C. sg 2.54.

STRUTT, John William. See RAYLEIGH, BARON.

STRUVE, Otto (1897–1963), Russian-born US astronomer known for work on stellar evolution (see STAR) and primarily for his contributions to astronomical SPECTROSCOPY, especially his discovery thereby of INTERSTELLAR MATTER (1938).

STRYCHNINE, poisonous ALKALOID from nux vomica seeds causing excessive SPINAL CORD stimulation. Death results from spinal CONVULSIONS and ASPHYXIA.

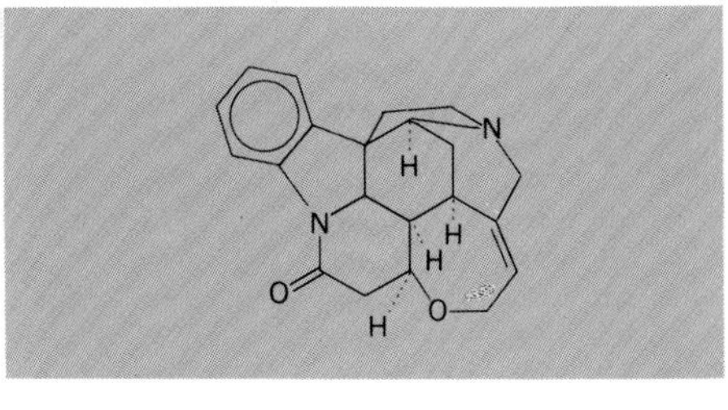

The complex structure of strychnine, first elucidated in 1947.

STUCCO, a fine, white MORTAR applied to interior walls and ceilings to form a smooth surface that may be painted (notably with frescoes), or to form modeled or molded relief decoration. In modern use, stucco is the common cement rendering applied to exterior walls.

STUDENT'S *t*-DISTRIBUTION, in STATISTICS, a way of testing how closely a model of a population corresponds to the results of sampling. *t* is given by

$$t = \frac{\sqrt{n}(\bar{x}-\mu)}{s}$$

where n is the number of items in the sample, s the sample STANDARD DEVIATION, $\bar{x}$ the sample mean (see MEAN, MEDIAN AND MODE) and μ the mean of the NORMAL DISTRIBUTION which is one's model.

STURGEONS, a family of fish that is a survivor of the fossil group CHONDROSTEI. Some species live in the sea but migrate to northern rivers to spawn. Their eggs are sold as CAVIAR.

STUTTER. See SPEECH AND SPEECH DISORDERS.

STY. See BOIL.

STYLE. See FLOWER.

STYRENE, or vinylbenzene, colorless liquid, an AROMATIC COMPOUND found in COAL TAR and ESSENTIAL OILS, and made by dehydration of ethylbenzene. It is polymerized to make PLASTICS and RUBBERS; especially **polystyrene**, a molding plastic which (expanded to a solid foam) is used for heat insulation. MW 104.2, mp −31°C, bp 145°C.

quark	symbol	charge	antiquark	charge
up	u	$+\frac{2}{3}$	$\bar{u}$	$-\frac{2}{3}$
down	d	$-\frac{1}{3}$	$\bar{d}$	$+\frac{1}{3}$
sideways or strange	s	$-\frac{1}{3}$	$\bar{s}$	$+\frac{1}{3}$
charmed	c	$+\frac{2}{3}$	$\bar{c}$	$-\frac{2}{3}$

In an attempt to explain the multiplicity of subatomic particles physicists have proposed that the hadrons should be considered as being made up of sub-subatomic particles called quarks. These quarks, of spin $\frac{1}{2}$, fractional electric charge and bayon number $\frac{1}{3}$ ($-\frac{1}{3}$ for antiquarks) are not known to have any observable properties which are not properties of the particles of which they are postulated to be constituent. The strange quark has stangeness −1 (+1 for $\bar{s}$) while all other quarks are S=0. A new quantum number, charm (C), is postulated in the modified quark theory. The charmed quark is given charm +1 (−1 for $\bar{c}$), all other quarks having C=0. According to the quark theory baryons are considered as comprising three quarks or three antiquarks and mesons as comprising one quark and one antiquark. Thus the proton comprises 2u+1d quarks and the pion (π^+), 1u+1$\bar{d}$.

SUBATOMIC PARTICLES, or **Elementary Particles**, small packets of matter-energy which are constituent of ATOMS or are produced in nuclear reactions or in interactions between other subatomic particles. The first such particle to be discovered was the (negative) ELECTRON (e^-), the constituent of CATHODE RAYS. Next, the **nucleons** were discovered; first the (positive) PROTON (p^+); then, in 1932, the (neutral) NEUTRON (n°). The same year saw the discovery of the first antiparticle, the positron (or antielectron, $\bar{e}^+$—see ANTIMATTER), and from that time the number of known subatomic particles, found in COSMIC RAYS or detected using particle ACCELERATORS, grew rapidly, until by the early 1970s about 100 were known or suspected. As yet no attempt to find theoretical order in this multitude of particles, many of which are highly unstable and have very short HALF-LIVES, has proved entirely successful. A first division of the particles classifies them according to whether they obey BOSE-EINSTEIN STATISTICS (bosons), or FERMI-DIRAC STATISTICS (fermions). Another division groups them into classons, leptons and hadrons. The **classons** are massless bosons which are associated with the fields known to classical physics: the familiar PHOTON associated with ELECTROMAGNETIC RADIATION and the as yet hypothetical graviton, the particle associated with GRAVITATION. The **leptons** are the electrons, the *neutrinos* and the *muons*. These fermions interact with the classical fields and the "weak force" involved in beta-decay. The neutrinos, of rest MASS zero, are products in various decay processes. The **hadrons**,

group	particle	symbol	spin	charge	mass (GeV)	antiparticle	charge
classons	graviton	g	2	0	0	the same	0
	photon	γ	1	0	0	the same	0
leptons	electron	e^-	$\frac{1}{2}$	−1	0.000 5	e^+ (position)	+1
	electron neutrino	ν_e	$\frac{1}{2}$	0	0	$\bar{\nu}_e$	0
	muon	μ^-	$\frac{1}{2}$	−1	0.105 7	μ^+	+1
	muon neutrino	ν_μ	$\frac{1}{2}$	0	0	$\bar{\nu}_\mu$	0
baryons (fermion hadrons)	*nucleons* proton	p	$\frac{1}{2}$	+1	0.938 3	$\bar{p}$	−1
	neutron	n	$\frac{1}{2}$	0	0.939 6	$\bar{n}$	0
	hyperons lambda	Λ	$\frac{1}{2}$	0	1.115 5	$\overline{\Lambda}$	0
	sigma	Σ^+	$\frac{1}{2}$	+1	1.189 5	$\overline{\Sigma^+}$	−1
		Σ^0	$\frac{1}{2}$	0	1.192 5	$\overline{\Sigma^0}$	0
		Σ^-	$\frac{1}{2}$	−1	1.197 4	$\overline{\Sigma^-}$	+1
	xi	Ξ^0	$\frac{1}{2}$	0	1.315	$\overline{\Xi^0}$	0
		Ξ^-	$\frac{1}{2}$	−1	1.321	$\overline{\Xi^-}$	+1
	omega	Ω^-	$\frac{3}{2}$	−1	1.672	$\overline{\Omega^-}$	+1
mesons (boson hadrons)	pion	π^+	0	+1	0.139 6	$\overline{\pi^+} = \pi^-$	−1
		π^0	0	0	0.130 0	the same	0
		π^-	0	−1	0.139 6	$\overline{\pi^-} = \pi^+$	+1
	kaons	K^+	0	+1	0.494	$\overline{K^+} = K^-$	−1
		K^0	0	0	0.498	$\overline{K^0}$	0
	eta	η	0	0	0.549	the same	0

Table of the principal stable and long-lived quasi-stable subatomic particles, also indicating their antiparticles.

including the mesons, nucleons and hyperons, interact additionally with the "strong force"—the intense force that holds the atomic nucleus together in spite of the mutual electric repulsion of its constituent protons. Boson hadrons are known as *mesons*; these were originally postulated as mediating the "strong force" in a similar way to that in which photons mediate the classical electromagnetic field. The mesons include the *pions* (pi mesons) and the heavier *kaons* (*K*-mesons). Fermion hadrons are known as *baryons*. These include the nucleons (protons and neutrons), and the heavier *hyperons*. The omega-minus particle (Ω^-) is a quasi-stable hyperon with a half-life of about 0.1ns. A recent attempt to explain the multiplicity of subatomic particles has involved postulating the existence of an order of yet smaller particles, called **quarks**, supposed to be constituent of all the conventional hadron particles.

SUBCONSCIOUS, the area between CONSCIOUSNESS and UNCONSCIOUSNESS; in PSYCHOANALYSIS, a rarely used synonym for the UNCONSCIOUS.

SUBLIMATION, in psychoanalysis, process whereby energies derived from instinctive DRIVES, particularly the sexual and aggressive (see SEX; AGGRESSION), are channelled into noninstinctive behavior, through INHIBITION or otherwise.

SUBLIMATION, transformation of a substance from the solid to the vapor state without its becoming liquid. All solids will sublime below their triple point (at which solid, liquid and vapor are all in EQUILIBRIUM), but in only a few cases—including DRY ICE, IODINE, NAPHTHALENE and SULFUR—is this at a high enough temperature and rate to be useful for purification. Freeze-drying (see DEHYDRATION) is by sublimation.

SUBLIMINAL PERCEPTION. See PERCEPTION; SUGGESTION.

SUBMARINE, a ship capable of underwater operation. The idea is an old one, but the first working craft was not built until 1620, by Cornelis Drebbel; it was a wooden frame covered with greased leather. The first submarine used in warfare was invented by David BUSHNELL (1776). It was a one-man, hand-powered, screw-driven vessel supposed to attach mines to enemy ships. In the Civil War the Confederate States produced several submarines. Propulsion, the major problem, was partly solved by the Rev. G. W. Garrett, who built a steam-powered submarine (1880). In the 1890s John P. HOLLAND and his rival Simon Lake designed vessels powered by gasoline engines on the surface and by electric motors when submerged, the forerunners of modern submarines. They were armed with TORPEDOES and guns. Great advances were made during WWI and WWII, which demonstrated the submarine's military effectiveness. The German U-boats were notably efficient, and introduced SNORKELS to hinder detection while recharging batteries. But none of these vessels could remain submerged for very long, and a true (long-term) submarine awaited the advent of nuclear power, independent of the oxygen of the air for propulsion. The first nuclear-powered submarine was the U.S.S. *Nautilus* (1955), which in 1958 made the first voyage under the polar ice-cap. The US, USSR, UK and France have nuclear submarine fleets fitted with ballistic missiles. Modern submarines are streamlined vessels, generally with a double hull, the inner being a pressure hull with fuel and ballast tanks between it and the outer hull. The submarine submerges by flooding its ballast tanks to reach neutral buoyancy, i.e., displacing its own weight of water (see ARCHIMEDES), and dives using its hydrofoil diving planes. Submarines are equipped with PERISCOPES and INERTIAL GUIDANCE systems. As well as their military uses, submarines are used for oceanographic research and exploration, salvage and rescue.

SUBMARINE CANYON, suboceanic canyon cutting across the CONTINENTAL SHELF, sometimes the continental slope. They are thought to be of comparatively recent origin.

SUBTRACTION, the inverse operation of ADDITION. It can be seen as the solution of an EQUATION since the difference $(a-b)$ may be expressed as $b+x=a$. The subtraction of negative numbers is equivalent to addition, since $a-(-b) \equiv a+b$.

SUBWAY, an underground railroad system designed for efficient urban and suburban passenger transport. The TUNNELS usually follow the lines of streets, for ease of construction by the cut-and-cover method in which an arched tunnel is built in an open trench, covered with earth and the street restored. Outlying parts of the system usually emerge to the surface. The first subway was built in London (1860–63) by the cut-and-cover method; it used steam trains and was a success despite fumes. A three-mile section of London subway was built (1886–90) using a shield developed by J. H. Greathead: this is a large cylindrical steel tube forced forward through the clay by hydraulic jacks; the clay is removed and the tunnel walls built. Deep tunnels are thus possible, and there is no surface disturbance. This London "tube" was the first to use electrically-powered trains, which soon replaced steam trains everywhere. Elevators were provided for the deep stations, later mostly replaced by escalators. Many cities throughout the world followed London's lead, notably Paris (the Métro, begun 1898) and New York (begun 1900). The New York subway, using the multiple-unit trains developed by Frank SPRAGUE, is now the largest in the world. The Moscow subway (begun 1931) is noted for its palatial marble stations. With increasing road traffic congestion in the 1960s, the value of subways was apparent, and many cities extended, improved and automated their systems; some introduced quieter rubber-tired trains running on concrete guideways.

SUCCESSION, the progressive change in a plant population during the development of vegetation, such as occurs in an abandoned field, which first becomes overgrown with weeds, then is progressively invaded by woody species until a climax vegetation of woodland develops; in all, this type of succession can take some 100 years. The term **sere** is often used to describe the sequence of plant communities; for example, lithosere describes the succession that starts on a bare rock, and hydrosere, one that starts with water.

SUCCULENTS, plants that have swollen leaves or stems and are thus adapted to living in arid regions. Cacti are the most familiar but representatives occur in other families, notably the Crassulaceae (stonecrops and houseleeks) and Aizoaceae (living stones, mesembryanthemum). Many succulents have attractive foliage and colorful, though often short-lived, flowers.

SUCROSE ($C_{12}H_{22}O_{11}$), or cane sugar, disaccharide CARBOHYDRATE, commercially obtained from sugar beet, sugarcane and Sweet sorghum. As table sugar, sucrose is the most important of the SUGARS. It comprises a GLUCOSE unit joined to a FRUCTOSE unit. Sucrose, glucose and fructose all exhibit OPTICAL ACTIVITY and when sucrose is hydrolyzed the rotation changes from right to left. This is called inversion and an equimolar mixture of glucose and fructose is called invert sugar. The ENZYME which hydrolyzes sucrose to glucose and fructose is called invertase.

SUEDE, LEATHER used with the buffed, napped surface outward for shoes and clothing.

SUFFICIENCY. See NECESSITY AND SUFFICIENCY.

SUFFOCATION. See ASPHYXIA.

SUGARS, sweet, soluble CARBOHYDRATES (of general formula $C_x(H_2O)_y$), comprising the monosaccharides and the disaccharides. **Monosaccharides** cannot be further degraded by HYDROLYSIS and contain a single chain of CARBON atoms. They normally have the suffix -ose and a prefix indicating the length of the carbon chain; thus trioses, tetroses, pentoses, hexoses and heptoses contain 3, 4, 5, 6 and 7 carbon atoms respectively. The most abundant natural monosaccharides are the hexoses, $C_6H_{12}O_6$ (including

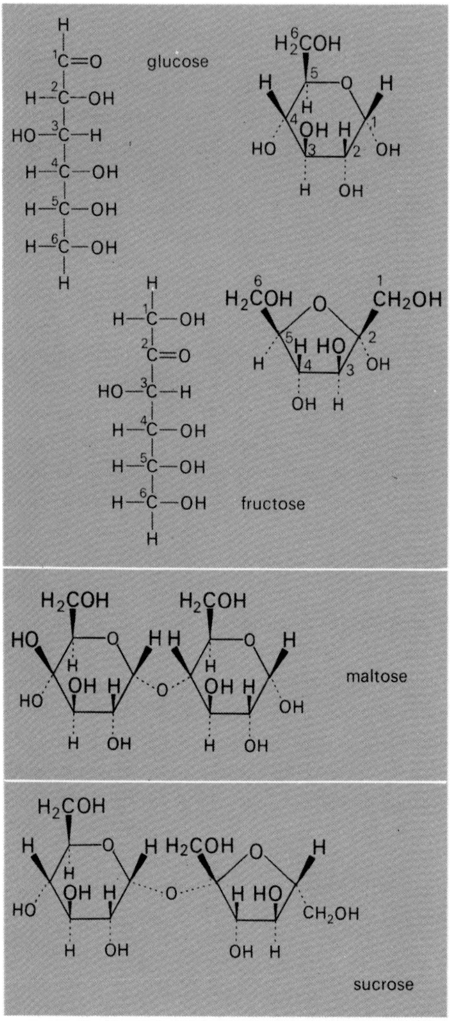

Monosaccharides glucose (a hexose) and fructose (a pentose), shown in both open-chain (*left*) and hemiacetal-ring (*right*) forms. Also the disaccharides maltose and sucrose.

GLUCOSE), and the pentoses, $C_5H_{10}O_5$ (including xylose). Many different isomers of these sugars are possible and often have names reflecting their source, or a property, e.g., FRUCTOSE is found in fruit, arabinose in gum arabic and the pentose, xylose, in wool. **Disaccharides** contain two monosaccharide units joined by an oxygen bridge. Their chemical and physical properties are similar to those of monosaccharides. The most important disaccharides are SUCROSE (cane sugar), LACTOSE and MALTOSE. (Table sugar consists of sucrose.) The most characteristic property of sugars is their sweetness. If we accord sucrose an arbitrary sweetness of 100, then glucose scores 74, fructose 173, lactose 16, maltose 33, xylose 40 (compare SACCHARIN 55 000). The sweetness of sugars is correlated with their solubility.

SUGGESTION, process whereby an individual loses his critical faculties and thus accepts IDEAS and beliefs that may be contrary to his own. People under HYPNOSIS are particularly suggestible (see also BRAINWASHING), as are those in a state of exhaustion. **Heterosuggestion** (dependent on an exterior source) is usually verbally derived, but may involve any of the SENSES. **Autosuggestion** implies that the individual himself is the source. **Mass suggestion** is one of the main aims of advertising, whereby a mass of people may be influenced in favor of a certain product; it may employ **subliminal suggestion** (illegal in most countries), where the stimuli are so brief or faint that they are registered only by the UNCONSCIOUS.

SULFA DRUGS, or **sulfonamides**, synthetic compounds (containing the $—SO_2NH_2$ group) that inhibit the multiplication of invading BACTERIA, thus allowing the body's cellular defense mechanisms to suppress infection. The first sulfa drug, sulfanilamide (Prontosil), was synthesized in 1908 and used widely as a dye, before, in 1935, DOMAGK reported its effectiveness against STREPTOCOCCI. Since then it has proved effective against several other bacteria including those causing SCARLET FEVER, certain VENERAL DISEASES and MENINGITIS. This and the many other sulfa drugs are now generally used in conjunction with ANTIBIOTICS.

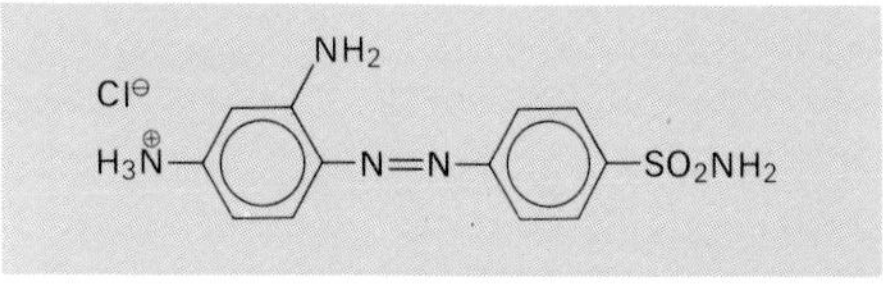

Sulfanilamide (Prontosil), the prototypical sulfa drug.

SULFATES, salts of SULFURIC ACID, containing the sulfate ion (SO_4^{2-}); formed by reaction of the acid with metals, their oxides or carbonates, or by oxidation of SULFIDES or sulfites (see SULFUR). Most sulfates are soluble in water, the main exceptions being calcium, strontium, barium and lead sulfates. They decompose at high temperatures to give sulfur trioxide and dioxide. Sulfates form LIGAND complexes and double salts (see ALUM). Many sulfate minerals occur in nature, often as evaporites or from oxidation of SULFIDES (see ANHYDRITE; BARITE; EPSOM SALTS; GYPSUM). **Bisulfates** contain the ion HSO_4^-; they are acid, and are converted to pyrosulfates ($S_2O_7^{2-}$) on heating. ESTERS of sulfuric acid $(RO)_2SO_2$ are also called sulfates.

SULFIDES, binary compounds of SULFUR. For organic sulfides see THIOETHERS. Nonmetal sulfides, formed by direct synthesis, include CARBON disulfide and several sulfides of nitrogen and phosphorus. Metal sulfides (S^{2-}) are mostly insoluble (except those of the ALKALI METALS and ALKALINE-EARTH METALS), and are prepared by precipitation with hydrogen sulfide. Soluble sulfides are readily hydrolyzed to the soluble bisulfides (HS^-), and are used as reducing agents and in making dyes and pesticides. Many sulfide minerals are important ores: see ARGENTITE; ARSENOPYRITE; BORNITE; CHALCOCITE; CHALCOPYRITE; CINNABAR; REALGAR; SPHALERITE; STIBNITE; WURTZITE.

Hydrogen Sulfide (H_2S), is a colorless, highly toxic gas with a foul odor of rotten eggs, occurring in volcanoes; a covalent HYDRIDE. It is obtained industrially as a by-product of petroleum refining, and prepared in the laboratory by reacting a sulfide with an acid. Hydrogen sulfide burns to give sulfur dioxide and water; in aqueous solution it is a very weak acid, forming sulfides with most metal salts. It is a good reducing agent. With sulfur, hydrogen sulfide gives hydrogen polysulfides (H_2S_n, $n = 2$–9), whose salts are also known. mp −85.5°C, bp −61°C.

SULFONAMIDES. See SULFA DRUGS.

SULFONIC ACIDS, organic compounds of general formula RSO_2OH; strong, water-soluble ACIDS often used as their sodium salts, sulfonates. AROMATIC sulfonic acids are made by **sulfonation** with fuming SULFURIC ACID. They are used to make detergents (see SOAPS AND DETERGENTS), DYES, SULFA DRUGS and ION-EXCHANGE resins. They are useful in synthesis, since the sulfonate group is readily replaced.

SULFUR (S), nonmetal in Group VIA of the PERIODIC TABLE. There are large deposits in Tex. and La., and in Japan, Sicily and Mexico; the American sulfur is extracted by the FRASCH PROCESS. It is also recovered from natural gas and petroleum. Combined sulfur occurs as SULFATES and SULFIDES. There are two main allotropes of sulfur (see ALLOTROPY): the yellow, brittle rhombic form is stable up to 95.6°C, above which monoclinic sulfur (almost colorless) is stable. Both forms are soluble in carbon disulfide; they consist of eight-membered rings S_8. Plastic sulphur is an amorphous form made by suddenly cooling boiling sulfur. Sulfur is reactive, combining with most other elements. It is used in gunpowder, matches, as a fungicide and insecticide, and to vulcanize rubber. AW 32.1, mp 113°C (rh), 119°C (mono), bp 445°C, sg 2.07 (rh, 20°C). **Sulfur dioxide** (SO_2) is a colorless, acrid gas, formed by combustion of sulfur. It is an oxidizing and reducing agent and is important as an intermediate in the manufacture of sulfur trioxide and SULFURIC ACID. It is also used in petroleum refining and as a refrigerant, disinfectant, preservative and bleach. It reacts with water to give sulfurous acid (H_2SO_3), which is corrosive. Thus sulfur dioxide in flue gases is a harmful cause of POLLUTION. mp −73°C, bp −10°C. **Sulfites** are salts containing the ion SO_3^{2-}, formed from sulfur dioxide and BASES; readily oxidized to SULFATES. Bisulfites are acid sulfites, containing the ion HSO_3^-. **Sulfur Trioxide** (SO_3) is a volatile liquid or solid formed by oxidation of sulfur dioxide (see CONTACT PROCESS). It reacts violently with water to give SULFURIC ACID. mp 17°C (α), bp 45°C (α). **Thiosulfates** are salts containing the ion $S_2O_3^{2-}$, usually prepared by dissolving sulfur in an aqueous sulfite solution. They are mild reducing agents, and form LIGAND complexes; in acid solution they decompose to give sulfur and sulfur dioxide. (For sodium thiosulfate, see SODIUM.)

SULFURIC ACID (H_2SO_4), an oily, colorless liquid, made in large quantities by the CONTACT PROCESS or the LEAD-CHAMBER PROCESS. It is an oxidizing agent, reacting with metals, sulfur and carbon on heating, and is a powerful dehydrating agent (see DEHYDRATION). In aqueous solution it is a strong ACID, and reacts with BASES and most metals to give SULFATES. Fuming sulfuric acid, or **oleum**, is 100% sulfuric acid containing dissolved sulfur trioxide; it is used to make SULFONIC ACIDS. Sulfuric acid is used to make fertilizers, paints, pigments, explosives, dyes and detergents, to refine petroleum and coal tar, and in lead-acid storage batteries. mp 10°C, bp 338°C (98%).

SULLIVAN, Harry Stack (1892–1949), US psychiatrist who made important contributions to SCHIZOPHRENIA studies and originated the idea that PSYCHIATRY depends on study of interpersonal relations (including that between therapist and patient).

SULPHUR. See SULFUR.

SUM, the result of ADDITION. If $a+b = c$, then c is the sum of a and b. The sum of two complex numbers (see IMAGINARY NUMBERS) $a+ib$ and $c+id$ is given by $(a+c)+i(b+d)$. (See also VECTOR ANALYSIS.)

SUMMATION, in mathematics, synonym for ADDITION or for the resulting SUM.

SUMNER, James Batcheller (1887–1955), US biochemist who shared with J. H. NORTHROP and W. M. STANLEY the 1946 Nobel Prize for Chemistry for his first crystallization of an ENZYME (urease, 1926), showing it was a PROTEIN.

SUN, the star about which the earth and the other planets of the SOLAR SYSTEM revolve. The sun is an incandescent ball of gases, by mass 69.5% hydrogen; 28% helium; 2.5% carbon, nitrogen, oxygen, sulfur, silicon, iron and magnesium altogether, and traces of other elements. It has a diameter of about 1 393Mm, and rotates more rapidly at the equator (24.65 days) than at the poles (about 34 days). Although the sun is entirely gaseous, its distance creates the optical illusion that it has a surface: this visible edge is called the PHOTOSPHERE. It is at a temperature of about 6000K, cool compared to the center of the sun (20 000 000K) or the corona (1 000 000K); the photospheres of other stars may be at temperatures of less than 2000K or more than 500 000K. Above the photosphere lies the **chromosphere**, an irregular layer of gases between 1.5Mm and 15Mm in depth. It is in the chromosphere that SUNSPOTS, FLARES and **prominences** occur: these last are great plumes of gas that surge out into the corona and occasionally off into space. The **corona** is the sparse outer atmosphere of the sun. During solar ECLIPSES it may be seen to extend several thousand megametres and as bright as the full moon, though in fact it extends to around the orbit of JUPITER. The earth lies within the corona, which at this distance from the sun is termed the SOLAR WIND. The sun is a very normal STAR, common in characteristics though rather smaller than average. It lies in one of the spiral arms of the MILKY WAY.

SUNBURN, burning effect on the SKIN following prolonged exposure to ULTRAVIOLET RADIATION from the sun, common in travelers from temperate zones to hot climates. First-degree BURNS may occur but usually only a delayed ERYTHEMA is seen with extreme skin sensitivity. Systemic disturbance occurs in severe cases. Fair-skinned persons are most susceptible.

SUNDIAL, ancient type of CLOCK, still used (though rarely) in its original form. It consists of a style parallel to the earth's axis that casts a shadow on the calibrated dial plate, which may be horizontal or vertical. It assumes that the sun's apparent motion lies always on the celestial equator (see CELESTIAL SPHERE). Sundials usually show local TIME but may be calibrated to show standard time.

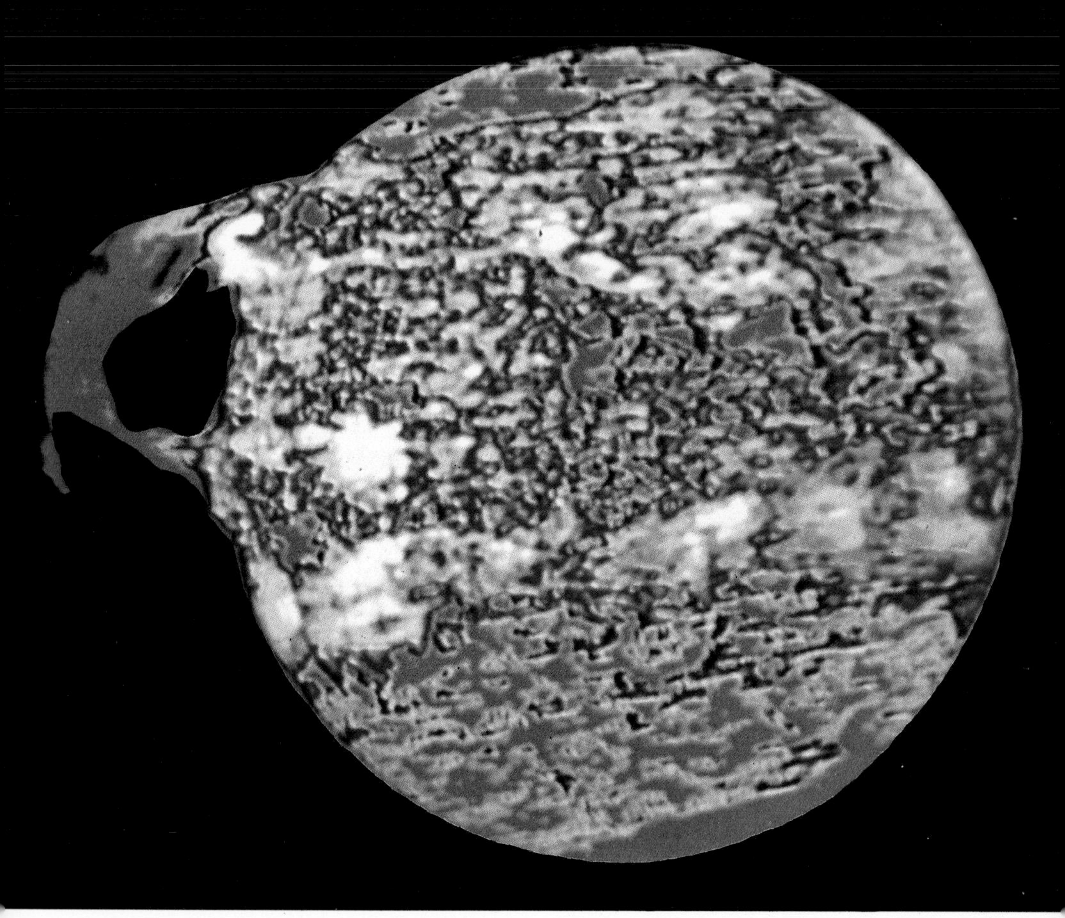

False-color spectroheliogram of the sun in the extreme ultraviolet showing a major solar eruption. In this photograph, taken from Skylab in December 1973, the least active regions of the chromosphere appear red, those displaying the most intense activity white.

SUNSPOTS, apparently dark spots visible on the face of the SUN. Vortices of gas associated with strong electromagnetic activity, their dark appearance is merely one of contrast with the surrounding photosphere. Single spots are known, but mostly they form in groups or pairs. They are never seen at the sun's poles or equator. Their cause is not certainly known. Their prevalence reaches a maximum about every 11 years.

SUNSTROKE, or **heatstroke**, rise in body TEMPERATURE and failure of sweating in hot climates, often following exertion. DELIRIUM, COMA and CONVULSIONS may develop suddenly and rapid cooling should be effected.

SUPERCONDUCTIVITY, a condition occurring in many metals, alloys, etc., at low temperatures, involving zero electrical RESISTANCE and perfect DIAMAGNETISM. In such a material an electric current will persist indefinitely without any driving voltage and applied MAGNETIC FIELDS are exactly cancelled out by the magnetization they produce. In **type I superconductors**, both these properties disappear abruptly when the temperature or applied magnetic field exceed critical values (typically 5K and 10^4A/m), but in **type II superconductors** the diamagnetism decay is spread over a range of field values. Large ELECTROMAGNETS sometimes use superconducting coils which will carry large currents without overheating, and the exclusion of fields by superconducting materials can be exploited to screen or direct magnetic fields. Superconductivity was discovered by H. KAMERLINGH-ONNES in 1911, and is due to an indirect interaction of pairs of ELECTRONS via local elastic deformations of the metal CRYSTAL.

SUPERCOOLING AND SUPERHEATING. A liquid cooled below its FREEZING POINT without the solid phase separating out is in a metastable supercooled state. Addition of a small amount of solid or shaking may cause the liquid to freeze. A liquid heated above its BOILING POINT or a saturated vapor heated after all traces of liquid have evaporated is superheated. This state is also metastable.

SUPEREGO, according to FREUD'S METAPSYCHOLOGY, the third and last part of the psychic apparatus to develop, a part of the EGO containing self-criticism, INHIBITIONS, etc. Unlike the conscience, its strictures may date from an earlier stage of the individual's development, even clashing with his current values.

SUPERFLUIDITY, the property whereby "superfluids" such as liquid HELIUM below 2.186K exhibit apparently frictionless flow. The effect requires QUANTUM MECHANICS for its explanation.

SUPERGIANT STAR. See STAR.

SUPERIORITY COMPLEX, overevaluation by an individual of his abilities, usually a DEFENSE MECHANISM countering an INFERIORITY COMPLEX.

SUPERNOVA, a NOVA which initially behaves like other novae but, after a few days at maximum brightness, increases to a far higher level of luminosity (a supernova in the ANDROMEDA galaxy, 1885, was one tenth as bright as the entire galaxy). It is thought that supernovae may be caused by the gravitational collapse of a star, or cloud of gas and dust, into a neutron STAR.

SUPEROXIDES, compounds of OXYGEN containing the ion O_2^-, resembling the PEROXIDES. Potassium superoxide is used in respirators. Rubidium and cesium superoxide are also known.

SUPERPHOSPHATE. See PHOSPHATES.

SUPERPOSITION, Principle of, law of STRATIGRAPHY first stated by William SMITH, that, when SEDIMENTARY ROCK strata are undisturbed, the younger ROCKS lie above the older.

SUPERSATURATION. See SATURATION.

SUPERSONICS, the study of fluid flow at velocities greater than that of SOUND, usually with reference to the supersonic flight of AIRPLANES and MISSILES when the relative velocity of the solid object and the air is greater than the local velocity of sound propagation.

SUPRARENAL GLANDS. See ADRENAL GLANDS.

SURDS, now little-used term for such irrational ROOTS as $\sqrt{3}$, $\sqrt{5}$, etc., or for their SUMS OF PRODUCTS; or for IRRATIONAL NUMBERS in general.

SURFACE INTEGRAL, the DOUBLE INTEGRAL (see also CALCULUS) obtained through integration of a FUNCTION $f(x, y)$ over a surface S whose vertical projection on the xy-plane is the region R. The surface integral gives the VOLUME between the surface and the xy-plane and is written

$$\iint_R f(x,y)\, dx\, dy.$$

In vector notation (see VECTOR ANALYSIS), the surface integral is generally written

$$\iint_S \mathbf{V} . d\mathbf{S},$$

which is a SCALAR, $d\mathbf{S}$ being a surface element treated as the vector $d\mathbf{x} \times d\mathbf{y}$ and $\mathbf{V}$ being a vector function.

SURFACE TENSION, FORCE existing in any boundary surface of a liquid such that the surface tends to assume the minimum possible area. It is defined as the force perpendicular to a line of unit length drawn on the surface. Surface tension arises from the cohesive forces between liquid molecules and makes a liquid surface behave as if it had an elastic membrane stretched over it. Thus, the weight of a needle floated on water makes a depression in the surface. Surface tension governs the wetting properties of liquids, CAPILLARITY and detergent action.

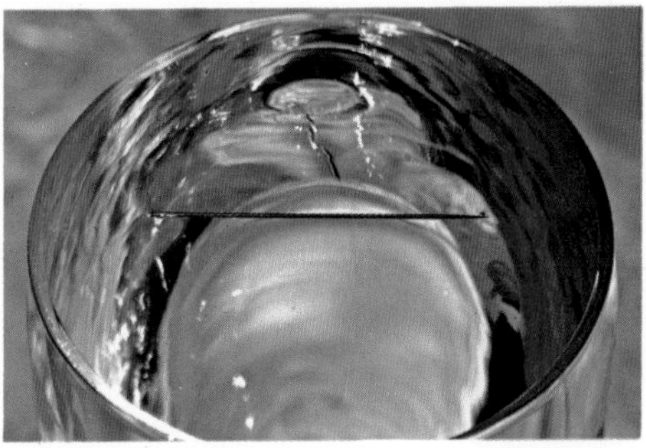

Surface tension effects allow adapted insects to "skate" on the surface of a pond, or a needle to be "floated" in a cup of water.

SURGERY, the branch of MEDICINE chiefly concerned with manual operations to remove or repair diseased, damaged or deformed body tissues. With time surgery has become more complex and has split up into a number of specialities. In 1970 ten surgical speciality boards existed in the US and Canada: general surgery; OPHTHALMOLOGY; otolaryngology; OBSTETRICS and GYNECOLOGY; ORTHOPEDICS; colon and rectal surgery; urology; PLASTIC SURGERY; neurosurgery, and thoracic (chest) surgery. **Otolaryngology** deals with the EAR, LARYNX (voicebox) and upper respiratory tract: tonsillectomy is one of its most common operations. **Colon and rectal surgery** deals with the large intestine. **Urological surgery** deals with the urinary system (KIDNEYS, ureters, BLADDER, urethra) and male reproductive system. **Neurosurgery** deals with the NERVOUS SYSTEM (BRAIN, SPINAL CORD, nerves); common operations include the removal of TUMORS, the repair of damage caused by severe injury, and the cutting of dorsal roots (rhizotomy) and certain parts of the spinal cord (cordotomy) to relieve unmanageable pain. **Thoracic surgery** deals with structures within the chest cavity. There are also a number of subspecialities; thus **cardiovascular surgery**, a subspeciality of thoracic surgery, deals with the heart and major blood vessels.

SURVEYING, the accurate measurement of distances and features on the earth's surface. The science began to attain modern accuracy in the 17th century with the introduction of Gunter's chain (1620)—66ft long and the standard for measuring distance until superseded by the steel or INVAR tape—and of the VERNIER SCALE, the telescopic sight and the spirit LEVEL. For making MAPS and charts, the LATITUDE and longitude of certain primary points are determined from astronomical observations. Geodetic surveying, for large areas, takes the earth's curvature into account (see GEODESY). After a base line of known length is established, the positions of other points are found by triangulation (measuring the angles of the point from each end of the base line) or by trilateration (measuring all the sides of the triangle formed by point and base line). Trigonometry, in particular the SINE RULE, yields the distances or angles not directly measured. A series of adjacent triangles is thus formed, each having one side in common with the next. Distances are measured by tape or electronically, sending a frequency-modulated light or microwave beam to the farther point and back, and measuring the phase shift. Angles are measured with the THEODOLITE or (vertically) the alidade. Vertical elevations are determined by LEVELS. Much modern surveying is done by PHOTOGRAMMETRY, using the STEREOSCOPE to determine contours. Surveying is important not only for mapmaking but also to chart land, to fix boundaries and to plan transportation routes, dams, etc.

SURVIVAL OF THE FITTEST, term first used by

Symbiosis, a mutually beneficial partnership between species. The oxpecker feeds on blood-sucking ticks and flies it finds on the buffalo's hide.

Herbert SPENCER in his *Principles of Biology* (1864) and adopted by Charles Darwin to describe his theory of EVOLUTION by NATURAL SELECTION.

SUSPENSION, system of macroscopic particles dispersed in a fluid in which settlement is hindered by intermolecular collisions and by the fluid's VISCOSITY. (See also COLLOID.)

SUTHERLAND, Earl Wilbur, Jr. (1915–1974), US physiologist awarded the 1971 Nobel Prize for Physiology or Medicine for demonstrating the role of cyclic adenosine 3′,5′-monophosphate in the way that HORMONES affect bodily organs.

SVEDBERG, Theodor (1884–1971), Swedish chemist awarded the 1926 Nobel Prize for Chemistry for inventing the ultracentrifuge, important in studies of COLLOIDS and large MOLECULES. (See also CENTRIFUGE.)

SWAMMERDAM, Jan (1637–1680), Dutch microscopist whose precision enabled him to make many discoveries, including red BLOOD cells (before 1658).

SWAMP, a poorly drained, low-lying area of land permanently saturated with water. Swamps usually develop where the surface is flat enough for rainwater runoff to be very slow, or where a lake basin has become filled in; vegetation helps retain the swampiness. **Marshes** have standing surface water, and are usually only temporary. (See also BOG; MUSKEG.)

SWEAT. See PERSPIRATION.

SWEETENING AGENTS, substances used to sweeten food and drink (see TASTE). The commonest are the SUGARS, especially SUCROSE (table sugar) and GLUCOSE, which are themselves FOODS. Artificial sweeteners, with no food value and up to several thousand times sweeter than sugar, are used by diabetics and in DIETETIC FOODS, toothpaste, etc. They include CYCLAMATE and SACCHARIN. Some have been banned because of possible harmful effects.

SWIM BLADDER. See AIR BLADDER.

SWINE FEVER, a BACTERIAL DISEASE of hogs causing FEVER, immobility and failure of appetite.

SYDENHAM, Thomas (1624–1689), "the English Hippocrates," who pioneered the use of QUININE for treating MALARIA and of LAUDANUM as an ANESTHETIC, wrote an important treatise on GOUT, and first described Sydenham's CHOREA (St. Vitus' Dance).

SYLLABARY, a set of written characters which represent the syllables of a language, rather than the PHONEMES (as do letters) or complete ideas or words (see IDEOGRAM). Wholly or partly syllabary scripts include the MINOAN LINEAR SCRIPTS, Japanese, and some CUNEIFORM scripts.

SYLLOGISM, the logical form of an argument consisting of three statements: two premises and a conclusion. The conclusion of a valid syllogism follows logically from the premises and is true if the premises are true. (See also LOGIC.)

SYLVITE, HALIDE mineral consisting of POTASSIUM chloride (KCl); white, cubic crystals. Occurs in West Germany and N.M. in EVAPORITE deposits.

SYMBIOSIS, the relationship between two organisms of different species in which mutual benefit is derived by both participants. The main types of symbiotic relationship are commensalism and mutualism. **Commensalism** implies eating at the same table, e.g., the Sea anemone that lives on the shell occupied by the Hermit crab: the anemone hides the crab but feeds on food scattered by the crab. **Mutualism** is more intimate, there being close physiological dependence between participants. An example is seen in bacteria that live in the gut of herbivorous mammals. Here the bacteria aid digestion of plant material.

SYMMETRY. A geometrical figure is symmetrical about a POINT (CENTER OF SYMMETRY), LINE (AXIS OF SYMMETRY) or PLANE (PLANE OF SYMMETRY) if, respectively, the point lies at the midpoint of any line drawn through it that cuts the figure in two places; any line drawn PERPENDICULAR to the axis of symmetry cuts the figure at equal distances on either side, or any line drawn perpendicular to the plane intersects (see INTERSECTION) the figure at equal distances on either side. Figures may be symmetrical about combinations or pluralities of centers, axes and planes. **Crystal Symmetry.** If a plane can be drawn through a CRYSTAL such that the halves of the crystal on either side of it are exact mirror images of each other, the plane is a plane of symmetry, denoted *m*. If an axis can be drawn through it such that, when the crystal is rotated through a certain ANGLE (60°, 90°, 120°, 180° or 360°) about the axis, it fills exactly the same space, the axis is a (rotation) axis of symmetry, denoted 6, 4, 3, 2 or 1 (called hexad, tetrad, triad, diad or identity axes respectively) depending on how many times the "symmetry operation" must be repeated to bring the crystal back to its original orientation. (All crystals have an infinite number of identity axes.) The crystal may also have an inversion axis of symmetry about which, after inversion, the crystal may be rotated through a certain angle to occupy exactly the original space. Inversion axes are denoted $\bar{6}$, $\bar{4}$, $\bar{3}$ and $\bar{2}$ (a $\bar{2}$ axis is equivalent to a plane of symmetry and this notation is thus not generally used); a $\bar{1}$ axis implies that the crystal may be inverted through the center to occupy the same space. Crystals are usually classified according to the symmetry which they display.

SYNAPSE, the point of connection between two nerves or between nerve and muscle. An electrical nerve impulse releases a chemical transmitter (often

ACETYLCHOLINE) which crosses a small gap and initiates electrical excitation (or inhibition) of the succeeding nerve or muscle. (See NERVOUS SYSTEM.)

SYNCHROCYCLOTRON, type of CYCLOTRON with one dee-shaped electrode where the accelerating electric field is frequency-modulated to give charged particles, such as PROTONS, high energies while their relativistic mass is large. Pulses of particles start at the center of the accelerator and move in circles of increasing radius, being accelerated as they enter and leave the dee.

SYNCHROTRON, particle accelerator for accelerating pulses of PROTONS, deuterons or ELECTRONS to high energies. The particles move in a circular path of almost constant radius, while the particles are accelerated by an alternating electric field. The path radius is kept constant by increasing the magnetic field while each pulse is accelerated.

SYNCLINE. See FOLD; GEOSYNCLINE.

SYNCOPE. See FAINTING.

SYNERGISM, the working together of two or more agencies (e.g., synergistic MUSCLES, or a chemical with a mechanical phenomenon, or even a chemist with a physicist) to greater effect than both would have working independently.

SYNGE, Richard Laurence Millington (1914–), awarded the 1952 Nobel Prize for Chemistry with A. J. P. MARTIN for developing paper CHROMATOGRAPHY, a tool of great biochemical importance.

SYNOVIAL FLUID, the small amount of fluid which lubricates JOINTS and the synovial sheaths of TENDONS. It contains hyaluronic acid which contributes to its lubricating properties.

SYNTAX. See GRAMMAR.

SYNTHETIC FIBER, man-made textile FIBER derived from artificial POLYMERS, as opposed to regenerated fibers (such as rayon) made from natural substances, or to natural fibers. Almost all types of long-chain polymer may be used: NYLON, the first to be discovered, is a polyamide; **Dacron** is a polyester, useful for nonstretch clothing. Other widely-used synthetic fibers include ORLON, POLYETHYLENE and FIBERGLASS. Polyurethane fibers are ELASTOMERS, used in stretch fabrics. To make the fibers, the polymer is usually converted to a liquid by melting or dissolving it; this is extruded through a spinneret with minute holes, and forms a filament as the solvent evaporates (dry spinning) or as it passes into a suitable chemical bath (wet spinning). The filaments are drawn (stretched) to increase strength by aligning the polymer molecules. They may then be used as such, or cut into short lengths which are twisted together, forming yarn.

SYPHILIS. See VENEREAL DISEASES.

SYRINGE, a simple PUMP for drawing in and ejecting liquids. The ear syringe is merely a tapering tube with a rubber bulb at one end. The hypodermic syringe, used to give INJECTIONS, has a cylindrical barrel containing a PISTON and with an attached hollow needle.

SYSTEMS ANALYSIS, management technique for determining the optional paths to a complex objective and evaluating the relative effectiveness and probable cost of each; or, the study of the general properties of controllable behavioral systems or, more broadly, of any system.

SZENT-GYÖRGYI VON NAGYRAPOLT, Albert (1893–), Hungarian-born US biochemist awarded the 1937 Nobel Prize for Physiology or Medicine for work on biological COMBUSTION processes, especially in relation to VITAMIN C.

SZILARD, Leo (1898–1964), Hungarian-born US physicist largely responsible, with FERMI, for the development of the US atom bomb (see MANHATTAN PROJECT). In 1945 he was a leader of the movement against using it. Later he made contributions in the field of molecular biology.

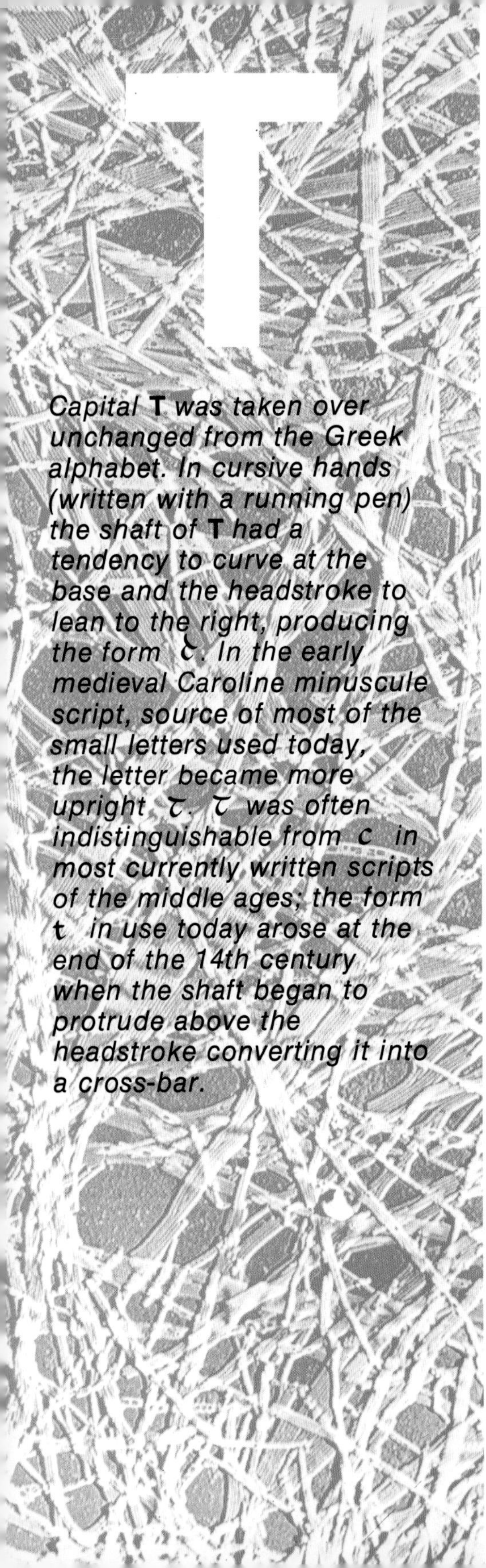

T

Capital **T** *was taken over unchanged from the Greek alphabet. In cursive hands (written with a running pen) the shaft of* **T** *had a tendency to curve at the base and the headstroke to lean to the right, producing the form* ƭ. *In the early medieval Caroline minuscule script, source of most of the small letters used today, the letter became more upright* ꞇ. ꞇ *was often indistinguishable from* c *in most currently written scripts of the middle ages; the form* t *in use today arose at the end of the 14th century when the shaft began to protrude above the headstroke converting it into a cross-bar.*

TABES DORSALIS, form of tertiary syphilis (see VENEREAL DISEASES) in which certain tracts in the SPINAL CORD—particularly those concerned with position sense—degenerate, leading to a characteristic high-stepping gait, sensory abnormalities and sometimes disorganization of JOINTS. Attacks of abdominal pain and abnormal pupil reactions are typical.

TABOO, tabu, tapu or kapu, Polynesian words meaning that which is forbidden. Negative taboos arise from fear of possible ill effects (e.g., incest); positive taboos from awe or reverence (e.g., approaching a god). In tribal society the TOTEM of each CLAN is often subject to taboo.

TABULA RASA (Latin: scraped tablet), philosophical term referring to the condition of the mind before it is modified by experience; often used by empiricists (see EMPIRICISM) to emphasize the dependency of knowledge on the senses.

TACHOMETER, instrument for measuring the angular VELOCITY of a rotating shaft. The simplest is a timed revolution counter. Other mechanical tachometers include the centrifugal tachometer, similar to the flyball GOVERNOR; the vibrating-reed tachometer, a group of reeds of different lengths which is held against the shaft housing so that the reed whose natural vibration frequency equals the rotation frequency of the shaft vibrates by resonance; and the velocity-head tachometer, in which a pump or fan on the shaft produces a measured air pressure. Electrical tachometers are usually electric GENERATORS or electric impulse counters. The eddy-current tachometer is used as a SPEEDOMETER.

TACHYCARDIA. See HEART.

TACONITE, an unleached, low-grade IRON ore. It consists of fine-grained FLINT containing HEMATITE, MAGNETITE and several silicates. Taconite must be concentrated by leaching or magnetic processes before smelting. The US has large quantities, including a huge deposit near Lake Superior.

TADPOLES, the larvae of frogs and toads. An aquatic larva is characteristic of all the AMPHIBIA but in salamanders and newts it is similar in appearance to the adult. In frogs and toads, the tadpole is globular with a long muscular tail. A full METAMORPHOSIS must be undergone to reach adult form.

TAIGA, Siberian forest region lying between the TUNDRA and the STEPPES. Conifers predominate, though birches are occasionally found. Much of the ground is swampy. The term is applied also to other, similar, N-Hemisphere FORESTS.

TALBOT, William Henry Fox (1800–1877), English scientist and inventor of the Calotype (Talbotype) method of PHOTOGRAPHY. In Calotype a latent image in silver iodide is developed in gallic acid and fixed in sodium thiosulfate giving a paper "negative." Thus, for the first time, any number of positive prints could be made from a single exposure by contact printing from the negative.

TALC, basic magnesium SILICATE mineral $Mg_3Si_4O_{10}(OH)_2$, occurring in METAMORPHIC ROCKS, chiefly in the US, USSR, France and Japan. It has a layer structure resembling that of MICA, and is extremely soft (see HARDNESS). Talc is used in ceramics, roof insulation, cosmetics, as an insecticide carrier and as a filler in paints, paper and rubber. Compacted talc forms SOAPSTONE.

TALISMAN. See AMULET.
TALUS, or **scree**, accumulation of rocky debris at the base of a cliff or steep mountain slope, the result of mechanical weathering (see EROSION) of the rocks above. A **breccia** is a SEDIMENTARY ROCK formation of consolidated talus.
TAMM, Igor Yevgenevich (1895–1971), Soviet physicist awarded, with P. A. CHERENKOV and I. M. FRANK, the 1958 Nobel Prize for Physics for work with Frank interpreting the Cerenkov effect (1937).
TANGENT. See TRIGONOMETRY.
TANGENT OF A CURVE, a LINE touching, but not intersecting (see INTERSECTION), a CURVE. The GRADIENT of the tangent is equal to the instantaneous gradient of the curve at the POINT of contact (see CALCULUS). (See also CIRCLE.)
TANH. See HYPERBOLIC FUNCTIONS.
TANK, armored combat vehicle, armed with guns or missiles, and self-propelled on caterpillar treads; the chief modern conventional ground assault weapon. Tanks were first built in 1915 by Britain and used from 1916 against Germany in WWI. These early tanks were very slow, and development between the wars greatly improved speed and firepower. The Spanish civil war and WWII showed the effectiveness of concentrated tank attacks. Amphibious and airborne tanks were developed. Heavy tanks proved cumbersome, and were generally abandoned in favor of the more maneuverable (though more vulnerable) light and medium tanks. Improved models are now used where heavy guns are needed. Light tanks (less than 25 tonnes) are used mainly for infantry support.
TANNING, the conversion of animal hide into LEATHER. After cleaning and soaking, a tanning agent is applied that converts the GELATIN of the hide into an insoluble material which cements the PROTEIN fibers together and makes them incorruptible. Until the end of the 19th century vegetable extracts containing TANNINS were used; then the process was greatly shortened by using CHROMIUM salts, and also FORMALDEHYDE and FORMIC ACID.
TANNINS, or **tannic acid**, a group of complex organic substances occurring in many plants, especially oak gallnuts, tea, and the bark of oak, mangrove and sumac, from which they are extracted by boiling in water. Tannins may be classified as hydrolyzable (yielding GALLIC ACID) or condensed. They are used for TANNING, for making DYES and INKS, and in medicine as an ASTRINGENT.
TANTALUM (Ta), hard, silvery-gray metal in Group VB of the PERIODIC TABLE; a TRANSITION ELEMENT. It is found in COLUMBITE with NIOBIUM, which it resembles closely. Tantalum oxide is separated by solvent extraction and reduced to the metal. It is highly inert, and is used in laboratory ware, capacitors, surgical instruments and as a "getter." AW 180.9, mp 2996°C, bp 5425°C, sg 16.6 (20°C).
TAPE RECORDER, instrument for SOUND RECORDING on magnetic tape, and subsequent playback. The tape, consisting of small magnetic particles of iron oxides on a thin plastic film base, is wound from the supply reel to the take-up reel by a rotating capstan which controls the speed. The tape passes in turn the erase head, which by applying an alternating field reduces the overall magnetization to zero; the recording head, and the playback head. Standard tape speeds are $1\frac{7}{8}$, $3\frac{3}{4}$, $7\frac{1}{2}$, 15 or 30in/s, the higher speeds being used for HIGH-FIDELITY reproduction. **Cassettes** contain thin tape handily packaged, running at $1\frac{7}{8}$in/s. The somewhat larger **cartridges** contain an endless loop of tape on a single reel. Most recorders use two, four or even more tracks side by side on the tape.
TAPEWORMS, intestinal parasites, so named because they are long and flat, forming the class Cestoda of the flatworm phylum PLATYHELMINTHES. A scolex, or head, only 1.5–2mm (about 0.06in) in diameter is attached to the gut and behind this the body consists of a ribbon of identical flat segments, or proglottids, each containing reproductive organs. These proglottids are budded off from behind the scolex. Mature proglottids containing eggs pass out with the feces where larval stages can infect intermediate hosts.
TAR, dark, odorous liquid obtained by destructive distillation of coal (see COAL TAR) or wood, especially from conifers. (See also CREOSOTE; PITCH.)
TARNISHING. See CORROSION.
TARTAGLIA, Niccolò (1499–1557), Italian Renaissance mathematician who discovered a method of solving cubic EQUATIONS. He is also known for an encyclopedic work on elementary mathematics and for his contributions to BALLISTICS.
TARTARIC ACID, or dihydroxybutanedioic acid (HOOC.CHOH.CHOH.COOH), a CARBOXYLIC ACID having three STEREOISOMERS, used in foods and SOFT DRINKS, as a metal cleaner, and in dyeing and photography. It is obtained from the lees of wine fermentation, in which it occurs as potassium hydrogen tartrate, known as **argol**, or **cream of tartar** when pure, used in BAKING POWDER and in ELECTROPLATING. From argol are made **Rochelle salt**, potassium sodium tartrate, used in making processed cheese, mirrors and cathartics; and **tartar emetic**, antimony potassium tartrate, used as an emetic, insecticide, and mordant in dyeing.
TASMANIANS, now extinct native population of Tasmania, perhaps once the native race of Australia and physically and culturally quite unlike the AUSTRALIAN ABORIGINES. The pure stock were extinguished 1804–1876, although a few halfbreeds still exist.
TASTE, special SENSE concerned with the differentiation of basic modalities of food or other substances in the mouth; receptors are distributed over the surface of the TONGUE and are able to distinguish salt, sweet, sour, bitter and possibly water as primary tastes. Much of what is colloquially termed taste is actually SMELL perception of odors reaching the olfactory EPITHELIUM via the naso-PHARYNX. Receptors for sweet are concentrated at the tip of the tongue, for salt and sour along the sides, with bitter mainly at the back. Taste nerve impulses pass via the BRAIN stem to the cortex.
TATUM, Edward Lawrie (1909–1975), US biochemist awarded the 1958 Nobel Prize for Physiology or Medicine with G. W. BEADLE and J. LEDERBERG for work with Beadle showing that individual GENES control production of particular ENZYMES (1937–40).
TAURUS (the Bull), a large constellation on the ECLIPTIC; the second sign of the ZODIAC. It contains the CRAB NEBULA, the GALACTIC CLUSTERS the Hyades and

PLEIADES, and the bright star Aldebaran.

TAUTOMERISM, the existence of two interconvertible ISOMERS of a compound, usually in labile EQUILIBRIUM, though in some cases isolable. It may be demonstrated by spectroscopy or by the exhibition of properties characteristic of both tautomers. Most tautomerism is by hydrogen transfer, as with carbonyl compounds, in which the keto form (—CH—C=O) is in equilibrium with the enol form (—C=C—OH). SUGARS display tautomerism between cyclic and straight-chain forms.

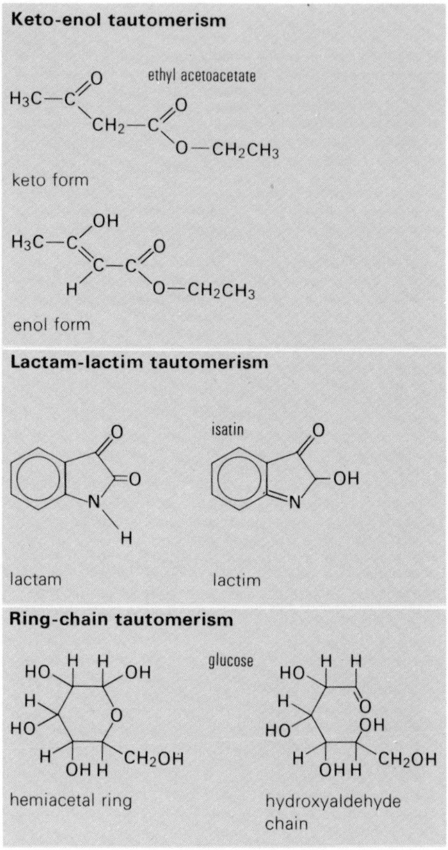

Three kinds of tautomerism.

TAXONOMY, the science of classifying PLANTS and ANIMALS. The theory of EVOLUTION states that organisms come into being as a result of gradual change and that closely related organisms are descended from a relatively recent common ancestor. One of the main aims of taxonomy is to reflect such changes in a classification of groups, or taxa, which are arranged in a hierarchy such that small taxa contain organisms that are closely related and larger taxa contain organisms that are more distantly related. Taxa commonly employed are (in their conventional typography and starting with the largest): Kingdom, Phylum, Class, Order, Family, *Genus* and *species*.

***t*-DISTRIBUTION.** See STUDENT'S *t*-DISTRIBUTION.

TEAR GAS, volatile substance that incapacitates for a time by powerfully irritating the eyes, provoking tears. Various halogenated organic compounds are used, including α-chloroacetophenone (**Mace gas** or CN), and the even more potent CS gas. They are packed in grenades and used for riot control. (See also CHEMICAL AND BIOLOGICAL WARFARE.)

TEARS, watery secretions of the lacrymal GLANDS situated over the EYES which provide continuous lubrication and protection of cornea and sclera. A constant flow runs across the surface of the eye to the nasolacrymal duct at the inner corner, where tears drain into the NOSE. Excess tears produced in states of high emotion and conjunctival or corneal irritation overflow over the lower eyelid.

TECHNETIUM (Tc), radioactive metal (see RADIOACTIVITY) in Group VIIB of the PERIODIC TABLE; a TRANSITION ELEMENT. It does not occur naturally, but was discovered in 1937 by Emilio SEGRÈ and Carlo Perrier in bombarded molybdenum—the first element to be made artificially. It is now recovered from the fission products of nuclear reactors. Technetium is chemically very like RHENIUM. AW 99, mp 2172°C, bp 4877°C, sg 11.5 (20°C).

TEETH, the specialized hard structures used for biting and chewing food. Their numbers vary in different species and at different ages, but in most cases an immature set of teeth (milk teeth) is replaced during growth by a permanent set. In man the latter consists of 32 teeth comprising 8 incisors, 4 canines, 8 premolars and 12 molars, of which the rearmost are the late-erupting wisdom teeth. ("Dentition" refers to the numbers and arrangement of the teeth in a species.) Each tooth consists of a crown, or part above the gum line, and a root, or insertion into the BONE of the jaw. The outer surface of the crowns are covered by a thin layer of enamel, the hardest animal tissue. This overlies the dentine, a substance similar to bone, and in the center of each tooth is the pulp which contains blood vessels and nerves. The **incisors** are developed for biting off food with a scissor action, while the **canines** are particularly developed in some species for maintaining a hold on an object. The **molars** and **premolars** are adapted for chewing and macerating food, which partly involves side-to-side movement of one jaw over the other. Maldevelopment and CARIES of teeth are the commonest problems encountered in DENTISTRY.

TEFLON, or polytetrafluoroethene (PTFE), a FLUOROCARBON plastic, inert and heat-resistant, used as a nonstick coating for cooking utensils, and for making chemical and electrical components.

TEKTITES, glassy objects, usually of less than 100mm diameter, found only in certain parts of the world. Most are rich in SILICA: they resemble OBSIDIAN, though have less water. Despite suggestions that they are of extraterrestrial, particularly lunar, origin, it seems most likely that they have resulted from meteoritic impacts in SEDIMENTARY ROCK in the remote past.

TELEGRAPH, electrical apparatus for sending coded messages. The term was first applied to Claude Chappe's SEMAPHORE. Experiments began on electric telegraphs after the discovery (1819) that a magnetic needle was deflected by a current in a nearby wire. In 1837 W. F. Cooke and Charles WHEATSTONE patented a system using six wires and five pointers which moved in pairs to indicate letters in a diamond-shaped array. It was used on English railroads. In the same year Samuel MORSE, in partnership with Alfred Vail, and helped by Joseph HENRY, patented a telegraph system using MORSE CODE in the US. The first intercity line was inaugurated in 1844. At first the receiver embossed or printed the code symbols but this was soon replaced by a sounding device. In 1858 Wheatstone invented a high-speed automatic Morse telegraph, using punched paper tape in transmission. The TELEX system, using teletypewriters, is now most popular. In 1872 Jean-Maurice-Émile Baudot invented a multiplexing system for sharing the time on each transmission line between several operators. Telegraph signals are now transmitted not only by wires and land lines but also by submarine cables and radio.

TELEMETRY, the transmission of data from distant automatic monitoring stations to a recording station for analysis. It is of immense importance in SPACE EXPLORATION. (See also RADIO.)

TELEOLOGY (from Greek *telos*, end), the study of an action, event, idea or thing with reference to its purpose or end. ARISTOTLE argued that to have a complete understanding of anything its "final cause," its purpose in existing, had to be taken into account. Many modern scientists have questioned this notion, preferring to consider biological processes purely in mechanistic terms. The teleological argument for God's existence is that from design.

TELEOSTEI, a group of fish of the ACTINOPTERYGII that includes the vast majority of bony fish alive today.

TELEPATHY. See ESP.

TELEPHONE, apparatus for transmission and reproduction of sound by means of frequency electric waves. Precursors in telecommunication included the megaphone, the speaking tube and the string telephone—all of which transmitted sound as such—and the TELEGRAPH, working by electrical impulses. Although the principles on which it is based had been known 40 years earlier, the telephone was not invented until 1876, when Alexander Graham BELL obtained his patent. Bell's transmitter worked by the voltage induced in a coil by a piece of iron attached to a vibrating diaphragm. The same apparatus, working in reverse, was used as a receiver. Modern receivers use the same principle, but it was soon found that a more sensitive transmitter was needed, and by 1878 the carbon MICROPHONE (invented by Thomas EDISON) was used. A battery-powered DC circuit connected microphone and receiver. In 1878 the first commercial exchange was opened in New Haven, Conn., and local telephone networks spread rapidly in the US and elsewhere. Technical improvements made for longer-distance transmission included the use of hard-drawn copper wire, underground dry-core CABLE, and two-wire circuits to avoid the cross-talk that occurred when the circuit was completed via ground. Distortion in long circuits was overcome by introducing loading coils at intervals to increase the INDUCTANCE. The introduction also of repeaters, or AMPLIFIERS, made long-distance telephone calls possible. Today, MICROWAVE and RADIO links, and telecommunications SATELLITES are used. Telephone subscribers are connected to a local exchange, these in turn being linked by trunk lines connecting a

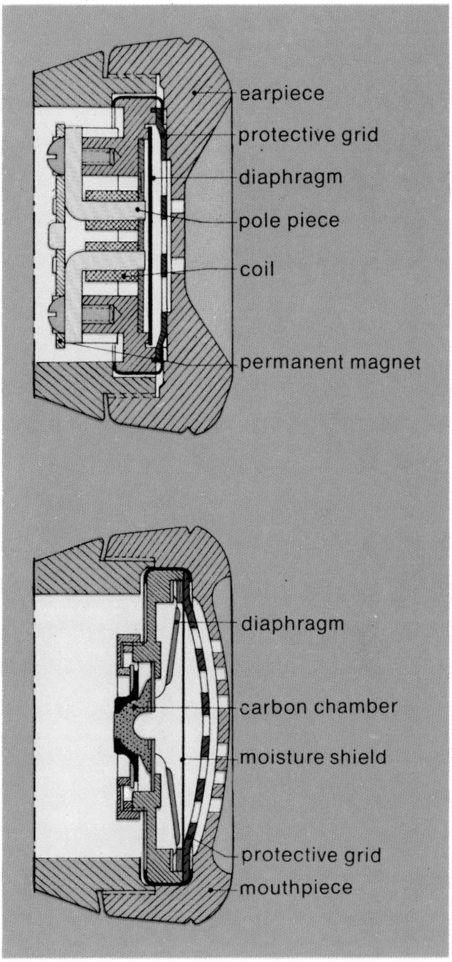

Sections through the ear- and mouthpieces of a typical telephone handset.

hierarchy of switching centers so that alternative routes may be used. When a call is dialed, each digit is coded as pulses or pairs of tones which work electromechanical or electronic switches. (See also PICTUREPHONE.)

TELEPHOTO, or **wirephoto**, an apparatus for facsimile transmission of news photographs, weather maps etc. over electric telecommunication channels. Photographic scanning is used, and the data, transmitted over intercity networks, are photographically reproduced, usually in HALFTONE.

TELESCOPE, Optical, instrument used to detect or examine distant objects. It consists of a series of lenses and mirrors capable of producing a magnified IMAGE and of collecting more light than the unaided eye. The refracting telescope essentially consists of a tube with a LENS system at each end. Light from a distant object first strikes the objective lens which produces an inverted image at its focal point. In the terrestrial telescope the second lens system, the eyepiece, produces a magnified, erect image of the focal image,

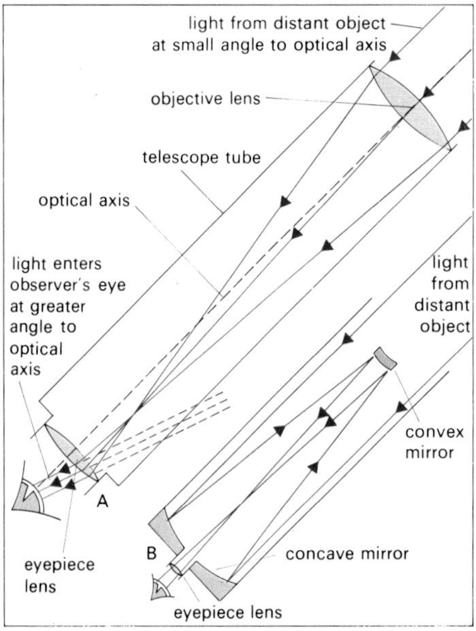

(A) The principle of the astronomical refracting telescope. The magnifying effect of the lens combination is achieved by increasing the angle subtended at the observer's eye by the object. The image is inverted (upside-down). (B) The Cassegrain reflecting telescope: the objective lens of the refracting telescope is replaced by a combination of mirrors.

but in instruments for astronomical use, where the image is usually recorded photographically, the image is not reinverted, thus reducing light losses. The reflecting telescope uses a concave MIRROR to gather and focus the incoming light, the focal image being viewed using many different combinations of lenses and mirrors in the various types of instrument, each seeking to reduce different optical ABERRATIONS. The size of a telescope is measured in terms of the diameter of its objective. Up to about 30cm diameter the resolving power (the ability to distinguish finely separated points) increases with size but for larger objectives the only gain is in light gathering. A 500cm telescope can thus detect much fainter sources but resolve no better than a 30cm instrument. Because mirrors can be supported more easily than large lenses, the largest astronomical telescopes are all reflectors. (See also ASTRONOMY; ASTROPHOTOGRAPHY; BINOCULARS; CLARK, ALVAN; GALILEO; OBSERVATORY; RADIO TELESCOPE; SCHMIDT.)

TELEVISION, the communication of moving pictures between distant points using wire or radio transmissions. In television broadcasting, centrally prepared programs are transmitted to a multitude of individual receivers, though closed-circuit industrial and education applications are of increasing importance. Often, a sound signal is transmitted together with the picture information. In outline, a television CAMERA is used to form an optical IMAGE of the scene to be transmitted and convert it into electrical signals. These are amplified and transmitted, either directly by cable (closed-circuit) or as radio waves, to a receiver where the scene is reconstituted as an optical image on the screen of a CATHODE-RAY TUBE. Today most television cameras are of the image orthicon or vidicon types, these having largely replaced the earlier iconoscope and orthicon designs. Since it is impossible to transmit a whole image at once, the image formed by the optical LENS system of the camera is scanned as a sequence of 525 horizontal lines, the varying light value along each being converted into a fluctuating electrical signal and the whole scan being repeated 30 times a second to allow an impression of motion to be conveyed without noticeable flicker. The viewer sees the image as a whole because of the persistence of VISION effect. In **color television**, the light entering the camera is analyzed into red, green and blue components—corresponding to the three primary COLORS of light—and electrical information concerning the saturation of each is superimposed on the ordinary luminance (brightness) monochrome signal. In the color receiver this information is recovered and used to control the three electron beams which, projected through a shadow mask (a screen containing some 200 000 minute, precisely positioned holes), excite the mosaic of red, green and blue PHOSPHOR dots which reproduce the color image. All three color television systems in use around the world allow monochrome receivers to work normally from the color transmissions.

Development of television. Early hopes of practical television date back to the early days of the electric TELEGRAPH, but their realization had to await several key developments. First was the discovery of the photoconductive properties of selenium (see

Different television standards are operative in different parts of the world. The number of frames per second is tied to the local electrical supply frequency (60Hz in North America; 50Hz elsewhere). The different color systems are the National Television System Committee (NTSC) system, the phase alternating line (PAL) system and the SECAM (*système en couleurs à mémoire*) system.

Television systems of the world

region	number of frames per second	number of lines per frame	color system
North America, Japan, South America	525	30	NTSC
France, Algeria	819, 625	25	SECAM
USSR, DDR, Hungary	625	25	SECAM
UK	625, 405	25	PAL
Rest of Europe, Africa, Australia	625	25	PAL

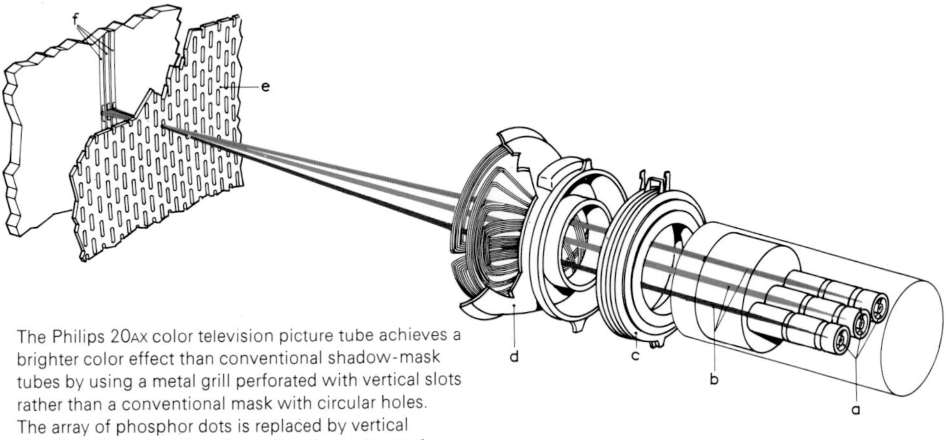

The Philips 20AX color television picture tube achieves a brighter color effect than conventional shadow-mask tubes by using a metal grill perforated with vertical slots rather than a conventional mask with circular holes. The array of phosphor dots is replaced by vertical phosphor lines, one line of each hue for each set of slots in the grill, while a single electron gun with three cathodes, horizontally aligned, substitutes for the three electron guns of the conventional system. Only a single system of focusing coils is required for the three electron beams and far more electrons are able to strike the viewing screen. (a) Cathodes; (b) electron beams for the three primary colors; (c) electromagnetic coils controlling focusing, color purity, etc.; (d) deflection yoke; (e) shadow-mask grill with vertical slots; (f) three of the vertical phosphor lines on the inside of the viewing screen.

PHOTOCONDUCTIVE DETECTOR), followed by the development of the cathode-ray tube (1897) and the ELECTRON TUBE (1904). The first practical television system, demonstrated in London in 1926 by J. L. BAIRD, used a mechanical scanning method devised by Paul Nipkow in 1884. Electronic scanning dates from 1923 when ZWORYKIN filed a patent for his iconoscope camera tube. Television broadcasting began in London in 1936 using a 405-line standard. In the US public broadcasting began in 1941, with regular color broadcasting in 1954. US television broadcasts are made in the VHF (Channels 2–13) and the UHF (Channels 14–83) regions of the RF spectrum (see RADIO). (See also ELECTRONICS; PICTUREPHONE; VIDEOTAPE.)

TELFORD, Thomas (1757–1834), Scottish civil engineer responsible for many British roads, harbors, canals and bridges. He is best known for his suspension BRIDGE over the Menai Straits, North Wales (1819–26); the construction of the Caledonian Canal, Scotland (1804–22); the Göta Canal, Sweden (1808–10), and for several Scottish harbors.

TELLER, Edward (1908–), Hungarian-born US nuclear physicist who worked with FERMI on nuclear FISSION at the start of the MANHATTAN PROJECT, but who is best known for his fundamental work on, and advocacy of, the HYDROGEN BOMB.

TELLURIUM (Te), silvery-white metalloid in Group VIA of the PERIODIC TABLE, occurring as heavy-metal tellurides, and extracted as a by-product of copper refining. Tellurium resembles SELENIUM in its chemistry, but is rather more metallic; tellurium (IV) compounds resemble PLATINUM (IV). It is added to metals to make them easier to machine and to lead for greater corrosion resistance; bismuth telluride is used in thermoelectric devices. AW 127.6, mp 450°C, bp 1390°C, sg 6.24 (20°C).

TEMIN, Howard Martin (1934–), US virologist who shared with R. DULESCO and D. BALTIMORE the 1975 Nobel Prize for Physiology or Medicine for their work on cancer-forming VIRUSES.

TEMPERATURE, the degree of hotness or coldness of a body, as measured quantitatively by THERMOMETERS. The various practical scales used are arbitrary: the FAHRENHEIT scale was originally based on the values 0°F for an equal ice-salt mixture, 32°F for the freezing point of water and 96°F for normal human body temperature. (See also CELSIUS, ANDERS.) Thermometer readings are arbitrary also because they depend on the particular physical properties of the thermometric fluid etc. There are now certain primary calibration points corresponding to the triple points, boiling points or freezing points of particular substances, whose values are fixed by convention. The thermodynamic, or ABSOLUTE, temperature scale, is not arbitrary; starting at ABSOLUTE ZERO and graduated in KELVINS, it is defined with respect to an ideal reversible heat engine working on a CARNOT cycle between two temperatures T_1 and T_2. If Q_1 is the heat received at the higher temperature T_1, and Q_2 the heat lost at the lower temperature T_2, then T_1/T_2 is defined equal to Q_1/Q_2. Such absolute temperature is independent of the properties of particular substances, and is a basic THERMODYNAMIC function, arising out of the zeroth law. It is an intensive property, unlike HEAT, which is an extensive property—that is, the temperature of a body is independent of its mass or nature; it is thus only indirectly related to the heat content (internal energy) of the body. Heat flows always from a higher temperature to a lower. On the molecular scale, temperature may be defined in terms of the statistical distribution of the kinetic energy of the molecules.

TEMPERATURE, Body. Animals fall into two classes: COLD-BLOODED ANIMALS, which have the same temperature as their surroundings, and WARM-BLOODED ANIMALS, which have an approximately constant temperature maintained by a "thermostat" in the brain. The normal temperature for most such animals lies between 95°F (35°C) and 104°F (40°C); it is greatly reduced during HIBERNATION. For man, the normal mouth temperature usually lies between 97°F (36°C) and 99°F (37.2°C), the average being about 98.6°F (37.0°C). It fluctuates daily, and in women monthly. The temperature setting is higher than normal in FEVER. When the body is too hot, the blood vessels near the skin expand to carry more blood and to lose heat by radiation and convection, and the sweat glands produce PERSPIRATION which cools by evaporation. When the body is too cold, the blood vessels near the skin contract, the metabolic rate increases and SHIVERING occurs to produce more heat. Fat under the skin, and body hair (FUR in other animals), help to keep heat in. If these defenses against cold prove inadequate, **hypothermia** results: body temperature falls, functions become sluggish, and death may result. Controlled cooling may be used in surgery to reduce the need for oxygen.

TEMPERATURE HUMIDITY INDEX (THI), formerly **discomfort index**, an empirical measure of the discomfort experienced in various warm weather conditions, and used to predict how much power will be needed to run air-conditioning systems. It is given by

$$THI = 0.4(T_1 + T_2) + 15$$

where T_1 is the dry-bulb temperature and T_2 the wet-bulb temperature in degrees Fahrenheit (see HYGROMETER). When the index is 70 almost everyone feels comfortable; at 80 or more, no one.

TEMPERATURE INVERSION. See INVERSION.

TEMPERING, heat-treatment process in METALLURGY, used to toughen an ALLOY, notably STEEL. The metal is heated slowly to the desired temperature, held there while stresses are relieved and excess solution precipitates out from the supersaturated solid solution, and then cooled, usually by rapid quenching. The temperature determines the properties produced, and may be chosen to retain hardness.

TENDON, fibrous structure formed at the ends of most MUSCLES, which transmits the force of contraction to the point of action (usually a BONE). They facilitate mechanical advantage and allow bulky power muscles to be situated away from small bones concerned with fine movements, as in the HANDS.

TENSILE STRENGTH, the resistance of a material to tensile stresses (those which tend to lengthen it). The tensile strength of a substance is the tensile force per unit area of cross-section which must be applied to break it.

TENSOR, an abstract quantity expressed in terms of definite components within a coordinate system, and whose components obey certain transformation laws when the tensor is considered in relation to a different coordinate system. **Tensor Analysis** has as its chief aim the inspection and mathematical formulation of laws whose validity is unaffected by transfer from one coordinate system to any other. (See also VECTOR.)

TERBIUM (Tb), one of the LANTHANUM SERIES. AW 158.9, mp 1360°C, bp 3041°C, sg 8.229 (25°C).

TERMAN, Lewis Madison (1877–1956), US psychologist best known for developing the STANFORD-BINET TEST.

TERPENES, HYDROCARBONS that are oligomers or POLYMERS of ISOPRENE, and their derivatives. Most are odorous liquids, reactive and unstable, and are found in ESSENTIAL OILS, especially TURPENTINE. They contain double bonds and usually one or more rings.

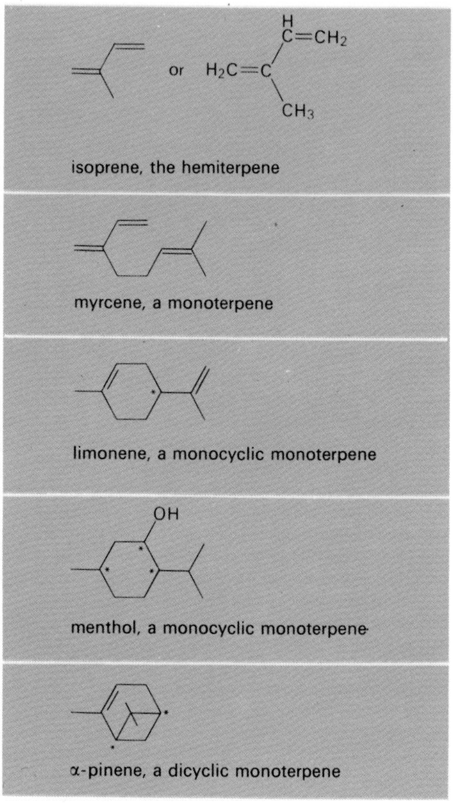

Isoprene and some monoterpenes. Centers of asymmetry are starred. Sesquiterpenes contain three isoprene units and diterpenes (including vitamin A) four. Tri- and teraterpenes are common in nature and various natural rubbers can be considered polyterpenes, containing several thousand isoprene units. Many synthetic rubbers imitate this structure.

They include CAMPHOR, CAROTENOIDS, MENTHOL and VITAMIN A. Latex is a polyterpene.

TERRA COTTA (Italian: baked earth), any fired earthenware product, especially one made from coarse, porous CLAY, red-brown in color and unglazed. Being cheap, hard and durable, it has been used from ancient times for building and roofing, and for molded architectural ornament and statuettes. Its use for sculpture and plaques was revived in the Renaissance and in the 18th century. (See also POTTERY AND PORCELAIN.)

TERRAMYCIN, or oxytetracycline. See TETRACYCLINES.

TERRITORIALITY, a behavioral drive causing animals to set up distinct territories defended against other members of the same species (conspecifics) for the purposes of establishing a breeding site, home range or feeding area. It is an important factor in the spacing out of animal populations. Territoriality is shown by animals of all kinds: birds, mammals, fishes and insects, and may involve displays or the scent-marking of boundaries. A territory may be held by individuals, pairs or even family groups.

TERTIARY, the period of the CENOZOIC before the advent of man, lasting from about 65 to 4 million years ago. Sometimes the Tertiary is regarded as synonymous with the Cenozoic, the QUATERNARY being merely a subperiod. (See also GEOLOGY.)

TESLA, Nikola (1856–1943), Croatian-born US electrical engineer whose discovery of the rotating magnetic field permitted his construction of the first AC INDUCTION MOTOR (c1888). Since it is easier to transmit AC than DC over long distances, this invention was of great importance.

TESTES, pair of male GONADS, which in humans lie in the scrotum suspended from the perineum below the PENIS. This position allows a lower temperature than in the ABDOMEN, thus favoring SPERM production, the principal function of the testes. ANDROGEN hormones (mainly TESTOSTERONE) are also secreted by the testes under the control of HYPOTHALAMUS and PITUITARY GLAND. The testes develop at the back of the abdomen and descend in the FETUS and infant. Failure to descend in childhood may require surgical correction.

TESTOSTERONE, ANDROGEN STEROID produced by the interstitial cells of the TESTES, and to a lesser extent by the ADRENAL GLAND cortex, under the control of LUTEINIZING HORMONE. It is responsible for most male sexual characteristics—VOICE change, HAIR distribution and sex-organ development.

TETANUS, or **lockjaw**, BACTERIAL DISEASE in which a TOXIN produced by anaerobic tetanus bacilli growing in contaminated wounds causes MUSCLE spasm due to nerve toxicity. Minor cuts may be infected with the bacteria which are common in soil. The first symptom may often be painful contraction of jaw and neck muscles; trunk muscles including those of RESPIRATION and muscles close to the site of injury are also frequently involved. Untreated, many cases are fatal, but ARTIFICIAL RESPIRATION, antiserum and PENICILLIN have improved the outlook. Regular VACCINATION and adequate wound cleansing are important in prevention.

TETANY, involuntary MUSCLE contractions, with excessive muscular irritability due to lack of ionic CALCIUM in the BLOOD and tissues. True hypocalcemia may be due to PARATHYROID GLAND insufficiency or pancreatitis, while ALKALOSIS (e.g., from overbreathing) may transiently reduce ionization of calcium compounds.

TETHYS, primeval sea that lay between the supercontinents GONDWANALAND and LAURASIA, separating what is now Africa from what is now S Eurasia. As the continents evolved toward their present form, the Tethys narrowed, leaving only the present Mediterranean. The sediments of the Tethys GEOSYNCLINE are to be found in folded MOUNTAIN ranges such as the Himalayas. (See also PLATE TECTONICS.)

TETRACYCLINES, broad-spectrum ANTIBIOTICS (including Aureomycin and Terramycin) which may be given by mouth. While useful in BRONCHITIS and other minor infections, they are especially valuable in disease due to RICKETTSIA and related organisms; they can also be used in ACNE. Staining of TEETH in children and deterioration in KIDNEY failure cases are important side effects.

TETRAETHYL LEAD. See LEAD.

TETRAHEDRON. See POLYHEDRON.

TEXAS FEVER, tick-borne disease of cattle caused by the protozoan *Babesia bigemina*.

TEXTILES, fabrics made from natural FIBERS or SYNTHETIC FIBERS, whether knitted, woven, bonded or felted. The fibers are prepared (WOOL preparation being typical) and spun (see SPINNING) into yarn. This is then formed into fabric by WEAVING or other methods. Finishing processes include BLEACHING; CALENDERING; MERCERIZING; dyeing (see DYES AND DYEING), brushing, sizing, fulling and tentering. Chemical processes are used to impart crease-resistance, fireproofing, stain-resistance, water-proofing, or nonshrink properties.

THALAMUS, two nuclei of the upper BRAIN stem involved in transmission of impulses to and from the cerebral cortex, especially in sensory pathways.

THALES (early 6th century BC), ancient Greek PRESOCRATIC philosopher, one of the Seven Sages. He is reputed to have invented geometry and to have attempted the first rational account of the universe, claiming that it originated from water.

THALIDOMIDE, mild SEDATIVE introduced in the late 1950s and withdrawn a few years later on finding that it was responsible for congenital deformities in children born to mothers who took the DRUG. This was due to an effect on the EMBRYO in early PREGNANCY, in particular causing defective limb bud formation.

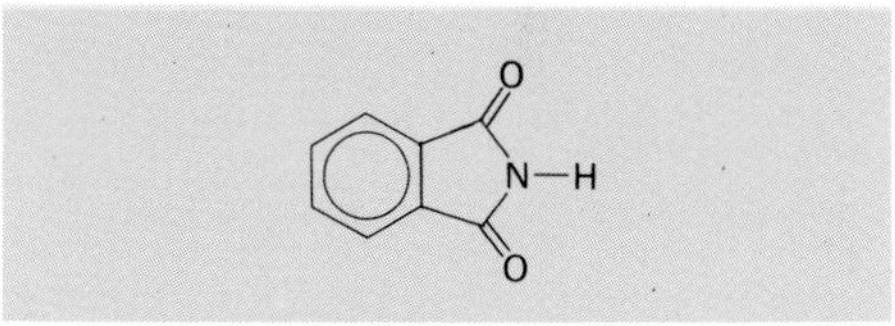

1,3-isoindoledione (phthalimide), marketed under the trade name Thalidomide.

THALLIUM (Tl), soft, bluish-gray metal in Group IIIA of the PERIODIC TABLE, resembling LEAD. Its extraction is the same as INDIUM's. It forms monovalent (thallous) compounds which may be oxidized to the less stable trivalent (thallic) compounds. Thallous sulfide is used in photocells, and mixed crystals of the bromide and iodide in infrared detectors. Thallium compounds are dangerously toxic. AW 204.4, mp 303.5°C, bp 1457°C, sg 11.85 (20°C).

THALLOPHYTA, a traditional but artificial division of the PLANT KINGDOM, which includes ALGAE and FUNGI. The thallophyta represents plants that lack specialized conducting tissues and which are not differentiated into true ROOTS, STEMS and LEAVES. However, since both the constituent groups are so diverse, it is clear they should be placed in separate divisions.

THEILER, Max (1899–1972), South-African-born US microbiologist awarded the 1951 Nobel Prize for Physiology or Medicine for his discovery that an attenuated strain of YELLOW FEVER could be prepared by infecting mice, so making possible the first yellow fever vaccine.

THEODOLITE, surveying instrument comprising a sighting TELESCOPE whose orientation with respect to two graduated angular scales, one horizontal, the other vertical, can be determined. It represents a development of the *transit*, which traditionally included only the horizontal scale.

THEOPHRASTUS (c370–c285 BC), Greek philosopher of the Peripatetic School, a pupil of both PLATO and ARISTOTLE, generally considered the father of modern BOTANY. His *Enquiry into Plants* deals with description and classification, his *Plant Etiology* with physiology and structure. But he is best known for *Characters*, a collection of 30 brief character sketches, often satirical.

THEORELL, Axel Hugo Teodor (1903–), Swedish biochemist awarded the 1955 Nobel Prize for Physiology or Medicine for his studies of ENZYME action, specifically the roles of enzymes in biological OXIDATION AND REDUCTION processes.

THEOREM, any statement logically implied by a set of postulates, or AXIOMS. Once its dependence on the axioms is proved, a newly discovered theorem becomes a potential stepping-stone to the discovery and proof of other, similarly dependent, theorems.

THERMAL ANALYSIS, or **thermoanalysis**, group of methods for detecting and studying physical and chemical changes in substances heated at a standard rate through a temperature range; sometimes used for chemical ANALYSIS. In **thermogravimetric analysis** (TGA), the sample is weighed in a thermobalance—a sensitive BALANCE with the sample pan inside a furnace—and its weight is plotted against temperature. Weight loss is due to giving off gases or vapors; weight gain to reaction with the atmosphere. In **differential thermal analysis** (DTA), the sample is heated simultaneously with an inert reference substance (usually aluminum oxide), and the temperature difference between them is plotted against temperature. This deviates from zero in one direction when an exothermic reaction occurs, and in the other direction when an endothermic reaction occurs (see THERMOCHEMISTRY).

THERMAL POLLUTION, the release of excessive waste heat into the environment, notably by pumping warm water from power plant cooling towers into rivers and lakes. This may kill off some living species, decrease the oxygen supply, and adversely affect reproduction.

THERMIONIC EMISSION, spontaneous emission of ELECTRONS from metal or oxide-coated metal surfaces at temperatures between 1000K and 3000K. It supplies the electrons in electron- and cathode-ray tubes.

THERMISTOR, a SEMICONDUCTOR device the electrical RESISTANCE of which falls rapidly as its TEMPERATURE rises. It is used as a sensor in electronic circuits measuring or regulating temperature and also in time-delay circuits.

THERMITE, mixture of powdered ALUMINUM and IRON oxide (Fe_3O_4) in equivalent amounts, used in WELDING and INCENDIARY BOMBS. On ignition with a barium peroxide or magnesium fuze, a violently exothermic OXIDATION reaction occurs, producing molten iron at 2500°C and alumina slag. It thus supplies both the heat and the metal for welding, and can be used to join large parts in a preheated refractory mold.

THERMOCHEMISTRY, branch of PHYSICAL CHEMISTRY that deals with HEAT changes accompanying chemical reactions. Practical thermochemistry is mainly by CALORIMETRY, which yields standard heats of reaction or enthalpy values ($\Delta H°$) (see THERMODYNAMICS). If this is negative, the reaction is termed *exothermic* (heat-producing); if positive, *endothermic* (heat-absorbing). **Hess' law**, or the law of constant heat summation, a corollary of the first law of THERMODYNAMICS, states that the overall heat change in a chemical reaction is the same whether it takes place in one or several steps. Thus, by algebraic addition of chemical equations and their $\Delta H°$ values, inaccessible heats of reaction may be calculated, including the **heat of formation** of a compound, which is the heat change when one mole of the compound is formed from its constituent elements in their standard states.

THERMOCOUPLE, an electric circuit involving two junctions between different METALS or SEMICONDUCTORS; if these are at different temperatures, a small ELECTROMOTIVE FORCE is generated in the circuit (Seebeck effect). Measurement of this emf provides a sensitive, if approximate THERMOMETER, typically for the range 70K–1000K, one junction being held at a fixed temperature and the other providing a compact and robust probe. Semiconductor thermocouples in particular can be run in reverse as small refrigerators. A number of thermocouples connected in series with one set of junctions blackened form a **thermopile**, measuring incident radiation through its heating effect on the blackened surface. **Thermoelectricity** embraces the Seebeck and other effects relating heat transfer, thermal gradients, ELECTRIC FIELDS and currents.

THERMODYNAMICS, division of PHYSICS concerned with the interconversion of HEAT, WORK and other forms of ENERGY, and with the states of physical systems. Being concerned only with bulk matter and energy, **classical thermodynamics** is independent of theories of their microscopic nature; its axioms are sturdily empirical, and from them theorems are derived with mathematical rigor. It is basic to ENGINEERING, parts of GEOLOGY, METALLURGY and PHYSICAL CHEMISTRY. Building on earlier studies of the thermodynamic functions TEMPERATURE and heat, Sadi CARNOT pioneered the science by his investigations of the cyclic heat ENGINE (1824), and in 1850 CLAUSIUS stated the first two laws. Thermodynamics was further developed by J. W. GIBBS, H. L. F. von HELMHOLTZ, Lord KELVIN and J. C. MAXWELL.

In thermodynamics, a *system* is any defined collection of matter: a *closed system* is one that cannot exchange matter with its surroundings; an *isolated system* can exchange neither matter nor energy. The *state* of a system is specified by determining all its properties such as pressure, volume, etc. A system in stable EQUILIBRIUM is said to be in an equilibrium state, and has an equation of state (e.g., the general GAS law) relating its properties. (See also PHASE EQUILIBRIA.) A *process* is a change from one state A to another B, the path being specified by all the intermediate states. A *state function* is a property or FUNCTION of properties which depends only on the state and not on the path by which the state was reached; a differential dX of a function X (not necessarily a state function) is termed a *perfect differential* if it can be integrated between two states to give a value $X_{AB} = \int_A^B dX$ which is independent of the

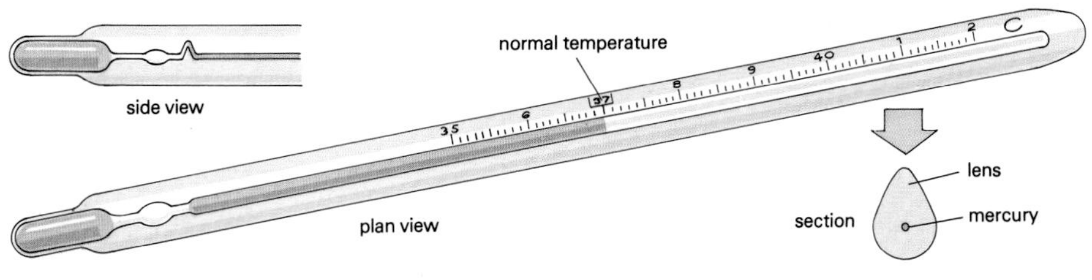

A clinical thermometer. When the bulb of the thermometer is placed under the tongue of the patient, the mercury in the bulb expands and forces a column of mercury up the scale tube. When the thermometer is removed for inspection, the mercury in the bulb contracts but that in the tube cannot flow back past the neck; the reading thus stays the same. The thermometer is reset by centrifugal shaking.

path from A to B. If this holds for all A and B, X must be a state function.

There are four basic laws of thermodynamics, all having many different formulations that can be shown to be equivalent. The **zeroth law** states that, if two systems are each in thermal equilibrium with a third system, then they are in thermal equilibrium with each other. This underlies the concept of temperature. The **first law** states that for any process the difference of the heat Q supplied to the system and the work W done by the system equals the change in the internal energy U: $\Delta U = Q - W$. U is a state function, though neither Q nor W separately is. Corollaries of the first law include the law of conservation of ENERGY, Hess' law (see THERMOCHEMISTRY), and the impossibility of PERPETUAL MOTION machines of the first kind. The **second law** (in Clausius' formulation) states that heat cannot be transferred from a colder to a hotter body without some other effect, i.e., without work being done. Corollaries include the impossibility of converting heat entirely into work without some other effect, and the impossibility of PERPETUAL MOTION machines of the second kind. It can be shown that there is a state function ENTROPY, S, defined by

$\Delta S = \int dQ/T$, where T is the absolute temperature.

The entropy change ΔS in an isolated system is zero for a reversible process and positive for all irreversible processes. Thus entropy tends to a maximum (see HEAT DEATH). It also follows that a heat ENGINE is most efficient when it works on a reversible CARNOT cycle between two temperatures T_1 (the heat source) and T_2 (the heat sink), the EFFICIENCY being $(T_1 - T_2)/T_2$. The **third law** states that the entropy of any finite system in an equilibrium state tends to a finite value (defined to be zero) as the temperature of the system tends to absolute zero. The equivalent NERNST heat theorem states that the entropy change for any reversible isothermal process tends to zero as the temperature tends to zero. Hence absolute entropies can be calculated from specific heat data. Other thermodynamic functions, useful for calculating equilibrium conditions under various constraints, are: **enthalpy** (or heat content) $H = U + pV$; the **Helmholtz free energy** $A = U - TS$; and the **Gibbs free energy** $G = H - TS$. The free energy represents the capacity of the system to perform useful work. **Quantum statistical thermodynamics**, based on QUANTUM MECHANICS, has arisen in the 20th century. It treats a system as an assembly of particles in quantum states. The entropy is given by $S = k \ln P$ where k is the BOLTZMANN constant and P the statistical probability of the state of the system. Thus entropy is a measure of the disorder of the system.

THERMOELECTRICITY. See THERMOCOUPLE.

THERMOGRAPH, any type of THERMOMETER that is self-registering, recording variations of temperature with time on a graph. A bimetallic strip is often used as the temperature-sensitive element, its deflection being recorded on a rotating drum via a system of levers. Thermographs are widely used in meteorology and atmospheric investigations.

THERMOLUMINESCENCE, emission of light from a steadily heated material that has previously been excited by exposure to radiation. It is a type of LUMINESCENCE and arises from electron displacements within the material's crystal lattice. A long time delay may elapse between excitation and subsequent light emission; this is utilized in thermoluminescent dating in archaeology.

THERMOMETER, instrument for measuring the relative degree of hotness of a substance (its TEMPERATURE) on some reproducible scale. Its operation depends upon a regular relationship between temperature and the change in size of a substance (as in the mercury-in-glass thermometer) or in some other physical property (as in the platinum resistance thermometer). The type of instrument used in a given application depends on the temperature range and accuracy required.

THERMONUCLEAR REACTIONS, the reactions used in nuclear FUSION devices such as the HYDROGEN BOMB.

THERMOPHILE, any plant that either requires a high temperature for growth or can tolerate high temperatures. Thermophilic BACTERIA require temperatures between 45°C and 65°C for optimum growth.

THERMOPILE. See THERMOCOUPLE.

THERMOS BOTTLE, trade name for VACUUM BOTTLE.

THERMOSTAT, device for maintaining a material or enclosure at a constant temperature by automatically regulating its HEAT supply. This is cut off if the TEMPERATURE rises and reconnected if it falls below that required. A thermostat comprises a sensor whose dimensions or physical properties change with temperature and a relay device which controls a switch or valve accordingly. **Bimetallic strips** are widely used in thermostats; they consist of two metals with widely different linear thermal coefficients fused together. As the temperature rises, the strip bends away from the side with the larger coefficient. This motion may be sufficient to control a heater directly.

THIAMINE, or aneurin, alternative name for VITAMIN B_1.

THIOETHERS, organic compounds structurally similar to ETHERS but in which sulfur substitutes for the oxygen bridging the two alkyl groups.

THIOSULFATES. See SULFUR.

THIRST AND HUNGER, complex specific sensations or desires for water and food respectively, which have a role in regulating their intake. Thirst is the end result of a mixture of physical and psychological effects including dry mouth, altered BLOOD mineral content, and the sight and sound of water; hunger, those of STOMACH contractions, low blood sugar levels, HABIT, and the SMELL and sight of food. Repleteness with either inhibits the sensation. Food and water intake are regulated by the HYPOTHALAMUS, and are closely related to the control of HORMONE secretion and other vegetative functions, being part of the system preserving the HOMEOSTASIS (constancy) of the body's internal environment. DRUGS, SMOKING, systemic disease and local BRAIN damage are among the many factors influencing thirst and hunger. Excessive thirst may be a symptom of DIABETES or KIDNEY failure (UREMIA), but organic excessive hunger is rare.

THOMPSON, Benjamin. See RUMFORD, COUNT.

THOMSEN, Christian Jürgensen (1788–1865), Danish archaeologist who devised a three-part classification of prehistoric technologies (since applied also to contemporary primitive cultures): STONE AGE; BRONZE AGE, and IRON AGE.

THOMSON, Sir Charles Wyville. See CHALLENGER EXPEDITION.

THOMSON, Sir George Paget (1892–1975), British physicist awarded with C. DAVISSON the 1937 Nobel Prize for Physics for showing that ELECTRONS can be diffracted, thus demonstrating their wave nature.

THOMSON, Sir Joseph John (1856–1940), British physicist generally regarded as the discoverer of the ELECTRON. It had already been shown that CATHODE RAYS could be deflected by a MAGNETIC FIELD; in 1897 Thomson showed that they could also be deflected by an ELECTRIC FIELD, and could thus be regarded as a stream of negatively charged particles. He showed their mass to be much smaller than that of the HYDROGEN atom—this was the first discovery of a SUBATOMIC PARTICLE. His model of the ATOM, though imperfect, provided a good basis for RUTHERFORD's more satisfactory later attempt. Thomson was awarded the 1906 Nobel Prize for Physics.

THOMSON, William. See KELVIN, BARON.

THORAX, the middle part of the body in insects, lying between the head and the ABDOMEN and bearing the wings and legs. It houses many of the viscera and is equivalent to the mammalian CHEST region.

THORIUM (Th), silvery-white radioactive metal, one of the ACTINIDES. Its chief ore is MONAZITE. Thorium is tetravalent, resembling ZIRCONIUM and HAFNIUM. The metal is used in magnesium ALLOYS and to produce uranium-233 for atomic fuel. The refractory thorium (IV) oxide was used to make incandescent gas mantles; it is added in small amounts to the tungsten filaments in electric lamps. AW 232.0, mp c1700°C, bp c4000°C, sg 11.66 (17°C).

THORNDIKE, Edward Lee (1874–1949), US psychologist whose system of psychology, **connectionism**, had a profound influence on US school education techniques, especially his discovery that the learning of one skill only slightly assists in the learning of another, even if related.

THROMBIN, and **thromboplastin.** See CLOTTING.

THROMBOSIS, the formation of clot (thrombus) in the HEART or BLOOD vessels. It commonly occurs in the legs and is associated with VARICOSE VEINS but is more serious if it occurs in the heart or in the brain arteries. Detachments from a thrombus in the legs may be carried to the lungs as an embolus (see EMBOLISM); this may have a fatal outcome if large vessels are occluded. The treatment includes ANTICOAGULANTS.

THRUSH, or **monilia** or **candidiasis**, mucous membrane infection with the fungus *Candida*, seen as multiple white spots, most often affecting the mouth or vagina. Patients with DIABETES or KIDNEY failure and those on STEROIDS and/or ANTIBIOTICS are particularly at risk. Antifungal antibiotics are required.

THRUST, in aerodynamics, the FORCE that propels an airplane or missile. The chemical energy converted in a rocket or jet engine exhausts a high-velocity gas stream whose MOMENTUM, according to Newton's third law of motion, produces the thrust force.

THULIUM (Tm), least abundant of the RARE EARTHS; one of the LANTHANUM SERIES. AW 168.9, mp 1545°C, bp 1727°C, sg 9.321 (25°C).

THUNDER, the acoustic shock wave caused by the sudden expansion of air heated by a LIGHTNING discharge. Thunder may be a sudden clap, or a rumble lasting several seconds if the lightning path is long and thus varies in distance from the hearer. It is audible up to about 15km away; the distance in kilometres can be roughly estimated as one-third the time in seconds between the lightning and thunder.

THUNDERSTORM, a STORM accompanied by THUNDER and LIGHTNING, heavy PRECIPITATION and SQUALLS. Usually short-lived, thunderstorms are generally associated with cumulonimbus CLOUDS.

THURSTONE, Louis Leon (1887–1955), US psychologist whose application of the techniques of STATISTICS to the results of PSYCHOLOGICAL TESTS permitted their more accurate interpretation and demonstrated that a plurality of factors contributed to an individual's score.

THYMINE. See NUCLEIC ACIDS; NUCLEOTIDES.

THYMUS, a ductless two-lobed gland lying just behind the breast bone and mainly composed of lymphoid cells (see LYMPH). It plays a part in setting up the body's IMMUNITY system. Autoimmunity is thought to result from its pathological activity. After PUBERTY it declines in size.

THYROID GLAND, a ductless two-lobed gland lying in front of the trachea in the neck. The principal HORMONES secreted by the thyroid are thyroxine and triiodothyroxine; these play a crucial role in regulating the rate at which cells oxidize fuels to release ENERGY, and strongly influence growth. The release of thyroid hormones is controlled by thyroid stimulating hormone (TSH) released by the PITUITARY GLAND when blood thyroid-hormone levels are low. Deficiency of thyroid hormones (hypothy-

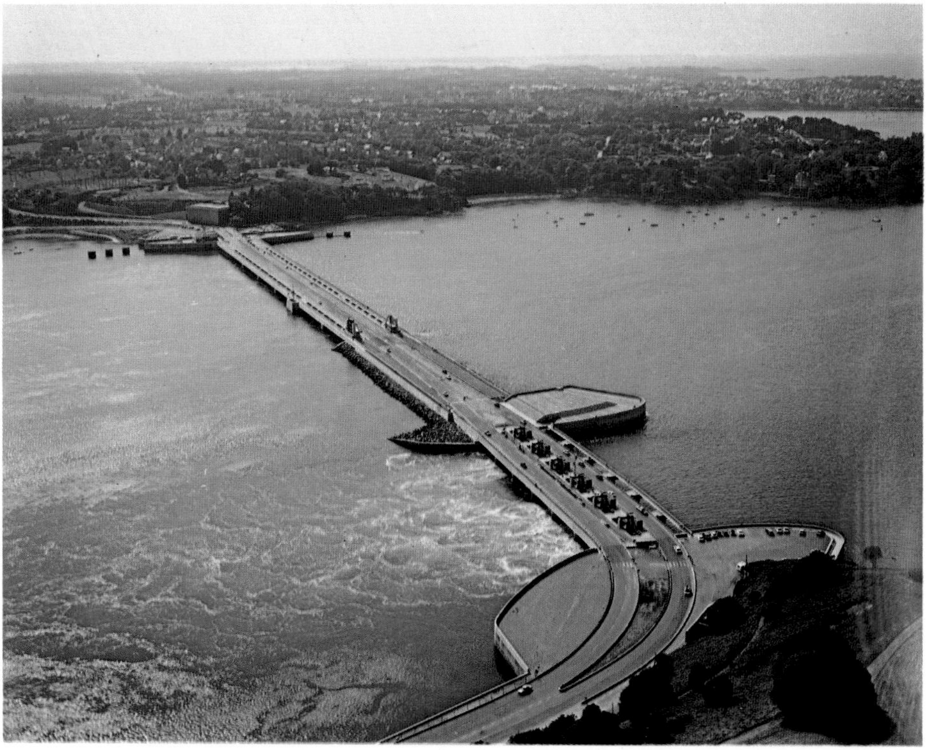

Tidal power: the barrage on the Rance river in northern France

roidism) in adults leads to **myxedema**, with mental dullness and cool, dry and puffy skin. Oversecretion of thyroid hormones (hyperthyroidism or thyrotoxicosis) produces nervousness, weight loss and increased heart rate. GOITER, an enlargement of the gland, may result when the diet is deficient in iodine. (See also CRETINISM.)

TIBIA, or shin bone, the principal BONE of the lower LEG, paralleled by the FIBULA.

TIC, a stereotyped movement, habit spasm or vocalization which occurs irregularly, but often more under stress, and which is outside voluntary control. Its cause is unknown. **Tic dolouroux** is a condition in which part of the FACE is abnormally sensitive, any TOUCH provoking intense PAIN.

TIDAL POWER, form of HYDROELECTRICITY produced by harnessing the ebb and flow of the TIDES. Barriers containing reversible TURBINES are built across an estuary or gulf where the tidal range is great. The Rance power plant in the Gulf of St. Malo, Brittany, the first to be built (1961–67), produces 240MW power, mostly at ebb tide.

TIDAL WAVE, obsolete term for TSUNAMI.

TIDES, the periodic rise and fall of land and water on the earth. Tidal motions are primarily exhibited by water: the motion of the land is barely detectable. As the earth-moon system rotates about its center of gravity, which is within the earth, the earth bulges in the direction of the moon and in the exactly opposite direction, owing to the resultant of the moon's gravitational attraction and the centrifugal forces resulting from the system's revolution. Toward the moon, the lunar attraction is added to a comparatively small centrifugal force; in the opposite direction it is subtracted from a much larger centrifugal force. As the moon orbits the earth in the same direction as the earth rotates, the bulge "travels" round the earth each lunar day (24.83h); hence most points on the earth have a high tide every 12.42h. The sun produces a similar though smaller tidal effect. Exceptionally high high tides occur at full and new moon (spring tides), particularly if the moon is at perigee (see ORBIT); exceptionally low high tides (neap tides) at first and third quarter. The friction of the tides causes the DAY to lengthen 0.001s per century.

TILE, thin slab of TERRA COTTA or other kinds of POTTERY AND PORCELAIN, used in building to cover surfaces. Roof tiles are commonly unglazed and functional; they are either flat, hooked over roof battens, or curved (often S-shaped) and cemented. Floor and structural tiles are hard and vitreous. Wall tiles, used from ancient times, are often decorated with bas-relief molding, painting and glazing. Seventeenth-century Delft tiles are famous. Plain glazed wall tiles are now commonly used in bathrooms etc. By analogy, squares of linoleum, vinyl polymers and cork are also called tiles.

TILL, or **boulder clay**, the unsorted material left behind on the land after the retreat of a GLACIER. (See also DRIFT; DRUMLIN; MORAINE.)

TILLITE. See CONGLOMERATE.

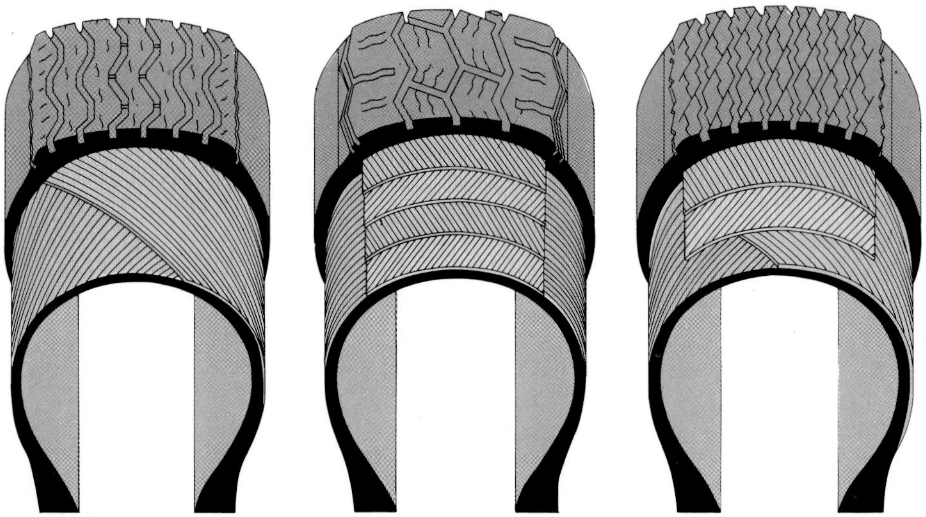

Three types of tire. *Left:* bias-ply or diagonal; *center:* radial-ply; *right:* bias-belted.

TIMBER. See FORESTRY.

TIME, a concept dealing with the order and duration of events. If two events occur nonsimultaneously at a point, they occur in a definite order with a time lapse between them. Two intervals of time are equal if a body in equilibrium moves over equal distances in each of them; such a body constitutes a clock. The sun provided man's earliest clock, the natural time interval being that between successive passages of the sun over the local meridian—the solar DAY. For many centuries the rotation of the earth provided a standard for time measurements, but in 1967 the SI UNIT of time, the SECOND was redefined in terms of the frequency associated with a cesium energy-level transition. In everyday life, we can still think of time in the way Newton did, ascribing a single universal time-order to events. We can neglect the very short time needed for light signals to reach us, and believe that all events have a unique chronological order. But when velocities close to that of light are involved, relativistic principles become important; simultaneity is no longer universal and the time scale in a moving framework is "dilated" with respect to one at rest—moving clocks appear to run slow (see RELATIVITY).

TIN (Sn), silvery-white metal in Group IVA of the PERIODIC TABLE, occurring as CASSITERITE in SE Asia, Bolivia, Zaire and Nigeria. The ore is reduced by smelting with coal. Tin exhibits ALLOTROPY: white (β) tin, the normal form, changes below 13.2°C to gray (α) tin, a powdery metalloid form resembling GERMANIUM, and known as "tin pest." Tin is unreactive, but dissolves in concentrated acids and alkalis, and is attacked by HALOGENS. It is used as a protective coating for steel, and in alloys including solder (see SOLDERING), BRONZE, PEWTER, BABBITT METAL and type metal. AW 118.7, mp 232°C, bp 2270°C, sg (β) 7.31. Tin forms organotin compounds, used as biocides, and also inorganic compounds: tin (II) and tin (IV) salts. **Tin (IV) Oxide** (SnO_2), white powder prepared by calcining CASSITERITE or burning finely divided tin; used in glazes and as an abrasive. subl 1800°C. **Tin (II) Chloride** ($SnCl_2$), white crystalline solid, prepared by dissolving tin in hydrochloric acid, used as a reducing agent, in tin-plating, and as a mordant for dyes. mp 246°C, bp 652°C.

TINBERGEN, Nikolaas (1907–), Dutch ethologist awarded with K. LORENZ and K. von FRISCH the 1973 Nobel Prize for Physiology or Medicine for their individual, major contributions to the science of ANIMAL BEHAVIOR.

TING, Samuel (1936–), US physicist who shared the 1976 Nobel Prize for physics with B. RICHTER for his independent discovery of the *J* particle (also known as the psi(3095) particle).

TIRE, ring-shaped cushion fitted onto a wheel rim as a shock absorber and to provide traction. The pneumatic tire (filled with compressed air) was patented in 1845 by R. W. Thomson, an English engineer, who used a leather tread and a rubber inner tube. Solid rubber tires were more popular, however, until the pneumatic tire was reinvented by John Boyd DUNLOP (1888), whose outer tube was of canvas covered by vulcanized rubber. The modern tubeless tire (without inner tube) dates from the 1950s. The basic structure of a tire comprises layers (plies) of rubberized fabric (usually polyester cord). The plies are combined with "beads"—inner circular wire reinforcements—and the outer tread and sidewalls on a tire-building drum. The tire is then shaped and vulcanized (see VULCANIZATION) in a heated mold under pressure, acquiring its tread design. Three types of tire are made: the bias-ply tire has the plies with cords running diagonally, alternately in opposite directions; the bias-belted tire is similar, with fiberglass belts between plies and tread; the radial-ply tire has the cords running parallel to the axle, and steel-mesh belts.

TISELIUS, Arne Wilhelm Kaurin (1902–1971), Swedish chemist awarded the 1948 Nobel Prize for Chemistry for his development of new techniques and equipment in order to apply ELECTROPHORESIS to the study of PROTEINS, notably those of the BLOOD.

TISSUES, similar CELLS grouped together in certain areas of the body of multicellular ANIMALS and PLANTS. These cells are usually specialized for a single function; thus MUSCLE cells contract but do not secrete; nerve cells conduct impulses but have little or no powers of contraction. The cells are held together by intercellular material such as COLLAGEN. Having become specialized for a single or at most a very narrow range of functions, they are dependent upon other parts of the organism for items such as food or oxygen. Groups of tissues, each with its own functions, make up ORGANS. **Connective tissue** refers to the material in which all the specialized body organs are embedded and supported. It includes ADIPOSE TISSUE and the material of LIGAMENTS and TENDONS. (See also HISTOLOGY.)

TITANIUM (Ti), silvery gray metal in Group IVB of the PERIODIC TABLE; a TRANSITION ELEMENT. Titanium occurs in RUTILE and in ILMENITE, from which it is extracted by conversion to titanium (IV) chloride and reduction by magnesium. The metal and its alloys are strong, light, and corrosion- and temperature-resistant, and, although expensive, are used for construction in the aerospace industry. Titanium is moderately reactive, forming tetravalent compounds, including titanates (TiO_3^{2-}), and less stable di- and trivalent compounds. **Titanium (IV) oxide** (TiO_2) is used as a white pigment in paints, ceramics, etc. **Titanium (IV) chloride** ($TiCl_4$) finds use as a catalyst. AW 47.9, mp 1660°C, bp 3287°C, sg 4.54.

TITCHENER, Edward Bradford (1867–1927), British-born US psychologist, a disciple of WUNDT, who played a large part in establishing experimental PSYCHOLOGY in the US, especially through his *Experimental Psychology* (4 vols., 1901–05).

TITRATION, common technique of VOLUMETRIC ANALYSIS in which a standard solution of one reagent is added little by little from a BURETTE to a second reagent whose amount is to be determined. The end point, at which an exactly equivalent amount of reagent has been added, may be determined by using an INDICATOR, or by measurements of color, resistance, current flow or potential, whose variation with added reagent changes abruptly when the end point is reached.

TNT, or **trinitrotoluene**, pale yellow crystalline solid made by NITRATION of TOLUENE. It is the most extensively used high EXPLOSIVE, being relatively insensitive to shock, especially when melted by steam heating and cast. MW 227.1, mp 82°C.

TOBACCO, dried and cured leaves of varieties of the tobacco plant (*Nicotiana tabacum*), used for smoking, chewing and as SNUFF. Native to America, tobacco was introduced to Europe by the Spanish in the 16th century and from there spread to Asia and Africa. Today the US remains the world's largest producer, followed by China, India and the USSR. Consumption is increasing despite the health hazards of SMOKING. Tobacco is grown in alluvial or sandy soils and may be harvested in about four months. Cultivation is dependant on hand labor. Family: Solanaceae.

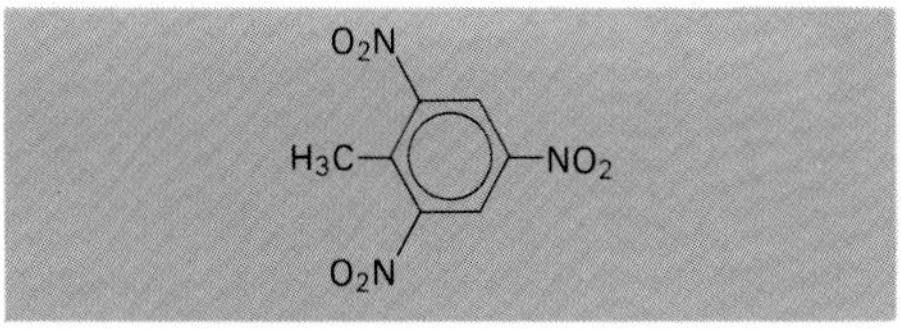

TNT (2,4,6-trinitrotoluene).

TOBACCO MOSAIC, VIRUS disease of plants, strains of which affect TOBACCO, tomatoes, beans and many decorative plants. It restricts growth and causes the leaves to develop a mottled or mosaic pattern. Apart from avoiding infection, no cure has been found for this the most studied of all viruses, having been the first virus to be isolated and the first to be purified. (See also PLANT DISEASES.)

TOCOPHEROL, or Vitamin E. See VITAMINS.

TODD, Alexander Robertus, Baron Todd of Trumpington (1907–), British organic chemist awarded the 1957 Nobel Prize for Chemistry for his work on the structure and synthesis of NUCLEOTIDES.

TOIT, Alexander Logie du (1878–1948), South African geologist, a disciple of WEGENER, whose work did much to validate Wegener's CONTINENTAL DRIFT theories and was to a great extent responsible for their eventual acceptance.

TOLUENE, or methylbenzene ($C_6H_5CH_3$), colorless liquid HYDROCARBON, an AROMATIC COMPOUND produced from COAL TAR and by catalytic reforming of PETROLEUM hydrocarbons. It is used as a solvent, in GASOLINE, and for making TNT, BENZALDEHYDE, BENZOIC ACID etc. MW 92.2, mp −95°C, bp 111°C.

TOMBAUGH, Clyde William (1906–), US astronomer who discovered the planet PLUTO (1930).

TOMONAGA, Shinichiro (1906–), Japanese physicist who shared with FEYNMAN and SCHWINGER the 1965 Nobel Prize for Physics for their independent work on quantum electrodynamics.

TON, name of various actual or nominal units of WEIGHT. The short ton commonly used in the US is 2 000lb, the long ton 2 240lb. The metric ton or **tonne (t)** is 1 000kg (2 204.62lb). The ton used for measuring ships' cargoes is 40cu ft in the US, 42cu ft in the UK. The register ton used for describing the capacity of merchant ships is 100cu ft; the displacement ton used for describing warships (equivalent to 35cu ft of sea water) refers to the weight of water displaced by the ship—and hence to the actual weight of the ship.

TONGUE, muscular organ in the floor of the mouth which is concerned with the formation of food boluses and self-cleansing of the mouth, TASTE sensation and VOICE production. Its mobility allows it to move substances around the mouth and to modulate sound production in speech. In certain animals, the tongue is extremely protrusile and is used to draw food into the mouth from a distance.

TONNE (t), 1 000kg, the metric TON.

TONSILLITIS, INFLAMMATION of the TONSILS due to VIRUS or BACTERIAL infection. It may follow sore throat or other pharyngeal disease or it may be a primary tonsil disease. Sore throat and red swollen tonsils, which may exude PUS or cause swallowing difficulty, are common; LYMPH nodes at the angle of the jaw are usually tender and swollen. QUINSY is a rare complication. ANTIBIOTIC treatment for the bacterial cause usually leads to a resolution but removal of the tonsils is needed in a few cases.

TONSILS, areas of LYMPH tissue aggregated at the sides of the PHARYNX. They provide a basic site of body

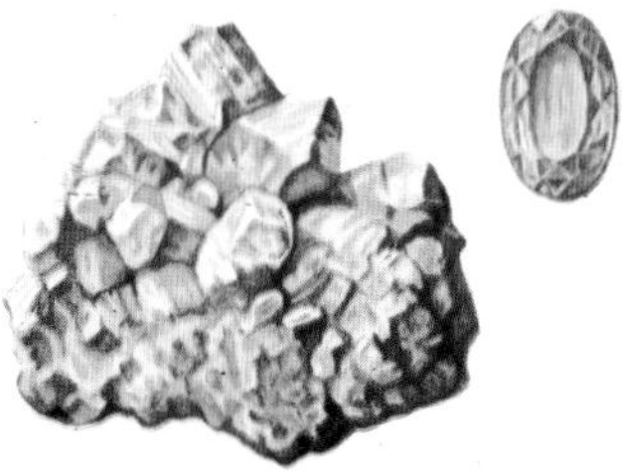

Topaz: crystals in pegmatite, and cut and polished as a gemstone.

defense against infection via the mouth or NOSE and are thus particularly susceptible to primary infection (TONSILLITIS). As with the ADENOIDS, they are particularly important in children first encountering infectious microorganisms in the environment.

TOOTH. See TEETH.

TOPAZ, aluminum SILICATE mineral of composition $Al_2SiO_4(F,OH)_2$, forming prismatic crystals (orthorhombic) which are variable and unstable in color, and valued as GEM stones. The best topazes come from Brazil, Siberia and the US.

TOPOLOGY, a branch of mathematics related to GEOMETRY and dealing with the positions and relative positions of features of a geometric figure. Topologists are primarily concerned with pattern; that is, those features which are not affected by changes of size, angle, etc. (See also FOUR-COLOR PROBLEM; KLEIN BOTTLE; MÖBIUS STRIP.)

TOPSOIL. See SOIL.

TORNADO, the most violent kind of STORM; an intense WHIRLWIND of small diameter, extending downward from a convective cloud in a severe THUNDERSTORM, and generally funnel-shaped. Air rises rapidly in the outer region of the funnel, but descends in its core, which is at very low pressure. The funnel is visible owing to the formation of cloud droplets by expansional cooling in this low pressure region. Very high winds spiral in toward the core. These, and explosions due to the low pressure, account for the almost total devastation and loss of life in the path of a tornado—which itself may move at up to 200m/s. Though generally rare, tornadoes occur worldwide, especially in the US and Australia in spring and early summer. (See also WATERSPOUT.)

TORPEDO, self-propelled streamlined missile that travels underwater, its explosive warhead detonating when it nears or strikes its target. The torpedo was invented by Robert Whitehead, a British engineer, in 1866. Modern torpedoes are launched by dropping from airplanes or by firing from ships or submarines. They are electrically driven by propellers and guided by rudders controlled by a GYROPILOT. Many can be set to home in acoustically on their target. Rocket-propelled torpedoes are fired as guided missiles, and convert into torpedoes when they enter the water near their target. Torpedoes are now chiefly antisubmarine weapons.

TORQUE, a measure of the effectiveness of a FORCE or MOMENT in setting a body in rotation. In mechanics, a torque is a twisting moment or couple which tends to twist a fixed object such as a shaft about a rotation axis. If the shaft starts to rotate, the POWER it transmits is given by the product of the rotational speed and the torque.

TORRICELLI, Evangelista (1608–1647), Italian physicist and mathematician, a one-time assistant of GALILEO, who improved the telescope and microscope and invented the (mercury) BAROMETER (1643).

TORSION, strain produced by a twisting motion about an axis (a *torque*), such as a couple applied perpendicular to a cylinder axis. The resistance of a bar of given material to torsion is a measure of its rigidity and elasticity.

TORUS, or **anchor ring**, doughnut-shaped topological space formed by rotating a CIRCLE about a straight LINE which lies in the same PLANE as the circle but nowhere intersects it. It has a VOLUME of π^2r^2d and a surface AREA of $4\pi^2rd$, where r is the radius of the circle and d the distance of its center from the line.

TOTEM, an object, animal or plant toward which a TRIBE, CLAN or other group feels a special affinity, often considering it as a mythical ancestor. Killing of the totemic animal or animals by members of the group is TABOO, except, with some peoples, ritually during religious ceremonies. **Totem poles**, on which are carved human and animal shapes representing the particular warrior's heritage, were at one time common among the AMERINDS.

TOUCH, the sensory system concerned with surface sensation, found in all external body surfaces including the SKIN and some mucous membranes. Touch sensation is crucial in the detection and recognition of objects at the body surface, including those explored by the limbs, and also in the protection of these surfaces from injury. Functional categories of touch sensation include light touch (including movement of HAIRS), heat, cold, pressure and pain sensation. These are to some degree physiologically distinct. Receptors for all the SENSES are particularly concentrated and developed over the FACE and HANDS. When the various types of skin receptor are stimulated, they activate nerve impulses in cutaneous nerves; these impulses pass via the SPINAL CORD and brain stem to the BRAIN, where coding and perception occur. With painful stimuli, REFLEX withdrawal movements may be induced at the segmental level.

TOURMALINE, borosilicate mineral of variable composition (in general $XY_3Al_6(BO_3)_3Si_6O_{18}(OH)_4$ where X = Na,Ca and Y = Al,Fe^{III}, Li, Mg), found in PEGMATITES as trigonal/hexagonal crystals used as GEM stones. Tourmaline crystals exhibit DOUBLE REFRACTION and PIEZOELECTRICITY, and hence are used in polarizers and pressure-sensing devices.

Tourmaline: crystals in calcite, and cut and polished as a gemstone.

TOWNES, Charles Hard (1915–), US physicist awarded the 1964 Nobel Prize for Physics with N. BASOV and A. PROKHOROV for independently working out the theory of the MASER and, later, the LASER. He built the first maser in 1951.

TOXIN, a poisonous substance produced by a living organism. Many microorganisms, animals and plants produce chemical substances which are poisonous to some other organism; the toxin may be released continuously into the immediate environment or released only when danger is imminent. Examples include FUNGI which secrete substances which destroy BACTERIA (as ANTIBIOTICS these are of great value to man) and poisonous spiders and snakes which deliver their toxin via fangs. In some organisms, the function of toxins is obscure, but in many others they play an important role in defense and in killing prey. The symptoms of many INFECTIOUS DISEASES in man (e.g., CHOLERA; DIPHTHERIA; TETANUS) are due to the release of toxins by the bacteria concerned. (See also ANTITOXINS.)

TOXOID. See ANTITOXINS.

TRACE ELEMENTS, minerals required in minute quantities in an adequate human diet (see NUTRITION) or for the optimum growth and yield of plants (see FERTILIZERS).

TRACER, Radioactive. See RADIOCHEMISTRY; RADIOISOTOPE.

TRACHEA, the route by which air reaches the LUNGS from the PHARYNX. Air is drawn in through the mouth or NOSE and passes via the LARYNX into the trachea, which then divides into the major BRONCHI. It may be seen below the Adam's apple. In tracheostomy, it is incised to bypass any obstruction to RESPIRATION.

TRACHEAE, respiratory pathways permeating the insect body, branching from surface spiracles.

TRACHEOPHYTA. See VASCULAR PLANTS.

TRACHOMA, INFECTIOUS DISEASE due to an organism (bedsonia) intermediate in size between BACTERIA and VIRUSES, the commonest cause of BLINDNESS in the world. It causes acute or chronic CONJUNCTIVITIS and corneal INFLAMMATION with secondary blood-vessel extension over the cornea resulting in loss of translucency. Eyelid deformity with secondary corneal damage is also common. It is transmitted by direct contact; early treatment with SULFA DRUGS or TETRACYCLINE may prevent permanent corneal damage.

TRACTOR, self-propelled motor vehicle similar in principle to the AUTOMOBILE, but designed for high power and low speed. Used in agriculture, construction etc., tractors may pull other vehicles or implements, and may carry bulldozer and digging attachments. In the early 20th century the tractor, powered by the internal-combustion engine, largely superseded the steam traction engine and stationary farm-machinery engines. Many tractors have four-wheel drive or endless crawler tracks.

TRACTRIX, the CURVE to which a straight line AB, of length a, is always a TANGENT at A as B moves along the x-axis from $-\infty$ to $+\infty$. The term is sometimes loosely used to describe the solid formed by rotation of the curve about the x-axis.

TRADE WINDS, persistent warm moist WINDS that blow westward from the high-pressure zones at about 30°N and S latitude toward the DOLDRUMS (intertropical convergence zone) at the equator. They are thus northeasterlies in the N Hemisphere and southeasterlies in the S Hemisphere. They are stronger and displaced toward the equator in winter.

TRANQUILIZERS, agents which induce a state of quietude in anxious or disturbed patients. Minor tranquilizers are SEDATIVES (e.g., benzodiazepines) valuable in the anxious. In psychosis (see MENTAL ILLNESS), especially schizophrenia and (hypo-)mania, major tranquillizers are required to suppress abnormal mental activity as well as to sedate; phenothiazines (e.g., chlorpromazine) are often used.

TRANSCENDENTAL NUMBERS, those numbers that cannot be expressed as the ROOTS of a polynomial EQUATION whose coefficients are INTEGERS. Such numbers are not RATIONAL NUMBERS; they are irrational, though not all IRRATIONAL NUMBERS are transcendental. The two best-known such numbers are e (see EXPONENTIAL) and π (see PI).

TRANSDUCERS, devices which convert power levels or signals carried in one energy mode to equivalent signals in another mode; e.g., electric MOTORS; MICROPHONES; LOUDSPEAKERS; TURBINES.

TRANSDUCTION, a special type of RECOMBINATION of genetic material which involves an infectious process rather than the fusion of GAMETES. It occurs in microorganisms where DNA from a particular strain of BACTERIA can transform some of the genetic characteristics of a second strain. Resistance to ANTIBIOTICS can be passed from one bacterium to another by transduction.

TRANSFERENCE, the coloration of an individual's observation of a person (or object) by his ASSOCIATION of that person with another. In PSYCHOANALYSIS, the term means solely the effects of this process on a patient's attitudes toward his analyst.

TRANSFER-RNA. See NUCLEIC ACIDS.

TRANSFINITE CARDINAL NUMBER, in SET THEORY, the number of elements of an infinite set. Though apparently meaningless, the concept of transfinite cardinal numbers is of use in the comparison of the number of elements in *different* infinite sets. For example, that of the set of all INTEGERS, aleph null ($\aleph_0$), is smaller than that of the set of POINTS on a straight LINE since, however the points are numbered, it is always possible to construct further points between them. (See INFINITY.)

TRANSFORMATION, term synonymous with MAPPING.

TRANSFORMER, a device for altering the voltage of an AC supply (see ELECTRICITY), used chiefly for converting the high voltage at which power is transmitted over distribution systems to the normal domestic supply voltage, and for obtaining from the latter voltages suitable for electronic equipment. It is

Part of the tractrix (locus of A) for line AB of length a.

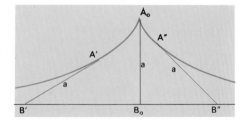

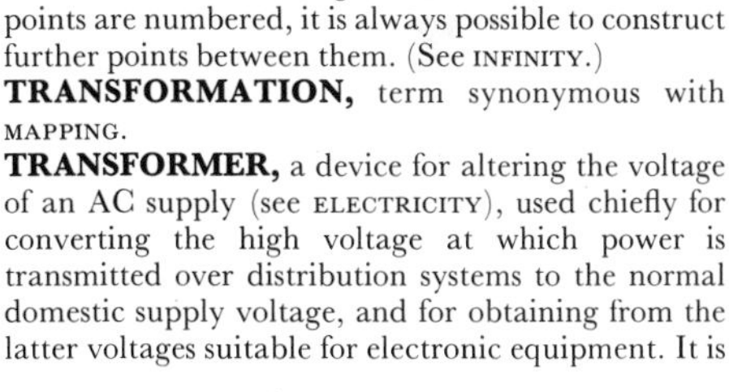

based on INDUCTION: the "primary" voltage applied to a coil wound on a closed loop of a ferromagnetic core creates a strong oscillating MAGNETIC FIELD which in turn induces in a "secondary" coil wound on the same core an AC voltage proportional to the number of turns in the secondary coil. The core is laminated to prevent the flow of "eddy" currents which would otherwise also be induced by the magnetic field and would waste some ENERGY as HEAT.

TRANSFUSION, Blood, means of BLOOD replacement in ANEMIA, SHOCK or HEMORRHAGE by intravenous infusion of blood from donors. It is the simplest and most important form of transplant, though, while of enormous value, it carries certain risks. Blood group compatibility based on ANTIBODY AND ANTIGEN reactions is of critical importance as incompatible transfusion may lead to life-threatening shock and KIDNEY failure. Infection (e.g., HEPATITIS) may be transmitted by blood, and FEVER or ALLERGY are common.

TRANSHUMANCE, the practice of some farming peoples, especially of mountainous areas, of moving their herds from one region to another to allow for different climatic conditions through the year: e.g., from mountain to valley in winter and the reverse in summer.

TRANSISTOR, electronic device made of semiconducting materials used in a circuit as an AMPLIFIER, RECTIFIER, detector or switch. Its functions are similar to those of an ELECTRON TUBE, but it has the advantage of being smaller, more durable and consuming less power. The early and somewhat unsuccessful point-contact transistor has been superseded by the junction transistor, invented in 1948 by BARDEEN, BRATTAIN and SHOCKLEY. The junction transistor is a layered device consisting of two p-n junctions (see SEMICONDUCTOR) joined back to back to give either a p-n-p or n-p-n transistor. The three layers are formed by controlled addition of impurities to a semiconductor crystal, usually SILICON or GERMANIUM. The thin central region (p-type in an n-p-n transistor and n-type in a p-n-p one) is known as

Transition elements

element	Z	atomic weight	outermost electrons	oxidation states†
scandium (Sc)	21	44.96	$3d^1 4s^2$	+III
titanium (Ti)	22	47.90	$3d^2 4s^2$	+II +III +IV
vanadium (V)	23	50.94	$3d^3 4s^2$	−I 0 +I +II +III +IV +V
chromium (Cr)	24	52.00	$3d^5 4s^1$	−II −I 0 +I +II +III +IV +V +VI
manganese (Mn)	25	54.94	$3d^5 4s^2$	−I 0 +I +II +III +IV +V +VI +VII
iron (Fe)	26	55.85	$3d^6 4s^2$	−II 0 +I +II +III +IV +V +VI
cobalt (Co)	27	58.93	$3d^7 4s^2$	−I 0 +I +II +III +IV
nickel (Ni)	28	58.70	$3d^8 4s^2$	−I 0 +I +II +III +IV
copper (Cu)	29	63.55	$3d^{10} 4s^1$	+I +II +III
zinc (Zn)	30	65.38	$3d^{10} 4s^2$	+II
yttrium (Y)	39	88.91	$4d^1 5s^2$	+III
zirconium (Zr)	40	91.22	$4d^2 5s^2$	+II +III +IV
niobium (Nb)	41	92.91	$4d^4 5s^1$	−I +I +II +III +IV +V
molybdenum (Mo)	42	95.94	$4d^5 5s^1$	−II 0 +I +II +III +IV +V +VI
technetium (Tc)	43	97*	$4d^5 5s^2$	−I 0 +I +II +III +IV +V +VI +VII
ruthenium (Ru)	44	101.07	$4d^7 5s^1$	−II 0 +II +III +IV +V +VI +VII +VIII
rhodium (Rh)	45	102.91	$4d^8 5s^1$	−I 0 +I +II +III +IV +V +VI
palladium (Pd)	46	106.4	$4d^{10}$	0 +II +IV
silver (Ag)	47	107.87	$4d^{10} 5s^1$	+I +II +III
cadmium (Cd)	48	112.41	$4d^{10} 5s^2$	+II
lanthanides‡	57–71	—	—	—
hafnium (Hf)	72	178.49	$4f^{14} 5d^2 6s^2$	+III +IV
tantalum (Ta)	73	180.95	$4f^{14} 5d^3 6s^2$	−I +I +III +IV +V
tungsten (W)	74	183.85	$4f^{14} 5d^4 6s^2$	−II 0 +I +II +III +IV +V +VI
rhenium (Re)	75	186.21	$4f^{14} 5d^5 6s^2$	−I 0 +I +II +III +IV +V +VI +VII
osmium (Os)	76	190.2	$4f^{14} 5d^6 6s^2$	0 +II +III +IV +V +VI +VII +VIII
iridium (Ir)	77	192.22	$4f^{14} 5d^7 6s^2$	0 +I +III +IV +V +VI
platinum (Pt)	78	195.09	$4f^{14} 5d^9 6s^1$	0 +II +IV +V +VI
gold (Au)	79	196.97	$4f^{14} 5d^{10} 6s^1$	+I +III
mercury (Hg)	80	200.59	$4f^{14} 5d^{10} 6s^2$	+I +II

* radioactive, the best known isotope.
† principal states in medium type, others may only occur in coordination compounds.
‡ see table under lanthanides.

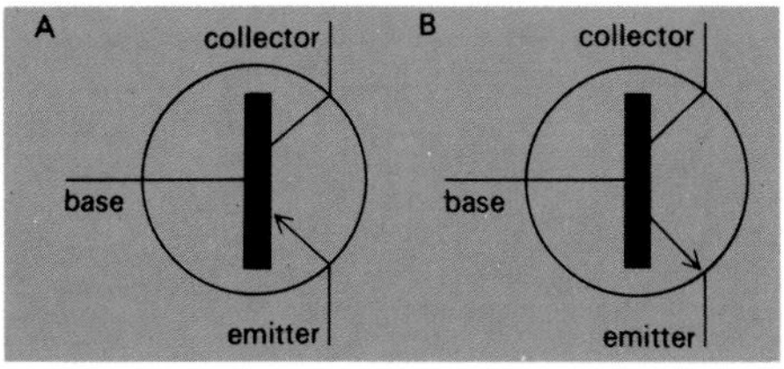

Graphic symbols for transistors: (A) with n-type base; (B) with p-type base.

the *base*, and the two outer regions (n-type semiconductor in an n-p-n transistor) are the *emitter* and *collector*, depending on the way an external voltage is connected. To act as an amplifier in a circuit, an n-p-n transistor needs a negative voltage to the collector and base. If the base is sufficiently thin, it attracts ELECTRONS from the emitter which then pass through it to the positively charged collector. By altering the bias applied to the base (which need only be a few volts), large changes in the current from the collector can be obtained and the device amplifies. A collector current up to a hundred times the base current can be obtained. This type of transistor is analogous to a TRIODE, the emitter and collector being equivalent to the CATHODE and ANODE respectively and the base to the control grid. The functioning of a p-n-p transistor is similar to the n-p-n type described, but the collector current is mainly holes rather than electrons. Transistors revolutionized the construction of electronic circuits, but are being replaced by INTEGRATED CIRCUITS in which they and other components are produced in a single semiconductor wafer.

TRANSIT, the passage of a star across an observer's meridian (the great circle on the CELESTIAL SPHERE passing through his ZENITH and the north point of his horizon). The term is also applied to the passage of the inferior planets, Mercury and Venus, across the disk of the sun.

TRANSITION ELEMENTS, the elements occupying the short groups in the PERIODIC TABLE— i.e., Groups IIIB to VIII, IB and IIB—in which the *d*-ORBITALS are being filled. The transition elements are all metals, and include most of the technologically important ones. In general they are dense, hard and of high melting point. Their electronic structures, with many loosely-bound unpaired *d*-electrons, account for their properties: they exhibit many different VALENCE states, form stable LIGAND complexes, mostly colored and paramagnetic, and are generally good catalysts. They form many stable ORGANOMETALLIC COMPOUNDS and carbonyls (compounds in which carbon monoxide, CO, acts as a ligand) with specially stable "push-pull" bonding. The second and third row transition elements are less reactive than the first row, and stable in higher valence states.

TRANSMISSION, in engineering, a device for transmitting and adapting power from its source to its point of application. Most act by changing the angular velocity of the power shaft, either by step-variable means—GEARS, as in automobiles, or CHAINS, as in bicycles—with fixed ratios and no slip, or by stepless means—belt-and-pulley systems or traction drives employing adjustable rolling contact—with continuously variable ratios but liable to slip. In an AUTOMOBILE with manual transmission, the flywheel on the engine crankshaft is connected to the gearbox via the **clutch**, two plates that are normally held tightly together by springs so that through friction they rotate together. When the clutch pedal is depressed, the plates are forced apart so that the engine is disengaged from the rest of the transmission. This is necessary when changing gear: sliding different sets of gears into engagement by means of a manual lever. Modern gearboxes have **syncromesh** in all forward gears: a coned clutch device that synchronizes the rotation of the gears before meshing. The gearbox is coupled to the final drive by a drive shaft with universal joints. A crown wheel and pinion, connected to the half-shafts of each drive wheel via a DIFFERENTIAL, complete the system. In **automatic transmission** there is no clutch pedal or gear lever; a fluid clutch (see FLUID COUPLING), combined with sets of epicyclic gears selected by a GOVERNOR according to the program set by the driver, provides a continuously variable torque ratio for maximum efficiency at all speeds.

TRANSPIRATION, the loss of water by EVAPORATION from the aerial parts of PLANTS. Considerable quantities of water are lost in this way, far more than is needed for the upward movement of solutes and for the internal metabolism of the plant alone. Transpiration is a necessary corollary of PHOTOSYNTHESIS, in that in order to obtain sufficient CARBON dioxide from the air, considerable areas of wet surface, from which high loss of water by evaporation is inevitable, have to be exposed. Plants have many means for reducing water loss, STOMATA playing an important part. XEROPHYTES in particular are adapted for minimizing transpiration. (See also WILT.)

TRANSPLANTS, organs that are removed from one person and surgically implanted in another to replace lost or diseased organs. Autotransplantation is the moving of an organ from one place to another within a person where the original site has been affected by local disease (e.g., skin grafting—see PLASTIC SURGERY). Blood TRANSFUSION was the first practical form of transplant. Here BLOOD cells and other components are transferred from one person to another. The nature of BLOOD allows free transfusion between those with compatible blood groups. The next, most important, and now most successful of organ transplants, was that of the KIDNEY. Here a single kidney is transplanted from a live donor who is a close relative or from a person who has recently suffered sudden DEATH (e.g., by traffic accident or irreversible BRAIN damage), into a person who suffers from chronic renal failure. The kidney is placed beneath the skin of the abdominal wall and plumbed into the major ARTERIES and VEINS in the PELVIS and into the BLADDER. High doses of STEROIDS and IMMUNITY suppressants are used to minimize the body's tendency to reject the foreign tissue of the graft. These doses are gradually reduced to lower maintenance levels, but may need to be increased again if rejection threatens. Here, tissue typing methods are used additionally to blood grouping to minimize rejection. HEART transplantation has been much publicized, but is limited to a few centers, many problems remaining. LIVER and LUNG transplants

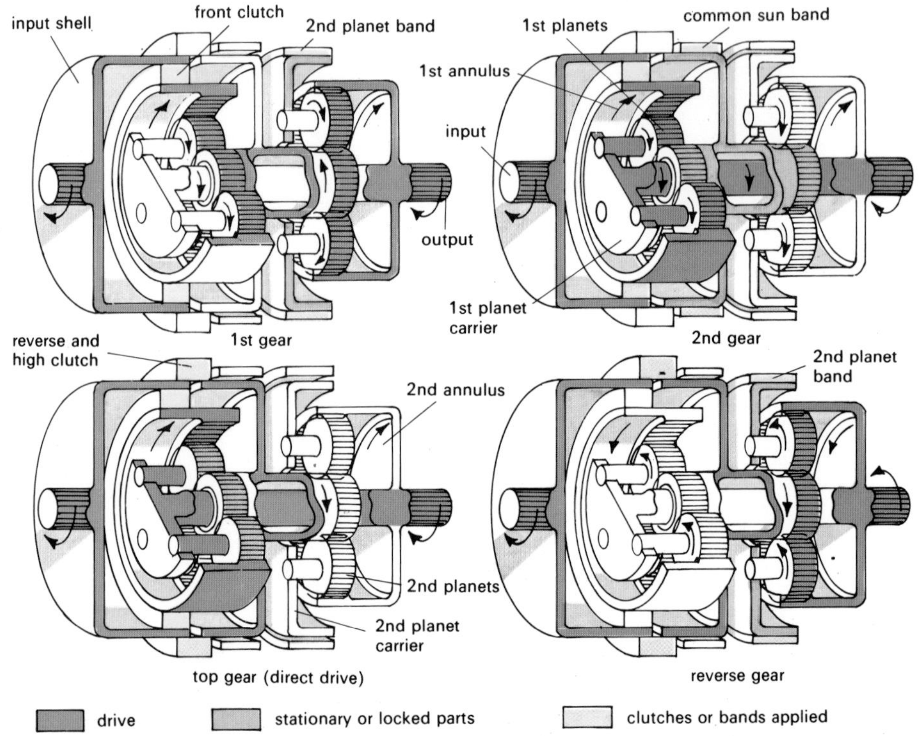

Three-speed automatic transmissions comprise a gear set (including two sun, planet and ring-gear (annulus) sets with a common sun), a fluid clutch ("converter coupling"—see illustration page 230) and a set of hydraulically operated clutches and band brakes for locking various of the component parts together to achieve different torque ratios. Drive from the engine is mediated through the fluid clutch to the input shell of the gear set and supplied by the output shaft to the differential drive (see illustration page 169). In first (lowest) gear, the front clutch is engaged, driving the annulus of the first sun-and-planet set and thus the first planets and the common sun of both sets. The second planet-carrier brake band being locked on, drive is conveyed to the output shaft via the second planets and annulus. In second gear, the common-sun brake band is locked on and drive is conveyed from the first annulus to the output shaft via the first planets and first planet carrier (which is connected directly to the output shaft); the second planets idle. Top gear involves direct drive from input to output shafts: both the front clutch and the reverse clutch are engaged, which has the effect of locking both sets of sun-and-planet gears. For reverse gear, the reverse clutch and second planet-carrier brake bands are engaged, drive transmission occurring through the second sun-and-planet set while the first idles.

have also been attempted although here too the difficulties are legion. Corneal grafting is a more widespread technique in which the cornea of the EYE of a recently dead person replaces that of a person with irreversible corneal damage leading to BLINDNESS. The lack of blood vessels in the cornea reduces the problem of rejection. Grafts from nonhuman animals are occasionally used (e.g., pig SKIN as temporary cover in extensive BURNS). Both animal and human heart valves are used in cardiac surgery.

TRANSURANIUM ELEMENTS, the elements with atomic numbers greater than that of URANIUM (92—see PERIODIC TABLE; ATOM). None occurs naturally: they are prepared by bombardment (usually with NEUTRONS or ALPHA PARTICLES) of suitably-chosen lighter ISOTOPES. All are radioactive (see RADIOACTIVITY), and those of higher atomic number tend to be less stable. Those so far discovered are the ACTINIDES from neptunium through lawrencium, RUTHERFORDIUM and HAHNIUM. Only neptunium and plutonium have been synthesized in large quantity; most of the others have been produced in weighable amounts, but some with very short HALF-LIVES can be studied only by special tracer methods.

TRAPEZIUM, and **trapezoid**. See QUADRILATERAL.

TRAUMA, any sudden wound to the body or mind, often closely followed by SHOCK. In PSYCHOANALYSIS, a trauma is viewed as an immediate cause of ANXIETY which may develop into NEUROSIS. An *infantile trauma* is one that occurred in childhood but which affects the adult.

TRAVERTINE, compact, banded LIMESTONE, usually light-colored, evaporated or deposited from hot springs; sometimes applied also to STALACTITES AND STALAGMITES. Taking a high polish, it is used for interior decoration. **Tufa**, or calcareous sinter, is a porous equivalent.

TREADMILL, vertically-mounted wheel turned by a man treading on pedals fixed to its rim, or a hollow cylinder turned by an animal walking along its internal circumference; an early power source, widely used for raising water and operating mills.

TREE, woody perennial PLANT with a well defined main stem, or trunk, which either dominates the form throughout the life cycle (giving a pyramidal shape) or is dominant only in the early stages later forking to form a number of equally important branches (giving a rounded or flattened form to the tree). It is often difficult to distinguish between a small tree and a SHRUB, but the former has a single trunk rising some distance from the ground before it branches while the latter produces several stems at, or close to, ground level. The trunk of a tree consists almost wholly of thick-walled water-conducting cells (xylem) which are renewed every year (see WOOD), giving rise to the familiar ANNUAL RINGS. The older wood in the center of the tree (the heartwood) is much denser and harder than the younger, outer sapwood. The outer skin or the BARK, insulates and protects the trunk and often shows characteristic cracks, or falls off leaving a smooth skin. Trees belong to the two most advanced groups of plants, the GYMNOSPERMS and the ANGIOSPERMS (the flowering plants). The former include the cone-bearing trees such as the pine, spruce and cedar; they are nearly all evergreens and mostly live in the cooler regions of the world. The angiosperms have broader leaves and much harder wood; in tropical climates they are mostly EVERGREEN, but in temperate regions they are DECIDUOUS. (See also FORESTRY; FORESTS.)

TREMATODA, parasitic flukes included in the PLATYHELMINTHES, that are similar to free-living flatworms (TURBELLARIA) but which have a thick cuticle, and suckers which enable them to cling to their hosts. Trematode worms are responsible for diseases such as SCHISTOSOMIASIS.

TRENCH FOOT. See IMMERSION FOOT.

TRENCH MOUTH, WWI name for VINCENT'S ANGINA.

TREPHINE, surgical instrument used for trepanning—making a circular hole in an internal organ in order to drain fluid, PUS, BLOOD or air. Trepanning the SKULL was successfully practiced in ancient times.

TREVITHICK, Richard (1771–1833), British mining engineer and inventor primarily remembered for his work improving the STEAM ENGINE and for building the first railroad LOCOMOTIVE (c1804).

TRIANGLE, a three-sided POLYGON. There are three main types of plane triangle: scalene, in which no side is equal in length to another; isosceles, in which two of the sides are equal in length; and equilateral, in which all three sides are equal in length. A right (or right-angled) triangle has one INTERIOR ANGLE equal to 90°, and may be either scalene or isosceles (see PYTHAGORAS' THEOREM). The "corners" of a triangle are termed vertices (singular, vertex). The sum of the angles of a plane triangle is 180°. A **spherical triangle** is an area, bounded by arcs of three great circles (see SPHERICAL GEOMETRY), of the surface of a SPHERE, each arc being less than 180°, each side and interior angle being termed an element. The sum of the three sides is never greater than 360°, the sum of the three angles always in the range 180°–540°. (See also COSINE RULE; SINE RULE.)

TRIASSIC, the first period of the MESOZOIC era, which lasted from about 225 to 190 million years ago. (See also GEOLOGY.)

TRIBE (from Latin *tribus*, a third, referring to the three peoples who founded Rome), a people with a common territory, customs and, usually, language or dialect. Tribes may be merely a few families or great in numbers; social structures and customs vary from one tribe to another.

TRICEPS, the major ARM muscle concerned with straightening the elbow. It has three heads or bellies at its upper end that have separate insertions.

TRICERATOPS, a DINOSAUR that grew to a length of 6m (20ft), with a head bearing a short horn on the snout and a larger horn over each eye. The rear part of the skull was extended backward to form a bony frill over the neck.

TRICHINOSIS, infestation with the larva of a worm (*Trichinella*), contracted from eating uncooked pork etc., causing a feverish illness. EDEMA around the eyes, MUSCLE pains and DIARRHEA occur early; later the LUNGS, HEART and BRAIN may be involved. It is avoided by the adequate cooking of pork. CHEMOTHERAPY may be helpful in severe cases.

TRIGONOMETRY, the branch of GEOMETRY that deals with the ratios of the sides of right-angled TRIANGLES, and the applications of these ratios. The principal ratios, when considering ANGLE A of triangle ABC whose sides are a, b and c, where b is the HYPOTENUSE, are:

name	*abbreviation*	*ratio*
tangent	tan A	$\frac{a}{c}$
sine	sin A	$\frac{a}{b}$
cosine	cos A	$\frac{c}{b}$
cotangent	cot A	$\frac{c}{a}$
cosecant	cosec A	$\frac{b}{a}$
secant	sec A	$\frac{b}{c}$

As can be seen, the cotangent is the RECIPROCAL of the tangent, the cosecant that of the sine, and the secant that of the cosine. (See COSINE RULE; SINE RULE.)

From these ratios are derived the **trigonometric functions**, setting y equal to tan x, sin x, etc. These FUNCTIONS are termed transcendental (nonalgebraic). Of particular importance is the sine wave, in terms of which many naturally-occurring WAVE MOTIONS, such as SOUND and LIGHT, are studied. (See also CALCULUS, SPHERICAL TRIGONOMETRY.)

TRIHEDRON, in DIFFERENTIAL GEOMETRY, a set of three mutually perpendicular (see ANGLE) VECTORS directed from a single point. If one considers a trihedron moving from point to point along, say, a CURVE, one has a handy coordinate system, using the three vectors as AXES, for study of the curve. This is termed a moving trihedron.

TRINITROTOLUENE. See TNT.

TRIODE, ELECTRON TUBE with positive ANODE, electron-emitting CATHODE and negatively biased control grid, used as an AMPLIFIER or OSCILLATOR.

TRIPLETS. See MULTIPLE BIRTHS.

TRISECTION OF AN ANGLE, the division of an ANGLE into three equal parts. Impossible using only straight edge and COMPASSES, it was one of the famous construction problems of classical GEOMETRY.

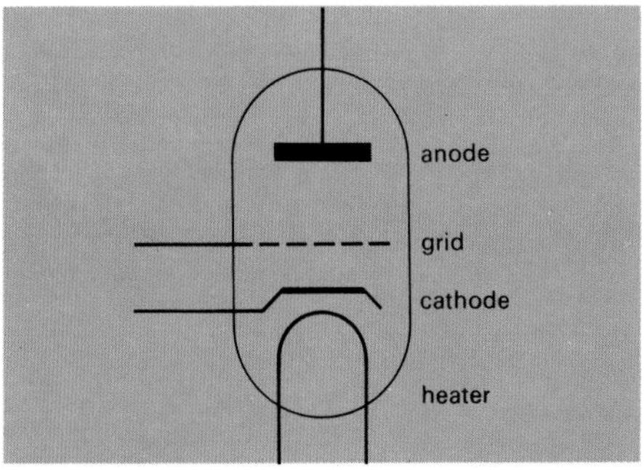

Graphic symbol for a triode.

TRITIUM (T or $_1H^3$), the heaviest ISOTOPE of HYDROGEN, whose nucleus has one PROTON and two NEUTRONS, produced by neutron irradiation of lithium (Li^6). Tritium is weakly radioactive and emits BETA RAYS. It is used for tracer studies, in luminous paints, and (together with DEUTERIUM) in HYDROGEN BOMBS. (See also FUSION, NUCLEAR). AW 3.0, mp −252.5°C, bp −248.1°C.

TROPICAL MEDICINE, branch of MEDICINE concerned with the particular diseases encountered in and sometimes imported from the tropics. These largely comprise INFECTIOUS DISEASES due to VIRUSES (e.g., YELLOW FEVER, SMALLPOX, lassa fever), BACTERIA (e.g., CHOLERA), protozoa (e.g., MALARIA, TRYPANOSOME diseases) and worms (e.g., FILARIASIS) which are generally restricted to tropical zones. The ANTIBIOTIC treatment and CHEMOTHERAPY of bacterial and PARASITIC DISEASES and their prevention with prophylactic DRUGS and by control of insect or other vectors are important aspects of this speciality. The diseases of MALNUTRITION—KWASHIORKOR, marasmus and the VITAMIN deficiency diseases of BERI-BERI, PELLAGRA etc.—often fall in the province of tropical disease as do SUNSTROKE and SNAKE BITES.

TROPICS, the lines of latitude lying about 23½° N (**Tropic of Cancer**) and S (**Tropic of Capricorn**) representing the farthest northerly and southerly latitudes where the sun is, at one time of the year, directly overhead at noon. This occurs at the time of summer SOLSTICE in each hemisphere. The term is used also of the area between the two tropics.

TROPISMS, movements of PLANTS in response to external directional stimuli. If a plant is laid on its side, the stem will soon start to bend upward again. This movement (geotropism) is a response to the force of gravity. The stem is said to be negatively geotropic. Roots are generally positively geotropic and grow downward. Phototropisms are bending movements in response to the direction of illumination. Stems are generally positively phototropic (bend toward the light). Most roots are negatively phototropic, although some appear unaffected by light. Some roots exhibit positive hydrotropism: they bend toward moisture. This response is more powerful than the response to gravity; roots can be deflected from their downward course if the plants are watered only on one side. Tropisms are controlled by differences in concentration of growth HORMONES. (See also AUXINS.)

TROPOSPHERE, the innermost zone of the earth's ATMOSPHERE, extending from the surface up to the tropopause.

TROY WEIGHT, a system of weights used for precious metals and stones, named for the French town of Troyes, famed for its medieval trade fairs. Today the price of GOLD is still quoted in dollars per troy ounce. The troy ounce, equivalent to 1.0971 ounces avoirdupois, is equal to the apothecaries' ounce. (See also APOTHECARIES' WEIGHTS; WEIGHTS AND MEASURES.)

Troy weight

Unit	Equivalent in same system	Approximate SI equivalent
pound (lb t)	12 ounces	0.373 242 kg
ounce (oz t)	20 pennyweight	31.103 g
pennyweight (dwt)	24 grains	1.555 g
grain (gr)	—	*64.798 91 mg

*exact equivalent

TRUCK, automotive vehicle used for transporting freight by road. The typical long-distance truck is an articulated vehicle comprising a two- or three-axled "truck tractor" coupled to a two-axled "semitrailer." A two- or three-axled "full trailer" may in addition be coupled to the semitrailer. Most trucks are powered by a DIESEL ENGINE, have a manual TRANSMISSION with perhaps as many as 16 forward gears, and have AIR BRAKES. In the US, trucks carry about 40% of all intercity freight (compared with the railroads' 30%).

TRUTH, philosophical concept deriving from the everyday notion of "telling the truth" as opposed to telling a deliberate falsehood. "Correspondence" theories of truth hold that a statement is true if it corresponds to the "facts" of experience. "Coherence" theories of truth contend, however, that facts are themselves statements of a kind whose truth cannot be tested by looking for further correspondences, but only by considering their logical coherence with other statements about supposed reality. "Pragmatic" theories of truth stress that the only usefully testable "truths" are those that enable us to anticipate or control the course of events. (See also EMPIRICISM; RATIONALISM; PRAGMATISM.)

TRUTH SERUM. See PENTOTHAL SODIUM.

TRUTH TABLE, in LOGIC, a convenient way of displaying the range of "truth values" (truth or falsity) of a compound statement as determined by the truth values of its simple component statements. For example, if the simple statements *p* and *q* are both true, then the compound statement *p and q* is true (1); but if either *p* or *q* or both *p* and *q* are false, then *p and q* is false (0). This is usually displayed in tabular form thus:

p	q	$p \cdot q$
1	1	1
1	0	0
0	1	0
0	0	0

The table can be extended to show the range of truth values for other combinations of *p* and *q*.

TRYPANOSOMES, PROTOZOA responsible for trypanosomiasis of the African (SLEEPING SICKNESS) and South American (CHAGAS' DISEASE) varieties, carried by the tsetse fly and certain bugs respectively. They are relatively insensitive to CHEMOTHERAPY in established cases; prevention is therefore important.

TRYPSIN, ENZYME catalyzing the breakdown of PROTEINS in the vertebrate DIGESTIVE SYSTEM.

TSIOLKOVSKY, Konstantin Eduardovich (1857–1935), Russian physicist who pioneered ROCKET science, but who is perhaps most important for his role in educating the Soviet government and people into acceptance of the future potential of SPACE EXPLORATION. He also built one of the first WIND TUNNELS (c1892). A large crater on the far side of the MOON is named for him; and the timing of Sputnik I's launch commemorated the 100th anniversary of his birth.

TSUNAMI, formerly called **tidal wave**, fast-moving ocean wave caused by submarine EARTHQUAKES, volcanic eruptions, etc., found mainly in the Pacific, and often taking a high toll of lives in affected coastal areas. In midocean, the wave height is usually under 1m, the distance between succeeding crests of the order of 200km, and the velocity about 750km/h. Near the coast, FRICTION with the sea bottom slows the wave, so that the distance between crests decreases, the wave height increasing to about 25m or more.

TUATARA. See RHYNCHOCEPHALIA.

TUBER, swollen underground stems and roots which are organs of perennation and vegetative propagation and contain stored food material. The potato is a stem tuber. It swells at the tip of a slender underground stem (or stolon) and gives rise to a new plant the following year. Dahlia tubers are swollen roots.

TUBERCULIN, PROTEIN derivative of the mycobacteria responsible for TUBERCULOSIS. This may be used in tests of cell-mediated IMMUNITY to tuberculosis, providing evidence of previous disease (often subclinical) or immunization (BCG). The substance was originally isolated by KOCH.

TUBERCULOSIS (TB), a group of INFECTIOUS DISEASES caused by the BACILLUS *Mycobacterium tuberculosis*, which kills some 3 million people every year throughout the world. TB may invade any organ but most commonly affects the respiratory system where it has been called consumption or phthisis (see also LUPUS VULGARIS and SCROFULA). In 1906 it killed 1 in every 500 persons in the US, but today it leads to only 1 in 30 000 deaths, because of effective drugs and better living conditions. The disease is spread in three ways: inoculation via cuts, etc.; inhalation of infected sputum, and ingestion of infected food. In pulmonary TB there are two stages of infection. In primary infection there are usually no significant symptoms: dormant small hard masses called tubercles are formed by the body's defenses. In postprimary infection the dormant BACTERIA are reactivated due to weakening of the body's defenses and clinical symptoms become evident. Symptoms include fatigue, weight loss, persistent cough with green or yellow sputum and possibly blood. Treatment nowadays is mainly by triple drug therapy with streptomycin, para-aminosalicylic acid (PAS) and isoniazid, together with rest. Recovery usually takes about two years.

The TUBERCULIN skin test can show whether a person has some IMMUNITY to the disease, though the detection of the disease in its early stages, when it is readily curable, is difficult. Control of the disease is accomplished by preventive measures such as X-RAY screening, BCG vaccination, isolation of infectious people and food sterilization.

TUFA. See TRAVERTINE.

TULAREMIA, or **Rabbit Fever**, INFECTIOUS DISEASE due to BACTERIA, causing FEVER, ulceration, LYMPH node enlargement and sometimes PNEUMONIA. It is carried by wild animals, particularly rabbits, and insects. ANTIBIOTICS are fully effective in treatment.

TULL, Jethro (1674–1741), British agricultural scientist who invented a horse-drawn hoe for cultivating the ground between the rows (after 1715).

TUMOR, strictly, any swelling on or in the body, but more usually used to refer only to an abnormal overgrowth of tissue (or **neoplasm**). These may be benign proliferations such as fibroids of the WOMB, or they may be forms of CANCER, LYMPHOMA or SARCOMA, which are generally malignant. The rate of growth, the tendency to spread locally and to distant sites via the BLOOD vessels and LYMPH system, and systemic effects determine the degree of malignancy of a given tumor. Tumors may present as a lump, by local compression effects (especially with BRAIN tumors), by bleeding (GASTROINTESTINAL TRACT tumors) or by systemic effects including ANEMIA, weight loss, false HORMONE actions, NEURITIS etc. Treatments include surgery, RADIATION THERAPY and CHEMOTHERAPY.

TUNDRA, the treeless plains of the Arctic Circle. For most of the year the temperature is less than 0°C, and even during the short summer it never rises above 10°C. The soil is a thin coating over PERMAFROST. Tundra vegetation includes lichens, mosses and stunted shrubs. Similar regions on high mountains (but generally without permafrost) are **alpine tundra**.

TUNG OIL, or China wood oil, important drying OIL and ingredient of many PAINTS and VARNISHES. It dries very rapidly, to form an intact and durable coating. Poisonous, it is extracted from the seeds of the tung oil tree (*Aleurites fordii*) and related species. Family: Euphorbiaceae.

TUNGSTEN (W), or **wolfram**, hard, silvery-gray metal in Group VIB of the PERIODIC TABLE; a TRANSITION ELEMENT. Its chief ores are SCHEELITE and WOLFRAMITE. The metal is produced by reduction of heated tungsten dioxide with hydrogen. Its main uses are in tungsten steel ALLOYS for high-temperature applications, and for the filaments of incandescent lamps. It is relatively inert, and resembles MOLYBDENUM. Cemented **tungsten carbide** (WC) is used in cutting tools. AW 183.9, mp 3410°C, bp 5660°C, sg 19.3 (20°C).

TUNICATES. See UROCHORDATA.

TUNING FORK, simple two-pronged instrument that emits a pure tone of fixed PITCH when struck. Made within a FREQUENCY range of 20 to 20 000Hz, they are used as acoustical frequency standards.

TUNNEL, underground passageway, usually designed to carry a highway or railroad, to serve as a conduit for water or sewage, or to provide access to an underground working face (see MINING). Although tunnels have been built since prehistoric times,

A tunneling shield breaks through into a 20m-deep station excavation during construction of London–Heathrow airport's subway link.

tunneling methods remained primitive and hazardous until the 19th century. Modern soft-ground tunneling was pioneered by Marc BRUNEL, who in 1824 invented the "tunneling shield"—a device subsequently improved (1869–86) by James Greathead. The Greathead shield is basically a large steel cylinder with a sharp cutting edge driven forward by hydraulic rams. Used in conjunction with a compressed-air atmosphere, it protects excavating workmen against cave-ins and water seepage. Tunneling through hard rock is facilitated by an array of pneumatic drills mounted on a "jumbo" carriage running on rails. Explosives are inserted in a pattern of holes drilled in the rock face and then detonated. Increasingly used today, however, are automatic tunneling machines called "moles," with cutting heads consisting of a rotating or oscillating wheel that digs, grinds or chisels away the working face. Another common tunnel-building method—used in constructing the New York SUBWAY—is "cut-and-cover," which involves excavating a trench, building the tunnel-lining and then covering it. The world's longest vehicular or railroad tunnel is the 19.8km (12.3mi) Simplon II in the Alps, completed in 1922.

TUNNEL DIODE, SEMICONDUCTOR device with a high impurity concentration and negative RESISTANCE over part of its operating range, used in amplifying, oscillating and switching circuits. Its operation depends on quantum-mechanical tunneling of charges through a narrow p-n junction at zero voltage which ceases at increased forward voltages.

TURBELLARIA, free-living, leaf-shaped PLATYHELMINTHES composed of three layers of cells. Turbellarians are found in ponds, rivers and the sea and grow to a length of 10–35mm (0.4–1.4in). They have a branched gut with a single external opening.

TURBINE, machine for directly converting the kinetic and/or thermal ENERGY of a flowing FLUID into useful rotational energy. The working fluid may be air, hot gas, steam or water. This either pushes against a set of blades mounted on the drive shaft (impulse turbines) or turns the shaft by reaction when the fluid is expelled from nozzles (or nozzle-shaped vanes) around its circumference (reaction turbines). Water turbines were the first to be developed. They now include the vast inward-flow reaction turbines used in the generation of HYDROELECTRICITY and the smaller-scale tangential-flow "Pelton wheel" impulse types used when exploiting a very great "head" of water. In the 1880s, Charles Algernon Parsons (1854–1931), a British engineer, designed the first successful steam turbines, having realized that the efficient use of high-pressure steam demanded that its energy be extracted in a multitude of small stages. Steam turbines thus consist of a series of vanes mounted on a rotating drum with stator vanes redirecting the steam in between the moving ones. They are commonly used as marine engines and in thermal and nuclear power plants. GAS TURBINES are not as yet widely used except in airplanes (see JET PROPULSION) and for peak-load electricity generation.

TURBOJET. See JET PROPULSION.

TURBULENCE, type of irregular flow of FLUIDS in which the motion at any point varies rapidly in magnitude and direction. The value of REYNOLDS NUMBER determines whether fluid flow is laminar (smooth and well-defined), or turbulent. Most natural fluid motion is turbulent.

TURPENTINE, exudate obtained from injured pine trees. This is distilled to give the RESIN, rosin, and the ESSENTIAL OIL, oil of turpentine (also known as turpentine). The chief constituent of oil of turpentine, used as a solvent and thinner for PAINTS and VARNISHES, is pinene (see TERPENES).

TURQUOISE, hydrated copper aluminum PHOSPHATE mineral, $CuAl_6(PO_4)_4(OH)_8.4H_2O$; used as a semiprecious GEM stone, blue in color. Deposited from water, it occurs in veinlets and as masses. The finest turquoise comes from Iran.

TWINS. See MULTIPLE BIRTHS; SIAMESE TWINS.

TYLOR, Sir Edward Burnett (1832–1917), British anthropologist whose work, culminating in *Primitive Culture* (1871), established him as a father of cultural ANTHROPOLOGY. In 1896 he was appointed the first Professor of Anthropology in the University of Oxford.

TYNDALL, John (1820–1893), British physicist who, through his studies of the scattering of light by colloidal particles or large molecules in SUSPENSION (the **Tyndall effect**), showed that the daytime sky is blue because of the Rayleigh SCATTERING of impingent sunlight by dust and other colloidal particles in the air (see COLLOID).

TYPE. See PRINTING.

TYPE METAL, various ALLOYS of lead, antimony and tin, easily cast to make printing type because they expand on solidifying.

TYPESETTING. See LINOTYPE; MONOTYPE; PRINTING.

TYPEWRITER, writing machine activated manually or electrically by means of a keyboard. Normally, when a key is depressed, a pivoted bar bearing a type character strikes an inked ribbon against a sheet of paper carried on a cylindrical rubber "platen," and the platen carriage automatically moves a space to the left. In some electric models all the type is carried on a single rotatable sphere that moves from left to right and strikes a fixed platen. The first efficient typewriter was developed in 1868 by C. L. SHOLES.

TYPHOID FEVER, INFECTIOUS DISEASE due to a SALMONELLA species causing FEVER, a characteristic rash, LYMPH node and SPLEEN enlargement, GASTROINTESTINAL TRACT disturbance with bleeding and ulceration, and usually marked malaise or prostration. It is contracted from other cases or from disease carriers, the latter often harboring asymptomatic infection in the GALL-BLADDER or urine, with contaminated food and water as major vectors. Carriers must be treated with ANTIBIOTICS (and have their gall-bladder removed if this site is the source); they must also stop handling food until they are free of the bacteria. VACCINATION may help protect high risk persons; antibiotics—chloramphenicol or cotrimoxazole—form the treatment of choice.

TYPHOON. See HURRICANE.

TYPHUS, INFECTIOUS DISEASE caused by RICKETTSIA and carried by lice, leading to a feverish illness with a rash. Severe HEADACHE typically precedes the rash, which may be erythematous or may progress to skin HEMORRHAGE; mild respiratory symptoms of cough and breathlessness are common. Death ensues in a high proportion of untreated adults, usually with profound SHOCK and KIDNEY failure. Recurrences may occur in untreated patients who recover from their first attack, often after many years (Brill-Zinsser disease). A similar disease due to a different but related organism is carried by fleas (Murine typhus). Chloramphenicol or TETRACYCLINES provide suitable ANTIBIOTIC therapy.

Type metal alloys

	lead	tin	antimony	copper
Electrotype	93	3	4	—
Linotype	86	3	11	—
Stereotype	81	6	13	—
Monotype	74	10	16	—
Foundry type	60	13	25	2

Percentage compositions of typical type metals. The hardest alloys are nearest the bottom.

TYRANNOSAURUS, a DINOSAUR with powerful hind limbs but tiny fore limbs that bore only two fingers. *Tyrannosaurus* was a carnivore with a large skull bearing long, pointed teeth.

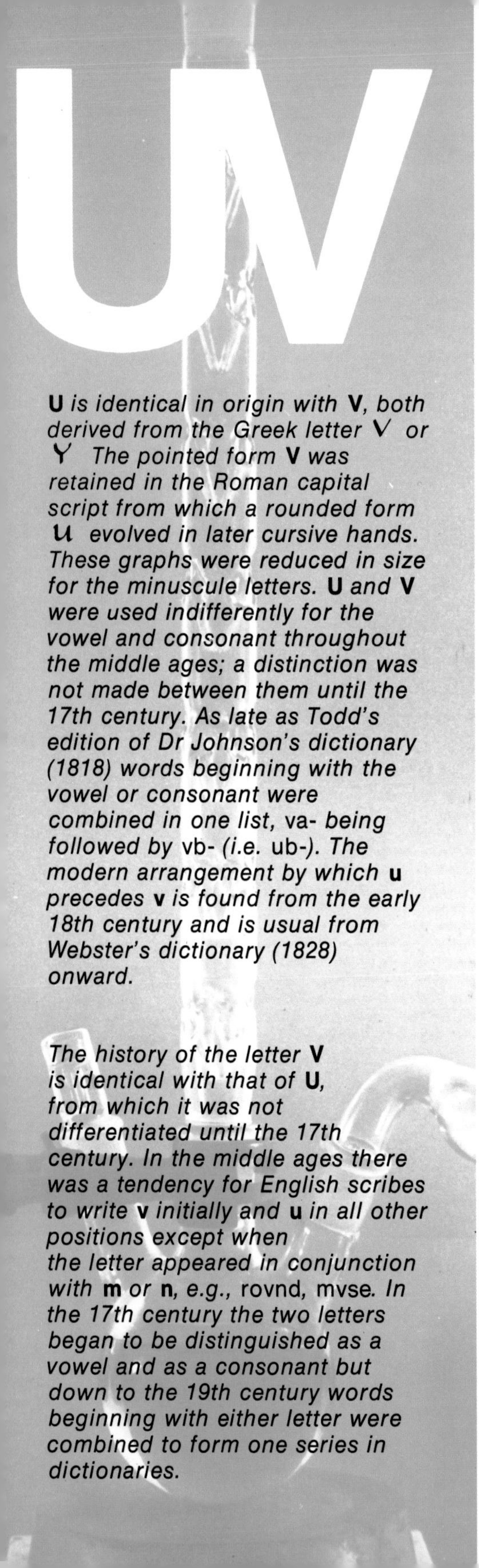

UV

U *is identical in origin with* **V**, *both derived from the Greek letter* V *or* Y *The pointed form* **V** *was retained in the Roman capital script from which a rounded form* u *evolved in later cursive hands. These graphs were reduced in size for the minuscule letters.* **U** *and* **V** *were used indifferently for the vowel and consonant throughout the middle ages; a distinction was not made between them until the 17th century. As late as Todd's edition of Dr Johnson's dictionary (1818) words beginning with the vowel or consonant were combined in one list,* va- *being followed by* vb- *(i.e.* ub-*). The modern arrangement by which* **u** *precedes* **v** *is found from the early 18th century and is usual from Webster's dictionary (1828) onward.*

The history of the letter **V** *is identical with that of* **U**, *from which it was not differentiated until the 17th century. In the middle ages there was a tendency for English scribes to write* **v** *initially and* **u** *in all other positions except when the letter appeared in conjunction with* **m** *or* **n**, *e.g.,* rovnd, mvse. *In the 17th century the two letters began to be distinguished as a vowel and as a consonant but down to the 19th century words beginning with either letter were combined to form one series in dictionaries.*

ULCER, pathological defect in skin or other epithelium, caused by inflammation secondary to infection, loss of blood supply, failure of venous return or cancer. Various skin lesions can cause ulcers, including infection, arterial disease, varicose veins and skin cancer. Aphthous ulcers in the mouth are painful epithelial ulcers of unknown origin. Peptic ulcers include gastric and duodenal ulcers, although the two have different causes; they may cause characteristic pain, acute hemorrhage, or lead to perforation and peritonitis. Severe scarring or edema around the pylorus may cause stenosis with vomiting and stomach distension. Antacids, rest, stopping smoking, and licorice derivatives may help peptic ulcer but surgery may also be needed.

ULTRACENTRIFUGE. See centrifuge.

ULTRA-HIGH-FREQUENCY WAVES (UHF). See electromagnetic radiation; radio.

ULTRAMICROSCOPE, a microscope for studying liquid suspensions of particles too small for direct microscopy (10nm–1 μm), using light scattered by the particles at right angles. It allows their number and position to be determined, their motion to be followed, and their size to be estimated, though no structural detail can be discerned.

ULTRASONICS, science of sound waves with frequencies above those that humans can hear (>20 kHz). With modern piezoelectric techniques, ultrasonic waves having frequencies above 24 kHz can readily be generated with high efficiency and intensity in solids and liquids, and exhibit the normal wave properties of reflection, refraction and diffraction. They can thus be used as investigative tools or for concentrating large amounts of mechanical energy. Low-power waves are used in thickness gauging and holography, high-power waves in surgery and for industrial homogenization, cleaning and machining.

ULTRAVIOLET RADIATION, electromagnetic radiation of wavelength between 0.1nm and 380nm, produced using gas discharge tubes. Although it constitutes 5% of the energy radiated by the sun, most falling on the earth is filtered out by atmospheric oxygen and ozone, thus protecting life on the surface from destruction by the solar ultraviolet light. This also means that air must be excluded from optical apparatus designed for ultraviolet light, similar strong absorbtion by glass necessitating that lenses and prisms be made of quartz or fluorite. Detection is photographic or by using fluorescent screens. The principal use is in fluorescent tubes (see lighting) but important medical applications include germicidal lamps, the treatment of rickets and some skin diseases and the vitamin-D enrichment of milk and eggs.

UMBILICAL CORD, long structure linking the developing embryo or fetus to the placenta through most of pregnancy. It consists of blood vessels taking blood to and from the placenta, and a gelatinous matrix. At birth the cord is clamped to prevent blood loss and is used to assist delivery of the placenta. It undergoes atrophy and becomes the **navel**.

UMBRA. See shadow; eclipse.

UNCERTAINTY PRINCIPLE, or **indeterminacy principle,** a restriction, first enunciated by W. K. Heisenberg in 1927, on the accuracy with which the position and momentum of an object can be established simultaneously: the product of the accuracies attainable in each cannot be less than the PLANCK CONSTANT. Relevant only near the atomic level, the principle arises from the wave nature of matter: a particle consists of a superposition of waves with slightly different speeds producing a localized disturbance of which neither the position nor the speed is precisely defined.

UNCONFORMITY, a surface between two contiguous rock strata representing a break in the normal succession; usually owing to EROSION having removed layers of rock before the deposition of the younger stratum. They may be parallel (strata parallel), angular (strata not parallel), heterolithic (sediment over intruded IGNEOUS ROCKS) or nondepositional (a genuine break in the deposition pattern). (See also SEDIMENTARY ROCKS.)

UNCONSCIOUS, that part of the mind in which take place events of which the individual is unaware; i.e., the part of the mind that is not the CONSCIOUS. Unconscious processes can, however, alter the behavior of the individual (see also DREAMS; INSTINCT). FREUD renamed the unconscious the ID. (See also COLLECTIVE UNCONSCIOUS.)

UNCONSCIOUSNESS, lack of awareness, the commonest example of which is sleep; or the lack of self-awareness displayed by most, if not all, animals (see CONSCIOUSNESS). (See also COMA.)

UNDULANT FEVER. See BRUCELLOSIS.

UNGULATES, general name for all hoofed mammals, including both the odd-toed and even-toed groups: PERISSODACTYLA and ARTIODACTYLA respectively.

UNIFIED FIELD THEORY, theory which tries to incorporate electromagnetic together with the strong and weak nuclear forces into the general theory of RELATIVITY. If successful, one set of equations would describe these fundamental force fields, including gravity, in terms of the geometry of space-time. Einstein made the first attempt to produce such a theory; he wanted to represent physical reality entirely in terms of fields, yet, in his general theory of relativity, particles still exist as SINGULARITIES—regions where field equations break down.

UNIFORMITARIANISM, the principle, due to J. HUTTON and C. LYELL, that the same agencies are at work in nature today, operating at the same intensities, as they have always done throughout geologic time. It was originally opposed to CATASTROPHISM.

UNION. See SET THEORY.

UNITS. See ABAMPERE; CGS UNITS; METRIC SYSTEM; MKSA UNITS; SI UNITS; STATAMPERE; WEIGHTS AND MEASURES.

UNITY, mathematical term for the number one.

UNIVERSAL, philosophical term referring to any possible attribute of more than one particular. Redness, for example, is the universal common to all red things. The question arises whether the general term naming a universal refers to an entity that exists independently of thought or is merely a principle of classification. (See also IDEALISM; REALISM; CONCEPTUALISM; NOMINALISM.)

UNIVERSE, the closed system of all that exists or happens (see also HEAT DEATH). In COSMOLOGY, the term is applied to our universe, i.e., all that we can observe and the presumably homogeneous and isotropic extension thereof, since it is conceived that there may have been prior universes and that there may be subsequent universes.

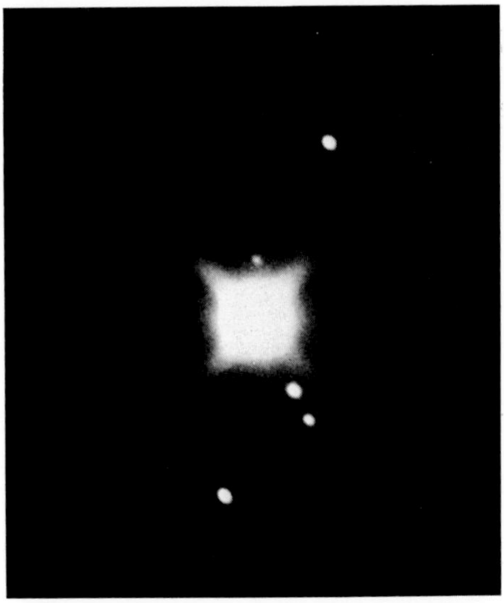

The planet Uranus with its moons (in increasing radius of orbit) Miranda, Ariel, Umbriel, Titania and Oberon.

URACIL. See NUCLEIC ACIDS; NUCLEOTIDES.

URANINITE. See PITCHBLENDE.

URANIUM (U), soft, silvery-white radioactive metal in the ACTINIDE series; the heaviest natural element. Uranium occurs widespread as PITCHBLENDE (uraninite), CARNOTITE and other ores, which are concentrated and converted to uranium (IV) fluoride, from which uranium is isolated by electrolysis or reduction with calcium or magnesium. The metal is reactive and electropositive, reacting with hot water and dissolving in acids. Its chief oxidation states are +4 and +6, and the uranyl (UO_2^{2+}) compounds are common. Uranium has three naturally-occurring ISOTOPES: U^{238} (HALF-LIFE 4.5×10^9yr), U^{235} (half-life 7.1×10^8yr) and U^{234} (half-life 2.5×10^5yr). More than 99% of natural uranium is U^{238}. The isotopes may be separated by fractional DIFFUSION of the volatile uranium (VI) fluoride. Neutron capture by U^{235} leads to nuclear FISSION, and a chain reaction can occur which is the basis of NUCLEAR REACTORS and of the ATOMIC BOMB. U^{238} also absorbs neutrons and is converted to an isotope of PLUTONIUM (Pu^{239}) which (like U^{235}) can be used as a nuclear fuel. Uranium is the starting material for the synthesis of the TRANSURANIUM ELEMENTS. Some of its compounds are used to color ceramics. AW 238.0, mp 1132°C, bp 3818°C, sg 19.05 (α).

URANUS, the third largest planet in the SOLAR SYSTEM and the seventh from the sun. Physically very similar to NEPTUNE, but rather larger (53 Mm ± 5% equatorial radius), it orbits the sun every 84.02 years at a mean distance of 19.2AU, rotating in 10.75h. The plane of its equator is tilted 98° to the plane of its orbit, such that the rotation of the planet and the revolution of its five moons, which orbit closely parallel to the equator, are retrograde (see RETROGRADE MOTION).

UREA, or carbamide, $CO(NH_2)_2$, the AMIDE of carbonic acid, a white crystalline solid, the end-product of protein METABOLISM in many animals, excreted in the URINE. Virtually the first organic compound to be synthesized—by Wöhler in 1828 from ammonium cyanate—it is now prepared by heating AMMONIA and CARBON dioxide under pressure. Urea's major uses are as a nitrogenous FERTILIZER, to make urea-formaldehyde resins (see PLASTICS), and to make BARBITURATES. MW 60.1, mp 135°C.

UREMIA, the syndrome of symptoms and biochemical disorders seen in KIDNEY failure, associated with a rise in blood UREA and other nitrogenous waste products of PROTEIN metabolism. Nausea, VOMITING, malaise, itching, pigmentation, ANEMIA and acute disorders of fluid and mineral balance are common presentations, but the manifestations depend on the type of disease, rate of waste buildup, etc. DIETARY FOODS may reduce uremic symptoms in chronic renal failure but dialysis or TRANSPLANTATION may be needed.

UREY, Harold Clayton (1893–), US chemist awarded the 1935 Nobel Prize for Chemistry for his discovery of DEUTERIUM, an isotope of HYDROGEN having one proton and one neutron in its nucleus, and who played a major role in the MANHATTAN PROJECT. He is also important as a cosmologist: his researches into geological dating using oxygen ISOTOPES enabled him to produce a model of the atmosphere of the primordial planet earth; and hence to formulate a theory of the planets' having originated as a gaseous disk about the sun (see SOLAR SYSTEM).

URIC ACID, or 2,6,8-trihydroxypurine, the end-product of protein METABOLISM in birds, invertebrates and snakes, and of PURINE metabolism in many insects, reptiles, birds, primates (including man) and the Dalmatian dog. Sufferers from GOUT have a high blood level of uric acid.

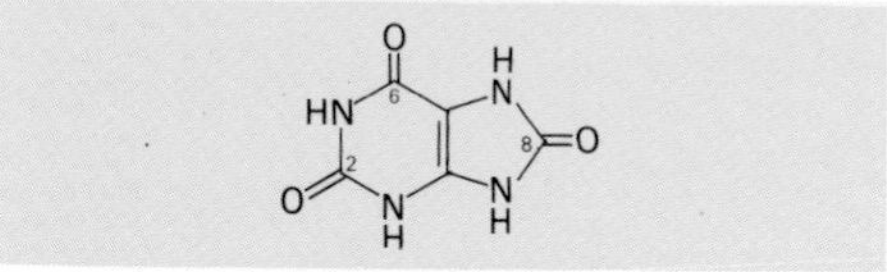

The lactan-type formula of uric acid; the name 2,6,8-trihydroxypurine refers to the lactim form (see Tautomerism).

URINE, waste product comprising a dilute solution of excess salts and unwanted nitrogenous material, such as UREA and deaminated PROTEIN, excreted by many animals. The wastes are filtered from the BLOOD in the KIDNEYS or equivalent structures and stored in the BLADDER till excreted. The passage of urine serves not only to eliminate wastes, but also provides a mechanism for maintaining the water and salt concentrations and pH of the blood. While all mammals excrete their nitrogenous wastes in urine, other groups—birds, insects and fishes—excrete them as AMMONIA or in solid crystals as URIC ACID.

UROCHORDATA, or **Tunicates**, a subphylum of the CHORDATA containing the sea squirts. They are marine and have sac-shaped bodies with two openings, a mouth and an atrial port. Water is sucked in through the mouth, filtered for food particles, and passed out through the atrial port.

URODELA, or **Caudata**, an order of the AMPHIBIA that includes forms with tails: the newts and salamanders. There are about 300 species found in temperate and tropical regions. They have fore and hind legs of more or less equal length, unlike the ANURA, and occupy a large range of habitats: aquatic, terrestrial and even arboreal.

URSA MAJOR. See GREAT BEAR.

URSA MINOR. See LITTLE DIPPER.

URTICARIA. See HIVES.

UTERUS. See WOMB.

UVULA, soft central portion of the soft PALATE which hangs at the back of the PHARYNX and forms part of the occluding mechanism which can functionally separate the nasopharynx from the oropharynx.

VACCINATION, method of inducing IMMUNITY to INFECTIOUS DISEASE due to BACTERIA or VIRUSES. Based on the knowledge that second attacks of diseases such as SMALLPOX were uncommon, early methods of protection consisted in inducing immunity by deliberate inoculation of material from a mild case. Starting from the observation that farm workers who had accidentally acquired cowpox by milking infected cows were resistant to smallpox, JENNER in the 1790s inoculated cowpox material into nonimmune persons who then showed resistance to smallpox. PASTEUR extended this work to experimental chicken CHOLERA, human ANTHRAX and RABIES. The term vaccination became general for all methods of inducing immunity by inoculation of products of the infectious organism. ANTITOXINS were soon developed in which specific immunity to disease TOXINS was induced. Vaccination leads to the formation of antibodies and the ability to produce large quantities rapidly at a later date (see ANTIBODIES AND ANTIGENS); this gives protection equivalent to that induced by an attack of the disease. It is occasionally followed by a reaction resembling a mild form of the disease, but rarely by the serious manifestations. Patients on STEROIDS, with immunity disorders or ECZEMA may suffer severe reactions and should not generally receive vaccinations.

VACUUM, any region of space devoid of ATOMS and MOLECULES. Such a region will neither conduct HEAT nor transmit SOUND waves. Because all materials which surround a space have a definite VAPOR PRESSURE, a perfect vacuum is an impossibility and the term is usually used to denote merely a space containing air or other gas at very low PRESSURE. Pressures less than 0.1μPa occur naturally about 800km above the earth's surface, though pressures as low as 0.01nPa can be attained in the laboratory. The low pressures required for many physics experiments are obtained using various designs of vacuum PUMP.

VACUUM BOTTLE, double-walled glass container designed by DEWAR in the late 19th century for storing liquified gases at low temperatures, but equally effective for storing hot substances. Heat transfer from the surroundings is minimized by silvering the glass walls to cut down heat RADIATION and evacuating the space between them to reduce the CONDUCTION of heat through them. A protective casing usually surrounds the glass walls.

VACUUM TUBE, an evacuated ELECTRON TUBE.

VALENCE, or **valency**, the combining power of an ELEMENT, expressed as the number of chemical BONDS which one atom of the element forms in a given compound. In general, the characteristic valence of an element in Group *N* of the PERIODIC TABLE is *N* or (8–*N*). (For **Oxidation Number**, see OXIDATION AND REDUCTION.)

VALLEY, long narrow depression in the earth's surface, usually formed by GLACIER or river EROSION. Young valleys are narrow, steep-sided and V-shaped; mature valleys, broader, with gentler slopes. Some, RIFT VALLEYS, are the result of collapse between FAULTS. **Hanging valleys**, of glacial origin, are side valleys whose floor is considerably higher than that of the main valley.

VALVE, mechanical device which, by opening and closing, enables the flow of fluid in a pipe or other vessel to be controlled. Common valve types are generally named after the shape or mode of operation of the movable element, e.g., cone, or needle, valve; gate valve; globe valve; poppet valve; and rotary plug cock. In the butterfly valve a disk pivots on one of its diameters. Self-acting valves include: safety valves, usually spring-loaded and designed to open at a predetermined pressure; nonreturn valves, which permit flow in one direction only; and float-operated valves, set to shut off a feeder pipe before a container overflows.

VALVE, Electronic. See ELECTRON TUBE.

VANADIUM (V), silvery-white, soft metal in Group VB of the PERIODIC TABLE; a TRANSITION ELEMENT. It is widespread, the most important ores being CARNOTITE and roscoelite; it is isolated by reduction of vanadium (V) oxide with calcium. Most is used in ALLOYS to make hard and wear-resistant STEELS. Vanadium is fairly unreactive. It forms compounds in oxidation states +2, +3, +4 and +5. Vanadium (V) oxide is used in ceramics and as a catalyst in the CONTACT PROCESS. AW 50.9, mp 1890°C, bp 3380°C, sg 5.96 (20°C).

VAN ALLEN, James Alfred (1914–), US physicist responsible for the discovery of the VAN ALLEN RADIATION BELTS (1958).

VAN ALLEN RADIATION BELTS, the belts of high-energy charged particles, mainly PROTONS and ELECTRONS, surrounding the earth, named for VAN ALLEN, who discovered them in 1958. They extend from a few hundred to about 50000km above the earth's surface, and radiate intensely enough that astronauts must be specially protected from them. The mechanisms responsible for their existence are similar to those involved in the production of the AURORA.

VAN DE GRAAFF GENERATOR. See ELECTROSTATIC GENERATOR.

VAN DER WAALS, Johannes Diderik (1837–1923), Dutch physicist who investigated the properties of real GASES. Noting that the KINETIC THEORY of gases assumed that the molecules had neither size nor interactive forces between them, in 1873 he proposed **Van der Waals' Equation**: $(P+a/V^2)(V-b)=RT$, (where P, V and T are pressure, volume and absolute temperature, R is the universal GAS constant, and a and b are constants whose values depend on the particular gas in question) in which allowance is made for both these factors. The weak attractive forces between molecules are therefore named **Van der Waals forces**. He received the 1910 Nobel Prize for Physics.

VAN HISE, Charles Richard (1857–1918), US geologist best known for his studying the PRECAMBRIAN rock formations in the Lake Superior region, with particular regard to their iron ore deposits.

The Van Allen radiation belts surround the earth like a large doughnut (*below*). The doughnut is centered on the earth's magnetic rather than rotation axis (*right*).

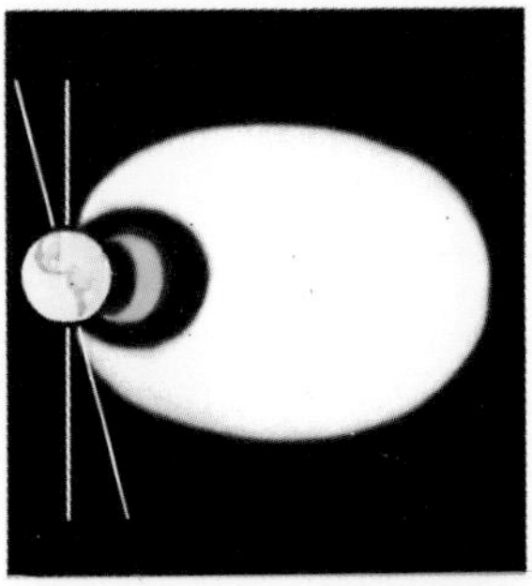

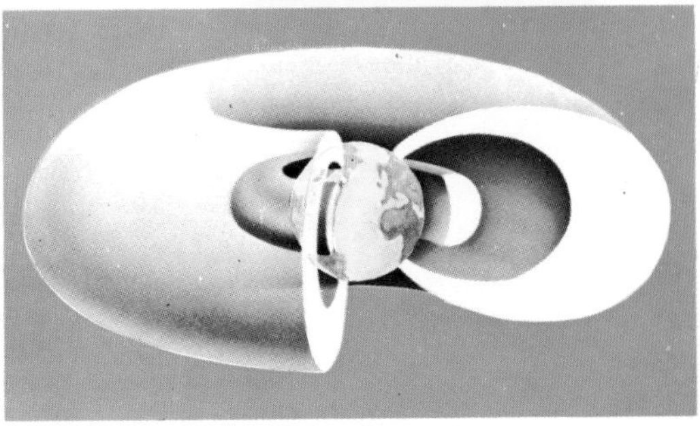

Graph of the Van der Waals equation for a real gas (red). If a quantity of gas is compressed at constant temperature under normal conditions, the red curve is not followed all the way: at a certain crtical pressure a change of state occurs—there is a sudden reduction in volume and the gas liquifies (broken blue) line. A whole family of related Van der Waals curves exist for different temperatures: in three dimensions these give rise to a pressure–volume–temperature surface. For a certain crtical temperature, there is no well in the curve, only a point of inflection. At and above this temperature, there is no definable phase transition between the gaseous and liquid states.

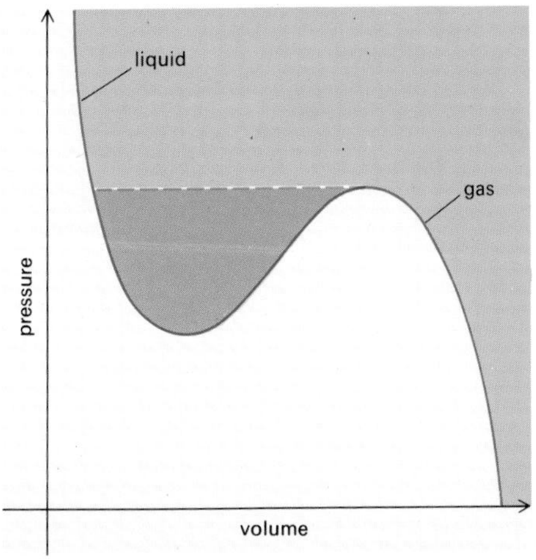

VAN 'T HOFF, Jacobus Henricus (1852–1911), Dutch physical chemist awarded in 1901 the first Nobel Prize for Chemistry for his work laying the foundations of STEREOCHEMISTRY. (See also OSMOSIS.)

VAN VLECK, John Hasbrouck (1899–), US physicist who shared the 1977 Nobel Prize for Physics with P. W. ANDERSON and N. F. MOTT for his work on the QUANTUM THEORY of magnetic materials. He also contributed to LIGAND field theory.

VAPOR, the gaseous state of a substance (usually one that is solid or liquid at room temperature). An isothermal increase in pressure can convert a vapor to a liquid. To convert a solid or liquid to vapor, heat is needed to overcome the cohesive forces between molecules and allow them to escape.

VAPOR LAMPS. See LIGHTING.

VAPOR PRESSURE, the pressure exerted by a VAPOR in EQUILIBRIUM with its liquid or solid. In an enclosed space this occurs when equal numbers of MOLECULES are entering and leaving the vapor, which is then saturated. For a pure substance, the saturated vapor pressure (SVP) depends on the temperature. The BOILING POINT of a liquid is reached when the SVP equals the external pressure.

VARIABLE, in mathematics, the algebraic (see ALGEBRA) quantity whose numerical value may change. The most-used symbols in elementary algebra are x and y. If one variable is a FUNCTION of the other, as in the EQUATION $y = ax^2 + bx + c$, where a, b, c are CONSTANTS, then y is termed the **dependent variable**, since its value depends on the value of the **independent variable** x. (See also PARAMETER.)

VARIABLE STARS, stars that vary in brightness. There are two main categories. **Extrinsic variables** are those whose variation in apparent brightness is caused by an external condition, as in the case of eclipsing binaries (see DOUBLE STAR). **Intrinsic variables** vary in absolute brightness owing to physical changes within them. They may vary either regularly or irregularly: NOVAE and SUPERNOVAE are irregular intrinsic variables, though some novae erupt in an approximate cycle. Pulsating variables, which vary in size, are the most common type of variable star: they include the RR Lyrae stars, with periods from 1.5h to little over a day, W Virginis stars and RV Tauri stars (all three types appearing principally in GLOBULAR CLUSTERS); long period and semiregular variables, which are red giants; and, possibly erroneously, the CEPHEID VARIABLES. Types of variable stars whose periods are known to have a relationship to their absolute brightness are especially important in that they can be used to determine large astronomical distances.

VARIANCE, in a sample (see STATISTICS) of items from a population, the average of the squares of the individual deviations of the items from the mean (see MEAN, MEDIAN AND MODE). The sample variance is usually denoted by s^2, the variance of the whole population by σ^2 (see also STANDARD DEVIATION).

VARIATION, diversity found in all natural populations of organisms. Variation is the result of differing effects of environmental factors and of differences in the genetic constitution of each individual. Genetic diversity is important because it provides variants, some of which may be more suited to prevailing conditions than others, which provide raw material for NATURAL SELECTION. (See also EVOLUTION.)

VARICELLA. See CHICKENPOX.

VARICOSE VEINS, enlarged or tortuous VEINS in the legs resulting from incompetent or damaged valves in the veins, with the pressure of the venous BLOOD causing venous distension and subsequent changes in the vein wall. Although unpleasant in appearance, they are more important for causing venous stagnation, with skin ECZEMA and ULCERS on the inside of the ankle, HEMORRHAGE and EDEMA. Treatment is by stripping or sclerosing injections.

VARNISH, solution of RESIN which dries to form a hard transparent film; widely applied to wood, metal and masonry to improve surface properties without changing appearance. There are two main types: "spirit varnishes," consisting of natural or synthetic resins dissolved in a volatile solvent such as alcohol; and "oleoresinous varnishes"—more resistant to heat and weather—which are mixtures of resins and drying oils dissolved in TURPENTINE or a petroleum oil. Lacquer, the original wood varnish, is the sap of the varnish tree.

VARVE, the layer of sediment deposited in the course of a single year, specifically in a lake formed of glacial meltwater. Characteristically, a varve has a SILT layer overlying a SAND layer. Study of varves is of great use in geological dating. (See also GLACIER; SEDIMENTATION.)

VASCULAR PLANTS, members of the division **Tracheophyta** of the PLANT KINGDOM in which the SPOROPHYTE is dominant and the GAMETOPHYTE is inconspicuous and dependent on the sporophyte. Vascular tissues, such as XYLEM and PHLOEM are present and the plants are differentiated into ROOTS, STEMS and LEAVES. Vascular plants include the CLUB MOSSES, HORSETAILS, FERNS, GYMNOSPERMS and ANGIOSPERMS.

VASCULAR SYSTEM, the BLOOD CIRCULATION system, comprising BLOOD, ARTERIES, CAPILLARIES, VEINS and the HEART; the LYMPH vessels form a further subdivision. Its function is to deliver nutrients (including OXYGEN) to, and remove wastes from all organs, and to transport HORMONES and the agents of body defense.

VASECTOMY, form of FAMILY PLANNING in males in which the *vas deferens* on each side is ligated and cut to prevent SPERM from reaching the seminal vesicles and hence the urethra of the PENIS. It does not affect ejaculation but causes permanent STERILIZATION.

VASELINE, or **petrolatum**, high-boiling HYDROCARBON residue from the distillation of PETROLEUM; a jelly used for LUBRICATION and as an emollient.

VASOCONSTRICTION, narrowing of BLOOD vessels, facilitating control of blood pressure and body TEMPERATURE.

VASODILATION, or vasodilatation, widening of BLOOD vessels, facilitating control of blood pressure and body TEMPERATURE.

VASOPRESSIN, or antidiuretic hormone (ADH), HORMONE produced by the HYPOTHALAMUS and posterior PITUITARY GLAND, which is a mild vasoconstrictor, but primarily inhibits diuresis or loss of water in URINE. It is a vital link in the system for preserving the HOMEOSTASIS or constancy of body fluids.

VEAL, pale, fine-textured, mild-flavored meat from calves (young cattle) slaughtered when less than 14

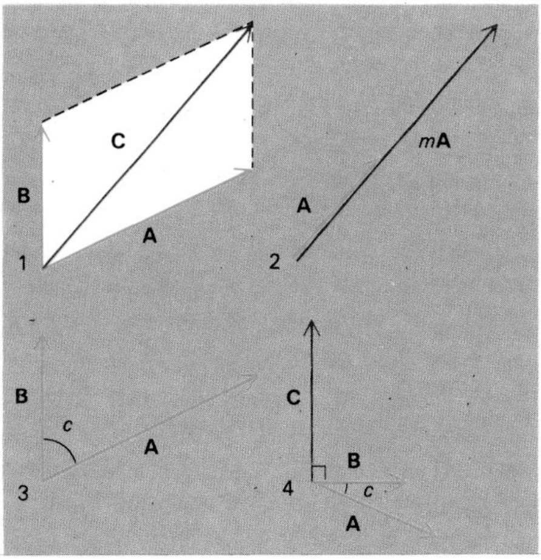

Vector algebra. (1) Addition of vectors **A** and **B** yields vector **C** according to the parallelogram rule. (2) Multiplication of vector **A** by a scalar m yields vector m**A**. (3) The scalar product of vectors **A** and **B**, **A.B** = AB cos c. (4) The vector product of vectors **A** and **B**, **A**×**B** = **C** where **C** = (AB sin c)**n** and **n** is a unit vector perpendicular to the plane containing **A** and **B**.

weeks old, generally factory-farmed and fed a high-protein, iron-deficient diet.

VEBLEN, Oswald (1880–1960), US mathematician best known for his contributions to projective and differential GEOMETRY and especially for his pioneering role in the development of TOPOLOGY, set out in *Analysis Situ* (1922).

VECTOR, a quantity having both magnitude and direction, unlike a SCALAR, which has only magnitude. One example of a vector quantity is VELOCITY. (See VECTOR ANALYSIS.)

VECTOR ANALYSIS, the application of the techniques of ANALYSIS and ALGEBRA to the study of VECTORS. Since vector quantities have both magnitude and direction, they may be represented geometrically by LINES with specified directions, whose lengths represent the magnitudes of the vectors concerned; the magnitude of a vector **V** is written V. The dot (or scalar) PRODUCT of two vectors **A** and **B** inclined at an angle c to each other is defined as $\mathbf{A}\,.\,\mathbf{B} = AB\cos c$. The cross (or vector) product of two vectors **A** and **B** inclined at an angle c to each other is defined as $\mathbf{A}\times\mathbf{B} = (AB\sin c)\,\mathbf{n}$ where **n** is a unit vector whose direction is PERPENDICULAR to the plane of **A** and **B** such that, looking along the direction of **n**, a clockwise turn of less than 180° is required to bring **A** into **B**.

VECTOR FIELD, a set of VECTORS such that every POINT in a particular region of space R is associated with a single vector; e.g., a MAGNETIC FIELD.

VECTOR SPACE, a set (see SET THEORY) of VECTORS together with a FIELD of SCALARS such that: the sum of any two members of the set is a vector also in the set, and multiplication of a member of the set by a member of the field produces a vector also in the set.

VEGA, Alpha Lyrae, the fourth brightest star in the night sky (absolute magnitude +0.5). It is 8pc distant, and 40 times as bright as the sun.

VEGETABLES, general term for plants whose leaves, flowers, roots, stems or fruits are edible. "Vegetable" is used for those plants that are eaten in main courses of meals, usually after cooking, while the general term "fruit" refers to plants used as appetizers or desserts, or eaten out of the hand. Thus, botanically a tomato is a fruit, but it is popularly considered a vegetable.

VEGETATION, characteristic plant life of a region. There are four basic types: DESERT; FOREST; GRASSLAND, and TUNDRA. (See also CLIMATE.)

VEIN, a mineral formation of far greater extent in two dimensions than in the third. Sheetlike **fissure veins** occur where fissures formed in the rock become filled with MINERAL. **Ladder veins** form in series of fractures in, e.g., DIKES. **Saddle-veins** are lens-shaped, concave below and convex above. Veins that contain economically important ORES are often termed **lodes**.

VEINS, thin-walled collapsible vessels which return BLOOD to the HEART from the tissue CAPILLARIES and provide a variable-sized pool of blood. They contain valves which prevent back-flow—especially in the legs. Blood drains from the major veins into the inferior or superior VENA CAVA. Blood in veins is at low pressure and depends for its return to the heart on intermittent muscle compression, combined with valve action.

VELD, or **veldt,** open GRASSLAND of South Africa, divided into three types: High Veld, around 1500m above sea level, which is similar to the PRAIRIES; Cape Middle Veld, somewhat lower, covered with scrub and occasional low ridges of hills; and Low Veld, under about 750m above sea level.

VELLUM. See PARCHMENT.

VELOCITY, the rate at which the position of a body changes, expressed with respect to a given direction. Velocity is thus a VECTOR quantity, of which the corresponding scalar is **speed**: the rate of change in the position of a body without respect to direction. Translational velocity, usually expressed in calculus as

$$\mathbf{v} = \frac{d\mathbf{s}}{dt},$$

refers to movement through space; angular or rotational velocity,

$$\omega = \frac{d\theta}{dt},$$

to rotation about a given axis. (See also ACCELERATION.)

VENA CAVA. The *superior vena cava* is a VEIN collecting BLOOD from the head, neck and arms, and delivering it to the right side of the HEART. The *inferior vena cava* performs the same function with blood from the legs and abdomen.

VENEER, thin slice of wood applied to the faces of cheap and unattractive wood. Low quality veneers are used for PLYWOOD, but normally high quality veneer, such as mahogany or walnut, is used to improve the finish of furniture.

VENEREAL DISEASES, those INFECTIOUS DISEASES transmitted mainly or exclusively by sexual contact, usually because the organism responsible is unable to survive outside the body and the close contact of genitalia provides the only means for transmitting viable organisms. **Gonorrhea** is an acute BACTERIAL

DISEASE which is frequently asymptomatic in females who therefore act as carriers, although they may suffer mild cervicitis or urethritis. In males it causes a painful urethritis with urethral discharge of PUS. ARTHRITIS, SEPTICEMIA and other systemic manifestations may also occur, and urethral stricture follow. Infection of an infant's eyes by mothers carrying the gonococcus causes neonatal OPTHALMIA, previously a common cause of childhood BLINDNESS. Gonorrhea is best treated with PENICILLIN. **Syphilis**, due to *Treponema pallidum*, a SPIROCHETE, is a disease with three stages. A painless genital ULCER or chancre—a highly infective lesion—develops in the weeks after contact; this is usually associated with LYMPH node enlargement. Secondary syphilis, starting weeks or months after infection, involves systemic disease with FEVER, malaise and a characteristic rash, mucous membrane lesions and occasionally MENINGITIS, HEPATITIS or other organ disease. If the disease is treated with a full course of penicillin in the early stages, its progression is prevented. Tertiary syphilis takes several forms; e.g., gummas—chronic granulomas affecting SKIN, EPITHELIUM, BONE or internal organs—may develop. Largely a disease of blood vessels, tertiary syphilis causes disease of the AORTA with aneurysm and aortic valve disease of the HEART, with incompetence. Syphilis of the NERVOUS SYSTEM may cause TABES DORSALIS, primary EYE disease, chronic meningitis, multifocal vascular disease resembling STROKE or general PARESIS with mental disturbance, personality change, failure of judgment and muscular weakness. Penicillin may only partially reverse late syphilis. Congenital syphilis is disease transmitted to the FETUS during PREGNANCY and leads to deformity and visceral disease. Other venereal diseases include Reiter's disease with arthritis, CONJUNCTIVITIS and urethritis (in males only); genital trichomonas; THRUSH; *Herpes simplex* virus, and "nonspecific urethritis." Tropical venereal diseases include chancroid; lymphogranuloma venereum, and granuloma inguinale.

VENN DIAGRAMS, a graphical way of representing concepts and relations occurring in SET THEORY.

VENOM. See SNAKE BITE.

VENTURI TUBE, short open-ended pipe with a central constriction used for measuring the flow rate of a FLUID. The fluid velocity increases and its pressure drops in the constriction. The fluid velocity is calculated from the pressure difference between the center and ends of the tube.

Venn diagram illustrating two sets, $S_1 = \{a,b,c,d,e\}$ and $S_2 = \{b,d,f,g\}$ for which the intersection of S_1 and S_2, $S_1 \cap S_2 = \{b,d\}$ and their union, $S_1 \cup S_2 = \{a,b,c,d,e,f,g\}$.

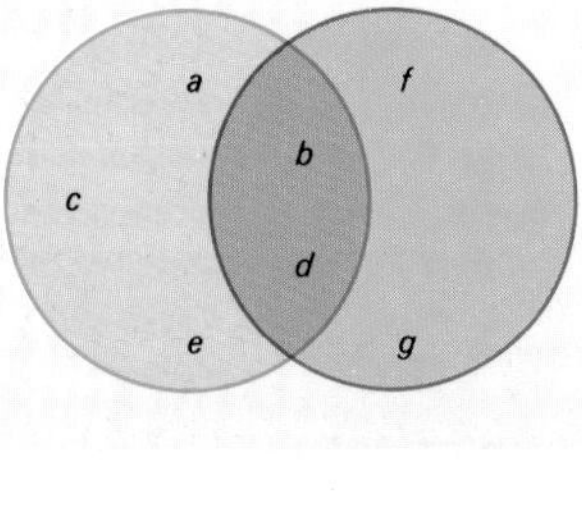

VENUS, the planet second from the sun, about the same size as the earth. Its face is completely obscured by dense clouds containing sulfuric acid, though the USSR's Venera-9 and Venera-10 (Oct. 1975) landers have provided photographs of the planet's rocky surface. Venus revolves about the sun at a mean distance of 0.72AU in 225 days, rotating on its axis in a retrograde direction (see RETROGRADE MOTION) in 243 days. Its diameter is 12.1Mm. Its atmosphere is 97% carbon dioxide. Its surface temperature is about 750K. Venus has no moons, and almost certainly does not support life.

VERD ANTIQUE, green mottled form of SERPENTINE used for indoor decoration; also a dark green PORPHYRY containing FELDSPAR. **Verde antico** is the green PATINA formed on bronze.

VERDIGRIS, blue-green powder, a basic copper (II) acetate, made by pickling copper in acetic acid and used as a pigment and a mordant in dyeing.

VERNAL EQUINOX. See EQUINOXES.

VERNIER SCALE, an auxiliary scale used in conjunction with the main scale on many instruments (in particular the vernier caliper), allowing greater precision of reading. The vernier scale is graduated such that nine graduations on the vernier scale equal ten on the main scale. By observing which vernier graduation nearest the zero on the vernier scale coincides with a graduation on the main scale, the precision with which a reading can be made is improved by a factor of 10.

VERRUCA. See WART.

VERTEBRAE, BONES forming the backbone or **spinal column**, which is the central pillar of the SKELETON of the group of animals, including man, called VERTEBRATES. Vertebrae exist for each segmental level of the body and are specialized to provide the trunk with both flexibility and strength. In the neck, cervical vertebrae are small and their JOINTS allow free movement to the head. The thoracic vertebrae provide the bases for the ribs. The lumbar spine consists of large vertebrae with long transverse processes that form the back of the ABDOMEN; the sacral and coccygeal vertebrae, which are fused in man, link the spine with the bony PELVIS. Within the vertebrae there is a continuous canal through which passes the SPINAL CORD; between them run the segmental nerves. Around the spinal column are the powerful spinal muscles and ligaments.

VERTEBRATES, animals with a backbone, forming a subphylum of the CHORDATA.

VERTEX. See ANGLE; CONE; TRIANGLE.

VERTICAL TAKEOFF AND LANDING AIRPLANE (VTOL), AIRPLANE that can lift-off and land vertically, such as a HELICOPTER; now chiefly associated with "jump-jet" systems in which the exhaust gases of horizontally-mounted jet engines can be deflected downward. The British Hawker "Harrier" military strike airplane, capable of supersonic speeds in level flight, is the best-known example.

VERTIGO, sensation of rotation in space resulting from functional (spinning of head with sudden stop) or organic disorders of the balance system of the EAR or its central mechanisms. It commonly induces nausea or VOMITING and may be suppressed by DRUGS.

VERY HIGH FREQUENCY (VHF). See

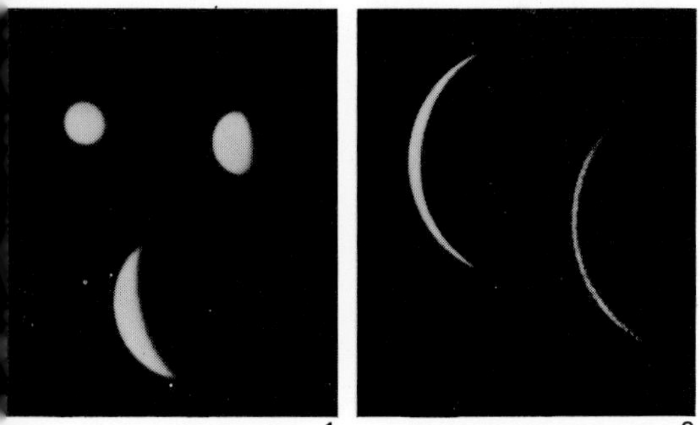

1 2

Seen from earth, Venus appears to have phases like those of the moon. At its full phase (1, top left) the planet is at its farthest from earth, and so seems small. It appears larger as it approaches, but the area illuminated by the sun is increasingly that turned away from us, and so we see a diminishing crescent. At its closest we see only a bright thin sickle shape (2, right)

ELECTROMAGNETIC RADIATION; RADIO.

VESALIUS, Andreas (1514–1564), Flemish biologist regarded as a father of modern ANATOMY. Initially a Galenist he became, after considerable experience of dissection, one of the leading figures in the revolt against GALEN. In his most important work, *On the Structure of the Human Body* (1543), he described several organs for the first time.

VESTIGIAL ORGAN, an anatomical structure which is nonfunctional and frequently underdeveloped in a modern species but which represents the remnant of an organ which in the remote past was fully functional in an ancestor species; e.g., the vermiform APPENDIX in man.

VETERINARY MEDICINE, the medical care of sick animals, sometimes including the delivery of their young. It is practiced separately from human MEDICINE since animal diseases differ largely from those affecting humans. Veterinarians treat domestic, farm, sport and zoo animals. Sometimes diseases can be controlled only through the slaughter of known or suspected carriers.

VIBRATION, periodic motion, such as that of a swinging PENDULUM or a struck TUNING FORK. The simplest and most regular type of vibration is SIMPLE HARMONIC MOTION. ENERGY from a vibration is propagated as a WAVE MOTION. Excess mechanical vibration, as with noise pollution, can do considerable damage to buildings.

Use of a vernier scale. The measuring point on the vernier scale is set exactly opposite the value to be measured on the main scale. The distance of the measuring point from the last scale division is then found by noting which division on the vernier scale most closely coincides with a division on the main scale. In this case, the reading is 81.7.

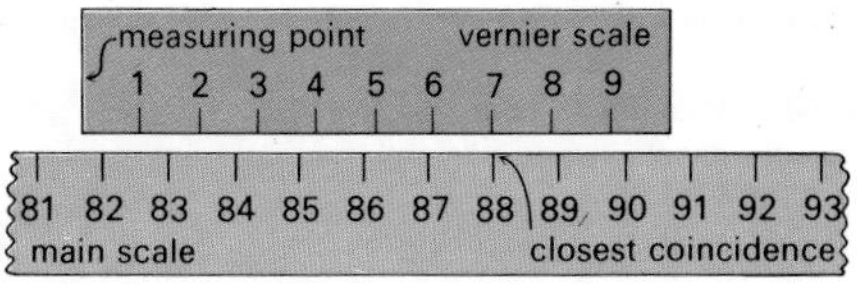

VIDEOTAPE, magnetic tape used to record TELEVISION programs. In order to record the vast amounts of information necessary to reconstruct a television picture, 2in-wide tape must be run through the tape heads at 15in/s. (See also TAPE RECORDER; SOUND RECORDING.)

VIKING PROGRAM, series of US unmanned space probes designed to land on and study Mars. Viking 1 landed on July 20, 1976, sending back TV pictures; its soil-analysis experiments yielded results which suggest the presence of life, but may represent only unusual chemical reactions. Viking 2 landed Sept. 3, 1976.

VILLANOVANS, early IRON AGE culture named for Villanova, Italy. They arrived in Italy from E Europe in the 10th–9th centuries BC, worked Tuscan iron and copper mines and were fine metalworkers.

VINCENT'S ANGINA, sore throat due to infection of the PHARYNX with a pair of BACTERIA, commonly seen in undernourished or debilitated persons. Only ANTIBIOTICS and local measures are required.

VINCULUM (from Latin *vincire*, to bind), symbol used in mathematics to indicate that two or more terms are to be treated as one: e.g., $a-\overline{b+c}$ meaning $a-(b+c)$. It is most commonly used with the ROOT symbol:

$$\sqrt[4]{\overline{6+10}}=\sqrt[4]{16}=2.$$

VINEGAR, a sour liquid used to flavor and preserve food. The taste comes from ACETIC ACID, which forms at least 4% of the total volume. Vinegar is made by FERMENTATION of wine, cider or any other alcohol solutions; BACTERIA (*Acetobacter* and *Acetomonas*) oxidize the alcohol (ETHANOL) to acetic acid. Malt vinegar is produced from alcohol fermented from potatoes or cereals.

VINYL COMPOUNDS, compounds containing the vinyl group, $CH_2{=}CH{-}$, formally derived from ethylene, but generally produced from ACETYLENE. They are polymerized (see POLYMERS) to form PLASTICS and RUBBERS. (See also ORLON.)

VIRAL DISEASES, generally INFECTIOUS DISEASES

The Hawker Siddeley Harrier fighter passes the exhaust gases of its Pegasus engine through four "vectored thrust" nozzles which direct thrust downward during takeoff and landing and backward during forward flight.

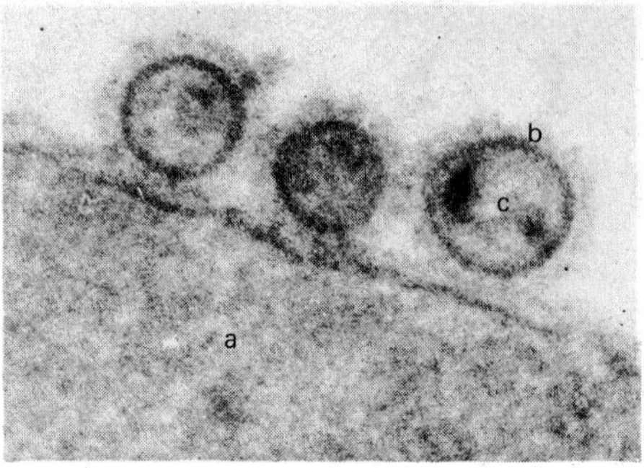

Microscopic section showing three influenza viruses on the outer membrane of red blood cell (a). Also visible are the protein shell of the virus (b) and its nucleus (c).

due to VIRUSES. The common COLD, INFLUENZA, CHICKENPOX, MEASLES and GERMAN MEASLES are common in childhood, while SMALLPOX and YELLOW FEVER are important tropical virus diseases. Viruses may also cause specific organ disease such as HEPATITIS, MENINGITIS, ENCEPHALITIS, MYOCARDITIS and pericarditis. Most virus diseases are self-limited and mild, but there are few specific drugs effective in cases of severe illness. Prevention by VACCINATION is therefore crucial.

VIRCHOW, Rudolf (1821–1902), Pomeranian-born German pathologist whose most important work was to apply knowledge concerning the CELL to PATHOLOGY, in course of which he was the first to document LEUKEMIA and EMBOLISM. He was also distinguished as an anthropologist and archaeologist.

VIRGO (the Virgin), a large constellation on the ECLIPTIC; the sixth sign of the ZODIAC. It contains the bright spectroscopic binary (see DOUBLE STAR) Spica and a cluster of galaxies.

VIRTANEN, Artturi Ilmari (1895–1973), Finnish biochemist awarded the 1945 Nobel Prize for Chemistry for his work on winter SILAGE. He showed that keeping the silage acid ($pH < 4$) stopped the FERMENTATION that would otherwise destroy it, without reducing its nutritive value or its palatability to animals (see PH).

VIRUS, submicroscopic parasitic microorganism comprising a PROTEIN or protein/lipid sheath containing nucleic acid (DNA or RNA). Viruses are inert outside living cells, but within appropriate cells they can replicate (using raw material parasitized from the cell) and give rise to the manifestations of the associated VIRAL DISEASE in the host organism. Various viruses infect animals, plants and BACTERIA (in which case they are BACTERIOPHAGES). Few drugs act specifically against viruses, although IMMUNITY can be induced in susceptible cells against particular viruses. Various pathogenic organisms formerly regarded as large viruses are now distinguished as *bedsonia*.

VISCOSITY, the property of a FLUID by which it resists shape change or relative motion within itself. All fluids are viscous, their viscosity arising from internal FRICTION between molecules which tends to oppose the development of velocity differences. The viscosity of liquids decreases as they are heated, but that of gases increases.

VISION, the special sense concerned with reception and interpretation of LIGHT stimuli reaching the EYE; the principal sense in man. Light reaches the CORNEA and then passes through this, the AQUEOUS HUMOR, the lens and the VITREOUS HUMOR before impinging on the RETINA. Here there are two basic types of receptor: **rods** concerned with light and dark distinction, and **cones**, with three subtypes corresponding to three primary visual COLORS: red, green and blue. Much of vision and most of the cones are located in the central area, the macula, of which the FOVEA is the central portion; gaze directed at objects brings their images into this area. When receptor cells are stimulated, impulses pass through two nerve cell relays in the retina before passing back toward the BRAIN in the optic nerve. Behind the eyes, information derived from left and right visual fields of either eye is collected together and passes back to the opposite cerebral hemisphere, which it reaches after one further relay. In the cortex are several areas concerned with visual perception and related phenomena. The basic receptor information is coded by nerve interconnections at the various relays in such a way that information about spatial interrelationships is derived with increasing specificity as higher levels are reached. Interference with any of the levels of the visual pathway may lead to visual symptoms and potentially to BLINDNESS.

VITALISM, the theory, dating from ARISTOTLE, that there is a distinguishing vital principle ("life force") in living organisms that is absent from nonliving objects.

VITAMINS, specific nutrient compounds which are essential for body growth or METABOLISM and which should be supplied by normal dietary foods. They are denoted by letters and are often divided into fat-soluble (A, D, E and K) and water-soluble (B and C) groups. **Vitamin A** , or **retinol**, is essential for the integrity of EPITHELIUM and its deficiency causes SKIN, EYE and mucous membrane lesions; it is also the precursor for RHODOPSIN, the retinal pigment. Vitamin-A excess causes an acute encephalopathy or chronic multisystem disease. Important members of the **vitamin B** group include thiamine (B_1), riboflavin (B_2), Niacin, Pyridoxine (B_6), Folic acid and cyanocobalamin (B_{12}). **Thiamine** acts as a coenzyme in CARBOHYDRATE metabolism and its deficiency, seen in rice-eating populations and alcoholics, causes BERIBERI and a characteristic encephalopathy. **Riboflavin** is also a coenzyme, active in oxidation reactions; its deficiency causes epithelial lesions. **Niacin** is a general term for nicotinic acid and nicotinamide, which are coenzymes in carbohydrate metabolism; their deficiency occurs in millet- or maize-dependent populations and leads to PELLAGRA. **Pyridoxine** provides an enzyme important in energy storage and its deficiency may cause nonspecific disease or ANEMIA. **Folic acid** is an essential cofactor in NUCLEIC ACID metabolism and its deficiency, which is not uncommon in PREGNANCY and with certain DRUGS, causes a characteristic anemia. **Cyanocobalamin** is essential for all cells, but the development of BLOOD cells and GASTROINTESTINAL-TRACT epithelium and NERVOUS SYSTEM function are particularly affected by its deficiency, which occurs in pernicious ANEMIA and in extreme vegetarians. Pantothenic acid, Biotin, Choline, Inositol and Para-aminobenzoic acid are

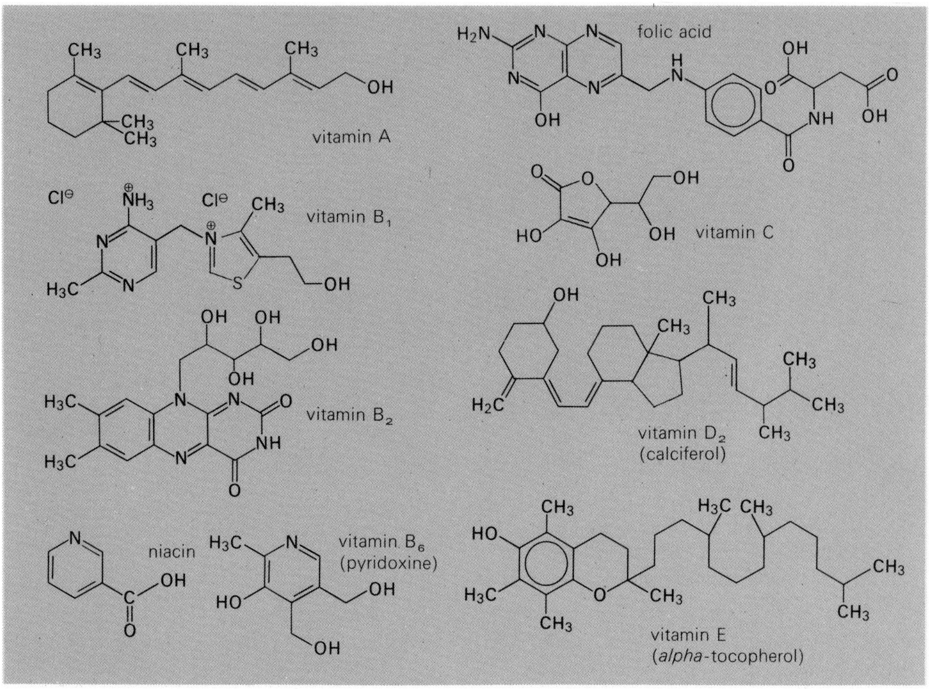

The structures of some vitamins. Closely related compounds often have similar metabolic effects, thus vitamin B_6 exists in aldehyde and amine forms as well as the alcohol shown here.

other members of the B group. **Vitamin C**, or **ascorbic acid**, is involved in many metabolic pathways and has an important role in healing, blood cell formation and bone and tissue growth; SCURVY is its deficiency disease. **Vitamin D**, or **calciferol**, is a crucial factor in CALCIUM metabolism, including the growth and structural maintenance of BONE; lack causes RICKETS, while overdosage also causes disease. **Vitamin E**, or **tocopherol**, appears to play a role in blood cell and nervous system tissues, but its deficiency is uncommon and its beneficial properties have probably been overstated. **Vitamin K** provides essential cofactors for production of certain CLOTTING factors in the LIVER; it is used to treat some clotting disorders, including that seen in premature infants. Vitamin A is derived from both animal and vegetable tissue and most B vitamins are found in green vegetables. though B_{12} is found only in animal food (e.g., liver). Citrus fruit are rich in vitamin C. Vitamin D is found in animal tissues, COD LIVER OIL providing a rich source. Vitamins E and K are found in most biological material.

VITREOUS HUMOR, the jelly-like substance forming much of the inner substrate of the EYE, through which light is transmitted from the lens to the RETINA.

VITRIOL, Oil of, obsolete term for SULFURIC ACID.

VIVISECTION, strictly, the dissection of living animals, usually in the course of physiological or pathological research; however, the use of the term is often extended to cover all animal experimentation. Although the practice remains the subject of considerable popular controversy, it is doubtful whether research, particularly medical, can be effectively carried on without a measure of vivisection.

VOICE, the sound emitted in speech (see SPEECH AND SPEECH DISORDERS), the method of communication exclusive to *Homo sapiens*. It is dependent for its generation upon the passage of air from the LUNGS through the TRACHEA, LARYNX, PHARYNX and mouth and its quality in each individual is largely determined by the shape and size of these structures and the resonance of the NOSE and nasal SINUSES. Phonation is the sounding of the elements of speech by the action of several small muscles on the vocal cords of the larynx; these regulate the air passing through and vibrate when tensed against this air stream. Articulation consists in the modulation of these sounds by the use of the TONGUE, TEETH and lips in different combinations. Vowels are produced mainly by phonation while consonants derive their characteristics principally from articulation.

VOLCANISM, or **vulcanicity**, the processes whereby MAGMA, a complex of molten silicates containing water and other volatiles in solution, rises toward the earth's surface, there forming IGNEOUS ROCKS. These may be extruded on the earth's surface (see LAVA; VOLCANO) or intruded into subsurface rock layers as, for example, DIKES, SILLS and LACCOLITHS. (See also FUMAROLE; GEYSER; HOT SPRINGS.)

VOLCANO, fissure or vent in the earth's crust through which MAGMA and associated material may be extruded onto the surface. This may occur with explosive force. The extruded magma, or LAVA, solidifies in various forms soon after exposure to the

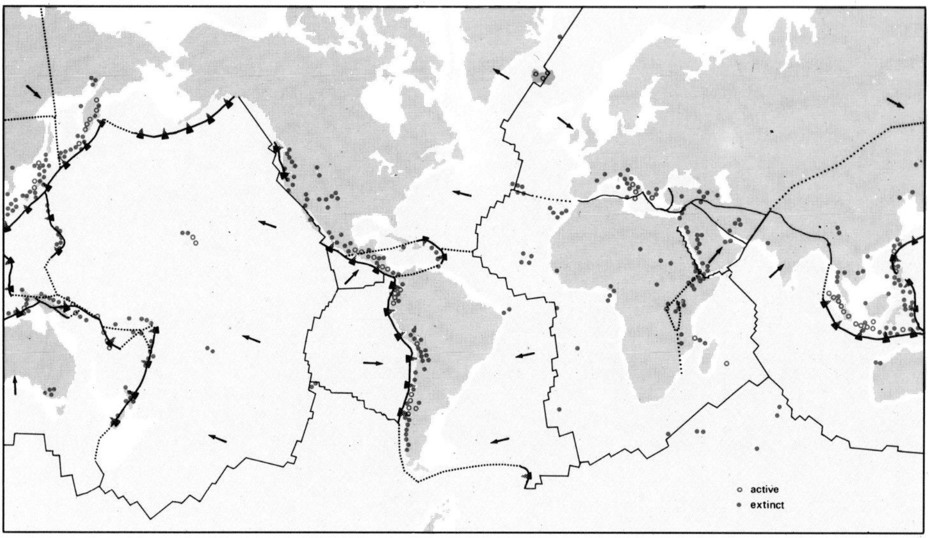

Above: Principal regions of volcanic activity. Note the close correlation between vulcanicity and plate margins.
Below: Cross section of a stratovolcano. (1) magma; (2) main conduit; (3) main outlet; (4) layers of ash and lava from previous eruptions; (5) large explosion crater, or caldera, partially filled with new volcanic matter; (6) branches of the main conduit; (7) parasitic cone formed by one of these.

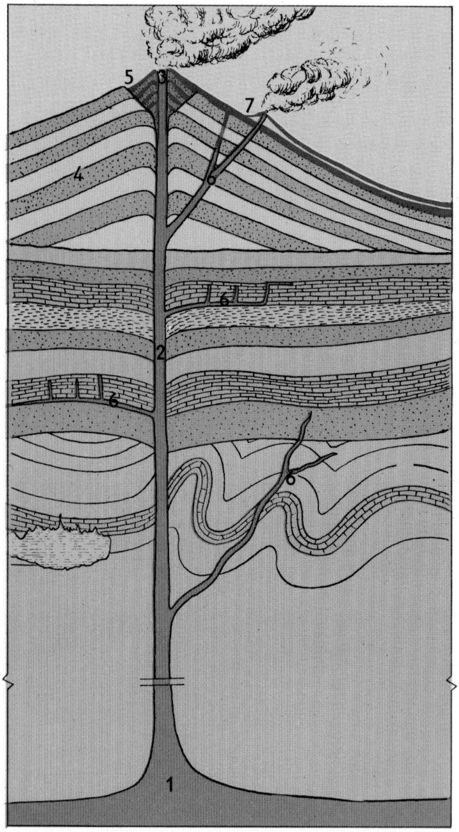

atmosphere. In particular it does so around the vent, building up the characteristic volcanic cone, at the top of which is a crater containing the main vent. There may be subsidiary vents forming "parasitic cones" in the slopes of the main cone. If the volcano is dormant or extinct the vents may be blocked with a *plug* (or *neck*) of solidified lava. On occasion these are left standing after the original cone has been eroded away. Volcanoes may be classified according to the violence of their eruptions. In order of increasing violence the main types are: Hawaiian, Strombolian, Vulcanian, Vesuvian, Peléan. Volcanoes are generally restricted to belts of seismic activity, particularly active plate margins (see PLATE TECTONICS). At mid-ocean ridges magma rises from deep in the mantle and is added to the receding edges of the plates (see SEA-FLOOR SPREADING). In MOUNTAIN regions, where plates are in collision, volatile matter ascends from the subducted edge of a plate, perhaps many km below the surface, bursting through the overlying plate in a series of volcanoes. (See also EARTHQUAKES.)

VOLT (V), the SI UNIT of electric POTENTIAL, potential difference, ELECTROMOTIVE FORCE, etc., defined such that ENERGY is dissipated at the rate of one WATT when a one-AMPERE current drops through a potential difference of one volt.

VOLTA, Alessandro Giuseppe Antonio Anastasio (1745–1827), French physicist who invented the voltaic pile (the first BATTERY) and thus provided science with its earliest continuous electric-current source. Volta's invention (c1800) demonstrated that "animal electricity" could be produced using solely inanimate materials, thus ending a long dispute with the supporters of GALVANI's view that it was a special property of animal matter.

VOLTMETER, an instrument used to estimate the difference in electrical POTENTIAL between different points in a circuit. Most consist of an AMMETER connected in series with a high RESISTANCE and calibrated in VOLTS. By OHM'S LAW the current flowing is proportional to the potential difference, though the instrument itself inevitably reduces the potential under test. Accurate determinations of potential difference must employ a POTENTIOMETER.

VOLUME, the amount of SPACE occupied by a three-dimensional object. (See MENSURATION.)

VOLUMETRIC ANALYSIS, method of quantitative chemical ANALYSIS in which quantities are measured in terms of volumes, either of solutions or of gases, using apparatus such as the BURETTE, the pipette (a calibrated tube, filled by suction, capable of delivering a known volume of liquid), and the volumetric (calibrated) flask. The chief technique of volumetric analysis is TITRATION; also important is the measurement in a gas burette of the gas produced in a reaction, the weight of one reactant being known.

VOMITING, the return of food or other substance (e.g., blood) from the STOMACH. It occurs by reverse PERISTALSIS after closure of the pyloric SPHINCTER and opening of the esophago-gastric junction. It may be induced by DRUGS, MOTION SICKNESS, GASTROENTERITIS or other infection, UREMIA, stomach or pyloric disorders. Morning vomiting may be a feature of early PREGNANCY. Drugs may be needed to control vomiting, and fluid and nutrient replacement may be needed.

VON NEUMANN, John (1903–1957), Hungarian-born US mathematician who contributed to QUANTUM MECHANICS, showing the equivalence of HEISENBERG'S matrix mechanics and SCHRÖDINGER'S wave mechanics. But his most important work was to formulate GAME THEORY, especially the minimax theorem. He also devised high-speed computers which contributed to the US development of the HYDROGEN BOMB.

VORTEX, a whirling mass of FLUID such as seen in a TORNADO, a WHIRLPOOL, a smoke ring or water running out of a bath. The term is used in HYDRODYNAMICS for a portion of fluid in which the individual particles have circular motions. A vortex may be produced when two adjacent fluid streams have different velocities or when a solid body moves

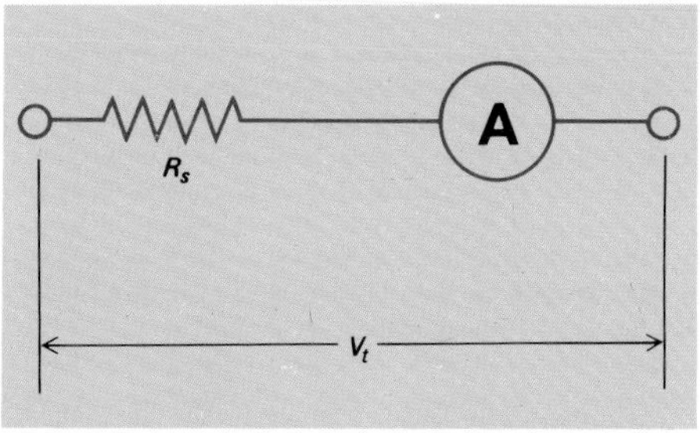

DC voltmeter circuit using a milliammeter (A) in series with a high resistance (R_s). If the resistance of the ammeter is known and the combined resistance of R_s and A is large compared with the resistance in the circuit under test, the ammeter reading allows a good estimate of the test voltage (V_t) to be made. For a particular ammeter–series resistance combination, the meter can be calibrated directly in volts.

through a fluid and vortex lines cannot begin or end inside the fluid.

VOWEL, speech element in whose production breath passes through the mouth with little or no restriction. The mouth, acting as a resonance chamber, alters configuration to form different vowel sounds, assisted by the tongue. In our alphabet there are five vowels (*a, e, i, o, u,* with, sometimes, *y*) but these, often compounded, may be used to represent many different vowel sounds.

VRIES, Hugo de (1848–1935), Dutch botanist who rediscovered the Mendelian laws of inheritance (see HEREDITY) and applied them to C. DARWIN'S theory of EVOLUTION in his *Mutation Theory* (1900–1903).

VTOL. See VERTICAL TAKEOFF AND LANDING AIRPLANE.

VULCANIZATION, process for enhancing the durability of RUBBER by heating it with SULFUR or sulfur compounds, usually in the presence of "accelerators" which speed the reaction. Vulcanization involves the creation of sulfur bridges between the long-chain rubber polymer molecules.

W

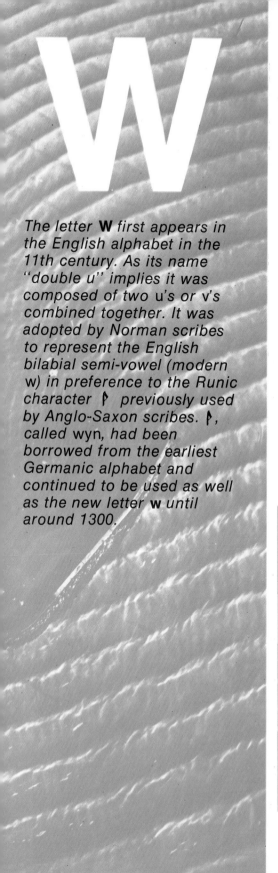

The letter **W** *first appears in the English alphabet in the 11th century. As its name "double u" implies it was composed of two u's or v's combined together. It was adopted by Norman scribes to represent the English bilabial semi-vowel (modern w) in preference to the Runic character ᚹ previously used by Anglo-Saxon scribes. ᚹ, called wyn, had been borrowed from the earliest Germanic alphabet and continued to be used as well as the new letter* **w** *until around 1300.*

WAALS, Johannes Diderik Van der. See VAN DER WAALS, JOHANNES DIDERIK.

WADDINGTON, Conrad Hall (1905–1975), British biologist and philosopher of science. His most important work was in embryology and genetics, and as a popular writer of books such as *Evolution of an Evolutionist* (1975).

WADI, or **arroyo,** stream-bed through which water flows only occasionally, found mainly in semiarid areas. Such **ephemeral streams** can swiftly erode quite deep, flat-bottomed gullies (see EROSION).

WAGNER VON JAUREGG (or Wagner-Jauregg), Julius (1857–1940), Austrian psychologist and neurologist awarded the 1927 Nobel Prize for Physiology or Medicine for his discovery that innoculation with MALARIA markedly helped sufferers from general PARESIS, hitherto fatal.

WAKSMAN, Selman Abraham (1888–1973), Russian-born US biochemist, microbiologist and soil scientist. His isolation of STREPTOMYCIN, the first specific antibiotic (a term he coined) against TUBERCULOSIS, won him the 1952 Nobel Prize for Physiology or Medicine.

Naturalist Alfred Russel Wallace was the first person to notice that a line could be drawn between the islands of the Malayan archipelago to the west of which few typically "Australasian" animals had naturally penetrated, the fauna largely comprising "Oriental" species. As later modified by T. H. Huxley, this "Wallace's line" divides the Australasian from the Oriental zoogeographic regions. To the east of Wallace's line is a further boundary, known as Weber's line. This represents the zone of approximate faunal balance: along this zone Australasian and Oriental species occur in roughly equal numbers. It is noteworthy that Wallace's line in its modified form approximates to the edge of the southeast Asian continental shelf.

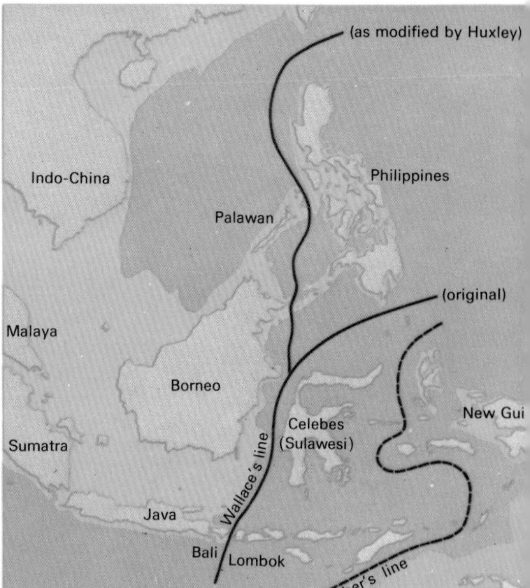

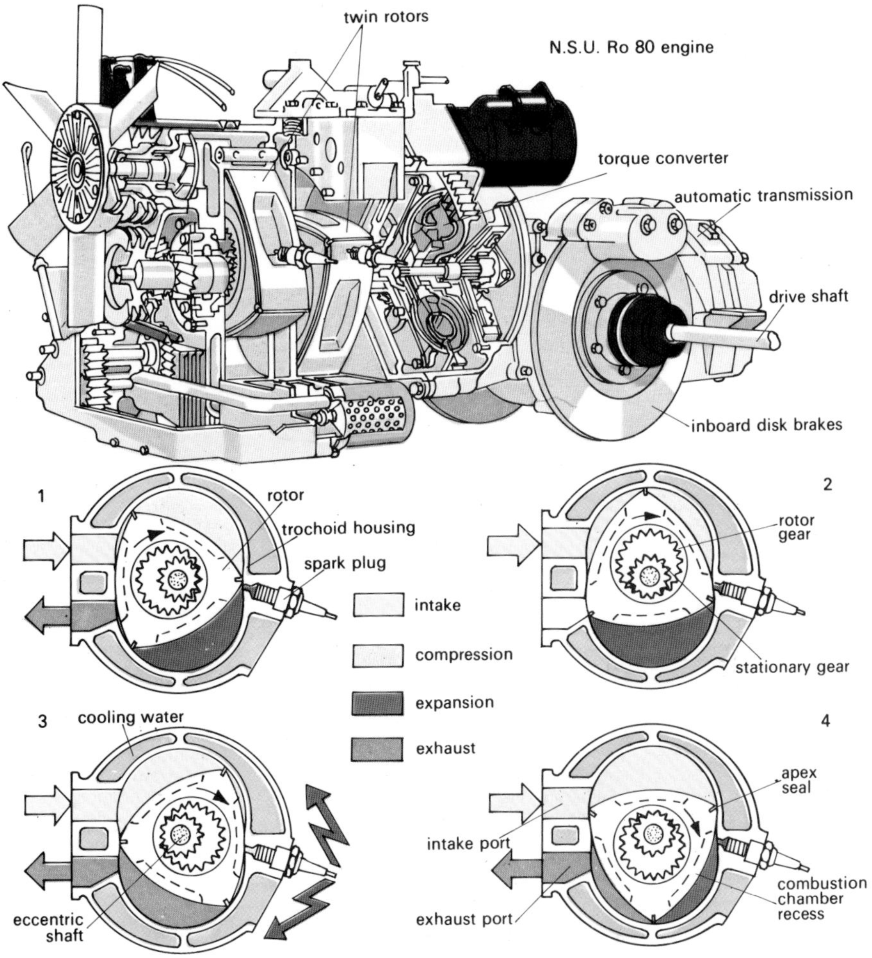

Top: cutaway of an automobile Wankel engine with automatic transmission, based on the NSU Ro 80 engine. This engine has two rotors and is mounted longitudinally, forward of the front (driving) wheels. *Above:* stages in the "working cycle" of a Wankel engine. Actually, each of the three combustion chambers on each rotor follows its own working cycle of intake, compression, combustion and exhaust, at phase angles 120° before and after its fellows.

WALD, George (1906–), US chemist and prominent pacifist whose work on the chemistry of VISION brought him a share, with HARTLINE and GRANIT, of the 1967 Nobel Prize for Physiology or Medicine.

WALKIE-TALKIE, portable two-way RADIO frequently used by policemen, sportsmen and others on the move to communicate over distances up to a few km. In the US, walkie-talkies operate on one or more of 23 channels lying between 26.960 and 27.255MHz.

WALLACE, Alfred Russel (1823–1913), British naturalist and socialist regarded as the father of ZOOGEOGRAPHY. His most striking work was his formulation, independently of C. DARWIN, of the theory of NATURAL SELECTION as a mechanism for the origin of species (see EVOLUTION). He and Darwin presented their results in a joint paper in 1858 before the Linnean Society.

WALLACH, Otto (1847–1931), Russian-born German experimental organic chemist awarded the 1910 Nobel Prize for Chemistry for his analysis of the structures of the TERPENES.

WALLIS, Sir Barnes Neville (1887–), British aeronautics engineer who designed the Wellington bomber and the skipping bomb used by the Dam Busters in WWII. He later devised the swing-wing principle used in some supersonic military airplanes.

WALLIS, John (1616–1703), British mathematician and cryptographer whose *Arithmetica infinitorum* (Arithmetic of Infinities, 1655) laid the foundations for much of modern algebra, introducing the concept of the LIMIT and the symbol ∞ for INFINITY. From this work were later developed CALCULUS and the BINOMIAL THEOREM.

WALTON, Ernest Thomas Sinton (1903–), Irish nuclear physicist who shared with COCKCROFT the 1951 Nobel Prize for Physics for their development of the first particle ACCELERATOR, using which they initiated the first nuclear FISSION reaction using nonradioactive substances.

WANKEL ENGINE, INTERNAL COMBUSTION ENGINE that produces rotary motion directly. Invented by the German engineer Felix Wankel (1902–), who completed his first design in 1954, it is now widely used in automobiles and airplanes. A triangular rotor with spring-loaded sealing plates at its apexes rotates eccentrically inside a cylinder, while the three combustion chambers formed between the sides of the rotor and the walls of the cylinder successively draw in, compress and ignite a fuel-and-air mixture. The Wankel engine is simpler in principle, more efficient and more powerful weight for weight, but more difficult to cool, than a conventional reciprocating engine.

WARBURG, Otto Heinrich (1883–1970), German biochemist awarded the 1931 Nobel Prize for Physiology or Medicine for his work elucidating the chemistry of cell RESPIRATION.

WARFARIN, a derivative of coumarin used as an ANTICOAGULANT and hence also to kill rodents by causing internal bleeding. However, warfarin-resistant races of rodents have developed in some areas.

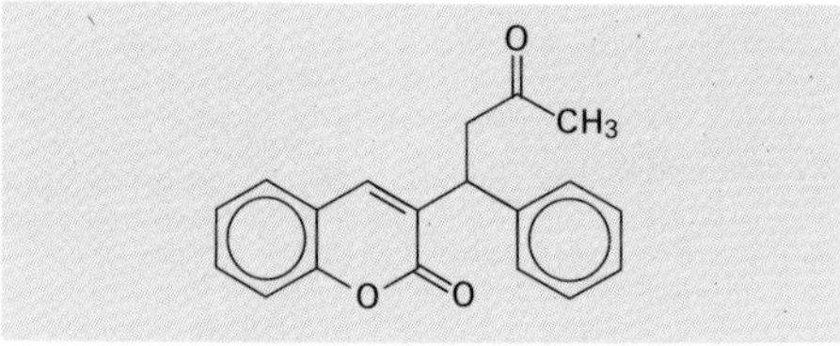

The structure of warfarin (4-hydroxy-3(1-phenyl-3-oxobutyl)-coumarin).

WARM-BLOODED ANIMALS, or Homoiotherms, animals whose body TEMPERATURE is not dependent on external temperature but is maintained at a constant level by internally-generated metabolic heat. This constant temperature enables the chemical processes of the body, many of them temperature-dependent, to be more efficient. Modern animals which have developed this homoiothermy are the mammals and birds, and it is now believed that pterodactyls, therapsids, and many other extinct reptiles may also have been warm-blooded.

WART, scaly excrescence on the SKIN caused by a VIRUS which may arise without warning and disappear equally suddenly. Numerous remedies have been suggested but local freezing or CAUTERIZATION are often effective. **Verrucas** are warts pushed into the soles of the feet by the weight of the body.

WASSERMANN TEST, screening test for syphilis (see VENEREAL DISEASES) based on a nonspecific serological reaction which is seen not only in syphilis but also in YAWS and diseases associated with immune disorders. More specific tests are available to discriminate between these.

WASTE DISPOSAL, disposal of such matter as animal excreta and the waste products of agricultural, industrial and domestic processes, where an unacceptable level of environmental POLLUTION would otherwise result. Where an ecological balance exists (see ECOLOGY), wastes are recycled naturally or by technological means (see RECYCLING) before accumulations affect the quality of life or disrupt the ecosystem. The most satisfactory waste-disposal methods are therefore probably those that involve recycling, as in manuring fields with dung, reclaiming metals from scrap or pulping waste paper for remanufacture. Recycling, however, may be inconvenient, uneconomic or not yet technologically feasible. Many popular waste-disposal methods consequently represent either an exchange of one form of environmental pollution for another less troublesome, at least in the short term—e.g., the dumping or burying of non-degradable garbage or toxic wastes—or a reducing of the rate at which pollutants accumulate—e.g., by compacting or incinerating bulk wastes before dumping. Urban wastes are generally disposed of by means of dumping, sanitary landfill, incineration and SEWAGE processing. Agricultural, mining and mineral-processing operations generate most solid wastes—and some of the most intractable waste-disposal problems: e.g., the "factory" farmer's problem of disposing of surplus organic wastes economically without resorting to incineration or dumping in rivers; the problems created by large mine dumps and open-cast excavations; and the culm-dumps that result from the processing of anthracite COAL. Another increasingly pressing waste-disposal problem is presented by radioactive wastes. Those with a "low level" of RADIOACTIVITY can be safely packaged and buried; but "high-level" wastes, produced in the course of reprocessing the fuel elements of NUCLEAR REACTORS, constitute a permanent hazard. Even the practice of encasing these wastes in thick concrete and dumping them on the ocean bottom is considered by many environmentalists to be an inadequate long-term solution (see also NUCLEAR ENERGY).

WATCH. See CLOCKS AND WATCHES.

WATER (H_2O), pale-blue odorless liquid which, including that trapped as ICE in icecaps and glaciers, covers about 74% of the earth's surface. Water is essential to LIFE, which began in the watery OCEANS; because of its unique chemical properties, it provides the medium for the reactions of the living CELL. Water is also man's most precious natural resource, which he must conserve and protect from POLLUTION (see also WATER SUPPLY.) Chemically, water can be viewed variously as a covalent HYDRIDE, an OXIDE, or a HYDROXIDE. It is a good solvent for many substances, especially ionic and polar compounds; it is ionizing and itself ionizes to give a low concentration of hydroxide and hydrogen ions (see PH). It is thus both a weak ACID and a weak BASE, and conducts electricity. It is a good, though labile, LIGAND, forming HYDRATES. Water is a polar molecule, and shows anomalies due to HYDROGEN BONDING, including contraction when heated from 0°C to 4°C. Formed when hydrogen or volatile hydrides are burned in oxygen, water oxidizes reactive metals to their ions, and reduces fluorine and chlorine. It converts basic oxides to hydroxides, and acidic oxides to OXY-ACIDS. (See also DEHYDRATION; HARD WATER; HEAVY WATER; HYDROLYSIS; POLYWATER; STEAM.) mp 0°C, bp 100°C, triple point 0.01°C, sg 1.0.

WATER CLOCK. See CLEPSYDRA.

WATER CYCLE. See HYDROLOGIC CYCLE.

WATERFALL, a vertical fall of water where a river flows from hard rock to one more easily eroded (see EROSION), or where there has been a rise of the land relative to sea level or blockage of a river by a landslide. Largest in the world is one of the Angel Falls, Venezuela (815m).

WATER GAP, a short, narrow gorge cut through a ridge or region of high ground by a stream or river (see EROSION). If the river no longer passes through it, the gorge is termed a **wind gap**.

WATER GAS, or blue gas (because of its blue flame), a FUEL GAS consisting of HYDROGEN and CARBON monoxide, made by blowing steam (alternately with air) over red-hot coke. Enriched with petroleum hydrocarbons, it becomes carbureted water gas.

WATER GLASS, aqueous solution of sodium SILICATE (Na_2SiO_3)—concentrated, syrupy and alkaline—made by fusing sodium carbonate with silica. It is used to preserve eggs, as a cement, in ore flotation, and for water- and fireproofing.

WATER POLLUTION. See POLLUTION; WATER; WATER SUPPLY.

WATER POWER. See HYDROELECTRICITY; TURBINE.

WATERPROOFING, any method of making an absorbent material, such as cloth, paper, leather, wood or masonry, resist penetration by water. Waterproofing is usually effected by the application of such agents as hardened oils, greases, waxes, plastics or rubber. Waterproofed TEXTILES are of two main kinds. Those that are said to be "impervious" to water are protected by a solid film which sheds water completely, but as they are also impervious to air they are uncomfortable to wear. "Water-repellent" fabrics, sometimes termed "showerproof," admit air but will stand up to exposure to water for a limited period only.

WATERSHED, a catchment basin (see DRAINAGE); or a DIVIDE.

Water droplets on a fabric surface made water-repellent using a silicone spray.

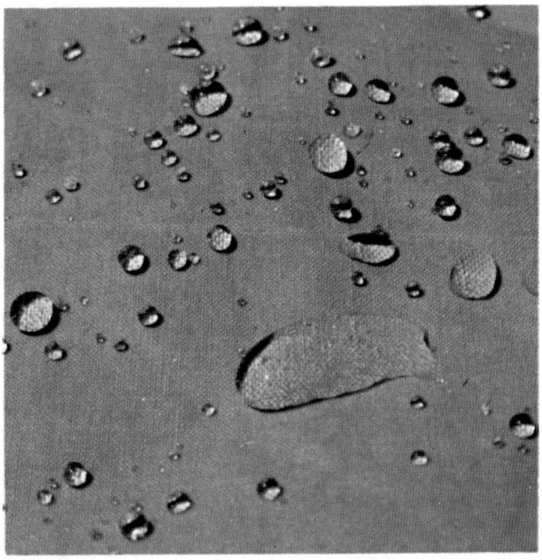

WATER SOFTENING. See HARD WATER.

WATERSPOUT, effect of a rotating column of air, or TORNADO, as it passes over water. A funnel-like CLOUD of condensed water vapor extends from a parent cumulonimbus cloud to the water surface, where it is surrounded by a sheath of spray.

WATER SUPPLY, available WATER resources and the means by which sufficient water of a suitable quality is supplied for agricultural, industrial, domestic and other purposes. Water precipitated over land (see HYDROLOGIC CYCLE) is available either as "surface water," in the form of rivers and lakes, usually supplemented by reservoirs, or as GROUND WATER, held underground—typically in an AQUIFER underlaid by impermeable rock—and brought to the surface by pumping or else rising as a spring or ARTESIAN WELL. Water may also be extracted from SEWAGE, purified and recycled (see RECYCLING).

WATER TABLE. See GROUNDWATER.

WATSON, James Dewey (1928–), US biochemist who shared with F. H. C. CRICK and M. H. F. WILKINS the 1962 Nobel Prize for Physiology or Medicine for his work with Crick establishing the "double helix" molecular model of DNA. His personalized account of the research, *The Double Helix* (1968), became a best-seller.

WATSON, John Broadus (1878–1958), US psychologist who founded BEHAVIORISM, a dominant school of US psychology from the 1920s to 1940s, and whose influence is still strong today.

WATSON-WATT, Sir Robert Alexander (1892–), British physicist largely responsible for the development of RADAR, patenting his first "radiolocator" in 1919. He perfected his equipment and techniques from 1935 through the years of WWII, his radar being largely responsible for the British victory of the Battle of Britain.

WATT (W), the unit of POWER in SI UNITS, defined as the power dissipated when ENERGY is utilized at a rate of one JOULE per second.

WATT, James (1736–1819), Scottish engineer and inventor. His first major invention was a STEAM ENGINE with a separate condenser and thus far greater efficiency. For the manufacture of such engines he entered partnership with John ROEBUCK and later (1775), more successfully, with Matthew BOULTON. Between 1775 and 1800 he invented the sun-and-planet gear wheel, the double-acting engine, a throttle valve, a pressure gauge and the centrifugal governor—as well as taking the first steps toward determining the chemical structure of water. He also coined the term HORSEPOWER and was a founder member of the LUNAR SOCIETY.

WAVEGUIDE, means for channeling high-frequency ELECTROMAGNETIC RADIATION by confining it within a tube whose walls are made of a conducting material (typically metal). Waveguides find their greatest use in MICROWAVE technology.

WAVE MECHANICS, branch of QUANTUM MECHANICS developed by SCHRÖDINGER which considers MATTER rather in terms of its wavelike properties (see WAVE MOTION) than as systems of particles. Thus an orbital ELECTRON is treated as a 3-dimensional system of standing waves represented by a *wave function*. In accordance with the UNCERTAINTY PRINCIPLE, it is not possible to pinpoint both the instantaneous position and velocity of the electron;

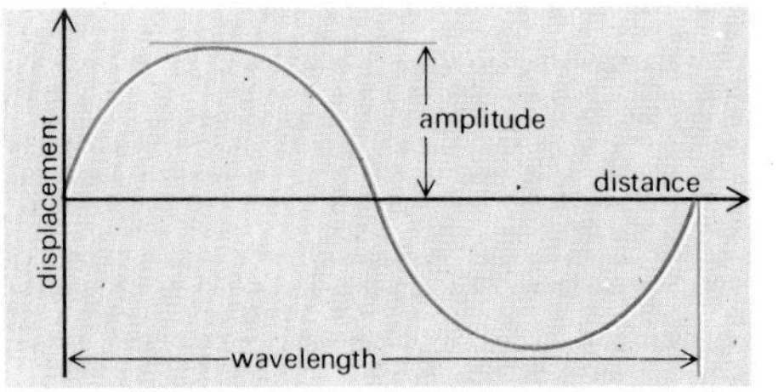

Amplitude and wavelength for a simple sine wave motion.

however, the square of the wave function yields a measure of the probability that the electron is at any given point in space-time. The pattern of such probabilities provides a model for the "shape" of the electron ORBITAL involved. Given wave functions can be obtained from the Schrödinger wave equation. Usually, and not unsurprisingly, this can only be solved for particular values of the ENERGY of the system concerned.

WAVE MOTION, a collective motion of a material or extended object, in which each part of the material oscillates about its undisturbed position, but the oscillations at different places are so timed as to create an illusion of crests and troughs running right through the material. Familiar examples are furnished by surface waves on water, or transverse waves on a stretched rope; SOUND is carried through air by a wave motion in which the air molecules oscillate parallel to the direction of propagation, and LIGHT or RADIO waves involve ELECTROMAGNETIC FIELDS oscillating perpendicular to it. The maximum displacement of the material from the undisturbed position is the *amplitude* of the wave, the separation of successive crests, the *wavelength*, and the number of crests passing a given place each second, the *frequency*. The product of the wavelength and the frequency gives the *velocity of propagation*. According to the direction and form of the local oscillations of the medium, different *polarizations* of the wave are distinguished (see POLARIZED LIGHT). *Standing waves* (apparently stationary waves, where the nodes and antinodes—points of zero and maximum amplitude—appear not to move) arise where identical waves traveling in opposite directions superpose. The characteristic properties of waves include propagation in straight lines; REFLECTION at plane surfaces; REFRACTION—a change in direction of a wave transmitted across a plane interface between two media; DIFFRACTION—diffuse SCATTERING by impenetrable objects of a size comparable with the wavelength; and INTERFERENCE—the cancellation of one wave by another wave half a wavelength out of step (or *phase*) so that the crests of one wave fall on the troughs of the other. If the *wave velocity* is the same for all wavelengths, then quite arbitrary forms of disturbance will travel as waves, and not simply regular successions of crests and troughs. When this is not the case, the wave is said to be *dispersive* and localized disturbances move at a speed (the *group velocity*) quite different from that of the individual crests, which can often be seen moving faster or slower within the disturbance "envelope," which becomes progressively broader as it moves. Waves carry ENERGY and MOMENTUM with them just like solid objects; the identity of the apparently irreconcilable wave and particle concepts of matter is a basic tenet of QUANTUM MECHANICS.

WAX, moldable water-repellent solid. There are several entirely different kinds. **Animal waxes** were the first known: *wool wax* when purified yields LANOLIN; *beeswax*, from the honeycomb, is used for some candles and as a sculpture medium (by carving or casting); *spermaceti wax*, from the sperm whale, is used in ointments and cosmetics. **Vegetable waxes**, like animal waxes, are mixtures of ESTERS of long-chain ALCOHOLS and CARBOXYLIC ACIDS. *Carnauba wax*, from the leaves of a Brazilian palm tree, is hard and lustrous, and is used to make polishes; *candelilla wax*, from a wild Mexican rush, is similar but more resinous; *Japan wax*, the coating of sumac berries, is fatty and soft but tough and kneadable. **Mineral waxes** include *montan wax*, extracted from lignite (see COAL), bituminous and resinous; *ozokerite*, an absorbent hydrocarbon wax obtained from wax shales, and *paraffin wax* or petroleum wax, the most important wax commercially: it is obtained from the residues of PETROLEUM refining by solvent extraction, and is used to make candles, to coat paper products, in the electrical industry, to waterproof leather and textiles, etc. Various **synthetic waxes** are made for special uses.

WEATHER, the hour-by-hour variations in the atmospheric conditions experienced at a given place. (See ATMOSPHERE; METEOROLOGY; WEATHER FORECASTING AND CONTROL.)

WEATHER FORECASTING AND CONTROL, the practical application of the knowledge gained through the study of METEOROLOGY. **Weather forecasting**, organized nationally by government agencies such as the US National Weather Service, is coordinated internationally by the World Meteorological Organization (WMO). There are three basic stages: observation; analysis, and forecasting. Observation involves round-the-clock weather watching and the gathering of meteorological data by land stations, weather ships, and by using RADIOSONDES and weather SATELLITES. In analysis, this information is coordinated at national centers, and plotted in terms of ISOBARS, FRONTS, etc., on synoptic charts (weather maps). Then, in forecasting, predictions of the future weather pattern are made by the "synoptic method" (in which the forecaster applies his experience of the evolution of past weather patterns to the current situation) and by "numerical forecasting" (which treats the ATMOSPHERE as a fluid of variable density and seeks to use hydrodynamic equations to determine its future parameters). These methods yield short- and medium-term forecasts—up to four days ahead. Long-range forecasting, a recent development, depends additionally on the statistical analysis of past weather records in attempting to discern the future weather trends over the next month or season. **Weather control**, or weather modification, is an altogether less reliable technology. Indeed, the natural variability of weather phenomena makes it difficult to assess the success of experimental procedures. To date, the best results have been obtained in the fields of CLOUD seeding and the dispersal of supercooled FOGS.

WEATHERING. See EROSION.

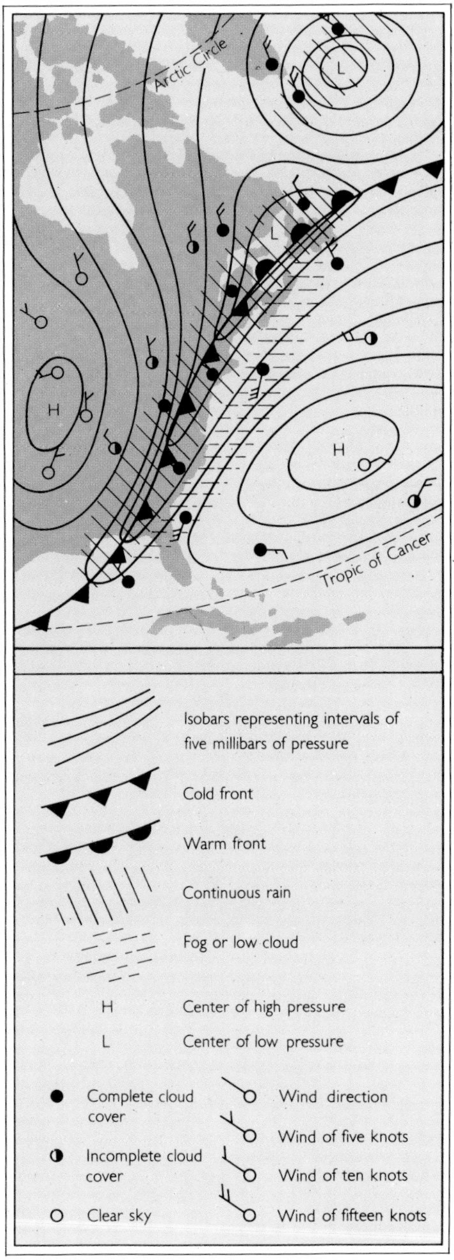

Weather forecasting: a portion of a synoptic chart showing a mid-latitude cyclone (depression) with an associated area of rain over the western Atlantic off the coast of Newfoundland. *Below:* some of the symbols used on such charts.

WEAVING, making a fabric by interlacing two or more sets of threads. In "plain" weave, one set of threads—the *warp*—extends along the length of the fabric; the other set—the *woof,* or *weft*—is at right angles to the warp and passes alternately over and under it. Other common weaves include "twill," "satin" and "pile." In basic twill, woof threads, stepped one warp thread further on with each line, pass over two warp threads, under one, then over two again, producing diagonal ridges, or wales, as in denim, flannel and gaberdine. In satin weave, a development of twill, long "float" threads passing under four warp threads give the fabric its characteristically smooth appearance. Pile fabrics, such as corduroy and velvet, have extra warp or weft threads woven into a ground weave in a series of loops that are then cut to produce the pile. Weaving is usually accomplished by means of a hand- or power-operated machine called a loom. Warp threads are stretched on a frame and passed through eyelets in vertical wires (heddles) supported on a frame (the harness). A space (the shed) between sets of warp threads is made by moving the heddles up or down, and a shuttle containing the woof thread is passed through the shed. A special comb (the reed) then pushes home the newly woven line. (See also TEXTILES.

WEBER (Wb), the unit of magnetic flux in SI UNITS, defined such that an ELECTROMOTIVE FORCE of one VOLT is induced in a single coil when the flux changes in the coil at the rate of one weber per second. (See INDUCTION, ELECTROMAGNETIC.)

WEBER, Wilhelm Eduard (1804–1891), German physicist best known for his work with GAUSS on GEOMAGNETISM.

WEDGE. See MACHINE.

WEEDKILLERS, or **herbicides,** chemical compounds used to kill weeds (plants growing where they are unwanted). Originally general herbicides were used—diesel oil, sulfuric acid, sodium arsenite, sodium chlorate, etc.—but these dangerous substances have been largely superseded since WWII by a vast host of selective weedkillers, complex organic compounds (also dangerous) which at a suitable dosage are much more toxic to the prevailing weeds than to the crop. Contact herbicides, including paraquat, kill only the parts of the plants on which they are sprayed. The use of weedkillers complements good agricultural management, mechanical destruction of weeds and biological control.

WEEK, an arbitrary division of time, through most of the Christian era, of duration seven days. In most European languages, the days of the week are named for the planets or deities which were considered to preside over them.

WEGENER, Alfred Lothar (1880–1930), German meteorologist, explorer and geologist. His *The Origin of Continents and Oceans* (1915) set forth "Wegener's hypothesis," the theory of CONTINENTAL DRIFT, whose developments were in succeeding decades to revolutionize man's view of the planet he lives on (see also PLATE TECTONICS).

WEIDENREICH, Franz (1873–1948), German-born US physical anthropologist and anatomist best known for his work on fossil remains of *Sinanthropus, Pithecanthropus* and *Meganthropus* (see PREHISTORIC MAN), and for his chronological arrangement of the various stages in man's evolution.

WEIGHT, the attractive FORCE experienced by an object in the presence of another massive body in accordance with the law of universal GRAVITATION.

Weights and Measures

US Customary System

Quantity	Unit	Symbol	Equivalent same system	Approximate SI equivalent
weight (avoirdupois)	ton (short)	—	20 hundredweight (short)	0.907 tonne
	hundredweight	cwt	100 pounds	45.359 kg
	pound	lb (lb av)	16 ounces (7 000 grains)	*0.453 592 37 kg
	ounce	oz (oz av)	16 drams	28.350 g
	dram	dr (dr av)	27.344 grains	1.772 g
	grain	gr	—	*64.798 91 mg
length	mile	mi	1 760 yards	*1.609 344 km
	yard	yd	3 feet	*0.914 4 m
	foot	ft	12 inches	*0.304 8 m
	inch	in	—	*25.4 mm
area	square mile	sq mi (mi^2)	640 acres	2.589 99 km^2
	acre	—	4 840 sq yards	0.404 69 ha
	square yard	sq yd (yd^2)	9 sq feet	0.836 13 m^2
	square foot	sq ft (ft^2)	144 sq inches	0.092 90 m^2
	square inch	sq in (in^2)	—	*645.16 mm^2
volume	cubic yard	cu yd (yd^3)	27 cu ft	0.764 555 m^3
	cubic foot	cu ft (ft^3)	1 728 cu inches	0.028 317 m^3
	cubic inch	cu in (in^3)	—	*16 387.064 mm^3
capacity (liquid measure)	gallon	gal	4 quarts	3.785 4 litre
	quart	qt	2 pints	0.946 3 litre
	pint	pt	4 gills	0.473 2 litre
	gill	gi	4 fluidounces	0.118 3 litre
	fluidounce	fl oz	8 fluidrams	29.573 ml
	fluidram	fl dr	60 minims	3.697 ml
	minim	min	—	0.061 610 ml
capacity (dry measure)	bushel	bu	4 pecks	35.238 litre
	peck	pk	8 quarts	8.810 litre
	quart	qt	2 pints	1.101 litre
	pint	pt	—	0.551 litre

British Imperial System (where different from US Customary System)

Quantity	Unit	Symbol	Equivalent in same system	Approximate SI and US equivalents
weight (avoirdupois)	ton (long)	—	20 hundredweight (long)	1.016 tonne 1.12 short tons
	hundredweight (long)	cwt	112 pounds	50.802 kg 1.12 short cwt
liquid and dry measure	bushel	bu	4 pecks	36.369 litre
	peck	pk	2 gallons	9.092 litre
	gallon	gal	4 quarts	4.546 litre 1.201 US gal
	quart	qt	2 pints	1.137 litre
	pint	pt	4 gills	568.260 ml
	gill	gi	5 fluidounces	142.065 ml
	fluidounce	fl oz	8 fluidrams	28.413 ml 0.961 US fl oz
	fluidram	fl dr	60 minims	3.551 6 ml
	minim	min	—	0.059 194 ml

*exact equivalent.

The weight of a body (measured in newtons) is given by the product of its MASS and the local ACCELERATION due to gravity (*g*). Weight differs from mass in being a VECTOR quantity.

WEIGHTS AND MEASURES, units of WEIGHT, LENGTH, AREA and VOLUME commonly used in the home, in commerce and in industry. Although like other early peoples the Hebrews used measures such as the foot, the cubit (the length of the human forearm) and the span, which could easily be realized in practice using parts of the body, in commerce they also used standard containers and weights. Later, weights were based on the quantity of precious metal in coins. During and after the Middle Ages, each region evolved its own system of weights and measures. In the 19th century these were standardized on a national basis, these national standards in turn being superseded by those of the METRIC SYSTEM. In the western world, only the British Empire and the US retained their own systems (the Imperial System and the US Customary System) into the mid-20th century. With the UK's adoption of the International System of Units (SI UNITS), the US has found itself alone in not using metric units, although, as has been the case since 1959, the US customary units are now defined in terms of their metric counterparts and not on the basis of independent standards. In the US the administration of weights and measures is coordinated by the National Bureau of Standards (NBS) who also publish the version of the International System used in this volume. (See also APOTHECARIES' WEIGHTS; TROY WEIGHT.)

WEISMANN, August (1834–1914), German biologist regarded as a father of modern GENETICS for his demolition of the theory that ACQUIRED CHARACTERISTICS could be inherited, and proposal that CHROMOSOMES are the basis of HEREDITY. He coupled this proposal with his belief in NATURAL SELECTION as the mechanism for EVOLUTION.

WELCH, William Henry (1850–1934), US pathologist and bacteriologist whose most significant achievements were in the field of medical education, playing a large part in the founding (1893) and development of the Johns Hopkins Medical School.

WELDING, bringing two pieces of metal together under conditions of heat or pressure or both, until they coalesce at the joint. The oldest method is forge welding, in which the surfaces to be joined are heated to welding temperature and then hammered together on an anvil. The most widely used method today is metal-arc welding: an ELECTRIC ARC is struck between an ELECTRODE and the workpieces to be joined, and molten metal from a "filler rod"—usually the electrode itself—is added. Gas welding, now largely displaced by metal-arc welding, is usually accomplished by means of an oxyacetylene torch, which delivers the necessary heat by burning ACETYLENE in a pure OXYGEN atmosphere. Sources of heat in other forms of welding include the electrical RESISTANCE of the joint (resistance welding), an electric arc at the joint (flash welding), a focused beam of ELECTRONS (electronbeam welding), pressure alone, usually well in excess of 1 400 000kPa (cold welding), and friction (friction welding). Some more recently applied heat sources include hot PLASMAS, LASERS, ULTRASONIC vibrations and explosive impacts.

WELL, man-made hole in the ground used to tap water, gas or minerals from the earth. Most modern wells are drilled and fitted with a lining, usually of steel, to forestall collapse. Though wells are sunk for NATURAL GAS and PETROLEUM oil, the commonest type yields water. Such wells may be horizontal or vertical, but all have their innermost end below the water table (see GROUNDWATER). If it should be below the permanent water table (the lowest annual level of the water table) the well will yield water throughout the year. Most wells require to be pumped, but some operate under natural pressure (see ARTESIAN WELL).

WELLER, Thomas Huckle (1915–), US bacteriologist and virologist who shared with J. F. ENDERS and F. C. ROBBINS the 1954 Nobel Prize for Physiology or Medicine for their cultivation of POLIOMYELITIS virus in non-nerve tissues.

WELLS, Horace (1815–1848), US dentist and pioneer of surgical ANESTHESIA, using (largely without success) nitrous oxide (see NITROGEN).

WELSBACH, Carl Auer von, Baron (1858–1929), Austrian chemist who invented the incandescent gas mantle (patented 1885) and the ALLOY, Auer metal, used to make lighter flints.

WEN, or sebaceous CYST, blocked SEBACEOUS GLAND, often over the scalp or forehead, which forms a cyst containing old sebum under the SKIN. It may become infected. Its excision is a simple procedure.

WERNER, Abraham Gottlob (1750–1817), Silesian-born mineralogist who taught for over 40 years at the Freiberg Mining Academy, disseminating the doctrines of NEPTUNISM.

WERNER, Alfred (1866–1919), French-born Swiss chemist awarded the 1913 Nobel Prize for Chemistry for his theory of coordination complexes. (See BOND, CHEMICAL; LIGANDS.)

WERTHEIMER, Max (1880–1943), Czechoslovakian-born US founder, with KOFFKA and KÖHLER, of the school of GESTALT PSYCHOLOGY.

WESTERLIES. See PREVAILING WESTERLIES.

WESTINGHOUSE, George (1846–1914), US engineer, inventor and businessman who pioneered the use of high-voltage AC electricity. In 1869 he founded the Westinghouse Air Brake Company to develop the AIR BRAKES he had invented for RAILROAD use. From 1883 he did pioneering work on the safe transmission of NATURAL GAS. In 1886 he founded the Westinghouse Electric Company, employing notably TESLA, to develop AC INDUCTION MOTORS and transmission equipment: this company was largely responsible for the acceptance of AC in preference to DC for most applications—in spite of opposition from the influential EDISON.

WEYL, Hermann (1885–1955), German mathematician and mathematical physicist noted for his contributions to the theories of RELATIVITY and QUANTUM MECHANICS.

WHEATSTONE, Sir Charles (1802–1875), British physicist and inventor who popularized the WHEATSTONE BRIDGE; and invented the electric TELEGRAPH (with the help of Joseph HENRY) before MORSE (1837), the STEREOSCOPE (1838) and the concertina (1829).

WHEATSTONE BRIDGE, an electric circuit used for comparing or measuring RESISTANCE. Four resistors, including the unknown one, are connected in a square, with a BATTERY between one pair of diagonally opposite corners and a sensitive

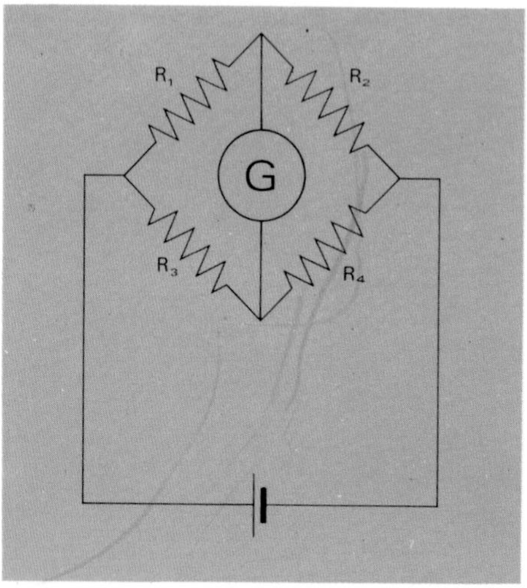

Wheatstone bridge circuit. When no current is flowing through the sensitive galvanometer (G), $R_1/R_2 = R_3/R_4$.

GALVANOMETER between the other. When no current flows through the meter, the products of opposite pairs of resistances are equal. Similar bridge circuits are used for IMPEDANCE measurement.

WHEEL, disk-like mechanical device mediating between rotary and linear motion, widely used to transmit POWER, store ENERGY (see FLYWHEEL) and to facilitate the movement of heavy objects. Wheels may be solid or spoked, flanged or unflanged, with or without TIRES. Most usefully, they are attached to an axle through the center. Indeed, the **wheel and axle** is one of the classic simple MACHINES, exemplified in the capstan, the WINCH and TRANSMISSION gears.

WHEWELL, William (1794–1866), English philosopher and Master of Trinity College, Cambridge (1841–66), renowned as the last polymath. His interests ranged from mineralogy to moral philosophy, but he is chiefly remembered for his *Philosophy of the Inductive Sciences* (1840), which reflected his study of KANT and led to a famous controversy with J. S. MILL.

WHIPPLE, George Hoyt (1878–), US pathologist awarded the 1934 Nobel Prize for Physiology or Medicine for his discovery that feeding raw liver to anemic dogs improved their condition: the successful work of G. MINOT and W. MURPHY (who shared the award with him) in finding a treatment for pernicious ANEMIA sprang directly from this.

WHIRLPOOL, a rotary current in water. Permanent whirlpools may arise in the ocean from the interactions of the TIDES (see OCEAN CURRENTS). They occur also in streams or rivers where two currents meet or the shape of the channel dictates. Short-lived whirlpools may be created by wind. (See also VORTEX; WHIRLWIND.)

WHIRLWIND, rotating column of air caused by a pocket of low atmospheric pressure formed—unlike a TORNADO—near ground level by surface heating. They are far less violent than tornadoes. Whirlwinds passing over dry dusty country are sometimes called "dust devils."

WHISKEY, strong spirituous DISTILLED LIQUOR, drunk mixed or neat, made from grain. When from Scotland or Canada, whisky is spelled without an "e". The ingredients and preparation vary. In the US corn and rye are commonly used: 51% corn for *bourbon whiskey* and 51% rye for *rye whiskey*. A grain mash is allowed to ferment, then distilled, diluted and left to age. Bourbon and rye whiskey stand in oak barrels for four years. *Canadian whisky* is made from corn, rye and malted (germinated) barley and aged for 4–12 years. *Irish whiskey* uses barley, wheat, oats and rye, and vessels called potstills for the distilling process. *Scotch whisky* is the finest form: the best types are pure barley malt or grain whiskies, but blended varieties are cheaper. The secret of its flavor is supposed to be the peat-flavored water of certain Scottish streams. Whiskey is one of the most popular of ALCOHOLIC BEVERAGES. In the US an average of 16 bottles per person are drunk every year.

WHITE, Paul Dudley (1886–1972), US physician regarded from the 1940s to 1960s as the world's leading cardiologist.

WHITE CORPUSCLES. See BLOOD.

WHITE GOLD, an ALLOY of GOLD with nickel and sometimes other NOBLE METALS, used in dentistry.

WHITEHEAD, Alfred North (1861–1947), English mathematician and philosopher. He was co-author with Bertrand RUSSELL of *Principia Mathematica* (1910–13), a major landmark in the philosophy of mathematics; and while teaching at Harvard University (from 1924) he developed a monumental system of metaphysics, most comprehensively expounded in his *Process and Reality* (1929).

WHITE LEAD. See CERUSSITE.

WHITEPRINT PROCESS. See OZALID PROCESS.

WHITEWASH, cheap nondurable PAINT composed of CHALK, a glue or casein binder, and water; the dry form is called **calcimine**.

WHITNEY, Eli (1765–1825), US inventor of the COTTON GIN (1793), from which he earned little because of patent infringements, and pioneer of MASS PRODUCTION. In 1798 he contracted with the US Government to make 10 000 muskets: he took 8 years to fulfil the 2-year contract, but showed that with unskilled labor muskets could be put together using parts that were precision-made and thus interchangeable, a benefit not only during production but also in later maintenance.

WHITTLE, Air Commodore Sir Frank (1907–), British aeronautical engineer who invented the first aircraft JET PROPULSION unit (patented 1937, first used in flight 1941).

WHITWORTH, Sir Joseph, Baronet (1803–1887), British engineer, metallurgist and philanthropist who brought radically improved accuracy to the manufacture of MACHINE TOOLS and who invented the Whitworth rifle, used by the British Army in the late 1860s and 1870s, which had a hexagonal barrel.

WHOOPING COUGH, or **pertussis**, BACTERIAL DISEASE of children causing upper respiratory symptoms with a characteristic whoop or inspiratory noise due to INFLAMMATION of the LARYNX. It is usually a relatively mild illness, except in the very young, but VACCINATION is widely practiced to prevent it.

WHORF, Benjamin Lee (1897–1941), US linguist best known for proposing the theory that a language's structure determines the thought processes of its speakers. (See also LINGUISTICS.)

WIELAND, Heinrich Otto (1877–1957), German organic chemist noted for his work on STEROIDS, especially his research on the BILE acids, which brought him the 1927 Nobel Prize for Chemistry.

WIEN, Wilhelm (1864–1928), Prussian-born German physicist best known for his work on BLACKBODY RADIATION, work which was later to be a foundation stone for Planck's QUANTUM THEORY.

WIENER, Norbert (1894–1964), US mathematician who created the discipline CYBERNETICS. His major book is *Cybernetics: Or Control and Communication in the Animal and the Machine* (1948).

WIGNER, Eugene Paul (1902–), Hungarian-born US physicist who shared with J. H. D. JENSEN and M. G. MAYER the 1963 Nobel Prize for Physics for his work in the field of nuclear physics. He also worked with FERMI on the MANHATTAN PROJECT, and received the 1960 Atoms for Peace Award.

WILEY, Harvey Washington (1844–1930), US chemist whose main achievements were in promoting pure food laws, being largely responsible for instituting the Pure Food and Drugs Act (1906).

WILKINS, Maurice Hugh Frederick (1916–), British biophysicist who shared with F. H. CRICK and J. D. WATSON the 1962 Nobel Prize for Physiology or Medicine for his X-RAY DIFFRACTION studies of DNA, work that was vital to the determination by Crick and Watson of DNA's molecular structure.

WILKINSON, Geoffrey (1921–), British inorganic chemist awarded with Ernst Otto FISCHER the 1973 Nobel Prize for Chemistry for their work on ORGANOMETALLIC COMPOUNDS.

WILLIAMS, Daniel Hale (1858–1931), US surgeon who carried out the first repair operation on the damaged outer surface of a human heart (1893).

WILL-O'-THE WISP, or **Jack O'Lantern,** or **Ignis Fatuus** (Latin: foolish fire), light seen at night over marshes, caused by SPONTANEOUS COMBUSTION of METHANE produced by putrefying matter. Luring travelers into danger, it was popularly regarded as a wandering damned spirit bearing its own hell-fire.

WILLSTÄTTER, Richard (1872–1942), German chemist awarded the 1915 Nobel Prize for Chemistry for his studies of the structure of CHLOROPHYLL.

WILSON, Charles Thomson Rees (1869–1959), British physicist awarded with A. H. COMPTON the 1927 Nobel Prize for Physics for his invention of the CLOUD CHAMBER (1911).

WILT, condition where plants droop and wither due to a lack of water in their CELLS. This can be caused by lack of available moisture, physiological disorders, or FUNGI or BACTERIA damaging water conducting tissues inside roots or stems.

WINCH, device facilitating the hoisting or hauling of loads. It comprises a rotatable drum around which is wound a rope or cable attached to the load. The drum is turned by means of a hand-operated crank or a motor. (See also CRANE; WINDLASS.)

WIND, body of air moving relative to the earth's surface. The world's major wind systems, or general winds, are set up to counter the equal heating of the earth's surface and modified by the rotation of the earth. Surface heating, at its greatest near the equator, creates an equatorial belt of low pressure (see DOLDRUMS) and a system of CONVECTION currents transporting heat toward the Poles (see HORSE LATITUDES). The earth's rotation deflects the currents of the N Hemisphere to the right and those of the S Hemisphere to the left of the directions in which they would otherwise blow, producing the NE and SE TRADE WINDS, the PREVAILING WESTERLIES and the Polar Easterlies. Other factors influencing general wind patterns are the different rates of heating and cooling of land and sea and the seasonal variations in surface heating. Mixing of air along the boundary between the Westerlies and the Polar Easterlies—the polar front—causes depressions in which winds follow circular paths, counterclockwise in the N Hemisphere and clockwise in the S Hemisphere (see CYCLONE). Superimposed on the general wind systems are local winds—winds, such as the CHINOOKS, caused by temperature differentials associated with local topographical features such as mountains and coastal belts, or winds associated with certain CLOUD systems. (See also ATMOSPHERE; BEAUFORT SCALE; HURRICANE; JET STREAM; MONSOON; TORNADO; WEATHER FORECASTING; WHIRLWIND.)

WINDAUS, Adolf (1876–1959), German organic chemist awarded the 1928 Nobel Prize for Chemistry for his work on the STEROLS, in course of which he determined the structure of CHOLESTEROL and showed that ULTRAVIOLET RADIATION transforms ergosterol into VITAMIN D_2.

WIND GAP. See WATER GAP.

WINDLASS, simple WINCH once widely used to draw water from wells. A load-bearing rope passes around a small-diameter drum which is turned by a hand-operated crank. A ship's capstan uses the same principle.

WINDMILL, machine that performs WORK by harnessing wind power. In the traditional windmill, the power applied to a horizontal shaft by four large radiating sails was transmitted to milling or pumping machinery housed in a sizable supporting structure. The windmill's modern cousin is the wind turbine, often seen in remote rural areas. Here a multibladed turbine wheel mounted on a steel derrick or mast and pointed into the wind by a "fantail" drives a pump or electric generator.

WINDPIPE, common name for the TRACHEA.

WIND TUNNEL, tunnel in which a controlled stream of air is produced in order to observe the effect on scale models or full-size components of airplanes, missiles, automobiles or such structures as bridges and skyscrapers. An important research tool in AERODYNAMICS, the wind tunnel enables a design to be accurately tested without the risks attached to full-scale trials. "Hypersonic" wind tunnels, operating on an impulse principle, can simulate the frictional effects of flight at over five times the speed of sound.

WINE, an ALCOHOLIC BEVERAGE made from fermented grape juice; wines made from other fruits are always named accordingly. **Table wines** are red, rosé or "white" in color; red wines are made from dark grapes, the skins being left in the fermenting mixture; white wines may be made from dark or pale grapes, the skins being removed. The grapes—normally varieties of *Vitis vinifera*—are allowed to ripen until they attain suitable sugar content—18% or more—and acidity (in cool years or northern areas sugar may have to be added). After crushing, they

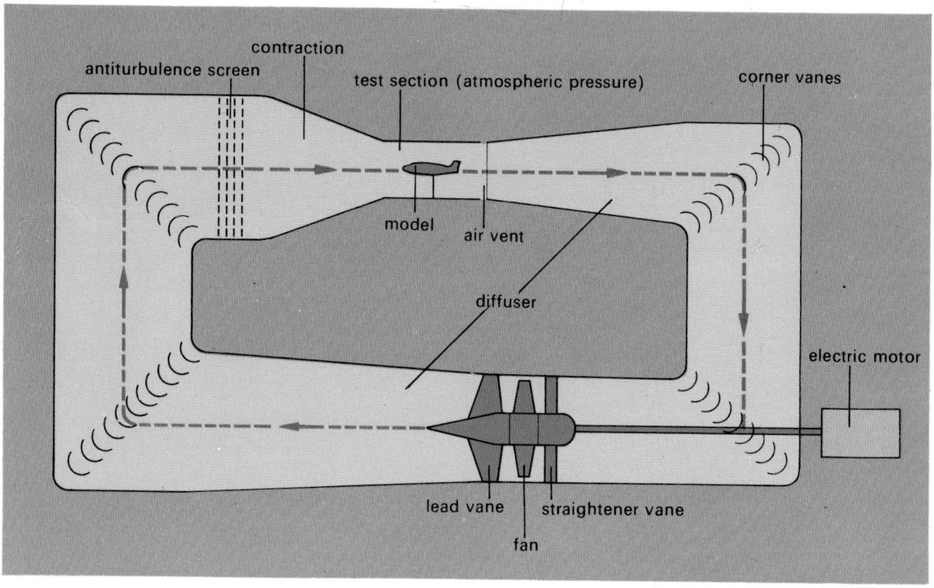

Schematic diagram of a return-flow low-velocity wind tunnel.

undergo FERMENTATION in large tanks, a small amount of sulfur dioxide being added to inhibit growth of wild yeasts and bacteria; the wine yeast used, *Saccharomyces cerevisiae*, is resistant to it. When the alcohol and sugar content is right, the wine is cellared, racked off the lees (from which argol is obtained—see TARTARIC ACID), clarified by FILTRATION or fining (adding absorbent substances such as BENTONITE, GELATIN and ISINGLASS), aged in the wood and bottled. Sweet wines contain residual sugar; dry wines little or none. The alcohol content of table wines varies from 8% to 14% by volume. **Sparkling wines**—notably champagne—are made by secondary fermentation under pressure, in bottles or in tanks. **Fortified wines**, or dessert wines—including sherry, port and madeira—have brandy added during or after fermentation, and contain about 20% alcohol. Vermouth is a fortified wine flavored with wormwood (or absinthe) and other herbs. Major wine-producing areas of the world include France, Germany, Spain and Portugal, Italy, and, in the US, Cal.

WINTERGREEN, *Gaultheria procumbens*, a North American shrub, the leaves of which yield an ESSENTIAL OIL called oil of wintergreen, containing methyl salicylate. The oil is used as an antiseptic and to treat rheumatism. The berries are eaten in pies. Family: Ericaceae.

WIRE, a length of metal that has been drawn out into a thread. Wire is usually flexible, circular in cross-section and uniform in diameter. Wire diameters generally range from about 0.001 to 0.5in (0.025–12.7mm). To manufacture wire, normally a hot-rolled metal rod pointed at one end is coated with a lubricant, threaded through a tungsten-carbide or diamond DIE and attached to a drum called a draw block. The draw block is rotated and wire of a diameter, or gauge, determined by the diameter of the die is drawn until the entire metal rod is reduced to wire. Steel, iron, aluminum, copper and bronze are the metals most widely used for wire making, although others, including gold, platinum and silver, are used as well. Copper and aluminum are preferred for electrical wiring, since they combine high ductility with low resistance to electric current. (See also BARBED WIRE; CABLE.)

WIREPHOTO. See TELEPHOTO.

WISHBONE. See COLLARBONE.

WITHERING, William (1741–1799), British physician, mineralogist and botanist, who first made use of DIGITALIS in the treatment of dropsy (see EDEMA). The mineral **witherite** ($BaCO_3$) is named for him.

WITTGENSTEIN, Ludwig (1889–1951), Austrian philosopher, whose two chief works, *Tractatus Logico-Philosophicus* (1921) and *Philosophical Investigations* (published posthumously in 1953), have profoundly influenced the course of much recent British and US philosophy. The *Tractatus* dwells on the logical nature and limits of language, understood as "picturing" reality. The *Investigations* rejects the assumption in the *Tractatus* that all representations must share a common logical form and instead relates the meanings of sentences to their uses in particular contexts: philosophical problems are attributed to misuses of language. Wittgenstein was professor of philosophy at Cambridge U., England (1929–47).

WÖHLER, Friedrich (1800–1882), German chemist who first synthesized an organic compound from inorganic material, UREA from ammonium cyanate (1828), and who, with LIEBIG, discovered the benzoyl radical.

WOLFF, Caspar Friedrich (1733–1794), German anatomist regarded as the father of modern EMBRYOLOGY for his demonstration that the organs and other bodily parts form from undifferentiated tissue in the fetus. It had earlier been thought that the fetus was a body in miniature.

WOLFRAM. See TUNGSTEN.

WOLFRAMITE, or iron manganese tungstate $(Fe, Mn)WO_4$, the chief ore of TUNGSTEN. It forms brown

..o black monoclinic crystals, and is found with TIN ores. China is the main producer.

WOLLASTON, William Hyde (1766–1828), British chemist and physicist who, through the process he devised for isolating PLATINUM in pure and malleable form, founded the technology of powder METALLURGY. He also discovered PALLADIUM (1803) and RHODIUM (1804), was the first to observe the FRAUNHOFER LINES (1802), and invented the CAMERA LUCIDA (1807) and the reflecting GONIOMETER (1809).

WOMB, or **uterus**, female reproductive organ which is specialized for IMPLANTATION of the EGG and development of the EMBRYO and FETUS during PREGNANCY. The regular turnover of its lining under the influence of ESTROGEN and PROGESTERONE is responsible for MENSTRUATION. Disorders of the womb include malformation, abnormal position and disorders of menstruation. Benign tumors or fibroids are a common cause of the latter. CANCER of the womb or its cervix is relatively common and may be detected by the use of regular PAPANICOLAOU tests. Removal of the womb for cancer, fibroids etc., is HYSTERECTOMY.

WOOD, the hard, dead tissue obtained from the trunks and branches of TREES and SHRUBS. Woody tissue is also found in some herbaceous PLANTS. Botanically, wood consists of *xylem* tissue which is responsible for the conduction of water around the plant. A living tree trunk is composed of (beginning from the center): the *pith* (remains of the primary growth); wood (xylem); CAMBIUM (a band of living cells that divide to produce new wood and phloem); PHLOEM (conducting nutrients made in the leaves), and the bark. The wood nearest the cambium is termed *sapwood* because it is capable of conducting water. However, the bulk of the wood is *heartwood* in which the xylem is impregnated with LIGNIN which gives the cells extra strength but prevents them from conducting water. In temperate regions, a tree's age can be found by counting its ANNUAL RINGS. Commercially, wood is divided into hardwood (from deciduous ANGIOSPERM trees) and softwood (from GYMNOSPERMS). (See also FORESTS; FORESTRY; PAPER.)

WOOD, Jethro (1774–1834), US inventor of an improved plow (patents 1814, 1819), many of whose features are incorporated in modern plows.

WOOD ALCOHOL. See METHANOL.

WOODWARD, Robert Burns (1917–), US chemist who has synthesized QUININE (1944), CHOLESTEROL (1951), CORTISONE (1951), STRYCHNINE (1954), LSD (1954), reserpine (1956; see RAUWOLFIA SERPENTINA), CHLOROPHYLL (1960) and VITAMIN B_{12} (1971). He received the 1965 Nobel Prize for Chemistry for his many organic syntheses.

WOOL, animal FIBER that forms the fleece, or protective coat, of sheep. Coarser than most vegetable or synthetic fibers, wool fibers are wavy (up to 10 waves/cm) and vary in color from the usual white to

Some of the many types of wood used in commerce: (1) Pitch pine; (2) redwood; (3) walnut; (4) Oregon pine; (5) coromandel; (6) oak; (7) sapele; (8) teak.

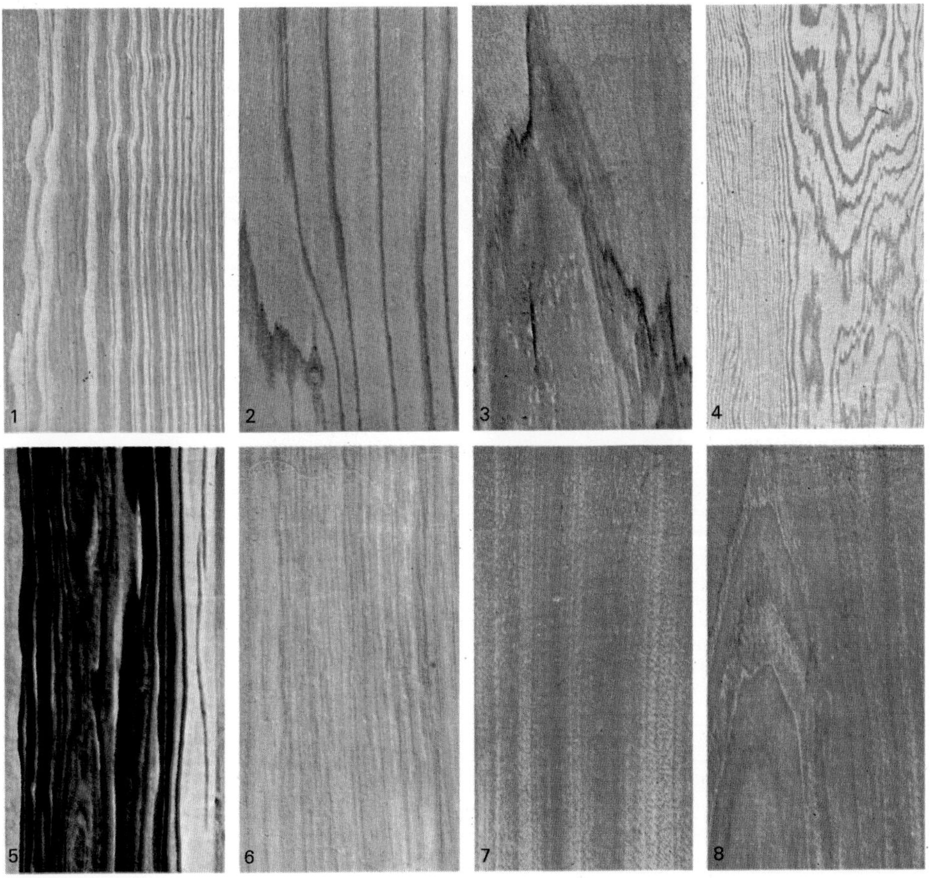

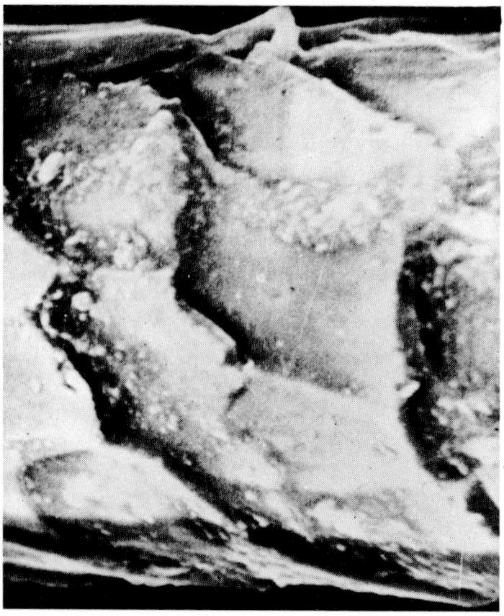

Scanning electron micrograph of a Romney Marsh wool fiber of 44s quality. This fiber is 36 μm wide.

brown or black. They are composed of the protein KERATIN, whose molecules are long, coiled chains, giving wool elasticity and resilience. Reactive side groups result in good affinity for DYES, and enable new, desirable properties to be chemically imparted. Wools lasts if well cared for, but is liable to be damaged by some insect larvae (which eat it), by heat, sunlight, alkalis and hot water. It chars and smolders when burned, but is not inflammable. Wool strongly absorbs moisture from the air. It is weakened when wet, and liable to form felt if mechanically agitated in water. Wool has been used from earliest times to make cloth. Sheep are shorn, usually annually, and the fleeces are cleaned—the wool WAX removed is the source of LANOLIN—and sorted, blended, carded (which disentangles the fibers and removes any foreign bodies) and combed if necessary to remove shorter fibers. A rope of woolen fibers, roving, is thus produced, and is spun (see SPINNING). The woolen yarn is woven into cloth, knitted, or made into carpets or blankets. The main producing countries are Australia, New Zealand, the USSR and India. Because the supply of new (virgin) wool is inadequate, inferior textiles are made of reprocessed wool which has been recovered as "shoddy"

WOOLLEY, Sir Charles Leonard (1880–1960), English archaeologist, best known for his excavations at Ur in Mesopotamia (Iraq) 1922–34, where his most spectacular discoveries were the royal tombs. He also worked in Syria and Turkey.

WORK, alternative name for ENERGY, used particularly in discussing mechanical processes. Work of one JOULE is done when a FORCE of one NEWTON acts through a distance of one METRE.

WORMS, True. See ANNELIDA.

WREN, Sir Christopher (1632–1723), greatest English architect. He had a brilliant early career as a mathematician and professor of astronomy, Oxford (1661–73), and was a founder-member of the ROYAL SOCIETY OF LONDON (president 1681–83). In 1663 he turned to architecture, in which he was largely self-taught. After the Great Fire of London (1666) Wren was appointed principal architect to rebuild London, where he was responsible for 52 churches, all of different design. His greatest building was the new St. Paul's Cathedral, noted for its monumental Baroque facade. He also worked at Greenwich (see GREENWICH OBSERVATORY), Oxford and Cambridge.

WRIGHT, Orville (1871–1948) and **Wilbur** (1867–1912), US aeronautical engineers who built the first successful powered heavier-than-air aircraft, flown first at Kitty Hawk, N.C., on Dec. 17, 1903, over distances of 120–852ft (37–260m). Their early experiments were with gliders, influenced by the work of Otto LILIENTHAL: Wilbur incorporated the AILERON (1899), a major step forward in their first man-carrying glider, flown at Kitty Hawk in 1900. In 1901 they built and experimented with a WIND TUNNEL, and their findings were used for their 1902 glider, by far the most advanced of its time. Following their first successful powered flight in 1903, they made further developments and by 1906 they were able to stay aloft for more than an hour. The American Wright Company for manufacturing airplanes was formed in 1909. (See also FLIGHT, HISTORY OF.)

WRIGHT, Sewall (1889–), US geneticist regarded as a founder of population GENETICS, the application of STATISTICS to the study of EVOLUTION.

WRIST, the flexible set of BONES linking the end of the ARM with the HAND. The small bones of the wrist allow movement in various directions and provide basic maneuverability for the hand itself.

WRITING, History of. Human communication has two primary forms: the transient, e.g. speech, SIGN LANGUAGE; and the permanent or semipermanent, of which the most important is writing. Forerunners of writing are the use of carved sticks or knotted cords to convey information; but the earliest form of writing was the PICTOGRAPHY of ancient Sumeria and Egypt. Originally the pictographs depicted objects, but some 5 000 years ago there developed IDEOGRAMS (representing ideas) and logograms (words). Sumerian CUNEIFORM and Egyptian HIEROGLYPHICS had complex word signs; as does Chinese to this day. The Hittites, Egyptians and Mesopotamians derived symbols for specific sounds; that is, phonetic writing. During the 2nd millennium BC the Semitic ALPHABETS emerged, and from these were derived the Greek and later Roman alphabets and so, in time, our own. (See also LANGUAGE; MINOAN LINEAR SCRIPTS; PUNCTUATION; RUNES; SHORTHAND; SYLLABARY.)

WROUGHT IRON, refined IRON produced by purifying PIG IRON in a "puddling" furnace; it contains a uniform distribution of ferrous silicate slag. Although it has useful mechanical properties, little wrought iron is now manufactured.

WUNDT, Wilhelm (1832–1920), German psychologist regarded as the father of experimental PSYCHOLOGY, opening the first psychological institute in 1879 and so ushering in the modern era of psychology.

WURTZITE, the high-temperature (α) form of zinc sulfide (ZnS); a minor ore of ZINC which inverts to SPHALERITE. It forms brown-black pyramidal crystals in the hexagonal system, and occurs in Bolivia, Mont. and Nev.

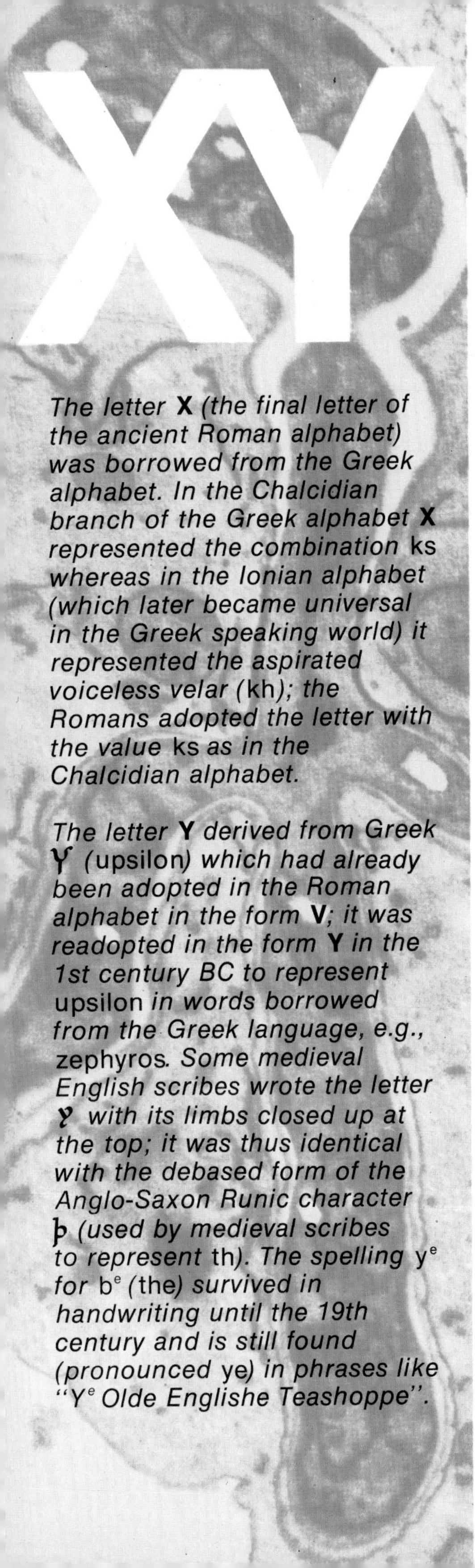

XY

The letter **X** *(the final letter of the ancient Roman alphabet) was borrowed from the Greek alphabet. In the Chalcidian branch of the Greek alphabet* **X** *represented the combination* ks *whereas in the Ionian alphabet (which later became universal in the Greek speaking world) it represented the aspirated voiceless velar (*kh*); the Romans adopted the letter with the value* ks *as in the Chalcidian alphabet.*

The letter **Y** *derived from Greek* ϒ *(*upsilon*) which had already been adopted in the Roman alphabet in the form* **V***; it was readopted in the form* **Y** *in the 1st century BC to represent* upsilon *in words borrowed from the Greek language, e.g.,* zephyros*. Some medieval English scribes wrote the letter* ƴ *with its limbs closed up at the top; it was thus identical with the debased form of the Anglo-Saxon Runic character* þ *(used by medieval scribes to represent* th*). The spelling* y^{e} *for* þ^{e} *(*the*) survived in handwriting until the 19th century and is still found (pronounced* ye*) in phrases like "Y^{e} Olde Englishe Teashoppe".*

X AND Y CHROMOSOMES, or **sex chromosomes**, the CHROMOSOMES which determine the sex of a person (as well as carrying some genetic information not related to sex determination). Sex chromosomes are inherited (see HEREDITY) in the same way as the other 22 human chromosome pairs, normal persons being either XX (female) or XY (male). The Y chromosome carries little genetic information and it is largely the properties of the X chromosome that determine "sex-linked" characteristics in males. Sex-linked characteristics include HEMOPHILIA and COLOR BLINDNESS, which are carried as recessive genes in females.

XANTHATE, an organic salt of general formula $ROCS_2$, formed by reacting an ALCOHOL with carbon disulfide and an alkali. Some are used as FLOTATION agents. CELLULOSE xanthate is an intermediate in making rayon and CELLOPHANE.

XANTHOPHYTA. See ALGAE.

XENOBIOLOGY. See EXOBIOLOGY.

XENON (Xe), one of the NOBLE GASES, used in flash lamps, electric-arc lamps, BUBBLE CHAMBERS and radiation counters; and also to produce ANESTHESIA. It forms several stable covalent compounds with fluorine and oxygen, showing oxidation numbers +2, +4, +6 and +8. They are reactive and highly oxidizing, and are unstable in acid solution. Most of the crystalline xenate (VIII) salts (XeO_6^{4-}) are stable. AW 131.3, mp −112°C, bp −107°C.

XENUSION, a famous fossil once generally claimed to belong to the ONYCHOPHORA and to be the missing link between the ANNELIDA and ARTHROPODA. Recent research has established beyond doubt that *Xenusion* is in fact a fossil colonial cnidarian, related to sea pens.

XEROGRAPHY, an electrostatic copying method. Light reflected from the original is focused onto an electrostatically charged (see ELECTRICITY), selenium-coated drum. Selenium is photoconductive (see PHOTOCONDUCTIVE DETECTOR), so where the light strikes the drum, the charge leaks away, leaving a reversed electrostatic image of the original on the drum. This is dusted with "toner," a dry ink powder which sticks only to the charged image. The toner is then transferred to a sheet of ordinary paper and fixed by applying heat. The paper thus carries a positive copy of the original. Repeated exposure on the rotating drum is needed to produce further copies. Other electrostatic copying processes form a positive print on paper specially coated with zinc oxide.

XEROPHYTE, a plant that has adaptations which enable it to live in dry habitats. These adaptations allow it to store water; cut down water loss by having small leaf area, and recover after wilting. Cacti are typical xerophytes.

X-RAY ASTRONOMY, the study of the X-RAYS emitted by celestial objects. Since the earth's atmosphere absorbs most X-radiation before it reaches the surface, observations are usually made from high altitude balloons, satellites and rockets. A number of celestial X-ray sources are known, including the sun and the CRAB NEBULA.

X-RAY DIFFRACTION, a technique for determining the structure of CRYSTALS through the way in which they scatter X-RAYS. Because of INTERFERENCE between the waves scattered by different ATOMS, the scattering occurs in directions characteristic of the spatial arrangement of the unit cells of the crystal,

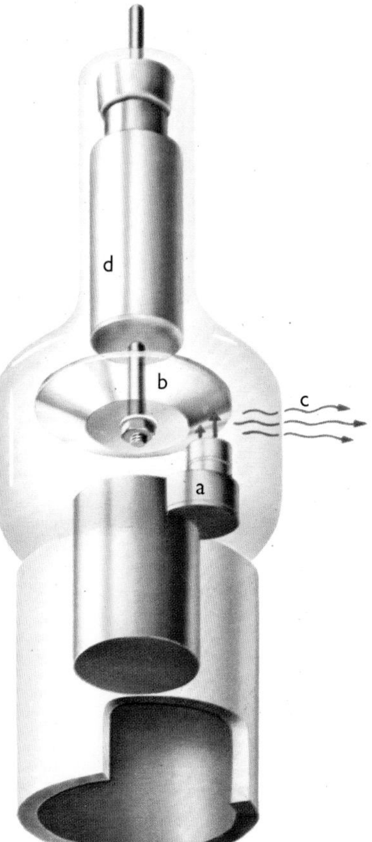

In the modern X-ray tube, the heated cathode (a) emits a stream of electrons which strikes the tungsten target anode (b), causing its atoms to emit X-rays (c). At low voltages most of the electrons dissipate their energy as heat and the anode must be cooled. In this case the electric motor (d) rotates the target, so that the electron beam always strikes a cool area.

while the relative intensity of the different beams reflects the structure of the unit cell itself. Unfortunately, more than one structure can often produce the same DIFFRACTION pattern. The technique is also used to study local ordering in "amorphous" solids and liquids.

X-RAYS, highly energetic, invisible ELECTROMAGNETIC RADIATION of wavelengths ranging between 0.1pm to 1nm. They are usually produced using an evacuated ELECTRON TUBE in which ELECTRONS are accelerated from a heated CATHODE toward a large tungsten or molybdenum ANODE by applying a POTENTIAL difference of perhaps 1MV. The electrons transfer their energy to the anode which then emits X-ray PHOTONS. X-rays are detected using PHOSPHOR screens (as in medical fluoroscopy), with GEIGER and SCINTILLATION COUNTERS and on photographic plates. X-rays were discovered by ROENTGEN in 1895, but because of their extremely short wavelength their wave nature was not firmly established until 1911, when von LAUE demonstrated that they could be diffracted from crystal LATTICES. X-rays find wide use in medicine both for diagnosis and treatment (see RADIOLOGY) and in engineering where RADIOGRAPHS are used to show up minute defects in structural members. X-ray tubes must always be carefully shielded because the radiation causes serious damage to living tissue.

XYLEM, the main water-conducting tissue in higher plants. (See WOOD.)

YALE, Linus (1821–1868), US inventor of the Yale lock (1861–65) and the dial combination lock (1862). (See LOCKS AND KEYS.)

YALLOW, Rosalyn Sussman (1921–) US medical physicist who was awarded half-share in the 1977 Nobel Prize for Physiology and Medicine (the other half was shared by R. GUILLEMIN and A. V. SCHALLY) for the development of radioimmunoassay techniques for peptide HORMONES.

YANG, Chen Ning (1922–), Chinese-born US physicist who shared with Tsung Dao LEE the 1957 Nobel Prize for Physics for their studies of violations of the conservation of PARITY.

YARD (yd), base unit of length in the US Customary and British Imperial unit systems. The International yard, defined as 0.9144m exactly, is generally used as a scientific standard. The yard of US commerce, defined as 3600/3937m, is marginally greater.

YAWS, a disease, caused by an organism related to that of syphilis (see VENEREAL DISEASES), common in the tropics. It occurs often in children and consists of a local lesion on the limbs; there is also mild systemic disease. Chronic destructive lesions of SKIN, BONE and CARTILAGE may develop later. The WASSERMANN TEST is positive as in syphilis and PENICILLINS are the treatment of choice.

Y CHROMOSOMES. See X AND Y CHROMOSOMES.

YEAR (yr), name of various units of time, all depending on the revolution of the earth about the sun. The *sidereal year* (365.256 36 mean solar DAYS) is the average time the earth takes to complete one revolution measured with respect to a fixed direction in space. The *tropical year* (365.242 20 mean solar days), the year measured by the changing SEASONS, is that in which the mean longitude of the sun moves through 360°. The *anomalistic year* (365.259 64 mean solar days) is the average interval between successive terrestrial perihelions (see ORBIT). The *civil year* is a period of variable duration, usually 365 or 366 days (leap year), depending on the type of CALENDAR in use.

YEASTS, single-celled plants classified with the FUNGI. Some cause diseases of the skin and mucous membranes (see THRUSH), while others, notably the strains of *Saccharomyces cerevisiae*, Bakers' yeast, are used in baking (see LEAVEN), brewing and wine-making. Yeasts employ either or both of two metabolic processes: FERMENTATION involves the anaerobic decomposition of hexose SUGARS to yield alcohol (ETHANOL) and CARBON dioxide; "respiration" involves the exothermic decomposition of various sugars in the presence of oxygen to give carbon dioxide and water. Yeasts are also grown as a source of food rich in B-complex VITAMINS.

YELLOW FEVER, INFECTIOUS DISEASE caused by a VIRUS carried by MOSQUITOES of the genus AËDES and occurring in tropical America and Africa. The disease consists of FEVER, headache, backache, prostration and VOMITING of sudden onset. PROTEIN loss in the URINE, KIDNEY failure, and LIVER disorder with JAUNDICE are also frequent. HEMORRHAGE from mucous membranes, especially in the GASTROINTESTINAL TRACT is also common. A moderate number of cases are fatal but a mild form of the disease is also recognized. VACCINATION to induce IMMUNITY is important and effective as no specific therapy is available; mosquito control provides a similarly important preventive measure.

YERKES, Robert Mearns (1876–1956), US pioneer of comparative (animal/human) PSYCHOLOGY and of intelligence testing. He initiated the first mass psychological testing program in WWI, involving nearly 1¾ million US army men, and in the 1920s and 1930s he was the world's foremost authority on PRIMATES.

YERSIN, Alexandre Émile John (1863–1943), Swiss bacteriologist who discovered, independently of KITASATO, the PLAGUE bacillus (1894), and developed a SERUM to combat it.

YOGHURT, or yogurt, semisolid cultured MILK food made by inocculating pasturized milk with a culture of *Streptococcus thermophilus* and *Lactobacillus bulgaricus* and incubating until the desired acidity is achieved. Various fruits can be added in packaging.

YOUNG, Thomas (1773–1829), British linguist, physician and physicist. His most significant achievement was by demonstrating optical INTERFERENCE, to resurrect the wave theory of LIGHT, which had been occulted by NEWTON's particle theory. He also suggested that the eye responded to mixtures of three primary colors (see VISION), and proposed the modulus of ELASTICITY known as *Young's modulus* (*E*—see MATERIALS, STRENGTH OF).

YTTERBIUM (Yb), one of the LANTHANUM SERIES. AW 173.0 mp 824°C, bp 1193°C, sg 6.972 (25°C).

YTTRIUM (Y), silvery-white RARE-EARTH metal. It is used in ALLOYS and as a "getter" to help evacuate ELECTRON TUBES. A red PHOSPHOR (yttrium oxide or vanadate excited by europium) is used in color televisions, and yttrium-iron GARNETS are used in RADAR and MICROWAVE devices. AW 88.9, mp 1523°C, bp 3337°C, sg 4.46 (25°C).

YUKAWA, Hideki (1907–   ), Japanese physicist who postulated the meson (see SUBATOMIC PARTICLES) as the agent bonding the atomic nucleus together. In fact, the mu-meson discovered shortly afterwards (in 1936) by C. D. ANDERSON, does not fulfil this role and Yukawa had to wait until C. F. POWELL discovered the pi-meson in 1947 for vindication of his theory. He received the 1949 Nobel Prize for Physics.

Z

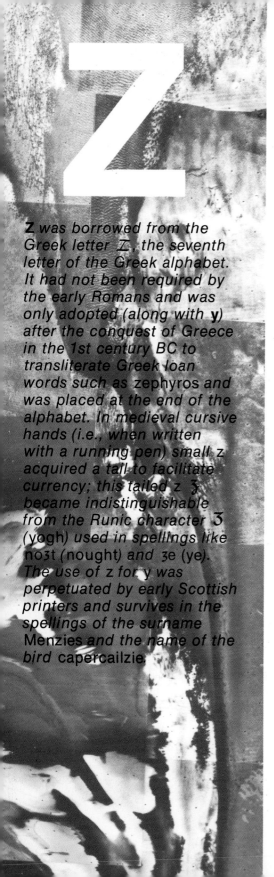

Z was borrowed from the Greek letter Ζ, the seventh letter of the Greek alphabet. It had not been required by the early Romans and was only adopted (along with **y**) after the conquest of Greece in the 1st century BC to transliterate Greek loan words such as zephyros and was placed at the end of the alphabet. In medieval cursive hands (i.e., when written with a running pen) small z acquired a tail to facilitate currency; this tailed z ʒ became indistinguishable from the Runic character Ȝ (yogh) used in spellings like noȝt (nought) and ȝe (ye). The use of z for y was perpetuated by early Scottish printers and survives in the spellings of the surname Menzies and the name of the bird capercailzie.

ZEEMAN, Pieter (1865–1943), Dutch physicist who shared with LORENTZ the 1902 Nobel Prize for Physics for his discovery of the **Zeeman effect**. This involves the splitting of spectral lines (see SPECTROSCOPY) in a MAGNETIC FIELD.

ZEISS, Carl (1816–1888), German optical manufacturer who founded a famous workshop at Jena in 1846. Realizing that optical technology had much to gain from scientific research, in the mid-1860s he formed a fruitful association with the physicist, Ernst ABBE.

ZENITH, in astronomy, the point on the CELESTIAL SPHERE directly above an observer and exactly 90° from the celestial horizon. It is directly opposite to the NADIR.

ZENO OF ELEA (c450 BC), Greek philosopher member of the Eleatic school. He is most important for his four paradoxes, best known of which are the "Achilles and tortoise" paradox (in a race, the tortoise is given a start: by the time Achilles reaches the point where the tortoise *was*, the tortoise has advanced—therefore Achilles can never overtake the tortoise) and the "arrow" paradox (at any instant in its flight an arrow is in only one place, and therefore at rest—therefore the arrow cannot move).

ZEOLITES, group of hydrated aluminosilicate minerals, mostly found in volcanic ROCKS and hydrothermal veins. Variable in form, they are light, with an open-framework structure which permits their use as "molecular sieves" and for ION EXCHANGE (especially for softening HARD WATER). They undergo reversible DEHYDRATION.

ZEPPELIN, Count Ferdinand Adolf August Heinrich von (1838–1917), German aeronautical engineer who designed and built almost a hundred powered BALLOONS (1900 on), called zeppelins for him.

ZERNIKE, Frits (1888–1966), Dutch physicist awarded the 1953 Nobel Prize for Physics for his development of phase-contrast microscopy, whereby CELLS may be studied without prior staining.

ZERO, in mathematics, a NUMBER smaller than any finite positive number, but larger than any finite negative number. It obeys

$$x \pm 0 = x$$
$$x \times 0 = 0$$
$$0/x = 0$$
$$x^0 = 1.$$

Division by zero is an undefined operation. Zero may be regarded as the identity element for ADDITION in the FIELD of real numbers.

ZERO-POINT ENERGY, the kinetic ENERGY which, in accordance with the UNCERTAINTY PRINCIPLE, is retained by a substance at ABSOLUTE ZERO. Each component oscillator retains one half quantum ($h\nu/2$) of energy.

ZIEGLER, Karl (1898–1973), German chemist who shared with NATTA the 1963 Nobel Prize for Chemistry for his work on ORGANOMETALLIC COMPOUNDS, in course of which he dramatically advanced POLYMER science and technology.

ZINC (Zn), bluish-white metal in Group IIB of the PERIODIC TABLE, an anomalous TRANSITION ELEMENT. It occurs naturally as SPHALERITE, SMITHSONITE, HEMIMORPHITE and WURTZITE, and is extracted by roasting to the oxide and reduction with carbon. It is used for GALVANIZING; as the cathode of dry cells, and in ALLOYS including BRASS. Zinc is a vital trace element, occurring in red BLOOD cells and in INSULIN. Chemically zinc is reactive, readily forming divalent ionic salts (Zn^{2+}), and zincates (ZnO_2^{2-}) in alkaline solution; it forms many stable LIGAND complexes. Zinc oxide and sulfide are used as white pigments. Zinc chloride is used as a FLUX, for fireproofing, in dentistry, and in the manufacture of BATTERIES and FUNGICIDES. AW 65.4, mp 420°C, bp 907°C, sg 7.133 (25°C).

Mid 17th-century illustration depicting the geocentric universe. The spheres of the planets lie nestled within that of the fixed stars. The signs of the zodiac are marked along the ecliptic. The intersection of the ecliptic with the celestial equator (equinoctial) facing us is the vernal equinox (first point of Aries), then and now in Pisces. At the bottom right is a diagram of Tycho Brahe's cosmology; Ptolemy's appears at the bottom left.

ZINJANTHROPUS. See PREHISTORIC MAN.

ZINSSER, Hans (1878–1940), US bacteriologist who directed the development and production of vaccines to counter certain strains of TYPHUS.

ZIRCON ($ZrSiO_4$), hard SILICATE mineral, a major ore of ZIRCONIUM, of widespread occurrence. It forms prismatic crystals in the tetragonal system, which when transparent are used as GEMS. They may be colorless, red, orange, yellow, green or blue, and have a high refractive index.

ZIRCONIUM (Zr), a silvery-white TRANSITION ELEMENT in Group IVB of the PERIODIC TABLE. It occurs naturally as baddeleyite (ZrO_2) and ZIRCON; the metal is extracted by reducing zirconium (IV) chloride with magnesium. It is corrosion-resistant at ordinary temperatures, owing to an inert oxide layer, but reactive at high temperatures. The metal is used in photographic flash bulbs and to clad uranium fuel elements in atomic reactors. The refractory oxide is used for ceramics, and other zirconium compounds

Zircon crystals, as they occur in limestone, and cut and polished as a gemstone.

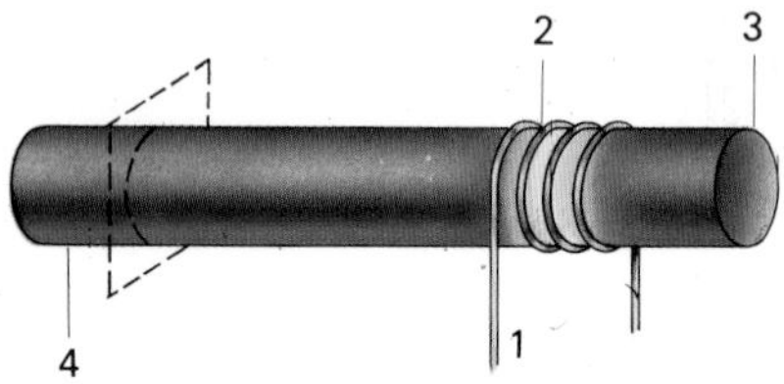

In zone refining a heating coil (1) with an associated section of molten material (2) is slowly moved from one end (3) to the other (4) of a bar of material. Any impurities in the bar tend to be carried in the molten section to the latter end which is then cut off and discarded.

are used in pharmaceuticals and as mordants in dyeing. AW 91.2, mp 1852°C, bp 3580°C, sg 6.49 (20°C).

ZODIAC, the band of the heavens whose outer limits lie 9° on each side of the ECLIPTIC. The 12 main CONSTELLATIONS near the ecliptic, corresponding to the 12 signs of the zodiac, are ARIES; TAURUS; GEMINI; CANCER; LEO; VIRGO; LIBRA; SCORPIO; SAGITTARIUS; CAPRICORNUS; AQUARIUS; PISCES. The orbits of all the planets except Pluto lie within the zodiac and their positions, as that of the sun, are important in ASTROLOGY. The 12 signs are each equivalent to 30° of arc along the zodiac.

ZODIACAL LIGHT, a hazy band of light, usually to be seen in the W after sunset or the E before sunrise, but in fact extending all along the ECLIPTIC. Associated is the **gegenschein**, a circular patch of light on the ecliptic directly opposite to the sun. Both phenomena are caused by the reflection of sunlight on clouds of interplanetary dust.

ZONE REFINING, purifying process, used particularly with METALS and SEMICONDUCTORS, in which a moving zone of an ingot of the impure material is melted, impurities tending to remain in the moving molten zone, while the purified material is left behind. Similar processes are used for preparing large uniform semiconducting crystals and for distributing a desired impurity uniformly through a pure ingot.

ZOO, or zoological garden, a collection of wild animal species preserved for public education, scientific research and the breeding of endangered species. The first modern zoo was that of the Royal Zoological Society at Regent's Park, London, established in 1826.

ZOOGEOGRAPHY, the study of the geographical distribution of animal species and populations. Physical barriers, such as wide oceans and mountain ranges, major climatic extremes, intense heat or cold, may prevent the spread of a species into new areas, or may separate two previously linked populations, allowing them to develop into distinct species. The presence of these barriers to movement and interbreeding, both now and in the past, are reflected in the distributions and later adaptive radiations of animal species, resulting in the zoogeographical distributions we find today. The major **zoogeographic regions** of the world are the Ethiopian (sub-Saharan Africa); the Oriental (India and SE Asia); the Australasian (including Australia, New Guinea and New Zealand); the Neotropical (Central and South America), and the Holarctic (the whole northerly region, often divided into the Nearctic—North America—and the Palearctic—most of Eurasia with N Africa).

ZOOIDS. See ENTOPROCTA; HEMICHORDATA; POLYZOA.

ZOOLOGY, the scientific study of animal life. Originally concerned with the classification of animal groups (see ANIMAL KINGDOM), comparative ANATOMY and PHYSIOLOGY, the science now embraces studies of EVOLUTION, GENETICS, EMBRYOLOGY, BIOCHEMISTRY, ANIMAL BEHAVIOR and ECOLOGY.

ZOOPLANKTON, the animal elements of the PLANKTON, made up of small marine animals, PROTOZOA, and, principally, the larvae of other marine creatures, mainly mollusks and crustacea.

ZSIGMONDY, Richard (1865–1929), Austrian-born German chemist awarded the 1925 Nobel Prize for Chemistry for his work on COLLOIDS.

ZWICKY, Fritz (1898–1974), Swiss-born US astronomer and astrophysicist best known for his studies of SUPERNOVAS, which he showed to be quite distinct from, and much rarer than, NOVAS. He also did pioneering work on JET PROPULSION.

ZWITTERION, complex ION carrying both a positive and a negative charge in different parts (owing to both an acidic and a basic FUNCTIONAL GROUP being present), and hence neutral overall. They usually show TAUTOMERISM with an uncharged molecule; e.g., glycine:

$$H_2NCH_2COOH \rightleftharpoons {}^{+}H_3NCH_2COO^{-}.$$

ZWORYKIN, Vladimir Kosma (1889–), Russian-born US electronic engineer regarded as the father of modern TELEVISION: his kinescope (patented 1924), little adapted, is our modern picture tube; and his iconoscope, though now obsolete, represents the basis of the first practical television camera. He has also made important contributions to the ELECTRON MICROSCOPE.

ZYGOMYCETES. See FUNGI.

ZYGOTE, CELL produced by the fusion of two GAMETES and which contains the diploid chromosome number (see MEIOSIS). The offspring is then produced by mitotic division (see MITOSIS) of the zygote to give 2, 4, 8, 16, 32 . . . 2^n cells.

ZYMASE, ENZYME complex found in YEASTS which catalyzes the alcoholic FERMENTATION of CARBOHYDRATES. It was first isolated by E. BUCHNER.